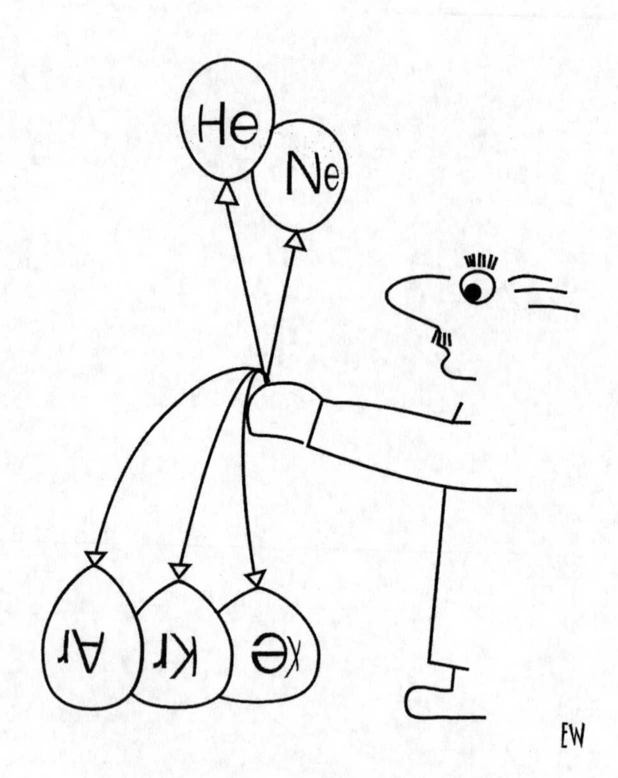

Separating the light from the heavy noble gases.

REVIEWS in MINERALOGY and GEOCHEMISTRY

Volume 47 2002

— NOBLE GASES — IN GEOCHEMISTRY AND COSMOCHEMISTRY

Editors:

Donald Porcelli *Department of Earth Sciences University of Oxford Oxford, United Kingdom*

Chris J. Ballentine *Department of Earth Sciences University of Manchester Manchester, United Kingdom*

Rainer Wieler *Institute for Isotope Geology and Mineral Resources E T H – Zürich Zürich, Switzerland*

Front cover: *Courtesy http://philip.greenspun.com*
Frontispiece: *Cartoon by Elisabeth Wilding*

Series Editors: ***Jodi J. Rosso & Paul H. Ribbe***

GEOCHEMICAL SOCIETY

MINERALOGICAL SOCIETY of AMERICA

REVIEWS IN MINERALOGY AND GEOCHEMISTRY

(Formerly: REVIEWS IN MINERALOGY)

ISSN 1529-6466

Volume 47

Noble Gases in Geochemistry and Cosmochemistry

ISBN 0-939950-59-6

** This volume is the ninth of a series of review volumes published jointly under the banner of the Mineralogical Society of America and the Geochemical Society. The newly titled *Reviews in Mineralogy and Geochemistry* has been numbered contiguously with the previous series, *Reviews in Mineralogy*.

Additional copies of this volume as well as others in this series may be obtained at moderate cost from:

THE MINERALOGICAL SOCIETY OF AMERICA
1015 EIGHTEENTH STREET, NW, SUITE 601
WASHINGTON, DC 20036 U.S.A.

FOREWORD

Noble Gases in Geochemistry and Cosmochemistry is the 47th volume in a series that was initiated by the Mineralogical Society of America (MSA) in 1974 under the title "Short Course Notes." Within a few years the series was re-named *Reviews in Mineralogy*, which from the mid-1980s was added to the 4000 or so scientific periodicals tracked by the Institute for Scientific Information in their *Journal Citation Reports*. Beginning in 2000 with Volume 39, the series was given a new title, *Reviews in Mineralogy and Geochemistry* (*RiMG*), in response to the partnership that was formed at that time between MSA and the Geochemical Society—*GS* in the mini-logo that now appears on the spine of all *RiMG* volumes:

G
MSA

The two societies manage their own short courses (if any) and edit their own volumes; Paul Ribbe is editor for MSA and Jodi Rosso for GS.

For this volume, both editors were involved. There was no short course, although an attempt was undertaken to get the volume printed in time for the V.M. Goldschmidt conference in Davos, Switzerland (mid-August 2002) at which there was a major symposium on noble gases. *Noble Gases in Geochemistry and Cosmochemistry* was edited by D. Porcelli, C. J. Ballentine, and R. Wieler, all of whom were at the Institute for Isotope Geology and Mineral Resources, ETH—Zürich when this project was begun [their present affiliations are in the heading of Chapter 1, page 1].

If all proceeds according to plan, this will be the first *RiMG* volume to be published in electronic format as well as in paper copy. Visit the MSA website, www.minsocam.org, for breaking news.

Jodi Rosso & Paul Ribbe
West Richland, Washington
Blacksburg, Virginia
July 18, 2002

PREFACE

Noble Gases – Noble Science

The scientific discoveries that have been made with noble gas geochemistry are of such a profound and fundamental nature that earth science textbooks should be full of examples. Surprisingly, this really is not so. The "first discoveries" include presolar components in our solar system, extinct radionuclides, primordial volatiles in the Earth, the degassing history of Mars, secular changes in the solar wind, reliable present day mantle degassing fluxes, the fluxes of extraterrestrial material to Earth, groundwater paleotemperatures and the ages of the oldest landscapes on Earth. Noble gas geochemistry has scored so many such "firsts" or "home runs" that it should permeate a lot of earth science thinking and teaching. Yet rather surprisingly it does not.

Noble gas geochemistry also is a broader and more versatile field than almost any other area of geochemistry. It pervades cosmochemistry, Earth sciences, ocean sciences, climate studies and environmental sciences. Yet most modern Earth, planetary and environmental science departments do not consider noble gas geochemistry to be at the top of their list in terms of hiring priorities these days. Furthermore, with the exception of Ar geochronologists, noble gas geochemists are a surprisingly rare breed.

Why is the above the case? Perhaps the reasons lie in the nature of the field itself.

First, although noble gas geochemists work on big problems, the context of their data is often woefully under-constrained so that it becomes hard to make progress beyond the first order fundamental discoveries. Noble gas data are often difficult to interpret. Although some concepts are straightforward and striking in their immediate implications (e.g. mantle ^{3}He in the oceans), others are to this day shrouded in lack of clarity. The simple reason for this is that in many situations it is only the noble gases that offer any real insights at all and the context of other constraints simply does not exist.

Some examples of the big issues being addressed by noble gases are as follows and I have deliberately posed these as major unresolved questions that only exist because noble gas geochemistry has opened windows through which to view large-scale issues and processes that otherwise would be obscure. (1) Is the presolar noble gas component present in a tiny fraction of submicroscopic meteoritic C or is it ubiquitously distributed? (2) How did solar noble gases get incorporated into the Earth? (3) How did solar noble gases survive the protracted accretion of the Earth via giant impacts? (4) What is the origin of the noble gas pattern in the Earth's atmosphere? (5) Why are the Earth and Mars almost opposites in terms of the relative isotopic differences between atmosphere and mantle? (6) What is the present source of Earth's primordial helium? Can we ignore the core? (7) What is the ^{20}Ne/^{22}Ne of the mantle, how was it acquired and why is it different from the atmosphere? (8) How does one reconcile the strong fractionation in terrestrial Xe with data for other noble gases? (9) How much radiogenic ^{40}Ar should the Earth have? How well do we know K/U? (10) Are the light isotopes of Xe the same in the mantle and the atmosphere? If not, why not? (11) How are noble gases transported through the creeping solid earth? (12) How does one explain the heat – helium paradox? (13) How incompatible are the noble gases during melting? (14) How are atmospheric components incorporated into volcanic samples? (15) How are the excess air components incorporated into groundwater? (16) Why are continental noble gas paleotemperature records offset from oceanic temperature records?

Noble gas data tell us that the Earth and solar system represent very complex environments. When we make our simple first order conclusions and models we are only at the tip of the iceberg of discoveries that are needed to arrive at a thorough understanding of the behavior of volatiles in the solar system. Who wants to hear that things are complicated? Who wants to hire in a field that will involve decades of data acquisition and analysis in order to sort out the solar system? Sadly, too few these days. This is the stuff of deep scientific giants and bold, technically difficult long-term research programs. It is not for those who prefer superficiality and quick, glamorous, slick answers. Noble gas geochemists work in many areas where progress is slow and difficult even though the issues are huge. This probably plays a part in the limited marketability of noble gas geochemistry to the non-specialist.

Second, noble gases is a technically difficult subject. That is, noble gas geochemists need to be adept at technique development and this has to include skills acquired through innovation in the lab. Nobody can learn this stuff merely with a book or practical guide. Reading "Zen and the Art of Motorcycle Maintenance" (by Robert Pirsig) would give you a clearer picture.

This magnificent MSA-GS volume is going to be enormously useful but on its own it won't make anybody into a noble gas geochemist. Although the mass spectrometry principles are not complex, the tricks involved in getting better data are often self taught or passed on by working with individuals who themselves are pushing the boundaries further. Furthermore, much of the exciting new science is linked with technical developments that allow us to move beyond the current measurement capabilities. Be they better crushing devices, laser resonance time of flight, multiple collection or compressor sources - the technical issues are central to progress.

Lastly, noble gas geochemists need a broad range of other skills in order to make progress. They have to be good at mass spectrometry as already stated. However, nowadays they also need to be able to understand fields as different as mantle geochemistry, stellar evolution, cosmochemistry, crustal fluids, oceanography and glaciology. They are kind of "Renaissance" individuals.

Therefore, if you are thinking broadly about hiring scientists who love science and stand a good chance of making a major difference to our understanding of the solar system, earth and its environment – I would recommend you hire a really good noble gas geochemist. However, the results may take a while. If you want somebody who will crank out papers at high speed and quickly increase the publication numbers of your department then you may need to think about somebody else. The two are not mutually exclusive but think hard about what is really important.

This volume is nothing short of a fantastic accomplishment. This is true in many ways. The editors are to be commended for working so efficiently to pull it off. However, there are others who also should be honored in this preface. Of course, the authors have, between them, written what is bound to be an essential handbook for all those who are interested in geochemistry. But they are building on the work of some remarkably clever and insightful scientists who had the original vision that has led to the big discoveries that I have already talked about at the start of this little section. In this case we are talking of amazing people like: Al Nier, who developed noble gas mass spectrometry and K-Ar geochronology; John Reynolds, who first discovered ^{129}Xe excesses in the Richardton meteorite, then wrote that astonishing paper entitled "The Age of the Elements" and had the clever of idea of doing stepwise degassing of irradiated samples to establish that the radiogenic xenon was derived from iodine sites; Igor Tolstikhin and Boris Mamyrin, who first discovered primordial ^{3}He; Brian Clarke, who developed the Clarke source and first showed ^{3}He variations in ocean waters; Harmon Craig, who first developed our understanding of primordial ^{3}He being released from the Earth into the oceans; Claude Allègre, who first measured Xe isotopic anomalies in MORB; Craig Merrihue and Grenville Turner, who first developed ^{40}Ar-^{39}Ar dating; Peter Signer, who first defined the Planetary noble gas component; Keith O'Nions and Ron Oxburgh, who formulated the heat-helium paradox; and Minoru Ozima and Frank Podosek, who established the key reference work for all those interested in noble gas geochemistry.

Many more could be named but it is time for me to stop and to say I am looking forward to seeing this book launched at the Goldschmidt Conference in Davos where we should be able to toast some of the above and their remarkable accomplishments. I encourage you to jump into these pages of deep science and "learn loads" about the truly enormous array of interesting research and discovery that has followed from noble gas geochemistry.

Alex N. Halliday
Zürich, May 2002

Table of Contents
Noble Gases in Geochemistry and Cosmochemistry

Reviews in Mineralogy and Geochemistry, Vol. 47

1 An Overview of Noble Gas Geochemistry and Cosmochemistry

Donald Porcelli, Chris J. Ballentine, Rainer Wieler

2 Noble Gases in the Solar System

Rainer Wieler

3 Noble Gases in Meteorites – Trapped Components

Ulrich Ott

4 Noble Gases in the Moon and Meteorites: Radiogenic Components and Early Volatile Chronologies

Timothy D. Swindle

5 Cosmic-Ray-Produced Noble Gases in Meteorites

Rainer Wieler

6 Martian Noble Gases

Timothy D. Swindle

7 Origin of Noble Gases in the Terrestrial Planets

Robert O. Pepin, Donald Porcelli

8 Noble Gas Isotope Geochemistry of Mid-Ocean Ridge and Ocean Island Basalts: Characterization of Mantle Source Reservoirs

David W. Graham

9 Noble Gases and Volatile Recycling at Subduction Zones

David R. Hilton, Tobias P. Fischer, Bernard Marty

10 Storage and Transport of Noble Gases in the Subcontinental Lithosphere

Tibor J. Dunai, Donald Porcelli

11 Models for the Distribution of Terrestrial Noble Gases and Evolution of the Atmosphere

D. Porcelli, C.J. Ballentine

12 Production, Release and Transport of Noble Gases in the Continental Crust

Chris J. Ballentine, Pete G. Burnard

13 Tracing Fluid Origin, Transport and Interaction in the Crust

Chris J. Ballentine, Ray Burgess, Bernard Marty

14 Noble Gases in Lakes and Ground Waters

Rolf Kipfer, Werner Aeschbach-Hertig, Frank Peeters, Marvin Stute

15 Noble Gases in Ocean Waters and Sediments

Peter Schlosser, Gisela Winckler

16 Cosmic-Ray-Produced Noble Gases in Terrestrial Rocks: Dating Tools for Surface Processes

Samuel Niedermann

17 K-Ar and Ar-Ar Dating

Simon Kelley

18 (U-Th)/He Dating: Techniques, Calibrations, and Applications

Kenneth A. Farley

1 An Overview of Noble Gas Geochemistry and Cosmochemistry

Donald Porcelli[1], Chris J. Ballentine[2] and Rainer Wieler

Institute for Isotope Geology and Mineral Resources
ETH Zürich, Sonneggstrasse 5
8092 Zürich, Switzerland

don.porcelli@earth.ox.ac.uk

[1] Now at: *Dept. of Earth Sciences, University of Oxford*
Parks Road, Oxford OX1 3PR, United Kingdom

[2] Now at: *Dept. of Earth Sciences, University of Manchester,*
Oxford Road, Manchester M13 9PL, United Kingdom

INTRODUCTION

A wealth of fundamental information regarding the Earth and solar system is based upon observations of the highly volatile elements He, Ne, Ar, Kr, and Xe. At first, this may seem surprising, considering that these elements are generally thought to reside almost entirely in the atmosphere, and so are considered strongly 'atmophile.' However, increasingly sophisticated analytical techniques have provided the means for precisely measuring their abundances in a wide range of geological and cosmochemical materials. Fittingly, these elements are known collectively as the rare gases, reflecting their general scarcity in geological materials. It is this feature that continues to provide challenges for analysts. These elements are also the noble gases, in tribute to their disdain for engaging in chemical consort with other species. Such behavior has been responsible for the early difficulties in their detection, and facilitates their continuing migration to the atmosphere. However, others refer to these as 'the inert gases,' which seems to imply that their behavior is dictated by a lack of interest in chemical reaction, a deficiency in chemical drive. Overall, the choice of appellation depends upon whether scarcity, nobility, or inertness is considered the most important characteristic. Regardless of their motivations, these noble gases can be profitably considered together, because physical and chemical properties vary systematically with atomic weight. However, much of the utility of noble gases is based on the widespread variations in their isotopic compositions. This is related to their overall depletion, which has made these elements vulnerable to isotopic modification from nuclear processes involving relatively more abundant parent elements. The wide applicability of noble gas systematics is due to the range of such processes. In cosmochemistry, fundamental contributions have been made to understanding the sources and distributions of volatiles throughout the solar system, to identifying the preservation of nucleosynthetic anomalies in meteorites, and to defining early solar system chronologies. Studies of the distribution of noble gases within the Earth are a critical component in studies of mantle geochemistry and the formation history of the atmosphere. Within the crust, noble gases have been key components in studies of crustal evolution, of flow patterns in hydrological systems and ocean basins, and in a range of dating techniques.

The present volume contains a series of focused reviews of noble gases across the solar system, in the Earth's mantle, in the crust, and as applied in geochronology. These are written by researchers closely associated with each field of research. Other books that are of interest in complementing these works include the earlier review of *He Isotopes in Nature* (Mamyrin and Tolstikhin 1984) and the recently published second edition of

1529-6466/00/0047-0001$0.500

Noble Gas Geochemistry (Ozima and Podosek 2001), which provide different perspectives, as well as detailed discussions of some background information that is not covered in the same detail in the more application-oriented papers of the present volume.

NOBLE GAS MASS SPECTROMETRY

The progress of noble gas geochemistry and cosmochemistry has been paced by the rate of developments in mass spectrometry. This has been driven by the need to attain increasingly greater precision to distinguish the often subtle variations in isotopic compositions, higher sensitivity to measure the low abundances found in many materials, and lower blanks to remove interference from atmospheric gases. The earliest measurements and ages, including U-He ages, did not involve the detection of isotopes, relying only on the total abundances of the elements present. Different isotopes of a single element were first detected for a noble gas, Ne (Thomson 1914), and following this Dempster (1918) and Aston (1919) produced the first focusing machines for relative isotope abundance measurements. The Nier design mass spectrometer (Nier 1947), including what has become the standard design for the electron bombardment source, provided precise compositions of the atmosphere. Reynolds (1956) introduced the high-sensitivity static mass spectrometer design that is essential to most modern noble gas studies. The high resolution required for precise $^3He/^4He$ measurements was achieved using a magnetic resonance mass spectrometer (Mamyrin et al. 1969; see Mamyrin and Tolstikhin 1984) and high-resolution sector mass spectrometer (Clarke et al. 1969).

Meteorite research has made particular demands upon noble gas analysis as it was found that there are noble gas compositional variations over very fine spatial scales. Separation of noble gas carrier phases using chemical procedures were developed, leading to the isolation of essentially pure exotic 'Ne-E' by Junck and Eberhardt (1979), then the discovery of gas-rich acid-resistant residues by Lewis et al. (1975), and culminating in the isolation of the first pure presolar grains (Lewis et al. 1987; Zinner et al. 1987). The sensitivity required to obtain high precision isotopic compositions of small abundance samples, especially for the heavy noble gases, was achieved through advances in pulse counting and developments in ultra-low blank extraction systems (Hohenberg 1980), design of the high transmission Baur-Signer source with rotational symmetry (Baur 1980), resonance ionisation mass spectrometry (Gilmour et al. 1994), and the recent construction of a compressor source with an increase in sensitivity for He and Ne by two orders of magnitude (Baur 1999). Another development has been in designing mass spectrometers for planetary probes, which have provided compositions from other planets (see Nier and Hayden 1971). Also, space missions have made measurements of the solar wind and cosmic rays (see Wieler 2002a, this volume). Important recent examples are the solar wind Ne composition determination by the SOHO mission (Kallenbach et al. 1997) and the Ne isotopic analyses during solar energetic particle events by ACE (Leske et al. 1999).

THE ATMOSPHERIC STANDARD

The standard for noble gas measurements, and the reference for discussions of data, is the composition of the terrestrial atmosphere. The abundances of the noble gases are in Table 1, along with those of other major and minor gases that are more abundant than Xe. With the exception of Ar, which is dominated by radiogenic ^{40}Ar, the noble gases are present as trace constituents. Isotopic compositions are provided in Table 2. Since air is the reference standard for laboratory analyses, measurements of other media are typically normalized to these values.

Table 1. Composition of the terrestrial atmosphere.

Constituent	*Volume Mixing Ratio*	*Total Inventory (cm^3STP)*
Dry Air	1	$(3.961\pm0.006)\times10^{24}$
N_2	7.81×10^{-1}	
O_2	2.09×10^{-1}	
Ar	$(9.34\pm0.01)\times10^{-3}$	$(3.700\pm0.004)\times10^{22}$
CO_2	3.7×10^{-4}	
Ne	$(1.818\pm0.004)\times10^{-5}$	$(7.202\pm0.016)\times10^{19}$
He	$(5.24\pm0.05)\times10^{-6}$	$(2.076\pm0.020)\times10^{19}$
CH_4	$1\text{-}2\times10^{-6}$	
Kr	$(1.14\pm0.01)\times10^{-6}$	$(4.516\pm0.040)\times10^{18}$
H_2	$4\text{-}10\times10^{-7}$	
N_2O	3×10^{-7}	
CO	$0.1\text{-}2\times10^{-7}$	
Xe	$(8.7\pm0.1)\times10^{-8}$	$(3.446\pm0.040)\times10^{17}$
Rn	$\sim6\times10^{-20}$	2×10^{5}

Based on dry tropospheric air. Water generally accounts for ≤4% of air. Other chemical constituents have mixing ratios less than Xe. Data from compilations by Lewis and Prinn (1984) and Ozima and Podosek (2001). CO_2 data from Keeling and Whorf (2000).

Table 2. Noble gas isotope composition of the atmosphere.

Isotope	*Relative abundances*	*Percent molar abundance*
^{3}He	$(1.399\pm0.013)\times10^{-6}$	0.000140
^{4}He	≡ 1	100
^{20}Ne	9.80±0.08	90.50
^{21}Ne	0.0290±0.0003	0.268
^{22}Ne	≡ 1	9.23
^{36}Ar	≡ 1	0.3364
^{38}Ar	0.1880±0.0004	0.0632
^{40}Ar	295.5±0.5	99.60
^{78}Kr	0.6087±0.0020	0.3469
^{80}Kr	3.9599±0.0020	2.2571
^{82}Kr	20.217±0.004	11.523
^{83}Kr	20.136±0.021	11.477
^{84}Kr	≡ 100	57.00
^{86}Kr	30.524±0.025	17.398
^{124}Xe	2.337±0.008	0.0951
^{126}Xe	2.180±0.011	0.0887
^{128}Xe	47.15±0.07	1.919
^{129}Xe	649.6±0.9	26.44
^{130}Xe	≡ 100	4.070
^{131}Xe	521.3±0.8	21.22
^{132}Xe	660.7±0.5	26.89
^{134}Xe	256.3±0.4	10.430
^{136}Xe	217.6±0.3	8.857

After compilation by Ozima and Podosek (2001). He: Mamyrin et al. (1970). Ne: Eberhardt et al. (1965). Ar: Nier (1950), Steiger and Jäger (1977). Kr, Xe: Basford et al. (1973).

PRODUCTION OF NOBLE GAS ISOTOPES

The dominant cause of the extensive isotopic variations seen in natural samples is production of noble gas nuclides by nuclear processes. The parents of simple decay schemes producing noble gases are listed in Table 3. ^{4}He is copiously produced by the ^{238}U, ^{235}U, and ^{232}Th decay series, each of which involves decay of various intermediate nuclides before producing stable Pb. Many of the transformations in these series involve alpha decay, producing alpha particles that require only the acquisition of electrons to become ^{4}He. Spontaneous fission, although accounting for a very small fraction of ^{238}U decay, produces Xe isotopes that can cause appreciable isotopic variations due to the generally low concentrations of Xe. The short-lived nuclides ^{129}I and ^{244}Pu also produce Xe (and Kr) isotopes, although these were only present within the Earth in significant amounts early in solar system history. The fission Xe isotope yield spectra for ^{238}U and ^{244}Pu decay are given in Table 4. Spontaneous fission of ^{238}U also produces Kr, but generally does not produce appreciable isotope variations. Some of the other reactions that produce noble gas nuclides within the Earth are listed in Table 5. There are also some radioactive noble gas isotopes of interest that are listed in Table 6. A comprehensive review of radiogenic, nucleogenic and fissiogenic noble gas isotope production is given by Ballentine and Burnard (2002, this volume).

The heaviest noble gas, Rn, is represented in the environment by only one isotope, ^{222}Rn. It is produced by ^{226}Ra (as part of the ^{238}U-^{206}Pb decay chain), and decays with a half-life of only 3.8 days. Its distribution therefore reflects that of its parent and short-term transport processes. Therefore, it is more fruitfully considered with the U-Th series nuclides, and is not included in this volume.

Table 3. Half-lives of parent nuclides for noble gases[a]

Nuclide	*Half-life*	*Daughter*	*Yield (atoms/decay)*	*Comments*
^{3}H	12.26 a	^{3}He	1	Continuously produced in atm.
^{238}U	4.468 Ga	^{4}He	8[b]	
		^{136}Xe	3.6×10^{-8} [c]	Spontaneous fission
			$(4.4 \pm 0.1) \times 10^{-8}$ [d]	
^{235}U	0.7038 Ga	^{4}He	7[b]	^{238}U/^{235}U = 137.88
^{232}Th	14.01 Ga	^{4}He	6[b]	Th/U = 3.8 in bulk Earth
		^{136}Xe	$<4.2 \times 10^{-11}$ [e]	No significant production in Earth
^{40}K	1.251 Ga	^{40}Ar	0.1048[a]	^{40}K = 0.01167% total K
^{244}Pu	80.0 Ma	^{136}Xe	7.00×10^{-5}	^{244}Pu/^{238}U = 6.8×10^{-3} at 4.56Ga[f]
^{129}I	15.7Ma	^{129}Xe	1	^{129}I/^{127}I = 1.1×10^{-4} at 4.56Ga[g]

Notes:

a. From data compilations of Blum (1995), Ozima and Podosek (2001), Pfennig et al. (1998)
b. Assuming secular equilibrium for entire decay series.
c. Eikenberg et al. (1993).
d. Ragettli et al. (1994).
e. Wieler and Eikenberg (1999) upper limit conservatively assuming 10% of fission events produce ^{136}Xe.
f. Hudson et al. (1989).
g. Hohenberg et al. (1967b).

Table 4. Xe Isotope spontaneous fission yield spectra.

Parent, *reference*	^{129}Xe	^{131}Xe	^{132}Xe	^{134}Xe	^{136}Xe
^{238}U					
Ragettli et al. (1994)	<0.06	8.3±0.06	57.7±0.1	82.8±0.1	≡100
Hebeda et al. (1987)	0.60±0.3	8.8±0.2	56.8±0.3	82.8±0.4	≡100
Eikenberg et al. (1993)	<0.11	8.3±0.2	57.0±0.3	81.8±0.3	≡100
Wetherill (1953)	<0.2	7.6±0.3	59.5±1.0	83.2±1.2	≡100
^{244}Pu					
Alexander et al. (1971)	-	25.1±2.2	87.6±3.1	92.1±2.7	≡100
Lewis et al.(1975)	4.8±5.5	24.6±2.0	88.5±3.0	93.9±0.8	≡100
Hudson et al. (1989)	-	24.8±1.5	89.3±1.3	93.0±0.5	≡100

Table 5. Some nuclear reactions producing noble gases.

Reaction	*Upper crust production ratios**
$^{6}Li(n,\alpha)^{3}H(\beta-)^{3}He$	$^{3}He/^{4}He = 1 \times 10^{-8}$
$^{17}O(\alpha,n)^{20}Ne$	$^{20}Ne/^{4}He = 4.4 \times 10^{-9}$
$^{18}O(\alpha,n)^{21}Ne$	$^{21}Ne/^{4}He = 4.5 \times 10^{-8}$
$^{24}Mg(n,\alpha)^{21}Ne$	$^{21}Ne/^{4}He = 1 \times 10^{-10}$
$^{25}Mg(n,\alpha)^{22}Ne$	*Combined crustal production:*
$^{19}F(\alpha,n)^{22}Na(\beta+)^{22}Ne$	$^{22}Ne/^{4}He = 9.1 \times 10^{-8}$
$^{35}Cl(n,\gamma)^{36}Cl(\beta-)^{36}Ar$	$^{40}Ar/^{36}Ar = 1.5 \times 10^{7}$

*See Ballentine and Burnard (2002, this volume) for details and production rates in different compositions.

Table 6. Radioactive noble gas nuclides.

Isotope	*Half-life*	*Source*
^{37}Ar	35.0 days	$^{40}Ca(n,\alpha)^{37}Ar$
^{39}Ar	269 years	$^{39}K(n,p)^{39}Ar$
^{81}Kr	2100 years	Spallation reactions
^{85}Kr	10.72 years	Spallation reactions
^{222}Rn	3.82 days	^{226}Ra decay

Table 7. Some physical and chemical properties of noble gases.

	Atomic No.	*Atomic wt.*[a]	*Atomic radius(Å)*[b]	*First ionization potential (eV)*[c]	*Second ionization potential (eV)*[c]
He	2	4.002602(2)	1.8	24.59	54.42
Ne	10	20.1797 (6)	1.6	21.56	40.96
Ar	18	39.948 (1)	1.9	15.76	27.63
Kr	36	83.80 (1)	2.0	14.00	24.36
Xe	54	131.29 (2)	2.2	12.13	21.21

Notes: a. Based on the atmospheric composition. b. Ozima and Podosek (2001) c. Lide (1995)

BEHAVIOR OF THE NOBLE GASES

A detailed treatment of noble gas behavior relevant to geochemistry and cosmochemistry is given by Ozima and Podosek (2001). A brief review is provided here only to set out some of the more commonly required data and to outline the nature of the processes that control the variations in the relative elemental abundances and isotopic compositions of the noble gases. Some basic elemental data are shown in Table 7.

Water, natural gas, and oil partitioning

Equilibration between noble gases in the atmosphere and the hydrosphere, including oceans and groundwaters, occurs according to the solubility of each element. Their solubilities in water, which increase with atomic weight, are well defined from laboratory determinations. There is also a temperature dependence on solubility that is different for each element, so that the elemental ratios of atmosphere-derived noble gases in solution then reflect the temperature at equilibration. This is the basis for determining past atmospheric temperatures from groundwaters. Noble gases in groundwaters that interact with other phases, such as natural gas or oil, will partition between these phases according to their relative solubilities, and in proportion to the volume of the phase and the conditions of equilibration. Partitioning histories can then be deduced from the present abundance patterns. The specific solubility data, partitioning models and applications are presented in Kipfer et al. (2002, this volume) for ground and surface waters and Ballentine et al (2002, this volume) for oil, gas, hydrothermal systems and deep groundwater.

Silicate melt solubilities

Noble gases can be lost from silicate melts according to melt solubilities. The clearest example of this is the degassing of mid-ocean ridge basalts that are exsolving CO_2 and so are progressively losing noble gases by partitioning into bubbles. The results are basaltic melts that typically have extensively fractionated noble gas elemental abundances, and the volatile composition of these melts are then preserved by quenching to glass during submarine eruptions (see Graham 2002, this volume). The solubility of noble gases in different melt compositions has been explored experimentally. However, while this information may provide some details about melt structure, it has not found widespread application in approaching geological problems largely because it is rare to find other magmatic compositions that have retained trapped noble gases that have not been lost by later processes of crystallization and weathering.

Crystal-melt partitioning

As expected for such unreactive species, the noble gases are generally considered to be highly incompatible during partial melting, greatly preferring less selective melt structures. Unfortunately, it has been notoriously difficult to experimentally determine partition coefficients. The circumstances in which this has been most extensively discussed is for production of mid-ocean ridge basalts, which clearly contain noble gases inherited from the upper mantle source region, which is dominated by olivine and pyroxenes. However, a generally recognized, consistent data set for the noble gases remains to be established. Other magma compositions have not been found to have appreciable quantities of noble gases incorporated from the melt source region, and so the partitioning behavior of noble gases in these systems have not been evaluated as systematically.

Iron-silicate partitioning

Early planetary differentiation included separation of Fe-Ni from mantle silicates to form the core. While a considerable fraction of some elements partitioned into core-forming material at this time, it is not known to what extent significant amounts of noble gases were transferred to the core in this way. The basic data for noble gas partitioning between silicates and metal at high temperatures and pressures is still scarce (see Porcelli and Ballentine 2002, this volume).

Adsorption

Noble gases may interact with surfaces by sorption processes, in defiance of their generally inert character. This is especially true for the heavier elements, and is a common source of experimental problems. There are some data available, which provides a sense of the magnitude of the quantities that can be sorbed (see Ozima and Podosek 2001 for compilation). However, this information has generally not been used in a rigorous fashion to evaluate natural concentrations, since it appears that incorporation of noble gases in geological materials is more variable than accountable by a single mechanism, and likely reflects a range of surface conditions and incorporation mechanisms that to date have defied simple quantification and often have not been of great scientific interest. Significant advances in this field would greatly contribute to our understanding of noble gas incorporation in pre-accretionary material, the mechanisms of atmospheric contamination of magmatic and other silicate samples, and the origin of excess atmospheric gases released from sedimentary rocks.

Diffusion

Noble gases can be transported and fractionated by diffusion, and the lighter noble gases are expected to be the more nimble. Diffusion coefficients for various media have been determined (see compilations by Ozima and Pososek 2001; Ballentine and Burnard 2002; Ballentine et al. 2002, both in this volume). This is most easily considered through relatively homogeneous materials, for example through single grains. Therefore, diffusion has been most extensively discussed for retention of noble gases within individual minerals, especially in the context of obtaining age information. Since many silicate phases may be open to noble gas loss by diffusion at elevated temperatures but not at lower temperatures, information on the thermal history of materials can be obtained (see Farley 2002). On a larger scale, different diffusive paths are available, such as diffusion along grain boundaries that might affect large-scale transport. However, obtaining coefficients for such pathways by experiment or field conditions is problematic, and so such a process is sometimes raised but cannot be quantitatively evaluated. Also, diffusion clearly provides the potential for generating isotopic variations as well, although this has not been widely documented. Diffusion as a process for transporting noble gases within the crust is discussed in detail by Ballentine and Burnard (2002, this volume) and its potential for isotopic fractionation by Ballentine et al. (2002, this volume).

Atmosphere losses

Losses of noble gases from the atmosphere to space can also modify noble gas compositions. Under present conditions, only He is sufficiently light to escape, and so the relative abundance of He is greatly depleted compared to the others. This is a problem of atmospheric physics, and noble gas geochemists appreciate the benefits of a He-depleted experimental background. However, during the early, more energetic conditions of the Earth, conditions may have favoured losses of all the noble gases by processes such as hydrodynamic entrainment during loss of major species, and so it is in the context of global budgets and planetary comparisons that such loss processes are important (see Pepin and Porcelli 2002, this volume).

NOBLE GASES IN COSMOCHEMICAL AND GEOCHEMICAL STUDIES

The reviews in this volume have been divided into the broad fields of cosmochemistry, mantle geochemistry, noble gases in the crust, and geochronology. Many of the fields of research covered by the noble gas reviews in this volume include acronyms that are repeatedly used and may be unfamiliar to some readers. A list of those commonly used here is presented in Appendix 1 (Table 8). Conversion factors for some commonly used units in noble gases are also provided there (Table 9).

Noble gases in cosmochemistry

Reservoirs. In contrast to their scarcity in much of the present solar system, the noble gases were quite abundant in the solar nebula, which was comprised of ~10% He. The largest existing reservoir of noble gases in the solar system is the Sun, which has presumably acquired its composition from the solar nebula without much modification. The abundances of many elements in the Sun have been obtained spectrometrically, but noble gases are an exception. Although He was first identified by Janssen and Lockyer in a spectrum taken during a solar eclipse in 1868, no noble gas lines that are useful for abundance determinations are present in photospheric spectra. While solar abundance estimates could be made from the pattern of abundances of other elements, solar elemental and isotopic compositions are obtained largely from measurements of atoms streaming out of the Sun. This solar wind is analyzed either directly in space or from targets such as lunar regolith and meteorites. Although some changes have occurred due to solar processes, the greatest difficulty in obtaining the composition of the Sun, and so the solar nebula, is correcting for fractionation effects in the solar wind. Nonetheless, the solar composition remains the best baseline for studies of the evolution of the solar system. However, the atmospheres of the giant planets, which presumably have also been acquired from the solar nebula and are accessible to more direct measurement through planetary probes such as Galileo (Mahaffy et al. 2000), are emerging as future baselines of the solar nebula. There are also considerable data for the Moon, which serves as an archive of solar history through the record of implanted solar wind. In comparison with the solar compositions, the terrestrial planets are highly depleted in rare gases, as first noted for the Earth by Aston (1924). There are also data from planetary probes for the atmospheres of Venus and Mars (Istomin et al. 1983; Nier and McElroy 1977), terrestrial planets that are highly depleted in noble gases, and some information for the exospheres of Mercury and the Moon. In this volume, Wieler (2002a) presents a systematic evaluation of noble gas data for the Sun and the planets, but also for more exotic subjects such as comets, interplanetary dust particles, and the interplanetary medium. Mars presents a special case, since there are not only data for the atmosphere, but also considerable information from Martian meteorites, and so has been reserved for separate treatment (see below). The masses of volatile reservoirs in the terrestrial planets are listed in Appendix 1 (Table 10).

Trapped noble gases. Gerling and Levskii (1956) first discovered trapped noble gases (neither cosmogenic nor radiogenic) in meteorites, and two different noble gas abundance patterns, the solar pattern, resembling 'cosmic' abundances, and the 'planetary' pattern resembling the terrestrial atmosphere, were recognized (Signer and Suess 1963). While the solar pattern is due to implantation of solar wind ions into asteroidal regoliths, it was later recognized that the 'planetary' pattern is a misnomer, as 'planetary' noble gases in meteorites represent various primordial components, many of them having been brought into the solar nebula by surviving presolar grains. The first clear detection of such components was by Black and Pepin (1969) of Ne highly enriched in ^{22}Ne (the product of short-lived ^{22}Na), and dubbed Ne-E (Black 1972), although earlier data by Reynolds and Turner (1964) contained evidence for Xe-HL enriched in both the heaviest and lightest isotopes relative to normal meteoritic Xe, and so was the first reported nucleosynthetic isotope anomaly of any element. Junck and Eberhardt (1979) first isolated the presolar grain carrier phases of Ne-E, which appear to be SiC (Zinner et al. 1987). The Xe component was found to be in nanometer-sized presolar diamonds (Lewis et al. 1987). Subsequent investigations have found many noble gas variations due to contributions from different environments that have not been completely homogenized in the solar nebula. These materials provide information on the stellar sources that contributed to the solar system and on subsequent processes within the nebula or on meteorite parent bodies. Today, the "noble gas alphabet" of cosmochemical component labels contains at least as many letters as the Latin alphabet, which is somewhat unfortunate even for the insider. The present knowledge of trapped noble gases in meteorites is reviewed by Ott (2002, this volume), who also promotes a nomenclature system that should help to reduce the number of letters.

Chronology. Brown (1947) suggested that isotopic variations in meteorites due to extinct radioactive isotopes present only early in Earth history might be detected in meteorites. This was followed by the landmark paper by Reynolds (1960) that not only found evidence for extinct ^{129}I and so provided the basis for I-Xe meteorite dating, but also used this to calculate the time interval since nucleosynthesis. I-Xe dating became an established method with the development of irradiation techniques that linked I and ^{129}Xe through step-heating (Jeffrey and Reynolds 1961), and the evidence that ^{129}I was uniformly mixed with ^{127}I in the early solar system (Hohenberg et al. 1967b). ^{244}Pu had also been suggested as a source of meteoritic Xe in meteorites (Kuroda 1960) and was indeed found to be present in many meteorites (Pepin 1964; Hohenberg et al. 1967a). These dating methods complemented those involving long-lived parents such as ^{40}K. The contributions that noble gases have made to solar system chronology are summarized by Swindle (2002a, this volume), who also looks at the chronological work done in lunar samples using radiogenic gases.

Cosmogenic nuclides. It was suggested by Bauer (1947) that unreasonably high U-^{4}He ages of iron meteorites could be due to cosmic ray production. This was confirmed with the measurement of very high ^{3}He/^{4}He in iron meteorites (Paneth et al. 1952). Since then, there has been extensive work on using cosmic-ray-produced nuclides in meteorites. As a nuclear product due to exposure, this is distinctive in not defining the formation of a material or closure to daughter nuclide loss, but rather the amount of sample exposure in space either before incorporation into a parent body, near a parent body surface, or during flight to the Earth. Cosmogenic nuclides provide valuable information on the orbital history of meteorites, and help to recognize common falls to Earth of different meteorite specimens or common ejection events from parent bodies. The field is reviewed by Wieler (2002b, this volume).

Mars. Mars presents a distinctive case for planetary studies, since meteorites sampling the planet are available. While it had been initially argued on circumstantial grounds that the SNC meteorites were of Martian origin (Wasson and Wetherill 1979), the first direct evidence to their identification was the discovery by Bogard and Johnson (1983) that noble gases in meteorite EETA 79001 were distinctive from other meteoritic compositions but strikingly similar to those of the Martian atmosphere, and so were likely to have been implanted by shock into the sample while on the planet. The SNC meteorites have since become the source of a wealth of information about Mars, including further high precision data of volatiles in the Martian atmosphere as well as the planetary interior. The noble gas evidence that has accumulated is described by Swindle (2002b, this volume).

Origins of noble gases on the terrestrial planets. Brown (1949) and Suess (1949) compared the terrestrial atmosphere noble gases with solar abundances, and suggested that the Earth suffered depletion by hydrodynamic outflow to space and fractionating processes. Reynolds (1960) further suggested that the difference between the light meteoritic Xe isotopes and those of the Earth's atmosphere may be due to strong mass fractionation of meteoritic Xe in a gravitational field, and that the heavier isotopes then required the addition of a component possibly due to fission. The subsequent study of Xe and other noble gases in the terrestrial atmosphere have been refinements to these statements. In addition, noble gas isotopes are the only evidence for the presence of short-lived nuclides in the Earth, and so also provide strong constraints on early Earth history. Excess ^{129}Xe in the atmosphere relative to the solar composition was clearly shown by Marti (1967), and used to infer a late 'formation age' for the Earth (Wetherill 1975). Although the identification of fissiogenic Xe from ^{244}Pu is more difficult, it appears to be present in the atmosphere as well (Pepin and Phinney 1978). The goal of terrestrial volatile evolution models is to explain the levels of depletion and isotopic characteristics of noble gases relative to the Sun on the terrestrial planets. While the atmospheres are most accessible, an understanding of the origin of noble gases must also consider noble gases captured and retained within the solid planet, and this can be done at least for the case of the Earth. An evaluation of these processes, including an accessible review of the principles underlying the fractionating hydrodynamic escape mechanisms that were likely to have operated early in Earth history, is provided by Pepin and Porcelli (2002, this volume).

Noble gases in the mantle

The convecting mantle. Early studies of noble gases in the mantle and the rate of degassing considered the rate at which ^{40}Ar, which is produced within the solid Earth, was released to the atmosphere over the history of the Earth (Turekian 1959). Similar calculations were also applied to ^{4}He. The calculated flux of ^{4}He into the oceans was expected to produce He concentrations above those of dissolved atmosphere, and such saturation anomalies were observed in deep Pacific Ocean waters (Suess and Wänke 1965). The discovery that such ^{4}He excesses correlated with ^{3}He excesses indicated that primordial (i.e., nonradiogenic) ^{3}He was also degassing (Clarke et al. 1969). Mantle He with high $^{3}He/^{4}He$ ratios was also found in volcanic gases (Mamyrin et al. 1969) and in mid-ocean ridge basalts (Krylov et al. 1974; Lupton and Craig 1975; Kurz et al. 1982), pointing to an incontrovertible link between magmatic activity and continued degassing of primordial volatiles from the solid Earth. Subsequent discoveries that the distribution of He isotopes in the mantle was not uniform and appeared to contain relatively gas-rich regions set the basis for mantle noble gas studies (Polyak et al. 1976; Craig and Lupton 1976; Kaneoka and Takaoka 1978). Noble gases, and especially He, thus joined the host of other trace element isotopic tracers that have been used to examine the creation and maintenance of mantle reservoirs, but are unique in tracing atmospheric degassing as well. Another major noble gas advance was the identification of nonatmospheric Ne with solar composition in ocean island basalts (Craig and Lupton 1976; Honda et al. 1991) and mid-ocean ridge basalts (Poreda and Radicati di Brozolo 1984; Sarda et al. 1988). Atmospheric Ne requires a separate early history, and is due not simply due to degassing of the mantle. Another significant aspect of noble gases in the Earth was the discovery of high $^{129}Xe/^{130}Xe$ ratios that reflected inputs from extinct ^{129}I first in well gases (Butler et al. 1963), then in mantle-derived materials (Kaneoka and Takeoka 1978; Staudacher and Allègre 1982). In this volume, Graham (2002) reviews the evidence for mantle variations and the relationships with other tracers.

Subduction zones. While oceanic basalts sample noble gases primarily of mantle origin, lavas and other magmatic phenomena associated with subduction zones are also characterised by noble gases principally derived from the mantle. At the same time as the initial discovery of primordial ^{3}He emanating from submarine oceanic ridges (Clarke et al., 1969), subaerial hydrothermal activity from the Kurile island arc was found to possess high $^{3}He/^{4}He$ ratios (Mamyrin et al. 1969; Baskov et al. 1973). Subsequent work has shown that mantle noble gases dominate in this tectonic environment (in the arc and back-arc regions) and are useful for recognising and quantifying recycling from the crust, atmosphere, and hydrosphere to the mantle. In this volume, Hilton et al. (2002) review the noble gas data and provide a comprehensive evaluation of volatile fluxes into and out of subduction zones—both on an arc-by-arc basis as well as on a global basis.

The lithosphere. The subcontinental lithosphere forms a reservoir separate from the underlying convecting mantle, and so has the potential for maintaining distinctive geochemical characteristics. It is sampled by xenoliths that are torn from the source region by rising magmas. High $^3He/^4He$ ratios were found in mantle-derived xenoliths (Tolstikhin et al. 1974) essentially concurrently with the initital investigations of mid-ocean ridges. Since then, substantial work has been done on mapping the isotopic signatures of the lithosphere at many locations, and examining the evolution and redistribution of volatiles. In this volume, Dunai and Porcelli (2002) evaluate the evidence contained in these samples for the conditions in the lithosphere.

Mantle models. Models for the formation of the atmosphere are clearly tied to the structure of the mantle. Brown (1949) and Rubey (1951) argued that the atmosphere is secondary, resulting from the degassing of the solid Earth. Turekian (1959) calculated the rate of degassing using the atmospheric abundance of ^{40}Ar. This was followed by a series of increasingly refined models based upon the noble gas characteristics in the mantle, starting with Ozima and Kudo (1972). Also of particular interest have been mantle variations in Xe isotopes, which provide constraints on early degassing. It is interesting that Xe is the only element that preserves variations due to short-lived nuclides present early in Earth history (of ^{129}I and ^{244}Pu). The delineation of multiple mantle reservoirs has added to the complexity of these models, and thus discussions of formation of the atmosphere, the distribution of noble gases within the Earth, and the evolution of the mantle are now irrevokably coupled. Porcelli and Ballentine (2002, this volume) evaluate constraints on the global parameters and the global transport processes involved, and then provide an up-to-date critical evaluation of the wide range of mantle models that are available.

Noble gases in surface reservoirs

The crust as a noble gas reservoir. Twenty-seven years after the discovery of the He spectra in the Sun, William Ramsay found the same spectral lines in gas released from cleveite treated with sulphuric acid (1895). It took a further thirteen years before He was shown to be a component in a nitrogen rich natural gas produced from Dexter county, Kansas, USA (Cady and McFarland 1908). By 1917 the US Bureau of Mines had set up the He Survey and Resource Evaluation Center in Amarillo, Texas, prompting other countries to produce works such as the 'Report on some sources of helium in the British Empire' (McLennan 1920). The US Bureau of Mines was nevertheless responsible for establishing the Cliffside strategic helium reserve near Amarillo of $1.1x10^9$ m^3 STP He extracted from local He-rich natural gases. Although the atmosphere is depleted in helium compared to the other noble gases, radiogenic He (largely ^4He) found within natural gases in the crust provides a resource capable of cheaply supplying the current commercial demand for this gas. The identification of other noble gas nuclides produced in the crust came with the mushrooming of the nuclear industry, both military and civil, that resulted in a huge database quantifying nuclear reactions and particle interaction probabilities. The now classic works on nuclear production of Ne and Ar isotopes (Wetherill 1954) and of ^3He (Morrison and Pine 1955) by natural nuclear reactions in the crust were directly derived from this new database. In the search to define and understand crustal noble gas compositions, some isotopic compositions were found that could not be accounted for by extant nuclear processes. For example, the discovery of excess ^{129}Xe in the Harding County (New Mexico, USA) CO_2-rich natural gases provided the first evidence for the terrestrial effects of an extinct (^{129}I) isotope (Butler et al. 1963). Radiogenic, nucleogenic and fissiogenic production of noble gases and their release from the crust is reviewed in this volume by Ballentine and Burnard (2002).

Deep crustal fluids. Recognition that the ^4He in natural gases is derived from U and

Th decay and the sudden interest in the late 1940s in uranium, resulted in several early studies that attempted to correlate He-rich natural gases with U mineralization. It was soon realised that He mobility and focusing mechanisms in the crust meant that helium as a uranium prospecting tool was fundamentally flawed. Perhaps the most important development at this stage was the realisation that equilibration between the groundwater and natural gas phases would result in almost quantitative transfer of noble gases dissolved in the groundwater into the gas phase (Goryunov and Kozlov 1940) and that groundwater movement could play a significant role in He transport. It was not for a further 20 years before the full use of noble gas partitioning between different phases to quantify gas/water volume ratios was explored in the western literature (Zartman et al. 1961). Although the concepts were well established by then, the noble gas solubility data in water under different temperature and salinities was not available, nor was there a comprehensive understanding of the different sources and transport mechanisms of groundwater, crustal and mantle-derived noble gases in the crustal fluid system. Indeed, it was not until Bosch and Mazor (1988) re-investigated the control of phase solubility on noble gas abundance patterns that the detail noble gases could provide about subsurface fluids again received attention. The recent applications of noble gases to study multi-phase crustal fluid systems, reviewed by Ballentine et al. (2002, this volume) has only become possible because of the huge advances in understanding noble gas physical and nuclear chemistry during the intervening time.

Lakes and groundwaters. The decay of tritium ($t_{1/2}$ = 12.43 yr) in the environment (Grosse et al. 1951; Kaufman and Libby 1954) to ^{3}He has provided an invaluable tool for the high sensitivity determination of tritium concentrations, and therefore a dating tool, in water. The contribution that noble gases could make to climate studies was recognized by Mazor (1972), who noted that the noble gas abundance patterns in groundwaters reflected the temperature of recharge; progressively older waters down gradient of an aquifer preserves a record of past climatic temperature variations. In this way, noble gases have allowed past climate in different regions of the Earth to be studied. More recent developments in the numerical handling of data and the combination of noble gas studies with other tracers has shed light on recharge conditions and groundwater source. As water resources have become more scarce and climate change more politically topical, less abundant or more short-lived tracers of groundwater age, such as ^{39}Ar, ^{85}Kr and ^{81}Kr, have become the focus of more research. In addition to the huge amount of work on groundwater systems, surface waters and lakes have also been the subject of intense research, utilising, for example, a combination of tritium dating techniques and the inert character of the crustal noble gases entering into lakes to determine turnover rates and the extent of communication between different water bodies at depth. These applications and the advances in numerical handling of data are reviewed in this volume by Kipfer et al. (2002).

Oceans and deep sea sediments. When extraterrestrial helium was discovered in deep sea sediments by Merrihue (1964), the interest this would generate in disciplines as far apart as understanding the origin of primitive noble gases in the mantle and quantifying deep sea sedimentation rates through to understanding the mechanisms of mass extinction, could not be foreseen. The patterns of ocean circulation and the distribution of chemical constituents in seawater have been studied using a wide variety of trace elements. This is possible whenever source inputs of the tracer can be defined and the ocean signature has not been homogenized due to long tracer residence times. In the early 1970s, noble gas solubilities in water and seawater were accurately determined by Weiss (1971), motivated by the possibility of using tritium tracing techniques, with tritium atmospheric noble gases, and the newly discovered mantle ^{3}He injected into the deep ocean (e.g., Clarke et al. 1969), to trace ocean circulation patterns and identify the exchange mechanisms between the atmosphere and oceans. The use of these techniques and their applications are described by Schlosser and Winkler (2002, this volume).

Noble gases and geochronology

Ar methods. Dating methods using noble gases reach across the entire history of isotope geochemistry. Rutherford (1906) first suggested obtaining ages from radioactive minerals using U-He and U-Pb methods. The first geochemical dates were obtained using the amount of ^{4}He that had accumulated in pitchblende (Strutt 1905). The next noble gas decay scheme was born when von Weizäcker (1937) famously deduced, partly by noting the atmospheric overabundance of ^{40}Ar relative that expected from the overall pattern of elemental abundances, that ^{40}Ar was produced by ^{40}K. Aldrich and Nier (1948) confirmed this with measurements of ^{40}Ar in potassium-bearing minerals, and this developed into a geological dating method. The K-Ar method was soon complemented by the Ar-Ar method that elegantly relates K in different sites within a sample directly with the ^{40}Ar produced by that K (Merrihue 1965; Merrihue and Turner 1966). There are several excellent textbooks and reviews available that cover the methodologies and widespread applications of geochronological methods, including Faure (1986), Dickin (1995), and McDougall and Harrison (2002). Kelley (2002, this volume) discusses the considerable progress that has been made recently in K-Ar and Ar-Ar dating with high sensitivity Ar measurement techniques and deserves a separate, focused review in this volume.

Thermal history dating. Considerable work has been done recently using the U-He method. Although this was the first dating method explored, the possibility of only partial He retention was recognized from the start (Rutherford 1906), and was soon documented (Strutt 1908). However, recent work has focused on minerals that do quantitatively retain He below certain temperatures. The accumulation of He, and therefore the U-He age, therefore does not correspond to the formation of the host mineral, but rather the timescale over which such low temperatures were attained. Thus an uplift history of a mineral that was previously resident at greater depths (and so higher temperatures) can be obtained. Such techniques have been applied to metamorphic studies in the past with Rb-Sr and K-Ar systems, but the newest advances in the U-He method extend to lower temperatures. Progress in this field, and its growing importance in tectonic studies, are described by Farley (2002, this volume).

Exposure age dating. Another application that has gained widespread use recently is exposure dating, which involves the generation of nuclides during exposure at the Earth's surface to irradiation from space. This follows the exposure age dating techniques developed earlier for meteorites, which are exposed to greater levels of radiation without the shielding of the Earth's atmosphere and magnetic field, but with more complex issues of exposure depth due to breakup of the parent body. Although the first determination of a cosmogenic nuclide (^{36}Cl) in a terrestrial sample was done soon after exposure dating had been established for meteorites (Davis and Schaeffer 1955), and the production of cosmogenic nuclides in the atmosphere (e.g., ^{14}C, ^{10}Be) has been has been used for some time in a range of geological applications, the exposure age dating of geological materials has developed more recently. Spallogenic Xe was reported in sedimentary barites (Srinivasan 1976), but has not been widely observed or utilized. The production of ^{3}He was first demonstrated at high elevations on Maui (Craig and Poreda 1986; Kurz 1986). The production of cosmogenic ^{21}Ne was also confirmed there (Marti and Craig 1987). The use of exposure dating on the Earth has expanded with progress over issues such as different rates of production in different target materials, at different locations and elevations, and at different times, and is now an important component in erosion and tectonic studies. The field is reviewed by Niedermann (2002, this volume).

END NOTE

The number of noble gas isotope geochemistry laboratories that are active throughout the world is testament to the continuing rewards achieved across a wide range of geochemical and cosmochemical fields. It is hoped that the papers included in this volume, in reviewing the main lines of noble gas research, indicate where future fruitful work lies, and provide a valuable resource not only for noble gas investigators, but also for those in related fields that would like both to appreciate further the contributions that noble gases have made and to understand where noble gases can be applied further.

ACKNOWLEDGMENTS

We thank all the authors who have contributed to this volume and are responsible for successfully creating a valuable reference over the full breadth of noble gas research. The work of a long list of external reviewers has greatly contributed to the high quality of the papers. The essential editorial work of Series Editors Jodi Rosso and Paul Ribbe is greatly appreciated. David Hilton and Bob Pepin kindly provided reviews of this introduction.

REFERENCES

Aldrich LT, Nier AO (1948) Argon 40 in potassium minerals. Phys Rev 74:876-877
Aston FW (1919) A positive ray spectrograph. Phil Mag ser VI, 38:707-714
Aston FW (1924) The rarity of the inert gases on the Earth. Nature 114:786
Ballentine CJ, Burnard P (2002) The crust as a noble gas reservoir. Rev Mineral Geochem 47:481-538
Ballentine CJ, Burgess R, Marty B (2002) Tracing fluid origin, transport and interaction in the crust. Rev Mineral Geochem 47:539-614
Basford JR, Dragon JC, Pepin RO, Coscio Jr MR, Murthy VR (1973) Krypton and xeon in lunar fines. Proc 4th Lunar Sci Conf 2:1915-1955
Baskov Y, Vetshteyn V, Surikov S, Tolstikhin I, Malyuk G, Mishina T (1973) Isotopic composition of H, O, C, Ar, and He in hot springs and gases in the Kurile-Kamchatka volcanic region as indicators of formation conditions. Geochem Intl 10:130-138
Bauer CA (1947) Production of helium in meteorites by cosmic radiation. Phys Rev 72:354-355
Baur H (1980) Numerical simulation and practical testing of an ion source with rotational symmetry for gas mass spectrometers. PhD dissertation, ETH-Zürich Nr 6596, 94 p (in German)
Baur H (1999) A noble-gas mass spectrometer compressor source with two orders of magnitude improvement in sensitivity. EOS, Trans Am Geophys Union 46:F1118
Black DC, Pepin RO (1972) On the origins of trapped helium, neon, and argon isotopic variations in meteorites-II. Carbonaceous meteorites. Geochim Cosmochim Acta 36:377-394
Black DC, Pepin RO (1969) Trapped neon in meteorites-II. Earth Planet Sci Lett 6:395-405
Blum JD (1995) Isotopic decay data. *In* Global Earth physics: a handbook of physical constants. Ahrens TJ (ed) American Geophysical Union, Washington DC, p 271-282
Bogard DD, Johnson P (1983) Martian gases in an Antarctic meteorite? Science 221: 651-654
Bosch A, Mazor E (1988) Natural gas association with water and oil depicted by atmospheric noble gases: case studies from the southern Mediterranean Coastal Plain. Earth Planet Sci Lett 87:338-346
Brown H (1947) An experimental method for the estimation of the age of the elements. Phys Rev 72:348
Brown H (1949) Rare gases and the formation of the Earth's atmosphere. *In* Atmospheres of the Earth and the planets. Kuiper GP (ed) University of Chicago Press, Chicago, p 258-266
Butler WA, Jeffery PM, Reynolds JH, Wasserburg GJ (1963) Isotopic variations in terrestrial xenon. J Geophys Res 68:3283-3291
Cady HP, McFarland DF (1908) Chemical composition of Kansas gas. Kansas Geol Surv 9: 228-284
Clarke WB, Beg MA, Craig H (1969) Excess ^{3}He in the sea: evidence for terrestrial primordial helium. Earth Planet Sci Lett 6:213-220
Craig H, Poreda RJ (1986) Cosmogenic ^{3}He in terrestrial rocks: the summit of lavas of Maui. Proc Natl Acad Sci USA 83:1970-1974
Craig H, Lupton JE (1976) Primordial neon, helium, and hydrogen in oceanic basalts. Earth Planet Sci Lett 31:369-385
Davis R, Schaeffer OA (1955) Chlorine-36 in nature. Ann New York Acad Sci 62:105-122
Dempster AJ (1918) A new method of positive ray analysis. Phys Rev 11: 316-325
Dickin AP (1995) Radiogenic Isotope Geology. Cambridge University Press, Cambridge

Dunai TJ, Porcelli D (2002) The storage and transport of noble gases in the subcontinental lithosphere. Rev Mineral Geochem 47:371-409

Eberhardt P, Eugster O, Marti K (1965) A redetermination of the isotopic composition of atmospheric neon. Z Naturforsch 20a:623-624

Eikenberg J, Signer P, Wieler R (1993) U-Xe, U-Kr, and U-Pb systematics for dating uranium minerals and investigations of the production of nucleogenic neon and argon. Geochim Cosmochim Acta 57: 1053-1069

Farley KA (2002) (U-Th)/He dating: techniques, calibrations, and applications. Rev Mineral Geochem 47:

Faure G (1986) Principles of Isotope Geology, 2nd edition. John Wiley, New York

Gerling EK, Levsky LK (1956) On the origin of inert gases in stone meteorites. Geokhimiya 7:59-64 (in Russian)

Gerling EK, Mamyrin BA, Tolstikhin IN, Takovleva SS (1971) Isotope composition of helium in rocks. Geokhimiya 5:608-617 (in Russian)

Gilmour JD, Lyon IC, Johnston WA, Turner G (1994) RELAX—An ultrasensitive, resonance ionization mass spectrometer for xenon. Rev Sci Instr 65:617-625

Goryunov MS, Kozlov AL (1940) A study of the geochemistry of helium-bearing gases and the conditions for the accumulation of helium in the earth's crust. State Sci-Tech Publ Co Oil and Solid Fuel Lit, Leningrad-Moscow (in Russian)

Graham DW (2002) Noble gas isotope geochemistry of mid-ocean ridge and ocean island basalts: characterization of mantle source reservoirs. Rev Mineral Geochem 47:247-318

Grosse AV, Johnston WM, Wolfgang RL, Libby WF (1951) Tritium in nature. Science 113:1-2

Hilton DR, Fischer TP, Marty B (2002) Noble gases and volatile recycling at subduction zones. Rev Mineral Geochem 47:319-370

Hohenberg CM (1980) High sensitivity pulse counting mass spectrometer for noble gas analysis. Rev Sci Instrum 51:1075-1082

Hohenberg CM, Munk MN, Reynolds JH (1967a) Spallation and fissiogenic xenon and krypton from stepwise heating of the Pasamonte achondrite; the case for extinct plutonium 244 in meteorites; relative ages of chondrites and achondrites. J Geophys Res 72: 3139-3177

Hohenberg CM, Podosek FA, Reynolds JH (1967b) Xenon-iodine dating: sharp isochronism in chondrites. Science 156:233-236

Honda M, McDougall I, Patterson DB, Doulgeris A, Clague DA (1991) Possible solar noble gas component in Hawaiian basalts. Nature 349:149-151

Hudson GB, Kennedy BM, Podosek FA, Hohenberg CM (1989) The early solar system abundance of ^{244}Pu as inferred from the St. Severin chondrite. Proc 19th Lunar Planet Sci Conf, p 547-557

Istomin VG, Gechnev KV, Kochnev VA (1983) Venera 13, Venera 14: mass spectrometry of the atmosphere. Kossmech Issled 21:410-420

Jeffery PM, Reynolds JH (1961) Origin of excess ^{129}Xe in stone meteorites. J Geophys Res 66:3582-3

Jungck MHA, Eberhardt P (1979) Neon-E in Orgueil density separates. Meteoritics 14:439-440

Kallenbach R, Ipavich FM, Bochsler P, Hefti S, Hovestadt D, Grünwaldt H, Hilchenbach M, Axford WI, Balsiger H, Bürgi A, Coplan MA, Galvin AB, Geiss J, Gliem F, Gloeckler G, Hsieh KC, Klecker B, Lee MA, Livi S, Managadze GG, Marsch E, Möbius E, Neugebauer M, Reiche K-U, Scholer M, Verigin MI, Wilken B, Wurz P (1997) Isotopic composition of solar wind neon measured by CELIAS/MTOF on board SOHO. J Geophys Res 102:26895-26904

Kaneoka I, Takaoka N (1978) Excess ^{129}Xe and high ^{3}He/^{4}He ratios in olivine phenocrysts of Kauho lava and xenolithic dunites from Hawaii. Earth Planet Sci Lett 39:382-386

Kaufman S, Libby WF (1954) The natural distribution of tritium. Phys Rev 93:1337-1344

Keeling CD, Whorf TP (2000) Atmospheric CO_2 records from sites in the SIO air sampling network. *In* Trends: a compendium of data on global change. Carbon Dioxide Information Analysis Center, Oak Ridge National Laboratory, Oak Ridge, TN

Kelley SP (2002) K-Ar and Ar-Ar dating. Rev Mineral Geochem 47:785-818

Kipfer R, Aeschbach-Hertig W, Peeters F, Stute M (2002) Noble gases in lakes and ground waters. Rev Mineral Geochem 47:615-700

Krylov AY, Mamyrin BA, Khabarin LA, Mazina TI, Silin YI (1974) Helium isotopes in ocean floor bedrock. Geochim Intl 11:839-843

Kuroda PK (1960) Nuclear fission in the early history of the Earth. Nature 187:36-38

Kurz MD (1986) Cosmogenic helium in a terrestrial igneous rock. Nature 320:435-439

Kurz MD, Jenkins WJ, Hart SR (1982) Helium isotopic systematics of oceanic islands and mantle heterogeneity. Nature 297:43-47

Leske RA, Mewaldt RA, Cohen CMS, Cummings AC, Stone EC, Wiedenbeck ME, Christian ER, von Rosenvinge TT (1999) Event-to-event variations in the isotopic composition of neon in solar energetic particle events. Geophys Res Lett 26:2693-2696

Lewis JS, Prinn RG (1984) Planets and their Atmospheres: Origin and Evolution. Academic Press, Orlando, Florida
Lewis RS, Srinivasan B, Anders E (1975) Host phase of a strange xenon component in Allende. Science 190:1251-1262
Lewis RS, Tang M, Wacker JF, Anders E, Steel E (1987) Interstellar diamonds in meteorites. Nature 326:160-162
Lide DR (1995) CRC Handbook of Chemistry and Physics. CRC Press, Boca Raton, Florida
Lupton JE, Craig H (1975) Excess ^{3}He in oceanic basalts: evidence for terrestrial primordial helium. Earth Planet Sci Lett 26:133-139
Mahaffy PR, Niemann HB, Alpert A, Atreya SK, Demick J, Donahue TM, Harpold DN, Owen TC (2000) Noble gas abundance and isotope ratios in the atmosphere of Jupiter from the Galileo Probe Mass Spectrometer. J Geophys Res E105:15061-15071
Mamyrin BA, Tolstikhin IN (1984) Helium Isotopes in Nature. Elsevier, Amsterdam.
Mamyrin BA, Tolstikhin IN, Anufriev GS, Kamensky IL (1969) Anomalous isotopic composition of helium in volcanic gases. Dokl Akad Nauk SSSR 184: 1197-1199 (in Russian)
Mamyrin BA, Anufriyev GS, Kamenskii IL, Tolstikhin IN (1970) Determination of the isotopic composition of atmospheric helium. Geochem Intl 7:498-505
Marti K (1967) Isotopic composition of trapped krypton and xenon in chondrites. Earth Planet Sci Lett 3:243-248
Marti K, Craig H (1987) Cosmic-ray-produced neon and helium in the summit lavas of Maui. Nature 325:335-337
Mazor E (1972) Paleotemperatures and other hydrological parameters deduced from noble gases dissolved in groundwaters; Jordan Rift Valley, Israel. Geochim Cosmochim Acta 36:1321-1336
McLennan JC (1920) Report on some sources of helium in the British Empire. Canada, Dept Mines, Mines Branch, Bull 31:1-72
McDougall I, Harrison MT (2002) Geochronology and Thermochronology by the ^{40}Ar/^{39}Ar Method. Oxford University Press, Oxford
Merrihue C (1964) Rare gas evidence for cosmic dust in modern Pacific red clay. Ann New York Acad Sci 119:351-367
Merrihue CM (1965) Trace element determinations and potassium-argon dating by mass spectroscopy of neutron-irradiated samples. Trans Am Geophys Union 46:125
Merrihue CM, Turner G (1966) Potassium-argon dating by activation with fast neutrons. J Geophys Res 71:2852-2857
Morrison P, Pine J (1955) Radiogenic origin of the helium isotopes in rocks. Ann New York Acad Sci 62:69-92
Niedermann S (2002) Cosmic-ray-produced noble gases in terrestrial rocks as a dating tool for surface processes. Rev Mineral Geochem 47:731-784
Nier AO (1947) A mass spectrometer for isotope and gas analysis. Rev Sci Instrum 18:398-411
Nier AO (1950) A redetermination of the relative abundances of the isotopes of carbon, nitrogen, oxygen, argon and potassium. Phys Rev 77:789-793
Nier AO, Hayden JL (1971) A miniature Mattauch-Herzog mass spectrometer for the investigation of planetary atmospheres. Intl J Mass Spectrom Ion Phys 6:339-346
Nier AO, McElroy MB (1977) Composition and structure of Mars' upper atmosphere: results from the neutral mass spectrometers on Viking 1 and 2. J Geophys Res 82:4341-4349
Ott U (2002) Noble gases in meteorites- trapped components. Rev Mineral Geochem 47:71-100
Ozima M, Kudo K (1972) Excess argon in submarine basalts and an earth-atmosphere evolution model. Nature Phys Sci 239:23-24
Ozima M, Podosek FA (2001) Noble Gas Geochemistry, 2nd edition. Cambridge University Press, Cambridge
Paneth FA, Reasbeck P, Mayne KI (1952) Helium 3 content and age of meteorites. Geochim Cosmochim Acta 2:300-303
Pepin RO (1964) Isotopic analyses of xenon. *In* The origin and evolution of atmospheres and oceans. Brancazio PJ, Cameron AGW (eds) Wiley, New York, p 191-234
Pepin RO, Phinney D (1978) Components of xenon in the solar system. Unpublished manuscript
Pepin RO, Porcelli D (2002) The origin of noble gases in the terrestrial planets. Rev Mineral Geochem 47: 191-246
Pfennig G, Klewe-Nebenius H, Seelann-Eggebert W (1998) Karlsruhe chart of the nuclides, 6th edition, revised reprint. Institut für Instrumentelle Analytik, Karlsruhe
Polak BG, Kononov VI, Tolstikhin IN, Mamyrin MA, Khabarin LV (1976) The helium isotopes in thermal fluids. *In* Thermal and chemical problems of thermal waters (Proc. Grenoble Symp, Aug. 1975). Johnson AI (ed) Intl Assoc Hydrol Sci Publ 119:15-29

Porcelli D, Ballentine CJ (2002) Models for the distribution of terrestrial noble gases and the evolution of the atmosphere. Rev Mineral Geochem 47:411-480

Poreda R, Radicati di Brozolo F (1984) Neon isotope variations in Mid-Atlantic Ridge basalts. Earth Planet Sci Lett 69:277-289

Ragettli RA, Hebeda EH, Signer P, Wieler R (1994) Uranium-xenon chronology: precise determination of $\lambda_{sf}*^{136}Y_{sf}$ for spontaneous fission of ^{238}U. Earth Planet Sci Lett 128:653-670

Ramsey W (1895) Discovery of helium. Chem News 71:151

Reynolds JH (1956) High sensitivity mass spectrometer for noble gas analysis. Rev Sci Instrum 27:928-934

Reynolds JH (1960) Determination of the age of the elements. Phys Rev Lett 4:8-10

Reynolds JH, Turner G (1964) Rare gases in the chondrite Renazzo. J Geophys Res 69:3263-3275

Rubey WW (1951) Geologic history of seawater. Geol Soc Am Bull 62:1111-1148

Rutherford E (1906) Radioactive Transformations. Charles Scribner's Sons, New York

Sarda P, Staudacher T, Allègre CJ (1988) Neon isotopes in submarine basalts. Earth Planet Sci Lett 91: 73-88

Schlosser P, Winckler G (2002) Noble gases in ocean waters and sediments. Rev Mineral Geochem 47:701-730

Srinivasan B (1976) Barites: anomalous xenon from spallation and neutron-induced reactions. Earth Planet Sci Lett 31:129-141

Staudacher T, Allègre CJ (1982) Terrestrial xenology. Earth Planet Sci Lett 60:389-406

Steiger RH, Jäger E (1977) Subcommission on geochronology: convention on the use of decay constants in geo- and cosmochronology. Earth Planet Sci Lett 36:359-362

Strutt RJ (1905) On the radio-active minerals. Proc Roy Soc Lond A76:88-101

Strutt RJ (1908) On the accumulation of the helium in geological time. Proc Roy Soc Lond A81:272-277

Suess HE (1949) The abundance of the noble gases on the earth and in the cosmos. J Geol 57:600-607 (in German)

Suess HE and Wänke H (1965) On the possibility of a helium flux through the ocean floor. *In* Progress in oceanography. Vol. 3. Sears M (ed) Pergamon Press, Oxford, p 347-353

Swindle TD (2002a) Noble gases in the Moon and meteorites: Radiogenic components and early volatile chronologies. Rev Mineral Geochem 47:101-124

Swindle TD (2002b) Martian noble gases. Rev Mineral Geochem 47:171-190

Thomson JJ (1914) Rays of positive electricity. Proc Roy Soc London A89:1-20

Tolstikhin IN, Mamyrin BA, Khabarin LB, Erlikh EN (1974) Isotope composition of helium in ultrabasic xenoliths from volcanic rocks of Kamchatka. Earth Planet Sci Lett 22:73-84

Turekian KK (1959) The terrestrial economy of helium and argon. Geochim Cosmochim Acta 17:37-43

von Weizäcker CF (1937) On the possibility of a dual beta-decay of potassium. Physikalische Z 38: 623-624 (in German)

Wasson JT, Wetherill GW (1979) Dynamical, chemical, and isotopic evidence regarding the formation locations of asteroids and meteorites. *In* Asteroids. Gehrels T (ed) University of Arizona Press, Tucson, p 926-974

Weiss RF (1971) Solubility of helium and neon in water and seawater. J Chem Eng Data 16:235-241

Wetherill GW (1954) Variations in the isotopic abundances of neon and argon extracted from radioactive minerals. Phys Rev 96:679-683

Wetherill GW (1975) Radiogenic chronology of the early solar system. Ann Rev Nuclear Sci 25:283-328

Wieler R (2002a) Noble gases in the solar system. Rev Mineral Geochem 47:21-70

Wieler R (2002b) Cosmic-ray-produced noble gases in meteorites. Rev Mineral Geochem 47:125-170

Wieler R, Eikenberg J (1999) An upper limit on the spontaneous fission decay constant of ^{232}Th derived from xenon in monazites with extremely high Th/U ratios. Geophys Res Lett 26:107-110

Zartman RE, Wasserburg GJ, Reynolds JH (1961) Helium, argon and carbon in some natural gases. J Geophys Res 66:277-306

Zinner E, Tang M, Anders E (1987) Large isotopic anomalies of Si, C, N, and noble gases in interstellar silicon carbide from the Murray meteorite. Nature 330:730-732

APPENDIX I

Table 8. Commonly used acronyms and abbreviations.

AAD	Australian-Antarctic Discordance
ACR	Anomalous Cosmic Rays
AGB	Asymptotic Giant Branch star
ASW	Air Saturated Water
BABB	Back-Arc Basin Basalt
CAI	Calcium-Aluminium-rich Inclusions in meteorites
CE	Closed-system Equilibration model (NGT)
CI chondrites	Ivuna-type (chemically primitive) Carbonaceous chondrites [a]
CME	Coronal Mass Ejection
COSPEC	Correlation Spectrometer for atmospheric SO_2 concentrations
CSSE	Closed Sytem Stepwise Etching (for noble gas extraction)
EUV	Extreme UltraViolet
FIP	First Ionisation Potential
FTPAZ	Fission Track Partial Annealing Zone
GCR	Galactic Cosmic Rays
HED	Howardite-Eucrite-Diogenite meteorites (derived from Vesta?)
HePRZ	Helium Partial Retention Zone
HFS	High Field Strength elements
ICPMS	Inductively Coupled Plasma Mass Spectrometer
IDP	Interplanetary Dust Particle
LIC	Local Interstellar Cloud
LIL	Large Ion Lithophile element
L-K correlation	Loihi-Kilauea Ne isotope correlation
LREE	Light Rare Earth Elements
MAR	Mid-Atlantic Ridge
MDD	Multi Diffusion Domain (for Ar)
MORB	Mid-Ocean Ridge Basalt
MR	Multi-step partial Re-equilibration model (NGT)
NGT	Noble Gas Temperature (from groundwater)
OCZ	Outer Convective Zone of the sun
ODP	Ocean Drilling Program
OIB	Ocean Island Basalt
PR	Partial Re-equilibration model (NGT)
Q gases	Primordial meteoritic noble gas component [a]
REE	Rare Earth Elements
R/R_A	$^3He/^4He$ ratio normalized to that of the atmosphere
R_A	Atmospheric $^3He/^4He$
R_C	$^3He/^4He$ ratio corrected for air, or air-saturated water derived, He
SCR	Solar Cosmic Rays
SEIR	South East Indian Ridge
SNC	Shergottite-Nakhlite-Chassignite (Martian) meteorites
UA	Unfractionated excess Air model (NGT)
UV	UltraViolet
$^{40}Ar^*$ or $^{40*}Ar$	Radiogenic Ar; similarly for other daughter isotopes

a. See Ott (2002) for discussion of other meteorite classes and meteoritic components.

Table 9. Conversion factors.

1 cm^3 STP	2.6868 x 10^{19} atoms
1 Tritium Unit	1 ^3H per 10^{18} atoms of H $\equiv 4.012 \times 10^{-14}$ cm^3STP^{3}He if all ^{3}H decayed
1 mole	6.02217×10^{23} atoms

Table 10. Volatile reservoirs on the terrestrial planets

	Mass (g)
Earth	5.98×10^{27}
Atmosphere	5.1×10^{21}
Hydrosphere	1.5×10^{24}
Continental Crust	2.5×10^{25}
Mantle	4.0×10^{27}
Core	1.96×10^{27}
Venus	4.87×10^{27}
Atmosphere	4.75×10^{23}
Mars	6.4×10^{26}
Atmosphere	2.7×10^{19}
Sun	1.99×10^{33}

2 Noble Gases in the Solar System

Rainer Wieler

Isotope Geology, ETH Zürich
Sonneggstrasse 5
CH-8092 Zürich, Switzerland

wieler@2erdw.ethz.ch

INTRODUCTION

This chapter provides an overview of available noble gas data for solar system bodies apart from the Earth, Mars, and asteroids. Besides the Sun, the Moon, and the giant planets, we will also discuss data for the tenuous atmospheres of Mercury and the Moon, comets, interplanetary dust particles and elementary particles in the interplanetary medium and beyond. In addition, we summarize the scarce data base for the Venusian atmosphere. The extensive meteorite data from Mars and asteroidal sources are discussed in chapters in this volume by Ott (2002), Swindle (2002a,b) and Wieler (2002). Data from the Venusian and Martian atmospheres are discussed in more detail in chapters by Pepin and Porcelli (2002) and Swindle (2002b). Where appropriate, we will also present some data for other highly volatile elements such as H or N.

The solar system formed from a molecular cloud fragment—traditionally called the solar nebula—that was rather well mixed. Therefore, isotopic abundances in almost all available solar system materials are very similar to each other, and elemental abundances in primitive meteorites are also similar to the values in the Sun. The major exceptions to this rule are the noble gases. Because they are chemically inert and volatile, they are very strongly depleted in solid matter. As a consequence, numerous noble gas "components" can be recognized throughout the solar system which are not necessarily related to the composition of the bulk nebula. Still, one major question in cosmochemistry is to what extent planetary bodies contain reservoirs that reflect the noble gas composition in the nebula or the presolar cloud.

To discuss this, we first need a proxy for the nebula composition. The obvious choice is the bulk Sun, except for He, which has been produced in the Sun throughout its history. We will discuss the large database that allows us to infer noble gas abundances and isotopic composition of the present Sun as well as the protosun. This will include data from samples from the lunar dust surface (regolith) and meteorites from asteroidal surfaces which represent an archive of the ancient solar wind, the particle radiation emitted by the Sun. Giant planets are the next choice for deducing the nebula composition, because they formed either from collapsing "subnebulae" or by gravitational attraction of gas from the nebula onto their early solid cores. Indeed, the $^{3}He/^{4}He$ ratio determined by the probe descending during the Galileo mission into Jupiter's atmosphere is presently considered to be the best value for the composition of protosolar He, before deuterium was converted to ^{3}He in the early Sun.

ANALYTICAL TECHNIQUES

Table 1 summarizes the techniques used to measure noble gases. By far the most important is mass spectrometry. Mass spectrometers in space are used, e.g., for solar wind and solar energetic particle measurements or atmospheric analyses on Venus, Moon, Mars and Jupiter, while mass spectrometers in the laboratory allow us to analyze extraterrestrial samples available on Earth, i.e., lunar samples, meteorites, interplanetary dust or solar corpuscular radiation trapped by foils exposed in space. Of course,

1529-6466/00/0047-0002$10.00

Table 1. Analytical techniques.

Techniques used for specified object	*Remarks and selected references*
Sun	
UV spectroscopy (flares, corona, emerging active regions)	No useful noble gas lines in photospheric spectra. Feldman and Widing 1990; Feldman 1998; Widing 1997.
Space-borne mass spectrometry (solar wind, SW; solar energetic particles, SEP)	Kallenbach et al. 1997; Leske et al. 1999; Gloeckler and Geiss 1998b; review by Wimmer-Schweingruber 1999.
Mass spectrometry in laboratory (solar wind collector foils etc.)	Apollo SW Composition experiment, MIR space station; upcoming Genesis mission. Geiss et al. 1972.
Mass spectrometry, laboratory (SW/SEP trapped in lunar regolith and meteorites)	Lunar regolith archive of ancient solar wind. Pepin et al. 1970a,b; Wieler et al. 1986; Marti 1969.
Helioseismology & solar models	Solar He abundance. Christensen-Dalsgaard 1998; Gough 1998.
Mercury	
UV-Spectroscopy (Mariner 10)	He, Ar in atmosphere (exosphere). Stern 1999b.
Venus	
Mass spectrometry, gas chromatography (Pioneer Venus, Venera 11-14)	Atmospheric analyses. Hoffman et al. 1980a,b; Donahue and Pollack 1983; Moroz 1983.
Moon	
Ion- and neutral mass spectrometry	He and Ar in lunar exosphere on ground, $^{36}Ar/^{40}Ar$. Stern 1999b.
UV spectroscopy	Upper limit for Kr, Xe from orbit. Feldman and Morrison 1991.
Mass spectrometry, laboratory (Apollo & Luna samples, lunar meteorites)	Ar-Ar ages, exposure ages, solar wind etc. Turner 1970; Culler et al. 2000; Wieler 1998.
Mars	
Mass spectrometry (Viking)	Atmosphere. Nier and McElroy 1977; Owen et al. 1977.
EUV spectroscopy	He in atmosphere. Krasnopolsky et al. 1994.
Mass spectrometry, laboratory (Martian meteorites)	Martian atmosphere, crust, mantle. Bogard et al. 1984; Becker and Pepin 1984; Swindle 2002b.
Asteroids	
Mass spectrometry, laboratory (meteorites)	Ott 2002; Swindle 2002a.
Jupiter and other giant planets	
Mass spectrometry (Galileo probe)	He-Ar, Xe elemental and isotopic abundance in outer Jovian atmosphere. Mahaffy et al. 2000.
In-situ interferometry (Galileo probe)	He abundance in outer Jovian atmosphere. v. Zahn et al. 1998.
Radio occultation & infrared spectroscopy (Voyager)	He abundance in giant planet outer atmospheres. Conrath et al. 1991; Conrath and Gautier 2000.
Comets	
EUV spectroscopy	Ar abundance, upper limits for He, Ne. Stern 1999a.
in-situ dust collection (see also IDPs)	Stardust mission
CI chondritic meteorites?	Lodders and Osborne 1999; Ehrenfreund et al. 2001.
Interplanetary dust particles, micrometeorites	
Mass spectrometry, laboratory	Dust from asteroids & comets, collected in near Earth space, ice, & sediments. Nier and Schlutter 1992; Olinger et al. 1990.
Heliospheric ions	
Mass spectrometry, space	Sources: Sun, Galaxy, local interstellar cloud. Mewaldt et al. 2001a; Binns et al. 2001; Gloeckler and Geiss 1998b.

laboratory analyses will usually yield data of higher precision than *in situ* measurements. Conversely, the latter offer the advantage of high temporal and spatial resolution, allowing one to study, e.g., solar wind emitted from different regions or during different regimes, correlations between noble gases and other elements, or short-term variations of fluxes or composition.

Noble gases are intrinsically difficult to detect by spectroscopy. For example, solar photospheric spectra, which form the basis for solar abundance values of most elements, do not contain lines from noble gases (except for He, but this line cannot be used for abundance determinations). Yet, ultraviolet spectroscopy is the only or the major source of information on noble gas abundances in the atmospheres of Mercury and comets. In the Extreme Ultraviolet (EUV), photon energies exceed bond energies of molecules and the first ionization potential of all elements except F, He, and Ne, so that only these elements are visible in this part of the spectrum (Krasnopolsky et al. 1997). Other techniques can be used to determine the abundance of He where this element is a major constituent. Studies of solar oscillations (helioseismology) allow a precise determination of the He abundance in the solar interior, and the interferometer on the Galileo probe yielded a precise value for the refractive index and hence the He abundance in the upper atmosphere of Jupiter (see respective sections of this chapter).

THE SUN

Solar noble gas abundances

Determining elemental and isotopic compositions of the elements in the Sun is a fundamental task in cosmochemistry and abundance tables are updated at regular intervals (Anders and Grevesse 1989; Palme and Beer 1993; Grevesse and Sauval 1998). For most elements, the compilations are based on photospheric spectroscopy on the one hand and meteorite data on the other. This reflects the very remarkable fact that in a very rare class of meteorites, the CI chondrites, relative elemental abundances are identical within uncertainties to photospheric values for essentially all but the most volatile elements, e.g., H, N, O, and the noble gases (a further exception is Li, which is destroyed in the sun, Grevesse and Sauval 1998). Anders and Grevesse (1989) define "solar abundances" as the best estimates obtained from photospheric analyses, if necessary augmented by data from solar wind, solar energetic particles, or astronomical observations, but not from meteorites. "Solar system abundances" are defined as best estimates for the entire solar system, and are mostly based on the CI chondrite data. Because photospheric and CI chondrite data agree so well with each other, the distinction between "solar" and "solar system" is often not made in practice, and actually, cosmochemists usually use the "solar abundances" obtained from CI chondrites, due to their higher precision. Since the elemental abundances in the sun are also remarkably similar to those in many other stars, the term "cosmic" abundances is also sometimes used for the former, although this appears to overestimate the role of our place in the universe.

For noble gases, the situation is different. Meteoritic noble gas abundances are by many orders of magnitude lower than any likely values in the primordial solar nebula. In addition, because no noble gas lines that could be used for abundance determinations are visible in photospheric spectra, among the noble gases only the "photospheric" abundances of Ne and Ar have been measured spectroscopically in active regions and the corona. This and other data sources that need to be considered for noble gases are explained in the following two subsections. Furthermore, all deuterium in the nascent Sun has been converted to ^{3}He, raising the $^{3}He/^{4}He$ ratio of the present-day sun by a factor of ~2.5-3 above the value of the proto-Sun. Both isotopes of helium are also produced in the present Sun, as intermediate (^{3}He) and end product (^{4}He), respectively, of hydrogen fusion. For these reasons, "solar" abundance compilations adopt best estimates of the protosolar ^{3}He/^{4}He ratio, as well as the protosolar ^{4}He abundance.

Elemental abundance values of noble gases in the sun recommended here are listed in Table 2, and isotopic compositions are given in Tables 3-5. In the next two subsections, we comment on these adopted values. Note that in Tables 7, 11, and 13 (found later in this chapter) elemental abundances in various reservoirs are stated relative to solar abundances. These are always the values as reported in the original references, i.e., they have not been renormalized to the solar abundances used here.

Elemental abundances. (A) Helium. Solar models are able to deduce accurate, although somewhat model-dependent, protosolar He values by fits to the observed present-day luminosity (e.g., Christensen-Dalsgaard 1998, see *Helium in the sun* section). The value stated in the body of Table 2 in three different notations is from the compilation by Grevesse and Sauval (1998), two other values for the initial He mass fraction are given in the Table caption (see also next section). In summary, the models provide the protosolar He abundance with an uncertainty of only a few percent.

(B) Neon. We adopt the Ne abundance recommended by Holweger (2001), which is based on EUV spectroscopy of emerging active regions (Widing 1997). Holweger (2001) points out that "emerging flux events most likely permit direct observation of unfractionated photospheric material," and prefers this data source over the solar particle measurements or extra-solar data used by Anders and Grevesse (1989). Holweger (2001) modified the original Widing (1997) Ne abundance value of [8.08] in the astrophysical notation (see Table 2) by also using ogygen (besides Mg) as a reference element combined with updated photospheric abundances of the reference elements. The Holweger (2001) Ne abundance is ~20% lower than the Widing (1997) value (adopted by Grevesse and Sauval 1998) and 23% lower than the Anders and Grevesse value, but all these estimates agree within stated uncertainties with each other.

(C) Ar. Grevesse and Sauval (1998) discuss two recent solar Ar abundance values. One is from coronal spectroscopy, yielding a photospheric Ar/Ca abundance ratio of 1.31±0.30, corresponding to an Ar abundance of [6.47±0.10] (Young et al. 1997). The second value is from solar energetic particles, yielding an Ar abundance of [6.39±0.027] (Reames 1998, value slightly updated and with a lower uncertainty than that stated in Grevesse and Sauval 1998). Grevesse and Sauval (1998) prefer the SEP value due to its lower uncertainty. In Table 2 we adopted the mean of both values, however, with a more conservative uncertainty of [0.1] dec (~25%) that would allow for some fractionation between coronal and photospheric Ar abundance. The Ar value in Table 2 is about 35% lower than that recommended by Anders and Grevesse (1989).

(D) Kr and Xe. The only available direct data of Kr and Xe in the Sun are from implanted solar wind (SW) atoms in lunar samples and gas-rich meteorites. These data are not very useful for abundance determinations, because both elements are enriched in the solar corpuscular radiation, as is discussed in the *Moon* section. The most reliable way to estimate the Kr and Xe abundances in the Sun is by interpolating concentrations of suitable isotopes in CI chondrites. Best suited are isotopes of elements close to Kr and Xe in the periodic system which are largely produced by s-process nucleosynthesis. Table 2 lists the values given by Palme and Beer (1993). Uncertainties of the s-process calculations are estimated to be 5-10%, to which another similar uncertainty has to be added for the abundances of the elements used for interpolation. The Kr and Xe estimates in Table 2 differ by ~15-20% from those adopted by Anders and Grevesse (1989), which

Table 2. Elemental abundances of noble gases in the Sun and the solar corpuscular radiation

	He* (*^{4}He*)	Ne (*^{20}Ne*)	Ar (*^{36}Ar*)	Kr (*^{84}Kr*)	Xe (*^{132}Xe*)	$^{4}He/^{20}Ne$	$^{20}Ne/^{36}Ar$	$^{36}Ar/^{84}Kr$	$^{84}Kr/^{132}Xe$
Sun									
refs. 1-3	$(2.73\pm0.13)\times10^{9}$	$(2.80\pm0.48)\times10^{6}$	$(7.51\pm1.94)\times10^{4}$	56±8	4.1±0.9	1050±190	41±12	1980±600	29±8
($Si=10^{6}$)	*$(2.73\pm13)\times10^{9}$*	*$(2.61\pm45)\times10^{6}$*	*$(6.35\pm1.64)\times10^{4}$*	*(32±5)*	*(1.09±0.25)*				
refs. 1-3	10.99±0.02	8.001±0.069	6.43±0.10	3.30±0.06	2.16±0.09				
(log- scale)[a]	*(10.99)*	*(7.97±0.07)*	*(6.36±0.10)*	*(3.05±0.06)*	*(1.58±0.09)*				
ref. 1 (N_{He}/N_{H})	0.098								
ref. 1 (Y_0)[b]	0.275±0.010								
Solar wind[4]						650±50	47±3	1670-2390	4.7-9.0
Solar energetic particles[5]									
gradual events	3.75×10^{8}	1.00×10^{6}	2.17×10^{4}			400±25	51±4		
$Si=10^{6}$;	(3.75×10^{8})	(9.34×10^{5})	(1.83×10^{4})						
[$O=10^{3}$]	[56,000]	[$^{20}Ne=142$]	[$^{36}Ar=2.8$]						
impuls. events [$O=10^{3}$]	[46,000]	[$^{20}Ne=388$]							

1: (He & Ar): Grevesse and Sauval 1998, for Ar see text; other recent protosolar He values from solar modeling: Y_0=0.2713 (Christensen-Dalsgaard 1998), and Y_0=0.2735 (Bahcall et al., 2001). **2** (Ne): Holweger 2001. **3**: (Kr & Xe): Palme and Beer 1993. All Si-normalized values assume H/Si=27900 (Anders and Grevesse 1989). **4**: Murer et al. 1997; Wieler and Baur 1995; Kr and Xe abundances show secular changes. **5**: Miller 1998; Reames 1998; Abundances in [ ] are reported relative to O as in orig. ref., because Si is enhanced in SEP due to FIP-effect. Miller 1998: gradual events $^{4}He/H$ ~0.1. Reames et al. 1994: impulsive events $^{4}He/H$ = 0.005-0.15.

*: protosolar He abundance (before onset of hydrogen burning)

a: The astronomic abundance scale is logarithmic, with the hydrogen abundance log N_H=12 are based on earlier s-process calculations and CI abundances. In summary, whereas the He abundance in the (proto)sun is known today with high precision, mainly thanks to solar modeling, the values for Ne-Xe are probably still uncertain by some 20%.

Isotopic abundances. Our knowledge of the isotopic composition of noble gases in the sun so far is exclusively based on analyses of the solar wind and solar energetic particles. These are discussed in detail below in this section and in the *Moon* section. We will also discuss estimates of isotopic fractionations between the sun and the solar corpuscular radiation and conclude that such fractionations are small but probably not entirely negligible. The He isotopic composition in the outer convective zone of the sun given in Table 3 has been corrected for various estimated fractionation effects. Also the solar Ne and Ar isotopic compositions in Table 4 (below) have been very slightly corrected, whereas the Kr and Xe compositions in Table 5 (below) are measured solar wind values. Alternative reservoirs that might conceivably contain isotopically unfractionated "solar" noble gases, notably the atmosphere of Jupiter (or the other Giant planets), have not been analyzed so far with the necessary precision to yield values of equal quality as the solar wind data (*Giant Planets* section). However, the $^{3}He/^{4}He$ in Jupiter's atmosphere is the currently most widely used estimate for the protosolar He isotopic composition (prior to deuterium-burning) and is thus given in Table 3.

Helium in the Sun

During its main-sequence lifetime over the past ~4.6 Gyr, the Sun has been generating its energy by the conversion of hydrogen into ^{4}He in its core where temperatures are high enough that atomic nuclei can overcome the repulsive force due to their electric charge. Hydrogen fusion as well as further stellar nucleosynthesis reactions are discussed in, e.g., Clayton (1968) and Rolfs and Rodney (1988). Briefly, the net reaction

$$^{4}H \rightarrow {}^{4}He + 2\,e^{+} + 2\,\nu$$

produces an energy of 26.73 MeV, part of which will be carried away by the neutrinos. As the probability of a direct fusion of four protons is essentially zero, the conversion of hydrogen into helium occurs in various chains of two-particle interactions. The most important ones are the three p-p chains (cf. Rolfs and Rodney 1988, Fig. 6.13). In all these reaction chains also ^{3}He is produced as an intermediate product and in two of them also Be, Li, and B nuclei are produced and destroyed again. In stars like our Sun, which also contain sizeable amounts of elements heavier than H and He, hydrogen fusion also occurs by cyclic reactions involving heavier nuclei, the most important one being the CNO bi-cycle (Rolfs and Rodney 1988, Fig. 6.22). The C, N, and O nuclei involved act as catalysts only, and the net reaction is again the conversion of four protons into a ^{4}He nucleus.

The details of these nucleosynthesis processes are major ingredients for solar models. Although they are conceptually simple (Gough 1998), the so-called standard solar models (e.g., Bahcall and Pinsonneault 1995; Bahcall et al. 2001) are thought to describe the Sun surprisingly well. As noted above, solar modeling provides a means for accurately inferring the He abundance of the Sun. The initial abundance of He, given as the mass fraction Y_0, is a free parameter in solar modeling. Because the luminosity of the Sun depends sensitively on the helium abundance, the value Y_0 can be chosen such that the model yields the correct present-day luminosity. The total initial mass fraction of the elements heavier than He (Z_0) is then obtained by adjusting the model to the correct present-day abundance ratio of heavy elements and hydrogen in the photosphere (Z_s/X_s = 0.0245). Christensen-Dalsgaard (1998) obtains $Y_0 = 0.2713$ and $Z_0 = 0.0196$, Bahcall et al. (2001) report for their standard solar model $Y_0 = 0.2735$. Figure 1 shows the hydrogen

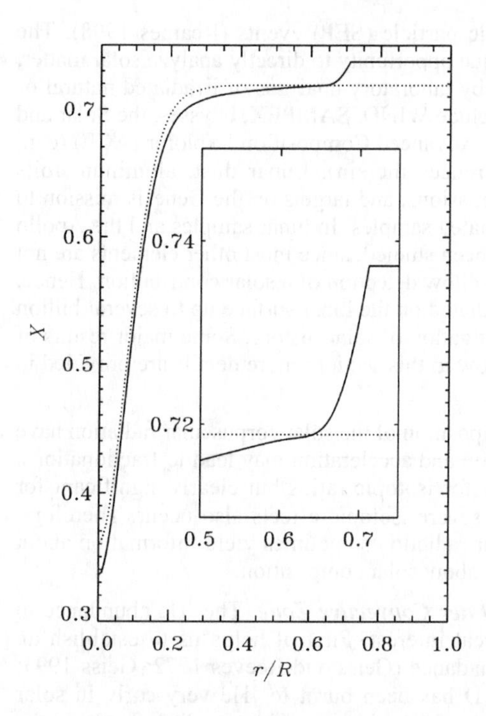

Figure 1. Hydrogen model abundance X (by mass) in the Sun as a function of distance from the center (solar surface at r/R = 1). The He abundance profile is essentially the complement of the H profile minus ~2% of heavy elements. Helioseismology provides evidence for gravitational settling of He, which is incorporated in the model represented by the solid line, where the He abundance increases abruptly below the outer convective zone at r/R = 0.7. The dotted line is for a model neglecting He settling. [Used by kind permission of Kluwer Academic Publishers, from Christensen-Dalsgaard (1998) *Space Science Reviews*, Vol. 85, Fig. 1, p 22.]

abundance in the present-day Sun as a function of distance from the solar center. The dotted line is for a model that neglects gravitational settling of helium and heavier elements towards the Sun's center. The helium profile essentially follows this dotted line, although lower by 2% to account for the heavy elements. Near the center, hydrogen is depleted due to nuclear burning. Outwards of about 0.25 solar radii, however, the hydrogen abundance stays almost constant and so therefore does the He abundance. In this model, the He/H ratio at the surface is very close to the initial ratio 4.6 Ga ago. A model that takes into account element settling below the Outer Convective Zone (e.g., Vauclair 1998) is shown by the solid line in Figure 1. In the Outer Convective Zone (OCZ) in the outermost ~30% of the solar radius, mixing leads to a constant H/He ratio, which is above the initial ratio, however. He is therefore depleted near the solar surface by ~10% relative to the value expected without gravitational settling. Helioseismology, the science of probing the Sun's interior by observing solar oscillations, allows a deduction of the He abundance in the outer part of the Sun (e.g., Gough et al. 1996) and provides clear experimental support for a He depletion near the surface. The He abundance in the convective zone determined by helioseismology is close to 0.25 (Christensen-Dalsgaard 1998), clearly lower than the initial value of ~0.275 predicted by standard solar models without He-settling, but in good agreement with models accounting for element settling. It may be worth noting here that the "solar neutrino problem," i.e., the unexpectedly low abundance of neutrinos from the solar interior detected on Earth, is no longer suspected to indicate severe flaws in standard solar models (e.g., Bahcall 2000).

Noble gases in the solar corpuscular radiation

The Sun emits the solar wind (SW), a continuous stream of particles with velocities mostly between ~300-800 km/s (Parker 1997). Ions of higher energy are emitted especially during so-called solar energetic particle (SEP) events (Reames 1998). The solar corpuscular radiation provides a unique opportunity to directly analyze solar matter, either by mass spectrometers in space or by laboratory analyses of irradiated natural or artificial targets. Recent space missions include WIND, SAMPEX, Ulysses, the Solar and Heliosperic Observatory (SOHO), and the Advanced Composition Explorer (ACE) (e.g., Wimmer-Schweingruber 2001, and references therein). Lunar dust, aluminum foils exposed on the Moon during the Apollo missions, and targets on the Genesis mission to be returned to Earth are examples of irradiated samples. In lunar samples and the Apollo foils almost exclusively noble gases have been studied, since most other elements are not scarce enough in the targets themselves to allow detection of a solar contribution. Hence, noble gases in dust samples that were irradiated on the lunar surface up to several billion years ago are very important for the investigation of solar history. Some major results of the lunar sample studies are presented below in this section; more details are provided in the *Moon* section.

Studies of elemental and isotopic composition of the solar corpuscular radiation have to account for the fact that particle selection and acceleration may lead to fractionations. In the solar wind, such effects are minor for isotopic ratios but clearly significant for some elemental ratios, whereas in SEPs severe isotopic effects also occur. Therefore, abundance studies in the solar corpuscular radiation sometimes yield information about fractionation processes rather than directly about solar composition.

Helium in the solar wind and the Outer Convective Zone. The ^{3}He abundance in the present and past solar wind is of great interest. First, it helps us to establish an estimate for the protosolar deuterium abundance (Geiss and Reeves 1972; Geiss 1993; Gloeckler and Geiss 2000), because all D has been burnt to ^{3}He very early in solar history. Second, ^{3}He builds up in the course of solar history at intermediate depths in the sun by incomplete hydrogen burning. The $^{3}He/^{4}He$ ratio in the solar wind and especially its long-term evolution are therefore very sensitive indicators of mixing of material into the outer convective zone (Bochsler et al. 1990).

Table 3 shows $^{3}He/^{4}He$ values in the solar wind obtained by different techniques. Although all values agree within ±7% with each other, reported uncertainties are often quite large. The recent SWICS Ulysses value is at the low end of those stated in Table 3. It represents a SW regime associated with coronal holes ("high-speed SW" of up to 800 km/s), which is thought to be less fractionated relative to the OCZ (see below) than the "interstream-SW" with speeds on the order of ~300-500 km/s. Figure 2 shows that the $^{3}He/^{4}He$ ratio in this "slow-SW" is probably some 5-10% higher than in the coronal-hole dominated SW (Bodmer and Bochsler 1998b; Gloeckler and Geiss 2000). The weak dependence of $^{3}He/^{4}He$ on the SW regime indicates, however, that overall the solar wind isotopic composition reflects the He composition at the solar surface quite accurately (Bodmer and Bochsler 1998b). These authors estimate the ($^{3}He/^{4}He$) ratio in coronal-hole SW to be only ~9% higher than the value in the OCZ, which they calculate as $(3.75\pm0.70) \times 10^{-4}$. Gloeckler and Geiss (2000) arrive at the same value but with a lower uncertainty (Table 3). The OCZ-estimate is also important in assessing the protosolar D/H ratio in the next section. The data from lunar samples will be discussed in more detail in the *Moon* section.

In summary, the He isotopic composition in the solar wind and hence in the OCZ is reasonably well known, but further improvements are nevertheless desirable, e.g., to reduce uncertainties in the value of the protosolar D/H ratio, as is discussed next.

The protosolar D/H ratio. The protosolar deuterium abundance is an upper limit to the primordial abundance of this isotope and provides a crucial observational test for the Big Bang cosmological model (Schramm 1993). For example, the primordial deuterium

Table 3. $^3He/^4He$ in the sun (in units of 10^{-4}).

	SW	OCZ
Protosun (Jupiter)[1]	1.66±0.05	
Solar wind and outer convective zone	**SW**	**OCZ**
Ulysses-SWICS[2] (coronal-hole SW)	4.08±0.77	3.75±0.70
Ulysses-SWICS[3]		3.75±0.27
ISEE-3[4]	4.37	
Apollo foils[5]	4.25±0.21	
lunar soil[6]		
last ~100 Myr	4.57±0.08	
last ~1000 Myr	4.18-4.63	
Solar energetic particles		
gradual events (CME)[7]	19±2	
impulsive events (flares)[8]	$(\sim1\text{-}330)\times10^3$	
SEP-He in lunar soil[6]	2.17±0.05	
Protosolar deuterium[9]		
$(D+^3He)/^4He$	$(3.60\pm0.38)\times10^{-4}$	
D/H	$(1.94\pm0.39)\times10^{-5}$	

1: Mahaffy et al. 1998; Jupiter value assumed to be identical to protosolar value (*Giant Planets* section). 2: Bodmer and Bochsler 1998b. 3: Gloeckler and Geiss 2000. 4: Bochsler 1984. 5: Geiss et al. 1972. 6: Benkert et al. 1993. 7: Mason et al. 1999, average value of 12 events. 8: Reames et al. 1994; Mason et al. 2000. 9: Gloeckler and Geiss 2000, deduced from $(^3He/^4He)_{OCZ}$ and $^3He/^4He$ in Jupiter (ref. 1, Table 7).

Stated uncertainties of SW values are not directly comparable with each other. Uncertainty in (2) mostly due to an assumed systematic uncertainty of detector efficiencies; uncertainties in (6) do not consider possible alterations in samples.

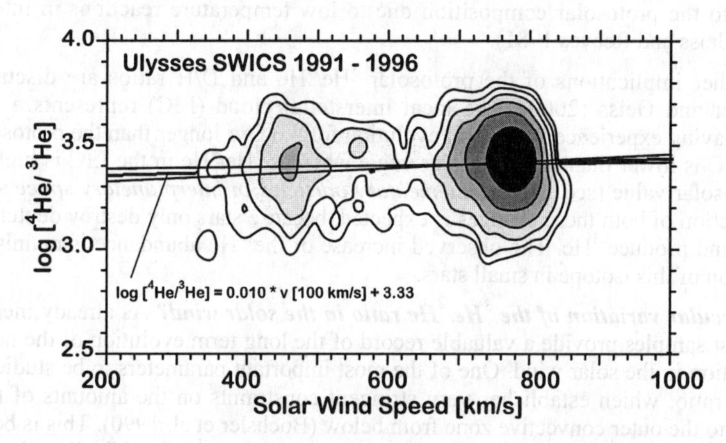

Figure 2. $^4He/^3He$ (log-scale) as a function of solar wind speed as determined with the Solar Wind Ion Composition Spectrometer on the Ulysses spacecraft. Contour lines indicate 2, 4, 8, 16, 32, and 64 cases per bin. The scatter is mostly (but probably not entirely) due to statistical fluctuations. A weak increase of $^4He/^3He$ with SW speed is indicated (400 km/s: $^4He/^3He = 2350$, i.e., $^3He/^4He = 4.26\times10^{-4}$; 800 km/s: $^4He/^3He = 2570$, i.e., $^3He/^4He = 3.89\times10^{-4}$). [Used by permission of Springer Verlag, from Bodmer and Bochsler (1998b), Astron. Astrophys., Vol. 337, Fig. 6, p 926.]

abundance constrains the baryonic density in the universe. Protosolar D is also important for tracing galactic evolution in the past 4.6 Gyr. To derive the $(D + {}^3He)/{}^4He$ ratio in the protosun from the present-day $({}^3He/{}^4He)_{SW}$ ratio, one needs to consider several effects which may alter the He isotopic composition in the solar wind. First, there is a probable mass fractionation between the SW and the OCZ as discussed in the previous paragraph. Furthermore, the He isotopic composition in the OCZ may itself be slightly different from the protosolar $(D + {}^3He)/{}^4He$ due to gravitational settling of He into deeper layers (see the subsection *Helium in the Sun* and Fig. 1) and a possible minor contribution of ^3He produced during the main sequence life of the Sun and mixed into the convective zone (Geiss 1993; Geiss and Gloeckler 1998). The latter effect is also discussed in the next subsection. Gloeckler and Geiss (2000) estimate that gravitational settling and mixing together lead to a modest enrichment of ^3He in the OCZ by 4±2% and calculate a protosolar $(D + {}^3He)/{}^4He$ ratio of $(3.60 \pm 0.38) \times 10^{-4}$ (Table 3). Assuming for the protosolar $^3He/^4He$ ratio the value of 1.66×10^{-4} measured in Jupiter (see the *Giant Planets* section) and protosolar He/H = 0.10, Gloeckler and Geiss (2000) obtain a protosolar (D/H) ratio of $(1.94 \pm 0.39) \times 10^{-5}$, close to the original estimate of $(2.5 \pm 0.5) \times 10^{-5}$ by Geiss and Reeves (1972). Alternatively, assuming He in the meteoritic "Phase-Q" ($^3He/^4He = 1.23 \times 10^{-4}$) to represent the protosolar composition, Busemann et al. (2001) derive protosolar D/H = $(2.4 \pm 0.7) \times 10^{-5}$.

It is not the place here to discuss all the major implications of this result, but a historical problem is worth mentioning, which was a major motivation for the Apollo solar wind composition experiment (cf. Geiss 1993). In the 1960s, the D/H ratio in terrestrial oceans (1.6×10^{-4}) was assumed to represent the "cosmic" or protosolar deuterium abundance. Based on this assumption the $^3He/^4He$ ratio in the solar wind was expected to be above 10^{-3}, but meteorites thought to contain implanted solar wind noble gases yielded lower values of around 4×10^{-4}. These were confirmed by all subsequent studies discussed above, and Geiss and Reeves (1972) and Black (1972) were therefore able to conclude that deuterium is enriched in sea water by almost an order of magnitude relative to the protosolar composition due to low temperature reactions in interstellar clouds (Geiss and Reeves 1981).

Further implications of the protosolar $^3He/^4He$ and D/H ratios are discussed by Gloeckler and Geiss (2000). The local interstellar cloud (LIC) represents a galactic sample having experienced chemical evolution for 4.6 Gyr longer than the protosun. D/H in the LIC is lower than the protosolar value whereas $^3He/^4He$ in the LIC is higher than the protosolar value (see also the *Elementary particles in interplanetary space* section). The direction of both these changes is expected, because stars only destroy deuterium but destroy and produce ^3He. The observed increase of the ^3He abundance is mainly due to production of this isotope in small stars.

A secular variation of the $^3He/^4He$ ratio in the solar wind? As already mentioned, lunar dust samples provide a valuable record of the long term evolution of the noble gas composition in the solar wind. One of the most important parameters to be studied is the $^3He/^4He$ ratio, which establishes very stringent constraints on the amounts of material mixed into the outer convective zone from below (Bochsler et al. 1990). This is because a ^3He-rich region gradually evolves at intermediate depths in the solar interior, where temperatures are high enough to produce ^3He in the p-p chains, but too low to further process this isotope. The maximum enrichment after 4.6 Gyr is about a factor of 30 at about 0.3 solar radii (Bochsler et al. 1990). Some non-standard solar models predict that a considerable fraction of this ^3He is mixed into the convective zone and should thus be visible in the solar wind. If so, the $^3He/^4He$ ratio in samples that trapped SW several Ga ago should be significantly lower than in more recently irradiated samples.

Data relevant to this problem are shown in Figure 9. This Figure will be discussed in more detail in the *Moon* section, so here we just note the most salient points. Most importantly, the isotopic composition of SW-He has definitely not changed by more than some 10% per Gyr, as shown by the bulk sample analyses compiled by Geiss (1973) shown as solid circles. This finding rules out non-standard solar models invoking strong internal mixing or high mass-loss after onset of hydrogen burning (Bochsler et al. 1990). These data also were one of the crucial observations allowing Geiss and Reeves (1972) to conclude that the He composition in the outer convective zone and hence the solar wind is close to the protosolar (D + ^{3}He)/^{4}He ratio (see above). Whether the lunar data actually indicate a small secular increase of the ^{3}He abundance in the convective zone is still unclear, as is discussed in the *Moon* section. New analyses, where the solar noble gases from the outermost dust grain layers are selectively sampled (solid squares in Fig. 9), apparently confirm a small secular change, but it seems probable that this is an experimental artifact. If so, corrected data (open squares) indicate an almost constant He isotopic composition in very ancient as well as recently irradiated samples.

He in solar energetic particles. SEPs provide another source of information on the composition of the solar corona. SEP events are basically classified into large gradual events and smaller impulsive events (e.g., Reames 1998, 2001). Rise and fall times of gradual events are on the order of days, while those of impulsive events are considerably shorter. SEPs were originally all thought to be accelerated in solar flares (intense abrupt brightenings on the solar surface with radiation emitted over the entire electromagnetic spectrum due to a sudden release of magnetic energy), but it has become clear now that in gradual events solar wind and coronal particles are accelerated at shock waves produced by "coronal mass ejections," and that flares are another manifestation of coronal mass ejections (e.g., Gosling 1993).

Typical He abundances and isotopic compositions are given in Tables 2 and 3. Large events show a modest, variable enrichment of ^{3}He, on average by about a factor of 5 over the solar wind value (Mason et al. 1999). In contrast, particle acceleration in impulsive solar flares yields a highly fractionated sample of the corona, e.g., ~250- to 80,000-fold enhancements in the ^{3}He/^{4}He ratio and enrichments of heavy elements (e.g., Fe/O) of an order of magnitude. These enhancements are believed to be due to resonant wave particle interactions in a solar flare (Reames 1998; Miller 1998). However, on average, impulsive events only contribute a minor fraction to the SEP population. Also, Mason et al. (1999) suggest that the ^{3}He-enrichment in gradual flares may be due to remnant flare material in the interplanetary medium.

Note that in the next subsection and in the *Moon* section we will discuss a noble gas component in lunar samples also dubbed SEP, as it is thought to be of solar origin and implanted at higher energies than the solar wind. This component seems, however, not to represent the same energy range as SEPs detected in space.

Ne-Xe in the solar wind and solar energetic particles. Being rarer than He, the heavier noble gases in the SW are more difficult to analyze directly by mass spectrometers in space. Only recently the first Ne and Ar isotopic data were obtained by the SOHO and WIND missions (Kallenbach et al. 1997; Wimmer-Schweingruber et al. 1998; Weygand et al. 2001). Hence, much of the current knowledge on the heavier noble gases in the SW is based on the analysis of lunar samples and gas-rich meteorites, as is discussed in some detail in the *Moon* section. Here we compare some of these data with results on the present-day solar wind based on *in situ* measurements and the Apollo foils.

Table 4 shows ^{20}Ne/^{22}Ne and ^{36}Ar/^{38}Ar ratios in the solar wind determined by the three techniques. As noted for He already, the stated error bars should be regarded with caution. In particular, the low nominal error bars of the lunar and meteoritic data ignore

Table 4. Isotopic abundances of Ne and Ar in the sun and the solar corpuscular radiation.

	$^{20}Ne/^{22}Ne$	$^{21}Ne/^{22}Ne$	$^{36}Ar/^{38}Ar$
Solar[a]	13.6±0.4	0.0326 ±0.0010	5.5±0.2
Solar wind			
Apollo foils[1]	13.7±0.3	0.0333 ±0.0044	5.3±0.3
SOHO[2]	13.8±0.7	0.0314 ±0.0080	5.55±0.65
WIND[3]	13.64±0.7		
Moon/meteorites (SW)[4]	13.8±0.1	0.0328 ±0.0005	5.58±0.03 5.77±0.07
SEP-component[4]	11.2±0.2	0.0295±0.0005	4.87±0.05
Solar energetic particles[5]	~5-20		

a: Solar wind composition corrected for an assumed fractionation relative to bulk sun (see text). Solar $^{40}Ar/^{38}Ar = 1.63\times10^{-3}$ (Anders and Grevesse 1989).

1: Geiss et al. 1972; Cerutti 1974. 2: Kallenbach et al. 1997; Weygand et al. 2001, mean of reported interstream and coronal hole values. 3: Wimmer-Schweingruber et al. 1998; 4: Benkert et al. 1993; Becker et al. 1998; Pepin et al. 1999. 5: Leske et al. 1999, 2001.

possible systematic uncertainties (see the *Moon* section). It is nevertheless gratifying that all three techniques yield very similar values, which suggests that the solar wind Ne and Ar isotopic composition is known to within a few percent. The lunar data also allow for a precise determination of the abundance of the rare isotope ^{21}Ne, which is not as well constrained by the other two methods.

Anders and Grevesse (1989) adopted for solar Ar the $^{36}Ar/^{38}Ar$ value 5.32 of the terrestrial atmosphere. The value measured in the Apollo foils and also the new SOHO value are within their somewhat limited precision consistent with this assumption. However, recent solar wind values derived from lunar and meteoritic samples with a considerably higher precision (Table 4) suggest that the $^{36}Ar/^{38}Ar$ ratio in the SW is several percent higher than the air value. As mentioned above, this suggestion relies on the assumption that the lunar and meteoritic values are not severely compromised by systematic errors. The good agreement of the lunar- and meteorite-derived $^{20}Ne/^{22}Ne$ ratio with the respective Apollo and SOHO values (Table 4) is a good argument that this is not the case. Two slightly different $(^{36}Ar/^{38}Ar)_{SW}$ values are given in Table 4. The lower one of 5.58 from an acid-etch study of the gas-rich meteorite Kapoeta is preferred by this author, whereas the higher one of 5.77 is adopted in the chapter by Pepin and Porcelli (2002, this volume). If taken at face value, the $^{36}Ar/^{38}Ar$ ratio in Jupiter's atmosphere of 5.6±0.25 (Table 7) appears to be another argument that SW-Ar is indeed isotopically slightly different from atmospheric Ar, but we discuss in the *Giant Planets* section that this argument is less straightforward than it may seem.

The Kr and Xe isotopic compositions in the solar wind have not yet been determined directly, neither in space nor with foils. On the other hand, lunar soil samples are well suited to study the isotopic composition of these elements in the solar wind, because the moon contains hardly any indigenous Kr and Xe. The good agreement of the He and Ne compositions deduced from regolithic samples with the values determined *in situ* or the Apollo solar wind composition experiment gives us confidence that the lunar data for the two heavy noble gases are also reliable. Frequently used recent values are given in Table 5. They are largely based on the first extraction steps of *in vacuo* etch runs, which should contain essentially the pure and unfractionated SW component, whereas earlier values, e.g., those adopted by Anders and Grevesse (1989), probably reflect the presence of isotopically heavier SEP-Kr and SEP-Xe in bulk sample analyses.

Relationships between Kr and Xe in the SW and other planetary reservoirs are discussed in several other chapters in this book (Swindle 2002b; Ott 2002; Pepin and Porcelli 2002). Figure 3 compares various important Xe components, all normalized to the composition of the terrestrial atmosphere. One of the biggest challenges in noble gas cosmochemistry is to explain the relationship between the Xe composition in the Earth's atmosphere and the SW. Atmospheric Xe is depleted in the lighter isotopes by some 4.2% per amu relative to solar Xe (Fig. 3), but a simple mass-dependent fractionation plus additions of radiogenic and fissiogenic isotopes to air-Xe would lead to an overabundance of the heaviest two Xe isotopes in the atmosphere. Therefore, in a widely circulated preprint that was to become the by far best-known piece of Samisdat literature in noble gas cosmochemistry, R.O. Pepin and D. Phinney postulated in 1978 that terrestrial atmospheric Xe as well as meteoritic trapped Xe derive from a primordial component different from solar Xe (Pepin 2000; see chapters in this volume by Pepin and Porcelli 2002 and Ott 2002). This component has been dubbed U-Xe (Table 5, Fig. 3). Relative to U-Xe, xenon in the Sun would be enriched by several percent in ^{134,136}Xe only. This conundrum remains unsolved, but it seems clear that it is not caused by a grossly wrong determination of the heavy isotope abundances of SW-Xe.

Atmospheric Kr is probably also slightly fractionated relative to SW-Kr. The composition adopted by Pepin et al. (1995; Table 5) yields a depletion of lighter isotopes of ~7.5‰ per amu, that of Wieler and Baur (1994) only some 2‰ per amu, with an uncertainty that would also allow atmospheric Kr to be identical to SW-Kr. More important than this slight discrepancy is the fact that, unlike Xe, atmospheric and solar wind Kr essentially are related to each other by a simple mass dependent isotopic fractionation.

The Ne isotopic composition in solar energetic particles has been measured in space since the late 1970s (Dietrich and Simpson 1979; Mewaldt et al. 1979). These first measurements by satellites in (gradual) solar energetic particle events suggested an enrichment of the heavier Ne isotope relative to the SW-Ne composition, but uncertainties were large. Precise data have become available in particular with the ACE mission (Leske et al. 1999, 2001). Unlike in the solar wind, the ^{20}Ne/^{22}Ne ratio in solar energetic particles is highly variable from event to event, from ~0.4 to ~1.5 times the SW value in the 18 events studied by Leske et al. (2001). This shows that SEPs are less well suited than the SW to determine isotopic abundances in the Sun. The Ne isotopic composition correlates strongly with elemental ratios such as Na/Mg (Fig. 4), suggesting that elemental and isotopic fractionations are both governed by the charge to mass ratio (apart from elemental fractionations due to the FIP effect, see below). The large event-to-event variability makes it difficult to determine the average Ne composition in SEPs (one event may dominate the SEP flux of an entire 11-year solar cycle). Therefore, it is not yet clear whether the average ^{20}Ne/^{22}Ne in SEPs and the SW differ from each other. Yet, the mean value determined by ACE up to the year 2000 is a few 10% lower than the SW ratio (Leske et al. 2001). This may be important in view of the second trapped solar Ne component (also dubbed SEP) discussed in more detail in the *Moon* section. In all five noble gases, this component is isotopically heavier than the respective SW values, e.g., ^{20}Ne/^{22}Ne = 11.2±0.2.

Isotopic fractionation Sun – solar wind. We noted above that the two He isotopes are expected to be only slightly fractionated between the outer convective zone and the

Table 5. Kr and Xe in the solar wind (^{84}Kr≡100; ^{130}Xe≡100)

	^{78}Kr	^{80}Kr	^{82}Kr	^{83}Kr	^{86}Kr			
Solar wind								
Ref. 1	0.6365±.0034	4.088±.014	20.482±.054	20.291±.026	30.24±.10			
Ref. 2			20.28±.10	20.04±.10	30.37±.13			
terrestrial atm.³	0.6087	3.960	20.217	20.136	30.524			
	^{124}Xe	^{126}Xe	^{128}Xe	^{129}Xe	^{131}Xe	^{132}Xe	^{134}Xe	^{136}Xe
Solar wind⁴	2.939±.070	2.549±.082	51.02±.54	627.3±4.2	498.0±1.7	602.0±3.3	220.68±.90	179.71±.55
U-Xe⁵	2.928±.010	2.534±.013	50.83±.06	628.6±.6	499.6±.6	604.7±.6	212.6±.4	165.7±.3
terrestrial atm.³	2.337	2.180	47.15	649.6	521.3	660.7	256.3	217.6

All values normalized to terrestrial atmospheric Kr and Xe composition by Basford et al. 1973.

1: Pepin et al. 1995, average of 14 analyses. **2**: Wieler and Baur 1994, in-vacuo etch of lunar soil 71501, ^{78,80}Kr used for spallation correction. **3**: Basford et al. 1973. **4**: Data of lunar soil 71501 by Wieler and Baur 1994, values according to Pepin et al. 1995. **5**: Pepin 2000.

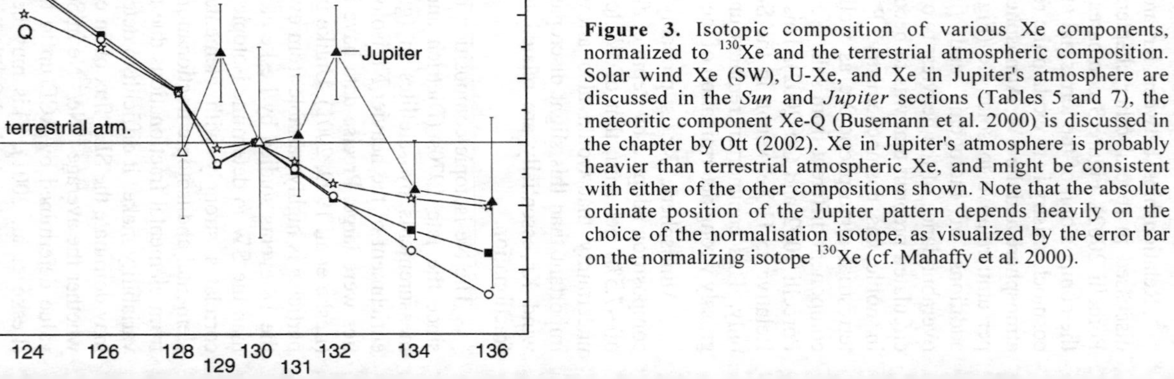

Figure 3. Isotopic composition of various Xe components, normalized to ^{130}Xe and the terrestrial atmospheric composition. Solar wind Xe (SW), U-Xe, and Xe in Jupiter's atmosphere are discussed in the *Sun* and *Jupiter* sections (Tables 5 and 7), the meteoritic component Xe-Q (Busemann et al. 2000) is discussed in the chapter by Ott (2002). Xe in Jupiter's atmosphere is probably heavier than terrestrial atmospheric Xe, and might be consistent with either of the other compositions shown. Note that the absolute ordinate position of the Jupiter pattern depends heavily on the choice of the normalisation isotope, as visualized by the error bar on the normalizing isotope ^{130}Xe (cf. Mahaffy et al. 2000).

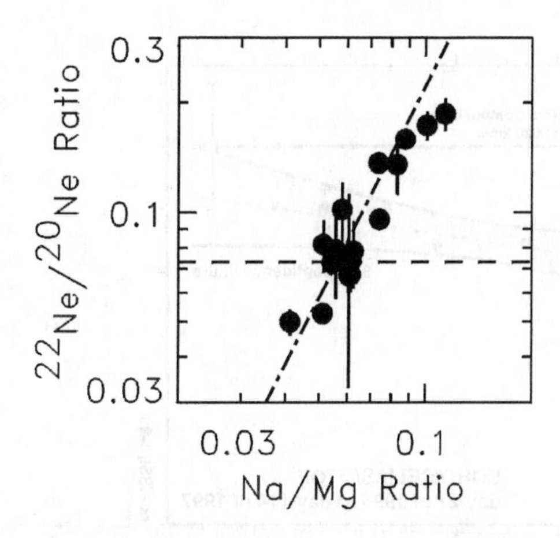

Figure 4. $^{22}Ne/^{20}Ne$ versus Na/Mg in 18 large (gradual) solar energetic particle events measured mostly by the Advanced Composition Explorer (Leske et al. 2001). The Ne isotopic composition varies by almost a factor of 4 between the different events, but correlates well with elemental ratios such as Na/Mg, indicating that both elemental and isotopic fractionations are governed by the ionic charge-to-mass ratio. The average Ne isotopic composition in SEP events may be different from the SW value, which is indicated by the dashed horizontal line. Figure courtesy of R.A. Leske.

solar wind, e.g., by ~9% in coronal-hole-dominated (fast) SW. Isotopes with a smaller relative mass difference are expected to show even less fractionation. Potential fractionation mechanisms in the solar wind are discussed by Bochsler (2000), Kallenbach et al. (1998b), and Kallenbach (2001). The $^{24}Mg/^{26}Mg$ ratio is expected to be enhanced in the SW by no more than 3% due to inefficient Coulomb coupling (Bodmer and Bochsler 1998a,b) and probably by 1-2% due to gravitational settling in the outer convective zone (Kallenbach et al. 1998b; Bochsler 2000). Precise Mg and Si data from the SOHO, ACE and WIND missions can be used to test these predictions (Kallenbach et al. 1998b; Kallenbach 2001). The SOHO data are shown in Figure 5. At a SW velocity of ~620 km/s the Si and Mg isotopic compositions are identical to the terrestrial or meteoritic values within a systematic uncertainty of the instrument calibration of ~1.5% per amu. At lower velocities, the light isotopes become slightly enriched, by (1.4±1.3)% per amu at 350 km/s (dashed line in Fig. 5 and its 2σ error). So, whereas the systematic uncertainties are still somewhat too large to firmly conclude whether Mg and Si are isotopically fractionated in the bulk solar wind, the results strongly indicate a slightly variable isotopic composition as a function of solar wind speed, with the heavy isotopes being less abundant in the slow wind. Kallenbach (2001) reports isotopic ratios of various elements in the bulk solar wind obtained as average values from various missions. The $^{24}Mg/^{26}Mg$ and $^{28}Si/^{30}Si$ ratios obtained from WIND and SOHO (R. Kallenbach, pers. comm.) are both larger than the solar system values, by (1±2.2)% and (1±1.4)%, respectively. Ne and Ar in the solar wind can be expected to be similarly fractionated as Mg and Si. This suggests that $^{20}Ne/^{22}Ne$ and $^{36}Ar/^{38}Ar$ values in the solar wind differ from solar values probably by between 1-3% (R. Kallenbach, pers. comm. 2001). The solar Ne and Ar isotopic compositions adopted here (Table 4) take this Mg-Si-derived fractionation and its uncertainty into account. The SW isotopic compositions used to derive the solar values are those based on lunar and meteorite samples. Table 4 illustrates that the Ne and Ar isotopic compositions of the bulk sun are probably less accurately known than the currently most precise solar wind data may suggest, due to the uncertainty of the fractionation correction sun-solar wind. Interestingly, the solar Ar composition is within uncertainty still equal to the terrestrial atmospheric value or that of primordial meteoritic Ar (Busemann et al. 2000; Ott 2002, this volume), as was already the case for the Anders and Grevesse (1989) value. Therefore, even though we are quite sure today that the atmospheric $^{36}Ar/^{38}Ar$ ratio is slightly different from the solar wind value, it is still uncertain whether terrestrial atmospheric Ar is fractionated relative to bulk solar Ar (see also the discussion of Ar in Jupiter in the *Giant Planets* section).

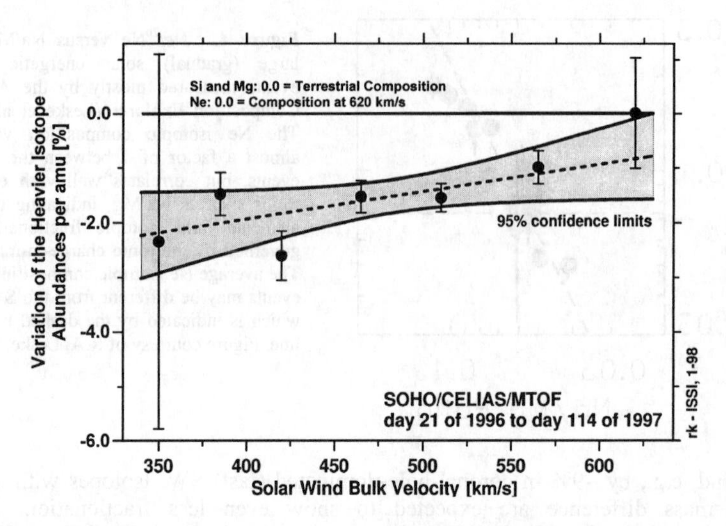

Figure 5. Isotopic compositions of Mg, Si, and Ne as a function of solar wind velocity, as measured by the SOHO mission. The ordinate shows the weighted average variations of the abundances of the heavier isotope of the three elements (per amu), relative to the abundances at 620 km/s. For Mg and Si, the isotopic compositions at 620 km/s are identical to the terrestrial values, although all Mg and Si data may be systematically offset by up to 1.5% per amu due to uncertainties in the instrument calibration. [Used by permission of Kluwer Academic Publishers, from Kallenbach et al. (1998b), *Space Science Reviews*, Vol. 85, Fig. 5, p 364.]

Elemental abundances of noble gases in the solar wind; the FIP effect in Kr and Xe. Elemental abundances of noble gases in the SW are given in Table 2 and discussed in the *Moon* section, as these data are largely derived from lunar (and meteoritic) regolith samples. Here we briefly mention the particularly interesting case of Kr and Xe. Both Kr/Ar and especially Xe/Ar in lunar samples are systematically higher than the inferred solar ratios. Very probably this is not due to fractionation during or after implantation but rather indicates an enrichment of Xe and Kr in the solar wind source region. Such enrichments are well known for elements with a first ionization potential (FIP) less than about 10 eV and the enrichment factor for Xe of ~4 is similar to that of low-FIP elements in the (slow) solar wind (see the *Moon* section for details).

THE GIANT PLANETS

The He abundances

The hydrogen- and helium-dominated atmospheres of the giant planets remind us strongly of the composition of the Sun. Therefore, the predominant view prior to the Voyager missions was that the He abundance in the atmospheres of the giant planets is identical to the protosolar value, and it was hoped that an accurate determination would be of importance for our understanding of the evolution of the universe. The first He abundance determination in the Jovian atmosphere by infrared spectroscopy onboard the Pioneer 10 and 11 missions (He mole fraction = 0.12±0.06; Orton and Ingersoll 1976) was too inaccurate to test this assumption. However, a few years later the Voyager data revealed a considerable variability in the He abundances of the four giant planets, indicating that Jupiter and, in particular, Saturn are depleted in He in their outer atmospheres. Therefore, rather than being of cosmological significance, an accurate measurement of the He abundance in the atmospheres of the giant planets is crucial for our understanding of the origin and evolution of the planets themselves. Although a recent reanalysis of the Voyager data for Saturn did not confirm the originally reported large He depletion (Conrath and Gautier 2000; Table 6 and Fig. 6), this view has been strengthened by the Galileo mission, which confirmed a modest depletion of He in Jupiter's atmosphere and revealed in addition a large depletion of Ne and considerable enrichments of Ar-Xe, as discussed below.

Tables 6 and 7 show the available noble gas data for the outer atmospheres of the giant planets. The data base is by far most extensive for Jupiter. The mass spectrometer on the Galileo probe descending into Jupiter's atmosphere yielded abundance values for all five noble gases and isotopic ratios for all except Kr (Mahaffy et al. 1998, 2000). Furthermore, a very accurate He abundance value was obtained with an interferometer on the Galileo probe measuring the refractive index of the Jovian atmosphere (von Zahn et al. 1998). For the other three giant planets, the only available noble gas data are relatively imprecise He abundance values, derived mostly by combining infrared spectroscopy with radio-occultation experiments onboard the Voyager spacecrafts. The occultation of the spacecraft's radio signal by a planet's atmosphere yielded the ratio T/m as a function of the atmospheric depth, where T is the temperature and m the mean molecular weight. The thermal emission spectrum depends on T and the opacity, which in turn is dominated by collisions between the major atmospheric constituents H_2 and He. Combining the data from the two experiments therefore yields both T and the He abundance in the upper atmosphere (e.g., Conrath et al. 1987).

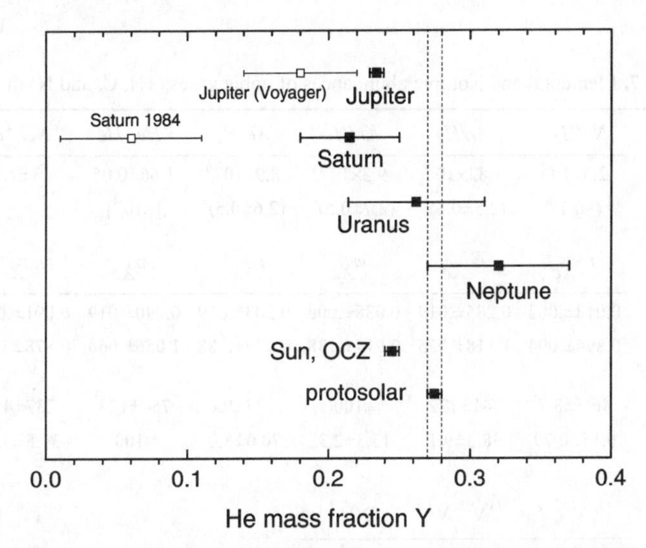

Figure 6. The He mass fraction Y in the atmospheres of the four giant planets. The Jovian value is from the interferometer on the Galileo probe, while all others (including a less precise earlier Jovian value) are from the Voyager missions (Table 6). The Saturn value represented by the filled symbol is the revised value of Conrath and Gautier (2000) based on a new analysis of the original Voyager data (labeled "Saturn 1984"). Also shown are the He abundances in the solar outer convective zone (OCZ) and the protosolar value. He in the atmosphere of Jupiter, and probably also in that of Saturn, is depleted relative to the protosolar abundance due to gravitational settling, although the depletion in Saturn's atmosphere is considerably less pronounced as believed based on the original Voyager analysis. The Uranus and Neptune data are consistent with the expectation that no gravitational settling of He occurred in these planets.

Table 6. He abundances in the giant planets.

	Mixing ratio $^4He/H_2$	*Mole fraction*[a]	*Mass fraction Y*[b]	*Rel. to proto-solar abund.*
Jupiter				
Galileo probe interferometer[1]	0.157±0.003	0.1359±0.0027	*0.234±0.005*	*0.85±0 .018*
Galileo probe mass spectrometer[2]	0.157±0.030	0.136±0.026	0.234±0.044	*0.85± 0.16*
Voyager, revised[3,5]	0.114±0.025	*0.104±0.023*	0.18±0.04	*0.65± 0.14*
Saturn[4]	0.11-0.16	*0.10-0.14*	0.18-0.25	*0.65-0.91*
	(0.034±0.024)		(0.06±0.05)	*(0.22± 0.18)*
Uranus[5]	*0.183±0.046*	0.152±0.033	0.262±0.048	*0.95±0 .17*
Neptune[6]	*0.244±0.055*	0.190±0.032	0.32±0.05	*1.16± 0.18*

Values in italics calculated here from original data.

a: mole fraction = $n_{He}/(n_{H2}+n_{He}+\_n_i)$; $\_n_i$= sum of all elements heavier than He.

b: Mass fraction Y = $m_{He}/(m_{H2}+m_{He}+\_m_i)$. Assumed total mass of elements heavier than He: $\_m_i = Z_0 = 0.0192$

1: von Zahn et al. 1998. **2**: Niemann et al. 1998. **3**: Conrath et al. 1984, Conrath and Gautier 2000, IR spectroscopy & radio occultation. **4**: Conrath and Gautier 2000, revised from Conrath et al. 1984 (in parentheses). **5**: Conrath et al. 1987. **6**: Conrath et al. 1991.

Table 7. Elemental and isotopic abundances of noble gases (+H, C, and N) in Jupiter.

	Ne/H₂	*Ar/H₂*	*Kr/H₂*	*Xe/H₂*	$^3He/^4He$	$^{20}Ne/^{22}Ne$	$^{36}Ar/^{38}Ar$
absolute	2.3×10^{-5}	1.82×10^{-5}	9.3×10^{-9}	8.9×10^{-10}	1.66±0.05	13±2.0	5.6±.25
(rel. to solar)	(~0.1)[a]	(2.5±0.5)[a]	(2.7±0.5)[a]	(2.6±0.5)[a]	[$\times10^{-4}$]		
	^{128}Xe	^{129}Xe	^{130}Xe	^{131}Xe	^{132}Xe	^{134}Xe	^{136}Xe
^iXe/total-Xe[b]	0.018±.002	0.285±.019	0.038±.006	0.203±.019	0.290±.019	0.091±.008	0.076±.011
rel. to nonrad. terrestrial Xe[c]	0.898±.094	1.118±.075	0.914±.138	0.929±.088	1.056±.068	0.878±.074	0.866±.126
rel. to ^{130}Xe[d]	46.4±8.7	741±122	≡100	527±93	754±123	237±41	197±41
rel. to ^{132}Xe[d]	6.15±0.79	98.3±9.1	13.3±2.3	70.0±8.0	≡100	31.5±3.4	26.1±4.1
other isotopic ratios	$^{13}C/^{12}C$[e]	$^{15}N/^{14}N$[f]	*D/H*[f]				
	0.0108 ±0.0005	0.0023 ±0.0003	$(2.6\pm0.7)\times10^{-5}$				

For He abundance see Table 6. All data from Galileo probe (noble gases: Mahaffy et al. 2000; $^3He/^4He$: Mahaffy et al. 1998).

a: Mixing ratios of Ne-Xe are also given relative to solar values as stated in original reference (in parentheses).

b: Sum of ^{128}Xe-^{136}Xe = 1.0. Uncertainties from P.R. Mahaffy, pers. comm. 2001 (uncertainties reported in Mahaffy et al. 2000 are relative values for normalization to nonrad. terrestrial Xe).

c: P.R. Mahaffy, pers. comm. 2001, nonradiogenic terrestrial Xe from Pepin 1991.

d: P.R. Mahaffy, pers. comm. 2001, uncertainties of normalizing isotope (16% for ^{130}Xe, 6.5% for ^{132}Xe) quadratically included in stated uncertainties of ratios.

e: Niemann et al. 1998. **f**: Owen et al. 2001.

The He abundances in Table 6 are given in three different ways, as mixing ratios (He atoms per H_2 molecule), as mole fractions (see also Table caption) and as mass fractions Y (in the astrophysical notation, X, Y, and Z denote the mass fractions of hydrogen, helium and the sum of all heavier elements, respectively). In Table 6, Y has been corrected for the presence of heavier elements by assuming a solar system Z_0 of 0.0192 (the protosolar value), which may not pertain to Jupiter itself (cf. von Zahn et al. 1998).

Figure 6 shows the He mass fractions of the giant planets compared with the protosolar value and that in the Sun's outer convective zone (see the *Sun* section). As noted above, the originally published value for Saturn ("Saturn 1984") is much lower than the solar He mass fraction. Following Stevenson and Salpeter (1977a,b), Conrath et al. (1984) interpreted this to indicate that He had become immiscible in metallic hydrogen at high pressure in Saturn's interior, leading to a migration of He droplets towards the planet's center and a corresponding He depletion in the observable atmosphere. For Jupiter the situation was less clear, but a relatively modest He depletion in Jupiter's atmosphere also appeared likely, in agreement with expectations by Stevenson and Salpeter (1977b). For Uranus and Neptune, the Voyager data were, and still are, consistent with the expectation that He has not segregated in these two smaller planets, although in view of the revision of the Saturn value (see the next but one paragraph), the accuracy of the Uranus and Neptune data appears unclear.

The accurate Galileo probe interferometer data has allowed confirmation of a He depletion in Jupiter's atmosphere. Actually, the Jovian He abundance is, within error, the same as the value in the solar convective zone, but this is accidental, because He settling also occurs in the Sun (*Sun* section). On the other hand, the He abundance in Jupiter is lower than the lower limit of the protosolar value by about 6 sigma. So, whereas the implication from the Voyager data of a considerable He depletion in Jupiter's atmosphere relative to the solar convective zone does not hold any more, the Galileo data nevertheless clearly indicate that He has segregated, or still does so, in the atmosphere and interior of Jupiter (vs. Zahn et al. 1998).

The Galileo probe results also prompted a reevaluation of the Saturn He data from Voyager (Conrath and Gautier 2000). The revised He abundance for Saturn is Y_s = 0.18-0.25, much larger than the original value of 0.06±0.05 and consistent with a recent model prediction of 0.11-0.25 (Guillot 1999). So, while it appears that Saturn's atmosphere is somewhat depleted in He, such a depletion is much less dramatic than was assumed after the Voyager missions.

The Ne-Xe abundances in Jupiter

The Galileo probe mass spectrometer also showed that the abundances of the four heavier noble gases in Jupiter's atmosphere relative to H are distinctly non-solar (Table 7). Ne is depleted by about an order of magnitude, whereas Ar, Kr, and Xe all are enriched by a factor of ~2.5. The striking underabundance of Ne is viewed as supporting evidence for He segregation in Jupiter's interior (Niemann et al. 1998; von Zahn et al. 1998), because Ne may well preferentially partition into segregating He drops. The similarity of the Ar, Kr, and Xe enhancement factors is remarkable, the more so given that carbon, sulfur, and probably nitrogen also are enriched by very similar factors (Owen et al. 1999). The authors conclude that these highly volatile elements were mainly supplied by very cold icy planetesimals, which would contain them in solar proportions relative to each other. Owen et al. (1999) suggest comets that formed in the Kuiper belt region outside Neptune as the most plausible candidates, because Kuiper belt objects are thought to have formed at ~30 K, low enough to trap N, Ar, Kr, and Xe in solar abundance relative to C (Owen and Bar Nun 1995), whereas comets now much farther out in the Oort cloud are thought to stem from the Uranus-Neptune region where formation temperatures were higher. Owen et al. (1999) conclude that the heavy element abundances in Jupiter's atmosphere argue against the conventional view that giant planet formation is related to the position of the "snowline" where temperatures in the solar nebula were just low enough (~160 K) for water ice to condense, since Ar and N_2 in such warm planetesimals would be depleted by about four orders of magnitude. He and Ne would not condense even in Kuiper belt objects, hence these gases in giant planets would not originate from planetesimals and the original He-Ne abundances in giant planets can be expected to be proto-solar (Niemann et al. 1998).

Isotopic ratios in Jupiter

Jupiter is the only giant planet for which isotopic ratios of noble gases are available, provided by the Galileo probe mass spectrometer (Mahaffy et al. 1998, 2000; Table 7). This instrument also yielded quite precise values for D/H, $^{13}C/^{12}C$ and $^{15}N/^{14}N$, which are also listed in Table 7.

The D/H and $^3He/^4He$ ratios in giant planet atmospheres may be expected to reflect the protosolar values, and, as discussed in the *Sun* section, the $^3He/^4He$ value of $(1.66\pm0.05) \times 10^{-4}$ deduced by the Galileo probe is now widely accepted as the best value for protosolar He. The Jovian value is similar to that of $(1.5\pm0.3) \times 10^{-4}$ often adopted for primordial He in meteorites, which led Niemann et al. (1998) to conclude that neither the trapping of He in the meteorites nor the raining out of He in Jupiter's interior have led to a large isotopic fractionation. However, primordial He in bulk meteorites is probably mainly contained in preserved interstellar nanodiamonds (Ott 2002) and may therefore not well represent He in the bulk solar nebula. A possibly better meteoritic candidate for solar nebula He is the component He-Q, for which Busemann et al. (2001) report $^3He/^4He = 1.23 \times 10^{-4}$. It is gratifying that this is also quite similar to the Galileo probe value, but the difference of some 25% suggests a fractionation, either in the meteorites or in Jupiter. Precise He isotopic measurements in other giant planets could rule on this.

The D/H ratio of $(2.6\pm0.7) \times 10^{-5}$ (Niemann et al. 1998, Table 7) agrees within error limits with the inferred but more precise value of $(1.94\pm0.39) \times 10^{-5}$ determined from He measurements in the solar wind and Jupiter (*Sun* section, Table 3). Implications of this result for galactic evolution and cosmology are discussed by Gloeckler and Geiss (2000) and Niemann et al. (1998) (see also the *Sun* section).

The isotopic compositions of Ne, Ar, and Xe as measured by the Galileo probe mass spectrometer (Table 7) are within error limits identical to solar wind values (Tables 4 and 5). The most precisely measured ratio, $^{36}Ar/^{38}Ar$, almost perfectly agrees with values determined for solar wind with lunar and meteoritic samples and the SOHO spacecraft. The interpretation of this similarity is not as straightforward as it may seem, however. First, within its uncertainty, the Jupiter value is also very close to the terrestrial atmospheric ratio or that of primordial meteoritic Ar (Busemann et al. 2000; see Ott 2002, this volume). Second, we noted in the *Sun* section that $^{36}Ar/^{38}Ar$ in the solar wind may be up to a few percent higher than the bulk solar value. Third, Ar trapped in water ice at ~30 K in the laboratory shows a depletion of the lighter isotope of ~1.4-1.9%/amu (Notesco et al. 1999). Therefore, if Jupiter's Ar has been delivered predominantly by cold icy planetesimals, as is indicated by its elemental abundance (see above), the source of this Ar might have a $^{36}Ar/^{38}Ar$ ratio of ~5.8±0.3 instead of the 5.6±0.25 now observed. In summary, even though quite precisely measured, the Ar isotopic composition in Jupiter appears only to constrain the protosolar $^{36}Ar/^{38}Ar$ ratio to the interval ~5.3-6.0 and hence cannot be used at present to decide whether this protosolar ratio is close to the primordial meteoritic (or terrestrial atmospheric) value of ~5.32, about 4% higher as suggested by the solar wind data at face value (see also the *Sun*

Table 8. Noble gases in the atmosphere of Venus.

radiogenic isotopes (mixing ratios)		
^{4}He (ppm)	12^{+24}_{-6}	upper atmosphere (ref. 1)
^{4}He (ppm)	0.6-12	mixed atmosphere, model-dependent extrapolation of upper-atm. data (ref. 3)
^{40}Ar (ppm)	31^{+20}_{-10}	ref. 2
^{129}Xe (ppb)	<9.5	ref. 4
Venus/Earth (abund. ratio)		
^{4}He	175-3700	ref. 3, He lost from terrestrial atmosphere
^{40}Ar	0.25	ref. 3
non-radiogenic isotopes (mixing ratios)		
^{20}Ne (ppm)	7	adopted value in ref. 3. Reported range (excl. uncertainties): 4.3-13 ppm (ref. 4).
^{36}Ar (ppm)	30^{+20}_{-10}	adopted value in ref. 3, uncertainty from ref. 2. Reported range: 21-48 ppm (ref. 4).
^{84}Kr (ppb)	25^{+13}_{-18}	adopted value in refs. 3 & 4
^{132}Xe (ppb)	<10	adopted value in refs. 3 & 4
Venus/Earth (abund. ratio)		
^{20}Ne	21	ref. 3
^{36}Ar	70	ref. 3
Kr	3	ref. 3
Xe	<35	ref. 3
Isotopic ratios		
^{3}He/^{4}He	$<3\times10^{-4}$	refs. 1 & 2
^{20}Ne/^{22}Ne[a]	12.15±0.4	ref. 5
^{20}Ne/^{22}Ne[b]	11.8±0.6	
^{21}Ne/^{22}Ne	<0.067	ref. 5
^{36}Ar/^{38}Ar	5.45±0.1	ref. 5
^{40}Ar/^{36}Ar	1.11±0.02	ref. 5

1: Hofmann et al. 1980a. **2**: Hofmann et al. 1980b. **3**: Donahue and Russell 1997. **4**: Donahue and Pollack 1983. **5**: Istomin et al. 1983.

a: Mean of reported final Venera 13 and Venera 14 data (12.2 and 12.1, respectively); see text for uncertainty adopted here. **b**: Value in ref. 5 corrected for mass discrimination (see text).

section), or even higher (if Jupiter's Ar would be fractionated upon being trapped in ice).

The ^{20}Ne/^{22}Ne ratio in Jupiter has a relatively large uncertainty, but the reported value is within ~6% identical to values deduced for the solar wind. As the difference between the solar Ne composition and that of atmospheric or meteoritic Ne is considerably larger than in the case of Ar (e.g., Ott 2002), the Ne data from the Galileo probe therefore allow us to conclude what was not possible for Ar: Ne in Jupiter has a solar-like isotopic composition that is distinct from primordial Ne in meteorites or Ne in the Earth's atmosphere. However, the uncertainties of the ^{20}Ne/^{22}Ne ratio are too large to decide whether or not the Ne depletion in Jupiter's atmosphere by an order of magnitude resulted in any isotopic fractionation.

The Xe isotopic composition measured by the Galileo probe is reported in Table 7 and shown in Figure 3 for $^{128-136}$Xe. Although the error bars are large, Jovian Xe is probably isotopically lighter than Xe in the terrestrial atmosphere (Mahaffy et al. 2000). Considering the five most abundant isotopes only (filled triangles in Fig. 3), these authors noted that the Jupiter data compare best with U-Xe, the inferred primordial Xe component (see the *Sun* section and chapter by Ott 2002). However, given the uncertainties of 10-15% for $^{134-136}$Xe, the data are also consistent with the composition of solar wind Xe (see also Pepin and Porcelli 2002).

The ^{15}N/^{14}N ratio measured by the Galileo probe is almost 40% lower than the terrestrial atmospheric value (Owen et al. 2001; Table 7). These authors consider the Jovian value of $(2.3\pm0.3) \times 10^{-3}$ to represent the protosolar N isotopic composition. This result is of potential importance in the dispute about the isotopic composition of nitrogen in the Sun and the solar wind. This issue is discussed in the *Moon* section, where noble gases and nitrogen trapped in dust from the lunar and asteroidal surfaces are presented. Further implications of the N composition in Jupiter are discussed by Owen et al. (2001).

VENUS

Compared to Mars, noble gas data from Venus are scarce and limited to the atmosphere. In this chapter we only briefly present the basic data and some first order implications. A more extensive discussion about the origin of noble gases in Venus and the other terrestrial planets is given by Pepin and Porcelli (2002) elsewhere in this volume.

Noble gases in the Venusian atmosphere have been analyzed by mass spectrometry and gas chromatography on board the Pioneer Venus and several Venera spacecraft (Hoffman et al. 1980a; Donahue and Pollack 1983; Istomin et al. 1983; Moroz 1983; Donahue and Russell 1997). The data and their sources are summarized in Table 8. In the early 1980s, greatly different Kr and Xe abundances were reported by the Pioneer Venus and the Venera teams (see Appendix in Donahue and Pollack 1983). This discrepancy was settled, when Venera 13 and 14 mass spectrometer data were found to be in agreement with the Pioneer Venus results (Donahue 1986). The early Venera results appear to have been compromised by contamination with calibration gas (Istomin et al. 1983). Abundances of the lighter noble gases are rather uncertain also (Donahue and Pollack 1983), especially ^{4}He, which was reliably measured only in the upper atmosphere. Extrapolation of this value to the mixed atmosphere is model-dependent, which results in an uncertainty of a factor of 20 (Donahue and Russell 1997).

A comment on the commonly used values of the isotopic compositions of Ne and Ar is in order here. A rather ill-constrained ^{20}Ne/^{22}Ne ratio of $14.3^{+5.7}_{-3.2}$ was measured by Pioneer Venus (Hoffman et al. 1980a). Istomin et al. (1982) reported a much more precise value of 11.8±0.7 from the Venera 13 mass spectrometer (see also Moroz 1983). This is the value adopted by Donahue (1986) and subsequently by most Western workers (usually citing Donahue 1986). However, in their final data evaluation, Istomin et al. (1983) slightly corrected their ^{20}Ne/^{22}Ne ratio from Venera 13 to a value of 12.2 and in addition reported a value of 12.1 from Venera 14. The first value in Table 8 is the mean of these two ratios. No uncertainties are given for the final Venera data, but it seems clear that they must be smaller than the ~6% reported for the preliminary evaluation of each of two Venera 13 cycles. This is also supported by the excellent agreement of the final Venera 13 and 14 values and the adopted uncertainty in Table 8 here seems thus to be rather conservative. In addition, however, Istomin et al. (1982, 1983) also reported a ^{20}Ne/^{22}Ne ratio for the terrestrial atmosphere of 10.07±0.35 measured in a preflight calibration of the Venera 13 instrument. This is about 3% higher than today's commonly adopted value of 9.80 by Eberhardt et al. (1965). Istomin et al. (1982, 1983) apparently did not correct their reported Venus value for this difference (T. M. Donahue, pers. comm. 2001). Doing so yields the value of 11.8±0.6 stated in Table 8 as $^{20}Ne/^{22}Ne$ ratio of Venus' atmosphere corrected for mass discrimination. Luckily, this value turns out to be essentially identical to the one widely used so far. Note also that the $^{36}Ar/^{38}Ar$ value of 5.45±0.1 by Istomin et al. (1983) reported in Table 8 is slightly different from, and has a considerably lower uncertainty than the value of 5.55±0.6 usually adopted by Western scientists (e.g., Donahue and Pollack 1983).

Despite the noted ambiguities, several important comparisons between noble gas element and isotope abundances in the atmospheres of Venus and the Earth are possible (Table 8), which are discussed in more detail by Pepin and Porcelli (2002, this volume):

(1) Amounts of non-radiogenic Ne and Ar in the atmosphere of Venus are several ten times larger than those for the Earth. Kr also may be somewhat more abundant in Venus than in the Earth. The large Ne and Ar overabundances in Venus are usually attributed to a solar wind component implanted into proto-Venusian dust or planetesimals (e.g., Bogard 1988).

(2) Radiogenic ^{40}Ar in Venus' atmosphere only amounts to about a quarter of that in the Earth's atmosphere. In contrast, there is between 175 to 3700 times more radiogenic ^{4}He in Venus' atmosphere as in that of the Earth. However, ^{4}He in the terrestrial atmosphere is readily lost to space and the ^{4}He outgassed from the Earth's interior throughout its lifetime is at least 3 times (perhaps 60 times) larger than the amount now present in the atmosphere of Venus (Donahue and Russell 1997). Therefore, the non-radiogenic gases suggest that outgassing rates on Venus are several times lower than on Earth (Krasnopolsky et al. 1994; Donahue and Russell 1997; Kaula 1999).

(3) The $^{20}Ne/^{22}Ne$ ratio in Venus is higher than the terrestrial atmospheric value. This is consistent with a considerable solar wind contribution to the Ne budget of Venus' atmosphere as indicated by the high Ne abundance. The $^{36}Ar/^{38}Ar$ ratio of Venus also may be slightly higher than in the terrestrial atmosphere and close to the solar wind value, but uncertainties are too large for any firm conclusion.

THE MOON

The Moon is highly depleted in volatile elements. It is therefore not surprising that no indigenous noble gases have been found in lunar rocks. On the other hand, the lunar regolith (the some 10 m thick dust layer on the lunar surface) contains large amounts of noble gases implanted by the solar wind. We presented in the *Sun* section some of the major findings from this precious archive of solar history. In this section we discuss the lunar data in more detail and will occasionally also consider work on solar-gas-rich meteorites, which are compacted samples from the regoliths of asteroidal parent bodies. Besides the solar noble gases, the grains contain other implanted noble gas components of lunar origin, such as degassed and re-implanted radiogenic ^{40}Ar, whose abundance yields a semi-quantitative estimate of the time when a sample trapped its noble gases at the lunar surface. Furthermore, we will discuss nitrogen in lunar samples and we will also mention the "cosmogenic" noble gases produced in lunar soils and rocks by interactions with the galactic and solar cosmic radiation. Noble gases in the tenuous lunar atmosphere are discussed in the following section and radiogenic ^{40}Ar as a dating tool for lunar soils and rocks is covered by Swindle (2002b) elsewhere in this volume.

Trapped solar (and other) noble gases in the lunar (and asteroidal) regoliths

The solar wind is implanted only a few ten nanometers deep into solid matter. Yet almost every lunar regolith grain from the sampled uppermost 2 meters contains noble gases from the solar wind (e.g., Bogard et al. 1973). This testifies to the intense stirring or "gardening" of the regolith by meteorite bombardment which has brought most grains at least once to the immediate surface. Some of the samples in our collection trapped their solar wind probably as early as 4 Ga ago, whereas other samples contain much more recently implanted SW. To make full use of the lunar archive of solar history, we therefore need to estimate about how long ago a sample collected its share of SW, as is discussed next.

The antiquity of lunar samples. The "antiquity" of a sample is the time in the past when it trapped its solar wind ions. There is no straightforward way to determine the antiquity of lunar samples (Kerridge et al. 1991b). A qualitative measure may be the depth within the regolith, as samples with an early exposure on the surface may on average be buried deeper today. Another rough indicator may be the cosmic-ray exposure age, because samples that have been in the regolith for a long time may have had their first exposure on the very top rather early. A more direct measure of antiquity relies on radiogenic nuclides that were trapped at the same time as the solar wind. In particular, the ^{40}Ar residing near grain-surfaces is widely used as a semi-quantitative antiquity estimate. This surface-correlated ^{40}Ar is radiogenic Ar from the decay of ^{40}K (half-life 1.25 Gyr) in the Moon and degassed to the atmosphere. Most probably it got ionized by solar UV in the lunar atmosphere and was then accelerated in the electromagnetic field induced by the solar wind and re-implanted into the regolith surface (Manka and Michel 1971). Such "parentless" ^{40}Ar is found in every lunar soil sample. Xenon-129 from decay of ^{129}I (15.7 Myr) and $^{131-136}$Xe from fission of ^{244}Pu (α-half-life 80 Myr) are other radiogenic nuclides for which a parentless component has been found on grain surfaces of a few, apparently very ancient samples. This is reviewed by e.g., Swindle et al. (1986).

To use parentless ^{40}Ar as a measure of antiquity, it first must be normalized to the length of time a sample has spent on the lunar surface. This is done with the concentration of solar ^{36}Ar. Next the ^{40}Ar/^{36}Ar ratio as a function of time needs to be determined. Figure 7 shows such a calibration by Eugster et al. (2001). The solid line is the calibration curve. On its low-age end, it is defined by sample 67601 whose antiquity can be constrained to $\leq$50 Ma before present, the age of the crater on whose rim the sample was taken. The three other solid symbols represent samples for which the time when they were irradiated on the regolith surface could be determined by the amount of Xe from the neutron-induced fission of ^{235}U (half-life 704 Myr). The open symbols represent less reliable antiquity estimates. Samples 14301 and 14318 are regolith breccias (former soils compacted by meteorite impact) containing also parentless ^{129}Xe and $^{131-136}$Xe. This confirms at least qualitatively their exceptional antiquity, although it is unclear whether their nominal I-Pu-Xe ages of only several 10 Myr younger than the 4.57 Gyr of the oldest meteorites (Swindle et al. 1986) should be equated with the time the parentless Xe (and the SW gases) were trapped at their present sites. Eugster and coworkers' antiquity estimate for these samples is several hundred Myr younger (Fig.7).

Quite remarkably, the calibration curve follows closely the exponential function describing the decay of ^{40}K (dashed line). This might suggest that the amount of degassing ^{40}Ar has remained proportional to its production rate over the last 4 Gyr (if the flux of ^{36}Ar from the solar wind has remained constant during this time), rather than to the total amount of ^{40}Ar accumulated in the lunar interior, as might be expected. Alternatively, the coincidence of the two curves might be fortuitous and the decreasing ^{40}Ar/^{36}Ar ratio be a result of a declining relative degassing rate because of secular cooling of the moon (see *Exospheres...* section). In any case, the calibration curve in Figure 7 should be taken with a grain of salt, because all calibration points have considerable uncertainties, and particularly because the antiquity of most samples is probably not very

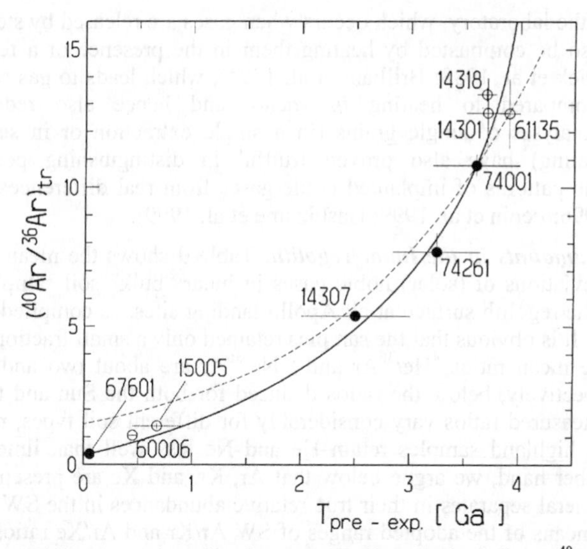

Figure 7. The solid line shows the calibration curve for the antiquity indicator $^{40}Ar/^{36}Ar$ for lunar soils, proposed by Eugster et al. (2001). The antiquity (abscissa) is the time when a sample has trapped its solar wind noble gases at the immediate lunar surface. The ordinate shows the ratio $(^{40}Ar/^{36}Ar)_{tr}$ of Ar trapped on grain surfaces. Filled and open symbols represent calibration data of higher and lower quality, respectively. The dashed line is an exponential curve illustrating the decay of ^{40}K, the parent isotope of ^{40}Ar. [Used by permission of The Meteoritical Society, from Eugster et al. (2001), Meteoritics Planet. Sci., Vol. 36, Fig. 4, p 1107.]

well defined, since soils may consist of grains with quite different regolith histories.

Unfortunately, no direct antiquity estimate exists for gas-rich meteorites. It is often believed that most of them collected very ancient solar wind from the early Sun, when collisions in the asteroid belt were much more numerous and hence regolith formation rates much higher than today. The recent discovery of halite grains with formation ages >4 Gyr in two gas-rich meteorites (Whitby et al. 2000; Bogard et al. 2001) supports the idea that at least some gas-rich meteorites indeed contain very old solar wind, because it is questionable whether fragile halite crystals could survive for long times in an active regolith. On the other hand, present-day asteroids are covered with thick, near the surface probably active regoliths. It appears therefore conceivable that surface dust gets compacted and ejected by recent collisions and hence that some gas-rich meteorites have an antiquity not much exceeding their cosmic ray exposure ages of between several and several 10 Myr (Wieler 2002, this volume).

Analytical techniques. Solar noble gas records in lunar and meteoritic samples are not easy to read. In lunar soils, most of the gas resides in reworked glassy particles (agglutinates) which do not conserve the original grain surfaces onto which implantation occurred. It is therefore preferable to study mineral separates or single mineral grains. Even minerals are not perfect recorders, however, as most of them lost part of their solar He and Ne by diffusion, and careful experiments are needed to recognize such distortions. A gas release technique that has proven particularly useful is etching of samples in vacuo in a device connected directly to a mass spectrometer (closed system stepped etching, CSSE; e.g., Wieler et al. 1986, Benkert et al. 1993). Gases are released at room temperature by slowly etching the carrier minerals by HF or HNO_3 in a device where vacuum-exposed parts consist of Au and Pt only. This minimizes noble gas fractionation in the laboratory, which occurs when gases are released by stepwise heating. Samples can also be combusted by heating them in the presence of a few millibars of oxygen (e.g., Frick et al. 1988; Brilliant et al. 1994), which leads to gas release at lower temperatures compared to heating *in vacuo*, and hence also reduces diffusive fractionation. Analyses of single grains (in a single extraction or in several steps by incremental heating) have also proven fruitful in distinguishing post-implantation alterations of the patterns of implanted noble gases from real differences between soils (Wieler et al. 1996; Pepin et al. 1999; Hashizume et al. 1999).

Noble gas amounts in the lunar regolith. Table 9 shows the mean concentrations and standard deviations of (solar) noble gases in lunar "bulk" soil samples (the <1mm fraction) from the regolith surface at all Apollo landing sites, as compiled by Fegley and Swindle (1993). It is obvious that the samples retained only a small fraction of the He and Ne, because the mean ratios $^{4}He/^{36}Ar$ and $^{20}Ne/^{36}Ar$ are about two and one orders of magnitude, respectively, below the ratios deduced for both the Sun and the solar wind. Note that the measured ratios vary considerably for different soil types, mainly because plagioclase-rich highland samples retain He and Ne less well than ilmenite-rich mare soils. On the other hand, we argue below that Ar, Kr, and Xe are present in lunar bulk samples and mineral separates in their true relative abundances in the SW (which is why in Table 9 the means of the adopted ranges of SW Ar/Kr and Ar/Xe ratios are similar to the mean measured ratios). It is therefore instructive to compare the amounts of solar Ar-Xe in the entire regolith with the expected fluences onto the lunar surface over a regolith lifetime of some 4 Gyr. Such an estimate is necessarily rather uncertain because only the topmost ~2.5 meters of the 8-15 m thick regolith have been sampled. Noble gas concentrations generally appear to decrease with depth down to 2.5 m, although the trends are not very clear (Fegley and Swindle 1993). Modifying an earlier estimate by Geiss (1973), Wieler (1997) deduced that about 2 Gyr of solar wind irradiation at present-day intensity are needed to account for all the Xe in the two drill cores at the Apollo 15 and 16 sites. Given the uncertain Xe concentration in the unsampled lower portion of the regolith, Wieler (1997) concluded that the Xe amounts in the core require an average flux of SW particles in the last 4 Gyr equal to or perhaps a few times higher than today. Unless the mean SW flux in the past was considerably higher than today, the data therefore indicate that all Xe ions that ever hit the Moon still are present in the regolith, though not necessarily in their original trapping sites. We will use this observation below to argue that the relative abundances of Ar, Kr, and Xe represent the true SW values. Note that such estimates need to consider the fact that the solar wind fluence varies with location on the Moon and is actually minimal around the near-side center from where the Apollo and Luna samples are from (e.g., Fegley and Swindle 1993).

Table 9. Noble gases in lunar samples

	^{3}He (10^{-6} cc/g)	$^{4}He^{a}$ (10^{-2} cc/g)	^{20}Ne (10^{-6} cc/g)	^{36}Ar (10^{-6} cc/g)	^{84}Kr (10^{-8} cc/g)	^{132}Xe (10^{-8} cc/g)
mean concentrations ($\pm 1\sigma$) at all Apollo sites[1]	31±25	7.8±6.3	1370±900	310±120	14.5±5.8	2.4±1.0
		$^{4}He/^{36}Ar$	$^{20}Ne/^{36}Ar$		$^{36}Ar/^{84}Kr$	$^{36}Ar/^{132}Xe$
mean at Apollo sites[1]		250	4.4		2140	12900
solar wind[2]		30500	47		1670-2390	9900-16800
sun[2]		43000	41		1980	57400

1: Fegley and Swindle 1993. **2**: see Table 2.

Helium-3 as a future resource? The inventory of volatile elements in the lunar regolith is also of interest in view of their potential use as resources (Fegley and Swindle 1993). Lunar ^{3}He is often proposed as fuel for future deuterium-^{3}He fusion reactors (e.g., Wittenberg et al. 1992; Kulcinski and Schmitt 1992). Fegley and Swindle (1993) estimate a total lunar ^{3}He inventory of roughly 10^{9} kg. Johnson et al. (1999) present a ^{3}He abundance map which is partly based on the titanium concentration determined by the Clementine mission, since the Ti concentration is a measure of the abundance of the He-retentive mineral ilmenite. Perhaps some 7×10^{6} kg of ^{3}He may be mineable from Mare Tranquilitatis alone (Wittenberg et al. 1992), compared to a few kilograms per year available from terrestrial natural gas wells, or to a potential ~100 kg from the decay of ^{3}H from current fission reactors (if collected). Proponents of this idea argue that such mining would only minimally affect the lunar environment and could be implemented without severely disrupting international order (Wittenberg et al. 1992). In contrast, others propose that large-scale human activities on the Moon should be avoided in order not to compromise its unique scientific record (e.g., Bochsler 1994).

Isotopic composition of solar noble gases in regolithic samples. In the *Sun* section we compared the isotopic compositions of He-Ar in the solar wind as derived from lunar and meteoritic samples with direct SW analyses and foil data. Here we discuss the lunar data in more detail. Figure 8 shows several typical Ne data sets, mainly obtained using *in vacuo* etching as the gas release technique. In such three-isotope diagrams with a common denominator, a mixture of two components is represented by a point on the straight line connecting the two end-member compositions and an n-component mixture plots within the respective polygon. Different noble gas components may thus be recognized either by measuring grain-size fractions of a sample or by degassing a sample in several steps. For many data sets in Figure 8 the points of the first gas extractions fall close to the composition labeled SW, representing SW-Ne from the Apollo foils (Table 4). Later extractions define a path pointing first towards the point labeled SEP (solar energetic particles) and then deviating towards the point GCR, representing Ne produced by galactic cosmic rays (see the *Cosmogenic noble gases in lunar samples* subsection). The data sources and further details are given by Wieler (1998).

The two major observations from Figure 8 to be discussed now are:

Well chosen lunar and meteoritic samples appear to conserve the correct isotopic composition, or nearly so, of SW-Ne in their very surface layers, and this composition can be deduced by suitable techniques.

The samples appear to contain a second trapped Ne component besides the SW, residing at larger depth than the few tens of nanometers typical for SW-Ne. Because of its presumed larger implantation energy, this second component has been dubbed SEP (Wieler et al. 1986), a term widely used today in noble gas cosmochemistry. However, this terminology has caused some confusion, because, as we will see below, SEP noble gases in extraterrestrial samples appear to be much too abundant to be due mainly to the rare Solar Energetic Particles that have energies of tens of MeV per amu and are widely studied by the Solar Physics community (*Sun* section). In the following, the reader should be aware of this distinction.

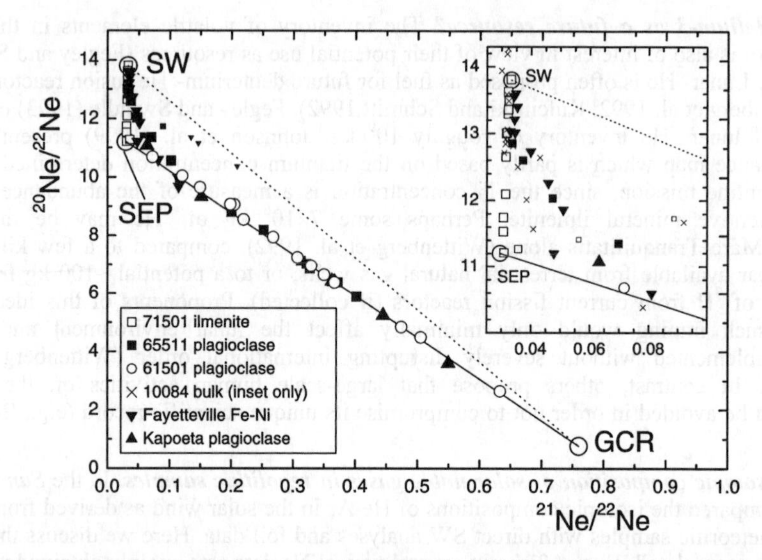

Figure 8. Ne data from from 4 lunar and 2 meteoritic samples containing solar noble gases. Gases were released by *in vacuo* etching, except for samples 10084 (stepwise heating) and 61501 (total fusion of aliquot samples previously etched off-line). Besides solar wind Ne (SW), all samples appear to contain a second solar component, labeled SEP. This is indicated e. g. by the many points on the mixing line SEP-GCR, where GCR represents the composition of Ne produced by Galactic Cosmic Rays. The ^{20}Ne/^{22}Ne ratios above the SW point in the first release fractions of 10084 (inset) indicate isotopic fractionation during gas extraction, which is not observed in the etch-experiments. Data sources listed in Wieler (1998). [Used by permission of Kluwer Academic Publishers, from Wieler (1998), Space Sci. Rev., Vol. 85, Fig. 1, p 304.]

The SW component. It is not straightforward to assume that initial etch fractions should represent the true solar wind composition, because gas losses may lead to fractionations and because of the slightly different penetration depths of different SW isotopes. Wieler (1998), however, points out that such effects can largely be controlled by careful data evaluation. Concerning the implantation effect, simulations have shown that the true isotopic ratio is conserved at a grain surface where ion-implantation and surface loss by sputtering are in equilibrium (R. H. Becker, pers. comm.. 1997; Wieler 1998). In the *Sun* section we also noted the excellent agreement between the isotopic composition of SW-Ne as determined from lunar and meteoritic samples on the one hand, and the foil- and SOHO-based values on the other. This is another argument that the SW noble gas compositions based on regolithic samples are reliable. Therefore, it is now widely agreed upon that the first steps of *in vacuo* etch runs yield the currently best values for the isotopic composition of Kr and Xe in the solar wind, i.e. the elements for which no direct determinations from spacecraft or foil experiments exist (Table 5). Possibly, the slightly higher ^{3}He/^{4}He ratio in the SW derived from lunar samples compared to space-probe values (Table 3) indicates a small, uncorrected cosmogenic ^{3}He contribution or another minor analytical bias, but this is not expected to be a significant problem for studying the possibility of secular variations of the He isotopic composition in the SW as discussed in one of the next subsections.

The SEP component. This component has remained enigmatic ever since it was proposed after the first lunar samples became available (Pepin et al. 1970a; Reynolds et al. 1970; Black 1972). On the one hand, a component with a well-defined isotopic composition different from that of the SW appears to be present near the grain surfaces in essentially every solar-gas-containing lunar or meteoritic sample studied carefully enough. On the other hand, this component appears to be too abundant by several orders of magnitude to be easily reconciled with the flux of particles above typical SW energies. This leads to the question of whether SEP in lunar soils may not be a discrete component, but rather is caused by some fractionating process such as diffusion or the deeper implantation of heavier isotopes.

In all data sets of Figure 8, later steps define essentially the same SEP-Ne component, sometimes directly as the composition of the last step, but in most cases as one end-member of the mixing line pointing towards the GCR point. The large number of points on this line (also displayed by many samples not shown here) is very remarkable and is the main evidence that lunar samples and meteorites contain a well-defined SEP component that is distinct from the SW. The SEP component has also been observed for the other noble gases (Benkert et al. 1993; Wieler and Baur 1994; Rao et al. 1993), and its isotopic composition for He-Ar is given in Tables 3 and 4. Note that the component called Ne-C (Black 1972) is identical to SEP-Ne, although the nominal compositions differ somewhat from each other.

The major problem in understanding the SEP component is that it is so prominent (e.g., Wieler 1998). SEP-Xe nominally represents roughly 20% of the total Xe in many lunar samples, while >0.1 MeV/amu particles are estimated to contribute only a small fraction (10^{-4} to 10^{-5}) to the total solar flux. For mineral grains, part of this SEP overabundance may be explained by loss of the SW component due for example to sputtering or grain abrasion. However, this explanation does not work for the lunar regolith as a whole, if it indeed retains essentially all the solar Xe atoms that ever hit the moon, as we discussed in the *Noble gas concentrations*... subsection. Even though the flux estimates above are plagued by many uncertainties, it seems that alternative explanations need to be considered. A potentially far-reaching hypothesis has been proposed by Wimmer-Schweingruber and Bochsler (2000, 2001), who postulate a galactic origin for SEP noble gases in regolithic samples. According to these authors, the long-term average flux of ions with energies exceeding solar wind values is dominated by accelerated pick-up ions (see the *Elementary Particles in Interplanetary Space* section), mostly from dense interstellar clouds. Accordingly, Wimmer-Schweingruber and Bochsler rename the SEP component HEP, for "Heliospheric Energetic Particles." If this hypothesis turns out to be correct, the lunar regolith might serve not only to study the ancient solar wind but also as an archive of our galactic environment. A possible problem with this view is the observed connection between the composition of SEP and SW. For all five noble gases, the isotopic ratios of SEP are related to the corresponding SW ratio by a factor equal to the square of the respective mass ratios (Wieler and Baur 1994), implying a close link.

Alternative explanations try to explain the SEP component as an artifact. Its discoverers cautioned that isotopic fractionation of a single SW component during laboratory degassing might mimic an isotopically heavier component in later heating steps (Pepin et al. 1970b, Hohenberg et al. 1970). This concern was eliminated by the *in vacuo* etch data, where gases are released at room temperature. Wieler (1998) also considered it very unlikely that either migration of solar wind gas into the grain interior or the slightly larger implantation depth of the heavier SW isotopes would lead to a uniform composition over a large depth range and in many different minerals (as indicated by the many points falling on the tie line SEP-GCR in Fig. 8). Recently, however, Mewaldt et al. (2001b) used new measurements of the fluences of solar particles over a wide energy range to model the depth dependence of the isotopic composition of implanted gases with a uniform source composition, and concluded that the isotope-dependent depth of implantation could in fact mimic an isotopically heavier component. If so, the SEP component would indeed be due mainly to solar particles of somewhat higher energy than the solar wind, but its inferred isotopic composition would be wrong. On the other hand, we noted in the *Sun* section that the solar energetic particles in the ~10 MeV/amu range measured by ACE appear to have a $^{20}Ne/^{22}Ne$ ratio consistent with the value of the SEP component in lunar samples (Leske 2001). Also, Mewaldt et al. (2001b) concluded that the present-day solar particle flux in the energy range above that of the SW falls short by some two orders of magnitude to explain the apparent abundance of the SEP component in lunar samples. In short, this component remains enigmatic. Note that the inference above that the outermost grain layers conserve basically the correct SW isotopic composition remains valid whether or not the SEP component actually has a distinct isotopic composition.

The $^3He/^4He$ ratio in lunar samples of different antiquity. We noted in the *Sun* section that lunar dust samples allow us to study the secular evolution of the $^3He/^4He$ ratio in the solar wind, which has important implications for solar models and estimates of the protosolar $^3He/^4He$ and D/H ratios. Here we discuss relevant data in more detail. We first refer again to the discussion of Table 3 where we noted that the $(^3He/^4He)_{SW}$ ratios in lunar samples are close to values deduced for the present-day SW, indicating that suitable analyses allow a reliable deduction of the isotopic composition of SW He in the past. Minor systematic uncertainties, e.g., due to the different penetration depths of the two isotopes, should not hinder deducing possible temporal trends. Figure 9 shows $(^3He/^4He)_{SW}$ values as a function of antiquity for different lunar samples. The first order observation is that all data fall in a narrow range, showing that mixing of ^3He-rich material into the convective zone during the past 4 Gyr must have been very limited (see

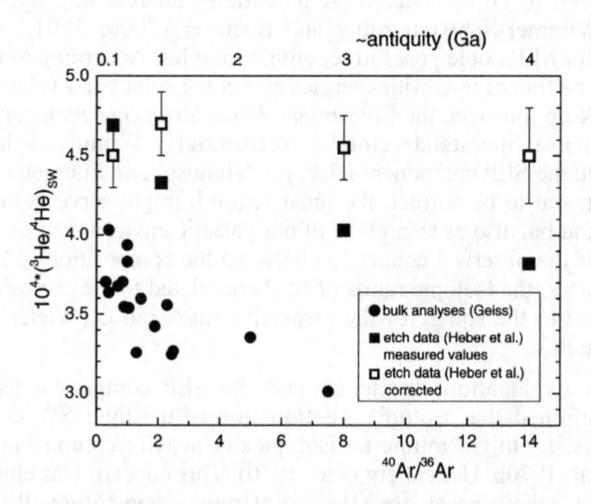

Figure 9. Isotopic composition of solar He in lunar samples as a function of the antiquity indicator $^{40}Ar/^{36}Ar$. The approximate antiquity (Fig. 7) is shown on the upper abscissa. Bulk sample data (Geiss 1973) suggested a secular increase of the ^3He abundance in the outer convective zone of some 10% per Gyr. High resolution *in vacuo* etch analyses of mineral separates (Heber et al. 2001b) appear to imply a somewhat smaller increase of ~5% per Gyr. However, correcting the He data with the Ne isotopic data (not shown) for a probable partial loss of the SW component in the older samples (see *Moon* Section) indicates that the He isotopic composition in the solar wind has actually remained essentially constant throughout the past ~4 Gyr (open squares).

the *Sun* section). However, a closer look suggests a small secular increase of the ^{3}He abundance in the SW and hence the convective zone. The bulk soil grain size fraction data compiled by Geiss (1973) suggest an increase of the ^{3}He abundance of some 30% over the past ~3 Gyr. Data obtained by *in vacuo* etching (Heber et al. 2001b), where noble gases from the outermost grain layers are selectively sampled, show higher ordinate values than the bulk data because the latter represent a mixture of SW-He and SEP-He. At face value, the etch data also suggest an increase of ^{3}He/^{4}He of about 20% over the past ~4 Gyr, slightly less than inferred earlier from the bulk data. However, it is probable that this is an experimental artifact, because samples with high antiquity might have lost relatively more of the outermost SW component, e.g., by diffusion or mechanical abrasion (Heber et al. 2001b). This is suggested by the correlation between the He and Ne isotopic compositions measured in the etch analyses (Heber et al. 2001b), since a secular change of the Ne composition in the solar wind would be unexpected. A correction of the measured ^{3}He/^{4}He values based on the ^{20}Ne/^{22}Ne ratios is given by the open squares in Figure 9. The corrected data are consistent with a constant He isotopic composition in the solar wind over most of solar history. This suggests that the present-day SW analyses can be used to deduce the protosolar (D+^{3}He)/^{4}He composition without needing to correct for main-sequence-produced ^{3}He admixed into the outer convective zone.

Elemental composition of solar noble gases in regolithic samples. Solar He and Ne are severely depleted relative to the heavier noble gases in all constituents of lunar samples due to diffusive losses. This is illustrated by the ^{4}He/^{36}Ar and ^{20}Ne/^{36}Ar ratios in bulk samples when compared to solar or solar wind abundances (Table 9). Depletion factors are strongly mineral-dependent, reflecting variable diffusivities (Signer et al. 1977). Also bulk samples of solar-gas-rich meteorites have lost part of their He and Ne. However, metallic Fe-Ni grains from regolithic chondrites appear to conserve the true relative abundances of solar He, Ne, and Ar, as judged from the flat depth profiles of He/Ar and Ne/Ar observed during *in vacuo* etch runs (Murer et al. 1997). These long-term averages also agree very well with the present-day values deduced by the Apollo Solar Wind Composition experiment (Table 2).

No Kr and Xe data are available yet for the modern solar wind, so lunar samples and a very few meteorites are the only source of information on the abundances of these elements in the solar wind. Table 2 shows that Ar/Kr and Kr/Xe ratios are both quite variable and below the values inferred for the bulk Sun. By analogy with He and Ne, this was first thought to indicate diffusive losses of Ar, and to a lesser extent also Kr and possibly Xe (e.g., Bogard et al. 1973). Kerridge (1980) noted, however, that Kr/Xe ratios do not scatter randomly but correlate with the antiquity of the samples, indicating a secular increase of the Kr/Xe ratio in the solar wind. This conclusion was later confirmed by Wieler and Baur (1995) and Wieler et al. (1996), who showed that all minerals in lunar soils as well as bulk samples retain the true Ar/Kr/Xe elemental ratios in the solar wind. The lunar regolith therefore retains an archive of a secularly variable enhancement of Xe and Kr in the solar wind (Wieler et al. 1996; Wieler 1998). Xenon is enhanced in the solar wind relative to Ar by a factor of about four, similar to the enhancements observed in the interstream (slow) SW for elements with a first ionization potential (FIP) <10 eV (von Steiger et al. 1997). The Kr enhancement is smaller and more variable. It may be unexpected that Xe and Kr display this "FIP-effect" given that their FIPs are above the 10 eV threshold for which no enhancement is observed for other elements, but models where the ionization time in the chromosphere is the actual parameter governing the FIP-effect predict Xe and Kr enrichments (Geiss et al. 1994; see also Wieler 1998). The few available data from meteorites indicate that the Xe and Kr enrichments in the solar wind are even more variable than inferred from lunar samples and that the FIP-effect may have existed since 4 Ga ago or even earlier (Heber et al. 2001a).

Nitrogen in lunar samples

Nitrogen in lunar soils, solar or not? Nitrogen belongs to the very few elements besides the noble gases which are rare enough in lunar rocks that it might be possible to detect the solar wind component in soil samples. However, unlike the noble gases, it has proven very difficult to deduce the isotopic composition of solar nitrogen from lunar samples and no consensus exists up to this day (e.g., Becker 2000). The crucial observations are:

1. Nitrogen concentrations are much larger in lunar soils than in lunar rocks (Mathew and Marti 2001) and in stepwise analyses most of the nitrogen in soils is released in the same fractions as solar wind implanted noble gases (e.g., Frick et al. 1988). This indicates that most of the nitrogen in lunar soils is a trapped component.
2. The isotopic composition of this trapped nitrogen varies by up to 35% between different bulk samples as well as between different extraction steps of individual samples (Kerridge 1975; Clayton and Thiemens 1980).
3. The abundance ratio $N/^{36}Ar$ in lunar samples is about an order of magnitude higher than the corresponding solar wind ratio (e.g., Kerridge 1993).

These observations generally have been interpreted to indicate a secular increase in the $^{15}N/^{14}N$ ratio in the solar wind from about 2.9×10^{-3} several Ga ago to perhaps up to 4.1×10^{-3} in the recent solar wind (e.g., Kerridge 1993). A major problem with this hypothesis is that no generally accepted mechanism for such a secular change has been proposed (Geiss and Bochsler 1982). This classical interpretation also implies that essentially all trapped nitrogen in the soils represents implanted solar wind atoms and so would require a preferential loss of solar wind ^{36}Ar relative to nitrogen, e.g., by diffusion. Wieler and Baur (1995) and Wieler et al. (1996) argued that this is unlikely, based on Ar, Kr, and Xe depth profiles in grains and single grain analyses, respectively.

Therefore, an alternative interpretation is that in most lunar soil grains most of the trapped nitrogen (on average some 90%) has a non-solar source, as suggested for example by the highly variable $N/^{36}Ar$ ratios and the concomitant nearly constant Ar/Kr/Xe ratios in single grains (Wieler et al. 1999). Recent ion-probe measurements on single grains thought to contain relatively little non-solar N yielded an upper limit of $< 2.79 \times 10^{-3}$ for the $^{15}N/^{14}N$ ratio in the solar wind (Hashizume et al. 2000). This value is close to the one inferred earlier for the ancient solar wind, but Hashizume and coworkers argue that the higher ratios observed in younger samples do not indicate a secular variation in the composition of the solar wind but rather a contribution of isotopically heavier non-solar nitrogen. The Hashizume et al. value is in good agreement with the $^{15}N/^{14}N$ ratio of $(2.3 \pm 0.3) \times 10^{-3}$ reported for the Jovian atmosphere from Galileo probe data (Owen et al. 2001; see Table 7). On the other hand, both of these values are in striking disagreement with the direct measurement by the SOHO spacecraft, which yielded a solar wind $^{15}N/^{14}N$ value of $\left(5.0^{+1.9}_{-1.1}\right) \times 10^{-3}$ (Kallenbach et al. 1998a). It is hoped that the upcoming Genesis mission will yield a precise and unambiguous value for the N composition in the present-day solar wind.

Indigenous lunar nitrogen. The relatively large amounts of trapped nitrogen in lunar soils mean that indigenous lunar nitrogen should be preferentially found in interior portions of rock samples (Murty and Goswami 1992; Mathew and Marti 2001) or in pyroclastic deposits (Kerridge et al. 1991a). However, the isotopic signature of the indigenous N component may well be affected by cosmic-ray-produced N even in such samples, and the low abundances of indigenous N, on the order of 1 ppm (Mathew and Marti 2001), also make the analyses prone to atmospheric contamination. Nevertheless, the indigenous component can be separated by a stepwise gas extraction protocol in suitable samples, and three studies all find the isotopic composition of indigenous lunar nitrogen to be enriched in the heavy isotope by 13-17‰ relative to the composition of the terrestrial atmosphere (Kerridge et al. 1991a; Murty and Goswami 1992; Mathew and Marti 2001). Mathew and Marti (2001) note the approximate consistency of this composition with values reported for the terrestrial mantle but caution that fractionation effects during the formation of the moon are difficult to assess.

Cosmogenic noble gases in lunar samples

Production systematics. The production of noble gases and radionuclides by galactic cosmic rays in meteorites and terrestrial surface samples, and their applications, are discussed in two separate chapters by Wieler (2002) and Niedermann (2002), respectively. Here we briefly review some of the major findings based on studies of cosmogenic noble gases in lunar samples (for lunar meteorites see Wieler 2002). For a more detailed discussion of cosmogenic nuclide production systematics the reader is also referred to Reedy and Arnold (1972), Hohenberg et al. (1978), Vogt et al. (1990), Masarik et al. (2001) and Leya et al. (2001). The last paper also provides production rates of cosmogenic Ne (as well as important radionuclides) from each major target element as a function of depth based on a physical model. Earlier data (including the other noble gases) are given by Hohenberg et al. (1978), Regnier et al. (1979), and Reedy (1981).

Figure 10 shows the production rate of ^{21}Ne for the average chemical composition of the Apollo 15 deep drill core calculated using two different physical models (Leya et al. 2001; Masarik et al. 2001; I. Leya and J. Masarik, pers. comm. 2001). The differences of some 15-20% between the two curves reflect the uncertainties to be expected for current model predictions of this kind. Near the surface, the production rate first increases with depth due to the build-up of the secondary neutron cascade (cf. Wieler 2002), but then decreases steeply and at 500 g/cm^2 it is already about an order of magnitude lower than the maximum value near the surface. This means that nominal exposure ages of lunar samples are inherently uncertain, unless shielding information is available. In addition, the regolith is constantly stirred by meteorite impacts. This "gardening" implies that samples usually did not acquire all their cosmogenic nuclides at the same depth. This is especially true of the soils, which consist of many individual grains. Exposure ages of lunar samples stated in the literature (in particular the commonly determined ^{21}Ne and ^{38}Ar ages) usually are calculated by adopting a production rate equal to that at the immediate surface or to the maximum production rate slightly below the surface. Therefore, such ages often are minimum values. As a further problem, plagioclase does not quantitatively retain cosmogenic Ne. This mainly affects calculated exposure ages of the plagioclase-rich lunar highland samples.

Shielding corrections commonly used for meteorites that are based on the ratio ^{22}Ne/^{21}Ne (Wieler 2002) are impractical for lunar samples, because this parameter becomes ambiguous at larger depths (Masarik et al. 2001). Moreover, in soils this ratio usually cannot be determined due to the large amounts of trapped solar Ne. A more suitable shielding index for lunar samples is the ratio ^{126}Xe/^{131}Xe (Eugster 1985). Shielding is also inherently corrected in the ^{81}Kr-Kr method, where the production rate of a stable Kr isotope in a sample with unknown shielding is determined via the concentration of radioactive ^{81}Kr, which is in steady state concentration due to its short half-life of 230,000 years (e.g., Marti et al. 1970). However, this method is also based on the assumption that the entire exposure of a sample occurred at the depth where the ^{81}Kr was acquired over the past one Myr or so. Nevertheless, lunar samples have frequently been dated by ^{81}Kr-Kr, because they are rich in the main target elements for Kr production (Sr, Y, and Zr) and because heterogeneous concentrations of these target

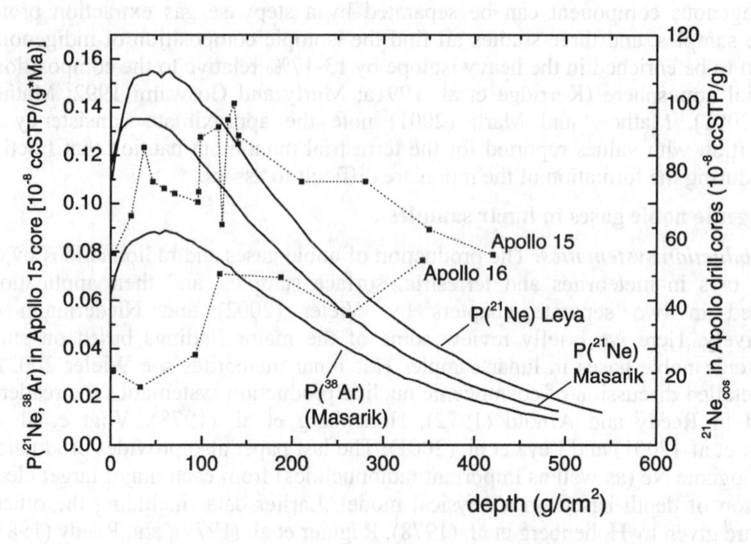

Figure 10. The solid lines show production rates of cosmogenic ^{21}Ne and ^{38}Ar as a function of depth in a lunar regolith with the chemical composition of the Apollo 15 Deep Drill Core (left ordinate scale). The three curves were calculated with the nuclide production models by Masarik et al. (2001) and Leya et al. (2001), respectively (data courtesy Jozef Masarik and Ingo Leya). The difference of ~15-20% between the two ^{21}Ne curves illustrates the accuracy of present-day physical model calculations. Also shown are measured concentrations of cosmogenic ^{21}Ne in the Apollo 15 and 16 deep drill cores (Bogard et al. 1973; right ordinate scale; depth scale converted assuming a bulk soil density of 1.75 g/cm^3 for Apollo 15 and of 1.5 and 1.65 g/cm^3 in the uppermost 80 cm and the lower portions, respectively, of the Apollo 16 core; Carrier et al. 1991, Fig. 9.14). Whereas the production rate decreases by about an order of magnitude along the topmost ~500 g/cm^2, the measured concentrations show no clear trend (Apollo 15) or even increase with depth (Apollo 16), indicating that the lunar regolith suffered a complex irradiation history.

elements are of little concern.

It is also important to note that, in addition to nuclides produced by galactic cosmic rays, isotopes produced by solar cosmic radiation (SCR) are common in lunar samples (e.g., Vogt et al. 1990), unlike in meteorites where the outermost few cm of the pre-atmospheric body—the penetration depth of SCR—are usually lost. SCR noble gas production rates are provided by Hohenberg et al. (1978), Regnier et al. (1979), and Reedy (1981). SCR production dominates over the galactic component in the first ~1-2 cm, but, because for most samples the regolith is mixed to considerably larger depths during the relevant exposure period, SCR contributions usually are neglected when exposure ages are calculated.

Cosmic-ray exposure ages of lunar samples. Despite the considerable uncertainties discussed above, cosmogenic nuclides have proven very valuable for deciphering lunar regolith dynamics and the near-surface history of samples. Numerous exposure ages have been reported, largely in the various Proceedings of the Lunar (and Planetary) Science Conferences in Houston. Soil exposure ages usually fall into a range of one to several hundred million years. As noted above, in many cases these ages should not be taken literally, because they may not date a real event that excavated a sample from at least several meters depth, i.e. from below the reach of galactic cosmic rays. In some cases, however, such an assignment is possible. Several relatively young craters have been dated by cosmogenic nuclides in samples from their rims or from rays originating from these craters, e.g., Tycho at ~100 Ma (Arvidson et al. 1976), North Ray Crater at ~50 Ma (Marti et al. 1973; Drozd et al. 1974), Cone Crater at ~26 Ma (Turner et al. 1971), Shorty Crater at 17 Ma (Eugster et al. 1979), and South Ray Crater at a mere 2 Ma (Drozd et al. 1974; Eugster 1999).

Profiles of cosmogenic nuclide concentrations along deep drill cores constrain the exposure history of the uppermost ~2.5 m of the regolith at the respective sites. Cosmogenic ^{21}Ne concentrations in the Apollo 15 and 16 deep drill cores (Bogard et al. 1973) vary by about a factor of 2-3 in each core (Fig. 10). Whereas the Apollo 15 core does not show a trend with depth, the uppermost ~60 cm at the Apollo 16 site appear to have experienced a lower fluence of GCR particles than the lower portions. By comparing the measured ^{21}Ne profiles with the production rate curves it becomes obvious that the regolith has not experienced a simple deposition and accumulation history. For example, nominal exposure ages of ~370 Myr and ~4.3 Gyr are obtained, respectively, for the uppermost and lowermost sample of the Apollo 15 core, assuming production rates at these depths. Bogard et al. (1973) discuss various scenarios for regolith accumulation at the two sites. A very fast accretion over a few Myr of previously unirradiated material followed by long residence times clearly does not explain either core, and also a very slow accretion with the topmost layers irriadiated only for a very short time can be rejected. The Apollo 15 core data are roughly consistent with an accumulation of unirradiated material during perhaps 600 Myr followed by a static irradiation of about 400 Myr. However, the relatively large variations of ^{21}Ne over narrow depth ranges indicate that the true history of the core material probably is more complicated, with variably pre-irradiated layers having accumulated on top of each other. The lower ^{21}Ne concentrations in the top samples of the Apollo 16 core may be due to a ~60 cm layer of relatively fresh material added relatively late, or may alternatively reflect a relatively slow continuous accretion followed by a period of irradiation in a static regolith.

Other cosmogenic nuclides. Cosmogenic radionuclides of various half lives have also been extensively studied in lunar samples. Measurements on the Apollo drill cores (e.g., Nishiizumi et al. 1976, 1984, and references therein) yield information on regolith mixing over times comparable to the half lives of the respective nuclide. The long exposure ages of lunar samples also lead to the accumulation of measurable amounts of stable cosmogenic non-noble-gas nuclides. Best known are several isotopes of Gd, Sm, and Cd, which have very high capture cross sections for slow (thermal and epithermal) neutrons, which are secondary products from the galactic cosmic rays. Recent such work is described, e.g., by Sands et al. (2001), who also give references to earlier studies. One of the model histories for the Apollo 15 deep drill core proposed by Russ et al. (1972) based on neutron stratigraphy is in agreement with the long accumulation/long static irradiation model discussed by Bogard et al. (1973).

THE EXOSPHERES OF THE MOON AND MERCURY

The Moon and Mercury are usually called atmosphereless bodies. However, they are actually surrounded by tenuous atmospheres with near-surface densities not unlike those of cometary comae (cf. Stern 1999b). Atoms or molecules in both atmospheres interact predominantly with the planet's surface rather than with each other. The atmospheres of the Moon and Mercury are therefore usually called "exospheres," although the behavior of an exosphere lying above a denser atmosphere is much simpler and better understood than that of an exosphere interacting with a solid planetary surface (Hunten and Sprague 1997). Reviews on the composition, origin, and character of the two atmospheres are provided by Hunten et al. (1988); Sprague and Hunten (1995); Hunten and Sprague

1997). Reviews on the composition, origin, and character of the two atmospheres are provided by Hunten et al. (1988); Sprague and Hunten (1995); Hunten and Sprague (1997); Killen and Ip (1999); and Stern (1999b). Here we present measured abundances of noble gases or upper limits thereof (Table 10).

Neutral He and Ar were detected near the lunar surface by the Lunar Atmospheric Composition Experiment (LACE) on the Apollo 17 mission. An early inference of ^{20}Ne in the lunar atmosphere based on the Suprathermal Ion Detector Experiment (SIDE) has not been confirmed and has been largely ignored (Stern 1999b). Also initial reports of Ne in the daytime atmosphere of the moon (Hoffman et al. 1973) appear to be consistent with contamination (Hodges et al. 1973; Stern 1999b), although the reported ^{20}Ne/^{22}Ne ratios of around 14 (Hoffman et al. 1973) are close to the solar wind ratio. The upper limits for Kr and Xe in Table 10 were obtained by the Apollo 17 UV spectrometer. However, from the abundance of ^{36}Ar and solar elemental ratios, true Kr and Xe abundances are expected to be considerably below these limits. Stern (1999b) reports a total He amount in the lunar atmosphere of 10^{30} atoms (7×10^6 g), Hodges (1975) gives a total Ar amount of 1.5×10^{28} atoms (1×10^6 g). Note that the corresponding He/Ar ratio is not entirely consistent with that derived from the respective column densities given in Table 10, which apparently reflects uncertainties in the assumptions involved in deriving the various values.

In Mercury's atmosphere, He has been detected by UV spectroscopy during the Mariner 10 flybys. This instrument only yielded an upper limit for ^{40}Ar.

Further species identified in the lunar and/or Hermean exosphere are hydrogen (Mercury only), sodium and potassium, and possibly (on Mercury only) oxygen (see cited review articles). For various other species upper limits have been reported. The non-detection of H on Moon is unexplained, and it is certainly possible that undetected

Table 10. Noble gases in the exospheres of Mercury and the Moon.

Species	*Mercury*	*Moon*	*Units and comments*
He[1,2]	6,000	2,000 (day) 40,000 (night)	Number densities at surface of planet (cm^{-3}) Nightly lunar maximum: 20,000-40,000 (ref 2)
He[3]	3×10^{11}	1×10^{11}	Column density (cm^{-2})
He[4]		9×10^{23}	loss rate to space (atoms/s). In addition, ~5×10^{23} He atoms/s probably lost from non-atmospheric component (see text)
He[4]		2×10^5	lifetime (s) against (thermal) loss
Ar[1,2,5]	<3×10^7	<100 (nightly min.) >35,000 (sun-rise max.)	Number densities (cm^{-3}). Lunar minimum and maximum correspond to detection limit & detector saturation, resp. (refs. 4, 2). Ar during lunar day not unambiguously detected (ref. 2)
Ar[3]	<2×10^{13}	2×10^{10}	Column density (cm^{-2})
Ar[4]		2×10^{21}	loss rate to space (atoms/s)
Ar[4]		~9×10^6	lifetime (s) against (photoionisation) loss, including time spent adsorbed on surface
^{40}Ar/^{36}Ar[4,2]		~10	measured prior to local sunrise
Kr[6]		<20,000	Number density at surface (cm^{-3})
Xe[6]		<2,000	Number density at surface (cm^{-3})

Cited references are not always original data source. **1**: Hunten and Sprague 1997. **2**: Stern 1999b. **3**: Killen and Ip 1999. **4**: Hodges 1975. **5**: Hoffman et al. 1973. **6**: Feldman and Morrison 1991, Apollo 17 UV spectrometer.

constituents in both atmospheres increase the observed total number densities of ~10^5 cm^{-3} by several orders of magnitude (Hunten and Sprague 1997).

Sources and sinks

The $^{40}Ar/^{36}Ar$ ratio measured by the Apollo 17 mass spectrometer is about 10 (Hofmann et al. 1973). This value has been determined from data taken shortly before local sunrise, when the ^{40}Ar signal had already strongly increased relative to its nighttime minimum, indicating that the analyzed ^{40}Ar is not to any large extent from terrestrial atmospheric contamination carried onto the Moon, but rather exospheric Ar frozen onto the soil during nighttime and released at the warming terminator. Therefore, the measured $^{40}Ar/^{36}Ar$ ratio of ~10 is probably the true exospheric value. It seems clear that the ^{36}Ar derives almost exclusively from the solar wind, having been implanted into the regolith and later released, for example by sputtering (Sprague and Hunten 1995; Hunten and Sprague 1997; Stern 1999b). The ^{40}Ar source is degassed radiogenic Ar from K decay in the lunar interior, either by direct outgassing or by loss from the surfaces of regolith grains into which it had been implanted by the Manka-Michel mechanism (see the *Moon* section). The LACE data suggest that the ^{40}Ar release from the interior is variable, perhaps due to tidally induced moonquakes (Stern 1999b). It also seems clear that the major source of ^4He (^3He has not been detected) in the lunar and Hermean atmospheres is solar wind He, with a possible minor contribution of radiogenic He from the lunar interior (see following paragraphs).

The primary loss process for exospheric Ar is photo-ionization and subsequent acceleration by the magnetic field of the solar wind, followed by re-implantation into the regolith or escape to space. Helium, on the other hand, is gravitationally bound neither on Mercury nor the Moon, and therefore is lost primarily by Jeans escape (Stern 1999b; Killen and Ip 1999). Hodges (1975) models the escape of lunar atmospheric helium and argon. Deduced loss rates and lifetimes are listed in Table 10. The average loss rate of ^{40}Ar of 2×10^{21} atoms/s corresponds to 8% of the present production rate from ^{40}K decay in the entire Moon. Less "new" radiogenic ^{40}Ar atoms would be needed if a considerable fraction is recycled between the atmosphere and regolith, but Hodges (1975) notes that the temporal variation of the exospheric ^{40}Ar abundance argues strongly against substantial recycling of ^{40}Ar. A fraction of 8% of the total freshly produced Ar being released from the lunar interior appears astonishingly high. Hodges (1975) argues that its most probable source is a semi-molten central region of the moon of about 750-km radius with a bulk lunar K abundance of 100 ppm. He suggests that the Ar is released to the surface possibly episodically, correlated with seismic processes or creating seismicity when gas under high pressure opens vents to the lunar surface, although it is not clear whether such a process would work. Alternatively, assuming that production and loss of ^{40}Ar are not in equilibrium, one may also compare the present loss rate with the total number of ~1.5×10^{40} atoms of ^{40}Ar ever produced in the moon. This yields a relative loss rate of ~1.3×10^{-19} s^{-1}, if nearly all of these atoms were still retained in the lunar interior, or less than 2% loss throughout lunar history. A problem with this interpretation may be that the $^{40}Ar/^{36}Ar$ ratio of trapped Ar in lunar soils has decreased throughout lunar history approximately exponentially, roughly consistent with the half-life of ^{40}K (see Fig. 7 and corresponding discussion), whereas this ratio might be expected to increase with time if the ^{40}Ar concentration in the atmosphere is roughly proportional to the ^{40}Ar accumu-lated in the lunar interior. Perhaps this indicates that the relative degassing rate of ^{40}Ar from the lunar interior was considerably higher in the past due to the secular cooling of the Moon.

Hodges (1975) also estimates a supply rate of about 1×10^{23} atoms of radiogenic ^4He from the lunar interior to the atmosphere, or about 10% of the total loss rate to space.

This estimate is not based directly on atmospheric He data but on the assumption that He is outgassed from the same volume as Ar and taking the present day production rate ratio of radiogenic ^{4}He to ^{40}Ar, which is more than twice as high as the integrated production ratio over 4.5 Gyr.

In addition, Hodges (1975) estimates that the total loss rate of atmospheric ^{4}He is only about 70% of the flux rate supplied by the solar wind, and suggests that this is due to an additional, non-atmospheric loss of He as sputtered ions or suprathermal atoms, which would not have been detected by the Apollo 17 mass spectrometer. Whereas we have discussed in the *Moon* section that most of the solar wind Kr and Xe ever having hit the lunar surface may still be retained in the regolith, the He/Xe ratio in lunar soils is about three orders of magnitude lower than the solar wind ratio (Table 9) so that indeed gain and loss of solar wind He must be approximately in equilibrium.

The fact that the ^{40}Ar/^{36}Ar ratio in the lunar exosphere is about an order of magnitude higher than in relatively recently irradiated lunar samples (*Moon* section) is perhaps surprising. Hodges et al. (1973) conclude that this suggests that the soil is not saturated with ^{36}Ar and hence most of the impinging solar ^{36}Ar is permanently trapped.

For non-condensable gases, the number density of a species is expected to be proportional to $T^{-5/2}$ (Hodges and Johnson 1968; see Stern 1999b). The ~20 times higher He density at night compared to the daytime value nicely obeys this relationship. In contrast, the Ar density in the lunar atmosphere is lower during the night than during the day and steeply increases at sunrise, because Ar gets adsorbed in the regolith at lunar night-side temperatures.

COMETS

Comets are considered to be chemically even more primitive than CI chondrites, the most primitive meteorites, because comets are enriched relative to CIs in the highly volatile elements H, N, and O (Altwegg et al. 1999). Therefore, it has been a long-standing goal to determine noble gas abundances in comets. Noble gases are a powerful tool for constraining the thermal history of cometary ices (Stern 1999a). Also the idea that comet-like icy planetesimals may have played a major role in the origin of planetary atmospheres (Owen and Bar-Nun 1995) can be tested with cometary noble gas data (Owen et al. 1999; see the *Giant Planets* section).

So far, only a single identification of a noble gas in a comet has been published. Stern et al. (2000) report a production rate ratio Ar/O of 0.0058±0.0017 in Hale-Bopp's coma (Table 11). Assuming this value to represent the abundance ratio Ar/O, the authors conclude that Ar is enriched in the coma by a factor of 1.8±0.96 relative to oxygen and solar values, indicating that the comet's nucleus never experienced temperatures higher than 35-40K. However, as C/O, Si/O, and S/O in comets are found to be solar, this selective enrichment of Ar is not comparable to the situation in Jupiter's atmosphere, where Ar is enriched by a factor of ~2.5 relative to H but is present in solar proportion relative to C, N, S, Kr, and Xe. Whereas the Jupiter data can be explained by trapping volatiles at low temperatures in planetesimals (*Giant Planets* section), the selective Ar enrichment in Hale-Bopp is enigmatic (T. Owen, pers. comm. 2001), also in view of the upper limit of Ar/O < 0.1 times solar set by Feldman et al. (2001) in comet C/2001 A2 (Linear) using the Far Ultraviolet Explorer.

Besides this Ar measurement, several significant upper limits have been reported for He and Ne in comets (Table 11). Stern et al. (1992) found that He is depleted in the coma of Comet Austin by more than four orders of magnitude relative to solar abundances (a not very stringent limit for Ar is also given in the same paper, corresponding to an

Table 11. Noble gases in comets

Comet		*rel. to solar ratio*[a]	*depletion factors*[a]
Austin (coma, subliming ices)[1]	He/O < 7.2×10^{-3}	<6.5×10^{-5}	>15,000
Hale-Bopp (gas)[2]	Ne/O < 6×10^{-3}	<0.04	>25
Hale-Bopp (bulk nucl.)[2]	Ne/O < 7×10^{-4}	<.0047	>200
Halley (coma)[3]	Ne/H_2O < 1.5×10^{-3}	<0.008	>125
Hale-Bopp (coma)[4]	Ar/O = 0.0058±0.0017	1.8±0.96	
Hyakutake			
ice[5]	Ne/O < 2.2×10^{-4}	<1.5×10^{-3}	>700
ice+dust[5]	Ne/O <5.8×10^{-5}	<3.9×10^{-4}	>2600

a: Ratios relative to solar abundances as stated in original reference.
1: Stern et al. 1992. **2**: Krasnopolsky et al. 1997. **3**: Geiss et al. 1999. **4**: Stern et al. 2000.
5: Krasnopolsky and Mumma 2001. Halley value from the Giotto ion mass spectrometer, other data from EUV spectroscopy.

enrichment factor <30). Krasnopolsky et al. (1997), Geiss et al. (1999), and Krasnopolsky and Mumma (2001) found Ne in the ices of Hale-Bopp, Halley, and Hyakutake to be depleted by factors of at least 25, 125, and 700, respectively. Krasnopolsky and Mumma (2001) even report a Ne depletion factor of >2600 for the sum of ice and dust in Hyakutake. These Ne depletions imply that the ices of these comets formed at, or later suffered, temperatures higher than ~25 K. Combining the inferences from Ne and Ar, Stern (1999a) and Stern et al. (2000) conclude that Hale-Bopp's ice formed at ~20-35 K (unless the Ar abundance in the solar nebula was much higher than assumed). This would indicate an origin of Hale Bopp in the Kuiper belt region beyond Neptune, rather than closer to the Sun as is usually inferred from the comet's recent home, the Oort cloud.

Other information about comets is potentially contained in interplanetary dust particles, which are discussed in the next section. Note that some workers also presume that the chemically most primitive meteorites, the CI- and CM carbonaceous chondrites, originate from comets (Lodders and Osborne 1999; Ehrenfreund et al. 2001). However, as these claims are controversial, we do not discuss noble gas data from CI/CM chondrites from a cometary perspective.

INTERPLANETARY DUST PARTICLES AND MICROMETEORITES

Interplanetary dust particles (IDP) are collected in the Earth's stratosphere and represent a wide variety of solar system materials, usually in the size range of a few to a few tens of microns (Brownlee 1985). Micrometeorites are extraterrestrial dust grains (typically 100-200 µm) collected in ice, mostly in Greenland and Antarctica (Engrand and Maurette 1998; Maurette et al. 2000). Many of these particles are probably fragments of comets, and hence are very valuable material from parent bodies that so far are probably otherwise unsampled (see the *Comets* section). Deep sea sediments are a further source of micrometeorites, from which usually the magnetic fraction is collected only. The "large" micrometeorites in the 100-micron range represent by far the major source of material falling onto Earth today. Noble gas analyses of micrometeorites have helped to prove their extraterrestrial origin (Merrihue 1964; Amari and Ozima 1988; Olinger et al. 1990).

Table 12 gives the ranges of concentrations or gas amounts and isotopic compositions observed in two studies of IDPs and micrometeorites. The major source of the noble gases in these particles is the solar wind, which is obvious both from the very high concentrations, which in IDPs often exceed values in lunar regolith samples, as well as the isotopic compositions. $^{20}Ne/^{22}Ne$ ratios usually fall between the values of the solar components SW and SEP (13.8 and 11.2, respectively, Table 4) and $^{3}He/^{4}He$ ratios in many cases also fall between SW and SEP values (Pepin et al. 2000). The solar gases were implanted rather recently while the particles were spiraling towards the Sun due to Poynting-Robertson drag, and the high solar noble gas abundances unfortunately swamp any primordial cometary noble gases that may be present. There may be a possible exception to this, however. Some IDPs analyzed at the University of Minnesota (Nier and Schlutter 1993; Pepin et al. 2000; Pepin et al. 2001) show intriguingly high $^{3}He/^{4}He$ ratios of up to 40 times the solar wind value. This does not appear to be due to helium from impulsive solar flares or pick-up ions from the interstellar medium with $^{3}He/^{4}He$ ratios of around unity (Pepin et al. 2001, see Tables 3 and 13). Further, modeling erosion and fragmentation of IDPs on their way to the terrestrial orbit yields lifetimes far shorter than those required to account for the ^{3}He excesses by cosmic-ray production (Pepin et al. 2001). It appears possible that the ^{3}He was produced by cosmic rays when the particles still were residing in their parent body regoliths, although corresponding ^{21}Ne excesses have not been verified so far. However, required regolith exposure times of in some cases >1Gyr appear very long, although they might be reasonable for regoliths in Kuiper-Belt comets (Pepin et al. 2001). Pepin and coworkers therefore also consider the possibility that the ^{3}He excesses are not cosmic-ray-produced but represent a ^{3}He-rich primordial component from an unknown source.

The fact that many IDPs and micrometeorites have retained part of their solar wind noble gases in their surface layers proves that these particles have not suffered very high temperatures during atmospheric entry. Nier and Schlutter (1992, 1993) used the release pattern of SW-He as a function of temperature to determine the maximum temperature during entry. Many particles probably were heated to less than 600°C. Nier and Schlutter suggested that the He release pattern might be useful for distinguishing asteroidal from cometary IDPs, because the latter will have higher entry velocities.

Table 12. Noble gases in interplanetary dust particles (IDPs) and micrometeorites.

	Interplanetary Dust Particles[1]	*Micrometeorites*[2]
mass	0.06-48 ng	0.2-21μg
^{4}He	3.0×10^{-12}-2.9×10^{-9} ccSTP *(1.7×10^{-4}-14.5) ccSTP/g*	
$^{3}He/^{4}He$	$(1.8\text{-}31)\times10^{-4}$	
^{20}Ne	1.6×10^{-13}-6.9×10^{-11} ccSTP *(2.4×10^{-5}-0.34) ccSTP/g*	$(59\text{-}5040)\times10^{-8}$ ccSTP/g
$^{20}Ne/^{22}Ne$	10.3-12.4	11.0-12.1
$^{21}Ne/^{22}Ne$	0.021-0.037	0.029-0.07
^{36}Ar	3.8×10^{-14}-2.3×10^{-12} ccSTP *(6.3×10^{-5}-1.2×10^{-2}) ccSTP/g*	
$^{36}Ar/^{38}Ar$	4.72-4.87	

1: Pepin et al. 2000. **2**: Olinger et al. 1990, unmelted particles from Greenland ice.

Olinger et al. (1990) report ^{21}Ne cosmic ray exposure ages of between <0.5 and 20 Myr for micrometeorites from Greenland, consistent with the range of exposure ages of deep sea spherules determined by ^{10}Be (Raisbeck and Yiou 1989). This spread of ages suggests a cometary origin of most of these particles (Raisbeck and Yiou 1989).

ELEMENTARY PARTICLES IN INTERPLANETARY SPACE

The solar system is filled with elementary particles of various origins. The dominant source is the Sun and we have extensively discussed in this chapter noble gases from the solar corpuscular radiation (*Sun* and *Moon* sections). Here we discuss galactic cosmic ray (GCR) particles, "pick-up" ions of interstellar or interplanetary origin, and "anomalous cosmic rays." A recent overview is provided in a conference proceedings volume (Wimmer-Schweingruber 2001). A summary of the noble gas data is given in Table 13.

Galactic cosmic rays

Noble gas elemental abundances for the GCR source as compiled by Meyer et al. (1997) are given in Table 13. The GCR source composition is deduced from the measured GCR composition by correcting for secondary contributions by spallation reactions, solar modulation effects, and energy loss during propagation. GCR source abundances of all elements show a similar general pattern to that of the solar corona, in particular an enhancement of elements with a relatively low first ionization potential. The noble gases appear to obey this pattern also, as the enrichment of the heavier gases appears to be somewhat larger than that of Ne. Also, He is depleted by about a factor of three relative to Ne, about the same as in the solar corona. However, Meyer et al. (1997) consider this overall similarity between GCR source abundances and solar coronal abundances to be coincidental. They propose that GCR originate mainly from dust grains accelerated by a supernova remnant blast wave and subsequently eroded by sputtering (see also Ellison et al. 1997). A smaller proportion only would originate from accelerated gas. Meyer et al. (1997) and Ellison et al. (1997) argue that GCR source abundances therefore reflect the volatility of the elements or their major chemical compounds rather than the first ionization potential (FIP) which is governing abundances in the corona (*Sun* section). The similar trends of coronal and GCR abundances is then a consequence of the fact that volatility and FIP of elements roughly correlate with each other.

The ACE mission provided a high-precision value of the ^{20}Ne/^{22}Ne ratio of galactic cosmic rays in the energy range 80-280 MeV/nucleon (Binns et al. 2001). The rare isotope ^{21}Ne is used to correct the other two Ne isotopes for secondary production by spallation. Strikingly, the ^{22}Ne/^{20}Ne ratio of the GCR source is 5.0±0.2 times larger than the solar wind value, and also considerably larger than the ratios of primordial meteoritic components (Ott 2002, this volume). However, this does not mean that the Ne isotopic composition of the solar system is highly unusual. Rather, the observations support models where GCR preferentially represent dust or gas from Wolf Rayet stars or supernovae from the central regions of the galaxy (Ellison et al. 1997; Binns et al. 2001).

The ^{3}He/^{4}He ratio in GCR is enhanced relative to the solar or protosolar value by several orders of magnitude (Table 13). This is because ^{3}He is almost entirely a secondary product believed to be due to break-up of primary ^{4}He (Reimer et al. 1998). The ^{3}He/^{4}He ratio in GCR is therefore an excellent probe of the propagation history of the galactic cosmic rays.

Interstellar and interplanetary pick-up ions

Interstellar pick-up ions are neutral particles from the local interstellar cloud (LIC) that entered the heliosphere, where they became ionized and "picked-up" by the solar wind (the heliosphere is the volume in space shaped by the solar wind and its interaction with the surrounding and flowing interstellar medium). Pick-up ions are almost entirely singly charged, which is the criterion for distinguishing them from the solar wind. They are dominated by elements with a high first ionization potential, because a larger fraction of these elements is present in neutral form in the interstellar medium and can therefore enter the heliosphere. Table 13 gives He and Ne abundances in the LIC inferred from analyses of pick-up ions (Gloeckler and Geiss 1998a). Most data were obtained by the SWICS instrument on the Ulysses mission. The necessary corrections for various fractionation effects are large and not very precise. Because the solar Ne abundance itself is still uncertain by some 20% (see the *Sun* section), the apparent ~40% over-abundance of Ne in the Sun, relative to the interstellar medium, needs to be taken cautiously. Yet pick-up ions are a promising new source of information on noble gas abundances in the galaxy, albeit the very local neighborhood.

Table 13. Noble gases in the interplanetary medium and the galaxy.

GCR source	*ratio relative to solar value*[a]	*uncertainty factor*[b]
He/H[1]	0.7	1.2
^{20}Ne/H[1,c]	2.4	1.35
Ar/H[1]	4.3	1.4
(Kr/H)[1,d]	4.7	1.6
(Xe/H)[1,d]	10.3	1.8
^{22}Ne/^{20}Ne[2]	5.0±0.2	
GCR		*absolute ratio*
^{3}He/^{4}He[3]		0.12-0.23
Local interstellar medium (LIC)[4]	*ratio relative to solar value*[a]	*absolute ratio*
^{4}He/H	1	0.10
^{20}Ne/H	0.61	7.5×10^{-5}
^{3}He/^{4}He[5]		$(2.48^{+.68}_{-.62})\times10^{-4}$
^{3}He/^{4}He[6]		$(1.70^{+.50}_{-.42})\times10^{-4}$
(protosolar[e])		$(1.66\pm0.05)\times10^{-4}$
Anomalous cosmic rays		
H/He[7]		$5.6^{+.6}_{-.5}$
C/He[7]		$(4.8\pm1.1)\times10^{-4}$
N/He[7]		0.011±0.001
O/He[7]		0.075±0.006
Ne/He[7]		0.0050±0.0004
^{22}Ne/^{20}Ne[8]		$0.077^{+.085}_{-.023}$
^{20}Ne/^{22}Ne[8]		$13.0^{+5.5}_{-6.8}$

a: ratios relative to solar abundances as stated in original references. **b:** nominal uncertainties given as factors. **c:** ^{22}Ne strongly enhanced, see text. **d:** uncertainty possibly underestimated. **e:** Jupiter value, see Table 7.

1: Meyer et al. 1997; Ellison et al. 1997. **2:** Binns et al. 2001. **3:** Reimer et al. 1998. **4:** Gloeckler and Geiss 1998a. **5:** Gloeckler and Geiss 1998b (SWICS-Ulysses). **6:** Salerno et al. 2001 (Mir, foils). **7:** Cummings and Stone 1996. **8:** Leske et al. 1996.

The ^{3}He/^{4}He ratio in the LIC has also been measured both by Ulysses (Gloeckler and Geiss 1998b) as well as in foils exposed on the MIR space station (Salerno et al. 2001). The two values agree within their rather large uncertainties (Table 13) and it appears that the ^{3}He abundance in the solar neighborhood today is larger than it was 4.6 Ga ago, as is expected (Gloeckler and Geiss 2000, see the *Sun* section).

Pick-up ions do not originate exclusively from the LIC, but also from within the solar system. (Geiss et al. 1995). Part of the C^{+} and O^{+} from this "Inner Source" is likely to derive from interstellar grains evaporated at distances of a few AU from the Sun (Geiss et al. 1996). However, Ne from the Inner Source probably represents predominantly solar wind particles implanted into interplanetary dust grains and released when the grains get close to the Sun at about 20 solar radii (Gloeckler and Geiss 2001).

Anomalous cosmic rays

Anomalous cosmic rays (ACR) are thought to be pick-up ions (both from the Interstellar and the Inner Source) that were carried by the solar wind out to the solar wind termination shock (the boundary where the SW becomes subsonic), where they are accelerated to energies on the order of ~5-200 MeV/nucleon (Jokipii 1990; Jokipii and Kóta 2000). Besides the three noble gases He, Ne, and Ar, also H, C, N, and O have been detected. Table 13 gives measured abundances of ACR noble gases (Cummings and Stone 1996), as well as the Ne isotopic composition (Leske et al. 1996). Within the large uncertainties, the latter is consistent with either solar or primordial meteoritic Ne. On the other hand, ACR-Ne is distinctly different from GCR-Ne as discussed above, which indicates again that GCRs include contributions from ^{22}Ne-rich sources (Leske et al. 1996).

ACKNOWLEDGMENTS

I thank Jozef Masarik and Ingo Leya for providing the data shown in Figure 10, Rick Leske for providing Figure 4 and Veronika Heber for allowing me to show her unpublished data in Figure 9. I appreciate discussions with Bob Binns, Henner Busemann, George Gloeckler and Don Porcelli. Very constructive reviews and comments by Henner Busemann, Uli Ott, Toby Owen, Bob Pepin, Don Porcelli, and Robert Wimmer-Schweingruber have been of great help.

REFERENCES

Altwegg K, Balsiger H, Geiss J (1999) Composition of the volatile material in Halley's coma from *in situ* measurements. Space Sci Rev 90:3-18

Amari S, Ozima M (1988) Extra-terrestrial noble gases in deep sea sediments. Geochim Cosmochim Acta 52:1087-1095

Anders E, Grevesse N (1989) Abundances of the elements: meteoritic and solar. Geochim Cosmochim Acta 53:197-214

Arvidson R, Drozd R, Guinness E, Hohenberg C, Morgan C, Morrison R, Oberbeck V (1976) Cosmic ray exposure ages of Apollo 17 samples and the age of Tycho. Proc Lunar Sci Conf 7th:2817-2832

Bahcall JN (2000) Solar neutrinos: an overview. Physica Scripta T85:63-70

Bahcall JN, Pinsonneault MH (1995) Solar models with helium and heavy-element diffusion. Rev Mod Phys 67:781-808

Bahcall JN, Pinsonneault MH, Basu S (2001) Solar models: Current epoch and time dependences, neutrinos, and helioseismological properties. Astrophys J 555:990-1012

Basford JR, Dragon JC, Pepin RO, Coscio Jr. MR, Murthy VR (1973) Krypton and Xenon in lunar fines. Proc Lunar Sci Conf 4th:1915-1955

Becker RH (2000) Nitrogen on the moon. Science 290:1110-1111

Becker RH, Pepin RO (1984) The case for a martian origin of the shergottites: nitrogen and noble gases in EETA 79001. Earth Planet Sci Lett 69:225-242

Becker RH, Schlutter DJ, Rider PE, Pepin RO (1998) An acid-etch study of the Kapoeta achondrite: Implications for the argon-36/argon-38 ratio in the solar wind. Meteoritics Planet Sci 33 (1):109-113

Benkert J-P, Baur H, Signer P, Wieler R (1993) He, Ne, and Ar from the solar wind and solar energetic particles in lunar ilmenites and pyroxenes. J Geophys Res E 98:13147-13162

Binns WR, Wiedenbeck ME, Christian ER, Cummings AC, George JS, Israel MH, Leske RA, Mewaldt RA, Stone EC, von Rosenvinge TT, Yanasak NE (2001) GCR neon abundances: comparison with Wolf-Rayet star models and meteoritic abundances. *In* Solar and galactic composition. Wimmer-Schweingruber RF (ed) AIP Conf Proc Vol 598, Am Inst Phys, Melville NY, p 257-262

Black DC (1972) On the origins of trapped helium, neon and argon isotopic variations in meteorites-I. Gas-rich meteorites, lunar soil and breccia. Geochim Cosmochim Acta 36:347-375

Bochsler P (1984) Helium and oxygen in the solar wind: dynamic properties and abundances of elements and helium isotopes as observed with the ISEE-3 plasma composition experiment. Habilitation Thesis, Univ Bern, Switzerland

Bochsler P (1994) Solar wind composition from the moon. Adv Space Res 14:161-173
Bochsler P (2000) Abundances and charge states of particles in the solar wind. Rev Geophys 38:247-266
Bochsler P, Geiss J, Maeder A (1990) The abundance of ^{3}He in the solar wind—a constraint for models of solar evolution. Solar Phys 128:203-215
Bodmer R, Bochsler P (1998a) Fractionation of minor ions in the solar wind acceleration process. Phys Chem Earth 23:687-692
Bodmer P, Bochsler P (1998b) The helium isotopic ratio in the solar wind and ion fractionation in the corona by inefficient Coulomb drag. Astron Astrophys 337:921-927
Bogard DD (1988) On the origin of Venus' atmosphere: possible contributions from simple component mixtures and fractionated solar wind. Icarus 74:3-20
Bogard DD, Nyquist LE, Hirsch WC, Moore DR (1973) Trapped solar and cosmogenic noble gas abundances in Apollo 15 and 16 deep drill samples. Earth Planet Sci Lett 21:52-69
Bogard DD, Nyquist LE, Johnson P (1984) Noble gas contents of shergotittes and implications for the Martian origin of SNC meteorites. Geochim Cosmochim Acta 48:1723-1739
Bogard DD, Garrison DH, Masarik J (2001) The Monahans chondrite and halite: Argon-39/argon-40 age, solar gases, cosmic-ray exposure ages, and parent body regolith neutron flux and thickness. Meteoritics Planet Sci 36:107-122
Brilliant DR, Franchi IA, Pillinger CT (1994) Nitrogen components in lunar soil 12023: Complex grains are not the carrier of isotopically light nitrogen. Meteoritics 29:718-723
Brownlee DE (1985) Cosmic dust: collection and research. Annu Rev Earth Planet Sci 13:147-173
Busemann H, Baur H, Wieler R (2000) Primordial noble gases in "phase Q" in carbonaceous and ordinary chondrites studied by closed-system stepped etching. Meteoritics Planet Sci 35:949-973
Busemann H, Baur H, Wieler R (2001) Helium isotopic ratios in carbonaceous chondrites: significant for the early solar nebula and circumstellar diamonds? Lunar Planet Sci, XXXII, Lunar & Planetary Institute, Houston, abstract # 1598, CD-ROM
Carrier WD, Olhoeft GR, Mendell W (1991) Physical properties of the lunar surface. *In* The Lunar Sourcebook. Heiken GH, Vanimann DT, French BM (eds) Cambridge University Press, Cambridge, p 475-594
Cerutti H (1974) Die Bestimmung des Argons im Sonnenwind aus Messungen an den Apollo-SWC-Folien. PhD Thesis, Univ Bern, Switzerland
Christensen-Dalsgaard J (1998) The 'standard' Sun—Modelling and helioseismology. Space Sci Rev 85: 19-36
Clayton DD. (1968) Priciples of stellar evolution and nucleosynthesis. McGraw-Hill, New York.
Clayton RN, Thiemens MH (1980) Lunar nitrogen: Evidence for secular change in the solar wind. *In* The Ancient Sun. Pepin RO, Eddy JA, Merrill RB (eds) Geochim Cosmochim Acta, suppl 13:463-473
Conrath BJ, Gautier D (2000) Saturn helium abundance: A reanalysis of Voyager measurements. Icarus 144:124-134
Conrath BJ, Gautier D, Hanel RA, Hornstein JS (1984) The helium abundance of Saturn from Voyager measurements. Astrophys J 282:807-815
Conrath B, Gautier D, Hanel R, Lindal G, Marten A (1987) The helium abundance of Uranus from Voyager measurements. J Geophys Res A 92:15,003-15,010
Conrath BJ, Gautier D, Lindal GF, Samuelson RE, Shaffer WA (1991) The helium abundance of Neptune from Voyager measurements. J Geophys Res 96, suppl.:18907-18919
Culler TS, Becker TA, Muller RA, Renne PR (2000) Lunar impact history from ^{40}Ar/^{39}Ar dating of glass spherules. Science 287:1785-1788
Cummings AC, Stone EC (1996) Composition of anomalous cosmic rays and implications for the heliosphere. Space Sci Rev 78:117-128.
Dietrich WF, Simpson JA (1979) The isotopic and elemental abundances of neon nuclei accelerated in solar flares. Astrophys J 231:L91-L95
Donahue TM (1986) Fractionation of noble gases by thermal escape from accreting planetesimals. Icarus 66:195-210
Donahue TM, Pollak JB (1983) Origin and evolution of the atmosphere of Venus. *In* Venus. Hunten DM, Colin L, Donahue TM, Moroz VI (eds) Univ Arizona Press, Tucson, p 1003-1036
Donahue TM, Russell CT (1997) The Venus atmosphere and ionosphere and their interaction with the solar wind: an overview. *In* Venus II. Bougher SW, Hunten DM, Phillips RJ (eds) Univ Arizona Press, Tucson, p 3-31
Drozd RJ, Hohenberg CM, Morgan CJ, Ralston CE (1974) Cosmic-ray exposure history at the Apollo 16 and other lunar sites: lunar surface dynamics. Geochim Cosmochim Acta 38:1625-1642
Eberhardt P, Eugster O, Marti K (1965) A redetermination of the isotopic composition of atmospheric Neon. Z Naturf 20a:623-624

Eberhardt P, Eugster O, Marti K (1965) A redetermination of the isotopic composition of atmospheric Neon. Z Naturf 20a:623-624

Ehrenfreund P, Glavin DP, Botta O, Cooper G, Bada JL (2001) Extraterrestrial amino acids in Orgueil and Ivuna: Tracing the parent body of CI type carbonaceous chondrites. Proc Natl Acad Sci USA 98: 2138-2141

Ellison DC, Drury LO'C, Meyer J-P (1997) Galactic cosmic rays from supernova remnants. II. Shock acceleration of gas and dust. Astrophys J 487:197-217

Engrand C, Maurette M (1998) Carbonaceous micrometeorites from Antarctica. Meteoritics Planet Sci 33:565-580

Eugster O (1985) Multistage exposure history of the 74261 soil constituents. Proc Lunar Planet Sci Conf 16th, D95-D102

Eugster O (1999) Chronology of dimict breccias and the age of South Ray crater at the Apollo 16 site. Meteoritics Planet Sci 34:385-391

Eugster O, Grögler N, Eberhardt P, Geiss J (1979) Double drive tube 74001/2: History of the black and orange glass; determination of a pre-exposure 3.7 AE ago by $^{136}Xe/^{235}U$ dating. Proc Lunar Planet Sci Conf 10th:1351-1379

Eugster O, Terribilini D, Polnau E, Kramers J (2001) The antiquity indicator argon-40/argon-36 for lunar surface samples calibrated by uranium-235-xenon-136 dating. Meteoritics Planet Sci 36:1097-1115

Fegley B, Swindle TD (1993) Lunar volatiles: implications for lunar resource utilization. *In* Resources of Near-Earth Space. Lewis J, Matthews MS, Guerrieri ML (eds), Univ Arizona Press, Tucson, p367-426

Feldman U (1998) FIP effect in the solar upper atmosphere: Spectroscopic results. Space Sci Rev 85: 227-240

Feldman U, Widing KG (1990) Photospheric abundances of oxygen, neon, and argon derived from the XUV spectrum of an impulsive flare. Astrophys J 363:292-298

Feldman PD, Morrison D (1991) The Apollo 17 ultraviolet spectrometer–lunar atmosphere measurements revisited. Geophys Res Lett 18:2105-2108

Feldman PD, Weaver HA, Burgh EB (2001) COMET C/2001 A2 (Linear). IAU Circ No. 7681:15, August 2001

Frick U, Becker RH, Pepin RO (1988) Solar wind record in the lunar regolith: Nitrogen and noble gases. Proc Lunar Planet Sci Conf 18th:87-120

Geiss J (1973) Solar wind composition and implications about the history of the solar system. Conf Paper 13th Intl Cosmic Ray Conf:3375-3398

Geiss J (1993) Primordial abundances of hydrogen and helium isotopes. In Origin and Evolution of the Elements. Prantzos N, Vangioni-Flam E, Cassé M (eds). Cambridge Univ Press, Cambridge, p 89-106

Geiss J, Bochsler P (1982) Nitrogen isotopes in the solar system. Geochim Cosmochim Acta 46:529-548

Geiss J, Bochsler P (1985) Ion composition in the solar wind in relation to solar abundances. *In* Proc Conf Isotopic Ratios in the Solar System. Cepadues-Editions, Toulouse, France, p 213-228

Geiss J, Gloeckler G (1998) Abundances of deuterium and helium-3 in the protosolar cloud. Space Sci Rev 84:239-250

Geiss J, Reeves H (1972) Cosmic and solar system abundances of deuterium and helium-3. Astron Astrophys 18:126-132

Geiss J, Reeves H (1981) Deuterium in the solar system. Astron Astrophys 93:189-199

Geiss J, Gloeckler G, von Steiger R (1994) Solar and heliospheric processes from solar wind composition measurements. Phil Trans Royal Soc London A 349:213-226

Geiss J, Gloeckler G, Fisk LA, von Steiger R (1995) C^+ pickup ions in the heliosphere and their origin. J Geophys Res 100:23,373-323,377

Geiss J, Gloeckler G, von Steiger R (1996) Origin of C^+ ions in the heliosphere. Space Sci Rev 78:43-52.

Geiss J, Altwegg K, Balsiger H, Graf S (1999) Rare atoms, molecules and radicals in the coma of P/Halley. Space Sci Rev 90:253-268

Geiss J, Bühler F, Cerutti H, Eberhardt P, Filleux C (1972) Solar wind composition experiment. Apollo 16 Prelim Sci Rep, NASA SP-315:14.11-14.10

Gloeckler G, Geiss J (1998a) Interstellar and inner source pickup ions observed with SWICS on Ulysses. Space Sci Rev 86:127-159

Gloeckler G, Geiss J (1998b) Measurement of the abundance of helium-3 in the Sun and in the local interstellar cloud with SWICS on Ulysses. Space Sci Rev 84:275-284

Gloeckler G, Geiss J (2000) Deuterium and helium-3 in the protosolar cloud. *In* The light elements and their evolution. da Silva L, Spite M, de Medeiros JR (eds) IUA Symposium 198:224-233

Gloeckler G, Geiss J (2001) Heliospheric and interstellar phenomena deduced from pickup ion distributions. Space Sci Rev 97:169-181

Gosling JT (1993) The solar flare myth. J Geophys Res A 98:18937-18949

Gough D (1998) On the composition of the solar interior—Rapporteur Paper I. Space Sci Rev 85:141-158

Gough DO, Leibacher JW, Scherrer PH, Toomre J (1996) Perspectives in helioseismology. Science 272:1281-1283

Grevesse N, Sauval AJ (1998) Standard solar composition. Space Sci Rev 85:161-174

Guillot T (1999) A comparison of the interiors of Jupiter and Saturn. Planet Space Sci 47:1183-1200

Hashizume K, Marty B, Wieler R (1999) Single grain analyses of the nitrogen isotopic composition in the lunar regolith—in search of the solar wind component. Lunar Planet Sci XXX, abstr #1567, Lunar Planet Institute, Houston (CD-ROM)

Hashizume K, Chaussidon M, Marty B, Robert F (2000) Solar wind record on the moon: Deciphering presolar from planetary nitrogen. Science 290:1142-1145

Heber VS, Baur H, Wieler R (2001a) Solar krypton and xenon in gas-rich meteorites: new insights into a unique archive of solar wind. *In* Solar and galactic composition. Wimmer-Schweingruber RF (ed) AIP Conf Proc Vol 598, Am Inst Phys, Melville, New York, p 387-392

Heber VS, Baur H, Wieler R (2001b) Is there evidence for a secular variation of helium isotopic composition in the solar wind? Meteoritics Planet Sci 36 (9, suppl.):A76-A77

Hodges RR (1975) Formation of the lunar atmosphere. Moon 14:139-157

Hodges Jr. RR, Johnson FS (1968) Lateral transport in planetary exospheres. J Geophys Res 73:7307-7317

Hodges RR, Hoffman JH, Johnson FS, Evans DE (1973) Composition and dynamics of lunar atmosphere. Proc Lunar Sci Conf 4th:2855-2864

Hoffman JH, Hodges RR, Johnson FS, Evans DE (1973) Lunar atmospheric composition results from Apollo 17. Proc Lunar Sci Conf 4th:2865-2875

Hoffman JH, Oyama VI, von Zahn U (1980a) Measurements of the Venus lower atmosphere composition: a comparison of results. J Geophys Res 85:7871-7881

Hoffman JH, Hodges RR, Donahue TM, McElroy MB (1980b) Composition of the Venus lower atmosphere from the Pioneer Venus mass spectrometer. J Geophys Res 85:7882-7890

Hohenberg CM, Davis PK, Kaiser WA, Lewis RS, Reynolds JH (1970) Trapped and cosmogenic rare gases from stepwise heating of Apollo 11 samples. Proc Apollo 11 Lunar Sci Conf:1283-1309

Hohenberg CM, Marti K, Podosek FA, Reedy RC, Shirck JR (1978) Comparisons between observed and predicted cosmogenic noble gases in lunar samples. Proc Lunar Planet Sci Conf 9th:2311-2344

Holweger H (2001) Photospheric abundances: problems, updates, implications. *In* Solar and galactic composition. Wimmer- Schweingruber RF (ed) AIP Conf Proc Vol 598, Am Inst Phys, Melville, New York, p 23-30

Hunten DM, Sprague AL (1997) Origin and character of the Lunar and Mercurian atmospheres. Adv Space Res 19:1551-1560

Hunten DM, Morgan TH, Shemansky DE (1988) The Mercury atmosphere. *In* Mercury. Vilas F, Chapman CR, Matthews MS (eds) Univ Arizona Press, Tucson, p 562-612

Istomin VG, Grechnev KV, Kochnev VA (1982) Mass spectrometry on the Venera 13 and Venera 14 landers: preliminary results. Sov Astron Lett 8: 211-215

Istomin VG, Gechnev KV, Kochnev VA (1983) Mass spectrometry of the atmosphere by Venera 13 and Venera 14. Cosmic Research 21:329-338

Johnson JR, Swindle TD, Lucey PG (1999) Estimated solar wind-implanted helium-3 distribution on the Moon. Geophys Res Lett 26:385-388

Jokipii JR (1990) The anomalous components of cosmic rays. *In* Physics of the outer heliosphere. Grzedzielski S, Page DE (eds) Pergamon, New York, p 169-178

Jokipii JR, Kóta J (2000) Galactic and anomalous cosmic rays in the heliosphere. Astrophys Space Sci 274:77-96

Kallenbach R (2001) Isotopic composition measured *in situ* in different solar wind regimes by CELIAS/MTOF on board SOHO. *In* Solar and galactic composition. Wimmer-Schweingruber RF (ed) AIP Conf Proc Vol 598, Am Inst Phys, Melville, New York, p 113-119

Kallenbach R, Ipavich FM, Bochsler P, Hefti S, Hovestadt D, Grünwaldt H, Hilchenbach M, Axford WI, Balsiger H, Bürgi A, Coplan MA, Galvin AB, Geiss J, Gliem F, Gloeckler G, Hsieh KC, Klecker B, Lee MA, Livi S, Managadze GG, Marsch E, Möbius E, Neugebauer M, Reiche K-U, Scholer M, Verigin MI, Wilken B, Wurz P (1997) Isotopic composition of solar wind neon measured by CELIAS/MTOF on board SOHO. J Geophys Res A 102:26895-26904

Kallenbach R, Geiss J, Ipavich FM, Gloeckler G, Bochsler P, Gliem F, Hefti S, Hilchenbach M, Hovestadt D (1998a) Isotopic composition of solar wind nitrogen: First *in situ* determination with the CELIAS/MTOF spectrometer on board SOHO. Astrophys J 507:L185-L188

Kallenbach R, Ipavich FM, Kucharek H, Bochsler P, Galvin AB, Geiss J, Gliem F, Gloeckler G, Grünwaldt H, Hefti S, Hilchenbach M, Hovestadt D (1998b) Fractionation of Si, Ne, and Mg isotopes in the solar wind as measured by SOHO/CELIAS/MTOF. Space Sci Rev 85:357-370

Kaula WM (1999) Constraints on Venus evolution from radiogenic argon. Icarus 139:32-39

Kerridge JF (1975) Solar nitrogen: evidence for a secular increase in the ratio of nitrogen-15 to nitrogen-14. Science 188:162-164

Kerridge JF (1980) Secular variations in composition of the solar wind: Evidence and causes. *In* The Ancient Sun. Pepin RO, Eddy JA, Merrill RB (eds). Pergamon, New York, p 475-489

Kerridge JF (1993) Long-term compositional variation in solar corpuscular radiation: evidence from nitrogen isotopes in the lunar regolith. Rev Geophys 31:423-437

Kerridge JF, Eugster O, Kim JS, Marti K (1991a) Nitrogen isotopes in the 74001/74002 double-drive tube from Shorty Crater, Apollo 17. Proc Lun Planet Sci 21:291-299

Kerridge JF, Signer P, Wieler R, Becker RH, Pepin RO (1991b) Long-term changes in composition of solar particles implanted in extraterrestrial materials. *In* Sonett CP, Giampapa MS, Matthews MS (eds) The Sun in Time. Univ Arizona Press, Tucson, pp 389-412

Killen RM, Ip W-H (1999) The surface-bounded atmospheres of Mercury and the Moon. Rev Geophys 37:361-406

Krasnopolsky VA, Mumma MJ (2001) Spectroscopy of comet Hyakutake at 80-700 Å: First detection of solar wind charge-transfer emissions. Astrophys J 549:629-634

Krasnopolsky VA, Bowyer S, Chakrabarti S, Gladstone GR, McDonald JS (1994) First measurement of helium on Mars: implications for the problem of radiogenic gases on the terrestrial planets. Icarus 109:337-351

Krasnopolsky VA, Mumma MJ, Abbott M, Flynn BC, Meech KJ, Yeomans BK, Feldman PD, Cosmovici CB (1997) Detection of soft x-rays and a sensitive search for noble gases in comet Hale-Bopp (C/1995 O1). Science 277:1488-1491

Kulcinski GL, Schmitt HH (1992) Fusion power from lunar resources. Fusion Technology 21:2221-2229

Leske RA, Mewaldt RA, Cummings AC, Cummings JR, Stone EC, von Rosenvinge TT (1996) The isotopic composition of anomalous cosmic rays from Sampex. Space Sci Rev 78:149-154.

Leske RA, Mewaldt RA, Cohen CMS, Cummings AC, Stone EC, Wiedenbeck ME, Christian ER, von Rosenvinge TT (1999) Event-to-event variations in the isotopic composition of neon in solar energetic particle events. Geophys Res Lett 26:2693-2696

Leske RA, Mewaldt RA, Cohen CMS, Christian ER, Cummings AC, Slocum PL, Stone EC, von Rosenvinge TT, Wiedenbeck ME (2001) Isotopic abundances in the solar corona as inferred from ACE measurements of solar energetic particles. *In* Solar and Galactic Composition. Wimmer-Schweingruber RF (ed) AIP Conf Proc Vol 598, Am Inst Phys, Melville NY, p 127-132

Leya I, Neumann S, Wieler R, Michel R (2001) The production of cosmogenic nuclides by GCR-particles for 2π exposure geometries. Meteoritics Planet Sci 36:1547-1561

Lodders K, Osborne R (1999) Perspectives on the comet-asteroid-meteorite link. Space Sci Rev 90:289-297

Mahaffy PR, Donahue TM, Atreya SK, Owen TC, Niemann HB (1998) Galileo probe measurements of D/H and $^3He/^4He$ in Jupiter's atmosphere. Space Sci Rev 84:251-263

Mahaffy PR, Niemann HB, Alpert A, Atreya SK, Demick J, Donahue TM, Harpold DN, Owen TC (2000) Noble gas abundance and isotope ratios in the atmosphere of Jupiter from the Galileo Probe Mass Spectrometer. J Geophys Res E 105:15061-15071

Manka RH, Michel FC (1971) Lunar atmosphere as a source of lunar surface elements. Proc Lunar Sci Conf 2nd:1717-1728

Marti K (1969) Solar-type xenon: a new isotopic composition of xenon in the Pesyanoe meteorite. Science 166:1263-1265

Marti K, Lugmair GW, Urey HC (1970) Solar wind gases, cosmic-ray spallation products and the irradiation history of Apollo 11 samples. Proc Apollo 11 Lunar Sci Conf:1357-1367

Marti K, Lightner BD, Osborn TW (1973) Krypton and xenon in some lunar samples and the age of North Ray Crater. Proc Lunar Sci Conf 4th:2037-2048

Masarik J, Nishiizumi K, Reedy RC (2001) Production rates of cosmogenic helium-3, neon-21, and neon-22 in ordinary chondrites and the lunar surface. Meteoritics Planet Sci 36:643-650

Mason GM, Mazur JE, Dwyer JR (1999) ^{3}He enhancements in large solar energetic particle events. Astrophys J Lett 525:L133-L136

Mason GM, Dwyer JR, Mazur JE (2000) New properties of ^{3}He-rich solar flares dedcued from low-energy particle spectra. Astrophys J Lett 545:L157-L160

Mathew KJ, Marti K (2001) Lunar nitrogen: indigenous signature and cosmic-ray production rate. Earth Planet Sci Lett 184:659-669

Maurette M, Duprat J, Engrand C, Gounelle M, Kurat G, Matrajt G, Toppani A (2000) Accretion of neon, organics, CO_2, nitrogen and water from large interplanetary dust particles on the early Earth. Planet Space Sci 48:1117-1137

Merrihue C (1964) Rare gas evidence for cosmic dust in modern pacific red clay. Ann New York Acad Sci 119:351-367

Mewaldt RA, Spalding JD, Stone EC, Vogt RE (1979) The isotopic composition of solar flare accelerated neon. Astrophys J 231:L97-L100

Mewaldt RA, Mason GM, Gloeckler G, Christian ER, Cohen CMS, Cummings AC, Davis AJ, Dwyer JR, Gold RE, Krimigis SM, Leske RA, Mazur JE, Stone EC, von Rosenvinge TT, Wiedenbeck ME, Zurbuchen TH (2001a) Long-term influences of energetic particles in the heliosphere. *In* Solar and galactic composition. Wimmer-Schreingruber RF (ed) Am Inst Phys Conf Proc 598:165-170

Mewaldt RA, Ogliore RC, Gloeckler G, Mason GM (2001b) A new look at Neon-C and SEP-Ne. *In* Solar and galactic composition. Wimmer-Schweingruber RF (ed) AIP Conf Proc Vol 598, Am Inst Phys, Melville, New York, p 393-398

Meyer JP, Drury LO'C, Ellison DC (1997) Galactic cosmic rays from supernova remnants. I. a cosmic ray composition controlled by volatility and mass to charge ratio. Astrophys J 487:182-196.

Miller JA (1998) Particle acceleration in impulsive solar flares. Space Sci Rev 86:79-105

Moroz VI (1983) Summary of preliminary results of the Venera 13 and Venera 14 missions. *In* Venus. Hunten DM, Colin L, Donahue TM,Moroz VI (eds) Univ Arizona Press, Tucson, p 45-68

Murer CA, Baur H, Signer P, Wieler R (1997) Helium, neon, and argon abundances in the solar wind: in vacuo etching of meteoritic iron-nickel. Geochim Cosmochim Acta 61:1303-1314

Murty SVS, Goswami JN (1992) Nitrogen, noble gases, and nuclear tracks in lunar meteorites MAC88104/105. Proc Lunar Planet Sci 22:225-237

Niedermann S (2002) Cosmic-ray-produced noble gases in terrestrial rocks: Dating tools for surface processes. Rev Mineral Geochem 47:731-784

Niemann HB, Atreya SK, Carignan GR, Donahue TM, Haberman JA, Harpold DN, Hartle RE, Hunten DM, Kasprzak WT, Mahaffy PR, Owen TC, Way SH (1998) The composition of the Jovian atmosphere as determined by the Galileo probe mass spectrometer. J Geophys Res E 103:22831-22845

Nier AO, McElroy MB (1977) Composition and structure of Mars' upper atmosphere: results from the neutral mass spectrometers on Viking 1 and 2. J Geophys Res 82:4341-4349

Nier AO, Schlutter DJ (1992) Extraction of helium from individual interplanetary dust particles by step-heating. Meteoritics 27:166-173

Nier AO, Schlutter DJ (1993) The thermal history of interplanetary dust particles collected in the Earth's stratosphere. Meteoritics 28:675-681

Nishiizumi K, Imamura M, Honda M, Russ GP III, Kohl CP, Arnold JR (1976) ^{53}Mn in the Apollo 15 and 16 drill stems: Evidence for surface mixing. Proc Lunar Sci Conf 7th:41-54

Nishiizumi K, Elmore D, Ma XZ, Arnold JR (1984) ^{10}Be and ^{36}Cl depth profiles in an Apollo 15 drill core. Earth Planet Sci Lett 70:157-163

Notesco G, Laufer D, Bar Nun A, Owen T (1999) An experimental study of the isotopic enrichment in Ar, Kr, and Xe when trapped in water ice. Icarus 142:298-300

Olinger CT, Maurette M, Walker RM, Hohenberg CM (1990) Neon measurements of individual Greenland sediment particles: proof of an extraterrestrial origin. Earth Planet Sci Lett 100:77-93

Orton GS, Ingersoll AP (1976) Pioneer 10 and 11 and ground-based infrared data on Jupiter: the thermal structure and He-H_2 ratio. *In* Jupiter. Gehrels T (ed) Univ Arizona Press, Tucson, p 206-215

Ott U (2002). Noble gases in meteorites-trapped components. Rev Mineral Geochem 47:71-100

Owen T, Bar-Nun A (1995) Comets, Impacts, and Atmospheres. Icarus 116:215-226

Owen T, Biemann K, Rushneck DR, Biller JE, Howarth DW, Lafleur AL (1977) The composition of the atmosphere at the surface of Mars. J Geophys Res 82:4635-4639

Owen T, Mahaffy P, Niemann HB, Atreya S, Donahue T, Bar Nun A, de Pater I (1999) A low-temperature origin for the planetesimals that formed Jupiter. Nature 402:269-270

Owen T, Mahaffy PR, Niemann HB, Atreya SK, Wong M (2001) Protosolar nitrogen. Astrophys J 553:L77-L79

Palme H, Beer H (1993) Abundances of the elements in the solar system. *In* Landolt-Börnstein, Group VI: Astronomy and Astrophysics, Voigt HH (ed), 3(a). Springer, Berlin, p 196-221

Parker EN (1997) Mass ejection and a brief history of the solar wind. *In* Cosmic Winds and the Heliosphere. Jokipii JR, Sonett CP, Giampapa MS (eds) Univ Arizona Press, Tucson, p 3-27

Pepin RO (1991) On the origin and early evolution of terrestrial planet atmospheres and meteoritic volatiles. Icarus 92:2-79

Pepin RO (2000) On the isotopic composition of primordial xenon in terrestrial planet atmospheres. Space Sci Rev 92:371-395

Pepin RO, Porcelli D (2002). Origin of noble gases in the terrestrial planets. Rev Mineral Geochem 47: 191-246

Pepin RO, Nyquist LE, Phinney D, Black DC (1970a) Isotopic composition of rare gases in lunar samples. Science 167:550-553

Pepin RO, Nyquist LE, Phinney D, Black DC (1970b) Rare gases in Apollo 11 lunar material. Proc Apollo 11 Lunar Sci Conf:1435-1454

Pepin RO, Becker RH, Rider PE (1995) Xenon and krypton isotopes in extraterrestrial regolith soils and in the solar wind. Geochim Cosmochim Acta 59:4997-5022
Pepin RO, Becker RH, Schlutter DJ (1999) Irradiation records in regolith materials. I: Isotopic compositions of solar-wind neon and argon in single lunar mineral grains. Geochim Cosmochim Acta 63:2145-2162
Pepin RO, Palma RL, Schlutter DJ (2000) Noble gases in interplanetary dust particles, I: The excess helium-3 problem and estimates of the relative fluxes of solar wind and solar energetic particles in interplanetary space. Meteoritics Planet Sci 35:495-504
Pepin RO, Palma RL, Schlutter DJ (2001) Noble gases in interplanetary dust particles, II: excess ^{3}He in cluster particles and modeling constraints on IDP exposures to cosmic ray irradiation. Meteoritics Planet Sci 36:1515-1534
Raisbeck GM, Yiou F (1989) Cosmic ray exposure ages of cosmic spherules. Meteoritics 24:318
Rao MN, Garrison DH, Bogard DD, Reedy RC (1993) Solar-flare-implanted He-4/He-3 and solar-proton-produced Ne and Ar concentration profiles preserved in lunar rock 61016. J Geophys Res A 98:7827-7835
Reames DV (1998) Solar energetic particles: Sampling coronal abundances. Space Sci Rev 85:327-340
Reames DV (2001) Energetic particle composition. *In* Solar and galactic composition. Wimmer-Schweingruber RF (ed) AIP Conf Proc Vol 598, Am Inst Phys, Melville NY, p 153-164
Reames DV, Meyer J-P, von Rosenvinge TT (1994) Energetic particle abundances in impulsive solar flare events. Astrophys J, suppl 90:649-667
Reedy RC (1981) Cosmic-ray-produced stable nuclides: Various production rates and their implications. Proc Lunar Planet Sci, 12B:1809-1823
Reedy RC, Arnold JR (1972) Interaction of solar and galactic cosmic-ray particles with the Moon. J Geophys Res 77:537-555
Regnier S, Hohenberg CM, Marti K, Reedy RC (1979) Predicted versus observed cosmic-ray produced noble gases in lunar samples: improved Kr production ratios. Proc Lunar Planet Sci Conf 10th:1565-1586
Reimer O, Menn W, Hof M, Simon M, Davis AJ, Labrador AW, Mewaldt RA, Schindler SM, Barbier LM, Christian ER, Krombel KE, Mitchell JW, Ormes JF, Streitmatter RE, Golden RL, Stochaj SJ, Webber WR, Rasmussen IL (1998) The cosmic-ray ^{3}He/^{4}He ratio from 200 MeV per nucleon^{-1} to 3.7 GeV per nucleon^{-1}. Astrophys J 496:490-502
Reynolds JH, Hohenberg CM, Lewis RS, Davis PK, Kaiser WA (1970) Isotopic analysis of rare gases from stepwise heating of lunar fines and rocks. Science 167:545-548
Rolfs CE, Rodney WS. (1988) Cauldrons in the cosmos. Univ Chicago Press, Chicago.
Russ GP, Burnett DS, Wasserburg GJ (1972) Lunar neutron stratigraphy. Earth Planet Sci Lett 15:172-186
Salerno E, Bühler F, Bochsler P, Busemann H, Eugster O, Zastenker GN, Agafonov YuN, Eismont NA (2001) Direct measurement of ^{3}He/^{4}He in the LISM with the COLLISA experiment. *In* Solar and galactic composition. Wimmer- Schweingruber RF (ed) AIP Conf Proc Vol 598, Am Inst Phys, Melville NY, p 275-280
Sands DG, De Laeter JR, Rosman KJR (2001) Measurements of neutron capture effects on Cd, Sm and Gd in lunar samples with implications for the neutron energy spectrum. Earth Planet Sci Lett 186:335-346
Schramm DN (1993) Primordial nucleosynthesis. *In* Origin and Evolution of the Elements. Prantzos N, Vangioni Flam E, Cassé M (eds) Cambridge Univ Press, Cambridge, p 112-131
Signer P, Baur H, Derksen U, Etique P, Funk H, Horn P, Wieler R (1977) Helium, neon, and argon records of lunar soil evolution. Proc Lunar Sci Conf 8th:3657-3683
Sprague AL, Hunten DM (1995) Mercury's atmosphere and its relation to the surface. *In* Volatiles in the Earth and solar system, Farley KA (ed). AIP Conf Proc 341, Am Inst Phys, p 200-208
Stern SA (1999a) Studies of comets in the ultraviolet: The past and the future. Space Sci Rev 90:355-361
Stern SA (1999b) The lunar atmosphere: History, status, current problems, and context. Rev Geophys 37:453-491
Stern SA, Green JC, Cash W, Cook TA (1992) Helium and argon abundance constraints and the thermal evolution of comet Austin (1989c1). Icarus 95:157-161
Stern SA, Slater DC, Festou MC, Parker JWm, Gladstone GR, A'Hearn MF, Wilkinson E (2000) The discovery of argon in comet C/1995 O1 (Hale-Bopp). Astrophys J 544:L169-L172
Stevenson DJ, Salpeter EE (1977a) The phase diagram and transport properties for hydrogen-helium fluid planets. Astrophys J suppl 35:221-237
Stevenson DJ, Salpeter EE (1977b) The dynamics and helium distribution in hydrogen-helium fluid planets. Astrophys J suppl 35:239-261
Swindle TD (2002a) Noble gases in the Moon and meteorites: Radiogenic components and early volatile chronologies. Rev Mineral Geochem 47:101-124
Swindle TD (2002b) Martian noble gases. Rev Mineral Geochem 47:171-190

Swindle TD, Caffee MW, Hohenberg CM, Taylor SR (1986) I-Pu-Xe dating and the relative ages of the Earth and Moon. *In* Origin of the Moon. Hartmann WK, Phillips RJ, Taylor GJ (eds) Lunar Planet. Institute, Houston, p 331-357

Turner G (1970) Argon-40/Argon-39 dating of lunar rock samples. Science 167:466-468

Turner G, Huneke JC, Podosek FA, Wasserburg GJ (1971) ^{40}Ar-^{39}Ar ages and cosmic ray exposure ages of Apollo 14 samples. Earth Planet Sci Lett 12:19-35

Vauclair S (1998) Element settling in the solar interior. Space Sci Rev 85:71-78

Vogt S, Herzog GF, Reedy RC (1990) Cosmogenic nuclides in extraterrestrial materials. Rev Geophys 28:253-275

von Steiger R, Geiss J, Gloeckler G (1997) Composition of the solar wind. *In* Cosmic winds and the heliosphere. Jokipii JR, Sonett CP, Giampapa MS (eds) Univ Arizona Press, Tucson, p 581-616

von Zahn U, Hunten DM, Lehmacher G (1998) Helium in Jupiter's atmosphere: Results from the Galileo probe helium interferometer experiment. J Geophys Res E 103:22815-22829

Weygand JM, Ipavich FM, Wurz P, Paquette JA, Bochsler P (2001) Determination of the ^{36}Ar/^{38}Ar isotopic abundance ratio of the solar wind using SOHO/CELIAS/MTOF. Geochim Cosmochim Acta, 65: 4589-4596

Whitby J, Burgess R, Turner G, Gilmour J, Bridges J (2000) Extinct ^{129}I in halite from a primitive meteorite: Evidence for evaporite formation in the early solar system. Science 288:1819-1821

Widing KG (1997) Emerging active regions on the sun and the photospheric abundance of neon. Astrophys J 480:400-405

Wieler R (1997) Why are SEP noble gases so abundant in extraterrestrial samples? Lunar Planet Sci XXVIII (Lunar Planet Inst, Houston, Texas) p1551-1552

Wieler R (1998) The solar noble gas record in lunar samples and meteorites. Space Sci Rev 85:303-314

Wieler R (2002) Cosmic-ray-produced noble gases in meteorites. Rev Mineral Geochem 47:125-170

Wieler R, Baur H (1994) Krypton and xenon from the solar wind and solar energetic particles in two lunar ilmenites of different antiquity. Meteoritics 29:570-580

Wieler R, Baur H (1995) Fractionation of Xe, Kr, and Ar in the solar corpuscular radiation deduced by closed system etching of lunar soils. Astrophys J 453:987-997

Wieler R, Baur H, Signer P (1986) Noble gases from solar energetic particles revealed by closed system stepwise etching of lunar soil minerals. Geochim Cosmochim Acta 50:1997-2017

Wieler R, Kehm K, Meshik AP, Hohenberg CM (1996) Secular changes in the xenon and krypton abundances in the solar wind recorded in single lunar grains. Nature 384:46-49

Wieler R, Humbert F, Marty B (1999) Evidence for a predominantly non-solar origin of nitrogen in the lunar regolith revealed by single grain analyses. Earth Planet Sci Lett 167:47-60

Wimmer-Schweingruber RF (ed) (2001) Solar and galactic composition. AIP Conf Proc Vol 598, Am Inst Phys, Melville, New York

Wimmer-Schweingruber RF, Bochsler P (2000) Is there a record of interstellar pick-up ions in lunar soils? *In* Acceleration and transport of energetic particles observed in the heliosphere. ACE 2000 Symp. Mewaldt RA, Jokipii JR, Lee MA, Möbius E, Zurbuchen TH (eds) Am Inst Phys Conf Proc 528:270-273

Wimmer-Schweingruber RF, Bochsler P (2001) Lunar soils: a long-term archive for the galactic environment of the heliosphere. *In* Solar and galactic composition. Wimmer-Schweingruber RF (ed) AIP Conf Proc Vol 598, Am Inst Phys, Melville, New York, p 399-404

Wimmer-Schweingruber RF, Bochsler P, Wurz P (1999) Isotopes in the solar wind: New results from ACE, SOHO, and WIND. Solar Wind 9:147-152

Wimmer-Schweingruber RF, Bochsler P, Kern O, Gloeckler G, Hamilton DC (1998) First determination of the silicon isotopic composition of the solar wind: WIND/MASS results. J Geophys Res 103: 20621-20630

Wittenberg LJ, Cameron EN, Kulcinski GL, Ott SH, Santarius JF, Sviatoslavsky GI, Sviatoslavsky IN, Thompson HE (1992) A review of ^{3}He resources and acquisition for use as fusion fuel. Fusion Technology 21:2230-2253

Young PR, Mason HE, Keenan FP, Widing KG (1997) The Ar/Ca relative abundance in solar coronal plasma. Astron Astrophys 323:243-249

3 Noble Gases in Meteorites – Trapped Components

Ulrich Ott
Max-Planck-Institut für Chemie (Otto-Hahn-Institut)
Becherweg 27
D-55128 Mainz, Germany
ott@mcph-mainz.mpg.de

OVERVIEW AND HISTORY

Fundamental insights in the field of cosmochemistry have come from the study of the abundances and isotopic compositions of trapped noble gases. A case in point is the discovery of presolar grains in primitive meteorites: isotope abundance anomalies in noble gases were known long before the discovery of presolar grains such as nanodiamonds, silicon carbide and graphite, and it was the search for the carrier phases of these anomalies that ultimately led to the identification and isolation of these types of grains (e.g., Anders and Zinner 1993; Ott 1993).

Starting with the work of John Reynolds and his colleagues at Berkeley, much of the work on trapped noble gases has centered on xenon, so much that an own term "xenology" was coined to describe this field of noble gas cosmochemistry (Reynolds 1963). The first of the so-called "gas-rich meteorites" were discovered by Russian workers in 1956 (Gerling and Levskii 1956), the same year that Reynolds published details of his innovative static noble gas mass spectrometer (Reynolds 1956). Soon thereafter it was discovered that the elemental abundance patterns observed in meteorites basically fell into one of two patterns: (a) the gas-rich meteorites with a "solar" pattern—high abundances of the lightest gases He and Ne compared to the heavy ones, the origin being implanted solar wind; and (b) the "planetary" pattern with strong elemental fractionation, i.e., strong enrichment of the heavy noble gases relative to the light ones, when compared to the solar abundance pattern (Signer and Suess 1963; Fig. 1). "Planetary" gases are most abundant in the most primitive carbonaceous chondrites (Marti 1967; Mazor et al. 1970). In this chapter I will deal exclusively with the type of noble gases in meteorites included in the "planetary component" as originally defined. Solar wind noble gases as well as truly planetary components such as terrestrial and Martian noble gases are being discussed in separate chapters in this book.

Detailed investigations of the "planetary component" have shown that, while there is some basic similarity to the elemental abundance pattern in the Earth's atmosphere as well as to that of Mars (Fig. 1) and while the term "planetary" still is used, "planetary" is actually a misnomer and should better be avoided given that there are not only variations in the elemental abundance patterns, but also really large differences in *isotopic* compositions. As detailed below, in itself "planetary" noble gases in meteorites are complex and consist of a number of different individual components characterized by unique isotopic (sometimes also elemental) compositions which are carried by specific host phases. Some—termed isotopically approximately normal—are not too different in isotopic composition from solar noble gases (Wieler 2002) and may be (but need not be) of "local" origin. Others—as pointed out in the beginning—are carried by presolar grains and show in their isotopic compositions abundance anomalies that reflect excess contributions from specific processes of nucleosynthesis in stars. A general overview is given in Table 1, where also alternative names for the various components—currently or previously used, often for neon or krypton/xenon only—are given. It is suggested that workers in the field should settle for an agreed set of names (suggested to be the one used here) in order not to confuse too much the non-specialists.

1529-6466/00/0047-0003$05.00

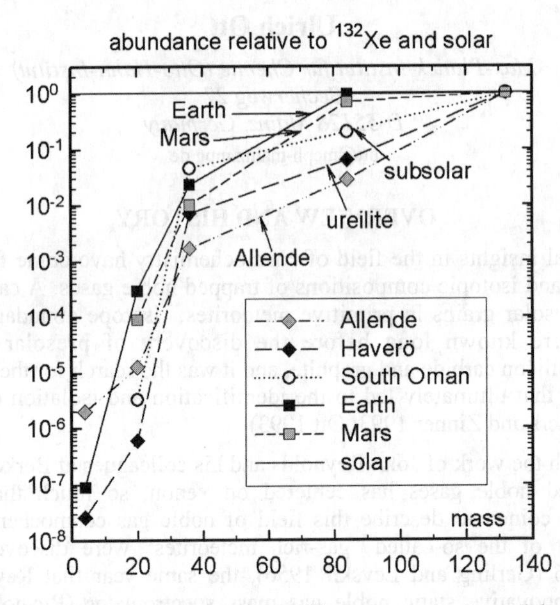

Figure 1. Basic "planetary" patterns as observed in various meteorites and the atmospheres of the Earth and Mars. Shown are abundance ratios relative to ^{132}Xe of ^{4}He, ^{20}Ne, ^{36}Ar and ^{84}Kr normalized to the solar ratios (Wieler 2002). In order to be able to show data for ^{4}He with minimal radiogenic contribution, for the carbonaceous chondrite Allende data for a carbon-rich acid residue (BA1, Ott et al. 1981) are shown, for the ureilite Haverö data for a handpicked carbon-rich sample (B-18-1, Göbel et al. 1978). Enstatite data (South Oman) are from Crabb and Anders (1981), those for Mars from Pepin (1991).

Table 1. Some properties of important trapped noble gas components—an overview (solar gases and Martian meteorites excluded).

Component	*Alternative Name(s)*	*Occurrence*	*Carrier Phase*	*Remarks*
Q gases	P1 (Xenon) OC-Xe	primitive meteorites	Q (definition)	dominates Ar, Kr, Xe
P3 gases	(Neon) Ne-A1	primitive meteorites	presolar diamond	isotopes ~normal
P6 gases	--	primitive meteorites	presolar diamond	~normal (?)
HL gases	(Neon) Ne-A2 (= HL + P6)	primitive meteorites	presolar diamond	nuclear component (r-process ?)
N component*	--	primitive meteorites	SiC	isotopes ~normal
G component*	(Ne, Kr, Xe) Ne-E(H), Kr-S, Xe-S	primitive meteorites	SiC	nuclear component (s-process)
R component	(Neon) Ne-E(L)	primitive meteorites	presolar graphite	radiogenic ^{22}Ne
ureilite gases	Kenna type	ureilites	diamond, graphite	isotopes ~normal
subsolar gases	--	enstatite chondrites	enstatite	less fractionated

* Similar, but probably discrete components are present in presolar graphite (see section *Isotopic compositions*). Additional components discussed in the text: sub-Q, noble gases in other achondrites, as well as in sulfide, metal and iron meteorite samples.

Probably the most abundant and widespread of the components in Table 1 is the "Q" component (Lewis et al. 1975; Wieler 1994; Ozima et al. 1998), also named P1 (Huss et al. 1996) or (for xenon) OC-Xe (Lavielle and Marti 1992). Others, as noted above, are carried by presolar grains, while additional ones, such as ureilite and subsolar noble gases (Göbel et al. 1978; Crabb and Anders 1981), are primarily found in special types of meteorites only.

The rest of this review is organized as follows. First a general overview is given of elemental and isotopic abundance patterns of major and/or well-defined components. This is followed by a chapter that gives a discussion of origin and history of these components and that also discusses some less common and/or less well-characterized components. After possible relationships are considered, a short summary is given.

ELEMENTAL ABUNDANCE PATTERNS

Elemental abundance ratios for various of the components listed in Table 1 are given in Table 2 and are shown in Figure 2. Unlike (in general) the isotopic compositions discussed in the following chapter, most of the elemental compositions investigated in detail show variations between different samples analyzed, in some cases at least due to metamorphism (cf. Huss and Lewis 1994a; Busemann et al. 2000). So the elemental ratios listed in Table 2, and shown in Figure 2, are kinds of "typical" or "average" values (Q, ureilites) or best estimates of primitive compositions (see footnote to Table 2 for details). For ureilites, where all ratios vary by more than an order of magnitude, it is especially difficult to define a representative composition. Only the subsolar composition is simply that found by Crabb and Anders (1981) in the E-chondrite South Oman. Where

Table 2. Elemental abundance ratios in various trapped components.

Component	$^{4}He/^{132}Xe$	$^{20}Ne/^{132}Xe$	$^{36}Ar/^{132}Xe$	$^{84}Kr/^{132}Xe$
Q (P1)	374 ± 72	3.2 ± 0.5	76 ± 7	0.81 ± 0.05
P3	67,000 ± 20,000	80 ± 53	400 ± 25	2.7 ± 0.4
HL	300,000 ± 30,000	485 ± 26	50 ± 20	0.48 ± 0.04
G	321,000 ± 45,000	138 ± 19	3.55 ± 0.50	0.465 ± 0.066
N	661,000 ± 93,000	826 ± 117	79 ± 11	1.468 ± 0.208
ureilite	64.5 ± 9.1	1.41 ± .20	409 ± 58	1.85 ± 0.26
subsolar	--	7.59 ± 1.07	2660 ± 376	5.86 ± 0.84

Variations occur within the individual elemental abundance patterns as observed in different samples (see discussion in text). Values listed here:

a) for Q (also named P1) the average deduced in the closed system etch analyses of HF/HCl residues by Busemann et al. (2000);

b) for P3 and HL: "best estimate of primitive component" of Huss and Lewis (1994a);

c) for G and N components as given by Lewis et al. (1994) and/or calculated based on their sample KJ; the "typical" G and N patterns listed and shown are based on their decomposition of Ne, Kr and Xe for the KJ sample, together with their reported $(^{4}He/^{22}Ne)_G = 193$ and $(^{4}He/^{20}Ne)_N = 800$. For the conversion of ^{22}Ne abundances into ^{20}Ne abundances values for $^{20}Ne/^{22}Ne$ in the G component of 0.0827 (Table 3) and of 8.4 in the N component were used. Ar was calculated by partitioning, following their approach, assuming $^{38}Ar/^{36}Ar$ as 0.660 and 0.1705 in G and N, respectively (Table 3).

d) for ureilites: data for a carbon-rich separate from Haverö (Göbel et al. 1978);

e) for the subsolar component: as measured in the enstatite chondrite South Oman by Crabb and Anders (1981).

Where no specific errors are given in the original literature, a 10% error in elemental abundances was assumed and propagated.

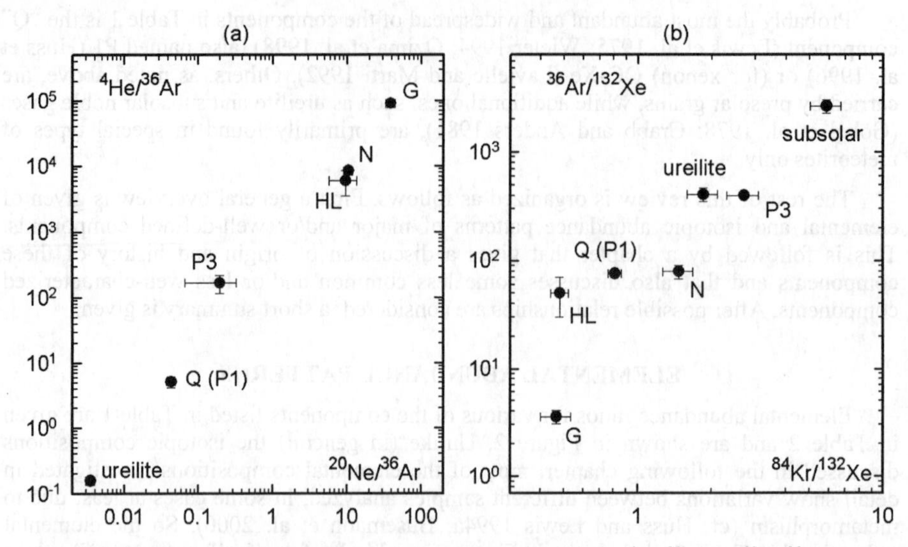

Figure 2. Elemental abundance ratios in major trapped components, $^4He/^{36}Ar$ vs. $^{20}Ne/^{36}Ar$ (a) and $^{36}Ar/^{132}Xe$ vs. $^{84}Kr/^{132}Xe$ (b). Variations occur in compositions as observed in different samples, for selection of values shown see footnote to Table 2 and text. Based on listed values or derived from data of Busemann et al. (2000), Huss and Lewis (1994a), Lewis et al. (1994), Göbel et al. (1978) and Crabb and Anders (1981). Where no errors are given in the original literature, an assumed 10% error in elemental abundances was propagated.

no explicit errors are reported in the original literature, errors have been assigned for this compilation assuming 10% uncertainties in elemental abundance determinations. Uncertainties due to model-dependent assumptions were not taken into account.

All patterns are fractionated, relative to the solar abundance pattern (cf. Figs. 1 and 2), to various degrees, however. Extremely strong is the fractionation relative to the heavy gases of He and Ne for the Q and ureilite abundance patterns. The similarity of the two patterns and the similarity of isotopic compositions (see following section) have led to speculation that these components are related, if not identical (see *Relations* section; Ott et al. 1984, 1985b; Busemann et al. 2000). Gases contained in diamonds and in SiC generally show a less fractionated pattern. Also less fractionated is the subsolar component found in enstatite meteorites which has been the reason for giving it its name. However, there appears to be a general break in fractionation patterns between light and heavy noble gases. The subsolar component, e.g., is less fractionated in Ar/Xe and Kr/Xe than the other components listed in Table 2, but Ne/Xe is among the most fractionated, while the G component in presolar SiC shows just the opposite behavior (Fig. 2). A similar effect—a break between light and heavy gases—has been also seen by Busemann et al. (2000) in the variations within the Q component alone. Possibly the dominant mechanism(s) causing fractionation among the heavy noble gases are not the same as those responsible for the fractionation between heavy and light ones (cf. also discussion in *Relations* section further below).

ISOTOPIC COMPOSITIONS

Given the evidence for variations in elemental composition within various components, isotopic compositions generally are more characteristic for the presence of a specific component. Isotopic compositions are listed in Tables 3-5 and are shown in Figures 3-5, 7 and 8, below. Comparison is made with solar compositions derived from solar wind measurements (Wieler 2002), either as measured (Kr, Xe) or corrected for a possible fractionation in the solar wind relative to the true solar composition (Ne, Ar). The individual trapped components are discussed below. Basically, there are two ways these compositions have been derived: a) consistent determinations in a variety of samples and/or release steps where they were assumed to occur in essentially "pure" form (e.g., Q, ureilites); b) in cases where predominantly mixtures of various components were measured, from observed mixing lines and assumptions on one isotopic ratio for the composition of the end-member of interest (e.g., diamond and SiC noble gases). As for the latter approach, the value chosen for the ratio on which to extrapolate the "pure" component is arbitrary in principle, although the choice usually is reasonable. This limitation should be kept in mind when assessing the meaning of the different compositions and of the differences between the various near-normal components. Only in the case of the G component hosted by presolar silicon carbide and (for Kr and Xe) corresponding to material synthesized in the s-process, the assumption involved in the decomposition (essentially zero ^{136}Xe/^{130}Xe in s-process-Xe, e.g.; see section *Origins and History*) is based on a solid and well-founded physical understanding. Generally also not taken into account are possible fractionation effects during trapping (for the case of the P3 and HL components discussed in the *Origins and History* section).

Helium

Compositions are listed in Table 3. Given the small grain size of most important carrier phases (e.g., Q and diamond) and almost homogeneous distribution in the matrix (cf. Banhart et al. 1998), these grains must have acquired basically the matrix concentration of spallogenic ^{3}He and radiogenic ^{4}He due to recoil out of other matrix material. With ^{3}He/^{4}He generally low (~10^{-4}) in trapped components, and on the other hand abundantly produced by cosmic ray spallation, ^{3}He generally is more strongly affected than ^{4}He. Precise compositions of trapped ^{3}He/^{4}He have been obtained only in a few cases, i.e., for the Q component (Busemann et al. 2000, 2001a) as well as the P3 and HL gases hosted by presolar diamond (Huss and Lewis 1994b), all lying in the range 1.2 to 1.7×10^{-4}, with P3 possibly characterized by a lower ratio of ~0.45×10^{-4} (Busemann et al. 2001a). Nevertheless, strictly speaking, even in these cases obtained ^{3}He/^{4}He ratios must be regarded as upper limits only. The exact size of the cosmogenic interference depends, of course, on the cosmic ray exposure age of the host meteorite and on the concentration of trapped He in the respective host phase, and hence must be minor in phases like diamond which contains trapped He in high concentration. A similar upper limit of 2.6×10^{-4} has been inferred for the N component in presolar silicon carbide (Lewis et al. 1994). The ^{3}He/^{4}He ratio in the Q-component is based on closed system etch analyses of HF/HCl-resistant residues (Busemann et al. 2000, 2001a), which selectively release Q gases as operationally defined (see section *Origins and History*). The value in Table 3 of 1.23×10^{-4} (Busemann et al. 2001a) was obtained in the analysis of a residue from the Isna CO3 chondrite, which has an extremely short cosmic ray exposure. The exact value for ^{3}He/^{4}He in Q has wider implications, e.g., in cosmology (Yang et al. 1984; Schramm and Turner 1998; Busemann et al. 2000, 2001a). This is, because with the Q component likely to be representative of the primordial noble gas component of the solar system (Wieler 1994; Ozima et al. 1998), $(^3\text{He}/^4\text{He})_Q$ should correspond to the protosolar value. Hence, in turn, from the difference between ^{3}He/^{4}He in Q and in the solar wind, it should be possible to determine the protosolar D/H ratio (see also Wieler 2002).

Table 3. Isotopic compositions of He, Ne, and Ar in various trapped components.

Component	*$^3He/^4He$ [10^{-4}]*	*$^{20}Ne/^{22}Ne$*	*$^{21}Ne/^{22}Ne$*	*$^{36}Ar/^{38}Ar$*	*References*
Q (P1)	1.23 ± 0.02	10.67 ± 0.02	0.0294 ± 0.0010	5.34 ± 0.02	[1,2,3]
		10.11 ± 0.02			[1]
P3	≤ 1.35 ± 0.10	8.910 ± 0.057	0.029 ± 0.001	5.26 ± 0.03	[4]
HL	≤ 1.70 ± 0.10	8.500 ± 0.057	0.036 ± 0.001	4.41 ± 0.06	[4]
G	--	< 0.1	< 0.0015	--	[5]
G (theory)	*~0*	*0.0827*	*0.00059*	*1.52 ± 0.20*	*[5]*
N	≤ 2.6	--	--	5.87 ± 0.07	[5]
R	--	< 0.01	< 0.0001		[6]
ureilite	--	10.70 ± 0.25	--	5.26 ± 0.06	[1,7]
		10.4 ± 0.3			[8]
subsolar	--	--	--	5.45 ± 0.04	[9]
	~2.1 × 10^{-4}	~11.65	--	~5.36	[10]

References: [1] Busemann et al. (2000); [2] Wieler et al. (1992); [3] Busemann et al. (2001a); [4] Huss and Lewis (1994b); [5] Lewis et al. (1994); [6] Amari et al. (1995); [7] Göbel et al. (1978); [8] Ott et al. (1985a); [9] Crabb and Anders (1981); [10] Busemann et al. (2001b).

Notes:

(a) He isotopic ratios can be generally regarded as upper limits only because of omnipresent spallogenic contributions.

(b) For the G component experimental upper limits for the Ne isotopic ratios are listed plus (in italics) the theoretical values used in the decomposition of gases measured in samples of presolar silicon carbide.

(c) for $^3He/^4He$ in P3 a value of (0.45 ± 0.04) × 10^{-4} has been suggested by Busemann et al. (2001a).

(d) Ratios measured in solar wind: $^3He/^4He = 4.1 \times 10^{-4}$; $^{20}Ne/^{22}Ne = 13.8$; $^{21}Ne/^{22}Ne = 0.0328$; $^{36}Ar/^{38}Ar = 5.58$.
Solar: $^3He/^4He$ (outer convective zone) = 3.75×10^{-4}; $^{20}Ne/^{22}Ne = 13.6$, $^{21}Ne/^{22}Ne = 0.0326$, $^{36}Ar/^{38}Ar = 5.5$ (Wieler 2002).
For more detailed discussion see text.

Neon

Compositions are also listed in Table 3; a selection of $^{20}Ne/^{22}Ne$ ratios, along with $^{36}Ar/^{38}Ar$ ratios is shown in Figure 3. Neon isotopic compositions, together with those in xenon, have turned out to be the most diagnostic ones that allow to identify and sort out the different components. Strongly standing out is the composition of Ne in the graphite (R) and in the SiC-G component, both almost pure ^{22}Ne, also termed Ne-E (Table 3; not shown in Fig. 3). In early work (cf. Swindle 1988) these components had been given the names Ne-E(L) and Ne-E(H), because the early separation attempts had shown them to be located in a carrier phase of low density and low release temperature (L; now known to be graphite) and in another carrier of high density and high release temperature (H; now known to be silicon carbide).

Strictly speaking, only very low upper limits on the abundance of ^{20}Ne and ^{21}Ne in the R and G components are known (Table 3). But it is generally accepted that Ne-R [= Ne-E(L)] is pure ^{22}Ne, the decay product of ^{22}Na ($T_{½}$ = 2.6 a), which in turn was trapped when the grains condensed. This apparently "radiogenic origin" led Amari et al. (1995) to re-name this component Ne-R. The G component is thought to be material from the He burning shell of Red Giant AGB stars (e.g., Gallino et al. 1990; see also section *Origins and History*) and theoretical values for the small, but finite amounts of ^{20}Ne and ^{21}Ne accompanying ^{22}Ne are listed in Table 3 (in italics) together with the experimentally determined upper limits.

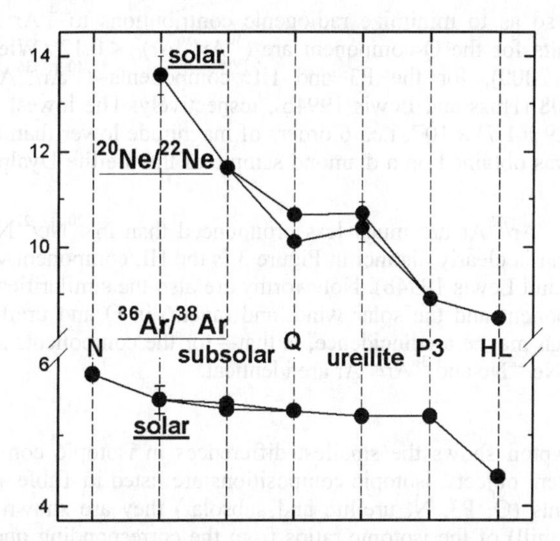

Figure 3. Isotopic ratios $^{20}Ne/^{22}Ne$ and $^{36}Ar/^{38}Ar$ as determined for various trapped noble gas components (Table 3). Not shown are Ne isotopic ratios for the nucleogenic G [Ne-E(H)] and R [Ne-E(L)] components, which are almost or (probably) entirely pure ^{22}Ne (see discussion in text). Also not shown is the (theoretical) value for Ar-G. HL may contain small contributions from P6. Obvious are the similarities between the Q and ureilite compositions as well as the similarity in $^{36}Ar/^{38}Ar$ between solar wind and the subsolar component. Both ratios (maybe by coincidence) decrease in the order from left to right.

Neon in the other components in Table 3/Figure 3 is characterized by higher $^{20}Ne/^{22}Ne$ ratios, although in all cases lower than in solar wind neon, occupying a range ~8 to ~11, which encompasses the ratio observed in the terrestrial atmosphere (9.8). Two values are listed for Ne in the Q component because one group of meteorites gives a $^{20}Ne/^{22}Ne$ ratio of ~10.7, while another group gives a lower ratio of ~10.1 (Busemann et al. 2000). Also for ureilite Ne two values are listed: 10.70 (Busemann et al. 2000), which is the average ratio for the ureilite carbon/diamond samples analyzed by Göbel et al. (1978), and a lower ratio of 10.4 reported by Ott et al. (1985a) for the Hajmah ureilite with very low spallation contributions. The similarity in isotopic composition between Q and ureilite neon strengthens the case for a close relationship between the two components as already evident in the elemental abundance pattern (Fig. 2).

Neon in presolar diamonds (P3, HL, P6) dominates the neon inventory of primitive meteorites, and formerly identified components Ne-A, or Ne-A1 and Ne-A2 (e.g., Swindle 1988) can be related to these newly identified primary components: Ne-A1 corresponds to P3, A2 to HL+P6, and A is probably a mixture of all three plus, in addition, contributions from Ne-G and/or Ne-R. Since these contributions are difficult to resolve, the value listed for HL in Table 3 and shown in Figure 3 actually contains possible contributions from the P6 component. In this sense it more closely corresponds to the old Ne-A2, but contributions from P6 in all likelihood are minor (Huss and Lewis 1994b).

Argon

Compositions for various components are listed in Table 3. Given are only $^{36}Ar/^{38}Ar$ ratios. Measured $^{40}Ar/^{36}Ar$ ratios in trapped-gas-rich samples are generally low (often <1) and actually reported values critically depend on blank corrections and/or purity of the analyzed samples so as to minimize radiogenic contributions to ^{40}Ar from K-bearing phases. Upper limits for the Q-component are $(^{40}Ar/^{36}Ar)_Q < 0.12$ (Wieler et al. 1992; Busemann et al. 2000), for the P3 and HL components $(^{40}Ar/^{36}Ar)_{P3} < 0.03$ and $(^{40}Ar/^{36}Ar)_{HL} < 0.08$ (Huss and Lewis 1994b), respectively. The lowest value measured for $^{40}Ar/^{36}Ar$ is $(2.9 \pm 1.7) \times 10^{-4}$, i.e., 6 orders of magnitude lower than in the terrestrial atmosphere, and was obtained on a diamond sample of the ureilite Dyalpur (Göbel et al. 1978).

Variations in $^{36}Ar/^{38}Ar$ are much less pronounced than in $^{20}Ne/^{22}Ne$ (Fig. 3). The only component that is clearly distinct in Figure 3 is the HL component with $^{36}Ar/^{38}Ar = 4.41 \pm 0.06$ (Huss and Lewis 1994b). Noteworthy are also the similarities of $^{36}Ar/^{38}Ar$ in the subsolar component and the solar wind, and (again) in Q and ureilite noble gases. Interesting, although maybe a coincidence, is that—for the components shown in Figure 3—the trends in $^{20}Ne/^{22}Ne$ and $^{36}Ar/^{38}Ar$ are identical.

Krypton

Generally, krypton shows the smallest differences in isotopic composition among different solar system objects. Isotopic compositions are listed in Table 4. For the "more normal" components (Q, P3, N, ureilite and subsolar) they are shown in Figure 4 as deviations (in per mill) of the isotopic ratios from the corresponding ones measured for the solar wind. Not shown in Figure 4 is the $^{78}Kr/^{84}Kr$ ratio for the P3 component (because of its large uncertainty) and that for the N component (no value reported). Again, as in the case of Ar, the subsolar component shows a marked similarity to the solar wind pattern. The others show very similar patterns, mass fractionated relative to the solar wind favoring the heavy isotopes by something like 0.8% per mass unit. The exception is at mass 86, which in the N component is depleted by about 6% and which also shows a larger spread among the other components.

Large anomalies are, of course, observed in the nucleosynthetic components carried by the presolar diamond, SiC and graphite grains (Table 4). Kr-G (s-process krypton) and Kr-N are carried by SiC grains, where they occur in roughly equal abundance, and have been studied extensively. The compositions listed in Table 4 are that derived by Lewis et al. (1994) on SiC by partitioning the measured compositions with the assumption of a $^{84}Kr/^{82}Kr$ ratio for the G component of 2.40, as based on astrophysical theoretical grounds. The inferred ratios $^{80}Kr/^{84}Kr$ and $^{86}Kr/^{84}Kr$ for the G component are variable and constitute a sensitive measure of physical conditions during their creation in the slow neutron capture process (s-process) of nucleosynthesis (section *Origins and History*). Similar krypton of s-process origin is present in presolar graphite (Amari et al. 1995), where even more extreme variations have been observed. This may be due to the presence in varying abundance ratios of two discrete components, Kr-S(H) and Kr-S(L) carried by graphite grains of different density.

In contrast to the approach for G and N in silicon carbide, where the reported compositions are extrapolations, the Kr composition for Kr-P3 and Kr-HL in presolar diamond (Huss and Lewis 1994b) is based on the most extreme *measured* compositions, and Kr-HL of the reported composition is unlikely to correspond to Kr as produced in a nucleosynthetic event. Defining a composition for the P6 component—which is released from diamonds at high temperature, mixed with HL gases—has turned out to be difficult. Two possible compositions have been suggested by Huss and Lewis (1994b) and are listed in Table 4: one assumes a "normal" $^{84}Kr/^{82}Kr$ ratio for P6 corresponding to the one assumed for the Kr-P3 component, the other ("exotic" P6) assumes a $^{84}Kr/^{82}Kr$ just beyond the most extreme measured data point in the Kr-HL/Kr-P6 mixtures. Details of these anomalies and their implications are discussed in the *Origins and History* section.

Table 4. Isotopic composition of Kr in various trapped components.

Component	$^{78}Kr/^{84}Kr$	$^{80}Kr/^{84}Kr$	$^{82}Kr/^{84}Kr$	$^{83}Kr/^{84}Kr$	$^{86}Kr/^{84}Kr$	*References*
Q (P1)	0.00603 ± 0.00003	0.03937 ± 0.00007	0.2018 ± 0.0002	0.2018 ± 0.0002	0.3095 ± 0.0005	[1]
P3*	0.0064 ± 0.0010	0.0395 ± 0.0004	≡ 0.2023	0.2032 ± 0.0005	0.3128 ± 0.0006	[2]
HL*	0.0042 ± 0.0010	0.0305 ± 0.0010	≡ 0.1590	0.1989 ± 0.0010	0.3623 ± 0.0018	[2]
P6*	0.0059	0.0381 ± 0.0008	≡ 0.2023	02013 ± 0.0010	0.3147 ± 0.0030	[2]
P6(exotic)*	0.0052	0.0354 ± 0.0004	≡ 0.1839	0.1999 ± 0.0005	0.3348 ± 0.0012	[2]
G*ˣ	--	0.0133 – 0.0183	≡ 0.4167	0.1192 ± 0.0054	0.454 – 1.176	[3]
N*	--	0.03962 ± 0.00040	0.2028 ± 0.0021	0.2018 ± 0.0032	0.2842 ± 0.0036	[3]
ureilite	0.00601 ± 0.00008	0.0399 ± 0.0005	0.2037 ± 0.0018	0.2026 ± 0.0015	0.3091 ± 0.0012	[4]
subsolar	0.00634 ± 0.00016	0.04050 ± 0.00029	0.2045 ± 0.0009	0.2027 ± 0.0007	0.3073 ± 0.0017	[5]

References : [1] Busemann et al. (2000); [2] Huss and Lewis (1994b), renormalized following Busemann et al. (2000); [3] Lewis et al. (1994), [4] Göbel et al. (1978), avg. of 7 ureilites, except for $^{78}K/^{84}Kr$ (ureilite Kenna; Wilkening and Marti 1976). [5] Crabb and Anders (1981), South Oman.

Notes: Ratios measured for solar wind Kr are $^{78}Kr/^{80}Kr/^{82}Kr/^{83}Kr/^{84}Kr/^{86}Kr$ = 0.006365/0.04088/0.20482/0.20291/ ≡ 1/0.3024 (Pepin et al., 1995; see also contribution by Wieler).

* compositions derived from mixing lines, involving assumptions about $^{84}Kr/^{82}Kr$ ratios in end-members (see text).

ˣ ratios $^{80}Kr/^{82}Kr$ and $^{86}Kr/^{82}Kr$ in the G component are variable. Ranges are for grain size fractions of presolar SiC. For graphite $^{86}Kr/^{84}Kr$ extends to even higher values.

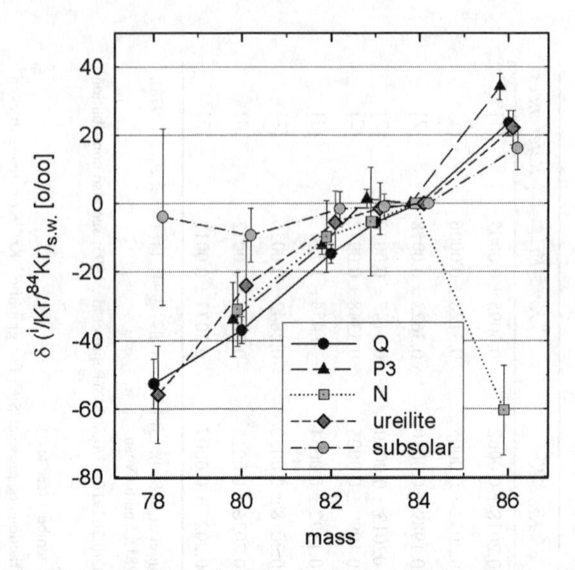

Figure 4. Kr isotopic compositions of various "approximately normal" trapped components: Q (P1), ureilite, subsolar, as well as P3 in presolar diamonds and N in presolar SiC and graphite (Table 4). More anomalous components are shown in Figure 7. Ratios are normalized to ^{84}Kr, and shown are deviations in per mill of the ^{i}Kr/^{84}Kr ratios from the corresponding ratios measured for the solar wind (Table 4). Not shown because of its large error is ^{78}Kr/^{84}Kr in the P3 component.

Xenon

Besides neon, xenon has turned out to be the element most diagnostic in its isotopic composition. An important role has been played by Xenon-HL, which was the first of the nucleosynthetic isotope anomalies to be discovered (Reynolds and Turner 1964). The HL component has received its name for the simultaneous overabundance of the heavy xenon isotopes (≡ Xe-H) and the light xenon isotopes (≡ Xe-L). Because the H part originally was more reliably determined, Xenon-HL was first believed to be associated with fission, possibly of a superheavy element (e.g., Anders et al. 1975), but in the end the search for its host phase led to the discovery of the existence of grains of presolar origin in primitive meteorites (Lewis et al. 1987).

Compositions are listed in Table 5. Again, as in the case of Kr, Xe-HL as listed is based on mixing lines, assuming ^{136}Xe/^{132}Xe in the HL component to have a value of 0.70, which is close to the most extreme measured value and the intersection of the P3-HL and P6-HL mixing lines (Huss and Lewis 1994b). To derive the other end composition of the mixing line observed in the low-temperature release, which corresponds to Xe-P3, a ^{136}Xe/^{132}Xe ratio of ~0.310, just below the lowest measured value was assumed (Table 5; Huss and Lewis 1994b). Again, for P6, the component with which Xe-HL mixes at higher release temperature, two compositions are listed, based on two different (normal vs. exotic) assumptions for ^{136}Xe/^{132}Xe in it.

The compositions of Xe-G and Xe-N present in presolar SiC and graphite grains—in analogy to Kr—have been partitioned based on mixing lines and theoretical guidelines, i.e., assumption of essentially no ^{136}Xe (^{136}Xe/^{130}Xe = 0.0071) for the G component (Lewis et al. 1994). For the N component, these authors chose the ratio ^{136}Xe/^{130}Xe to be equal to 1.305 times that measured in the ureilite Kenna. This leads (by definition) to

Table 5. Isotopic composition of Xe in various trapped components.

Component	$^{124}Xe/^{132}Xe$	$^{126}Xe/^{132}Xe$	$^{128}Xe/^{132}Xe$	$^{129}Xe/^{132}Xe$	$^{130}Xe/^{132}Xe$	$^{131}Xe/^{132}Xe$	$^{134}Xe/^{132}Xe$	$^{136}Xe/^{132}Xe$	*References*
Q (P1)	0.00455 (2)	0.000406 (2)	0.0822 (2)	1.042 (2)	0.1619 (3)	0.8185 (9)	0.3780 (11)	0.3164 (8)	[1]
P3*	0.00446 (6)	0.00400 (4)	0.0806 (2)	1.042 (4)	0.1589 (2)	0.8247 (19)	0.3767 (10)	≡ 0.3096	[2]
HL*	0.00839 (9)	0.00564 (8)	0.0905 (6)	1.056 (2)	0.1542 (3)	0.8457 (13)	0.6356 (13)	≡ 0.6991	[2]
P6*	0.00433 (25)	0.00440 (28)	0.0890 (20)	1.114 (8)	0.1658 (11)	0.8229 (47	0.3288 (50)	≡ 0.3096	[2]
P6(exotic)*	0.00679 (8)	0.00516 (8)	0.0899 (5)	1.078 (2)	0.1587 (3)	0.8370 (13)	0.5176 (13)	≡ 0.5493	[2]
G^x	≡ 0	≡ 0	0.2159 (23)	0.1182 (112)	0.4826 (42)	0.1858 (117)	0.0222 (53)	≡ 0.00343	[3]
N^x	0.00470 (13)	0.00357 (18)	0.0785 (11)	1.000 (14)	0.1603 (13)	0.8109 (130)	0.4183 (59)	0.4006 (32)	[3]]
Ureilite	0.00463 (6)	0.00416 (4)	0.0827 (5)	1.035 (5)	0.1627 (5)	0.8195 (13)	0.3776 (12)	0.3152 (19)	[4]
Subsolar	0.00490 (16)	0.00432 (14)	0.0843 (8)	--	0.1649 (10)	0.8301 (34)	0.3765 (25)	0.3095 (20)	[5]

Uncertainties in the last digits are given in parentheses.

References: [1] Busemann et al. (2000); [2] Huss and Lewis (1994b), renormalized following Busemann et al. (2000); [3] Lewis et al. (1994); [4] Göbel et al. (1978), avg. of 7 ureilites; [5] Crabb and Anders (1981), South Oman.

Notes: Ratios measured for solar wind Xe are $^{124}Xe^{126}Xe/^{128}Xe/^{129}Xe/^{130}Xe/^{131}Xe/^{132}Xe/^{134}Xe/^{136}Xe$ = 0.004882/0.004234/0.08475/1.0420/0.1661/0.8272/≡1/0.3666/0.2985 (Pepin et al. 1995; see also Wieler 2002).

* compositions derived from mixing lines, involving assumptions about $^{136}Xe/^{132}Xe$ ratios in endmembers (see text). Two possible compositions for P6 are listed based on different assumptions.

^{x} compositions derived from mixing lines, involving assumptions about $^{136}Xe/^{130}Xe$ ratios in endmembers (see text). ^{124}Xe and ^{126}Xe derived in this approach may be influenced by spallation contributions. For these isotopes the theoretically expected value (≡ 0) is listed.

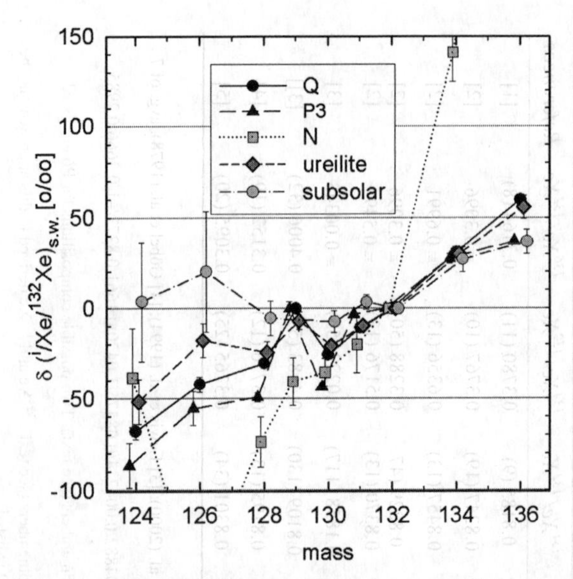

Figure 5. Xe isotopic compositions in various trapped components: Q (P1), ureilite, subsolar as well as P3 in presolar diamonds and N in presolar SiC and graphite (Table 5). More anomalous components are shown in Figure 8. Ratios are normalized to ^{132}Xe, and shown are deviations in per mill of the ^{i}Xe/^{132}Xe ratios from the corresponding ratios measured for the solar wind (Table 5; errors for solar wind not included). Values for ^{126}Xe and ^{136}Xe in the N component are off scale at −156 and +342.

excess ^{136}Xe (and also to excess ^{134}Xe), but a Kenna-like (ureilite-like) composition for the lighter isotopes in Xe-N (Fig. 5). A shortcoming of this treatment—when it comes to ^{124}Xe and ^{126}Xe—is that cosmogenic contributions were ignored. But there must have been such contributions produced during exposure to the cosmic radiation while the SiC grains resided in the interstellar medium and, as demonstrated by Ott and Begemann (2000), there is actually evidence for their presence in the SiC data of Lewis et al. (1994). For this reason, for ^{124}Xe and ^{126}Xe in the G component instead of the Lewis et al. (1994) values the theoretically expected value of zero is given in Table 5; it is not clear, however, how the composition derived for Xe-N has been affected by ignoring the spallation contributions.

As for Kr, the nucleosynthetic components in Xe will be discussed in more detail in the *Origins and History* section, while the more normal Xe compositions (Q, P3, N, ureilite and subsolar) are shown in comparison with Xe determined for the solar wind in Figure 5. Again, as in Kr, a general characteristic is an overall relative depletion of the light isotopes relative to the heavy ones when compared to solar wind. At the lighter isotopes there is some scatter, relative uncertainties are large there, however. More interesting is that there seem to be differences not only in ^{136}Xe/^{132}Xe but also in ^{130}Xe/^{132}Xe. Noteworthy again is the strong similarity between Q-Xe and that found in ureilites and the fact that except for ^{134}Xe and ^{136}Xe, the subsolar pattern is within (unfortunately rather large) errors identical with solar wind.

ORIGINS AND HISTORY

The exact origin and the history for most of the trapped components discussed here still present a puzzle. This even holds for the Q component which is generally regarded as representative for the most important noble gas reservoir in the Solar System outside of the Sun (e.g., Wieler 1994; Ozima et al. 1998). Somewhat clearer are our ideas regarding the components hosted by grains of presolar origin because of the unique nucleosynthetic information recorded in their isotopic signatures which points to the stellar site from which they originate.

Q(P1)-gases

For the heavy noble gases Ar, Kr and Xe, the Q component is the one that dominates the noble gas inventory of primitive meteorites. For example, in the extreme case of the Orgueil CI carbonaceous chondrite (no thermal metamorphism) about 95% of all ^{132}Xe belongs to the Q(P1) component (Huss and Lewis 1995; Huss et al. 1996; Fig. 6). The definition of the Q phase as noble gas host phase and that of Q gas as a gas component is operational (Lewis et al. 1975): Q is that phase which together with the bulk of noble gases survives treatment with hydrochloric and hydrofluoric acid (which leaves typically on the order of a percent of a meteorite), and from which, during following treatment with nitric acid, essentially all of the "Q-gases" are lost, with little loss of mass. Recent work has shown that there are alternatives to the HF/HCl treatment and that samples closely corresponding to the HF/HCl residues can also be prepared—relying on the small size of the carrier grains—by purely physical means, however with much lower yield (Matsuda et al. 1999). After initial controversies following its original discovery (Lewis et al. 1975), it now seems well established that Q is a carbonaceous phase (e.g., Ott et al. 1981). Primarily because of their rather normal isotopic composition, it is generally assumed (but in no way certain) that Q-gases (and hence Q) are of "local," possibly Solar System, origin, unlike the gases hosted by the presolar circumstellar phases being discussed further below.

Within the Q component as operationally defined variations in composition have been observed, partly at least connected with metamorphic history (Huss et al. 1996; Busemann et al. 2000). The most clear-cut trend is the decrease of the abundance of Q gases with increasing metamorphic grade of the host meteorites, coupled with an increase in median release temperature (Huss et al. 1996). As far as isotopic compositions are

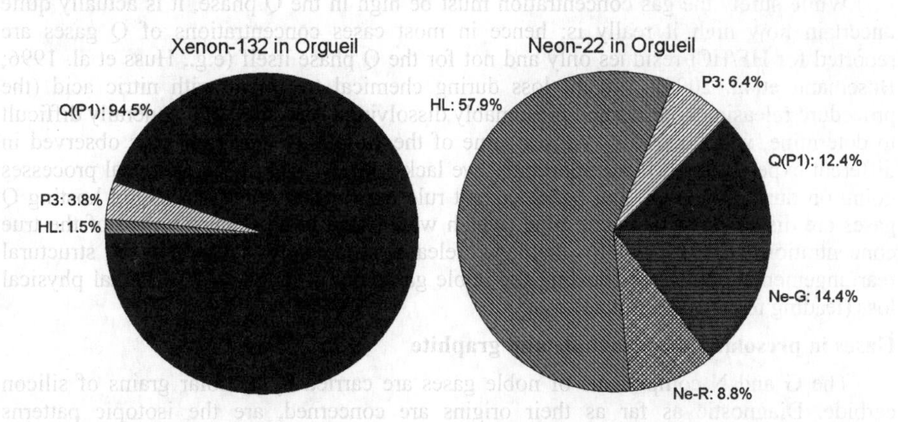

Figure 6. Distribution of ^{132}Xe and ^{22}Ne among the various components in the Orgueil CI carbonaceous chondrite. Because in primitive meteorites virtually all (non-solar) trapped gases are contained in acid-resistant phases (e.g., Ott et al. 1981), where a better separation of components is achieved during stepwise heating, the numbers shown are based on such residue data (Huss and Lewis 1995): for neon and xenon in the Q (P1) component on the HF/HCl residue data, for the Xe-P3 and Xe-HL components on the etched residue data. Shown are only components which contribute more than 1% to the total.

concerned, there may be variations caused by re-trapping of other components (primarily HL gases) by the Q phase in meteorites of higher metamorphic grade (Huss et al. 1996). However, this interpretation, which is an attempt to explain observations from stepwise heating experiments, is probably not unique. The only clear-cut variation seen during closed system etching (i.e., definitely in Q as defined) is in neon: ratios for $^{20}Ne/^{22}Ne$ in Q from different meteorites group around two values, ~10.1 and ~10.7 (section *Isotopic compositions* and Table 3; Busemann et al. 2000). Also the elemental ratios vary: for example, $^{36}Ar/^{132}Xe$ and $^{84}Kr/^{132}Xe$ ratios vary by about ±50% around the average listed in Table 2, and even larger variations occur in the abundance ratio of light to heavy elements (Huss et al. 1996; Busemann et al. 2000). Busemann et al. (2000) suggest the existence of two types of subcarriers "Q1" and "Q2" with slightly different chemical properties. In their interpretation variations in Ar/Xe and Kr/Xe reflect both thermal metamorphism and aqueous alteration of the carriers, while the relatively high He/Ar and Ne/Ar in meteorites that suffered strong aqueous alteration indicate that the subcarriers differ in how susceptible they are to aqueous alteration.

In some way the most puzzling property of Q is the apparent high concentration of noble gases. There have been a variety of attempts (e.g., Yang and Anders 1982; Wacker et al. 1985; Zadnik et al. 1985; Nichols et al. 1992; Sandford et al. 1998) to simulate gas trapping by Q, but in no case have inferred distribution coefficients—at least for gases sited as retentively as Q-gases—approached those inferred from the gas abundance in Q and pressures expected in the solar nebula during trapping. If any, the currently most widely accepted model in order to explain the Q phenomenon is that of Wacker (1989) who suggests that Q gases are physically adsorbed on interior surfaces formed by a pore labyrinth within amorphous carbon. Adsorption/desorption times in this model are controlled by choke points that restrict the movement of noble gases within the labyrinth. Another interesting and new mechanism currently being explored (Hohenberg et al. 2002) involves impingement of low energy noble gases onto a growing surface accompanied by the formation of chemical bonds—under conditions where not sufficient chemically more active elements are present—that last long enough so that the noble gases are buried by the growing surface ("active capture and anomalous adsorption").

While surely the gas concentration must be high in the Q phase, it is actually quite uncertain how high it really is; hence in most cases concentrations of Q gases are reported for HF/HCl residues only and not for the Q phase itself (e.g., Huss et al. 1996; Busemann et al. 2000). Weight loss during chemical treatment with nitric acid (the procedure releasing Q gases and presumably dissolving Q) is small and generally difficult to determine, which certainly causes some of the variability in weight loss observed in different experiments. Most importantly, we lack a knowledge of the chemical processes going on during the etch, and hence cannot rule out (a) that other phases not hosting Q gases are dissolved at the same time (which would lead to an underestimate of the true concentration in Q) nor (b) that gas release is actually caused by a structural rearrangement of the phase hosting the noble gases not accompanied by actual physical loss (leading to an overestimate).

Gases in presolar silicon carbide and graphite

The G and N components of noble gases are carried by presolar grains of silicon carbide. Diagnostic as far as their origins are concerned, are the isotopic patterns observed in the G component for Ne and Kr/Xe: Ne-G [= Ne-E(H)] is dominated by ^{22}Ne, with only small amounts of $^{20,21}Ne$; while Kr-G and Xe-G show large overabundances of the isotopes made in the slow neutron capture process (s-process) of nucleosynthesis (Kr-S, Xe-S, Tables 4, 5; Figs. 7, 8). Graphite is characterized by the occurrence of Ne-R [= Ne(E)-L], generally considered as monoisotopic ^{22}Ne from the decay of ^{22}Na

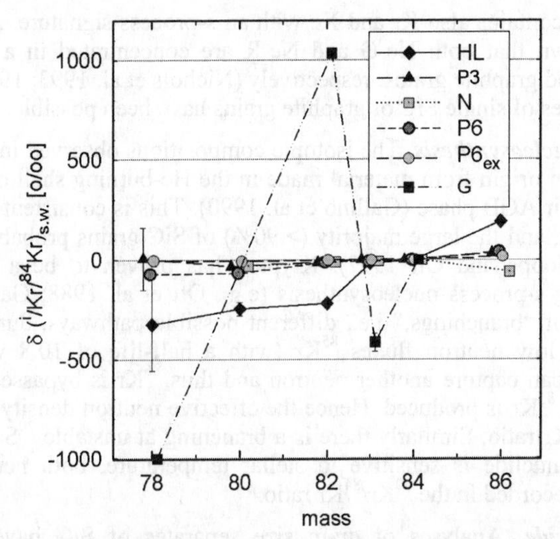

Figure 7. Kr isotopic compositions in gas components hosted by presolar grains: P3, P6, HL in presolar diamonds, G and N in presolar SiC and graphite (Table 4). Ratios are normalized to ^{84}Kr, and shown are deviations in per mill of the ^{i}Kr/^{84}Kr ratios from the corresponding ratios measured for the solar wind (Table 4). Derived compositions generally involve assumption of a value for one characteristic ratio (see text). Two compositions are shown for the P6 component, which involve different assumptions about ^{82}Kr/^{84}Kr in P6. For ^{78}Kr/^{84}Kr in these no error has been given (and hence no error is shown). No experimental data have been reported for ^{78}Kr/^{84}Kr in Kr-G and Kr-N, and for the G component the theoretically expected value of –1000 per mill is plotted. Ratios ^{80}Kr/^{84}Kr and ^{86}Kr/^{84}Kr in Kr-G are variable and not shown. For discussion see text.

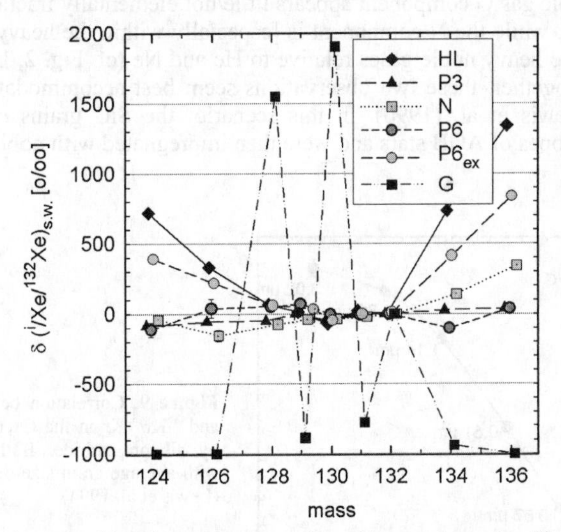

Figure 8. Xe isotopic compositions in gas components hosted by presolar grains: P3, P6, HL in presolar diamonds, G and N in presolar SiC and graphite (Table 5). Ratios are normalized to ^{132}Xe, and shown are deviations in per mill of the ^{i}Xe/^{132}Xe ratios from the corresponding ratios measured for the solar wind (Table 5; errors for solar wind not included). Derived compositions generally involve assumption of a value for one characteristic ratio (see text). Two compositions are shown for the P6 component, which involve different assumptions about ^{136}Xe/^{132}Xe in P6. For ^{124}Xe/^{132}Xe and ^{126}Xe/^{132}Xe in the G component the theoretically expected values of –1000 per mill are shown. For discussion see text.

($T_{½}$ = 2.6 a) and contains also Kr and Xe with an s-process signature. Analyses of single grains have shown that both Ne-G and Ne-R are concentrated in a small fraction of silicon carbide and graphite grains, respectively (Nichols et al. 1993; 1994). So far, for Kr and Xe no analyses of single SiC or graphite grains have been possible.

S-process nucleosynthesis. The isotopic compositions observed in the G component clearly point to an origin from material made in the He-burning shell of carbon-rich Red Giant stars in their AGB phase (Gallino et al. 1990). This is consistent with observations in other elements, and the large majority (> 90%) of SiC grains probably come from this stellar source (Hoppe and Ott 1997). Krypton has proven to be a diagnostic of the conditions during s-process nucleosynthesis (e.g., Ott et al. 1988; Gallino et al. 1990). This is because of "branchings," i.e., different possible pathways, during the s-process. For example, at low neutron fluxes ^{85}Kr (with a half-life of 10.8 years), completely decays before it can capture another neutron and thus ^{86}Kr is bypassed; while at higher neutron densities ^{86}Kr is produced. Hence the effective neutron density is recorded in the observed ^{86}Kr/^{84}Kr ratio. Similarly there is a branching at unstable ^{79}Se, and because the half-life of this nuclide is sensitive to stellar temperature, both neutron density and temperature are recorded in the ^{80}Kr/^{84}Kr ratio.

Silicon carbide. Analyses of grain size separates of SiC have shown that the sensitive Kr isotopic ratios vary as a function of size of the SiC grains (Lewis et al. 1994). These variations occur coupled together with variations in the elemental composition, where especially the abundance ratio Ne/Xe in the G component (Ne-E(H)/Xe-S) sensitively depends on grain size (Fig. 9; Lewis et al. 1994; cf. also Ott and Merchel 2000).

Trends observed in noble gas isotopic patterns are, however, contrary to the trends observed in other elements, such as, e.g., Ba (Gallino et al. 1993; Lewis et al. 1994). In addition, the noble gas G component appears little/not elementally fractionated relative to its stellar source, while the N component is [especially within the heavy noble gases and in the ratio of the heavy noble gases relative to He and Ne (cf. Fig. 2; Lewis et al. 1990, 1994)]. Taken together, these two observations seem best accommodated by a scenario suggested by Lewis et al. (1990). In this scenario, the SiC grains condensed in the expanding envelopes of AGB stars and were then impregnated with noble gas ions from a

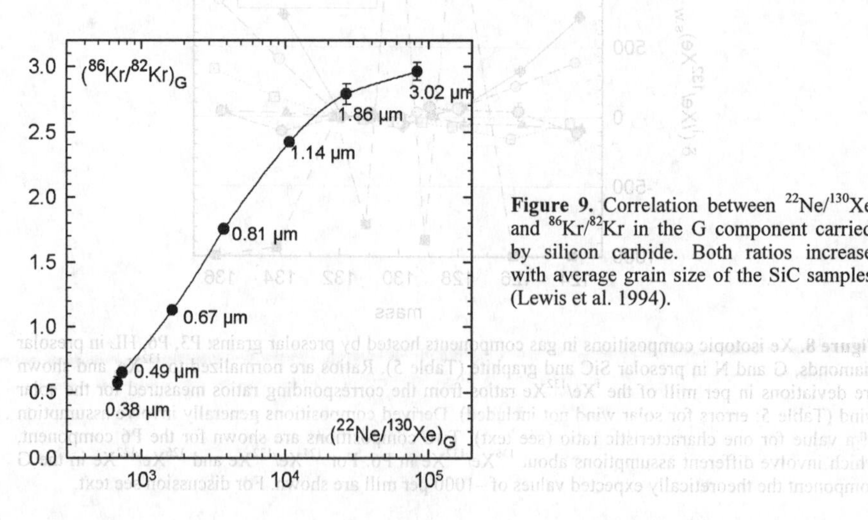

Figure 9. Correlation between ^{22}Ne/^{130}Xe and ^{86}Kr/^{82}Kr in the G component carried by silicon carbide. Both ratios increase with average grain size of the SiC samples (Lewis et al. 1994).

stellar wind. Ion implantation would thus dominate the noble gas pattern, while chemically active elements such as Ba would have also been taken up during condensation and their inventory would be dominated by this source. The scenario is consistent with the elemental abundance ratios that point to a > $10^4\times$ higher Ba/Xe ratio in the grains as compared to the source. To also accommodate the elemental trends within the noble gases, two wind components might be required: a minor component from a fully ionized region, contributing most of He, Ne and Ar, and a component from a cooler, only partially ionized region, relatively enriched in Kr and Xe because of their lower ionization potentials.

Graphite. In contrast to silicon carbide which is dominated by grains originating from a single source, AGB stars, graphite grains have noticeable contributions from a number of different sources (e.g., Zinner 1998), and this also shows in the noble gases (Amari et al. 1995). There are obviously contributions from AGB stars as for the SiC grains, which is shown by the signature of the s-process in the Kr and Xe isotopic compositions. Compared to SiC, the Kr isotopic ratios sensitive to physical conditions during the process, show even more extreme values. For example, (^{86}Kr/^{84}Kr) in s-process Kr in graphite ranges up to ~2.2 (Amari et al. 1995) as compared to the maximum of ~1.2 observed in SiC (Lewis et al. 1994; Table 4; cf. also Fig. 9) and the distribution appears bimodal rather than continuous. In addition, there is the "R component" [Amari et al. 1995; previously Ne-E(L)], which appears to be monoisotopic ^{22}Ne, presumably from the decay of ^{22}Na (half-life 2.6 a) and trapped by a sub-fraction of the graphite grains before decay (Nichols et al. 1994; Amari et al. 1995). Given the evidence for ion implantation as an important trapping mechanism in the case of presolar diamond (see below) and silicon carbide grains, it appears likely that this applies also to graphite. Elemental abundance patterns for Ar, Kr, Xe in graphite are similar to those in SiC, but Ne is low by one to two orders of magnitude, possibly due to diffusive loss (Amari et al. 1995).

Gases trapped in presolar diamond (HL, P3 and P6 components)

Among the gases that are contained in the presolar phases, diamond noble gases (P3, HL, P6) are by far the most abundant. Remarkably, they even make a significant contribution to the whole noble gas inventory of primitive meteorites: for the heavy noble gases like Xe, whose abundance is dominated by the Q component, diamonds contribute typically on the order of 10%, while for He and Ne the gases carried by the diamonds even dominate the whole inventory (Fig. 6; Huss and Lewis 1995). As a consequence, the original "planetary" Xe (AVCC = average carbonaceous chondrite Xe; e.g., Eugster et al. 1967) actually is a mixture of Q-Xe and diamond Xe, while the composition of the "classical" Ne-A component (^{20}Ne/^{22}Ne = 8.2±0.4; Black 1972), which is the Ne of the "planetary" component in primitive meteorites, is essentially that of the Ne in the HL component carried by the diamonds.

Presolar diamonds carry a total of three isotopically different noble gas components: P3, with an approximately normal isotopic composition of the heavy noble gases Kr and Xe, the isotopically anomalous HL component, and the less abundant and less well-characterized P6 component (Figs. 7, 8). The most complete characterization of the isotopic compositions has been given by Huss and Lewis (1994b), and the compositions listed in Tables 4 and 5 are based on their data. The compositions of P3 and HL are based on mixing lines defined by gases released at low temperature (<1235°C), while compositions of gas released at higher temperatures are interpreted as a mixture of HL and P6. The decomposition of the components involves assumptions about the end-member compositions of a given isotopic ratio in order to derive the full set, e.g., ^{136}Xe/^{132}Xe = 0.3096 and 0.6991 in P3 and HL, respectively (renormalized from 0.310

and 0.700 following Busemann et al. 2000). For HL this is the intersection of the P3-HL and P6-HL mixing lines, but no definite constraints exist on the other end of those lines. For P6, Huss and Lewis (1994b) give two sets of ratios: one assuming P6 to be a "normal component" ($^{136}Xe/^{132}Xe = 0.3096$ as in P3), and another assuming P6 to be "exotic," with the ratio just slightly lower than in the temperature steps with the highest P6/HL ratio ($^{136}Xe/^{132}Xe = 0.5493$; Table 5).

From an astrophysical point of view the most interesting component is the HL component of likely supernova origin, in which the Xe isotopes produced solely by the p-process (^{124}Xe, ^{126}Xe; Xe-L) and those produced only in the r-process (^{134}Xe, ^{136}Xe; Xe-H), are strongly enriched relative to those of intermediate mass that have contributions from the s-process (Fig. 8). Prominent among the various puzzles surrounding the HL component is the close association of the excesses in the light and heavy Xe isotopes, and so far all reports of observed separations between Xe-H and Xe-L have not been confirmed by later experiments trying to reproduce the results (cf. Huss and Lewis 1994a). In the latest such report (Meshik et al. 1999) variable enhancements of light and heavy isotopes seemed to have been seen in experiments involving gas extraction using laser light of different wavelengths, but it was found later that there had been problems with the blank corrections in the experiment. Currently it appears an open question whether a definite separation will be achieved using this approach (Meshik et al. 2001).

Another puzzle is the connection between the P3 and HL components. P3 is thermally significantly more labile than HL and P6 (peak release of P3 in pyrolysis of Orgueil nanodiamonds at ~490°C; Huss and Lewis 1994a; cf. Fig. 10), and its abundance relative to that of HL—as well as, on the higher temperature side, the ratio P6/HL—may be used to estimate metamorphic temperatures (Huss and Lewis 1994a; 1995). In this context, it is important to note that the diamonds are only nm-sized (average size 2.6 nm; Daulton et al. 1996), with the average diamond consisting of approximately thousand carbon atoms only, and that gases such as Xe occur in an abundance that there is only one Xe atom per about a million diamond grains. Hence there may be a variety of subpopulations among the diamonds, and the majority may not even be of presolar origin, an idea supported by their isotopically normal $^{12}C/^{13}C$ ratio (cf. Anders and Zinner 1993).

There are indications that the noble gases were introduced into the nanodiamonds by ion implantation. Among other things, they do not contain excessive abundances of ^{129}Xe

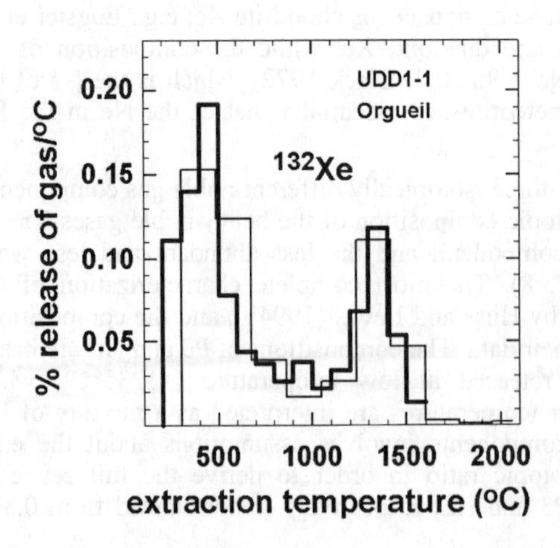

Figure 10. Comparison of thermal release of ^{132}Xe from Orgueil nanodiamonds (Huss and Lewis 1994b) with those from artificial nanodiamonds implanted with noble gas ions (according to Koscheev et al. 2001; sample UDD-1-1). Both show two release peaks. Shown is the percent release per °C temperature interval in stepwise degassing.

which would have been expected, if incorporation was by trapping during grain formation in an early stage after the explosion of a supernova. This is because at this time ^{129}I (a r-process only isotope) would have been alive, and because chemically active elements would be expected to have been more effectively trapped than chemically inert noble gases (Lewis and Anders 1981). Other evidence for an ion implantation process comes from combustion experiments on different grain size fractions of diamond (Verchovsky et al. 1998). Building on this evidence, Koscheev et al. (2001) studied the ion implantation process by implanting ~1 keV ions (He, Ar, Kr, Xe) into synthetic nanodiamonds of similar size and found a bimodal release similar to that in the meteoritic diamonds (Fig. 10). They infer from their observations that the P3 component, besides the evident low temperature part, must have also a high temperature part and that s-process-only isotopes such as ^{130}Xe observed in the (high-temperature) release and commonly assigned to the HL component (e.g., Huss and Lewis 1994b) may actually belong to the high temperature part of P3. As a consequence, the HL composition in Table 5 and shown in Figure 8 would need to be modified by extrapolating the mixing lines to a ^{130}Xe-free composition (Koscheev et al. 2001). In addition, the isotopic fractionation observed in the high-temperature portion of ion-implanted gases needs to be taken into account (Huss et al. 2000; Koscheev et al. 2001). Of course, similar corrections as for xenon would have to be made to the other noble gases as well.

A ^{130}Xe-free composition for the HL component is also more in line with astrophysical scenarios. Since relative excesses in ^{124}Xe and ^{126}Xe are not identical and nor are the relative excesses of the two r-only isotopes (^{134}Xe and ^{136}Xe), the nucleosynthetic sources cannot be the p- and r-processes proper. Two explanations which are currently debated for the heavy (the H) part of Xenon-HL involve a neutron burst intermediate between s- and r-process (Howard et al. 1992; Meyer et al. 2000) and a regular r-process augmented by separation of stable Xe isotopes and radioactive precursors on a short time scale after the explosion of a supernova (Ott 1996; Richter et al. 1998).

Ureilite gases

The trapped noble gases in ureilites are part of the properties that make this class of achondrites "enigmatic," as they often have been called because they combine properties of primitive and not-so-primitive meteorites. Among the "primitive" ones is the high abundance of noble gases that approach or in some cases even surpass those found in the most primitive chondrites (Göbel et al. 1978). This, as well as the similarity in isotopic composition and elemental abundance pattern to the Q component has repeatedly been cited as evidence for a close relationship between those two components (Ott et al. 1984, 1985b; Busemann et al. 2000).

In most ureilites the noble gases are hosted by diamonds, where they seem to be correlated with nitrogen characterized by δ^{15}N ~ -100‰ (Yamamoto et al. 1998; Rai et al. 2000), with evidence that some may reside in a different carbonaceous phase (Göbel et al. 1978). The ureilite diamonds, unlike the nanodiamonds in primitive chondrites discussed earlier, are of solar system, not of presolar, origin and most likely produced from graphitic carbon due to shock transformation (Vdovykin 1970). In fact, in the few ureilites that are less shocked (e.g., ALH78019; Wacker 1986) trapped noble gases seem to be carried by graphite, and it is generally assumed that in all ureilites trapped noble gases were originally carried by graphite or another carbonaceous phase (phase Q?) and that they were retained during shock transformation into diamond.

Subsolar and sub-Q gases

Subsolar gases are characterized by high Ar/Xe and Kr/Xe ratios when compared to the "typical" trapped noble gas pattern (excepting solar wind) as shown in Figure 2. They are typical for enstatite chondrites, primarily such of higher petrologic type (Crabb and Anders 1981, 1982; Patzer and Schultz 2000). Unlike Q, the major host phase of Ar/Kr/Xe in ordinary and carbonaceous chondrites (Fig. 6), the phase that hosts the subsolar gases is largely soluble in HF/HCl (although apparently not completely so; Busemann et al. 2001b). Most likely it is the major mineral in enstatite chondrites, enstatite, or some phase closely associated with it (Crabb and Anders 1981, 1982). Besides in enstatite chondrites, the subsolar (or a similar) component appears to be present in at least some ordinary chondrites, as shown by the fact that the Ar/Xe ratio often is substantially higher in bulk meteorites as compared to Q-dominated acid-resistant residues (Alaerts et al. 1979; Schelhaas et al. 1990). Isotopically, subsolar gases, where they have been determined, are similar to solar wind noble gases (Crabb and Anders 1981; Busemann et al. 2001b; section *Isotopic Compositions*; Figs. 3-5), although the errors may be somewhat large for a firm conclusion.

While subsolar gases show up mostly in enstatite chondrites of higher types, there is evidence in the more primitive E3 chondrites for still another component named sub-Q because of the very low $^{36}Ar/^{132}Xe$ (23±4) and $^{84}Kr/^{132}Xe$ (0.68±0.34) abundance ratios (Patzer and Schultz 2000). While sub-Q has been interpreted as a separate component (Patzer and Schultz 1999, 2000), it is worthwhile to remember that there is quite significant variation within the Q component proper (Busemann et al. 2000), and so the so-called sub-Q gases may just be an extreme example of that.

Sitings

Important information regarding origin, relations, and history of the various trapped noble gas components may be obtained from a knowledge about where they—and their carrier phases—occur within the host meteorites. To obtain this is most straightforward where the nature of the carrier phases is known and where analytical techniques with high spatial resolution (SEM, TEM) can be applied in the search for these phases. Both silicon carbide grains (carriers of the G and N components) as well as grains of nanodiamond (P3, HL, P6) have been identified in the matrices of carbonaceous chondrites (Alexander et al. 1990; Banhart et al. 1998), and it is reasonable to assume on that basis that this is where the bulk of those noble gases occurs. The situation is different where the carriers are unknown or ill-defined. A useful approach in this case is search for noble gases *in situ*. Using this approach, Nakamura et al. (1999), using *in situ* laser beam extraction with a spatial resolution of 50-100 µm, have determined that in CM chondrites Q-type noble gases are concentrated in fine-grained accretionary rims around chondrules. Similar results were obtained by Vogel et al. (2000, 2001), who measured noble gases in small hand-picked samples in different units from primitive chondrites. Chondrules were found to be gas-poor, consistent with earlier measurements performed on larger samples with different objectives (e.g., Swindle et al. 1983). The recent observation by Okazaki et al. (2001) of chondrules rich in subsolar noble gases in an enstatite meteorite shows, however, that there are exceptions to this rule.

Other (lesser) components

Besides the components discussed in detail above, a number of lesser and/or less-well characterized components have been identified, primarily for xenon, which cannot be exhaustively discussed here; for several a confirmation of their existence would be welcome. One of these components, found in sulfides of the Allende meteorite, is characterized by overabundances of ^{124}Xe in gas fractions released in the 1400°C – 1500°C range, without corresponding overabundances in the other light isotopes as expected for Xe-L or spallation xenon (Lewis et al. 1979). Others are connected with achondrites other than ureilites. Among these achondrites are the brachinites and related meteorites which contain trapped gases with isotopic similarities to Q or ureilites, and an elemental pattern more like ureilites (Ott et al. 1993; Weigel et al. 1997). Of interest are also lodranites and diogenites (Michel and Eugster 1994; Weigel and Eugster 1994) which may contain especially primitive Xe (see section *Relations*). Still others are connected with iron meteorites, some of which contain inclusions quite rich in trapped noble gases with compositions similar to those of the "normal" components discussed above (Bogard et al. 1971; Niemeyer 1979). Noteworthy in addition is also the occurrence of solar-type noble gases in some iron and stony-iron meteorites (Becker and Pepin 1984; Mathew and Begemann 1997). A few specific examples are briefly discussed below.

Xenon in chondritic metal. Marti et al. (1989) have identified a xenon component (FVM-Xe) in a metal separate of the Forest Vale (H4) chondrite which appears to be distinct from xenon identified in other solar system reservoirs. It is characterized by relative abundances of the heaviest isotopes with unusually high ^{134}Xe. A possible explanation is recoil of fission fragments into the metal grains, possibly from ^{244}Pu, ^{248}Cm or neutron induced fission of ^{235}U (Marti et al. 1989). This suggestion is consistent with the observed grain size dependence of FVM which favors a near-surface location.

Xenon and krypton in silicate-graphite inclusions of the El Taco iron meteorite. In their analysis of noble gases in silicate-graphite inclusions of the El Taco IAB iron meteorite Mathew and Begemann (1995) found, besides a trapped component of Ar, Kr, Xe in the silicates with Xe isotopically close to ureilite xenon, a different component in the high temperature release of noble gases from graphite and schreibersite [$(Fe,Ni)_3P$]. Xe in graphite-schreibersite is unlike any of the well-established types of Xe and can be generated as a mixture of mass fractionated U-Xe (the putative primary component of "planetary" Xe; Pepin et al. 1995; Pepin 2000; Pepin and Porcelli 2002; see discussion in *Relations* section) and ^{244}Pu fission Xe, where the fission component must have been added prior to incorporation of Xe into the El Taco parent body. Most likely the mass fractionation event occurred after mixing of U-Xe with fission Xe.

Xenon in acid residues of iron meteorites. Murty et al. (1983) analyzed noble gases in acid-resistant residues from the Canyon Diablo and Campo del Cielo (El Taco) meteorites and found in the nonmagnetic fractions xenon that was air-like except for ^{124}Xe and ^{126}Xe. The relative abundance of these isotopes is lower than in air (in which their abundance already is the lowest among well known solar system reservoirs) by up to 15%. Murty et al. (1983) suggest that this xenon may constitute presolar nebular matter which was trapped in micro-inclusions of iron meteorites.

RELATIONS

Given the number of components discussed in the preceding sections, it is an obvious question whether all of them are really distinct, independent components or if some of them are related. This question has been addressed already in several instances, such as a) regarding a possible relation between Q and ureilite noble gases—primarily based on the similarity of elemental and isotopic patterns, and b) the question of the sub-Q gases which may be an extreme case of the variations in elemental abundance pattern observed among Q gases proper. Two more fundamental questions are briefly discussed below.

Relationship between Q and solar gases

Given that more than 99% of the mass in the solar system is contained in the Sun and that the Sun, in contrast to the rocky material that constitutes the meteorites, has retained its full complement of volatile elements, it is clear that the Sun is the primary reservoir of noble gases in the Solar System. Hence solar gases must be the benchmark against which other components must be measured—bearing in mind that the solar wind upon which most of our information is based is fractionated relative to the true solar composition to some extent (e.g., Wieler 2002). On the other hand, the Q component is widespread among primitive solar system materials which suggests that it may be a distinct component established as such in the gas phase of the solar nebula prior to incorporation into planetary solids (Ozima et al. 1998). Obviously, a central question that arises is that of a relation between these two important noble gas components.

Key to understanding any relationship—if one exists—should be the understanding of the trapping process. This has been studied phenomenologically in a variety of experiments, however with moderate success only (see discussion of Q in *Origins and History* section). While there is no shortage of ideas (popular, e.g., are adsorption in labyrinth pores, or on presolar grain mantles; cf. Wacker 1989; Huss et al. 1996; see also Hohenberg et al. 2002; and review by Swindle 1988), an alternative approach is to look at what the key physical parameters may be and in which way they may have entered. An important observation in this context may be that generally light isotopes are depleted in the Q component relative to heavy ones when compared to the solar wind composition, and also light elements relative to heavy elements. For example, $^{20}Ne/^{22}Ne$ in Q as most commonly measured (Busemann et al. 2000) is ~10.7, ~22% lower than the solar value of ~13.6 (Table 3), while $^{136}Xe/^{132}Xe$ is higher by ~6% (0.3164 vs. ~0.2985; Table 5), and $^{20}Ne/^{132}Xe$ (Table 2) is lower by about 5 orders of magnitude. Ozima et al. (1998), therefore, have suggested that Q may be related to solar type noble gases via a process in which mass enters as a major factor in generating the observed elemental and isotopic pattern, e.g., in the form of some type of Rayleigh distillation. They have derived an empirical relationship

$$\log \{(n_k/n_j)_Q / (N_k/N_j)_{sw}\} = \{F_{He}\} \{(m_{He} / m_k)^{½} - (m_{He} / m_j)^{½}\} \quad (1)$$

where n and N are present and initial abundances in Q and in the solar wind of two gases/isotopes (k,j), and F_{He} the depletion factor ($\equiv n_{He}/N_{He}$) of ^4He which is used as a reference.

Relevant elemental abundance ratios as given in Table 2 are plotted in Figure 11a. As required by relation (1) the data points plotting on the left side indeed follow approximately the straight line anchored at the origin and fitted to those data points. (Note that the uncertainties shown of ~20% are arbitrary and for illustrative purposes only). The fit, however, is not really convincing, and, in addition, data points more to the right, where ^4He is involved and which are characterized by larger relative mass differences, do not fall on the same line. In addition, data for isotopic rather than elemental ratios show a vastly different behavior indicative of much less mass-dependent fractionation than inferred from the elemental abundance ratios (Fig. 2 in Ozima et al. 1998).

So mass may be an important factor, but overall the situation is likely to be more complex. For example, ionization efficiency could also have played a role, given (a) the evidence in several cases for ion implantation as the process by which noble gases were introduced into their host phases (e.g., SiC, diamond; section *Origins and History*), and (b) the observation of Göbel et al. (1978) that the elemental fractionation pattern for ureilite gases—which, as often stressed, are compositionally similar to Q gases—correlates with ionization energy. This fact is demonstrated in Figure 11b, which is the same as Figure 11a, except that differences in ionization energy rather than mass are plotted on the abscissa.

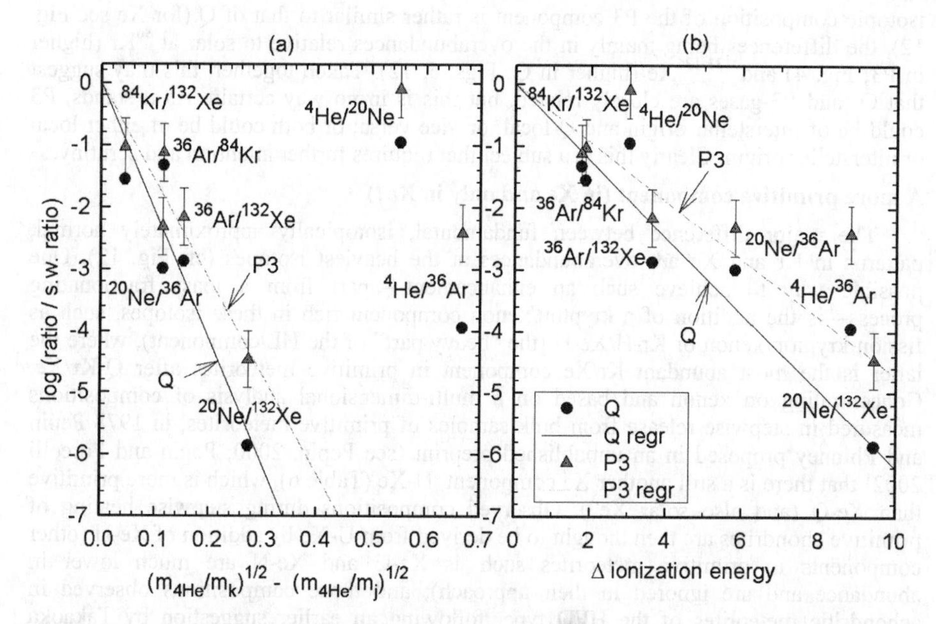

Figure 11. (a) Mass fractionating depletions relative to the solar composition for the Q and P3 components vs. a measure of relative mass difference. Following Ozima et al. (1998), log $[(n_k/n_j) / (N_k/N_j)]$ is plotted vs. $[(m_{4He}/m_k]^{1/2} - (m_{4He}/m_j)^{1/2}]$, where n_k/n_j and N_k/N_j are currently observed and "original" (assumed solar wind) abundance ratios and m_k, m_j, the corresponding atomic masses. Data taken from the values in Table 2 are plotted for various nuclide pairs as indicated. Arbitrary errors of ~20% are shown to help judging the quality of the correlation lines (anchored to the origin) to which the data points on the left may conform. As indicated by the correlations, these data points are consistent with the fractionation equation proposed by Ozima et al. (1998), while the data points plotting on the right and involving ^{4}He (not included in the fit) appear to be not. See text for discussion. (b) as in (a), but plotted vs. differences in ionization energy. Ratios involving ^{4}He again are not included in the fit, which is anchored at the origin.

Having a mass-dependent relationship between solar and Q gases as suggested by Ozima et al. (1998) [or some other kind of relationship] leaves, of course, the question how other "normal components" fit into this picture. As pointed out earlier, ureilite and sub-Q gases might easily fit in, since they may be variants of, if not identical to, Q gases. The normal component (N) observed in presolar silicon carbide grains most likely constitutes material from the envelopes of Red Giant stars, and although approximately normal in composition there is no reason why there should be a simple relationship to normal solar system reservoirs. This leaves the normal P3 component found in nanodiamonds as the other major component for which it might be of interest to search for a relation. This is in spite of the fact that the diamonds are "presolar," because we do not know what fraction of the diamonds really is. Given the normal isotopic composition of carbon and possibly also nitrogen—if the recent determination of $^{14}N/^{15}N$ for the Jovian atmosphere (Owen et al. 2001) and inferences from recent measurements on lunar soil grains (Hashizume et al. 2000) give the true protosolar nitrogen isotopic ratio—the majority of the nanodiamonds may be of "local" origin, and with it the P3 noble gases.

As also shown in Figures 11a/b, elemental abundance ratios in P3 may be related to those in the solar wind in a similar fashion as those in Q, with slightly different fractionation factors. Furthermore, for Kr and Xe (although definitely not for Ne), the isotopic composition of the P3 component is rather similar to that of Q (for Xe see Fig. 12), the differences being mainly in the overabundances relative to solar at ^{86}Kr (higher in P3; Fig. 4) and $^{131-136}$Xe (higher in Q; Figs. 5, 12). Taken together, this may suggest that Q- and P3-gases are closely related, but this is in no way certain. As it stands, P3 could be of interstellar origin and Q local, or vice versa; or both could be of either local or interstellar origin. Clearly this is a subject that requires further attention and scrutiny.

A more primitive component (in Xe and only in Xe?)

The major difference between fundamental, isotopically approximately normal, patterns in Kr and Xe are overabundances at the heaviest isotopes (cf. Fig. 12). One possible way to achieve such an enhancement—apart from a mass fractionating process—is the addition of a krypton/xenon component rich in these isotopes, such as fission krypton/xenon or Kr-H/Xe-H (the "heavy part" of the HL component), where the latter is the most abundant Kr/Xe component in primitive meteorites after Q-Kr/Xe. Concentrating on xenon and based on a multi-dimensional analysis of compositions measured in stepwise release from bulk samples of primitive meteorites, in 1978 Pepin and Phinney proposed in an unpublished preprint (see Pepin, 2000; Pepin and Porcelli 2002) that there is a still another Xe component, U-Xe (Table 6), which is more primitive than Xe-Q (and also solar Xe?). Observed compositions during stepwise heating of primitive chondrites are then thought to be derived from U-Xe by addition of Xe-H (other components in primitive meteorites such as Xe-G and Xe-N are much lower in abundance and are ignored in their approach); and those compositions observed in achondritic meteorites of the HED type, following an earlier suggestion by Takaoka (1972), are thought to be derived by the addition to U-Xe of ^{244}Pu-fission Xe.

Compared with solar xenon, U-Xe agrees, within the respective uncertainties, in all isotopes but the two heaviest ones, ^{134}Xe and ^{136}Xe (Table 6; Fig. 12) which constitutes something of a puzzle. One piece of the puzzle is that this suggests solar Xe to contain a not so small extra fraction of a nuclear component, Xe-H, amounting to some 8% of ^{136}Xe in the Sun (Pepin et al. 1995). Also puzzling appears that in primitive meteorites Xe-H and Xe-L are carried by presolar nanodiamonds, where both always appear to occur in constant abundance ratio, but that according to the evaluation by Pepin and Phinney (Pepin 2000; Pepin and Porcelli 2002) Xe-H and Xe-L appear to be decoupled: extra Xe-H seems to be present in the solar wind, unaccompanied by Xe-L. Not so puzzling, on the other hand, is that no analogous components have been detected in Ar and Kr: this may simply be due to the fact that for these the H component makes a smaller contribution and is less outstanding in its isotopic composition (Figs. 2-5). It is interesting, nevertheless, that a similar relationship as there exists between U-Xe and solar Xe may also exist between U-Xe and Xe-N, the Xe component deriving from the envelope of AGB stars and found in presolar SiC grains (Pepin et al. 1995), and that this relation may even extend to the P6 composition found in presolar diamonds (Russell et al. 1997).

So far U-Xe may have been directly observed in a low temperature release step from the non-colloidal part of an acid-resistant residue from the Murray meteorite (Niemeyer and Zaikowski 1980), and it has also been reported as being present in two achondritic meteorites (Tatahouine and Lodran; Michel and Eugster 1994; Weigel and Eugster 1994). However, the latter finding could not be confirmed in more recent analyses of the same meteorites (Busemann and Eugster 2000). Clearly, one of the major challenges for the future, apart from further efforts to establish the separability of the H and L parts of the HL component, is to establish whether U-Xe actually exists as a noble gas component. Among other things, an exercise of the kind that led Pepin and Phinney to postulate the existence of U-Xe, but based on modern data, would be an important step. If the existence of U-Xe can be confirmed, this may lead to a more unified view of the

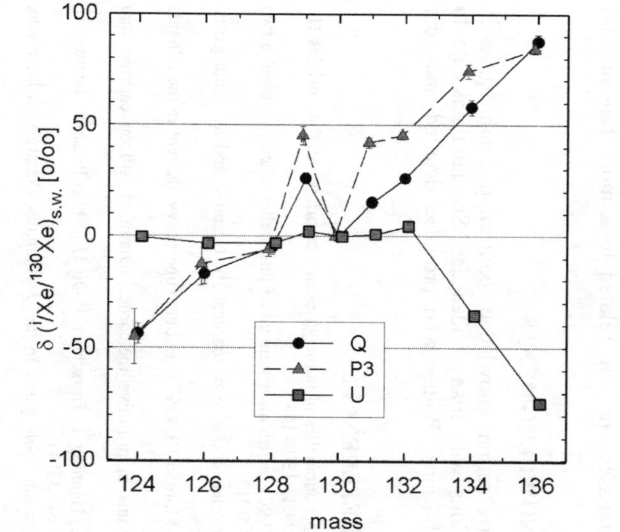

Figure 12. Relative deviations of Q-Xe, Xe-P3 and the putative primitive U-Xe from solar wind xenon. Contrary to Figures 5 and 8, normalization is to ^{130}Xe. Errors in the composition of U-Xe and solar wind xenon have not been included.

Table 6. Isotopic composition of Xe in the solar wind and in primitive trapped components.

Component	*^{124}Xe/^{132}Xe*	*^{126}Xe/^{132}Xe*	*^{128}Xe/^{132}Xe*	*^{129}Xe/^{132}Xe*	*^{130}Xe/^{132}Xe*	*^{131}Xe/^{132}Xe*	*^{134}Xe/^{132}Xe*	*^{136}Xe/^{132}Xe*	*References*
solar wind	0.004882	0.004234	0.08475	1.0420	0.1661	0.8272	0.3666	0.2985	[1], [2]
U-Xe	0.004873	0.004201	0.08411	1.0395	0.1653	0.8243	0.3520	0.2750	[1]
Q-Xe	0.00455 (2)	0.00406 (2)	0.0822 (2)	1.042 (2)	0.1619 (3)	0.8185 (9)	0.3780 (11)	0.3164 (8)	[3]
Xe-P3	0.00446 (6)	0.00400 (4)	0.0806 (2)	1.042 (4)	0.1589 (2)	0.8247 (19)	0.3767 (10)	≡ 0.3096	[4]

Uncertainties in the last digits for the Q and P3 components are given in parentheses.

References: [1] Pepin et al. (1995); [2] Wieler (2002); [3] Busemann et al. (2000); [4] Huss and Lewis (1994b), renormalized following Busemann et al. (2000).

multitude of different components of trapped noble gases in meteorites and the Solar System in general.

CONCLUDING REMARKS

Meteorites contain a variety of trapped noble gas components other than the solar wind. In many cases their compositions have not been measured in "pure" form, but have been determined from measured (i.e., not pure) compositions based on correlation / mixing lines involving assumptions about one isotopic ratio in the endmember. Some of the components established in this way are of nucleosynthetic origin and carried by presolar grains (diamond, SiC, graphite). Their isotopic compositions are testament to the nuclear processes by which elements are made in the interior of the stars around which those carrier grains formed.

Others are more likely to be of "local" origin, as inferred by normal or near-normal isotopes, with their original (pre-metamorphic) compositions established in the solar nebula or the presolar cloud from which the solar system formed. They are most abundant in the most primitive meteorites, where the majority of heavy gases (Ar, Kr, Xe) is carried by the Q phase, which is insoluble in HF/HCl and from which gases are released during treatment with HNO_3. Similar gas components are, however, also found in some achondritic meteorites (e.g., ureilites), and also in nanodiamonds of presumed presolar origin (P3). It will be a major challenge in future work to establish in more detail if/what type of relations exist and to determine whether, and if yes, how these components, including solar noble gases, may be related to a more fundamental component.

ACKNOWLEDGMENTS

My ideas about how to synthesize current knowledge about trapped noble gases in meteorites were shaped by discussions with many colleagues. Special thanks go to Henner Busemann, Gary Huss and Rainer Wieler who provided detailed and very constructive reviews.

REFERENCES

Alaerts L, Lewis RS, Anders E (1979) Isotopic anomalies of noble gases in meteorites and their origins-III. LL chondrites. Geochim Cosmochim Acta 43:1399-1415

Alexander CMO'D, Swan P, Walker RM (1990) *In situ* measurement of interstellar silicon carbide in two CM chondrite meteorites. Nature 348:715-717

Amari S, Lewis RS, Anders E (1995) Interstellar grains in meteorites: III. Graphite and its noble gases. Geochim Cosmochim Acta 59:1411-1426

Anders E, Higuchi H, Gros J, Takahashi H, Morgan JW (1975) Extinct superheavy element in the Allende meteorite. Science 190:1262-1271

Anders E, Zinner E (1993) Interstellar grains in primitive meteorites: Diamond, silicon carbide, and graphite. Meteoritics 28:490-514

Banhart F, Lyutovich Y, Braatz A, Jäger C, Henning T, Dorschner J, Ott U (1998) Presolar diamond in unprocessed Allende. Meteoritics Planet Sci 33:A12

Becker RH, Pepin RO (1984) Solar composition noble gases in the Washington County iron meteorite. Earth Planet Sci Lett 70:1-10

Black DC (1972) On the origins of trapped helium, neon and argon isotopic variations in meteorites—II. Carbonaceous meteorites. Geochim Cosmochim Acta 36:377-394

Bogard DD, Huneke JC, Burnett DS, Wasserburg GJ (1971) Xe and Kr analyses of silicate inclusions in iron meteorites. Geochim Cosmochim Acta 35:1231-1254

Busemann H, Baur H, Wieler R (2000) Primordial noble gases in "phase Q" in carbonaceous and ordinary chondrites studied by closed-system stepped etching. Meteoritics Planet Sci 35:949-973

Busemann H, Eugster O (2000) Primordial noble gases in Lodran metal separates and the Tatahouine diogenite. Lunar Planet Sci XXXI, Abstr #1642 (CD-ROM)

Busemann H, Baur H, Wieler R (2001a) Helium isotopic ratios in carbonaceous chondrites: significant for the early solar nebula and circumstellar diamonds? Lunar Planet. Sci. XXIII, Abstr #1598 (CD-ROM)

Busemann H, Baur H, Wieler R (2001b) Subsolar noble gases in an acid-resistant residue of the EH5 chondrite St. Mark's. Meteoritics Planet Sci 36:A34

Crabb J, Anders E (1981) Noble gases in E-chondrites. Geochim Cosmochim Acta 45:2443-2464

Crabb J, Anders E (1982) On the siting of noble gases in E-chondrites. Geochim Cosmochim Acta 46:2351-2361

Daulton TL, Eisenhour DD, Bernatowicz TJ, Lewis RS, Buseck PR (1996) Genesis of presolar diamonds: comparative high-resolution transmission electron microscopy study of meteoritic and terrestrial nano-diamonds. Geochim Cosmochim Acta 60:4853-4872

Eugster O, Eberhardt P, Geiss J (1967) Krypton and xenon isotopic composition in three carbonaceous chondrites. Earth Planet Sci Lett 3:249-257

Gallino R, Busso M, Picchio G, Raiteri CM (1990) On the astrophysical interpretation of isotope anomalies in meteoritic SiC grains. Nature 348:298-302

Gallino R, Raiteri CM, Busso M (1993) Carbon stars and isotopic Ba anomalies in meteoritic SiC grains. Astrophys J 410:400-411

Gerling EK, Levskii LK (1956) On the origin of the rare gases in stony meteorites. Doklady Akad Nauk USSR (Geochemistry) 110:750

Göbel R, Ott U, Begemann F (1978) On trapped noble gases in ureilites. J Geophys Res 83:855-867

Hashizume K, Chaussidon M, Marty B, Robert F (2000) Solar wind record on the Moon: Deciphering presolar from planetary nitrogen. Science 290:1142-1145

Hohenberg CM, Thonnard N and Meshik A (2002) Active capture and anomalous adsorption: new mechanisms for the incorporation of heavy noble gases. Meteoritics Planet Sci 37:257-267

Hoppe P, Ott U (1997) Mainstream silicon carbide grains from meteorites. *In* Astrophysical Implications of the Laboratory Study of Presolar Materials. Bernatowicz TJ, Zinner E (eds) AIP Conf Proc 402, Am Inst Phys, Woodbury, New York, p 27-58

Howard WM, Meyer BS, Clayton DD (1992) Heavy-element abundances from a neutron burst that produces Xe-H. Meteoritics 27:404-412

Huss GR, Lewis RS (1994a) Noble gases in presolar diamonds II: Component abundances reflect thermal processing. Meteoritics 29:811-829

Huss GR, Lewis RS (1994b) Noble gases in presolar diamonds I: Three distinct components and their implications for diamond origins. Meteoritics 29:791-810

Huss GR, Lewis RS (1995) Presolar diamond, SiC, and graphite in primitive chondrites: Abundances as a function of meteorite class and petrologic type. Geochim Cosmochim Acta 59:115-160

Huss GR, Lewis RS, Hemkin S (1996) The "normal planetary" noble gas component in primitive chondrites: Compositions, carrier, and metamorphic history. Geochim Cosmochim Acta 60:3311-3340

Huss GR, Ott U, Koscheev AP (2000) Implications of ion-implantation experiments for understanding noble gases in presolar diamonds. Meteoritics Planet Sci 35:A79-A80

Koscheev AP, Gromov MD, Mohapatra RK, Ott U (2001) History of trace gases in presolar diamonds as inferred from ion implantation experiments. Nature 412:615-617

Lavielle B, Marti K (1992) Trapped xenon in ordinary chondrites. J Geophys Res E 97:20875-20881

Lewis RS, Amari S, Anders E (1990) Meteoritic silicon carbide: pristine material from carbon stars. Nature 348:293-298

Lewis RS, Amari S, Anders E (1994) Interstellar grains in meteorites: II. SiC and its noble gases. Geochim Cosmochim Acta 58:471-494

Lewis RS, Anders E (1981) Isotopically anomalous xenon in meteorites: A new clue to its origin. Astrophys J 247:1122-1124

Lewis RS, Hertogen J, Alaerts L, Anders E (1979) Isotopic anomalies in meteorites and their origins–V. Search for fission fragment recoils in Allende sulfides. Geochim Cosmochim Acta 43:1743-1752

Lewis RS, Srinivasan B, Anders E (1975) Host phase of a strange xenon component in Allende. Science 190:1251-1262

Lewis RS, Tang M, Wacker JF, Anders E, Steel E (1987) Interstellar diamonds in meteorites. Nature 326:160-162

Marti K (1967) Trapped xenon and the classification of chondrites. Earth Planet Sci Lett 2:193-196

Marti K, Kim JS, Lavielle B, Pellas P, Perron C (1989) Xenon in chondritic metal. Z Naturforsch 44a: 963-967

Mathew KJ, Begemann F (1995) Isotopic composition of xenon and krypton in silicate-graphite inclusions of the El Taco, Campo del Cielo, IAB iron meteorite. Geochim Cosmochim Acta 59:4729-4746

Mathew KJ, Begemann F (1997) Solar-like noble trapped noble gases in the Brenham pallasite. J Geophys Res E102:11015-11026

Matsuda J-I, Amari S, Nagao K (1999) Purely physical separation of a small fraction of the Allende meteorite that is highly enriched in noble gases. Meteoritics Planet Sci 34:129-136
Mazor E, Heymann D, Anders E (1970) Noble gases in carbonaceous chondrites. Geochim Cosmochim Acta 34:781-824
Meshik AP, Pravdivtseva OV, Hohenberg CM (1999) Separation of Xe-H and Xe-L by selective laser absorption in Murchison diamonds. Lunar Planet Sci XXX, Abstr #1621 (CD-ROM)
Meshik AP, Pravdivtseva OV, Hohenberg CM (2001) Selective laser extraction of Xe-H from Xe-HL in meteoritic nanodiamonds: real effect or experimental artifact? Lunar Planet Sci XXXII, Abstr #2158 (CD-ROM)
Meyer BS, Clayton DD, The L-S (2000) Molybdenum isotopes from a neutron burst. Lunar Planet Sci XXXI, Abstr #1458 (CD-ROM)
Michel Th, Eugster O (1994) Primitive xenon in diogenites and plutonium-244-fission xenon ages of a diogenite, a howardite, and eucrites. Meteoritics 29:593-606
Murty SVS, Goel PS, Minh DVu, Shukolyukov YuA (1983) Nitrogen and xenon in acid residues of iron meteorites. Geochim Cosmochim Acta 47:1061-1068
Nakamura T, Nagao K, Metzler K, Takaoka N (1999) Microdistribution of primordial noble gases in CM chondrites determined by *in situ* laser microprobe analysis: decipherment of nebular processes. Geochim Cosmochim Acta 63:241-255.
Nichols RH Jr, Amari S, Hohenberg CM, Hoppe P, Lewis RS (1993) ^{20,22}Ne-E(H) and ^{4}He measured in single interstellar SiC grains of known C-isotopic composition. Meteoritics 28:410-411
Nichols RH Jr, Kehm K, Brazzle R, Amari S, Hohenberg CM, Lewis RS (1994) Ne, C, N, O, Mg, and Si isotopes in single interstellar graphite grains: multiple stellar sources for Neon-E(L). Meteoritics 29:510-511
Nichols RH Jr, Nuth JA III, Hohenberg CM, Olinger CT, Moore MH (1992) Trapping of noble gases in proton-irradiated silicate smokes. Meteoritics 27:555-559
Niemeyer S (1979) I-Xe dating of silicate and troilite from IAB iron meteorites. Geochim Cosmochim Acta 43:843-860
Niemeyer S, Zaikowski A (1980) I-Xe age and trapped Xe components of the Murray (C-2) chondrite. Earth Planet Sci Lett 48:335-347
Okazaki R, Takaoka N, Nagao K, Sekiya M, Nakamura T (2001) Noble-gas-rich chondrules in an enstatite meteorite. Nature 412:795-798
Ott U (1993) Interstellar grains in meteorites. Nature 364:25-33
Ott U (1996) Interstellar diamond xenon and timescales of supernova ejecta. Astrophys J 463:344-348
Ott U, Begemann F (2000) Spallation recoil and age of presolar grains. Meteoritics Planet Sci 35:53-63
Ott U, Begemann F, Yang J, Epstein S (1988) S-process krypton of variable isotopic composition in the Murchison meteorite. Nature 332:700-702
Ott U, Kronenbitter J, Flores J, Chang S (1984) Colloidally separated samples from Allende residues: Noble gases, carbon and an ESCA-study. Geochim Cosmochim Acta 48:267-280
Ott U, Löhr HP, Begemann F (1985a) Trapped neon in ureilites – a new component. *In* Rapports Isotopiques Dans le Systeme Solaire (Isotopic Ratios in the Solar System), Centre National d'Etudes Spatiales, Paris, p 129-136
Ott U. Löhr HP, Begemann F (1985b) Trapped noble gases in 5 more ureilites and the possible role of Q. Lunar Planet Sci 16:639-640
Ott U, Löhr HP, Begemann F (1993) Noble gases in Yamato-75097 inclusion: similarities to brachinites (only?) 18th Symp Antarct Meteorites, NIPR, Tokyo, pp 236-239
Ott U, Mack R, Chang S (1981) Noble-gas-rich separates from the Allende meteorite. Geochim Cosmochim Acta 45:1751-1788
Ott U, Merchel S (2000) Noble gases and the not so unusual size of presolar SiC in Murchison. Lunar Planet Sci XXXI, Abstr #1356 (CD-ROM)
Owen T, Mahaffy PR, Niemann HB, Atreya S, Wong M (2001) Protosolar nitrogen. Astrophys J 553:L77-L79
Ozima M, Wieler R, Marty B, Podosek FA (1998) Comparative studies of solar, Q-gases and terrestrial noble gases, and implications on the evolution of the solar nebula. Geochim Cosmochim Acta 62: 301-314
Patzer A, Schultz L (1999) Trapped noble gases in enstatite chondrites: a "sub-Q" component in EH3s? Meteoritics Planet Sci 34:A89-A90
Patzer A, Schultz L (2000) New noble gas data of enstatite chondrites: another piece to the puzzle. Meteoritics Planet Sci 35:A125
Pepin RO (1991) On the origin and early evolution of terrestrial planet atmospheres and meteoritic volatiles. Icarus 92:2-79

Pepin RO (2000) On the isotopic composition of primordial xenon in terrestrial planet atmospheres. Space Sci Rev 92:371-395

Pepin RO, Becker RH, Rider PE (1995) Xenon and krypton isotopes in extraterrestrial regolith soils and in the solar wind. Geochim Cosmochim Acta 59:4997-5022

Pepin RO, Porcelli D (2002) Origin of noble gases in the terrestrial planets. Rev Mineral Geochem 47: 191-246

Rai VK, Murty SVS, Ott U (2000) Lewis Cliff 85328: a monomict ureilite containing both heavy and light nitrogen. Meteoritics Planet Sci 35:A132-A133

Reynolds JH (1956) High-sensitivity mass spectrometer for noble gas analysis. Rev Sci Instr 27:928-934

Reynolds JH (1963) Xenology. J Geophys Res 68:2939-2956

Reynolds JH, Turner G (1964) Rare gases in the chondrite Renazzo. J Geophys Res 69:3263-3281

Richter S, Ott U, Begemann F (1998) Tellurium in pre-solar diamonds as an indicator for rapid separation of supernova ejecta. Nature 391:261-263

Russell SS, Ott U, Alexander CMO'D, Zinner EK, Arden JW, Pillinger CT (1997) Presolar silicon carbide from the Indarch (EH4) meteorite: comparison with silicon carbide populations from other meteorite classes. Meteoritics Planet Sci 32:719-732

Sandford SA, Bernstein MP, Swindle TD (1998) The trapping of noble gases by the irradiation and warming of interstellar ice analogs. Meteoritics Planet. Sci 33: A135.

Schelhaas N, Ott U, Begemann F (1990) Trapped noble gases in unequilibrated ordinary chondrites. Geochim Cosmochim Acta 54:2869-2882

Schramm DN and Turner MS (1998) Big-bang nucleosynthesis enters the precision era. Rev Mod Phys 70:303-318

Signer P, Suess HE (1963) Rare gases in the sun, in the atmosphere, and in meteorites. *In* Earth Science and Meteorites. Geiss J, Goldberg ED (eds) North Holland, Amsterdam, p 241-272

Swindle T (1988) Trapped noble gases in meteorites. *In* Meteorites and the Early Solar System. Kerridge JF, Matthews MS (eds) University of Arizona Press, Tucson, p 535-564

Swindle TD, Caffee MW, Hohenberg CM, Lindstrom MM (1983) I-Xe studies of individual Allende chondrules. Geochim Cosmochim Acta 47:2157-2177

Takaoka N (1972) An interpretation of general anomalies of xenon and the isotopic composition of primitive xenon. Mass Spectr 20:287-302

Verchovsky AB, Fisenko AB, Semjonova LF, Wright IP, Lee MR, Pillinger CT (1998) C, N, and noble gas isotopes in grain size separates of presolar diamonds from Efremovka. Science 281:1165-1168

Vdovykin GP (1970) Ureilites. Space Sci Rev 10:483-510

Vogel N, Baur H, Bischoff A, Semenenko VP, Wieler R (2000) Microdistribution of light noble gases in primitive chondrites and implications for their accretionary history. Meteoritics Planet Sci 35: A165-A166

Vogel N, Baur H, Bischoff A, Wieler R (2001) Contrasts in chondrites—microdistribution of noble gases in Allende, Leoville, and Krymka. Meteoritics Planet Sci 36:A216

Wacker JF (1986) Noble gases in the diamond-free ureilite, ALHA 78019: The roles of shock and nebular processes. Geochim Cosmochim Acta 50:633-642

Wacker JF (1989) Laboratory simulation of meteoritic noble gases. III. Sorption of neon, argon, krypton, and xenon on carbon: Elemental fractionation. Geochim Cosmochim Acta 53:1421-1433

Wacker JF, Zadnik MG, Anders E (1985) Laboratory simulation of meteoritic noble gases. I. Sorption of xenon on carbon: Trapping experiments. Geochim Cosmochim Acta 49:1035-1048

Weigel A, Eugster O (1994) Primitive trapped Xe in Lodran minerals and further evidence from EET84302 and Gibson for break-up of the lodranite parent asteroid ~4 Ma ago. Lunar Planet Sci 25:1479-1480

Weigel A, Eugster O, Koeberl C, Krähenbühl U (1997) Differentiated achondrites Asuka 881371, an angrite, and Divnoe: Noble gases, ages, chemical composition, and relation to other meteorites. Geochim Cosmochim Acta 61:239-248

Wieler R (2002) Noble gases in the solar system. Rev Mineral Geochem 47:21-70

Wieler R (1994) "Q-gases" as "local" primordial noble gas component in primitive meteorites. *In* Noble Gas Geochemistry and Cosmochemistry. Matsuda J (ed) Terra Scientific Publishing, Tokyo, p 31-41

Wieler R, Anders E, Baur H, Lewis RS, Signer P (1992) Characterization of Q-gases and other noble gas components in the Murchison meteorite. Geochim Cosmochim Acta 56:2907-2921

Wilkening LL, Marti K (1976) Rare gases and fossil particle tracks in the Kenna ureilite. Geochim Cosmochim Acta 40:1465-1473

Yamamoto T, Hashizume K, Matsuda J-I, Kase T (1998) Multiple nitrogen isotopic components coexisting in ureilites. Meteoritics Planet Sci 33:857-870

Yang J, Anders E (1982) Sorption of noble gases by solids, with reference to meteorites. III. Sulfides, spinels, and other substances; on the origin of planetary gases. Geochim Cosmochim Acta 46:877-892

Yang J, Turner MS, Steigmann G, Schramm DN and Olive KA (1984) Primordial nucleosynthesis: a critical comparison of theory and observation. Astrophys J 281:493-511

Zadnik G, Wacker JF, Lewis RS (1985) Laboratory simulation of meteoritic noble gases. II. Sorption of xenon on carbon: Etching and heating experiments. Geochim Cosmochim Acta 49:1049-1059

Zinner E (1998) Stellar nucleosynthesis and the isotopic composition of presolar grains from primitive meteorites. Ann Rev Earth Planet Sci 26:147-188

4 Noble Gases in the Moon and Meteorites: Radiogenic Components and Early Volatile Chronologies

Timothy D. Swindle

Lunar and Planetary Laboratory
University of Arizona
Tucson, Arizona 85721

tswindle@u.arizona.edu

INTRODUCTION

One of the reasons noble gases make such good tracers of processes occurring in rocks is their scarcity. Thus, a process that converts a small fraction of a relatively rare element into a noble gas can have a large effect on the noble gas. Radioactive decay can be such a process. In a meteorite which has formed very early in the solar system's history and been little altered since, the effect can be even more dramatic. Furthermore, since radioactive decay proceeds at a known rate, radiogenic noble gases, those produced by decay, have been crucial in deciphering the chronology of the solar system.

Several "radionuclides," radioactive isotopes that decay to noble gases, are listed in Table 1. In a radioactive decay, the radioactive isotope is referred to as the "parent" isotope, while the decay product, the noble gas isotope, is referred to as the "daughter" isotope. In some cases listed, there is more than one mode of decay possible. In other cases, a single decay starts a chain that will ultimately produce several noble gas atoms. To take both into account, the yield (the number of noble gas atoms produced for each parent atom) is also given. Finally, radionuclides that fission may produce any of the several different isotopes of Xe in a characteristic spectrum (Table 2). All of the systems listed, with the exception of the decay of U and Th to Xe, have been exploited in extraterrestrial samples at one time or another.

Table 1: Radiogenic noble gas isotopes and their parents

Parent isotope	*Half-life (years)*	*Daughter isotope(s)*	*Yield[c] (atom/atom)*	*Proxy[a]*	*Abundance[b]*
^{22}Na	2.602	^{22}Ne	1		
^{36}Cl	3.01×10^{5}	^{36}Ar	1	^{37}Cl→^{38}Ar	^{36}Cl/^{35}Cl ~ 10^{-6}
^{40}K	1.277×10^{9}	^{40}Ar	0.1048	^{39}K→^{39}Ar	0.0117
^{129}I	1.57×10^{7}	^{129}Xe	1	^{127}I→^{128}Xe	^{129}I/^{127}I ~ 10^{-4}
^{232}Th	1.405×10^{10}	^{4}He	6		100
^{235}U	7.038×10^{8}	^{4}He	7		0.72
^{238}U	4.468×10^{9}	^{4}He	8		99.28
^{238}U	4.468×10^{9}	$^{131, 132, 134, 136}$Xe	3.50×10^{-8}	^{235}U→$^{131-136}$Xe	99.28
^{244}Pu	8.08×10^{7}	$^{131, 132, 134, 136}$Xe	7.00×10^{-5}	^{235}U→$^{131-136}$Xe	^{244}Pu/^{238}U~5×10^{-3}

[a]Nuclear reaction(s) producing noble gas isotope(s) from stable or long-lived isotope during an irradiation in a nuclear reactor. [b]For parent isotopes with half-lives less than 10^8 years, typical or suspected abundance at the time of formation of the solar system. For longer-lived parent isotopes, current abundance (%) of element. [c]For actinides, ^{4}He yields are the number of atoms produced per decay chain, fission Xe yields are branching ratio for ^{136}Xe. Yields for other isotopes are given in Table 2.

1529-6466/00/0047-0004$05.00

Table 2: Xenon fission spectra of actinides

Actinide	^{131}Xe	^{132}Xe	^{134}Xe	^{136}Xe	*Reference*
^{238}U	0.078	0.595	0.832	1.0	(Wetherill1953)
$^{235}U^{a}$	0.669	1.0	1.841	1.475-2.5	(Hohenberg & Kennedy1981; Hyde1971)
^{244}Pu	0.246	0.885	0.939	1.0	(Lewis1975)

[a]The lower number for the abundance of ^{136}Xe is for low-flux environments. Typical irradiation experiments give values more like 2.5, because the half-life of ^{135}Xe is long enough (9.14 hours), and the neutron-capture cross-section high enough to produce a significant amount of ^{136}Xe (Hohenberg and Kennedy1981).

A typical analysis of radiogenic noble gases involves trying to find the ratio of the daughter isotope to some other, stable, isotope of the parent element, for example, the ratio ^{129}Xe*/^{127}I or ^{40}Ar*/^{40}K, where the asterisk refers to radiogenic gas. For a long-lived radionuclide like ^{40}K, the result is usually quoted as an age, the length of time it would have taken for that much of the radiogenic gas to build up. For a short-lived radionuclide, where all the radionuclide has long since decayed, the result is often converted into the relative abundance of the parent isotope at the time of isotopic closure, such as ^{129}I/^{127}I. There is also chronological information in this ratio, although there are other complications, such as the possibility of isotopic inhomogeneity, the question of when and how the radionuclide was created, and the calibration of what the ratio was at some known time, details that will be discussed later in this chapter for several radionuclides.

Isotopic closure will not necessarily be the same thing as formation of the host rock. Since noble gases diffuse more readily than other radiogenic isotopes, they are uniquely suited to studies of thermal processes. However, one of the things that has become increasingly apparent is that it is crucial to determine what sample is required to date a particular process (or, to determine what process a particular sample is dating).

In this chapter, we will first consider the long-lived nuclides, such as ^{40}K and the actinides U and Th, and then the short-lived "extinct" radionuclides.

LONG-LIVED NUCLIDES: CHRONOLOGY OF SOLAR SYSTEM EVOLUTION

Of the parent isotopes listed in Table 1, ^{40}K is by far the most abundant. Only 0.017% of the potassium in a modern rock is ^{40}K, but potassium is at least a minor element in many rocks, and the half-life of ^{40}K (1280 Ma) is relatively long. Because ^{40}K is so abundant, it is not surprising that the ^{40}K-^{40}Ar system is the most widely used in meteorites as well as terrestrial samples. Although the ^{40}K-^{40}Ar system can provide some useful information about crystallization ages, in most cases in meteorites and lunar samples it is most useful for piecing together the thermal history, since it is more sensitive to low-temperature heating events than the Rb-Sr, Sm-Nd or even U,Th-Pb systems. For example, metamorphic heating of chondrites or heating by the deposition of warm ejecta by an impact event can frequently be dated using the K-Ar system.

The actinides, Th and U, also decay by α decay en route to the stable Pb isotopes. For each actinide nucleus, several α particles (^{4}He nuclei) are produced. However, the same problems that make it difficult to use U,Th-He dating in terrestrial studies (Farley 2002, this volume), the recoil upon creation and the extremely low temperature at which it is lost, make it difficult to apply to meteorites as well. So the K-Ar system, particularly the ^{40}Ar-^{39}Ar version, is used far more often than is the U-Th-He system, and hence will be mentioned far more often in applications.

Another chapter in this book (Kelley 2002) contains details about the history of the technique and pertinent technical details about its application. These will not be repeated here, except to note that the most common application of the technique for meteorite and lunar studies is step-heating ^{40}Ar-^{39}Ar dating. In this application, a sample is irradiated with neutrons, converting some ^{39}K into ^{39}Ar through neutron capture. The sample is then heated to progressively higher temperatures, and the gas released at each temperature is analyzed. This yields an "age spectrum" of apparent age vs. temperature (Kelley 2002).

Solar system impact history

The dominant process occurring on most solid surfaces in the solar system for the last 4.5 Ga has been impact cratering. When Galileo turned his new telescope to the Moon in 1609, what he noted most was the myriad of near-perfect circles. Spacecraft views of asteroids, planetary moons, and even planets have looked much the same, except for cases like Venus or Jupiter's moons Io and Europa, where there has been enough internal heating to cause very recent resurfacing. Our understanding of how the cratering process works has grown in recent decades, spurred in part by the observations of other solar system bodies and by a recognition that craters might be important to Earth's history, and aided by nuclear weapons tests (which provide the closest analog to an impact) and the development of sophisticated hydrocodes (Melosh 1989). Recognition of the inevitability of the cratering process has led to the use of relative crater densities to determine the relative chronology of surfaces on the Moon, Mars or any other heavily cratered body. However, crater densities determine relative, not absolute, chronology. Crater densities can be turned into absolute chronology only if the cratering rate is known, a difficult proposition at best (Strom et al. 1992).

Although an impact-cratering event carries a huge amount of energy and can disrupt a vast area of a planet's surface, it will not cause isotopic resetting of every rock that it moves. In fact, a shock event will not even cause isotopic resetting of every rock in which it leaves mineralogical evidence. The shock wave caused by the impact can move through fast enough that it may disrupt mineral grains without giving radiogenic isotopes an opportunity to move and equilibrate. Even if radiogenic isotopes are mobilized, it may simply be enough to disturb a system without resetting it. Hence, finding the age of a crater is often difficult. In a few cases on the Moon, there are craters young enough ($<$100 Ma) that their ages can be determined by the clustering of cosmic ray exposure ages of rocks they have uncovered (Arvidson et al. 1975). In most cases, an impact event has to be dated through the collateral heating that it produces. Since the ^{40}Ar-^{39}Ar system is the most sensitive to heating of the major chronology systems, it will be reset in many more rocks than systems such as Rb-Sr or Sm-Nd. Bogard has written two good reviews of this application of ^{40}Ar-^{39}Ar studies, a decade and a half apart (Bogard 1979, 1995).

For an impact event to be able to reset the ^{40}Ar-^{39}Ar system, the rock has to be at a high enough temperature for a long enough time. There are many cases in meteorite and lunar samples, as for terrestrial samples (Kelley 2002, this volume), where samples have apparently not reached high enough temperatures to completely degas—mineral sites that degas at low-temperatures yield lower apparent ages, and the age spectrum can be interpreted in terms of diffusive loss. On the other hand, even if a sample is melted, the ^{40}Ar-^{39}Ar system will not be reset if it cools quickly. This has been shown dramatically in the cases of the meteorites Peace River (McConville et al. 1988) and Chico (Bogard et al. 1995), both of which have glassy areas that were melted in a shock event, but then quickly cooled. In each case, detailed ^{40}Ar-^{39}Ar studies, comparing multiple samples, have shown that some samples have retained ("inherited") a fraction of the ^{40}Ar that they contained before the impact, and yield age spectra that would be very difficult to interpret without multiple samples. Hence, in trying to find the age of a crater, one would like to find an impact melt that cooled slowly enough to fully degas (for example, one that cooled slowly enough to recrystallize). Such slow cooling requires being near the center of a fairly large (perhaps 100 km in diameter) crater. In many cases, we will be limited to imperfect samples, either because we are trying to date a smaller crater or because of the fundamental limitation of the available samples, so it becomes necessary to interpret a complicated age spectrum (Kelley 2002).

Lunar impact history: a cataclysm? Given its proximity to the Earth, the Moon's impact history presumably is indicative of the impact history of the Earth as well. The majority of published ^{40}Ar-^{39}Ar ages of lunar rocks come from the 1970s, most of them published in the annual *Proceedings of the Lunar Science Conference*. However, these early studies raised questions that are still being sorted out by detailed studies with more advanced technology.

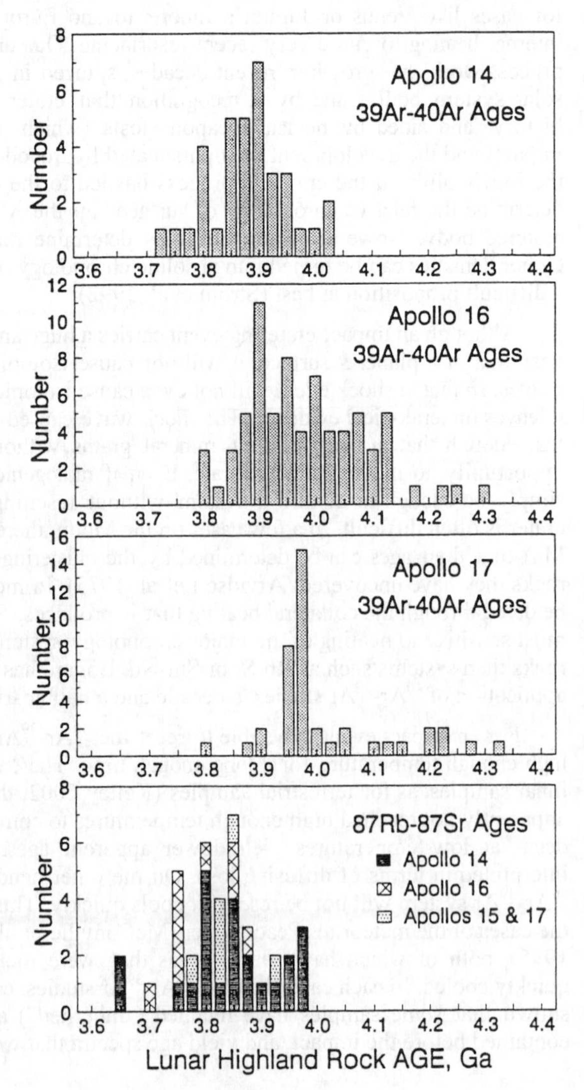

Figure 1. Histogram of apparent ^{40}Ar-^{39}Ar ages of lunar samples from three Apollo landing sites and Rb-Sr ages from four sites. Note the prominence of ages between 3.8 and 4.0 Ga. [Used by permission of the editor of *Meteoritics*, from Bogard (1995), Fig. 6, p. 256].

The most notable aspect of ^{40}Ar-^{39}Ar ages of lunar impact melts is that the majority of them cluster at an age just younger than 4.0 Ga (Fig. 1), first noted by Turner et al. (1973). A similar effect is seen in the U-Th-Pb and Rb-Sr systems (Fig. 1), which led Tera et al. (1974) to suggest that a "lunar cataclysm" occurred 3.8 to 4.0 Ga ago. No one suggested a good reason why such a cataclysm should have occurred, and many lunar geologists quickly argued that instead of a "cataclysm," the isotopic systems might be recording the end of an epoch of "heavy bombardment," (Baldwin 1974; Hartmann 1975). It has also been suggested that the apparent "cataclysm" might be reflecting a single event, the impact that formed Imbrium. Imbrium is one of the most recent of the basins, craters several hundred kilometers in diameter that dominate lunar geology. The returned lunar samples all come from a relative small area on the Near Side of the Moon, and that region can be all be seen to have suffered some effects of the Imbrium impact, so there simply might not be any material surviving with a record of previous impacts. On the other hand, attempts to date other basins based on impact melts that could have come from them still suggested that the basins had all formed in a relatively short period of time (see Dalrymple and Ryder (1996) and Stöffler and Ryder (2001) for discussion of many isotopic attempts to find the ages of basins).

Two studies that appeared less than a year apart led to detailed experimental studies that have lent strong support to the idea of a cataclysm. Although the idea still cannot be considered verified, some of the earlier questions have been answered, and means have been found to address a new set of questions.

First, G. Ryder revived the cataclysm hypothesis (Ryder 1990). He looked at the petrology of samples that had been dated, and argued that none of the ages older than 4.0 Ga were actually dating impacts. Instead, he suggested that those ages all came from samples that were either igneous rocks, or rocks that had been at most partially reset by impacts. This led to a detailed study by Ryder and Dalrymple of clasts that definitely were impact melts. With the advances in techniques that had occurred, they were able to date much smaller samples than had been possible in the 1970s. In a series of studies (Dalrymple and Ryder 1993; Dalrymple and Ryder 1996) they found many impact melts 3.8 to 4.0 Ga old. They suggested that most of the visible basins may have formed in as little as 55 to 60 Ma, implying a very high flux of very large objects for a very short period of time. They found several other samples that were older than 4.0 Ga, but none of the old samples were impact melts. Since some older samples were observed, they argued that the lack of old impact melts was not a result of all old rocks having been reset, but rather the result of a cataclysm.

Another insight came from G.J. Taylor (1991), who suggested a way to test the influence of Imbrium. In a study of a lunar meteorite, MacAlpine Hills 88105, he pointed out that the meteorite's bulk chemistry was inconsistent with an origin in the part of the Near Side from which samples had been returned, and that it had impact melt clasts with compositions different from anything expected from impacts on the Near Side. Hence, he suggested that if those clasts could be dated, the ages would assuredly not be reflecting Imbrium. Cohen et al. (2000) determined ages from 31 impact-melt clasts in MacAlpine Hills 88105 and three other lunar meteorites, and still failed to see any ages older than 4.0 Ga. The source of the cataclysm is still unexplained, but its reality now seems more likely.

In reviewing impact ages of extraterrestrial bodies, Bogard (1995) pointed out that there may also be evidence for a cataclysm in ^{40}Ar-^{39}Ar ages of other samples from the inner solar system. He noted that several of the eucrites and diogenites, differentiated meteorites that apparently come from the outer portion of the Main Belt asteroid Vesta (Drake 2001), show impact ages of 3.5 to 4.0 Ga. The same is true of many of the mesosiderites and IIE iron meteorites, which are both types of metal-rich meteorites that presumably come from near the cores of differentiated asteroids. Furthermore, the only Martian meteorite old enough to have been in existence during the cataclysm, Allan Hills 84001, has also apparently been reset by impact about 3.9 to 4.0 Ga ago. Although there is not a strong signal of a 3.9 to 4.0 Ga impact among the most common meteorites, the chondrites, K-Ar ages have been determined for far more chondrites, and many of these do appear to have lost Ar within the last 4 Ga (Heymann 1967).

Lunar samples have also been used in an attempt to determine the frequency of smaller cratering events on the Moon and, by extension, the Earth. Culler et al. (2000)

analyzed many samples from an Apollo 14 "soil" (lunar regolith) sample. In contrast to Dalrymple and Ryder (1993, 1996) and Cohen et al. (2000), who restricted themselves to impact melts that had cooled slowly enough to recrystallize, Culler et al. (2000) chose to analyze glass spherules, which must have cooled quickly, reasoning that this would increase their chances of sampling smaller craters, which in turn would increase their chances of sampling many different craters. Many of their samples had ages less than 1 Ga (some, though gave ages approaching 4.0 Ga), and while there are certainly some ages that appear more often than others, the meaning of the age spectrum, (e.g., whether there is a periodicity), remains to be seen (Culler et al. 2000). In addition, the use of glass spherules also makes such a study more vulnerable to samples that have not been completely degassed in the impact event, as shown by the Peace River data (McConville et al. 1988) discussed above.

Many of the advances in the last decade of the 20th Century were possible because of advances in techniques. Dalrymple and Ryder (1993, 1996), Cohen et al. (2000), and Culler et al. (2000) all used laser extraction techniques that allowed them to use smaller samples (typically no more than a few hundred micrograms, sometimes as small as 1 μg). This in turn, enabled them to determine ages of samples that would have been completely undatable in the 1970s. Lasers were used for ^{40}Ar-^{39}Ar experiments as early as 1973 (Megrue 1973). However, the early experiments used the lasers as microprobes, basically finding the K-Ar ages of small regions within a sample. More recent experiments have used lasers in step-heating experiments as a heating device that generates extremely little background interference. The latter approach has proven to be more widely applicable.

Ordinary chondrites. The ordinary chondrites, the most common meteorites, do not show a strong signature of a cataclysm at 3.9 Ga. Bogard (1995), in pointing this out, suggested that it was a selection effect, that chondritic asteroids that suffered large impacts in the cataclysm were simply destroyed, so we preferentially sample those that were not affected. There are, however, more recent impact events that are clearly recorded in ordinary chondrites.

The most prominent signature is of one or more events that affected, perhaps destroyed, the L (low-iron) chondrite parent body less than 1 Ga ago (Fig. 2). This can be seen even in the simplest of noble gas radiometric ages, K-Ar and U-Th-He ages. Because L chondrites all have roughly the same K and actinide contents, the distribution of ^{40}Ar and ^{4}He, respectively, are equivalent to age distributions. Several authors (Heymann 1967; Wasson and Wang 1991; Zähringer 1968) noted a preponderance of ages of roughly 350-500 Ma, although it is impossible to be more precise on this basis.

More detailed ^{40}Ar-^{39}Ar ages have only slightly clarified the situation. Bogard and Hirsch (1980) analyzed seven severely shocked chondrites and found most showed either well-defined plateaus at 500-600 Ma or age minima at 700-1000 Ma. In a more detailed study of a single sample, with multiple smaller samples from various locations within a shock-melted L chondrite, Chico, Bogard et al. (1995) documented the prevalence of partial resetting in this event—the lowest ^{40}Ar-^{39}Ar ages clustered around 550 Ma, but even that is higher than the Rb-Sr age of about 470 Ma, so they suggested that at least 1-2% of the radiogenic ^{40}Ar was retained in all of the samples. In a similar study of another shock-melted L chondrite, Cat Mountain, Kring et al. (1996) argued that Cat Mountain suffered a shock event 800-900 Ma ago, an age also shown by a few other L chondrites. Whether this truly represents a second shock event, or simply reflects partial resetting 400-500 Ma ago, will require future studies to resolve.

Metamorphic heating of chondritic meteorites. In terrestrial studies, one of the most common uses of the K-Ar system is to date metamorphic events (Kelley 2002; McDougall and Harrison 1999). In meteorites, the most prominent metamorphic situation is the heating of ordinary chondrites. Terrestrial studies have demonstrated that the "age" of a metamorphosed sample depends on the closure temperature, which in turn depends on both the time-temperature history of the sample and the specific mineral under consideration. In addition, whether a particular question can be answered depends on the precision of the system being considered. The story of ordinary chondrite metamorphism provides examples of all of these facets.

Ordinary chondrites are divided into chemical classes, based primarily on their iron abundance (H – high; L – low; LL – very low). Each chemical class is further divided into metamorphic classes or petrographic grades (3 to 6 from least to most metamorphosed), based on, among other things, the degree of equilibration of mafic silicates and the degree of recrystallization of textural components (Dodd 1981; Wasson 1985).

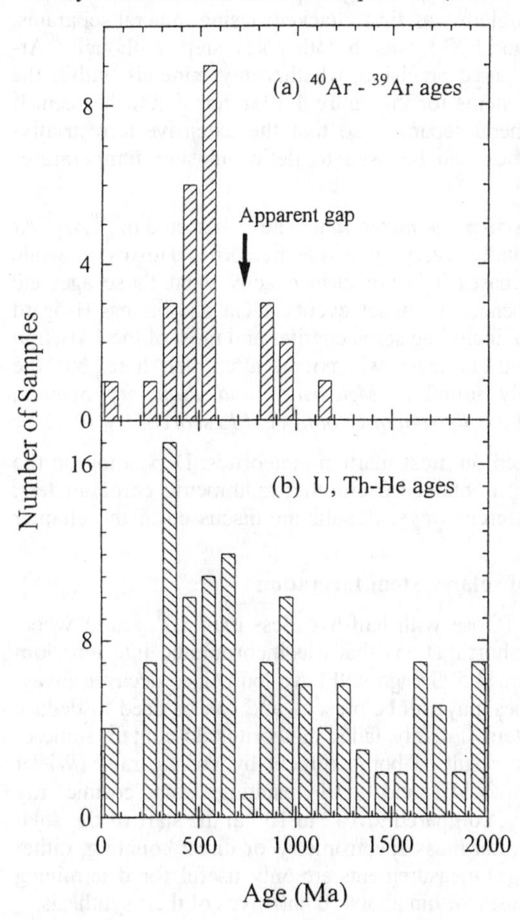

Figure 2. Histogram of reported recent shock-reset ages of L chondrite meteorites. Note the apparent presence of one peak at about 500 Ma and the possibility of another near 900 or 1000 Ma. [Used by permission of the editor of the *Journal of Geophysical Research*, from Kring et al. (1996), Fig. 12, p. 29,367].

The question is how and where that metamorphism occurred. Is it prograde or retrograde (i.e., is it reflecting a reheating, or does it merely reflect slow cooling from some very high initial temperature)? Did it occur within an internally heated, stratified parent body, or within pieces of a rubble pile that were heated, broken apart and reassembled? Within a stratified parent body, the expectation would be that the more equilibrated (higher petrographic grade) meteorites would have come from closer to the center of the body, so they would have stayed hot longer (younger ages) and would have cooled more slowly.

For many years, the chronological and cooling rate data did not give a coherent picture. Cooling rates based on the size and composition of metal grains typically showed no correlation with petrographic grade (Taylor et al. 1987), nor did I-Xe ages ((Jordan et al. 1980; Swindle and Podosek 1988); note that these experiments were on whole-rock samples, rather than the mineral separates that have proven more amenable to interpretation, as discussed below). On the other hand, cooling rates based on damage tracks from fissioning ^{244}Pu, really cooling ages based on that extinct radionuclide, did seem to show the expected correlation (Pellas and Störzer 1981). A ^{40}Ar-^{39}Ar study of 13 ordinary chondrites, meanwhile, concluded that the ages were nearly indistinguishable at about 4.5 Ga old (Turner et al. 1978). However, these typically had uncertainties of 30 Ma or more, comparable to the expected variations, and 10 of 13 were of a single one of the three chemical classes (H). When all available analyses were considered, a correlation was suggested, particularly for the LL chemical class (Lipschutz et al. 1989).

Later radiometric studies finally began to show a correlation. The most telling evidence came from Göpel et al. (1994), who found a correlation in Pb-Pb ages from apatites. Later, Brazzle et al. (1999) found that the I-Xe ages of apatites (in some cases, aliquots of the samples analyzed by Göpel) and feldspars also showed a correlation. The I-Xe work will be discussed more in the next section.

In this case, the crucial data appears to require higher precision than the ^{40}Ar-^{39}Ar technique was capable of when the problem was first attacked. Using mineral separates, instead of whole rock samples (Pb-Pb and I-Xe), was the other key step. Although ^{40}Ar-^{39}Ar experiments are capable of the required precision, whether any minerals within the meteorites will have remained closed systems for the entire 4.5 Ga remains to be seen. If so, the logical next step is to use mineral separates so that the extensive terrestrially-derived database on diffusion properties can be used to define closure temperatures (McDougall and Harrison 1999).

Other applications of the K-Ar system to meteorites. The K-Ar and/or ^{40}Ar-^{39}Ar techniques have been applied to virtually every type of meteorite. However, while precise ages can be obtained, in many cases it is not clear exactly what those ages are dating. While many of them show evidence of impact events 4 Ga ago or less (Bogard 1995), others give ages of nearly 4.5 Ga, including some eucrites and most of the IAB iron meteorites (Niemeyer 1980). A detailed summary will not be attempted here, but the individual studies are most commonly found in *Meteoritics and Planetary Science*, *Geochimica et Cosmochimica Acta*, and the *Journal of Geophysical Research*.

^{40}Ar-^{39}Ar dating has been performed on most martian meteorites. Depending on the meteorite, and on what results have been obtained from other radiometric chronometers, the results have been interpreted in different ways. Results are discussed in the chapter on martian samples (Swindle 2002).

Extinct radionuclides: Chronology of solar system formation

There are several cases in Table 1 (those with half-lives less than 10^8 years) where the half-life of the parent isotope is so short that any that was incorporated into a rock at the time of formation of the solar system 4.5 Ga ago will have long since decayed away. However, the noble gas daughter isotopes may still be present, and can be used to deduce the very early history of the solar system. Freshly fallen meteorites do contain some of these "short-lived radioactivities," as a result of bombardment by cosmic rays (Wieler 2002, this volume). But ratios like $^{36}Cl/Cl$ and $^{129}I/I$ resulting from cosmic ray bombardment are typically 10^{-10} to 10^{-15}, compared to 10^{-4} to 10^{-6} at the start of the solar system, so they are measured by accelerator mass spectrometry or direct counting, rather than by measuring noble gases. Noble gas measurements are only useful for determining the abundances of short-lived radionuclides within about 10 half-lives of their synthesis.

Extinct radionuclides have a huge advantage in precision over long-lived radionuclides. For example, a change in a factor of two in the abundance of ^{129}I requires only one half-life, 15.7 Ma, while the same length of time will cause a change of less than 1% in the abundance of ^{40}K, since it is only about 1% of a half-life. On the other hand, since the extinct radionuclide is, by definition, completely decayed away, it is necessary to determine its abundance at some particular time to determine ages, a problem for all extinct radionuclides.

The relative amount of attention given to extinct radionuclides compared to the ^{40}Ar-^{39}Ar system in this chapter does not reflect the relative number of studies. However, the greater applicability of the ^{40}Ar-^{39}Ar system means that there are more reviews available for that technique (Bogard 1979, 1995; Kelley 2002; McDougall and Harrison 1999).

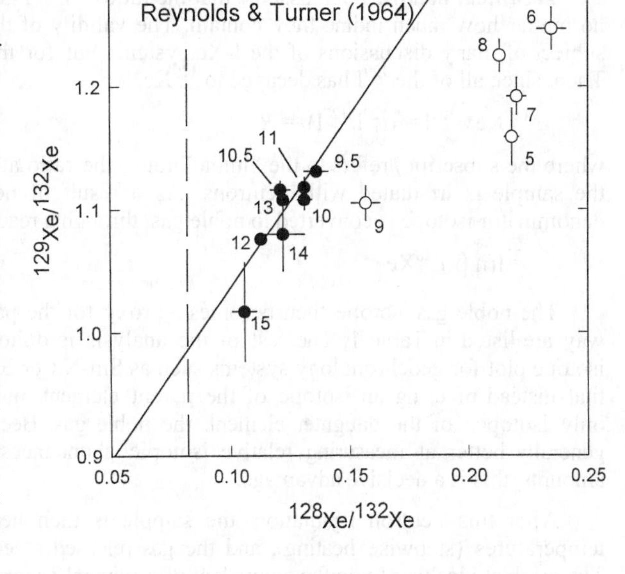

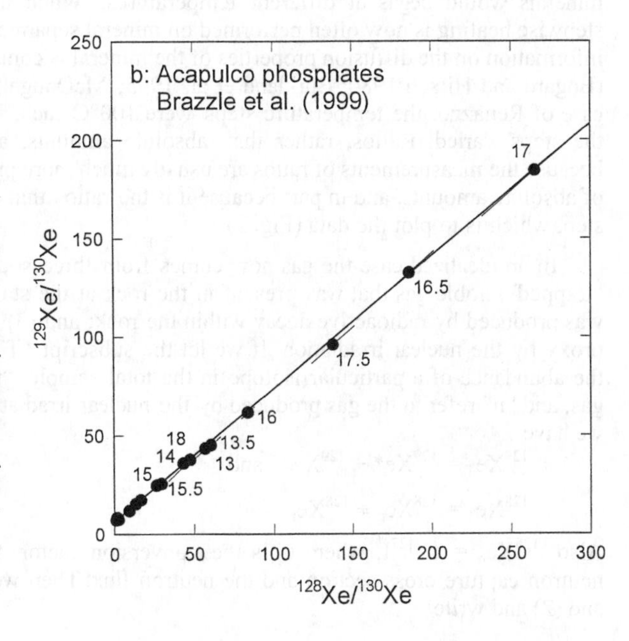

Figure 3. Three-isotope plots for I-Xe experiments. Data in (a) are from Reynolds and Turner (1964). Labels next to points are extraction tempeatures (in hundreds of degrees Celsius). Solid points are those defining the correlation line. Dashed line is at a ^{128}Xe/^{132}Xe ratio typical of chondritic meteorites. Data in (b) are from Brazzle et al. (1999). The extraction temperatures in (b) are actually the temperature of a heating coil containing the sample, rather than the sample itself. In this case, even the low temperature extractions fall on the correlation line. A different denominator is used because the ^{132}Xe in the Acapulco phosphates is dominated by fission-produced gas. For the sake of conversion, the trapped ^{130}Xe/^{132}Xe ratio is typically about 0.16 (Ott 2002, this volume). The equivalent dashed line would be indistinguishable from the ordinate, because the I/Xe ratios are so high.

Iodine-xenon

The first extinct radionuclide studied, and by far the most commonly studied of the extinct radionuclides, is ^{129}I. Most chondrite meteorites contain ^{129}Xe that appears to be a result of the decay of ^{129}I (as do the atmospheres of Earth and Mars). We can illustrate the basic technique commonly used in studies of many of the short-lived radiogenic noble gases in meteorites with two examples (Fig. 3). Figure 3a is data from one of the first I-Xe analyses, a study of the carbonaceous chondrite Renazzo. Figure 3b is data from an important recent analysis, of a phosphate mineral separate from the meteorite Acapulco. The data from Renazzo are less precise, and more difficult to interpret, than more recent data, and the data from Acapulco are among the most precise and simplest to interpret. Between them, they illustrate several of the basic principles that will be discussed. We will then compare that to a typical ^{40}Ar-^{39}Ar analysis, which has been described elsewhere in this volume (Kelley 2002), and then briefly consider the complications involved in analyzing other systems.

Technical details. First, assume that the ratio $^{129}I/^{127}I$ is the same in all mineral sites, no matter how much iodine they contain. The validity of that assumption has been the subject of many discussions of the I-Xe system, but for the moment, we will make it. Then, since all of the ^{129}I has decayed to $^{129}Xe^*$,

$$^{129}Xe^*/^{127}I = (^{129}I/^{127}I)_I = R \quad (1)$$

where the subscript $_I$ refers to the "initial" ratio, the ratio at the time of formation. Next, the sample is irradiated with neutrons. As a result of neutron capture, some of the denominator isotope is converted to noble gas, through a reaction like

$$^{127}I(n,\beta^-)^{128}Xe$$

The noble gas isotope then becomes a proxy for the parent. The isotopes used this way are listed in Table 1. The rest of the analysis is quite similar to a standard three-isotope plot for geochronology systems such as Sm-Nd or Rb-Sr. The major difference is that instead of using an isotope of the parent element, noble gas-based techniques use only isotopes of the daughter element, the noble gas. Because mass spectrometers are generally better at measuring relative isotopic abundances than at measuring absolute amounts, this is a decisive advantage.

After this neutron irradiation, the sample is then heated to progressively higher temperatures (stepwise heating), and the gas released at each temperature is analyzed. The original idea was to do the equivalent of a mineral separation, assuming that different minerals would degas at different temperatures. While that is true to some extent, stepwise heating is now often performed on mineral separates with the understanding that information on the diffusion properties of the mineral is contained in the degassing results (Bogard and Hirsch 1980; Burkland et al. 1995; McDougall and Harrison 1999). In the case of Renazzo, the temperature steps were 100°C each, in the Acapulco phosphates, the steps varied. Ratios, rather than absolute amounts, are usually reported, in part because the measurements of ratios are usually much more precise than the measurements of absolute amounts, and in part because it is the ratios that will be important in the next step, which is to plot the data (Fig. 3).

In an idealized case the gas now comes from three sources: (1) the background (or "trapped") noble gas that was present in the rock at the start; (2) the radiogenic gas that was produced by radioactive decay within the rock; and (3) the gas produced as a parent proxy by the nuclear irradiation. If we let the subscript "T" refer to an isotopic ratio or the abundance of a particular isotope in the total sample, "t" refer to the trapped noble gas, and "n" refer to the gas produced by the nuclear irradiation, then for ^{129}Xe and ^{128}Xe we have

$$^{129}Xe_T = {}^{129}Xe^* + {}^{129}Xe_t \quad \text{and} \quad (2a)$$

$$^{128}Xe_T = {}^{128}Xe_n + {}^{128}Xe_t\,. \quad (2b)$$

Also, $^{128}Xe_n = C\ ^{127}I$, where C is the conversion factor, which is determined by the neutron capture cross-section and the neutron flux. Then we can combine Equations (1) and (2) and write

$$^{129}\text{Xe}^* = R\ ^{127}\text{I} = CR\ ^{128}\text{Xe}_n\,. \tag{3}$$

The presence of the trapped Xe means that the total ^{129}Xe/^{128}Xe ratio will differ from one mineral to another, because the ratio of I to Xe in the minerals will differ.

Some other Xe isotopes, notably ^{130}Xe and ^{132}Xe, may be entirely "trapped" Xe (in reality, a fraction of these may have been produced by spallation, in the former case, or fission, in the latter case, but those are often second-order complications). If we measure, for example, ^{129}Xe/^{130}Xe and ^{128}Xe/^{130}Xe, then we can use Equations (2a) and (2b) to write

$$(^{129}\text{Xe}/^{130}\text{Xe})_T = {}^{129}\text{Xe}^*/^{130}\text{Xe} + {}^{129}\text{Xe}_t/^{130}\text{Xe} \quad \text{and} \tag{4a}$$

$$(^{128}\text{Xe}/^{130}\text{Xe})_T = {}^{128}\text{Xe}_n/^{130}\text{Xe} + {}^{128}\text{Xe}_t/^{130}\text{Xe}. \tag{4b}$$

But since $^{129}\text{Xe}^* = CR\ ^{128}\text{Xe}_n$, we can substitute in Equation (4a), and write

$$(^{129}\text{Xe}/^{130}\text{Xe})_T = CR\ ^{128}\text{Xe}_n/^{130}\text{Xe} + {}^{129}\text{Xe}_t/^{130}\text{Xe}\,, \tag{5a}$$

we can also rewrite Equation (4b) as

$$^{128}\text{Xe}_n/^{130}\text{Xe} = (^{128}\text{Xe}/^{130}\text{Xe})_T - {}^{128}\text{Xe}_t/^{130}\text{Xe}\,. \tag{5b}$$

Making one last substitution of Equation (5b) into Equation (5a), we have

$$(^{129}\text{Xe}/^{130}\text{Xe})_T = CR\,[(^{128}\text{Xe}/^{130}\text{Xe})_T - {}^{128}\text{Xe}_t/^{130}\text{Xe}] + {}^{129}\text{Xe}_t/^{130}\text{Xe}\,. \tag{6}$$

This is the equation of a line involving the two measured ratios. Much of the data in Figure 3, where the two ratios are plotted, clearly do fall along a line, which is referred to as an "isochron", since all the points falling along the line have the same ("iso") initial ratio or age ("chron"). We will return later to the significance of the Renazzo points that deviate from the isochron. Note that if the I/Xe ratio is higher, as is the case for most of the data from the Acapulco phosphates, the trapped Xe is less significant. However, there are also cases where the isochron is defined by fewer points than the Renazzo plot in Figure 3.

The slope is *CR*, the product of the initial iodine isotope ratio and the conversion factor. The real point of the experiment is to determine *R*, the initial iodine isotopic composition. To do so, however, it is necessary to know *C*, the efficiency with which ^{127}I is converted to ^{128}Xe. This is normally done by including an "irradiation monitor" in the nuclear irradiation, a sample with a known $(^{129}\text{I}/^{127}\text{I})_0$ ratio. The data from the monitor are then plotted on the same type of isochron plot, but for the monitor, the unknown is *C*, the conversion factor for that particular experiment. Then, knowing the conversion factor, *R* can be determined for the other samples irradiated at the same time.

It might seem that the best monitors of the conversion factors would be relatively pure samples, such as potassium iodide (KI) for either K-Ar or I-Xe experiments. In fact, such samples are difficult to analyze, because they contain so much more neutron-derived gas than the unknown samples, and have so much iodine that they may produce self-shielding effects in the reactor (Hohenberg et al. 2000). Instead, the monitors are usually samples with known ratios similar to those expected in the experiment. For meteorites, the most commonly used monitors in recent years have been the meteorites Bjurbole and Shallowater (Table 3). Other meteorites have been used as monitors, as well, but only for these two have enough samples been analyzed to demonstrate that the $(^{129}\text{I}/^{127}\text{I})_0$ ratio is uniform. In experiments where other monitors have been used, the data should be approached with caution (Hohenberg et al. 1981, 2000; Swindle and Podosek 1988).

Table 3: Iodine-xenon monitors and calibrations.

Sample	*$^{129}I/^{127}I$*	*Relative age (Ma)*	*Absolute age (Ma) relative to Acapulco*	*Ste. Marguerite*
Shallowater	≡1.072	≡0	4566	4564
Bjurbole	1.095±29	-0.46±0.15	4566	4564
Acapulco phosphates	0.731	+8.8±0.2	≡4557	4555
Ste. Marguerite feldspars	1.105	-0.7±0.4	4567	≡4565

From Brazzle et al. (1999), Gilmour and Saxton (2001).
Relative ages are positive for later ages.

Another technique that might seem sufficient for calculating the conversion factors would be to simply calculate the conversion efficiency of ^{127}I to ^{128}Xe from the neutron flux and the cross-section of the reaction. However, just as in the case of ^{39}Ar production, much of the production comes not from the thermal neutrons but from resonances at higher energies, and the details of the energy spectrum can be a function of the geometry of the irradiation package, as well as self-shielding.

The other piece of information available in a plot such as Figure 3 is the initial isotopic ratio of the daughter element, in this case $(^{129}Xe/^{130}Xe)_t$, the $^{129}Xe/^{130}Xe$ ratio before iodine decay began. The initial $^{129}Xe/^{130}Xe$ ratio, however, is not the intercept of the isochron with the y-axis, as it would be for Rb-Sr or Sm-Nd. Instead, Equation (6) shows that the ordinate-intercept is $(^{129}Xe/^{130}Xe)_t - CR(^{128}Xe/^{130}Xe)_t$. A better way of thinking of this is that if the I/Xe ratio is zero, $(^{128}Xe/^{130}Xe) = (^{128}Xe/^{130}Xe)_t$, which is usually well known (Ott 2002, this volume). Thus, $(^{129}Xe/^{130}Xe)_t$ is the intercept of the isochron with a vertical line at $(^{128}Xe/^{130}Xe)_t$, as given by the dotted line in Figure 3. If the I/Xe ratio is high, as for the Acapulco phosphates, $(^{129}Xe/^{130}Xe)_t$ is poorly defined.

All the data will be colinear in Figure 3 only if all the mineral sites that are degassed started out with the same $^{129}I/^{127}I$ ratio and the same $^{129}Xe/^{130}Xe$ ratio, and there were no subsequent events that affected different sites differently. The latter assumption is the one mostly likely to be invalid. In the common case of a later event that leads to partial resetting, some of the points will fall off the line. For example, in Figure 3a, it is the data points from the higher temperature steps that define the line. Gas released at lower temperatures comes, by definition, from sites that are easier to degas, so it is quite likely that the Renazzo meteorite has been partially degassed. In fact, those are also the sites that would be easier to contaminate with terrestrial iodine, which would produce the same effect, so it is often impossible to tell which has happened. In some cases, though, the pattern of apparent initial iodine isotopic ratios is consistent enough with the pattern expected from diffusive loss that it is possible to determine apparent thermal histories.

The initial $^{129}Xe/^{130}Xe$ or $^{129}Xe/^{132}Xe$ ratio, as well, potentially contains information about the evolution of the system before decay began (Swindle 1998; Swindle et al. 1991b), just as the initial isotopic compositions of Sr or Nd do. However, it is not commonly used, and Gilmour et al. (2001) suggest that it should not be used, in part because some of the initial Xe isotopic ratios derived (such as the one for Renazzo in Fig. 3a) are less than 1, less than any suspected trapped Xe component (Ott 2002). They argue that plots such as Figure 3 are more complicated than described above, and that the trapped Xe must be accompanied by some trapped iodine as well.

Comparison with ^{40}Ar-^{39}Ar. There are several differences between I-Xe dating and ^{40}Ar-^{39}Ar dating. The most fundamental difference is that a ^{40}Ar-^{39}Ar age is intrinsically an absolute age, while I-Xe ages are not. Traditionally, I-Xe ages have been reported as either initial $^{129}I/^{127}I$ ratios or as ages relative to other I-Xe ages. Since it is often not obvious what event last reset the xenon isotopic composition, and most absolute age systems do not have the resolution that I-Xe does, calibration is difficult. However, the Pb-Pb system can now produce 1-2 Ma scale precision, and has been applied to phosphates in meteorites that also have enough iodine to produce precise I-Xe ages. The most common intercalibration currently in use (Table 3, below) involves the assumption that the I-Xe and Pb-Pb ages of Acapulco phosphates date the same event, and hence they represent the same age (Brazzle et al. 1999; Nichols et al. 1994). To try to better calibrate I-Xe with chronometry based on the extinct radionuclide ^{53}Mn, Gilmour and Saxton (2001) have suggested an alternate calibration, assuming that the Pb-Pb and Mn-Cr ages given by the meteorite LEW86010 represent the same event and that the Mn-Cr and I-Xe ages of feldspars from the ordinary chondrite Ste. Marguerite give the same age. The difference in absolute I-Xe ages between the two calibration schemes is less than the uncertainty in the Pb-Pb age of Acapulco phosphates, so they are not really in conflict with one another.

To determine an age from an initial iodine isotopic composition, we can rewrite the equation of radioactive decay as

$$(^{129}I/^{127}I)\,(t_2) = (^{129}I/^{127}I)\,(t_1) \times \exp[-\lambda\,(t_2 - t_1)]\,,$$

where t_1 and t_2 are two different times, and λ is the decay constant for ^{129}I. From this, it is possible to solve for $(t_2 - t_1)$, the time difference. Without the calibration to give the absolute age of one of those times, it is only possible to give relative ages. Incidentally, when the first I-Xe experiments were performed, the best determination of the half-life of ^{129}I was 17.2±0.9 Ma (Katcoff et al. 1951), which was used in early I-Xe experiments, and then was often used for consistency even after a more precise determination of 15.7±0.4 Ma became available (Emery et al. 1972). At present 15.7 Ma is most commonly used (Table 1), even though 17 Ma may actually be more accurate (Holden 1990). Relative ages are proportional to the half-life, so conversion from one to the other simply requires multiplying by the ratio of the adopted half-lives.

Another fundamental difference is that while potassium is a minor element, and there are minerals that can be identified that contain relatively large amounts of potassium (e.g., orthoclase or biotite), iodine is a trace element, and is seldom a major part of a mineral. Hence it is usually not known what mineral contains the bulk of the iodine in an extraterrestrial sample. A notable exception is the mineral sodalite in the Allende meteorite, which contains the bulk of the iodine in that meteorite (Kirschbaum 1988). Although the siting of the iodine has been identified in a few other cases, discussed below, it may be that most of the iodine is along grain boundaries and in interstitial sites.

Although the early ^{40}Ar-^{39}Ar and I-Xe experiments were one and the same (consisting of xenon and argon analyses of the same irradiated samples), they are optimized in slightly different conditions. For ^{127}I, there is a large capture cross section from thermal neutrons, while the highest neutron-capture cross section of ^{39}K is at higher energies, so different reactors are preferred for the two types of experiments (see McDougall and Harrison (1999) for a list of reactors commonly used for ^{40}Ar-^{39}Ar dating). In addition, since Ar diffuses more readily than Xe, the crucial temperature steps in a ^{40}Ar-^{39}Ar stepwise heating experiment are often at lower temperatures than the crucial steps in an I-Xe experiment.

There are also several differences between the ways I-Xe and ^{40}Ar-^{39}Ar data are usually presented. I-Xe data are usually presented in isochron form, ^{40}Ar-^{39}Ar data usually are not. Instead, an initial argon isotopic ratio (in this case $^{40}Ar/^{36}Ar$) is assumed, and an apparent age (really, an apparent $^{40}Ar^*/K$ ratio) calculated. Next, the data are presented as an age spectrum plot (Kelley 2002). In this representation, if all the mineral sites formed at the same time with the same $^{40}Ar/^{36}Ar$ ratio, they will all have the same apparent age, and the result will be a flat series of points across the plot. Or, if some low temperature sites have been degassed more recently, the result will be a series of steps up to what appears to be a plateau, in which case the value of the plateau is usually interpreted to be the age of resetting, and the lower temperature steps may be used to define the thermal history. In some cases, including the studies of the shock-melted meteorites discussed above and studies of the shergottite martian meteorites that will be discussed in the chapter on martian noble gases (Swindle 2002), there can be a combination of partial degassing and even injection of gas from the atmosphere or surrounding rock that makes the plateau plot difficult or impossible to interpret. Similar problems can occur with the interpretation of the I-Xe system, particularly of shocked meteorites (Hohenberg et al. 2000). It should be mentioned that in cases where the I/Xe ratio is high, so the assumed xenon isotopic composition has little affect on the apparent age, I-Xe experiments are also often presented as age spectrum plots.

Finally, it should be noted that just as a typical ^{40}Ar-^{39}Ar irradiation also produces argon from calcium and chlorine, a typical I-Xe irradiation also produces xenon from tellurium, barium and uranium (Turner 1965) and krypton from bromine. Conversion factors for many or all of these reactions have been measured or calculated by several authors, most recently by Irwin and co-workers for use in studies of fluid inclusions (Irwin and Reynolds 1995; Irwin and Roedder 1995). Many of these elements are not routinely determined by other techniques, so direct comparisons with other techniques are rare (but see (Garrison et al. 2000; Swindle et al. 1991b)).

Applications. The iodine-xenon system has been reviewed by Swindle and Podosek (1988), and, more recently, by Gilmour (2000). In this chapter, we will highlight some of the areas in which I-Xe has been, or is likely to be, applied the most.

The first I-Xe experiments, like the Renazzo experiment described above, were on whole-rock samples, and served primarily to roughly establish the initial $^{129}I/^{127}I$ ratios typical of the early solar system. This, combined with nucleosynthetic calculations, made it possible to make estimates of the time between the synthesis of ^{129}I and the formation of solids in the solar system (e.g., Hohenberg et al. 1967). Initial results suggested a difference of approximately 100 Ma, and provided the first experimental constraint of this important parameter. More recently, other extinct radionuclides with even shorter half-lives have been documented within meteorites. They do not necessarily refer to the same nucleosynthetic event as ^{129}I, so the ~100 Ma timescale for ^{129}I still appears to be correct (Cameron 1993).

Later, as more I-Xe experiments were performed, they were used in an attempt to develop higher precision chronology within the early solar nebula. For example, the I-Xe system was applied to ordinary chondrites of different classes in an attempt to determine the details of chondrite metamorphism, as discussed above. A lack of correlation with petrographic grade was taken as an indication as either that metamorphism occurred in rubble piles, so cooling rates and times did not correlate with petrographic grade (Taylor et al. 1987) or that ^{129}I was distributed inhomogeneously in the early solar nebula (Clayton 1980; Clayton et al. 1977), so that the I-Xe system could not be used as a chronometer. In fact, the biggest problem was probably the fact that whole-rock samples were being analyzed, and different rocks had different amounts of various iodine carriers that required different time-temperature histories to reset. Burkland et al. (1995) showed that a single chondrite, Bjurbole, contained at least three different iodine carriers, some of which would not have been reset by the metamorphism that chondrite experienced. When samples of feldspar and the phosphate apatite, both secondary minerals, were analyzed (Brazzle et al. 1999), petrographic grade did correlate with apparent age (Fig. 4). Feldspar and apatite are presumably not the only minerals that contains enough iodine use for I-Xe, and that could be reset or created at temperatures low enough to be dating some aspect of chondrite metamorphism, but they are the only ones so far applied.

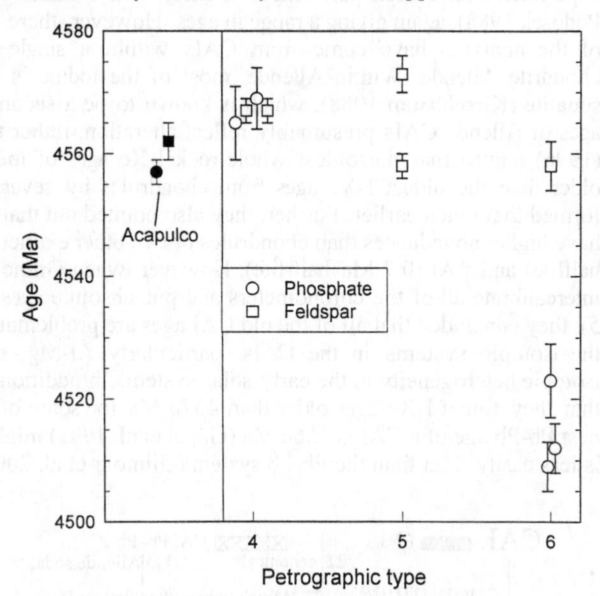

Figure 4. Correlation of I-Xe ages with petrographic type for phosphates and feldspars from ordinary chondrites. Data are from Brazzle et al. (1999). Note that type 6 (more equilibrated) chondrites have later ages than type 4 chondrites. The very old age for one of the type 5 chondrites may only be apparent, an artifact of a shock event (Hohenberg et al. 2000). The absolute ages are calibrated by assuming the I-Xe and U-Th-Pb systems for Acapulco phosphates reflect the same time.

Another process that the I-Xe system should be effective at dating is aqueous alteration, since iodine is extremely mobile in water. Several I-Xe experiments on objects that have been affected by aqueous alteration have been interpreted in this fashion (Krot et al. 1999; Swindle et al. 1991a; Whitby et al. 2000). The biggest problem is separating out the details of the complicated histories that these objects have experienced.

Iodine-xenon experiments have also been applied, but only with minimal success, to dating the formation of chondrules and calcium-aluminum-rich inclusions (Fig. 5). These two types of objects make up large portions of many meteorites, and are believed to be among the earliest solids formed in the solar nebula.

The formation of chondrules, the spherical objects that make up the bulk of many of the most common, "chondritic," meteorites, is particularly contentious. Two conferences more than a decade apart (Hewins et al. 1996; King 1983) each failed to reach consensus on the mechanism of the origin of chondrules, although formation from molten droplets of rock is a feature of most of models of their origin. Clearly, a chronology of chondrule origin would be desirable. I-Xe experiments have been performed on nearly 100 individual chondrules (see Swindle et al. 1996; Whitby et al. 2002). A wide variety of ages has been observed, of which the oldest must represent a minimum time of the formation of the first chondrules. Much of the variation probably represents secondary processing, rather than differences in formation ages (Swindle et al. 1991a,b). But even when meteorites with very simple histories are considered, there is still a range in ages. For example, a 2.6 Ma range in I-Xe ages in enstatite chondrites has been interpreted as representing the minimum duration of chondrule formation (Whitby et al. 2002).

Calcium-aluminum-rich inclusions (CAIs) are among the most refractory objects within meteorites (MacPherson et al. 1988), and hence are often thought to have been among the first objects formed in a solar system that was cooling from high temperatures. They are primarily found in meteorites that have not been heavily metamorphosed. I-Xe experiments have been performed on more than a dozen individual CAIs (Swindle and Podosek 1988), again giving a range in ages. However, there is a problem with this. Most of the analyses have come from CAIs within a single meteorite, the carbonaceous chondrite Allende. Within Allende, most of the iodine is contained within the mineral sodalite (Kirschbaum 1988), which is known to be a secondary phase. Hence, the I-Xe ages of Allende CAIs presumably reflect alteration, rather than formation. Swindle et al. (1996) argued that the oldest whole-rock I-Xe ages of meteorites containing CAIs are older than the oldest I-Xe ages from chondrules by several Ma, suggesting that CAIs formed that much earlier. Further, they also pointed out that it had been argued that CAIs have higher abundances than chondrules of two other extinct radionuclides, ^{53}Mn (3.7 Ma halflife) and ^{26}Al (0.7 Ma half-life). However, when Gilmour and Saxton (2001) tried to intercalibrate all of the chronometers and put absolute ages on the various samples (Fig. 5), they concluded that all of the old CAI ages are problematical. In fact, they suggest that the isotopic systems in the CAIs, particularly Al-Mg, may have been disturbed by isotopic heterogeneity in the early solar system. In addition, they suggested that the fact that they found I-Xe ages older than 4570 Ma for some other samples, even though the best Pb-Pb age of a CAI is 4566 Ma (Göpel et al. 1991) might mean that the I-Xe system is less easily reset than the Pb-Pb system (Gilmour et al. 2000).

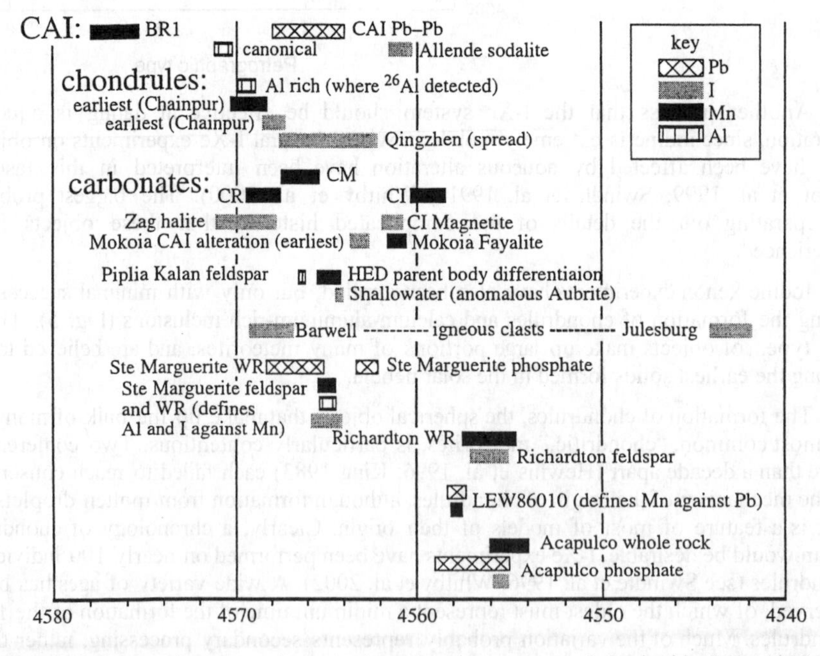

Figure 5. Summary of a variety of ages in the early solar system, as determined by I-Xe, Al-Mg, Mn-Cr and Pb-Pb. Absolute I-Xe are generated by assuming Acapulco phosphate (near bottom of figure) became closed to the I-Xe and U-Th-Pb systems at the same time, although an alternative calibration point is to assume that Ste. Marguerite phosphates are the same age in both the I-Xe and Pb-Pb systems. [Used by permission of the editor of *Philosophical Transactions: Mathematical, Physical and Engineering Sciences*, from Gilmour and Saxton (2001), Fig. 2, p. 2044.]

This ties into another unsolved problem: what is the oldest age in the solar system? In the I-Xe system, this translates into the highest $^{129}I/^{127}I$ ratio. Several different authors (Gilmour 2000; Gilmour and Saxton 2001; Swindle and Podosek 1988; Swindle et al. 1996) have all discussed this question and come to different conclusions, so it is probably safest to say that the answer is not yet known, although current arguments range between 1.3×10^{-4} and 1.45×10^{-4} for the $^{129}I/^{127}I$ ratio, corresponding to absolute ages between 4570 and 4580 Ma. There are several factors that can lead to a $^{129}I/^{127}I$ ratio that appears higher than it is, including an improperly calibrated monitor (Hohenberg et al. 1981, 2000) or a complicated history involving shock (Caffee et al. 1982) or aqueous alteration (Turner et al. 2000). Finally, of course, the intercalibration between I-Xe and Pb-Pb is not yet certain. Since the I-Xe system is, by nature, not an absolute chronometer, absolute I-Xe ages are derived by normalization of the two systems. Acapulco phosphate has a Pb-Pb age with about 2 Ma uncertainty, introducting this uncertainty into all absolute I-Xe ages calibrated in this fashion (Brazzle et al. 1999), while I-Xe ages calibrated via Mn-Cr (Gilmour and Saxton 2001) have a smaller statistical uncertainty, but the calibration involves an extra step. However, with increasing analyses of mineral separates for I-Xe making it both easier to intercalibrate systems and to understand the processes being dated, this problem is becoming more tractable.

^{244}Plutonium

With its relatively long 82 Ma halflife, it is not surprising that ^{244}Pu was present in the early solar system. In fact, the presence of ^{244}Pu in the early solar system was postulated at nearly the same time ^{129}I was discovered (Kuroda 1960), although it took another decade for its existence to be definitively proven.

One difficulty is that there is no stable isotope of plutonium with which to compare its abundance. To really quantify its abundance, it is necessary to consider the amount of ^{244}Pu relative to an isotope of a similar element. The definition of “similar” depends on the problem to be addressed. In studies of nucleosynthesis, the “similar” element used is usually uranium, another actinide, which is produced in the same stellar environments. In studies of the history of specific meteorite parent bodies, the “similar” element is more commonly a light rare earth element (LREE) like neodymium, since the geochemical behavior of plutonium is apparently most similar to that of the LREE. We will discuss the details of the experimental technique of each approach below.

The other difficulty is that unlike ^{129}I or ^{40}Ar, each of which decays to a single specific noble gas isotope, the fission of ^{244}Pu can produce any of several Xe isotopes. However, the spectrum of the Xe isotopes produced makes it possible to distinguish the actinides (Table 2). In fact, the original suggestion of ^{244}Pu was made on the basis of its fission spectrum inferred from the composition of the terrestrial atmosphere (Kuroda 1960). Subsequently, the predicted spectrum was indeed measured on artificial ^{244}Pu (Alexander et al. 1971). Only the heaviest xenon isotopes are produced in fission. Since stable nuclei become progressively more neutron-rich at higher atomic number, fission produces two neutron-rich fragments, each of which then undergoes a series of β^- decays until it reaches stability. Hence fission will ultimately produce only those isotopes which are not shielded by stable isotopes to their lower right in the chart of the nuclides. In the case of xenon, the unshielded isotopes are at masses 129, 131, 132, 134 and 136. Fission typically produces unequal sized fragments, and ^{136}Xe is near one of the peaks of production for actinide fission (^{129}Xe is far enough from the peak that there is usually too little production to be usefully measured). The other peak is near krypton, but only ^{86}Kr is both close to the production peak and unshielded, so there is no krypton isotopic spectrum that can be used to diagnose actinide fission.

Nucleosynthesis studies. For nucleosynthesis studies, a neutron irradiation, much the same as the ones for ^{40}Ar-^{39}Ar or I-Xe, is done. But the reaction of interest is neutron-introduced fission of ^{235}U. As noted in Table 2, this produces a different isotopic spectrum than ^{244}Pu decay. Since there are four isotopes affected, and only two spectra, it is usually possible to determine the ratio of ^{244}Pu fission to ^{235}U fission and hence, with the help of a monitor, the $^{244}Pu/^{238}U$ ratio (Hudson et al. 1989). The more common uranium isotope, ^{238}U, has a far lower cross-section for neutron-induced fission, but can be used as a reference isotope because $^{235}U/^{238}U$ is constant in all samples measured so far. Natural samples do contain some Xe from spontaneous fission of ^{238}U, but in most meteorites, the amount is negligible compared to the amount of ^{244}Pu fission Xe, since the fission branching ratio of ^{244}Pu is so much higher (Table 1).

For example, Figure 6 shows a three-isotope plot involving ^{130}Xe, ^{134}Xe, and ^{136}Xe. Points are colinear, suggesting a two-component mixture. In this case, the two likely components are trapped Xe (to the upper right) and a fission component (in the lower left). Note that the fission component is neither ^{244}Pu fission nor neutron-induced fission of ^{235}U, but rather a mixture of the two. Knowing the compositions of the two possible fission components, it is possible to determine how much of each isotope is a result of each actinide. If, for example, we determine the relative amount of ^{136}Xe from each, then

$$[^{136}\mathrm{Xe}]_{244} = B_{244} \times y_{244,136} \times [^{244}\mathrm{Pu}] \tag{7a}$$

$$[^{136}\mathrm{Xe}]_{235} = [\mathrm{n}] \times B_{235,\mathrm{n}} \times y_{235,136} \times (^{235}\mathrm{U}/^{238}\mathrm{U}) \times [^{238}\mathrm{U}] \tag{7b}$$

where the B's represent branching ratios (for fission, instead of alpha decay), the y's

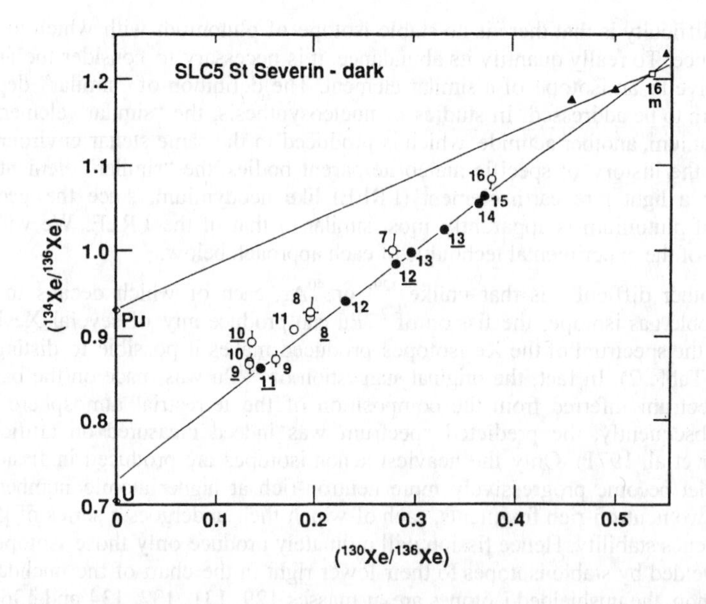

Figure 6. Three-isotope plot for an irradiated sample of the St. Severin meteorite, demonstrating the presence of ^{244}Pu-derived fission Xe. Triangles near the upper right-hand corner represent plausible trapped (fission-free) compositions. "Pu" and "U" represent the fission composition of Xe derived from ^{244}Pu and ^{235}U (for this particular experiment). Labels next to points are temperatures in hundreds of degrees Celsius. Underlined numbers are 50°C higher (e.g., 13 represents 1350°C). [Adapted by permission of the editor of the *Proceedings of the 19th Lunar and Planetary Science Conference*, from Hudson et al. (1989), Fig. 4, p. 551.]

represent the yield of ^{136}Xe in each of the fission processes, [n] is the total neutron fluence, the isotopic values in square brackets represent the absolute amounts of those isotopes, and the U ratio is the measured ratio. Simply taking a ratio of Equation (7a) to (7b), it is clear that the ratio of ^{244}Pu/^{238}U is proportional to the ratio that can be measured.

The initial ^{244}Pu/^{238}U ratio of the solar system is valuable for studies of the history of the material that became a part of the solar system. The most massive stable nucleus is ^{209}Bi, which contains far less nucleons than ^{244}Pu, ^{238}U or ^{235}U. Hence, these actinides can only be synthesized by rapid neutron capture ("r-process"). Their relative rates of production can be predicted from theoretical considerations, so their abundances at the beginning of the solar system can be used to infer the time between cessation of r-process activity and the formation of the solar system (Hudson et al. 1989; Jones 1982). As mentioned above, it is particularly useful when combined with data on ^{129}I, another r-process nuclide (Cameron 1993), and confirms the ~100 Ma timescale suggested by ^{129}I.

Pu-Xe dating. Decay of ^{244}Pu can be used as a chronometer of the first 100 Ma for some specific meteorite parent bodies. Both Pu and the LREE tend to be concentrated in refractory minerals like phosphates. Which LREE is the best proxy for Pu? Various authors have suggested Nd (Lugmair and Marti 1977), Sm (Jones and Burnett 1987), or Pr or Ce (Boynton 1978). There are no neutron-induced reactions that produce a rare gas from any of the LREE, but all of these, particularly Nd, do produce the light xenon isotopes like ^{124}Xe and ^{126}Xe through cosmic-ray-induced spallation reactions (Wieler 2002, this volume). In fact, in many cases, the LREE (and presumably Pu), are probably not fractionated much from each other. Hence, if the cosmic ray dose (i.e., the cosmic ray exposure age) is known, and the production rate of isotopes like ^{124}Xe and ^{126}Xe is also known, then the abundance of the LREE can be calculated. Then the ratio of $^{136}Xe_{244}$ (Pu-derived fission ^{136}Xe) to $[^{126}Xe]_{spall.}$ (LREE-derived spallation ^{126}Xe) is proportional to the ratio of ^{244}Pu to LREE, since

$$[^{126}Xe]_{spall.} = P_{LREE,126} \times [T_{exposure}] \times [LREE] \quad (8)$$

where the P is the production rate of ^{126}Xe from LREE and $T_{exposure}$ is the time of exposure to cosmic rays. Once again, $^{136}Xe_{244}$ is as given in Equation (7a).

Varieties of this technique, Pu-Xe dating, have been applied to igneous meteorites like the eucrites (Marti et al. 1977, 1979; Miura et al. 1998; Shukolyukov and Begemann 1996a), which have enough actinides and LREE to make the measurement possible, and have histories which extend over several hundred Ma, so that the measurement is useful. One of the most interesting results to come from this is the fact that in some cases, the Pu-Xe ages are older than Pb-Pb ages on the same meteorite, suggesting that the Pu-Xe system is more resistant to resetting than the Pb-Pb system (Shukolyukov and Begemann 1996a), consistent with the suggestion that the I-Xe system might be more retentive than Pb-Pb (Gilmour et al. 2000).

The Pu-Xe technique does have its drawbacks. The determination of LREE abundances is dependent on the determination of the cosmic-ray exposure age, although by analyzing Kr at the same time, it is possible to determine precise ^{81}Kr-Kr exposure ages (Miura et al. 1998; Shukolyukov and Begemann 1996b). Another potential problem with the spallation-produced Xe is that Ba also produces Xe isotopes through spallation. Two approaches to dealing with Ba have been tried. Marti et al. (1979) used the isotopic composition of the spallation Xe to determine the Ba/LREE ratio. Other authors (Miura et al. 1998; Shukolyukov and Begemann 1996a) have used measurements of Ba and LREE determined by other techniques on other samples of the same meteorite. More problematical, the correlation of Pu with Nd (or any other LREE) is unlikely to be perfect

(Crozaz et al. 1989). As a result of this, plus the inherent uncertainties resulting from the determination of the correlation, the best uncertainties achieved are roughly 20 Ma.

Other fissioning nuclides?

Only a few years after the presence of ^{244}Pu had been suggested, and well before it had been demonstrated definitively, a different fission-like Xe spectrum had been identified in Renazzo (Reynolds and Turner 1964) and other carbonaceous chondrite meteorites. The isotopic spectrum did not match any known fissioning nuclide, and there were other properties that were also inconsistent with actinide fission. Most prominently, the light Xe isotopes such as ^{124}Xe and ^{126}Xe, which could not be produced by fission, were also enriched. The component was dubbed "carbonaceous chondrite fission" (or "CCF") Xe, even though there were questions from the start about whether it was truly a fission component. There were other actinides that were potential progenitors, such as ^{247}Cm, with a half-life of 15.6 Ma, but these were ruled out on various grounds. If it was a fission component, the most likely progenitor seemed to be a "superheavy" element, with a mass number of roughly 115, lying in a predicted island of stability far beyond the elements which either occur in nature or had been synthesized (Anders et al. 1975). In fact, "CCF" Xe was not a fission component at all, but rather a nucleosynthetic effect, now known as "H"-Xe, contained in pre-solar grains (Ott 2002, this volume). In a classic example of dogged pursuit of a scientific problem, the same laboratory that put together the most convincing story for a superheavy progenitor was also the one that laid that theory to rest after a decade of progressively more detailed studies (Lewis et al. 1987).

Other radiogenic noble gases

Two other short-lived radionuclides may have produced measurable amounts of radiogenic noble gases in meteorites. ^{22}Sodium has a half-life of only 2.6 years (Table 1). Remarkably, there are meteorites that contain evidence of ^{22}Na decay. However, ^{22}Na is a special case—the samples in which evidence of its decay is found are presolar grains, and the decay produces the component known as Ne-E(L) or Ne-R (Ott 2002). So while the ^{22}Na was incorporated into a solid within a few half-lives of synthesis, its existence puts constraints on the formation of grains around supernovae rather than on the timing of any process within the solar system.

Another radionuclide, ^{36}Cl, has a half-life (300 ka) that is not as short as that of ^{22}Na, but is still shorter than ^{129}I or ^{244}Pu. Hence, it would be expected to have been less abundant in the early solar system. Furthermore, it is more difficult to determine its abundance relative to other Cl isotopes. The problem is that while there is a stable chlorine isotope which will produce a noble gas through a neutron capture (^{37}Cl producing ^{38}Ar), the only Ar isotope left that could be used as the denominator in a plot such as Figure 3 is ^{40}Ar, and the change in the ^{36}Ar/^{40}Ar ratio is most likely dominated by K decay rather than Cl decay. Clayton (1977) suggested a way to get around that problem by using only the ^{36}Ar/^{38}Ar ratio and comparing two identical samples, one irradiated and one unirradiated, but this has not been successfully applied.

The only meteorite in which evidence of live ^{36}Cl has been found is Efremovka, a carbonaceous chondrite that has suffered less processing than most meteorites. Murty et al. (1997) have reported the presence of excess ^{36}Ar, presumably from the decay of ^{36}Cl, in a sample of matrix from Efremovka, but no one has successfully measured the ^{36}Cl/^{35}Cl ratio. Murty et al. suggested a ^{36}Cl/^{35}Cl ratio of a little more than 10^{-6}, based on the excess ^{36}Ar and the Cl abundance. Since the half-life of ^{36}Cl is so short, this would be one of the shortest-lived radionuclides yet identified in the solar system. Since evidence of a radionuclide generally disappears within a few half-lives, the very presence of ^{36}Cl would require it to have been synthesized no more than 1 to 2 million years before incorporation into solar system solids. However, another study (Swindle et al. 2001) failed to find any evidence for ^{36}Cl in a similar Efremovka matrix sample, so its existence remains only a tantalizing possibility at this point.

With ^{36}Cl, the roster of potential noble-gas-producing radionuclides is probably exhausted. However, with the exception of ^{36}Cl, studies of radiogenic noble gases have moved from the age of discovery and confirmation to the age of detailed investigation of what they have to say about solar system history.

ACKNOWLEDGMENTS

Constructive reviews and comments by J. Gilmour, D. Garrison, C. Hohenberg, U. Ott and R. Wieler are gratefully acknowledged. This work was supported in part by NASA Grant NAG 5-4767.

REFERENCES

Alexander EC, Lewis RS, Reynolds JH, Michel ML (1971) Plutonium-244: Confirmation as an extinct radioactivity. Science 172:837-840

Anders E, Higuchi H, Gros J, Takahashi H, Morgan JW (1975) Extinct superheavy element in the Allende meteorite. Science 190:1262-1271

Arvidson R, Crozaz G, Drozd R, Hohenberg C, Morgan C (1975) Cosmic ray exposure ages of features and events at the Apollo landing sites. The Moon 13:259-276

Baldwin RB (1974) Was there a "Terminal Lunar Cataclysm" $3.9\text{-}4.0 \times 10^9$ years ago? Icarus 23:157-166

Bogard DD (1979) Chronology of asteroid collisions as recorded in meteorites. *In* Asteroids. Gehrels T (ed) University of Arizona Press, Tucson, p 558-578

Bogard DD (1995) Impact ages of meteorites: A synthesis. Meteoritics 30:244-268

Bogard DD, Hirsch WC (1980) ^{40}Ar/^{39}Ar dating, Ar diffusion properties, and cooling rate determinations of severely shocked chondrites. Geochim Cosmochim Acta 44:1667-1682

Bogard DD, Garrison DH, Norman M, Scott ERD, Keil K (1995) ^{39}Ar/^{40}Ar age and petrology of Chico: Large-scale melting on the L chondrite parent body. Geochim Cosmochim Acta 59:1383-1399

Boynton WV (1978) Fractionation in the solar nebula II. Condensation of Th, U, Pu, and Cm. Earth Planet Sci Lett 40:63-70

Brazzle RH, Pravdivtseva OV, Meshik AP, Hohenberg CM (1999) Verification and interpretation of the I-Xe chronometer. Geochim Cosmochim Acta 63:739-760

Burkland MK, Swindle TD, Baldwin SL (1995) Diffusion studies of the I-Xe system in the meteorite Bjurbole. Geochim Cosmochim Acta 59:2085-2094

Caffee MW, Hohenberg CM, Horz F, Hudson B, Kennedy BM, Podosek FA, Swindle TD (1982) Shock disturbance of the I-Xe system. J Geophys Res 87:A318-A330

Cameron AGW (1993) Nucleosynthesis and star formation. *In* Protostars and Planets III. Levy EH, Lunine JI (eds) University of Arizona, Tucson, p 47-73

Clayton DD (1977) Interstellar potassium and argon. Earth Planet Sci Lett 36:381-390

Clayton DD (1980) Chemical and isotopic fractionation by grain-size separates. Earth Planet Sci Lett 47:199-210

Clayton DD, Dwek E, Woosley SE (1977) Isotopic anomalies and proton irradiation in the early solar system. Astrophys J 214:300-315

Cohen BA, Swindle TD, Kring DA (2000) Support for the lunar cataclysm hypothesis from lunar meteorite impact melt ages. Science 290:1754-1756

Crozaz G, Pellas P, Bourot-Denise M, de Chazal SM, Fieni C, Lundberg LL, Zinner E (1989) Plutonium, uranium and rare earths in the phosphates of ordinary chondrites—the quest for a chronometer. Earth Planet Sci Lett 93:157-169

Culler TS, Becker TA, Muller RA, Renne PR (2000) Lunar impact history from ^{40}Ar/^{39}Ar dating of glass spherules. Science 287:1785-1788

Dalrymple GB, Ryder G (1993) ^{40}Ar/^{39}Ar age spectra of Apollo 15 impact melt rocks by laser step-heating and their bearing on the history of lunar basin formation. J Geophys Res 98:13,085-13,095

Dalrymple GB, Ryder G (1996) Argon-40/argon-39 age spectra of Apollo 17 highlands breccia samples by laser step heating and the age of the Serenitatis basin. J Geophys Res 101:26,069-26,084

Dodd RT (1981) Meteorites, a petrologic-chemical synthesis. Cambridge University Press, New York.

Drake MJ (2001) The eucrite/Vesta story. Meteoritic Planet Sci 36:501-513

Emery JF, Reynolds SA, Wyatt EL, Gleason GI (1972) Half-lives of radionuclides - IV. Nucl Sci Eng 48:319-323

Farley KA (2002) (U-Th)/He dating techniques, calibrations, and applications. Rev Mineral Geochem 47:819-845

Garrison D, Hamlin S, Bogard D (2000) Chlorine abundances in meteorites. Meteoritic Planet Sci 35:419-429

Gilmour JD (2000) The extinct radionuclide timescale of the Solar System. Space Sci Rev 192: 123-132.

Gilmour JD, Saxton JM (2001) A time-scale of formation of the first solids. Phil Trans R Soc London A 359:2037-2048

Gilmour JD, Whitby JA, Turner G, Bridges JC, Hutchison R (2000) The iodine-xenon system in clasts and chondrules from ordinary chondrites: implications for early Solar System chronology. Meteoritic Planet Sci 35:445-455

Gilmour JD, Whitby JA, Turner G (2001) Negative correlation of iodine-129/iodine-127 and xenon-129/xenon-132: product of closed-system evolution or evidence of a mixed component. Meteoritic Planet Sci 36:1283-1286

Göpel C, Manhes G, Allègre CJ (1991) Constraints on the time-scale of accretion and thermal evolution of chondrite parent bodies by precise U-Pb dating of phosphates. Meteoritics 26:338

Göpel C, Manhes G, Allègre CJ (1994) U-Pb systematics of phosphates from equilibrated ordinary chondrites. Earth Planet Sci Lett 121:153-171

Hartmann WK (1975) Lunar "cataclysm": A misconception. Icarus 24:181-187

Hewins RH, Jones RH, Scott ERD (eds) (1996) Chondrules and the protoplanetary disk. Cambridge University Press, New York.

Heymann D (1967) On the origin of hypersthene chondrites: Ages and shock effects of black chondrites. Icarus 6:189-221

Hohenberg CM, Kennedy BM (1981) I-Xe dating: inter-comparisons of neutron irradiations and reproducibility of the Bjurbole standard. Geochim Cosmochim Acta 45:251-256

Hohenberg CM, Podosek FA, Reynolds JH (1967) Xenon-iodine dating: Sharp isochronism in chondrites. Science 156:233-236

Hohenberg CM, Hudson B, Kennedy BM, Podosek FA (1981) Noble gas retention chronologies for the St. Severin meteorite. Geochim Cosmochim Acta 45:535-546

Hohenberg CM, Pravdivtseva O, Meshik A (2000) Reexamination of anomalous I-Xe ages: Orgueil and Murchison magnetites and Allegan feldspar. Geochim Cosmochim Acta 64:4257-4262

Holden NE (1990) Total half-lives for selected nuclides. Pure Appl Chem 62:942-958.

Hudson GB, Kennedy BM, Podosek FA, Hohenberg CM (1989) The early solar system abundance of ^{244}Pu as inferred from the St. Severin chondrite. Proc Lunar Sci Conf 19:547-557

Hyde EK (1971) The nuclear properties of the heavy elements. Prentice-Hall, Englewood Cliffs, New Jersey

Irwin JJ, Reynolds JH (1995) Multiple stages of fluid trapping in the Stripa granite indicated by laser microprobe analysis of Cl, Br, I, K, U and nucleogenic plus radiogenic Ar, Kr, and Xe in fluid inclusions. Geochim Cosmochim Acta 59:355-369

Irwin JJ, Roedder E (1995) Diverse origins of fluid in magmatic inclusions at Bingham (Utah, USA), Butte (Montana, USA), St. Austell (Cornwall, UK), and Ascension Island (mid-Atlantic, UK), indicated by laser microprobe analysis of Cl, K, Br, I, Ba+Te, U, Ar, Kr, and Xe. Geochim Cosmochim Acta 59:295-312

Jones JH (1982) The geochemical coherence of Pu and Nd and the ^{244}Pu/^{238}U ratio of the early solar system. Geochim Cosmochim Acta 46:1793-1804

Jones JH, Burnett DS (1987) Experimental geochemistry of Pu and Sm and the thermodynamics of trace element partitioning. Geochim Cosmochim Acta 51:769-782

Jordan J, Kirsten T, Richter H (1980) ^{129}I/^{127}I: A puzzling early solar system chronometer. Z Naturforsch 35a:145-170

Katcoff S, Schaeffer OA, Hastings JM (1951) Half-life of I^{129} and the age of the elements. Phys Rev 82:688-690

Kelley S (2002) K-Ar and Ar-Ar dating. Rev Mineral Geochem 47:785-818

King EA. (ed) (1983) Chondrules and their origins. Lunar and Planetary Institute, Houston, Texas

Kirschbaum C (1988) Carrier phases for iodine in the Allende meteorite and their associated ^{129}Xe$_r$/^{127}I ratios: A laser microprobe study. Geochim Cosmochim Acta 52:679-699

Kring DA, Swindle TD, Britt DT, Grier JA (1996) Cat Mountain: A meteoritic sample of an impact-melted asteroid regolith. J Geophys Res 101:29,353-329,371

Krot AN, Brearley AJ, Ulyanov AA, Biryukov VV, Swindle TD, Keil K, Mittlefehldt DW, Scott ERD, Clayton RN, Mayeda TK (1999) Mineralogy, petrography, bulk chemical, iodine-xenon, and oxygen-isotopic compositions of dark inclusions in the reduced CV3 chondrite Efremovka. Meteoritic Planet Sci 34:67-89

Kuroda PK (1960) Nuclear fission in the early history of the earth. Nature 187:36-40

Lewis RS (1975) Rare gases in separated whitlockite from the St. Severin chondrite: Xenon and krypton from fission of extinct ^{244}Pu. Geochim Cosmochim Acta 39:417-432

Lewis RS, Ming T, Wacker JF, Steele IM (1987) Interstellar diamonds in meteorites. Nature 326:160-162

Lipschutz ME, Gaffey ME, Pellas P (1989) Meteorite parent bodies: nature, number, size and relation to present-day asteroids. *In* Asteroids II. Binzel RP, Gehrels T, Matthews MS (eds) University of Arizona, Tucson, p 740-788

Lugmair GW, Marti K (1977) Sm-Nd-Pu timepieces in the Angra dos Reis meteorite. Earth Planet Sci Lett 35:273-284

MacPherson GJ, Wark DA, Armstrong JT (1988) Primitive material surviving in chondrites: refractory inclusions. *In* Meteorites and the Early Solar System. Kerridge JF, Matthews MS (eds) University of Arizona Press, Tucson, p 746-817

Marti K, Lugmair GW, Scheinin NB (1977) Sm-Nd-Pu systematics in the early solar system. Lunar Planet Sci VIII:619-621

Marti K, Kurtz JP, Regnier S (1979) Pu-Nd-Xe dating: a stepwise approach. Meteoritics 14:482-483

McConville PS, Kelley S, Turner G (1988) Laser probe ^{40}Ar-^{39}Ar studies of the Peace River shocked L6 chondrite. Geochim Cosmochim Acta 52:2487-2499

McDougall I, Harrison TM (1999) Geochronology and thermochronology by the ^{40}Ar/^{39}Ar method. Oxford University Press, New York.

Megrue GH (1973) Spatial distribution of ^{40}Ar/^{30}Ar ages in lunar breccia 14301. J Geophys Res 78:3216-3221

Melosh HJ (1989) Impact cratering: a geologic process. Oxford University Press, New York.

Miura YN, Nagao K, Sugiura N, Fujitani T, Warren PH (1998) Noble gases, ^{81}Kr-Kr exposure ages and ^{244}Pu-Xe ages of six eucrites, Bereba, Binda, Camel Donga, Juvinas, Millbillillie, and Stannern. Geochim Cosmochim Acta 62:2369-2387

Murty SVS, Goswami JN, Shukolyukov YA (1997) Excess ^{36}Ar in the Efremovka meteorite: a strong hint for the presence of ^{36}Cl in the solar system. Astrophys J 475:L65-L68

Nichols RHJr, Hohenberg CM, Kehm K, Kim Y, Marti K (1994) I-Xe studies of the Acapulco meteorite: absolute I-Xe ages of individual phosphate grains and the Bjurbole standard. Geochim Cosmochim Acta 58:2553-2561

Niemeyer S (1980) I-Xe and ^{40}Ar-^{39}Ar dating of silicate from Weekeroo Station and Netschaevo IIE iron meteorites. Geochim Cosmochim Acta 44:33-44

Ott U (2002) Noble gases in meteorites—Trapped components. Rev Mineral Geochem 47:71-100

Pellas P, Störzer D (1981) ^{244}Pu fission track thermometry and its applications to stony meteorites. Proc R Soc London A 324:253-270

Reynolds JH, Turner G (1964) Rare gases in the chondrite Renazzo. J Geophys Res 69:3263-3281

Ryder G (1990) Lunar samples, lunar accretion and the early bombardment of the Moon. EOS, Trans Am Geophys Union 71:322-323

Shukolyukov A, Begemann F (1996a) Pu-Xe dating of eucrites. Geochim Cosmochim Acta 60:2453-2471

Shukolyukov A, Begemann F (1996b) Cosmogenic and fissiogenic noble gases and ^{81}Kr-Kr exposure age clusters of eucrites. Meteoritic Planet Sci 31:60-72

Stöffler D, Ryder G (2001) Stratigraphy and isotope ages of lunar geologic units: chronological standard for the inner solar system. Space Sci Rev 96:9-54

Strom RG, Croft SK, Barlow NG (1992) The Martian impact cratering record. *In* Mars. Kieffer HH, Jakosky BM, Snyder CW, Matthews MS (eds) University of Arizona Press, Tucson, p 383-423

Swindle TD (1998) Implications of iodine-xenon studies for the timing and location of secondary alteration. Meteoritic Planet Sci 33:1147-1155

Swindle TD (2002) Martian noble gases. Rev Mineral Geochem 47:171-190

Swindle TD, Podosek FA (1988) Iodine-xenon dating. *In* Meteorites and the Early Solar System. Kerridge JF, Matthews MS (eds) University of Arizona Press, Tucson, p 1127-1146

Swindle TD, Grossman JN, Garrison DH, Olinger CT (1991a) Iodine-xenon, chemical and petrographic studies of Semarkona chondrules. Geochim Cosmochim Acta 55:3723-3734

Swindle TD, Caffee MW, Hohenberg CM, Lindstrom MM, Taylor GJ (1991b) Iodine-xenon studies of petrographically and chemically characterized Chainpur chondrules. Geochim Cosmochim Acta 55:861-880

Swindle TD, Davis AM, Hohenberg CM, MacPherson GJ, Nyquist LE (1996) Formation times of chondrules and Ca-Al-rich inclusions: Constraints from short-lived radionuclides. *In* Chondrules and the protoplanetary disk. Hewins RH, Jones RH, Scott ERD (eds) Cambridge University Press, New York, p 77-86

Swindle TD, Olson EK, Bart GB (2001) Searching for evidence of extinct ^{36}Cl in Efremovka. Meteoritic Planet Sci 36:A201
Taylor GJ (1991) Impact melts in the MAC 88105 lunar meteorite: Inferences for the lunar magma ocean hypothesis and the diversity of basaltic impact melts. Geochim Cosmochim Acta 55:3031-3036
Taylor GJ, Maggiore P, Scott ERD, Rubin AF, Keil K (1987) Original structures, and fragmentation and reassembly histories of asteroids: Evidence from meteorites. Icarus 69:1-13
Tera F, Papanastassiou DA, Wasserburg GJ (1974) Isotopic evidence for a terminal lunar cataclysm. Earth Planet Sci Lett 22:1-21
Turner G (1965) Extinct iodine 129 and trace elements in chondrites. J Geophys Res 70:5433-5445
Turner G, Cadogan PH, Yonge CJ (1973) Argon selenochronology. Proc Lunar Sci Conf 4th, p 1889-1914
Turner G, Enright MC, Cadogan PH (1978) The early history of chondrite parent bodies inferred from ^{40}Ar-^{39}Ar ages. Proc Lunar Sci Conf 9th, p 989-1025
Turner G, Gilmour JD, Whitby JA (2000) High iodine-129/iodine-127 ratios in primary and secondary minerals: Chronology or fluid processes. Meteoritic Planet Sci 35:A160-A161
Wasson JT (1985) Meteorites: Their record of early solar system history. W. H. Freeman, New York.
Wasson JT, Wang S (1991) The histories of ordinary chondrite parent bodies: U,Th-He age distributions. Meteoritics 26:161-167
Wetherill GW (1953) Spontaneous fission yields from uranium and thorium. Phys Rev 92:907-912
Whitby JA, Burgess R, Turner G, Gilmour JD, Bridges J (2000) Extinct ^{129}I in halite from a primitive meteorite: Evidence for evaporite formation in the early solar system. Science 288:1819-1821
Whitby JA, Gilmour JD, Turner G, Prinz M, Ash RD (2002) I-Xe dating of chondrules from the Qingzhen and Kota Kota enstatite chondrites. Geochim Cosmochim Acta 66:347-359
Wieler R (2002) Cosmic-ray-produced noble gases in meteorites. Rev Mineral Geochem 47:125-170
Zähringer J (1968) Rare gases in stony meteorites. Geochim Cosmochim Acta 32:209-237

5 Cosmic-Ray-Produced Noble Gases in Meteorites

Rainer Wieler
Isotope Geology, ETH-Zürich
Sonneggstrasse 5
CH-8092 Zürich, Switzerland
wieler@2erdw.ethz.ch

INTRODUCTION

Energetic particles of the galactic cosmic radiation (GCR) have a mean penetration depth in rock of about 50 cm, comparable to the typical size of a meteorite. GCR-induced effects therefore provide a means to study the history of meteorites as small objects in space or in the top few meters of their parent body. These effects include cosmic ray tracks, i.e., the radiation damage trails in a crystal lattice produced by heavy ions in the GCR (Fleischer et al. 1975), and thermoluminescence, i.e., the light emitted by a heated sample which had been irradiated by energetic particles (Benoit and Sears 1997). By far most important, however, are "cosmogenic" nuclides, produced by interactions of primary and secondary cosmic ray particles with target atoms.

This chapter concentrates on the cosmogenic noble gas nuclides in meteorites. A separate chapter in this book by Niedermann (2002) discusses cosmogenic noble gases in terrestrial rocks, which are by now a major tool in quantitative geomorphology. Cosmogenic noble gases in lunar samples are briefly presented in the chapter by Wieler (2002) on noble gases in the solar system. Here and in the other chapters mentioned, cosmogenic radionuclides such as ^{10}Be, ^{26}Al, or ^{36}Cl are often also considered, as comprehensive studies usually require combining noble gas and radionuclide analyses. Not further discussed here are cosmic-ray-induced shifts of the abundances of stable isotopes of a few elements besides noble gases. The most important of these are Gd, Sm, and Cd, which have isotopes with extremely high cross sections for the capture of slow (thermal or epithermal) neutrons produced as secondary cosmic ray particles by interactions of the primary GCR protons with target atoms (Eugster et al. 1970; Hidaka et al. 1999; Sands et al. 2001). Because the flux of thermal and epithermal neutrons peaks at greater depths than that of more energetic particles, neutron-induced shifts are mainly produced at relatively high shielding and are thus particularly useful to study the exposure history of large meteorites or a previous exposure stage near the surface of a meteorite parent body (see *Cosmogenic noble gases produced by capture of low-energy neutrons* subsection).

Sometimes, cosmic-ray-induced isotope shifts are also a nuisance, because they may compromise the precise determinations of isotopic compositions of elements of interest. This is often a major problem in the case of "trapped" noble gases (Ott 2002, this volume), but may also require attention when e.g., small excesses or deficits of daughter isotopes of an extinct radionuclide present in the early solar system are to be determined (e.g., Leya et al. 2000c).

In the following, we will first discuss the production systematics of cosmogenic nuclides in meteorites. Next, we will present the distributions of cosmic ray exposure ages of different meteorite classes and discuss what they tell us about meteorite transport from the various parent bodies to Earth. We will also consider so-called "complex" or "pre-exposure" histories, which is when a meteorite also contains cosmogenic nuclides acquired before it became the body with the size and shape it had immediately before it fell to Earth. Some of these meteorites represent material that once was within the top

1529-6466/00/0047-0005$05.00

few meters of the surface of their parent body and thus contain information on asteroidal or planetary surface processes. Other meteorites with complex exposure histories may even record an intense irradiation by energetic particles from the early Sun. We will also see how cosmogenic nuclides are used to study the history of the particles that produced them, e.g., possible temporal variations in the fluxes of the cosmic radiation. In the last section, cosmogenic noble gases produced by energetic particles from the Sun (in rather recent times) are briefly mentioned. Earlier reviews on cosmogenic nuclides in extraterrestrial samples include Anders (1962), Reedy et al. (1983), Vogt et al. (1990), Marti and Graf (1992), and chapters 6 and 9 in Tuniz et al. (1998).

THE PRODUCTION OF COSMOGENIC NUCLIDES IN METEORITES

Fundamentals

In this section we discuss how cosmogenic nuclides are produced in solid matter in space and how production rates can be determined for a given sample. The major challenge is that production rates depend on the size of a preatmospheric meteorite (meteoroid) and the position of a given sample within this body, collectively somewhat loosely called the "shielding" of the sample. Since a meteoroid may lose well over 90% of its preatmospheric mass when passing through the Earth's atmosphere, and moreover may disintegrate into many fragments, the shielding is a priori unknown and has to be assumed or deduced by some means.

The shielding dependence of production rates is illustrated in Figure 1, which shows the production rates of ^{21}Ne, P(^{21}Ne), in chondrites and iron meteorites, as well as the

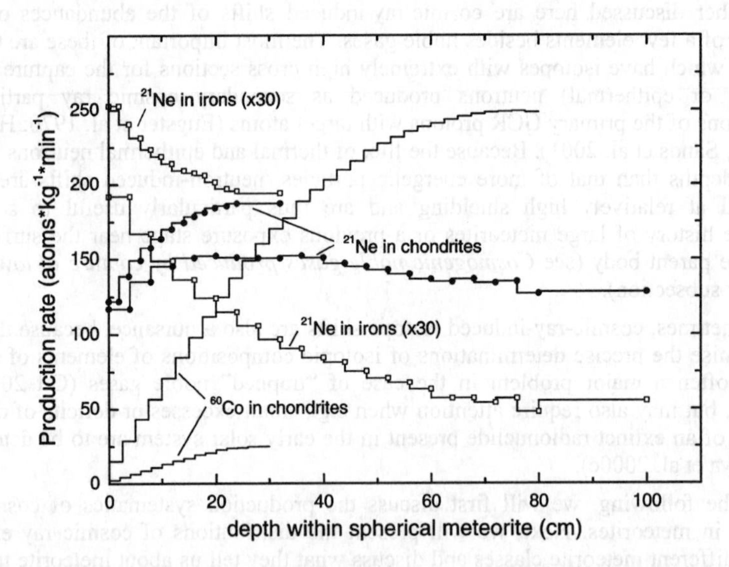

Figure 1. Production rates by galactic cosmic rays of ^{60}Co in chondrites and ^{21}Ne in chondrites and iron meteorites. Two curves are shown for each nuclide-target combination, representing spherical meteorites of two different radii (30 and 80 cm for ^{60}Co, 30 and 100 cm for ^{21}Ne). P(^{21}Ne) in iron meteorites has been multiplied by a factor of 30. The ^{60}Co curves are based on the nuclide production model by Masarik and Reedy (1994), assuming a Co concentration of 690 μg/g, and the ^{21}Ne curves are calculated with the model by Leya et al. (2000a). The figure illustrates that depth-and size-dependencies of production rates may be highly dissimilar for different nuclides or different nuclide-target combinations.

production rate of the radionuclide ^{60}Co (half-life 5.3 yr) in chondrites. The two curves for each nuclide correspond to spherical meteorites with two different radii (equal to the largest depth shown) and the production rates are calculated by models to be discussed in more detail below (Leya et al. 2000a; Masarik and Reedy 1994). Cosmogenic nuclides in meteorites are produced by protons and alpha-particles of the primary GCR (see also Niedermann 2002) as well as the resulting cascade of secondary protons and neutrons that develops within the meteoroid. The relative contributions from primaries and secondaries can be highly different for various target-nuclide combinations, which is the reason for the widely different shapes of the various curves. The production rate of ^{21}Ne in iron meteorites is highest at the surface and decreases more or less exponentially with depth. This is because ^{21}Ne from iron is a so-called "high-energy product," made preferentially by high energy particles (mostly primary protons), because the mass difference between the target elements (Fe and Ni) and the product ^{21}Ne is large. In contrast, P(^{21}Ne) in chondrites increases from the surface up to ~20- to 40-cm depth, and then decreases less steeply than in iron meteorites. This increase is due to a large contribution by secondary neutrons and protons with lower energies than primary protons, as there is a relatively small mass difference between ^{21}Ne and the main target elements for this nuclide in rocks (Mg, Al, and Si). The much higher absolute production rates in chondrites compared to iron meteorites (whose ordinate values in Figure 1 have been multiplied by a factor of 30) illustrate that production rates also strongly depend on the elemental composition of the target. Generally, production rates are higher when the mass difference between target and product is small, e.g., P(^{21}Ne) from Mg is ~2-3 times higher than P(^{21}Ne) from Si and several ten times higher than P(^{21}Ne) from Fe. Note also that production rates at the surface are higher for smaller bodies (especially smaller iron meteorites) because they see a higher flux of particles from the "opposite" side of the meteoroid.

Whereas the increase of the ^{21}Ne production rate with depth near the surface is modest, and P(^{21}Ne) is somewhat smaller at all depths in the larger body in Figure 1 compared to the smaller one, the variations of P(^{60}Co) with meteoroid size and sample depth are dramatic, e.g., P(^{60}Co) in the center of the R = 80 cm meteoroid is about 10 times higher than in the center of the R = 30 cm body. This is because this nuclide is mostly produced when ^{59}Co captures secondary cosmic ray neutrons that had been slowed down to more or less thermal energies. The flux of these thermal neutrons increases with depth and is highest in the center of a meteorite of roughly one meter radius. Cosmogenic nuclide production by neutron-capture is further discussed in a separate section below.

In summary, Figure 1 illustrates that it is far from trivial to obtain a precise set of production rates for a given sample. Luckily, the preatmospheric radius of stony meteorites rarely exceeded a meter and mostly was less than 40 cm or so. For such bodies, P(^{21}Ne) does not change by more than a factor of ~2 and ~1.5, respectively, at all depths (see also Fig. 4). Hence, an approximate exposure age can normally be estimated even when shielding information is lacking. As an uncertainty of the order of 50% usually is not satisfying, however, either a correction for shielding or a method that largely self-corrects for shielding is required. This will be discussed in the next section. Here we end with an introduction of the most important shielding-sensitive ratios of pairs of cosmogenic noble gas nuclides.

Shielding parameters. The most widely used shielding parameter for stony meteorites is the ratio ^{22}Ne/^{21}Ne of the cosmogenic component (note that we will always talk here about the cosmic-ray-produced fraction of a nuclide, which often has been obtained from measured values after substantial correction for other contributions, such as primordial noble gases; see *Isotopic abundances of cosmogenic noble gases* subsection). Although P(^{22}Ne) also increases with depth near the surface, it increases somewhat less than P(^{21}Ne), because with a higher flux of secondary neutrons ^{21}Ne production by the reaction ^{24}Mg(n,α)^{21}Ne becomes more important. Therefore, ^{22}Ne/^{21}Ne is largest at the surface of a meteoroid. This is shown in Figure 6, which will be discussed in more detail below. This ratio is also larger in the center of a small meteoroid than in the center of a larger one with the same (relative) abundances of the main target elements producing cosmogenic Ne (mainly Mg, Al, and Si). The ^{22}Ne/^{21}Ne ratio can therefore be used to correct P(^{21}Ne) for shielding (Nishiizumi et al. 1980; Eugster 1988). However, we will see in the next section that this is possible only within certain limits, as the relation between P(^{21}Ne) and ^{22}Ne/^{21}Ne becomes ambiguous at larger shielding. The ratio ^{22}Ne/^{21}Ne also depends somewhat on the Mg/(Mg + Al + Si) ratio. Another often measured shielding-sensitive ratio is ^{3}He/^{21}Ne, which in chondrites positively correlates with ^{22}Ne/^{21}Ne (Eberhardt et al. 1966). However, as cosmogenic ^{3}He and its radioactive precursor ^{3}H ($T_{1/2}$ = 12.3 yr) are prone to diffusive loss, the ^{3}He/^{21}Ne ratio is actually often used instead to recognize meteoroids that may have been heated, e.g., on orbits that brought them relatively close to the Sun. Other shielding indicators in stony meteorites are the ratios ^{78}Kr/^{83}Kr (Eugster et al. 1969; Regnier et al. 1979; Eugster 1988; Lavielle and Marti 1988; Lavielle et al. 1997), and ^{124,126,128}Xe/^{131}Xe (Eugster and Michel 1995; Lavielle et al. 1997), which all correlate positively with ^{22}Ne/^{21}Ne and which also correlate well among one another. The Kr and Xe ratios may be useful in cases where (^{22}Ne/^{21}Ne)$_{cos}$ cannot be determined. In iron meteorites, ^{4}He/^{21}Ne is a useful shielding parameter (Signer and Nier 1960, 1962; Voshage 1978, 1984).

Production systematics

Production systematics of cosmogenic nuclides are studied by different approaches:

(1) survey analyses of many meteorites, e.g., of a specific class;

(2) systematic analyses of several samples from the same meteorite, possibly from different fragments;

(3) nuclide production models that are based on meteorite data, thin-target cross section analyses, or experiments that simulate the cosmic radiation by bombarding thick targets with energetic protons.

In all such studies, multi-nuclide analyses are the rule, and the various approaches are often also combined. Note that a commonly used simplification is to assume that meteoroids had spherical shapes.

Multi-meteorite studies. Perhaps the most straightforward way to determine "average" production rates for meteorites with a certain chemical composition is to analyse a radioactive-stable nuclide pair in a number of meteorites (Herzog and Anders 1971; Nishiizumi et al. 1980; Moniot et al. 1983). Some of these need to have exposure ages comparable to the half-life of the involved radionuclide. Others must have been exposed for much longer, so that the radionuclide activity (number of decays per unit of mass per unit of time) is saturated, i.e., for each freshly made nuclide on average one produced earlier decays. Preferentially, the production rates of the radioactive and stable nuclide should show a similar shielding dependence. Frequently used pairs are ^{26}Al-^{21}Ne, ^{10}Be-^{21}Ne, and ^{53}Mn-^{21}Ne. If all the samples studied are from average-sized meteorites, their radionuclide activities will all have similar values at saturation, and the mean saturation activity is equal to the mean production rate of a radionuclide in the subject meteorite class. The exposure ages of the meteorites in which the radionuclide has not yet reached its saturation level, and hence also the production rate of the stable nuclide can then be determined according to the following equations:

$$T_{exp} = \frac{1}{\lambda} \ln \frac{A_{sat}}{A_{sat} - A_{meas}}, \text{ where } A_{sat} > A_{meas} \quad (1)$$

$$P_{stable} = \frac{C_{stable}}{T_{exp}} \quad (2)$$

Here T_{exp} is the exposure age, λ is the decay constant of the radionuclide, and A_{sat} and A_{meas} are the saturation activity and the measured activity of the radionuclide at the time of fall, respectively. P_{stable} and C_{stable} are the production rate and the measured concentration of the stable nuclide, respectively. This method requires that all meteorites considered had a simple exposure history. The production rates deduced in this way are basically the mean values over the last few half-lives of the radionuclide. We will discuss in section *The Cosmic Ray Flux in Time* that this approach can be used to study possible variations of the GCR flux with time.

A further example of the multi-meteorite approach is the work by Eugster (1988), who presents production rates for cosmogenic ^{3}He, ^{21}Ne, ^{38}Ar, ^{83}Kr, and ^{126}Xe as a function of ^{22}Ne/^{21}Ne for various chondrite classes. The data are based on noble gas analyses of a considerable number of ordinary chondrites whose exposure ages were determined by the ^{81}Kr-Kr method, which is largely self-correcting for shielding (see the respective subsection below). The production rate equations and the correction factors F for different chemical compositions are given in Table 1. These factors are based on earlier work in which the relative production rates of each nuclide from various elements or element combinations had been determined either by analysing different minerals from the same meteorite (e.g., Bogard and Cressy 1973) or from model calculations that are discussed below (e.g., Hohenberg et al. 1978). The Eugster formalism has the virtue of yielding a unique production rate for a given value of the routinely determined shielding parameter ^{22}Ne/^{21}Ne. However, as the actual relationship between ^{22}Ne/^{21}Ne and production rate is ambiguous for relatively high shielding, for chondrites the formulas in Table 1 should only be applied for ^{22}Ne/^{21}Ne ratios larger than about 1.08-1.10. They become unreliable for ^{22}Ne/^{21}Ne ratios below about 1.07, as they imply increasing production to very large depth, which is clearly not the case (these limiting ^{22}Ne/^{21}Ne ratios are slightly different for meteorites with higher or lower than chondritic Mg/(Mg + Al + Si) ratios). Furthermore, for very large shielding above some 50 cm or so the trend of ^{22}Ne/^{21}Ne with shielding reverses, as is shown by both model calculations

Table 1. Shielding-corrected production rates in chondrites (Eugster 1988)

	F factors for various chemical classes							
	CI	*CM*	*CO*	*CV*	*H*	*L/LL*	*EH*	*EL*
$P(^{3}He)=F(2.09-0.43\times^{22}Ne/^{21}Ne)$	1.01	1.00	0.99	0.99	0.98	1.00	0.97	1.00
$P(^{21}Ne)=1.61F/(21.77\times^{22}Ne/^{21}Ne-19.32)$	0.67	0.79	0.96	0.96	0.93	1.00	0.78	0.96
$P(^{38}Ar)=F(0.112-0.063\times^{22}Ne/^{21}Ne)$	0.75	0.88	1.03	1.10	1.08	1.00	0.98	0.89
$P(^{83}Kr)=0.0196F/(0.62\times^{22}Ne/^{21}Ne-0.53)$	0.71	0.94	1.02	1.13	1.00	1.00	0.75	0.80
$P(^{126}Xe)=F(0.0174-0.0094\times^{22}Ne/^{21}Ne)$	0.66	0.93	1.18	1.40	1.00	1.00	0.72	0.72

Production rates P in [10^{-8} cc STP/(g × Myr)] for He, Ne and Ar, and in [10^{-12} cc STP/(g × Myr)] for Kr and Xe (1 cc STP=2.687 × 10^{19} atoms). All values from Eugster (1988), except P(^{38}Ar) for which a ~12% lower value than that given by Eugster (1988) is adopted (Schultz et al. 1991; Graf and Marti 1995). ^{22}Ne/^{21}Ne is the ratio of the cosmogenic component. The shielding correction proposed here is approximately valid only for ^{22}Ne/^{21}Ne > 1.08-1.10, and may fail completely for the rare very large meteorites (see text).

and analyses of the very large Gold Basin meteorite (Masarik et al. 2001; Wieler et al. 2000b; see following subsection). However, such large meteorites are probably very rare, and the $^{22}Ne/^{21}Ne$ shielding parameter in practice is useful in a majority of cases. Note also that more recent studies indicate that P(^{38}Ar) by Eugster (1988) appears to be about 12% too high (Schultz et al. 1991; Graf and Marti 1995). In Table 1 we adopted the lower P(^{38}Ar) values, although the issue is not settled (Eugster et al. 1998b).

Systematic studies on single meteorites and semiempirical models. An example of the second approach to determine production systematics is the study of cosmogenic He, Ne, and Ar, some radionuclides and nuclear tracks in the large chondrite Knyahinya (Graf et al. 1990b) along with follow-up studies that included cosmogenic Kr and Xe (Lavielle et al. 1997) as well as further radionuclides (Reedy et al. 1993; Jull et al. 1994). The main mass of Knyahinya was split in two almost equally sized fragments upon its fall. Fortunately the cut passed very close to the preatmospheric center of the meteorite and the slab included an edge where only a very few cm has been ablated. This happy accident allowed the determination of concentration gradients of the various cosmogenic nuclides as a function of preatmospheric depth. The data were also used to determine the parameters of a semiempiric model of cosmogenic nuclide production (Graf et al. 1990a). This model, based on earlier work by Signer and Nier (1960, 1962) and others, describes production of each nuclide by two depth- and size-dependent terms with only one free parameter each, both of which are independent of shielding. Based on the Knyahinya data and an independently determined exposure age of this meteorite, the model is quite successful in predicting cosmogenic noble gas and radionuclide production rates as a function of depth in chondrites of different sizes (see also figures in Wieler et al. 1996). Similar semiempirical models have been used to describe nuclide production in stony-iron or iron meteorites (e.g., Honda 1988). Voshage (1978, 1984) combined noble gas data with determinations of the very long-lived cosmogenic radionuclide ^{40}K ($T_{1/2}$ = 1.25 Gyr) in iron meteorites (see also *The Cosmic Ray Flux in Time* section).

Many further comprehensive studies on single meteorites have either contributed to our understanding of cosmogenic nuclide production systematics or the exposure histories of the studied meteorites. Some examples are the investigations on Grant (Signer and Nier 1960, 1962; Graf et al. 1987), Keyes (Wright et al. 1973), St. Severin (Schultz and Signer 1976; Lavielle and Marti 1988), Jilin (Begemann et al. 1996; Heusser et al. 1996 and references therein), Chico (Garrison et al. 1992), Bur Gheluai (Vogt et al. 1993); Canyon Diablo (Heymann et al. 1966; Michlovich et al. 1994), and Gold Basin (Kring et al. 2001; Wieler et al. 2000b).

Physical models. Much progress has also been made in modeling cosmogenic nuclide production in meteorites based essentially only on physical principles and without using any free parameters except for the absolute cosmic ray flux. These models simulate all the processes relevant for particle production and transport using Monte Carlo techniques (Masarik and Reedy 1994; Masarik et al. 2001; Leya et al. 2000a, Leya et al. 2001b and references therein). The basic equation underlying these models is (e.g., Leya et al. 2000a).

$$P_j(R,d,M) = \sum_{i=1}^{N} c_i \frac{N_A}{A_i} \sum_{k=1}^{3} \int_0^{\infty} \sigma_{j,i,k}(E) \times J_k(E,R,d,M)dE \tag{3}$$

where $P_j(R,d,M)$ is the production rate of nuclide j as a function of the meteoroid radius R, sample depth d, and the solar modulation parameter M (this parameter is a measure for the reduction of the GCR particle flux by the Sun's magnetic field; see also Niedermann 2002). The first sum goes over all target elements i, while the index k in the second sum represents the reaction particle type (primary or secondary proton, secondary neutron). N_A is Avogadro's number, A_i the mass number (in amu) of the target element i, c_i the abundance of element i (in g/g). $\sigma_{j,i,k}(E)$ is the excitation function for the production of nuclide j from element i by reactions induced by particles of type k, and $J_k(E,R,d,M)$ is the differential flux density of particle type k, which depends on the energy E of the reacting particles (as well as on R, d, and M). Note that the primary GCR intensity is assumed to be constant in time (see also *The Cosmic Ray Flux in Time* section). Note also that the models by Leya and coworkers and Masarik and coworkers take into account primary and secondary alpha particles only approximatively, by multiplying production rates from Equation (3) by a factor of 1.55 (the 12% alpha particles in the GCR have 55% as much mass—and hence energy—as the 87% GCR protons).

Equation (3) shows that the most critical ingredients for such models are the numerous excitation functions needed. The data base for proton-induced reactions is fairly complete for most relevant noble gas nuclides as well as for most of the important radionuclides. An important exception is ^{36,38}Ar production from Ca, for which cross sections are still based only on theoretical nuclear models (e.g., Blann 1971). Measured cross sections for neutron-induced reactions are, however, very scarce (e.g., Sisterson et al. 2001). This is a serious problem because secondary neutrons usually dominate the cosmogenic nuclide production. Leya (1997) and Leya and Michel (1997) therefore derived excitation functions for neutron-induced reactions with the data from five thick-target simulation experiments, where stony or iron spheres of various radii and filled with a large number of target foils were isotropically irradiated by energetic protons (Michel et al. 1986; Leya et al. 2000b and references therein). Neutron excitation functions were derived for all reactions where the proton-induced production could be reliably calculated, based on their known excitation functions, and so subtracted.

Figures 2-6 present some important results from the models by Leya et al. (2000a) and Masarik et al. (2001). Figure 2 shows that both models do reproduce the measured ^{21}Ne depth profile in Knyahinya well, and hence can be expected to reliably predict nuclide production in meteorites of a wide range of sizes. Remarkably, secondary neutrons contribute about two thirds to the total ^{21}Ne production at the surface and this fraction increases to 85% near the center. This illustrates the importance of reliable neutron cross section data. Figures 3 and 4 show the ^{3}He and ^{21}Ne production rates, respectively, in the two most abundant meteorite classes, the H and L chondrites, as a function of depth and size. As noted above, for average-sized meteorites ($R < 40$ cm), production rates vary within only about a factor of 1.5. On the other hand, for the Gold Basin chondrite with its preatmospheric radius of perhaps 3 m, nuclide concentrations vary by more than an order of magnitude (Kring et al. 2001; Wieler et al. 2000b). This meteorite is almost represented by the lines denoting an infinite radius (2π).

Figure 5 shows P(^{21}Ne) values versus the ^{22}Ne/^{21}Ne ratio for objects of various sizes and samples of various depths according to calculations by Leya et al. (2001a). Also shown is the empirical curve according to Eugster (1988) that is given in Table 1. The Figure shows that the relation between P(^{21}Ne) and ^{22}Ne/^{21}Ne is more or less unique only for ^{22}Ne/^{21}Ne ≥ 1.13, i.e., for meteorites with radii less than some 15 cm or for near-surface samples of somewhat larger bodies. In more heavily shielded samples, P(^{21}Ne) may differ by up to a factor of 2 at a given ^{22}Ne/^{21}Ne. However, this large spread is only observed for objects with radii above ~85 cm (filled symbols), which fortunately are rare. For more common sizes and mean values of ^{22}Ne/^{21}Ne of 1.08-1.14, the model predictions agree quite well with the Eugster (1988) calibration curve. At lower shielding, the empirical calibration curve runs parallel to the model results, although with an offset of up to 30%. In summary, Figure 5 indicates that the shielding-corrected P(^{21}Ne) values according to Eugster (1988) are usually correct to within some $\pm$ 15% for average

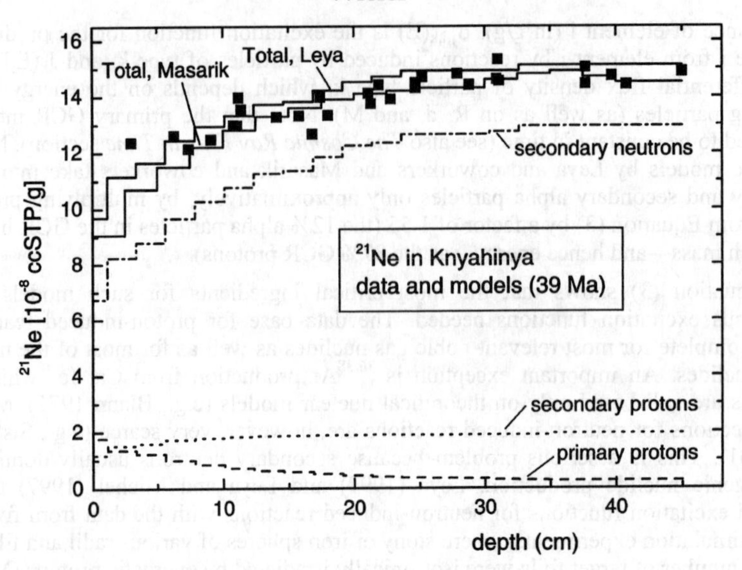

Figure 2. Comparison of measured (squares) and modelled (solid lines) ^{21}Ne concentrations in the L/LL chondrite Knyahinya. The measured data are from Graf et al. (1990b), and the two model curves from Leya et al. (2000a) and Masarik et al. (2001). Both models assume a radius of Knyahinya of 45 cm and an exposure age of 39 Myr. Also shown are the individual contributions by primary protons and secondary protons and neutrons (Leya et al. 2000a). Note that even at the surface, most of the ^{21}Ne is produced by secondary particles.

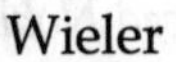

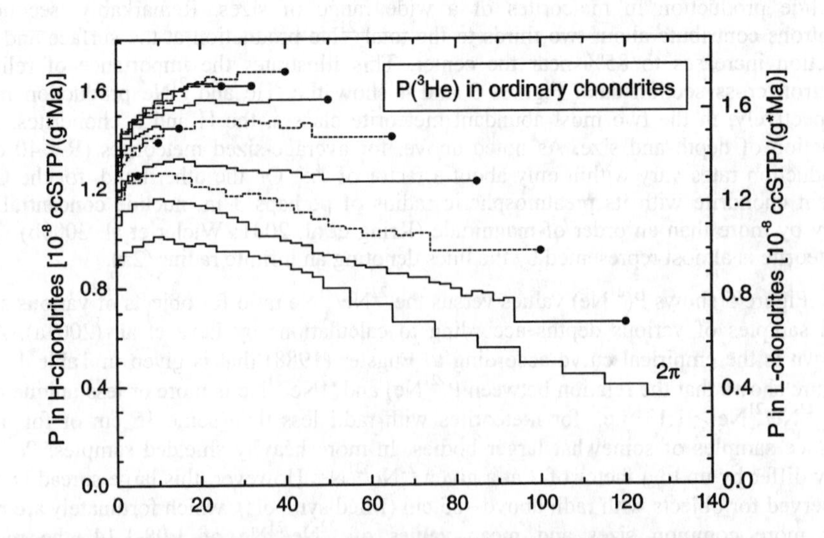

Figure 3. GCR-induced production rates of ^{3}He in ordinary chondrites of classes H (left ordinate) and L (right ordinate). For LL chondrites, all values are 2% larger than those of L chondrites. The curves represent meteoroids with radii of between 5 and 120 cm. Also included is a depth profile on an infinitely large flat body (2π). Model calculations are from Leya et al. (2000a) and updates (I. Leya, pers. comm. 2001).

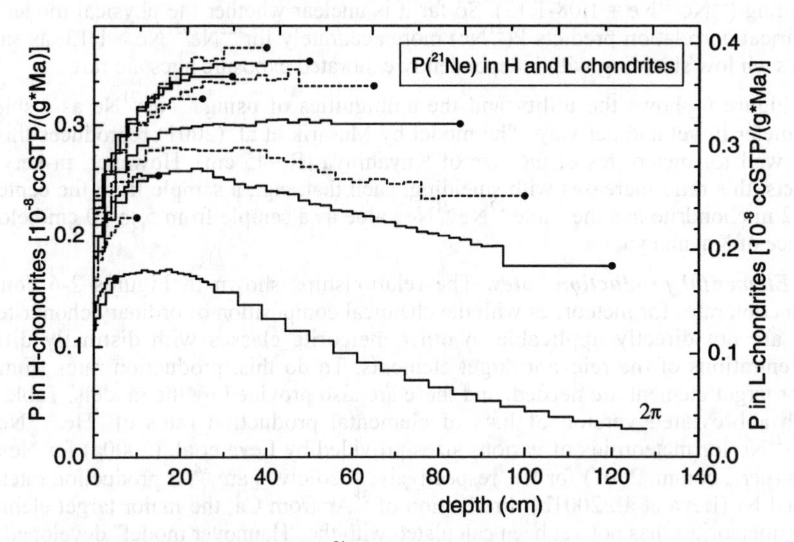

Figure 4. Same as Fig. 3 but for ^{21}Ne. For LL chondrites, the L-scale on the right ordinate has to be multiplied by 1.02. Data from Leya et al. (2000a).

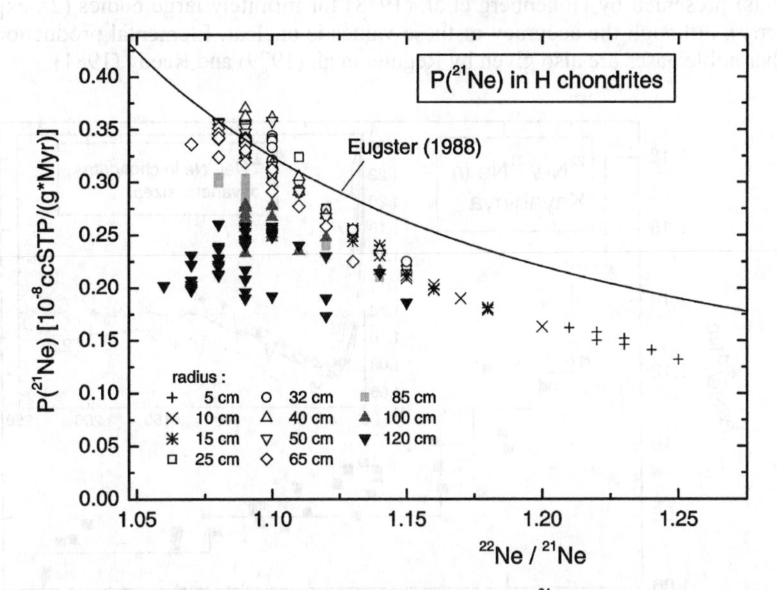

Figure 5. The symbols show the GCR production rate of ^{21}Ne as a function of the shielding parameter ^{22}Ne/^{21}Ne in H chondrites of various radii and samples from variable depths, calculated according to Leya et al. (2001a). Also shown is the empirical relation between P(^{21}Ne) and ^{22}Ne/^{21}Ne according to Eugster (1988). The model data show that only for ^{22}Ne/^{21}Ne $\geq$ 1.13 there exists an essentially unequivocal P(^{21}Ne) value. At lower ^{22}Ne/^{21}Ne values, P(^{21}Ne) varies by up to twofold at a given ^{22}Ne/^{21}Ne. Nevertheless, for meteorite radii not larger than ~50 cm, the Eugster (1988) relation is in agreement with the model results to within ±15% for 1.08 < ^{22}Ne/^{21}Ne < 1.14. It is unclear whether the physical model or the empirical correlation describes the data better for ^{22}Ne/^{21}Ne > 1.13.

shielding ($^{22}Ne/^{21}Ne$ = 1.08-1.13). So far it is unclear whether the physical model or the empirical correlation predicts P(^{21}Ne) more accurately for $^{22}Ne/^{21}Ne$ > 1.13, as samples with such low shielding with independently calibrated exposure ages are rare.

Figure 6 shows the utility and the ambiguities of using $^{22}Ne/^{21}Ne$ as a shielding parameter in yet another way. The model by Masarik et al. (2001) reproduces this ratio very well for meteorites of the size of Knyahinya (R ~45 cm). However, in very large objects, this ratio increases with shielding, such that, e.g., a sample from the center of a R = 2 m chondrite has the same $^{22}Ne/^{21}Ne$ value as a sample from 5 to 10 cm below the surface of Knyahinya.

Elemental production rates. The relationships shown in Figures 2-6 constrain production rates for meteorites with the chemical composition of ordinary chondrites, but they are not directly applicable to other meteorite classes with distinctly different concentrations of the relevant target elements. To do this, production rates from each major target element are needed, and these are also provided by the models. Table 2 is a much abbreviated version of lists of elemental production rates of ^{3}He, ^{21}Ne, and $^{22}Ne/^{21}Ne$ for meteoroids of various sizes provided by Leya et al. (2000a) for Ne and I. Leya (pers. comm. 2001) for He, respectively. Also given are ^{38}Ar production rates from Fe and Ni (Leya et al. 2001b). Production of ^{38}Ar from Ca, the major target element in stony meteorites, has not yet been calculated with the "Hannover model" developed by R. Michel, I. Leya and coworkers, because of the lack of experimentally determined cross sections. We therefore reproduce in Table 2 the older ^{38}Ar values from Ca (and other elements) presented by Hohenberg et al. (1978) for infinitely large bodies (2π exposure geometry), although the accuracy of these values is unclear. Elemental production rates for other noble gases are also given by Regnier et al. (1979) and Reedy (1981).

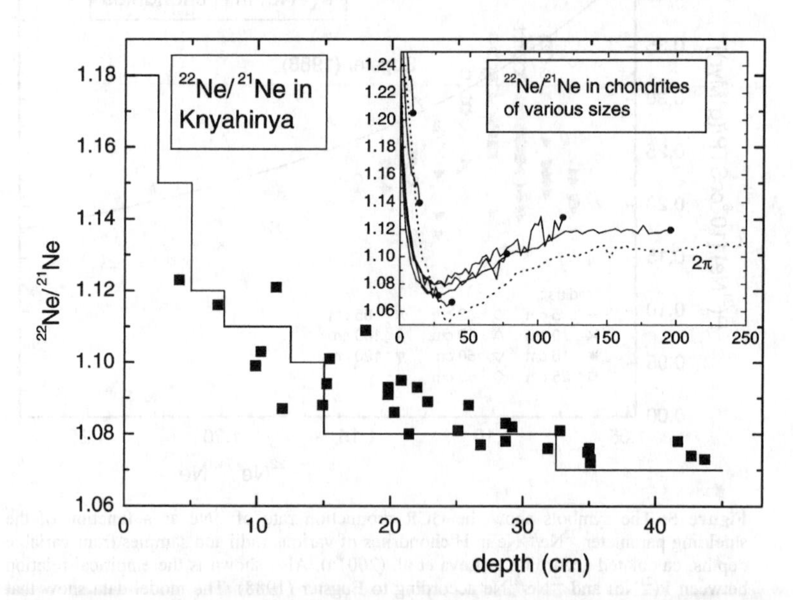

Figure 6. In the main panel, the $^{22}Ne/^{21}Ne$ ratio predicted by the model of Masarik et al. (2001) is compared with data from Knyahinya samples from known preatmospheric depths (Graf et al. 1990b). The inset shows the modelled ratios for various meteorite radii and a 2π irradiation. At depths below ~50 cm, $^{22}Ne/^{21}Ne$ increases again. Therefore, the same $^{22}Ne/^{21}Ne$ ratio corresponds to widely different depths, which illustrates the ambiguity of this ratio as a shielding index in very large meteorites.

Table 2. Elemental production rates of cosmogenic He, Ne, and Ar in meteorites.

radius/ depth	^{3}He					^{21}Ne					^{22}Ne/^{21}Ne				^{38}Ar			
	O	Mg	Al	Si	Fe	Mg	Al	Si	Ca	Fe	Mg	Al	Si	Ca	Ca	Ti	Fe	Ni
5,s	148	113	124	138	86.6	53.2	32.6	25.2	8.57	2.03	1.290	1.188	1.197	1.256			7.47	6.54
5,c	166	124	130	143	89.0	69.0	37.8	28.8	8.77	1.99	1.200	1.224	1.197	1.236			7.43	6.58
10,s	163	125	130	142	88.2	70.2	37.4	28.8	8.69	1.97	1.191	1.228	1.190	1.234			7.39	6.50
10,c	190	144	139	149	90.8	98.7	45.9	34.9	8.89	1.86	1.091	1.265	1.187	1.197			7.23	6.41
15,s	170	132	131	143	87.3	80.4	39.6	30.6	8.53	1.88	1.152	1.260	1.187	1.223			7.19	6.33
15,c	202	158	139	150	87.2	121.0	49.9	38.0	8.40	1.66	1.054	1.325	1.185	1.175			6.66	5.89
25,s	186	148	135	147	86.8	102.7	44.7	34.6	8.36	1.75	1.095	1.309	1.184	1.200			6.78	6.01
25,c	232	188	149	158	88.2	159.2	59.3	45.5	8.32	1.46	1.020	1.377	1.188	1.127			6.17	5.52
40,s	188	152	132	144	82.8	111.7	45.5	35.6	7.88	1.60	1.076	1.339	1.176	1.185			6.33	5.60
40,c	246	205	148	156	82.7	184.3	65.0	51.2	7.63	1.14	1.002	1.413	1.151	1.075			5.32	4.79
65,s	173	144	120	133	74.5	108.0	41.8	33.2	7.02	1.41	1.071	1.369	1.174	1.179			5.60	4.95
65,c	210	180	119	127	62.0	171.7	53.6	43.4	5.44	0.69	0.969	1.470	1.159	1.022			3.63	3.22
100,s	161	135	112	123	68.1	104.3	39.2	30.9	6.33	1.26	1.066	1.388	1.189	1.180			5.08	4.47
100,c	138	130	78.5	86.2	38.7	120.6	36.5	30.2	3.20	0.37	1.020	1.537	1.150	0.994			2.00	1.80
2π, 0	118	97	90	100	58	59.9	24.7	19.9	5.40	1.31	1.061	1.190	1.224	1.235	*61*	*11.8*	4.42	3.96
2π, 40	146	113	86	90	49	93.1	34.3	27.1	4.50	0.73	0.927	1.175	1.196	1.100	*113*	*11.1*	3.14	2.83
2π, 100	127	98	66	70	35	83.1	28.9	23.7	3.06	0.37	0.896	1.188	1.163	1.027	*103*	*7.48*	1.96	1.75
2π, 200	83	64	40	43	20	52.7	17.7	15.0	1.63	0.16	0.905	1.208	1.145	0.978	*72*	*3.91*	0.97	0.78
2π, 500	15	12	7	8	3	9.9	3.0	2.8	0.22	0.01	0.917	1.255	1.116	0.908	*12.0*	*.42*	0.12	0.10

Production rates [in 10^{-10} cc STP/(g×Ma)] according to model calculations (1 cc STP=2.687×10^{19} atoms). Given are values for the surface (subscript s) and the center (subscipt c), respectively, of meteorites with radii between 5 and 100 cm. The lowermost 5 rows give the values for an infinitely large body (2π irradiation) at depths between 0 and 500 g/cm^2. Most data are from I. Leya and coworkers (^{3}He: I. Leya, pers. comm. 2001; ^{21}Ne (4π): Leya et al. 2000a; Ne (2π): Leya et al. 2001b; ^{38}Ar from Fe and Ni: Leya et al. 2001b). ^{38}Ar from Ca and Ti (2π) values are from Hohenberg et al. 1978. These values should not be compared straightforwardly with the respective ^{38}Ar data from Fe and Ni by Leya et al. 2001b, because the latter values are considerably higher than P(^{38}Ar) from Fe and Ni also reported by Hohenberg et al. 1978. ^{22}Ne/^{21}Ne from Fe as target not given, because P(^{22}Ne) from Fe reported by Leya et al. (2000a, 2001b) is unreliable. This table is a summary of much more extended data sets given in the original publications.

The matrix effect. When calculating a production rate for a sample with a given chemical composition by multiplying the elemental production rates times the fractional abundance of the respective target element and summing over all relevant elements, one ignores a possible dependence of elemental production rates on the overall chemical composition of a meteorite. This so-called matrix effect can arise because particle fluxes and spectra are a function of the chemical composition of the target. For example, the matrix effect is expected to be important for silicate inclusions in stony-iron meteorites, where the high Fe concentrations lead to an enhanced flux of secondary neutrons relative to stony meteorites (Begemann and Schultz 1988; Masarik and Reedy 1994). The matrix effect will be particularly pronounced for nuclides largely produced by relatively low-energy neutrons, e.g., ^{21}Ne from ^{24}Mg and ^{38}Ar from Ca. Albrecht et al. (2000) showed that ^{21}Ne production rates in silicates from mesosiderites on average are some 60% higher than would be expected for samples with the same composition but in a silicate-only matrix. This result is in agreement with the model calculations by Masarik and Reedy (1994). For stony meteorites of variable composition the matrix effect is generally expected to be < 10%, however, an exception being ^{38}Ar production from Ca, which is several ten percent lower in eucrites and aubrites compared to chondrites (Masarik and Reedy 1994).

Production rate ratios P(^{10}Be)/P(^{21}Ne) and P(^{26}Al)/P(^{21}Ne). Because of the ambiguity of the parameter ^{22}Ne/^{21}Ne, other methods for deducing shielding-corrected production rates are often useful. These methods operate with a radioactive and a stable nuclide pair that have production rate ratios that can be assumed to depend only weakly on the irradiation conditions. We already mentioned the systems ^{10}Be-^{21}Ne, ^{26}Al-^{21}Ne, and ^{53}Mn-^{21}Ne, which can be used not only to deduce average production rates as shown above, but in principle also to derive a shielding-corrected ^{21}Ne production rate for an individual sample.

Based on the Knyahinya data, Graf et al. (1990a) proposed the following relations for L chondrites (in atoms/atom):

$$\frac{P(^{10}\mathrm{Be})}{P(^{21}\mathrm{Ne})} = (0.140 \pm 0.001) + (0.02 \pm 0.05) \times \left(\frac{^{22}\mathrm{Ne}}{^{21}\mathrm{Ne}} - 1.11 \right) \tag{4}$$

$$\frac{P(^{26}\mathrm{Al})}{P(^{21}\mathrm{Ne})} = (0.37 \pm 0.02) - (0.4 \pm 0.7) \times \left(\frac{^{22}\mathrm{Ne}}{^{21}\mathrm{Ne}} - 1.11 \right) \tag{5}$$

Note that the two production rate ratios P(^{10}Be)/P(^{21}Ne) and P(^{26}Al)/P(^{21}Ne) are given as functions of ^{22}Ne/^{21}Ne and are thus basically also subject to the reservations on the reliability of this shielding parameter. However, the dependence of P(^{10}Be)/P(^{21}Ne) on ^{22}Ne/^{21}Ne in Equation (4) is actually very weak, which reflects the fact that the ratio ^{10}Be/^{21}Ne is nearly constant over the entire range of shielding represented by the Knyahinya samples. The pair ^{10}Be-^{21}Ne therefore offers a way to determine shielding-corrected exposure ages in chondrites (Eqn. 4), if ^{10}Be can be assumed to be saturated, i.e., for exposure ages above ~6 Myr. A word of caution is necessary, however. In contrast to the model of Graf et al. (1990a), that by Leya et al. (2000a) predicts P(^{10}Be)/P(^{21}Ne) in chondrites to depend substantially on the ratio ^{22}Ne/^{21}Ne. If so, ^{10}Be-^{21}Ne ages are correct only to within the limits imposed by the reliability of ^{22}Ne/^{21}Ne as shielding indicator. Leya et al. (2000a) estimate that for ^{22}Ne/^{21}Ne < 1.10 ^{10}Be-^{21}Ne ages are ambiguous to within ±15%.

In iron meteorites this ambiguity does not appear to exist. Both data and models show that the ratio P(^{10}Be)/P(^{21}Ne) is constant over a wide range of shielding. Lavielle et al. (1999a) derive a $P(^{10}Be)/P(^{21}Ne)$ value of 0.55, in perfect agreement with the model-based value by Leya et al. (2000a), which is essentially independent on shielding for all meteorites with radii not larger than one meter. Graf et al. (1987) had deduced a $P(^{10}Be)/P(^{21}Ne)$ ratio of 0.772±0.05 with data from the Grant iron meteorite, by adopting an exposure age derived with the ^{40}K-^{41}K method. This ratio is 40% higher than that published by Lavielle et al. (1999a), which is based on the ^{36}Cl-^{36}Ar age of Grant and another iron meteorite. As Cl-Ar and K-K ages of iron meteorites differ systematically by ~40%, probably because the cosmic ray intensity has changed with time (see *The Cosmic Ray Flux in Time* section), the production rate ratios $P(^{21}Ne)/P(^{10}Be)$ given by Graf et al. (1987), Lavielle et al. (1999a), and Leya et al. (2000a) therefore all agree very well. If we adopt the ^{36}Cl-^{36}Ar age scale, a ^{10}Be-^{21}Ne age of iron meteorites can therefore be calculated by

$$T_{exp} = \left(\frac{^{21}Ne}{P(^{10}Be)} \right) \times 0.55 \tag{6}$$

where ^{21}Ne is the concentration in atoms per unit of mass and $P(^{10}Be)$ the production rate in atoms per unit of mass per unit of time.

In iron meteorites, the $P(^{26}Al)/P(^{21}Ne)$ ratio also appears to be rather insensitive to shielding. Leya et al. (2000a) predict an essentially constant value of 0.45, somewhat higher that that of 0.38 proposed by Hampel and Schaeffer (1979), which is based on the ^{36}Cl-^{36}Ar age scale. The mean value of 0.56 by Graf et al. (1987) (based on the ^{40}K-^{41}K age scale) is again 25-47% higher than the other two values. However, in contrast to the model prediction (Leya et al. 2000a), the measured $P(^{26}Al)/P(^{21}Ne)$ in Grant increases somewhat with increasing shielding (Graf et al. 1987).

The ^{36}Cl-^{36}Ar, ^{129}I-^{129}Xe, and ^{81}Kr-Kr methods. These methods do not involve $^{22}Ne/^{21}Ne$, and are thus self-correcting for shielding in a more strict sense than the ^{10}Be-^{21}Ne method in chondrites. The ^{36}Cl-^{36}Ar pair has the advantage that in metallic Fe-Ni samples most of the ^{36}Ar is produced through its radioactive precursor ^{36}Cl, and hence the production rate ratio $P(^{36}Ar)/P(^{36}Cl)$ is undoubtedly essentially shielding-independent. One advantage of the ^{81}Kr-Kr technique is that both radioactive and stable nuclide are isotopes of the same element krypton, so only an isotopic ratio measurement is needed. Note, however, that these methods also will yield erroneous ages if the assumption of a single stage exposure history of the studied sample is violated.

Lavielle et al. (1999a) and Graf et al. (2001) give values for $P(^{36}Cl)/[\gamma P(^{36}Cl) + P(^{36}Ar_{direct})]$ of 0.84±0.04 for Fe and 0.78±0.03 for Ni. These values should be shielding-independent for production by both protons and neutrons. The coefficient $\gamma = 0.981$ is the branching ratio for the decay of ^{36}Cl into ^{36}Ar. With an average Ni concentration of 9% in chondritic metal, the ^{36}Cl-^{36}Ar age can be calculated by

$$T_{exp} = 0.363 * \left(\frac{^{36}Ar_{cos}}{^{36}Cl} \right) \tag{7}$$

if [atoms per gram] are used as units for the concentrations of cosmogenic (subscript cos) ^{36}Ar and ^{36}Cl, or

$$T_{exp} = 427 * \left(\frac{^{36}Ar_{cos}}{^{36}Cl} \right) \tag{8}$$

using [10^{-8} ccSTP/g] for ^{36}Ar and [dpm/kg] for ^{36}Cl.

Begemann et al. (1976), Leya et al. (2000a) and Albrecht et al. (2000) give coefficients of 425, 433 and 430, respectively, for Equation (8). Whereas ^{36}Cl-^{36}Ar ages are thus basically very reliable, they are quite difficult to determine, because very high-purity metal separates are required to avoid contamination with ^{36}Cl and ^{36}Ar from silicate impurities (e.g., Graf et al. 2001). Also, ^{36}Cl and ^{36}Ar have to be analyzed on two different, often small aliquots. Therefore, rather few ^{36}Cl-^{36}Ar exposure ages have been reported so far, although the method is becoming increasingly popular.

Another potentially interesting method where the stable cosmogenic noble gas nuclide is predominantly produced through a radioactive precursor is the live-^{129}I-^{129}Xe method (Marti et al. 1986). All cosmogenic ^{129}Xe produced from Te will have gone through ^{129}I (half-life 15.7 Myr). For samples where ^{129}Xe contributions from Ba can be corrected for (using ^{124,126}Xe) and for which ^{129}Xe contributions from extinct ^{129}I from the early solar system are of no concern, this system is ideal in being self-correcting for both shielding and variable Te abundance. Marti et al. (1986) determined the exposure age of the big Cape York iron meteorite with the live-^{129}I-^{129}Xe couple in troilite inclusions. Otherwise, the method has not been widely applied so far, but as ^{129}I analyses by accelerator mass spectrometry are becoming routine, it should prove very useful to study cosmic ray flux variations over timescales of a few ten million years.

The ^{81}Kr-Kr technique was proposed by Marti (1967). Cosmogenic Kr is mainly produced from the target elements Rb, Sr, Y, and Zr. ^{81}Kr has a half-life of 229,000 yrs. Marti and Lugmair (1971) observed that the isotopic ratios of cosmogenic Kr vary systematically in Apollo 12 lunar rocks, due to variable shielding. They observed the following relations (all Kr concentrations or isotopic ratios refer to the spallation Kr component):

$$P(^{81}Kr)/P(^{83}Kr) = 0.95\left(\frac{^{80}Kr + ^{82}Kr}{2\times ^{83}Kr}\right) \tag{9}$$

and

$$P(^{81}Kr)/P(^{83}Kr) = 1.262\frac{^{78}Kr}{^{83}Kr} + 0.381 \tag{10}$$

These relations allow determination of a shielding-corrected exposure age based on a single Kr analysis The factor 0.95 in Equation (9) represents the isobaric fraction yield of ^{81}Kr (Marti 1967). Note that this value has recently been redetermined to ~0.92 for chondritic abundances of the relevant target elements (B. Lavielle, pers. comm. 2001). The second relation is insensitive to ^{80}Kr and ^{82}Kr from neutron capture on Br, which is present in relatively large or Br-rich meteorites (see next section). Note, that these relations are strictly valid only for samples which have the same relative abundances of the major target elements as Apollo 12 rocks. For example, Apollo 11 samples, which have ~2-3 times higher Zr/Sr ratios than Apollo 12 samples, have a P(^{81}Kr)/P(^{83}Kr) that is slightly higher at a given ^{78}Kr/^{83}Kr value than predicted by Equation (10) (Lugmair and Marti 1971). However, the difference is only about 1-3%. Moreover, Finkel et al. (1978) showed that the Kr spallation systematics derived from Apollo 12 samples are also valid for the two chondrites San Juan Capistrano and St. Severin, apparently because Apollo 12 rocks and chondrites have rather similar Zr/Sr ratios (although very different Rb/Sr ratios, Rb contributing substantially to Kr production in chondrites but not in lunar samples). Equations (9) and (10) are thus also widely used today for meteorites. Eugster (1988) presented a correlation between P(^{81}Kr)/P(^{83}Kr) and ^{22}Ne/^{21}Ne for ordinary chondrites, but from the only available ^{81}Kr depth profile in Knyahinya it is unclear whether this relation holds for low shielding (Lavielle et al. 1997). The ^{81}Kr-Kr method has the disadvantage of often not being very precise, both because of the low abundance of ^{81}Kr and because large corrections for non-cosmogenic Kr are often required.

An important caveat is that all procedures for deducing production rates discussed above assume a single-stage exposure. In other words, it is assumed that a meteorite was ejected from its parent body from at least several meters depth, where it had been completely shielded from the GCR, and then travelled to Earth without ever changing its shape by a further collision or by space erosion (see *Complex Exposure Histories* section). Only in this case is the present-day activity of a radionuclide a good measure for the production rate of a stable nuclide, as the radionuclide concentration depends only on the shielding of the sample during the past few half-lives, whereas the stable nuclide has been accumulating during the entire exposure of the sample to cosmic rays. However, we will see in the *Complex Exposure Histories* section that for a fairly high fraction of all meteorites this requirement of a single exposure stage is probably not fulfilled. Complex exposures may well go unnoticed, e.g., if a single noble gas analysis on a single sample has been carried out only.

A factor further limiting the accuracy of production rates is a possible change in the mean GCR flux over the timescales of interest. We will discuss in section *The Cosmic Ray Flux in Time* that such variations on a million-year scale appear to be modest but that the mean flux during the last few or few ten million years probably has been some 40 to 50% higher than the mean over the past 1-2 Gyr. The resulting additional uncertainty of exposure ages is often only a minor concern, however. This is because one is often more interested in whether exposure ages of different meteorite classes cluster at certain values rather than in the absolute position of such clusters. It needs, however, to be taken into account if ages based on production rates derived from different nuclide systems are compared with each other, particularly for iron meteorites (see the respective subsection below).

In summary, production rates of cosmogenic nuclides in meteorites have quite a variable level of accuracy and reliability. In general, the more nuclides that are determined on the same sample and the more samples that are measured from the same meteorite, the higher will be our confidence in a stated exposure age. A single age determination will usually not be more accurate than to within 15-20%, with the probable exception of ^{36}Cl-^{36}Ar ages, which may have a precision of better than 10%. Uncertainties may even be considerably larger than 20%, e.g., if an unusually heavy shielding is not recognized. On the other hand, we will see below that the accuracy that is presently achieved generally is good enough to study the history of meteorites as small bodies in space and collisional processes on parent bodies.

Cosmogenic noble gases produced by capture of low-energy neutrons

So far, we have mainly discussed production of cosmogenic nuclides by neutrons or protons with energies high enough (tens of MeVs or higher) to break up a target nucleus. The resulting cosmogenic nuclide (commonly referred to as “spallation product”) usually has a lower mass than the original target. In contrast to this, target nuclei may also capture secondary cosmic ray neutrons which have already been slowed down to much lower energy without having yet interacted with a nucleus. These are so-called thermal (<0.6 eV) or epithermal (up to hundreds of eV) neutrons. We mentioned in the *Introduction* that several isotopes of Cd, Sm and Gd have a cross section high enough so that capture of thermal or epithermal neutrons induces measurable shifts in their isotopic composition in samples exposed to GCR. We also mentioned ^{60}Co, a nuclide predominantly produced by thermal neutron capture on ^{59}Co. Here we discuss how several stable noble gas nuclides are also efficiently produced by slow neutron-capture reactions so that sometimes a sizeable fraction of the respective nuclide may be produced by this pathway, although this is unfortunately often difficult to quantify. The most important of these nuclides are ^{36}Ar, ^{80,82}Kr, and ^{128,131}Xe.

The flux of thermal and epithermal neutrons peaks at larger depths (in large objects roughly around 200 g/cm^2) than the depth of maximum production of nuclides made by particles of higher energy (Eberhardt et al. 1963; Lingenfelter et al. 1972; Spergel et al. 1986; Nishiizumi et al. 1997b). Nuclides produced by thermal or epithermal neutrons are therefore potentially very useful for studying cosmic ray interactions at relatively high shielding. For example, high levels of neutron-capture products in a meteorite otherwise thought to have been of "average" preatmospheric size would reveal a complex exposure history, since the neutron-capture nuclides would have been acquired on the parent body or in a larger precursor meteoroid (see *Complex Exposure Histories* section). Neutron-capture products have also been widely used to study the GCR irradiation history of the lunar regolith (see Wieler et al. 2002). Bogard et al. (1995) give an overview of applications of neutron-capture-produced nuclides in extraterrestrial samples.

The noble gases are themselves much too rare to serve as significant neutron-capture targets. For this reason, neutron-capture-produced noble gas isotopes must have a radioactive precursor formed from another element. For example, neutron-capture-produced ^{36}Ar (^{36}Ar$_n$) results from the reaction ^{35}Cl(n,γ)^{36}Cl(β$^-$)^{36}Ar. Similarly, ^{80}Kr and ^{82}Kr are produced from ^{79}Br and ^{81}Br through ^{80,82}Br (Lugmair and Marti 1971). The most important neutron-capture-produced isotope of Xe is ^{131}Xe from ^{130}Ba (with intermediate products ^{131}Ba and ^{131}Cs). This is a particularly interesting system because Ba is also the major target element for the production of cosmogenic Xe by high-energy reactions, so ratios such as ^{131}Xe/^{126}Xe provide a measure for the depth of irradiation virtually independent of the chemical composition of a sample (Eberhardt et al. 1971). ^{128}Xe from ^{127}I may also be detectable, e.g., in calcium-aluminum-rich inclusions in Allende (Göbel et al. 1982) or near the surface of lunar dust grains where volatile iodine had been concentrated (Pepin et al. 1995). These examples show that sometimes care needs to be taken to distinguish neutron-capture products from other contributions like potential nucleosynthetic anomalies or implanted solar wind Xe.

A major problem limiting the applicability of neutron-capture-produced noble gases is that other noble gas components often hamper a quantitative determination of the neutron-capture component. This is a particular problem in the case of ^{36}Ar. Essentially every meteorite contains cosmogenic Ar produced by high-energy spallation (^{36}Ar/^{38}Ar ~ 0.65), and most stony meteorites contain primordial Ar (^{36}Ar/^{38}Ar ~ 5.3) or some atmospheric Ar with essentially the same ^{36}Ar/^{38}Ar ratio as the primordial component. As long as only these components have to be taken into account, the concentrations of cosmogenic ^{36}Ar and ^{38}Ar can be deduced unambiguously, but this cannot be done any more when ^{36}Ar$_n$ may also possibly be present. To firmly prove the presence or absence of ^{36}Ar$_n$ in a meteorite is therefore generally quite a formidable task, involving e.g., Ar release in several temperature steps and combining such measurements with analyses of other neutron-capture nuclides (Bogard et al. 1995) and/or analyses of multiple samples from the same meteorite (e.g., Welten et al. 2001a). Even then, it may still be difficult to quantitatively determine the concentration of the neutron-capture-produced ^{36}Ar, so that only lower and upper limits may be estimated. In practice, this often leads researchers to neglect the possible presence of ^{36}Ar$_n$ even in cases where this may not be warranted. Bogard et al. (1995) showed that ^{36}Ar$_n$ accounts for perhaps more than half of the total (cosmogenic and trapped) ^{36}Ar in the large chondrite Chico, and they discuss observations indicating the presence of ^{36}Ar$_n$ in several other large meteorites or meteorites with a known complex exposure history. They note that this is not a surprise but rather is to be expected. Therefore, neglecting this component may often compromise the determination of conventional ^{38}Ar exposure ages. This is because the measured ^{38}Ar essentially must always be corrected for primordial or atmospheric Ar, as just mentioned, and this correction will be too large if the $^{36}Ar_n$ is erroneously viewed to be primordial or atmospheric (e.g., Welten et al. 2001a). It is unclear how often this will happen, because—as we will see below—the fraction of meteorites with a complex exposure history is not well constrained and because it is also not well constrained how many meteorites with a large preatmospheric size go unnoticed as such. However, Bogard et al. (1995) note from a data survey that higher $^{36}Ar/^{38}Ar$ ratios are in general due to a higher contribution of primordial or atmospheric Ar rather than to a possible widespread presence of $^{36}Ar_n$, so the problem does not seem to be an ubiquitous one.

As Kr and Xe have more isotopes than Ar, an unambiguous determination of the neutron-capture component is easier in principle. Nevertheless, precise corrections for primordial or atmospheric Kr and Xe are often also difficult.

Isotopic abundances of cosmogenic noble gases

In order to deduce the cosmic-ray-produced fraction of a noble gas nuclide in a meteorite, it is commonly necessary to correct for other noble gas components, mostly trapped primordial gases or atmospheric contamination. To this end, the isotopic composition—or at least some crucial isotopic ratios—of the various components should be well constrained. Trapped components are discussed in previous chapters by Ott (2002) and Wieler (2002). Here we discuss the isotopic compositions of cosmogenic noble gases, which are summarized in Tables 3 and 4. As we have seen above, this composition will depend on the shielding and the chemical composition of a sample. Therefore, the values in Tables 3 and 4 should only serve as guidelines and the best values to be adopted may need to be evaluated from case to case.

Helium. Iron meteorites often contain essentially pure cosmogenic He, i.e., negligible radiogenic or atmospheric ^{4}He. The range of $(^4He/^3He)_{cos}$ given in Table 3 has been determined from the noble gas compilation by Schultz and Franke (2000), using relatively He-rich samples (^{3}He $> 200 \times 10^{-8}$ cc/g) of all iron meteorite classes. $(^4He/^3He)_{cos}$ correlates with the widely used shielding parameter $^4He/^{21}Ne$, increasing with shielding from ~3 to ~4.5. However, this correlation is often disturbed in meteorites that suffered ^{3}H losses.

In most stony meteorites, ^{4}He is overwhelmingly radiogenic, such that $(^4He/^3He)_{cos}$ is difficult to measure directly (and of rather little practical concern). The two values given in Table 3 have both been derived from a suite of meteorites with low and presumably rather constant radiogenic ^{4}He. These are L chondrites which record a major collision on their parent body some 500 Myr ago that led to a

Table 3. Isotopic composition of cosmogenic He, Ne, and Ar.

	$^4He/^3He$	Corresponding $^4He/^{21}Ne$ range
iron meteorites	3.2-4.4	200-440
stony meteorites	5.2±0.3[1]	
	6.1±0.3[2]	
	$^{22}Ne/^{21}Ne$	Corresponding $^{20}Ne/^{21}Ne$ range
chondrites	1.05-1.25	0.88-0.98[a]
	$^{36}Ar/^{38}Ar$	Corresponding $^4He/^{21}Ne$ range
iron meteorites	0.60-0.665 (mean 0.63)	450-200
stony meteorites	~0.65	

1: Heymann (1967). 2: Alexeev (1998); other values see text.
a: range somewhat uncertain, see text.

complete loss of the radiogenic ^4He accumulated up to that time (Swindle 2002b). Within their limits of uncertainty, the L chondrite values should also hold for other stony meteorite classes. This is also indicated by some stony meteorites, in particularly diogenites, which have measured $^4He/^3He$ ratios as low as ~5 due to very low U and Th concentrations (Welten et al. 1997).

Neon. We have seen above that $(^{22}Ne/^{21}Ne)_{cos}$ in stony meteorites is quite variable, which makes it useful as a shielding parameter. The range stated in Table 3 is for chondrites (see Fig. 6), and will be different for stony meteorites with rather different relative Mg, Al, and Si abundances. The differences can be estimated by means of Table 2. The $(^{20}Ne/^{21}Ne)_{cos}$ ratio in chondrites is difficult to evaluate because at least minor contributions of primordial or atmospheric ^{20}Ne are ubiquitous. Consideration of only ordinary chondrites of high exposure ages and high petrographic types 5 and 6 (i.e., little or no primordial Ne; Schultz and Franke 2000) suggests that $(^{20}Ne/^{21}Ne)_{cos}$ increases with increasing $(^{22}Ne/^{21}Ne)_{cos}$ (Table 3).

Argon. Iron meteorites again often allow direct measurement of the $(^{36}Ar/^{38}Ar)_{cos}$ ratio. Iron meteorites with relatively high $^{38}Ar_{cos}$ show an inverse correlation between $^{36}Ar/^{38}Ar$ and the shielding parameter $^4He/^{21}Ne$ (Table 3). $(^{36}Ar/^{38}Ar)_{cos}$ decreases by up to ~10% with higher shielding. The average value is ~0.63. Determining $(^{36}Ar/^{38}Ar)_{cos}$ in stony meteorites is again compromised by the common presence of primordial or atmospheric Ar and sometimes also by neutron-capture-produced ^{36}Ar. Yet, the lowest measured $^{36}Ar/^{38}Ar$ ratios in achondrites poor in primordial gases and with relatively high exposure ages are ~0.63, similar to the mean value of iron meteorites (Schultz and Franke 2000). More achondrite values cluster around ~0.65, however, which is a widely adopted $(^{36}Ar/^{38}Ar)_{cos}$ value for stony meteorites. It appears rather surprising that ^{36}Ar and ^{38}Ar are produced in almost equal proportion from the target elements Ca and Fe, respectively.

Krypton. The average isotopic composition of cosmogenic Kr in lunar soils and in chondrites is given in Table 4. These data are from compilations by Pepin et al. (1995) and Lavielle and Marti (1988). When correcting a measured Kr composition for a trapped component, the assumption is commonly made that the cosmic-ray-produced contribution on ^{86}Kr can be neglected, so that the measured ^{86}Kr can be assumed to be entirely trapped. Note that $^{80,82}Kr$ in different meteorites may contain variable contributions from

Table 4. Isotopic composition of cosmogenic Kr and Xe (relative to ^{83}Kr, $^{126}Xe \equiv 1$)

	^{78}Kr	^{80}Kr	^{82}Kr	^{84}Kr	^{86}Kr		
average, lunar bulk soils[1]	0.20(2)	0.54(7)	0.72(5)	0.32(10)	≡0		
chondrites[2]	0.18(4)	0.60(8)	0.76(8)	0.67(15)	≡0		
	^{124}Xe	^{128}Xe	^{129}Xe	^{130}Xe	^{131}Xe	^{132}Xe	^{134}Xe	^{136}Xe
average, lunar bulk samples[1]	0.56(3)	1.48(6)	1.64(15)	0.95(7)	5.30(42)	0.77(20)	0.05(3)	≡0
Chondritic Ba/REE[3]	0.595(10)	1.52(10)		0.98(06)	3.77(16)	0.83(06)	0.044(15)	≡0
meteorites[4]	0.55(10)	1.40(15)	1.6(2)	1.0(2)	2.5(5)	1.0(2)	0	≡0

Uncertainties in ( ) given in units of the last digit. **1**: Pepin et al. 1995. **2**: Lavielle & Marti 1988, mean and standard deviation from 11 chondrites. **3**: Hohenberg et al. 1981, based on Angra dos Reis achondrite data. **4**: Kim & Marti 1992.

Table 5. Exposure age ranges and clusters of meteorite classes

Class	*Range (Myr)*	*Clusters (Ma)*	*Comments*
Chondrites			
H[1]	1-80	**7.6** & 33	all petrographic types
		7.0?	H5 a.m. falls only
		24	H6 only
L[2]	1-70	**40**	mainly L5 & L6
		28	mainly L5 & L6 & ^{40}Ar-poor
		15 & 5	^{40}Ar-poor
LL[3]	0.03-70	**15**	mainly LL5 & LL6 & ^{40}Ar-rich
		10?	LL6 only
		28?	mainly LL3
		40?	mainly LL4
EH & EL[4]	0.07-66	25? & 8? & 3.5?	clusters need confirmation
CO, CV, CK[5]	0.15-63	9?	CV, CK only
		29?	CO, CV, CK
CM & CI[6]	0.05-7	**0.22**	CM only (CI poor statistics)
R[7]	0.2-50		many ages between ~7-40 Ma
Other meteorites			
HED[a,8]	5-76	**21** & 38	H&E&D
		12?	Eucrites only?
		50?	diogenites only?
Aubrites[9]	~17-130	45-80??	cluster unclear (see text)
Lodranites-acapulcoites[10]	4-10		similar as 7 Ma peaks for H chondrites
Ureilites[11]	0.1-34		
Iron meteorites[b]	10-2300		
IVA[12]		255	^{36}Cl-^{36}Ar ages
		207	
IIIAB[13]		460	
Mesosiderites[14]	10-180		^{36}Cl-^{36}Ar ages
Lunar[15]	<0.01-8		hardly any source crater pairing
Martian[15]	0.7-20		4-9 events on Mars

Particularly conspicuous exposure age clusters in bold face, uncertain ones marked by ? **a**: Howardites-Eucrites-Diogenites. **b:** Ages of clusters of iron meteorites refer to ^{36}Cl age-scale (see text). In the ^{40}K-scale, the double peak at 255 & 207 Ma in group IVA corresponds to a single peak at 375 Ma, whereas the IIIAB peak at 460 Ma is shifted to 650 Ma (Voshage 1978, 1984). Oldest stated age of 2300 Myr (Deep Springs) refers to ^{40}K-scale.

References: **1**: Graf and Marti 1995; Graf et al. 2001. **2**: Marti and Graf 1992. **3**: Graf and Marti 1994; lowest age from ref. 4. **4**: Patzer and Schultz 2001. **5**: Scherer and Schultz 2000. **6**: Nishiizumi et al. 1993; Caffee and Nishiizumi 1997; K. Nishiizumi, pers. comm. 2000. **7**: Schultz and Weber 2001. **8**: Eugster and Michel 1995; Welten et al. 1997; Welten et al. 2001b. **9**: Eberhardt et al. 1965. **10**: Terribilini et al. 2000; **11**: Scherer et al. 1998. **12**: Lavielle et al. 2001. **13**: B. Lavielle 2001, pers. comm. **14**: Begemann et al. 1976. **15**: Wieler and Graf 2001. **16**: Nyquist et al. 2001.

neutron-capture on Br.

Xenon. The average isotopic composition of cosmogenic Xe in lunar samples and meteorites is also given in Table 4. The lunar data are again from Pepin et al. (1995).

Hohenberg et al. (1981) studied Xe in different mineral fractions of the Angra dos Reis achondrite and were able to derive separately the isotopic composition of the cosmogenic Xe produced by Ba and rare earth elements. The respective entry in Table 4 is for a chondritic value of (La + Ce + Nd)/Ba = 0.52. The value for meteorites given by Kim and Marti (1992) is modified from earlier data on chondrites and eucrites.

EXPOSURE AGE DISTRIBUTIONS OF METEORITES

Even though a single exposure age value may be equivocal, as we noted above, the impressive data base on exposure age distributions of individual meteorite classes available today provides crucial information on how meteorites are delivered to Earth. Exposure ages tell us mainly how long meteorites have been travelling as meter-sized objects in interplanetary space and peaks in exposure age histograms indicate common impact events that led to the ejection of many meteorites from a common parent body.

Most published exposure ages are based on noble gas analyses, which are compiled by Ludolf Schultz and coworkers at the Max Planck Institut für Kosmochemie in Mainz, Germany. In the year 2000, the compilation contained more than 6000 entries from more than 1500 meteorites (Schultz and Franke 2000). From these data, exposure ages for many hundreds of meteorites can be calculated. For meteorites with short exposure ages of a very few million years or less, radionuclide data are often also considered, in particular for carbonaceous chondrites of the classes CM and CI. A compilation of radionuclide data available up to the late 1980s is given by Nishiizumi (1987). Radionuclide-based exposure ages are reliable as long as the measured activity is sufficiently smaller than the saturation activity, so that the uncertainty of the latter is not a crucial factor in the calculation, i.e., for exposure ages up to perhaps two half-lives of the nuclide under consideration. For meteorite finds it must also be verified that an activity below saturation is not due to a long terrestrial age (see *Terrestrial ages* subsection).

A potential problem when studying exposure age distributions is a bias introduced by unrecognized "pairings," i.e., meteorite fragments from a common fall. This is a major concern especially for meteorites found in deserts. In Antarctica, sometimes dozens or hundreds of fragments recovered by organized searches belong to the same shower of a meteorite broken up in the atmosphere. For meteorites that are observed to fall, pairing is usually readily recognized, however. Exposure age histograms shown below are thus based either mainly on observed falls, or are corrected for pairing to the extent possible.

Figures 7-10 show exposure age distributions of most of the more common meteorite classes, and Table 5 summarizes the figures. We note the following first-order observations and conclusions:

(1) Most stony meteorites have nominal exposure ages ranging between a few and about 70 Myr. Apart from possible misassignments due to unrecognized complex exposures or grossly overestimated production rates, this interval thus indicates the range of times stony meteorites spend in space as small objects after ejection from their "immediate" parent body. Remarkable exceptions are the carbonacous chondrites of types CM and CI, which mostly show exposure ages below 1 Myr, and the aubrites, which on average have the longest exposure ages of all stony meteorites.

(2) Iron meteorites mostly have considerably longer exposure ages than stony meteorites, often amounting to several hundred million years. The iron meteorite Deep Springs has an exposure age of about 2.3 Gyr and hence has lived as a small object in interplanetary space for just about half the solar system lifetime. The highest exposure ages of stony-iron meteorites are in between the highest values of stony and iron meteorites, respectively. Because CI chondrites, and presumably also CMs, are more fragile than other stony meteorites, which in turn are more fragile than iron meteorites and probably also stony irons, it is certainly reasonable to conclude that mechanical strength influences the mean lifetime against collisional

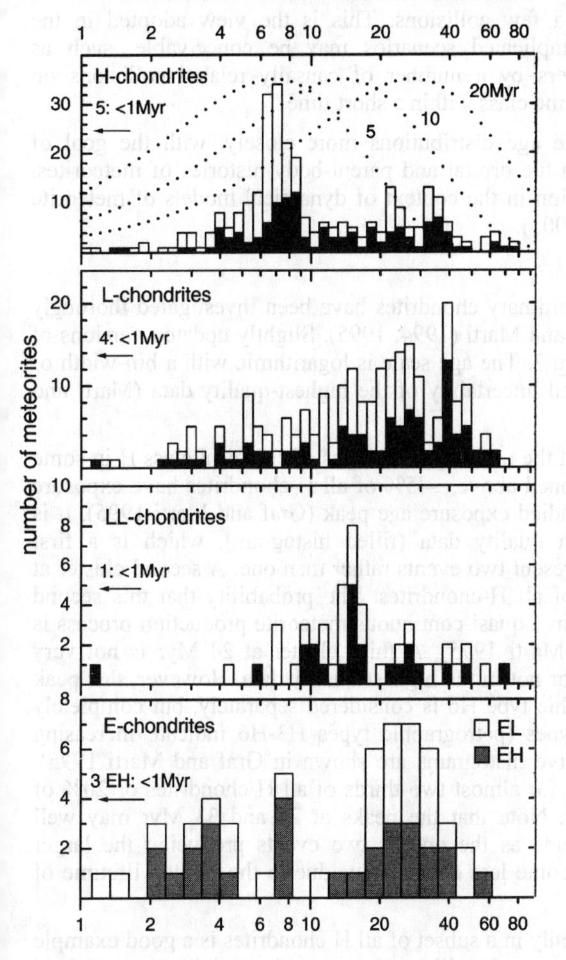

Figure 7. Exposure age distributions of ordinary chondrites (H, L, LL) and enstatite chondrites (EL, EH). The histograms for the ordinary chondrites are from the compilations by Marti and Graf (1992), Graf and Marti (1994, 1995), and updates (Th. Graf, pers. comm. 2001). The time resolution on the logarithmic abscissa scale is 10%, corresponding to the estimated accuracy of the highest quality data (class A of Graf and Marti, 1992; shown as filled bars in the top three panels). The E-chondrite histograms are from Patzer and Schultz (2001), with a time resolution of 20%. Some meteorites with nominal exposure ages < 1 Myr are off-scale and indicated by arrows. The dotted lines in the top panel represent expected exposure age distributions assuming continuous delivery and mean orbital lifetimes of 5, 10, and 20 Myr, respectively (Graf and Marti 1995).

destruction and thus is an important factor in determining meteorite exposure ages. Evidence for this is provided by the observation that chondrites with long exposure ages have lower $^{22}Ne/^{21}Ne$ ratios and hence larger masses than those with shorter ages (Loeken et al. 1992). On the other hand, this cannot be the whole story, as aubrites are more fragile than most other stony meteorite classes, and yet show unusually long exposure ages.

(3) Exposure age distributions of many, if not most, meteorite classes show one or more

clusters. In some cases, these clusters are very distinct, as the 7-Myr peak representing almost half of the H chondrites, the 15-Myr peak comprising a third of the LL chondrites and the 21-Myr peak that includes half of the HED meteorites. In other cases, peaks may require quite detailed data analysis to be recognized, as we will explain below, and sometimes it is unclear at all whether a couple of meteorites with similar exposure ages represent a statistically significant cluster. The most straightforward interpretation of these clusters is that each of them is the result of a—presumably very large—collision on a single immediate parent body. Each confirmed cluster thus indicates that a sizeable fraction of all meteorites of the respective class results from a few collisions. This is the view adopted in the following, although more complicated scenarios may be conceivable, such as producing exposure age clusters by a number of causally related collisions on different parent bodies of the same class within a short time.

We will next inspect exposure age distributions more closely, with the goal of obtaining additional information on the orbital and parent-body histories of meteorites. Later we will discuss this information in the context of dynamical models of meteorite delivery (see also Wieler and Graf 2001).

Undifferentiated meteorites

Exposure age distributions of ordinary chondrites have been investigated thorougly by Marti and Graf (1992) and Graf and Marti (1994, 1995). Slightly updated versions of their histograms are shown in Figure 7. The age scale is logarithmic with a bin-width of 10%, corresponding to the presumed uncertainty of the highest-quality data (Marti and Graf 1992).

H-chondrites. We consider first the ordinary chondrites of chemical class H in some detail (uppermost panel). As mentioned above, ~45% of all H-chondrites have exposure ages of ~7 Myr. This is the best studied exposure age peak (Graf and Marti 1995). It is rather broad, even for the highest quality data (filled histogram), which is a first indication that it might actually represent two events rather than one. A second cluster at 33 Myr comprises a further 10% of all H-chondrites. The probability that this second peak is just a statistical fluctuation in a quasi-continuous meteorite production process is estimated to be < 0.5% (Graf and Marti 1995). A third cluster at 24 Myr is not very conspicuous in Figure 7, in particular not in the highest quality data. However, this peak is rather prominent if the petrographic type H6 is considered separately, but completely absent in the other petrographic types (petrographic types H3-H6 indicate increasing thermal metamorphism; the respective histograms are shown in Graf and Marti 1995). Hence, three or four events account for almost two-thirds of all H chondrites or 20% of all stony meteorites that fall today. Note that the peaks at 24 and 33 Myr may well represent events of similar magnitude as the one or two events producing the larger 7-Myr peak, as older peaks will become less conspicuous due to the limited lifetime of meteorites discussed below.

That the 24-Myr peak appears only in a subset of all H chondrites is a good example of how detailed examinations may reveal collisional events that might go unnoticed otherwise. Unfortunately, most meteorite classes besides H and L chondrites are not numerous enough to allow such detailed assessments. The individual "7-Myr" peaks of types H4 and H5 are also displaced by roughly 1 Myr relative to each other, the H5s being older. Graf and Marti (1995) furthermore point out that H5 chondrites more often than other H chondrites have lost their cosmogenic tritium (and hence ^{3}He) due to heating during close passage by the Sun and that H5s do not show the excess of afternoon falls relative to morning falls that is typical for the other ordinary chondrite classes. These observations indicate that a subgroup of the H5 chondrites experienced a special orbital evolution. It is possible that these are the meteorites formed in the older of the two presumed 7-Myr collisions, happening on a different parent body than that yielding the other 7-Myr H chondrites. This second collision and that at 33 Myr sampled several or all petrographic types of H chondrites, indicating that a structure according to petrographic type, perhaps layered with type 3 being near the top (Pellas and Fiéni 1988), is partly preserved on at least one of the immediate H chondrite parent bodies. The 24-Myr event may have happened on a part of this body dominated by H6 material or on a separate H6-rich immediate parent body.

In summary, three or four large collisions on one to four asteroids are responsible for two thirds of all H chondrites. It is often assumed that all H chondrites ultimately derive from a single asteroid, having broken up and partly reassembled early in solar system history (Keil et al. 1994). In this scenario, the parent bodies from which meteorites get delivered to Earth today would have been formed in these early or in later collisions. Cosmic ray exposure age studies cannot decide whether these immediate parent bodies—if more than one—indeed derive from a common ancestor.

Given that the majority of the H chondrites can be traced back to a very few collisions, it is reasonable to ask whether the remaining H chondrites which do not obviously belong to one of the major peaks were formed either by many smaller events as a quasi-continuous "background," or in a few other rather large collisions. It is also possible that the respective ages are wrong, and we will see in the *Complex Exposure Histories* section that this is probably indeed so for part, but not all, of the "non-peak" meteorites. Continuous background production has been modelled by Graf and Marti (1995). The dotted lines in the uppermost panel of Figure 7 show their expected exposure age distributions (not to scale) for constant production and three different mean lifetimes against either ejection from the solar system or collisions with a planet or the Sun (the apparent maxima of these curves at ~7-30 Myr are the result of the logarithmic scale on the abscissa). The observed distribution shows the same sharp drop-off at >60 Myr as the 10-Myr-lifetime curve and this model distribution may also be consistent with the small but discernible fraction of meteorites with exposure ages <4Myr. However, Graf and Marti (1995) point out that such young ages may often be flawed, so that many of the meteorites nominally younger than 4 Myr were actually created in one of the 7-Myr events. They concluded that the dotted lines do not represent a possible background population well, unless either minimum transfer times from the asteroid belt are several million years for all but possibly the fastest few meteorites, or the backgound was absent in the past few Myr. In conclusion, it is possible that many of the meteorites outside one of the large peaks are the result of a small number of additional rather large collisions (Graf and Marti 1995). We will further discuss this in the Ex*posure ages and dynamical models of meteorite delivery* subsection.

L and LL chondrites. The histograms of these two classes have many similarities with that of the H chondrites: one very conspicuous peak each (at 40 Myr for the L chondrites and 15 Myr for the LLs), no ages larger than some 70 Myr, and (for LLs) very few ages below 5 Myr. Note that the few L chondrites below 2 Myr are strongly shocked and possibly have an atypical collisional history (Graf and Marti 1994). Therefore the observation of a scarcity of very low-exposure ages to some extent also holds for L chondrites. As with the H chondrites, the L chondrites also have minor peaks (at 28, 15, and 5 Myr) that become prominent only in a subgroup, here the ^{40}Ar-poor L chondrites (Marti and Graf 1992). Many L chondrites show deficits in radiogenic ^{4}He and ^{40}Ar due to a major collision on their parent body roughly 500 Myr ago (Bogard 1995; Swindle 2002a). Interestingly, the LL chondrites show a minor peak at 28 Myr and possibly another one at 40 Myr, both at positions also occupied by L chondrite peaks (Graf and

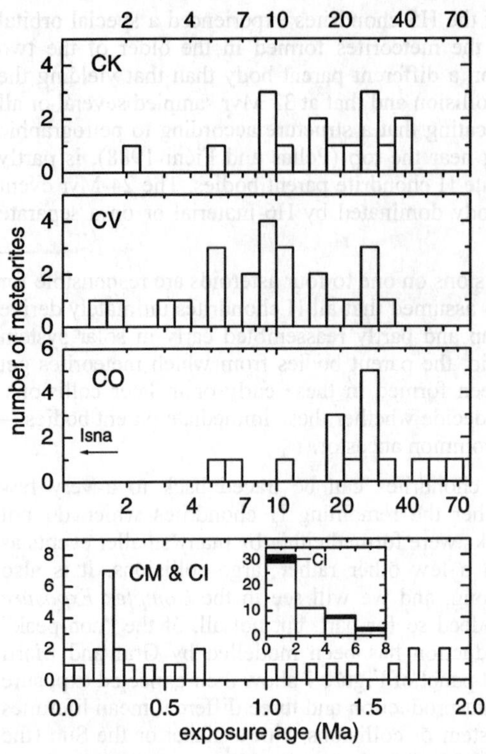

Figure 8. Exposure age distributions of carbonaceous chondrites. The top 3 panels (CK, CV, and CO chondrites) display data from Scherer and Schultz (2000) on a logarithmic time scale with 20% resolution. The bottom panel shows the histogram for the CM chondrites (including the few CI chondrite ages in the inset) according to Nishiizumi et al. (1993), Caffee and Nishiizumi (1997), and K. Nishiizumi (pers. comm., 2001). Note the widely different (linear) age scale in this panel.

Marti 1994). The LL chondrite Galim/a has an extraordinarily low-exposure age of 30,000 years only (Patzer and Schultz 2001).

Given these similarities in the exposure-age distributions, the conclusions for L and LL chondrites are similar to those for the H chondrites: a very few collisions are responsible for a large fraction of the L and LL chondrites, and it is conceivable that most of the others were also created in a few additional events. If a quasi-continuous background exists, transfer times are mostly at least several million years.

Enstatite chondrites. A comprehensive study of exposure ages of enstatite chondrites by Patzer and Schultz (2001) indicates clusters at 3.5, 8, and 25 Myr (Fig. 7), although the authors caution that these peaks need to be confirmed, since both the age uncertainties for enstatite chondrites are larger than for ordinary chondrites, and the data base for E-chondrites still is comparatively small. Keil (1989) suggests that the two subgroups EH and EL derive from two different parent bodies. It is therefore instructive to compare the exposure age distributions of EH and EL chondrites, even though this is hampered by somewhat poor statistics. Patzer and Schultz (2001) note that the two distributions are not distinguishable from each other, which would be remarkable if the two classes derive from different parent bodies. Yet, the suspected peaks at 3.5 and 8 Myr do not show up in the EL chondrites. On the other hand, the most clear-cut peak at 25 Myr in the lowermost panel of Figure 7 is indeed seen in both types, EH and EL (although it is somewhat irritating that only very few of the 29 E chondrites compiled by Okazaki et al. 2000 have exposure ages around 25 Myr). We further note that, very similar to ordinary chondrites, ages >60 Myr are absent and quite few ages are below 2

Myr. A notable exception is the EH3 chondrite Galim/b, which seems to be an inclusion in the LL-chondritic polymict breccia Galim/a with an exposure age of only ~70,000 years (Patzer and Schultz 2001).

Rumuruti chondrites. Schultz and Weber (2001) report an age range of 0.2-50 Myr for 18 members of this class. Many ages fall between 7 and 40 Myr, but Schultz and Weber (2001) note that the remarkably low age of about 200,000 yr of the R chondrite Northwest Africa 053 demonstrates that short transfer times from the Asteroid belt are possible. The number of R chondrites is too low to expect clear-cut exposure age clusters.

Carbonaceous chondrites. Carbonaceous chondrites are in many respects the most primitive meteorites. The exposure age distributions of the various chemical classes reveal a clear dichotomy (Fig. 8). The CV, CO, and CK histograms essentially look very similar to those of the other stony meteorite classes, with most ages falling in the range of between a few Myr up to 40-60 Myr (Scherer and Schultz 2000). The very low age of only 0.15 Myr of the CO chondrite Isna is remarkable. On the other hand, the chemically most primitive classes CI and CM have strikingly low-exposure ages of less than 7 Myr, the majority of them falling between a mere 50 kyr and 2 Myr (Nishiizumi et al. 1993; Caffee and Nishiizumi 1997; Eugster et al. 1998a; K. Nishiizumi, pers. comm. 2001, see Wieler and Graf 2001). CV and CK chondrites possibly show a peak in their exposure age distribution at ~9 Myr and all three classes CV, CK, and CO may display a cluster at ~29 Myr (Scherer and Schultz 2000). If these common events could be verified, this would support the idea that these classes are closely linked to each other (Kallemeyn et al. 1991). CI meteorites are much too scarce to reveal any possible clustering, but CM chondrites show a very distinct peak at 0.2-0.25 Myr.

Why does the majority of CM and CI chondrites have such low-exposure ages in a range almost unoccupied by all other stony meteorite classes? We noted above that one reason may be the mechanical weakness of these meteorites and their parent bodies, which may strongly reduce their mean lifetime against collisional destruction. This would imply than CM (and CI?) chondrite meteoroids are abundantly produced. However, as also noted above, the high exposure ages of the fragile aubrites illustrate that other factors must be involved, such as mean transfer times and dynamical lifetimes, as discussed below. Perhaps the distinct 0.2- to 0.25-Myr peak is due to a collision on an immediate CM parent body with an orbit crossing that of the Earth (Caffee and Nishiizumi 1997). Besides the one distinct peak, the exposure ages of CM chondrites are distributed rather evenly, perhaps suggesting that a comparatively large fraction of them derives from relatively frequent smaller collisions.

Differentiated meteorites

HED meteorites. The Howardite-Eucrite-Diogenite (HED) clan is the largest class of achondrites. It is of particular importance here because it is believed that we know the parent body: the asteroid Vesta, or perhaps some of the smaller "Vestoids" thought to have been spalled off Vesta in a very large collision (Binzel and Xu 1993). The HED exposure age distribution (Fig. 9) is again strikingly similar to those of ordinary chondrites: all ages fall between 5-80 Myr, and distinct peaks emerge at 20-25 and 35-42 Myr (Eugster and Michel 1995; Welten et. al. 1997). Welten et al. (2001b) propose that there is an additional peak at ~50 Myr for the diogenites, although this is not resolved in Figure 9. Possibly as few as 5 or 6 events can explain most of the HED exposure ages (Eugster and Michel 1995; Welten et al. 1997). Based on this low number, Welten et al. (1997) and Bottke et al. (2000) favor Vesta rather than the Vestoids as the major HED source.

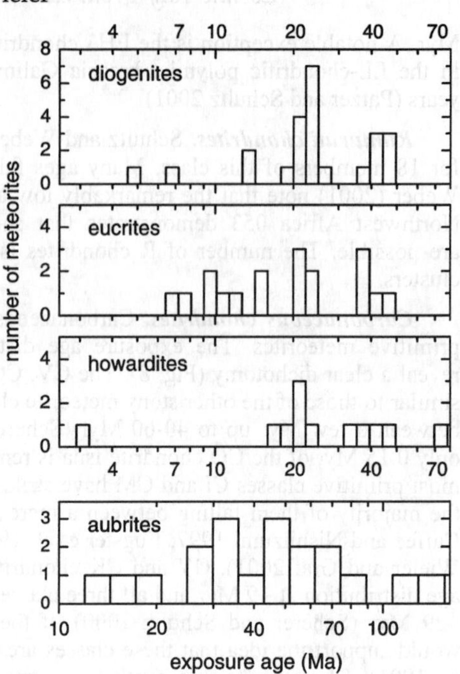

Figure 9. Exposure age distributions of HED meteorites (howardites, eucrites, diogenites) and aubrites or enstatite achondrites. The HED data are from the compilation by Welten et al. (1997), with only the (shielding corrected) ^{81}Kr-Kr ages being displayed for the eucrites. Included are 4 new diogenite ages from Welten et al. (2001b). In addition, the ~3 Myr age of the howardite Kapoeta has been added (Caffee and Nishiizumi 2001). For data sources of aubrites see text. Note the different abscissa scales between the lowermost panel and the others.

Aubrites (enstatite achondrites). This class is similar in many respects to enstatite chondrites, although the two classes are thought to derive from different parent bodies (Keil 1989). In the bottom panel of Figure 9, the exposure ages of aubrites are shown. These ages are more uncertain than those of chondrites, because production rate systematics have been less well studied for this rare class. The ages shown here rely on published cosmogenic Ne data (Schultz and Franke 2000) and the production model by Leya et al. (2000a), or, in a few cases, on ^{81}Kr-Kr analyses (Miura et al. 2000). In addition, we include ages given for eight aubrites from Antarctica by Lorenzetti et al. (2001). In the first systematic study of the cosmic ray record of aubrites, Eberhardt et al. (1965) already noted that aubrites have unusually long exposure ages among stony meteorites. This has become even more accentuated as the production rates we adopt here are lower than those used by Eberhardt and coworkers. Note on the other hand that the exposure age of the Norton County aubrite of 240-280 Myr reported by Begemann et al. (1957) has later been revised downwards and is shown here as only 130 Myr. Yet, it is quite remarkable that this meteorite with the first ever published cosmic ray exposure age still holds the record among stones, within uncertainties together with Mayo Belwa, another aubrite (Lorenzetti et al. 2001). Eberhardt et al. (1965) noted a cluster at 40-50 Myr, but uncertainties in production rates are too large to confirm this cluster in Figure 9 (at roughly 50-60 Myr with present-day production rates). Nevertheless, exposure ages of aubrites clearly are higher than those of most enstatite chondrites, consistent with the conclusion that aubrites derive from a separate parent body. The long interplanetary journeys of the fragile aubrites also indicate a distinctly different orbital evolution compared to other meteorite classes, resulting in longer survival times against ejection from the solar system or collision with a planet, as already pointed out by Eberhardt and coworkers.

Acapulcoites and lodranites. These meteorite classes, also called *primitive* *achondrites*, are believed to derive from the same parent body and are residues from partial melting of chondritic precursors (McCoy et al. 1997). Exposure ages of both classes show a remarkably tight clustering between 4 and 10 Myr, in the same range as the prominent 7 Myr cluster of the H chondrites (Weigel et al. 1999; Terribilini et al. 2000a; Ma et al. 2001). Terribilini et al. conclude that this may be due to one impact (with subsequent secondary break-ups of a larger meteoroid) or several impacts closely spaced in time. They point out that the coincidence of exposure age clusters of acapulcoites/lodranites and H chondrites might suggest an enhanced collisional activity in the asteroid belt some 7 Myr ago.

Ureilites. Goodrich (1992) and Scherer et al. (1998) compiled exposure ages of ureilites. Values range between 0.1 and 34 Myr. Again, the majority of the ureilites, i.e., >70% of the 22 meteorites compiled by Scherer et al. (1998), have ages above 3 Myr. No exposure age clusters are observed. However, since many ureilites contain relatively large amounts of primordial Ne, a shielding correction via $^{22}Ne/^{21}Ne$ is often not possible. Therefore, exposure ages of ureilites often have a quite high uncertainty, and so clusters, even if present, might be less easily recognized than in other meteorite classes. Mean activities of ^{10}Be and ^{26}Al in ureilites are ~20% lower than expected, indicating that many ureilites had a small preatmospheric size (Aylmer et al. 1990).

Iron meteorites and stony-irons. Reliable exposure ages for iron meteorites are often even more difficult to obtain than for stony meteorites, because quite a few iron meteorites had preatmospheric sizes of one to several meters, so that large production rate variations due to variable shielding are common. As a somewhat extreme example, ^{3}He concentrations in fragments of the R ~15 m Canyon Diablo meteorite vary by a factor of 10,000 (Heymann et al. 1966), and even in the 13 most gas-rich samples analyzed by these authors, He, Ne, and Ar concentrations varied by ~20 to 35 times. Reliable corrections for production rates over a wide range of shielding are therefore a prerequisite in exposure age studies of iron meteorites. Considerable efforts are being spent to achieve this, because iron meteorites often have much higher exposure ages than stony meteorites and therefore are particularly important for studying possible long term variations of the flux of the galactic cosmic radiation. Three main methods are He-Ne-Ar, ^{40}K-^{41}K, and ^{36}Cl-^{36}Ar. The relationship between noble gas production rates in iron meteorites and shielding parameters such as $^{4}He/^{21}Ne$ has been described by Signer and Nier (1960, 1962) using a semiempirical model. The ^{40}K-^{41}K method, involving the very long-lived radionuclide ^{40}K (half-life 1.25 Gyr) has been developed by H. Voshage (Voshage 1978 1984, and references therein) and the ^{36}Cl-^{36}Ar method has recently systematically been applied to iron meteorites by Lavielle et al. (1999a, 2001). These latter studies are also discussed in section *The Cosmic Ray Flux in Time*.

Figure 10 shows the exposure age distribution of iron meteorites. Ages >200 Myr are based on the ^{40}K-^{41}K method (Voshage 1978), while younger ages are based on ^{38}Ar (Lavielle et al. 1985), because few K-ages exist for the age range 0-200 Myr. The Figure may thus well be biased somewhat due to sample selection, possibly underestimating the fraction of iron meteorites with relatively low-exposure ages. The abscissa may also be compromised by a possible GCR intensity change as discussed in section *The Cosmic Ray Flux in Time*. Nevertheless, it is clear that the majority of iron meteorites has exposure ages exceeding the highest known age of a stony meteorite of some 130 Myr or so. Two clusters of exposure ages have been recognized for a long time, indicating major collisions on the IIIAB and IVA iron meteorite parent bodies about 650 Myr and 375 Myr ago, respectively, using the ^{40}K-age scale (Voshage 1978; Lavielle et al. 1985). Lavielle et al. (2001) proposed that the 375 Myr peak actually represents two events at 255 and 207 Myr ago in the ^{36}Cl-scale, whereas the 650-Myr peak corresponds to an

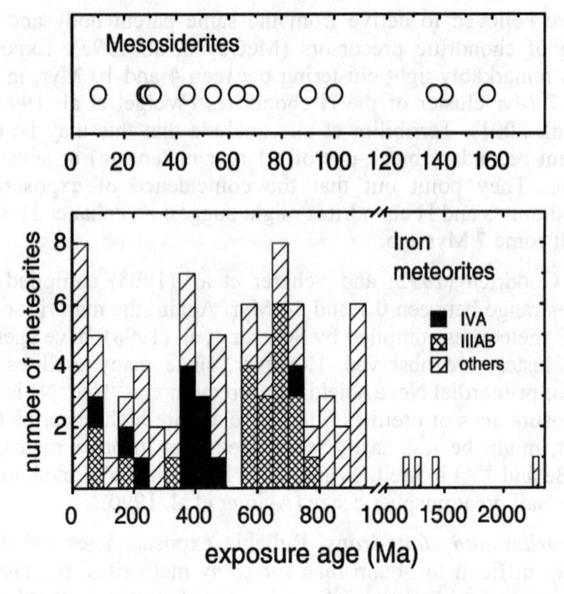

Figure 10. Exposure age distributions of iron meteorites and mesosiderite stony-irons. Iron meteorite data > 200 Myr were obtained by the ^{40}K-^{41}K method (Voshage 1978), ages < 200 Myr are based on ^{38}Ar (Lavielle et al. 1985). Iron meteorite groups IVA and IIIAB show exposure age clusters around 375 and 650 Myr (^{40}K-scale) and are shown separately. The 375 Myr cluster appears to be composed of two independent peaks not shown here (at 207 and 255 Myr in the ^{36}Cl-scale, see text). Mesosiderite ages (^{36}Cl-^{36}Ar) are from Albrecht et al. (2000), as updated from Begemann et al. (1976). Note the widely different abscissa scales of the two panels.

event at 460 Myr in this scale (Lavielle 2001, pers. comm). In contrast to other iron meteorites, the two classes IIA and IIE show exposure age ranges not unlike those of stony meteorites. Ages of IIA and IIE irons may be as low as a few Myr (Olsen et al. 1994), with the most reliable among the lowest values being around 10 Myr (e.g., Rafrüti: Terribilini et al. 2000b; Kodaikanal, Braunau, and Watson: Lavielle 2001, pers. comm.).

A study by Begemann et al. (1976), updated in Albrecht et al. (2000), determined ^{36}Cl-^{36}Ar ages of 16 stony-iron meteorites, mostly mesosiderites (Fig. 10). The values fall between ~10-180 Myr (10-160 Myr for mesosiderites alone), with about a third of the values being larger than 100 Myr. The age range for stony-irons is thus broadly in between those of stones and most irons. Welten et al. (2001c) note a hint for an exposure age cluster around 60-70 Myr.

The comparatively very old exposure ages of iron meteorites, and perhaps also the rather high ages of stony irons, are probably partly the result of their mechanical strength and hence long survival times against collisional destructions. Long collisional lifetimes could also explain why iron meteorites probably derive from considerably more parent bodies than stony meteorites. Petrographic and chemical evidence suggests that there are 90 or so different original iron meteorite parents (Wasson 1995), and a large number of them are sampled only by relatively small meteorites (Wasson 1990).

Lunar and Martian meteorites. At the time of this writing, our collections contain about 20 individual meteorites (corrected for pairings) from the Moon and 18 from Mars. A list of the Martian meteorites is given by Swindle (2002b), who also briefly discusses their cosmic ray record. Exposure ages of Martian meteorites are also compiled and discussed in detail by Nyquist et al. (2001). Cosmogenic noble gases and radionuclides are a major tool for deciding whether different meteorite specimens, e.g., found in a desert nearby to each other, belong to the same fall or whether different meteorites from any find location were ejected by the same event (source crater pairing).

Quite as expected, most meteorites from the nearby Moon have had considerably shorter transit times than the meteorites from the asteroid belt discussed so far. Most exposure ages of lunar meteorites are below 1 Myr, and many even below 0.1 Myr. Very remarkable is the short transit time of Dhofar 026 of at most 4 Kyr (Nishiizumi and Caffee 2001). The highest age found so far is ~8 Myr (Polnau and Eugster 1998; Nishiizumi et al. 1999). Very few, if any, lunar meteorite falls are source crater paired (Warren 1994). Essentially all lunar meteorites have a complex exposure history, i.e. they also contain cosmogenic nuclides acquired on the Moon (e.g., Warren 1994; Polnau and Eugster 1998). This means that most lunar meteorites were ejected in relatively minor cratering events from within the uppermost very few meters of the lunar surface, where they had previously spent up to several hundred million years. Nishiizumi et al. (1999) note a correlation between the depth of ejection and the transit time to Earth, with meteorites from smaller events arriving earlier.

Martian meteorites had travelled for considerably longer than lunar meteorites, with transit times of between 0.7 and 20 Myr (Nyquist et al. 2001). Unlike for lunar meteorites, source crater pairing is common for Martian meteorites. The number of individual ejection events on Mars needed to account for the available meteorites is controversial, however (see also Swindle 2002b). Based on crystallisation ages, four to five impacts were proposed by Nyquist et al. (1998) (see also Nyquist et al. 2001). On the other hand, exposure ages suggest 7-9 events (Nyquist et al. 2001, Nishiizumi et al. 2000; K. Nishiizumi, pers. comm. 2000). Nyquist et al. (2001) discuss scenarios involving secondary collisions to reconcile the two lines of evidence. Since, unlike lunar meteorites, none of the Martian meteorites show signs of a complex exposure (Warren 1994; Nyquist et al 2001; however see also Hidaka et al. 2001), the bodies originally launched from Mars would have to have been quite large. The apparently single-stage exposure history also indicates that all Martian meteorites were launched from at least a few meters below the Martian surface (Warren 1994). A caveat to this conclusion is that a parent body exposure stage is more difficult to detect in meteorites having had a longer subsequent exposure during their transit to Earth. However, Martian meteorites on Earth are usually considerably larger than their lunar counterparts and hence probably also had a larger preatmospheric size, which also suggests that they stem from larger events than lunar meteorites. This is indeed expected given the different escape velocities from the two planets.

Micrometeorites and interplanetary dust particles (IDPs). The noble gas record in micrometeorites and IDPs is discussed by Wieler (2002). Cosmic ray exposure ages of micrometeorites from Greenland (Olinger et al. 1990) and of deep-sea spherules (Raisbeck and Yiou 1989) are between < 0.5 and 26 Myr, suggesting a cometary origin of most of these particles (Raisbeck and Yiou 1989). Cosmic ray exposure ages of IDPs are uncertain. Pepin et al. (2001) report large ^{3}He excesses in some IDPs which, if cosmic-ray-produced, would imply GCR ages of up to more than one Gyr, much larger than orbital lifetimes of small particles. The authors consider that this might be due to very long exposures in the regoliths of Kuiper Belt comets, but they also discuss the possibility that the excess ^{3}He is primordial (see Wieler 2002).

Exposure ages and dynamical models of meteorite delivery

Here we discuss constraints imposed by the observed distributions of exposure ages on ideas and models of how meteorites get delivered to Earth from the asteroid belt, the Moon and Mars (see also Wieler and Graf 2001).

Stony meteorites from the asteroid belt. It has been known for a long time that ~15% of all meteorites falling today were produced in one—or more probably two—collisions manifested in the prominent 7 Myr peak in the exposure age histogram of the H chondrites. It has often been thought, however, that this was an exceptional event and that a quasi continuous production of meteorites in many smaller events is the rule. We have seen above that to some extent this discussion goes on today, as it remains open whether most of the meteorites not obviously belonging to a major peak in an exposure age distribution should be assigned to the "background" or to one of a few additional events, and how many of even these meteorites should be shifted to one of the large peaks. Furthermore, for all but the most common meteorite classes, even the quite impressive data base available today still often does not allow an unambiguous recognition of peaks. Nevertheless, exposure age peaks are identified or suspected in the histograms of many classes, which allows us to reiterate a conclusion already made by Anders (1964): a large part of the meteorites were produced in a few distinct major collisions. The quantification of this statement is somewhat difficult, but we note that more than 50% of all H chondrites derive from a few events, whereas small iron meteorites appear to be due mostly to a quasi continuous production.

It also has long since been recognized that orbital resonances with planets are a major factor determining the delivery of meteorites from the asteroid belt (Wetherill and Williams 1979; Wisdom 1983; Gladman et al. 1997; Farinella et al. 2001). Until recently, a common assumption was that the velocity of meteorites ejected from their parent asteroids needs to be sufficiently high to place them into a resonance, e.g., the 3:1 Kirkwood gap in the main asteroid belt, where an object orbits three times around the Sun in the time it takes Jupiter to complete one orbit. This has led to the hypothesis that only large collisions on parent bodies close to a resonance could produce large numbers of meteorites on Earth. This would qualitatively explain the distinct peaks in exposure age histograms, but it might be difficult to account for the observed flux of meteorites with the limited number of viable parent bodies (see discussion by G.W. Wetherill in Greenberg and Nolan 1989). Recent dynamical models now indicate that objects placed into a so-called chaotic zone in a resonance will have a much shorter dynamical lifetime than has previously been thought. Within a very few million years they will either collide with the Sun or be ejected from the solar system (Gladman et al. 1997). As these times are considerably less than typical exposure ages of almost all meteorite classes, in this scenario most meteorites need to spend most of their lifetime as meter-sized objects in the main belt, slowly drifting into a resonance (Morbidelli and Gladman 1998). It has recently been realized that the asymmetric radiation of thermal energy, the so-called Yarkovsky effect, can provide enough momentum to allow meter-sized objects to drift into a resonance on timescales comparable with meteorite exposure ages (e.g., Bottke et al. 2000). The Yarkovsky effect appears to be able to explain also the paucity of meteorites with exposure ages less than a few million years (Bottke et al. 2000). If so, asteroids throughout large parts of the main belt may be potential meteorite parent bodies. This, however, is only compatible with the marked peaks in the exposure age distributions, if a very few ones among these many potential parent asteroids contribute a major part of all meteorites. These selected bodies presumably should be among the largest asteroids. Wieler and Graf (2001) discuss some evidence that this might indeed be the case. On the other hand, the original H chondrite parent body or parent bodies, responsible for a large fraction of all meteorites today, were rather small, with a radius of only about 100 km (Pellas and Fiéni 1988). Bottke et al. (2000) also discuss this problem. Clearly, exposure age distributions with improved temporal resolution will be needed to further rule on the importance of the Yarkovsky effect as meteorite delivery mechanism.

Even though meteorites may be brought to Earth quickly from within a resonance in the main belt, it is not clear whether the 0.2-Myr peak of the CM chondrites can be explained this way. This would require a very large collision on a CM parent body in a position where the time needed to drift into a resonance would be extremely short and from which we would see today perhaps just some forerunners. It appears more likely that the 0.2 Myr CM chondrites are the products of a collision on a parent body already on an orbit crossing that of the Earth (Caffee and Nishiizumi 1997).

Iron meteorites. Meter-sized iron meteorites have a slower Yarkovsky drift rate than stones of the same size (Farinella et al. 2001), which is in qualitative agreement with the much older exposure ages of iron meteorites (see also Bottke et al. 2000; Morbidelli and Gladman 1998). The old ages lead to the expectation that many more collisions are recorded in the iron meteorite histograms compared to those of stony meteorites, consistent with the rareness of peaks in the former. The comparatively large number of iron meteorite parent bodies can be explained if iron meteorites are able to drift into a resonance from even larger portions of the belt than stones, because of their longer collisional lifetimes due to their mechanical strength. Wasson (1990) proposed that small iron meteorites move faster through the belt than large ones due to their higher ejection velocity, which would account for the high number of parent bodies sampled by small iron meteorites.

Lunar and Martian meteorites. Gladman et al. (1996) calculate that most lunar meteorites that end up on Earth do so within several 10^4 years. Fragments escaping this fate will reach orbits outside the Earth's immediate influence after about a million years. The calculations are thus essentially in agreement with the observations. Gladman (1997) calculates transfer times from Mars in agreement with the observed exposure ages for material ejected just slightly above escape velocity. Martian meteorites travel much longer than lunar ones simply because the orbits of Martian ejecta will not cross Earth's orbit initially. Given the closeness of the Moon, it appears astonishing that we do not have many more lunar than Martian meteorites. Gladman (1997) notes, however, that the higher fraction of escaped lunar meteorites that end up on Earth and the higher number of fragments per impact on the Moon may be largely offset by higher cratering rates on Mars and the fact that many Martian meteorites are source-crater-paired. It is thus not quite clear whether or not the similar numbers of lunar and Martian meteorites actually constitutes a puzzle. If it does so, this may be explained by mechanical properties of the two planetary surfaces. The low porosities of lunar meteorites suggest that compaction of loose regolith and simultaneous ejection may not work so that only the small fraction of already coherent rocks on the lunar surface may have a chance to end up as meteorites (Warren 2001).

COMPLEX EXPOSURE HISTORIES

The term *complex exposure* is commonly used to denote a situation where an entire meteorite acquired part of its cosmogenic nuclides detectable today in a larger body than the one immediately prior to atmospheric entry. A complex exposure history thus indicates a break-up of a precursor meteoroid or an ejection of a meteoroid from a near-surface location of a parent asteroid or planet. In the first case we talk about two different so-called 4π exposure stages, in that during both stages each sample saw cosmic rays from all directions in space, while in the second case a 2π irradiation is followed by the 4π exposure. It may also occur, however, that only certain parts of a meteorite (e.g., clasts or individual mineral grains) show excesses of cosmogenic nuclides relative to other fractions of the meteorite. This is commonly called *pre-compaction* exposure, and the precompaction stage may have occurred early in solar system history or later in a dust and gravel layer on the parent body surface, the *regolith.*

Thus, meteorites with a complex exposure history may help us to constrain collisional dynamics in the asteroid belt, the dynamics of asteroidal regoliths, and perhaps the energetic particle environment in the early solar system. On the other hand, we mentioned above that complex exposure histories often hinder our efforts to determine exposure age distributions.

Sometimes, a complex exposure history is quite easily recognized, for example if cosmogenic noble gases indicate an exposure during several ten million years in an average-sized meteorite, whereas the ^{10}Be activity is lower than the saturation value of medium-sized meteorites. Often, however, analyses of many different nuclides in different samples, if possible from well defined positions relative to each other, are required to unambiguously decide whether a meteorite had a simple or complex history. Nuclides such as ^{60}Co with its steeply rising production rate with preatmospheric depth (e.g., Heusser et al. 1996; see Fig. 1) or nuclear tracks with their steeply declining production rate with preatmospheric depth (Bhattacharya et al. 1973) are particularly useful but such data are rarely available (and ^{60}Co has decayed completely a few decades after the fall of a meteorite). It should thus not come as a surprise that only for few meteorites a complex exposure history has been unambiguously documented or at least found to be very likely. Vogt et al. (1993) and Herzog et al. (1997) list just 15 stony meteorites with a probable or certain complex history. It is even more difficult to exclude a complex exposure history (a prominent example is Knyahinya, where from numerous analyses a single stage lasting ~40 Myr has been deduced; Graf et al. 1990b). Therefore, the fraction of meteorites with a complex history cannot be estimated with confidence. Wieler and Graf (2001) guess that it could be around 30% for stony meteorites. Similarly, Lavielle et al. (1985, 2000) note that about one-fourth of all iron meteorites show indications of a complex history.

Some meteorites with a well documented complex history are Jilin, Bur Gheluai, Tsarev, Torino, Gold Basin and QUE93021 (Begemann, et al. 1996; Vogt et al. 1993; Wieler et al. 1996; Welten et al. 2001d,e). The second stages lasted on the order of 1-15 Myr and the first stages sometimes up to a hundred or a few hundred million years. The size of the first stage body is often unconstrained, but very long first stages seem to require an asteroidal parent body. Meteorites with nominally short exposure ages appear to often show a discernible complex history (Herzog et al. 1997; Merchel et al. 1999). This is not too surprising, because a first stage contribution is easier to detect if the subsequent second stage lasted only briefly. In these examples, first stages typically lasted on the order of 10 Myr, second stages only around a few hundred thousand years. This does not mean, however, that very low 4π exposure ages are more common than it may appear from the exposure age histograms, because the first stage may already have occurred in a meteorite-sized body rather than on a parent asteroid. This is illustrated by the H-chondrite Jilin, which is a good example of how complex histories smear out peaks in exposure age histograms. Jilin was probably ejected in one of the 7-Myr events and broken up in a second collision 0.3 to 0-6 Myr ago, whereas calculated single stage ages are between 1-5 Myr only, depending on the sample analysed (Begemann et al. 1996; Heusser et al. 1996). We mentioned above that a large part of the few H chondrites with <7-Myr nominal ages may have left their parent body in one of the 7-Myr events. On the other hand, it is worth repeating here that the mere existence of peaks in exposure age histograms is compelling evidence that the single stage assumption yields correct ages for a large part of all meteorites. Hence, while unrecognized complex exposures do compromise exposure age histograms to some extent, this even strengthens the conclusion that a large fraction of all meteorites is produced in a very limited number of collisions in the asteroid belt.

Pre-compaction exposures. When parts of a meteorite show excesses of cosmogenic nuclides with respect to other parts of the same meteorite, and when these excesses cannot be explained by higher production rates due to higher concentrations of major target elements or favourable shielding, then the excesses must have been acquired before the various constituents of the meteorite were assembled as they are observed today. Such pre-compaction exposures are often recognized in so-called gas-rich meteorites, which contain noble gases implanted by the solar wind and represent compacted dust and pebbles from the surface regolith of an asteroid. Cosmogenic noble gas excesses have also been reported in single mineral grains or single chondrules from meteorites with no obvious signs of a parent-body-regolith exposure. These excesses may indicate an irradiation by an intense Sun in the early solar system. It might also be feasible to determine the lifetime of presolar grains before they were incorporated into meteorite parent bodies. These three topics are discussed in the following.

Gas-rich meteorites usually are a mixture of compacted solar-gas-bearing dust (matrix) and solar-gas-free cm-sized pebbles (inclusions). Often the matrix is darker than the inclusions. In many meteorites studied in detail, some of the inclusions contain excesses of cosmogenic noble gases, corresponding to 2π exposures on the order of millions to tens of millions of years (e.g., Wieler et al. 1989b and references therein; Pun et al. 1998). Commonly this is interpreted as the time a pebble spent near the surface of its parent body regolith, or, more exactly, as the differences of the regolith exposure durations of different pebbles. Compared with exposure ages in the lunar regolith of a few hundred million years, meteoritic regoliths are thus less mature, i.e., they had less time to evolve under the influence of (micro)meteorite bombardment and solar and galactic particle irradiation. Some of the inclusions are xenolithic and hence must ultimately derive from another parent body than their host meteorite (e.g., St. Mesmin Schultz and Signer 1977; Fayetteville: Wieler et al. 1989b; Kapoeta: Pedroni 1989; Pun et al. 1998). The respective excesses may therefore at least partly have been acquired on the grandparent body or perhaps while the future xenoliths were on their way to their foster parent bodies. Caffee and Nishiizumi (2001) studied ^{10}Be in clasts of the gas-rich howardite Kapoeta known to contain excess ^{21}Ne. They did not find ^{10}Be concentrations above those expected to have been acquired during ~3 Myr of meteoroid exposure, which indicates that the parent body exposure that produced the ^{21}Ne excesses had not occurred immediately prior to the ejection of the Kapoeta meteoroid. Regolith compaction and meteorite ejection therefore did not happen at the same time, at least not in the case of Kapoeta.

Matrix samples of gas-rich meteorites often also contain excess cosmogenic noble gases (Wieler et al. 1989a, Pedroni 1989; Assonov et al. 1996; Bogard et al. 2001), corresponding again to parent body exposure ages of a few 10^6 to a few 10^7 years. In some meteorites, the concentrations of solar and cosmogenic noble gases correlate well with each other, as may be expected (Wieler et al. 1989a; Assonov et al. 1996).

The group led by Charles Hohenberg in St. Louis has extensively studied cosmogenic ^{21}Ne in single mineral grains or batches of grains from several meteorites (Caffee et al. 1983, 1987; Hohenberg et al. 1990; Woolum and Hohenberg 1993). The grains were selected according to the presence or absence of solar-flare tracks, acquired while a grain was within the top few mm of the parent body surface. The track-rich grains almost all contained large excesses of cosmogenic Ne, whereas the track-poor grains did not. The excesses correspond to parent body exposure times to galactic cosmic rays of several 10^7 to several 10^8 years. Because this appeared too long, it was proposed that the cosmogenic Ne in the track-rich grains was due to an irradiation with a high flux of energetic particles from the early Sun, which was far more active then than today (Caffee et al. 1983, 1987). In these early studies, mostly gas-rich meteorites were studied, which led Wieler et al. (1989a, 2000a) to propose the alternative explanation that the correlation of solar flare tracks with cosmogenic Ne excesses was produced by mixing of parcels of mature parent body regolith, which had been in the top meter or so for tens of millions of years, with less mature regolith. In this view, the Ne excesses are due to GCR production in the parent body regolith.

Whereas the long parent-body GCR irradiation scenario appears quite straightforward to explain the single grain data from gas-rich meteorites, even larger ^{21}Ne excesses were subsequently found in individual track-rich olivine grains from several carbonaceous chondrites of type CM (Hohenberg et al. 1990). The implied regolith GCR exposure ages of at least ~150 Myr, but more probably ~300 Myr, appear unreasonably long (Woolum and Hohenberg 1993). Therefore, the CM chondrite data are the most promising candidates for revealing evidence of an early active Sun. On the other hand, the solar flare track densities in CMs might be expected to be higher than observed if the cosmogenic Ne excesses were due to solar cosmic rays (Woolum and Hohenberg 1993; Wieler et al. 2000a).

Polnau et al. (1999, 2001) reported for several meteorites systematically higher concentrations of cosmogenic noble gases in chondrules than in matrix samples of the same meteorite. The excesses would correspond to 4π ages on the order of 0.1-3 Myr. The authors exclude the possibility that this is due to preferential gas loss from the fine-grained matrix, or to systematic errors in the corrections for variable target element chemistry, and conclude that the chondrules contain noble gases from a precompaction exposure. Only one of the meteorites studied contains solar noble gases (Polnau et al. 1999) and is therefore expected to show effects of pre-exposure in the parent body regolith. Polnau et al. (2001) therefore suggest that the chondrules in the other meteorites may have experienced exposure in a very high flux of solar energetic particles in the early solar system. This conclusion deserves further verification, especially by meteorites with very short exposure ages.

Lewis et al. (1994) calculated the pre-compaction ages of presolar SiC grains in primitive meteorites based on their concentrations of cosmogenic ^{21}Ne (after correcting for the ^{21}Ne produced in the recent irradiation of the host meteoroids). These ages are important for understanding the provenance of the grains and their history before incorporation into larger solid bodies in the early solar system (see Ott and Begemann 2000). The presolar ages derived by Lewis et al. (1994) of between ~10 and ~130 Myr were surprisingly low, with larger grains showing higher ages. Later, Ott and Begemann (2000) invalidated these ages when they showed that the loss of cosmogenic ^{21}Ne from micron-sized SiC grains due to recoil is much larger than was assumed by Lewis et al. (1994). In fact, the recoil loss rates of ^{21}Ne are so large as to inhibit a meaningful correction, except perhaps for the largest grains. Cosmogenic ^{126}Xe may be more promising (Ott and Begemann 2000; Ott et al. 2001), because of its considerably lower recoil range compared to ^{21}Ne.

Space erosion. It is conceivable that meteoroids change their shape not only by a macroscopic collision but by continuous erosion by micrometeorites in space. This space erosion has occasionally been invoked to explain discrepant exposure ages based on various nuclide systems. However, no generally accepted space erosion rates appear to have been derived and space erosion is not routinely taken into account when deducing exposure ages of meteorites. Voshage (1984) estimated space erosion rates on the order of 0.1 mm/Myr for iron meteorites by comparing the minimum depths of iron meteorite samples with variable exposure ages, but he cautioned that this might not be a well constrained estimate, as simulation experiments yielded about 5 times lower values (Schaeffer et al. 1981). Space erosion rates on the lunar surface are usually assumed to be on the order of ~1 mm/Myr (e.g., Kohl et al. 1978). If these values apply to stony meteorites also, detectable effects would hardly be expected or would be limited to meteorites with exposure ages of at least several 10^7 Myr. Schaeffer et al. (1981) deduced a value of 0.65 mm/Myr for stony meteorites. However, Welten et al. (2001b) observed systematic differences between $^{10}Be/^{21}Ne$ and ^{21}Ne ages for diogenites with high exposure ages, whereas both methods yield consistent ages for younger diogenites. This might be explained by a space erosion rate of ~5 mm/Myr. Welten and coworkers caution that their observation might instead indicate a combination of (microscopic) space erosion and repeated chipping-off of larger fragments. On the other hand, if meteorites indeed spend most of their lifetime as meter-sized objects in the main asteroid belt as is indicated by modern dynamical models (see *Exposure ages and dynamical models of meteorite delivery* subsection), space erosion rates might be larger than commonly assumed due to a higher flux of micrometeorites. It therefore appears advisable to reconsider the issue of space erosion. Perhaps, discrepancies between noble gas and radionuclide data are not always due to a complex exposure history in the classical sense of collisional fragmentation (Welten et al. 2001b).

THE COSMIC RAY FLUX IN TIME

Concentrations of cosmogenic nuclides in meteorites—as well as in lunar and terrestrial samples—are widely used to study variations in the flux of galactic and solar cosmic radiation over time scales in the range of years to billions of years. GCR flux variations may occur due to variable modulations by the Sun or variations in the intensity of the flux beyond the heliosphere, the region of influence of the Sun. Changes of the interstellar GCR flux may occur, e.g., when the Sun passes through a spiral arm in the Galaxy, where the number of nearby supernovae is higher. In terrestrial samples, variations of the geomagnetic field intensity and several other parameters also have to be considered (Damon and Sonett 1991; Niedermann 2002). This makes meteorites particularly attractive for studying flux variations. On the other hand, radionuclide concentrations in terrestrial reservoirs often provide much higher time resolution, because they record instantaneous, and not integrated, production. Examples are ^{10}Be in sediments or ice cores and ^{14}C in tree rings, which allow, e.g., the study of solar activity during periods of reduced sunspot numbers such as during the Maunder Minimum (Beer et al. 1991; Damon and Sonett 1991).

In meteorites, short-term cosmic ray flux variations can be recognized by measuring the activity of a radionuclide with a short half-life in meteorites with a known fall date. Longer-term variations are studied by comparing apparent production rates of stable (noble gas) nuclides based on various pairs of a stable and a radioactive nuclide, or by comparing experimental and model data (see *Production systematics* subsection).

Short-term variations. Evans et al. (1982) showed that in freshly fallen meteorites between 1967 and 1978 the shielding- and target-chemistry-corrected activities of four radionuclides with half-lives between 16 days and 2.6 years were 2.5-3 times lower at solar maximum than at solar minimum. These data showed that solar modulation of the galactic cosmic radiation by the 11-year solar cycle strongly influences production rates of cosmogenic nuclides. Evans et al. (1982) also noted that their results may suggest variations of solar-modulated cosmic ray fluxes on much longer timescales. A similar variation with the solar cycle between 1965 and 1993 was observed by Bhandari et al. (1994) for the ratio $^{22}Na/^{26}Al$, which is rather insensitive to shielding. This result also implies that a careful correction is needed when applying the couple ^{22}Na-^{22}Ne to determine exposure ages. The radionuclide ^{44}Ti ($T_{1/2} \approx 60$ yr) has been extensively studied by G. Bonino and coworkers in meteorites that fell during the past two hundred years (e.g., Bonino et al. 1999). They find that shielding- and chemistry-corrected ^{44}Ti production rates vary by about 50%. The highest values are observed about 35 years after the two Gleissberg minima of solar activity at ~1800 and ~1900, a phase lag arising because the ^{44}Ti activity at the time of fall represents the time-integrated production rate, not the instantaneous one. Bonino and coworkers note that the ^{44}Ti variations are larger than expected, probably implying that the solar magnetic field during the Gleissberg minima was lower than during recent 11 year cycle minima.

Longer-term variations. Whereas short-term variations are well established and the Sun is known to be their major reason, it is less clear whether meteorites record cosmic ray intensity variations on a million to billion year time scale. The ^{21}Ne production rates derived from the nuclide couple ^{26}Al-^{21}Ne often turned out to be about 50% higher than ^{21}Ne production rates based on other radio-nuclides, in particular ^{22}Na ($T_{1/2} = 2.6$ yr), ^{81}Kr (2.3×10^5 yr), ^{10}Be (1.51×10^6 yr), and ^{53}Mn (3.7×10^6 yr). Such calibrations are discussed by Nishiizumi et al. (1980), Moniot et al. (1983) and earlier studies referenced therein (see *Production systematics* subsection). It is very unlikely that this is due to an incorrectly determined half-life of ^{26}Al. One other possible reason is a higher GCR flux during a time interval for which ^{26}Al is more sensitive than any of the other nuclides studied. As some of these have longer and others shorter half-lives than ^{26}Al, this higher flux would have prevailed perhaps around a few hundred thousand years or so but would have returned to "normal" levels during the more recent periods mainly recorded by ^{81}Kr and ^{22}Na. This is certainly not impossible. Nevertheless, at present a cosmic ray flux change is not the preferred explanation for the aberrant ^{26}Al-based calibration of production rates. A popular explanation is that a large fraction of the meteorites studied with the ^{26}Al-^{21}Ne couple suffered a complex exposure history and hence contain a discernible amount of ^{21}Ne from a first irradiation stage (e.g. Moniot et al. 1983; Vogt et al. 1993). A contribution from an earlier exposure stage contributes more to the total inventory of a stable cosmogenic nuclide if the second stage lasted only briefly (see *Complex Exposure Histories* section). It is therefore to be expected that ^{21}Ne from a previous exposure stage is more easily seen in meteorites with an exposure age short enough that ^{26}Al is not yet in saturation than in meteorites with longer exposures, where ^{10}Be or ^{53}Mn (but not ^{26}Al) are undersaturated. However, according to this logic, complex exposures should compromise nuclide couples involving ^{81}Kr or ^{22}Na even more than those with ^{26}Al. This explanation therefore fails to explain the agreement of the ^{21}Ne production rates based on either ^{81}Kr or ^{22}Na with the values based on ^{10}Be or ^{53}Mn. Vogt et al. (1993) suggest that improper shielding corrections may compromise the ^{26}Al-^{21}Ne system more than the other nuclide couples discussed here.

Fortunately, there is an independent way to use radionuclide concentrations in meteorites to constrain possible temporal variations of the GCR flux. This is by comparing measured nuclide activities in meteorites having a known single-stage exposure history with predicted activites based on data from simulated GCR irradiations and physical nuclide production models (see *Production systematics* subsection). The model by Leya et al. (2000a) does reproduce measured depth profiles of concentrations of ^{10}Be, ^{26}Al, ^{36}Cl, and ^{53}Mn in the chondrite Knyahinya simultaneously using the same assumed value of the GCR flux. This means that the mean cosmic ray flux has been constant to within the combined uncertainties of analyses and model predictions over a few half-lives of the involved radionuclides, i.e., over perhaps 7 to 10 Myr. The range of uncertainty of this statement is somewhat difficult to quantify but should not exceed 30%. Note, however, that the present-day activity of a given nuclide is much more sensitive to a more recent production rate change than to one of the same magnitude and duration that occurred several half-lives ago. Therefore, if the GCR intensity were higher or lower by 30% earlier than say some 6 Myr ago, this would not be recognized even in the ^{53}Mn activity with present-day precision. To extend the time scale for investigations of the past GCR flux, nuclides with a longer half-life than ^{53}Mn are thus essential. The prime candidate is ^{129}I (15.7 Myr), although only very few data are available so far. Preliminary work by Schnabel et al. (2001) suggests that any change in the GCR flux larger than 35% and 50% over the last 15 and 20 Myr, respectively, is unlikely (see also the discussion of the live ^{129}I-^{129}Xe method in the *Production systematics* subsection).

Beyond ^{129}I, only one radionuclide has been used to study the GCR flux, which is ^{40}K (1.25 Gyr) in iron meteorites (Voshage 1978, 1984 and references therein). This nuclide offers the most firm evidence that the cosmic ray flux may be variable over time scales much longer than those governed by solar activity changes (up to thousands of years), as revealed by meteorites (see above) and ^{10}Be in terrestrial archives (e.g., Beer et al. 1991). Voshage (1984) concluded that the allowed range of ^{21}Ne production rates as a function of ^{4}He/^{21}Ne according to the Signer-Nier model is independent from the ^{40}K-^{41}K ages, indicating that the long-term (~100 Myr or so) average cosmic ray intensity has been constant to within 10-20% over the last 1 Gyr (the exposure age range covered by most iron meteorites). Conversely, by comparing ^{40}K-^{41}K ages with ^{36}Cl-^{36}Ar ages, he concluded that the GCR intensity over the past few Myr (the past few half-lives of ^{36}Cl) was higher than the long-term average by some 50%. Hampel and Schaeffer (1979) arrived at a similar conclusion by comparing ^{26}Al-^{21}Ne ages of iron meteorites with ^{36}Cl-^{36}Ar, ^{39}Ar-^{38}Ar and ^{40}K-^{41}K ages. The ^{36}Cl-^{36}Ar method has recently been systematically applied to iron meteorites by B. Lavielle and coworkers (Lavielle et al. 1999a). These authors confirm a discrepancy between nominal ^{40}K-^{41}K ages and ages based on ^{36}Cl-^{36}Ar or other pairs involving radionuclides with a much shorter half-life than ^{40}K, such as ^{10}Be and ^{53}Mn. In agreement with the conclusions drawn by Voshage, a recent increase of the cosmic ray intensity by some 40% was found to be the most straightforward explanation for this observation, but Lavielle et al. (1999a) caution that the production rate systematics of potassium isotopes may not be known accurately enough to firmly conclude this.

In summary, meteorites provide evidence for GCR flux variations of up to a few times governed by solar activity changes on times scales of 1-100 years. On time scales of up to 5-20 Myr no clear evidence for changes has been found, but in the last 5-20 Myr, and possibly longer, the mean GCR intensity presumably was some 40-50% higher than the average over a billion years. Perhaps this may be explained by the position of the Sun in the Orion arm of the galaxy relatively close to the galactic midplane (Schaeffer 1975). Nothing is known whether such fluctuations on a roughly ten-Myr scale occur regularly, but on a hundred-Myr scale, iron meteorites indicate no discernible GCR flux change.

TERRESTRIAL AGES

Terrestrial ages of meteorites that were not observed to fall are determined with cosmogenic radionuclides, whose production is essentially stopped when the meteorite becomes shielded by terrestrial atmosphere and magnetic field (Jull 2001). Noble gases are used in most such studies only to verify the assumption that the meteorite's exposure age in space was long enough for the radionuclide activity to have reached saturation. However, the radioactive noble gas nuclide ^{81}Kr can also be used to determine terrestrial ages (Freundel et al. 1986; Miura et al. 1993). The most reliable (shielding-corrected) terrestrial ages are obtained by a combined analysis of two or more radionuclides with different half-lives (e.g., ^{36}Cl-^{10}Be, ^{14}C-^{10}Be, ^{41}Ca-^{36}Cl; Nishiizumi et al. 1997a; Lavielle et al. 1999a; Kring et al. 2001; Welten et al. 2001a). The time of fall of a meteorite is important for studying infall rates, weathering rates, and possible meteorite concentration mechanisms, particularly in areas such as Antarctica or hot deserts where meteorites are actively searched today (Jull 2001). For exposure age determinations using a stable-radioactive nuclide pair, terrestrial ages are also required to correct measured activites of the radionuclide to the time of fall.

COSMOGENIC NOBLE GASES PRODUCED BY SOLAR COSMIC RAYS (SCR)

SCR protons and alpha-particles penetrate only a very few cm into solid matter. Therefore, SCR-produced cosmogenic nuclides are normally not expected to be observed in meteorites, because their outermost few cm usually were ablated in the Earth's atmosphere. SCR nuclide concentration profiles therefore mostly have been calculated for the Moon, i.e., for samples with lunar chemical composition irradiated at 1 AU from the Sun. Most of the recent work has been done by R.C. Reedy and coworkers (e.g., Rao et al. 1994; Reedy 1998a,b) and R. Michel and coworkers (Michel et al. 1996; Neumann 1999). The lunar data also allow one to study temporal variations of SCR fluxes (Reedy and Marti 1991; Reedy 1998b).

In chondrites there is indeed hardly any firm evidence for the presence of SCR noble gases (Garrison et al. 1995). The most promising candidates appear to be small meteorites which hopefully were ablated only little. Indeed a few small meteorites show high levels of ^{26}Al indicative of SCR production (e.g., Nishiizumi et al. 1990), but none of them unequivocally contains SCR-produced ^{21}Ne. Pätsch (2000) reported hints for SCR Ne contributions near the postatmospheric surface for one out of five small Antarctic chondrites studied. It is possible that SCR noble gases may be found near the edges of clasts in meteoritic regolith breccias, i.e., from near-surface samples of pebbles once irradiated in an asteroidal regolith. The few systematic such studies so far have not yielded evidence for SCR noble gases, however (e.g., Wieler et al. 1989b).

One class of martian meteorites, the shergottites, are a very remarkable exception, as all of them show signs of the presence of SCR-Ne in the form of relatively low ^{21}Ne/^{22}Ne ratios (Garrison et al. 1995). In contrast, no representative of the other types of martian meteorites (Chassigny, Nakhlites, ALH84001) shows SCR-Ne. Garrison et al. (1995) suggest that Shergottites may have had atypical orbital parameters compared to other meteorite classes, leading either to a higher production of SCR nuclides relatively close to the Sun or to an atypically low ablation due to low-entry velocities.

Garrison et al. (1995) and Weigel et al. (1999) conclude that SCR-Ne is also present or likely to be present in some acapulcoites and lodranites. For some reason, members of these meteorite classes often had a small preatmospheric size. It should be noted, however, that these conclusions are based on high ^{22}Ne/^{21}Ne ratios of samples without a documented relative position to each other (this is also true for most of the shergottites). A verification by means of depth profiles of ^{21}Ne, ^{22}Ne/^{21}Ne as well as ^{26}Al would be highly desirable.

SCR noble gases have also been found in one iron meteorite (Lavielle et al. 1999b). Arlington has a highly unusual shape that led to a partial preservation of its preatmospheric surface.

ACKNOWLEDGMENTS

I thank Thomas Graf, Bernard Lavielle and Ingo Leya for numerous discussions and suggestions. Ingo Leya and Jozef Masarik kindly provided the data shown in Figures 1-6, Thomas Graf those in Figure 7. Kuni Nishiizumi is acknowledged for his most recent data and compilation of exposure ages of CM and CI chondrites. Very helpful reviews by

Gregory Herzog, Bernard Lavielle, Ingo Leya, Don Porcelli, and Kees Welten are gratefully acknowledged.

REFERENCES

Albrecht A, Schnabel C, Vogt S, Xue S, Herzog GF, Begemann F, Weber HW, Middleton R, Fink D, Klein J (2000) Light noble gases and cosmogenic radionuclides in Estherville, Budulan, and other mesosiderites: Implications for exposure histories and production rates. Meteoritics Planet Sci 35: 975-986

Alexeev VA (1998) Parent bodies of L and H chondrites: Times of catastrophic events. Meteoritics Planet Sci 33 (1):145-152

Anders E (1962) Meteorite ages. Rev Mod Phys 34:287-325

Anders E (1964) Origin, age, and composition of meteorites. Space Sci Rev 3:583-714

Assonov SS, Ivanova MA, Yasevich AN, Shukolyukov YuA (1996) Dengli H3.8 chondrite noble gases: Evidence of the regolith nature of brecciation and of two-stage cosmic irradiation. Geochem Intl 34:825-833

Aylmer D, Vogt S, Herzog GF, Klein J, Fink D, Middleton R (1990) Low ^{10}Be and ^{26}Al contents of ureilites: Production at meteoroid surfaces. Geochim Cosmochim Acta 54:1775-1784

Beer J, Raisbeck GM, Yiou F (1991) Time variations of Be-10 and solar activity. *In* The Sun in Time. Sonett CP, Giampapa MS, Matthews MS (eds) Univ. Arizona Press, Tucson, p 343-359

Begemann F, Schultz L (1988) The influence of bulk chemical composition on the production rate of cosmogenic nuclides in meteorites. Lunar Planet Sci IXX, Lunar Planet Institute, Houston, p 51-52

Begemann F, Geiss J, Hess DC (1957) Radiation age of a meteorite from cosmic-ray-produced He^3 and H^3. Phys Rev 107:540-542

Begemann F, Weber HW, Vilcsek E, Hintenberger H (1976) Rare gases and ^{36}Cl in stony-iron meteorites: cosmogenic elemental production rates, exposure ages, diffusion losses and thermal histories. Geochim Cosmochim Acta 40:353-368

Begemann F, Fan CY, Weber HW, Wang XB (1996) Light noble gases in Jilin: More of the same and something new. Meteoritics Planet Sci 31:667-674.

Benoit PH, Sears DWG (1997) The orbits of meteorites from natural thermoluminescence. Icarus 125: 281-287.

Bhandari N, Bonino G, Cini Castagnoli G (1994) The 11-year solar cycle variation of cosmogenic isotope production rates in chondrites. Meteoritics 29:443-444

Bhattacharya SK, Goswami JN, Lal D (1973) Semiempirical rates of formation of cosmic ray tracks in spherical objects exposed in space: preatmospheric and postatmospheric depth profiles. J Geophys Res 78:8356-8363

Binzel RP, Xu S (1993) Chips off of Asteroid-4 Vesta—evidence for the parent body of basaltic achondrite meteorites. Science 260:186-191

Blann M (1971) Hybrid model for pre-equilibrium decay in nuclear reactions. Phys Rev Lett 27: 337 - 340

Bogard DD (1995) Impact ages of meteorites: A synthesis. Meteoritics 30:244-268

Bogard DD, Cressy PJ (1973) Spallation production of ^{3}He, ^{21}Ne, and ^{38}Ar from target elements in the Bruderheim chondrite. Geochim Cosmochim Acta 37:527-546

Bogard DD, Nyquist LE, Bansal BM, Garrison DH, Wiesmann H, Herzog GF, Albrecht AA, Vogt S, Klein J (1995) Neutron-capture ^{36}Cl, ^{41}Ca, ^{36}Ar, and ^{150}Sm in large chondrites: Evidence for high fluences of thermalized neutrons. J Geophys Res E100:9401-9416

Bogard DD, Garrison DH, Masarik J (2001) The Monahans chondrite and halite: Argon-39/argon-40 age, solar gases, cosmic ray exposure ages, and parent body regolith neutron flux and thickness. Meteoritics Planet Sci 36:107-122

Bonino G, Cini Castagnoli G, Bhandari N, Della Monica P, Taricco C (1999) Galactic cosmic ray variations in the last two centuries recorded by cosmogenic ^{44}Ti in meteorites. Adv Space Res 23: 607-610

Bottke WF, Rubincam DP, Burns JA (2000) Dynamical evolution of main belt meteoroids: Numerical simulations incorporating planetary perturbations and Yarkovsky thermal forces. Icarus 145:301-331

Caffee MW, Nishiizumi K (1997) Exposure ages of carbonaceous chondrites: II. Meteoritics Planet Sci 32:A26

Caffee MW, Nishiizumi K (2001) Exposure history of separated phases from the Kapoeta meteorite. Meteoritics Planet Sci 36:429-437

Caffee MW, Goswami JN, Hohenberg CM, Swindle TD (1983) Cosmogenic neon from precompaction irradiation of Kapoeta and Murchison. Proc Lunar Planet Sci Conf 14th:B267-B273

Caffee MW, Hohenberg CM, Swindle TD, Goswami JN (1987) Evidence in meteorites for an active early Sun. Astrophys J 313:L31-L35

Damon PE, Sonett CP (1991) Solar and terrestrial components of the atmospheric ^{14}C variation spectrum. *In* The Sun in Time. Sonett CP, Giampapa MS, Matthews MS (eds) Univ. Arizona Press, Tucson, p 360-388

Eberhardt P, Geiss J, Lutz H (1963) Neutrons in meteorites. *In* Earth Science and Meteoritics. Geiss J, Goldberg ED (eds) North Holland, Amsterdam, p 143-168

Eberhardt P, Eugster O, Geiss J (1965) Radiation ages of aubrites. J Geophys Res 70:4427-4434

Eberhardt P, Eugster O, Geiss J, Marti K (1966) Rare gas measurements in 30 stone meteorites. Z Naturf 21a:414-426

Eberhardt P, Geiss J, Graf H (1971) On the origin of excess ^{131}Xe in lunar rocks. Earth Planet Sci Lett 12:260-262

Eugster O (1988) Cosmic ray production rates for ^{3}He, ^{21}Ne, ^{38}Ar, ^{83}Kr, and ^{126}Xe in chondrites based on ^{81}Kr-Kr exposure ages. Geochim Cosmochim Acta 52:1649-1662

Eugster O, Michel T (1995) Common asteroid break-up events of eucrites, diogenites, and howardites and cosmic ray production rates for noble gases in achondrites. Geochim Cosmochim Acta 59:177-199

Eugster O, Eberhardt P, Geiss J (1969) Isotopic analyses of krypton and xenon in fourteen stone meteorites. J Geophys Res 74:3874-3896

Eugster O, Tera F, Burnett DS, Wasserburg GJ (1970) Isotopic composition of gadolinium and neutron-capture effects in some meteorites. J Geophys Res 75:2753-2768

Eugster O, Eberhardt P, Thalmann C, Weigel A (1998a) Neon-E in CM-2 chondrite LEW90500 and collisional history of CM-2 chondrites, Maralinga, and other CK chondrites. Geochim Cosmochim Acta 62:2573-2582

Eugster O, Polnau E, Terribilini D (1998b) Cosmic ray- and gas retention ages of newly recovered and of unusual chondrites. Earth Planet Sci Lett 164:511-519

Evans JC, Reeves JH, Rancitelli LA, Bogard DD (1982) Cosmogenic nuclides in recently fallen meteorites: evidence for galactic cosmic ray variations during the period 1967-1978. J Geophys Res 87:5577-5591

Farinella P, Vokrouhlicky D, Morbidelli A (2001) Delivery of material from the asteroid belt. *In* Accretion of extraterrestrial matter throughout Earth's history. Peucker-Ehrenbrink B, Schmitz B (eds) Kluwer, New York, p 31-49

Finkel RC, Kohl CP, Marti M, Martinek B, Rancitelli L (1978) The cosmic ray record in the San Juan Capistrano meteorite. Geochim Cosmochim Acta 42:241-250

Fleischer RL, Price PB, Walker RM. (1975) Nuclear tracks in solids. Univ. California Press, Berkeley

Freundel M, Schultz L, Reedy RC (1986) Terrestrial ^{81}Kr-Kr ages of Antarctic meteorites. Geochim Cosmochim Acta 50:2663-2673

Garrison DH, Bogard DD, Albrecht AA, Vogt S, Herzog GF, Klein J, Fink D, Dezfouly-Arjomandy B, Middleton R (1992) Cosmogenic nuclides in core samples of the Chico L6-chondrite—evidence for irradiation under high shielding. Meteoritics 27:371-381

Garrison DH, Rao MN, Bogard DD (1995) Solar-proton-produced neon in shergottite meteorites and implications for their origin. Meteoritics 30:738-747

Gladman B (1997) Destination: Earth. Martian meteorite delivery. Icarus 130:228-246

Gladman BJ, Burns JA, Duncan M, Lee P, Levison HF (1996) The exchange of impact ejecta between terrestrial planets. Science 271:1387-1392

Gladman BJ, Migliorini F, Morbidelli A, Zappala V, Michel P, Cellino A, Froeschle C, Levison HF, Bailey M, Duncan M (1997) Dynamical lifetimes of objects injected into asteroid belt resonances. Science 277:197-201

Göbel R, Bergmana F, Ott U (1982) On neutron-induced and other noble gases in Allende inclusions. Geochim Cosmochim Acta 46:1777-1792

Goodrich CA (1992) Ureilites—A critical review. Meteoritics 27:327-352

Graf T, Marti K (1994) Collisional records in LL-chondrites. Meteoritics 29:643-648

Graf T, Marti K (1995) Collisional history of H chondrites. J Geophys Res E100:21247-21263

Graf T, Vogt S, Bonani G, Herpers U, Signer P, Suter M, Wieler R, Wölfli W (1987) Depth dependence of ^{10}Be and ^{26}Al production in the iron meteorite Grant. Nucl Instr Methods B29:262-265

Graf T, Baur H, Signer P (1990a) A model for the production of cosmogenic nuclides in chondrites. Geochim Cosmochim Acta 54:2521-2534

Graf T, Signer P, Wieler R, Herpers U, Sarafin R, Vogt S, Fiéni C, Pellas P, Bonani G, Suter M, Wölfli W (1990b) Cosmogenic nuclides and nuclear tracks in the chondrite Knyahinya. Geochim Cosmochim Acta 54:2511-2520

Graf T, Caffee MW, Marti K, Nishiizumi K, Ponganis KV (2001) Dating collisional events: ^{36}Cl-^{36}Ar exposure ages of H-chondritic metals. Icarus 150:181-188

Greenberg R, Nolan MC (1989) Delivery of asteroids and meteorites to the inner solar system. *In* Asteroids II. Binzel RP, Gehrels T, Matthews MS (eds) Univ. Arizona Press, Tucson, p 778-804

Hampel W, Schaeffer OA (1979) ^{26}Al in iron meteorites and the constancy of cosmic ray intensity in the past. Earth Planet Sci Lett 42:348-358
Herzog GF, Anders E (1971) Absolute scale for radiation ages of stony meteorites. Geochim Cosmochim Acta 35:605-611
Herzog GF, Vogt S, Albrecht A, Xue S, Fink D, Klein J, Middleton R, Weber HW, Schultz L (1997) Complex exposure histories for meteorites with short exposure ages. Meteoritics Planet Sci 32: 413-422.
Heusser G, Ouyang Z, Oehm J, Yi W (1996) Aluminum-26, sodium-22 and cobalt-60 in two drill cores and some other samples of the Jilin chondrite. Meteoritics Planet Sci 31:657-665.
Heymann D (1967) On the origin of hypersthene chondrites: ages and shock effects of black chondrites. Icarus 6:189-221
Heymann D, Lipschutz ME, Nielsen B, Anders E (1966) Canyon Diablo meteorite: metallographic and mass spectrometric study of 56 fragments. J Geophys Res 71:619-641
Hidaka H, Ebihara M, Yoneda S (1999) High fluences of neutrons determined from Sm and Gd isotopic compositions in aubrites. Earth Planet Sci Lett 173:41-51
Hidaka H, Yoneda S, Nishiizumi K (2001) Neutron capture effects on Sm and Gd isotopes in Martian meteorites. Meteoritics Planet Sci 36:A80-A81
Hohenberg CM, Marti K, Podosek FA, Reedy RC, Shirck JR (1978) Comparisons between observed and predicted cosmogenic noble gases in lunar samples. Proc Lunar Planet Sci Conf 9th:2311-2344
Hohenberg CM, Hudson B, Kennedy BM, Podosek FA (1981) Xenon spallation systematics in Angra dos Reis. Geochim Cosmochim Acta 45:1909-1915
Hohenberg CM, Nichols RH, Olinger CT, Goswami JN (1990) Cosmogenic neon from individual grains of CM meteorites: Extremely long pre-compaction exposure histories or an enhanced early particle flux. Geochim Cosmochim Acta 54:2133-2140
Honda M (1988) Statistical estimation of the production of cosmic-ray-induced nuclides in meteorites. Meteoritics 23:3-12
Jull AJT (2001) Terrestrial ages of meteorites. *In* Accretion of extraterrestrial matter throughout Earth's history. Peucker-Ehrenbrink B, Schmitz B (eds). Kluwer, New York, p 241-266
Jull AJT, Donahue DJ, Reedy RC, Masarik J (1994) A Carbon-14 depth profile in the L5 chondrite Knyahinya. Meteoritics 29:649-651
Kallemeyn GW, Rubin AE, Wasson JT (1991) The compositional classification of chondrites: V. The Karoonda (CK) group of carbonaceous chondrites. Geochim Cosmochim Acta 55:881-892
Keil K (1989) Enstatite meteorites and their parent bodies. Meteoritics 24:195-208
Keil K, Haack H, Scott ERD (1994) Catastrophic fragmentation of asteroids: Evidence from meteorites. Planet Space Sci 42:1109-1122
Kim JS, Marti K (1992) Solar-type xenon: isotopic abundances in Pesyanoe. Proc Lunar Planet Sci 22: 145-151
Kring DA, Jull AJT, McHargue LR, Bland PA, Hill DH, Berry FJ (2001) Gold Basin meteorite strewn field, Mojave Desert, northwestern Arizona: Relic of a small late Pleistocene impact event. Meteoritics Planet Sci 36:1057-1066
Kohl CP, Murrell MT, Russ GP, Arnold JR (1978) Evidence for the constancy of the solar cosmic ray flux over the past ten million years: ^{53}Mn and ^{26}Al measurements. Proc Lunar Planet Sci Conf 9th: 2299-2310
Lavielle B, Marti K (1988) Cosmic ray produced Kr in St. Severin core AIII. Proc Lunar Planet Sci Conf 18th:565-572
Lavielle B, Regnier S, Marti K (1985) Ages d'exposition des météorites de fer: histoires multiples et variations d'intensité du rayonnement cosmique. *In* Isotopic ratios in the solar system, Cepadues-Editions, Toulouse, France:15-20
Lavielle B, Toe S, Gilabert E (1997) Noble Gas Measurements In the L/LL5 Chondrite Knyahinya. Meteoritics Planet Sci 32:97-107.
Lavielle B, Marti K, Jeannot JP, Nishiizumi K, Caffee M (1999a) The Cl-36-Ar-36-K-40-K-41 records and cosmic ray production rates in iron meteorites. Earth Planet Sci Lett 170:93-104
Lavielle B, Gilabert E, Marti K (1999b) Records of solar energetic particles in the iron meteorite Arlington. Meteoritics Planet Sci 34:A72-A73
Lavielle B, Gilabert E, Marti K, Nishiizumi K, Caffee MW (2000) Collisional history in irons: interpreting the cosmic ray record. Meteoritics Planet Sci 35:A96
Lavielle B, Caffee MW, Gilabert E, Marti K, Nishiizumi K, Ponganis KV (2001) Irradiation records in Group IVA irons. Meteoritics Planet Sci 36:A110-A111
Lewis RS, Amari S, Anders E (1994) Interstellar grains in meteorites: II. SiC and its noble gases. Geochim Cosmochim Acta 58:471-494

Leya I (1997) Modellrechnungen zur Beschreibung der Wechselwirkungen galaktischer kosmischer Teilchenstrahlung mit Stein- und Eisenmeteoroiden. PhD thesis, University of Hannover
Leya I, Michel R (1997) Determination of neutron cross sections for nuclide production at intermediate energies by deconvolution of thick-target production rates. *In* Proc Int Conf Nuclear Data for Science and Technology. Reffo G (ed) Trieste, Italy, p 1463-1467
Leya I, Lange HJ, Neumann S, Wieler R, Michel R (2000a) The production of cosmogenic nuclides in stony meteoroids by galactic cosmic ray particles. Meteoritics Planet Sci 35:259-286
Leya I, Lange HJ, Lüpke M, Neupert U, Daunke R, Fanenbruck O, Michel R, Rösel R, Meltzow B, Schiekel T, Sudbrock F, Herpers U, Filges D, Bonani G, Dittrich Hannen B, Suter M, Kubik PW, Synal HA (2000b) Simulation of the interaction of galactic cosmic ray protons with meteoroids: On the production of radionuclides in thick gabbro and iron targets irradiated isotropically with 1.6 GeV protons. Meteoritics Planet Sci 35:287-318
Leya I, Wieler R, Halliday AN (2000c) Cosmic ray production of tungsten isotopes in lunar samples and meteorites and its implications for Hf-W cosmochemistry. Earth Planet Sci Lett 175:1-12
Leya I, Graf T, Nishiizumi K, Wieler R (2001a) Cosmic-ray production rates of helium, neon and argon isotopes in H chondrites based on chlorine-36/argon-36 ages. Meteoritics Planet Sci 36:963-973
Leya I, Neumann S, Wieler R, Michel R (2001b) The production of cosmogenic nuclides by GCR-particles for 2π exposure geometries. Meteoritics Planet Sci 36:1547-1561
Lingenfelter RE, Canfield EH, Hampel VE (1972) The lunar neutron flux revisited. Earth Planet Sci Lett 16:355-369
Loeken T, Scherer P, Weber HW, Schultz L (1992) Noble gases in eighteen stone meteorites. Chem Erde 52:249-259
Lorenzetti S, Eugster O, Burbine T, McCoy TJ, Marti K (2001) Break-up events on the aubrite parent body. Meteoritics Planet Sci 36:A116-A117
Lugmair GW, Marti K (1971) Neutron capture effects in lunar gadolinium and the irradiation histories of some lunar rocks. Earth Planet Sci Lett 13:32-42
Ma P, Herzog GF, Faestermann T, Knie K, G. Korschinek (2001) ^{53}Mn activities and ^{26}Al-^{53}Mn exposure ages of lodranites and acapulcoites. Meteoritics Planet Sci 36:A120
Marti K (1967) Mass-spectrometric detection of cosmic-ray-produced Kr^{81} in meteorites and the possibility of Kr-Kr dating. Phys Rev Lett 18:264-266
Marti K, Lugmair GW (1971) Kr^{81}-Kr and K-Ar^{40} ages, cosmic ray spallation products, and neutron effects in lunar samples from Oceanus Procellarum. Proc Lunar Sci Conf 2nd:1591-1605
Marti K, Graf T (1992) Cosmic ray exposure history of ordinary chondrites. Ann Rev Earth Planet Sci 20:221-243
Marti K, Murty SVS, Nishiizumi K. (1986) Neutron exposure age of the Cape York iron meteorite. Lunar Planet Sci XVII, Lunar Planet Institute, Houston:516-517.
Masarik J, Reedy RC (1994) Effects of bulk composition on nuclide production processes in meteorites. Geochim Cosmochim Acta 58:5307-5317
Masarik J, Nishiizumi K, Reedy RC (2001) Production rates of cosmogenic ^{3}He, ^{21}Ne, and ^{22}Ne in ordinary chondrites and the lunar surface. Meteoritics Planet Sci 36:643-650
McCoy TJ, Keil K, Clayton RN, Mayeda TK, Bogard DD, Garrison DH, Wieler R (1997) A petrologic and isotopic study of lodranites: evidence for early formation as partial melt residues from heterogeneous precursors. Geochim Cosmochim Acta 61:623-637
Merchel S, Altmaier M, Faestermann T, Herpers U, Knie K, Korschinek G, Kubik PW, Neumann S, Michel R, Suter M (1999) Saharan meteorites with short or complex exposure histories. *In* Workshop on extraterrestrial materials from cold and hot deserts. Lunar Planet. Institute, Houston, Contrib No. 997:53-56
Michel R, Dragovitsch P, Englert P, Peiffer F, Stück R, Theis S, Begemann F, Weber HW, Signer P, Wieler R, Filges D, Cloth P (1986) On the depth dependence of spallation reactions in a spherical thick diorite target homogeneously irradiated by 600 MeV protons. Simulation of production of cosmogenic nuclides in small meteorites. Nucl Instr Methods B16:61-82
Michel R, Leya I, Borges L (1996) Production of cosmogenic nuclides. *In* Meteoroids: accelerator experiments and model calculations to decipher the cosmic ray record in extraterrestrial matter. Nucl Instr Methods B 113:434-444
Michlovich ES, Vogt S, Masarik J, Reedy RC, Elmore D, Lipschutz ME (1994) Aluminum 26, ^{10}Be, and ^{36}Cl depth profiles in the Canyon Diablo iron meteorite. J Geophys Res 99:23187-23194
Miura Y, Nagao K, Fujitani T (1993) ^{81}Kr terrestrial ages and grouping of Yamato eucrites based on noble gas and chemical compositions. Geochim Cosmochim Acta 57:1857-1866
Miura YN, Nagao K, Okazaki R (2000) Noble gas studies of aubrites. Meteoritics Planet Sci 35:A112

Moniot RK, Kruse TH, Tuniz C, Savin W, Hall GS, Milazzo T, Pal D, Herzog GF (1983) The ^{21}Ne production rate in stony meteorites estimated from ^{10}Be and other radionuclides. Geochim Cosmochim Acta 47:1887-1895

Morbidelli A, Gladman B (1998) Orbital and temporal distributions of meteorites originating in the asteroid belt. Meteoritics Planet Sci 33:999-1016

Neumann S (1999) Aktivierungsquerschnitte mit Neutronen mittlerer Energien und die Produktion kosmogener Nuklide in extraterrestrischer Materie. PhD dissertation, University of Hannover

Niedermann S (2002) Cosmic-ray-produced noble gases in terrestrial rocks as a dating tool for surface processes. Rev Mineral Geochem 47:731-784

Nishiizumi K (1987) ^{53}Mn, ^{26}Al, ^{10}Be and ^{36}Cl in meteorites: data compilation. Nucl Tracks Radiat Meas 13:209-273

Nishiizumi K, Caffee MW (2001) Exposure histories of lunar meteorites Dhofar 025, 026, and Northwest Africa 482. Meteoritics Planet Sci 36:A148-A149

Nishiizumi K, Regnier S, Marti K (1980) Cosmic ray exposure ages of chondrites, pre-irradiation and constancy of cosmic ray flux in the past. Earth Planet Sci Lett 50:156-170

Nishiizumi K, Nagai H, Imamura M, Honda M, Kobayashi K, Kubik PW, Sharma P, Wieler R, Signer P, Goswami JN, Reedy RC, Arnold JR (1990) Solar cosmic ray produced nuclides in Salem meteorite. Meteoritics 25:392

Nishiizumi K, Arnold JR, Caffee MW, Finkel RC, Southon JR, Nagai H, Honda M, Imamura M, Kobayashi K, Sharma P (1993) Exposure ages of carbonaceous chondrites—I. Lunar Planet Sci XXIV; Lunar Planet Institute, Houston:1085-1086

Nishiizumi K, Caffee MW, Jeannot J-P, Lavielle B, Honda M (1997a) A systematic study of the cosmic-ray-exposure history of iron meteorites: Beryllium-10-chlorine-36/beryllium-10 terrestrial ages. Meteoritics Planet Sci 32:A100

Nishiizumi K, Fink D, Klein J, Middleton R, Masarik J, Reedy RC, Arnold JR (1997b) Depth profile of ^{41}Ca in an Apollo 15 drill core and the low-energy neutron flux in the Moon. Earth Planet Sci Lett 148:545-552

Nishiizumi K, Masarik J, Caffee MW, Jull AJT (1999) Exposure histories of pair lunar meteorites EET 96008 and EET 87521. Lunar Planet Sci Conf XXX, Lunar Planet Institute, Houston, TX, Abstr #1980, CD-ROM

Nishiizumi K, Caffee MW, Masarik J (2000) Cosmogenic radionuclides in the Los Angeles Martian meteorite. Meteoritics Planet Sci 35:A120

Nyquist LE, Borg LE, Shih CY (1998) The Shergottite age paradox and the relative probabilities for Martian meteorites of differing ages. J Geophys Res E103:31445-31455

Nyquist LE, Bogard DD, Shih C-Y, Greshake A, Stöffler D, Eugster O (2001) Ages and geologic histories of martian meteorites. Space Sci Rev 96:105-164

Olinger CT, Maurette M, Walker RM, Hohenberg CM (1990) Neon measurements of individual Greenland sediment particles: proof of an extraterrestrial origin. Earth Planet Sci Lett 100:77-93

Olsen E, Davis A, Clarke RS, Schultz L, Weber HW, Clayton R, Mayeda T, Jarosewich E, Sylvester P, Grossman L, Wang MS, Lipschutz ME, Steele IM, Schwade J (1994) Watson—a new link in the IIE iron chain. Meteoritics 29:200-213

Ott U (2002) Noble gases in meteorites—trapped components. Rev Mineral Geochem 47:71-100

Ott U, Begemann F (2000) Spallation recoil and age of presolar grains in meteorites. Meteoritics Planet Sci 35:53-63

Ott U, Altmeier M, Herpers U, Kuhnhenn J, Merchel S, Michel R, Mohapatra RK (2001) Update on recoil loss of spallation products from presolar grains. Meteoritics Planet Sci 36:A155-A156

Okazaki R, Takaoka N, Nakamura T, Nagao K (2000) Cosmic ray exposure ages of enstatite chondrites. Antarct Meteorite Res (Natl Inst Polar Research, Tokyo) 13:153-169

Pätsch M. (2000) Produkte der Solaren Kosmischen Strahlung in Meteoriten: Aufbau und Erprobung einer Laser-Extraktionsanlage für Edelgase. PhD Thesis, Johannes Gutenberg Universität, Mainz.

Patzer A, Schultz L (2001) Noble gases in enstatite chondrites I: exposure ages, pairing, and weathering effects. Meteoritics Planet Sci 36:947-961

Pedroni A (1989) Die korpuskulare Bestrahlung der Oberflächen von Asteroiden; eine Studie der Edelgase in den Meteoriten Kapoeta und Fayetteville. PhD dissertation, ETH Zürich, Nr. 8880.

Pepin RO, Becker RH, Rider PE (1995) Xenon and krypton isotopes in extraterrestrial regolith soils and in the solar wind. Geochim Cosmochim Acta 59:4997-5022

Pellas P, Fiéni C (1988) Thermal histories of ordinary chondrite parent asteroids. Lunar Planet Sci XIX, Lunar Planetary Institute, Houston, p 915-916

Polnau E, Eugster O (1998) Cosmic ray produced, radiogenic, and solar noble gases in lunar meteorites Queen Alexandra Range 94269 and 94281. Meteoritics Planet Sci 33:313-319

Polnau E, Eugster O, Krähenbühl U, Marti K (1999) Evidence for a precompaction exposure to cosmic rays in a chondrule from the H6 chondrite ALH76008. Geochim Cosmochim Acta 63:925-933
Polnau E, Eugster O, Burger M, Krähenbühl U, Marti K (2001) Precompaction exposure of chondrules and implications. Geochim Cosmochim Acta 65:1849-1866
Pun A, Keil K, Taylor GJ, Wieler R (1998) The Kapoeta howardite: Implications for the regolith evolution of the HED parent body. Meteoritics Planet Sci 33:835-851
Raisbeck GM, Yiou F (1989) Cosmic ray exposure ages of cosmic spherules. Meteoritics 24:318
Rao MN, Garrison DH, Bogard DD, Reedy RC (1994) Determination of the flux-distribution and energy-distribution of energetic solar protons in the past 2-Myr using lunar rock- 68815. Geochim Cosmochim Acta 58:4231-4245
Reedy RC (1981) Cosmic-ray-produced stable nuclides: Various production rates and their implications. Proc Lunar Planet Sci, 12B:1809-1823
Reedy RC (1998a) Studies of modern and ancient solar energetic particles. Proceedings Indian Acad Sci–Earth Planetary Sci 107:433-440
Reedy RC (1998b) Variations in solar-proton fluxes over the last million years. Meteoritics Planet Sci 33:A127-A127
Reedy RC, Marti M (1991) Solar-cosmic ray fluxes during the last ten million years. *In* The Sun in time. Sonett CP, Giampapa MS, Matthews MS (eds) Univ Arizona Press, Tucson, p 260-287
Reedy RC, Arnold JR, Lal D (1983) Cosmic ray record in solar system matter. Ann Rev Nucl Part Sci 33:505-537
Reedy RC, Masarik J, Nishiizumi K, Arnold JR, Finkel RC, Caffee MW, Southon J, Jull AJT, Donahue DJ (1993) Cosmogenic-radionuclide profiles in Knyahinya: new measurements and models. Lunar Planet Sci XXIV; Lunar Planet Institute, Houston, p 1195-1196
Regnier S, Hohenberg CM, Marti K, Reedy RC (1979) Predicted versus observed cosmic ray produced noble gases in lunar samples: Improved Kr production ratios. Proc Lunar Planet Sci Conf 10th: 1565-1586
Sands DG, De Laeter JR, Rosman KJR (2001) Measurements of neutron capture effects on Cd, Sm and Gd in lunar samples with implications for the neutron energy spectrum. Earth Planet Sci Lett 186:335-346
Schaeffer OA (1975) Constancy of galactic cosmic rays in time and space. 14th Intl Cosmic Ray Conf: 3508-3520
Schaeffer OA, Nagel K, Fechtig H, Neukum G (1981) Space erosion of meteorites and the secular variation of cosmic rays (over 10^9 years). Planet Space Sci 29:1109-1118
Scherer P, Schultz L (2000) Noble gas record, collisional history, and pairing of CV, CO, CK, and other carbonaceous chondrites. Meteoritics Planet Sci 35:145-153
Scherer P, Zipfel J, Schultz L (1998) Noble gases in two new ureilites from the Saharan desert. Lunar Planet Sci XXIX, Lunar Planet Institute, Houston, TX:Abstr #1383 (CD ROM)
Schnabel C, Leya I, Wieler R, Herd RK, Synal H-A, Krähenbühl U, Herzog GF (2001) ^{129}I in Knyahinya and Abee and a first estimate of GCR constancy over 20 Myr. Meteoritics Planet Sci 36:A183-A184
Schultz L, Signer P (1976) Depth dependence of spallogenic helium, neon, and argon in the St. Severin chondrite. Earth Planet Sci Lett 30:191-199
Schultz L, Signer P (1977) Noble gases in the St. Mesmin chondrite: Implications to the irradiation history of a brecciated meteorite. Earth Planet Sci Lett 36:363-371
Schultz L, Franke L (2000) Helium, neon, and argon in meteorites-a data collection, update 2000. Max-Planck-Institut für Chemie, Mainz, CD-ROM (Schultz@mpch-mainz.mpg.de)
Schultz L, Weber HW (2001) The irradiation history of Rumuruti-chondrites. 26th Symp Antarctic Meteorites, Natl Inst Polar Research, Tokyo, p 128-130
Schultz L, Weber HW, Begemann F (1991) Noble gases in H-chondrites and potential differences between Antarctic and non-Antarctic meteorites. Geochim Cosmochim Acta 55:59-66
Signer P, Nier AO (1960) The distribution of cosmic-ray-produced rare gases in iron meteorites. J Geophys Res 65:2947-2964
Signer P, Nier AO (1962) The measurement and interpretation of rare gas concentrations in iron meteorites. *In* Researches in meteorites. Moore CB (ed) John Wiley & Sons, New York, p 7-35
Sisterson JM, Jones DTL, Binns PJ, Langen K, Schroeder I, Buthelezi Z, Latti E, Brooks FD, Buffler A, Allie MS, Herbert MS, Nchodu MR, Makupula S, Ullmann J, Reedy RC (2001) Production of ^{22}Na and other radionuclides by neutrons in Al, SiO_2, Si, Ti, Fe and Ni targets: implications for cosmic ray studies. Lunar Planet Sci, XXXII, Lunar & Planetary Institute, Houston, Abstr #1302,CD-ROM
Spergel MS, Reedy RC, Lazareth OW, Levy PW, Slatest LA (1986) Cosmogenic neutron-capture-produced nuclides in stony meteorites. Proc Lunar Planet Sci Conf 16th:D483-D494
Swindle TD (2002a) Noble gases in the Moon and meteorites: Radiogenic components and early volatile chronologies. Rev Mineral Geochem 47:101-124
Swindle TD (2002b) Martian noble gases. Rev Mineral Geochem 47:171-190

Terribilini D, Eugster O, Herzog GF, Schnabel C (2000a) Evidence for common breakup events of the acapulcoites-lodranites and chondrites. Meteoritics Planet Sci 35:1043-1050

Terribilini D, Eugster O, Mittlefehldt DW, Diamond LW, Vogt S, Wang D (2000b) Mineralogical and chemical composition and cosmic ray exposure history of two mesosiderites and two iron meteorites. Meteoritics Planet Sci 35:617-628

Tuniz C, Bird JR, Fink D, Herzog GF. (1998) Accelerator Mass Spectrometry. CRC Press, Boca Raton, Florida

Vogt S, Herzog GF, Reedy RC (1990) Cosmogenic nuclides in extraterrestrial materials. Rev Geophys 28:253-275

Vogt SK, Aylmer D, Herzog GF, Wieler R, Signer P, Pellas P, Fiéni C, Tuniz C, Jull AJT, Fink D, Klein J, Middleton R (1993) On the Bur Gheluai H5 chondrite and other meteorites with complex exposure histories. Meteoritics 28:71-85

Voshage H (1978) Investigations on cosmic-ray-produced nuclides in iron meteorites, 2. New results on $^{41}K/^{40}K$-$^{4}He/^{21}Ne$ exposure ages and the interpretations of age distributions. Earth Planet Sci Lett 40:83-90

Voshage H (1984) Investigations of cosmic-ray-produced nuclides in iron meteorites, 6. The Signer-Nier model and the history of the cosmic radiation. Earth Planet Sci Lett 71:181-194

Warren PH (1994) Lunar and Martian meteorite delivery services. Icarus 111:338-363

Warren PH (2001) Porosities of lunar meteorites: Strength, porosity, and petrologic screening during the meteorite delivery process. J Geophys Res E106:10101-10111

Wasson JT (1990) Ungrouped iron meteorites in Antarctica: origin of anomalously high abundance. Science 249:900-902

Wasson JT (1995) Sampling the asteroid belt: how biases make it difficult to establish meteorite-asteroid connections. Meteoritics 30:595

Weigel A, Eugster O, Koeberl C, Michel R, Krähenbühl U, Neumann S (1999) Relationships among lodranites and acapulcoites: Noble gas isotopic abundances, chemical composition, cosmic ray exposure ages, and solar cosmic ray effects. Geochim Cosmochim Acta 63:175-192

Welten KC, Lindner L, van der Borg K, Loeken T, Scherer P, Schultz L (1997) Cosmic ray exposure ages of diogenites and the recent collisional history of the howardite, eucrite and diogenite parent body/bodies. Meteoritics Planet Sci 32:891-902

Welten KC, Nishiizumi K, Masarik J, Caffee MW, Jull AJT, Klandrud SE, Wieler R (2001a) Cosmic ray exposure history of two Frontier Mountain H-chondrite showers from spallation and neutron-capture products. Meteoritics Planet Sci 36:301-317

Welten KC, Nishiizumi K, Caffee MW, Schultz L (2001b) Update on exposure ages of diogenites: the impact history of the HED parent body and evidence of space erosion and/or collisional disruption of stony meteoroids. Meteoritics Planet Sci 36:A223

Welten KC, Bland PA, Russell SS, Grady MM, Caffee MW, Masarik J, Jull AJT, Weber HW, Schultz L (2001c) Exposure age, terrestrial age and pre-atmospheric radius of the Chinguetti mesosiderite: Not part of a much larger mass. Meteoritics Planet Sci 36:939-946

Welten KC, Nishiizumi K, Caffee MW (2001d) The search for meteorites with complex exposure histories among ordinary chondrites with low $^{3}He/^{21}Ne$ ratios. Lunar Planet Sci XXXII, Abstr #2148 (CD-ROM).

Welten KC, Nishiizumi K, Caffee MW, Masarik J, Wieler R (2001e) A complex exposure history of the Gold Basin L4-chondrite shower from cosmogenic radionuclides and noble gases. Lunar Planet Sci XXXII, Abstr #2110 (CD-ROM).

Wetherill GW, Williams JG (1979) Origin of differentiated meteorites. *In* Origin and distribution of the elements. Ahrens LH (ed) Pergamon, Oxford, p19-31

Wieler R (2002) Noble gases in the solar system. Rev Mineral Geochem 47:21-70

Wieler R, Graf T (2001) Cosmic ray exposure history of meteorites. *In* Accretion of extraterrestrial matter throughout Earth's history. Peucker-Ehrenbrink B, Schmitz B (eds) Kluwer, New York, p 221-240

Wieler R, Baur H, Pedroni A, Signer P, Pellas P (1989a) Exposure history of the regolithic chondrite Fayetteville: I. Solar-gas-rich matrix. Geochim Cosmochim Acta 53:1441-1448

Wieler R, Graf T, Pedroni A, Signer P, Pellas P, Fiéni C, Suter M, Vogt S, Clayton RN, Laul JC (1989b) Exposure history of the regolithic chondrite Fayetteville: II. Solar-gas-free light inclusions. Geochim Cosmochim Acta 53:1449-1459

Wieler R, Graf T, Signer P, Vogt S, Herzog GF, Tuniz C, Fink D, Fifield LK, Klein J, Middleton R, Jull AJT, Pellas P, Masarik J, Dreibus G (1996) Exposure history of the Torino meteorite. Meteoritics Planet Sci 31:265-272

Wieler R, Pedroni A, Leya I (2000a) Cosmogenic neon in mineral separates from Kapoeta: No evidence for an irradiation of its parent body regolith by an early active Sun. Meteoritics Planet Sci 35:251-257

Wieler R, Baur H, Jull AJT, Klandrud SE, Kring DA, Leya I, McHargue LR (2000b) Cosmogenic helium, neon, and argon in the large Gold Basin chondrite. Meteoritics Planet Sci 35:A169-A170
Wisdom J (1983) Chaotic behaviour and the origin of the 3/1 Kirkwood gap. Icarus 56:51-74
Woolum DS, Hohenberg C (1993) Energetic particle environment in the early solar system—extremely long pre-compaction meteoritic ages or an enhanced early particle flux. *In* Protostars and Planets III. Levy EH, Lunine JI (eds) Univ Arizona Press, Tucson, p 903-919
Wright RJ, Simms LA, Reynolds MA, Bogard DD (1973) Depth variation of cosmogenic noble gases in the ~120-kg Keyes chondrite. J Geophys Res 78:1308-1318

Martian Noble Gases

Timothy D. Swindle
Lunar and Planetary Laboratory
University of Arizona
Tucson, Arizona 85721
tswindle@u.arizona.edu

STUDYING MARS AS ANOTHER PLANET

Noble gas studies have been done on well-documented samples from two very different solar system bodies, the Earth and Moon. There is one other planet, Mars, from which we have samples, but those samples are nearly 20 (at the start of 2002) Martian meteorites, rocks from unknown locales on Mars. Even from those, though, we have learned enough to realize that Mars is a fascinating compromise between the Earth and Moon.

Like the Earth, Mars has a true atmosphere and interior volatile reservoirs, and it has been generating volcanic rocks throughout most of its history. Like the Moon, Mars has a surface that is exposed to enough galactic cosmic rays to generate abundant spallation reactions in surface rocks, and impact is likely one of the most common sources of heating for metamorphism.

In reviewing noble gas studies of Mars, we will touch on topics that are covered for the Earth, Moon or more common meteorites in several other chapters. In most cases, Mars provides another place to apply the techniques developed for the Earth and Moon, and gives intriguingly different results.

DISCOVERY OF MARTIAN METEORITES

The first measurement of noble gas on Mars came not from a Martian meteorite, but from a spacecraft. The Viking landers carried mass spectrometers that were able to measure the composition of noble gases in the Martian atmosphere well enough to determine that the $^{129}Xe/^{132}Xe$ and $^{40}Ar/^{36}Ar$ ratios are considerably higher than in the terrestrial atmosphere (Owen et al. 1977). On the other hand, the relative elemental abundances of the non-radiogenic noble gases are very similar to the terrestrial atmosphere, although the absolute abundances are all roughly a factor of 100 lower than in the Earth's atmosphere. The $^{15}N/^{14}N$ ratio is also higher than that in the terrestrial atmosphere. In fact, it is also higher than virtually any meteorite measurement, high enough that it was quickly concluded that the nitrogen isotopic composition is a result of some atmospheric loss process (McElroy et al. 1976), a theme that will be repeated in discussing the noble gases. Despite having uncertainties of tens of percent, these measurements were the key to demonstrating that some meteorites are Martian.

There was a small group of differentiated meteorites, the "SNC" meteorites that defied understanding. There was one key hint that they could be Martian (McSween and Stolper 1980; Wasson and Wetherill 1979; Wood and Ashwal 1981): they were clearly igneous, but had ages of 1300 Ma or less. Because the length of time that a body can maintain enough internal heat to generate igneous activity depends on its size, and the youngest known lunar basalts had ages of more than 3 Ga, this fact alone suggested a planet-sized body, of which Mars was the most likely candidate. However, the idea was controversial, in part because no one had figured out how to get a rock off Mars without pulverizing it.

1529-6466/00/0047-0006$05.00

In trying to perform $^{40}Ar/^{39}Ar$ dating on the Antarctic SNC meteorite Elephant Moraine ("EETA") 79001, Bogard and co-workers found apparent ages of more than 5000 Ma, older than the solar system itself, on some samples. They surmised that the reason was that the samples violated the assumption that all ^{40}Ar was radiogenic. In fact, a more detailed analysis of an unirradiated sample showed that the $^{40}Ar/^{36}Ar$ and $^{129}Xe/^{132}Xe$ ratios, particularly within glassy portions, were within the range of the Viking measurements, as were the noble gas elemental abundances (Bogard and Johnson 1983). Becker and Pepin (1984) found that the $^{15}N/^{14}N$ ratio in EETA 79001, though not as high as the Viking measurement, was also consistent with a contribution from the Martian atmosphere. In fact, not just the isotopic composition and the relative abundances matched the Martian atmosphere. The absolute abundances (in terms of atoms per cm^3) of six different gases, over a range of eight orders of magnitude, were the same in the EETA 79001 glass as in the Martian atmosphere (Pepin 1985; Fig. 1).

How did a nearly pure sample of Martian atmosphere end up inside EETA 79001? The trapping mechanism was apparently the shock event that ejected the rock from Mars. Simulation experiments, in which (terrestrial) basaltic samples were shocked in the presence of an atmosphere, showed that shock can produce glass that contains a sample of the ambient noble gas, with little or no elemental or isotopic fractionation (Bogard et al. 1986; Wiens and Pepin 1988). Furthermore, the Martian atmospheric signature is emphasized in EETA 79001. The rock is remarkably young (crystallization age of <200 Ma), so there is little radiogenic ^{40}Ar, and has a remarkably short exposure age (<1 Ma), so there is little cosmogenic noble gas of any kind.

However, there are several other meteorites, the SNC meteorites, which appear to be close relatives of EETA 79001, so these other meteorites probably come from Mars as well (McSween 1994). Hence, the noble gases within them can be used to study volatile reservoirs and transport processes on Mars. A listing of the Martian meteorites described in the scientific literature as of the time of publication is given in Table 1. McSween has written three valuable reviews of Martian meteorites as the topic has evolved (McSween 1985, 1994, 2002), and the Johnson Space Center has produced a useful compendium of information about them (Meyer 1998). More recently, Treiman et al. (2000) have reviewed the arguments that these meteorites are indeed from Mars, and Volume 96 of *Space Science Reviews*, stemming from a 2000 workshop on the Chronology and Evolution of Mars, contains several reviews of specific aspects of Martian meteorites, several of which will be referenced below.

Most of the Martian meteorites are described as shergottites. The type meteorite is Shergotty, the "S" of SNC. However, there are significant differences among the shergottites. There are really two basic types, lherzolites, which appear to be plutonic, and basalts, surface flows. The two types are probably closely related, because many of their formation ages are indistinguishable. In addition, EETA 79001 actually contains an igneous contact between two lithologies, one basaltic and one containing xenoliths that are lherzolitic. It is in the shergottites, particularly within glassy shock-produced inclusions, that the strongest evidence for Martian atmosphere is found (Becker and Pepin 1984; Bogard and Johnson 1983; Bogard and Garrison 1998; Bogard et al. 1984; Marti et al. 1995; Swindle et al. 1986). Some of the questions surrounding the ages of the shergottites will be discussed later.

There are then four meteorites, Chassigny and three nakhlites, which are 1300 Ma old. A fourth nakhlite, Northwest Africa 817, was only discovered in 2001, but also appears to be about 1300 Ma old, as discussed below.

Chassigny, the "C" of SNC, is a dunite, an olivine-rich rock that potentially comes from the Martian mantle. Chassigny contains a high abundance of noble gas that is not at all like the Martian atmosphere, and so is probably a sample of noble gas from the Martian interior (Mathew and Marti 2001; Ott 1988).

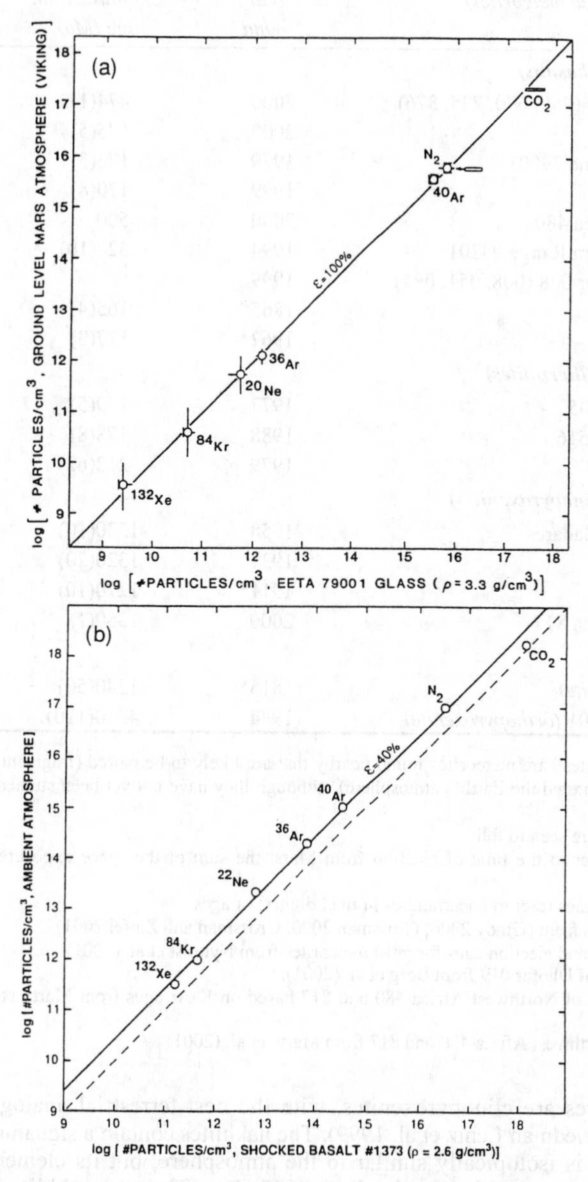

Figure 1. (a) Log-log plot of number of particles per cubic centimeter measured by Viking in the martian atmosphere vs. number of particles per cubic centimeter in EETA 79001 glass (Pepin 1985). More recent measurements of glass in martian meteorites (Bogard and Garrison 1998) suggest differences of up to a factor of two relative to the Viking measurement, but the overall good agreement remains. (b) Plot of number of particles per cubic centimeter in a shocked basalt sample, compared with the ambient atmosphere in which the sample was shocked (Wiens and Pepin 1988). The heavy line represents an emplacement efficiency of 40%. [Used by permission of University of Arizona Press, from Pepin and Carr (1988), Fig. 1, p. 128]

Table 1. Martian meteorites.

Name (associated meteorites)	*Year found*	*Crystallization age (Ma)*	*Ejection age (Ma)*
Shergottites (basalts)			
Dar al Gani 476 (489, 670, 735, 876)	2000	474(11)	1.24(12)
Dhofar 019	2000	575(5)	19.8(2.3)
Elephant Moraine 79001	1979	173(3)	0.73(15)
Los Angeles	1999	170(8)	3.10(20)
Northwest Africa 480	2000	<500	2.4(2)
Queen Alexandra Range 94201	1994	327(10)	2.71(20)
Sayh al Uhaymir 008 (008, 051, 094)	1999		1.5(3)
Shergotty	1865*	165(4)	2.73(20)
Zagami	1962*	177(3)	2.92(15)
Shergottites (lherzolites)			
Allan Hills 77005	1977	179(5)	3.06(20)
Lewis Cliffs 88516	1988	178(8)	3.94(40)
Yamato 793605	1979	212(62)	4.70(50)
Nakhlites (clinopyroxenites)			
Governador Valadares	1958	1330(10)	10.0(2.1)
Lafayette	1931	1320(20)	11.9(2.2)
Nakhla	1911*	1270(10)	10.75(40)
Northwest Africa 817	2000	1350(?)	9.7(1.1)
Other			
Chassigny *(dunite)*	1815*	1340(50)	11.3(6)
Allan Hills 84001 *(orthopyroxenite)*	1984	4510(110)	15.0(8)

"Associated meteorites" are meteorites found nearby that are likely to be paired (fragments of the same meteoroid that entered the Earth's atmosphere), although they have not yet been studied in sufficient detail to know.

* Meteorites that were seen to fall.

"Ejection ages" refer to the time of ejection from Mars, the sum of the space exposure age and the terrestrial age.

Numbers in parentheses refer to uncertainties in final digit(s) of ages.

Names and fall dates from (Grady 2000; Grossman 2000; Grossman and Zipfel 2001).

Crystallization ages and ejection ages for most meteorites from Nyquist et al. (2001).

Crystallization age of Dhofar 019 from Borg et al. (2001).

Crystallization ages of Northwest Africa 480 and 817 based on K-Ar ages from Marty et al. (2001) (see text).

Ejection ages of Northwest Africa 480 and 817 from Marty et al. (2001).

The nakhlites are clinopyroxenites, with the best terrestrial analogs being near-surface flows (Friedman Lentz et al. 1999). The nakhlites contain a signature of a trapped component that is isotopically similar to the atmosphere, but its elemental ratios are distinct, a puzzle that will be also be discussed below. The type nakhlite is Nakhla (the "N" of SNC), a meteorite that has the distinction of reputedly having killed a dog when it fell in Egypt in 1911 (Grady 2000). Two others (Lafayette and Governador Valadares) are finds that have been studied for at least 40 years. A fourth nakhlite, NWA 817, was only reported in 2001, so little information is available on it. The only information pertinent to its crystallization age is a K-Ar age of 1350 Ma (Marty et al. 2001), in agreement with bulk Ar-Ar ages for the other nakhlites and Chassigny (Bogard and

Husain 1977; Podosek 1973).

Only one of Martian meteorites, the orthopyroxenite Allan Hills (ALH) 84001, is older than 1300 Ma, even though much of the surface of Mars is thought to be relatively old (Nyquist et al. 1998; Tanaka et al. 1992). ALH84001 has become best known for claims of evidence of life (McKay et al. 1996). Although these are not widely accepted (e.g. Sears et al. (1997)), there are other reasons why a sample of the ancient crust is of interest, including the possibility of finding a sample of paleoatmosphere, discussed below. ALH84001 is also the only Martian meteorite that does not fall into one of the SNC classes. Although there are some respects in which ALH84001's noble gases are similar to those of the SNC meteorites, the identification of it as Martian does not rely on noble gases. ALH84001 has an oxygen isotopic composition that is matched only by the SNC meteorites. It also has some key elemental ratios, such as Fe/Mn, that match the SNCs. Hence, if the SNC meteorites are Martian, so is ALH84001.

Table 2. Relative elemental abundances.

	$^{20}Ne/^{132}Xe$	$^{36}Ar/^{132}Xe$	$^{84}Kr/^{132}Xe$
Solar[a]	2.35×10^{6}	57,400	29
CI chondrites	44	105	1.3
Earth	703	1347	27.8
Mars atmosphere	212	580	20.3
Mars atmosphere[b]	–	900±100	20.5±2.5
Mars interior[c]	–	19	1.2

Atom/atom ratios. All data from Pepin (1991), unless otherwise marked.
[a](Wieler 2002b, this volume); [b](Bogard and Garrison 1998); [c](Ott 1988)

MARTIAN ATMOSPHERE

A key use of Martian meteorites has been in providing data for the isotopic and elemental abundances of the noble gases at far higher precision than was possible for Viking (Bogard and Garrison 1998; Garrison and Bogard 1998; Marti et al. 1995; Swindle et al. 1986) (Tables 2 and 3). With this, it is then possible to make models of the origin and evolution of the Martian atmosphere, and to compare these with comparable models for Earth (Pepin 1994; Pepin and Porcelli 2002, this volume). Bogard et al. (2001) have recently reviewed the origin and evolution of volatiles, including noble gases, on Mars.

Martian atmospheric noble gases are remarkably similar to terrestrial noble gases in two puzzling properties:

1. First, the relative elemental abundances of Ne, Ar, Kr and Xe are very similar for the two planets (Fig. 2). When compared to solar wind, the Ne/Ar/Kr ratios appear to be fractionated in favor of the heavy elements. However, the Xe/Kr ratio is nearly the same as the solar ratio. Alternatively, if the elemental abundances are compared to the "Q" gases trapped in chondritic meteorites (Ott 2002, this volume), the Ne/Ar/Kr ratios are similar, but the Xe/Kr ratio is too low. In the case of the Earth, the unusual Xe/Kr ratio led to unsuccessful searches for "missing Xe" (Porcelli and Ballentine 2002, this volume). Whatever the cause on Earth, the fact that the Martian ratios are the same suggests that the answer may not be a planet-specific process, unless it occurred to the same extent on both Earth and Mars.

Table 3a. Neon, argon and krypton isotopic composition of martian atmosphere.

	$^{20}Ne/^{22}Ne$	$^{36}Ar/^{38}Ar$	$^{40}Ar/^{36}Ar$	$^{78}Kr/^{84}Kr$	$^{80}Kr/^{84}Kr$	$^{82}Kr/^{84}Kr$	$^{83}Kr/^{84}Kr$	$^{86}Kr/^{84}Kr$
Atmosphere[a]	10.1(7)	4.1(2)	≈2400	0.00637(36)	≡0.0409	0.2054(20)	0.2034(18)	0.3006(27)
Atmosphere[b,c]	7-10	≤4	≈1900	0.093(2)	0.0432(2)	0.2099(6)	0.2058(8)	0.2975(7)
Solar[a,d]	13.7	5.8	<1	0.006365(34)	0.04088(14)	0.20482(54)	0.20291(26)	0.3024(10)
Earth[a]	9.80(8)	5.320	295.5	0.006087(28)	0.03960(15)	0.20217(65)	0.20136(61)	0.30524(65)

Table 3b. Xenon isotopic composition of martian atmosphere and interior.

	$^{124}Xe/^{132}Xe$	$^{126}Xe/^{132}Xe$	$^{128}Xe/^{132}Xe$	$^{129}Xe/^{132}Xe$	$^{130}Xe/^{132}Xe$	$^{131}Xe/^{132}Xe$	$^{134}Xe/^{132}Xe$	$^{136}Xe/^{132}Xe$
Atmosphere[e]	0.0038(2)	0.0033(2)	0.0735(9)	2.40(2)	0.1543(9)	0.7929(27)	0.4007(15)	0.3514(15)
Atmosphere[b]			0.0751(33)	2.590(31)	0.1561(22)	0.810(12)	0.4043(59)	0.3533(47)
Interior[f]	0.00477(18)	0.00432(2)	0.0846(9)	1.080(7)	0.1656(11)	0.8241(6)	0.3650(34)	0.2982(13)
Solar[d]	0.00490(4)	0.00423(14)	0.0847(10)	1.04(9)	0.1661(9)	0.8272(59)	0.3666(25)	0.2985(19)
Earth[a]	0.003537(11)	0.003300(17)	0.07136(9)	0.9832(12)	0.15136(12)	0.7890(11)	0.3879(6)	0.3294(4)

Atom/atom ratios. Numbers in parentheses are uncertainties in last digits, as given in references.

References:

[a](Pepin 1991) and references therein

[b](Garrison and Bogard 1998), 1550°C extraction of EETA 79001,8A for Kr and Xe

[c](Bogard 1997)

[d](Pepin et al. 1995), measured solar wind composition for Kr and Xe

[e](Swindle et al. 1986), "SPB-Xe"

[f](Mathew and Marti 2001), "Chass-S" (500°C extraction of Chassigny Ol)

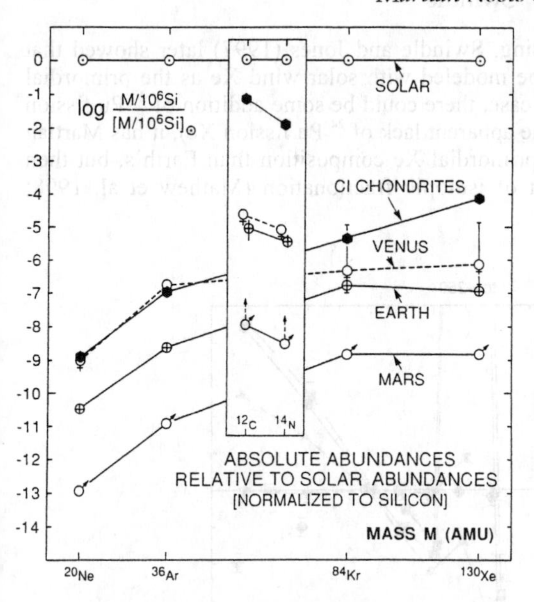

Figure 2. Relative elemental abundances of the four heavier noble gases in the Martian atmosphere, compared with terrestrial atmosphere and various meteorites, all plotted relative to solar abundances. The atmospheric abundances of C and N are also shown. The Martian atmosphere, like the terrestrial, has a noble gas pattern similar to CI chondrites, except that the Xe/Kr ratio is much lower, similar to the solar ratio. However, the Martian atmosphere contains approximately two orders of magnitude less noble gas than the terrestrial atmosphere. [Used by permission of Univ. of Arizona Press, from Pepin and Carr (1988), Fig. 1, p. 128.]

2. Second, there is a similarity in the overall Xe isotope pattern, before addition of radiogenic or fission-derived components (Fig. 3). As in the case of Earth, there appears to be an overall fractionation, relative to primordial reservoirs, in favor of the heavy isotopes. Furthermore, the fractionation on the two planets is to a similar degree. By far the best-developed model for this fractionation is that of hydrodynamic escape (Hunten et al. 1987; Pepin 1994; Pepin and Porcelli 2002, this volume). However, since hydrodynamic escape is a planet-specific process, it would take a remarkable coincidence for it to produce almost exactly the same effect on two very different planets. There have been some models that have tried to account for the similarity of the two planets' Xe isotope patterns in a pre-planetary process, such as diffusive separation in the gravitational field of a porous planetesimal (Zahnle et al. 1990) or fractionation during incorporation into ice that became a part of comets that later provided the planets' noble gases (Owen et al. 1992), but each of these models has problems serious enough to make equal degrees of hydrodynamic escape still the most attractive possibility.

On the other hand, there are also some key differences between Martian and terrestrial noble gases. Despite the glib description of the similarity in Xe isotopic patterns, the two planets' patterns differ in detail.

For example, terrestrial Xe has typically been modeled as starting with a primordial "U-Xe" composition (very similar to solar wind Xe), and then experiencing (a) isotopic fractionation, (b) addition of radiogenic ^{129}Xe, and (c) addition of heavy Xe isotopes derived from fission of ^{244}Pu (Pepin 2000). Martian Xe doesn't quite work that way. Swindle et al. (1986) first pointed out that U-Xe (Pepin 2000) will not work as the primordial Martian Xe. Instead, they found that except for an enrichment of radiogenic ^{129}Xe, Martian atmospheric Xe could be modeled as isotopically fractionated "AVCC Xe" (at the time considered a primitive component, now considered a mixture of Q-Xe and Xe from presolar diamonds (Ott 2002, this volume)). In that model there would be no need for any ^{244}Pu fission Xe. In fact, there would be no room for any. This would be strange: because ^{244}Pu has a longer half-life than ^{129}I, the ^{129}I would decay away first, yet it is the

^{244}Pu whose decay products are missing. Swindle and Jones (1997) later showed that Martian atmospheric Xe could also be modeled with solar wind Xe as the primordial composition, instead of U-Xe. In this case, there could be some addition of ^{244}Pu fission Xe. While this solves one problem (the apparent lack of ^{244}Pu fission Xe), it has Martian Xe starting with a slightly different primordial Xe composition than Earth's, but then undergoing almost the same amount of isotopic fractionation (Mathew et al. 1998; Swindle and Jones 1997).

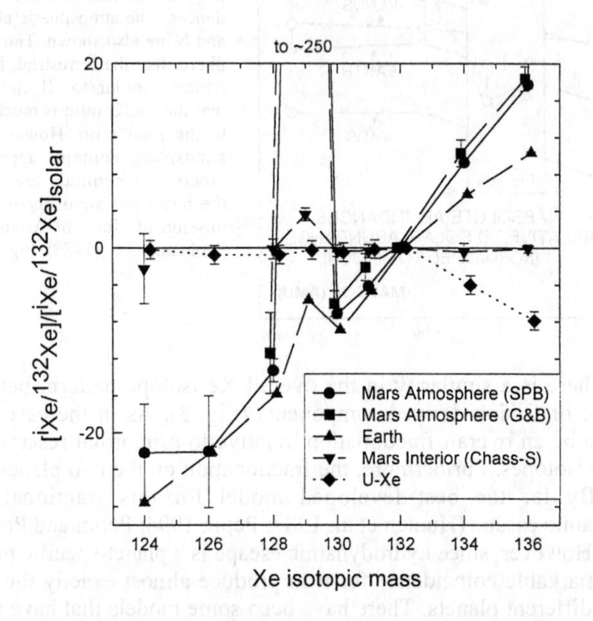

Figure 3. Xenon isotopic composition of the Martian atmosphere (represented by the "SPB-Xe" composition (Swindle et al. 1986) and the 1550°C extraction from glass 8A of EETA 79001 (Garrison and Bogard 1998)), the Martian interior (represented by the "Chass-S" composition (Mathew and Marti 2001), their spallation-corrected 500°C extraction of Chassigny olivine), the terrestrial atmosphere, and "U-Xe" (Pepin 2000), compared to measured solar wind Xe (Pepin et al. 1995). Deviations are given in percent. The interior (Chass-S) is identical to the solar wind, within uncertainties, except for a slight increase at ^{129}Xe. The Martian atmosphere has a light-isotope fractionation pattern that is quite similar to the terrestrial atmosphere, but has a much larger radiogenic addition at ^{129}Xe and is clearly distinct from the Earth at the heaviest isotopes. Implications of various models of the heavy isotopes of Xe in the terrestrial and Martian atmospheres are discussed in the text.

Another difference between Martian and terrestrial noble gases is that both the ^{40}Ar/^{36}Ar and ^{129}Xe/^{132}Xe ratios are more radiogenic (higher) on Mars (Table 3). The difference clearly involves some combination of the timing and extent of outgassing of radiogenic noble gas and the timing and extent of atmospheric loss. The fact that Mars has two orders of magnitude less noble gas in its atmosphere suggests that it has experienced considerably more atmospheric loss. This is generally attributed to its smaller gravitational potential. For example, Mars is far more likely to lose gas by stripping as a result of an impact (Melosh and Vickery 1989). Since ^{129}Xe is the result of the decay of ^{129}I, which has a half-life of only 16 Ma, the higher ^{129}Xe/^{132}Xe requires either very early atmospheric loss or storage of the bulk of the radiogenic ^{129}Xe within the planet until after the atmospheric loss. Musselwhite et al. (1991) suggested that iodine's solubility in water could lead to iodine dissolving in water, then being sequestered in minerals that might hold the radiogenic ^{129}Xe during Xe loss, if Mars had liquid water very early in its history. However, this still would not explain a lack of ^{244}Pu fission Xe. Clearly, understanding exactly how much (if any) ^{244}Pu fission Xe the Martian atmosphere contains would help in understanding the history of the Martian atmosphere.

Martian krypton also differs isotopically from terrestrial krypton. For most krypton isotopes, the Martian atmospheric ratios are indistinguishable from those of the solar wind, within the uncertainties in our knowledge of the two compositions, or perhaps isotopically fractionated in the opposite direction from terrestrial krypton (Garrison and Bogard 1998; see Fig. 4). Meanwhile, ^{80}Kr and ^{82}Kr are elevated in most samples of shergottite glass, presumably reflecting neutron capture on ^{79}Br and ^{81}Br, respectively. Whether this occurred within the rock or is a feature of the Martian atmosphere remains to be determined.

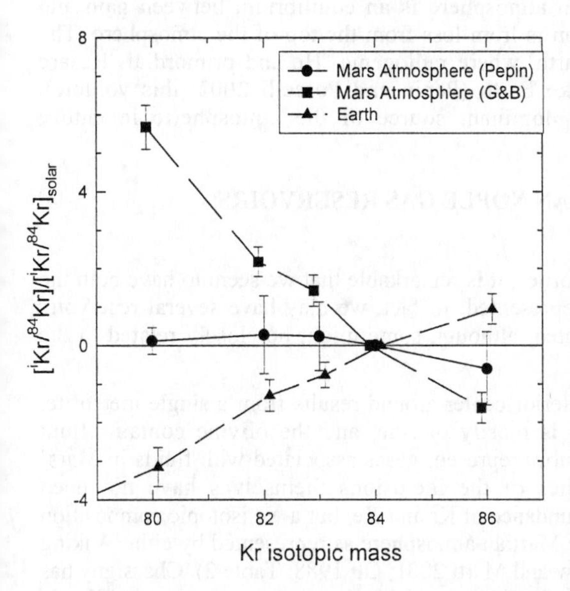

Figure 4. Krypton isotopic composition of various measurements of the Martian atmosphere (represented by the "Pepin Mars" composition (Pepin 1991) and the 1550°C extraction from glass 8A of EETA 79001 (Garrison and Bogard 1998)) and the terrestrial atmosphere, compared to measured solar wind Kr (Pepin et al. 1995). Deviations are given in percent. The EETA 79001, 8A glass appears to be fractionated in favor of the lighter isotopes by as much as 1%/amu, while the "Pepin Mars" composition is virtually identical to the solar wind. The terrestrial atmosphere, on the other hand, is fractionated in favor of the heavier isotopes. For more detailed discussion (and plots) see Garrison and Bogard (1998).

Another difference between terrestrial and Martian noble gases is in the ^{36}Ar/^{38}Ar ratio. The terrestrial ratio, 5.3, is within about 10% of the ratios in the solar wind and in "Q" gases (Ott 2002; Wieler 2002b, both this volume), but the Martian ratio is much lower, probably 4.0 or less (Bogard 1997). The best model for the difference is solar-wind-induced sputtering (Jakosky et al. 1994): neutral atoms in the upper atmosphere become ionized and picked up by the magnetic field of the flowing solar wind. However, because they are charged, they have a gyroradius, so some impact the exobase, the top of the atmosphere. Through momentum transfer, these can impart enough kinetic energy to atmospheric atoms to cause loss. Isotopic fractionation occurs because this process happens above the homopause, the height to which the atmosphere is isotopically well-mixed, so lighter species are preferentially removed. This is clearly a planet-specific process. Incidentally, the Viking measurement of the Martian ^{36}Ar/^{38}Ar ratio was at one point reported as "within 10% of terrestrial" (Owen et al. 1977), although this was subsequently corrected (Owen 1986).

Jakosky et al (1994) pointed out that if there is enough loss of Ar to change the ^{36}Ar/^{38}Ar ratio, there should be enough loss of Ne that its abundance would represent an equilibrium between outgassing and atmospheric loss. If so, the fact that the Ne/Ar ratio in the Martian atmosphere is very similar to that of the terrestrial atmosphere becomes a coincidence. Furthermore, the $^{20}Ne/^{22}Ne$ ratio in the Martian atmosphere appears to be somewhat similar to the terrestrial atmosphere (Garrison and Bogard 1998; Swindle et al. 1986; Wiens et al. 1986; Table 3). However, the isotopic composition of Ne is not as well defined as that of the heavier elements. Because Ne is produced so readily by spallation reactions induced by galactic cosmic rays (Wieler 2002a, this volume), there is significant spallation-produced Ne in even the glassy inclusions in shergottites, even in EETA79001, with its exposure age of less than 1 Ma.

Martian atmospheric Ne has proven difficult to measure in Martian meteorites, but Martian atmospheric He has proven impossible. However, it has been possible to make telescopic measurements of at least the He elemental abundance (Krasnopolsky et al. 1994). As in the case of the Earth, the abundance, and presumably the isotopic composition, of He in the Martian atmosphere is an equilibrium between gain and attrition. Like the Earth, the attrition is from loss from the top of the atmosphere. The situation may be similar to the Earth, where radiogenic ^{4}He and primordial ^{3}He are degassing from the interior. Unlike Earth (Pepin and Porcelli 2002, this volume), cosmogenic ^{3}He is probably the dominant source of the atmospheric inventory (Krasnopolsky et al. 1994).

OTHER MARTIAN NOBLE GAS RESERVOIRS

Martian interior (Chassigny)

With less than two dozen meteorites, it is remarkable that we seem to have both the atmosphere and the interior well represented. In fact, we may have several reservoirs other than the atmosphere represented, although some might be closely related to the atmosphere.

Most models of the Martian interior center around results from a single meteorite, the dunite Chassigny. Chassigny is mostly olivine, and the olivine contains fluid inclusions, so the noble gases plausibly represent gases associated with fluids in Mars' interior, although noble gas studies of the inclusions themselves have not been performed. Chassigny has a high abundance of Kr and Xe, but a Xe isotopic composition (Fig. 3) that is far different from the Martian atmosphere as represented by either Viking or the shergottite meteorites (Mathew and Marti 2001; Ott 1988; Table 2). Chassigny has a $^{129}Xe/^{132}Xe$ ratio of approximately 1, so it contains virtually no radiogenic ^{129}Xe. In addition, it has a Kr/Xe ratio far lower than the terrestrial or Martian atmospheres (Table 2). Ott, who performed the first detailed measurements of Chassigny Xe, noted that Martian meteorites that do not seem to contain pure atmospheric noble gases might contain mixtures of Chassigny (interior) noble gas and Martian atmospheric noble gases (Ott 1988). Ott's plot of $^{129}Xe/^{132}Xe$ vs. $^{84}Kr/^{132}Xe$ (Fig. 5) has proven a useful tool for separating out various Martian and terrestrial components, since the Martian interior (Chassigny), the Martian atmosphere, and the terrestrial atmosphere are clearly separated.

Although Chassigny contains enough cosmic-ray-spallation produced Xe to make it difficult to determine the underlying composition with precision, it appears to be indistinguishable from solar wind Xe (Mathew and Marti 2001; Ott 1988). On the basis of stepwise heating results on two different Chassigny samples, Mathew and Marti (2001) have argued that Chassigny might contain gas from not one, but two, Martian interior reservoirs, one of them containing a contribution from Pu fission.

Unlike the case for Earth, where the noble gases in the interior (such as MORB and OIB samples (Graham 2002, this volume)) are more radiogenic than the atmosphere, the

Martian interior, as represented by Chassigny, is much less radiogenic. Hence Chassigny must sample an undegassed interior reservoir.

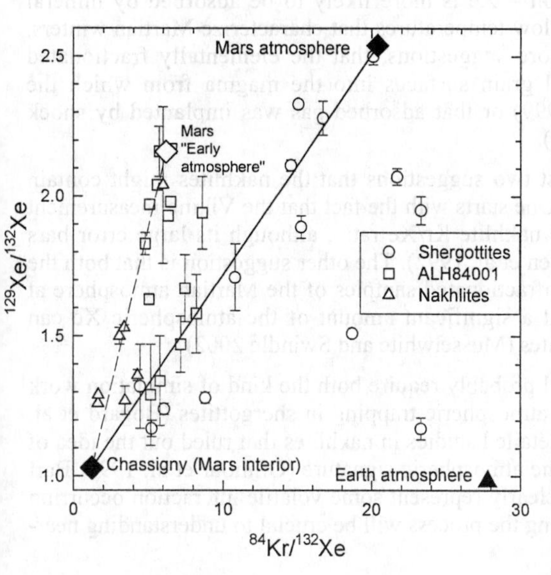

Figure 5. Plot of $^{129}Xe/^{132}Xe$ vs. $^{84}Kr/^{132}Xe$ for various Martian meteorite samples, the Martian atmosphere as measured by Martian meteorites, and the Martian interior as measured by Chassigny [Data from compilation of (Bogard and Garrison 1998)]. Simple two-component mixtures of Martian atmosphere and Martian interior should fall along the solid line. Terrestrial contamination would pull data points to the right (below the line). Shergottite samples fall on or to the right of the line. However, ALH84001 and most samples from nakhlites plot above the line, suggesting that the incorporation process(es) has resulted in elemental fractionation, preferentially incorporating Xe. Alternatively, many points fall close to the dashed line, a mixing line between Chassigny and the Mars "Early atmosphere" composition derived by Mathew and Marti (2001), and so could represent the paleoatmosphere. Various models for this effect are discussed in the text.

While Chassigny provides good samples of krypton and xenon from the Martian interior, its cosmic-ray-exposure age is high enough (11 Ma, Table 1), and the abundances of Ne and Ar low enough, that those gases are dominated by cosmogenic gas plus radiogenic ^{40}Ar. Although it can be used to determine the elemental abundance of Ar (Ott 1988), Chassigny cannot be used to determine the primordial Martian isotopic composition of either Ne or Ar, or even the elemental abundance of Ne.

Other reservoirs I: Nakhlite (and ALH84001) Xe

Although some samples of Martian meteorites appear to contain nearly pure atmosphere, and Chassigny seems to contain rather pure Martian interior noble gases, many Martian meteorites seem to contain gas that is similar to the atmosphere, but not identical.

In the nakhlites, the gas has $^{129}Xe/^{132}Xe > 1$ (greater than 2 in many samples of Nakhla), but with a Kr/Xe ratio far lower than would be expected of a mixture of Martian atmosphere (EETA 79001 glass) and interior (Chassigny), as shown in Figure 5. ALH84001 plots with the nakhlites. An obvious possibility is that the nakhlites (and ALH84001) contain elementally fractionated Martian atmosphere. There are several processes in which Xe is preferentially retained relative to Kr, and it seems that most of them have been suggested as the explanation for the atmospheric-like component in the nakhlites.

One possibility is solubility—Xe is more soluble in water than Kr, and nakhlites contain alteration products that seem to require the presence of liquid water to form (Treiman et al. 1993). Drake et al (1994) suggested that these alteration products contained the elementally fractionated atmosphere and reported an analysis of a single sample of weathering products consist with that possibility. Garrison and Bogard (1998) suggested a similar mechanism. However, studies of mineral separates (Gilmour et al.

1999; Bart et al. 2001) showed that the bulk of the Kr and Xe in Nakhla is in the major minerals or in the mesostasis, rather than the weathering products.

Another possibility is adsorption – Xe is more likely to be adsorbed by mineral surfaces than Kr, particularly at the low temperatures that characterize Martian winters. This idea is at the heart of two more suggestions, that the elementally fractionated atmosphere was released from soil grain surfaces into the magma from which the nakhlites formed (Gilmour et al. 1999) or that adsorbed gas was implanted by shock (Gilmour et al. 2000; Bart et al. 2001).

Finally, there have been at least two suggestions that the nakhlites might contain *un*fractionated Martian atmosphere. One starts with the fact that the Viking measurement is actually more consistent with the nakhlite Kr/Xe ratio, although its large error bars encompass the shergottite value (Owen et al. 1992). The other suggestion is that both the shergottites and nakhlites contain unfractionated samples of the Martian atmosphere at the time they were ejected, but that a significant amount of the atmospheric Xe can sometimes be trapped in polar clathrates (Musselwhite and Swindle 2002)

Sorting out the true cause(s) will probably require both the kind of simulation work that demonstrated the possibility of atmospheric trapping in shergottites (Bogard et al. 1986; Wiens 1988) and the kind of detailed studies in nakhlites that ruled out the idea of the alteration components carrying the atmospheric signature (Gilmour et al. 1999; Bart et al. 2001). Since these meteorites clearly represent some volatile interaction occurring near the Martian surface, understanding the process will be crucial to understanding near-surface volatile reservoirs.

Other reservoirs II: Shergottite Ar

Just as the nakhlites contain noble gases that do almost, but not quite, match the atmosphere, so do the shergottites. In this case, however, the evidence is in the $^{40}Ar/^{36}Ar$ ratio of the gas trapped within them.

Recall that Bogard and Johnson (1983) discovered the Martian atmosphere trapped within EETA 79001 while attempting ^{40}Ar-^{39}Ar dating. Bogard and co-workers later performed ^{40}Ar-^{39}Ar experiments on most of the shergottites known at the time (Bogard and Garrison 1999), and found that atmospheric contamination seemed to be the rule, rather than the exception. If the shergottites all incorporated Martian atmospheric argon with the same isotopic composition, then it should be possible to simply take the amount of ^{36}Ar trapped in the rock, and subtract an appropriate amount of ^{40}Ar:

$$^{40}Ar^* = {}^{40}Ar_T - [{}^{36}Ar_t \times ({}^{40}Ar/{}^{36}Ar)_{atm}] \,, \qquad (1)$$

where $^{40}Ar^*$ is the radiogenic (decay-produced) argon, and the subscripts T, t, and atm refer to total, trapped, and atmospheric, respectively. Bogard and Garrison (1999) found that if they used the atmospheric argon isotopic composition, they got different apparent ages for different meteorites. In general, they did not obtain the ages obtained by other techniques, and in most cases they calculated negative apparent ages. Therefore, they concluded that the shergottites had incorporated Ar with variable $^{40}Ar/^{36}Ar$ ratios. The most intriguing possibility is one or more near-surface volatile reservoirs that were available either when the rocks formed or when they were ejected from Mars. However, the same effect could also result from a combination of Martian atmospheric argon, incorporated terrestrial atmospheric argon (most of the meteorites involved have spent $>10^4$ years in Antarctica), and "inherited" ^{40}Ar, radiogenic ^{40}Ar that was incompletely degassed when the rock formed.

Paleoatmosphere

On Earth, after decades of searching, it is not clear that anyone has ever found a sample of paleoatmosphere with noble gas isotopic ratios distinct from the modern ratios. For Mars, with only 20 samples to work with, a strong case can be made that paleoatmosphere has been found.

To preserve a sample of paleoatmosphere, a rock would have to have trapped some atmosphere long enough ago that the noble gas isotopes in the atmosphere had been able to evolve significantly in the interim. On Earth, that is difficult. Most rocks are not that old. Many rocks that are old enough (e.g., Archean) are not the kinds of rocks that would have trapped atmosphere within them, since virtually the only mechanisms on Earth whereby atmospheric trapping occurs are those operating during sedimentation. Finally those rocks that might have had an opportunity to trap an isotopically distinct atmosphere have usually been metamorphosed or have suffered through other processes that could have caused them to lose that atmosphere. On Mars, ALH84001 is a rock that is more than 4 Ga old (old enough), it apparently suffered several shock events early in the history of Mars (which could have implanted some atmospheric gas), but it has apparently been altered little in the last 4 Ga, since Mars lacks plate tectonics and, to a large extent, the action of liquid water.

But does it have paleoatmosphere? The identification has largely been made based on expectations of what the Martian atmosphere might have been like 4 Ga ago. A definitive identification would require some demonstration that the identified component was implanted 4 Ga ago, which would be difficult.

The highest $^{129}Xe/^{132}Xe$ in ALH84001 is about 2.16, 10% lower than measured in the glasses in the shergottites. In addition, the $^{129}Xe/^{136}Xe/^{132}Xe$ systematics are not consistent with a combination of current Martian atmosphere, a Martian interior component like that identified in Chassigny, and terrestrial contamination. On this basis, the Xe has been argued to be an ancient component, rather than modern atmosphere (Gilmour et al. 1998; Mathew and Marti 2001; Murty and Mohapatra 1997).

In addition, the $^{15}N/^{14}N$ and $^{40}Ar/^{36}Ar$ ratios in ALH84001 are not as high as in the modern Martian atmosphere. Since those ratios are expected to rise over time as a result of atmospheric loss and outgassing, respectively, that could mean that ALH84001 has trapped Martian paleoatmosphere that is not as evolved (Grady et al. 1998; Mathew and Marti 2001; Murty and Mohapatra 1997). Unlike the case for Xe, different authors have come up with different isotopic ratios for the "unevolved" N and Ar. For $\delta^{15}N$, the values obtained have been +7‰ (Mathew and Marti 2001) or ≤+200‰ (Grady et al. 1998) or ≥+46‰ and <+620 (Murty and Mohapatra 1997). Grady et al. (1998) concluded $^{40}Ar/^{36}Ar$ was ≤300, while Murty and Mohapatra (1997) suggested ≤1400 (although a small fraction of their data suggested ~300). Finally, Mathew and Marti (2001) concluded that $^{36}Ar/^{38}Ar$ was ≥5.0, while Murty and Mohapatra (1997) assumed $^{36}Ar/^{38}Ar$ was 4.1.

Identifying Martian paleoatmosphere is fascinating in its own right, but also could put tight constraints on Martian atmospheric evolution. For example, based on Xe isotope systematics in ALH84001, Mathew and Marti have argued that not only had the Martian atmosphere not acquired its full complement of ^{129}Xe by 4 Ga ago, the isotopic fractionation that is required (see Fig. 3 and discussion above) had not yet occurred then either (Mathew and Marti 2001). Such a late fractionation would be very difficult to explain by atmospheric loss through hydrodynamic escape (Hunten et al. 1987; Pepin 1994; Pepin and Porcelli 2002, this volume).

MARTIAN CHRONOLOGY

As they are with terrestrial, lunar and asteroidal meteorite samples, noble gases have been crucial to deciphering the chronology of the Martian meteorites. However, just as in those other samples, noble gases are not the only chronometers that have been applied. Most of the information about crystallization ages has come from other chronometers. Nyquist et al. (2001) have reviewed the subject, and their best estimates of the formation ages of the Martian meteorites are listed in Table 1 and plotted in Figure 6.

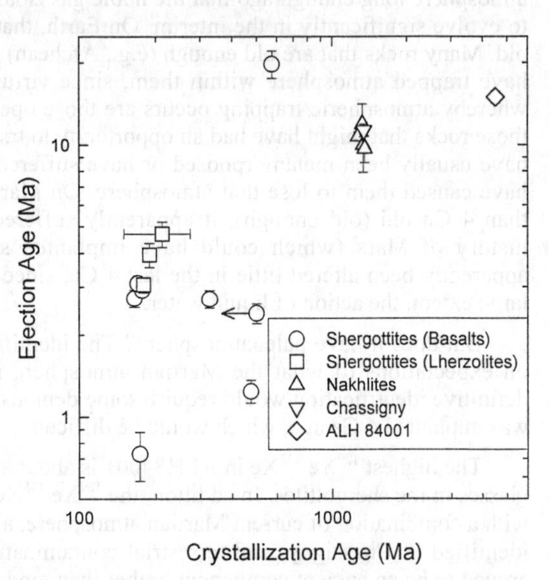

Figure 6. Plot of crystallization ages vs. ejection ages (cosmic-ray-exposure age plus terrestrial age) for Martian meteorites for which both ages have been measured (Table 1). Meteorites for which the ejection ages differ presumably came from distinct ejection events. Meteorites could have different crystallization ages, but the same ejection ages if an impact ejected material from, for example, distinct lava flows.

^{40}Ar-^{39}Ar dating

The high ^{40}Ar/^{36}Ar ratio of the Martian atmosphere (Table 2) is clearly a problem for ^{40}Ar-^{39}Ar dating. It is a worse problem than for terrestrial samples for two reasons. First, the higher ratio means that the correction that must be applied (Eqn. 1) is larger for Martian samples than for terrestrial samples. In addition, since the Martian meteorites have all had enough cosmic-ray exposure to create a significant amount of cosmogenic ^{36}Ar, there is an added uncertainty at the start of the correction process.

Because of incorporated Martian atmosphere, ^{40}Ar-^{39}Ar analyses of shergottites have yielded more information about trapped argon components (see above) than about rock ages. However, ^{40}Ar-^{39}Ar analyses of whole rock samples of three nakhlites and Chassigny have given very simple apparent age spectra, with virtually all apparent ages concordant with each other and with other chronometers at about 1300 Ma (Bogard and Husain 1977; Podosek 1973). This suggests that there is little, if any, Martian atmospheric argon incorporated into these rocks. Furthermore, whole-rock K-Ar ages of these rocks are also approximately 1300 Ma (Swindle 2001), as is that of newly discovered nakhlite NWA 817 (Marty et al. 2001).

K-Ar dating on a component within nakhlites has been used to argue for the presence of liquid water on Mars within the last 1000 Ma. The nakhlites contain evidence of aqueous alteration on Mars in the form of patches and coatings on olivine grains of a fine-grained material known as "iddingsite" (Treiman et al. 1993). Because it comes in small patches (the largest sample yet separated is 23 µg) and because its fine-grained nature means that ^{39}Ar recoil is a significant problem (see (Kelley 2002, this volume)), standard ^{40}Ar-^{39}Ar dating would be difficult to apply. However, when K-Ar dating was performed by measuring the potassium through neutron activation and the ^{40}Ar through mass spectrometry, ages of 0 to 670 Ma were determined on 12 iddingsite samples from the nakhlite Lafayette (Swindle et al. 2000). The oldest ages matched a two-point "isochron" from a Rb-Sr leaching experiment (Shih et al. 1998), and so may represent the beginning of iddingsite formation. The K-Ar ages certainly demonstrate that the aqueous alteration occurred on Mars, since they are much older than the meteorite's terrestrial residence time.

ALH84001 has a crystallization age of about 4.5 Ga (Table 1), nearly as old as the planet itself. Since the K-Ar system is so susceptible to resetting by heating, it is probably not surprising that ^{40}Ar-^{39}Ar ages on ALH84001 have turned out to be less than that. Instead, the ages are about 4 Ga, with the exact result depending on the assumptions made about the presence and composition of incorporated Martian atmospheric argon (Ash et al. 1996; Bogard and Garrison 1999; Ilg et al. 1997; Turner et al. 1997). That number presumably represents heating from a shock event—ALH84001 suffered through several shock events, whose results include some heating (Kring et al. 1998; Treiman 1998). One of the most intriguing things about the age of about 4 Ga, assuming it represents a shock event, is that it is very close to the age postulated for a lunar cataclysm (Bogard 1995; Swindle 2002, this volume).

An intriguing question is whether incorporated atmospheric argon will be a problem for rocks returned from Mars or analyzed *in situ* on the Martian surface (Swindle 2001). There are some reasons to think it will not be. The partial pressure of ^{40}Ar in the Martian atmosphere is roughly two orders of magnitude lower than in Earth's. Furthermore, most of the Martian atmosphere incorporated in the shergottites, the only Martian meteorites where the incorporation process is truly understood, is the result of shock, presumably the shock that ejected the rocks from Mars and sent them on their way to Earth. Rocks analyzed in situ, and even carefully handled returned samples, will not have the added confusion of the possibility of terrestrial contamination, nor will they have been shocked. However, the results of Bogard and Garrison (1999) suggests that there is some magmatic argon in shergottites, and even a little argon incorporated some other way (magmatically, for example) could make analyses of returned samples, and particularly *in situ* analyses, difficult to interpret.

Cosmic-ray exposure ages

The cosmic-ray exposure ages of Martian meteorites also tell some fascinating tales. The calculation of cosmic-ray exposure ages in meteorites is discussed in Wieler (2002a) and those of Martian meteorites are included in Table 1 and Figure 6.

When the SNC meteorites were first shown to be Martian, the question was how any rocks could reach Earth from Mars as meteorites. However, studies of the impact cratering process suggested a way to produce Martian meteorites (Melosh 1984). In an impact, there is a zone near the surface where the rock can be "spalled" off and accelerated to speeds approaching the speed of the incoming impactor without being shocked so severely that it would be crushed. Another suggestion was that oblique impacts could accelerate material off the planet in the downrange direction (Nyquist 1983), perhaps as a result of entrainment in a vapor plume (O'Keefe and Ahrens 1986). Furthermore, analytical calculations, first using Monte Carlo methods (Wetherill 1984) then using direct integrations of orbits (Gladman et al. 1996) indicated that a reasonable fraction of the rocks that are ejected from Mars could make it to Earth.

Then came the basic question, still not quite settled, of how many ejection events were required to yield the Martian meteorites that we have. At first, given the perceived difficulty of ejecting any meteorites from Mars, scenarios were suggested with as few impacts as possible.

One version suggested that the 180 Ma age-signature present in the shergottites really represented an ejection event, and that the exposure ages really represented break-up events (Vickery 1987). There would have to be one break-up event roughly 11 Ma ago to produce the nakhlites and Chassigny, another one about 3 Ma ago to produce most of the shergottites, and a third one less than 1 Ma ago to produce EETA79001.

However, the isotopic ratios of cosmogenic Ne in meteorites are depth sensitive (Wieler 2002a), and these suggest that the Martian meteorites have only been exposed as small meteoroids (Garrison et al. 1995). In addition, there are only hints of excesses of noble gases that could be produced by neutron capture at depth, such as ^{80}Kr, ^{82}Kr and ^{131}Xe. Finally, studies of cosmogenic radionuclides have found no evidence for multiple stage exposure [e.g. Nishiizumi et al. 1986, 1999, 2000). Hence, a few smaller impacts became the preferred scenario. However, as the number of Martian meteorites grew, so did the number of required events.

The nakhlites and Chassigny could still come from a single event, given their similar crystallization ages and indistinguishable cosmic-ray exposure ages. However, they are unlikely to come from the same event that ejected the shergottites (Treiman 1995).

Eugster et al. (1997) argued for at least three shergottite ejection events. Using the best production rates available, they concluded that two of the lherzolitic shergottites have ejection ages (the sum of the cosmic-ray-exposure age and the terrestrial exposure age) of roughly 4 Ma, distinct from most of the basaltic shergottites, which have 3 Ma ejection ages. They also argued for a separate event for EETA 79001.

Nyquist et al. (1998), trying to minimize the number of required ejection events of very young rocks, suggested that all the shergottites could come from a single event about 3 Ma ago, with EETA 79001 released in a later breakup. They suggested that the apparent difference in ejection ages between the basaltic and lherzolitic shergottites could be the result of incorrect corrections for the very distinct elemental compositions of the two types, even though Eugster et al. were clearly using the best production rates available.

The shergottites, however, can no longer be attributed to just one or two craters. Dhofar 019 has an exposure age of nearly 20 Ma (Table 1), EETA 79001 still has an age less than 1 Ma, and Dar al Gani 476 (Table 1) has an exposure age between those of EETA 79001 and the other basaltic shergottites. Hence, there seem to be five or more shergottite ejection events.

ALH84001 requires a separate event, because of its longer exposure age. Actually, since models of the distribution of ages of surface units on Mars predict that most of the surface should be older than 1300 Ma (Nyquist et al. 1998; Tanaka et al. 1992), a naive interpretation would be that there should be more events in terrain older than that of the nakhlites and Chassigny than in terrain that young or younger, but just the opposite is observed. This presumably means that either the estimates of the ages of terrains on Mars are wrong, or that the likelihood of a rock becoming a meteorite depends on the age of the terrain, perhaps because the ejection process is more efficient for younger, more consolidated rocks (Head and Melosh 1999).

Note that the same evidence that none of the meteorites were exposed as parts of a larger meteoroid means that there is no evidence for any of the Martian meteorites having had any measurable exposure on Mars. In the spall models of meteorite ejection, most of the material would come from rather close to the surface. Furthermore, *all* lunar meteorites have abundant evidence for pre-ejection exposure within the top one to two meters. Hence the question of why no pre-ejection exposure of Martian meteorites is observed also remains an unsolved puzzle (Nyquist et al. 2001).

In summary, as a result of the Martian meteorites, we know more about the noble gases on Mars than about those on any planet besides Earth. As our knowledge of Martian noble gases grows, we continue to find results that are unexpected, and, at the moment, inexplicable, providing a valuable check for models that we devise based on our more detailed knowledge of the Earth-Moon system.

ACKNOWLEDGMENTS

Constructive reviews by D. Bogard, R. Pepin, and R. Wieler are gratefully acknowledged. This work was supposed in part by NASA Grant NAG 5-4767.

REFERENCES

Ash RD, Knott SF, Turner G (1996) A 4-Gyr shock age for a Martian meteorite and implications for the cratering history of Mars. Nature 380:57-59

Bart GD, Swindle TD, Olson EK, Treiman AH (2001) Xenon and krypton in Nakhla mineral separates. Lunar Planet Sci XXXII, Abstr #1363 (CD-ROM)

Becker RH, Pepin RO (1984) The case for a Martian origin of the shergottites: nitrogen and noble gases in EETA 79001. Earth Planet Sci Lett 69:225-242

Bogard DD (1995) Impact ages of meteorites: A synthesis. Meteoritics 30:244-268

Bogard DD (1997) A reappraisal of the Martian $^{36}Ar/^{38}Ar$ ratio. J Geophys Res 102:1653-1661

Bogard DD, Husain L (1977) A new 1.3 Aeon-young achondrite. Geophys Res Lett 4:69-72

Bogard DD, Johnson P (1983) Martian gases in an Antarctic meteorite? Science 221:651-654

Bogard DD, Garrison DH (1998) Relative abundances of argon, krypton, and xenon in the Martian atmosphere as measured in Martian meteorites. Geochim Cosmochim Acta 62:1829-1835

Bogard DD, Garrison DH (1999) Argon-39-argon-40 "ages" and trapped argon in Martian shergottites, Chassigny, and Allan Hills 84001. Meteoritic Planet Sci 34:451-473

Bogard DD, Nyquist LE, Johnson P (1984) Noble gas contents of shergottites and implications for the Martian origin of SNC meteorites. Geochim Cosmochim Acta 48:1723-1739

Bogard DD, Horz F, Johnson P (1986) Shock-implanted noble gases: An experimental study with implications for the origin of Martian gases in shergottite meteorites. J Geophys Res 91:E99-E114

Bogard DD, Clayton RN, Marti K, Owen T, Turner G (2001) Martian volatiles: isotopic composition, origin, and evolution. Space Sci Rev 96:425-458

Borg LE, Nyquist LE, Reese Y, Wiesmann H, Shih C-Y, Ivanova M, Nazarov MA, Taylor LA (2001) The age of Dhofar 019 and its relationship to the other Martian meteorites. Lunar Planet Sci XXXII, Abstr #1144 (CD-ROM)

Drake MJ, Swindle TD, Owen T, Musselwhite DS (1994) Fractionated Martian atmosphere in the nakhlites. Meteoritics 29:854-859

Eugster O, Weigel A, Polnau E (1997) Ejection times of Martian meteorites. Geochim Cosmochim Acta 61:2749-2757

Friedman Lentz RC, Taylor GJ, Treiman AH (1999) Formation of a Martian pyroxenite: A comparative study of the nakhlite meteorites and Theo's Flow. Meteoritic Planet Sci 34:919-932

Garrison DH, Bogard DD (1998) Isotopic composition of trapped and cosmogenic noble gases in several Martian meteorites. Meteoritic Planet Sci 33:721-736

Garrison DH, Rao MN, Bogard DD (1995) Solar-proton-produced neon in shergottite meteorites and implications for their origin. Meteoritics 30:738-747

Gilmour JD, Whitby JA, Turner G (1998) Xenon isotopes in irradiated ALH84001: Evidence for shock-induced trapping of ancient Martian atmosphere. Geochim Cosmochim Acta 62:2555-2571

Gilmour JD, Whitby JA, Turner G (1999) Martian atmospheric xenon contents of Nakhla mineral separates: Implications for the origin of elemental mass fractionation. Earth Planet Sci Lett 166: 139-147

Gilmour JD, Whitby JA, Turner G (2000) Extraterrestrial xenon components in Nakhla. Lunar Planet Sci XXXI, Abstr #1513 (CD-ROM)

Gladman BJ, Burns JA, Duncan M, Lee P, Levison HF (1996) The exchange of impact ejecta between terrestrial planets. Science 271:1387-1392

Grady MM (2000) Catalogue of Meteorites: with special reference to those represented in the collection of The Natural History Museum, London. Cambridge University Press, Cambridge

Grady MM, Wright IP, Pillinger CT (1998) A nitrogen and argon stable isotope study of Allan Hills 84001: Implications for the evolution of the Martian atmosphere. Meteoritic Planet Sci 33:795-802
Graham DW (2002) Noble gas isotope geochemistry of mid-ocean ridge and ocean island basalts: Characterization of mantle source reservoirs. Rev Mineral Geochem 47:247-318
Grossman JN (2000) The Meteoritical Bulletin, No. 84, 2000 August. Meteoritic Planet Sci 35:A199-A225
Grossman JN, Zipfel J (2001) The Meteoritical Bulletin, No. 85, 2001 September. Meteoritic Planet Sci 36:A293-A322
Head JN, Melosh HJ (1999) Effects of layering on spall velocity: Numerical simulations. Lunar Planet Sci XXX, Abstr #1761 (CD-ROM)
Hunten DM, Pepin RO, Walker JGC (1987) Mass fractionation in hydrodynamic escape. Icarus 69:532-549
Ilg S, Jessberger EK, El Goresy A (1997) Argon-40/argon-39 laser extraction dating of individual maskelynites in SNC pyroxenite Allan Hills 84001. Meteoritic Planet Sci 32:A65
Jakosky BM, Pepin RO, Johnson RE, Fox JL (1994) Mars atmospheric loss and isotopic fractionation by solar-wind-induced sputtering and photochemical escape. Icarus 111:271-288
Kelley S (2002) K-Ar and Ar-Ar dating. Rev Mineral Geochem 47:785-818
Krasnopolsky VA, Bowyer S, Chakrabarti S, Gladstone GR, McDonald JS (1994) First measurement of helium on Mars: Implications for the problem of radiogenic gases on the terrestrial planets. Icarus 109:337-351
Kring DA, Swindle TD, Gleason JD, Grier JA (1998) Formation and relative ages of maskelynite and carbonate in ALH84001. Geochim Cosmochim Acta 62:2155-2166
Marti K, Kim JS, Thakur AN, McCoy TJ, Keil K (1995) Signatures of the Martian atmosphere in glass of the Zagami meteorite. Science 267:1981-1984
Marty B, Marti K, Barrat JA, Birck JL, Blichert-Toft J, Chaussidon M, Deloule E, Gillet P, Gopel C, Jambon A, Manhes G, Sautter V (2001) Noble gases in new SNC meteorites NWA 817 and NWA 480. Meteoritic Planet Sci 36:A122-A123
Mathew KJ, Marti K (2001) Early evolution of Martian volatiles: Nitrogen and noble gas components in ALH84001 and Chassigny. J Geophys Res 106:1401-1422
Mathew KJ, Kim JS, Marti K (1998) Martian atmospheric and indigenous components of xenon and nitrogen in the Shergotty, Nakhla, and Chassigny group meteorites. Meteoritic Planet Sci 33:655-664
McElroy MB, Yung YL, Nier AO (1976) Isotopic composition of nitrogen: Implications for the past history of Mars' atmosphere. Science 194:70-72
McKay DS, Gibson EKJr, Thomas-Keprta KL, Vali H, Romanek CS, Clemett SJ, Chillier XDF, Maechling CR, Zare RN (1996) Search for past life on Mars: Possible relic biogenic activity in Martian meteorite ALH84001. Science 273:924-930
McSween HY Jr (1985) SNC meteorites: Clues to Martian petrologic evolution? Rev Geophys 23:391-416
McSween HY Jr (1994) What we have learned about Mars from SNC meteorites. Meteoritics 29:757-779
McSween HY Jr (2002) The rocks of Mars from far and near. Meteoritic Planet Sci 37:7-25
McSween HY Jr, Stolper EM (1980) Basaltic meteorites. Sci Am 242:54-63
Melosh HJ (1984) Impact ejection, spallation and the origin of meteorites. Icarus 59:234-260
Melosh HJ, Vickery AM (1989) Impact erosion of the primordial atmosphere of Mars. Nature 338:487-489
Meyer C. (1998) Mars meteorite compendium 1998:273. Office of the Curator, Earth Science and Solar System Exploration Division, Johnson Space Center, Houston, Texas
Murty SVS, Mohapatra RK (1997) Nitrogen and heavy noble gases in ALH84001: Signatures of ancient Martian atmosphere. Geochim Cosmochim Acta 61:5417-5428
Musselwhite DS, Swindle TD (2002) Is polar clathrate storage fractionation of the Martian atmosphere the cause of the nakhlite Kr to Xe ratio? Icarus 154:207-216
Musselwhite DS, Drake MJ, Swindle TD (1991) Early outgassing of Mars: IInferences from the geochemistry of iodine and xenon. Nature 352:697-699
Nishiizumi K, Klein J, Middleton R, Elmore D, Kubik PW, Arnold JR (1986) Exposure history of the shergottites. Geochim Cosmochim Acta 50:1017-1023
Nishiizumi K, Masarik J, Welten KC, Caffee MW, Jull AJT, Klandrud SE (1999) Exposure history of the new Martian meteorite Dar Al Gani 476. Lunar Planet Sci XXX, Abstr #1566 (CD-ROM)
Nishiizumi K, Caffee MW, Masarik J (2000) Cosmogenic radionuclides in the Los Angeles Martian meteorite. Meteoritic Planet Sci 35:A120
Nyquist LE (1983) Do oblique impacts produce Martian meteorites? J Geophys Res 88:A785-A798
Nyquist LE, Borg LE, Shih C-Y (1998) The shergottite age paradox and the relative probabilities for Martian meteorites of differing ages. J Geophys Res 103:31445-31555
Nyquist LE, Bogard DD, Shih C-Y, Greshake A, Stoffler D, Eugster O (2001) Ages and geologic histories of Martian meteorites. Space Sci Rev 96:105-164
O'Keefe JD, Ahrens TJ (1986) Oblique impact: a process for obtaining meteorite samples from other planets. Science 234:346-349

Ott U (1988) Noble gases in SNC meteorites: Shergotty, Nakhla, Chassigny. Geochim Cosmochim Acta 52:1937-1948
Ott U (2002) Noble gases in meteorites—trapped components. Rev Mineral Geochem 47:71-100
Owen T. (1986) Update of the Anders-Owen model for Martian volatiles *In* Carr M, James P, Leovy C, Pepin R (eds) Workshop on the Evolution of the Martian Atmosphere LPI Tech Report 86-07:31-32. Lunar Planet Inst, Houston, Texas
Owen T, Biemann K, Rushneck DR, Howarth DW, Lafleur AL (1977) The composition of the atmosphere at the surface of Mars. J Geophys Res 82:4635-4640
Owen T, Bar-Nun A, Kleinfeld I (1992) Possible cometary origin of heavy noble gases in the atmospheres of Venus, Earth and Mars. Nature 358:43-46
Pepin RO (1985) Evidence of Martian origins. Nature 317:473-475
Pepin RO (1991) On the origin and early evolution of terrestrial planet atmospheres and meteoritic volatiles. Icarus 92:2-79
Pepin RO (1994) Evolution of the Martian atmosphere. Icarus 111:289-304
Pepin RO (2000) On the isotopic composition of primordial xenon in terrestrial planet atmospheres. Space Sci Rev 92:371-395
Pepin RO, Carr MH (1988) Major issues and outstanding questions. *In* Mars. Kieffer HH, Jakosky BM, Snyder CW, Matthews MS (eds) Univ. of Arizona Press, Tucson, p 120-143
Pepin RO, Porcelli D (2002) The origin of noble gases in the terrestrial planets. Rev Mineral Geochem 47: 191-246
Pepin RO, Becker RH, Rider PE (1995) Xenon and krypton isotopes in extraterrestrial regolith soils and in the solar wind. Geochim Cosmochim Acta 59:4997-5022
Podosek FA (1973) Thermal history of the nakhlites by the ^{40}Ar-^{39}Ar method. Earth Planet Sci Lett 19: 135-144
Porcelli D, Ballentine CJ (2002) Models for the distribution of terrestrial noble gases and the evolution of the atmosphere. Rev Mineral Geochem 47:411-480
Sears D, Scott E, Warren P (1997) The legacy of Allan Hills 84001. Meteoritic Planet Sci 33:545-546
Shih C-Y, Nyquist LE, Reese Y, Wiesmann H (1998) The chronology of the nakhlite, Lafayette: Rb-Sr and Sm-Nd isotopic ages. Lunar Planet Sci XXIV, Abstr #1145 (CD-ROM)
Swindle TD (2001) Could in situ dating work on Mars? Lunar Planet Sci XXXII, Abstr #1492 (CD-ROM)
Swindle TD (2002) Noble gases in the Moon and meteorites: Radiogenic components and early volatile chronologies. Rev Mineral Geochem 47:101-124
Swindle TD, Jones JH (1997) Xenon isotopic composition of the primordial Martian atmosphere: Contributions from solar and fission components. J Geophys Res 102:1671-1678
Swindle TD, Caffee MW, Hohenberg CM (1986) Xenon and other noble gases in shergottites. Geochim Cosmochim Acta 50:1001-1015
Swindle TD, Treiman AH, Lindstrom DL, Burkland MK, Cohen BA, Grier JA, Li B, Olson EK (2000) Noble gas studies of iddingsite from the Lafayette meteorite: Evidence for the timing of liquid water on Mars. Meteoritic Planet Sci 35:107-115
Tanaka KL, Scott CH, Greeley R (1992) Global stratigraphy. *In* Mars. Kieffer HH, Jakosky BM, Snyder CW, Matthews MS (eds) Univ of Arizona Press, Tucson, p 345-382
Treiman AH (1995) Multiple source areas for Martian meteorites. J Geophys Res 100:5329-5340
Treiman AH (1998) The history of Allan Hills 84001 revised: Multiple shock events. Meteoritic Planet Sci 33:753-764
Treiman AH, Barrett RA, Gooding JL (1993) Preterrestrial aqueous alteration of the Lafayette (SNC) meteorite. Meteoritics 28:86-97
Treiman AH, Gleason JD, Bogard DD (2000) The SNC meteorites are from Mars. Planet Space Sci 48:1213-1230
Turner G, Knott SF, Ash RD, Gilmour JD (1997) Ar-Ar chronology of the Martian meteorite ALH84001: Evidence for the timing of the early bombardment of Mars. Geochim Cosmochim Acta 61:3835-3850
Vickery AM (1987) The large crater origin of the SNC meteorites. Science 237:738-743
Wasson JT, Wetherill GW (1979) Dynamical, chemical and isotopic evidence regarding the formation locations of asteroids and meteorites. *In* Asteroids. Gehrels T (ed) Univ. of Arizona Press, Tucson, p 926-974
Wetherill GW (1984) Orbital evolution of impact ejecta from Mars. Meteoritics 19:1-13
Wieler R (2002a) Cosmic-ray-produced noble gases in meteorites. Rev Mineral Geochem 47:125-170
Wieler R (2002b) Noble gases in the solar system. Rev Mineral Geochem 47:21-70
Wiens RC (1988) On the siting of gases shock-emplaced from internal cavities in basalt. Geochim Cosmochim Acta 52:2775-2783

Wiens RC, Pepin RO (1988) Laboratory shock emplacement of noble gases, nitrogen, and carbon dioxide into basalt, and implications for trapped gases in shergottite EETA 79001. Geochim Cosmochim Acta 52:295-307

Wiens RC, Becker RH, Pepin RO (1986) The case for a Martian origin of the shergottites, II, Trapped and indigenous gas components in EETA 79001 glass. Earth Planet Sci Lett 77:149-158

Wood CA, Ashwal LD (1981) SNC meteorites: Igneous rocks from Mars? Proc Lunar Sci Conf 12: 1359-1375

Zahnle K, Pollack JB, Kasting JF (1990) Xenon fractionation in porous planetesimals. Geochim Cosmochim Acta 54:2577-2586

7 Origin of Noble Gases in the Terrestrial Planets

Robert O. Pepin
School of Physics and Astronomy
University of Minnesota
116 Church Street S.E.
Minneapolis, Minnesota 55455
pepin001@tc.umn.edu

Donald Porcelli
Department of Earth Science
University of Oxford, Parks Road
Oxford, OX1 3PR, United Kingdom

INTRODUCTION

Identifying the mechanisms that drove the evolution of planetary volatiles from primordial to present-day compositions is one of the classic challenges in the planetary sciences. The field bristles with models of one type or another, but none are without difficulties. Efforts to understand the histories of volatile species in the atmospheres and interiors of the terrestrial planets have concentrated on their noble gases, free of the entanglements of chemical interaction, as evolutionary tracers. The elemental and isotopic compositions of these minor constituents are rich in clues to the chemical characteristics of their source reservoirs, the physics of their evolution, and the nature of the astrophysical and planetary environments in which they evolved. Most workers would agree that atmospheric mass distributions of the nonradiogenic noble gases were probably established very early, through the action of processes operating before, during, or shortly after planetary accretion. But beyond this there are no certainties as yet on the specifics of sources or mechanisms.

The question of past and present volatile inventories on and in the terrestrial planets is intrinsically interesting in the more general context of the evolution of the planets themselves. For each individual body it crosscuts issues relating to atmospheric origin and compositional history, emergence of a coupled atmosphere-surface system and climate evolution, surface morphology and its record of past geological processing, and planetary rheology, differentiation, and degassing history. But the volatile problem extends well beyond its relationship to the initial state of a particular planet and the mechanisms that drove it down its specific evolutionary track. Attempts to decipher planetary histories in the broader context of the evolution of the solar system as a whole are focusing more and more attention on the sources and processing of volatiles in the primordial solar accretion disk, in primitive meteorites, and in the terrestrial planets as a class.

One of the salient characteristics of the composition of the Earth is the depletion in volatiles compared to the parental solar nebula relative abundances. This is most pronounced in the noble gases (Fig. 1). However, the acquisition by the planet of these unreactive elements at even these levels pose considerable problems. A comparison between noble gases on the terrestrial planets and other solar system objects reveals significant differences in both elemental ratios and isotopic compositions and indicate that complex processes were involved in sequestering planetary volatiles from the nebula, as well as providing important indications of the sources and evolutionary history of planetary volatiles. Also, noble gas isotope inventories that are produced by nuclear

1529-6466/00/0047-0007$10.00

processes within the planets are clearly identified against the strong depletion of primordial nuclides, and provide both chronological information about volatile retention on the planets and the extent of mantle degassing. While considerations of the origin of planetary noble gases have been predominantly focused on those presently found in the atmosphere, noble gases still within the Earth provide further constraints about volatile trapping within the planet during formation. A wide range of noble gas information for the Earth's mantle has been obtained from mid-ocean ridge basalts (MORB) and ocean island basalts (OIB). This indicates that separate reservoirs within the Earth were established early in Earth history that have characteristics that are distinct from those of the atmosphere, and demand inclusion in comprehensive models of Earth's volatile history. Planetary probes sent into the atmospheres of Venus and Mars have provided data to formulate model histories for these planets as well. Some clues to the interior composition of Mars are even available from Martian meteorites. Therefore, the evolution of Earth volatiles can be considered within the context of terrestrial planet formation across the solar system.

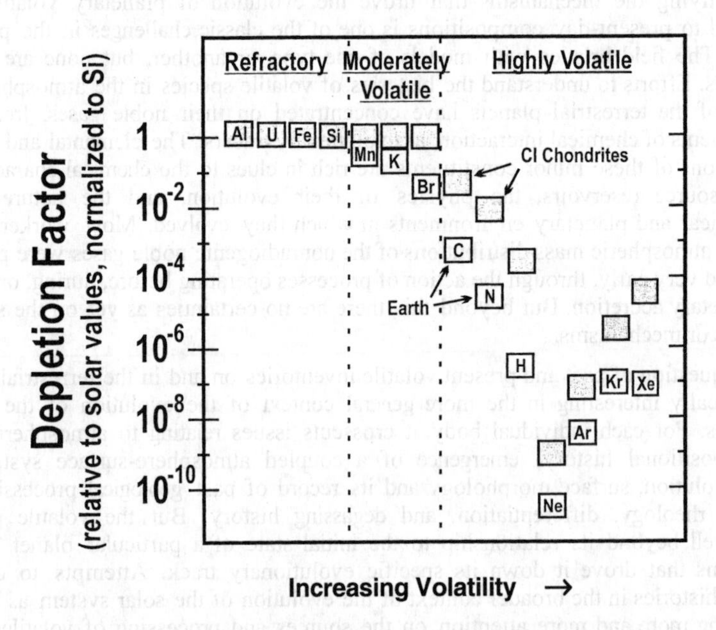

Figure 1. Compared to solar relative abundances, elemental abundances on the Earth show a general trend of increasing depletion with volatility, with noble gases amongst the most highly depleted. CI chondrites are not depleted in moderately volatile elements, but also show strong depletion in highly volatile elements. Data from Anders and Grevesse (1989), Wasson and Kallemeyn (1988), and McDonough and Sun (1995).

The origins of the noble gas features of the terrestrial planet atmospheres and interiors have defied simple explanations, and so a number of different models have been proposed for volatile sources and modification processes. These have often expanded in complexity as additional constraints have emerged, but all models still have problems. The relevant data for noble gases in the atmospheres and interiors of the terrestrial planets, and the constraints these provide, are summarized below. The presently observed volatile states of the planets are the result of acquisition processes that reflect the sources of noble gases and associated fractionations, followed by losses from the planets that included modification of the composition of residue species. Acquisition and loss processes are discussed separately below, although models that are consistent with the available data include elements of both aspects.

The discussions here are necessarily based on conclusions of more detailed evaluations of available data. More information can be found in this volume in the reviews of terrestrial noble gases (Graham 2002), of models for terrestrial degassing (Porcelli and Ballentine 2002), of solar and planetary characteristics (Wieler 2002), of Mars (Swindle 2002), and of meteorites (Ott 2002). For earlier reviews or alternative viewpoints on the origin of planetary volatiles, see, e.g., Lewis and Prinn (1984), Ahrens et al. (1989), Pepin (1991, 1992, 1994, 1997), Jakosky et al. (1994), Hutchins and Jakosky (1996), and Ozima and Podosek (2001).

CHARACTERISTICS OF PLANETARY NOBLE GASES

The primary references for comparing noble gases are those of the solar nebula, generally represented now by solar gases. Compared to these noble gases, those in the atmospheres of the terrestrial planets show strong elemental and isotopic fractionations. These provide the primary constraints on models of the origin of planetary noble gases, and are reviewed below.

Planetary noble gas abundance patterns

Noble gas abundances in terrestrial planetary atmospheres relative to the solar composition (which represents that of the primordial solar nebula) are shown in Figure 2, along with the pattern exhibited by trapped noble gases in bulk chondrites. Helium is lost from the planetary atmospheres to space, and is not included here. All objects have a striking depletion in noble gases relative to solar abundances (normalized to Si). Chondrites display a regular pattern across the noble gases, with the lightest noble gases displaying the greatest depletions. Although the mechanisms of noble gas trapping are not understood (as discussed further below), this pattern seems consistent with the generally greater retentivity of the heavier noble gases. The Ne/Ar ratios of the terrestrial planets and chondrites are similar, with normalized abundances varying by 10^3, from gas-rich Venus to gas-poor Mars. The heavier noble gases display greater variations in relative abundances. While the Kr/Ar ratios of chondrites and the Earth are similar, the terrestrial Xe/Kr ratio is much lower. The difference was initially thought to be due to the sequestration of terrestrial Xe in other terrestrial reservoirs. However, investigations of possible reservoirs of Xe, such as shales or glacial ice, failed to find this "missing Xe" (Bernatowicz et al. 1984, 1985; Wacker and Anders 1984; Matsuda and Matsubara 1989). The lower terrestrial Xe/Kr ratio has since usually been considered a feature of the Earth (see, however, Ozima and Podosek 1999 and Rayleigh distillation section below). Although relative abundances in chondrites were historically called the "planetary pattern" based on similarities with the Earth's atmosphere, and this in turn drove the search for the "missing Xe", it is now clear that these trapped meteoritic gases do not have such direct relevance to the planets, and so are no longer properly termed "planetary". However, the reason why the Earth does not exhibit a systematic depletion pattern relative to solar gases (Fig. 2) still requires explanation.

The pattern exhibited by Mars closely follows that of the Earth, with a near-solar Xe/Kr ratio. The available data for Venus likewise suggest solar-like Xe/Kr, although uncertainties are large. Despite these uncertainties, however, it seems clear that Venus differs markedly from the other two terrestrial planets in having an Ar/Kr ratio that is also close to solar.

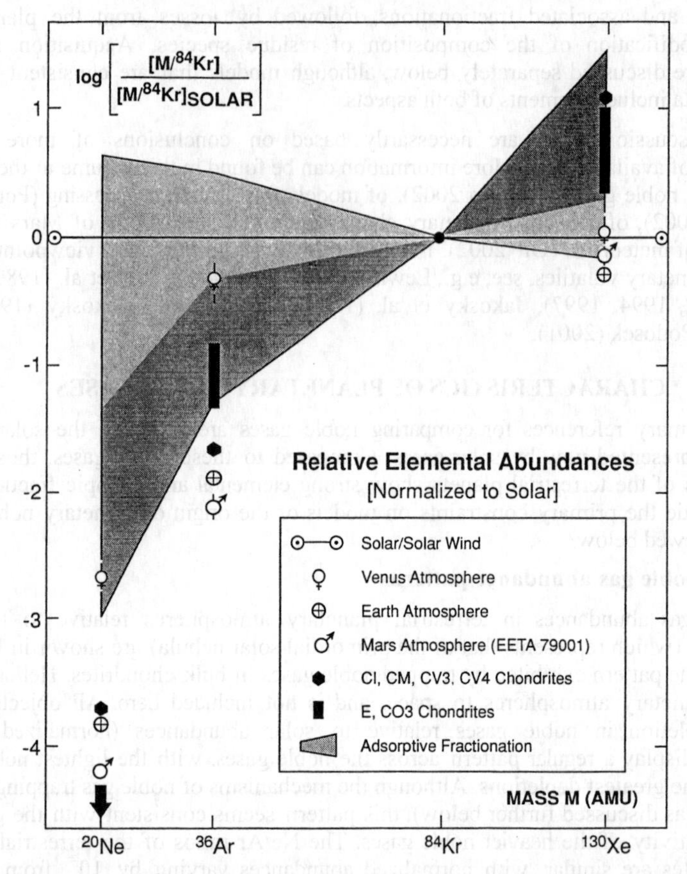

Figure 2. Noble gas M/^{84}Kr abundance ratios in terrestrial planet atmospheres and volatile-rich meteorites, plotted with respect to solar relative abundances and compared to the range of elemental fractionations (with respect to ambient gas-phase abundance ratios) determined from laboratory adsorptive experiments and from analyses of natural sedimentary materials (Pepin 1991). These fractionations generally fall in the darker shaded area of the figure, except for a number of measurements on carbon black (Wacker 1989) displaying the smaller or reversed patterns within the lighter shading. Data from references cited in the text and in Pepin (1991).

Rayleigh distillation and planetary noble gases. It has been suggested that the abundance pattern of trapped "Q-phase" noble gases—the most abundant and widely distributed primordial component in meteorites (Wieler 1994; Busemann et al. 2000; Ott 2000)—displays characteristics that are compatible with derivation by Rayleigh distillation of solar gases (Pepin 1991; Ozima et al. 1998). The Ozima et al. model, which assumes distillation of the entire nebular gas phase, was extended to the derivation of the terrestrial noble gases by Ozima and Podosek (1999) and Ozima and Igarashi (2000). Their arguments, if valid, would suggest that a distillation process other than hydrodynamic escape should be seriously considered in scenarios for the origin of terrestrial noble gases. If a well-mixed nebular reservoir that initially contains solar gases is progressively depleted by relative escape rates that are inversely proportional to the square root of mass, as assumed in the model, then a linear relationship is expected between the parameters shown in Figure 3. Several observations can be made here. There are two relative abundance ratios, ^{20}Ne/^{36}Ar and ^{36}Ar/^{84}Kr, that fall on a straight line that

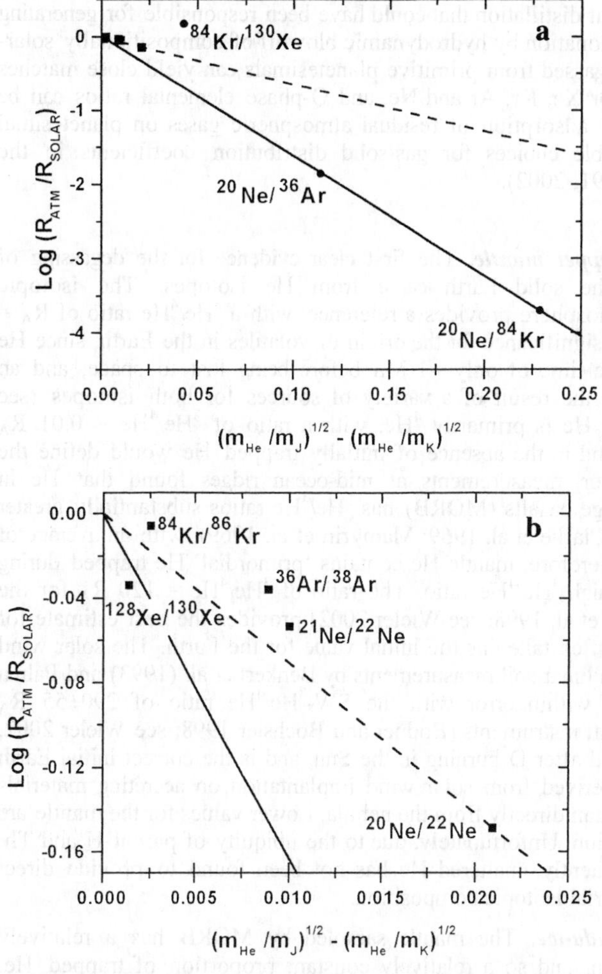

Figure 3.
(a) Rayleigh distillation of initially solar noble gases in the nebula, with the escape rate of each species inversely proportional to the square-root of its mass, will produce fractionations between any isotopes j and k that fall along a straight line in this plot. The solid line is fit through two of the three independent terrestrial noble gas ratios representing relative elemental abundances.
(b) Isotopic ratios fit poorly around a line (dashed). Note that the solid line for elemental ratios and the dashed line for isotope ratios do not fit all the data, and do not coincide with one another, indicating that this mechanism has not controlled any of the observed terrestrial noble gas characteristics.

passes through the origin. Note that other ratios that involve these same elements, such as ^{20}Ne/^{84}Kr, are not independent and so necessarily fall on the same line (Ozima and Podosek 1999). The ^{20}Ne/^{130}Xe ratio does not fall on the correlation; Ozima and Podosek (1999) argue that Xe has been either preferentially lost or hidden, although it is not clear how Xe could have been lost preferentially to the lighter noble gases, nor why any hidden Xe has not been found. Isotopic ratios of the atmosphere also do not fall on the correlation. While these may have been affected by other modification processes, strong isotopic fractionations generated by mass-dependent loss will necessarily modify elemental ratios. For Mars, a similar pattern is evident. The relative abundances of Ne, Ar, and Kr fall on the correlation, while neither Xe nor any of the isotopic compositions do. In sum, only 2 out of 3 elemental ratios and none of the isotopic compositions fall are compatible with the nebular Rayleigh distillation hypothesis; moreover there is no known astrophysical mechanism for the required distillation of the entire nebula. Therefore, adopting nebular Rayleigh distillation as an alternative to local processing by mechanisms such as hydrodynamic loss (see below) from individual planets leads to more features that require separate explanation than those that are easily explained. There are also mechanisms other than nebular distillation that could have been responsible for generating the meteoritic Q-gases. Fractionation by hydrodynamic blowoff of compositionally solar-like transient atmospheres degassed from primitive planetesimals can yield close matches to meteoritic isotopic data for Xe, Kr, Ar and Ne, and Q-phase elemental ratios can be reproduced in this model by adsorption of residual atmospheric gases on planetesimal surface grains with reasonable choices for gas-solid distribution coefficients in the adsorptive process (Pepin 1991, 2002).

Terrestrial noble gases

Primordial He in the upper mantle. The first clear evidence for the degassing of primordial volatiles from the solid Earth came from He isotopes. The isotopic composition of He in the atmosphere provides a reference, with a ^{3}He/^{4}He ratio of $R_A = 1.4 \times 10^{-6}$. However, it has no significance for the origin of volatiles in the Earth, since He has an atmospheric residence time of only ~1 Ma before being lost to space, and an isotopic composition that is the result of a variety of sources for both isotopes (see Torgersen 1989). Radiogenic He is primarily ^{4}He, with a ratio of ^{3}He/^{4}He ~ 0.01 R_A (Morrison and Pine 1955), and in the absence of initially trapped He would define the mantle composition. However, measurements at mid-ocean ridges found that He in mantle-derived mid-ocean ridge basalts (MORB) has ^{3}He/^{4}He ratios substantially greater than that of the atmosphere (Clarke et al. 1969; Mamyrin et al. 1969), with an average of 8 R_A (see Graham 2002). Therefore, mantle He contains 'primordial' He trapped during Earth formation and with a high ^{3}He/^{4}He ratio. The ratio of ^{3}He/^{4}He = 120 R_A for the Jupiter atmosphere (Mahaffy et al. 1998; see Wieler 2002) provides the best estimate for the solar nebula ratio and is often taken as the initial value for the Earth. The solar wind value of 330R_A deduced from lunar soil measurements by Benkert et al. (1993) and Palma et al. (2002)—which agrees within error with the SW-^{3}He/^{4}He ratio of 290±55 R_A measured by SOHO spacecraft instruments (Bodner and Bochsler 1998; see Wieler 2002, this volume)—was established after D burning in the Sun, and is the correct initial Earth value if terrestrial He was derived from solar wind implantation on accreting materials (Podosek et al. 2000) rather than directly from the nebula. Lower values for the mantle are due to radiogenic He production. Unfortunately, due to the ubiquity of parent U and Th, initially trapped and subsequently unaltered He has not been found to provide direct evidence for the initial terrestrial isotopic composition.

Upper mantle He abundance. The mantle sampled by MORB has a relatively uniform isotopic composition, and so a relatively constant proportion of trapped He. Early models for the distribution of noble gases in the mantle identified this reservoir as the upper mantle above the 670-km seismic discontinuity (e.g., Allègre et al. 1983). However, there is now evidence that there is no boundary to convection at this depth (e.g., Helffrich and Wood 2001; van Keken et al. 2002), although the extent of the MORB source reservoir is still debated (see Porcelli and Ballentine 2002, this volume, for discussion). The concentrations in the upper mantle reservoir are highly depleted due to degassing to the atmosphere. At present, available data suggests that there is (1.2-4.6) × 10^9 atoms ^{3}He/g and (1.4-5.4) × 10^9 atoms ^{20}Ne/g (see Porcelli and Ballentine 2002, this volume) in the MORB source. The amount that was originally incorporated there can be obtained only if the amount that has degassed and the reservoir volume are assumed. For example, if all of the atmospheric ^{20}Ne were degassed from 25% of the mantle (i.e., the volume above a depth of 670 km), then there was originally 1.75 × 10^{12} atoms ^{20}Ne/g in that volume. However, this does not consider losses to space, nor sources that added volatiles directly to the atmosphere.

Primordial He in the deep Earth. The ^{3}He/^{4}He ratios measured in ocean island basalts (OIB) are more variable, with values both below and above those of MORB.

$^3He/^4He$ values below MORB are often associated with basalts containing Pb isotopes more radiogenic than MORB, and probably include recycled components (e.g., Kurz et al. 1982; Hanyu and Kaneoka 1998). More significant for the origin of noble gases in the Earth are the OIB with higher values that require at least one mantle component with a time-integrated $^3He/(U+Th)$ ratio greater than that of the MORB-source mantle (Kurz et al. 1982; Allègre et al. 1983). This component must have a ratio at least as high as the values of 32-38 R_A measured in Loihi Seamount, the youngest volcano in the Hawaiian Island chain (Kurz et al. 1982), and Iceland (Hilton et al. 1999). It is likely that this component has been stored somewhere below the upper mantle in order to have avoided degassing and homogenization, although the nature of the reservoir is currently debated (see Porcelli and Ballentine 2002, this volume). It has been argued that the high $^3He/^4He$ component represents mantle that has remained undegassed since Earth formation. Assuming that this contains a bulk silicate Earth U concentration of 21 ppb and Th/U = 3.8, along with an initial value of $^3He/^4He$ = 120-330 R_A and a present value of 38 R_A, then the reservoir has (6.1-7.9) $\times 10^{10}$ atoms ^3He/g. If losses of He had occurred, the initial He concentration of this reservoir would have been greater. This provides an important constraint on the amount of He that must have been trapped within the mantle. It should be emphasized that this calculation does not make any assumptions regarding the volume or location of this reservoir in the mantle, issues that are presently contentious (see Porcelli and Ballentine 2002). Nevertheless, the presence of such a reservoir anywhere in the mantle requires that such concentrations were trapped in at least some portion of the mantle.

One alternative interpretation of the high $^3He/^4He$ ratios is that it represents He from the core. While this idea remains speculative, partitioning sufficient He into the core also requires high He concentrations into the mantle during core formation (Porcelli and Halliday 2001). The amount of He is uncertain and depend upon the conditions of core formation, but concentrations of He trapped within the mantle as high as (0.6-70) $\times 10^{11}$ atoms ^3He may be required (Porcelli and Halliday 2001). It has also been suggested that the source has a high $^3He/(U+Th)$ ratio due to greater depletion of U and Th than He (Graham et al. 1990; Helffrich and Wood 2001). This is possible if U and Th are more incompatible than He, which has not been demonstrated experimentally. Other difficulties in using this as a basis for a model for all the noble gases also have not been worked out (see Porcelli and Ballentine 2002, this volume, for further discussion). Alternatively, it has been argued that high $^3He/^4He$ ratios are from a reservoir that is depleted, rather than gas-rich, due to preferential removal of U relative to He (Graham et al. 1990; Anderson 1998; Helffrich and Wood 2001). This model is difficult to assess (see Porcelli and Ballentine 2002), but likely also requires high He concentrations as a starting condition so that sufficient He remains after depletion.

Atmospheric Ne. Ne isotopic compositions for the atmosphere and solar system reservoirs are summarized in Pepin (1991) and are shown in Figure 4. The greatest differences between solar system bodies are seen in the $^{20}Ne/^{22}Ne$ ratios. The value of $^{20}Ne/^{22}Ne = 13.8 \pm 0.1$ (Benkert et al. 1993; Pepin et al. 1995) derived for the solar wind is believed to represent the initial solar nebula composition. Meteorites have a variety of components. Ne isotope ratios in bulk CI chondrites scatter around an average $^{20}Ne/^{22}Ne$ of 8.9 ± 1.3. The meteoritic Q-component (Wieler 1994), sited in the surfaces of carbonaceous phases in primitive meteorites and possibly derived by fractionating processes operating on parent bodies (Pepin, 1991) or in the solar nebula itself (Ozima et al. 1998), and thought to be relatively free of 'exotic' components, has $^{20}Ne/^{22}Ne$ = 10.1 to 10.7 (Busemann et al. 2000). The terrestrial atmospheric ratio of 9.8 can be derived either from mixing meteorite and solar components or by fractionation of solar Ne. The difference between the solar (0.033) and atmospheric (0.029) $^{21}Ne/^{22}Ne$ ratios is consistent with fractionation of solar Ne and addition of radiogenic ^{21}Ne (see Porcelli and Ballentine 2002, this volume).

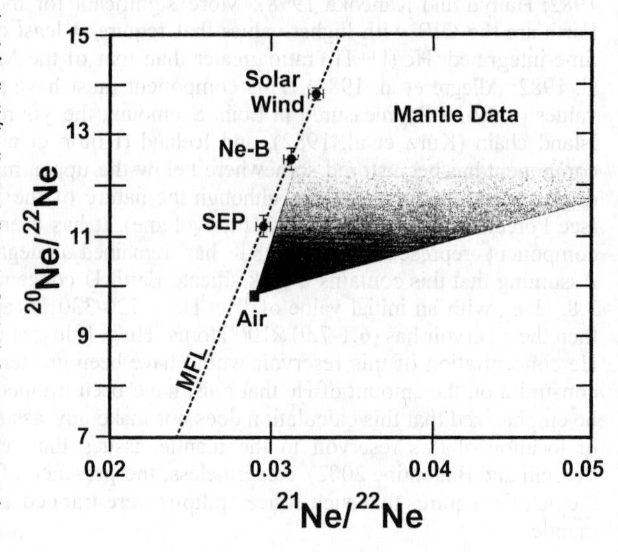

Figure 4. The Ne isotope composition of terrestrial reservoirs, compared to Ne from solar wind, solar energetic particle (SEP), and Ne-B (a meteoritic composition from irradiation of solar wind and SEP). The atmosphere can be derived from fractionation of solar Ne, with the addition of nucleogenic ^{21}Ne produced within the Earth. Mantle compositions have greater ^{20}Ne/^{22}Ne ratios, due to the presence of unfractionated solar Ne within the Earth, and greater ^{21}Ne/^{22}Ne ratios due to addition of nucleogenic ^{21}Ne. See compilations by Busemann et al. (2000) for extraterrestrial data and Graham (2002, this volume) for mantle data. MFL is the fractionation line for mass-dependent Rayleigh distillation losses.

Solar Ne within the Earth. Samples from the Earth's mantle have measured ^{20}Ne/^{22}Ne ratios that are greater than that of the atmosphere, and extend toward the solar value (Fig. 4). Since these isotopes are not produced in significant quantities in the Earth, this is unequivocal evidence for storage in the Earth of at least one nonradiogenic mantle component that is distinctive from the atmosphere and was trapped during formation of the Earth. This component must have a ^{20}Ne/^{22}Ne ratio at least as high as the highest measured mantle value, with lower measured ratios due to addition of contaminant air Ne. A few of the mantle ^{20}Ne/^{22}Ne ratios approach the solar wind value of 13.8, which is also assumed to represent the solar nebula, although most are lower. The solar wind/nebular ratio has commonly been taken to be that of the mantle. Recently, however, it has been suggested that OIB and MORB data indicate that the isotopically light component in the mantle is actually Ne-B (Trieloff et al. 2000, 2002; Ozima and Igarashi 2000), which is a mixture of solar wind and solar energetic particle (SEP) Ne. Ne-B has a ^{20}Ne/^{22}Ne ratio of about 12.5 (Black 1972), significantly below the solar value. The few measured mantle values above this (Sarda et al. 1988; Hiyagon et al. 1992) were insufficiently precise to firmly establish the presence of Ne with a higher value. This raises the possibility that mantle Ne was supplied by planetary accretion of planetesimals carrying Ne-B (Trieloff et al. 2000, 2002). However, recent analyses for Icelandic OIB samples found ^{20}Ne/^{22}Ne = 13.75±0.32 (Harrison et al. 1999), demanding the presence of solar Ne in the mantle, although the validity of this value is debated (Trieloff et al. 2000, 2001; Ballentine et al. 2001). It should be noted that it is possible that isotopic fractionation processes have modified Ne in mantle-derived samples during transport and degassing, but there is no information available to rigorously evaluate this.

The only other solar system composition with ^{20}Ne/^{22}Ne ratios greater than that of the atmosphere is Q-Ne with ^{20}Ne/^{22}Ne = 10.7 (Fig. 4). Recently, Ozima and Igarashi (2000) argued that there is a preferential grouping of both MORB and OIB data at the intermediate value of ^{20}Ne/^{22}Ne = 10.8, and suggested that this was due to the presence of Q-Ne in the mantle. This would demand mechanisms for the trapping and preservation of yet another interior component. Moreover, Q gases have $^3He/^{22}Ne \sim 0.14$ (Busemann et al. 2000); all MORB and OIB $^3He/^{22}Ne$ values are at least an order of magnitude greater (see below), and so this cannot be a Ne component in the mantle unless prior strong elemental fractionation has occurred. It should be noted that the statistical arguments for such a component are weak and do not account for the common observation that intermediate $^{20}Ne/^{22}Ne$ ratios commonly reflect some atmospheric contamination that is only partly resolved by step-heating. Without additional support, this hypothesis does not warrant further consideration.

Mantle He/Ne ratios. The MORB He and Ne isotopic compositions can be used to calculate the $^3He/^{22}Ne$ ratio of the source region prior to any recent fractionations created during transport and eruption. Since ^4He and ^{21}Ne production rates are directly coupled, $^3He/^4He$ and $^{21}Ne/^{22}Ne$ isotope variations should be correlated (see Porcelli and Ballentine 2002, this volume). A review of available Hawaii and MORB data (Honda and McDougall 1998) found a mantle average of $^3He/^{22}Ne = 7.7$. This is greater than the most recent estimate of 1.9 for the solar nebula (see discussion in Porcelli and Pepin 2000), and so requires some elemental fractionation of incorporated light solar noble gases. There is evidence that this value may not be uniform in the mantle, although it has not been conclusively established that more than one component was trapped in early Earth history and preserved since then (see Porcelli and Ballentine 2002 for further discussion).

Ar in the atmosphere. The initial $^{40}Ar/^{36}Ar$ ratio in the solar system is $\sim 10^{-4}$ to 10^{-3} (Begemann et al. 1976). The atmosphere has $^{40}Ar/^{36}Ar = 296$, so that essentially all the ^{40}Ar has been produced by ^{40}K decay. The atmospheric $^{38}Ar/^{36}Ar$ ratio of 0.188 is similar to that found in CI chondrites of 0.189±0.002 (Mazor et al. 1970) but substantially higher than the current estimate of 0.173 for the solar wind (Pepin et al. 1999; Palma et al. 2002). Atmospheric Ar derived from a solar source therefore must have been fractionated.

Radiogenic ^{40}Ar. A measure of the extent of mantle degassing can be obtained from the K-^{40}Ar budget. The concentration of K in the bulk silicate Earth is generally estimated to be 270 ppm K, which has produced 2.4×10^{42} atoms ^{40}Ar. Therefore, 41% of the ^{40}Ar that has been produced is now in the atmosphere (Allègre et al. 1986, 1996; Turcotte and Schubert 1988), and a significant reservoir of ^{40}Ar remains in the Earth. It should be noted that there has been some debate regarding the terrestrial K content (Albarède 1998; see Porcelli and Ballentine 2002, this volume, for discussion), and a downward revision could require a substantially greater degassing efficiency for the Earth. It has sometimes been assumed that the mantle reservoir that is rich in ^{40}Ar is the same as that with high $^3He/^4He$ and so is rich in ^3He, although this is not the case of all mantle models (see Porcelli and Ballentine 2002).

Solar Ar in the mantle? Measurements of MORB and OIB $^{38}Ar/^{36}Ar$ ratios typically are atmospheric within error, but have been of low precision due to the low abundance of these isotopes (Fig. 5). While some high precision analyses of MORB and OIB samples have found $^{38}Ar/^{36}Ar$ ratios lower than that of the atmosphere (Valbracht et al. 1997; Niedermann et al. 1997), others did not (Kunz 1999). Pepin (1998) argued that the low values are due to mantle components with solar light isotopes. The unambiguous identification of solar Ar in the mantle clearly would have important implications for the origin of terrestrial noble gases, and additional analyses are required to firmly establish whether there are mantle samples that contain non-atmospheric $^{38}Ar/^{36}Ar$ ratios not masked by air Ar contamination. However, atmospheric ratios in the mantle could be explained either by early trapping of Ar that had been fractionated or subduction of atmospheric Ar (see Porcelli and Wasserburg 1995b).

Kr isotopes. Solar, bulk meteorite, and terrestrial Kr isotopic compositions (Fig. 6) seem to be generally related to one another by mass fractionation (see Eugster et al. 1967;

Pepin 1991), with the atmosphere depleted in light isotopes by ~0.8%/amu relative to the solar composition (Pepin et al. 1995). Atmospheric Kr ("Air Kr") is precisely related to the solar composition by mass fractionation; a similarly good fit to terrestrial isotope ratios is generated by mixing of severely fractionated solar Kr with degassed solar Kr, as discussed later in the section "Hydrodynamic escape". Meteoritic Kr is also generally consistent within error with fractionated solar Kr, but with a possible excess in ^{86}Kr. Mars Kr, in contrast, is solar-like (Wieler 2002; Swindle 2002, both this volume).

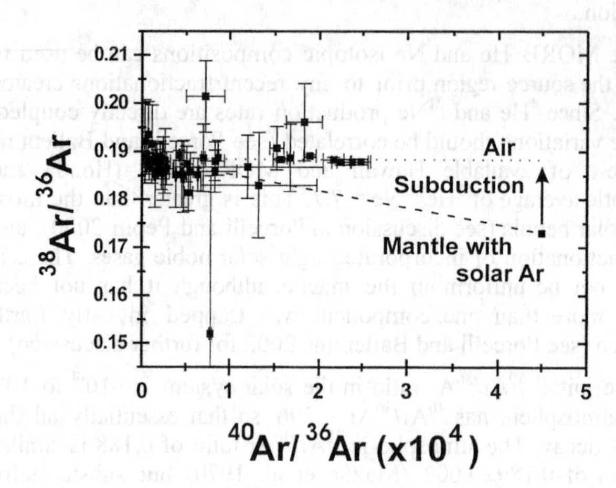

Figure 5. The $^{38}Ar/^{36}Ar$ ratio of mantle-derived volcanics are generally within error of the atmosphere, while $^{40}Ar/^{36}Ar$ ratios are much greater due to addition of radiogenic ^{40}Ar. A mantle component with a solar $^{38}Ar/^{36}Ar$ ratio falls outside the array of data, indicating that nonradiogenic mantle Ar was initially trapped after fractionation from the solar composition, or is dominated by Ar subducted after atmospheric fractionation. Data are from Sarda et al. (1985), Hiyagon et al. (1992), Honda et al. (1993), Niedermann et al. (1997), and Kunz (1999).

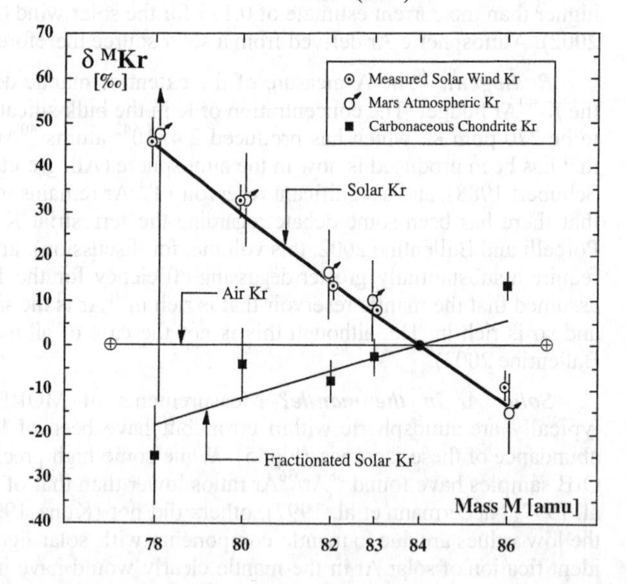

Figure 6. Krypton isotopic compositions in solar-system volatile reservoirs, plotted as permil deviations from the $^{M}Kr/^{84}Kr$ ratios in terrestrial air (Basford et al. 1973). The heavy "Solar Kr" curve represents a smooth fit to the measured solar-wind isotope ratios. Measured solar wind Kr from Wieler and Baur (1994) and Pepin et al. (1995); Mars Kr from Pepin (1991); carbonaceous chondrite Kr from Krummenacher et al. (1962), Eugster et al. (1967), and Marti (1967), all renormalized to Basford et al.'s air composition.

Nonradiogenic Xe isotopes in the atmosphere. The isotopic compositions of Xe components in the solar system have been more difficult to unravel. There is no suitable widespread solar system Xe component to use as a reference composition (see Ott 2002, this volume). The light isotopes of atmospheric Xe are related to both bulk chondritic and solar Xe by fractionation of ~4.2% per amu (first noted by Krummenacher et al. 1962), with a clear radiogenic excess of ~7% in ^{129}Xe from decay of ^{129}I. However, these components cannot be precisely related to the composition of the heavy Xe isotopes in the atmosphere directly or by simple processes (such as mass fractionation) and addition of expected fissiogenic components. This holds using any common fractionation function (e.g., Fick's law, hydrodynamic escape), and so these compositions cannot serve as the primordial terrestrial composition (see Pepin 2000). This is shown in Figure 7, where unfractionated SW-Xe isotope ratios are plotted in Figure 7a and its fractionated residue in Figure 7b, both normalized to corresponding ratios in the present atmosphere. It is evident that when solar Xe is fractionated to match the light isotopes of the atmosphere, the ^{136}Xe/^{130}Xe ratio is well above atmospheric. Therefore fractionated SW-Xe contributes more ^{136}Xe relative to ^{130}Xe than the atmosphere actually contains. Subtraction of a fission component from air-Xe worsens the discrepancy. Consequently SW-Xe is ruled out as the primordial terrestrial composition. This exclusion clearly applies with even greater force to the meteoritic AVCC-Xe and Xe-Q compositions, since both are considerably richer in the heavy isotopes than SW-Xe (Table 1; Ott 2002).

Table 1. Isotopic compositions of solar-system xenon components discussed in this chapter.

Component	^{124}Xe	^{126}Xe	^{128}Xe	^{129}Xe	^{130}Xe	^{131}Xe	^{132}Xe	^{134}Xe	^{136}Xe
Air-Xe$^{(a)}$	2.337(07)	2.180(11)	47.15(05)	649.6(06)	≡100	521.3(06)	660.7(05)	256.3(04)	217.6(02)
U-Xe$^{(b)}$	2.928(10)	2.534(13)	50.83(06)	628.6(06)	≡100	499.6(06)	604.7(06)	212.6(04)	165.7(03)
NEA-Xe$^{(c)}$	2.337(07)	2.180(11)	47.15(05)	605.3(29)	≡100	518.7(07)	651.8(13)	247.0(13)	207.5(13)
AVCC-Xe$^{(d)}$	2.851(51)	2.512(40)	50.73(38)	653.3$^{(d)}$	≡100	504.3(28)	615.0(27)	235.9(13)	198.8(12)
Xe-Q$^{(e)}$	2.810(13)	2.506(12)	50.77(16)	643.6(17)	≡100	505.6(11)	617.7(11)	233.5(08)	195.4(06)
SW$_1$-Xe$^{(f)}$	2.939(70)	2.549(82)	51.02(54)	627.3(42)	≡100	498.0(17)	602.0(33)	220.7(09)	179.7(06)
SW$_2$-Xe$^{(g)}$	2.928(10)	2.534(13)	50.83(06)	628.6(06)	≡100	499.6(06)	604.7(06)	220.7(09)	179.7(06)
Mars-Xe$^{(h)}$	2.46(13)	2.14(13)	47.63(65)	1555(16)	≡100	513.9(35)	648.1(38)	259.7(18)	227.7(16)
Mars-Xe$^{(i)}$	2.30(26)	≡2.10	47.3(14)	1556(08)	≡100	518.0(34)	652.9(42)	258.5(22)	226.9(16)

All isotope ratios are referenced to ^{130}Xe = 100. Uncertainties (±1σ) in the last two digits of the listed relative isotopic abundances are indicated in parentheses.

(a) Terrestrial atmospheric Xe (Basford et al. 1973).

(b) Pepin (2000).

(c) Nonradiogenic Earth Atmosphere Xe (Pepin 1991).

(d) Average Carbonaceous Chondrite Xe (Pepin 1991). The listed value for ^{129}Xe is an average for CI and CM carbonaceous chondrites; the approximate range is from 636 to 671 (Mazor et al. 1970).

(e) Busemann et al. (2000).

(f) Measured SW values in acid-etching experiments on lunar ilmenites (Wieler and Baur 1994; Pepin et al. 1995).

(g) Adopted SW values: assumes that the SW-Xe composition is identical to U-Xe from ^{124}Xe through ^{132}Xe, with ^{134}Xe and ^{136}Xe equal to Wieler and Baur's (1994) SW$_1$-Xe measured values.

(h) Swindle and Jones (1997).

(i) Mathew et al. (1998).

A suitable initial composition for the atmosphere has been deduced from meteorites. Multi-dimensional isotopic correlations of chondrite data have been used to constrain a range of compositions that, when mass-fractionated, yields the light-isotope ratios of terrestrial Xe. In order to match the terrestrial heavy Xe isotope ratios, addition of radiogenic ^{129}I and a heavy Xe isotope component is required. Constraining the composition of the heavy isotope component to known fission spectra then defines the "U-Xe" composition and identifies ^{244}Pu-derived fissiogenic Xe as the heavy isotope component (see "Primordial Xe" section). This is compatible with meteorite data that indicates that the amount of ^{244}Pu present in the early solar system (Hudson et al. 1989) will produce Xe that will dominate over the production by ^{238}U (see Porcelli and Ballentine 2002) and be the correctly identified parent of fissiogenic Xe. Therefore,

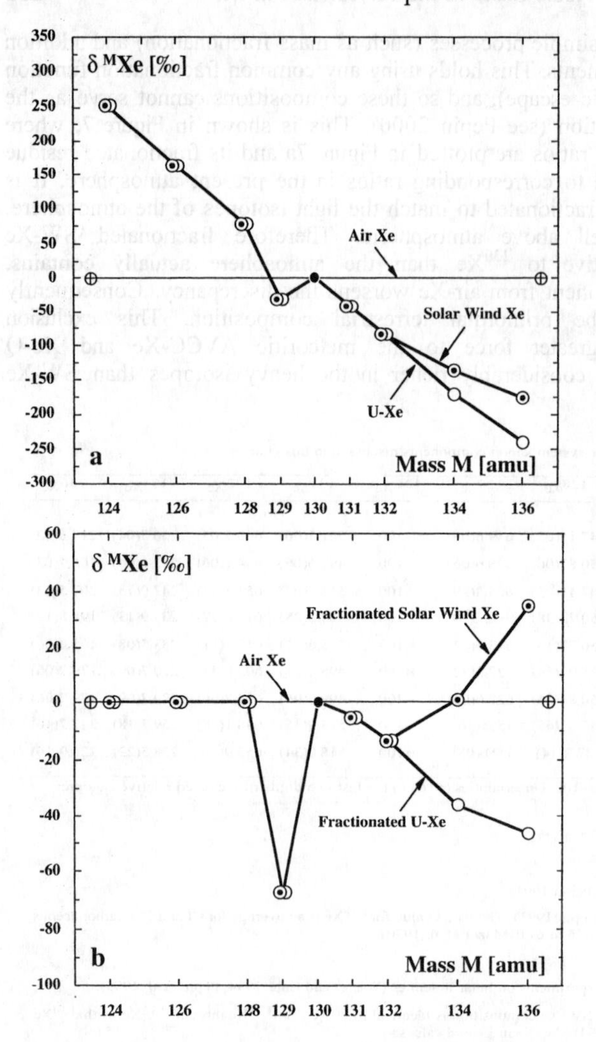

Figure 7. (a) Relationships of unfractionated SW_2-Xe and U-Xe to terrestrial atmospheric Xe, plotted as permil differences from the $^{M}Xe/^{130}Xe$ ratios in terrestrial air (Table 1). (b) As in Figure 6a, but now after hydrodynamic escape fractionation of SW_2-Xe and U-Xe to the extent required to match their $^{124-128}Xe/^{130}Xe$ ratios to the corresponding Air-Xe values. Fractionated SW-$^{136}Xe/^{130}Xe$ is elevated above the air ratio by ~10 times the 1σ uncertainty in the SW ratio measurement.

fractionation of U-Xe, followed by addition of radiogenic ^{129}Xe and fissiogenic Xe, can produce the Xe atmospheric composition. The fact that U-Xe is underabundant in the heaviest isotopes compared to solar wind Xe (Fig. 7, "Primordial Xe" section) suggests that there is a heavy isotope component in the sun but not in the early Earth. This is puzzling, since solar wind Xe, presumably reflecting the Xe composition in the accretion disk, would arguably be the more plausible contributor to primordial planetary inventories. However, due to the relationship between solar Xe and the atmosphere (Fig. 7), such a problematic relationship will be evident in any composition for the Earth. Note that Igarashi (1995) obtained an alternative composition for atmospheric nonradiogenic Xe. However, this composition implies that the atmosphere contains in addition a heavy isotope component that has relative proportions of Xe isotopes that do not match the spectrum of either ^{244}Pu or ^{238}U, and so cannot be used to determine the fissiogenic Xe abundance of the atmosphere. While Igarashi (1995) suggested that this heavy component matched fractionated U-derived Xe, mass balance constraints make a terrestrial source for this implausible (Ballentine and Burnard 2002, this volume). Also note that the substantially different U-Xe and fission Xe compositions obtained by Igarashi (1995) appear to reflect the use of a meteoritic database consisting largely of measurements on chemically separated phases that do not preserve mass balance (see discussion in Pepin 2000).

Nonradiogenic Xe isotopes in the mantle. The measured ratios of the nonradiogenic isotopes in MORB are indistinguishable from those of the atmosphere. However, some CO_2 well gases with presumably mantle-derived Xe have higher $^{124\text{-}128}Xe/^{130}Xe$ ratios (Phinney et al. 1978; Caffee et al. 1999) that can be explained either by a mixture of ~10% solar or U-Xe and ~90% atmospheric Xe, or a mantle component that has not been fractionated relative to solar Xe to the same extent as atmospheric Xe. Due to the lower precision of MORB data and the pervasiveness of atmospheric contamination in MORB samples, it has not been possible to confirm that this feature is present throughout the upper mantle.

Atmospheric Xe closure ages. The fractionated U-Xe ratios of $^{129}Xe/^{130}Xe = 6.053$ and $^{136}Xe/^{130}Xe = 2.075$ are the present best estimates of the isotopic composition of nonradiogenic terrestrial Xe. Therefore, 6.8±0.3% of atmospheric ^{129}Xe ($^{129*}Xe_{atm} = 1.7 \times 10^{35}$ atoms) and 4.65±0.50% of atmospheric ^{136}Xe ($^{136*}Xe_{atm} = 3.81 \times 10^{34}$ atoms) are radiogenic. The $^{129*}Xe_{atm}$ is only 0.8% of the total ^{129}Xe produced since 4.57 Ga; such a low value cannot be accounted for by incomplete degassing of the mantle nor from any uncertainties in the estimated amount of $^{129*}Xe$, and requires losses to space. Xe losses from the Earth to space must have occurred during early Earth history, when such heavy species could have been lost either from protoplanetary materials or from the growing Earth. Full accretion of the Earth may have occurred over ~100 Ma (Wetherill 1975). Over this time, almost all of the ^{129}I, and some of the ^{244}Pu, had decayed to daughter Xe isotopes that could have been lost to space. A 'closure age' of the Earth could be calculated by assuming that essentially complete loss of radiogenic and fissiogenic Xe occurred until closure time t, followed by no further loss. In this case (Wetherill 1975; Pepin and Phinney 1976), the relative proportions of radiogenic ^{129}Xe and fissiogenic ^{136}Xe in the atmosphere can be used to obtain a closure age of 80 Ma (see discussion in Porcelli and Ballentine 2002, this volume). The implication is that losses occurred over an extended period as the Earth accreted. If atmospheric Xe loss occurred during a massive Moon-forming impact, then the closure period corresponds to the time after an instantaneous catastrophic loss event. A more complicated loss history will result in some revision in the closure time, but this is unlikely to be significant. Note that if Pu-derived ^{136}Xe was lost over about one half-life (80 Ma) of ^{244}Pu, then ~40% of that produced subsequently in the Earth is in the atmosphere. This is compatible with the ^{40}Ar budget (see Porcelli and Ballentine 2002) and provides further support for the chosen nonradiogenic atmosphere composition.

Loss of nonradiogenic Xe from the atmosphere. None of the closure age calculations explicitly consider how nonradiogenic nuclides are retained. If the Earth had completely lost all Xe during the first ~100 Ma, there would be no nonradiogenic Xe present today. The nonradiogenic Xe budget can be examined with a simple two-stage model for the deep mantle reservoir. The reservoir begins with initial concentrations $^{i}C_{P0}$ and bulk silicate Earth parent element concentrations. At first, the reservoir behaves as a closed system, retaining all radiogenic $^{129*}Xe$ and fissiogenic $^{136*Pu}Xe$. At some time t, all Xe isotopes are instantaneously depleted, leaving only a fraction f. This is followed by complete retention to the present day. Then (Porcelli et al. 2001)

$$^{130}Xe_{Atm} = {}^{130}Xe_{Atm_o} f$$

and
$$\frac{^{129*}Xe}{^{136*Pu}Xe} = \frac{^{129}I_o[f+(1-f)e^{\lambda_{129}t}]}{^{244}Pu_o\,^{136}Y_{244}[f+(1-f)e^{\lambda_{244}t}]} \tag{1}$$

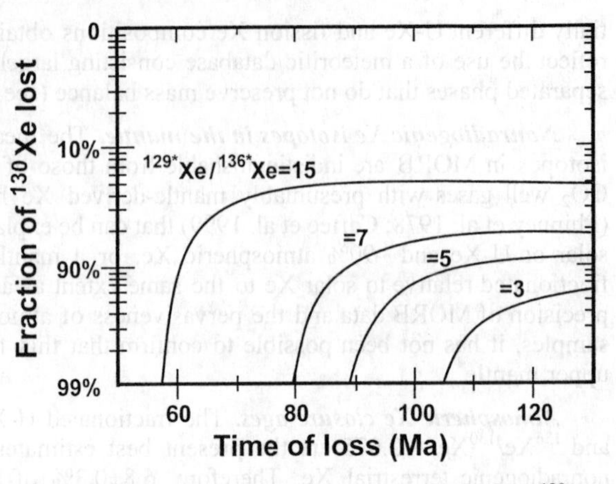

Figure 8. The relationship between the time at which an instantaneous loss of Xe occurred and the fraction of gas lost. While $^{129*}Xe/^{136*Pu}Xe$ = 4.4 for the atmosphere, estimates for the deep mantle vary between 3.5 and 10 (Porcelli et al. 2001), and require that losses are late (>60 Myr) and substantial (>97%) from throughout the planet.

where $^{136}Y_{244}$ is the fission yield of ^{136}Xe. Figure 8 shows the fraction f of retained ^{130}Xe, plotted as a function of the time of loss, for various present values of $^{129*}Xe/^{136*Pu}Xe$. It can be seen that for each value of $^{129*}Xe/^{136*Pu}Xe$, there is a minimum time of loss, which corresponds to very large depletions; late losses correspond to the maximum fraction of Xe retained. For the atmosphere, $^{129*}Xe/^{136*Pu}Xe = 4.4$, and so the minimum time of loss is 95 Ma, with losses >99%. Therefore, the Earth or Earth-forming materials initially contained at least 10^2 times the ^{130}Xe presently seen in the atmosphere. Loss of Xe during

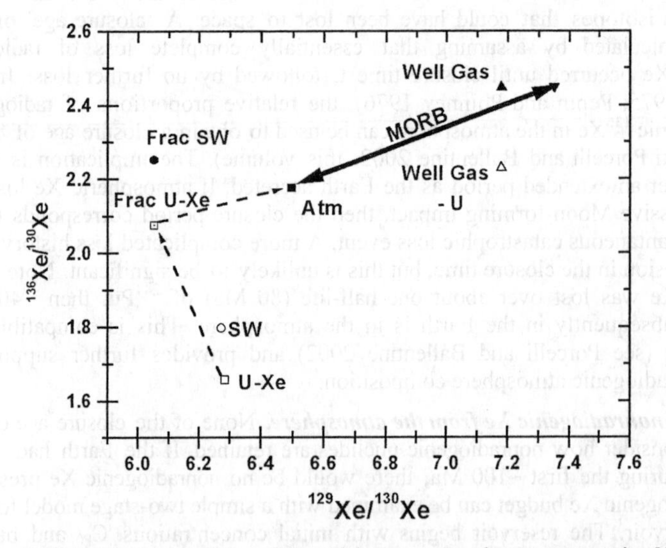

Figure 9. Xe isotope compositions of terrestrial precursors and present reservoirs. U-Xe, fractionated to match the light Xe isotopes (Fig. 7), provides an initial composition of the atmosphere, to which radiogenic ^{129}Xe and fissiogenic ^{136}Xe have been added. Similar fractionation of solar wind (SW) Xe produces Xe that is too heavy to supply the atmosphere. MORB typically have $^{129}Xe/^{130}Xe$ and $^{136}Xe/^{130}Xe$ ratios that are greater than the atmosphere due to radiogenic and fissiogenic additions and are correlated, with the range likely due to variable atmospheric contamination of samples. The most precise measurement of mantle Xe is for CO_2 well gas. When uranogenic Xe is subtracted, the well gas falls substantially below the MORB data, and indicating that the ratio of ^{244}Pu-derived ^{136}Xe to I-derived ^{129}Xe is lower than that of the atmosphere.

a single, brief event may have been driven by a Moon-forming impact. While more complicated loss histories can be constructed, the general conclusion that very large losses of Xe occurred is inescapable.

Radiogenic Xe in the mantle. Data for MORB indicate that the mantle has ^{129}Xe/^{130}Xe and ^{136}Xe/^{130}Xe ratios greater than those of the atmosphere (Fig. 9). There is considerable uncertainty and debate regarding how much of the ^{136}Xe excess is due to ^{244}Pu, rather than ^{238}U, decay. The most recent data suggest that ~30% is due to ^{244}Pu decay (Kunz et al. 1998), with considerable uncertainty due to scatter in the data. This is also consistent with measurements of mantle-derived Xe in well gases that indicate <20% is Pu-derived (Phinney et al. 1978; see further discussion in Porcelli and Ballentine 2002). It has been shown that these higher ratios cannot be due to early degassing of the mantle, but rather reflect the composition of a deep mantle reservoir that has evolved independently and where Xe produced by ^{129}I and ^{244}Pu is stored (Ozima et al. 1985; Porcelli and Ballentine 2002). This reservoir has been explicitly equated with a gas-rich source of high ^{3}He/^{4}He mantle (Porcelli and Wasserburg 1995a). While models for the distribution of noble gases within the mantle are still debated, the greatest obstacle to consensus has been how to maintain two separate mantle reservoirs (for MORB and OIB sources) throughout Earth history, and so it is more likely that the long-term storage of high ^{3}He/^{4}He ratios as well as high ^{129}Xe/^{130}Xe and ^{136}Xe/^{130}Xe occur in the same reservoir, rather than necessitating the maintenance of three mantle reservoirs.

Xe loss from the mantle. The ^{129*}Xe/^{136*Pu}Xe ratio of the mantle is not substantially different from that of the atmosphere, indicating that substantial losses also occurred from within the Earth. A value of ^{129*}Xe/^{136*Pu}Xe = 10 can be obtained from MORB data, although with considerable uncertainty in the amount of Pu-derived Xe as discussed above. In this case, the minimum time of loss from the deep mantle is ~70 Ma and a maximum fraction retained is f ~0.02 (Fig. 8). Therefore, at least 98% of the Xe was lost. Although further refinements can be made for calculating these values by considering the possible amounts of subducted Xe that is in the mantle (Porcelli et al. 2001), the results are qualitatively the same. If the Xe now seen in the upper mantle has been stored since early Earth history in an undegassed reservoir along with the high OIB ^{3}He/^{4}He ratios, then this reservoir initially contained = 10^2 times more Xe, and so also He, than it has at present. Taking the calculated closed system ^{3}He concentration of (6.1-7.9) × 10^{10} atoms ^{3}He/g, then = ~7 × 10^{12} atoms ^{3}He/g must have been initially trapped. Using ^{3}He/^{22}Ne = 7.7, then at least 10^{12} atoms ^{22}Ne/g was trapped. This may be compared to the value obtained by dividing the atmospheric ^{22}Ne inventory into the mass of the Earth of 3 × 10^{10} atoms/g. The initial gas-rich state of the Earth appears to have involved all reservoirs.

Noble gases on Venus

There are limited probe data available for the isotopic composition of the Venus atmosphere, and of course none available for noble gases in the Venus interior. Wieler (2002, this volume, p. 41: his Table 8 and references therein) provides the most recent compilation and assessment of the available data.

Nonradiogenic noble gases. For Ne, both of the two ^{20}Ne/^{22}Ne ratios derived from different data bases (11.8±0.6 and 12.15±0.40: see Wieler's (2002) Table 8) require some fractionation relative to the solar value, although not quite to the same extent as that of the terrestrial atmosphere. The constraint on the ^{21}Ne/^{22}Ne ratio of <0.067 cannot be used to further limit the source of Ne. The ^{36}Ar/^{38}Ar value of 5.45±0.10 measured by Venera spacecraft instruments is nominally somewhat above the terrestrial ratio but is essentially indistinguishable from it within error.

Radiogenic nuclides. The measured $^{40}Ar/^{36}Ar$ ratio is 1.11±0.02, substantially less radiogenic than the terrestrial atmosphere. For a mixing ratio of 21-48 ppm (Donahue and Pollack 1983), the atmosphere contains $(1.8\text{-}4.4) \times 10^{41}$ atoms ^{40}Ar. Divided by the mass of the planet, this corresponds to $(3.6\text{-}9.0) \times 10^{13}$ atoms $^{40}Ar/g$. This is 0.2-0.5 times the value for the Earth. However, Venus appears to be deficient in K. Data for the K/U ratio of the surface indicate that K/U = 7220±1220 (Kaula 1999), or 0.57±0.10 times that of the value of 1.27×10^4 commonly taken for the Earth. Assuming that Venus has the same U concentration as the total Earth of 14 ppb, then 12-28% of the ^{40}Ar produced in Venus is now in the atmosphere. This indicates that a substantial inventory of ^{40}Ar remains within the planet, possibly also accompanied by up to an equivalent fraction of nonradiogenic noble gases. In contrast, at least 40% of terrestrial ^{40}Ar is in the atmosphere. The fraction of ^{40}Ar in the Venus atmosphere has been related to tectonic activity and heat loss on the planet (Sasaki and Tajika 1995; Turcotte and Schubert 1988; Kaula 1999).

Noble gases on Mars

While there are probe data for the atmosphere of Mars, high precision information has also been obtained from Martian meteorites. These have also provided some information about the interior source regions of Martian volcanics. The most recent review is by Swindle (2002, this volume).

Nonradiogenic noble gases. The $^{20}Ne/^{22}Ne$ ratio of the Martian atmosphere is not well determined. The estimate of 10.1±0.7 from SNC meteorite data (Pepin 1991) is within error of the terrestrial atmosphere, and thus substantially lower than the solar value. The $^{38}Ar/^{36}Ar$ ratio is highly fractionated relative to the solar Ar ratio of 0.17; SNC meteorite analyses yield values ranging from 0.24±0.01 (Pepin 1991) to 0.26-0.30 (Bogard 1997; Garrison and Bogard 1998). The Kr isotopic composition of the atmosphere of Mars (Pepin 1991; Garrison and Bogard 1998) appears to be essentially indistinguishable from that of solar wind Kr, as seen in Figure 6, although it is possible that it is slightly fractionated to an isotopically somewhat lighter composition (see Garrison and Bogard 1998; Swindle 2002, this volume).

Primordial Xe (i.e., those isotopes unaffected by fissiogenic or radiogenic inputs) in the Martian atmosphere is greatly fractionated relative to the solar and chondrites compositions, with preferential enrichment of the heavy isotopes (Swindle et al. 1986; Mathew et al. 1998; Table 1). However, there are some uncertainties with constraining the precise composition of this Xe and this affects not only identification of the initial source of Martian Xe but also determination of the precise amounts of radiogenic and fissiogenic Xe in the atmosphere, particularly the latter (see "Primordial Xe" section and discussions below of radiogenic ^{129}Xe and fissiogenic ^{136}Xe budgets).

Radiogenic ^{40}Ar. The atmospheric $^{40}Ar/^{36}Ar$ ratio has been measured by Viking to be 3000±400 (Pepin and Carr 1992), although a lower value of ~1800 has been deduced from meteorite data (Pepin and Carr 1992; Bogard 1997; Bogard et al. 2001). Based on the Viking data, there are $(7.0\pm1.4) \times 10^{39}$ atoms ^{40}Ar in the atmosphere. The Mars mantle has been estimated to have 305 ppm K (Wänke and Dreibus 1988), so that 3.3×10^{41} atoms ^{40}Ar have been produced in Mars. Therefore, only 2% of Martian ^{40}Ar has degassed to the atmosphere, and most of the planet interior has retained the ^{40}Ar produced throughout its history. The history of degassing of ^{40}Ar from the interior has been discussed in several studies (Volkov and Frenkel 1993; Sasaki and Tajika 1995; Hutchins and Jakoski 1996; Tajika and Sasaki 1996).

The radiogenic ^{129}Xe budget. Martian atmospheric Xe clearly contains a considerable fraction of radiogenic ^{129}Xe. It has been estimated that the silicate portion of

Mars contains 32 ppb iodine (Wänke and Dreibus 1988). Assuming that $^{129}I/^{127}I = 1.1 \times 10^{-4}$ at 4.57 Ga (Hohenberg et al. 1967), then 8.44×10^{36} atoms ^{129}Xe have been produced in Mars or precursor materials. Using fractionated CI chondrite Xe or solar Xe for the nonradiogenic light Xe isotope composition (see discussion in "Primordial Xe"), the atmosphere is calculated to contain only 0.092% of what has been produced. Assuming there is none remaining in the planet, this corresponds to a closure age of 160 Ma. However, if only 2% has degassed (like ^{40}Ar) to the atmosphere, then a closure age of 70 Ma is obtained. This value is similar to that of the Earth, and suggests that there may also have been losses of volatiles from Mars over the same extended period of accretion.

The fissiogenic ^{136}Xe budget. The amount of ^{136}Xe produced in Mars or accreting materials, assuming that the silicate portion of Mars has 16 ppb of ^{238}U at present (Wänke and Dreibus 1988) and initially had $^{244}Pu/^{238}U = 0.0068$ (Hudson et al. 1989), is 1.9×10^{34} atoms ^{136}Xe from ^{244}Pu and 7.2×10^{32} atoms ^{136}Xe from ^{238}U. In contrast, there is a total of 2.8×10^{33} atoms ^{136}Xe in the atmosphere. Up to ~5% of the atmospheric ^{136}Xe may be Pu-derived (see "Primordial Xe" below); if so, and the closure age for Mars is 70 Ma, then 1-2% of the ^{244}Pu produced in the solid planet has degassed. This is consistent with the ^{129}Xe and ^{40}Ar budgets. As noted in the section *Primordial Xe*, plutogenic $^{136*}Xe$ could comprise much less that 5% of the total atmospheric inventory, requiring even less planetary degassing and greater very early isolation of interior volatiles from the atmosphere. Reports of significant abundances of ^{244}Pu fission Xe in several SNC meteorites do in fact point strongly to its efficient retention in the Martian crust (Marty and Marti 2002; Mathew and Marti 2002). Further discussion of the abundances of daughter Xe isotopes in the atmosphere is provided by Swindle and Jones (1997). Musselwhite et al. (1991) and Musselwhite and Drake (2000) discuss the distribution of I and the degassing of radiogenic ^{129}Xe.

Martian interior gases. Martian meteorites contain components other than those derived directly from the atmosphere (see detailed discussion by Swindle 2002, this volume). While some appear to be derived from the atmosphere by secondary processes such as alteration, noble gases in the dunite meteorite Chassigny appear to represent a distinct interior reservoir. Information on the relative abundances of the heavier noble gases (Ott 1988) suggests that the $^{84}Kr/^{132}Xe$ ratio (1.2) is lower than both the Martian atmosphere (20) and solar (16.9) values, but is similar to that of CI chondrites. If this is truly a source feature, it indicates that heavy noble gases trapped within the planet suffered substantially different elemental fractionation than the atmosphere. The interior $^{84}Kr/^{36}Ar$ ratio of 0.06 is much higher than the solar value of 2.8×10^{-4}, but it is close to the atmospheric value of 0.02 and so does not display the same contrast as the Kr/Xe ratio. It is not possible at present to conclusively determine whether the measured elemental abundance ratios reflect an interior reservoir that was initially different from atmospheric noble gases, rather than due either to planetary processing or transport and incorporation into the samples.

The isotopic composition of the Martian interior is only available for Xe. The nonradiogenic isotope ratios found in Chassigny (a dunite that is possibly mantle-derived) appear to be indistinguishable from solar values (Ott 1988; Mathew and Marti 2001), and so do not exhibit the strong fractionation seen in the atmosphere. The relative abundances of ^{129}Xe and ^{136}Xe are also close to solar, and so there is little scope for radiogenic additions, indicating that this reservoir had a high Xe/Pu ratio, at least during the lifetime of ^{244}Pu. In contrast, this solar Xe appears to be accompanied by a Kr/Xe ratio that is fractionated with respect to the solar value. Data from other meteorites indicate that there are other interior Martian reservoirs that contain solar Xe but with resolvable fissiogenic contributions (e.g., Nakhla: Mathew and Marti 2002), and so have had lower Xe/Pu ratios. Therefore the interior appears to be relatively undegassed, with high ratios of primordial Xe to parent elements.

ACQUISITION OF PLANETARY NOBLE GASES

The key diagnostic volatiles for tracing atmospheric origin and nonbiogenic evolution are the noble gases, nitrogen as N_2, and carbon as CO_2. Signatures of origin would be expected to survive most clearly in the chemically inert noble gases. Their record is complex, however, and not readily interpreted. As discussed in the preceding sections, absolute abundances of noble gases in the atmospheres of Earth, Venus and Mars and in the carbonaceous chondrites display highly variable depletions with respect to solar abundances, and isotopic patterns in each of these volatile reservoirs are generally distinct from each other and from inferred compositions in the sun and primitive solar nebula. Models of origin based on seemingly straightforward clues in one subset of this extensive data base have historically encountered inconsistencies in another. A celebrated example of this kind of difficulty involves neon-argon distributions. Measured $^{20}Ne/^{36}Ar$ elemental ratios are roughly the same, within a factor two or so, on Earth, Venus, and Mars, and in bulk samples of the primitive CI carbonaceous chondrites. This approximate concordance has prompted much discussion in the modeling literature (see, for example, the review by Donahue and Pollack 1983), usually with the view that it implies accretion of common parental material as the source of at least these two gases in inner solar system bodies. Isotope systematics, however, argue against this interpretation since the required isotopic uniformity is absent: $^{20}Ne/^{22}Ne$ on Venus is higher and $^{36}Ar/^{38}Ar$ on Mars much lower than the corresponding ratios on the Earth and in CI meteorites (Pepin 1991; Pepin and Carr 1992).

Prior to the mid-1980s, most attempts to account for the origin of terrestrial planet atmospheres focused, as in the example above, on deriving the elemental abundance patterns of atmospheric noble gases from primordial meteoritic or nebular reservoirs. One common approach, termed "gas-poor" models by Donahue and Pollack (1983), was to regard an atmosphere as a mixture of gases acquired by accretion on an initially volatile-poor planet of a veneer of known (or occasionally hypothetical) volatile-rich meteoritic or cometary materials or carriers of adsorbed nebular gases. Another class of "gas-rich" models postulated the initial presence and subsequent dissipation of gravitationally captured or impact-degassed primordial atmospheres on the planets themselves. With a few exceptions, comparatively little attention was paid to isotopic distributions, in particular those of the heavy species krypton and xenon, or to astrophysical environments in which evolutionary processing might plausibly have occurred.

The successes and problems of many of these "gas-poor" and "gas-rich" models are discussed by Donahue and Pollack (1983). A point to be emphasized here is that the gas-poor theories are intrinsically unable to account for the range of isotopic variability seen in planetary atmospheres. The processes of sorption and mixing invoked in these models do not fractionate isotopes (although mixing of isotopically different components can certainly generate variable compositions), yet there is strong evidence that such fractionating mechanisms have been at work. For this reason, gas-rich theories that assume the presence of primordial atmospheres of whatever origin on growing or fully accreted planetary bodies, or on (or in) large preplanetary planetesimals, appear fundamentally more attractive. They offer the potential for isotopic fractionation in the process(es) that subsequently dissipated these atmospheres, and the possibility that the variable distributions of noble gas abundances and isotopes seen in present-day planetary atmospheres may be understood as reflecting different degrees of processing on the individual bodies.

Early post-nebular solar wind

Noble gases implanted by low-energy solar wind irradiation are typically retained in solar-like elemental abundance proportions in lunar and meteoritic dust grains, particularly the three heaviest species. Implantation typically extends into the upper few tens of nanometers of the irradiated materials, and so the amounts of noble gases than can be accumulated in this way depend upon their surface areas. The irradiation of dust in the solar nebula will clearly contribute the highest concentrations of noble gases to the subsequently accumulated planetary bodies. An available analog is the fine material found in the lunar regolith. Accretion of planetesimals containing ~25-40 wt % of solar-wind-implanted dust (an enormous fraction, probably attainable only in small objects with thick regoliths of irradiated dust grains) loaded to lunar regolith levels in Xe, Kr and Ar but depleted ~100-fold in ^{20}Ne, could account for the absolute noble gas abundances in Venus's atmosphere. Neon (and helium) can be preferentially lost from irradiated grains by subsequent heating, or depleted even at comparatively low temperatures by rapid diffusion from certain mineral structures, notably plagioclase (e.g., Frick et al. 1988). However, the presence of substantial dust in the nebular disk prior to aggregation also greatly dampens penetration of solar wind out to much of the planet-forming region, while clearance of this dust results in larger targets and therefore smaller fractions by weight of irradiated material.

This general kind of solar wind source for noble gases on the terrestrial planets has been proposed in various contexts by Wetherill (1981), Donahue et al (1981), and McElroy and Prather (1981). Wetherill's (1981) model provides the large ^{36}Ar abundance on Venus by accretion of material irradiated in the inner edge of the nebular disk. He perceived the major difficulty of the model to be that of confining accretion of solar-wind-rich materials largely to just this planet, given the likelihood that gravitational scattering would tend to disperse it throughout the inner solar system. But it is important to note that this may no longer be a fatal objection. As described later in the *Hydrodynamic escape* section, Earth could have acquired a compositionally Venus-like primary atmosphere as well, later fractionated by partial loss, and there is evidence that solar-composition noble gases may have dominated the primordial atmosphere on Mars. The principal problem now with this hypothesis is that Xe in the atmospheres of the early Earth and Mars appears to have differed in isotopic composition—SW-Xe or CI-Xe on Mars, but U-Xe on Earth (see *Primordial Xe* below). Nevertheless the possibility that a solar-wind source of this kind contributed to some degree to noble gas inventories in the interiors or primordial atmospheres of the terrestrial planets should be taken seriously. It could turn out on compositional grounds to be more plausible than either the gravitational capture or icy planetesimal sources discussed below. Moreover Sasaki's (1991) arguments for off-disk penetration of an early and intense solar wind flux into a post-nebular environment rich in fine collisional dust may imply the existence of an ancient reservoir of abundant and heavily loaded carriers of solar-like noble gases. Such a source has been considered for providing the noble gases presently found within the Earth's mantle. Podosek et al. (2000) argued that the present concentrations of Ne estimated for a deep mantle gas-rich reservoir could have been derived from irradiated, km-sized planetesimals (assuming that sufficient turnover of the surfaces occurred so the process was not limited by grain saturation effects). This process would not have been limited by self-shielding by solid material across the accretionary disk due to removal of dust into larger bodies, and thus requires that a substantial fraction of the present mass of the Earth remained as small dispersed planetesimals until after nebula gas had cleared. Also, the gases must be retained in growing planetesimals and ultimately into the growing Earth, without being lost due to impacts or melting. This model cannot be assessed further without greater understanding of the chronology of accretion and gas dispersal, but it remains as a possible explanation for the origin of noble gases within the mantle. It should be noted that a definitive identification of trapped primordial Ne in the mantle as Ne-B (see *Solar Ne within the Earth* section) would provide strong support for this mechanism.

Adsorption on pre-planetary grains or protoplanetary bodies

Laboratory studies have shown that noble gases exposed to particular kinds of finely divided solid materials are adsorbed on or within the surfaces of individual grains. Adsorption is most efficient for various forms of carbon, where it appears to reflect intrinsic structural properties (Frick et al. 1979; Niemeyer and Marti, 1981; Wacker et al. 1985; Zadnik et al. 1985; Wacker, 1989), but has also been experimentally demonstrated for other minerals (Yang et al. 1982; Yang and Anders 1982a,b) and for polymineralic meteorite powder (Fanale and Cannon 1972). Moreover the process occurs naturally in sedimentary materials (Canalas et al. 1968; Fanale and Cannon 1971; Phinney 1972; Podosek et al. 1980, 1981; Bernatowicz et al. 1984). Adsorbed gases on these substrates generally display fractionation patterns, relative to ambient gas-phase abundances, in which heavier elements are enriched. For the most part these elemental fractionations are remarkably uniform, considering the wide range of experimental and natural conditions under which they are produced. Although occasional isotopic effects have been reported in natural samples (Phinney 1972), they are not observed in equilibrium adsorption in the laboratory (Bernatowicz and Podosek 1986).

The decline in solar-normalized noble gas abundance ratios from Kr to Ne in planetary atmospheres and meteorites is qualitatively similar to many of the adsorption fractionation patterns seen in the laboratory and in terrestrial sedimentary rocks, as shown in Figure 2. This has led to a suspicion that adsorption of nebular gases on meteoritic and protoplanetary materials has played a role in establishing these ratios. Adsorption of isotopically solar noble gases, however, cannot by itself generate the fractionated isotopic patterns seen in planets, and laboratory estimates of single-stage gas/solid partition coefficients are too low to account for planetary noble gas abundances by adsorption on free-floating nebular dust grains. Nevertheless adsorption or solution from a gas phase could well have been an important mechanism for supplying primordial volatiles to planets, if it occurred from a reservoir well above nebular pressures—for example, as discussed in the following section, by gravitational capture of atmospheres on protoplanetary or planetary bodies that had grown to the critical size for the process to be effective.

Gravitational capture

Capture by growing planets. A potentially powerful noble gas source for the terrestrial planets is direct capture of co-accreting primordial atmospheres from the nebula during planetary growth. Substantial gravitational capture of ambient gases requires the growth of protoplanets to appreciable masses (at least to ~Mercury-Mars size) prior to dissipation of the nebular gas phase. The extent to which this would have occurred hinges on the relative timing of nebular dissipation versus planetary accretion. Current estimates of timescales for loss of circumstellar dust and gas, from observation and theory, are on the order of ~10 Ma or perhaps somewhat less (Podosek and Cassen 1994), much shorter than the ~100 Ma or more for planetary growth to full mass in the standard model of planetary accumulation (Wetherill 1986, 1990a). Nebular lifetimes inferred from astronomical observation, however, are based solely on evolution of their fine dust component. Measurements of molecular line emissions that could at least set upper limits on the longevity of the gas-phase component in "naked" T-Tauri disks are challenging, and data are sparse (Strom et al. 1993). There is no reason to believe that disappearance of micrometer-size dust from infrared detectability, say by accretion into larger grains, would necessarily be coincident with gas loss. According to the standard model, terrestrial planet growth to roughly 80% of final masses occurred within ~20 Ma (Wetherill 1986). So if the observed dust clearing signals the beginning of planet-building, and a significant remnant of gas survived in the inner solar accretion disk for another 10-20 Ma or so, substantial gravitational capture would have taken place, at least on proto-Earth and proto-Venus.

It would appear that current observations and modeling cannot rule firmly for or against gravitational capture of massive solar-composition primary atmospheres on Earth and Venus. The extent to which this occurred, creating tenuous or dense, massive atmospheres, depends upon the mass of the protoplanets at the time of nebular gas dispersal. If it did occur, subsequent evolution from primary to present-day noble gas inventories and compositions must have involved loss processes that fractionated both elements and isotopes to generate the presently observed compositions (Fig. 2). There is presently no reason to assume elemental fractionation in the capture process itself, although it may be that this possibility needs more theoretical study.

Capture by planetary embryos. While the atmospheres captured by these small bodies may not provide sufficient terrestrial noble gas abundances in themselves, sufficient quantities of gases may have accumulated within the protoplanetary bodies by gas adsorption on surface materials followed by burial below the surface during continuing accretion (Pepin 1991). The process may have played an important role in creating internal volatile reservoirs for later outgassing of secondary atmospheres on the terrestrial planets, especially for the heavy noble gases. Protoplanetary cores, growing through sizes of a few lunar masses in the presence of nebular gas at temperatures and pressures estimated from astronomical observations and from accretion disk models, would have gravitationally captured atmospheres that may have then been incorporated into the solid body. Interaction of the gases with the surface is governed by the pressure at the base of the atmosphere, which depends on the thermal structure of the atmosphere. This in turn is a sensitive function of atmospheric opacity, which is difficult to estimate, although amplifications of surface pressure by ~4-6 orders of magnitude above that of the ambient nebula are likely (Pepin 1991). Adsorption and occlusion of these surface gases on and within growing planetary embryos would appear to be a natural consequence of their accumulation, in the presence of nebular gas, to bodies of up to ~Mercury-size within <1 Ma (Wetherill 1990b; Wetherill and Stewart 1993).

Another consideration bearing on the probable volatile-rich nature of planetary embryos is that impact velocities of materials accreting to form these small bodies are generally too low to promote efficient degassing of the impactors themselves. Consequently their volatiles also tend to be buried within the growing embryos (Tyburczy et al. 1986). If noble gases were acquired by these mechanisms, atmospheric formation would then occur by subsequent degassing and isotopic fractionation during loss to space. Noble gases trapped within the Earth and incorporated into the present deep mantle would exhibit solar isotopic compositions but elemental fractionation; for example, adsorption in laboratory experiments strongly fractionates ambient Ne and Ar (in favor of Ar) by factors of ~10 to >50.

Evidence of such a deep Earth abundance pattern coupled with solar isotopic compositions would be a strong indicator that adsorption could have played a role early in the Earth's accretional history (Pepin 1991, 1998). Although there is evidence that solar-like light noble gas isotopic compositions exist, the abundance pattern of trapped noble gases in the Earth cannot be easily constrained, since there is the possibility that subduction of heavy noble gases has since altered the pattern. With a Mars-size terrestrial embryo and the high value of 2×10^{-5} for the Ne Henry's constant assumed in Pepin's

(1991) treatment of adsorption, the amount of adsorbed Ne that could presently be stored in a gas-rich lower mantle may be achieved. However it is too low to account for an initial deep Earth abundance of 10^{12} atoms ^{22}Ne/g (see "Xe loss from the mantle" above). This suggests that if adsorption had been responsible for providing the heavier noble gases, another process, such as dissolution into surface magma, would be required to increase the ^{20}Ne inventory above the level supplied by adsorption.

Gravitational capture and dissolution into molten planets. If the Earth reached sufficient size in the presence of the solar nebula, a massive atmosphere of solar gases would have been gravitationally captured and supported by the luminosity provided by the growing Earth, and the underlying planet would have melted by accretional energy and the blanketing effect of the atmosphere (Hayashi et al. 1979). Gases from this atmosphere would have been sequestered within the molten Earth by dissolution at the surface and downward mixing (Mizuno et al. 1980). This mechanism can provide solar noble gases into the deep Earth with relative elemental abundances that have been fractionated according to differences in Henry's constants for solubility in silicates (with depletion of heavy noble gases). Initial calculations found that at least an order of magnitude more Ne than presently found in the deep mantle could be dissolved into the Earth unless the atmosphere began to escape when the Earth was only partially assembled (Mizuno et al. 1980; Mizuno and Wetherill 1984; Sasaki and Nakazawa 1990; Sasaki 1999). As noted above, the present abundances may be 10^{-2} times that of initial concentrations, and Porcelli et al. (2001) and Woolum et al. (1999) considered the conditions required to dissolve sufficient Ne to account for the initial deep mantle inventory prior to losses at ~10^8 a. If it is assumed that equilibration of the atmosphere with a thoroughly molten mantle was rapid, and uniform concentrations were maintained throughout the mantle by vigorous convection, then the initial abundances of gases retained in any mantle layer reflect surface noble gas partial pressures when that layer solidified. The depth of at which solidification occurs is determined by the surface temperature and the efficiency of convection in the molten mantle. Therefore, initial distributions of retained noble gases would be determined by the history of surface pressure and temperature during mantle cooling and solidification, i.e., the coupled cooling of Earth and atmosphere. For typical solubility coefficients (e.g., Lux 1987; Carroll and Stolper 1993; Shibata et al. 1998), a total surface pressure of the order of 100 atm under an atmosphere of solar composition was required to establish the initial deep mantle Ne concentration (Porcelli et al. 2001), along with surface temperatures high enough to melt the deep mantle (~4000 K). Therefore, for this model of the origin of deep He and Ne to be viable, two conditions must be shown to be achievable: (1) the surface pressures necessary to dissolve the required abundances of noble gases are achieved, and (2) the surface temperatures sufficient to melt to lower mantle depths are reached. The dense atmosphere is a balance between the gravitational attraction of the nebula-derived gases and expansion due to the Earth's luminosity (energy released by accreting planetesimals and the cooling Earth). The nebular temperature and pressure provide boundary conditions, and the atmospheric opacity (which controls the rate of energy loss to space) is a critical parameter. High luminosities (or low opacities) increase the surface temperature but lower the pressure, while decreasing nebular pressures will generate lower surface pressures and temperatures. Therefore, the temperature and pressure at the base of the atmosphere evolved as the energy released by accretion declined with time once planet assembly approached completion, and as the nebular pressure declined during nebula dispersal. Woolum et al. (1999) demonstrated that the necessary conditions were met under a range of parameter values for both convective and radiative atmospheric structures. However, the complexities in determining the effects of the composition of the lower atmosphere (which was probably strongly contaminated by evaporated terrestrial material that would have had a marked effect on opacity) and the transition from optically thick to thin atmospheric conditions presently remain to be resolved.

It should be noted that not all situations facilitate the dissolution of atmospheric gases. At low nebular pressures and high initial luminosities, it is possible that the initial cooling phases occurred when the entire atmosphere was optically thin; in some such cases, rapid magma solidification occurs without the incorporation of significant concentrations of atmospheric gases. However, in the presence of a massive atmosphere that promotes gas dissolution, the mantle cooling time is greatly extended over that which would otherwise occur (Tonks and Melosh 1990). Solidification times then commonly exceed a million years, allowing substantial time for downward convection of noble gases. Overall, it appears that sufficient He and Ne (along with associated Ar, Kr, and Xe) can be incorporated into the Earth over a wide range of conditions by this mechanism.

As discussed above, the plausibility of this gravitational capture mechanism depends upon whether a sufficient mass of the Earth accretes prior to dispersal of nebular gases, and further work is required to determine if dispersal of the nebula can extend over such time periods.

Accretion of volatile-rich planetesimals

Comet accretion models. Noble gases, as well as water, carbon, and nitrogen, could have been supplied to the inner planets by accretion of volatile-rich icy comets scattered inward from the outer solar system. Although noble gas isotopic distributions in comets are unknown, solar isotopic compositions would be expected in cometary gases acquired from the nebula. There is experimental evidence that the relative elemental abundances of heavier species (Xe, Kr, and Ar) trapped in water ice at plausible comet formation temperatures (~30 K) approximately reflect those of the ambient gas phase, and trapped noble gas abundances per gram of water are substantial (Bar-Nun et al. 1985; Owen et al. 1991). The compositional characteristics of an icy planetesimal source in comet accretion models for Venus, discussed below, require occlusion of nebular noble gases with approximately unfractionated elemental ratios for Ar:Kr:Xe but much lower Ne. Thermodynamic modeling suggests that noble gases incorporated in clathrates do indeed have low Ne/Ar ratios, but do not reflect ambient gas-phase compositions for Xe/Ar and Kr/Ar and instead are strongly enriched in the heavier species (Lunine and Stevenson 1985). Clathrated gases therefore appear unlikely to be the source of atmospheric noble gases, at least for Venus, and one must appeal to physical adsorption on ice.

Incorporation of a few percent or less by mass of icy cometary matter into an accreting terrestrial planet could potentially have supplied heavy noble gases of solar composition to its primary atmosphere, in addition to enough hydrogen (as H_2O) to fuel the subsequent hydrodynamic escape episode that generated the observed isotopic fractionations (see below). This kind of source is particularly attractive in that the low Ne/Ar ratio expected for trapping of ambient gases in amorphous ice at temperatures >20-25 K (Owen et al. 1991; Owen and Bar-Nun, 1995ab) would provide a natural explanation for the underabundance of Ne in the otherwise solar-like Venus pattern (Figs. 2 and 12). A cometary carrier of primordial Venus volatiles with just these characteristics has been proposed (Owen 1987, Hunten et al. 1988, Owen et al. 1991; Owen and Bar-Nun 1995a,b).

At likely nebular temperatures and pressures at its radial distance, Mars is too small to have condensed a dense early atmosphere from the nebula even in the limiting case of isothermal capture (Hunten 1979; Pepin 1991). Therefore, regardless of the plausibility of gravitational capture as a noble gas source for primary atmospheres on Venus and Earth, some other way is needed to supply Mars. An early inward flux of icy planetesimals would have contributed to all three of the terrestrial planets, even if it dominated the noble gas budget only on Mars. Accretion of a substantial mass of volatile-rich bodies resembling the present-day carbonaceous chondrites may be another possibility. Swindle et al. (1986) pointed out that the isotopic composition of Martian atmospheric Xe calculated from their SNC meteorite measurements was compatible with that of mass-fractionated CI-Xe, consistent with Dreibus and Wänke's (1985, 1987, 1989) SNC-based geochemical model of Mars' bulk composition which calls for a ~40% mass fraction of volatile-rich, oxidized CI-like material in the planet. However subsequent expansion and recalculation of Xe data from the SNC meteorites now suggest that solar-wind Xe is a viable alternative to CI-Xe as the principal atmospheric constituent on early Mars (see "Primordial Xenon" section below), raising the challenging question of how Mars could have acquired a large solar Xe component while Earth apparently did not.

A number of arguments can be made in favor of cometary carriers for inner planet volatiles. As noted above, reduction of the Ne/Ar ratio relative to the solar ratio, resembling the elemental pattern on Venus, is likely in such ices. Modeling discussed below indicates that a source of this nature could have supplied essentially identical primary atmospheres to both Venus and Earth if an initially Venus-like atmosphere on Earth were later elementally fractionated in hydrodynamic escape powered by a giant Moon-forming impact.

Accretion of icy cometary matter has been widely viewed as a plausible source for Earth's water. The D/H ratio in seawater, however, is substantially lower—by a factor of ~2—than in the few comets where D/H has been measured (Laufer et al. 1999). A significant contribution of terrestrial water by comets would still be permitted if their high D/H ratio were appropriately lowered by accretion of additional, deuterium-poor materials. Suggested possibilities for low D/H carriers include rocky planetary accretional components (Laufer et al. 1999), or a high influx during the heavy bombardment epoch of interplanetary dust particles heavily loaded with implanted solar-wind hydrogen (Pavlov et al. 1999).

Origins of volatile species on the terrestrial planets were modeled by Owen, Bar-Nun and co-workers as having resulted from accretion, in variable planet-specific proportions, of rocky materials as well as three types of comets. These formed at different heliocentric distances and thus at different nebular temperatures, leading to distinctive elemental fractionation patterns in volatiles trapped in their ices from ambient nebular gases (e.g., Owen at al. 1991, 1992; Owen and Bar-Nun 1995a,b). Present planetary inventories in such models are generated by simple mixing, in the sense that volatile abundances delivered by early cometary and rocky planetesimal bombardment are not subsequently altered by losses to space—except on Mars, where non-fractionating impact erosion has depleted volatile abundances. Suitable mixing of rocky-component volatiles with those contributed by comets formed at specific temperatures (~30 K for Venus, as noted above) can arguably reproduce relative elemental abundances of the noble gases (e.g., Fig. 3 in Owen and Bar-Nun 1995b) as well as C and N on the terrestrial planets (Owen and Bar-Nun 1995a).

The principal difficulty encountered by these mixing models in their present form is their inability to account for differences in nonradiogenic noble gas isotopic distributions between Earth and Mars, and between both of these and solar compositions (Swindle 2002; Wieler 2002), which imply processing of primordial atmospheres by isotopically fractionating mechanisms. As noted above, solar isotope ratios would be expected in the ambient nebular gases surrounding accreting cometary matter. Recent experiments specifically designed to investigate isotopic fractionation in the gas trapping process showed maximum heavy isotope enrichments which are too small to explain planetary isotopic offsets from solar-like ratios (Notesco et al. 1999)—in particular, those displayed by $^{36}Ar/^{38}Ar$ on Mars and by Xe on both planets. Also, models that rely on supply of noble gases from material trapped in the outer parts of the solar system cannot explain the abundances of mantle noble gases, since these materials are expected to be provided as a 'late veneer' when accreting bodies are supplied from a wider swathe of the nebula, and are more likely to devolatilize upon impact due to the size of the proto-Earth (see "Losses during accretion" section).

Xe fractionation in porous pre-planetary planetesimals. Nonradiogenic xenon isotope ratios on Earth and Mars are grossly mass-fractionated relative to the solar Xe composition. There are two current models for the origin of planetary Xe from solar Xe. The first involves driving initially solar-like Xe compositions to their present isotopically heavy states by fractionating losses to space by hydrodynamic escape from the planets themselves. This occurs after the acquisition of noble gas inventories by unrelated process, and is discussed below. The other possibility, originally suggested by Ozima and Nakazawa (1980) (see also Ozima and Igarashi 1989) and redeveloped and extended by Zahnle et al. (1990b), is fractionation of nebular Xe into the terrestrial composition by gravitational isotopic separation in large (~Ceres-size) porous planetesimals which have now vanished from the solar system. Nebular gases are gravitationally segregated within interconnected pore spaces due to the dependence of scale height on species mass. The process terminates when growth inhibits further diffusive equilibration and loss to space. Later accretion of these bodies supplies Earth-like Xe to all three terrestrial planets. The conceptual impetus for attributing planetary Xe to a common pre-planetary source lies in the view that fractionating processes operating on individual bodies as dissimilar as Earth and Mars, and perhaps driven by power sources as different as short-term decay of energy deposited in a giant impact on Earth and long-term exposure to EUV radiation on Mars, would be unlikely to result in the qualitatively similar nonradiogenic Xe compositions in their current atmospheres (e.g., Zahnle et al. 1990b)—a similarity that, in hydrodynamic escape modeling, must be regarded as coincidental.

Isotopic fractionation of nebular Xe to obtain the terrestrial atmospheric composition can be achieved in this model, although generation of the atmospheric Xe abundance requires retention of Xe within the planetesimal by adsorption as well (Zahnle et al. 1990b). Fractionations of Kr and Ar isotopes in this process are given by Ozima and Zahnle (1993) as functions of planetesimal radius. However, while these are in the right directions, it is not clear that terrestrial isotopic compositions of all three noble gases can be generated from solar compositions, for any distribution of planetesimal masses accreted by the Earth. The highly fractionated Kr and lighter noble gases must be mixed at some point with other, unfractionated components—and, in the case of Ne, further processed by fractionating escape to space (Ozima and Zahnle 1993)—to generate the isotopic distributions displayed by these species in the present-day terrestrial atmosphere. In this respect, post-accretional atmospheric evolution on Earth is modeled in much the same way as in the hydrodynamic escape model described below, with the early, isotopically fractionated noble gas inventory deposited by accretion of porous planetesimals rather than being generated on the planet itself by hydrodynamic escape of a primary atmosphere. A consequence of this mechanism is that the atmospheric noble gas characteristics are established in accreting materials, so that noble gases presently within the deep Earth are predicted to have the same characteristics. Since this is the only model to predict isotopic ratios for primordial (i.e., not recycled) noble gases still trapped within the Earth that are fractionated relative to solar, further constraints on mantle noble gases may speak to its applicability.

This kind of model has not been applied in any detail to Mars. A problem that confronts it on that planet is the observation that Xe on Mars and Earth, while displaying comparable light-isotope ratios, is compositionally different at the heaviest isotopes in ways that cannot be explained by variable additions of ^{244}Pu or ^{238}U fission Xe to either or both atmospheres. As discussed below, it appears that Xe compositions on the two planets reflect mass fractionation of two isotopically distinct primordial starting compositions, in conflict with the hypothesis of a common planetesimal source. Venus is the key. Even a moderately accurate isotopic analysis of Venusian Xe, where no data presently exist and where the nonradiogenic Xe compositions predicted by the two models—Earth-like if supplied by porous planetesimals, and solar-like if unfractionated by hydrodynamic escape—are very different, should rule definitively between them—or create problems for both.

LOSSES AND MODIFICATIONS OF PLANETARY NOBLE GASES

As discussed above, primary volatile sources in current models for the history of terrestrial planet atmospheres may include the solar nebula, solar wind, comets, one or more known meteorite classes, or other bodies carrying different volatile distributions —with the assumption that these represent materials that either are no longer extant in the solar system or exist but are unsampled. Some of these appear likely to preserve their compositional signatures during final addition to the planets. Such presumably nonfractionating processes of incorporation into the Earth and addition to the atmosphere include impact-degassing of meteorites or cometary matter accreted by planets, direct gravitational capture of ambient nebular gases, planetary outgassing, and ejection of gases by impact of accreting bodies into preexisting atmospheres. Other mechanisms fractionate either elements alone, or both elements and isotopes. Examples are implantation and diffusion of solar wind gases in dust grains prior to accretion, adsorption of gravitationally condensed nebular gases on protoplanetary or planetary surface materials, solution of ambient volatiles into the melted surfaces of planets during accretion, partitioning in planetary interiors, and loss of atmospheric constituents from planetesimals or planetary bodies by thermal evaporation (Jeans escape) or hydrodynamic escape or by nonthermal processes such as sputtering (see below).

None of the acquisition models described above can explain all of the presently observed noble gas characteristics in the terrestrial planets, indicating that subsequent loss processes have caused further modifications.

Losses during accretion

As growth of a protoplanet proceeds with increasing accretional energy, shock-induced devolatilization of the accreting materials occurs and volatile species are transferred into the growing atmosphere. Data summarized by Ahrens et al. (1989) indicate that efficient loss of CO_2 and H_2O from accreting solids on impact occurs when its mass approaches that of Mars. Above this size (on Earth and Venus) degassing would also be driven by extensive melting due to deposition of accretional energy (Safronov 1978), and further promoted by a radiative blanketing effect if a water-rich atmosphere is accumulated (see Abe and Matsui 1986). The overall extent of this degassing depends upon the depth and duration of melting, the rate of convection, and the efficiency of degassing of material at the protoplanetary surface. While this process will facilitate transfer of noble gases to the atmosphere, it will limit how much can be buried in the growing planet.

Loss of atmospheric gases to space can occur by impact erosion, when a sufficient transfer of energy from accreting bodies occurs so that a substantial portion of the protoplanetary atmosphere reaches escape velocity (see Cameron 1983; Ahrens 1993). For smaller accreting bodies, the maximum fraction of the atmosphere that can be expelled is ~6×10^{-4} (Vickery and Melosh 1990), equivalent to the total above the plane tangent to the planetary surface at the impact location. However, atmospheric loss may be much greater for very large impacts by bodies exceeding lunar size (Chen and Ahrens 1997). It should be emphasized that these impact-driven losses are not expected to generate elemental or isotopic fractionations in the noble gases, and contribute only to the overall depletion of these species.

A Moon-forming collision of an approximately Mars-sized body with Earth (e.g., Hartmann and Davis 1975; Cameron and Ward 1976) would clearly result in catastrophic loss of volatiles from the pre-existing atmosphere and may have caused substantial loss of deep-Earth noble gases as well. Detailed studies of atmospheric structure and dynamics during and following such an impact have not been carried out, and clearly would be a challenging task. A central issue is whether any remnant of the original atmosphere could have survived the event. Ahrens (1990, 1993), for example, argues that virtually complete expulsion might have occurred by direct ejection from the impacted hemisphere and by shock-induced outward ramming of the antipodal planetary surface. However a plausible alternative is that mechanical losses were incomplete, and may have been followed by additional losses driven by thermal processes. If so it may prove difficult to blow off the entire atmosphere, in particular its heavier constituents. Retention of a fraction of the primary pre-impact atmosphere is an important requirement in the model of terrestrial volatile evolution discussed below, where further hydrodynamic loss and fractionation of a post-impact atmospheric residue is responsible for setting the present abundance and isotopic composition of Xe.

Sputtering

Probable operation, on Mars, of an atmospheric loss mechanism that results in fractionation of elements and isotopes in the residual atmosphere was demonstrated by Luhmann et al. (1992) and Zhang et al. (1993). Oxygen atoms in the Martian exosphere, ionized by solar EUV radiation and accelerated in the motional $\mathbf{v} \times \mathbf{B}$ electric field of the solar wind, can impact species near the top of the atmosphere (the "exobase") with enough energy transfer to eject them from the planet's gravitational field. The exobase is defined as that atmospheric altitude where, in this case, an arriving oxygen ion would traverse one mean free path in overlying exospheric material before encountering it (Chamberlain and Hunten 1987). Loss rates of the dominant atmospheric constituent at the exobase (CO_2 on Mars) in this sputtering process depend on the magnitudes of the EUV flux and the solar wind velocity, and so estimates of how both electromagnetic and corpuscular radiation have evolved over solar history are needed in order to calculate sputtering losses in past epochs (Zhang et al. 1993) —note that this same requirement for knowledge of flux history applies to EUV-driven hydrodynamic escape as well (see next section). Escape fluxes of sputtered trace constituents in the atmosphere—e.g., noble gases and nitrogen—are proportional to their exobase mixing ratios with CO_2 (Jakosky et al. 1994).

Sputtering losses are greatly attenuated by the presence of a planetary magnetic field, most importantly because it deflects the solar wind around the planet and shields atoms photoionized in the outer atmosphere from the solar wind electric field that would otherwise accelerate some of them downward toward the exobase (Hutchins et al. 1997). For this reason the process would not have been important on Earth for as long as the core dynamo has existed, and it seems unlikely that the bulk composition of the massive Venusian atmosphere could have been substantially affected by sputtering loss with or without the protection of a paleofield. Escape of sputtered species is also impeded by the higher gravity of these two planets. In contrast, a thin, magnetically unshielded, and more weakly bound Martian atmosphere is particularly vulnerable to sputtering erosion.

Efficient operation of this fractionating loss mechanism over time on Mars is therefore centrally linked both to atmospheric pressure history and to the timing of the disappearance of the Martian paleomagnetic field (Hutchins et al. 1997; Connerney et al. 1999).

Fractionation from the sputtering process is governed primarily by the barometric law. An atmosphere is compositionally well-mixed up to an altitude called the homopause, on present-day Mars at ~120 km; above that point, it gravitationally separates according to $N(\Delta z) = N_H \exp[-mg\Delta z/kT]$ where N_H is the abundance of a particular species of mass m at the homopause, g is the gravitational acceleration, k is Boltzmann's constant, and Δz and T are the atmospheric altitude and temperature above the homopause. It follows directly that the abundance ratio of two species of masses m_1 and m_2 at Δz, relative to their ratio at the homopause, is $R = \exp[-\Delta mg\Delta z/kT]$, where $\Delta m = m_1 - m_2$ (Jakosky et al. 1994). If m_1 is less than m_2, R is greater than 1; sputtering loss from the exobase therefore preferentially removes lighter species, leaving the residual atmosphere enriched in heavier constituents. These atmospheric sputtering fractionations are thus seen to be generated by loss from a fractionated "target" (the exobase). Depletions of lighter species are further augmented in the escape process itself since ejection efficiency from the exobase (atoms ejected per incident ion) increases with decreasing atomic mass (Jakosky et al. 1994).

The exponential barometric decline of abundances with increasing mass generates an enormous range of elemental fractionations at the exobase; using Jakosky et al.'s (1994) estimate of ~0.4 km/K for Martian $\Delta z/T$, $N(\Delta z)/N_H$ is approximately 3×10^{-2} for ^{20}Ne and 2×10^{-3} for ^{36}Ar, but 3×10^{-7} and 10^{-10} for ^{84}Kr and ^{130}Xe respectively. Sputtering loss of the two heaviest noble gases is consequently extremely small, and isotopic fractionation in the process has negligible influence on the composition of their total atmospheric inventories.

Hydrodynamic escape

Much of the evolutionary modeling over the past 15 years or so has focused on fractionation from primordial source compositions during thermally driven atmospheric escape. In the first of these, Donahue (1986) considered fractionation effects resulting from classical Jeans escape of pure, solar-composition noble gas atmospheres from large planetesimals. Their atmospheres were assumed to derive from outgassing of nebular gases previously adsorbed on the surfaces of preplanetesimal dust grains, and were subsequently fractionated to different degrees by losses that depend on the rates of planetesimal growth. These bodies later accumulated in various proportions to form the terrestrial planets and their atmospheres, and in this respect the approach is similar to that taken in the porous planetesimal model discussed above although fractionation is attributed to entirely different mechanisms in the two cases. Donahue's model can account reasonably well for relative Ne:Ar:Kr elemental abundances and for Ne isotopic compositions in the atmospheres of Venus, Earth and Mars. Predicted $^{36}Ar/^{38}Ar$ ratios, however, are much lower than observed, and variations in Kr and Xe elemental and isotopic compositions in different planetary reservoirs cannot be explained since Jeans escape of such heavy species from the parent planetesimals would have been essentially nil.

These problems are proving to be more tractable in the context of a related thermal loss mechanism, hydrodynamic escape (Zahnle and Kasting 1986; Hunten et al. 1987, 1988, 1989); Sasaki and Nakazawa 1988, 1990; Zahnle et al. 1990a; Pepin 1991, 1994, 1997, 2000). Here atmospheric loss is assumed to occur from much larger bodies, partially or fully accreted planets. Their hydrogen-rich primordial atmospheres are heated at high altitudes, after the nebula has dissipated, by intense far-ultraviolet radiation from the young sun or alternatively, in the case of the Earth, by energy deposited in a large Moon-forming impact event. Under these conditions hydrogen escape fluxes can be large enough to exert upward drag forces on heavier atmospheric constituents sufficient to lift them out of the atmosphere. Lighter species are entrained and lost with the outflowing hydrogen more readily than heavier ones, leading to mass fractionation of the residual atmosphere. Hydrogen escape fluxes high enough to sweep out and fractionate atmospheric species as massive as Xe require energy inputs roughly two to three orders of magnitude greater than presently supplied to planetary exospheres by solar extreme ultraviolet (EUV) radiation —implying large but not prohibitive enhancements of surface activity on the early Sun. Hydrodynamic escape is particularly attractive for its ability to generate, in an astrophysically plausible environment, large isotopic fractionations with respect to solar Xe of the type displayed both by terrestrial xenon—first observed and attributed to an (unknown) fractionation process 40 years ago (Krummenacher et al. 1962)—and by Xe in the atmosphere of Mars (Swindle et al. 1986; Swindle 2002, this volume).

The potential power of the hydrodynamic loss mechanism, given adequate supplies of hydrogen and energy, to replicate observed isotopic distributions was well illustrated by Hunten et al.'s (1987) applications of the process to several specific cases, including the derivation of terrestrial Xe from solar Xe; by Sasaki and Nakazawa's (1988, 1990) independent treatment of the terrestrial Xe problem; and by Zahnle et al.'s (1990a) consideration of Ne and Ar compositions on Earth and Mars. Pepin (1991, 1994, 1997) assessed its more general viability as an actual instrument of volatile evolution by examining the consequences of hydrodynamic escape for the full range of elemental and isotopic mass distributions found in contemporary planetary atmospheres, and exploring the astrophysical and planetary conditions under which the process could account for this data base.

Historians of the field will recognize that it has come full circle with the recent emphasis on fractionation of atmospheric noble gases by escape from gravitational potential wells. This was just the approach taken by Brown (1949) and Suess (1949) in their pioneering attempts to account for the differences between the elemental abundance patterns of solar and terrestrial noble gases. A quote from a review of this early work by Signer and Suess (1963) makes the point: "All but a fraction of about 10^{-7} of the noble gases left the earth's gravitational field during an early stage of evolution by hydrodynamic outflow without undergoing separation. The rest, however, underwent some process by which a fractionation took place, shifting the abundance ratio (of Ne/Xe) by more than a factor of 10^4. A simple explanation for the fractionation of the noble gases is selective loss from a gravitational field during a limited period of time (Suess 1949)." At that time the theoretical framework of the hydrodynamic escape mechanism had not been developed, and the appeal was to classical Jeans escape, which required the fractionating loss to take place from small preplanetary planetesimals. Nevertheless this perception of evolutionary processing has a distinctly modern ring.

Theory. Theories of mass fractionation in hydrodynamic escape of gases from planetary atmospheres have been constructed by Zahnle and Kasting (1986) and by Hunten et al. (1987). Hunten et al.'s (1987) relatively simple analytic approach assumes the presence of an isothermal atmosphere consisting of an abundant light gaseous species and minor amounts of heavier components. Energy input at high altitudes drives thermal loss of the light constituent with escape flux F_1 particles cm^{-2} sec^{-1}. The escaping gas exerts upward drag forces on heavier species. For a given F_1 the drag is sufficient to lift all constituents with masses m_2 less than a critical mass m_c out of the atmosphere. Analytic expression of the theory (Hunten et al.1987; Pepin 1991, 1997), for an atmosphere containing inventory N_1 of the abundant light species with mass m_1 (constituent 1) and inventories N_2 of trace species with masses $m_2 > m_1$ (constituents 2), leads to escape fluxes F_2 of the minor, heavier constituents given by

$$F_2 = F_1 \frac{N_2}{N_1}\left[\frac{m_c - m_2}{m_c - m_1}\right] \tag{2}$$

for $m_C \geq m_2$. The critical or "crossover mass" m_C in Equation (2), representing the smallest mass for which the escape flux F_2 of a particular mass m_2 in constituent 2 is zero, is defined (Hunten et al. 1987) as

$$m_c = m_1 + \frac{kTF_1}{bgX_1} \tag{3}$$

where k = Boltzmann's constant, T = atmospheric temperature, g = gravitational acceleration, X_1 is the mole fraction of the abundant constituent 1 (assumed to remain near unity throughout the escape episode), and b is the diffusion parameter (the product of diffusion coefficient and total number density) of mass m_2 in the constituent 1 gas. Note from Equation (3) that a large F_1 yields a large crossover mass m_C; for F_1 high enough that $m_C >> m_2 > m_1$, losses described by Equation (2) are maximum (atmospheric "blowoff") and species in constituent 2 escape close to the ratios N_2/N_1 of their atmospheric inventories—i.e., without fractionation. Maximum fractionation occurs when F_1 drops to the level where m_C approaches m_2.

Constituent 1 is usually assumed to be molecular hydrogen. Constituent 2 is taken to be a multicomponent noble gas mixture—excluding He, which is only weakly bound in terrestrial planet atmospheres and escapes readily by other processes. The diffusion parameter b in Equation (3), and thus the crossover mass m_C, differs for different elements S in constituent 2. If $m_C(S)$ is known (or assumed) for one element, say Xe, then Equation (3) for both Xe and S yields

$$m_c(S) = m_1 + [m_c(Xe) - m_1]\left[\frac{b(Xe)}{b(S)}\right] \tag{4}$$

for the S crossover mass, where S represents Kr, Ar or Ne. Values of b(S) for noble gas diffusion in H_2 at various temperatures are known (Mason and Marrero 1970; Zahnle and Kasting 1986). They increase from Xe to Ne by a factor of approximately 2 that varies only slightly with temperature, and are identical or nearly so for isotopes of a given element.

The energy required for escape of a particle with mass m_1 from its local gravitational field, at radial distance $r \geq r_S$ from a body of mass M and radius r_S, is Gm_1M/r ergs per particle. If the global mean solar EUV input at heliocentric distance R and time t is $\phi(R,t)$ ergs cm^{-2} sec^{-1}, the energy-limited escape flux is

$$F_1(R,t) = \frac{\phi(R,t)\varepsilon}{Gm_1M/r} = \frac{r\phi(R,t)\varepsilon}{Gm_1M} \quad \text{particles cm}^{-2}\text{sec}^{-1} \tag{5}$$

where ε is the fraction of incident EUV energy flux converted to thermal escape energy of m_1. Energy input ϕ may be expressed in terms of the current mean EUV flux at Earth $\phi_{\oplus}(t_p)$, heliocentric distance R, and the ratio Φ_{EUV} of the flux at past time t to that at present time t_p. Defining $\phi(R,t_p) = (R_{\oplus}/R)^2\phi_{\oplus}(t_p)$ and $\Phi_{EUV}(t) = \phi(R,t)/\phi(R,t_p)$ yields $\phi(R,t) = (R_{\oplus}/R)^2\phi_{\oplus}(t_p)\Phi_{EUV}(t)$. Substituting this into Equation (5) and combining Equations (5) and (3), with $g = GM/r^2$ and M expressed in terms of r_S and planetary density ρ, yields

$$m_c(S) = m_1 + \left(\frac{3}{4\pi G\rho}\right)^2 \left(\frac{1}{r_S}\right)^3 \left(\frac{R_\oplus}{R}\right)^2 \left(\frac{kT}{X_1}\right) \left(\frac{\phi_\oplus(t_p)\mathcal{E}}{b(S)m_1}\right) \left(\frac{r}{r_S}\right)^3 \Phi_{EUV}(t) \tag{6}$$

relating crossover masses of minor species S to the EUV flux irradiating a planetary atmosphere at heliocentric distance R. It can be seen here that once the history of the driving energy source for loss, i.e., the EUV flux, is known, and assumptions are made about the initial inventory and ongoing supply of the major volatile species, H_2, the losses of each species can be calculated.

Information on what $\Phi_{EUV}(t)$ might have been in the early solar system comes from astronomical observations of radiation from young solar-type stars at various stages of pre- and early-main-sequence evolution. Since early solar EUV radiation could not have penetrated a full gaseous nebula to planetary distances, the applicable time dependence of stellar activity in the present model is that following dissipation of the dense accretion disks surrounding the classical T-Tauri stars at stellar ages of up to ~10 Ma (Walter et al. 1988; Strom et al. 1988; Walter and Barry 1991; Podosek and Cassen 1994). Among present observational data, soft (~3 to 60 Å) X-ray fluxes are most likely to be representative of at least the short-wavelength coronal component ($\lambda < 700$ Å) of the EUV spectrum. Figure 10 shows X-ray luminosities L_X, relative to that of the present sun, from observations of T-Tauri stars without disks (the "Weak T-Tauri Stars": Walter et al. 1988) and older solar-type main-sequence field stars (Simon et al. 1985), plotted against estimates of stellar ages (Walter and Barry 1991). Although considerable scatter from a single functional dependence of Φ_X on age is evident, most of the data for t between ~50 and 200 Ma do indicate a decline by factors of ~5-10 from levels at ~20 Ma.

In current escape models (Pepin 1991, 1994, 1997), Φ_{EUV} is assumed to fall off exponentially, with a mean decay time τ of 90 Ma, along the dashed curve in Figure 10; the solid portion of the curve identifies the period of interest for EUV-driven loss of planetary atmospheres. An alternative, power law function for $\Phi_X(t)$ of the type suggested by Feigelson and Kriss (1989) is also plotted. Models of hydrodynamic escape analogous to those for exponential flux decay have been constructed for this kind of power law dependence (Pepin 1989). Results are similar enough to indicate that the particular mathematical form of the decay in stellar flux through the period of atmospheric evolution is not centrally important. Decline of the EUV flux over the first few hundred Ma of solar history, by roughly the factor suggested by present astronomical data, is a crucial requirement of the modeling because, from Equation (6), it results in declining crossover masses to the level where fractionations described by Equation (2) and calculated below are large enough to match Xe isotopic distributions in current atmospheres.

In the case where energy deposited in an atmosphere by EUV radiation or from some other source declines exponentially with time, the flux F_1 of constituent 1 is given by

$$F_1 = F_1^o \exp\left[\frac{-(t-t_o)}{\tau}\right] \tag{7}$$

where F_1^0 is the initial escape flux at t_0 and τ is the mean decay time of the energy source. Equation (3) then requires the crossover mass $m_C(S)$ to decrease as well, from its initial value $m_C^0(S)$ at time t_0 and flux F_1^0 to its final value $m_C^f(S)$ at time t_f when escape either terminates due to hydrogen depletion or undergoes a transition to diffusion-limited hydrogen escape through a atmosphere of increasing molecular weight; in this latter case only Ne and perhaps Ar will continue to be lost (see Zahnle et al. 1990a). If $m_C(S)$ declines through the mass m_2 of a trace species S, loss of m_2 ends when $m_C(S) = m_2$ but escape of lighter isotopes or elements continues for as long as the $m_C(S)$ for each specific constituent is greater than its particular mass.

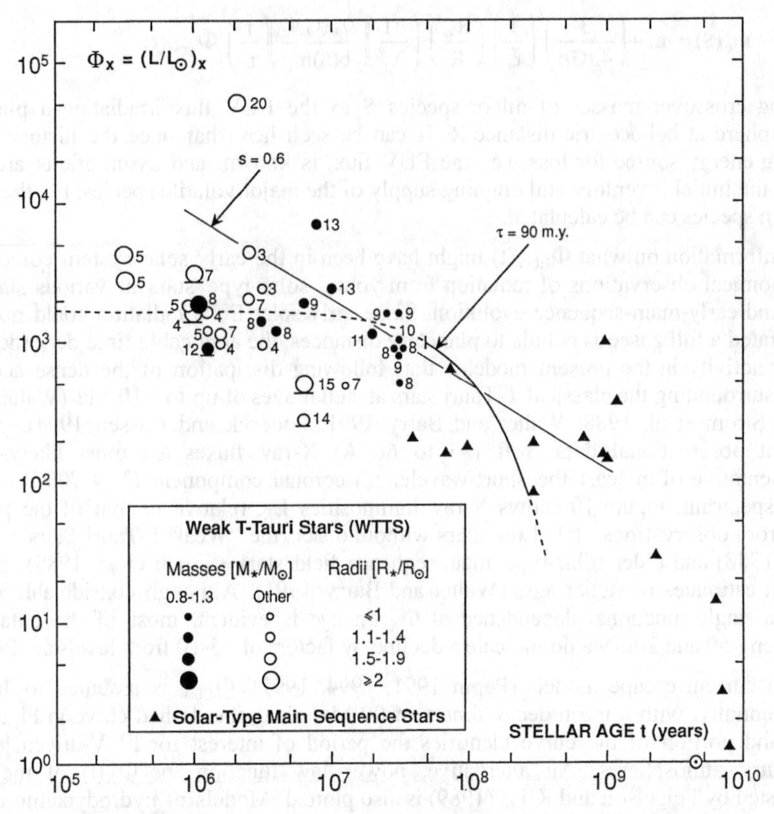

Figure 10. Observed X-ray luminosities L_X relative to the present sun vs. stellar ages for "weak-lined" T-Tauri stars (WTTS) and solar-type main sequence stars (MSS) (Pepin 1991). All WTTS data are from Walter et al. (1988), MSS data from Simon et al. (1985) and Walter and Barry (1990). Numbers adjacent to the plotted WTTS points are stellar mass estimates in tenths of a solar mass. Underlined symbols identify measurements that yielded only lower limits on luminosity. In the models explored in the text, both soft X-ray (Φ_X) and EUV (Φ_{EUV}) luminosity enhancements are taken to decline exponentially with a mean decay time τ = 90 Ma over the period of interest for planetary atmospheric evolution (indicated by the solid portion of the plotted curve). A $\Phi_X \propto (1/t)^s$ power law decay with s = 0.6 (Feigelson and Kriss 1989) is shown for comparison.

Fractionating effects of the escape process can now be calculated analytically if specific assumptions are made about the time dependence of the major (constituent 1) inventory N_1—that it is either replenished as fast as it escapes (constant inventory model), or is lost without replenishment along with the minor atmospheric species (Rayleigh fractionation model). In both cases the inventories N_2 of minor species, here the noble gases, are assumed to be in the atmosphere at t_0 and are lost without replenishment during the escape episode. For Rayleigh fractionation, adopted for this discussion, Equations (2), (3), and (7) and the definitions $F_1 = -dN_1/dt$ and $F_2 = -dN_2/dt$ may be combined and integrated, in the limit of $X_1 \equiv 1$, $m_C^0 > m_2 > m_1$, and $m_C^f \geq m_2$, to yield

$$\ln\frac{N_2^f}{N_2^o} = \ln\frac{N_1^f}{N_1^o} + \frac{\alpha_o}{1\ \alpha_o}\left[\frac{m_2 - m_1}{m_c^o - m_1}\right]\ln\left[\frac{N_1^f}{N_1^o}\frac{m_c^o - m_1}{m_2 - m_1}\right] \tag{8}$$

for the fractional depletion of a particular minor species of mass m_2 (Hunten et al. 1987; Pepin 1991). N_1^0, N_1^f and N_2^0, N_2^f are the initial and final inventories of the major and minor constituents. The dimensionless parameter α_o is defined as $\alpha_o = F_1^0 \tau/N_1^0$. The fractional retention of constituent 1 at t_f is given by

$$\frac{N_1^f}{N_1^o} = 1 - \alpha_o \left[\frac{m_c^o - m_c^f}{m_c^o - m_1} \right] \tag{9}$$

(Pepin 1991). From the definitions of m_c (Eqn. 3) and α_o, the initial inventory of constituent 1 is

$$N_1^o = \frac{bgX_1(m_c^o - m_1)\tau}{kT\alpha_o} \text{ particles cm}^{-2}. \tag{10}$$

The total duration $t_f - t_o$ of the escape episode is related to mean energy decay time and the initial and final crossover masses by

$$t_f - t_o = -\tau \ln\left[\frac{m_c^f - m_1}{m_c^o - m_1} \right]. \tag{11}$$

Note that equations containing t involve only time differences, not their absolute values (e.g., Eqns. 7 and 11). Consequently the fractionations of initial isotopic and elemental ratios generated by hydrodynamic escape do not depend on any specific choice for t_0, the time in solar evolutionary history when atmospheric escape begins. Permitted values of t_0 are constrained, however, by the solar EUV flux needed to drive an escape episode. For example, as noted in the following section, EUV-driven Xe loss from Earth requires a flux that exceeds the present solar level by a factor of ~450, and thus a t_0 no later than ~100 Ma if the flux history follows the τ = 90 Ma exponential in Figure 10. It is assumed that t_0, whatever its value, marks the time at which dust and gas in the nebular midplane had cleared to the extent that solar EUV radiation could penetrate to planetary distances, so that EUV-driven atmospheric loss would not have occurred prior to t_0 (Prinn and Fegley 1989).

Isotopic and elemental fractionations generated in constituent 2 residues are calculated from separate solutions of Equation (8) for particular choices of mass m_2 and a reference mass $[m_2]_{ref}$. Then $(N_2/N_2^0)/([N_2/N_2^0]_{ref}) = \{N_2/[N_2]_{ref}\}/\{N_2^0/[N_2^0]_{ref}\}$ is the corresponding fractionation factor. For example, if S is xenon and the inventories $N_2(S)$ and $[N_2(S)]_{ref}$ are respectively taken to represent ^{124}Xe and ^{130}Xe, the isotopic fractionation is $(^{124}Xe/^{124}Xe^0)/(^{130}Xe/^{130}Xe^0) = (^{124}Xe/^{130}Xe)/(^{124}Xe/^{130}Xe)^0$ where $(^{124}Xe/^{130}Xe)^0$ is the original (unfractionated) ratio.

Adjustable modeling parameters include the mass m_1 of the abundant light constituent (taken to be H_2 in most models); atmospheric temperature T; initial crossover mass m_C^0 for one noble gas—for example, for m_C^0 (Xe) as in Equation (4)—and its lower value m_C^f for the same species when the loss episode terminates (crossover masses for the remaining trace gases are fixed by Eqn. 4), which with m_1, T, and b(S) sets the initial value of F_1 via Equation (3); the decay constant τ for the thermal energy supply; the parameter $\alpha_0 = F_1^0\tau/N_1^0$ which for a given τ sets the initial ratio of escape flux to inventory for H_2; and the isotopic and elemental ratios of noble gases in the initial primordial atmosphere. It's important to note that allowed parameter space is not unconstrained. In Equation (9), for example, the light constituent abundance N_1 cannot be negative; N_1^f/N_1^0 must therefore be ≥ 0, which limits the permitted combinations of α_0 and m_C^f values for a particular choice of m_C^0.

With suitable choices of parameter values, hydrodynamic escape of hydrogen-rich primary atmospheres and outgassed volatiles from the terrestrial planets (plus sputtering losses on Mars), operating in an astrophysical environment for the early solar system inferred from observation of young star-forming regions in the galaxy and from models of planetary accretion, can account for most of the known details of noble gas distributions in their present-day atmospheres. The question, of course, is whether the required values or ranges of values required for the free modeling parameters reflect realistic astrophysical and planetary conditions in the early history of the solar system. The conditions required on each planet for obtaining the observed noble gas characteristics are described in the following sections.

Application to Earth. Current modeling requires the Earth to have acquired two isotopically primordial volatile reservoirs during accretion, one in the planet's interior, perhaps populated by a combination of nebular gases occluded in planetary embryo materials and dissolved in molten surface materials, and the other coaccreted as a primary atmosphere degassed from impacting planetesimals or gravitationally captured from ambient nebular gases during later planetary growth. These "isotopically primordial" reservoirs are characterized by isotope ratios measured in the solar wind, with the important exception of Xe; here the U-Xe composition, which differs substantially from solar wind Xe at the two heaviest isotopes, is required (see Fig. 7 and discussion in the "Terrestrial noble gases" section). Evolutionary processing proceeds in two stages. In the first stage, substantial depletion of Xe from the primary atmosphere occurs, driven by deposition of atmospheric energy. This sets the Xe inventory and generates the extensive Xe isotopic fractionation that is presently observed in the terrestrial atmosphere. However, the other, lighter noble gases are greatly overdepleted at the end of the first stage. Subsequently, mixing of the fractionated atmospheric residue with species degassed from the second, interior reservoir compensates for the overfractionation of Kr, Ar, and Ne elemental and isotopic abundances in the first stage, and produces the presently observed distributions of these gases. Modeling must therefore aim to define not only the hydrodynamic escape parameters that generate terrestrial Xe, but also the relative elemental abundances in both the primary atmospheric and interior reservoirs required to produce the final composition of the other noble gases.

In the original formulation of the model (Hunten et al. 1987; Pepin 1991), hydrodynamic losses of primary atmospheric volatiles are driven by intense EUV radiation from the young evolving Sun. Hydrogen outflow fluxes strong enough to enable Xe escape from Earth, and fractionation to its present isotopic composition, required atmospheric H_2 inventories equivalent to water abundances of up to a few wt.% of the planet's mass, and early solar EUV fluxes up to ~450× present levels—large but, as seen in Figure 10, not unrealistic enhancements if nebular dust and gas had dissipated to levels low enough for solar EUV radiation to penetrate the midplane to planetary distances within 100 Ma or so.

Energy sources other than solar EUV absorption may have powered atmospheric escape. Benz and Cameron (1990) suggested that hydrodynamic loss driven by thermal energy deposited in a giant Moon-forming impact could have generated the well-known fractionation signature in terrestrial Xe. Their model of the event calls for rapid invasion of the pre-existing primary atmosphere by extremely hot (~16,000 K) dissociated rock and iron vapor, emplacement of an orbiting rock-vapor disk with an inner edge at an altitude comparable to the atmospheric scale height at this temperature, and longer-term heating of the top of the atmosphere by reaccretion of dissipating disk material.

As yet there are no detailed theoretical calculations supporting the proposition that the physical conditions and energy-decay timescale needed to implement hydrodynamic escape could have been met in the aftermath of a giant impact on Earth. If a short post-impact escape episode did in fact occur, resulting in Rayleigh fractionation of whatever remnant of the primary atmosphere survived direct—and presumably non-fractionating—ejection in the impact event, required atmospheric H_2 inventories would be reduced by an order of magnitude or more compared to models in which losses are driven only by solar EUV radiation (Pepin 1997); this is in part a consequence of the scaling of N_l^0 with the energy decay constant τ in Equation (10), where, as noted below, τ for post-impact energy dissipation is assumed to be small. EUV-powered escape is still needed at some time following the impact event to account for Ne isotopic distributions (see below), but flux intensities for Ne-only loss are also about an order of magnitude lower than those required to lift heavier noble gas species out of the atmosphere. With appropriate choices for free parameters, both this and the original EUV-only model (Pepin 1991) can generate close matches to noble gas elemental and isotopic distributions in the contemporary terrestrial atmosphere.

The interplay of atmospheric escape fractionation and subsequent degassing required to achieve these matches is illustrated for the hybrid giant impact (GI)-EUV model in Figure 11. The mean decay time τ of energy deposited by the impact is presumably small, reflecting relatively rapid energy dissipation following the event: here τ is arbitrarily chosen to be 1000 years, and with this and other model parameter values, Equation (11) yields an escape episode that runs its course in a comparable time. Primordial atmospheric U-Xe is fractionated by GI-driven hydrodynamic escape to its present abundance and nonradiogenic isotope ratios. About 85% of the initial ^{130}Xe inventory is lost from the planet. The resulting "nonradiogenic Earth atmosphere" is used in Figure 11 as the Xe reference composition. Post-escape isotopic evolution of atmospheric Xe is largely restricted to degassing of radiogenic ^{129}Xe, and of $^{131\text{-}136}Xe$ generated primarily by spontaneous fission of ^{244}Pu, from the upper mantle and crust (see Fig. 13, below). This constraint on "pollution" of the nonradiogenic Xe isotopic distribution generated by hydrodynamic escape, either by subsequent degassing of solar Xe from the interior reservoir or by substantial addition of isotopically different Xe carried in later-accreting material, is central to the modeling (see discussion below in *Outstanding Issues*).

Increasingly severe fractionations of the lighter noble gases from their primordial isotopic compositions are imposed by the parameter values for GI-driven Xe escape. Residual Kr and Ar are both isotopically heavy (Fig. 11) and strongly depleted relative to the present atmosphere; only 6% and 0.8% respectively of the initial ^{84}Kr and ^{36}Ar inventories (and 0.4% of the ^{20}Ne) survive the event. Subsequent mixing of these residues of the primary atmosphere with solar-composition Kr and Ar degassed from the interior reservoir raises their abundances to present-day values, and yields generally good ($\geq 1\sigma$) matches to contemporary isotopic compositions except for a ~2‰ (3σ) excess of ^{86}Kr.

Neon is a special case in that $^{20}Ne/^{22}Ne$ and $^{21}Ne/^{22}Ne$ ratios in the fractionated residual atmosphere are substantially higher than at present, and later addition of outgassed solar Ne elevates them still more (Fig. 11). Here an episode of solar EUV energy deposition driving hydrodynamic escape of Ne at some time after GI fractionation and outgassing is needed to generate the contemporary $^{20}Ne/^{22}Ne$ ratio. But, as noted above, the EUV-driven H_2 escape flux must now be only intense enough to lift Ne, but not the heavier noble gases, out of the atmosphere. The waning solar EUV flux (Ayres 1997) may still have been sufficiently high (~60× present levels) to drive Ne-only escape at solar ages up to ~250 Ma (Fig. 10), with the actual timing —somewhere in the interval between Ne degassing and ~250 Ma—determined by the timescale for sufficient reduction of EUV dust-gas opacity in the nebular midplane (Prinn and Fegley 1989). Note from Figure 11 that EUV escape fractionation yielding a match to the present $^{20}Ne/^{22}Ne$ ratio

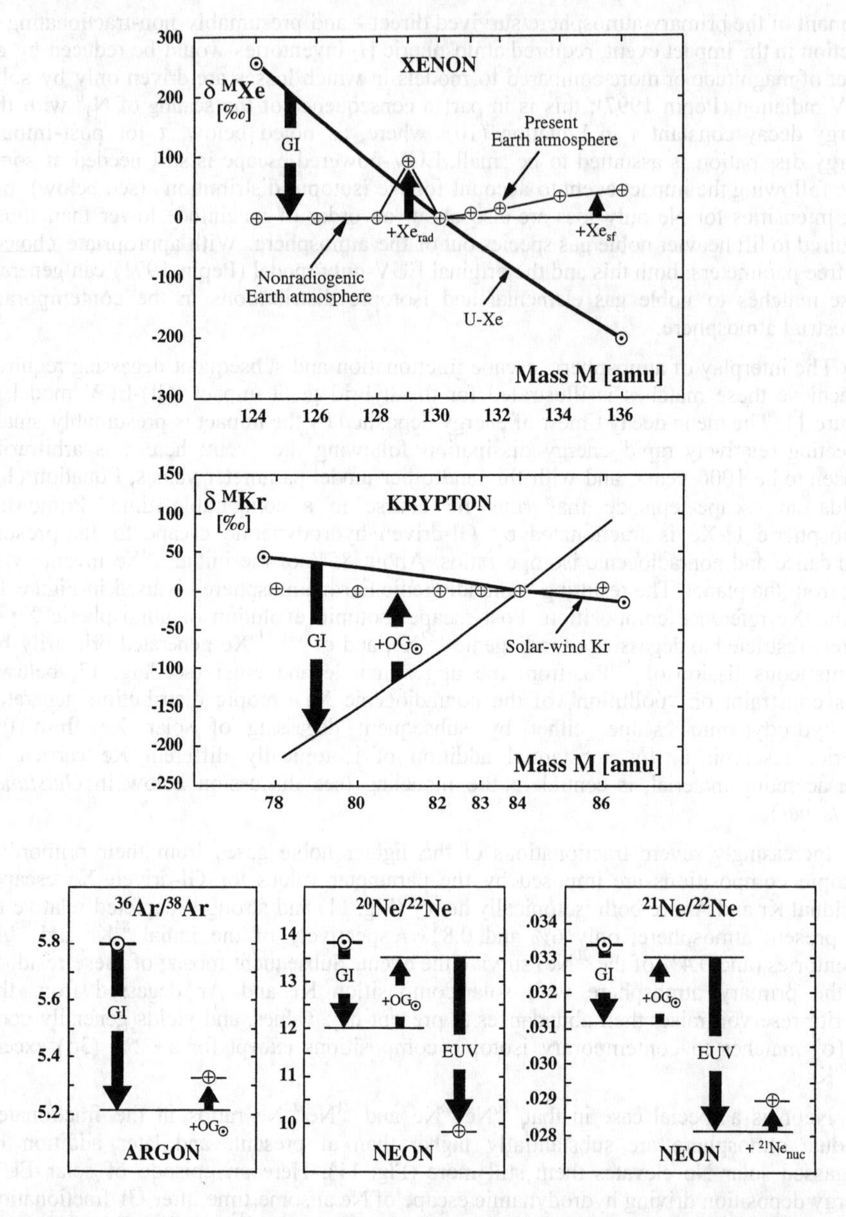

Figure 11. Evolution of terrestrial Xe, Kr, Ar and Ne from primordial atmospheric distributions to present-day compositions by Giant Impact (GI)-driven fractionation, addition of outgassed (OG) solar Kr, Ar and Ne, a later stage of solar EUV-powered Ne escape fractionation, and subsequent outgassing of radiogenic and fissiogenic Xe and nucleogenic ^{21}Ne (after Pepin 1997). The δ^MXe and δ^MKr representations are defined by $\delta^M = 1000\,[(R/R_{ref}) - 1]$ ‰; here R = ^{M}Xe/^{130}Xe and ^{M}Kr/^{84}Kr respectively, and the references R_{ref} are the corresponding isotope ratios in the nonradiogenic Earth atmosphere for Xe (Pepin 1991) and the present Earth atmosphere (Basford et al. 1973) for Kr. Xe data from Table 1; solar and atmospheric Kr from Wieler (2002, Table 5, refs. 2 and 4); solar ^{36}Ar/^{38}Ar = 5.80 ± 0.06, ^{20}Ne/^{22}Ne = 13.84 ± 0.04, and ^{21}Ne/^{22}Ne = 0.0334 ± 0.0003 from Palma et al. (2002).

overfractionates $^{21}Ne/^{22}Ne$ by a few percent, implying a ^{21}Ne deficit in the modeling. The atmospheric abundance of ^{21}Ne, however, is subsequently augmented by outgassing over geologic time of a nucleogenic $^{21}Ne_n$ component generated primarily by $^{18}O(\alpha,n)^{21}Ne_n$ reactions in the Earth's crust and mantle (Wetherill 1954). Estimates of the fractional abundance of $^{21}Ne_n$ in the present atmosphere are consistent with the amount needed to elevate the initially lower $^{21}Ne/^{22}Ne$ ratio in Figure 11 to its present value by post-escape degassing (see Porcelli and Ballentine 2002, this volume).

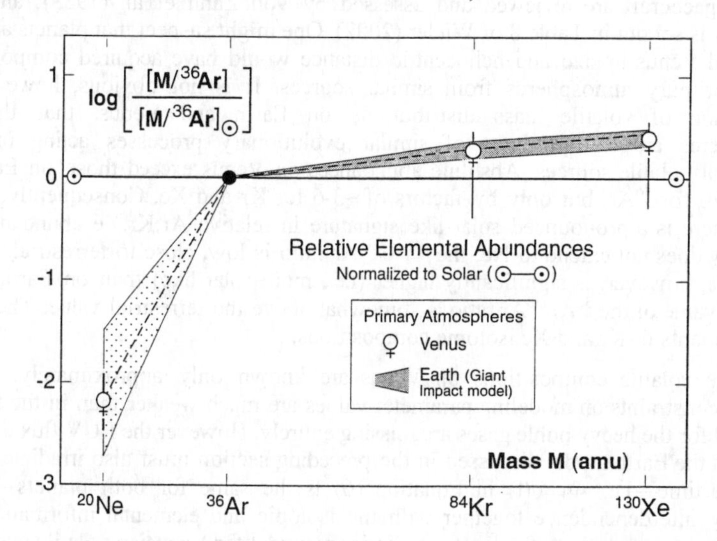

Figure 12. Modeling results for the elemental compositions of primary atmospheres on Earth and Venus, plotted relative to solar abundance ratios. The baseline model for impact-driven escape from Earth (dashed line) assumes H_2 loss from a 2000-K atmosphere. Different assumptions for atmospheric temperature and the identity of the abundant escaping constituent generate the shaded spread in Kr/Ar and Xe/Ar ratios originally present on Earth (see text). The stippled region represents the factor of ~5 range in initial terrestrial $^{20}Ne/^{36}Ar$, on either side of the Venus ratio, for which the Ne isotopic matches shown in Figure 1 can be generated by EUV-driven loss. All elemental data are from Pepin (1997).

Primordial heavy noble gas abundances, calculated from present inventories backward through the degassing and escape stages of evolution, yield the dashed-line relative abundance pattern shown in Figure 12 for the $^{84}Kr/^{36}Ar$ and $^{130}Xe/^{36}Ar$ ratios present in the pre-impact Earth atmosphere. Estimating the initial ^{20}Ne abundance on Earth is complicated by the terminal EUV-driven Ne fractionation stage. In calculating the Ne isotopic evolution shown in Figure 11 and the dashed-line elemental pattern in Figure 12, initial ^{20}Ne was assumed for illustration to be present in the same ratio relative to ^{36}Ar as in Venus' primary atmosphere (see below). Solutions yielding final Ne isotope ratios identical to those in Figure 11 exist, with differing EUV fractionation parameters and amounts of degassed $^{21}Ne_n$, for primordial atmospheric $^{20}Ne/^{36}Ar$ ranging from ~1/5 to 5 times the Venus ratio (stippled area in Fig. 12).

These modeling results assume an atmospheric temperature T = 2000 K and H_2 as the abundant hydrodynamically escaping species. Neither choice would be appropriate for escape from a very hot dissociated atmosphere, an environment likely to have characterized early stages of giant-impact-driven loss. However the results are not sensitive to either temperature or the exact identity of the carrier gas. The shaded area in

Figure 12 represents the relatively small spread in initial elemental composition introduced by assuming atmospheric temperatures ranging from 300 K to 10,000 K and, at temperatures ≥ 4000 K, H or O rather than H_2 as the dominant light atmospheric constituent.

Application to Venus. Data from in situ compositional measurements of the Venus atmosphere by mass spectrometers and gas chromatographs on the Pioneer Venus and Venera spacecraft are reviewed and assessed by von Zahn et al. (1983); an updated summary is set out in Table 8 of Wieler (2002). One might suspect that planets as alike as Earth and Venus in size and heliocentric distance would have acquired compositionally similar primary atmospheres from similar sources. It is not obvious, however, from comparison of volatile mass distributions on Earth and Venus, that these two atmospheres are end products of similar evolutionary processes acting on similar primordial volatile sources. Absolute abundances on Venus exceed those on Earth by a factor >70 for ^{36}Ar, but only by factors of ~3-6 for Kr and Xe. Consequently, as noted above, there is a pronounced solar-like signature in relative Ar:Kr:Xe abundances. This similarity does not extend to Ne: the ^{20}Ne/^{36}Ar ratio is low, close to terrestrial. Venusian ^{20}Ne/^{22}Ne, however, is significantly higher (i.e., more solar-like) than on Earth, and the nominal value of the ^{36}Ar/^{38}Ar ratio is somewhat above the terrestrial value. There are no measurements of Kr and Xe isotopic compositions.

Since volatile compositions on Venus are known only approximately, if at all, isotopic constraints on modeling parameter values are much weaker than in the terrestrial case, and for the heavy noble gases are missing entirely. However the EUV flux driving Ne escape in the Earth model discussed in the preceding section must also irradiate Venus at the same time—i.e., $\Phi_{EUV}(t)$ in Equation (6) is the same for both planets—and this planetary interdependence together with the isotopic and elemental information we do have allows construction of at least a preliminary model for Venusian volatile evolution. If EUV heating efficiencies and H_2 escape altitudes are assumed to be similar, it is seen from Equation (6) that the ratio of the crossover mass $m_C(S)$ on Venus to that on Earth for the same species is a function only of relative planetary radii, densities, heliocentric distances, and atmospheric temperatures. It turns out that the relatively weak solar EUV flux needed for Ne-only loss from Earth is still strong enough at Venus' orbital position to drive outflow of Kr and lighter gases from this smaller and less dense planet. But $m_C{}^0$(Xe) on Venus falls below the mass of the lightest Xe isotope, and so Venusian Xe is not lost and its nonradiogenic isotopic composition, presently unknown, is predicted to be unaltered from its primordial composition.

Results of this model of EUV-driven loss of an isotopically solar and elementally near-solar primordial atmosphere from Venus are sensitive to only one of the few remaining adjustable modeling parameters—α_O, which fixes the initial H_2 inventory $N_1{}^0$ via Equation (10). A single value of this parameter generates matches well within their 1σ uncertainties to the very limited isotopic data we currently have for Venus' present-day noble gases (^{20}Ne/^{22}Ne and ^{36}Ar/^{38}Ar only: see Table 8 in Wieler 2002, this volume). It is interesting that there is no discernible evidence in the data on hand that outgassing has played a significant role in establishing contemporary noble gas inventories. Fractionating loss of the primary atmosphere, governed by the parameters above, generates by itself approximate matches to observed compositions. Thus, in contrast to the case for Earth, the presence of an outgassed component on Venus is not required. This is not to say that the planet could not have degassed at some time during or after this stage of atmospheric evolution. But outgassed species would comprise only modest fractions of the large present-day Venusian atmospheric inventories even if bulk planetary concentrations were comparable to those on Earth. In fact, the atmospheric abundance of ^{40}Ar suggests that outgassing rates on Venus have been substantially less than terrestrial rates (see "Noble gases on Venus" section).

Elemental ratios characterizing the primary (pre-loss) Venus atmosphere in this evolutionary model are plotted in Figure 12. Although these computed primordial ^{84}Kr/^{36}Ar and ^{130}Xe/^{36}Ar ratios inherit the large uncertainties associated with measurements of present Venusian abundances, their nominal values fall squarely within the range of estimates calculated above for Earth's pre-impact primary atmosphere—a strong implication that noble gases on both planets could have evolved, clearly in quite different ways, from the same primordial distributions in the same types of primary planetary reservoirs. This result seems reasonably robust, provided of course that fractionating, GI-driven hydrodynamic escape actually did occur on Earth. However a central test which the Venus model must eventually confront is whether its predicted Xe and Kr isotopic compositions, respectively unfractionated and slightly fractionated with respect to primordial atmospheric compositions, are in accord with future measurements. Venus appears to represent an extreme in the range of atmospheric compositional patterns displayed by the terrestrial planets, and, as discussed below in "Outstanding Issues", this and other central questions could be answered, or at least more quantitatively addressed, if we can eventually manage to revisit the planet for moderately accurate in situ measurements of all noble gas isotopic distributions, or, even better, return an atmospheric sample to Earth for laboratory study.

Application to Mars: losses by hydrodynamic escape and sputtering. Information about volatile abundances and compositions on and in Mars come from in situ measurements by Viking spacecraft instruments and, by a great stroke of fortune, from high-precision laboratory analyses of Martian atmospheric gases trapped in the SNC meteorites (Swindle 2002, this volume). Early modeling of the history of noble gases on Mars based on these data suggested that they could have evolved from primordial to present-day distributions through two early episodes of hydrodynamic atmospheric escape (Pepin 1991). This model, however, did not address a number of processes that now appear germane to Martian atmospheric history. One, gas loss and fractionation by sputtering, was later proposed to be the dominant mechanism governing atmospheric CO_2 evolution on Mars over the past ~3-4 Ga (Luhmann et al. 1992; Zhang et al. 1993). Another, atmospheric erosion (Melosh and Vickery 1989), appears increasingly important (Chyba 1990, 1991; Zahnle 1993). In the absence at that time of a plausible mechanism, the possibility of isotopic evolution of noble gases heavier than Ne after the termination of hydrodynamic escape was not considered by Pepin (1991), and only qualitative attention was paid to the eroding effects of impact on abundances of all atmophilic species prior to the end of heavy bombardment ~3.8 to 3.7 Ga ago.

The more recent models of Martian atmospheric evolution constructed by Jakosky et al. (1994) and Pepin (1994) incorporate the sputtering loss mechanism proposed by Luhmann et al. (1992) and Zhang et al. (1993), and explore its consequences for elemental and isotopic fractionation of the noble gas and nitrogen in the residual atmosphere. They divide Martian atmospheric history into early and late evolutionary periods, the first characterized by an initial episode of hydrodynamic escape that sets the Martian Xe inventory (discussed later in the "Primordial Xenon" section), followed by high CO_2 pressures and a possible greenhouse, and the second by either a sudden transition to a low pressure environment similar to present-day conditions on the planet—perhaps initiated by abrupt polar CO_2 condensation ~3.7 Ga ago (Gierasch and Toon 1973; Haberle et al. 1992, 1994)—or by a more uniform decline to present pressure. Jakosky et al. (1994) focused on the late evolutionary epoch, and showed that contemporary Ne, Ar and N_2 abundances and isotope ratios—including the uniquely low Martian ^{36}Ar/^{38}Ar ratio

(Swindle 2002, this volume)—could have been generated by sputtering losses from an atmosphere that was continuously replenished by degassing of meteoritic (CI) N_2 and isotopically solar Ne and Ar, with the rate vs. time dependence of degassing chosen to be similar to estimates of volcanic flux vs. time over this period.

The second of these models (Pepin 1994) included some thoughts on early evolution of volatile distribution on Mars. The EUV-powered hydrodynamic escape episode driving Ne-only loss from Earth, and loss of Kr and lighter gases from Venus, would have been intense enough on Mars to lift all the noble gases out of its primordial atmosphere. Early in this pre-3.7-Ga epoch, Xe isotopes were therefore assumed to have been hydrodynamically fractionated to their present composition, with corresponding depletions and fractionations of lighter primordial atmospheric constituents. Subsequent CO_2 pressure and isotopic history was dictated by the interplay of estimated losses to impact erosion, sputtering, and carbonate precipitation, additions by outgassing and carbonate recycling, and perhaps also by feedback stabilization under greenhouse conditions. In a subsequent model of the early Martian atmosphere, Carr (1999) examined the influences of these same mechanisms in controlling CO_2 pressure history, and was led to similar results and conclusions. It should be stressed, however, that since almost nothing is actually known about the values of the parameters governing these various processes, models of this epoch are no more than qualitative illustrations of how they might have driven early atmospheric behavior.

Pepin's (1994) treatment of post-3.7-Ga evolution of Martian CO_2, N_2, and the noble gases, although differing somewhat in detail from that of Jakosky et al. (1994) and extended to include the radiogenic isotopes ^{40}Ar and ^{129}Xe, generated results close to those derived by Jakosky et al. Both models assume that early and late evolutionary stages on Mars were separated by atmospheric CO_2 pressure collapse near 3.7 Ga. Sputtering loss of an atmospheric species relative to that of CO_2 is directly proportional to its exobase mixing ratio with CO_2, and so sputtering fractionation of the atmospheric noble gas inventory is generally modest in a pre-3.7-Ga atmosphere dominated by CO_2 (Jakosky et al. 1994). Pressure collapse of the major atmospheric constituent abruptly increases the mixing ratios of pre-existing Ar, Ne and N_2 at the exobase, and since escape fluxes are proportional to exobase mixing ratios (see "Sputtering" section above), CO_2 collapse triggers their rapid removal by sputtering. This has the interesting consequence that no isotopic memory of their earlier processing survives. Current abundances and isotopic compositions are entirely determined by the action of sputtering and photochemical escape on gases supplied by outgassing during the late evolutionary epoch, and final distributions of the light noble gases and nitrogen are therefore decoupled from whatever their elemental and isotopic inventories might have been in the pre-3.7 Ga atmosphere. The present atmospheric Kr inventory (Fig. 6) also derives almost completely from solar-like Kr degassed during this period, which overwhelms the fractionated component inherited across the collapse episode —as noted above in the "Sputtering" section, both Kr and Xe are too massive to be sensibly affected by sputtering loss and fractionation during the late evolutionary stage. Consequently, among current observables, only the Xe isotopes and δ^{13}C survive as isotopic tracers of atmospheric history prior to its transition to low pressure. The assumption that early hydrodynamic escape fractionated the nonradiogenic Xe isotopes to at least approximately their present composition severely limits subsequent additions of unfractionated Xe to the atmospheric inventory by outgassing (consistent with the low degree of planetary degassing deduced above from radiogenic noble gas isotopes) or late-stage veneer accretion. This constraint, which threads through all hydrodynamic loss models for the terrestrial planets, is perhaps their most vulnerable characteristic (see "Outstanding Issues" below).

Jakosky and Jones (1997) reviewed progress up to that time in understanding some of the aspects of Martian atmospheric evolution. There have been several developments since the work of Jakosky et al. (1994) and Pepin (1994). Hutchins and Jakosky (1996) revisited the late evolution sputtering-degassing models to investigate in more detail the parameters controlling the evolution of Ne and Ar abundances and isotopes (including radiogenic ^{40}Ar), in particular those relating to Martian degassing history. Their interesting conclusion was that the outgassing flux of Ar and Ne attributable to degassing during epochs of volcanic activity would have been ~1-3 orders of magnitude too low to appropriately balance sputtering losses, and thus another major source of juvenile volatiles must have contributed to the atmosphere over geologic time, perhaps via input from gas-enriched hydrothermal systems. Hutchins et al. (1997) explored the important question of the extent to which a Martian paleomagnetic dipole field would have throttled sputtering losses by deflecting the solar wind around the upper atmosphere, and calculated the conditions, as functions of the time when paleomagnetic suppression of the sputtering mechanism ended, under which the combination of sputtering and degassing would still have generated present-day Ar and Ne distributions. The possible influence of an early global magnetic field on atmospheric evolution became more than a purely theoretical consideration with the discovery by Mars Global Surveyor of large-scale remnant magnetic lineations in the old Martian southern highlands (Connerney et al. 1999). The question of the existence of an active dynamo in Mars appears to be no longer if, but when.

Primordial xenon

Attempts to derive the compositions of the contemporary atmospheres on Earth and Mars from primitive nebular, planetesimal, or planetary source reservoirs have had to confront, in one way or another, an apparent decoupling of the evolutionary histories of Xe from those of Kr and the lighter noble gases. On Mars, for example, Xe is regarded as a highly fractionated residue of a likely CI- or solar-like progenitor composition; Kr, in contrast, appears to be unfractionated from an isotopically solar composition (Swindle 2002, this volume). This implies spatially and/or temporally separate provenances for these two species, processing mechanisms operating at different times, or both. In the models discussed in this section, atmospheric Xe inventories on both planets are considered to be fractionated relicts of their primary atmospheric Xe, while most of the Kr and lighter noble gases are products of planetary outgassing, compositionally modified on Mars by sputtering losses during the late evolutionary epoch.

Xenon therefore plays a crucial role in models of atmospheric evolution in which noble gases are fractionated from their initial compositions to isotopically heavier distributions by early hydrodynamic escape—it is the only observable among the noble gases that preserves a signature, albeit fractionated, of primordial atmospheric composition. With the assumption that nonradiogenic Xe isotope ratios in present-day atmospheres on Earth and Mars were generated in this way, backward modeling from these ratios through the fractionating process can in principle identify likely parental Xe compositions and thus the probable sources of noble gases in pre-escape atmospheres.

Current results of this exercise, using the Xe component compositions listed in Table 1, are reviewed in Pepin (2000). Applied to Earth, a modeling procedure involving derivation of fractionation relationships between nonradiogenic terrestrial and meteoritic Xe compositions simultaneously identifies a composition named U-Xe as primordial Xe, and establishes the presence of an atmospheric Xe component due principally to fission of extinct ^{244}Pu, as noted earlier in this chapter. Hydrodynamic escape of U-Xe leaves its fractionated residue on Earth. Evolution of atmospheric Xe from this early composition to its present isotopic state by subsequent degassing of fission and radiogenic components from the crust and mantle is shown schematically in Figure 13. To fill in the deficits at isotopes ^{129}Xe and $^{131-136}Xe$, Pu-Xe must comprise 4.65±0.30% of atmospheric ^{136}Xe, and 6.8±0.5% of the present abundance of ^{129}Xe is from decay of extinct ^{129}I. (A more accurate re-calculation of U-Xe composition by Pepin (2000) is responsible for the increase in the Pu-Xe contribution from Pepin's (1997) estimate of 3.9%).

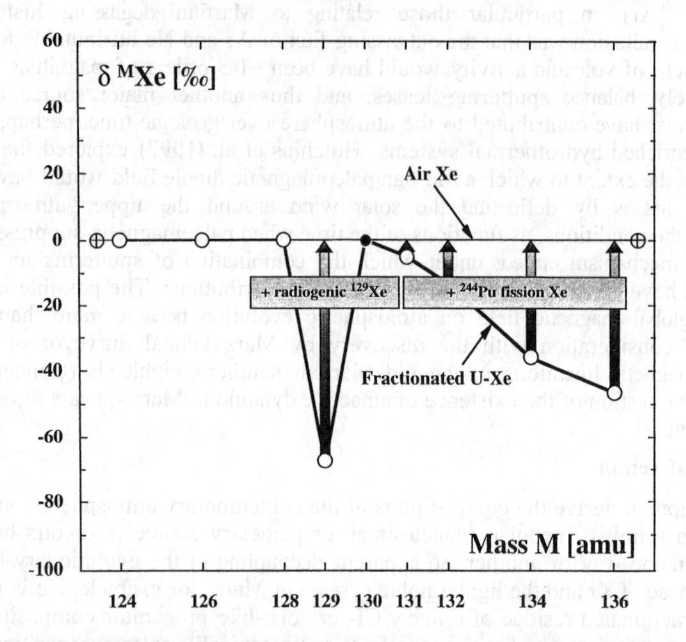

Figure 13. Post-escape evolution of Xe in the Earth's atmosphere (Pepin 2000). Escape-fractionated U-Xe defines an initially nonradiogenic terrestrial Xe composition (NEA-Xe, Table 1) to which radiogenic and fissiogenic components generated by decay of ^{129}I and ^{244}Pu in the crust and mantle were subsequently added by planetary outgassing.

The derived U-Xe composition is identical to that measured for solar-wind (SW) Xe except for relative underabundances of the two heaviest isotopes—an unexpected difference since the modeling otherwise points to solar wind compositions for the lighter noble gases in the primordial terrestrial atmosphere. However, as pointed out earlier (see Fig. 7 and associated discussion), SW-Xe cannot serve as primordial terrestrial Xe because its required fractionation generates a large overabundance of ^{136}Xe in the present atmosphere. Fractionated U-Xe ratios fall below the current atmosphere at $^{131-136}Xe$, defining a nonradiogenic terrestrial Xe spectrum to which a fissiogenic component is later added (Fig. 13); this property of the U-Xe composition is intrinsic to its derivation.

In contrast to Earth, Martian Xe apparently did not evolve from a U-Xe progenitor. Modeling derivation of primordial Xe composition on Mars is based on analyses of atmospheric gases trapped in glassy phases of SNC meteorites (Swindle 2002, this volume). Present ambiguities in this data base are such that two different solar-system Xe compositions, carbonaceous chondrite (CI)-Xe and SW-Xe, are possible candidates—but not U-Xe. Exclusion of U-Xe as the dominant primordial atmospheric inventory on Mars, despite the implication of the terrestrial modeling that it was a major component of the nebular gas phase, requires that accretion of CI- or SW-Xe-rich materials from sources more localized in space or time has overwhelmed the isotopic signature of its presence.

Swindle et al. (1986) pointed out that nonradiogenic Xe trapped in the glassy lithology of the SNC meteorite EET79001 strongly resembles mass-fractionated CI-Xe. Pepin (1991, 1994) chose CI-Xe to represent the primordial Xe composition on Mars in modeling the isotopic evolution of its atmosphere, in part based on Swindle et al.'s observation but also because, as noted earlier, a meteoritic source is consistent with models of the bulk chemical composition of the planet. The fact that fractionated CI-Xe *by itself* provides an excellent match to the Mars atmospheric composition, in particular at the four heaviest isotopes, has the interesting and somewhat unsettling consequence that additional contributions to the Xe inventory from degassed fission Xe are either very minor or absent altogether. And yet the present atmosphere is clearly heavily enriched in ^{129}Xe from extinct ^{129}I decay (Swindle 2002). The presence of radiogenic ^{129}Xe but apparently little if any ^{244}Pu-fission Xe, a situation quite unlike that on Earth, has been awkward to reconcile with models of Martian geochemical evolution and degassing history. Swindle and Jones (1997) considered this problem in detail, and constructed an alternative model of atmospheric evolution on Mars—using an atmospheric Xe composition close to that measured by Swindle et al. (1986)—with the specific objective of accommodating, if possible, a Pu fission Xe component. They chose SW-Xe rather than CI-Xe as primordial Xe, and demonstrated that an appropriate fractionation of this composition fell below measured Martian atmospheric $^{131-136}$Xe/^{130}Xe ratios by amounts completely consistent within error with the presence of an additional component with Pu-Xe fission yield ratios; moreover the calculated fissiogenic ^{136}Xe abundance was ~5% of total ^{136}Xe, similar to the fraction derived above for Earth.

Mathew et al. (1998) revised the earlier data base for Martian atmospheric Xe by including measurements of Xe composition in the Zagami shergotite and recalculating corrections for spallation Xe produced during the space exposures of SNC meteorites to cosmic-ray irradiation. The average of the data sets they judged to be of highest quality, plotted in Figure 14 referenced to SW-Xe, differs from Swindle et al.'s (1986) composition by <10‰ except for a ~60‰ decrease in δ^{124}Xe; their data are seen in the figure to be almost perfectly fit by hydrodynamic escape fractionation of SW-Xe, and thus could be consistent with no ^{244}Pu fission Xe contribution at all. However the second curve in Figure 14, which represents a somewhat less fractionated SW-Xe composition along the lines suggested by Swindle and Jones (1997), yields heavy-isotope residuals in excellent accord with addition of a Pu-Xe component at close to the 5% level at ^{136}Xe, although agreement with Mathew et al.'s (1998) light isotope ratios is somewhat degraded. Other than the criterion of quality of fit to measured isotopic distributions, there is nothing in the modeling of Martian Xe evolution by hydrodynamic escape that points to one or the other of these two fractionations as preferable.

It is important to note that if we return to the earlier assumption that CI-Xe represents the Martian primordial composition and renormalize Mathew et al.'s (1998) atmospheric ratios to CI-Xe, fractionated CI-Xe is found to be a significantly poorer fit to the data, particularly at the three lightest isotopes. The fractionation curve that best matches these light isotope ratios passes well above the heavy isotope data field, thus excluding CI-Xe as a candidate for primordial Martian Xe if the Mathew et al. composition really represents Mars. On the other hand, the original Swindle et al. (1986) data are compatible within error with fractionated CI-Xe at all isotopes. Until we know the actual Xe composition on Mars more precisely, either possibility is arguably viable. The key discriminators in ruling between these two primordial Xe candidates —and between the two curves in Figure 14, one of which allows a Pu-Xe component and the other not— are clearly the light isotope ratios, unfortunately both the most difficult to measure and the most subject to spallogenic perturbations.

The composition of primordial Xe on Mars and the presence or absence of ^{244}Pu fission Xe in the present atmosphere are important issues in the context of the provenance of accretional materials, the timing of planetary growth, and the subsequent geochemical and outgassing histories of the planet (e.g., see the earlier discussion of the Martian fissiogenic Xe budget). At the moment, the central question of the presence or absence of Pu-Xe is plagued by apparent coincidences. The excellent match of fractionated SW-Xe alone to the atmospheric data in Figure 14 is presumably fortuitous if Pu-Xe is present; and if it is absent, the fact that a weaker fractionation generates heavy-isotope residuals in good accord with ^{244}Pu fission yields must likewise be accidental.

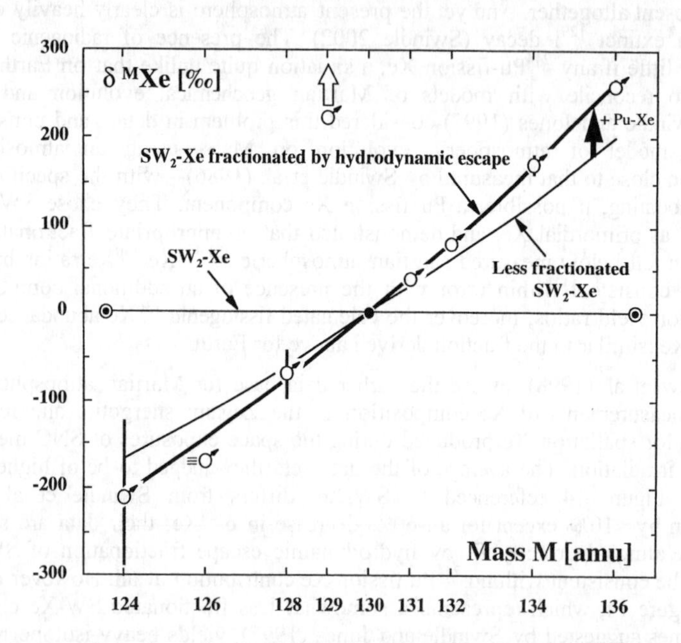

Figure 14. Mathew et al.'s (1998) Martian atmospheric composition plotted relative to SW_2-Xe (data from Table 1). The heavy curve demonstrates that SW_2-Xe in the primordial Martian atmosphere can be fractionated by escape to a close fit to all atmospheric isotope ratios except δ^{129}Xe (off-scale at +1480‰). The light curve represents a less severely fractionated SW_2-Xe composition which allows an additional component, isotopically consistent with ^{244}Pu-Xe at M = 131-136, in the present atmosphere (as suggested by Swindle and Jones 1997), but at the expense of a poorer fit to the nominal values of Mathew et al.'s (1998) light isotope ratios.

There remains the question of how and when Mars might have acquired solar wind-like Xe if that was indeed the primordial planetary composition. One possibility is that its accretional feedstock was ancient enough to have trapped ambient nebular gases during the early sun-building stage of disk evolution—thus sampling the same Xe that now comprises the solar inventory—and somehow eluded being swept into the sun. In this view Mars is an "older" planet than Earth. Alternatively, perhaps it is "younger" in the sense that a significant fraction of its pre-accretional materials were exposed, after formation of the Earth, to direct solar-wind irradiation in a waning dusty nebula which was thin enough to permit wind penetration to planetary distances but still opaque enough through the midplane to attenuate the solar EUV radiation that later powered hydrodynamic escape of Mars' primordial atmosphere.

OUTSTANDING ISSUES

Some progress has been made in the long-standing problem of understanding the sources of volatiles on and in the terrestrial planets, and the processes that could plausibly have driven their initial atmospheres down planet-specific evolutionary tracks to the amazingly divergent compositional states we observe today on Earth, Mars and Venus. Some of the ideas and models described in this chapter, although perhaps conceptually attractive, will turn out to be irrelevant in the sense that nature chose different paths. Others could be headed in the right direction but lack the data needed for meaningful experimental and observational tests of their predictive power. In this respect it is likely that much of our sharpening of volatile origin and evolution issues will emerge, at least in the near term, from studies of the terrestrial atmosphere-interior system where the data base is relatively large and growing, especially for volatile distributions in the Earth's mantle. Further progress toward the ultimate objective of understanding all three of the terrestrial planets as a linked class of objects clearly hinges on how much more we can learn about volatile abundances and compositions of noble gases and other atmophilic species on Venus and Mars. Venus, at the moment, is characterized only by the immensely valuable but still incomplete and relatively imprecise reconnaissance data from the Pioneer Venus and Venera spacecraft missions of the late 1970s. The combination of Viking in-situ measurements and the happy advent of the SNC meteorites has given us a much more quantitative view of the present state and possible history of Martian volatiles. Current modeling, however, is pushing beyond the edge of what is actually known about Mars, and now requires data constraints at levels of precision probably attainable only by laboratory analyses—in particular, in the context of this chapter, of a returned atmospheric sample. For Venus, on the other hand, the uncertainties noted in the modeling discussion—including the central issue of whether Venusian Xe does or does not resemble U-Xe—could be addressed, at least to first order, by in-situ measurements at precisions within the capabilities of current spacecraft instrumentation.

In this concluding section we summarize our views of the status of some of the current efforts to decipher volatile evolutionary history. This is by no means a comprehensive overview of the field. It concentrates instead on areas that appear to the authors to have the greatest potential for illuminating the general problem.

Acquisition of noble gases by planetary interiors. The evidence for primordial noble gases trapped deep within the Earth with solar (implanted or captured) Ne isotope characteristics, is unambiguous. Supplying these volatiles into the growing Earth will likely be related to the processes generating the initial atmosphere, although further surface additions may be possible. The capture of a primordial atmosphere by gravitational attraction of nebular gas is inescapable if the growing planet reaches sufficient mass in the presence of the nebula, and so the first criteria for this mechanism is firmly establishing if such a nebular history occurred. How much gas was then trapped in the Earth is a more complex issue requiring a substantial modeling effort. Parameters that need further consideration include the structure of the atmosphere, how long will the underlying mantle be molten and to what depth, and how this process is affected by continuing accretion. Whether or not this supplies deep mantle noble gases, a shallower reservoir may be more readily created that has supplied noble gases to the atmosphere and may no longer be represented in the mantle. Under conditions in which melting of the underlying planet is not achieved, adsorption of noble gases, enhanced by the increased pressure of the gravitationally focused nebular gases, may be an important source of noble gases that may become buried during accretion. Overall, it appears that there are various ways in which gravitational attraction of nebula gases can lead to acquisition of both atmospheric and interior noble gases, and the specific mechanisms that dominated depend upon the extent of aggregation by the time of nebula dispersal, and the temperature/pressure histories of the protoplanetesimals.

Burial of material bearing solar noble gases implanted by radiation also remains an option for the source of mantle noble gases. This requires the opposite conditions of gravitational capture; clearance of the solar nebula prior to substantial aggregation of solids to allow penetration of solar wind to where the terrestrial planets accumulate. However large protoplanetary objects, carrying irradiated regolith materials that comprise small fractions of their total masses, will be inefficient in supplying solar wind gases to planets, and may have already acquired noble gas budgets dominated by gravitational capture mechanisms. Here again, the history of the solar nebula is the deciding factor. One important constraint may come from resolving whether the mantle contains Ne with the isotopic composition of the solar nebula, rather than of implanted Ne-B.

Atmospheric origin and evolution. Hydrodynamic escape models, with appropriate numerical choices for free parameters and of primordial atmospheric and planetary interior compositions, are capable of replicating details of contemporary isotopic distributions as subtle as the need to degas nucleogenic ^{21}Ne from the terrestrial crust in an amount close to that predicted by independent calculations. One should recognize, however, that the model is highly parameterized and intrinsically multi-stage, requiring both escape fractionation and subsequent mixing with species degassed from planetary interiors in specific elemental proportions to generate contemporary elemental and isotopic abundances (e.g., Fig. 11). As promising as the modeling results may appear, they nevertheless should not obscure the fact that this evolutionary sequencing, and the parameter values controlling it, are subject to important uncertainties of one kind or another. For example, it seems clear that some degree of hydrodynamic loss and fractionation of planetary atmospheres would have been inevitable if the required conditions for energy source, hydrogen supply, and, in the case of solar EUV-driven escape, midplane transparency were even partially met. But the question of what species would actually be lost is another matter. The simple analytic theory outlined in this chapter assumes a hydrogen-dominated atmosphere and energy-limited escape, both arguably reasonable suppositions of primordial atmospheric conditions. However such atmospheres might also have contained substantial amounts of a heavy constituent, say CO_2, and in this case the escape flux of H_2 would have been limited by its ability to diffuse through the CO_2. Zahnle et al. (1990a) and Ozima and Zahnle (1993) have shown that only Ne and some Ar would be hydrodynamically lost under these conditions, with obvious implications for the possibility of Xe fractionation by this mechanism. A more complete short-term atmospheric blowoff, including heavier species, could have been driven by a very large deposition of collisional energy, but as noted above it has not been demonstrated that post-giant impact conditions would have led naturally to generation of large isotopic fractionations by hydrodynamic escape. Nor can one convincingly defend the implicit assumption that Venus' atmosphere was not also altered to some unknown degree by large-scale impact. Note also that the extents of Xe fractionation from primordial to present composition are similar on both Earth and Mars despite the much smaller mass of Mars, the apparent differences —U-Xe versus solar wind Xe— in their sources of precusor Xe, the much greater overall depletion of Martian noble gases, and the possibility that escape episodes were powered by distinctly different energy sources —EUV radiation on Mars versus giant impact on Earth. Is this just coincidence, or the expression of some more fundamental fractionating process that left similar signatures on all three of the terrestrial planet atmospheres?

Atmospheric Xe inventories in hydrodynamic escape modeling are considered to be fractionated relicts of their primary atmospheric Xe, while most of the Kr and lighter noble gases are products of planetary outgassing. This effective decoupling of the heaviest noble gas from lighter species follows naturally from the processes intrinsic to the model, in all respects but one. Isotope mixing systematics impose strict upper limits on the allowed levels of "contamination" of residual primary Xe by later addition of isotopically unfractionated Xe degassed from the interior or supplied by subsequent accretion of noble gas carrier materials, a constraint that applies with equal force to the model in which pre-fractionated Xe is delivered to planets by porous planetesimals. Estimates of these limits for Earth and Mars (Pepin 1991, 1994) fall well below the amounts of Xe that ordinarily would be expected to accompany the outgassed Kr components, and so one of two possible assumptions must be made about the Xe initially present in their interior reservoirs. Either it was largely sequestered within the planets by some mechanism and thus was never substantially transported to the atmosphere, or, following the fractional degassing models of Zhang and Zindler (1989) and Tolstikhin and O'Nions (1994), Xe was preferentially outgassed well before the bulk of the lighter noble gases, and most of it was already present in the primary atmospheres prior to Xe-fractionating hydrodynamic loss. Pepin (1991) originally proposed Xe retention deep in the planet by preferential partitioning into solid iron phases under high pressure. However Caldwell et al. (1997) have since shown experimentally that Xe does not alloy with iron even at pressures as high as that at the terrestrial core-mantle boundary. Jephcoat (1998) suggested an alternative sequestering process in which Xe atoms condense into initially nanometer-size solids in the deep mantle and segregate toward the core, although here the question arises as to whether separated phases of an element of such low abundance would be likely to form.

Whatever the mechanism, models that appeal to early fractionation of Xe from primordial to present-day planetary compositions, by hydrodynamic escape or in porous pre-planetary planetesimals, are viable only to the extent that the ancient fractionated Xe signatures can be preserved against major compositional perturbation by later additions of degassed or accreted noble gas components carrying isotopically different Xe. That this kind of isotopic pollution has not occurred is one of the more crucial assumptions of the modeling. In this context, it is clear that a version of the classical "missing Xe" problem is still with us—no longer driven, as it was originally, by the strikingly lower Xe/Kr elemental ratio in the terrestrial atmosphere compared to the chondritic meteorites, but now, in escape modeling, centered on the Xe isotopes and the postulated deficit in Xe degassed from planetary interiors.

The importance of Venus. High noble gas abundances and, within their substantial uncertainties, solar-like elemental ratios (except for Ne/Ar) suggest that at least the heavier noble gases on Venus are not greatly evolved from their primordial states (Cameron 1983; Pepin 1991). Neon and Ar isotope ratios also appear to be displaced toward solar values compared to their terrestrial counterparts. Venus therefore occupies a unique position among the triad of terrestrial planets in that its atmosphere may have been altered from its initial compositional configuration by planet-specific fractionating loss mechanisms to a much smaller extent than the highly processed atmospheres on Earth and Mars.

If this is the case, atmospheric compositions on Venus are enormously important in the context of models for the origin and evolution of terrestrial planet volatiles, particular in the case of Xe. The general similarity of nonradiogenic Xe isotope ratios on Earth and Mars is the strongest argument in favor of the fractionation of Xe on common pre-planetary carriers rather than on the planets themselves, although the correspondence does not appear to be exact. The ability of Venusian Xe to rule between the predictions of this and the hydrodynamic escape model was pointed out previously. Moreover, if Xe on Venus turns out to be isotopically lighter than terrestrial Xe, the extent to which it differs from solar composition could discriminate between hydrodynamic escape driven on all three planets entirely by intense solar EUV radiation, which would result in loss and fractionation of Venusian Xe (Pepin 1991), and a hybrid giant impact—EUV energy supply on Earth, which would not.

An equally significant question is whether nonradiogenic Xe on Venus, if solar-like, is compositionally closer to SW-Xe or to U-Xe, or to modestly fractionated derivatives of one or the other. One might expect that Earth and its sister planet would both have acquired their primordial gases from similar source(s) at about the same time, and that the U-Xe required for Earth in the modeling would also have been present in Venus' early atmosphere. If it was, the mystery of Mars' apparently different primordial Xe composition—SW-Xe or perhaps CI-Xe—is deepened; but if Venusian Xe more closely resembles SW-Xe, then Earth is the modeling anomaly.

The issue of what the composition of nebular gases actually was at the time(s) of planetary formation is clearly a central one for models which propose direct (gravitational condensation) or indirect (cometary carriers) nebular sources of primordial planetary volatiles. The solar wind has been adopted as a proxy of early nebular composition throughout the modeling discussed above, with the obvious exception of U-Xe. Xenon in Jupiter's atmosphere is of great importance in this context since it is one step closer to representing the ancient nebula than the solar wind, where the possibility of isotopic fractionation in processes transporting and releasing bulk solar Xe to and from the corona cannot be completely disregarded. Results from the Galileo Probe mass spectrometer (Table 1; Mahaffy et al. 2000) have narrowed the range of possible nebular Xe compositions at the time of Jovian atmospheric formation. Data scatter is substantial, as

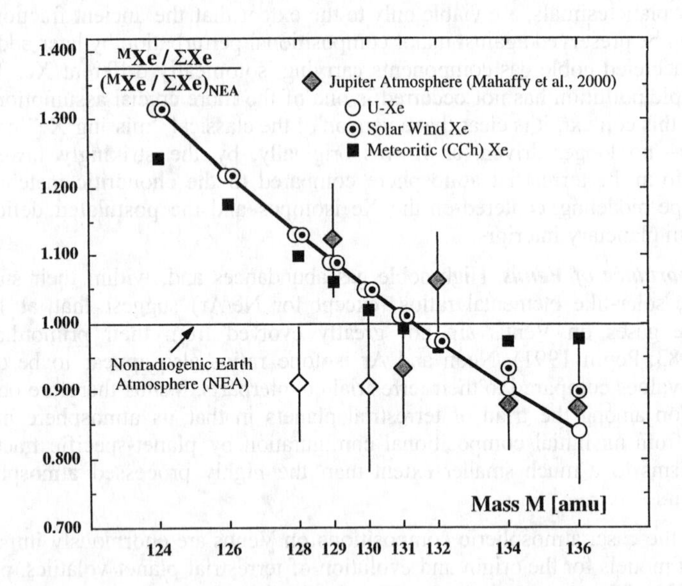

Figure 15. Jupiter atmospheric Xe composition measured by the Galileo Probe mass spectrometer (Mahaffy et al. 2000), calculated as the abundance of each isotope M divided by the total abundance for all M and plotted relative to the NEA-Xe composition (Table 1) represented in the same way. The five most abundant Jovian Xe isotopes are indicated by the shaded symbols. U-Xe, SW_2-Xe, and AVCC-Xe compositions in Table 1 are shown for comparison in the same representation. The data indicate deficits at the two heaviest isotopes relative to NEA-Xe and AVCC-Xe, but uncertainties are too large to rule between SW-Xe and U-Xe as the Jovian composition.

seen in Figure 15, but relative abundances of the most abundant isotopes (^{129}Xe and $^{131-136}$Xe) fall approximately along a fractionation curve with respect to nonradiogenic terrestrial Xe. The best-fit fractionation is most consistent with U-Xe at $^{134-136}$Xe although SW-Xe is allowed within 1σ uncertainties. The major-isotope pattern is incompatible with terrestrial Xe, suggesting that cometary ices acquiring their noble gases from the early nebula did not deliver Earth's present-day Xe composition to the planet.

Coupled histories of atmospheric and interior planetary volatiles. Highly detailed models of noble gas sources and evolution have been developed separately and almost independently for the atmosphere and interior of the Earth (e.g., Pepin 1991, 1997, 2000; Porcelli and Wasserburg 1995a,b,c; see Porcelli and Ballentine 2002). But the origin and history of atmospheric noble gases are not independent of the sources, distributions, and transport histories of noble gases within a planet —these two volatile systems must clearly be linked in nature through their primordial inventories and the processes of degassing and subduction. Indeed, many models of the atmosphere require degassing of some portion of the planetary interior, and mantle models include degassing of the atmosphere. The question, so far largely unexplored, is whether the requirements implicit in each of these models for the dynamical and compositional history of the other reservoir are compatible, and, if not, what the inconsistencies are and how might they be addressed. Integration of these historically separate approaches into a single consistent view of coupled volatile evolution in planetary atmospheres and interiors would be a major advance in the modeling research described in this chapter.

ACKNOWLEDGMENTS

Preparation of this chapter was supported at the University of Minnesota by Grant NAG5-7094 from the NASA Cosmochemistry Program. Reviews by R. Wieler and C. Ballentine are appreciated.

REFERENCES

Abe Y, Matsui T (1986) Early evolution of the Earth: accretion, atmosphere formation and thermal history. Proc 17th Lunar Planet Sci Conf, J Geophys Res 91:E291-E302

Ahrens TJ (1990) Earth accretion. *In* Origin of the Earth. Newsom HE, Jones JH (eds), Oxford University Press, New York, p 211-227

Ahrens TJ (1993) Impact erosion of terrestrial planetary atmospheres. Ann Rev Earth Planet Sci 21: 525-555

Ahrens TJ, O'Keefe JD, Lange MA (1989) Formation of atmospheres during accretion of the terrestrial planets. *In* Origin and Evolution of Planetary and Satellite Atmospheres. Atreya SK, Pollack JB, Matthews MS (eds) University of Arizona Press, Tucson, Arizona, p 328-385

Albarède (1998) Time-dependent models of U-Th-He and K-Ar evolution and the layering of mantle convection. Chem Geol 145:413-429

Allègre CJ, Staudacher T, Sarda P, Kurz M (1983) Constraints on evolution of Earth's mantle from rare gas systematics. Nature 303:762-766

Allègre CJ, Staudacher T, Sarda P (1986) Rare gas systematics: formation of the atmosphere, evolution and structure of the Earth's mantle. Earth Planet Sci Lett 81: 127-150

Allègre CJ, Hofmann A, O'Nions RK (1996) The argon constraints on mantle structure. Geophys Res Lett 23:3555-3557

Anders E, Grevesse N (1989) solar-system abundances of the elements. Geochim Cosmochim Acta 53: 197-214

Anderson DL (1998a) The helium paradoxes. Proc Natl Acad Sci USA 95:4822-4827

Ayres TR (1997) Evolution of the solar ionizing flux. J Geophys Res 102:1641-1651

Ballentine CJ, Burnard PG (2002) Production and release of noble gases in the continental crust. Rev Mineral Geochem 47:481-538

Ballentine CJ, Porcelli D, Wieler R (2001) Technical comment on 'Noble gases in mantle plumes' by Trieloff et al. (2000). Science 291:2269a

Bar-Nun A, Herman G, Laufer D, Rappaport ML (1985) Trapping and release of gases by water ice and implications for icy bodies. Icarus 63:317-332

Basford JR, Dragon JC, Pepin RO, Coscio MR Jr, Murthy VR (1973) Krypton and xenon in lunar fines. Proc Lunar Sci Conf 4th, p 1915-1955
Begemann F, Weber HW, Hintenberger H (1976) On the primordial abundance of argon-40. Astrophys J 203:L155-L157
Benz W, Cameron AGW (1990) Terrestrial effects of the giant impact. *In* Origin of the Earth. Newsom HE, Jones JH (eds) Oxford University Press, New York, p 61-67
Bernatowicz TJ, Kennedy BM, Podosek FA (1985) Xe in glacial ice and the atmospheric inventory of noble gases. Geochim Cosmochim Acta 49:2561-2564
Bernatowicz TJ, Podosek FA (1986) Adsorption and isotopic fractionation of Xe. Geochim Cosmochim Acta 50:1503-1507
Bernatowicz TJ, Podosek FA, Honda M, Kramer FE (1984) The atmospheric inventory of xenon and noble gases in shales: The plastic bag experiment. J Geophys Res 89:4597-4611
Bogard DD (1997) A reappraisal of the Martian Ar-36/Ar-38 ratio. J Geophys Res 102:1653-1661
Bogard DD, Clayton RN, Marti K, Owen T, Turner G (2001) Martian volatiles: isotopic composition, origin, and evolution. Space Sci Rev 96:425-458
Brown H (1949) Rare gases and formation of the earth's atmosphere. *In* The Atmospheres of the Earth and Planets. Kuiper GP (ed), University of Chicago Press, Chicago, p 258-266
Busemann H, Baur H, Wieler R (2000) Primordial noble gases in "phase Q" in carbonaceous and ordinary chondrites studied by closed-system stepped etching. Meteorit Planet Sci 35:949-973
Caffee MW, Hudson GU, Velsko C, Huss GR, Alexander Jr EC, Chivas AR (1999) Primordial noble cases from Earth's mantle: identification of a primitive volatile component. Science 285:2115-2118
Caldwell WA, Nguyen JH, Pfrommer BG, Mauri F, Louie SG, Jeanloz R (1997) Structure, bonding, and geochemistry of xenon at high pressures. Science 227:930-933
Cameron AGW (1983) Origin of the atmospheres of the terrestrial planets. Icarus 56:195-201
Cameron AGW, Ward WR (1976) The origin of the Moon. Lunar Science VII, Lunar and Planetary Institute, Houston, p 120-122
Canalas R, Alexander EC Jr, Manuel OK (1968) Terrestrial abundance of noble gases. J Geophys Res 73:3331-3334
Carr MH (1999) Retention of an atmosphere on early Mars. J Geophys Res 104:21897-21909
Carroll MR, Stolper EM (1993) Noble gas solubilities in silicate melts and glasses: new experimental results for argon and the relationship between solubility and ionic porosity. Geochim Cosmochim Acta 57:5039-5051
Chen GQ, Ahrens TJ (1997) Erosion of terrestrial planet atmosphere by surface motion after a large impact. Phys Earth Planet Int 100:21-26
Chyba CF (1990) Impact delivery and erosion of planetary oceans in the early inner solar system. Nature 343:129-133
Chyba CF (1991) Terrestrial mantle siderophiles and the lunar impact record. Icarus 92:217-235
Clarke WB, Bet MA, Craig H (1969) Excess ^{3}He in the sea: evidence for terrestrial primordial helium. Earth Planet Sci Lett 6:213-220
Connerney JEP, Acuña MH, Wasilewski PJ, Ness NF, Rème H, Mazelle C, Vignes D, Lin RP, Mitchell DL, Cloutier PA (1999) Magnetic lineations in the ancient crust of Mars. Science 284:794-798
Donahue TM (1986) Fractionation of noble gases by thermal escape from accreting planetesimals. Icarus 66:195-210
Donahue TM, Pollack JB (1983) Origin and evolution of the atmosphere of Venus. *In* Venus. Hunten D, Colin L, Donahue T, Moroz V (eds) University of Arizona Press, Tucson, Arizona, p 1003-1036
Donahue TM, Russell CT (1997) The Venus atmosphere and ionosphere and their interaction with the solar wind: an overview. *In* Venus II. Bougher SW, Hunten DM, Phillips RJ (eds) University of Arizona Press, Tucson, Arizona, p 3-31
Donahue TM, Hoffman JH, Hodges RR Jr (1981) Krypton and xenon in the atmosphere of Venus. Geophys Res Lett 8:513-516
Dreibus G, Wänke H (1985) Mars, a volatile-rich planet. Meteoritics 20:367-381
Dreibus G, Wänke H (1987) Volatiles on Earth and Mars: A comparison. Icarus 71:225-240
Dreibus G, Wänke H (1989) Supply and loss of volatile constituents during the accretion of terrestrial planets. *In* Origin and Evolution of Planetary and Satellite Atmospheres. Atreya SK, Pollack JB, Matthews MS (eds) University of Arizona Press, Tucson, Arizona, p 268-288
Eugster O, Eberhardt P, Geiss J (1967) The isotopic composition of krypton in unequilibrated and gas-rich chondrites. Earth Planet Sci Lett 2: 385-393
Fanale FP, Cannon WA (1971) Physical adsorption of rare gas on terrigenous sediments. Earth Planet Sci Lett 11:362-368
Fanale FP, Cannon WA (1972) Origin of planetary primordial rare gas: The possible role of adsorption. Geochim Cosmochim Acta 36:319-328

Feigelson ED, Kriss GA (1989) Soft X-ray observations of pre-main-sequence stars in the Chamaeleon dark cloud. Astrophys J 338:262-276
Frick U, Mack R, Chang S (1979) Noble gas trapping and fractionation during synthesis of carbonaceous matter. Proc 10th Lunar Planet Sci Conf, p 1961-1973
Frick U, Becker RH, Pepin RO (1988) Solar wind record in the lunar regolith: Nitrogen and noble gases. Proc 18th Lunar Planet Sci Conf, p 87-120
Garrison DH, Bogard DD (1998) Isotopic composition of trapped and cosmogenic noble gases in several Martian meteorites. Meteorit Planet Sci 33:721-736
Gierasch PJ, Toon OB (1973) Atmospheric pressure variation and the climate of Mars. J Atmos Sci 30:1502-1508
Graham DW, Lupton F, Albarède F, Condomines M (1990) Extreme temporal homogeneity of helium isotopes at Piton de la Fournaise, Réunion Island. Nature 347:545-548
Graham DW (2002) Noble gases in MORB and OIB: observational constraints for the characterization of mantle source reservoirs. Rev Mineral Geochem 47:247-318
Graham DW, Lupton F, Albarède F, Condomines M (1990) Extreme temporal homogeneity of helium isotopes at Piton de la Fournaise, Rèunion Island. Nature 347:545-548
Haberle RM, Tyler D, McKay CP, Davis WL (1992) Evolution of Mars' atmosphere: Where has the CO_2 gone? Bull Am Astron Soc 24:1015-1016
Haberle RM, Tyler D, McKay CP, Davis WL (1994) A model for the evolution of CO_2 on Mars. Icarus 109:102-120
Hanyu T, Kaneoka I (1998) Open system behavior of helium in case of the HIMU source area. Geophys Res Lett 25:687-690
Harrison D, Burnard P, Turner G (1999) Noble gas behaviour and composition in the mantle: Constraints from the Iceland Plume. Earth Planet Sci Lett 171:199-207
Hartmann WK, Davis DR (1975) Satellite-sized planetesimals and lunar origin. Icarus 24:504-515
Hayashi C, Nakazawa K, Mizuno H (1979) Earth's melting due to the blanketing effect of the primordial dense atmosphere. Earth Planet Sci Lett 43:22-28
Helffrich GR, Wood BJ (2001) The Earth's mantle. Nature 412:501-507
Hilton DR, Grönvold K, MacPherson CG, Castillo PR (1999) Extreme $^3He/^4He$ ratios in northwest Iceland: constraining the common component in mantle plumes. Earth Planet Sci Lett 173:53-60
Hiyagon H, Ozima M, Marty B, Zashu S, Sakai H (1992) Noble gases in submarine glasses from mid-oceanic ridges and Loihi Seamount: constraints on the early history of the Earth. Geochim Cosmochim Acta 56:1301-1316
Hoffman JH, Hodges RR, Donahue TM, McElroy MB (1980) Composition of the Venus lower atmosphere from the Pioneer Venus mass spectrometer. J Geophys Res 85:7882-7890
Hohenberg CM, Podosek FA, Reynolds JH (1967) Xenon-iodine dating: sharp isochronism in chondrites. Science 156:233-236
Honda M, McDougall I (1998) Primordial helium and neon in the Earth- a speculation on early degassing. Geophys Res Lett 25:1951-1954
Honda M, McDougall I, Patterson DB, Doulgeris A, Clague DA (1993) Noble gases in submarine pillow basalt glasses from Loihi and Kilauea, Hawaii—a solar component in the Earth. Geochim Cosmochim Acta 57:859-874
Hudson GB, Kennedy BM, Podosek FA, Hohenberg CM (1989) The early solar system abundance of ^{244}Pu as inferred from the St. Severin chondrite. Proc 19th Lunar Planet Sci Conf, p 547-557
Hunten DM (1979) Capture of Phobos and Deimos by protoatmospheric drag. Icarus 37:113-123
Hunten DM, Pepin RO, Walker JCG (1987) Mass fractionation in hydrodynamic escape. Icarus 69:532-549
Hunten DM, Pepin RO, Owen TC (1988) Planetary Atmospheres. *In* Meteorites and the Early Solar System. Kerridge JF, Matthews MS (eds) University of Arizona Press, Tucson, Arizona, p 565-591
Hunten DM, Donahue TM, Walker JCG, Kasting JF (1989) Escape of atmospheres and loss of water. *In* Origin and Evolution of Planetary and Satellite Atmospheres. Atreya SK, Pollack JB, Matthews MS (eds) University of Arizona Press, Tucson, Arizona, p 386-422
Hutchins KS, Jakosky BM (1996) Evolution of martian atmospheric argon: Implications for sources of volatiles. J Geophys Res 101:14933-14949
Hutchins KS, Jakosky BM, Luhmann JG (1997) Impact of a paleomagnetic field on sputtering loss of martian atmospheric argon and neon. J Geophys Res 102:9183-9189
Igarashi G (1995) Primitive Xe in the Earth. *In* Volatiles in the Earth and Solar System. Farley KA (ed), Am Inst Phys Conf Proc 341:70-80
Jakosky BM, Jones JH (1997) The history of martian volatiles. Rev Geophys 35:1-16
Jakosky BM, Pepin RO, Johnson RE, Fox JL (1994) Mars atmospheric loss and isotopic fractionation by solar-wind-induced sputtering and photochemical escape. Icarus 111:271-288
Jephcoat AP (1998) Rare-gas solids in the Earth's deep interior. Nature 393:355-358

Kaula WM (1999) Constraints on Venus evolution from radiogenic argon. Icarus 139:32-39
Krummenacher D, Merrihue CM, Pepin RO, Reynolds JH (1962) Meteoritic krypton and barium versus the general isotopic anomalies in xenon. Geochim Cosmochim Acta 26:231-249
Kunz J (1999) Is there solar argon in the Earth's mantle? Nature 399:649-650
Kunz J, Staudacher T, Allègre CJ (1998) Plutonium-fission xenon found in Earth's mantle. Science 280:877-880
Kurz MD, Jenkins WJ, Hart SR (1982) Helium isotopic systematics of oceanic islands and mantle heterogeneity. Nature 297:43-46
Laufer D, Notesco G, Bar-Nun A (1999) From the interstellar medium to Earth's oceans via comets—An isotopic study of HDO/H_2O. Icarus 140:446-450
Lee D-C, Halliday AN, Snyder GA, Taylor LA (1997) Age and origin of the moon. Science 278:1098-1103
Lewis JS, Prinn RG (1984) Planets and their Atmospheres: Origin and Evolution. Academic Press, Orlando, Florida
Luhmann JG, Johnson RE, Zhang MHG (1992) Evolutionary impact of sputtering of the martian atmosphere by O^+ pickup ions. Geophys Res Lett 19:2151-2154
Lunine JI, Stevenson DJ (1985) Thermodynamics of clathrate hydrate at low and high pressures with application to the outer solar system. Astrophys J Suppl Series 58:493-531
Lux G (1987) The behavior of noble gases in silicate liquids: Solution, diffusion, bubbles and surface effects, with applications to natural samples. Geochim Cosmochim Acta 51:1549-1560
Mahaffy PR, Donahue TM, Atreya SK, Owen TC, Niemann HB (1998) Galileo probe measurements of D/H and $^3He/^4He$ in Jupiter's atmosphere. Space Sci Rev 84:251-263
Mahaffy PR, Niemann HB, Alpert A, Atreya SK, Demick J, Donahue TM, Harpold DN, Owen TC (2000) Noble gas abundance and isotope ratios in the atmosphere of Jupiter from the Galileo probe mass spectrometer. J Geophys Res 105:15061-15071
Mamyrin BA, Tolstikhin IN, Anufriev GS, Kamenskiy IL (1969) Anomalous isotopic composition of helium in volcanic gases. Dokl Adad Nauk SSR 184:1197-1199 (in Russian)
Marti K (1967) Isotopic composition of trapped krypton and xenon in chondrites. Earth Planet Sci Lett 3:243-248
Marty B, Marti K (2002) Signatures of early differentiation on Mars. Earth Planet Sci Lett 196:251-263
Mason EA, Marrero TR (1970) The diffusion of atoms and molecules. Advan At Mol Phys 6:155-232
Mathew KJ, Marti K (2001) Early evolution of Martian volatiles: Nitrogen and noble gas components in ALH84001 and Chassigny. J Geophys Res 106:1401-1422
Mathew KJ, Marti K (2002) Martian atmospheric and interior volatiles in the meteorite Nakhla. Earth Planet Sci Lett 199:7-20
Mathew KJ, Kim JS, Marti K (1998) Martian atmospheric and indigenous components of xenon and nitrogen in the Shergotty, Nakhla, and Chassigny group meteorites. Meteorit Planet Sci 33:655-664
Matsuda J, Matsubara K (1989) Noble gases in silica and their implication for the terrestrial "missing" Xe. Geophys Res Lett 16:81-84
Mazor E, Heymann D, Anders E (1970) Noble gases in carbonaceous chondrites. Geochim Cosmochim Acta 34:781-824
McDonough WF, Sun S-s (1995) The composition of the Earth. Chem Geol 120:223-253
McElroy MB, Prather MJ (1981) Noble gases in the terrestrial planets. Nature 293:535-539
Melosh HJ, Vickery AM (1989) Impact erosion of the primordial atmosphere of Mars. Nature 338:487-489
Mizuno H, Wetherill GW (1984) Grain abundance in the primordial atmosphere of the Earth. Icarus 59: 74-86
Mizuno H, Nakazawa K, Hayashi C (1980) Dissolution of the primordial rare gases into the molten Earth's material. Earth Planet Sci Lett 50:202-210
Morrison P, Pine J (1955) Radiogenic origin of the helium isotopes in rock. Ann NY Acad Sci 62:69-92
Musselwhite DS, Drake MJ (2000) Early outgassing of Mars: implications from experimentally determined solubility of iodine in silicate magmas. Icarus 148:160-175
Musselwhite DS, Drake MJ, Swindle TD (1991) Early outgassing of Mars supported by differential water solubility of iodine and xenon. Nature 352:697-699
Niedermann S, Bach W, Erzinger J (1997) Noble gas evidence for a lower mantle component in MORBs from the southern East Pacific Rise: decoupling of helium and neon isotope systematics. Geochim Cosmochim Acta 61:2697-2715
Niemeyer S, Marti K (1981) Noble gas trapping by laboratory carbon condensates. Proc 12th Lunar Planet Sci Conf, p 1177-1188
Notesco G, Laufer D, Bar-Nun A, Owen T (1999) An experimental study of isotopic enrichments in Ar, Kr, and Xe when trapped in water ice. Icarus 142, 298-300
Ott U (1988) Noble gases in SNC meteorites: Shergotty, Nakhla, Chassigny. Geochim Cosmochim Acta 52:1937-1948

Ott U (2002) Noble gases in meteorites- trapped components. Rev Mineral Geochem 47:71-100
Owen T (1987) Could icy impacts reconcile Venus with Earth and Mars? *In* Mars: Evolution of its Climate and Atmosphere. Baker V, Carr M, Fanale F, Greeley R, Haberle R, Leovy C, Maxwell T (eds) LPI Tech Report 87-01, Lunar Planet Inst, Houston, Texas, p 91
Owen T, Bar-Nun A (1995a) Comets, impacts and atmospheres. Icarus 116:215-226
Owen T, Bar-Nun A (1995b) Comets, impacts and atmospheres II. Isotopes and noble gases. *In* Volatiles in the Earth and Solar System. Farley KA (ed), Am Inst Phys Conf Proc 341:123-138
Owen T, Bar-Nun A, Kleinfeld I (1991) Noble gases in terrestrial planets: Evidence for cometary impacts? *In* Comets in the Post-Halley Era. Vol. 1. Newburn Jr RL, Neugebauer M, Rahe J (eds) Kluwer Academic Publications, Dordrecht, The Netherlands, p 429-437
Owen T, Bar-Nun A, Kleinfeld I (1992) Possible cometary origin of heavy noble gases in the atmospheres of Venus, Earth and Mars. Nature 358:43-46
Ozima M, Igarashi G (1989) Terrestrial noble gases: Constraints and implications on atmospheric evolution. *In* Origin and Evolution of Planetary and Satellite Atmospheres. Atreya SK, Pollack JB, Matthews MS (eds) University of Arizona Press, Tucson, Arizona, p 306-327
Ozima M, Igarashi G (2000) The primordial noble gases in the Earth: a key constraint on Earth evolution models. Earth Planet Sci Lett 176:219-232
Ozima M, Nakazawa K (1980) Origin of rare gases in the Earth. Nature 284:313-316
Ozima M, Podosek FA (1999) Formation age of Earth from $^{129}I/^{127}I$ and $^{244}Pu/^{238}U$ systematics and the missing Xe. J Geophys Res 104:25493-25499
Ozima M, Podosek FA (2001) Noble Gas Geochemistry 2nd edn. Cambridge, Cambridge University Press, Cambridge
Ozima M, Zahnle K (1993) Mantle degassing and atmospheric evolution: Noble gas view. Geochem J 27:185-200
Ozima M, Wieler R, Marty B, Podosek FA (1998) Comparative studies of solar, Q-gases and terrestrial noble gases, and implications on the evolution of the solar nebula. Geochim Cosmochim Acta 62:301-314
Palma RL, Becker RH, Pepin RO, Schlutter DJ (2002) Irradiation records in regolith materials, II: Solar-wind and solar-energetic-particle components in helium, neon, and argon extracted from single lunar mineral grains and from the Kapoeta howardite by stepwise pulse-heating. Geochim Cosmochim Acta (in press)
Pavlov AA, Pavlov AK, Kasting JF (1999) Irradiated interplanetary dust particles as a possible solution for the deuterium/hydrogen paradox of Earth's oceans. J Geophys Res 104:30725-30728
Pepin RO (1989) On the relationship between early solar activity and the evolution of terrestrial planet atmospheres. *In* The Formation and Evolution of Planetary Systems. Weaver HA, Danly L (eds) Space Tel Sci Inst Symp Series #3, Cambridge University Press, New York p 55-74
Pepin RO (1991) On the origin and early evolution of terrestrial planet atmospheres and meteoritic volatiles. Icarus 92:2-79
Pepin RO (1992). Origin of noble gases in the terrestrial planets. Ann Rev Earth Planet Sci 20: 389-430
Pepin RO (1994) Evolution of the martian atmosphere. Icarus 111:289-304
Pepin RO (1997) Evolution of Earth's noble gases: Consequences of assuming hydrodynamic loss driven by giant impact. Icarus 126:148-156
Pepin RO (1998) Isotopic evidence for a solar argon component in the Earth's mantle. Nature 394:664-667
Pepin RO (2000) On the isotopic composition of primordial xenon in terrestrial planet atmospheres. Space Sci Rev 92:371-395
Pepin RO (2002) Generation of meteoritic noble gas isotope ratios by hydrodynamic escape of transient degassed atmospheres on planetesimals. Meteorit Planet Sci in press (abstract)
Pepin RO, Carr MH (1991) Major issues and outstanding questions. *In* Mars. Kieffer HH, Jakosky BM, Snyder CW, Matthews MS (eds) University of Arizona Press, Tucson, Arizona, p 120-143
Pepin RO, Phinney D (1976) The formation interval of the Earth. Lunar Sci VII:682-684
Pepin RO, Becker RH, Rider PE (1995) Xenon and krypton isotopes in extraterrestrial regolith soils and in the solar wind. Geochim Cosmochim Acta 59:4997-5022
Pepin RO, Becker RH, Schlutter DJ (1999) Irradiation records in regolith materials, I: Isotopic compositions of solar-wind neon and argon in single lunar mineral grains. Geochim Cosmochim Acta 63:2145-2162
Phinney D (1972) ^{36}Ar, Kr, and Xe in terrestrial materials. Earth Planet Sci Lett 16:413-420
Phinney D, Tennyson J, Frick U (1978) Xenon in CO_2 well gas revisited. J Geophys Res 83:2313-2319
Podosek FA, Cassen P (1994) Theoretical, observational, and isotopic estimates of the lifetime of the solar nebula. Meteoritics 29:6-25
Podosek FA, Honda M, Ozima M (1980) Sedimentary noble gases. Geochim Cosmochim Acta 44:1875-1884

Podosek FA, Bernatowicz TJ, Kramer FE (1981) Adsorption of xenon and krypton on shales. Geochim Cosmochim Acta 45:2401-2415
Podosek FA, Woolum DS, Cassen P, Nichols RH (2000) Solar gases in the Earth by solar wind irradiation? J Conf Abstr 5:804
Porcelli D, Ballentine CJ (2002) Models for the distribution of terrestrial noble gases and the evolution of the atmosphere. Rev Mineral Geochem 47:411-480
Porcelli D, Halliday AN (2001) The core as a possible source of mantle helium. Earth Planet Sci Lett 192:45-56
Porcelli D, Pepin RO (2000) Rare gas constraints on early Earth history. *In* Origin of the Earth and Moon. Canup RM, Righter K (eds) University of Arizona Press, Tucson, Arizona, p 435-458
Porcelli D, Wasserburg GJ (1995a) Mass transfer of xenon through a steady-state upper mantle. Geochim Cosmochim Acta 59:1991-2007
Porcelli D, Wasserburg GJ (1995b) Mass transfer of helium, neon, argon, and xenon through a steady-state upper mantle. Geochim Cosmochim Acta 59:4921-4937
Porcelli D, Wasserburg GJ (1995c) A unified model for terrestrial rare gases. *In* Volatiles in the Earth and Solar System. Farley KA (ed), Am Inst Phys Conf Proc 341:56-69
Porcelli DR, Woolum D, Cassen P (2001) Deep Earth rare gases: initial inventories, capture from the solar nebula, and losses during Moon formation. Earth Planet Sci Lett 193:237-251
Prinn RG, Fegley B Jr (1989) Solar nebula chemistry: Origin of planetary, satellite, and cometary volatiles. *In* Origin and Evolution of Planetary and Satellite Atmospheres. Atreya SK, Pollack JB, Matthews MS (eds) University of Arizona Press, Tucson, Arizona, p 78-136
Safronov VS (1978) The heating of the Earth during its formation. Icarus 33:1-12
Sarda P, Staudacher T, Allègre CJ (1985) $^{40}Ar/^{36}Ar$ in MORB glasses: constraints on atmosphere and mantle evolution. Earth Planet Sci Lett 72:357-375
Sarda P, Staudacher T, Allègre CJ (1988) Neon isotopes in submarine basalts. Earth Planet Sci Lett 91: 73-88
Sasaki S (1991) Off-disk penetration of ancient solar wind. Icarus 91:29-38
Sasaki S (1999) Presence of a primary solar-type atmosphere around the Earth: evidence of dissolved noble gas. Planet Space Sci 47:1423-1431
Sasaki S, Nakazawa K (1988) Origin of isotopic fractionation of terrestrial Xe: hydrodynamic fractionation during escape of the primordial H_2-He atmosphere. Earth Planet Sci Lett 89:323-334
Sasaki S, Nakazawa K (1990) Did a primary solar-type atmosphere exist around the proto-Earth? Icarus 85:21-42
Sasaki S, Tajika E (1995) Degassing history and evolution of volcanic activities of terrestrial planets based on radiogenic noble gas degassing models. *In* Volatiles in the Earth and Solar System. Farley KA (ed) Am Inst Phys Conf Proc 341:186-199
Shibata T, Takahashi E, Matsuda J (1998) Solubility of neon, argon, krypton, and xenon in binary and ternary silicate systems: A new view on noble gas solubility. Geochim Cosmochim Acta 62: 1241-1253
Signer P, Suess HE (1963) Rare gases in the sun, in the atmosphere, and in meteorites. *In* Earth Science and Meteoritics. Geiss J, Goldberg ED (eds) North-Holland, Amsterdam p 241-272
Simon T, Herbig G, Boesgaard AM (1985) The evolution of chromospheric activity and the spin-down of solar-type stars. Astrophys J 293:551-574
Strom SE, Strom KM, Edwards S (1988) Energetic winds and circumstellar disks associated with low mass young stellar objects. *In* NATO Advanced Study Institute: Galactic and Extragalactic Star Formation. Pudritz R (ed) D. Reidel, Dordrecht, The Netherlands p 53-68
Strom SE, Edwards S, Skrutskie MF (1993) Evolutionary timescales for circumstellar disks associated with intermediate- and solar-type stars. *In* Protostars and Planets III. Levy EH, Lunine JI (eds) University of Arizona Press, Tucson, Arizona, p 837-866
Suess HE (1949) The abundances of noble gases in the Earth and Cosmos. J Geol 57:600-607 (in German)
Swindle TD (2002) Martian noble gases. Rev Mineral Geochem 47:171-190
Swindle TD, Jones JH (1997) The xenon isotopic composition of the primordial Martian atmosphere: Contributions from solar and fission components. J Geophys Res 102:1671-1678
Swindle TD, Caffee MW, Hohenberg CM (1986) Xenon and other noble gases in shergottites. Geochim Cosmochim Acta 50:1001-1015
Tajika E, Sasaki S (1996) Magma generation on Mars constrained from an Ar-40 degassing model. J Geophys Res 101:7543-7554
Tolstikhin IN, O'Nions RK (1994) The Earth's missing xenon: A combination of early degassing and of rare gas loss from the atmosphere. Chem Geol 115:1-6
Tonks WB, Melosh HJ (1990) The physics of crystal settling and suspension in a turbulent magma ocean. *In* Origin of the Earth. Newsom HE, Jones JH (eds) Oxford University Press, Oxford, p 151-171

Torgersen T (1989) Terrestrial helium degassing fluxes and the atmospheric helium budget: implications with respect to the degassing processes of continental crust. Chem Geol 79:1-14

Trieloff M, Kunz J, Clague DA, Harrison D, Allègre CJ (2000) The nature of pristine noble gases in mantle plumes. Science 288:1036-1038

Trieloff M, Kunz J, Clague DA, Harrison D, Allègre CJ (2001) Reply to comment on noble gases in mantle plumes. Science 291:2269a

Trieloff M, Kunz J, Allègre CJ (2002) Noble gas systematics of the Réunion mantle plume source and the origin of primordial noble gases in Earth's mantle. Earth Planet Sci Lett 200:297-313

Turcotte DL, Schubert G (1988) Tectonic implications of radiogenic noble gases in planetary atmospheres. Icarus 74:36-46

Tyburczy JA, Frisch B, Ahrens TJ (1986) Shock-induced volatile loss from a carbonaceous chondrite: implications for planetary accretion. Earth Planet Sci Lett 80:201-207

Valbracht PJ, Staudacher T, Malahoff A, Allègre CJ (1997) Noble gas systematics of deep rift zone glasses from Loihi Seamount, Hawaii. Earth Planet Sci Lett 150: 399-411

van Keken PE, Hauri EH, Ballentine CJ (2002) Mixing in the mantle and the creation, preservation and destruction of mantle heterogeneity. Ann Rev Earth Planet Sci 30:493-525

Vickery AM, Melosh HJ (1990) Atmospheric erosion and impactor retention in large impacts with application to mass extinctions. *In* Global Catastrophes in Earth History. Sharpton VL, Ward PD (eds) Geol Soc Am Spec Paper 247:289-300

Volkov VP, Frenkel MY (1993) The modelling of Venus degassing in terms of K-Ar system. Earth Moon Planets 62:117-129

von Zahn U, Kumar S, Niemann H, Prinn R (1983) Composition of the Venus atmosphere. *In* Venus. Hunten D, Colin L, Donahue T, Moroz V (eds) University of Arizona Press, Tucson, Arizona, p 299-430

Wacker JF (1989) Laboratory simulation of meteoritic noble gases. III. Sorption of neon, argon, krypton and xenon on carbon: Elemental fractionation. Geochim Cosmochim Acta 53:1421-1433

Wacker JF, Anders E (1984) Trapping of xenon in ice: implications for the origin of the Earth's noble gases. Geochim Cosmochim Acta 48:2373-2380

Wacker JF, Zadnik MG, Anders E (1985) Laboratory simulation of meteoritic noble gases. I. Sorption of xenon on carbon: Trapping experiments. Geochim Cosmochim Acta 49:1035-1048

Walter FM, Barry DC (1991) Pre- and main sequence evolution of solar activity. *In* The Sun in Time. Sonett CP, Giampapa MS, Matthews MS (eds) University of Arizona Press, Tucson, Arizona, p 633-657

Walter FM, Brown A, Mathieu RD, Myers PC, Vrba FJ (1988) X-ray sources in regions of star formation. III. Naked T Tauri stars associated with the Taurus-Auriga complex. Astron J 96:297-325

Wänke H, Dreibus G (1988) Chemical composition and accretion history of terrestrial planets. Phil Trans R Soc London A325:545-557

Wasson JT, Kallemeyn GW (1988) Composition of chondrites. Phil Trans Roy Soc Lond A325:535-544

Wetherill GW (1954) Variations in the isotopic abundances of neon and argon extracted from radioactive minerals. Phys Rev 96:679-683

Wetherill GW (1975) Radiometric chronology of the early solar system. Ann Rev Nucl Sci 25:283-328

Wetherill GW (1981) Solar wind origin of ^{36}Ar on Venus. Icarus 46:70-80

Wetherill GW (1986) Accumulation of the terrestrial planets and implications concerning lunar origin. *In* Origin of the Moon. Hartmann WK, Phillips RJ, Taylor GJ (eds) Lunar and Planetary Institute, Houston, p 519-550

Wetherill GW (1990a) Formation of the Earth. Ann Rev Earth Planet Sci 18:205-256

Wetherill GW (1990b) Calculation of mass and velocity distributions of terrestrial and lunar impactors by use of theory of planetary accumulation. *In* Abstracts for the International Workshop on Meteorite Impact on the Early Earth, LPI Contrib 746. Lunar and Planetary Institute, Houston, Texas, p 54-55

Wetherill GW, Stewart GR (1993) Formation of planetary embryos: Effects of fragmentation, low relative velocity. and independent variation of eccentricity and inclination. Icarus 106:190-209

Wieler R (1994) "Q-gases" as "local" primordial noble gas component in primitive meteorites. *In* Noble gas geochemistry and cosmochemistry. Matsuda J (ed) Terra Scientific Publishing, Tokyo, p 31-41

Wieler R (2002) Noble gases in the solar system. Rev Mineral Geochem 47:21-70:

Wieler R, Baur H (1994) Krypton and xenon from the solar wind and solar energetic particles in two lunar ilmenites of different antiquity. Meteoritics 29:570-580

Woolum DS, Cassen P, Porcelli D, Wasserburg GJ (1999) Incorporation of solar noble gases from a nebula-derived atmosphere during magma ocean cooling. Lunar and Planetary Science XXX, Lunar and Planetary Institute, Houston abstract #1518 (CD-ROM)

Yang J, Anders E (1982a) Sorption of noble gases by solids, with reference to meteorites. II. Chromite and carbon. Geochim Cosmochim Acta 46:861-875

Yang J, Anders E (1982b) Sorption of noble gases by solids, with reference to meteorites. III. Sulfides, spinels, and other substances; on the origin of planetary gases. Geochim Cosmochim Acta 46:877-892
Yang J, Lewis RS, Anders E (1982) Sorption of noble gases by solids, with reference to meteorites. I. Magnetite and carbon. Geochim Cosmochim Acta 46:841-860
Zadnik MG, Wacker JF, Lewis RS (1985) Laboratory simulation of meteoritic noble gases. II. Sorption of xenon on carbon: Etching and heating experiments. Geochim Cosmochim Acta 49:1049-1059
Zahnle KJ (1993) Xenological constraints on the impact erosion of the early martian atmosphere. J Geophys Res 98:10899-10913
Zahnle KJ, Kasting JF (1986) Mass fractionation during transonic escape and implications for loss of water from Mars and Venus. Icarus 68:462-480
Zahnle KJ, Kasting JF, Pollack JB (1990a) Mass fractionation of noble gases in diffusion-limited hydrodynamic hydrogen escape. Icarus 84:502-527
Zahnle KJ, Pollack JB, Kasting JF (1990b) Xenon fractionation in porous planetesimals. Geochim Cosmochim Acta 54:2577-2586
Zhang MHG, Luhmann JG, Bougher SW, Nagy AF (1993) The ancient oxygen exosphere of Mars: Implications for atmospheric evolution. J Geophys Res 98:10915-10923
Zhang Y, Zindler A (1989) Noble gas constraints on the evolution of the Earth's atmosphere. J Geophys Res 94:13719-13737

8 Noble Gas Isotope Geochemistry of Mid-Ocean Ridge and Ocean Island Basalts: Characterization of Mantle Source Reservoirs

David W. Graham

College of Oceanic & Atmospheric Sciences
Oregon State University
Corvallis, Oregon 97331
dgraham@coas.oregonstate.edu

INTRODUCTION

The study of noble gases in oceanic basalts is central to understanding chemical heterogeneity of the Earth's mantle and origin of the atmosphere. In the terrestrial environment the abundances of noble gases are quite low because they were excluded from solid materials during planetary formation in the inner solar system. This low background inventory helps to make the noble gases excellent tracers of mantle reservoirs. In this context, mid-ocean ridge and ocean island basalts provide valuable windows into the Earth's mantle. These oceanic basalts are not prone to the degree of contamination often observed in continental lavas that results from their passage through thick continental lithosphere and crust. Mid-ocean ridge basalts (MORBs) form by partial melting as the ascending mantle beneath spreading ridges reaches its solidus temperature, and MORBs are generally accepted to represent a broad sampling of the convecting upper mantle. Ocean island basalts (OIBs) represent melting 'anomalies' that are generally related to mantle upwelling. The extent to which ocean islands are derived from a thermal boundary layer in the deep mantle (e.g., as a mantle plume) or from chemical heterogeneities embedded within the mantle convective flow (e.g., as a mantle 'blob') has been debated for decades, and is not currently resolved. The isotope compositions of noble gases in oceanic basalts bear significantly on such debates over the chemical structure of the mantle. When oceanic basalts erupt as submarine lavas, their quenched rims of glass may contain high volatile abundances (especially when they are deeply erupted under elevated hydrostatic pressure), providing the best available opportunity for precisely characterizing the noble gas composition of the Earth's mantle. In favorable cases, inclusions of melt or fluids trapped within magmatic phenocrysts and mantle xenoliths can also be precisely analyzed for noble gas composition.

Measurable changes in the isotope compositions of noble gases are closely related to the geochemical processes controlling the distribution of K, U and Th, the major heat producing nuclides in the Earth. The isotopic makeup of every noble gas is modified by the radioactive decay of one or more of these heat-producing nuclides (Table 1). (Radon is an unusual case because it has no stable isotopes, and its geochemical distribution is completely controlled by radioactive decay of U and Th). The geochemical distribution of He is directly related to α-particle-production by U and Th. The Ne isotope composition in terrestrial systems is modified by nucleogenic processes in which ^{21}Ne is dominantly produced when neutrons or α-particles collide with Mg and O atoms. The radioactive decay of ^{40}K generally controls the Ar isotope composition. Small amounts of ^{84}Kr and ^{86}Kr have been produced over geological time by the spontaneous fission of ^{238}U. Production of ^{131,132,134,136}Xe has occurred over geological time by the spontaneous fission of ^{238}U and extinct ^{244}Pu (half-life $t_{1/2}$ = 82 Ma), while ^{129}Xe production occurred by radioactive decay of extinct ^{129}I ($t_{1/2}$ = 17 Ma).

1529-6466/00/0047-0008$10.00

Table 1. Some important production pathways for noble gas isotopes.

Isotope ratio	*Process*	*Half-Life (Ma)*	*Energy Released (MeV/atom)*
LONG-LIVED RADIOACTIVITY			
	$^{238}U \text{ ---->} \ ^{206}Pb+8\ ^{4}He+6\beta^{-}$	4,468	47.4
$^{3}He/^{4}He$	$^{235}U \text{ ---->} \ ^{207}Pb+7\ ^{4}He+4\beta^{-}$	704	45.2
	$^{232}Th \text{---->} \ ^{208}Pb+6\ ^{4}He+4\beta^{-}$	14,010	39.8
$^{40}Ar/^{36}Ar$	$^{40}K \text{--}^{ec}\text{->} \ 0.105\ ^{40}Ar$	1,250	0.71
$^{136,134}Xe/^{130}Xe$	^{238}U Fission	4,468	
EXTINCT RADIOACTIVITY			
$^{129}Xe/^{130}Xe$	$^{129}I \text{ ---->} \ ^{129}Xe+\beta^{-}$	16	
$^{136,134}Xe/^{130}Xe$	^{244}Pu Fission	82	
NUCLEOGENIC REACTIONS FROM U AND TH DECAY			
$^{21}Ne/^{22}Ne$	$^{18}O(\alpha,n) \text{ --->} \ ^{21}Ne$		
	$^{24}Mg(n,\alpha) \text{ --->} \ ^{21}Ne$		

One of the most significant observations is the ubiquitous presence of 'excess' ^{3}He in mantle-derived rocks from ocean ridges and islands, indicating that primordial volatiles are still escaping from the Earth's interior. The highest $^{3}He/^{4}He$ ratios, along with $^{20}Ne/^{22}Ne$ and $^{21}Ne/^{22}Ne$ ratios that approach solar values, are found at ocean islands, most notably Iceland and Hawaii. The He-Ne isotope systematics of these ocean island basalts are currently the strongest geochemical evidence that some portions of the mantle have remained relatively undegassed over geologic time. These are striking findings, because they directly conflict with some geophysical and geochemical models that argue that the mantle convects as a single system and that no primordial or undegassed material remains in the Earth's interior.

The high $^{40}Ar/^{36}Ar$ and non-atmospheric $^{129}Xe/^{130}Xe$ in MORBs also provide fundamental clues to ancient planetary outgassing. Volatile loss from the Earth's interior over time, during ancient formation of the ocean and atmosphere and/or by continuing depletion through partial melting and magma generation, produces a range of parent/daughter ratios and is the primary cause for variable ratios of radiogenic to non-radiogenic noble gas species. Plate tectonic recycling also plays a significant role for the budget of the lithophile parental nuclides U, Th and K. In contrast, the importance of subduction in the mantle budgets of heavier noble gases such as Ar and Xe is still debatable (Staudacher and Allègre 1988; Porcelli and Wasserburg 1995a,b; Sarda et al. 1999a; Burnard 1999b).

Several books (Alexander and Ozima 1978; Ozima and Podosek 2002; Mamyrin and Tolstikhin 1984; Matsuda 1994) and review articles (Craig and Lupton 1981; Lupton 1983; Farley and Neroda 1998; Ozima 1994; McDougall and Honda 1998) currently provide a comprehensive background to terrestrial noble gas geochemistry. The aim of this chapter is to provide an up-to-date overview of key observations on noble gas isotopes in ocean ridge and island basalts that bear on models for the composition and evolution of the Earth's mantle (e.g., see the chapter by Porcelli and Ballentine 2002, this volume).

BACKGROUND

Noble gas chemical behavior

As a group, the noble gases are chemically inert, exhibiting only weak van der Waals type interactions. This serves to produce systematic and predictable variations within the group as a whole, often resulting in a coherent light-to-heavy noble gas trend that reflects the physical processes at hand, such as vapor/liquid/solid partitioning, molecular diffusion, and adsorption. Igneous processes such as partial melting, crystal fractionation, and magmatic degassing of major volatile species (CO_2 and H_2O) should lead to systematic elemental fractionations of the noble gases that are related to their variation in atomic size and the extent to which their electron cloud may be polarized (Keevil 1940). The elemental abundances of noble gases in oceanic basalts may therefore provide clues about petrogenetic processes (e.g., Dymond and Hogan 1978; Batiza et al. 1979).

Our ability to characterize noble gas reservoirs in the mantle depends to some extent on understanding vapor/liquid and solid/liquid partitioning. For example, if the amount of gas loss from a magma is known to be small (or zero), then measured noble gas concentrations in a basalt could be used in combination with vapor/melt and mineral/melt partition coefficients to estimate mantle source abundances. A meaningful determination of the relative elemental abundances in the mantle source requires a 'correction' of the measured concentrations in igneous samples for any fractionation or contamination that occurred during petrogenesis. This is a formidable task, and it can only be carried out effectively by using isotopic relationships among noble gas species. Argon provides a clear example. $^{40}Ar/^{36}Ar$ ratios >30,000 have been measured in some MORBs. In the case of a MORB sample with $^{40}Ar/^{36}Ar = 5,000$ (a seemingly high value because it is more than an order of magnitude above the air ratio of 296) it is still possible that >80% of the measured Ar is derived from atmospheric contamination, and it would be erroneous to draw conclusions about noble gas abundance ratios in the mantle source region based on the concentration data alone. Fortunately, much of this problem can be circumvented through the ratios of certain radiogenic species (e.g., $^{4}He/^{40}Ar^*$). This approach is discussed in the section on *Coupled Radiogenic/Nucleogenic Production*.

Much less uncertainty and fewer assumptions are involved when isotopic ratios (e.g., $^{3}He/^{4}He$ vs. $^{40}Ar/^{36}Ar$) are compared directly. Therefore, wherever possible, this review focuses on isotopic relationships rather than on elemental abundances alone. The isotopic approach allows one to best discern the effects of atmospheric contamination, which can be quite large, and it circumvents some of the ambiguities that result from using the concentrations measured in rocks to interpret the inter-elemental fractionations that occur during igneous processes. The observational data upon which the review is based are presented in diagrams that form an integral part of the chapter. The key points provided by the figures are described in the review, but a detailed description of what they readily show for all the individual localities is not always given, in order to maintain a focus on the general characterization of mantle source reservoirs.

Vapor-melt partitioning. Noble gas solubility in basaltic melt decreases with increasing atomic mass, and is directly related to the atomic radius of the gas (Jambon et al. 1986; Lux 1987; Broadhurst et al. 1992; Shibata et al. 1998). At 1400°C the experimentally determined values for mid-ocean ridge basalt are 56, 25, 5.9, 3.0 and 1.7×10^{-3} std cm^3/g-bar for He, Ne, Ar, Kr and Xe, respectively (Jambon et al. 1986). Degassing of basaltic melts will therefore lead to significant fractionation in the relative abundance of the noble gases, with the exsolved (volatile) phase enriched in the heavier noble gases, such as Ar and Xe, compared to the lighter species, such as He and Ne. The residual melt will show the opposite effect, with a preferential depletion in the heavier species compared to the lighter ones. The low solubilities lead to elevated concentrations in the vesicle phase of basalts. Mid-ocean ridge basalt glasses generally exhibit equilibrium vesicle/melt partitioning for helium (Kurz and Jenkins 1981) and for argon (Marty and Ozima 1986).

Noble gas solubility is only weakly dependent on temperature (Lux 1987) but it depends on melt composition. The compositional control is well described by the 'ionic porosity' (Carroll and Stolper 1993; Carroll and Draper 1994) or by the ratio of non-bridging oxygens to silicon (Shibata et al. 1998). The ionic porosity is the difference between the unit cell volume of a material and the calculated volume of the anions and cations, and provides an integrated measure of the interstitial sites available within the melt structure. The logarithm of noble gas solubility is linearly correlated with ionic porosity. The solubility also shows an increasing sensitivity to small changes in ionic porosity as the size of the gas atom increases. This observation led Carroll and Stolper (1993) to suggest that, as melt structure becomes more tightly packed, the availability of interstitial sites capable of accommodating the larger atoms decreases dramatically.

Noble gas solubility also depends on H_2O and CO_2 content of silicate melts (Paonita et al. 2000; Nuccio and Paonita 2000). Paonita et al. (2000) applied a novel method to study these effects, by adding a known amount of He-bearing glass to their experimental runs. The solubility of He was determined over a range of mixing proportions of H_2O and CO_2 in a rhyolite and in a trachybasalt. The He solubility is strongly influenced by H_2O content, showing about a factor of 3 increase with the addition of 3 wt % H_2O, apparently because the addition of water increases the availability of sites in the melt that accommodate noble gases. Solubility also increases exponentially with atomic size due to the addition of H_2O (Nuccio and Paonita 2000), so while Xe is less soluble than He in anhydrous melts, Xe and He solubilities are nearly the same when several percent H_2O is dissolved. The effect of CO_2 is more uncertain but it appears to be the opposite of H_2O, showing a decrease in the solubility of He by a factor of ~1.5 with addition of 0.05 wt % CO_2 (Nuccio and Paonita 2000). The major volatile composition of the melt therefore affects the relative degassing behavior of the noble gases. During extended degassing, a CO_2–rich anhydrous magma will retain its dissolved He more efficiently than a H_2O–rich magma (Paonita et al. 2000). Variations in the amount of H_2O and CO_2 that were initially present in variably degassed melts may be partly responsible for the observed range of $CO_2/^3He$ in oceanic basalts (see the *Helium* subsection, *Relation to major volatiles*).

The effect of pressure on noble gas solubility in silicate melts is currently an active area of investigation. Chamorro-Perez et al. (1998) reported an order of magnitude decrease for Ar solubility in an olivine melt near 5 GPa, corresponding to mantle depths near 150 km. They inferred that melt densification makes it impossible to accommodate Ar in interstitial sites near this pressure. This is a surprising result, because it would imply that Ar is moderately compatible at depth, and partial melting would not be an effective means of mantle degassing. However, more recent data indicate that the Ar clinopyroxene/silicate melt partition coefficient is relatively constant at $\sim 4 \times 10^{-4}$ for pressures up to at least 8 GPa (Chamorro et al. 2002), so there is no structural change in the melt over that pressure range. Recent work on Ar and Xe solubility in synthetic melts also does not show a decrease in solubility even at pressures of 11 GPa. Instead, it appears that Ar solubility increases to about 6 GPa, above which it reaches a threshold concentration of 0.8 wt % (Schmidt and Keppler 2002). An important consideration in applying the experimental results for solubility and partitioning should be the very low abundance of noble gases in the mantle.

Crystal-melt partitioning. Compared to noble gas solubility in melts, much less is known about their solubility in minerals. Several different approaches can be used to determine a crystal/melt partition coefficient D (where D = weight concentration in crystal/weight concentration in melt) but each one has its limitations. The presently available data are contradictory and the D values range by orders of magnitude, from $<10^{-4}$ to >1. A thoughtful summary of the issues involved in these estimates is presented by Carroll and Draper (1994).

A few experimental studies of noble gas mineral/melt partitioning have been carried out (Hiyagon and Ozima 1982, 1986; Broadhurst et al. 1990, 1992). These studies show a very wide range in partition coefficients for some minerals, and their applicability to mantle melting is difficult to evaluate. Hiyagon and Ozima (1982, 1986) powdered their run products to physically separate the glass from the crystals, so there are questions of atmospheric adsorption, the extent to which all adhering glass could be removed from crystals, and the possible presence of fluid inclusions. They obtained olivine/melt values of $D_{He} \leq 0.07$, $D_{Ar} = 0.05\text{-}0.15$ and $D_{Xe} \leq 0.3$. Broadhurst et al. (1990, 1992) used separate containers for minerals and melts and were extremely careful to use inclusion-free starting materials, but they also obtained a wide range of D values. They observed a weak increase in D from He through Xe for each of the minerals studied (forsterite, diopside, anorthite and spinel) and suggested that this was related to the increasing polarizability with atomic number. They also suggested that the wide range in partition coefficients is related to the number of lattice defects, because variation in the density of interstitial sites in the minerals on the scale needed to explain the range seemed unreasonable. All of these experimental studies obtained D values indicating that the noble gases, especially the heavier species, are more compatible than is usually assumed. Given the experimental difficulties, these D values should be considered as upper limits.

A second approach for obtaining crystal/melt partition coefficients is to use naturally occurring glass-mineral pairs. This approach has been used to study partitioning between olivine and basalt melt (Marty and Lussiez 1993; Kurz 1993; Valbracht et al. 1994). In two studies that used similar approaches to analyze olivine-rich basalts from the Mid-Atlantic Ridge, Marty and Lussiez obtained a value of $D_{He} \leq 0.008$ and $D_{Ar} \leq 0.003$, while Kurz (1993) obtained $D_{He} \leq 0.0058$. These investigators analyzed olivine and glass from the samples both by crushing in vacuum, to release He trapped in inclusions and bubbles, and by fusion, to release the trapped plus the dissolved components. Marty and Lussiez (1993) showed that the olivine and glass have the same $^3He/^4He$ ratio, and appear to be in chemical equilibrium based on their Fe/Mg ratio, although this was questioned by Hiyagon (1994a). The He and Ar released from this glass is dominated by the vesicle gas fraction (Marty and Lussiez 1993), and the He and Ar released from the olivine by crushing appears to be dominated by gas in shrinkage bubbles associated with trapped melt inclusions (Marty and Lussiez 1994). The latter observation would support the notion that the olivine crystals grew before the magma was vapor-saturated. If this was not true then the investigated partitioning involves three phases and is much more complicated. The presence of melt and fluid inclusions in crystals and the possibility of vesicle loss from a magma following crystallization are limitations to using natural samples, and led Marty and Lussiez (1993) and Kurz (1993) to present their D values as minima.

In situ laser analysis is becoming increasingly important in addressing crystal/melt partitioning of noble gases. In a preliminary study using a UV laser ablation microprobe, Brooker et al. (1998) obtained a range of Ar partition coefficients from 0.013-0.14 for olivine and 0.0016-0.589 for clinopyroxene. The low values are probably more realistic and less affected by possible adsorption effects, early partial melting of the crystals and the presence of fluid inclusions.

Although the partitioning behavior of the noble gases between minerals and melts is currently poorly understood and further work is needed, the least ambiguous results indicate that the noble gases have D values below 1. Given the low abundances of the noble gases in the mantle, the experimental data give some indication that mineral defects may play an important role. Both the experiments and the naturally occurring mineral/melt pair studies indicate that He behaves as a highly incompatible element. U and Th are also known to be highly incompatible, having bulk D values on the order of 10^{-3} or less, so the behavior of He relative to U and Th still needs to be established. Given that the lowest D value determined for He is likely to be a maximum, the large differences in $^3He/^4He$ between the mantle sources for MORBs and OIBs can be taken to reflect differences in their degassing history.

Mantle structure and noble gases

The chemical structure of the Earth's mantle is directly related to the style of mantle convection, and the debate over whole mantle vs. layered mantle convection has gone on for decades. There are several scales of mantle convection indicated by seismic tomography, gravity and geochemistry (Anderson 1998b). The largest scale is controlled by the pattern of cooling associated with subducting plates, while the smaller scales (400-1000 km) are probably controlled by the depth of phase transitions and the thickness of the upper mantle low viscosity region. Seismic evidence now clearly shows that some subducting slabs penetrate below the 660 km seismic discontinuity (van der Hilst et al. 1997), so this depth can no longer be viewed as a strong barrier to mass transport between the upper and lower mantle. Some investigators take this as sufficient evidence for whole mantle convection. In this case, the isotopic differences between depleted MORBs and enriched OIBs might be explained by a mantle that contains large scale blobs of chemically enriched or primitive material (e.g., Manga 1996; Becker et al. 1999). The origin of such blobs is unclear, and detecting them (if they exist) is currently beyond the resolution capabilities of seismology. Alternatively, the mantle may have some form of layered structure. If so, then buoyant upwellings produced at thermal or chemical boundary layers (mantle plumes) will be an important mechanism by which deep material is brought close to the Earth's surface, where it partially melts to form ocean island basalt magmas. Mantle plumes have been implicated in the origin of many ocean islands since the discovery of plate tectonics (Morgan 1971). Bathymetric tracks of volcanoes such as the Hawaiian-Emperor chain indicate that these plume sources move laterally much more slowly than the plates, so the depth of origin must lie below the convecting upper mantle, although the exact depth is currently not well constrained. Such plumes most likely originate from boundary layers, perhaps as deep as the core-mantle boundary, and they are expected to entrain small amounts of material from the underlying reservoir. In actuality there is a slow relative motion between the hotspots on the Earth's surface that is explained by advection of the plumes by large-scale mid-mantle flow. This mid-mantle flow is generally toward ridges and opposite in direction to the flow field in the upper mantle as indicated by plate motion directions (Steinberger and O'Connell 1998). In some cases, most notably East Africa, the upwelling also appears to be much broader than expected for a narrow plume conduit (Ritsema et al. 1999).

Obviously much less is known about the deep mantle than the upper mantle. The style and vigor of mantle convection are described by the Rayleigh number (Ra). Ra is estimated to be 10^7 for the uppermost mantle; the estimate for the deep mantle is more uncertain but it may be several orders of magnitude smaller, mostly due to its higher viscosity (Tackley 1998). Consequently, the mixing time for the upper mantle is relatively rapid compared to the deep mantle, and compositional heterogeneities have a greater likelihood of surviving deeper in the Earth's interior (Gurnis and Davies 1986). Nevertheless, state-of-the-art coupled convection-degassing models that incorporate a high viscosity lower mantle, a phase transition at the base of the upper mantle, and temperature- and pressure–dependent rheology, currently fail to produce an isolated, high

$^3He/^4He$ region in the Earth (van Keken and Ballentine 1998, 1999).

Recent seismic tomography seems to indicate that some type of boundary may be present at a depth near 1700 km in the lower mantle (van der Hilst and Kárason 1999; Kellogg et al. 1999). Below this depth the mantle may be denser than the overlying mantle due to differences in bulk composition. However, the deeper material could contain more of its original inventory of heat producing elements, so it is hot and only slightly more dense than mantle at the same depth lying on the adiabat (the pressure-temperature path for material that expands or contracts without gaining or losing heat). This configuration is dynamically stable but it may lead to the development of significant topography on the surface of this layer. The resulting small density differences (~0.5%) would be sufficient to inhibit mixing, but the layer's surface could respond dramatically to down-going slabs and rising thermal plumes (Kellogg et al. 1999). This model is consistent with laboratory studies of thermochemical convection in a chemically heterogeneous fluid that has a stratified density and viscosity structure (Davaille 1999). These lab experiments show that hot domes oscillate vertically through the fluid while thin plumes rise from their upper surface. There are other 'layered' models. For example, the 'perisphere' model places all OIB sources in a thin, shallow enriched layer beneath the lithosphere (Anderson 1995). It accounts for the presence of enriched basalts in continental rifts, but it is difficult to reconcile with the voluminous volcanism at some ocean islands. Many of the proposed models do not readily satisfy the noble gas observations that support the preservation of relict primitive mantle. Some contrasting ideas on mantle convection and its structure and evolution that consider the noble gas constraints are given in papers by Kellogg (1992), Davies and Richards (1992), Albarède (1998), Coltice and Ricard (1999) and Tackley (2000). Two books on mantle dynamics are also now available (Davies 1999; Schubert et al. 2001).

Intriguingly, there is no compelling evidence for the survival of primitive mantle based on refractory element ratios or the isotopes of Sr, Nd and Pb in ocean island basalts (Hofmann 1997). Hofmann et al. (1986) showed that certain elemental ratios, such as Ce/Pb and Nb/U, had uniform and similar values in both MORBs and OIBs. These Ce/Pb and Nb/U ratios are distinct from values for either the primitive mantle or the continental crust, indicating that the chemical signature of crust extraction over geologic time has mostly been homogenized throughout the mantle. Therefore, the observed Sr, Nd and Pb isotopic differences between MORBs and OIBs must be the result of more recent processes. Hofmann and White (1980) convincingly established that the extreme isotopic compositions of OIBs are probably controlled by plate-tectonic recycling, in which OIBs are produced by the heating and melting of subducted slabs. Zindler and Hart (1986b) and Weaver (1991) also demonstrated that the OIBs with the most radiogenic Pb isotopes were probably dominated by recycled oceanic crust, while other enriched OIB end-members probably contain a small percentage of recycled sediments (terrigenous vs. pelagic) mixed into their mantle source region.

In the layered mantle model, OIB magmas from intraplate hotspots with high $^3He/^4He$ come from deep mantle source regions that are convectively isolated from the upper mantle source of MORBs. These deep mantle regions contain some proportion of relatively undifferentiated, primitive mantle. OIBs that do not have elevated $^3He/^4He$ may originate from shallower source regions. The upwelling, deep OIB mantle source also supplies material to the MORB mantle. The steady-state model for the upper mantle is a current paradigm in noble gas geochemistry (e.g., Allègre et al. 1986/1987; O'Nions 1987; Kellogg and Wasserburg 1990; O'Nions and Tolstikhin 1994, 1996; Porcelli and Wasserburg 1995a,b). This steady-state is reached through a balance of deep mantle input from below, slab subduction from above, and radiogenic production. Further discussion of the steady-state model is provided by Porcelli and Ballentine (2002, this volume).

HELIUM

Significance

Helium isotope measurements in ocean ridge and island basalts provide some of the most basic geochemical information on mantle source reservoirs. More helium isotope analyses have been performed for oceanic volcanic rocks than for other noble gas species, and helium isotopes have played a leading role in the study of mantle heterogeneity. Helium isotope analyses are readily performed by modern mass spectrometers because there is a general absence of atmospheric contamination in samples due to the low concentration of helium in air (5.24 parts per million by volume at standard temperature and pressure). There are 2 naturally occurring isotopes of helium. ^3He is much less abundant than ^4He; for example, the atmospheric $^3He/^4He$ ratio (R_A) is 1.39×10^{-6} (Mamyrin et al. 1970; Clarke et al. 1976). Nearly all of the terrestrial ^4He has been produced as α-particles from the radioactive decay of ^{238}U, ^{235}U and ^{232}Th over geological time, while nearly all of the ^3He is primordial. Because helium undergoes gravitational escape from the thermosphere and has an atmospheric residence time of 1 to 10 million years, it is not recycled by plate tectonics to the Earth's interior, and this makes the $^3He/^4He$ ratio unique among isotopic tracers of mantle sources involved in volcanism (Lupton 1983). By far the most important terrestrial source of ^3He is degassing from the Earth's interior. Excess ^3He in volcanic rocks was first reported by Krylov et al. (1974) and Lupton and Craig (1975). The presence of this ^3He in mantle-derived materials has profound implications; it means that the Earth is still outgassing volatiles that were trapped at the time of its accretion more than 4500 Ma ago.

Other sources of ^3He to the atmosphere include the auroral precipitation of solar wind, direct accretion from cosmic rays, and the flux of cosmic dust and meteorites (Lupton 1983). Small, but measurable amounts of ^3He are produced in rocks at the Earth's surface by high energy cosmic rays, predominantly from spallation of O, Si, Mg and Fe atoms, providing a means for determining surface exposure ages and erosion rates (Kurz 1986). Very small amounts of ^3He are also produced during radioactive decay of U and Th as a result of neutron interactions with Li, by the reaction $^6Li(n,\alpha)^3H\rightarrow{}^3He$ (Morrison and Pine 1955). The neutrons are produced by α-particle interactions on target elements such as Mg, Si and O in the host rock. This results in a low $^3He/^4He$ production ratio (typically $<0.02\ R_A$), even in crustal rocks that are relatively enriched in Li. Consequently, the terrestrial $^3He/^4He$ ratio varies by several orders of magnitude, from high values ($>10^{-5}$) in mantle-derived lavas and fluids, to low values ($\sim10^{-8}$) in continental regions due to increased amounts of radiogenic ^4He. The general pattern in oceanic basalts is one in which MORBs show a relatively small range in $^3He/^4He$ ($8.75\pm2.14\ R_A$; Table 2), while OIBs are much more variable, and extend to values that are higher than the MORB mean by more than a factor of 4.

Table 2. Helium isotopes in MORB glasses.

Location	*n*	*Median*	*Mean*	*Standard deviation*	*Skewness*
Atlantic	236	8.08	9.58	2.94	1.10
Pacific	245	8.14	8.13	0.98	0.04
Indian	177	8.24	8.49	1.62	1.73
All	658	8.11	8.75	2.14	1.92

Radiogenic production

The amount of ^{4}He produced in a closed system, at secular equilibrium and over time t is given by

$$^4\text{He}^* = {}^{238}\text{U}\{8\,(e^{\lambda_{238}t} - 1) + (7/137.88)\,(e^{\lambda_{235}t}-1) + 6\,\kappa\,(e^{\lambda_{232}t}-1)\} \quad (1)$$

where ^{238}U is the amount present, λ_{238}, λ_{235} and λ_{232} are the decay constants for ^{238}U, ^{235}U and ^{232}Th (0.155125, 0.98485 and 0.049475 Gy^{-1}, respectively), 1/137.88 is the present ratio of ^{235}U/^{238}U and κ is ^{232}Th/^{238}U. The range of κ values in the Earth's mantle is about a factor of 2, from κ of ~2 to ~4 (Allègre et al. 1986; O'Nions and McKenzie 1993). By convention, the helium isotope ratio is often expressed as ^{3}He/^{4}He, i.e., as the non-radiogenic to radiogenic isotope, which causes some inconvenience when formulating changes due to radioactive ingrowth. The ^{3}He/^{4}He ratio in igneous samples, R, is usually expressed relative to the atmospheric ratio, i.e., as R/R_A. Marine air provides a convenient and useful standard given its isotopic homogeneity for ^{3}He/^{4}He.

Mid-ocean ridge basalts

Global variability. Helium isotope variations along mid-ocean ridges have been extensively studied, and the variability along ridges is important to understand in the context of convective mixing and melt generation in the upper mantle. The difference in ^{3}He/^{4}He ratio commonly observed between MORBs and OIBs is accepted by most investigators as evidence for two distinct mantle source regions. This viewpoint has been challenged by Anderson (2000a,b; 2001) on the basis of a statistical comparison of MORB and OIB data sets. The choice of basalt samples included in these data sets directly impacts the accuracy of any conclusions about the mantle, so it is important to discuss this choice in some detail and to make it wisely. Anderson (2001) chose to include lavas from Iceland as mid-ocean ridge basalts, on the grounds that the Mid-Atlantic Ridge passes through its center. He also included near-axis seamounts and back-arc basin basalts. The inclusion of Iceland certainly affects the mean and variance in any such comparison, because the ^{3}He/^{4}He ratios near its center are higher than those observed anywhere along the global ocean ridge system. The inclusion of seamounts and back-arc basin basalts affects the variance much more than the mean, due to the presence of lavas that experienced shallow-level degassing and significant lowering of their ^{3}He/^{4}He ratio by interaction with seawater or altered wallrock (e.g., Graham et al. 1988; Staudacher and Allègre 1989; Hilton et al. 1993). Anderson did not carry out a similar data compilation of the OIB data set, but used a mean value for 23 ocean islands summarized by Allègre et al. (1995) based on 276 individual OIB analyses. The OIB data included lavas from very different stages of evolution in the volcanic history of the islands, and the sources of the original data were not given in detail. Anderson (2001) compared the mean value for these 23 islands with the mean for ~500 individual MORB glasses, so the two data sets were also not treated in the same way. In this section I have made a first-order attempt to objectively estimate the mean ^{3}He/^{4}He, its variance and skewness, for mid-ocean ridges from the Atlantic, Pacific and Indian Oceans, and for all ridges collectively. I arrive at different values for the mean and standard deviation compared to Anderson's work (Table 2).

Mid-ocean ridge basalts display a relatively narrow range of ^{3}He/^{4}He, while ocean-island basalts are more variable and often extend to higher values. The full range in MORBs is between 1 and 18 R_A. This range excludes subaerial lavas from Iceland, but not submarine lavas from the ridges to its south or north (Reykjanes or Kolbeinsey). In making this comparison, I also have not included seamounts or back-arc basin basalts, because their clear tectonic association with ridge mantle is questionable. I have included only one analysis from any individual sampling locality (i.e., one sample per dredge or rock core), as a very coarse attempt to avoid sampling bias from the few areas that have been studied in considerably more detail. I have also excluded a handful of glass samples with $^3He/^4He < 2.5\ R_A$, which, given their differentiated and highly degassed nature ([He] $<3\times10^{-8}$ std cm^3/g) and their elevated chlorine contents, have clearly been influenced by contamination with seawater or altered crust (Michael and Cornell 1998). These sample types occur along propagating rifts or overlapping spreading centers, often areas having a low magma budget. This may lead to a prolonged residence in the crust, during which the magma can undergo significant gas loss and interaction with altered wallrock. Including these degassed, low $^3He/^4He$ samples does not lead to a very different estimate of population means, but it does lead to higher estimates of variance. No other filtering has been done, and it should be emphasized that the data used represent more than 95% of the original collection of $^3He/^4He$ analyses for submarine glasses sampled along mid-ocean ridges. The spatial distribution of the ~660 MORB localities that have been studied is shown in Figure 1.

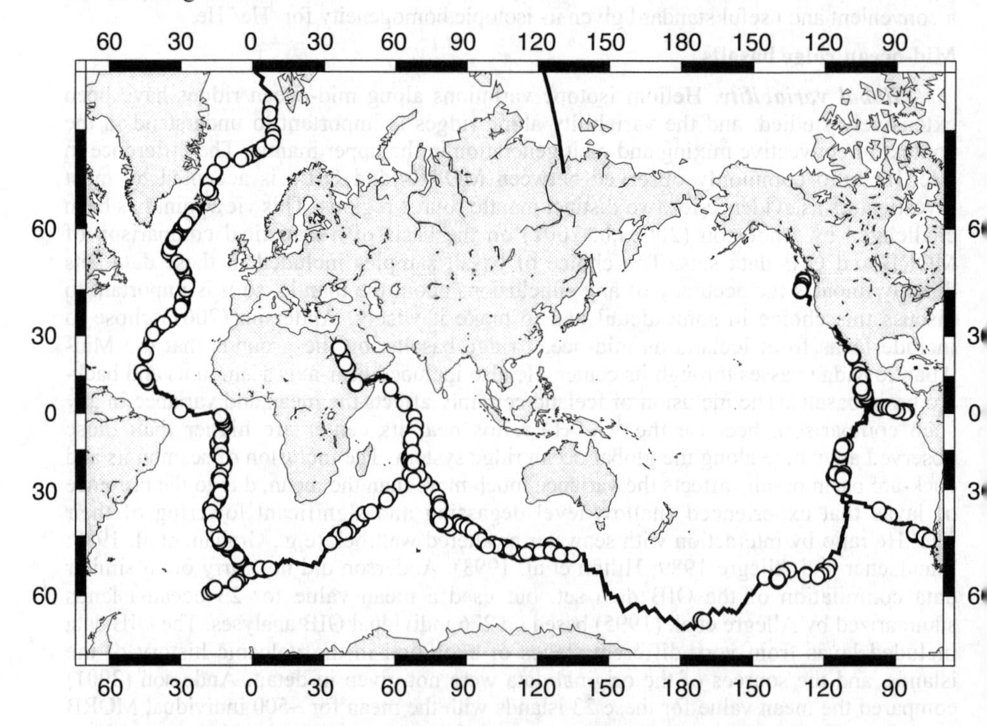

Figure 1. Location map of ocean ridge basalt glasses analyzed for $^3He/^4He$. Only submarine samples are included; the total number of sampling localities is 658. Data sources are Kurz (1982), Kurz and Jenkins (1981), Kurz et al. (1982b, 1998), Ozima and Zashu (1983), Rison and Craig (1984), Poreda et al. (1986, 1993), Graham et al. (1988, 1992b, 1996a,b; 1999, 2001 and unpublished data) Staudacher et al. (1989), Mahoney et al (1989, 1994), Hiyagon et al. (1992), Marty et al. (1993b) Lupton et al. (1993), Moreira et al. (1995, 1996), Taylor et al. (1997), Niedermann et al. (1997), Niedermann and Bach (1998), Schilling et al. (1998), Sturm et al. (1999), Sarda et al. (2000), Hilton et al. (2000b).

The mean and standard deviation for the respective populations from each ocean basin, and for all MORBs, are presented in Table 2 and displayed in Figure 2. For all MORB samples considered in the manner just described (n = 658), the mean is 8.75 R_A and the standard deviation is 2.14 R_A. MORB samples with the highest $^3He/^4He$ are from ridge sections that display other anomalous geochemical features, quite often where the sub-ridge mantle is influenced by nearby ocean islands. The range in MORBs is usually stated to be about 7 to 9 R_A, and this qualitatively describes most of the data when spreading ridges adjacent to ocean island hotspots are excluded. In only a handful of cases do ridge localities away from ocean islands show values outside of the 7 to 9 R_A range for a significant distance along-axis (≥50 km). These areas also show isotopic anomalies in other systems such as Sr, Nd and Pb. The cases are the super fast spreading section of the southern East Pacific Rise ($^3He/^4He$ up to 10.9 R_A; Mahoney et al. 1994), the Australian-Antarctic Discordance ($^3He/^4He$ down to 6.2 R_A; Graham et al. 2001), the 33°S isotope anomaly on the Mid-Atlantic Ridge (down to 6.2 R_A; Graham et al. 1996a), the ultraslow-spreading Southwest Indian Ridge from 9°-24°E (down to 6.3 R_A; Georgen et al. 2001) and the southern Chile Rise (down to 3.5 R_A; Sturm et al. 1999). In this last case, low Ce/Pb ratios indicate that crustal recycling has occurred into the MORB mantle source from a nearby subduction zone (Klein and Karsten 1995). Intriguingly, all of the other localities having low $^3He/^4He$ appear to show a limiting value near 6 R_A.

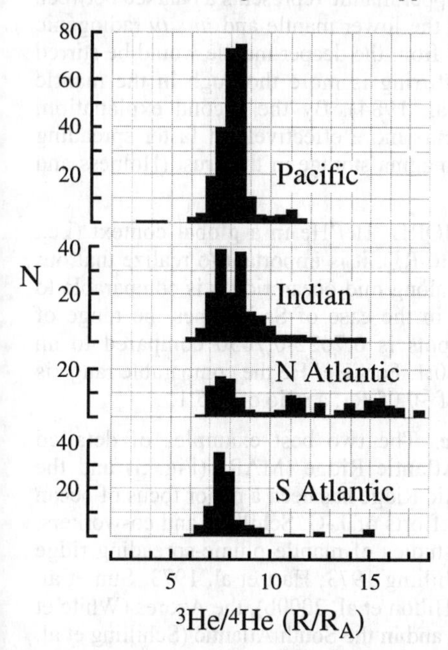

Figure 2. Histogram of MORB $^3He/^4He$ by ocean basin. For the few cases where several analyses have been performed at one sampling locality, only one sample was included as an attempt to avoid over sampling of unusually high or low $^3He/^4He$ localities. Data sources are given in the caption to Figure 1.

High $^3He/^4He$ localities other than Iceland seem to show an upper limiting value of ~15 R_A. These localities include the east and west rifts of the Easter Microplate (11.7 R_A; Poreda et al. 1993), the Southeast Indian Ridge near Amsterdam and St. Paul islands (14.1 R_A; Graham et al. 1999), the Gulf of Tadjoura near Afar (14.7 R_A; Marty et al. 1993b), the Shona and Discovery sections of the southern Mid-Atlantic Ridge (12.3 and 15.2 R_A, respectively; Moreira et al. 1995; Sarda et al. 2000), the Southwest Indian Ridge near Bouvet Island (14.9 R_A; Kurz et al. 1998) and the Manus Basin back-arc spreading center (15.1 R_A; Shaw et al. 2001). In each of these areas, the high $^3He/^4He$ signal can be attributed to introduction of high $^3He/^4He$ material from a nearby mantle hotspot. The apparent upper limit to the measured $^3He/^4He$ in these cases is usually presumed to stem from dilution with ambient upper mantle having $^3He/^4He$ between ~7 and 9 R_A.

No relationship between He isotope composition and ridge spreading rate is evident. However, the *variance* of $^3He/^4He$ along ridges may be inversely related to spreading rate

(Graham 1994, Allègre et al. 1995). There are several caveats that go with such an inferred simple relationship, however. First, an inverse relationship only applies when the high $^3He/^4He$ localities discussed above are excluded, because these localities naturally display the highest variance and they occur across most of the range in seafloor spreading rates. An inverse relationship also only holds when the length of ridges used in the comparison are relatively long, generally more than a few hundred kilometers. Finally, the proposed relationship has recently been called into question by Georgen et al. (2001), because a relatively narrow $^3He/^4He$ range is observed for basalts along the ultra-slow spreading Southwest Indian Ridge. Despite all these qualifications, there does appear to be some systematic behavior for most other sections of the ocean ridge system. An inverse relationship between variance and spreading rate may reflect either variations in the efficiency of upper mantle mixing, or differences in the degree of magma homogenization during MORB petrogenesis. By the first explanation, faster spreading ridges are ultimately a consequence of faster turnover rates in the upper mantle. The implicit assumption here is that $^3He/^4He$ in the upper mantle represents a balance between input of relatively high $^3He/^4He$ material from the lower mantle and *in situ* radiogenic addition of ^4He. Blobs of OIB material injected from the deeper mantle would be stirred into the upper mantle, and presumably this stirring is more thorough in the mantle beneath faster spreading ridges (Allègre et al. 1984). By the second explanation, "averaging" of small-scale heterogeneities occurs more effectively at faster spreading ridges, either during partial melting or during magma storage in the crust (Holness and Richter 1984, Batiza et al. 1984).

Despite the relatively narrow range of MORB $^3He/^4He$ in a global context (i.e., considering that OIBs extend to values above 30 R_A), it is important to realize that our ability to measure variations in $^3He/^4He$ signal along mid-ocean ridges is comparable to that for other isotopic systems. For example, in the case of Sr isotopes the range of $^{87}Sr/^{86}Sr$ in typical MORB away from hotspots is 0.7022-0.7030 compared to an analytical uncertainty of ~0.00001, a ratio of ~80:1. For $^3He/^4He$ the comparable range is 6.2-9.5 R_A relative to an analytical uncertainty of ~0.05 R_A, a ratio of ~65:1.

Regional variability—Mid-Atlantic Ridge. The two best examples of detailed regional variations in $^3He/^4He$ are the Mid-Atlantic Ridge (MAR) (Fig. 3) and the Southeast Indian Ridge (Fig. 4). The Mid-Atlantic Ridge has been a major focus of ocean ridge geochemical studies, largely through the efforts of J.-G. Schilling and co-workers. It has been especially targeted for numerous studies of mantle plume-spreading ridge interactions, particularly near Iceland (e.g., Schilling 1973; Hart et al. 1973; Sun et al. 1975; Schilling et al. 1998; Poreda et al. 1986; Hilton et al. 2000b), the Azores (White et al. 1976; Kurz et al. 1982b; Moreira et al. 1999) and in the South Atlantic (Schilling et al. 1985; Hanan et al. 1986; Graham et al. 1992b; Moreira et al. 1995; Kurz et al. 1998; Douglass et al. 1999; Sarda et al. 2000). The variation of $^3He/^4He$ along the Mid-Atlantic Ridge from north of Iceland to south of the Bouvet triple junction is shown in Figure 3. Very high $^3He/^4He$ ratios, near 30 R_A, are present in subaerial lavas centered on the Iceland hotspot. Northward and southward, along submarine portions of the oceanic ridge, $^3He/^4He$ ratios up to 18 R_A are present. The spatial extent of high $^3He/^4He$ ratios, when measured from the center of Iceland to the south along the Reykjanes Ridge, is about twice as long as it is to the north along the Kolbeinsey Ridge (Fig. 3). This has been interpreted as reflecting a more prominent southerly flow of plume material in the upper mantle (Schilling et al. 1998). It has also been noted that where plumes or blobs affect spreading ridges, the wavelength of He isotope variation is sometimes less than that of other tracers such as Sr and Pb isotopes (e.g., Poreda et al. 1993; Mahoney et al. 1994; Taylor et al. 1997; Schilling et al. 1998). This has been attributed to a relatively deep degassing process within the rising mantle plume, causing a strong $^3He/^4He$ peak near the plume center, while relatively degassed plume material that is still effectively traced by Sr, Nd or Pb isotopes becomes laterally dispersed at shallower depths (e.g., Schilling et al. 1998; Breddam et al. 2000).

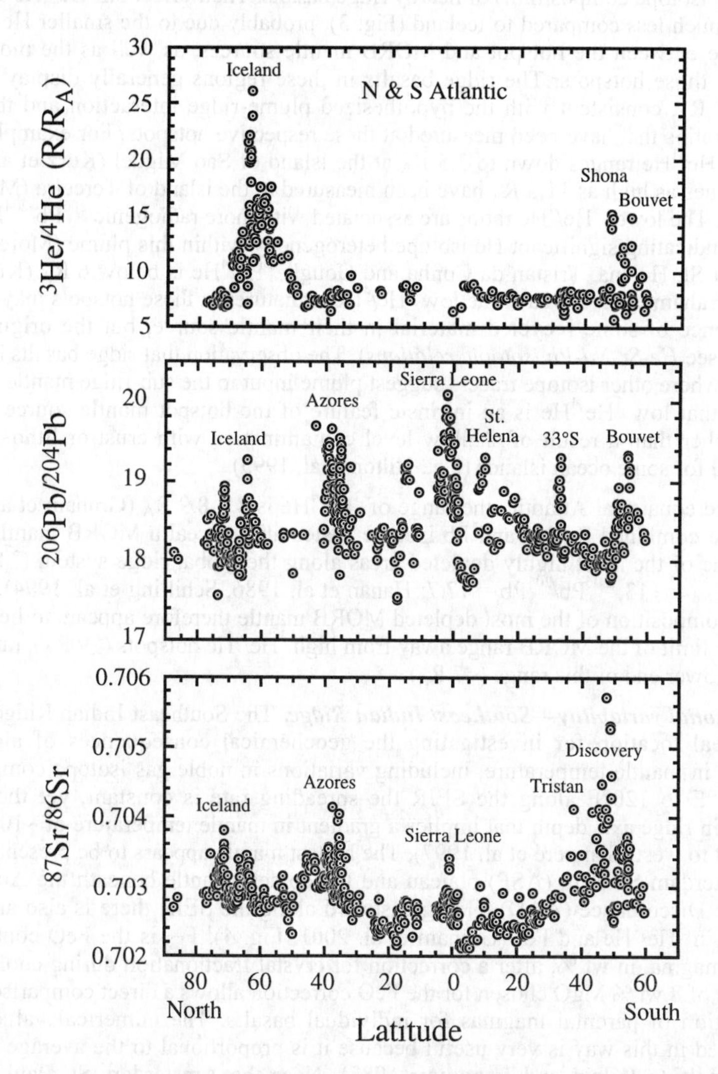

Figure 3. Variations in $^3He/^4He$, $^{206}Pb/^{204}Pb$ and $^{87}Sr/^{86}Sr$ along the axis of the Mid-Atlantic Ridge. Data sources are: MORB $^3He/^4He$ – Kurz et al. (1982b, 1998), Poreda et al. (1986), Graham et al. (1992b, 1996a,b and unpubl. data), Staudacher et al. (1989), Moreira et al. (1995), Taylor et al. (1997), Schilling et al. (1998), Sarda et al. (2000), Hilton et al. (2000b). Iceland $^3He/^4He$ – Condomines et al. (1983), Kurz et al. (1985), Dixon et al. (2000), Breddam et al. (2000), Moreira et al. (2001). Additional MORB Sr and Pb isotopes – Hart et al. (1973), Sun et al. (1975), White et al. (1976), Dickey et al. (1977), White and Schilling (1978), Dupré and Allègre (1980), Dupré et al. (1981), Machado et al. (1982), Hamelin et al. (1984), le Roex et al. (1987), Dosso et al. (1991, 1993, 1999), Hanan et al. (1986, 1994), Castillo and Batiza (1989), Mertz et al. (1991), Shirey et al. (1987), Schilling et al. (1994, 1998), Fontignie and Schilling (1996), Mertz and Haase (1997), Kurz et al. (1998), Douglass et al. (1999). Iceland Sr and Pb – Sun and Jahn (1975), O'Nions et al. (1976), Zindler et al. (1979), Elliott et al. (1991), Hémond et al. (1993), Hanan and Schilling (1997).

Other island hotspots, such as the Azores in the North Atlantic, and St. Helena, Tristan da Cunha and Gough Island in the South Atlantic, also show a strong effect on the Sr and Pb isotope compositions of nearby ridge basalts. Their effect on $^3He/^4He$ along the ridge is much less compared to Iceland (Fig. 3), probably due to the smaller He isotopic difference between the hotspot and MORB mantle sources, as well as the more distal nature of those hotspots. The ridge basalts in these regions generally display $^3He/^4He$ ratios <7 R_A, consistent with the hypothesized plume-ridge interaction and the lower $^3He/^4He$ ratios that have been measured at these respective hotspots. For example, in the Azores, $^3He/^4He$ ranges down to 3.5 R_A at the island of Sao Miguel (Kurz et al. 1990), while values as high as 11.3 R_A have been measured at the island of Terceira (Moreira et al. 1999). The lower $^3He/^4He$ ratios are associated with more radiogenic $^{206}Pb/^{204}Pb$ in the Azores, indicating significant He isotope heterogeneity within this plume (Moreira et al. 1999). At St. Helena, Tristan da Cunha and Gough, $^3He/^4He$ is below 6 R_A (Kurz et al. 1982a; Graham et al. 1992a). The low $^3He/^4He$ signatures at these hotspots may indicate the presence of some recycled material in their mantle source, but the origin is still debated (see *He-Sr-Nd-Pb isotopic relations*). The observation that ridge basalts have low $^3He/^4He$ where other isotope tracers suggest plume input to the sub-ridge mantle supports the idea that low $^3He/^4He$ is an intrinsic feature of the hotspot mantle source in these areas, rather than a result of shallow level contamination with crust or lithosphere as suggested for some ocean islands (e.g., Hilton et al. 1995).

In the equatorial Atlantic, the range of $^3He/^4He$ is 8.6-8.9 R_A (Graham et al. 1992), where the combined Sr, Nd and Pb isotope systematics reveal a MORB mantle source that is one of the most highly depleted areas along the global ridge system ($^{87}Sr/^{86}Sr$ = 0.7021, ε_{Nd} = +13, $^{206}Pb/^{204}Pb$ = 17.7; Hanan et al. 1986, Schilling et al. 1994). The He isotope composition of the most depleted MORB mantle therefore appears to lie close to the upper limit of the MORB range away from high $^3He/^4He$ hotspots (≥9 R_A), rather than near the lower end of this range (<7 R_A).

Regional variability—Southeast Indian Ridge. The Southeast Indian Ridge (SEIR) is an ideal location for investigating the geochemical consequences of along-axis variation in mantle temperature, including variations in noble gas isotope composition. From 88°E to 120°E along the SEIR the spreading rate is constant, yet there is an increase in ridge axis depth that implies a gradient in mantle temperature of ~100-150°C from east to west (Sempéré et al. 1997). The hottest mantle appears to be present beneath the Amsterdam-St. Paul (ASP) plateau and the coldest mantle beneath the Australian-Antarctic Discordance (AAD). Going eastward along the SEIR there is also an overall decrease in $^3He/^4He$ and Fe_8 (Graham et al. 2001; Fig. 4). Fe_8 is the FeO content of a basaltic magma, in wt %, after a correction for crystal fractionation during cooling. The reference of 8 wt % MgO chosen for the FeO correction allows a direct comparison of the composition of parental magmas for individual basalts. The numerical value of Fe_8 determined in this way is very useful because it is proportional to the average depth of mantle melting (Klein and Langmuir 1987). Near the Amsterdam-St. Paul plateau, $^3He/^4He$ ratios extend above 14 R_A due to the presence of the ASP hotspot. In the AAD, $^3He/^4He$ ratios are as low as 6.2 R_A. The good covariation of $^3He/^4He$ and Fe_8 over much of the SEIR indicates that higher $^3He/^4He$ may be associated with a higher mean pressure of melting in the underlying Indian Ocean mantle. This type of covariation appears best explained by melting of a heterogeneous mantle, in which blobs or veins of lithologically distinct mantle are embedded within a peridotite matrix. These types of heterogeneities could occur, for example, as veins of garnet pyroxenite that have low $^3He/^4He$, if they originated from recycled crust or lithosphere that was enriched in U and Th. Such recycled components should have very low $^3He/^4He$, <1 R_A, and only a small contribution to the He sampled by the basalts is required. Garnet pyroxenite will melt

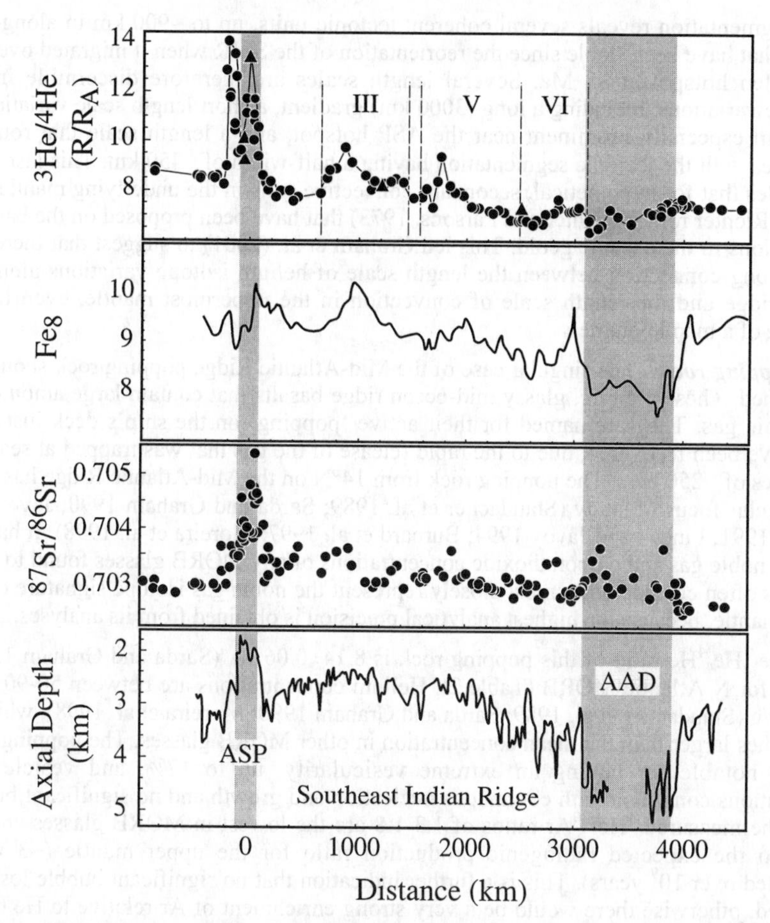

Figure 4. $^3He/^4He$, Fe8, $^{87}Sr/^{86}Sr$ and ridge axis water depth vs. distance along the Southeast Indian Ridge between the Rodrigues Triple Junction and the Australian-Antarctic Discordance. Data are from Graham et al. (1999, 2001), Dosso et al. (1988), Mahoney et al. (2002), and Nicolaysen et al. (2002). The line shown for Fe_8 is the 5-point running mean. Triangles represent lavas from seamounts and off-axis lava fields. The Amsterdam-St. Paul Plateau and AAD regions are depicted as shaded bands. Dashed lines and roman numerals delineate the regional tectonic segmentation from satellite gravity analysis of Small et al. (1999). Some of the geochemical structure roughly coincides with this tectonic segmentation.

preferentially to surrounding mantle peridotite because it has a lower solidus temperature and a smaller melting interval. However, the pyroxenite contribution to a magma will be less for melts that begin to form deeper, because the pyroxenitic melts become more effectively diluted by partial melts of peridotite (Hirschmann and Stolper 1996). Therefore, hotter mantle begins to melt deeper and, if pyroxenite is present, these two effects may lead to aggregated MORB melts that have higher values of Fe_8 and $^3He/^4He$.

There are also several peaks in $^3He/^4He$ and Fe_8, ranging between 200 and 500 km in basal width, superposed on the overall gradients along the SEIR (Fig. 4). These peaks are most prominent near the ASP plateau, and at along-axis distances of 910, 1800 and ~2850 km, respectively. The last three peaks occur near the centers of the regional segmentation of the ridge defined by fracture zone anomalies in satellite gravity data.

This segmentation reveals several coherent tectonic units, up to ~900 km in along-axis width, that have been stable since the reorientation of the SEIR when it migrated over the Kerguelen hotspot at 38 Ma. Several length scales are therefore discernible in the $^3He/^4He$ variations, including a long (3000 km) gradient, a short length scale variation of ~150 km especially prominent near the ASP hotspot, and a length scale that roughly coincides with the tectonic segmentation having a half-width of ~450 km. This last scale resembles that for hypothetical, secondary convective cells in the underlying mantle (so-called "Richter rolls"; Richter and Parsons, 1973) that have been proposed on the basis of undulations in the oceanic geoid. This led Graham et al. (2001) to suggest that there can be a strong connection between the length scale of helium isotope variations along an ocean ridge and the length scale of convection in the uppermost mantle, even in the absence of a mantle plume.

Popping rocks. The singular case of the Mid-Atlantic Ridge popping rock should be mentioned. These are rare, glassy mid-ocean ridge basalts that contain large amounts of magmatic gas. They are named for their active 'popping' on the ship's deck just after they have been recovered, due to the rapid release of the gas that was trapped at seafloor pressures of >250 bars. The popping rock from 14°N on the Mid-Atlantic Ridge has been a particular focus of study (Staudacher et al. 1989; Sarda and Graham 1990; Javoy and Pineau 1991; Pineau and Javoy 1994; Burnard et al. 1997; Moreira et al. 1998). It has the highest noble gas and carbon dioxide concentrations of any MORB glasses found to date, and it is often considered to most closely represent the noble gas isotope signature of the upper mantle, because the highest analytical precision is obtained from its analyses.

The $^3He/^4He$ ratio of this popping rock is 8.14±0.06 R_A (Sarda and Graham 1990), typical for N Atlantic MORB (Table 2). Helium concentrations are between $50\text{-}90\times10^{-6}$ std cm^3/g (Staudacher et al. 1989; Sarda and Graham 1990; Moreira et al. 1998), which is 5-10 times larger than the mean concentration in other MORB glasses. The popping rock is also notable for having an extreme vesicularity, up to 17%, and vesicle size distributions consistent with continuous nucleation and growth and no significant bubble loss. The measured $^4He/^{40}Ar$ ratios of 1.3-1.8 are the lowest in MORB glasses and are close to the expected radiogenic production ratio for the upper mantle (~3 when integrated over 10^9 years). This is a further indication that no significant bubble loss has occurred, otherwise there would be a very strong enrichment of Ar relative to He based on vapor/melt partitioning predicted from experimentally measured solubilities (see *Argon-Helium Systematics*). The popping rock is usually taken as an analog for a relatively undegassed MORB magma. The degassing flux of He at ridge crests, computed from the MORB helium deficit relative to the popping rock and from the crustal production rate by seafloor spreading ($2\text{-}4\times10^{16}$ g y^{-1}), is similar to that derived from the ^3He inventory of the deep ocean (Sarda and Graham 1990; see below).

Ridge helium flux. Because He is lost from the atmosphere, it is not possible to determine its integrated degassing history over geological time. However, an 'instantaneous' picture of the degassing at ridge crests can be obtained from the excess ^3He in the abyssal ocean and the upwelling rate as estimated from a ^{14}C box-model age (~1000 years for the deep Pacific; Craig et al. 1975). The ^3He flux estimated in this way is ~4±1 atoms cm^{-2} sec^{-1} (~1100 mol y^{-1}). In principle, this also provides a means to determine the flux of other elements at ridge crests using elemental/^3He ratios in hydrothermal vent fluids and MORB glasses (e.g., Edmond et al. 1982; Des Marais and Moore 1984; Marty and Jambon 1987; Sarda and Graham 1990; Graham and Sarda 1991), and to corroborate estimates for convective heat loss at ridge crests using vent fluid ^3He-temperature relationships (e.g., Jenkins et al. 1978; Lupton et al. 1989).

The value of the deep ocean He flux allows limits to be placed on the amount of U and Th in that part of the mantle which is being outgassed at ridges. If the production rate of ^4He by radioactive decay equals its loss rate from a mantle reservoir, then the reservoir is at steady-state and the short-term approximation applies (i.e., $t << 1/\lambda$ and $e^{\lambda t}-1 \approx \lambda t$). If the reservoir has a homogeneous $^3He/^4He$ ratio (a good approximation for the upper mantle MORB source), then the accompanying flux of ^3He is given by

$$F_3 = \{8\lambda_{238} + (7/137.88)\lambda_{235} + 6\kappa\lambda_{232}\} ([U]/238)\, M\, (^3He/^4He) \quad (2)$$

where M is the mass of the reservoir. Estimates for upper mantle κ are ~2.5 based on U-series studies of MORBs (e.g., Allègre et al. 1986; O'Nions and McKenzie 1993). The [U] of the upper mantle is ~5 ppb (5×10^{-9} g/g rock) based on models for the transfer of heat producing elements to the continental crust over time (O'Nions et al. 1979). Assuming the mantle above the 660 km seismic discontinuity ($M = 1.1\times10^{27}$ g) has $^3He/^4He = 1.2\times10^{-5}$ (the mean MORB value of 8.7 R_A from Table 2) and these U and Th contents, the calculated ^3He flux is ~520 mol y^{-1}. Obviously a larger proportion of depleted mantle ($M = 2.2\times10^{27}$ g or ~50% of the whole mantle) could bring this calculation in line with the observed flux of 1100 mol y^{-1} at ridge crests. However, the total oceanic heat flux of 20×10^{19} cal y^{-1} needs to be considered as well. The ^4He/heat production ratio at secular equilibrium is ~4×10^{12} atom ^4He/cal. A ^3He flux of 1100 mol y^{-1} corresponds to a ^4He flux of 9.3×10^7 mol y^{-1}, and a heat flux of 1.4×10^{19} cal y^{-1}. Therefore, the amount of radiogenic production supplying the ^4He flux at ridges supplies a maximum of 7% of the observed heat flux (O'Nions and Oxburgh 1983; van Keken et al. 2001). Much of the ^4He flux at ridges appears to be sustained by radiogenic production in the upper mantle, but most of the ^3He and heat fluxes appear to ultimately derive from either a deep mantle reservoir (O'Nions and Oxburgh 1983) or perhaps from the core (Porcelli and Halliday 2001). This concept has directly led to steady-state models for the upper mantle (Kellogg and Wasserburg 1990; O'Nions and Tolstikhin 1994, 1996; Porcelli and Wasserburg 1995a,b; Porcelli and Ballentine 2002) that critically depend on the noble gases measured in MORBs and OIBs.

Cosmic dust. Anderson (1993) argued that subduction of interplanetary dust particles (IDPs) in marine sediments could have supplied a significant amount of ^3He to the mantle over geologic time. However, extraordinarily high fluxes of IDPs in the Archean would be required to account for the amount of ^3He that is degassing at ocean ridges today (see above discussion). The terrestrial supply of ^3He from cosmic dust also depends critically on the size distribution of IDPs that survive atmospheric entry, because this controls the amount of ^3He retained during frictional heating and ablation (Trull 1994; Farley et al. 1997). Furthermore, the Anderson model requires insignificant diffusive loss of ^3He from IDPs during sediment subduction, despite experimental evidence that such loss is complete at subduction zone depths <25 km and temperatures <200°C (Hiyagon 1994b). Therefore, compared to volcanic degassing, this source of ^3He is unimportant from a mantle geochemistry viewpoint.

Relation to major volatiles. The major volatile species in oceanic basalts are carbon dioxide and water. Carbon dioxide has a lower solubility than water in basaltic melts, and it often reaches saturation levels even when eruption occurs at water depths of 4 to 5 km, which makes CO_2 the dominant gas species contained in vesicles. Because noble gases have relatively low solubilities they also tend to be concentrated in vesicles. Based on a global survey of MORB glasses, the helium in vesicles appears to be in Henry's Law equilibrium with the melt (Kurz and Jenkins 1981). This is not generally true of CO_2, which can be saturated or even over-saturated at the pressure corresponding to the eruption depth (e.g., Stolper and Holloway 1988: Dixon et al. 1988; Kingsley and Schilling 1995). This has been attributed to very rapid rise of magma, from depths in the crust at which it was last saturated with CO_2, to the site of eruption and quenching on the seafloor.

Since the seminal paper by Marty and Jambon (1987), the C/^{3}He ratio has been a parameter of considerable interest in chemical geodynamics of the mantle. The CO_2/^{3}He ratio of MORBs varies over a relatively narrow range. Marty and Jambon (1987) suggested that the best estimate of the upper mantle ratio was 2×10^9, on the basis of a small data set for MORB glasses. More detailed work by Marty and Zimmerman (1999) revealed that the range of values is from ~3×10^8 to ~8×10^9, but the mean is still similar to the original estimate. In this latter study the measured CO_2/^{3}He ratio was corrected for potential atmospheric contamination and gas loss, using the He and Ar concentrations and isotope compositions measured in the same samples along with experimentally determined He and Ar solubilities for basalt melts, and assuming that the magma originally had the radiogenic production ratio of ^{4}He/^{40}Ar that is appropriate for the mantle (see the section *Argon-Helium Systematics* for more discussion of this approach). Ocean ridge basalts from more enriched mantle sources, as indicated by ^{87}Sr/^{86}Sr or K/Ti ratios, generally have higher CO_2/^{3}He (Kingsley and Schilling 1995; Marty and Zimmerman 1999). These variations in CO_2/^{3}He probably reflect heterogeneities in mantle-source composition, perhaps as a result of subduction zone recycling of surface-derived carbon.

Overview. The amplitude and length scales of ^{3}He/^{4}He variations along ridges appear to be fundamentally related to the scales of solid state mixing in the uppermost parts of the mantle, and have been especially useful for investigating chemical geodynamics of mantle plume-spreading ridge interactions. The helium isotope composition of the MORB mantle source away from the influence of hotspots and plumes is best described by a relatively narrow range in ^{3}He/^{4}He, with >90% of those ridge basalts lying between 6.5-9.5 R_A. The median ^{3}He/^{4}He ratio is similar for MORBs from each ocean basin (8.08 to 8.24 R_A; Table 2). Values above 11 R_A are observed near the Iceland hotspot (up to 18 R_A along the Reykjanes Ridge and 13 R_A along the Kolbeinsey Ridge; Fig. 3), along the Easter Microplate (12 R_A), near the Shona-Discovery-Bouvet hotspots (up to 15 R_A) in the South Atlantic, in the Red Sea and Gulf of Tadjoura (15 R_A) near the Ethiopian mantle plume, and along the Southeast Indian Ridge near the Amsterdam-St. Paul plateau (14 R_A; Fig. 4). In the singular case of the super-fast spreading region of the southern East Pacific Rise, relatively high ^{3}He/^{4}He ratios (up to 11 R_A) are observed between 16 and 18°S despite the absence of a nearby ocean island hotspot. In this region other isotopic and geochemical anomalies are also observed, suggesting the presence of either an (as yet) undetected hotspot, or a mantle 'blob' that was embedded in the upper mantle flow (Mahoney et al. 1994). The lowest ^{3}He/^{4}He ratios along the global ridge system are also found in areas of geochemical anomalies, such as along the southern Chile Rise (down to 3.5 R_A), within the Australian-Antarctic Discordance (6.2 R_A), and at 33°S on the Mid-Atlantic Ridge (6.2 R_A), the latter being another region suggested to be affected by a mantle 'blob' or heterogeneity (Michael et al. 1994). In the singular case of Axial Volcano on the Juan de Fuca Ridge, ^{3}He/^{4}He lies within the range of ~8-9 R_A (Lupton et al. 1993), despite bathymetric, morphological, and isotopic evidence for the presence of a relatively weak hotspot (e.g., Eaby et al. 1984; Hegner and Tatsumoto 1989).

The fact that the range of ^{3}He/^{4}He variation along ridges is small in comparison to the global range including ocean islands certainly implies that mixing rates in the upper mantle MORB source are relatively rapid. Nevertheless, the range in MORBs is significant, and it results from a combination of (1) input of high ^{3}He/^{4}He material from deeper in the mantle, (2) radiogenic ingrowth in the upper mantle, (3) input of subducted crust and lithosphere that is enriched in U and Th (perhaps directly to the upper mantle or perhaps though mantle plumes), and (4) partial melting of upper mantle that is chemically or mineralogically heterogeneous.

A comparison of observed and calculated helium and heat fluxes at ocean ridges suggests that a deep mantle reservoir supplies most of the ^{3}He and heat to the upper mantle (O'Nions and Oxburgh 1983; Kellogg and Wasserburg 1990). The involvement of such a deep reservoir in the upper mantle heat and mass balance appears to make it inescapable that some form of stratification is present within the mantle, and that the mantle source of some ocean islands, such as Hawaii and Iceland, lies below the mantle source for ocean ridges.

Ocean island basalts

The highest magmatic ^{3}He/^{4}He ratios are found at ocean island localities such as Hawaii and Iceland, where they extend to values above 30 R_A. (The highest values to date are 35-43 R_A, found in the Miocene alkali basalts of northwest Iceland; Breddam and Kurz 2001). Other prominent, high ^{3}He/^{4}He localities include Galápagos, Samoa, Réunion, Easter, Juan Fernandez, Yellowstone and the Ethiopian Rift. The presence of such high ^{3}He/^{4}He ratios at these sites of extensive volcanism is consistent with the existence of mantle plumes or thermal upwellings from regions deep in the Earth. Parts of these deep regions may have remained more effectively isolated over geological time. They are thereby less degassed, and have lower time-integrated (U+Th)/^{3}He compared to the shallower mantle source regions for MORB. (Further support for deep, relatively undegassed regions of the mantle comes from the Ne isotope composition of MORBs and OIBs, as discussed in the section *Ne Isotopes*). Significant spatial and temporal variability in ^{3}He/^{4}He also occurs at ocean islands. This variability may be related to distance from the center of the mantle upwelling beneath an island, or to the stage of a volcano's evolution (e.g., seamount, shield or post-erosional). This variability can often be accounted for by variability in mixing between plume-derived material and material derived from the upper mantle, or by isotopic heterogeneity within the plume itself.

Global variability—relation to geologic and tectonic conditions. One might expect some systematic, global relationships between ^{3}He/^{4}He and other geological parameters. For example, interaction between a plume-derived magma and oceanic lithosphere may serve to lower ^{3}He/^{4}He ratios by shallow-level addition of radiogenic He (Hilton et al. 1995). As a consequence, the ^{3}He/^{4}He variability and the transition to lower ^{3}He/^{4}He at some localities might then be related to the speed of the overlying plate or to the magma supply rate (e.g., Kaneoka and Takaoka 1991). At the present time it is difficult to generalize about these issues, given the limited number of well studied ocean island hotspots and the fact that each one is somewhat unique in its setting. Key geological parameters of interest include the depth of plume origin, the plume mass flux, the extent and depth of partial melting, the proximity of a hotspot to a plate boundary (e.g., spreading ridge, continental margin, island arc), and the velocity and age (thickness) of the overlying plate. Some of these parameters, such as the depth of plume origin, are not readily quantified; others, such as depth and extent of partial melting, are relatively uncertain or show a large range. The ^{3}He/^{4}He ratios observed at ocean islands are compared on the basis of plume flux, lithosphere age and plate speed in Figure 5a-c.

The Hawaiian hotspot represents, by far, the largest plume mass flux (Fig. 5a). It is situated beneath a fast moving plate (100 mm/y) and old (mid-Cretaceous) lithosphere far from any plate boundaries. The highest ^{3}He/^{4}He ratios at Hawaii, up to 35 R_A, are found at Loihi Seamount, the youngest active volcano in the island chain (Kurz et al. 1982a, 1983; Rison and Craig 1983; Kaneoka et al. 1983; Hiyagon et al. 1992; Honda et al. 1993b; Valbracht et al. 1997; Hilton et al. 1997). High ^{3}He/^{4}He ratios persist throughout most of the tholeiitic shield building stage at Hawaiian volcanoes, and tend to show a rapid decrease near the end of this stage to values near 8 R_A characteristic of the ambient upper mantle (e.g., Kurz 1993; Kurz et al. 1983, 1987, 1996; Kurz and Kammer 1991; Rison and Craig 1983; Kaneoka 1987; Kaneoka and Takaoka 1978, 1980; Vance et al. 1989; Honda et al. 1993b; Roden et al. 1994; Mukhopadhyay et al. 1996).

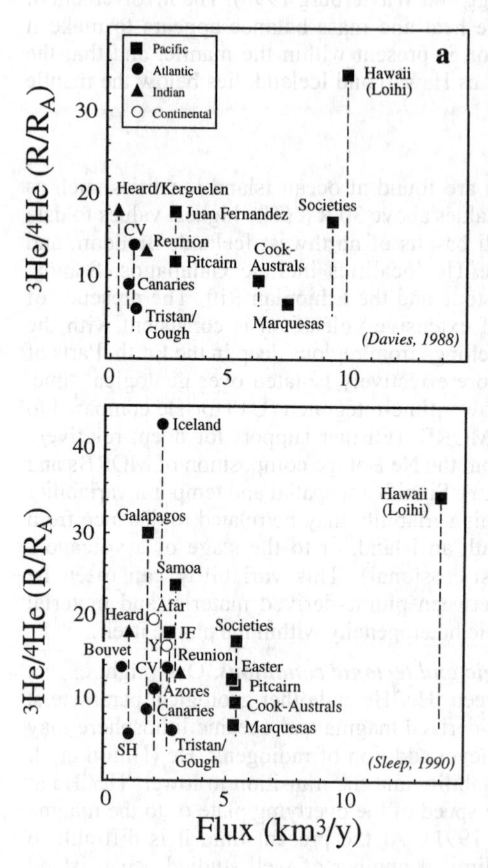

Figure 5. $^3He/^4He$ vs. (a–left) buoyancy flux. (b–opposite page, upper) Lithosphere age. (c–opp. page, lower) Plate speed. Points show the highest $^3He/^4He$ ratios at each island with dashed lines extending to the lowest values. Only crushing analyses are used to avoid post-eruptive addition of cosmogenic ^3He and radiogenic ^4He. Buoyancy flux is taken directly from Davies (1988) or calculated from Sleep (1990) assuming an excess plume temperature of 200°C. He data are from Kurz et al. (1982a, 1983) for Gough, Tristan da Cunha and Hawaii; Hilton et al. (1995) for Heard; Condomines et al. (1983), Kurz et al. (1985), Hilton et al. (1998b, 1999), Dixon et al. (2000), Breddam et al. (2000), Moreira et al. (2001) and Breddam and Kurz (2001) for Iceland; Kaneoka et al. (1986), Graham et al. (1990), Staudacher et al. (1990) and Marty et al. (1993a) for Réunion; Farley et al. (1992, 1993) for Samoa and Juan Fernandez; Hanyu et al. (1999) and Scarsi (2000) for Cook-Australs; Staudacher et al. (1989) and Scarsi (2000) for Societies; Graham et al. (1993) and Kurz and Geist (1999) for Galápagos; Graham et al. (1992a) for St. Helena; Poreda et al. (1993) for Easter; Kurz et al. (1998) for Bouvet; Christensen et al. (2001) for Cape Verde Islands; Graham et al. (1999) for Amsterdam/St. Paul; Moreira et al. (1999) for Azores; Hilton et al. (2000a) and Graham et al. (1996c) for Canary Islands; Desonie et al. (1991) for the Marquesas. For Easter and Amsterdam/St. Paul, the data are for values measured at the nearby spreading center because island data are not yet available. Two continental plumes (Y-Yellowstone and Afar/Ethiopia) shown for comparison: Afar data from Marty et al. (1996) and Scarsi and Craig (1996); 'Y' data from Craig et al. (1978).

The Iceland hotspot is ridge-centered, has a moderate plume flux (Fig. 5a), and shows a large spatial variability in basalt $^3He/^4He$, from high ratios (>40 R_A) to values lower than those typically found along spreading ridges (down to 5 R_A: Condomines et al. 1983; Kurz et al. 1985; Poreda et al. 1992; Burnard et al. 1994b; Marty et al. 1991; Hilton et al. 1990, 1998, 1999; Harrison et al. 1999; Breddam et al. 2000; Dixon et al. 2000; Moreira et al. 2001; Breddam and Kurz 2001). Much lower values (down to 0.07 R_A) are found in silicic and highly evolved volcanic rocks in Iceland, but lower crustal melting is involved in their petrogenesis and their $^3He/^4He$ ratios do not carry useful information about the underlying mantle (Condomines et al. 1983). The very high $^3He/^4He$ values are known to have persisted since the inception of the Iceland plume (Graham et al. 1998; Marty et al. 1998; Stuart et al. 2000; Kirstein and Timmerman 2000). The spatial and temporal variations at Hawaii and Iceland are usually interpreted to result from mixing, between a plume source and ambient upper mantle or lithosphere.

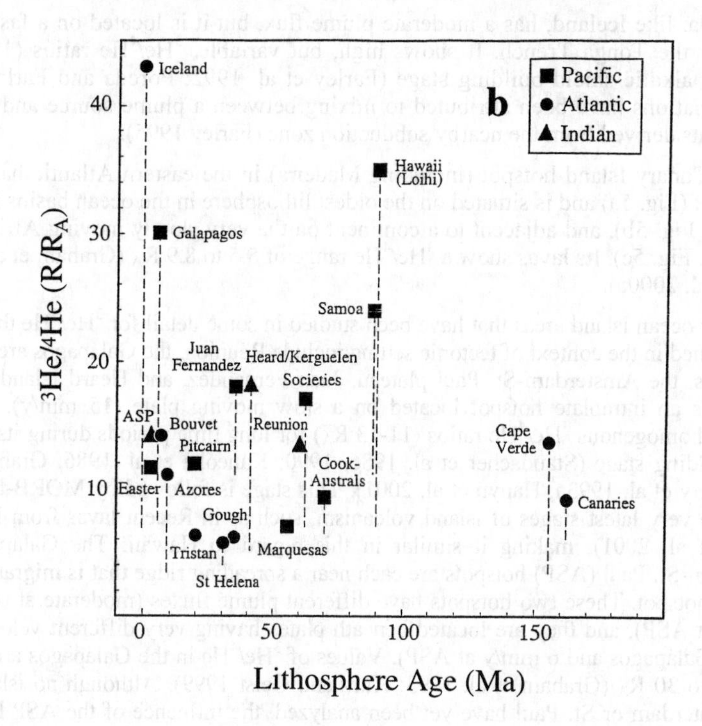

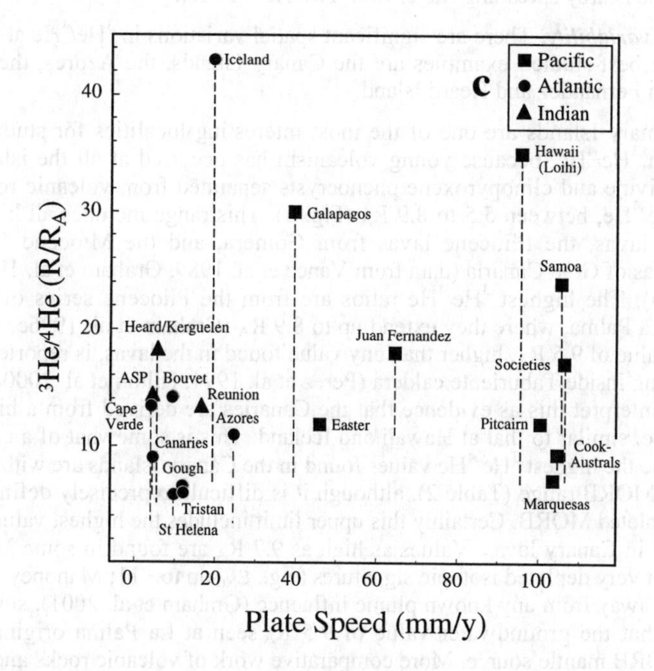

Figure 5, continued. $^3He/^4He$ vs. (b–upper) lithosphere age and (c–lower) plate speed.

Samoa, like Iceland, has a moderate plume flux, but it is located on a fast moving plate near the Tonga Trench. It shows high, but variable, ^{3}He/^{4}He ratios (11-24 R_A) during its alkalic shield-building stage (Farley et al. 1992; Poreda and Farley 1992). These variations have been attributed to mixing between a plume source and recycled components derived from the nearby subduction zone (Farley 1995).

The Canary Island hotspot (including Madeira) in the eastern Atlantic has a small plume flux (Fig. 5a) and is situated on the oldest lithosphere in the ocean basins (Jurassic, >170 Ma; Fig. 5b), and adjacent to a continent on the very slowly moving African plate (≤5 mm/y; Fig. 5c). Its lavas show a ^{3}He/^{4}He range of 5.5 to 8.9 R_A (Graham et al. 1996c; Hilton et al. 2000a).

Other ocean island areas that have been studied in some detail for ^{3}He/^{4}He that should be mentioned in the context of tectonic setting include Réunion, the Galápagos archipelago, the Azores, the Amsterdam-St. Paul plateau, Juan Fernandez, and Heard Island (Fig. 5). Réunion is an intraplate hotspot located on a slow moving plate (15 mm/y). It shows relatively homogenous ^{3}He/^{4}He ratios (11-13 R_A) for long time periods during its volcanic shield-building stage (Staudacher et al. 1986, 1990; Kaneoka et al. 1986; Graham et al. 1990; Marty et al. 1993a; Hanyu et al. 2001). This stage is followed by MORB-like ratios during the very latest stages of island volcanism, such as in Recent lavas from Mauritius (Hanyu et al. 2001), making it similar in this regard to Hawaii. The Galápagos and Amsterdam-St. Paul (ASP) hotspots are each near a spreading ridge that is migrating away from the hotspot. These two hotspots have different plume fluxes (moderate at Galápagos and low at ASP), and they are located beneath plates having very different velocities (40 mm/y at Galápagos and 6 mm/y at ASP). Values of ^{3}He/^{4}He in the Galápagos archipelago range up to 30 R_A (Graham et al. 1993; Kurz and Geist 1999). Although no island lavas from Amsterdam or St. Paul have yet been analyzed, the influence of the ASP hotspot is evident on the nearby spreading ridge, with ^{3}He/^{4}He > 14 R_A.

Spatial variability. There are significant spatial variations in ^{3}He/^{4}He at some ocean islands. The best-studied examples are the Canary Islands, the Azores, the Galápagos Islands, Juan Fernandez and Heard Island.

The Canary Islands are one of the most interesting localities for studying spatial variations in ^{3}He/^{4}He, because young volcanism has occurred at all the islands except Gomera. Olivine and clinopyroxene phenocrysts separated from volcanic rocks show a range in ^{3}He/^{4}He, between 5.5 to 8.9 R_A (Fig. 6). This range includes all historical and Quaternary lavas, the Pliocene lavas from Gomera, and the Miocene (shield) and Pliocene lavas of Gran Canaria (data from Vance et al. 1989; Graham et al. 1996c; Hilton et al. 2000a). The highest ^{3}He/^{4}He ratios are from the Pliocene series of Taburiente caldera on La Palma, where they extend up to 8.9 R_A (Graham et al. 1996c; Hilton et al. 2000a). A value of 9.5 R_A, higher than any value found in the lavas, is reported for a cold mineral spring inside Taburiente caldera (Perez et al. 1994; Hilton et al. 2000a). Hilton et al. (2000a) interpret this as evidence that the Canaries are derived from a high-^{3}He/^{4}He plume source, similar to that at Hawaii and Iceland. This is somewhat of a controversial point because the highest ^{3}He/^{4}He values found in the Canary Islands are within the upper limit of the MORB range (Table 2), although it is difficult to precisely define a limiting value for depleted MORB. Certainly this upper limit includes the highest values of 8.9 R_A as measured in Canary lavas. Values as high as 9.7 R_A are found in some MORB glass samples with very depleted isotopic signatures (e.g., ε_{Nd} up to +11; Mahoney et al. 1994), and in areas away from any known plume influence (Graham et al. 2001), suggesting the possibility that the groundwater value of 9.5 R_A seen at La Palma originates from a depleted MORB mantle source. More comparative work of volcanic rocks and associated fluids is needed to resolve this issue.

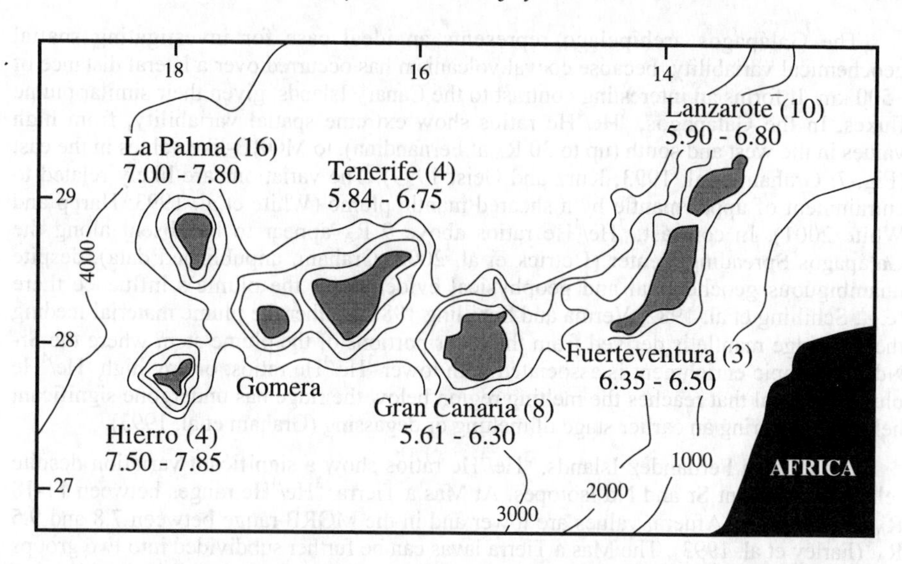

Figure 6. Map of the range in $^3He/^4He$ ratios for phenocrysts and xenoliths from historical and Quaternary lavas of the Canary Islands. Data are from Vance et al. (1989), Graham et al. (1996c) and Hilton et al. (2000a). $^3He/^4He$ ratios are systematically higher in young lavas from the western islands of La Palma and Hierro compared to the eastern islands. Pliocene lavas from La Palma extend to even higher values (8.9 R_A; Graham et al., 1996c, Hilton et al. 2000a). No Quaternary lavas are present on Gomera, but Pliocene lavas there show the full range seen in the Canary Island volcanic rocks (5.5-8.6 R_A; n = 4, Graham et al. 1996c).

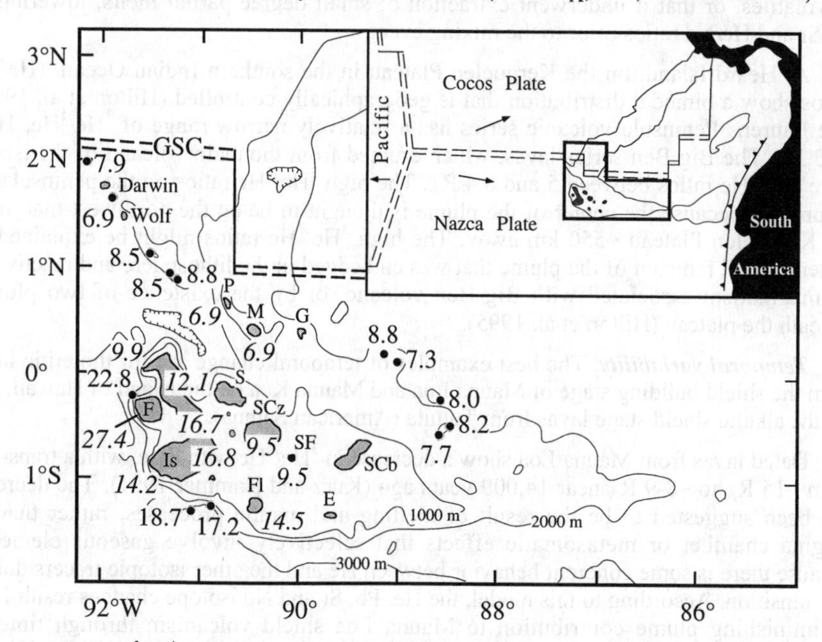

Figure 7. Map of $^3He/^4He$ ratios in young lavas and fumaroles from the Galápagos Archipelago. Data are from Graham et al. (1993), Kurz and Geist (1999), Goff et al. (2000). There is a systematic spatial variation in $^3He/^4He$ similar to variations for other isotopic systems in the region (e.g., White et al. 1993; Harpp and White 2001). The highest $^3He/^4He$ ratios occur in the western islands of Fernandina and Isabela, while MORB-like values are found to the east near the center of the archipelago.

The Galápagos archipelago represents an ideal case for investigating spatial geochemical variability, because coeval volcanism has occurred over a lateral distance of ~500 km. It forms an interesting contrast to the Canary Islands, given their similar plume fluxes. In the Galápagos, $^3He/^4He$ ratios show extreme spatial variability, from high values in the west and south (up to 30 R_A at Fernandina), to MORB-like values in the east (Fig. 7; Graham et al. 1993; Kurz and Geist 1999). The variations are likely related to entrainment of upper mantle by a sheared mantle plume (White et al. 1993; Harpp and White 2001). In contrast, $^3He/^4He$ ratios above 9 R_A appear to be absent along the Galápagos Spreading Center (Detrick et al. 2002; Graham, unpublished data), despite unambiguous geochemical and geophysical evidence for the plume's influence there (e.g., Schilling et al. 1982; Verma and Schilling 1982). Either the plume material feeding the sub-ridge mantle is derived from the outer portions of the plume stem where the Sr-Nd-Pb isotopic enrichment is associated with lower $^3He/^4He$ ratios, or any high $^3He/^4He$ plume material that reaches the melting region below the ridge has undergone significant helium loss during an earlier stage of melting or degassing (Graham et al. 1993).

In the Juan Fernandez Islands, $^3He/^4He$ ratios show a significant variation despite relatively constant Sr and Nd isotopes. At Mas a Tierra, $^3He/^4He$ ranges between 11-18 R_A, while at Mas Afuera, values are lower and in the MORB range between 7.8 and 9.5 R_A. (Farley et al. 1993). The Mas a Tierra lavas can be further subdivided into two groups that show no overlap in He isotope composition; the alkalic and tholeiitic shield group has $^3He/^4He$ = 14.5-18.0 R_A, and the post-shield basanite group has $^3He/^4He$ = 11.2-13.6 R_A. Farley et al. (1993) interpret the $^3He/^4He$ variations as the result of mixing between plume and upper mantle sources. The relatively constant $^{87}Sr/^{86}Sr$ and $^{143}Nd/^{144}Nd$ at the islands then requires that the plume component initially had a heterogeneous distribution of volatiles, or that it underwent extraction of small degree partial melts, lowering its He/Sr and He/Nd ratios prior to the mixing event.

At Heard Island, on the Kerguelen Plateau in the southern Indian Ocean, $^3He/^4He$ ratios show a bimodal distribution that is geographically controlled (Hilton et al. 1995). The Laurens Peninsula volcanic series has a relatively narrow range of $^3He/^4He$, 16.2-18.3 R_A. The Big Ben series lavas, which erupted from the main volcano of the island, have $^3He/^4He$ ratios between 5 and 8.4 R_A. The high $^3He/^4He$ ratios on the peninsula are surprising, because the center of the plume is thought to be on the northwest margin of the Kerguelen Plateau ~550 km away. The high $^3He/^4He$ ratios might be explained by either a fossil remnant of the plume that was embedded in the lithosphere and reactivated by magmatism associated with Big Ben volcano, or by the existence of two plumes beneath the plateau (Hilton et al. 1995).

Temporal variability. The best examples of temporal change are for tholeiitic lavas from the shield building stage of Mauna Loa and Mauna Kea on the island of Hawaii, and for the alkalic shield stage lavas from Tutuila (American Samoa).

Dated lavas from Mauna Loa show a decrease in $^3He/^4He$ with time, with a transition from ~15 R_A to ~8-9 R_A near 14,000 years ago (Kurz and Kammer 1991). The decrease has been suggested to be the result of melting and mantle processes, rather than of magma chamber or metasomatic effects that selectively involve gaseous elements, because there is some coherent behavior between He and the other isotopic tracers during the transition. According to this model, the He, Pb, Sr and Nd isotope changes result from a diminishing plume contribution to Mauna Loa shield volcanism through time. A subsequent He isotope investigation of Mauna Kea volcanism, through the Hawaii Scientific Drilling Project (HSDP), shows a similar change near the end of the shield building stage at Mauna Kea (Fig. 8), also with He and Nd isotope evidence for less plume involvement through time (Kurz et al. 1996; Lassiter et al. 1996). Lavas from the

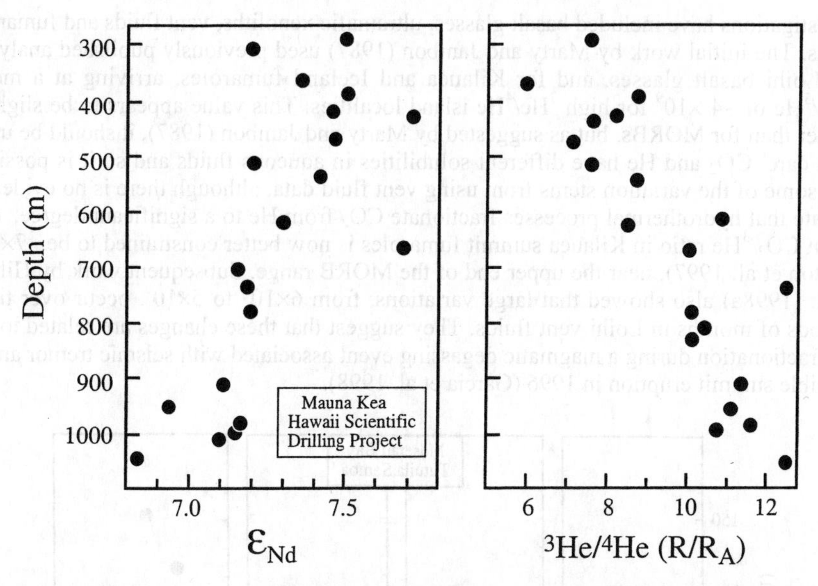

Figure 8. $^3He/^4He$ and ε_{Nd} in basaltic lavas from Mauna Kea sampled by the Hawaii Scientific Drilling Project (HSDP). Data are from Kurz et al. (1996) and Lassiter et al. (1996). There is a covariation in He and Nd isotopes during the end of the tholeiitic shield-building stage in lavas from Mauna Kea. The temporal trend is similar to that for dated lavas from Mauna Loa (Kurz et al. 1987; Kurz and Kammer 1991), in which MORB-like compositions dominate at the very end of the shield-building stage.

upper section of the HSDP core, between 240-270 m, were erupted from Mauna Loa and have $^3He/^4He$ between 13.9–15.8 R$_A$, similar to values found in Mauna Loa subaerial lavas that are >30,000 years old. The upper section of the Mauna Kea sequence, between 290-620 m, has $^3He/^4He$ between 6.8-7.0 R$_A$, similar to MORB values (Fig. 8). A single sample from this section has $^3He/^4He$ of 6 R$_A$, among the lowest values found at Hawaiian volcanoes. Some later-stage radiogenic He addition may be involved, but its origin is unclear. At depths below 620 m, higher $^3He/^4He$ values are found, ranging up to 12.5 R$_A$. The transition from high $^3He/^4He$ to MORB–like values occurs within the end of the tholeiitic shield stage of Mauna Kea (Kurz et al. 1996).

In contrast, the stratigraphic section from Tutuila (American Samoa) shows a large variation in $^3He/^4He$, between 11-24 R$_A$ (Fig. 9). There are large excursions in isotope composition over relatively short time periods, and no systematic trend though time (Farley et al. 1992). The variations appear to represent mixing of two isotopically extreme mantle components. The highly enriched component is characterized by radiogenic $^{87}Sr/^{86}Sr$ and low $^3He/^4He$, and may originate from recycled crustal material injected into the rising plume from the nearby subduction zone (Farley et al. 1992; Farley 1995). The high $^3He/^4He$ component has intermediate Sr, Nd and Pb isotope ratios compared to the full range of OIBs and MORBs. The rapid temporal variations in He, Sr, Nd and Pb isotopes at Samoa, without a systematic trend, indicate that the hypothesized mixing process must be erratic. This might occur when a deep rising plume intercepts and melts recycled crustal material trapped at a boundary layer in the mantle (Farley et al. 1992).

Relation to major volatiles. Studies of the $CO_2/^3He$ ratio have received the most attention in terms of the relations between helium and major volatiles at ocean islands.

Investigations have included basalt glasses, ultramafic xenoliths, vent fluids and fumarole gases. The initial work by Marty and Jambon (1987) used previously published analyses for Loihi basalt glasses, and for Kilauea and Iceland fumaroles, arriving at a mean $CO_2/^3He$ of ~4 $\times 10^9$ for high $^3He/^4He$ island localities. This value appears to be slightly higher than for MORBs, but as suggested by Marty and Jambon (1987), it should be used with care. CO_2 and He have different solubilities in aqueous fluids and so it is possible that some of the variation stems from using vent fluid data, although there is no evidence to date that hydrothermal processes fractionate CO_2 from He to a significant degree. The mean $CO_2/^3He$ ratio in Kilauea summit fumaroles is now better constrained to be ~7×10^9 (Hilton et al. 1997), near the upper end of the MORB range. Subsequent work by Hilton et al. (1998a) also showed that large variations, from 6×10^8 to 5×10^9, occur over time periods of months in Loihi vent fluids. They suggest that these changes are related to C-He fractionation during a magmatic degassing event associated with seismic tremor and a possible summit eruption in 1996 (Garcia et al. 1998).

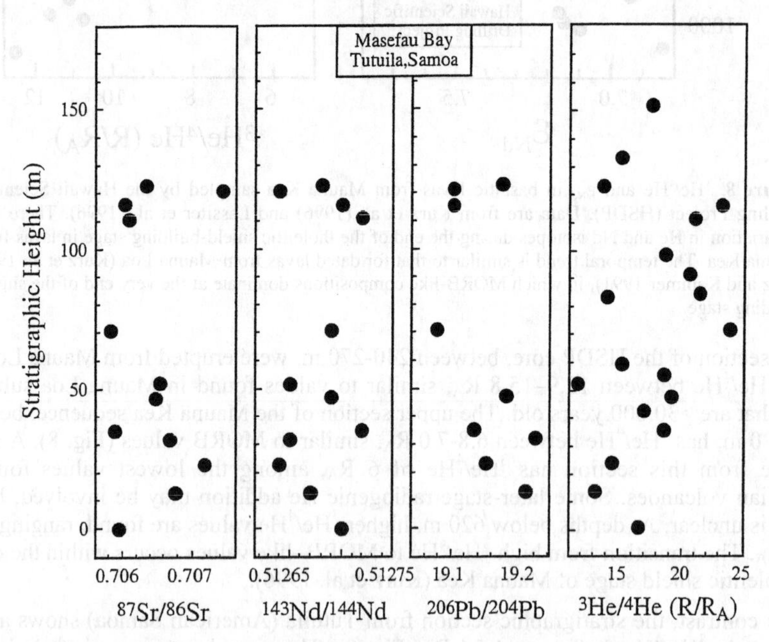

Figure 9. Stratigraphic section of the variations in $^3He/^4He$, $^{87}Sr/^{86}Sr$, $^{143}Nd/^{144}Nd$ and $^{206}Pb/^{204}Pb$ at Masefau Bay, island of Tutuila (Farley et al. 1992). The data are for the alkalic basalt shield-building stage which is characteristic of the Samoan Islands.

Considerable insight into OIB C-He relations has been gained from studies of ocean island xenoliths. Of special note are the laser extraction studies performed on ultramafic xenoliths from Réunion and Samoa by Burnard et al. (1994a, 1998). Much of the range in $CO_2/^3He$ at ocean islands (~1-20×10^9; Trull et al. 1993) can be observed at microscopic scales in a single magmatic cumulate nodule of dunite from Réunion ($CO_2/^3He$ = 1-10×10^9; Burnard et al. 1994a). The $CO_2/^3He$ ratio does not show any systematic variation as a function of fluid inclusion morphology or generation (e.g., primary vs. secondary). The observed variations may be related to the presence of immiscible melts having very different C solubilities, because there are rare examples of highly alkaline glassy inclusions preserved in Réunion lavas. The Samoan xenolith studied by Burnard et al. (1998) using laser extraction is even more complex than the Réunion xenolith, having numerous stages of melt and fluid inclusion growth. Very high $CO_2/^3He$ ratios were measured in this sample, between $11\text{-}41\times10^9$, but this sample may have suffered considerable diffusive loss of He, because many of the Samoan xenoliths have anomalously low $^4He/^{40}Ar^*$ ratios (~0.1) compared to the expected mantle production ratio (~1.5-4.0; see *Argon-Helium Systematics* for further discussion on using $^4He/^{40}Ar^*$).

The range of values of $CO_2/^3He$ for MORBs and OIBs is strikingly similar to the range of chondritic values estimated by Marty and Jambon (1987). Trull et al. (1993) showed that, while there was overlap between the $CO_2/^3He$ ratios of MORBs and OIBs, the basalts and xenoliths from high $^3He/^4He$ ocean islands also have considerably higher $CO_2/^4He$ ratios (the OIB range is $4\text{-}40\times10^4$ compared to the MORB range of $0.5\text{-}7\times10^4$). Trull et al. (1993) suggested that the roughly similar $CO_2/^3He$, but lower $^3He/^4He$ and $CO_2/^4He$ for MORBs compared to OIBs, was produced by radiogenic ^4He production in the upper mantle, consistent with the steady-state upper mantle model for helium proposed by Kellogg and Wasserburg (1990). Preferential carbon recycling to the upper mantle would lead to a higher $CO_2/^3He$ in MORBs compared to OIBs. The similarity of $CO_2/^3He$ between mid-ocean ridges and ocean islands having high $^3He/^4He$ appears to require carbon recycling to the OIB mantle source (Trull et al. 1993).

He-Sr-Nd-Pb isotope relations. Helium isotopes for MORBs and OIBs are plotted against $^{87}Sr/^{86}Sr$, $^{143}Nd/^{144}Nd$, $^{206}Pb/^{204}Pb$ and $^{208}Pb^*/^{206}Pb^*$ in Figure 10a-d. The highest $^3He/^4He$ ratios occur at intermediate values of Sr, Nd and Pb isotopes (Kurz et al. 1982a; Farley et al. 1992; Graham et al. 1993; Hanan and Graham 1996; Hilton et al. 1999). There is some covariation between He and the other isotope systems, particularly when data from individual localities are considered alone; but there are no simple global systematics. Mixing between MORB and OIB 'components' is usually invoked to explain much of the isotopic variation in oceanic basalts. However, the relationships between OIB mantle components and mantle reservoirs, if there are any, is unclear. At least 5 mantle components, including the MORB mantle, can be identified from the multi-isotopic approach (White 1985; Zindler and Hart 1986b). Some of the helium isotope variation in OIBs, and its relation to Pb, Nd and Sr isotope variations, appears to be adequately explained by mixing between high-$^3He/^4He$ plume material and MORB mantle. Ancient melting events are primarily responsible for the evolution of the MORB mantle to low $^{87}Sr/^{86}Sr$ and high $^{143}Nd/^{144}Nd$ ratios. In contrast, recycling of crust or lithosphere appears to play a dominant role in the Sr-Nd-Pb isotopic signatures of many OIBs, especially for the Kerguelen and Society island sources having $^{87}Sr/^{86}Sr>0.705$ (White 1985).

It is important to continually bear in mind that crustal material should have extremely low $^3He/^4He$ (~0.1 R_A) if it is degassed and more than 10-100 million years old. Yet the lowest $^3He/^4He$ ratios observed at ocean island localities suggested to contain such recycled components are about 3-5 R_A, or 300-500 times larger than the expected value. If recycled material is present in an OIB mantle source region, then the fraction of He from such a component must be extremely low for any of the island systems studied so far (e.g., Farley 1995).

The islands of Tristan da Cunha and Gough in the South Atlantic are notable because they have Pb, Nd and Sr isotope characteristics similar to hypothetical values for the bulk silicate Earth. Kurz et al. (1982a) showed that $^3He/^4He$ was 5-6 R_A at these islands, lower than MORB values (Fig. 10a), indicating that a primitive mantle reservoir is not involved. On this basis, Kurz et al. (1982a) suggested that recycled crust was present in the Tristan and Gough mantle source regions.

In the Azores, the $^3He/^4He$ ratio ranges between 3.5 and 11.3 R_A (Kurz et al. 1990; Moreira et al. 1999), and there is good covariation of $^3He/^4He$ with Pb isotopes at the

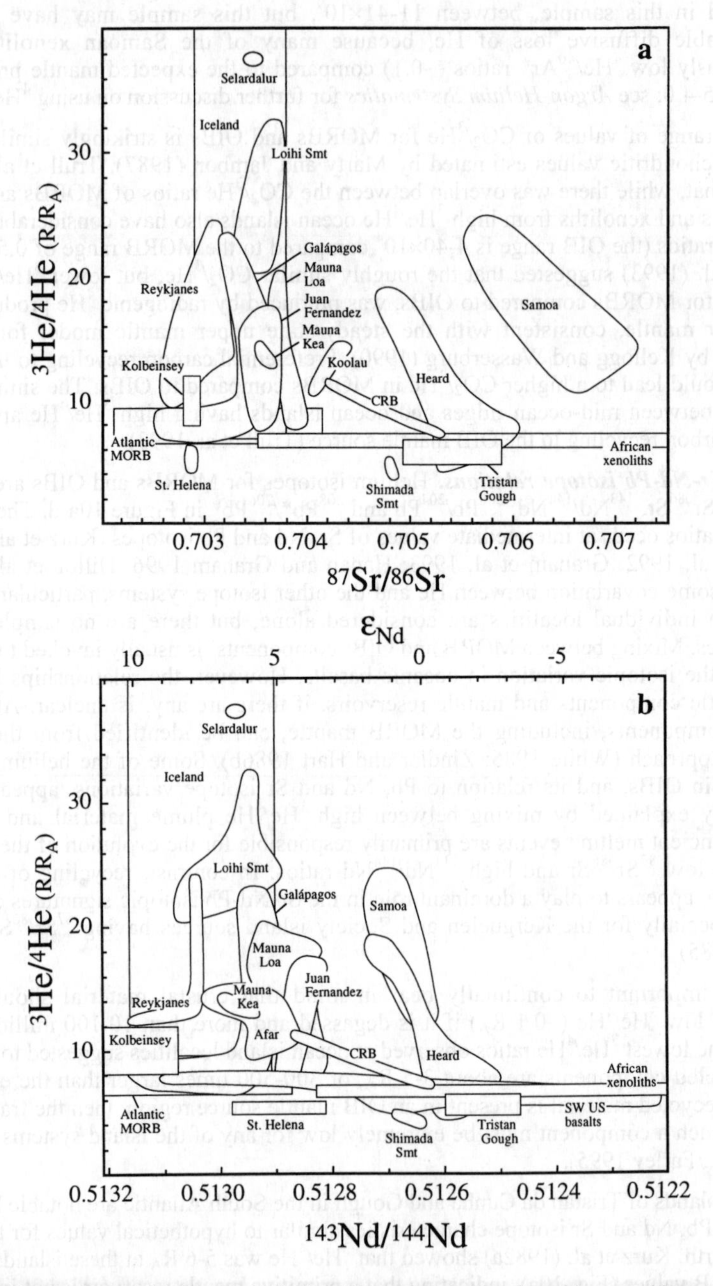

Figure 10. (a) He-Sr, (b) He-Nd, (c,d) He-Pb isotopic relations for mid-ocean ridge and ocean island basalts (after Graham et al. 1998 and references therein). The fields shown encompass paired analyses of the same samples in all cases, except that where boxes are shown they encompass the range of values at those localities. Selardalur, in northwest Iceland, is currently the locality having the highest measured $^3He/^4He$ ratio in a lava for which Sr, Nd and Pb isotope data are also available (Hilton et al. 1999).

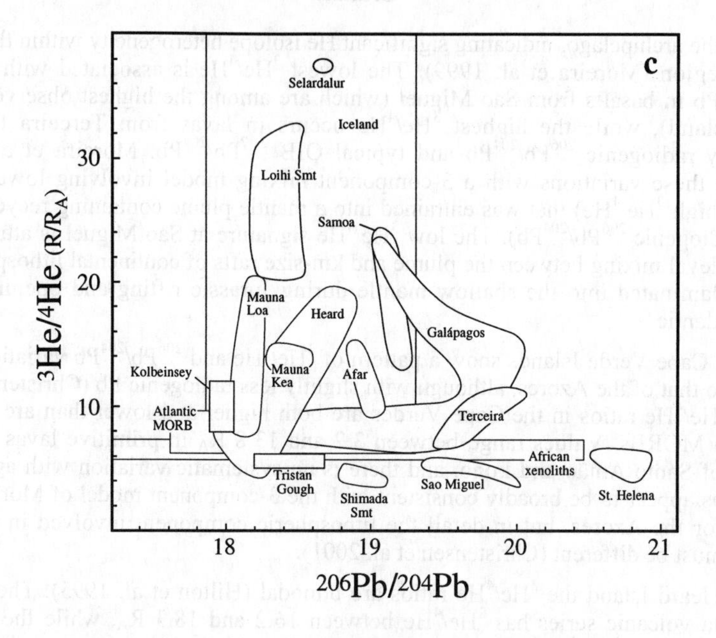

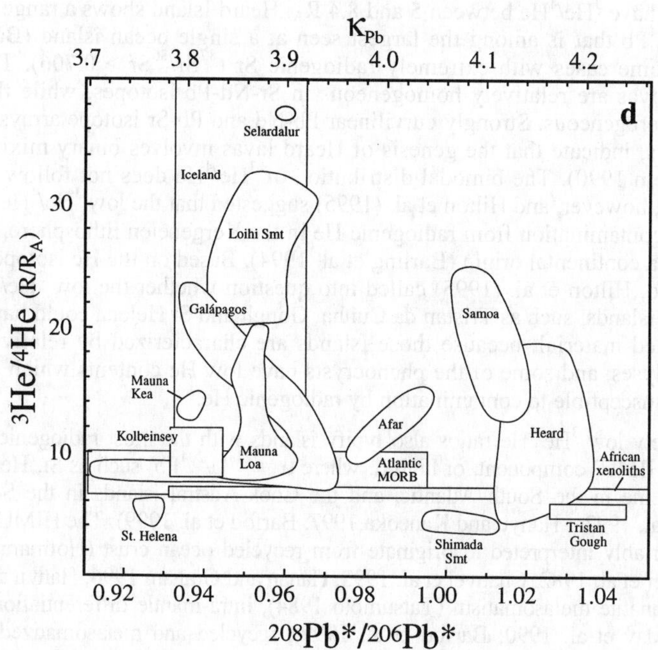

Figure 10, continued. (c,d) He-Pb isotopic relations for mid-ocean ridge and ocean island basalts. The ^{208}Pb*/^{206}Pb* is the isotopic ratio corrected for primordial Pb, and is related to the time-integrated Th/U ratio, designated κ_{Pb} (Allègre et al. 1986).

scale of the archipelago, indicating significant He isotope heterogeneity within the mantle source region (Moreira et al. 1999). The lowest $^3He/^4He$ is associated with elevated $^{207}Pb/^{204}Pb$ in basalts from Sao Miguel (which are among the highest observed at any ocean island), while the highest $^3He/^4He$ occurs in lavas from Terceira that have relatively radiogenic $^{206}Pb/^{204}Pb$ and typical OIB $^{207}Pb/^{204}Pb$. Moreira et al. (1999) interpret these variations with a 3-component mixing model involving lower mantle (having high $^3He/^4He$) that was entrained into a mantle plume containing recycled crust (with radiogenic $^{206}Pb/^{204}Pb$). The low $^3He/^4He$ signature at Sao Miguel is attributed to shallow level mixing between the plume and km-size rafts of continental lithosphere that were delaminated into the shallow mantle during Jurassic rifting and opening of the North Atlantic.

The Cape Verde Islands show a pattern of $^3He/^4He$ and $^{206}Pb/^{204}Pb$ variation that is similar to that of the Azores, although with slightly less radiogenic Pb (Christensen et al. 2001). $^3He/^4He$ ratios in the Cape Verdes are both higher and lower than are typically found in MORBs. Values range between 3.2 and 13.8 R_A in primitive lavas from the islands of Santo Antão and Fogo, and there is no systematic variation with age. These variations appear to be broadly consistent with the 3-component model of Moreira et al. (1999) for the Azores, but in detail the lithospheric component involved in the Cape Verdes must be different (Christensen et al. 2001).

At Heard Island the $^3He/^4He$ ratios are bimodal (Hilton et al. 1995). The Laurens Peninsula volcanic series has $^3He/^4He$ between 16.2 and 18.3 R_A, while the Big Ben series lavas have $^3He/^4He$ between 5 and 8.4 R_A. Heard Island shows a range in $^{87}Sr/^{86}Sr$ and $^{206}Pb/^{204}Pb$ that is among the largest seen at a single ocean island (Barling et al. 1994), in some cases with extremely radiogenic Sr ($^{87}Sr/^{86}Sr > 0.706$). The Laurens Peninsula lavas are relatively homogeneous in Sr-Nd-Pb isotopes, while the Big Ben series is heterogeneous. Strongly curvilinear Pb-Nd and Pb-Sr isotope arrays, and linear Pb-Pb arrays, indicate that the genesis of Heard lavas involves binary mixing (Barling and Goldstein 1990). The bimodal distribution of $^3He/^4He$ does not follow this simple relationship, however, and Hilton et al. (1995) suggested that the low $^3He/^4He$ ratios were derived by contamination from radiogenic He in the Kergeuelen lithosphere, which may partly have a continental origin (Barling et al. 1994). Based on the He isotope results for Heard Island, Hilton et al. (1995) called into question whether the low $^3He/^4He$ seen at other ocean islands, such as Tristan da Cunha, Gough and St Helena could have an origin from recycled material, because those islands are characterized by relatively few He isotope analyses, and some of the phenocrysts have low He contents which could make them more susceptible to contamination by radiogenic He.

Relatively low $^3He/^4He$ ratios also typify islands with the most radiogenic $^{206}Pb/^{204}Pb$ ratios (the 'HIMU' component, or high μ, where $\mu = {}^{238}U/^{204}Pb$) such as St. Helena and the Cameroon Line in the South Atlantic, and the Cook-Austral islands in the South Pacific (Graham et al. 1992a; Hanyu and Kaneoka 1997; Barfod et al. 1999). The HIMU component has been variably interpreted to originate from recycled ocean crust (Hofmann and White 1982; Zindler et al. 1982; Chauvel et al. 1992; Hanan and Graham 1996; Hanyu and Kaneoka 1997), carbonatite metasomatism (Tatsumoto 1984), intra-mantle differentiation by silicate melts (Halliday et al. 1990; Barfod et al. 1999), recycled and metasomatized continental lithosphere (McKenzie and O'Nions 1995), and recycled oceanic lithosphere (Moreira and Kurz 2001). The significance of low $^3He/^4He$ values at these ocean island localities is further complicated, because low ratios might also be produced by interaction with the oceanic lithosphere through which OIB magmas ascend, or by prolonged storage of degassed magma at crustal levels (Zindler and Hart 1986a; Graham et al. 1988). Thorough studies of "low-$^3He/^4He$" islands are needed to resolve many of these issues.

At the present time, the myriad of mantle components inferred from the Sr-Nd-Pb isotopes cannot be adequately placed into a reservoir framework within the mantle. Many of the components appear to be ubiquitous. For example, the enriched type of mantle present at some ocean islands also appears to comprise a worldwide pollutant of the MORB mantle (Hanan and Graham 1996). Furthermore, while it is obvious that the MORB mantle is well characterized for $^3He/^4He$, it is currently not possible to identify a single value of $^3He/^4He$ (or even a narrow range of values) that characterize each of the OIB isotopic end-members. One possible exception to this is the relatively uniform and low $^3He/^4He$ ratios (<7 R_A) of HIMU lavas, which suggests that this may be a general characteristic of this mantle source type (Graham et al. 1992; Hanyu and Kaneoka 1997).

In a global context, there is a convergence of individual ocean island isotopic arrays on a roughly similar composition that lies internal to the OIB end-members (Hart et al. 1992; Farley et al. 1992; Hauri et al. 1994; Hanan and Graham 1996). This convergence implies the presence of a distinct mantle reservoir with approximately the following characteristics; $^{87}Sr/^{86}Sr$ = 0.703-0.704, $^{143}Nd/^{144}Nd$ = 0.51285-0.51295 (ε_{Nd} = 4-6) and $^{206}Pb/^{204}Pb$ = 18.5-19.5. This is a large range. Nevertheless, $^3He/^4He$ of individual OIB suites often shows an increasing trend toward these intermediate compositions (Farley et al. 1992; Graham et al. 1993). Indeed, the highest measured $^3He/^4He$ in a lava for which Sr, Nd and Pb isotope data are also available, from Selardalur in northwest Iceland (Hilton et al. 1999), lies in this range (see Fig. 10). From the MORB viewpoint, the patterns of He isotope convergence appear to be different for the Mid-Atlantic Ridge (geochemical anomalies at 14°N and 33°S) compared to the East Pacific Rise (geochemical anomalies at 17°S and the Easter Microplate). In the Atlantic cases an increasingly radiogenic Pb signature, indicating more of the common material, is accompanied by a stronger radiogenic He signal (lower $^3He/^4He$), while in the Pacific the opposite appears to be true. Hanan and Graham (1996) inferred from this contrast that this material originates from a boundary layer within the mantle, below which the high $^3He/^4He$ reservoir is located. The common material with low $^3He/^4He$ was taken to originate from recycled material trapped in this layer, while the high $^3He/^4He$ in the Pacific cases implies a slightly larger mass transfer in that region from the underlying reservoir. This could further imply that the regional distribution of recycled crust trapped in the mantle accounts for the contrast in He and Pb isotope covariations.

Subduction of crust and lithosphere is a major mechanism for creating mantle heterogeneity. It is possible that recycled materials form distinct reservoirs that contribute to the large-scale structure of mantle, such as trapped megaliths (Ringwood 1982) or as large, coherent downgoing slabs that are impeded at deep boundary layers (Kellogg et al. 1999). As yet, however, no unequivocal evidence for this notion exists in the global isotope systematics. In fact, where the mixing relationships are best defined for individual island localities, through strongly hyperbolic arrays in Sr-Nd-Pb isotope relations such as those observed at Heard Island (Barling and Goldstein 1990), the individual island components do not correspond to the extreme mantle end-members defined from the global data set. This indicates that OIB source reservoirs often have intermediate isotopic compositions that were created by prior differentiation or mixing events (Barling and Goldstein 1990; Hart et al. 1992; Hanan and Graham 1996).

The comparisons in Figure 10 do indicate that some OIB source (deep mantle) regions have $^3He/^4He \geq 40$ R_A. These mantle regions also have κ_{Pb} values near 3.9-4.0 in some cases (Fig. 10d), consistent with a long-term history of quasi-isolation. Such κ_{Pb} values are present in the mantle source of high-$^3He/^4He$ tholeiitic basalts from the shield stage of many Hawaiian volcanoes, where a remarkable and internally consistent relationship between $^3He/^4He$ and $^{208}Pb/^{206}Pb$ was discovered (Eiler et al. 1998). For these Hawaiian shield lavas, Eiler et al. (1998) showed that it is possible to predict the $^3He/^4He$ ratio to ±2 R_A based only on Pb isotope systematics. This also suggests that some of the observed isotopic variations at ocean islands are related to internal heterogeneity within their mantle source regions.

He-Os isotope relations. An emerging area of investigation is the relationship between $^3He/^4He$ and $^{186}Os/^{188}Os$ in ocean island basalts. ^{186}Os is produced by long-lived radioactive decay of ^{190}Pt, and the Earth's outer core is thought to be enriched in Pt/Os, making the $^{186}Os/^{188}Os$ a potential tracer for the involvement of very deep Earth material in the origin of some OIBs (Walker et al. 1995; Brandon et al. 1998). Os isotopes in olivine-rich basalts from Hawaii can be explained by the involvement of ~1% of outer core material admixed with the Hawaiian mantle source, and there is a correlation showing that higher $^3He/^4He$ is associated with higher $^{186}Os/^{188}Os$ in Hawaiian picrites (Brandon et al. 1999). However, a reconnaissance sampling of picrites from Iceland shows that elevated $^{186}Os/^{188}Os$ is not associated with high $^3He/^4He$ (Brandon et al. 2001). More work on the He-Os isotope systematics will provide important constraints on possible involvement of material from the core-mantle boundary in the noble gas signature of ocean island basalts.

Overview. There does not seem to be any direct relationship between plate speed or plume flux and the value of $^3He/^4He$ measured at ocean islands. It does seem possible that somewhat more variable $^3He/^4He$ ratios may be found on old oceanic lithosphere or near continental margins/subduction zones, but this cannot be firmly concluded based on the available evidence (Fig. 5). Buoyancy flux, plate speed and lithosphere age may serve to modulate the $^3He/^4He$ signal at ocean islands, but it is clear that they do not control it.

The high $^3He/^4He$ ratios at localities such as Hawaii, Iceland, Galápagos and Samoa are consistent with the idea that some deep Earth regions have remained more effectively isolated and are less degassed than the upper mantle MORB source. The relations of Sr, Nd and Pb with He isotopes clearly identify two distinct reservoirs, namely, the depleted MORB mantle and a high $^3He/^4He$ mantle source for ocean islands such as Hawaii and Iceland. OIBs with the highest $^3He/^4He$ ratios show more enriched Sr and Nd isotope compositions compared to MORBs. Nevertheless, these OIBs are not characterized by primitive or 'bulk Earth' Sr and Nd isotope compositions ($^{87}Sr/^{86}Sr$ = 0.705, $^{143}Nd/^{144}Nd$ = 0.51264). Instead, they have depleted isotope signatures, indicating some earlier episode of melting in the history of their mantle sources. These isotope systematics might be explained by mixing of a very small amount (less than a few percent) of primitive, gas-rich mantle with differentiated mantle reservoirs. However, any self-consistent and detailed mass balance model that attempts to fully account for all the isotopic observations is likely to be non-unique, due to the range of crustal and lithospheric materials that have been recycled to the mantle.

At regional and local scales, the relationship of $^3He/^4He$ to other isotopic tracers is often consistently explained in terms of mantle or magma mixing processes. There are some cases that may need to be explained by plume heterogeneity in $^3He/^4He$, by magma degassing and radiogenic ingrowth of He, or by wallrock reaction/metasomatic effects during magma transport, but detailed studies have yet to be carried out that firmly establish cause and effects. Perhaps of most significance to characterizing reservoirs within the Earth, it is unclear whether the intermediate $^3He/^4He$ ratios (~10 to 30 R_A) often found at ocean islands are always due to mixing between high $^3He/^4He$ plume material and upper mantle or recycled components. The best example is along the Réunion hotspot track, where values of 11-13 R_A appear to persist for 65 Ma (Staudacher et al. 1990; Graham et al. 1990; Marty et al. 1993a; Basu et al. 1993; Hanyu et al. 2001). The constancy of intermediate $^3He/^4He$ for such long time periods is not readily explained by mixing, and suggests the possibility for deep mantle reservoirs having variable $^3He/^4He$.

NEON

Significance

Early neon isotope measurements indicated that the Ne isotope composition of the Earth's mantle differs from that of the atmosphere (Craig and Lupton 1976; Poreda and Radicati di Brozolo 1984), but confidence in the early results was limited by the low analytical precision. Once some of the analytical obstacles were overcome, the work by Honda et al. (1987) and Ozima and Zashu (1988) on diamonds, and by Sarda et al. (1988) on submarine volcanic glasses from ridges and islands, showed conclusively that the mantle is characterized by elevated ratios of $^{20}Ne/^{22}Ne$ and $^{21}Ne/^{22}Ne$ compared to the atmosphere. The relatively high $^{20}Ne/^{22}Ne$ in MORBs and OIBs indicates that the connection between the Earth's atmosphere and mantle is not a simple, complementary relationship. The different correlations between $^{20}Ne/^{22}Ne$ and $^{21}Ne/^{22}Ne$ in MORBs and OIBs each pass through the composition of air (Fig. 11), due to the ubiquitous presence of an atmospheric component. Key questions arise from the Ne isotope data, especially about the origin of the atmosphere. The chapters by Porcelli and Ballentine (2002) and Pepin and Porcelli (2002) discuss these questions in more detail. Of special note is the observation that the $^{20}Ne/^{22}Ne$ ratio in mantle-derived rocks approaches the solar-like composition while in the atmosphere it is much lower. This suggests that Ne in the Earth's atmosphere has been substantially fractionated from its terrestrial primordial composition (e.g., Porcelli and Pepin 2000) or that it was added as a late veneer (Marty 1989).

Precise Ne isotope analyses of mantle-derived rocks are still somewhat limited in number. The inherently low abundance of Ne in mantle melts makes atmospheric contamination a potential problem, and corrections for isobaric interference during the mass spectrometer analysis (e.g., doubly charged species $^{40}Ar^{++}$ and $^{12}C^{16}O_2^{++}$) always need to be determined. In general, the most reliable Ne isotope analyses today are produced by the stepwise release of gases (as they are for Ar and Xe isotopes), either by incremental crushing or incremental heating of a sample in vacuum.

Nucleogenic production

MORBs and OIBs are systematically different in their Ne isotope composition (Sarda et al. 1988; Honda et al. 1991, 1993a,b; Moreira et al. 1995; Moreira and Allègre 1998). This difference results from differences in the dilution, by primordial Ne, of the nucleogenic ^{21}Ne that is produced in their mantle sources. Even a small amount of nucleogenic production can markedly shift the mantle $^{21}Ne/^{22}Ne$ ratio because ^{21}Ne is so scarce. The nucleogenic production rate of Ne depends on several factors, including the elemental and isotopic compositions of the target elements within the source rocks, the neutron and α-particle energy spectrum, the neutron/α-particle production ratio, and the Ne isotope yields from the nuclear reaction (e.g., Kyser and Rison 1982; Yatsevich and Honda 1997; Leya and Wieler 1999). The most important pathways for ^{21}Ne production, often referred to as the Wetherill reactions (Wetherill 1954), are $^{18}O(\alpha,n)^{21}Ne$ and $^{24}Mg(n,\alpha)^{21}Ne$, with the former reaction being the dominant one for the Earth's mantle (Yatsevich and Honda 1997). In these reactions the α-particles and neutrons interact with target atoms of ^{18}O and ^{24}Mg. The α-particles are derived from U and Th decay, while the neutrons are produced from particle collisions with other elements such as oxygen. The production of radiogenic ^4He (α-particles) and nucleogenic ^{21}Ne in the mantle is therefore strongly coupled.

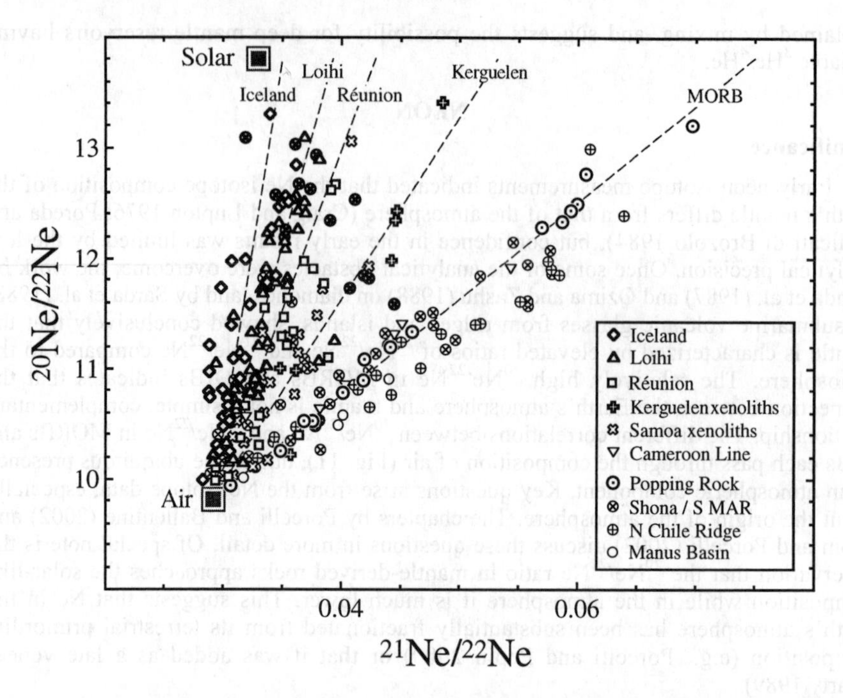

Figure 11. The Ne three-isotope diagram ($^{20}Ne/^{22}Ne$ vs. $^{21}Ne/^{22}Ne$). Data sources are MORB - Sarda et al. (1988, 2000), Hiyagon et al. (1992), Moreira et al. (1995, 1996, 1998), Moreira and Allègre (2002), Niedermann et al. (1997), Niedermann and Bach (1998), Shaw et al. (2001); OIB- Sarda et al. (1988), Poreda and Farley (1992), Staudacher et al. (1990), Hiyagon et al. (1992), Honda et al. (1991, 1993), Valbracht et al. (1996, 1997), Dixon et al. (2000), Barfod et al. (1999), Trieloff et al. (2000), Moreira et al. (2001) and Hanyu et al. (2001). Other MORBs and OIBs are omitted for clarity. Plotted points include individual analysis steps (crushing or heating), but only when the measured $^{20}Ne/^{22}Ne$ was >2-sigma above the air ratio. Darker symbols for the southern MAR data set crudely form a vector between the MORB line and solar compositions; these are lavas from the Shona and Discovery hotspot-influenced sections of the ridge, where $^{3}He/^{4}He$ ratios range up to 15 R_A (Moreira et al. 1995; Sarda et al. 2000); see text for further discussion. Dashed lines depict mixing between air and mantle Ne for Iceland, Loihi, Réunion, Kerguelen and MORB (MAR popping rock), drawn through the highest $^{20}Ne/^{22}Ne$ or the most precisely determined compositions for each locality. The extrapolated $^{21}Ne/^{22}Ne$ ratio, corresponding to solar $^{20}Ne/^{22}Ne$ at each locality, is 0.035 for Iceland, 0.039 for Loihi, 0.043 for Réunion, 0.053 for Kerguelen and 0.075 for MORB.

A good estimate of the $^{21}Ne/^{22}Ne$ ratio of the mantle source can be determined from the realization that Ne in a volcanic rock is a binary mixture between air and mantle (e.g., Moreira et al. 1995). This estimate is similar to that determined by a ternary deconvolution using $^{20}Ne/^{22}Ne$ and $^{21}Ne/^{22}Ne$ ratios for air, primordial (here assumed to be solar for illustration) and nucleogenic end-members (e.g., Honda et al. 1993a), because the nucleogenic ^{22}Ne production is minor compared to ^{21}Ne. The extrapolated $^{21}Ne/^{22}Ne$ corresponding to a solar $^{20}Ne/^{22}Ne$ ratio, $^{21}Ne/^{22}Ne_E$, can be determined for a line that passes through the air and any data point on the Ne three-isotope diagram, as illustrated in Figure 12. Its numerical value is given by

$$^{21}Ne/^{22}Ne_E = (^{21}Ne/^{22}Ne_M - {}^{21}Ne/^{22}Ne_A)/f_{22} + {}^{21}Ne/^{22}Ne_A \quad (3)$$

where E, S, A and M refer to extrapolated, solar, air and measured values, respectively, and where f_{22} is the proportion of mantle-derived Ne in a sample, i.e.,

$$f_{22} = \frac{{}^{20}Ne/{}^{22}Ne_M - {}^{20}Ne/{}^{22}Ne_A}{{}^{20}Ne/{}^{22}Ne_S - {}^{20}Ne/{}^{22}Ne_A}.$$

(Note: ${}^{20}Ne/{}^{22}Ne_S = 13.8$, ${}^{21}Ne/{}^{22}Ne_S = 0.0328$, ${}^{20}Ne/{}^{22}Ne_A = 9.8$ and ${}^{21}Ne/{}^{22}Ne_A = 0.029$).

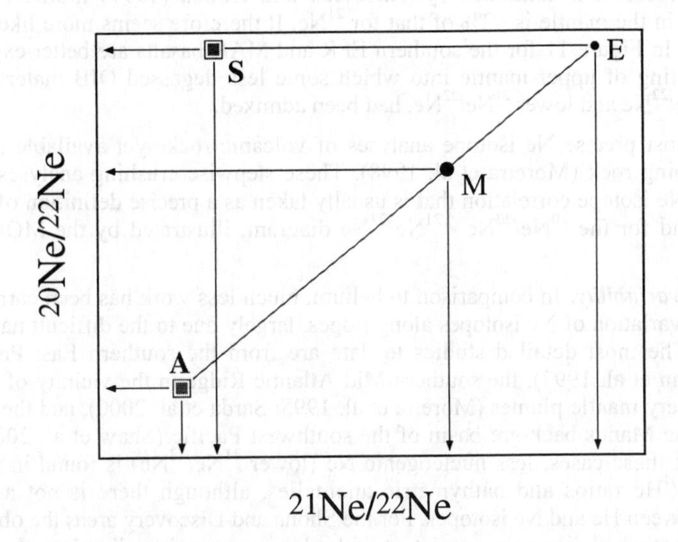

Figure 12. Schematic illustration for determining ${}^{21}Ne/{}^{22}Ne_E$, the extrapolated ${}^{21}Ne/{}^{22}Ne$ ratio of the mantle. S, A, M and E correspond to solar, air, measured and extrapolated values.

Mid-ocean ridge basalts

In a diagram of ${}^{20}Ne/{}^{22}Ne$ vs. ${}^{21}Ne/{}^{22}Ne$, ocean ridge basalts form a well defined array that passes through the composition of air and extends to values of ${}^{20}Ne/{}^{22}Ne \geq 12.5$ and ${}^{21}Ne/{}^{22}Ne \geq 0.07$ (Fig. 11). This array is usually interpreted as a mixing line between upper mantle Ne and air Ne. How and when this contamination by air occurs is conjectural. Some investigators propose that it takes place during magma ascent and eruption (e.g., Farley and Poreda 1993), or that it was introduced by ancient subduction (e.g., Sarda et al. 2000); others argue that it is largely an artifact introduced during sample preparation (e.g., Ballentine and Barfod 2000). The variability of repeat analyses by stepwise heating makes it evident that some allowance should be made for the effects of air contamination in every analysis, and the measured Ne isotope ratio should always be taken as a minimum estimate for the mantle source composition.

Global variability. In general, the Ne isotope measurements have relatively large and correlated analytical uncertainties. The scatter about the MORB Ne isotope array therefore makes it difficult to determine whether the upper mantle is characterized by a single Ne isotope composition. It seems likely that the whole mantle is characterized by a uniform value of ${}^{20}Ne/{}^{22}Ne$, but this is an unproven assumption. Because there are no known reactions that produce ${}^{20}Ne$ to a significant extent, variable ${}^{20}Ne/{}^{22}Ne$ in the mantle seems implausible unless nucleogenic production of ${}^{22}Ne$ is significant and spatially variable. Two MORB suites, from the southern East Pacific Rise (Niedermann et al. 1997) and the southern Mid-Atlantic Ridge (Moreira et al. 1995), each show a decrease in ${}^{20}Ne/{}^{22}Ne$ with increasing ${}^{21}Ne/{}^{22}Ne$, with a vector that points approximately to the solar Ne isotope composition (especially for the southern Mid-Atlantic Ridge samples with higher $^3He/^4He$ ratios near the Shona hotspot; see Fig. 11). Based on this evidence, Niedermann et al. (1997) speculated that the mantle could vary spatially in the production of ^{22}Ne (and consequently in its $^{20}Ne/^{22}Ne$ ratio), due to a heterogeneous distribution of fluorine and ^{22}Ne production via the reaction $^{19}F(\alpha,n)^{22}Na(\beta^+)^{22}Ne$. However, the recent estimates by Yatsevich and Honda (1997) indicate that ^{22}Ne production in the mantle is <2% of that for ^{21}Ne. It therefore seems more likely that the data trends in Figure 11 for the southern EPR and MAR basalts are better explained by partial melting of upper mantle into which some less degassed OIB material, having higher $^{20}Ne/^{22}Ne$ and lower $^{21}Ne/^{22}Ne$, had been admixed.

The most precise Ne isotope analyses of volcanic rocks yet available are for the MAR popping rock (Moreira et al. 1998). These stepwise crushing analyses display a very tight Ne isotope correlation that is usually taken as a precise definition of the upper mantle trend for the $^{20}Ne/^{22}Ne$ - $^{21}Ne/^{22}Ne$ diagram, illustrated by the MORB line in Figure 11.

Local variability. In comparison to helium, much less work has been carried out on the spatial variation of Ne isotopes along ridges, largely due to the difficult nature of the analyses. The most detailed studies to date are from the southern East Pacific Rise (Niedermann et al. 1997), the southern Mid-Atlantic Ridge in the vicinity of the Shona and Discovery mantle plumes (Moreira et al. 1995; Sarda et al. 2000), and the spreading center in the Manus back-arc basin of the southwest Pacific (Shaw et al. 2001). In the first two of these cases, less nucleogenic Ne (lower $^{21}Ne/^{22}Ne$) is found in regions of higher $^3He/^4He$ ratios and bathymetric anomalies, although there is not a strict 1:1 relation between He and Ne isotopes. For the Shona and Discovery areas the observed Ne and He isotope variations are consistent with plume material feeding into the sub-ridge mantle from nearby hotspot sources, as suggested from other geophysical and geochemical data (Small 1995; Douglass et al. 1999). There are no prominent ocean island hotspots or seamounts near the southern EPR between 14-19°S, but the bathymetric and geochemical anomalies along this section of the ridge also suggest an origin from hotspot-derived material, perhaps as a mantle blob that was entrained into the upper mantle flow (Mahoney et al. 1994).

There is evidence that some MORB suites show a different Ne isotope trend from the popping rocks. Basalts from the northern Chile Ridge lie to the right of the well-defined popping rock line in the Ne three-isotope diagram (Fig. 11), indicating a stronger component of nucleogenic Ne in their mantle source. Niedermann and Bach (1998) suggest that this could be the result of a prior melting event that fractionated the U/Ne ratio of the residual mantle. Assuming Ne is more incompatible than U, the mantle residue from an earlier melting event would acquire a higher value of $^{21}Ne/^{22}Ne$ through time compared to more fertile mantle. The time scale required is 10-100 Ma, consistent with the formation history of the northern Chile Ridge and westward mantle transport from a region of earlier melting beneath the East Pacific Rise. Some of these northern Chile Ridge basalts also have $^3He/^4He$ ratios at the low end of the MORB range (<7 R_A), which is consistent with a model residual mantle having elevated $U/^3He$. However, a few of the lavas have $^3He/^4He$ above 8 R_A, suggesting that some $^3He/^{22}Ne$ fractionation may also have occurred during the proposed earlier melting event.

Significant decoupling of He and Ne isotopes does appear to occur in some cases, and is best exemplified by the spreading ridge basalts from the Manus Basin (Shaw et al. 2001). Manus Basin basalts have $^3He/^4He$ ratios up to 15 R_A, yet they show a Ne isotope slope that is less than the MORB trend as defined by the popping rocks (Fig. 11). Shaw et al. (2001) considered several possible explanations for the discrepancy, ruling out crustal contamination, fluid addition from the subducting plate, or addition of Ne from ancient slabs that were recycled to the mantle. They conclude that either He was fractionated from Ne in the back-arc mantle by melting and degassing during the last 10 Ma, or that the Earth's mantle has maintained some record of primordial heterogeneity in its ^{3}He/^{22}Ne ratio.

Ocean island basalts

In comparison to MORBs, some OIBs show a much steeper correlation in the Ne three-isotope diagram (Fig. 11). This is especially true for the cases of Hawaii, Iceland and Réunion (Fig. 11; Honda et al. 1991; Valbracht et al. 1997; Trieloff et al. 2000; Dixon et al. 2000; Moreira et al. 2001; Hanyu et al. 2001). These steep trends reveal that these OIB mantle sources have less nucleogenic Ne (lower ^{21}Ne/^{22}Ne) than the MORB mantle source. Because MORBs also show more radiogenic He than OIBs from Hawaii, Iceland and Réunion, the mantle source for MORBs must be characterized by lower time-integrated ^{3}He/(U+Th) and ^{22}Ne/(U+Th) than the mantle source for those OIBs. Because the MORB mantle source is also trace element-depleted compared to the source for many OIBs, the OIB mantle source, at least for cases such as Hawaii, Iceland and Réunion, must be less degassed (i.e., it has higher ^{3}He and ^{22}Ne concentrations) than the upper mantle source of MORBs.

Because nucleogenic ^{21}Ne production is coupled to radiogenic ^{4}He production, one expects *a priori* that OIBs with a steeper slope on the Ne three-isotope diagram will have higher ^{3}He/^{4}He ratios. If He and Ne were perfectly coupled in the Earth (i.e., if no He/Ne fractionation occurred), the ^{3}He/^{4}He ratio at any locality could be predicted from the slope on the Ne isotope diagram, assuming the initial He and Ne isotope compositions are known (e.g., Honda et al. 1993a). A detailed comparison of the Ne isotope correlations for Iceland, Loihi, Réunion and the MAR popping rock is insightful in this context (see Fig. 11). The three ocean islands are characterized by relatively high ^{3}He/^{4}He (>30 R_A for some basalts from Iceland and Hawaii, and ~13 R_A for Réunion), while the popping rock, taken to exemplify the MORB mantle, has a ^{3}He/^{4}He ratio of 8 R_A. The isotope characteristics of the Réunion island source (^{3}He/^{4}He = 13 R_A and an intermediate slope on the Ne three-isotope diagram and ^{87}Sr/^{86}Sr > 0.704) cannot be explained by mixing of either recycled crust or MORB mantle with a high ^{3}He/^{4}He-source such as that for Loihi or Iceland. Hanyu et al. (2001) suggest that the Réunion plume source is therefore distinct from the Iceland and Loihi sources, and that the mantle contains more than one relatively undegassed reservoir, each having elevated but different ^{3}He/^{4}He ratios. This is an intriguing hypothesis that warrants further consideration, especially in light of developing models for deep mantle stratification that involve large domains having a complex topography that responds to convection and pressure from subducting slabs (e.g., Kellogg et al. 1999). Another possibility to bear in mind, however, is that some decoupling of He and Ne isotopes may occur if the mantle He/Ne ratio is fractionated during melting processes.

Solar hypothesis. The steep OIB correlations on the three-Ne isotope diagram, especially for basalts from Iceland and Loihi Seamount, trend toward Ne isotope compositions resembling either solar wind or the solar component commonly found in meteorites (Ne-B). It is not possible at the present time to distinguish unambiguously between these two possibilities, and the implications for early Earth evolution are quite different. The highest values of ^{20}Ne/^{22}Ne yet measured are ~12.5 for terrestrial lavas from several localities, including Loihi Seamount, Iceland and the popping rocks. This led Trieloff et al. (2000) to suggest that the Earth's initial Ne is best described by Ne-B, a component produced in meteorites by irradiation from the solar wind. Following this reasoning, Trieloff et al. (2000) argued that the Earth acquired its Ne while the young sun was still very active, during which time small planetesimals were thoroughly irradiated prior to their assembly to form the Earth. In contrast, Ballentine et al. (2001) argued that, because all basalt analyses can be expected to contain some air neon, a solar Ne isotopic composition ($^{20}Ne/^{22}Ne$ of ~13.8) best represents the Ne trapped within the Earth. This conclusion is based on the correlation between Ne and Ar isotopes seen in step crushing analyses of the MAR popping rock (Moreira et al. 1998; see Fig. 14, below) and its extrapolation to the very high $^{40}Ar/^{36}Ar$ measured by laser extraction from individual bubbles in the same lavas (Burnard et al. 1997), for which precise Ne isotope measurements were not possible. Following this reasoning, the Earth would have solar Ne isotope composition, and gravitational capture of a dense primitive atmosphere within the solar nebula during Earth accretion is a logical model (Mizuno et al. 1980).

ARGON

Significance

Argon isotopes can provide powerful constraints on the formation of the atmosphere, and on geodynamics and isotopic evolution of the mantle. Unlike He which is lost from the atmosphere, Ar has accumulated over Earth history. The primordial $^{40}Ar/^{36}Ar$ ratio is also extremely low, so the amount of ^{40}Ar initially present in the Earth can be reasonably taken as zero. Transport of ^{40}Ar to the atmosphere involves volcanic degassing, hydrothermal circulation through the crust, and erosion of continental crust that releases radiogenic Ar generated by decay of ^{40}K. The half-life of ^{40}K (1.25 Gy) is relatively short compared to the age of Earth, making the abundance of Ar in the atmosphere along with the $^{40}Ar/^{36}Ar$ of the atmosphere and upper mantle useful parameters for unraveling the history of degassing and plate tectonic recycling (e.g., Turekian 1959; Hamano and Ozima 1978; Fisher 1978; Sleep 1979; O'Nions et al. 1979; Hart et al. 1979; Sarda et al. 1985; Allègre et al. 1986/1987; Turner 1989; Albarède 1998; Coltice et al. 2000; Porcelli and Ballentine 2002). As one example, in the limiting case where all the ^{40}Ar generated in the crust has been released to the atmosphere, 30% of the Ar in the atmosphere would be from the crust and 70% from the mantle (Turcotte and Schubert 1988). Under these circumstances, only one-quarter of the whole mantle would be outgassed. As another example, under assumptions of a layered mantle (in which the 660 km seismic discontinuity represents the boundary) with known K content, about one-half of the terrestrial ^{40}Ar inventory is still retained in the deep Earth (Allègre et al. 1996). These conclusions depend directly on the K concentration of the Earth; a lower K/U ratio than commonly assumed leads to the conclusion of a more degassed deep Earth (Davies 1999).

Radiogenic production

In a closed system, the amount of radiogenic ^{40}Ar produced over time t is given by

$$^{40}Ar^* = a \cdot b \cdot K\,(e^{\lambda_{40} t} - 1) \tag{4}$$

where a is the natural abundance of ^{40}K (0.0117%), b is the branching ratio of ^{40}K decay to ^{40}Ar by electron capture (0.1048), K is the amount of potassium present, and λ_{40} is the total decay constant (0.554 Gy^{-1}).

Atmospheric contamination

Obtaining meaningful Ar isotope analyses on basalts is analytically challenging, because contamination by atmospheric components is notorious. Potential sources of this contamination include seawater, altered wall rock, or air itself. Basalt glasses may show complicated variations within an individual glassy rim, with lower $^{40}Ar/^{36}Ar$ in both the outermost glassy part as well as in the more crystalline interior (Kumagai and Kaneoka 1998). The atmospheric contaminant in MORBs appears to be significantly less in vesicles, because all measurable ^{36}Ar in the popping rock is accounted for by the amount released during the analysis of vesicle-free glass (Burnard et al. 1997), and there is no correlation between ^{3}He and ^{36}Ar from basalt crushing analyses (Ballentine and Barfod 2000). This suggests that for MORBs the contamination may not be occurring by crustal assimilation, but rather by adsorption or entrapment following eruption. In the case of many ocean islands, phenocrysts are separated from subaerially erupted lavas for noble gas analysis. Such samples usually show a large range in measured ^{40}Ar/^{36}Ar due to variable proportions of atmospheric and magmatic components (Farley and Craig 1994). The air component is not readily removed by physical or chemical treatment. It is also present to a large extent in fluid inclusions, along with the magmatic gas, as revealed by laser fusion analyses (Farley and Craig 1994; Burnard et al. 1994a). This indicates that the contamination may also occur by assimilation of crust, or by direct addition of seawater or air to a magma (Farley and Craig 1994). Argon isotope ratios in OIBs that resemble air are most likely due to contamination, and should not be taken as evidence for a deep mantle reservoir having near-atmospheric Ar isotope composition (Patterson et al. 1990). In fact, where OIBs have been analyzed that have high ^{3}He/^{4}He, they can range to moderately high ^{40}Ar/^{36}Ar ratios, up to ~8000 (Farley and Craig 1994; Burnard et al. 1994a,b; Trieloff et al. 2000).

Ballentine and Barfod (2000) reviewed the effects of atmospheric contamination in MORBs and OIBs in detail. Intriguingly, they showed that the modern air contaminants released during crushing are preferentially located in sites that are more easily ruptured than those containing magmatic gas. This suggests that much of the contamination occurs through the annealing of microfractures, following air entrapment that occurs either during sample recovery from seafloor depths or during sample preparation in the laboratory. This contaminant gas should, in principle, be separable from magmatic gas by stepwise crushing or heating, although a diagnostic test of when it is completely removed remains to be fully developed. Measured ^{40}Ar/^{36}Ar ratios should always be considered minimum values in young basalts.

It is commonly assumed that all ^{36}Ar present in a MORB or OIB sample could be derived from air contamination. This is a reasonable assumption provided that the ^{40}Ar/^{36}Ar is >3000, as seems likely for mantle source regions. This assumption allows a minimum estimate for the amount of magmatic ^{40}Ar present in a sample. This amount of radiogenic Ar derived from the mantle, denoted ^{40}Ar*, is given by

$$^{40}Ar^* = (^{40}Ar/^{36}Ar_M - ^{40}Ar/^{36}Ar_A)\ ^{36}Ar_M \tag{5}$$

where M designates the measured ratio and A the air ratio ($^{40}Ar/^{36}Ar_A = 295.5$). There is no straightforward way to estimate a sample's complement of primordial ^{36}Ar derived from the mantle. For example, if one assumes a mantle ^{40}Ar/^{36}Ar ratio (such as $^{40}Ar^*/^{36}Ar_P = 40{,}000$, where p refers to primordial Ar), any correlation involving $^{36}Ar_P$ (such as $^{3}He/^{36}Ar_P$) will be inherently the same as one involving ^{40}Ar*, since both are directly correlated by the assumed $^{40}Ar^*/^{36}Ar_P$ ratio.

Mid-ocean ridge basalts

Global variability. The mantle ^{40}Ar/^{36}Ar ratio is variable due to time-integrated variations in K/^{36}Ar. MORBs show ^{40}Ar/^{36}Ar ratios that range up to 40,000. The very highest values have been measured using a laser to rupture individual vesicles in the MAR popping rock (Burnard et al. 1997). Values up to 28,000 have been measured for these same samples by incrementally heating or crushing bulk samples of the glass in vacuum (Staudacher et al. 1989; Moreira et al. 1998). Very high ratios, with relatively large uncertainties, have also been measured in glassy basalts from the Mid-Atlantic Ridge (^{40}Ar/^{36}Ar = 42,400±9700; Marty and Humbert 1997) and from 10°S on the southern East Pacific Rise ($^{40}Ar/^{36}Ar$>35,000, limited by the absence of detectable ^{36}Ar; Fisher 1994). Ratios above 25,000 have also been measured in MORB glasses from the MAR near the Kane transform (Sarda et al. 1985), from the South Atlantic (Sarda et al. 2000), and from the N Chile Ridge in the eastern Pacific (Niedermann and Bach 1998). It is tempting to suggest that the whole upper mantle might be characterized by a $^{40}Ar/^{36}Ar$ value near 30,000 based on this similarity. However, individual analyses of the popping rock show the highest measured $^{40}Ar/^{36}Ar$ (>20,000) during the last two steps of crushing which release the lowest amounts of ^{36}Ar (<1.5×10^{-10} std cm^3; Moreira et al. 1998), indicating that even in these most gas-rich samples, atmospheric contamination is significant for Ar isotope analyses. The value of 40,000 measured by Burnard et al. (1997) using a laser probe should probably be taken as a minimum estimate for the upper mantle.

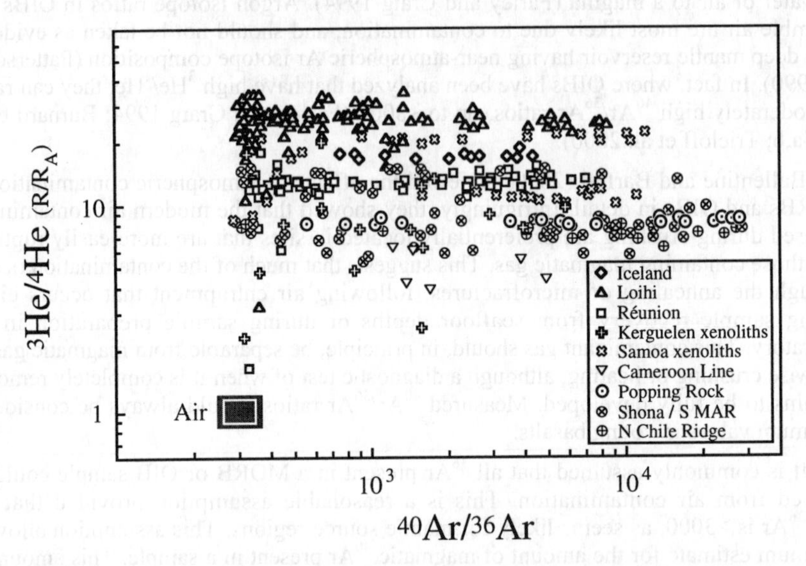

Figure 13. $^3He/^4He$ vs. $^{40}Ar/^{36}Ar$. Plotted points include individual analysis steps (crushing or heating), but only when the measured $^{40}Ar/^{36}Ar$ was >2σ above the air ratio. Data sources are given in the caption to Figure 11.

There is no systematic behavior between $^{40}Ar/^{36}Ar$ and $^3He/^4He$ for the upper mantle based on the MORB analyses. $^{40}Ar/^{36}Ar$ ratios for MORBs are quite variable, ranging by 2 orders of magnitude, while $^3He/^4He$ varies by about a factor of 2 (Fig. 13). For the case of the popping rock, the highest $^{40}Ar/^{36}Ar$ ratios are also associated with the highest measured $^{20}Ne/^{22}Ne$ (Fig. 14), consistent with the lowest degree of atmospheric contamination in those analyses. Moreira et al. (1998) inferred that the maximum $^{40}Ar/^{36}Ar$ of the MORB mantle is 44,000, based on extrapolation of the popping rock $^{40}Ar/^{36}Ar$-$^{20}Ne/^{22}Ne$ trend to solar $^{20}Ne/^{22}Ne$ = 13.8 (Fig. 14). High $^{40}Ar/^{36}Ar$ ratios are also found in other samples with lower $^{20}Ne/^{22}Ne$, such as from the N Chile Ridge and southern MAR. Much of the variation observed in Figure 14 must be due to air contamination. Different degrees of air contamination would only produce a single trajectory however, so the scatter in the data indicates a significant variation in the Ne/Ar ratio for either, or both, of the mixing end-members (i.e., the atmosphere-derived contaminant and the magmatic gas). For example, to account for the data scatter, any binary mixing between magmatic gas and unfractionated air would require that the

$^{22}Ne/^{36}Ar$ ratio of the magmatic gas ranges from an order of magnitude larger to an order of magnitude smaller than the air ratio. From a simultaneous consideration of Ne-Ar and Ne-Xe isotope systematics, Harrison et al. (2002) showed that contamination usually involves a fractionated atmospheric component having $^{22}Ne/^{36}Ar$ significantly less than air, such as might be expected if water-rich or adsorbed components play a role. This seriously complicates a full interpretation of even the best available heavy noble gas analyses. Measured $^{40}Ar/^{36}Ar$ ratios should be considered minimum estimates for their mantle source.

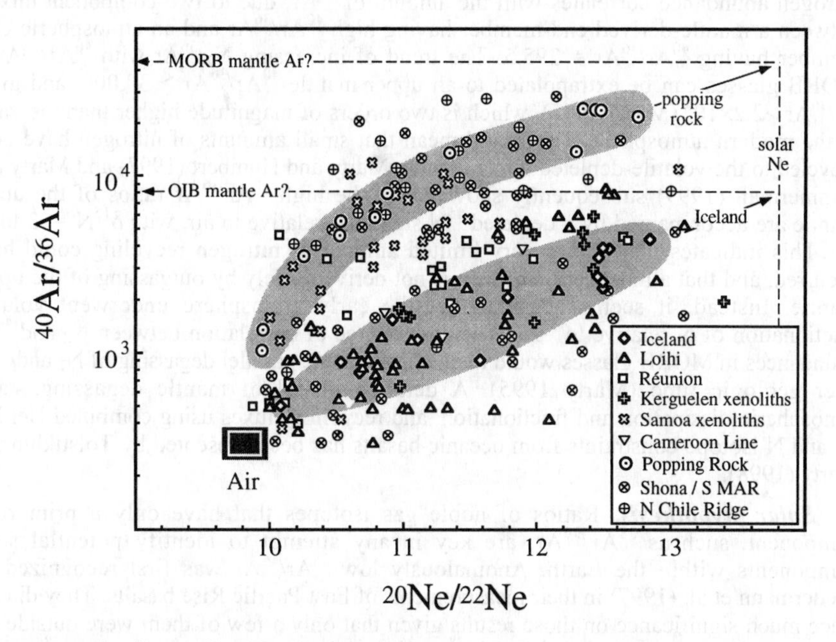

Figure 14. $^{40}Ar/^{36}Ar$ vs. $^{20}Ne/^{22}Ne$. Data sources are given in the caption to Figure 11. Plotted points include individual analysis steps (crushing or heating), but only when the measured Ne and Ar isotope ratios were $>2\sigma$ above the air ratio.

Regional variability. Given the strong potential effect of contamination by atmospheric components, it is difficult to quantitatively assess the spatial variability in $^{40}Ar/^{36}Ar$ along ridges. Sarda et al. (1999a) showed that the maximum $^{40}Ar/^{36}Ar$ ratio along the Mid-Atlantic Ridge, measured by step heating of basalt glasses, varies systematically with indicators of mantle heterogeneity such as Pb isotopes, with one end-member defined by the Azores hotspot. The correlations hold for the North and South Atlantic considered as a whole. Sarda et al. (1999a) reasoned that this ruled out shallow level contamination processes, and favored mixing that involved depleted upper mantle and recycled materials having low $^{40}Ar/^{36}Ar$ and radiogenic Pb. This would require that the recycled material has a $K/^{36}Ar$ ratio less than upper mantle values, i.e., that atmospheric noble gases are recycled to the mantle by subduction. Burnard (1999b) challenged this explanation however, and argued that shallow level contamination could explain the regional correlation between Ar and Pb isotopes. Radiogenic Pb signatures are typically found along axial topographic highs, where ridges are often influenced by nearby island hotspots. Low $^{40}Ar/^{36}Ar$ is also found along shallower ridge sections where degassing at shallower levels can lead to an increased susceptibility to contamination, and this can produce a correlation between $^{40}Ar/^{36}Ar$ and axial depth (Burnard 1999b;

Burnard et al. 2002). Sarda et al. (1999b) suggested that such a process should produce erratic results rather than a systematic trend. Whether the regional variation in Ar isotopes along ocean ridges carries information about variations in the upper mantle is still debated. Chlorine contents in suites of MORB glasses studied for Ar isotopes could potentially resolve this debate (e.g., Michael and Cornell 1998).

Relation to other volatiles. The $N_2/^{36}Ar$ ratio varies by several orders of magnitude in MORB glasses, and is correlated with $^{40}Ar/^{36}Ar$ (Marty 1995). This means that the nitrogen abundance correlates with the amount of ^{40}Ar, due to two-component mixing between a mantle-derived end-member having high $^{40}Ar/^{36}Ar$ and an atmospheric end-member having $^{40}Ar/^{36}Ar = 295.5$. The trend of increasing $N_2/^{36}Ar$ with $^{40}Ar/^{36}Ar$ in MORB glasses can be extrapolated to an upper mantle $^{40}Ar/^{36}Ar > 30,000$, and gives $N_2/^{36}Ar > 2.2 \times 10^6$ (Marty 1995), which is two orders of magnitude higher than the value in the modern atmosphere. This could mean that small amounts of nitrogen have been recycled to the volatile-depleted upper mantle. Marty and Humbert (1997) and Marty and Zimmerman (1999) subsequently showed that the high $^{40}Ar/^{36}Ar$ ratios of the upper mantle are accompanied by a depleted ^{15}N signature relative to air, with $\delta^{15}N$ = -3 to –5 ‰. This indicates that only a very limited amount of nitrogen recycling could have occurred, and that atmospheric nitrogen is not derived solely by outgassing of the upper mantle. Instead, it seems possible that the early atmosphere underwent volatile fractionation of N relative to ^{36}Ar. The high degree of correlation between N_2 and ^{40}Ar abundances in MORB glasses would then indicate some parallel degassing of N_2 and ^{40}Ar over geologic time (Marty 1995). A detailed model of mantle degassing, early atmospheric dissipation and fractionation, and recycling fluxes using combined He, Ne, Ar and N isotope constraints from oceanic basalts has been presented by Tolstikhin and Marty (1998).

Solar hypothesis. Ratios of noble gas isotopes that have only a primordial component, such as $^{38}Ar/^{36}Ar$, are key in any attempt to identify potential solar components within the Earth. Anomalously low $^{38}Ar/^{36}Ar$ was first recognized by Niedermann et al. (1997) in their measurements of East Pacific Rise basalts. They did not place much significance on those results given that only a few of them were outside the 2-sigma uncertainty of the air ratio (air $^{38}Ar/^{36}Ar = 0.188$). Low ratios, down to $^{38}Ar/^{36}Ar$ =0.185, were also measured for the deeply erupted glassy basalts at Loihi Seamount (Valbracht et al. 1997). The low $^{38}Ar/^{36}Ar$ ratios at Loihi were correlated with non-atmospheric and elevated $^{20}Ne/^{22}Ne$, providing preliminary evidence for a solar Ar component in the mantle. (The mean value of $^{38}Ar/^{36}Ar$ is 0.175 in solar wind). Subsequently, Pepin (1998) took the results from both the EPR and Loihi studies as promising evidence for a solar-like Ar component in the mantle, and discussed some of the implications for planetary accretion and atmospheric origin. The low $^{38}Ar/^{36}Ar$ ratios in the EPR samples are not correlated with high $^{40}Ar/^{36}Ar$, however, as would be expected from binary mixing between air and mantle components (Kunz 1999). Indeed, some of those low $^{38}Ar/^{36}Ar$ ratios were measured simultaneously with atmospheric $^{40}Ar/^{36}Ar$ ratios. It should also be recognized that very small analytical artifacts in the original studies might be the cause for the low $^{38}Ar/^{36}Ar$ ratios. Mass fractionation effects during gas extraction or analysis must be completely removed, and this is especially difficult to verify in the presence of relatively large amounts of ^{40}Ar. At this point in time it seems *possible* that a solar Ar component is present within the mantle, but the analytical evidence is weak and it remains unproven.

Ocean island basalts

The highest $^{40}Ar/^{36}Ar$ ratios at ocean islands are found in xenoliths from Samoa, where values range up to 21,500 (Farley et al. 1994). The origin of the high $^{40}Ar/^{36}Ar$ in these xenoliths is unclear, however, because the fluid inclusions show strong metasomatic overprints that occurred in the upper mantle (Burnard et al. 1998). The highest $^{40}Ar/^{36}Ar$ ratios obtained on ocean island basalt glasses are from Loihi Seamount, where they range up to 8300 (Trieloff et al. 2000). In the most detailed study of Ar isotopes at a single ocean island, Farley and Craig (1994) obtained a maximum $^{40}Ar/^{36}Ar$ of 8000 for olivine phenocrysts from Juan Fernandez, a high $^{3}He/^{4}He$ hotspot. Bulk crushing and laser fusion of individual crystals were carried out for forty splits of olivine from a single lava, and a variety of chemical and physical pre-treatments were attempted.

The relation between $^{40}Ar/^{36}Ar$ and $^{3}He/^{4}He$ or $^{20}Ne/^{22}Ne$ in ocean island basalts and xenoliths is generally poor (Figs. 13 and 14). The apparent difference in $^{40}Ar/^{36}Ar$–$^{3}He/^{4}He$ systematics between MORBs and OIBs was suggested by Kaneoka (1983, 1985) to indicate a layered mantle structure, in which OIB sources have higher $^{3}He/^{4}He$ and $^{40}Ar/^{36}Ar$ ratios that are nearly the same as the atmosphere. More recent data showing $^{40}Ar/^{36}Ar$ up to ~8000 in OIBs now makes a model involving OIB source reservoirs of near atmospheric composition untenable. High $^{40}Ar/^{36}Ar$ values, particularly in the Loihi glasses (Trieloff et al. 2000), are associated with $^{20}Ne/^{22}Ne$ ratios comparable to the highest values obtained in terrestrial basalts, evidence that these Loihi analyses are not very strongly affected by air contamination. The mantle source for OIBs from localities such as Iceland, Hawaii and Réunion clearly has a $^{40}Ar/^{36}Ar$ ratio higher than the atmosphere (Trieloff et al. 2000; Burnard et al. 1994b). It is not possible to assume that all air contamination has been eliminated even in these first rate analyses. A ratio of $^{40}Ar/^{36}Ar \geq 8000$ is arguably the best estimate for the mantle source of these OIBs (Fig. 14), although it is not currently possible to rule out the presence of a deep OIB source reservoir, having somewhat lower $^{40}Ar/^{36}Ar$, that generates OIB magma that subsequently undergoes slight contamination by volatiles derived from the shallower MORB mantle.

KRYPTON

Krypton isotopes in oceanic basalts have not been very diagnostic of mantle processes. The Kr isotopic composition of oceanic basalts is typically the same as modern air. Minor production of ^{83}Kr, ^{84}Kr and ^{86}Kr occurs from the spontaneous fission of ^{238}U, but it has not yet been detected in young oceanic basalts, due to the small fission yields (<1% for ^{86}Kr, ≤0.1% for ^{83}Kr and ^{84}Kr) in the mass region where Kr nuclides are present in relatively high abundance (natural abundance of 11.5%, 57% and 17.3%, respectively, for ^{83}Kr, ^{84}Kr and ^{86}Kr). Even if air contamination during sample analysis could be totally eliminated, it is unclear whether such small contributions could be resolved.

XENON

Significance

Xenon isotopes in oceanic basalts provide the most basic evidence on early differentiation of our planet and on formation of the atmosphere. Several Xe isotopes are produced by extinct radioactivity, providing a method to quantify events at the very start of Earth history near 4500 million years ago to a high precision, within several tens of millions of years. The difference between $^{129}Xe/^{130}Xe$ of the mantle and atmosphere is fundamental evidence for their early separation (e.g., Thompson 1980; Staudacher and Allègre 1982). It is thought that formation of the atmosphere involved some early degassing of the mantle. Whether the Earth ever had a primary atmosphere is still an open question, and ultimately depends upon whether the inner planets formed in the presence of a solar nebula gas phase. The timing of these early events, the degree to which Xe isotope differences reflect variations in I/Xe and Pu/Xe of source reservoirs, and the extent to which the mantle and atmosphere are complementary, all remain points of active investigation. For example, it has become widely recognized that the Moon may have formed by impact of a Mars-sized body on the early Earth (Hartmann 1984). The impact hypothesis is obviously relevant to models of atmosphere formation, because much of the original solid body would be left molten, and many of the lighter gas species originally present would escape the gravitational field (Pepin 1997). The Xe isotope systematics of oceanic basalts and the atmosphere indicate isotopic closure of Xe in the terrestrial system within ~10^8 years of the onset of accretion (Staudacher and Allègre 1982; Ozima et al. 1985; Podosek and Ozima 2000; Porcelli and Pepin 2000).

Radiogenic production

Several of the heavier Xe isotopes have contributions from spontaneous fission, both from decay of long-lived ^{238}U and from extinct ^{244}Pu. ^{129}Xe was also produced by the extinct β^- decay of ^{129}I (Table 1). The relative fission yields for ^{238}U are ^{129}Xe/^{131}Xe/^{132}Xe/^{134}Xe/^{136}Xe = <0.002/0.076/0.595/0.832/1, with the absolute yield of ^{136}Xe, $^{136}y = 6.3 \pm 0.4\%$ (Ozima and Podosek 2002; Porcelli et al. 2002). As an example, for a closed system the amount of ^{136}Xe produced by decay of ^{238}U is

$$^{136}\mathrm{Xe} = (\lambda_{sf}/\lambda_{238})\,^{136}y\,^{238}\mathrm{U}(e^{\lambda_{238}t} - 1) \tag{6}$$

where $\lambda_{sf}/\lambda_{238}$ is the ratio of fission to total decay (5.45×10^{-7}). For the case of ^{244}Pu, the relative fission yields are ^{129}Xe/^{131}Xe/^{132}Xe/^{134}Xe/^{136}Xe = 0.048/0.246/0.885/0.939/1, with the absolute yield of ^{136}Xe = 5.6±0.6% (Lewis 1975). The amount of ^{136}Xe produced by extinct ^{244}Pu is

$$^{136}\mathrm{Xe} = (\lambda_{sf}/\lambda_{244})\,^{136}y\,^{244}\mathrm{Pu}_0\, e^{-\lambda_{244}(T_0 - t)} \tag{7}$$

where λ_{244} is the decay constant for ^{244}Pu (8.45 Gy^{-1}), $\lambda_{sf}/\lambda_{244} = 1.25 \times 10^{-3}$, T_0 is the time at which the system became closed having an amount of plutonium $^{244}\mathrm{Pu}_0$, and $(T_0 - t)$ is referred to as the formation interval. ^{129}Xe was also produced from I^{129}, so

$$^{129}\mathrm{Xe} = {}^{129}\mathrm{I}_0 e^{-\lambda_{129}(T_0 - t)} \tag{8}$$

where λ_{129} is the decay constant for ^{129}I (40.8 Gy^{-1}).

The above considerations show that several ages may be calculated from Xe isotope systematics, provided that certain key parameters are specified. These ages may be computed as I-Xe or I-Pu-Xe ages. The different methods usually involve a comparison between planetary materials, such as meteorites and the Earth, or between different terrestrial reservoirs, such as the atmosphere and the upper mantle. The key parameters that need to be specified are the initial amounts of ^{129}I and ^{244}Pu, and the abundance of I in the Earth today. These are usually treated in terms of the ratios $(^{129}\mathrm{I}/^{127}\mathrm{I})_0$, $(^{244}\mathrm{Pu}/^{238}\mathrm{U})_0$, and $(^{127}\mathrm{I}/^{238}\mathrm{U})$, respectively. The I-Xe and I-Pu-Xe ages of the atmosphere, computed from differences between the ^{129}Xe/^{130}Xe and ^{136}Xe/^{130}Xe measured in air and in mid-ocean ridge basalts, indicate formation intervals of ~50-100 Ma (Staudacher and Allègre 1982; Ozima et al. 1985; Marty 1989). These short time scales indicate that the atmosphere and upper mantle have remained effectively separated for most of Earth's ~4500 Ma history. A comparison of the isotopic compositions of terrestrial, lunar and meteoritic Xe also provides a high resolution chronometer of early solar system events. I-Pu-Xe ages computed in this way indicate that the Earth-Moon system became closed to Xe loss ~100 Ma after the beginning of solar system formation (Wetherill 1975; Bernatowicz and Podosek 1978; Swindle et al. 1984; Lin and Manuel 1987; Ozima and Podosek 1999; Podosek and Ozima 2000).

Mid-ocean ridge basalts

One of the most remarkable findings in noble gas geochemistry to date has been the observed 'excess' ^{129}Xe in some mid-ocean ridge basalt glasses. Anomalous ^{129}Xe/^{130}Xe in MORBs was first reported by Staudacher and Allègre (1982), who measured ratios up to 7% higher than the modern atmosphere (air ^{129}Xe/^{130}Xe = 6.48). These anomalies are present in MORBs from all the major ocean basins. Small anomalies also occur in ^{134}Xe/^{130}Xe and ^{136}Xe/^{130}Xe (air ^{136}Xe/^{130}Xe = 2.17). These enrichments have been verified in subsequent MORB studies (e.g., Marty 1989; Kunz et al. 1998; Moreira et al. 1998; Sarda et al. 2000). The largest MORB ^{129}Xe/^{130}Xe ratios and ^{136}Xe/^{130}Xe, and the most precisely determined, are for the MAR popping rock, which shows ratios up to 7.73 and 2.57, respectively (Fig. 15). Moreira et al. (1998) infer that the maximum ^{129}Xe/^{130}Xe of the MORB mantle is 8.2, based on extrapolation of the popping rock ^{129}Xe/^{130}Xe-^{20}Ne/^{22}Ne trend to a solar ^{20}Ne/^{22}Ne = 13.8 (Fig. 16). Based on the ^{136}Xe/^{130}Xe-^{129}Xe/^{130}Xe correlation evident in Figure 15, this extrapolation gives ^{136}Xe/^{130}Xe = 2.7 for the upper mantle.

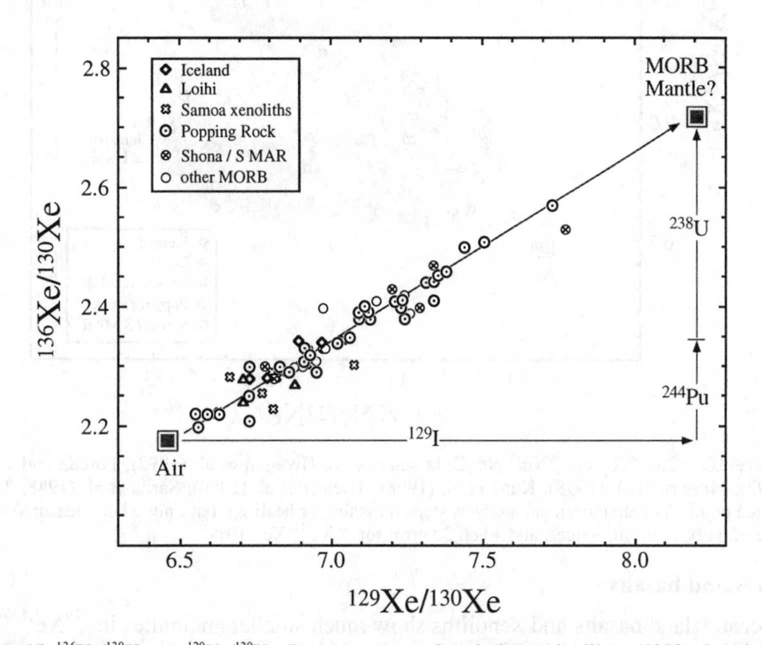

Figure 15. ^{136}Xe/^{130}Xe vs. ^{129}Xe/^{130}Xe. Data sources are Staudacher and Allègre (1982), Staudacher et al. (1989), Marty (1989), Hiyagon et al. (1992), Poreda and Farley (1992), Moreira et al. (1998), Kunz et al. (1998), Trieloff et al. (2000), Sarda et al. (2000). The lines show trends produced by decay of the indicated parental nuclides. Plotted points include individual analysis steps (crushing or heating), but only when both measured Xe isotope ratios were >2σ above the air values, and when 2σ errors were <10%. A proportion of 32% for fissiogenic ^{136}Xe that is ^{244}Pu-derived, as suggested by Kunz et al. (1998), is illustrated.

Excesses of ^{129}Xe in MORBs are derived from extinct radioactivity of ^{129}I, while those of ^{131,132,134,136}Xe are derived both from fission of ^{238}U and extinct ^{244}Pu. The correlation of MORB ^{136}Xe excesses with those of ^{129}Xe are best explained in terms of mixing between air and a depleted mantle component (Fig. 15). Spontaneous fission of ^{238}U and ^{244}Pu produces a vertical shift on Figure 15, while decay of ^{129}I produces a horizontal one. Some proportion of the fissiogenic component may be ^{244}Pu-derived, but the value of this proportion is sometimes a controversial point, and quantifying it requires precise analyses of $^{131,132,134,136}Xe/^{130}Xe$ in order to resolve small differences in the fissiogenic excesses (see the previous discussion of the production yields). Kunz et al. (1998) showed that when the most precise analyses are considered, or when the correlation lines are error weighted, the proportion of MORB fissiogenic ^{136}Xe that is ^{244}Pu-derived is 32±10%. Although the exact time scale is somewhat model dependent, this value leads to a refined I-Pu-Xe estimate for closure of the terrestrial Xe system that is ~50-70 Ma after solar system formation (Kunz et al. 1998).

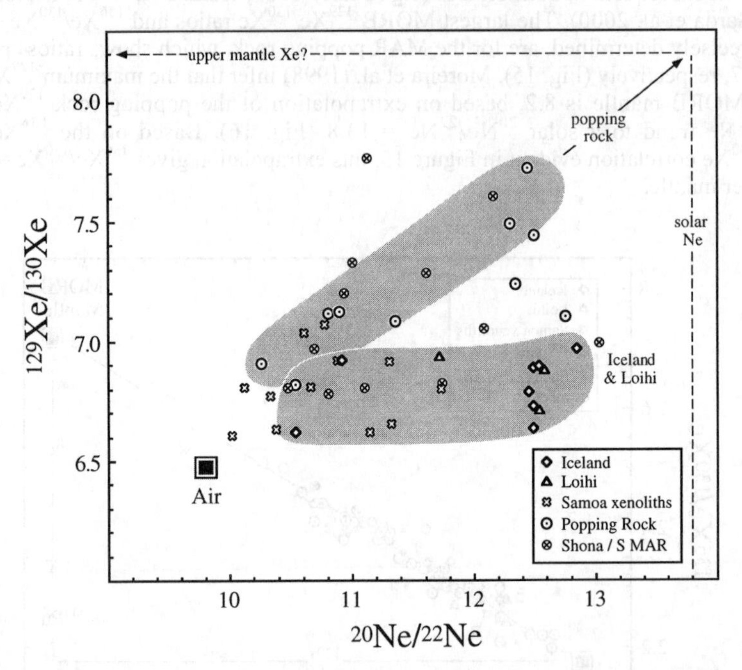

Figure 16. $^{129}Xe/^{130}Xe$ vs. $^{20}Ne/^{22}Ne$. Data sources are Hiyagon et al. (1992), Poreda and Farley (1992), Moreira et al. (1998), Kunz et al. (1998), Trieloff et al. (2000), Sarda et al. (1988, 2000). Plotted points include individual analysis steps (crushing or heating), but only when measured ratios were >2σ above the air values, and when 2σ error for $^{129}Xe/^{130}Xe$ <10%.

Ocean island basalts

Ocean island basalts and xenoliths show much smaller anomalies in $^{129}Xe/^{130}Xe$ and $^{136}Xe/^{130}Xe$ ratios than MORBs (Fig. 15). The small excesses in radiogenic and fissiogenic Xe appear to be correlated, but it is impossible to establish this firmly given typical analytical errors of ≥10% in $^{136}Xe/^{130}Xe$. It is also possible that Xe in OIB samples is sometimes contaminated by MORB-type Xe, as suggested for the Samoan xenoliths (Poreda and Farley 1992). If Xe isotope excesses are indigenous to the OIB sources, the relative fissiogenic contributions from ^{238}U and ^{244}Pu could, in principal, be different from those for the MORB mantle. One might expect from the steady-state mantle model (Porcelli and Wasserburg 1995b) that OIBs with high $^{3}He/^{4}He$ would have lower $^{136}Xe/^{130}Xe$ compared to MORBs, due to an enhanced ^{238}U-fissiogenic Xe component in the more degassed upper mantle. At the present time, a systematic, shallower trend of OIBs compared to MORBs is not clearly resolvable in Figure 15. A shallower slope for Iceland and Loihi is evident on the $^{129}Xe/^{130}Xe$–$^{20}Ne/^{22}Ne$ diagram (Fig. 16). This trend is poorly defined, but when extrapolated to solar $^{20}Ne/^{22}Ne$ = 13.8

there is the possibility that $^{129}Xe/^{130}Xe$ could be as high as ~7.2. This could be evidence for excess ^{129}Xe in those OIB mantle source regions. In contrast, if Ne-B ($^{20}Ne/^{22}Ne$ = 12.5) is the suitable choice (Trieloff et al. 2000; see discussion of the *Solar Hypothesis* for Ne isotopes) then no ^{129}Xe anomaly is apparent. Since OIBs usually erupt at shallower depths, they also degas more extensively and are more susceptible than MORB glasses to atmospheric contamination. This accounts for much of the observed scatter in Figures 15 and 16, and also makes it difficult to precisely define different trends for OIBs.

Solar hypothesis. Ratios involving the lighter Xe isotopes $^{124,126,128}Xe$ may be diagnostic of solar Xe. These nuclides are present in relatively low abundance (0.09-1.9%) and are quite difficult to measure. Consequently, no excesses of these isotopes have yet been reported in basalts. However, excesses have been found in CO_2 well gases from the southwestern US and S Australia (Caffee et al. 1999). These gases often contain a high abundance of noble gases, helping overcome some of the analytical difficulties. Excesses of $^{124,126,128}Xe$ in the well gases are correlated with excess ^{129}Xe and with $^{20}Ne/^{22}Ne$. The ultimate origin of such well gases is usually taken to be the Earth's mantle, with some additional crustal (radiogenic and nucleogenic) components (Staudacher 1987).

COUPLED RADIOGENIC/NUCLEOGENIC PRODUCTION

If the parental nuclide abundance ratios (such as K/U, Th/U, Pu/U) are known, the inter-elemental radiogenic production ratios (e.g., $^{4}He/^{40}Ar$, $^{4}He/^{21}Ne$, $^{4}He/^{136}Xe$ and ^{4}He/heat) may be predicted for all the noble gases (predictions involving Ne are possible when matrix composition is known). This is a very powerful approach to understanding Earth processes which is still being developed. The major limitation has been the few high quality measurements of all the noble gas isotopes in individual rock samples. This coupled approach provides some of the most powerful constraints on chemical heterogeneity of the mantle, the role of partial melting, and the significance of gas loss during magma ascent to the Earth's surface. Several examples are discussed below. With special treatment of the He-Ne systematics, measured concentrations of He and Ne are not needed to infer the ratio of the primordial nuclides $^{3}He/^{22}Ne$ in the mantle source, provided that the assumption of an initial isotope composition of the solid Earth, usually taken to be solar, is valid. In other treatments, the measured noble gas concentrations, after correction for atmospheric contamination, enter into estimates of the mantle source ratios (e.g., $^{3}He/^{22}Ne_S$, $^{4}He/^{21}Ne^*$, $^{4}He/^{40}Ar^*$). The ratios estimated in this way can be compared diagrammatically, and they reveal a wealth of information about physical processes in the mantle and in magma generated from it.

Neon-helium systematics

Time-integrated $^{3}He/^{22}Ne$ ratio. *A priori* one might expect the $^{3}He/^{22}Ne$ ratio to be constant over time. The time-integrated ratio (designated $\langle ^{3}He/^{22}Ne \rangle$) in a mantle reservoir may be inferred directly from the isotope compositions of He and Ne (Honda and McDougall 1998), without relying on measured He and Ne concentrations. This requires assumptions for the initial He and Ne isotope compositions for the Earth, such as solar values. The time-integrated ratio is given by

$$\langle ^{3}He/^{22}Ne \rangle = \frac{(^{21}Ne/^{22}Ne_E - {}^{21}Ne/^{22}Ne_S)}{(^{4}He/^{3}He_M - {}^{4}He/^{3}He_S)\, J_{Ne/He}} \tag{9}$$

where $J_{Ne/He} = 4.5\times10^{-8}$ is the nucleogenic ^{21}Ne to radiogenic ^{4}He production ratio of the mantle (Yatsevich and Honda 1997).

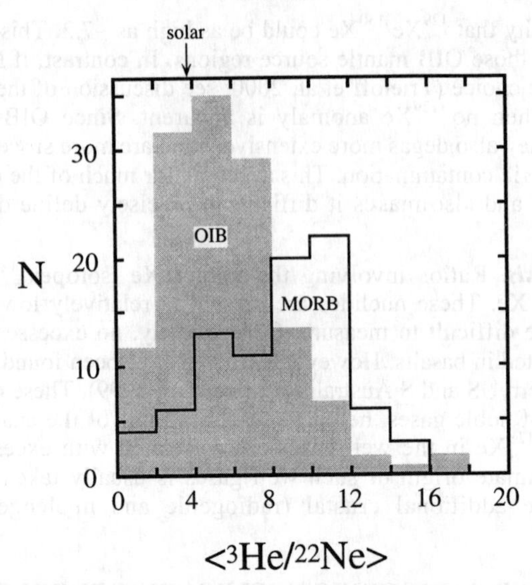

Figure 17. Histogram of <^{3}He/^{22}Ne> for MORBs and OIBs inferred from the He-Ne isotope systematics. Data sources are given in the caption to Figure 11.

Histograms for <^{3}He/^{22}Ne> are compared in Figure 17. These values were estimated assuming solar Ne isotope composition (^{20}Ne/^{22}Ne$_S$ = 13.8, ^{21}Ne/^{22}Ne$_S$ = 0.0328) and ^{3}He/^{4}He = 150 R$_A$. This choice of initial values is internally self-consistent with solar He and Ne isotope compositions deduced from analyses of the noble gases trapped in extraterrestrial materials, in particular the atmosphere of Jupiter (^{3}He/^{4}He = 120 R$_A$; Mahaffy et al. 2000). A self-consistent choice involving Ne-B, as suggested by Trieloff et al. (2000), would require using ^{3}He/^{4}He = 280 R$_A$, similar to the value for solar energetic particles or solar wind-implanted components.

Mean values of <^{3}He/^{22}Ne> for MORBs and OIBs are 8.8±3.5 (1 sd, n = 85) and 6.0±3.6 (1 sd, n = 169), respectively. These computed results are for crushing and step heating analyses for which ^{20}Ne/^{22}Ne was >2σ above the air value. The calculated ranges in <^{3}He/^{22}Ne> are large and there are sampling biases; e.g., individual samples, such as popping rocks and Loihi basalts, are included several times because they have been analyzed by different investigators using a variety of methods. This makes it difficult to say with certainty whether any apparent difference between MORBs and OIBs is real. For comparison, the ratio of ^{3}He/^{22}Ne in the modern-day solar wind is ~3.8 (Benkert et al. 1993). The mantle seems to have a slightly higher average ratio than this value, which might indicate that the ^{3}He/^{22}Ne ratio has been fractionated in the silicate Earth, with He showing a relative enrichment. Because helium is about twice as soluble as Ne in silicate melts (Jambon et al. 1986; Lux 1987), one speculative scenario proposed to account for such a fractionation is solubility controlled degassing of an early terrestrial magma ocean (Honda and McDougall 1998).

^{4}He/^{3}He-^{21}Ne/^{22}Ne relationships. MORBs and OIBs show a generally systematic relationship between measured He isotope composition and the ^{21}Ne/^{22}Ne ratio of the mantle source (^{21}Ne/^{22}Ne$_E$), calculated using Equation (3) and assuming the solar hypothesis is appropriate (Fig. 18). In this approach, it is more expedient to use the ^{4}He/^{3}He ratio, because it facilitates a diagrammatic comparison of hypothetical mixing and radiogenic evolution lines, and both of these processes are easily parameterized in terms of the ^{3}He/^{22}Ne ratio.

Higher values of $^3He/^4He$ (low $^4He/^3He$) at localities such as Loihi and Iceland are accompanied by lower values of $^{21}Ne/^{22}Ne_E$ (Fig. 18). The He and Ne isotopic differences between the MORB and OIB mantle sources require their source regions to have remained separated for long time periods (>10^9 years). Some of the observed He-Ne isotope variation suggests that mixing, either between mantle reservoirs or between different magmas, may have occurred relatively recently. For example, the trend may be produced by interaction of OIB source material, derived from the deep mantle and having solar-like $^{21}Ne/^{22}Ne$ and low $^4He/^3He$, with MORB source material, derived from the upper mantle and having $^4He/^3He$ = 84,600 ($^3He/^4He$ = 8.5 R_A) and $^{21}Ne/^{22}Ne$ = 0.075 (Moreira and Allègre 1995; Moreira et al. 2001). If r is defined as the concentration ratio $[^3He/^{22}Ne]_{MORB}/[^3He/^{22}Ne]_{OIB}$, where the subscripts designate the end-member components on Figure 18, a mixing line which best describes the data from Iceland would have r between 1 and 10. Compared to the OIB end-member, the MORB end-member must be enriched in He relative to Ne in this instance. This could indicate that any such mixing beneath Iceland is between a MORB magma that was previously degassed (leading to a higher He/Ne ratio as predicted from solubility considerations), and an undegassed, plume-like magma (Moreira et al. 2001). However, the observed scatter in Figure 18 for oceanic basalts also indicates a significant variation in the $^3He/^{22}Ne$ ratio from one location to another, and this might be due to differences in the type of mixing. For example, the Loihi data would be better described by r values <1, which might indicate that any mixing process occurs deeper beneath Hawaii, where magma degassing is less significant.

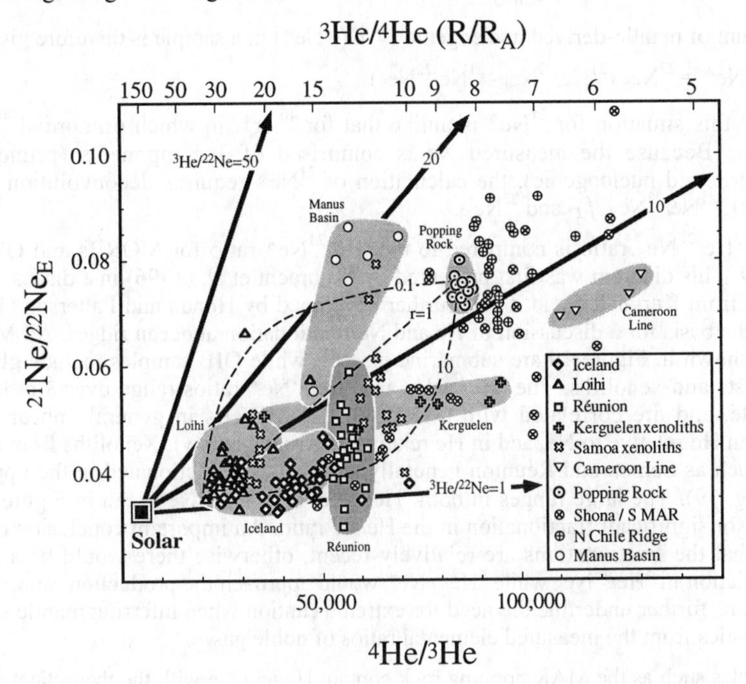

Figure 18. $^4He/^3He$ vs. $^{21}Ne/^{22}Ne_E$. Data sources are given in the caption to Figure 11. Two hypothetical mixing curves are shown by the dashed lines, for values of r = 1 and r = 10, where r = $[^3He/^{22}Ne]_{MORB}/[^3He/^{22}Ne]_{OIB}$. The OIB end-member has $^3He/^4He$ = 35 R_A and $^{21}Ne/^{22}Ne_E$ = 0.035, and the MORB end-member has $^3He/^4He$ = 8.5 R_A and $^{21}Ne/^{22}Ne_E$ = 0.075 for this illustration. Closed system evolution lines from solar He and Ne isotope compositions are also illustrated, for $^3He/^{22}Ne$ ratios of 1, 10, 20 and 50.

An alternative possibility is that the Earth's mantle is characterized by heterogeneity in $^3He/^{22}Ne$ ratio. If the initial Earth is taken to have solar-like Ne and He (e.g., $^3He/^4He$ = 150 R_A or $^4He/^3He$ = 4800), then the MORB end-member shown on Figure 18 would have $^3He/^{22}Ne$ = 12, while the OIB end-members ($^3He/^4He$ = 35 R_A or $^4He/^3He$ = 20,500), represented by Iceland ($^{21}Ne/^{22}Ne$ = 0.035) and Loihi ($^{21}Ne/^{22}Ne$ = 0.042), would have $^3He/^{22}Ne$ of 3 and 12, respectively (calculated in an analogous way to Eqn. 9). If heterogeneity accounts for the full variation observed in oceanic basalts, then $^3He/^{22}Ne$ must vary by about a factor of 20 throughout the mantle (Fig. 18).

Measured ^{3}***He/***^{22}***Ne***$_S$ ***and*** ^{4}***He/***^{21}***Ne****. Because ^{22}Ne in young oceanic basalts carries an insignificant nucleogenic component, the amount of mantle-derived ^{22}Ne is

$$^{22}Ne_S = {}^{22}Ne_M\, f_{22} \tag{10}$$

where $^{22}Ne_M$ is a sample's measured amount of ^{22}Ne and f_{22} is the proportion of mantle-derived Ne present, i.e.,

$$f_{22} = \frac{^{20}Ne/^{22}Ne_M - {}^{20}Ne/^{22}Ne_A}{^{20}Ne/^{22}Ne_S - {}^{20}Ne/^{22}Ne_A}\ .$$

The ratio of primordial ^3He to solar ^{22}Ne is

$$^3He/^{22}Ne_S = \frac{(^3He/^4He_M)\,[^4He]_M}{[^{22}Ne]_M\, f_{22}} \tag{11}$$

The amount of mantle-derived nucleogenic ^{21}Ne ($^{21}Ne^*$) in a sample is therefore given by

$$^{21}Ne^* = {}^{22}Ne_S\,(^{21}Ne/^{22}Ne_E - {}^{21}Ne/^{22}Ne_S) \tag{12}$$

Note that this situation for $^{21}Ne^*$ is unlike that for $^{40}Ar^*$, in which primordial ^{40}Ar is negligible. Because the measured Ne is comprised of 3 components (primordial, atmospheric and nucleogenic), the calculation of $^{21}Ne^*$ requires deconvolution using $^{21}Ne/^{22}Ne_E$, $^{21}Ne/^{22}Ne_S$, f_{22} and $^{22}Ne_M$.

The $^3He/^{22}Ne_S$ ratio is compared to the $^4He/^{21}Ne^*$ ratio for MORBs and OIBs in Figure 19. This diagram was first presented by Valbracht et al. (1996) in a discussion of xenoliths from Kerguelen and it was further developed by Honda and Patterson (1999), forming the basis for a discussion of He and Ne fractionation at ocean ridges. All MORB samples shown in Figure 19 are submarine glasses, while OIB samples include glasses, phenocrysts and xenoliths. The $^3He/^{22}Ne_S$ and $^4He/^{21}Ne^*$ ratios range over 3 orders of magnitude, and are correlated with a slope of ~1. MORBs in general appear to be enriched in He relative to Ne, and in He relative to Ar (see below). Xenoliths from ocean islands such as Samoa and Réunion generally appear to be fractionated in the opposite sense (Fig. 19). The large ranges in both $^3He/^{22}Ne_S$ and $^4He/^{21}Ne^*$ seen in Figure 19 is evidence for significant fractionation in the He/Ne ratio. An important conclusion of this work is that the fractionations are relatively recent, otherwise there would be a much larger variation in $^3He/^{22}Ne_S$ while $^4He/^{21}Ne^*$ would approach the production ratio. These observations further underline the need for extreme caution when inferring mantle source characteristics from the measured elemental ratios of noble gases.

Samples such as the MAR popping rock contain He and Ne with the theoretical ratios. Other MORBs generally do not, indicating that the helium enrichment characteristic of the MORB data set as a whole is probably not due to an enrichment in the upper mantle. It is also notable that a few samples from high $^3He/^4He$ localities do not show a significant fractionation from expected mantle ratios. These samples are the deeply erupted glassy basalts from Loihi seamount (Hiyagon et al. 1992; Valbracht et al. 1997; Trieloff et al.

2000) and several subglacial basalt glasses from Iceland (Dixon et al. 2000; Trieloff et al. 2000; Moreira et al. 2001). With the exception of the above three sample suites, the thousand-fold variation in He/Ne observed for other mantle-derived samples makes it clear that any attempt to estimate representative noble gas concentrations for their mantle sources is nearly futile. In light of this, the so-called "helium paradox" (i.e., the observation that OIB magmas with high $^3He/^4He$ often have lower helium contents than more deeply erupted MORB magmas; Anderson 1998a), is a manifestation of shallow level processes and has little bearing on mantle source characteristics.

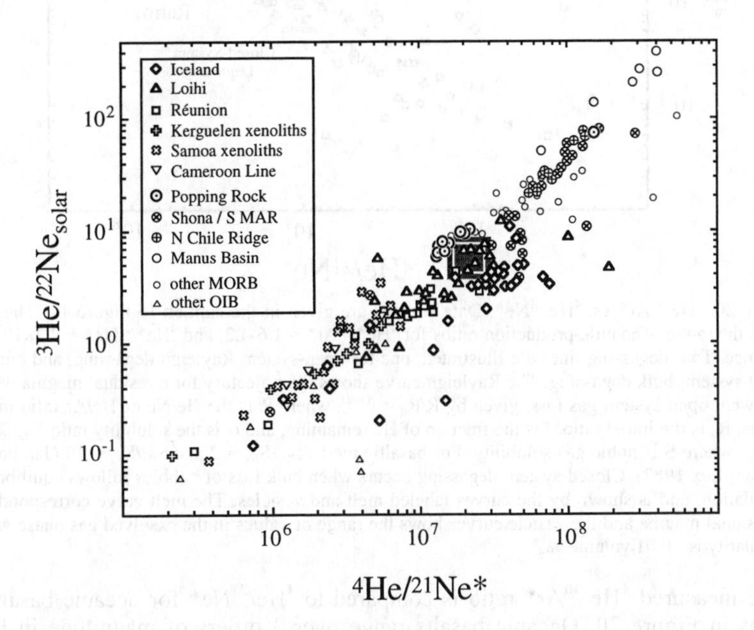

Figure 19. $^3He/^{22}Ne_S$ vs. $^4He/^{21}Ne^*$. Data sources are given in the caption to Figure 11. In Figures 19-22, individual analyses carried out by stepwise crushing are plotted, but for incremental heating analyses only the total gas contents are shown in order to avoid possible inter-elemental fractionation during heating. The box shows theoretical values for the mantle. The mantle production ratio for $^4He^*/^{21}Ne^* = 2.2\times10^7$ (Yatsevich and Honda 1997). The range of $^3He/^{22}Ne_S$ shown by the box is between 4, the value for modern solar wind, and 9, the mean value for the time-integrated $^3He/^{22}Ne$ ratio of the upper mantle (Fig. 17).

Argon-helium systematics

The $^4He/^{40}Ar^*$ ratio can be computed using the measured He content and $^{40}Ar^*$ calculated from Equation (5). This ratio may then be compared to theoretical values estimated for the mantle using Equations (1) and (4).

$$\frac{^4He^*}{^{40}Ar^*} = \frac{8\,(e^{\lambda_{238}t} - 1) + (7/137.88)\,(e^{\lambda_{235}t} - 1) + 6\,\kappa\,(e^{\lambda_{232}t} - 1)}{(K/U)\,1.23\times10^{-5}\,(e^{\lambda_{40}t} - 1)} \qquad (13)$$

The theoretical $^4He^*/^{40}Ar^*$ ratio varies between 1.6 (for t = 4500 Ma) and 4.2 (for $t \ll 1/\lambda_{235}$) for typical mantle values of $\kappa = 3$ and K/U (atomic) = 7.73×10^4 (K/U weight ratio = 1.3×10^4) (Jochum et al. 1983).

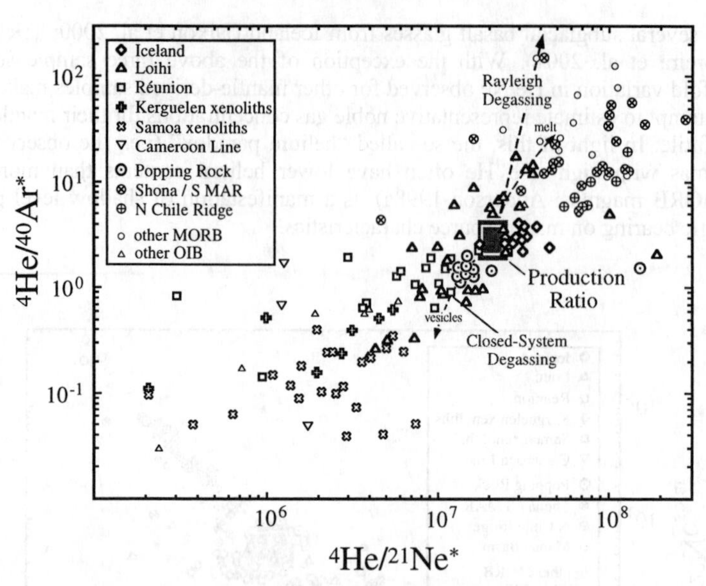

Figure 20. $^4He/^{40}Ar^*$ vs. $^4He/^{21}Ne^*$. Data sources are given in the caption to Figure 11. The box shows the range of mantle production ratios for $^4He^*/^{40}Ar^* = 1.6$-4.2, and $^4He^*/^{21}Ne^* = 2.2 \times 10^7$ for reference. Two degassing lines are illustrated, one for open-system Rayleigh degassing, and one for closed-system, bulk degassing. The Rayleigh curve shows the trajectory for a residual magma which underwent open system gas loss, given by $R/R_0 = f^{(1-\alpha)}$, where R is the He/Ne or He/Ar ratio in the magma, R_0 is the initial ratio, f is the fraction of He remaining, and α is the solubility ratio S_{He}/S_{Ne} or S_{He}/S_{Ar}, where S is noble gas solubility. For basaltic melts, $S_{He}/S_{Ne} = 2$ and $S_{He}/S_{Ar} = 10$ (Jambon et al. 1986; Lux 1987). Closed-system degassing occurs when bulk loss of bubbles follows equilibrium vesiculation, and is shown by the curves labeled melt and vesicles. The melt curve corresponds to the residual magma and the vesicle curve shows the range of values in the exsolved gas phase when vesicularity is >0.01 volume %.

The measured $^4He/^{40}Ar^*$ ratio is compared to $^4He/^{21}Ne^*$ for oceanic basalts and xenoliths in Figure 20. Oceanic basalts range over 3 orders of magnitude in He/Ar, similar to variations observed from the He-Ne systematics just considered. Nearly all MORB glasses are enriched in ^4He relative to $^{40}Ar^*$, ranging upward from the production ratio by more than a factor of 50. The popping rocks have $^4He/^{40}Ar^*$ ratios of 1.2-1.8, similar to the expected production ratio for a 4.5 Gy evolution, but about a factor of 2 lower than the value expected for an upper mantle mixing time of 10^9 years. A few of the most deeply erupted and least degassed Loihi seamount glasses analyzed by Trieloff et al. (2000) approach the theoretical ratios. Notably, the basalt glasses from Iceland reported by Trieloff et al. (2000) have also preserved the $^4He/^{40}Ar^*$ ratio of their mantle source.

A similar comparison may be made between $^4He/^{40}Ar^*$ and $^3He/^{22}Ne_S$ (Fig. 21). Again the data range is large and the ratios are broadly correlated. There is overlap between OIBs and MORBs near the expected mantle values, which includes the popping rock, Loihi Seamount and Iceland glasses. Some MORB and Loihi glasses trend at high angles away from the mantle values due to gas loss effects. In contrast, xenoliths from Kerguelen, Samoa and Réunion show a less steep slope, and trend to lower $^3He/^{22}Ne_S$ and $^4He/^{40}Ar^*$ ratios. Xenoliths from ocean islands often originate as fragments of oceanic lithosphere, and their shallower trend on Figure 21 may indicate that their trapped fluid inclusions represent gases extracted from ascending and fractionating magmas (Moreira and Sarda 2000).

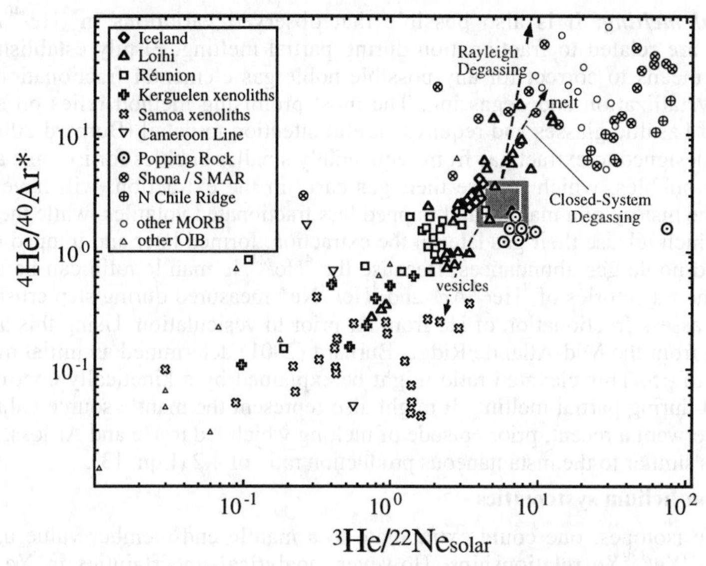

Figure 21. $^4He/^{40}Ar^*$ vs. $^3He/^{22}Ne_S$. Data sources are given in the caption to Figure 11. The box shows the range of $^4He^*/^{40}Ar^*$ = 1.6-4.2 and $^3He/^{22}Ne_S$ = 4-9 for reference. The two degassing curves are described in the caption to Figure 21.

Magma degassing. Figures 20 and 21 are especially diagnostic for investigating the physics of magma degassing. Moreira and Sarda (2000) and Sarda and Moreira (2002) suggest that the He-Ne-Ar covariations are produced by differences in the style of degassing for different tectonic regimes. Along mid-ocean ridges they propose that degassing is predominantly through bubble loss following closed system, equilibrium vesiculation, whereas beneath oceanic islands the degassing may largely occur through Rayleigh distillation (open system, fractional degassing), during which vesiculating bubbles are continuously lost (Figs. 20 and 21). Their model accurately reproduces the observed trends, but it requires the solubility of helium to be a factor of 4 larger than its experimentally determined value, while the solubility of Ar must remain unchanged (i.e., $S_{He} = 40S_{Ar}$ in their model compared to $S_{He} = 10S_{Ar}$ measured experimentally). The possibility that different degassing mechanisms operate beneath islands and ridges has important implications for using noble gases in basalts to infer mantle characteristics. Rayleigh degassing would lead to a much more effective gas loss, especially for the heavier noble gases, making it more difficult to detect Xe isotope anomalies in OIBs.

In situ laser analyses of two MORB glasses from the Mid-Atlantic Ridge by Burnard (1999a) reveal significant differences in He/Ar and Ar/CO_2 for individual vesicles within a sample. The observed fractionations occur during bubble growth and are consistent with a solubility-controlled Rayleigh-type degassing. Two studies of MORB glasses have also shown that $^4He/^{40}Ar^*$ is broadly correlated with axial depth and lava MgO content (Marty and Zimmerman 1999; Burnard et al. 2002). This suggests that the amount of gas loss from a MORB magma is related to confining pressure and/or the degree of crystallization of the melt. Because He solubility is an order of magnitude larger than that for Ar, the He/Ar ratio of gas bubbles will be about 10 times less than that of the parental melt. Consequently, any gas loss leads to elevated $^4He/^{40}Ar^*$ in the residual magma. Variation in $^4He/^{40}Ar^*$ for MORB suites from the East Pacific Rise and Mid-Atlantic Ridge (Marty and Zimmerman 1999) and from the Southeast Indian Ridge (Burnard et al. 2002) are consistent with such a degassing process accompanying fractional crystallization at 2-6 km depth.

Partial melting. It is also possible that observed variations in $^4He/^{40}Ar^*$ and $^4He/^{21}Ne^*$ are related to fractionation during partial melting. Firmly establishing this requires a means to correct for any possible noble gas elemental fractionation during magma crystallization and degassing. The most promising method relies on stepwise crushing of basaltic glasses and requires careful attention to detail (Burnard 2001). This method is designed to extract gas from sequentially smaller bubbles for isotopic analysis. The larger bubbles, which release their gas early in the extraction, will have formed earlier in the history of a magma and trapped less fractionated volatiles, while the smaller bubbles, which release their gas later in the extraction, formed later and trapped gas with fractionated noble gas abundances. Because the $^4He/^{21}Ne$ mantle ratio can be assumed constant, the trajectories of $^4He/^{40}Ar^*$ and $^4He/^{21}Ne^*$ measured during step crushing can be used to assess fractionation of He from Ar prior to vesiculation. Using this approach on a basalt from the Mid-Atlantic Ridge, Burnard (2001) determined an initial magmatic $^4He/^{40}Ar^*$ of ~5. This elevated ratio might be explained by a kinetically controlled He enrichment during partial melting. It might also represent the mantle source value, if the source underwent a recent, prior episode of melting which led to He and Ar loss, because the value is similar to the instantaneous production ratio of 4.2 (Eqn. 13).

Xenon-neon-helium systematics

For Xe isotopes, one could extrapolate to a mantle end-member value using the $^{20}Ne/^{22}Ne$-$^{136}Xe/^{130}Xe$ relationships. However, analytical uncertainties in Xe isotope analyses are sufficiently large to preclude doing this in a meaningful way at the present time. A simpler way to crudely estimate the amount of radiogenic ^{136}Xe in a sample is to assume that all ^{130}Xe derives from air, analogous to the way in which ^{36}Ar is used to calculate $^{40}Ar^*$.

$$^{136}Xe^* = (^{136}Xe/^{130}Xe_M - {}^{136}Xe/^{130}Xe_A)\ ^{130}Xe_M \tag{14}$$

where $^{136}Xe/^{130}Xe_A = 2.176$.

The theoretical $^4He^*/^{136}Xe^*$ ratio, calculated from Equations (1) and (6), is 4.5×10^8. A comparison of measured $^4He/^{136}Xe^*$ with measured $^4He/^{21}Ne^*$ is made in Figure 22.

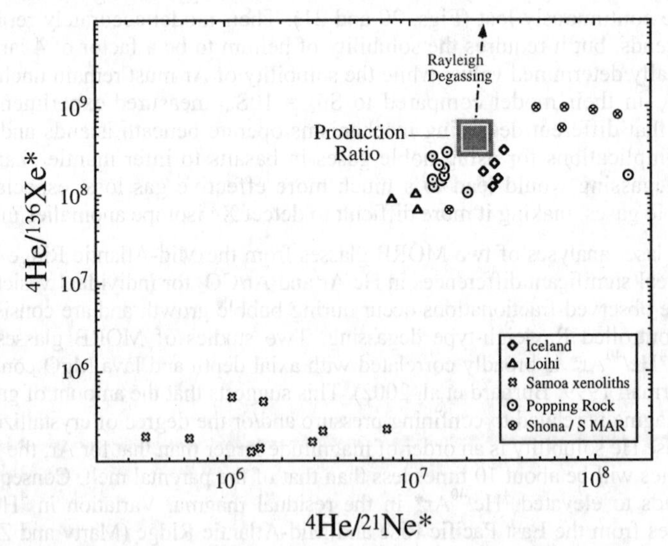

Figure 22. $^4He/^{136}Xe^*$ vs. $^4He/^{21}Ne^*$. Data sources are given in the caption to Figure 11. The box shows mantle production ratios for $^4He^*/^{136}Xe^* = 4.5\times10^8$ and $^4He^*/^{21}Ne^* = 2.2\times10^7$ for reference.

There is much less Xe isotope data available than for the other noble gases. Nevertheless, it is remarkable that once again the Iceland basalt glasses analyzed by Trieloff et al. (2000) and the MAR popping rock show values similar to the expected production ratios for the mantle.

MANTLE ABUNDANCE PATTERNS OF NOBLE GASES

The relative abundances of primordial noble gases can be calculated from the considerations in the previous sections. The exercise will be carried out here for MORB mantle represented by the popping rock, and for OIB mantle represented by Loihi and Iceland, because basalt glasses from these localities show the least fractionation of their radiogenic production ratios from expected mantle source values.

For the purposes of this calculation the following will be assumed:

1. $^3He/^4He$ is 8.5 R_A in the MORB mantle ($^4He/^3He$ = 84,600) and 40 R_A in the OIB mantle ($^4He/^3He$ = 18,000).
2. $^{40}Ar/^{36}Ar$ is 40,000 in the MORB mantle and 8,000 in the OIB mantle (Fig. 14).
3. $^{129}Xe/^{130}Xe$ is 8.2 in the MORB mantle and 7.2 in the OIB mantle (Fig. 16). Corresponding $^{136}Xe/^{130}Xe$ ratios are 2.70 and 2.40, respectively (Fig. 15). This gives $^{136}Xe^*/^{130}Xe = (^{136}Xe/^{130}Xe_M - ^{136}Xe/^{130}Xe_A) = 0.52$ in the MORB mantle and 0.22 in the OIB mantle (Eqn. 14).
4. $^{21}Ne/^{22}Ne_E$ (Eqn. 3) is 0.075 in the MORB mantle and 0.035 in the OIB mantle (Fig. 11).
5. The radiogenic production ratios are $^4He^*/^{21}Ne^* = 2.2\times10^7$, $^4He^*/^{40}Ar^* = 2.5$ and $^4He^*/^{136}Xe^* = 4.5\times10^8$ (which assumes all $^{136}Xe^*$ is derived from fission of ^{238}U).
6. Initial isotope ratios for the Earth are $^3He/^4He$ = 150 R_A ($^4He/^3He_0$ = 4800) and $^{21}Ne/^{22}Ne_S = 0.0328$.

The terrestrial primordial ratios are then computed from:

$$^3He/^{36}Ar=(^4He^*/^{40}Ar^*)(^{40}Ar/^{36}Ar)/\{^4He/^3He - {}^4He/^3He_0\}$$

$$^{22}Ne/^{36}Ar=(^4He^*/^{40}Ar^*)(^{40}Ar/^{36}Ar)/\{(^4He^*/^{21}Ne^*)(^{21}Ne/^{22}Ne_E - {}^{21}Ne/^{22}Ne_S)\}$$

$$^{130}Xe/^{36}Ar=(^4He^*/^{40}Ar^*)(^{40}Ar/^{36}Ar)/\{(^4He^*/^{136}Xe^*)(^{136}Xe/^{130}Xe_M - {}^{136}Xe/^{130}Xe_A)\}$$

The ratios of $^3He/^{36}Ar$ are between 1.2 and 1.5. A meaningful comparison with the atmospheric value (2.3×10^{-7}) is not possible due to the loss of ^3He to space. It is also not possible to use this approach for comparing Kr because isotope differences from atmospheric values have not been reported in oceanic basalts. Values for $^{22}Ne/^{36}Ar$ are 0.10-0.43, which is greater than the atmospheric value (0.053). For $^{130}Xe/^{36}Ar$, the OIB value is 2.0×10^{-4}, while for the popping rock it is 4.3×10^{-4}. These values are ~2 to 3 times the atmospheric $^{130}Xe/^{36}Ar$ ratio (1.13×10^{-4}).

The approach leads to some interesting first-order comparisons, but it is important to emphasize that these abundances are relative ones, and therefore do not constrain the amount of noble gases in different parts of the mantle. The abundances, normalized to solar ratios (Ozima and Podosek 2002) and to ^{36}Ar = 1, are shown in Figure 23. There are three basic observations. First, the MORB and OIB mantles show relatively flat $^3He/^{22}Ne$. In detail, compared to the solar ratio, the MORB mantle appears to be slightly enriched in $^3He/^{22}Ne$ by about a factor of 3 (consistent with the position of the popping rock in Fig. 19). Second, ^{22}Ne is depleted, relative to ^{36}Ar, by about a factor of 2 to 8 in the atmosphere compared to the mantle. Third, ^{130}Xe is enriched by about a factor of 2 to 4, relative to ^{36}Ar, in the mantle compared to the atmosphere. This might be evidence that there is a small subduction component of atmospheric Xe in the mantle (e.g., Porcelli and Wasserburg, 1995b). Obviously there are considerable uncertainties in any approach such as the one outlined here, and the accuracy of these mantle abundance patterns depends critically on the assumptions about the Ar and Xe isotope compositions and the appropriate radiogenic production ratios.

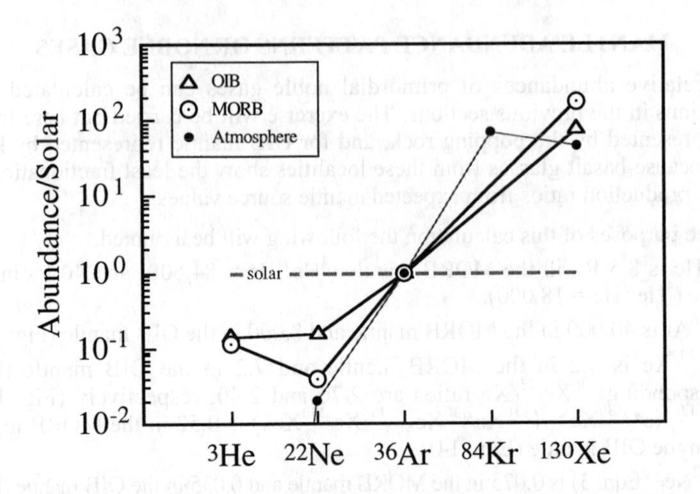

Figure 23. Abundance pattern of primordial noble gases in model mantle sources for MORBs and OIBs, calculated as outlined in the text. The ratios are normalized to the solar pattern and to $^{36}Ar = 1$. The pattern for the atmosphere is shown for comparison.

PRINCIPAL OBSERVATIONS

In attempting to characterize mantle reservoir compositions, any model of the geochemical structure of the Earth's mantle should include or allow for the following basic observations.

1. The range of $^{3}He/^{4}He$ in MORBs away from the influence of hotspots or mantle plumes is restricted to 6.5-9.5 R_A, with roughly a normal distribution. Lower (more radiogenic) values usually occur along ridge sections where other geochemical anomalies indicate the presence of compositionally distinct mantle 'blobs' that were entrained into the convective flow beneath the ridge. The higher values in this range of 6.5-9.5 R_A are usually associated with the most trace-element-depleted mantle sources. Fine-scale structure of $^{3}He/^{4}He$ along ridges may be related to the manner in which mantle heterogeneity is sampled by secondary convection and melting in the uppermost mantle.
2. $^{3}He/^{4}He$ in OIBs ranges to values >40 R_A, supporting the existence of mantle plumes from deep in the Earth. Best model cases are Iceland and Hawaii. Temporal and spatial variations at ocean island hotspots may be weakly associated with variations in plume flux and associated melting processes, but there is no systematic global behavior between plume flux and $^{3}He/^{4}He$ signal. Some ocean island hotspots, such as Réunion, may have relatively constant and intermediate $^{3}He/^{4}He$ ratios indicating a duration of 10^{7}-10^{8} years. This suggests the possibility of deep mantle reservoirs having variable $^{3}He/^{4}He$.
3. The mantle is characterized by elevated ratios of $^{20}Ne/^{22}Ne$ and $^{21}Ne/^{22}Ne$ compared to the atmosphere. Mid-ocean ridge basalts and ocean island basalts show distinct $^{20}Ne/^{22}Ne$-$^{21}Ne/^{22}Ne$ trends that result from differences in the dilution, by primordial Ne, of the nucleogenic ^{21}Ne that is produced in their respective mantle sources. The

MORB trend passes through the composition of air and extends to values of $^{20}Ne/^{22}Ne \geq 12.5$ and $^{21}Ne/^{22}Ne \geq 0.07$. OIB trends are much steeper; in the case of Iceland they pass through the composition of air and extend to values of $^{20}Ne/^{22}Ne \geq 12.5$ and $^{21}Ne/^{22}Ne \leq 0.035$. It seems likely that the whole mantle is characterized by a uniform value of $^{20}Ne/^{22}Ne$, but this has not been proven.

4. MORB and OIB samples show a range of measured $^{40}Ar/^{36}Ar$ that is mostly due to variable proportions of atmospheric and magmatic components. The highest $^{40}Ar/^{36}Ar$ ratio measured in a MORB is ~40,000. The correlation between $^{40}Ar/^{36}Ar$ and $^{20}Ne/^{22}Ne$ for the MAR popping rock, when extrapolated to a solar $^{20}Ne/^{22}Ne$ = 13.8, indicates a possible upper limit of 44,000 for $^{40}Ar/^{36}Ar$ in the upper mantle. High $^{3}He/^{4}He$ ocean islands range to moderately high $^{40}Ar/^{36}Ar$ ratios, up to ~8000. This may be the minimum value appropriate for the mantle source of OIBs.

5. $^{129}Xe/^{130}Xe$ anomalies are present in MORBs from all the major ocean basins. Small anomalies also occur in $^{131,132,134,136}Xe/^{130}Xe$ ratios, and these are correlated with the $^{129}Xe/^{130}Xe$ anomalies. The MORB excesses in ^{129}Xe and $^{131,132,134,136}Xe$ are derived from extinct radioactivity of ^{129}I, and fission of ^{238}U and extinct ^{244}Pu. The largest MORB $^{129}Xe/^{130}Xe$ and $^{136}Xe/^{130}Xe$ ratios, and the most precisely determined, are for the popping rock, which range up to 7.73 and 2.57, respectively. The maximum $^{129}Xe/^{130}Xe$ of the MORB mantle may be near 8.2, based on extrapolation of the popping rock Xe-Ne isotope trend to solar $^{20}Ne/^{22}Ne$ = 13.8 (Fig. 16). Ocean island basalts and xenoliths show smaller Xe isotope anomalies than MORBs. These small excesses in radiogenic and fissiogenic Xe appear to be correlated, but given the analytical uncertainties this is not firmly established. It is also possible that the OIBs have been contaminated by MORB-type Xe during magma ascent.

6. Coupled He-Ne isotope systematics reveal that the highest values of $^{3}He/^{4}He$ at ocean island localities are accompanied by the lowest values of $^{21}Ne/^{22}Ne_E$, the $^{21}Ne/^{22}Ne$ ratio corrected for air contamination and extrapolated to solar $^{20}Ne/^{22}Ne$. The general $^{3}He/^{4}He$-$^{21}Ne/^{22}Ne_E$ behavior (Fig. 18) suggests that either mixing between OIB and MORB components has occurred, or that the Earth's mantle is characterized by heterogeneity in $^{3}He/^{22}Ne$ ratio that is about a factor of 20. This heterogeneity may derive from fractionation processes associated with melting and degassing, or it may be primordial heterogeneity preserved since planetary accretion. The mean time-integrated $^{3}He/^{22}Ne$ ratio, inferred from coupled He-Ne isotope systematics, is similar for MORBs and OIBs (~9 and ~6, respectively) and appears to be higher than the ratio in the solar wind (~4) (Fig. 17).

7. Relatively recent fractionations dominate the measured ratios of $^{4}He/^{40}Ar^{*}$, $^{3}He/^{22}Ne_S$, $^{4}He/^{21}Ne^{*}$ and $^{4}He/^{136}Xe^{*}$ in MORBs and OIBs. These fractionations are controlled by magma degassing and by partial melting effects. Notable exceptions are deeply erupted basalts from Loihi Seamount and the suite of basalt glasses from Iceland studied by Trieloff et al. (2000), and the Mid-Atlantic Ridge popping rock studied by Moreira et al. (1998). These samples contain noble gases with radiogenic abundance ratios resembling values expected for the mantle. Abundance ratios of the primordial noble gases ($^{3}He/^{36}Ar$, $^{22}Ne/^{36}Ar$ and $^{130}Xe/^{36}Ar$) in the mantle sources of these basalt suites are significantly fractionated from solar ratios. The $^{3}He/^{22}Ne$ ratio is an exception, and is within a factor of three of the solar ratio.

SUMMARY

Variations in the isotope compositions of noble gases are related to processes controlling the distribution of K, U and Th, the major heat producing nuclides in the Earth. Mid-ocean ridge basalts have $^{3}He/^{4}He$ ratios between 3 and 15 R_A, while ocean island basalts have ratios between 3.5 and 43 R_A. The ubiquitous presence of 'excess' ^3He in mantle-derived rocks establishes that primordial volatiles are still escaping from the Earth's interior. In addition, $^3He/^4He$ ratios above 15 R_A are present only at ocean island localities, indicating a lower time-integrated $(U+Th)/^3He$ in their mantle source regions. The whole mantle is also characterized by elevated ratios of $^{20}Ne/^{22}Ne$ and $^{21}Ne/^{22}Ne$ compared to the atmosphere. The $^{20}Ne/^{22}Ne$ ratio appears to be relatively uniform throughout the mantle, resembling either solar wind or the solar component trapped in meteorites and the atmosphere of Jupiter. The $^{21}Ne/^{22}Ne$ ratio is variable in the mantle due to nucleogenic production of ^{21}Ne in reservoirs having different $(U+Th)/^{22}Ne$ ratios. This nucleogenic contribution is relatively larger in the MORB mantle source compared to some OIB sources (most notably at Iceland, Hawaii and Réunion), leading to higher $^{21}Ne/^{22}Ne$ in MORBs. The mantle $^{40}Ar/^{36}Ar$ ratio is variable due to variations in $K/^{36}Ar$, with MORBs displaying the highest $^{40}Ar/^{36}Ar$ ratios, up to ~40,000. The upper mantle source of MORBs is therefore characterized by higher radiogenic/primordial noble gas isotope ratios due to its degassed nature and consequent higher parent/daughter ratios (e.g., $(U+Th)/^3He$ and $K/^{36}Ar$). The relatively undegassed nature of the mantle source for some ocean islands is also clearly recognized from $^3He/^4He$, $^{20}Ne/^{22}Ne$-$^{21}Ne/^{22}Ne$ and $^{40}Ar/^{36}Ar$ in oceanic basalts. The contrast between MORB and OIB noble gas isotope compositions is the most fundamental geochemical evidence for some mode of stratification within the Earth's mantle. Ocean island basalts from localities such as Hawaii, Iceland and Réunion have a deeper mantle source than mid-ocean ridge basalts.

The near solar-like Ne isotope compositions, the high $^{40}Ar/^{36}Ar$ ratios, and the non-atmospheric $^{129}Xe/^{130}Xe$ and $^{136}Xe/^{130}Xe$ in oceanic basalts also provide basic clues on the origin and evolution of the atmosphere. The atmosphere appears to have 'formed' within the first 50-70 Ma of Earth's history, probably coincident with the period of terrestrial core formation (Halliday and Lee 1999).

The time-integrated ratio of the primordial nuclides ^3He and ^{22}Ne in the mantle may be inferred from coupled He-Ne isotope systematics. This time-integrated ratio, $\langle ^3He/^{22}Ne \rangle$, is independent of measured concentrations but rests on assumptions for the initial isotope composition of the solid Earth, usually taken to be solar. MORBs and OIBs show considerable overlap using this approach, and on average have $\langle ^3He/^{22}Ne \rangle$ ratios between 6 and 9, seemingly higher than the solar ratio of 4. Other methods for comparing relative abundance ratios in basalts depend on measured noble gas concentrations, and are subject to more uncertainty from elemental fractionations that occur both in nature and during sample preparation or analysis. Key ratios estimated in this way include $^3He/^{22}Ne_{solar}$, $^4He/^{21}Ne^*$, $^4He/^{40}Ar^*$ and $^4He/^{136}Xe^*$ (where the * designates a radiogenic or nucleogenic component after correction for air contamination). Abundance ratios computed in this way may be compared to expected mantle source values, assuming radioactive equilibrium and known values of K/U and Th/U in the mantle source. Differences from the expected mantle values provide information about the style of magma degassing (e.g., bulk vs. fractional) and about crystal/magma partitioning during partial melting.

Much of our knowledge about the noble gas geochemistry of the mantle relies on the analyses of a limited number of samples from three key areas; the popping rock from 14°N on the Mid-Atlantic Ridge, submarine basalt glasses from Loihi Seamount, and subglacial basalt glasses from Iceland. It is amazing in retrospect how much information about the Earth's mantle is contained in the noble gas isotope compositions of these few samples. We should anticipate that much more remains to be learned, as analytical techniques improve and as other key sample localities are discovered.

ACKNOWLEDGMENTS

I thank Don Porcelli for encouraging me to undertake this review, and for his patience in extracting it. The work began during an extended visit to Ken Farley's lab at CalTech, where Pete Burnard graciously shared many ideas with me on noble gas geochemistry and analytical techniques. Discussions with many other people, including Don Anderson, Barry Hanan, Bernard Marty, Manuel Moreira, Don Porcelli and Philippe Sarda were also instrumental in summarizing many of the issues mentioned. John Lupton made helpful comments on several versions of the manuscript. Pete Burnard, Steve Goldstein, Dave Hilton and Don Porcelli provided detailed and constructive reviews. The work was supported by the Marine Geology & Geophysics program of the National Science Foundation.

REFERENCES

Albarède F (1998) Time-dependent models of U-Th-He and K-Ar evolution and the layering of mantle convection. Chem Geol 145:413-429

Alexander EC, Ozima M (1978) Terrestrial Rare Gases. Adv Earth Planet Sci, vol 3. Center for Academic Publications, Japan, Tokyo

Allègre CJ, Turcotte DL (1985) Geodynamic mixing in the mesosphere boundary layer and the origin of oceanic islands. Geophys Res Lett 12:207-210

Allègre CJ, Dupré B, Lewin E (1986) Thorium/uranium ratio of the Earth Chem Geol 56:219-227

Allègre CJ, Hamelin B, Dupré B (1984) Statistical analysis of isotopic ratios in MORB: the mantle blob cluster model and the convective regime of the mantle. Earth Planet Sci Lett 71:71-84

Allègre CJ, Hofmann AW, O'Nions RK (1996) The argon constraints on mantle structure. Geophys Res Lett 23:3555-3557

Allègre CJ, Moreira M, Staudacher T (1995) $^4He/^3He$ dispersion and mantle convection. Geophys Res Lett 22:2325-2328

Allègre CJ, Staudacher T, Sarda P (1986/87) Rare gas systematics: formation of the atmosphere, evolution and structure of the Earth's mantle. Earth Planet Sci Lett 81:127-150

Allègre CJ, Staudacher T, Sarda P, Kurz M (1983) Constraints on evolution of Earth's mantle from rare gas systematics. Nature 303:762-766

Anderson DL (1993) Helium-3 from the mantle: primordial signal or cosmic dust? Science 261:170-176

Anderson DL (1995) Lithosphere, asthenosphere and perisphere. Rev Geophys 33:125-149

Anderson DL (1998a) The helium paradoxes. Proc Natl Acad Sci 95:4822-4827

Anderson DL (1998b) A model to explain the various paradoxes associated with mantle noble gas geochemistry. Proc Natl Acad Sci 95:9087-9092

Anderson DL (2000a) The statistics of helium isotopes along the global spreading ridge system and the central limit theorem. Geophys Res Lett 27:2401-2404

Anderson DL (2000b) The statistics and distribution of helium in the mantle. Intl Geol Rev 42:289-311

Anderson DL (2001) A statistical test of the two reservoir model for helium isotopes. Earth Planet Sci Lett 193:77-82

Ballentine CJ, Barfod DN (2000) The origin of air-like noble gases in MORB and OIB. Earth Planet Sci Lett 180:39-48

Ballentine CJ, Porcelli D, Wieler R (2001) Noble gases in mantle plumes. Science 291:2269a

Barfod DN, Ballentine CJ, Halliday AN, Fitton JG (1999) Noble gases in the Cameroon line and the He, Ne and Ar isotopic compositions of high μ (HIMU) mantle. J Geophys Res 104:29509-29527

Barling J, Goldstein SL (1990) Extreme isotope variations in Heard Island lavas and the nature of mantle reservoirs. Nature 348:59-62

Barling J, Goldstein SL, Nicholls IA (1994) Geochemistry of Heard Island (Southern Indian Ocean): characterization of an enriched mantle component and implications for enrichment of the sub-Indian ocean mantle. J Petrol 35:1017-1053

Basu AR, Renne PR, Das Gupta DK, Teichmann F, Poreda RJ (1993) Early and late alkali igneous pulses and a high ^3He plume origin for the Deccan flood basalts. Science 261:902-906

Batiza R (1984) Inverse relationship between Sr isotope diversity and rate of oceanic volcanism has implications for mantle heterogenity. Nature 309:440-441

Batiza R, Bernatowicz TJ, Hohenberg CM, Podosek FA (1979) Relations of noble gases to petrogenesis and magmatic evolution of some differentiated volcanic rocks. Contrib Mineral Petrol 69:301-314

Becker TW, Kellogg JB, O'Connell RJ (1999) Thermal constraints on the survival of primitive blobs in the lower mantle. Earth Planet Sci Lett 171:351-365

Benkert J-P, Baur H, Signer P, Wieler R (1993) He, Ne and Ar from the solar wind and solar energetic particles in lunar ilmenites and pyroxenes. J Geophys Res 98:13147-13162

Bernatowicz TJ, Podosek FA (1978) Nuclear components in the atmosphere. *In* Alexander EC, Ozima M (eds) Terrestrial Rare Gases. Center for Academic Publications Japan, Tokyo, p 99-135

Brandon AD, Graham D, Gautason B (2001) ^{187}Os-^{186}Os and He isotope systematics of Iceland picrites. EOS Trans Am Geophys Union 82:F1306

Brandon AD, Norman MD, Walker RJ, Morgan JW (1999) ^{186}Os-^{187}Os systematics of Hawaiian picrites. Earth Planet Sci Lett 174:25-42

Brandon AD, Walker RJ, Morgan JW, Norman MD, Prichard HM (1998) Coupled ^{186}Os and ^{187}Os evidence for core-mantle interaction. Science 280:1570-1573

Breddam K, Kurz MD (2001) Helium isotope signatures of Icelandic alkaline lavas. EOS Trans Am Geophys Union 82:F1315

Breddam K, Kurz MD, Storey M (2000) Mapping out the conduit of the Iceland mantle plume with helium isotopes. Earth Planet Sci Lett 176:45-55

Broadhurst CL, Drake MJ, Hagee BE, Bernatowicz TJ (1990) Solubility and partitioning of Ar in anorthite, diopside, forsterite, spinel, and synthetic basaltic liquids. Geochim Cosmochim Acta 54:299-309

Broadhurst CL, Drake MJ, Hagee BE, Bernatowicz TJ (1992) Solubility and partitioning of Ne, Ar, Kr and Xe in minerals and synthetic basaltic liquids. Geochim Cosmochim Acta 56:709-723

Brooker RA, Wartho J-A, Carroll MR, Kelley SP, Draper DS (1998) Preliminary UVLAMP determinations of argon partition coefficients for olivine and clinopyroxene grown from silicate melts. Chem Geol 147:185-200

Burnard PG (1999a) The bubble-by-bubble volatile evolution of two mid-ocean ridge basalts. Earth Planet Sci Lett 174:199-211

Burnard PG (1999b) Origin of argon-lead isotopic correlation in basalts. Science 286:871a

Burnard PG (2001) Correction for volatile fractionation in ascending magmas: noble gas abundances in primary mantle melts. Geochim Cosmochim Acta 65:2605-2614

Burnard PG, Farley KA, Turner G (1998) Multiple fluid pulses in a Samoan harzburgite. Chem Geol 147:99-114

Burnard PG, Graham DW, Farley KA (2002) Mechanisms of magmatic gas loss along the Southeast Indian Ridge and the Amsterdam-St. Paul Plateau. Earth Planet Sci Lett (in press)

Burnard PG, Graham DW, Turner G (1997) Vesicle-specific noble gas analyses of "popping rock": Implications for primordial noble gases gases in Earth. Science 276:568-571

Burnard PG, Stuart F, Turner G (1994a) C-He-Ar variations within a dunite nodule as a function of fluid inclusion morphology. Earth Planet Sci Lett 128:243-258

Burnard PG, Stuart FM, Turner G, Oskarsson N (1994b) Air contamination of basaltic magmas: implications for high ^{3}He/^{4}He mantle Ar isotopic composition. J Geophys Res 99:17709-17715

Caffee MW, Hudson GB, Veisko C, Huss GR, Alexander ECJ, Chivas AR (1999) Primordial noble gases from Earth's mantle: identification of a primitive volatile component. Science 285:2115-2118

Carroll MR, Draper DS (1994) Noble gases as trace elements in magmatic processes. Chem Geol 117: 37-56

Carroll MR, Stolper EM (1993) Noble gas solubilities in silicate melts and glasses: new experimental results for argon and the relationship between solubility and ionic porosity. Geochim Cosmochim Acta 57:5039-5051

Castillo P, Batiza R (1989) Strontium, neodymium and lead isotope constraints on near-ridge seamount production beneath the South Atlantic. Nature 342:262-265

Chamorro EM, Brooker RA, Wartho JA, Wood BK, Kelley SP, Blundy SP (2002) Ar and K partitioning between clinopyroxene and silicate melt to 8 GPa. Geochim Cosmochim Acta 66:507-519

Chamorro-Pérez E, Gillet P, Jambon A, Badro J, McMillan P (1998) Low argon solubility in silicate melts at high pressure. Nature 393

Chauvel C, Hofmann AW, Vidal P (1992) HIMU-EM: the French Polynesian connection. Earth Planet Sci Lett 110:99-119

Christensen BP, Holm PM, Jambon A, Wilson JR (2001) Helium, argon and lead isotopic composition of volcanics from Santo Antão and Fogo, Cape Verde Islands. Chem Geol 178:127-142

Clarke WB, Jenkins WJ, Top Z (1976) Determination of tritium by mass spectrometric measurement of ^{3}He. Intl J Appl Rad Isotopes 27:515-522

Coltice N, Ricard Y (1999) Geochemical observations and one layer mantle convection. Earth Planet Sci Lett 174:125-137

Coltice N, Albarède F, Gillet P (2000) ^{40}K-^{40}Ar constraints on recycling continental crust into the mantle. Science 288:845-847

Condomines M, Gronvold K, Hooker PJ, Muehlenbachs K, O'Nions RK, Oskarsson N, Oxburgh ER (1983) Helium, oxygen, strontium and neodymium isotopic relationships in Icelandic volcanics. Earth Planet Sci Lett 66:125-136

Craig H, Lupton JE (1976) Primordial neon, helium, and hydrogen in oceanic basalts. Earth Planet Sci Lett 31:369-385

Craig H, Lupton JE (1981) Helium-3 and mantle volatiles in the ocean and the oceanic crust. *In* Emiliani C (ed) The Oceanic Lithosphere, vol 7. John Wiley & Sons, Inc, New York, p 391-428

Craig H, Clarke WB, Beg MA (1975) Excess ^{3}He in deep water on the East Pacific Rise. Earth Planet Sci Lett 26:125-132

Craig H, Lupton JE, Welhan JA, Poreda R (1978) Helium isotope ratios in Yellowstone and Lassen Park volcanic gases. Geophys Res Lett 5:897-900

Davaille A (1999) Simultaneous generation of hotspots and superswells by convection in a heterogeneous planetary mantle. Nature 402:756-760

Davies GF (1988) Ocean bathymetry and mantle convection 1. Large-scale flow and hotspots. J Geophys Res 93:10467-10480

Davies GF (1999a) Geophysically constrained mantle mass flows and ^{40}Ar budget: a degassed lower mantle? Earth Planet Sci Lett 166:149-162

Davies GF (1999b) Dynamic Earth. Cambridge University Press. Cambridge

Davies GF, Richards MA (1992) Mantle convection. J Geol 100:151-206

Des Marais DJ, Moore JG (1984) Carbon and its isotopes in mid-oceanic basalt glasses. Earth Planet Sci Lett 69:43-57

Desonie DL, Duncan RA, Kurz MD (1991) Helium isotopic composition of isotopically diverse basalts from hotspot volcanic lineaments in French Polynesia. EOS Trans Am Geophys Union 72:536

Detrick RS, Sinton JM, Ito G, Canales JP, Behn M, Blacic T, Cushman B, Dixon JE, Graham DW, Mahoney J (2002) Correlated geophysical, geochemical and volcanological manifestations of plume-ridge interaction along the Galápagos Spreading Center. Geochem Geophys Geosys in press

Dickey JSJ, Frey FA, Hart SR, Watson EB, Thompson G (1977) Geochemistry and petrology of dredged basalts from the Bouvet triple junction, South Atlantic. Geochim Cosmochim Acta 41:1105-1118

Dixon ET, Honda M, McDougall I, Campbell IH, Sigurdsson I (2000) Preservation of near-solar neon isotopic ratios in Icelandic basalts. Earth Planet Sci Lett 180:309-324

Dixon JE, Stolper E, Delaney JR (1988) Infra-red spectroscopic measurements of CO_2 and H_2O in Juan de Fuca Ridge basaltic glasses. Earth Planet Sci Lett 90:87-104

Dosso L, Bougault H, Joron J-L (1993) Geochemical morphology of the North Mid-Atlantic Ridge, 10°-24°N: trace element-isotope complementarity. Earth Planet Sci Lett 120:443-462

Dosso L, Bougault H, Beuzart P, Calvez J-Y, Joron J-L (1988) The geochemical structure of the South-East Indian Ridge. Earth Planet Sci Lett 88:47-59

Dosso L, Hanan BB, Bougault H, Schilling J-G, Joron J-L (1991) Sr-Nd-Pb geochemical morphology between 10° and 17°N on the Mid-Atlantic Ridge: a new MORB isotope signature. Earth Planet Sci Lett 106:29-43

Dosso L, Bougault H, Langmuir C, Bollinger C, Bonnier O, Etoubleau J (1999) The age and distribution of mantle heterogeneity along the Mid-Atlantic Ridge (31-41°N). Earth Planet Sci Lett 170:269-286

Douglass J, Schilling J-G, Fontignie D (1999) Plume-ridge interactions of the Discovery and Shona mantle plumes with the southern Mid-Atlantic Ridge (40°-55°S). J Geophys Res 104:2941-2962

Dupré B, Allégre CJ (1980) Pb-Sr-Nd isotopic correlation and the chemistry of the North Atlantic mantle. Nature 286:17-22

Dupré B, Lambret B, Rousseau D, Allègre CJ (1981) Limitations on the scale of mantle heterogeneities under oceanic ridges. Nature 294:552-554

Dymond J, Hogan L (1978) Factors controlling the noble gas abundance patterns of deep-sea basalts. Earth Planet Sci Lett 38:117-128

Eaby J, Clague DA, Delaney JR (1984) Sr isotopic variations along the Juan de Fuca Ridge. J Geophys Res 89:7883-7890

Edmond JM, Von Damm KL, McDuff RE, Measures CI (1982) Chemistry of hot springs on the East Pacific Rise and their effluent dispersal. Nature 297:187-191

Eiler J, Farley KA, Stolper EM (1998) Correlated He and Pb isotope variations in Hawaiian lavas. Geochim Cosmochim Acta 62:1977-1984

Elliott T, Hawkesworth CJ, Grönvold K (1991) Dynamic melting of the Iceland plume. Nature 351: 201-206

Farley KA (1995) Rapid cycling of subducted sediments into the Samoan mantle plume. Geology 23: 531-534

Farley KA, Craig H (1994) Atmospheric argon contamination of ocean island basalt olivine phenocrysts. Geochim Cosmochim Acta 58:2509-2517

Farley KA, Neroda E (1998) Noble gases in the Earth's mantle. Ann Rev Earth Planet Sci 26:189-218
Farley KA, Poreda RJ (1993) Mantle neon and atmospheric contamination. Earth Planet Sci Lett 114: 325-339
Farley KA, Basu AR, Craig H (1993) He, Sr and Nd isotopic variations in lavas from the Juan Fernandez archipelago. Contrib Mineral Petrol 115:75-87
Farley KA, Love SG, Patterson DB (1997) Atmospheric entry heating and helium retentivity of interplanetary dust particles. Geochim Cosmochim Acta 61:2309-2316
Farley KA, Natland JH, Craig H (1992) Binary mixing of enriched and undegassed (primitive?) mantle components (He, Sr, Nd,Pb) in Samoan lavas. Earth Planet Sci Lett 111:183-199
Farley KA, Poreda RJ, Onstott TC (1994) Noble gases in deformed xenoliths from an ocean island: charcterization of a metasomatic fluid. *In* Matsuda J (ed) Noble Gas Geochemistry and Cosmochemistry. Terra Scientific, Tokyo, p 159-178
Fisher DE (1978) Terrestrial potassium and argon abundances as limits to models of atmospheric evolution. *In* Alexander EC, Ozima M (ed) Terrestrial Rare Gases. Center for Academic Publications Japan, Tokyo, p 173-183
Fisher DE (1994) Mantle and atmospheric-like argon in vesicles of MORB glasses. Earth Planet Sci Lett 123:199-204
Fisher DE (1997) Helium, argon and xenon in crushed and melted MORB. Geochim Cosmochim Acta 14:3003-3012
Fontignie D, Schilling J-G (1996) Mantle heterogeneities beneath the South Atlantic: a Nd-Sr-Pb isotope study along the Mid-Atlantic Ridge (3°S-46°S). Earth Planet Sci Lett 142:209-221
Garcia MO, Rubin KH, Norman MD, Rhodes JM, Graham DW, Muenow DW, Spencer K (1998) Petrology and geochronology of basalt breccia from the 1996 earthquake swarm of Loihi seamount, Hawaii: magmatic history of 1996 eruption. Bull Volc 59:577-592
Georgen JE, Kurz MD, Dick HJB, Lin J (2001) Helium isotope systematics of the western Southwest Indian Ridge: effects of plume influence, spreading rate and source heterogeneity. EOS Trans Am Geophys Union 82:F1169
Goff F, McMurtry GM, Counce D, Simac JA, Roldán-Manzo AR, Hilton DR (2000) Contrasting hydrothermal activity at Sierra Negra and Alcedo volcanoes, Galapagos Archipelago, Ecuador. Bull Volc 62:34-52
Graham DW (1994) Helium isotope variability along mid-ocean ridges: mantle heterogeneity and melt generation effects. Mineral Mag 58A:347-348.
Graham DW, Sarda P (1991) Reply to comment by T. M. Gerlach on "Mid-ocean ridge popping rocks: implications for degassing at ridge crests". Earth Planet Sci Lett 105:568-573
Graham DW, Michael PJ, Hanan BB (1996a) Helium-carbon dioxide relationships in MORB glasses from the Mid-Atlantic Ridge at 33°S. EOS Trans Am Geophys Union 77:F830
Graham DW, Castillo P, Lupton JE, Batiza R (1996b) Correlated helium and strontium isotope ratios in South Atlantic near-ridge seamounts and implications for mantle dynamics. Earth Planet Sci Lett 144:491-503
Graham DW, Christie DM, Harpp KS, Lupton JE (1993) Mantle plume helium in submarine basalts from the Galápagos platform. Science 262:2023-2026
Graham DW, Hoernle KA, Lupton JE, Schmincke H-U (1996c) Helium isotope variations in volcanic rocks from the Canary Islands and Madeira. *In* Bohrson WA, Davidson J, Wolff JA (eds) Shallow Level Processes in Ocean Island Magmatism: Distinguishing Mantle and Crustal Signatures, Chapman Conference, Puerto de la Cruz, Tenerife. American Geophys Union, p 13-14
Graham DW, Humphris SE, Jenkins WJ, Kurz MD (1992a) Helium isotope geochemistry of some volcanic rocks from Saint Helena. Earth Planet Sci Lett 110:121-131
Graham DW, Johnson KTM, Priebe LD, Lupton JE (1999) Hotspot-ridge interaction along the Southeast Indian Ridge near Amsterdam and St. Paul Islands: helium isotope evidence. Earth Planet Sci Lett 167:297-310
Graham DW, Lupton JE, Albarède F, Condomines M (1990) Extreme temporal homogeneity of helium isotopes at Piton de la Fournaise, Réunion Island. Nature 347:545-548
Graham DW, Lupton JE, Spera FJ, Christie DM (2001) Upper mantle dynamics revealed by helium isotope variations along the Southeast Indian Ridge. Nature 409:701-703
Graham DW, Jenkins WJ, Schilling J-G, Thompson G, Kurz MD, Humphris SE (1992b) Helium isotope geochemistry of mid-ocean ridge basalts from the South Atlantic. Earth Planet Sci Lett 110:133-147
Graham DW, Larsen LM, Hanan BB, Storey M, Pedersen AK, Lupton JE (1998) Helium isotope composition of the early Iceland mantle plume inferred from the Tertiary picrites of West Greenland. Earth Planet Sci Lett 160:241-255

Graham DW, Zindler A, Kurz MD, Jenkins WJ, Batiza R, Staudigel H (1988) He, Pb, Sr and Nd isotope constraints on magma genesis and mantle heterogeneity beneath young Pacific seamounts. Contrib Mineral Petrol 99:446-463
Gurnis M, Davies GF (1986) The effect of depth-dependent viscosity on convective mixing in the mantle and the possible survival of primitive mantle. Geophys Res Lett 13:541-544
Halliday AN, Lee D-C (1999) Tungsten isotopes and the early development of the Earth and Moon. Geochim Cosmochim Acta 63:4157-4179
Halliday AN, Davidson JP, Holden P, DeWolf C, Lee D-C, Fitton JG (1990) Trace-element fractionation in plumes and the origin of HIMU mantle beneath the Cameroon line. Nature 347:523-528
Hamano Y, Ozima M (1978) Earth-atmosphere evolution model based on Ar isotopic data. *In* Alexander EC, Ozima M (eds) Terrestrial Rare Gases. Center for Academic Publications Japan, Tokyo, p 155-171
Hamelin B, Dupré B, Allègre C-J (1984) Lead-strontium isotopic variations along the East Pacific Rise and the Mid-Atlantic Ridge: a comparative study. Earth Planet Sci Lett 67:340-350
Hanan BB, Graham DW (1996) Lead and helium isotope evidence from oceanic basalts for a common deep source of mantle plumes. Science 272:991-995
Hanan BB, Schilling J-G (1997) The dynamic evolution of the Iceland mantle plume: the lead isotope perspective. Earth Planet Sci Lett 151:43-60
Hanan BB, Graham DW, Michael PJ (1994) Highly correlated lead, strontium and helium isotopes in Mid-Atlantic Ridge basalts from a dynamically evolving spreading centre at 31-34°S. Mineral Mag 58A:370-371
Hanan BB, Kingsley RH, Schilling J-G (1986) Pb isotope evidence in the South Atlantic for migrating ridge-hotspot interactions. Nature 322:137-144
Hanyu T, Kaneoka I (1997) The uniform and low $^3He/^4He$ ratios of HIMU basalts as evidence for their origin as recycled materials. Nature 390:273-276
Hanyu T, Kaneoka I, Nagao K (1999) Noble gas study of HIMU and EM ocean island basalts in the Polynesian region. Geochim Cosmochim Acta 63:1181-1201
Hanyu T, Dunai TJ, Davies GR, Kaneoka I, Nohda S, Uto K (2001) Noble gas study of the Reunion hotspot: evidence for distinct less-degassed mantle sources. Earth Planet Sci Lett 193:83-98
Harpp KS, White WM (2001) Tracing a mantle plume: isotopic and trace element variations of Galápagos seamounts. Geochem Geophys Geosys Paper 2000GC000137
Harrison D, Burnard P, Turner G (1999) Noble gas behaviour and composition in the mantle: constraints from the Iceland plume. Earth Planet Sci Lett 171:199-207
Harrison D, Burnard P, Trieloff M, Turner G (2002) Resolving atmospheric contaminants in mantle noble gas analyses. submitted
Hart R, Dymond J, Hogan L (1979) Preferential formation of the atmosphere-sialic crust system from the upper mantle. Nature 278:156-159
Hart SR, Schilling J-G, Powell JL (1973) Basalts from Iceland and along the Reykjanes Ridge: Sr isotope geochemistry. Nature 246:104-107
Hart SR, Hauri EH, Oschmann LA, Whitehead JA (1992) Mantle plumes and entrainment: isotopic evidence. Science 256:517-520
Hartmann WK (1984) Moon origin: the impact-trigger hypothesis. *In* Hartmann WK, Phillips RJ, Taylor GJ (eds) Origin of the Moon. Lunar and Planetary Institute, Houston, p 579-608
Hauri EH, Whitehead JA, Hart SR (1994) Fluid dynamic and geochemical aspects of entrainment in mantle plumes. J Geophys Res 99:24275-24300
Hegner E, Tatsumoto M (1989) Pb, Sr and Nd isotopes in seamount basalts from the Juan de Fuca Ridge and Kodiak-Bowie seamount chain, northeast Pacific. J Geophys Res 94:17839-17846
Hémond C, Arndt NT, Lichtenstein U, Hofmann AW (1993) The heterogeneous Iceland plume: Nd-Sr-O isotopes and trace element constraints. J Geophys Res 98:15833-15850
Hilton DL, Hammerschmidt K, Loock G, Friedrichsen H (1993) Helium and argon isotope systematics of the central Lau Basin and Valu Fa Ridge: evidence of crust/mantle interactions in a back-arc basin. Geochim Cosmochim Acta 57:2819-2841
Hilton DR, Barling J, Wheller GE (1995) Effect of shallow-level contamination on the helium isotope systematics of ocean-island lavas. Nature 373:330-333
Hilton DR, Macpherson CG, Elliott TR (2000a) Helium isotope ratios in mafic phenocrysts and geothermal fluids from La Palma, the Canary Islands (Spain): implications for HIMU mantle sources. Geochim Cosmochim Acta 64:2119-2132
Hilton DR, McMurtry GM, Goff F (1998a) Large variations in vent fluid $CO_2/^3He$ ratios signal rapid chnages in magma chemistry at Loihi Seamount, Hawaii. Nature 396:359-362
Hilton DR, McMurtry GM, Kreulen R (1997) Evidence for extensive degassing of the Hawaiian mantle plume from helium-carbon relationships at Kilauea volcano. Geophys Res Lett 24:3065-3068

Hilton DR, Gronvold K, Macpherson CG, Castillo PR (1999) Extreme ^{3}He/^{4}He ratios in northwest Iceland: constraining the common component in mantle plumes. Earth Planet Sci Lett 173:53-60

Hilton DR, Grönvold K, O'Nions RK, Oxburgh R (1990) Regional distribution of ^{3}He anomalies in the Icelandic crust. Chem Geol 88:53-67

Hilton DR, Gronvold K, Sveinbjornsdottir AE, Hammerschmidt K (1998b) Helium isotope evidence for off-axis degassing of the Icelandic hotspot. Chem Geol 149:173-187

Hilton DR, Thirlwall MF, Taylor RN, Murton BJ, Nichols A (2000b) Controls on magmatic degassing along the Reykjanes Ridge with implications for the helium paradox. Earth Planet Sci Lett 183:43-50

Hirschmann MM, Stolper EM (1996) A possible role for garnet pyroxenite in the origin of the "garnet signature" in MORB. Contrib Mineral Petrol 124:185-208

Hiyagon H (1994a) Retention of solar helium and neon in IDPs in deep sea sediment. Science 263: 1257-1259

Hiyagon H (1994b) Constraints on rare gas partition coefficients from analysis of olivine-glass from a picritic mid-ocean ridge basalt-Comments. Chem Geol 112:119-122

Hiyagon H, Ozima M (1982) Noble gas distribution between basalt melt and crystals. Earth Planet Sci Lett 58:255-264

Hiyagon H, Ozima M (1986) Partition of noble gases between olivine and basalt melt. Geochim Cosmochim Acta 50:2045-2057

Hiyagon H, Ozima M, Marty B, Zashu S, Sakai H (1992) Noble gases in submarine glasses from mid-ocean ridges and Loihi seamount: constraints on the early history of the Earth. Geochim Cosmochim Acta 56:1301-1316

Hofmann AW (1997) Mantle geochemistry: the message from oceanic volcanism. Nature 385:219-229

Hofmann AW, White WM (1982) Mantle plumes from ancient oceanic crust. Earth Planet Sci Lett 57: 421-436

Hofmann AW, Jochum KP, Seufert M, White WM (1986) Nb and Pb in oceanic basalts: new constraints on mantle evolution. Earth Planet Sci Lett 79:33-45

Holness MB, Richter FM (1989) Possible effects of spreading rate on MORB isotopic and rare earth composition arising from melting of a heterogeneous source. J Geol 97:247-260

Honda M, McDougall I (1998) Primordial helium and neon in the Earth-a speculation on early degassing. Geophys Res Lett 25:1951-1954

Honda M, Patterson DB (1999) Systematic elemental fractionation of mantle-derived helium, neon, and argon in mid-oceanic ridge glasses. Geochim Cosmochim Acta 63:2863-2874

Honda M, McDougall I, Patterson D (1993a) Solar noble gases in the Earth: the systematics of helium-neon isotopes in mantle derived samples. Lithos 30:257-265

Honda M, Reynolds J, Roedder E, Epstein S (1987) Noble gases in diamonds: occurrences of solar like helium and neon. J Geophys Res 92:12507-12521

Honda M, McDougall I, Patterson DB, Doulgeris A, Clague DA (1991) Possible solar noble-gas component in Hawaiian basalts. Nature 349:149-151

Honda M, McDougall I, Patterson DB, Doulgeris A, Clague DA (1993b) Noble gases in submarine pillow basalt glasses from Loihi and Kilauea, Hawaii: a solar component in the Earth. Geochim Cosmochim Acta 57:859-874

Jambon W, Weber HW, Braun O (1986) Solubility of He, Ne, Ar, Kr and Xe in a basalt melt in the range 1250-1600°C. Geochemical Implications. Geochim Cosmochim Acta 50:401-408

Javoy M, Pineau F (1991) The volatiles record of a "popping" rock from the Mid-Altantic ridge at 14°N: chemical and isotopic composition of gas trapped in the vesicles. Earth Planet Sci Lett 107:598-611

Jenkins WJ, Edmond JM, Corliss JB (1978) Excess ^{3}He and ^{4}He in Galapagos submarine hydrothermal waters. Nature 272:156-158

Jochum KP, Hofmann AW, Ito E, Seufert HM, White WM (1983) K, U and Th in mid-ocean ridge basalt glasses and heat production, K/U and K/Rb in the mantle. Nature 306:431-435

Kaneoka I (1983) Noble gas constraints on the layered structure of the mantle. Nature 302:698-700

Kaneoka I (1987) Constraints on the characteristics of magma sources for Hawaiian volcanoes based on noble gas systematics. *In* Decker RW, Wright TL, Stauffer PH (eds) Volcanism in Hawaii, US Geol Survey Prof Paper 1350. U S Government Printing Office, Washington, p 745-757

Kaneoka I, Takaoka N (1978) Excess ^{129}Xe and high ^{3}He/^{4}He ratios in olivine phenocrysts of Kapuho lava and xenolithic dunites from Hawaii. Earth Planet Sci Lett 39:382-386

Kaneoka I, Takaoka N (1980) Rare gas isotopes in Hawaiian ultramafic nodules and volcanic rocks : constraints on genetic relationships. Science 208:1366-1368

Kaneoka I, Takaoka N (1985) Noble-gas state in the Earth's interior-some constraints on the present state. Chem Geol (Isotope Geosci) 52:75-95

Kaneoka I, Takaoka N (1991) Evolution of the lithosphere and its interaction with the underlying mantle as inferred from noble gas isotopes. Austraian J Earth Sci 38:559-567

Kaneoka I, Takaoka N, Clague DA (1983) Noble gas systematics for coexisting glass and olivine crystals in basalts and dunite xenoliths from Loihi Seamount. Earth Planet Sci Lett 66:427-437
Kaneoka I, Takaoka N, Upton BGJ (1986) Noble gas systematics in basalts and a dunite nodule from Réunion and Grand Comore islands, Indian Ocean. Chem Geol (Isotope Geosci) 59:35-42
Keevil NB (1940) Interatomic forces and helium in rocks. Proc Am Acad Arts Sci 73:311-359
Kellogg LH (1992) Mixing in the mantle. Ann Rev Earth Planet Sci 20:365-388
Kellogg LH, Wasserburg GJ (1990) The role of plumes in mantle helium fluxes. Earth Planet Sci Lett 99:276-289
Kellogg LH, Hager B, van der Hilst R (1999) Compositional stratification in the deep mantle. Science 283:1881-1884
Kingsley RH, Schilling J-G (1995) Carbon in mid-Atlantic ridge basalt glasses from 28°N to 63°N: Evidence for a carbon-enriched Azores mantle plume. Earth Planet Sci Lett 129:31-53
Kirstein LA, Timmerman MJ (2000) Evidence of the proto-Iceland plume in northwestern Ireland at 42 Ma from helium isotopes. J Geol Soc Lond 157:923-927
Klein EM, Karsten J (1995) Ocean-ridge basalts with convergent margin geochemical affinities from the Chile Ridge. Nature 374:52-57
Klein EM, Langmuir CH (1987) Global correlations of ocean ridge basalt chemistry with axial depth and crustal thickness. J Geophys Res 92: 8089-8115
Krylov A, Mamyrin BA, Khabarin LA, Mazina TI, Silin YI (1974) Helium isotopes in ocean floor bedrock. Geochem Intl 11:839-844
Kumagai H, Kaneoka I (1998) Variations in noble gas abundances and isotope ratios in a single MORB pillow. Geohys Res Lett 25:3891-3894
Kunz J (1999) Is there solar argon in the Earth's mantle? Nature 399:649-650
Kunz J, Staudacher T, Allègre CJ (1998) Plutonium-fission xenon found in Earth's mantle. Science 280:877-880
Kurz MD (1982) Helium isotopic geochemistry of oceanic volcanic rocks: implications for mantle heterogeneity and degassing. PhD dissertation MIT/WHOI
Kurz MD (1986) *In situ* production of terrestrial cosmogenic helium and some applications to geochronology. Geochim Cosmochim Acta 50:2855-2862
Kurz MD (1993) Mantle heterogeneity beneath oceanic islands: some inferences from isotopes. Phil Trans R Soc Lond A 342:91-103
Kurz MD, Geist D (1999) Dynamics of the Galapagos hotspot from helium isotope geochemistry. Geochim Cosmochim Acta 63:4139-4156
Kurz MD, Jenkins WJ (1981) The distribution of helium in oceanic basalt glasses. Earth Planet Sci Lett 53:41-54
Kurz MD, Kammer DP (1991) Isotopic evolution of Mauna Loa volcano. Earth Planet Sci Lett 103: 257-269
Kurz MD, Jenkins WJ, Hart SR (1982a) Helium isotopic systematics of oceanic islands and mantle heterogeneity. Nature 297:43-46
Kurz MD, le Roex AP, Dick HJB (1998) Isotope geochemistry of the oceanic mantle near the Bouvet triple junction. Geochim Cosmochim Acta 62:841-852
Kurz MD, Meyer PS, Sigurdsson H (1985) Helium isotopic systematics within the neovolcanic zones of Iceland. Earth Planet Sci Lett 74:291-305
Kurz MD, Garcia MO, Frey FA, O'Brien PA (1987) Temporal helium isotopic variations within Hawaiian volcanoes: basalts from Mauna Loa and Haleakala. Geochim Cosmochim Acta 51:2905-2914
Kurz MD, Jenkins WJ, Hart SR, Clague D (1983) Helium isotopic variations in the volcanic rocks from Loihi Seamount and the island of Hawaii. Earth Planet Sci Lett 66:388-406
Kurz MD, Jenkins WJ, Schilling J-G, Hart SR (1982b) Helium isotopic variations in the mantle beneath the central North Atlantic Ocean. Earth Planet Sci Lett 58:1-14
Kurz MD, Kammer DP, Gulessarian A, Moore RB (1990) Helium isotopes in dated alkali basalts from Sao Miguel, Azores. EOS Trans Am Geophys Union 71:657
Kurz MD, Kenna TC, Lassiter JK, DePaolo DJ (1996) Helium isotopic evolution of Mauna Kea Volcano: first results from the 1-km drill core. J Geophys Res 101:11781-11792
Kyser TK, Rison W (1982) Systematics of rare gas isotopes in basic lavas and ultramafic xenoliths. J Geophys Res 87:5611-5630
Lassiter JC, DePaolo DJ, Tatsumoto M (1996) Isotopic evolution of Mauna Kea volcano: results from the initial phase of the Hawaii Scientific Drilling Project. J Geophys Res 101:11769-11780
le Roex AP, Dick HJB, Gulen L, Reid AM, Erlank AJ (1987) Local and regional hetrogeneity in MORB from the Mid-Atlantic Ridge between 54.5°S and 51°S: evidence for geochemical enrichment. Geochim Cosmochim Acta 51:541-555

Lewis RS (1975) Rare gases in separated whitlockite from the St. Severin chondrite: xenon and krypton from fission of extinct ^{244}Pu. Geochim Cosmochim Acta 39:417-432

Leya I, Wieler R (1999) Nucleogenic production of Ne isotopes in the Earth's crust and upper mantle induced by alpha particles from the decay of U and Th. J Geophys Res 104:15439-15450

Lin WJ, Manuel OK (1987) Xenon decay products of extinct radionuclides in the Navajo, New Mexico well gas. Geochem J 21:197-207

Lupton JE (1983) Terrestrial inert gases: isotope tracer studies and clues to primordial components in the mantle. Ann Rev Earth Planet Sci 11:371-414

Lupton JE, Craig H (1975) Excess ^{3}He in oceanic basalts: evidence for terrestrial primordial helium. Earth Planet Sci Lett 26:133-139

Lupton JE, Baker ET, Massoth GJ (1989) Variable ^{3}He/heat ratios in submarine hydrothermal systems: evidence from two plumes over the Juan de Fuca Ridge. Nature 337:161-164

Lupton JE, Graham DW, Delaney JR, Johnson HP (1993) Helium isotope variations in Juan de Fuca Ridge basalts. Geophys Res Lett 20:1851-1854

Lux G (1987) The behavior of noble gases in silicate liquids: solution, diffusion, bubbles and surface effects, with applications to natural samples. Geochim Cosmochim Acta 51:1549-1560

Machado N, Ludden JN, Brooks C, Thompson G (1982) Fine-scale isotopic heterogeneity in the sub-Atlantic mantle. Nature 295:226-228

Mahaffy PR, Niemann HB, Alpert A, Atreya SK, Demick J, Donahue TM, Harpold DN, Owen TC (2000) Noble gas abundance and isotope ratios in the atmosphere of Jupiter from the Galileo probe mass spectrometer. J Geophys Res 105:15061-15071

Mahoney JJ, Graham DW, Christie DM, Johnson KTM, Hall LS, VonderHaar DL (2002) Between a hot spot and cold spot: asthenospheric flow and geochemical evolution in the Southeast Indian Ridge mantle, 86°-118°E. J Petrol in press

Mahoney JJ, Sinton JM, Kurz MD, Macdougall JD, Spencer KJ, Lugmair GW (1994) Isotope and trace element characteristics of a super-fast spreading ridge: East Pacific Rise, 13-23°S. Earth Planet Sci Lett 121:173-193

Mahoney JJ, Natland JH, White WM, Poreda R, Bloomer SH, Fisher RL, Baxter AN (1989) Isotopic and geochemical provinces of the western Indian Ocean spreading centers. J Geophys Res 94:4033-4052

Mamyrin BA, Tolstikhin IN (1984) Helium Isotopes in Nature. Elsevier. Amsterdam

Mamyrin BA, Anufriev GS, Kamenskii IL, Tolstikhin IN (1970) Determination of the isotopic composition of atmospheric helium. Geochem Intl 7:498-505

Manga M (1996) Mixing of heterogeneities in the mantle: effect of viscosity differences. Geophys Res Lett 23:403-406

Marty B (1989) Neon and xenon isotopes in MORB: implications for the earth-atmosphere evolution. Earth Planet Sci Lett 94:45-56

Marty B (1995) Nitrogen content of the mantle inferred from N_2-Ar correlation in oceanic basalts. Nature 377:326-329

Marty B, Humbert F (1997) Nitrogen and argon isotopes in oceanic basalts. Earth Planet Sci Lett 152: 101-112

Marty B, Jambon A (1987) C/^{3}He in volatile fluxes from the solid Earth: implications for carbon geodynamics. Earth Planet Sci Lett 83:16-26

Marty B, Lussiez P (1993) Constraints on rare gas partition coefficients from analysis of olivine-glass from a picritic mid-ocean ridge basalt. Chem Geol 106:1-7

Marty B, Lussiez P (1994) Constraints on rare gas partition coefficients from analysis of olivine-glass from a picritic mid-ocean ridge basalt -Reply. Chem Geol 112:122-127

Marty B, Ozima M (1986) Noble gas distribution in oceanic basalt glasses. Geochim Cosmochim Acta 50:1093-1098

Marty B, Zimmerman L (1999) Volatiles (He, C, N, Ar) in mid-ocean ridge basalts: assessment of shallow-level fractionation and characterization of source composition. Geochim Cosmochim Acta 63:3619-3633

Marty B, Pik R, Yirgu G (1996) Helium isotopic variations in Ethiopian plume lavas: nature of magmatic sources and limit on lower mantle contribution. Earth Planet Sci Lett 144:223-237

Marty B, Upton BGJ, Ellam RM (1998) Helium isotopes in early Tertiary basalts, northeast Greenland: evidence for 58 Ma plume activity in the North Atlantic-Iceland volcanic province. Geology 26: 407-410

Marty B, Meynier V, Nicolini E, Greisshaber E, Toutain JP (1993a) Geochemistry of gas emanations: a case study of the Réunion hot spot, Indian Ocean. Appl Geochem 8:141-152

Marty B, Appora I, Barrat J-A, Deniel C, Vellutini P, Vidal P (1993b) He, Ar, Sr, Nd and Pb isotopes in volcanic rocks from Afar: evidence for a primitive mantle component and constraints on magmatic sources. Geochem J 27:223-232

Marty B, Gunnlaugsson E, Jambon A, Oskarsson N, Ozima M, Pineau F, Torssander P (1991) Gas geochemistry of geothermal fluids, the Hengill area, southwest rift zone of Iceland. Chem Geol 91: 207-225

Matsuda J (1994) Noble Gas Geochemistry and Cosmochemistry. Terra Scientific Publishing Co, Tokyo

McDougall I, Honda M (1998) Primordial solar noble-gas component in the Earth: consequences for the origin and evolution of the Earth and its atmosphere. *In* Jackson I (ed) The Earth's mantle: composition, structure and evolution. Cambridge University Press, Cambridge, p 159-187

McKenzie D, O'Nions RK (1995) The source regions of ocean island basalts. J Petrol 36:133-159

Mertz DF, Haase KM (1997) The radiogenic isotope composition of the high-latitude North Atlantic mantle. Geology 25:411-414

Mertz DF, Devey CW, Todt W, Stoffers P, Hofmann AW (1991) Sr-Nd-Pb isotope evidence against plume-asthenosphere mixing north of Iceland. Earth Planet Sci Lett 107:243-255

Michael PJ, Cornell WC (1998) Influence of spreading rate and magma supply on crystallization and assimilation beneath mid-ocean ridges: evidence from chlorine and major element chemistry of mid-ocean ridge basalts. J Geophys Res 103:18325-18356

Michael PJ, Forsyth DW, Blackman DK, Fox PJ, Hanan BB, Harding AJ, Macdonald KC, Neumann GA, Orcutt JA, Tolstoy M, Weiland CM (1994) Mantle control of a dynamically evolving spreading center: Mid-Atlantic Ridge 31-34°S. Earth Planet Sci Lett 121:451-468

Mizuno H, Nakazawa K, Hayashi C (1980) Dissolution of the primordial rare gases into the molten Earth's material. Earth Planet Sci Lett 50:202-210

Moreira M, Allègre C-J (2002) Rare gas systematics on Mid Atlantic Ridge (37-40°N). Earth Planet Sci Lett 198:401-416

Moreira M, Allègre CJ (1998) Helium-neon systematics and the structure of the mantle. Chem Geol 147:53-59

Moreira M, Kurz MD (2001) Subducted oceanic lithosphere and the origin of the 'high μ' basalt helium isotopic signature. Earth Planet Sci Lett 189:49-57

Moreira M, Sarda P (2000) Noble gas constraints on degassing processes. Earth Planet Sci Lett 176: 375-386

Moreira M, Kunz J, Allègre C (1998) Rare gas systematics in popping rock: isotopic and elemental compositions in the upper mantle. Science 279:1178-1181

Moreira M, Valbracht PJ, Staudacher T, Allègre CJ (1996) Rare gas systematics in Red Sea ridge basalts. Geophys Res Lett 23:2453-2456

Moreira M, Doucelance R, Kurz MD, Dupré B, Allègre CJ (1999) Helium and lead isotope geochemistry of the Azores Archipelago. Earth Planet Sci Lett 169:489-205

Moreira M, Gautheron C, Breddam K, Curtice J, Kurz MD (2001) Solar neon in the Icelandic mantle: new evidence for an undegassed lower mantle. Earth Planet Sci Lett 185:15-23

Moreira M, Staudacher T, Sarda P, Schilling J-G, Allègre CJ (1995) A primitive plume neon component in MORB: the Shona ridge anomaly, South Atlantic (51-52°S). Earth Planet Sci Lett 133:367-377

Morgan WJ (1971) Convection plumes in the lower mantle. Nature 230:42-43

Morrison P, Pine J (1955) Radiogenic origin of the helium isotopes in rock. Annals NY Acad Sci 62:71-92

Mukhopadhyay S, Farley K, Bogue S (1996) Loihi-like $^3He/^4He$ ratios in shield and caldera-filling lavas from Kauai. EOS Trans Am Geophys Union 77:F811

Newsom HE, Taylor SR (1989) Geochemical implications of the formation of the moon by a single giant impact. Nature 338:29-34

Nicolaysen KE, Johnson KTM, Graham DW, Mahoney JJ, Frey FA (2002) Pb, Sr and Nd isotopes in Southeast Indian Ridge MORB near the Amsterdam and St. Paul hotspots. in prep.

Niedermann S, Bach W (1998) Anomalously nucleogenic neon in North Chile Ridge basalt glasses, suggesting a previously degassed mantle source. Earth Planet Sci Lett 160:447-462

Niedermann S, Bach W, Erzinger J (1997) Noble gas evidence for a lower mantle component in MORBs from the southern East Pacific Rise: decoupling of helium and neon isotope systematics. Geochim Cosmochim Acta 61:2697-2715

Nuccio PM, Paonita A (2000) Investigation of the noble gas solubility in H_2O-CO_2 bearing silicate liquids at moderate pressure II: the extended ionic porosity (EIP) model. Earth Planet Sci Lett 183:499-512

O'Nions RK (1987) Relationships between chemical and convective layering in the Earth. J Geol Soc Lond 144:259-274

O'Nions RK, McKenzie D (1993) Estimates of mantle thorium/uranium ratios from Th, U and Pb isotopic abundances in basaltic melts. Phil Trans R Soc Lond A342:65-74

O'Nions RK, Oxburgh ER (1983) Heat and helium in the Earth. Nature 306:429-431

O'Nions RK, Tolstikhin IN (1994) Behaviour and residence times of lithophile and rare gas tracers in the upper mantle. Earth Planet Sci Lett 124:131-138

O'Nions RK, Tolstikhin IN (1996) Limits on the mass flux between lower and upper mantle and stability of layering. Earth Planet Sci Lett 139:213-222
O'Nions RK, Evensen NM, Hamilton PJ (1979) Geochemical modeling of mantle differentiation and crustal growth. J Geophys Res 84:6091-6101
O'Nions RK, Pankhurst PJ, Grönvold K (1976) Nature and development of basalt magma sources beneath Iceland and the Reykjanes Ridge. J Petrol 17:315-338
Ozima M, Podosek FA (1999) Formation age of the Earth from $^{129}I/^{127}I$ and $^{244}Pu/^{238}U$ systematics and the missing Xe. J Geophys Res 104:25493-25499
Ozima M, Podosek FA (2002) Noble Gas Geochemistry. Cambridge University Press. Cambridge, UK
Ozima M, Zashu S (1983) Noble gases in submarine pillow volcanic glasses. Earth Planet Sci Lett 62: 24-40
Ozima M, Zashu S (1988) Solar-type Ne in Zaire cubic diamonds. Geochim Cosmochim Acta 52:19-25
Ozima M, Podosek FA, Igarashi G (1985) Terrestrial xenon isotope constraints on the early history of the Earth. Nature 315:471-474
Paonita A, Gigli G, Gozzi D, Nuccio PM, Trigila R (2000) Investigation of the He solubility in H_2O - CO_2 bearing silicate liquids at moderate pressure: a new experimental method. Earth Planet Sci Lett 181:595-604
Patterson DB, Honda M, Mcdougall I (1990) Atmospheric contamination: a possible source for heavy noble gases in basalts from Loihi seamount, Hawaii. Geophys Res Lett 17:705-708
Pepin RO (1997) Evolution of Earth's noble gases: consequences of assuming hydrodynamic loss driven by giant impact. Icarus 126:148-156
Pepin RO (1998) Isotopic evidence for a solar argon component in the Earth's mantle. Nature 394:664-667
Pepin RO, Porcelli D (2002) Origin of noble gases in the terrestrial planets. Rev Mineral Geochem 47: 191-246
Perez NM, Nakai S, Wakita H, Sano Y, Williams SN (1994) $^3He/^4He$ isotopic ratios in volcanic-hydrothermal discharges from the Canary Islands, Spain: implications on the origin of the volcanic activity. Mineral Mag 58:709-710
Pineau F, Javoy M (1994) Strong degassing at ridge crests: the behaviour of dissolved carbon and water in basalt glasses at 14°N, Mid-Atlantic Ridge. Earth Planet Sci Lett 123:179-198
Podosek FA, Ozima M (2000) The xenon age of the Earth. *In* Canup RM, Righter K (eds) Origin of the Earth and Moon. The University of Arizona Press, Tucson, p 63-72
Porcelli D, Ballentine CJ (2002) Models for the distribution of terrestrial noble gases and the evolution of the atmosphere. Rev Mineral Geochem 47:411-480
Porcelli D, Halliday AN (2001) The core as a possible source of mantle helium. Earth Planet Sci Lett 192:45-56
Porcelli D, Pepin RO (2000) Rare gas constraints on early Earth history. *In* Canup RM, Righter K (eds) Origin of the Earth and Moon. The University of Arizona Press, Tucson, p 435-458
Porcelli D, Wasserburg GJ (1995a) Mass transfer of helium, neon, argon, and xenon through a steady-state upper mantle. Geochim Cosmochim Acta 59:4921-4937
Porcelli D, Wasserburg GJ (1995b) Mass transfer of xenon through a steady-state upper mantle. Geochim Cosmochim Acta 59:1991-2007
Porcelli D, Ballentine CJ, Wieler R (2002) General Introduction. Rev Mineral Geochem 47:1-18
Poreda R, Radicati di Brozolo F (1984) Neon isotope variations in Mid-Atlantic Ridge basalts. Earth Planet Sci Lett 69:277-289
Poreda RJ, Farley KA (1992) Rare gases in Samoan xenoliths. Earth Planet Sci Lett 113:129-144
Poreda RJ, Schilling J-G, Craig H (1986) Helium and hydrogen isotopes in ocean-ridge basalts north and south of Iceland. Earth Planet Sci Lett 78:1-17
Poreda RJ, Schilling J-G, Craig H (1993) Helium isotope ratios in Easter Microplate basalts. Earth Planet Sci Lett 119:319-329
Poreda RJ, Craig H, Arnorsson S, Welhan JA (1992) Helium isotopes in Icelandic geothermal systems: I. ^3He, gas chemistry and ^{13}C relations. Geochim Cosmochim Acta 56:4221-4228
Richter FM, Parsons B (1973) On the interaction of two scales of convection in the mantle. J Geophys Res 80:2529-2541
Ringwood AE (1982) Phase transformations and differentiation in subducted lithosphere: implications for mantle dynamics, basalt petrogenesis, and crustal evolution. J Geol 90:611-643
Rison W, Craig H (1983) Helium isotopes and mantle volatiles in Loihi Seamount and Hawaiian Island basalts and xenoliths. Earth Planet Sci Lett 66:407-426
Rison W, Craig H (1984) Helium isotope variations along the Galapagos Spreading Center. EOS Trans Am Geophys Union 65:1139-1140
Ritsema J, van Heijst HJ, Woodhouse JH (1999) Complex shear wave velocity structure imaged beneath Africa and Iceland. Science 286:1925-1928

Roden MF, Trull T, Hart SR, Frey FA (1994) New He, Nd, Pb and Sr isotopic constraints on the constitution of the Hawaiian plume: results from Koolau Volcano, Oahu, Hawaii, USA. Geochim Cosmochim Acta 58:1431-1440

Sarda P, Graham D (1990) Mid-ocean ridge popping rocks: implications for degassing at ridge crests. Earth Planet Sci Lett 97:268-289

Sarda P, Moreira M (2002) Vesiculation and vesicle loss in mid-ocean ridge basalt glasses: He, Ne, Ar elemental fractionation and pressure influence. Geochim Cosmochim Acta 66:1449-1458

Sarda P, Moreira M, Staudacher T (1999a) Argon-lead isotopic correlation in Mid-Atlantic Ridge basalts. Science 283:666-668

Sarda P, Moreira M, Staudacher T (1999b) Response: Origin of argon-lead isotopic correlation in basalts. Science 286:871a

Sarda P, Staudacher T, Allègre CJ (1985) $^{40}Ar/^{36}Ar$ in MORB glasses: constraints on atmosphere and mantle evolution. Earth Planet Sci Lett 72:357-375

Sarda P, Staudacher T, Allègre CJ (1988) Neon isotopes in submarine basalts. Earth Planet Sci Lett 91: 73-88

Sarda P, Moreira M, Staudacher T, Schilling J-G, Allègre CJ (2000) Rare gas systematics on the southernmost Mid-Atlantic Ridge: constraints on the lower mantle and the Dupal source. J Geophys Res 105:5973-5996

Scarsi P (2000) Fractional extraction of helium by crushing of olivine and clinopyroxene phenocrysts: effects on the $^3He/^4He$ measured ratio. Geochim Cosmochim Acta 64:3751-3762

Scarsi P, Craig H (1996) Helium isotope ratios in Ethiopian Rift basalts. Earth Planet Sci Lett 144: 505-516

Schilling J-G (1973) Iceland mantle plume: Geochemical evidence along Reykjanes Ridge. Nature 242: 565-571

Schilling J-G, Kingsley RH, Devine JD (1982) Galapagos hot spot-spreading center system 1. spatial petrological and geochemical variations (83°W-101°W). J Geophys Res 87:5593-5610

Schilling J-G, Thompson G, Kingsley RH, Humphris SE (1985) Hotspot-migrating ridge interaction in the South Atlantic: geochemical evidence. Nature 313:187-191

Schilling J-G, Hanan BB, McCully B, Kingsley RH, Fontignie D (1994) Influence of the Sierra Leone mantle plume on the equatorial Mid-Atlantic Ridge: a Nd-Sr-Pb isotopic study. J Geophys Res 99:12005-12028

Schilling J-G, Kingsley R, Fontignie D, Poreda R, Xue S (1998) Dispersion of the Jan Mayen and Iceland mantle plumes in the Arctic: a He-Pb-Nd-Sr isotope tracer study of basalts from the Kolbeinsey, Mohns and Knipovich Ridges. J Geophys Res 104:10543-10569

Schmidt BC, Keppler H (2002) Experimental evidence for high noble gas solubilities in silicate melts under mantle pressures. Earth Planet Sci Lett 195:277-290

Schubert G, Turcotte DL, Olson P (2001) Mantle Convection in the Earth and Planets. Cambridge University Press, Cambridge, UK

Sempéré J-C, Cochran JR, Team SS (1997) The Southeast Indian Ridge between 88°E and 118°E: variations in crustal accretion at constant spreading rate. J Geophys Res 102:15489-15505

Shaw AM, Hilton DR, Macpherson CG, Sinton JM (2001) Nucleogenic neon in high $^3He/^4He$ lavas from the Manus back-arc basin: a new perspective on He-Ne decoupling. Earth Planet Sci Lett 194:53-66

Shibata T, Takahashi E, Matsuda J-I (1998) Solubility of neon, argon, krypton, and xenon in binary and ternary silicate systems: a new view on noble gas solubility. Geochim Cosmochim Acta 62:1241-1253

Shirey SB, Bender JF, Langmuir CH (1987) Three-component isotopic heterogeneity near the Oceanographer transform, Mid-Atlantic Ridge. Nature 325:217-223

Sleep NH (1979) Thermal history and degassing of the Earth: some simple calculations. J Geol 87:671-686

Sleep NH (1990) Hotspots and mantle plumes: some phenomenology. J Geophys Res 95:6715-6736

Small C (1995) Observations of ridge-hotspot interactions in the Southern Ocean. J Geophys Res 100:17931-17946

Small C, Cochran JR, Sempéré J-C, Christie DM (1999) The structure and segmentation of the Southeast Indian Ridge. Mar Geol 161:1-12

Staudacher T (1987) Upper mantle origin for Harding County well gases. Nature 325:605-607

Staudacher T, Allègre CJ (1982) Terrestrial xenology. Earth Planet Sci Lett 60:389-406

Staudacher T, Allègre CJ (1988) Recycling of oceanic crust and sediments: The noble gas subduction barrier. Earth Planet Sci Lett 89:173-183

Staudacher T, Allègre CJ (1989) Noble gases in glass samples from Tahiti: Teahitia, Rocard and Mehetia. Earth Planet Sci Lett 93:210-222

Staudacher T, Kurz MR, Allègre CJ (1986) New noble-gas data on glass samples from Loihi seamount and Hualalai and on dunite samples from Loihi and Réunion Island. Chem Geol 56:193-205

Staudacher T, Sarda P, Allègre CJ (1990) Noble gas systematics of Réunion Island. Chem Geol 89:1-17

Staudacher T, Sarda P, Richardson SH, Allègre CJ, Sagna I, Dmitriev LV (1989) Noble gases in basalt glasses from a Mid-Atlantic Ridge topographic high at 14°N: Geodynamic consequences. Earth Planet Sci Lett 96:119-133

Steinberger B, O'Connell RJ (1998) Advection of plumes in mantle flow: implications for hotspot motion, mantle viscosity and plume distribution. Geophys J Intl 132:412-434

Stolper EM, Holloway JR (1988) Experimental determination of the solubility of carbon dioxide in molten basalt at low pressure. Earth Planet Sci Lett 87:397-408

Stuart FM, Ellam RM, J. HP, Fitton JG, Bell BR (2000) Constraints on mantle plumes from the helium isotope composition of basalts from the British Tertiary Igneous Province. Earth Planet Sci Lett 177:273-285

Sturm ME, Klein E, M., Graham DW, Karsten J (1999) Age constraints on crustal recycling to the mantle beneath the southern Chile Ridge: He-Pb-Sr-Nd isotope systematics. J Geophys Res 104:5097-5114

Sun S-S, Jahn BM (1975) Lead and strontium isotopes in post-glacial basalts from Iceland. Nature 255:527

Sun S-S, Tatsumoto M, Schilling J-G (1975) Mantle plume mixing along the Reykjanes Ridge axis: Lead isotopic evidence. Science 190:143-147

Swindle TD, Caffee MW, Hohenberg CM, Taylor SR (1984) I-Pu-Xe dating and the relative ages of the Earth and Moon. *In* Hartmann WK, Phillips RJ, Taylor GJ (eds) Origin of the Moon. Lunar and Planetary Institute, Houston, p 331-357

Tackley PJ (1998) Three-dimensional simulations of mantle convection with a thermo-chemical basal boundary layer: D" ? *In* Gurnis M, Wysession ME, Knittle E, Buffett BA (eds) The Core-Mantle Boundary Region, 28. American Geophysical Union, Washington, DC, p 231-253

Tatsumoto M, Unruh DM, Stille P, Fujimaki H (1984) Pb, Sr and Nd isotopes in oceanic island basalts. *In* Proc 27th Intl Geol Congr 2, Geochemistry and Cosmochemistry, p 485-501

Taylor RN, Thirlwall MF, Murton BJ, Hilton DR, Gee MAM (1997) Isotopic constraints on the influence of the Icelandic plume. Earth Planet Sci Lett 148:E1-E8

Thompson L (1980) ^{129}Xe on the outgassing of the atmosphere. J Geophys Res 85:4374-4378

Tolstikhin IN, Marty B (1998) The evolution of terrestrial volatiles: a view from helium, neon, argon and nitrogen isotope modeling. Chem Geol 147:27-52

Trieloff M, Kunz J, Clague DA, Harrison D, Allègre CJ (2000) The nature of pristine noble gases in mantle plumes. Science 288:1036-1038

Trull TW (1994) Influx and age constraints on the recycled cosmic dust explanation for high ^{3}He/^{4}He ratios at hotspot volcanos. *In* Matsuda J (ed) Noble Gas Geochemistry and Cosmochemistry. Terra Scientific, Tokyo, p 77-88

Trull TW, Nadeau S, Pineau F, Polvé M, Javoy M (1993) C-He systematics in hotspot xenoliths: implications for mantle carbon contents and carbon recycling. Earth Planet Sci Lett 118:43-64

Turcotte DL, Schubert G (1988) Tectonic implications of radiogenic noble gases in planetary atmospheres. Icarus 74:36-46

Turekian KK (1959) The terrestrial economy of helium and argon. Geochim Cosmochim Acta 17:37f

Turner G (1989) The outgassing history of the Earth's atmosphere. J Geol Soc 146:147-154

Valbracht PJ, Staudacher TJ, Malahoff A, Allègre CJ (1997) Noble gas systematics of deep rift zone glasses from Loihi Seamount, Hawaii. Earth Planet Sci Lett 150:399-411

Valbracht PJ, Honda M, Staudigel H, McDougall I, Trost AP (1994) Noble gas partitioning in natural samples: results from coexisting glass and olivine phenocrysts in four Hawaiian submarine basalts. *In* Matsuda J (ed) Noble Gas Geochemistry and Cosmochemistry. Terra Scientific, Tokyo, p 373-381

Valbracht PJ, Honda M, Matsumoto T, Mattielli N, McDougall I, Ragettli R, Weis D (1996) Helium, neon and argon isotope systematics in Kerguelen ultramafic xenoliths: implications for mantle source signatures. Earth Planet Sci Lett 138:29-38

van der Hilst R, Kárason H (1999) Compositional heterogeneity on the bottom 1000 kilometers of Earth's mantle: toward a hybrid convection model. Science 283:1885-1888

van der Hilst RD, Widiyantoro S, Engdahl ER (1997) Evidence for deep mantle circulation from global tomography. Nature 386:578-584

van Keken PE, Ballentine CJ (1998) Whole-mantle versus layered mantle convection and the role of a high-viscosity lower mantle in terrestrial volatile evolution. Earth Planet Sci Lett 156:19-32

van Keken PE, Ballentine CJ (1999) Dynamical models of mantle volatile evolution and the role of phase transitions and temperature-dependent rheology. J Geophys Res 104:7137-7151

van Keken PE, Ballentine CJ, Porcelli D (2001) A dynamical investigation of the heat and helium imbalance. Earth Planet Sci Lett 188:421-434

Vance D, Stone JOH, O'Nions RK (1989) He, Sr, and Nd isotopes in xenoliths from Hawaii and other oceanic islands. Earth Planet Sci Lett 96:147-160

Verma SP, Schilling J-G (1982) Galapagos hot spot-spreading center system 2. ^{87}Sr/^{86}Sr and large ion lithophile element variations (85°W-101°W). J Geophys Res 87:10838-10856

Walker RJ, Morgan JW, Horan MF (1995) Osmium-187 enrichment in some plumes: Evidence for core-mantle interaction ? Science 269:819-822

Weaver BL (1991) Trace element evidence for the origin of ocean-island basalts. Geology 19:123-126

Wetherill GW (1954) Variations in the isotopic abundances of neon and argon extracted from radioactive minerals. Phys Rev 96:679-683

Wetherill GW (1975) Radiometric chronology of the early solar system. Ann Rev Nucl Sci 25:283-328

White WM (1985) Sources of oceanic basalts: radiogenic isotopic evidence. Geology 13:115-118

White WM, Schilling J-G (1978) The nature and origin of geochemical variation in Mid-Atlantic Ridge basalts from the central North Atlantic. Geochim Cosmochim Acta 42:1501-1516

White WM, McBirney AR, Duncan AR (1993) Petrology and geochemistry of the Galápagos islands: portrait of a pathological mantle plume. J Geophys Res 98:19533-19563

White WM, Schilling J-G, Hart SR (1976) Strontium isotope geochemistry of the central North Atlantic: evidence for the Azores mantle plume. Nature 263:659-

Yatsevich I, Honda M (1997) Production of nucleogenic neon in the Earth from natural radioactive decay. J Geophys Res 102:10291-10298

Zindler A, Hart SR (1986a) Helium: problematic primordial signals. Earth Planet Sci Lett 79:1-8

Zindler A, Hart SR (1986b) Chemical geodynamics. Ann Rev Earth Planet Sci 14:493-571

Zindler A, Jagoutz E, Goldstein SL (1982) Nd, Sr and Pb isotope systematics in a three-component mantle: a new perspective. Nature 298:519-523

Zindler A, Hart SR, Frey FA, Jakobsson SP (1979) Nd and Sr isotope ratios and rare earth element abundances in Reykjanes Peninsula basalts: Evidence for mantle heterogeneity beneath Iceland. Earth Planet Sci Lett 45:249-262

9 Noble Gases and Volatile Recycling at Subduction Zones

David R. Hilton

Geosciences Research Division, Vaughan Hall
Scripps Institution of Oceanography
La Jolla, California 92093
drhilton@ucsd.edu

Tobias P. Fischer

Department of Earth and Planetary Sciences
Northrop Hall, University of New Mexico
Albuquerque, New Mexico 87131

Bernard Marty

Centre de Recherches Petrographiques et Geochimiques (CRPG)
15 Rue Notre-Dame des Pauvres, B.P. 20
54501 Vandoeuvre les Nancy Cedex, France

INTRODUCTION

Volatiles are lost from the Earth's mantle to the atmosphere, hydrosphere and crust through a combination of subaerial and submarine volcanic and magmatic activity. These volatiles can be primordial in origin, trapped in the mantle since planetary accretion, produced *in situ,* or they may be recycled—re-injected into the mantle via material originally at the surface through the subduction process. Quantifying the absolute and relative contributions of these various volatile sources bears fundamental information on a number of issues in the Earth Sciences ranging from the evolution of the atmosphere and hydrosphere to the nature and scale of chemical heterogeneity in the Earth's mantle.

Noble gases have a pivotal role to play in addressing the volatile mass balance between the Earth's interior and exterior reservoirs. The primordial isotope ^3He provides an unambiguous measure of the juvenile volatile flux from the mantle (Craig et al. 1975). As such, it provides a means to calibrate other volatiles of geological and geochemical interest. A prime example is the CO_2 flux at mid-ocean ridges (MOR): by combining estimates of the ^3He flux at MOR with measurements of the $CO_2/^3He$ ratio in oceanic basalts, Marty and Jambon (1987) derived an estimate of the CO_2 flux from the (upper) mantle.

The approach of using ratios (involving noble gas isotopes) has also been extended to island arcs. Marty et al. (1989) found significantly higher $CO_2/^3He$ ratios in arc-related geothermal fluids than observed at mid-ocean ridges, consistent with addition of slab-derived CO_2 to the mantle wedge. Sano and Williams (1996) scaled the CO_2 flux to ^3He, showing that the output of CO_2 at subduction zones was comparable in magnitude to that at spreading ridges. Therefore, for CO_2 at least, subduction zones also represent a major conduit for the loss of volatiles from the solid Earth. However, the process of subduction is also the principal mechanism by which materials (including volatiles) are returned to the mantle. For this reason, constraining volatile fluxes and inventories at subduction zones is crucial to understanding volatile budgets on the Earth and the nature of recycling between the mantle and the atmosphere, hydrosphere and crust.

In this contribution, we focus on the noble gas systematics of subduction zones. First, we review the various methodologies of sampling noble gases (and other volatiles) using fluids and rocks in both the subaerial and submarine environments. The aim is to give the

1529-6466/00/0047-0009$05.00

reader background details on how the noble gas database has been accumulated and some of the issues of concern regarding data integrity. We continue by cataloguing the noble gas systematics of both arcs and back-arcs (the output regions)—pointing out similarities and differences in the worldwide record. We highlight some of the major controversies in the interpretation of the database. Finally, we concentrate on the issue of volatile mass balance at subduction zones, emphasising the role of the noble gases (especially ^3He) in constraining both the output fluxes of various volatile species and their provenance—both at individual arcs as well as on a global scale. We discuss implications of the volatile mass balance (or imbalance in some cases) for the recycling history of terrestrial volatiles between the Earth's internal and external reservoirs.

SAMPLING FOR NOBLE GASES

Noble gases are found in all types of fluids and rock samples. For example, fumarolic gas discharges, bubbling hot springs, groundwaters, and natural gases are prime sampling media for noble gases, as are submarine glasses and various minerals which can be crushed or melted *in vacuo* to release their trapped volatiles. In this section, we review the principal means of sampling noble gases at subduction zones.

Volcanic and geothermal fluids

Volcanic and geothermal gas discharges are widely exploited for sampling magmatic volatiles. Various sampling techniques and strategies have been developed depending upon the circumstances of the gas discharge—particularly the temperature. In all cases, precautions are undertaken to avoid, or at least minimize, atmospheric contamination.

High temperature fumaroles on passively degassing subaerial volcanoes provide the opportunity to sample volatiles released directly from a magma body. Lower temperature gases associated with bubbling hot springs, usually located on the flanks of volcanoes, allow for the sampling of noble gases released via hydrothermal systems. To facilitate the transfer of high temperature (>400°C) gases into sampling containers, a silica glass tube (~1-inch diameter) is normally inserted into a gas vent. In the case of vents with discharge temperatures ranging from the boiling point of water to ~400°C, a titanium tube (~1-inch diameter) is used. Gases bubbling into hot springs are sampled by placing a plastic funnel under water—if possible, at the bottom of the spring. The silica/titanium tube or the funnel is connected via a glass connector and silicone or Tygon® tubing to a sampling container, which is used for sample storage and transfer to the analytical facility. The following types of sampling containers are in common usage:

- *Lead-glass flask* (~50 cm^3 volume) with either one or two vacuum stopcocks. Lead-borate glass is used because of its low diffusivity for helium compared to Pyrex (helium diffuses 5 orders of magnitude more slowly through lead-borate glass than through Pyrex at 25°C; Norton 1957). Bottles with two stopcocks (one on each end) allow for the flushing of the bottle with sample prior to collection. The inlet section of a single stopcock flask (previously evacuated to UHV) is usually configured in a "Y" shape to allow flushing of air from the connecting tubes prior to opening the valve and admitting the gas to the flask.
- The second type is the *"Giggenbach" bottle* used by most gas geochemists to obtain the total chemistry of gas discharges. The sampling bottle is an approximately 200 cm^3 evacuated glass flask (Pyrex or lead glass) equipped with a Teflon stopcock. The flask contains 50-80 cm^3 of 4-6 N NaOH solution (Giggenbach and Goguel 1989). The solution absorbs the reactive gases (CO_2, SO_2, H_2S, HCl, HF) allowing build up of a large partial pressure of the remaining (non-reactive) gases (N_2, H_2, O_2, CO, hydrocarbons and noble gases) in the headspace volume. An upside-down position of

the bottle during storage, with the solution covering the Teflon stop-cock, helps prevent possible air contamination of the sample during periods of storage. Gas splits for noble gas analyses are taken from the headspace of the sample bottle.

- The third sampling container consists of a *copper tube* (~10 cm^3) that is crimped using a cold-welder sealing device or pinched off using refrigeration clamps after collection of the sample (see Weiss 1968; Kennedy et al. 1985). The clamps are designed to seal the tube by cold welding and are left in place until sample extraction. Copper is virtually impervious to helium, facilitating secure sample storage; however, interaction with high temperature, acid gases results in the formation of a mixture of copper sulphide, copper sulphate and copper chloride, which can impede the formation of a leak-tight cold seal (F. Goff written comm., 2001).

During sample collection, it is critical to avoid air contamination. Usually this is not a problem at high flow-rate fumaroles as the sampling tube, silicone rubber tubing and sampling bottles are rapidly flushed with the discharging gas. At lower flow-rate fumaroles and bubbling hot springs, more care must be taken to effectively flush the sampling system. This is best achieved by leading a tube from the out-flowing side of the sample bottle into water, allowing the gas to flow through the sample system and to bubble into water. Alternatively, a vacuum hand-pump may be used to facilitate flushing. Depending on the vigor of the gas discharges, flushing of the sampling system can take between ~5 and 40 minutes. Silicone tubes should be kept as short as practical in order to minimize the volume that needs to be flushed with gas prior to sample collection.

Gases from bubbling hot springs can also be sampled using the water-displacement method as described by Craig (1953), Mazor and Wasserburg (1965) and Kennedy et al. (1988). A funnel is submerged in the spring and the entire sampling apparatus is purged with spring fluids using a hand-pump. The funnel is then inverted over the upwelling gas bubbles and the gas is allowed to displace all the water in the apparatus—with the exception of the copper-tube sampler, which is connected to the flow-through sampling line via a "Y" connection. After a steady flow of gas is attained through the sampling line, the liquid from the copper-tube sampler is displaced. This technique maximizes the volume ratio of the displacing gas to that of the displaced liquid and minimizes the effect of back solution of gas into cooler fluid as it is displaced. While filling the copper-tube with gas, the outside of the tube may be cooled with water (or snow/ice) in order to lower the vapor pressure of the water. Holding the tube vertically during sampling allows for the condensed phase to flow out and a steady flow of condensed liquid and non-condensable gas is maintained. This technique, therefore, facilitates collection of non-condensable gas at ambient pressure and cooler than ambient temperatures (see Kennedy et al. 1988).

Hot spring waters and groundwaters

Noble gases dissolved in groundwaters and hot spring waters can be sampled in a fashion similar to that of gas discharges from fumaroles. The most common sampling device is a copper tube, either cold welded or crimped shut using refrigeration clamps. A Tygon® tube is connected to one end of the copper tube and inserted as deeply as possible into the mouth of the spring. A second Tygon® tube is attached to the other end of the copper tube with the discharging water flowing through the tube assembly. After flushing the system for a few minutes and tapping the copper tube lightly, e.g., with a wrench or screwdriver to release air bubbles from the walls of the copper tube, the tube is crimped shut. To avoid air contamination, the out-flowing end of the tube should be crimped first followed by the up-stream end.

Water samples can also be collected using evacuated flasks or "Giggenbach" bottles (without the caustic solution). As the spring water fills the evacuated bottles, dissolved gases will ex-solve and create a head-space in the bottle. Volatiles of interest, including the noble gases, can be withdrawn from the head-space volume for analysis.

Water samples from deep hot spring pools can be sampled at depth by use of a sampling device attached to an extension (aluminum) pole (Kennedy et al. 1988). This technique minimizes air contamination due to entrainment of circulating air-saturated waters from the surface of the pool. A copper tube is fastened to the aluminum pole with a Tygon® tube attached to the bottom end of the copper tube. A second Tygon® tube is attached to the top end of the copper tube and to the pole, and left open to the atmosphere. A spring-loaded Nylon clamp is attached to the pole below the copper tube and triggered by a stainless steel cable that allows for the release of the clamp and the pinching of the Tygon® tube. A second Nylon clamp is attached to the pole above the copper tube. For sample collection, the sampling device is kept vertical and lowered into the pool. At depth, the spring-loaded clamps are released, pinching the Tygon® tubes. The device is brought back to the surface, maintaining the hydrostatic head of the sampling depth. At the surface, the copper tube is pinched shut using a cold welding device or refrigeration clamps.

Groundwater pumped from wells is normally sampled using copper tubes that are connected to the well by Tygon® tubing. Caution must be exercised to sufficiently flush the sample device prior to crimping the copper tube. A "Y" connection may be used to release excess water pressure during and after pinching of the copper tube.

Geothermal wells

Gases from geothermal wells are collected in the same fashion as fumarole discharges. At geothermal wells, however, it is necessary to use a steam separator in order to efficiently sample the non-condensable gases, which partition into the steam phase. The collected sample then consists of steam plus gas.

Natural gases

Natural gases are generally CH_4-rich and at high pressure, making the sampling devices described above impractical. Natural gases are collected into 500 cm^3 or 1000 cm^3 "Whitey" stainless steel gas cylinders sealed at both ends with high-pressure valves. The pre-evacuated cylinders are either attached directly to the well-head or to the vapor-side of a gas-liquid separator. Once attached, the cylinders are opened, flushed with sample gas and sealed. In the laboratory, the cylinders are attached to a vacuum line and aliquots of the sample are transferred into copper tubes (Hiyagon and Kennedy 1992).

The partitioning of noble gases from oil or water into the gas phase is a function of the solubility of the noble gases in the liquid phase, which, among other factors, depends on the atomic mass of the gas. Therefore, fractionation may occur if samples are collected from a separator. According to Ballentine et al. (1996), between 82% and 96% of He, Ne, and Ar, is transferred into the gas phase at the separator. This confirms that noble gases strongly partition into the gas phase, resulting in minimal isotopic fractionation during the steam separation process. Gas fields with high gas/(water+oil) ratios further reduce noble gas (elemental) fractionation during collection of the samples at the separator (Torgersen and Kennedy 1999).

Mafic phenocrysts and xenoliths

Mafic minerals contained in volcanic rocks (phenocrysts) or in xenoliths are widely exploited in noble gas studies as they frequently contain fluid and/or melt inclusions that trap noble gases. Olivines and pyroxenes are the most commonly utilized minerals. Whole-rock samples are first crushed to 0.5 to 2 mm or larger, depending on the size of the crystals. The olivines and pyroxenes are then separated using a Frantz-Isodynamic magnet separator, followed by hand-picking under a binocular microscope to remove any adhering matrix. Mineral separates are then cleaned in distilled water, methanol or acetone. After cleaning and drying, the separates can be processed for noble gases. The two extraction techniques generally used to decrepitate inclusions in the samples are vacuum crushing and vacuum melting. Both methods, however, integrate volatiles from all types and sizes of inclusions in the sample. The relatively recent application of lasers to noble gas studies of mafic crystals (e.g., Burnard et al. 1994) offers the exciting possibility that individual inclusions, or trails of inclusions, can be targeted to reveal spatial and/or temporal variations preserved within single crystals.

There are two general categories of crushing devices whereby crushing takes place either on-line or off-line (see Hilton et al. 1999 for a description of the two types of crusher and a comparison of He isotope results produced by both). In the case of on-line devices, vacuum is maintained between the crusher and the mass spectrometer inlet line during the crushing or pulverizing of the sample. This is not the case for off-line crushers. Indeed, many samples processed for He-isotopes prior to the early-mid 1990s were processed by off-line "ball mills": in these cases, careful monitoring of possible air leakage (e.g., by neon analysis) was deemed essential to ensure the integrity of the helium isotope results (e.g., Hilton et al. 1992). A further potential problem with off-line ball-mills relates to their exceptional efficiency in crushing samples. Prolonged crushing times of mafic crystals can lead to the release of He components extraneous to the magmatic system under scrutiny, e.g., cosmogenic and/or radiogenic He sited in the mineral matrix (Hilton et al. 1993a; Scarsi 2000). Nowadays, most investigators use on-line crushing devices, with many adopting the added precaution of minimizing crushing times (Hilton et al. 1993a, 1999).

Submarine glasses

Submarine glasses from a variety of tectonic environments (including arcs and back-arcs) are a widely exploited medium for obtaining noble gas isotopic compositions and abundances. Rapid quenching of lavas as they are extruded onto the seafloor traps volatile phases in the glassy rinds. The normal procedure (e.g., Hilton et al. 1993b) involves selecting pristine, glass chips and shards by hand-picking under a binocular microscope. Only glass pieces with no visible signs of alteration are selected for analysis. The glasses are then cleaned ultrasonically with acetone and/or distilled water to remove any adhering matrix and/or alteration phases. Sizes of glass chips generally range from 1- to 5-mm diameter. As in the case of mafic minerals, samples can be processed by vacuum melting and/or crushing.

Problematic issues of noble gas analysis

Given the wide range of sampling media available for noble gas studies and the various means of extracting trapped gases for analysis, it is instructive to scrutinize results to test the integrity of both a particular sampling medium as well as the methodology or experimental approach.

Mafic minerals (olivine and clinopyroxene) are widely used to gain insight into the helium isotope systematics of the magmatic source region. However, there are processes that may compromise the integrity of $^{3}He/^{4}He$ results obtained using mafic crystals. These principally relate to addition of extraneous (radiogenic) helium through interaction with pre-existing crust or wallrock. The following observations are pertinent in helping to recognize these effects:

1. A correlation between measured $^{3}He/^{4}He$ ratio and helium content for co-genetic suites of samples may reveal late-stage addition of crustal (radiogenic) helium to helium-poor samples. Hilton et al. (1995) noted that He-poor crystals in some ocean-island phenocrysts had low $^{3}He/^{4}He$ ratios consistent with addition of radiogenic He. Due to

the higher diffusivity of helium in pyroxenes (Trull and Kurz 1993), this observation is more likely observed in pyroxenes than in olivines.

2. A correlation between $^3He/^4He$ ratios and an index of magmatic differentiation. For example, Gasparon et al. (1994) found lower $^3He/^4He$ ratios in more highly evolved phenocrysts (phenocrysts with a lower Mg #) from the Sunda arc, Indonesia.
3. The occurrence of isotopic disequilibrium between coexisting phenocryst pairs. Marty et al. (1994) reported higher $^3He/^4He$ ratios in olivine phenocrysts compared to pyroxene or hornblende for Mt. Etna volcano, an observation that is consistent with addition of late-stage, radiogenic or atmospheric helium. The same effect is seen in arc-related phenocrysts from the Lesser Antilles (Van Soest et al. 2002).

Submarine glasses can also exhibit noble gas variations depending on the method chosen for gas extraction. Crushing releases volatiles sited in vesicles whereas melting integrates vesicle gas and gas dissolved in the glass phase. Depending upon the degree of vesicularity, noble gases can fractionate from each other as a function of their relative solubilities in silicate melt (Jambon et al. 1986; Carroll and Webster 1994). A direct corollary of partitioning of noble gases between vesicle and melt is that the melt phase can become extremely depleted in noble gases. As a consequence, the isotopic composition of this residual gas (in the magma/glass) may become susceptible to modification by addition of volatiles resulting from radiogenic and nucleogenic reactions (e.g., magma aging; Zindler and Hart 1986). Alternatively, extraneous volatiles could be added to volatile-poor magma immediately prior to eruption thereby modifying a magmatic noble gas signature by one characteristic of crustal and/or seawater/atmospheric contamination (see Hilton et al. 1993a).

When comparing different sampling media directly, large differences in $^3He/^4He$ ratios have been observed. For example, at Cerro Negro volcano, Nicaragua, olivine phenocrysts erupted in 1992 have $^3He/^4He$ ratios of 3.5±0.5 R_A (Fischer et al. 1999a) whereas fumarole gases collected days after the eruption have ratios of 6.8 R_A (Sano and Marty 1995). An unreasonably high residence time of approximately 1 Ma is needed to lower $^3He/^4He$ ratios in the phenocrysts from 6.8 to 3.5 R_A assuming a magma U-content of 0.3 ppm and a He-content of 1.5×10^{-9} cm^3 STP/g olivine (and a He partition coefficient of 0.008; Marty and Lussiez 1993). This time period is much longer than the age of the Cerro Negro volcanic system (Fischer et al. 1999a). For this active volcanic system, therefore, the noble gas ratios of fumarolic discharges may be more representative of the present-day mantle source, with the 1992 activity erupting olivine phenocrysts having trapped helium from an earlier magmatic episode.

NOBLE GAS SYSTEMATICS OF ARC-RELATED VOLCANISM

The study of noble gases in arcs began with the seminal work of Mamyrin et al. (1969) who reported $^3He/^4He$ ratios greater than air for geothermal fluids from the Kurile Islands. This early observation of primordial (mantle-derived) ^3He in arc volcanoes has been confirmed by numerous subsequent studies (e.g., Craig et al. 1978; Sano and Wakita 1985; Poreda and Craig 1989) which have repeatedly emphasized the role of the mantle wedge in dominating the helium budget in the great majority of cases (the Banda arc of Indonesia being a notable exception; see below). It was also realized through noble gas studies that other contributors to the arc volatile inventory (the subducting slab and/or arc crust) could be traced through either the relative abundance of a major volatile phase to a noble gas or by the isotopic composition of a particular noble gas. An example of the former would be the high N_2/Ar ratios of island arcs tracing addition of slab-derived sedimentary N_2 to the source region (Matsuo et al. 1978; Kita et al. 1993): in the latter case, it has been argued that low $^3He/^4He$ ratios in the Andes indicate (upper) crustal additions to the volatile budget (Hilton et al. 1993a).

In this section, we review noble gas systematics of arc-related volcanism worldwide. Helium isotope studies dominate because most arc products are erupted subaerially, and air contamination is a relatively minor (correctable) problem for helium: this is not the case for Ne-Ar-Kr-Xe isotope systematics. Consequently, this section is weighted towards reporting observations of helium isotope variations in arc-related minerals and fluids. However, we summarize also the available database for neon, argon and xenon isotopes (to date Kr shows only air-like isotopic compositions). Finally, we consider the limited database of the relative abundances of the noble gases in arc-related products.

Helium isotope systematics of arc-related volcanism

At present, there are close to 1000 reported $^3He/^4He$ ratios from a variety of sampling media associated with arc-related volcanism. Table 1 summarizes the helium isotope database representing 26 individual arc segments worldwide (see Fig. 1 for locations). In this compilation, we focus our attention on the active arc only as detailed across-arc studies (e.g., Sano and Wakita 1985) have long established the link between the presence of mantle-derived helium and the locus of magmatic activity. Furthermore, we include in Table 1 all analyses irrespective of sampling medium—the only exception is duplicate analyses of the same sample (in which case the highest value has been selected). We have made no attempt to choose 'representative' values for a particular volcanic system: all analyses irrespective of distance from volcano summit (geothermal fluids) or age of lava flow (phenocrysts) are included. However, it should be noted that there is evidence of a geographic control on geothermal fluid $^3He/^4He$ ratios located around some individual volcanic centers: the highest values are found close to the eruptive vents and lower, more radiogenic values occur away from the volcanic centers (Sano et al. 1984; Williams et al. 1987; Marty et al. 1989; Hilton et al. 1993a; Van Soest et al. 1998). To facilitate discussion, we tabulate the range of $^3He/^4He$ ratios (maximum and minimum values) together with the average value (with 1 standard deviation from the mean). With reference to Table 1, the following points are emphasized:

1. The highest arc-related $^3He/^4He$ ratios fall within the range normally associated with depleted (N-type) MORB mantle (i.e., 8 ± 1 R_A; see Farley and Neroda 1998 and Graham 2002, this volume). The highest values (8.8 to 8.9 R_A) are reported for 4 arc segments: the southern Lesser Antilles (Pedroni et al. 1999), the Colombian Andes (Sano et al. 1997) and the Sunda arc system in both Sumatra (Gasparon et al. 1994) and Bali (Hilton and Craig 1989).
2. In nine of the arc segments, the highest $^3He/^4He$ ratio fails to reach the lower limit of the MORB range, i.e., the highest value is <7 R_A. These segments are: four regions of the Andes (central and northern Chile, Peru and Equador), Kamchatka, the Kuriles and Taiwan (where the total number of samples is low), and the eastern Sunda/Banda arc and the Campanian Magmatic Province of Italy (which both have a unique tectonic arrangement of plates; see below).
3. The lowest arc-related $^3He/^4He$ ratio (0.01 R_A) is found at the transition between the east Sunda and Banda arcs in Indonesia (Hilton et al. 1992). Other arc segments with low $^3He/^4He$ ratios (<1 R_A) are: the Chilean Andes (Hilton et al. 1993a), Colombia (Williams et al. 1987) and the Campanian Magmatic Province of Italy (Graham et al. 1993). Therefore, in a relatively small number of samples from these localities, radiogenic helium (with $^3He/^4He$ ratios ~0.05 R_A; see Andrews 1985) dominates over any magmatic helium input.

Table 1. Summary of helium isotope variations in arc-related volcanics and geothermal fluids worldwide.

Location (arc segment)	*R/R_A (max)*	*R/R_A (min)*	*R/R_A* (mean) (± 1 S.D.)*	*N*[$]	*Crustal thickness*[1] *(km) + type*[4]	*Conv. rate*[2] *(mm/yr)*	*Sediment subducted*[3] *(km³/Ma)*	*Slab age (Ma)*[4]	*Refs*[5]
N. Lesser Antilles (Saba – Dominica)									
	8.60	3.31	6.80± 1.33	73	30-35 O	20	19a	68	1-4
S. Lesser Antilles (Martinique – Grenada)									
	8.90	1.53	4.93± 1.85	82	30-35 O	20	19a	68	2-4
Andes – Central Chile									
	6.84	0.18	3.90± 1.74	18	40 C	70	74a	48	5
Andes – Northern Chile									
	6.02	0.82	2.39± 1.43	12	70 C	80	29b	82	5
Andes - Peru									
	-	-	2.8	1	70 C			45	6
Andes - Equador									
	4.80	2.10	3.22± 0.98	5	66 C			32	7
Andes - Columbia									
	8.84	0.90	5.66± 2.41	45	66 C			15± 10	6, 8-10
Central America									
	7.93	2.13	5.97± 1.44	31	42 C	65	23b	23± 5	11-14
Mexico									
	7.34	5.10	6.82± 0.97	5	30 C	70	35a	14 - 15	11, 15
Cascades									
	8.19	1.25	6.07± 1.93	19	25-35 C	35	34a	8	11, 16-1
Alaska									
	7.89	6.03	7.06± 0.75	8	36 C	70	177a	46	11
Aleutian Islands									
	8.01	1.60	5.74± 1.54	23	18-25 O	70	177a	54	11
Kamchatka									
	6.33	4.88	5.76± 0.54	7	25-45 C	90		90± 15	11, 18
Kurile Islands									
	6.76	5.70	6.33± 0.43	7	15-30 C	90	48b	119	11, 19
NE Japan									
	8.40	1.70	5.10± 1.68	129	27-36 C	105	29a	130	11, 20-2
SW Japan (Nankai)***									
	7.13	3.98	5.61± 1.07	16	25 C	30	18a	21	6, 20, 28
Taiwan (Ryukyu)									
	6.15	4.35	5.06± 0.65	6	35 C	60	19b	49	11
Mariana Islands									
	7.65	5.30	7.04± 0.88	6	15-18 O	90	52b	155	11, 16
Philippines									
	7.60	6.90	7.34± 0.24	10	-T	90	33b	-	11, 29
Indonesia – East Sunda/Banda arcs									
	4.50	0.01	2.18± 1.23	28	<15 O	70-80			11, 30-3
Indonesia - Sumatra									
	8.80	2.08	6.15± 1.74	20	- C	50	96a	55	32-33
Indonesia – Java-Flores									
	8.80	4.50	6.69± 1.03	26	25-30 T	75	37a	138	7, 30, 32
New Zealand (Taupo Volcanic Zone)									
	8.23	1.45	5.99± 1.38	145	36 C	70	24b	98	36-43
Tabar (New Ireland arc)									
	7.54	2.03	6.18± 1.37	53					44
Italy (Aeolian arc)**									
	8.00	1.23	5.32± 1.03	103	20	9			20, 45-5
Italy (Campanian Magmatic Provence)									
	4.62	0.76	2.52± 0.63	52		9			45-46,51-
GRAND TOTAL									
	8.90	0.01	5.37± 1.87	922					1-53

Footnotes to Table 1

* Includes only air- or ASW-corrected $^3He/^4He$ data except where no He/Ne ratios are given (see Appendix in Hilton 1996 for details of the correction procedure).

$ N = number of individual results incorporated into mean. In the case of duplicate analysis, the higher number is selected. Includes time-series results from single locality.

** Includes Mt. Etna on Sicily (see Schiano et al., 2001).

*** Xenolith samples from NE Kyushu in SW Japan (Sumino et al., 2000; Ikeda et al., 2001) with 'hotspot' $^3He/^4He$ ratios (>>$8R_A$) are not obviously arc-related and are not included in this compilation.

1. Crustal thickness (in km) from Gill (1981)
2. Orthogonal convergence rate from Von Huene and Scholl (1991)
3. Estimated (solid) fraction subducted (x 1000 km^3/Myr) from Von Huene and Scholl (1991) a = Type 1 (accretionary prism forms); b = Type 2 (no net accretion) – from Von Huene and Scholl (1991).
4. Age of slab (Myr) and type of crust from Jarrad (1986). Types: O = oceanic; C = continental; T = transitional.

Helium isotope data sources: 1= Chiodini et al. (1996); 2 = van Soest et al. (1998); 3 = Pedroni et al. (1999); 4 = van Soest et al. (2002); 5 = Hilton et al. (1993a); 6 = Sano and Marty (1995); 7 = Fischer and Sano (unpubl.); 8 = Williams et al. (1987); 9 = Sano et al., (1997); 10 = Lewicki et al. (2000); 11 = Poreda and Craig (1989); 12 = Hilton, Goff and McMurtry (unpubl.); 13 = Janik et al. (1992); 14 = Sano and Williams (1996); 15 = Taran et al. (1998); 16 = Craig et al. (1978); 17 = Dodson and Brandon (1999); 18 = Fischer et al. (1999a); 19 = Fischer et al. (1998); 20 = Marty et al. (1989); 21 = Igarashi et al. (1992); 22-26 = Sano et al (1984; 1986; 1988; 1991; 1994); 27 = Porcelli et al. (1992); 28 = Craig and Horibe (1994); 29 = Giggenbach and Poreda (1993); 30 = Hilton and Craig (1989); 31 = Hilton et al. (1992); 32 = Gasparon et al (1994); 33 = Hilton and Fischer (unpubl); 34 = Allard, (1983); 35 = Sano et al. (2001); 36 = Torgersen et al. (1982); 37 = Hulston et al. (1986); 38 = Sano et al. (1987); 39-40 = Hulston and Lupton (1989; 1996); 41 = Marty and Giggenbach (1990); 42 = Giggenbach et al. (1993); 43 = Patterson et al. (1994); 44 = Patterson et al. (1997); 45 = Sano et al. (1989); 46 = Marty et al. (1994); 47 = Tedesco et al. (1995); 48 = Tedesco and Nagao (1996); 49 = Allard et al. (1997); 50 = Nakai et al. (1997); 51 = Tedesco et al. (1990); 52 = Graham et al. (1993); 53 = Tedesco et al. (1998).

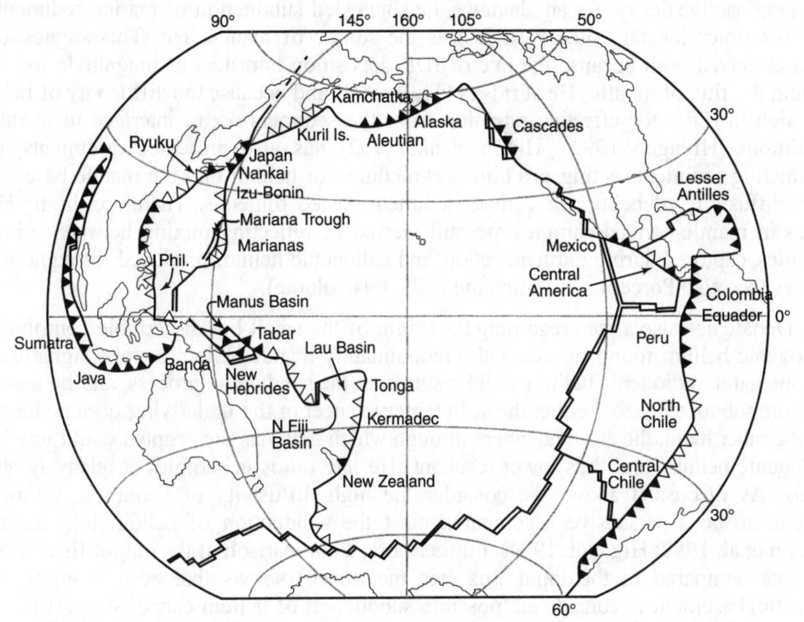

Figure 1. Locations of arc and back-arc systems for which helium isotope data are available (Note: Italy is omitted). Trenches are marked by barbed lines (filled barbs = accretionary trenches; open barbs = no accretionary prism). Modified from Von Huene and Scholl (1991).

4. The mean $^3He/^4He$ ratio of arc-related volcanism is 5.4±1.9 R_A (all arc segments). Although the significance of the mean value is unclear—it does not possess the significance of an individual maximum or minimum ratio for example, it is noteworthy that only the east Sunda/Banda arc, the Campanian Magmatic Province and three segments of the Andes (northern Chile, Peru and Equador) have average $^3He/^4He$ ratios which are more than one standard deviation displaced from the global mean. These are the segments that have the lowest absolute $^3He/^4He$ ratios (see above point).

The general picture that emerges from the helium isotope compilation is that in most cases arc-related volcanism samples helium with an isotopic composition close to that found in MORB, i.e., that emitted at divergent plate boundaries. This observation was made very early in the history of helium isotope studies of arc-related volcanism (e.g., Craig et al. 1978), and still stands today in light of the much more extensive database (Table 1). The first-order implication of this observation is that the helium in the majority of arcs is predominantly of mantle derivation, and presumably derived from the mantle wedge. This point can be illustrated by taking the mean global $^3He/^4He$ ratio calculated in Table 1 (5.4 R_A) and assuming that it represents a two-component mixture of MORB-type (8 R_A) and radiogenic (0.05 R_A) helium. In this case, approximately 67% (or two-thirds of the helium) is MORB-type helium and derived from the mantle wedge. Clearly, the proportion increases (up to 100% in some cases) for those samples with $^3He/^4He$ ratios higher than the mean value.

Considerable debate has arisen as to the (ultimate) origin of the high $^3He/^4He$ ratios typical of mantle-derived material in arcs and elsewhere. Anderson (1993) questioned the widely held notion that high $^3He/^4He$ ratios result from retention of primordial helium in the mantle and/or core with time-integrated lowering of the ratio by addition of ^4He produced from radioactive decay. As an alterative, he suggested subduction of marine sediments rich in IDPs (interplanetary dust particles) as the source of mantle ^3He. This suggestion has been criticized both because the rate of IDP deposition is orders of magnitude too low to sustain the flux of mantle ^3He at ridges (Trull 1994) and because the diffusivity of helium is too high in IDPs for effective retention from the sediment-ocean interface to mantle P-T conditions (Hiyagon 1994). Hilton et al. (1992) has also presented arguments against subducting sediments acting as a transport medium for helium into the mantle based on the high diffusivity of helium in various sediment-hosted minerals. Therefore, high $^3He/^4He$ ratios in mantle-derived samples are still viewed as reflecting mixing between primordial volatiles, captured during Earth accretion, and radiogenic helium produced throughout Earth history (see also Porcelli and Ballentine 2002, this volume).

Debate has also arisen regarding the origin of the small but discernible contribution of radiogenic helium found in arcs with predominantly mantle helium isotope signatures. On the one hand, radiogenic helium could result from the subduction process, and be associated with the subducted slab—either the sedimentary veneer or the underlying oceanic basement. On the other hand, the arc lithosphere through which magmas are erupted could provide the radiogenic helium, and thus lower resultant $^3He/^4He$ ratios in samples at relatively shallow levels. As discussed above, we consider the high diffusivity of helium in sedimentary material to be a persuasive argument against the subduction of helium into the mantle (Hilton et al. 1992; Higayon 1994). Furthermore, a comparison of the output flux of helium via arcs compared to the input flux (via the trench) shows that even when crystalline (basaltic) basement is considered, possible subduction of helium cannot support the return flux from the mantle (see section: Volatile mass balance at subduction zones). Circumstantial evidence also tends to implicate crustal contamination. For example, arc segments with significant volumes of subducting sediments (Alaska, the Aleutian Islands, Java, New Zealand; Table 1) and/or old oceanic basement (western Pacific arcs: Japan, the

Marianas, Java; Table 1) have $^3He/^4He$ ratios that fall within the canonical MORB-He isotope range of 8±1 R_A suggesting little or no slab influence. In contrast, $^3He/^4He$ ratios <7 R_A are observed more commonly in those arc segments which erupt through thickened crust of continental affinity (e.g., Taiwan, Kamchatka, the Andes—with the exception of Colombia; Table 1). For these reasons, we favor the hypothesis that the arc crust and/or lithosphere are responsible for the addition of radiogenic helium (see also discussion in Porcelli et al. 1992 and Hilton et al. 1993a).

In addition to the majority of arcs with predominantly mantle-derived helium, there are 3 segments which stand out as emitting helium with $^3He/^4He$ values consistently lower than the MORB range: the east Sunda/Banda arc of eastern Indonesia, the Campanian Magmatic Province of Italy and the Andes—particularly the Chilean segments. These arc systems are clearly unusual, at least in terms of their helium isotope systematics, and merit further comment.

East Sunda/Banda Arc. The eastern section of the Sunda arc and the contiguous Banda arc was the first arc system where predominantly radiogenic helium was found to characterize volcanic emissions (Poreda and Craig 1989). In this region, $^3He/^4He$ ratios reflect the overwhelming influence of radiogenic helium in the source region—a remarkable departure from observations at other western Pacific arcs. The transition from normal 'arc-like' ratios (6-8 R_A) to the predominantly radiogenic helium values (≤1-2 R_A) is centered on the island of Flores in the eastern Sunda arc and is extremely sharp (Hilton and Craig 1989, Hilton et al. 1992). It coincides with the change in composition of the subducting slab from normal oceanic lithosphere in the Sunda arc located to the west, to crust of continental affinity in the east—the leading edge of the Australian continental margin as it is subducted beneath the Banda Sea. The transition in type of subducting crust is traced with remarkable clarity in He-isotope space whereas other isotopic systems (e.g., $^{87}Sr/^{86}Sr$) display only a broad gradual change through the transition region (Hilton and Craig 1989). In this unusual tectonic environment, therefore, we see one of the rare examples where the effect of radiogenic helium associated with the subducting slab is discernible through the MORB-like helium signal normally associated with the mantle wedge.

Southern Italy. There is a pronounced northward decrease in $^3He/^4He$ ratios between the Aeolian Islands of southern Italy and the region of Mt. Vesuvius known as the Campanian Magmatic Province or the Neopolitan volcanic region. Mt. Etna on Sicily can be included in this trend as recent work (e.g., Schiano et al. 2001) indicates a progressive transition from a plume-related to a typical island-arc source for this volcano. The trend or 'step function' in the helium isotope systematics of southern Italy was first pointed out by Sano et al. (1989) and ascribed to shallow crustal contamination effects. However, subsequent work (Marty et al. 1994) has shown the combined He-O-Sr-C isotope systematics are consistent with the progressive involvement of Africa continental crust which has been subducted beneath the southern Tyrrhenian Sea. The influence of the African plate increases in a northerly direction (Marty et al. 1994)—towards Mt. Vesuvius (see data of Graham et al. 1993) and possibly as far north as the Roman and South Tuscany volcanic provinces (see Tedesco 1997).

The Andes. In the Chilean Andes, high (5-7 R_A) and low (<1 R_A) helium isotope ratios occur in both the Central and Southern Volcanic Zones (Hilton et al. 1993a). In contrast to the east Sunda/Banda arcs, there appears to be no spatial control on the distribution of $^3He/^4He$ ratios, nor any correlation between $^3He/^4He$ and Sr (or Pb) isotopes—an observation used to rule out a slab origin for the radiogenic helium isotope signal. Instead, it has been suggested that the principal control on the measured $^3He/^4He$ values is a combination of near surface magmatic degassing and crustal contamination of degassed magmas. It is noteworthy, however, that the highest $^3He/^4He$ ratios in each volcanic zone fall short of the MORB range (6.9 R_A in the Southern Zone and 6.0 R_A in the Central Zone) leading to the suggestion (Hilton et al. 1993a) that the magmatic base-line in $^3He/^4He$ ratios is set by assimilation processes in the lowermost crust (see Hildreth and Moorbath 1988) prior to transfer of magma towards the surface.

Neon and argon isotope systematics of arc-related volcanism

In contrast to the extensive helium isotope database described above, there are relatively few argon (n ~ 260) and even fewer neon (n < 100) isotope analyses available for arc-related samples. The available data are summarized in Table 2 (neon) and Table 3 (argon), adopting the same format as the helium isotope database. In the cases of neon and argon, however, reported isotope results invariably reflect varying degrees of contamination with air as a consequence of both the low intrinsic concentrations of these volatiles in natural samples and their greater abundance (relative to helium) in air. Therefore, more significance could be given to the extreme (higher) values as they represent samples least affected by air contamination.

Table 2. Summary of neon isotope variations in arc-related lavas and geothermal fluids.

Location		*maximum*	*minimum*	*N*	*mean (± 1 s.d.)*	*Refs.*[5]
New Zealand	$^{20}Ne/^{22}Ne$	10.59	9.69	9	10.02±0.25	[1]
	$^{21}Ne/^{22}Ne$	0.0554	0.0284	9	0.0332±0.0085	[1]
Central America	$^{20}Ne/^{22}Ne$	10.10	9.56	11	9.79±0.17	[2]
	$^{21}Ne/^{22}Ne$	0.0298	0.0286	11	0.0292±0.0004	[2]
Lesser Antilles[1]	$^{20}Ne/^{22}Ne$	10.55	9.56	65	9.94±0.17	[3]
	$^{21}Ne/^{22}Ne$	n.d.	n.d.	0		
Italy, Aeolian arc[2]	$^{20}Ne/^{22}Ne$	10.69	9.41	20	9.91±0.33	[4-5]
	$^{21}Ne/^{22}Ne$	0.0397	0.0290	20	0.0323±0.0033	[4-5]
Italy, Vesuvius[3]	$^{20}Ne/^{22}Ne$	9.95	8.88	11	9.50±0.33	[6]
	$^{21}Ne/^{22}Ne$	0.0451	0.0243	11	0.0337±0.0067	[6]
Japan[4]	$^{20}Ne/^{22}Ne$	10.36	9.73	38	9.89±0.14	[7]
	$^{21}Ne/^{22}Ne$	0.0306	0.0282	34	0.0292±0.0005	[7]
Summary	**$^{20}Ne/^{22}Ne$**	**10.69**	**8.88**	**154**	**9.89±0.24**	[1-7]
	$^{21}Ne/^{22}Ne$	**0.0554**	**0.0243**	**85**	**0.0309±0.0043**	[1-2, 4-7]

Footnotes to Table 2

1. Includes both northern and southern segments
2. Etna and Volcano only
3. Campanian Magmatic Provence
4. Includes data from NE and SW (Nankai) Japan.
5. Neon isotope data sources: 1 = Patterson et al. 1994, 2 = Kennedy et al. 1991, 3 = Pedroni et al. 1999, 4 = Nakai et al. 1997, 5 = Tedesco and Nagao 1996, 6 = Tedesco et al. 1998, 7 = Nagao et al. 1981.

n.d. = not determined.

Table 3. Summary of argon isotope variations ($^{40}Ar/^{36}Ar$) in arc-related lavas and geothermal fluids.

Location	*maximum*	*minimum*	*N*	*mean (± 1 s.d.)*	*Refs.*[6]
New Zealand	1160	296	22	431±209	[1-3]
Central America	330	290	17	300±9	[4-5]
Lesser Antilles (North)[1]	317	293	21	300±7	[6]
Lesser Antilles (South)[2]	435	286	46	302±21	[6]
Italy, Aeolian arc[3]	2082	288	116	503±381	[4, 7-10]
Italy, Vesuvius[4]	301	295	11	298±2	[11]
Japan[5]	325	287	28	304±9	[12]
Summary	**2082**	**286**	**261**	**401±278**	[1-12]

Footnotes to Table 3

1. Islands from Dominica to Nevis
2. Islands from Grenada to Martinique
3. Etna and Volcano only
4. Campanian Magmatic Province
5. Includes data from NE and SW (Nankai) Japan.

Argon isotope data sources: 1 = Torgersen et al. 1982, 2 = Marty and Giggenbach 1990, 3 = Patterson et al. 1994, 4 = Staudacher and Allegre 1988, 5 = Kennedy et al. 1991, 6 = Pedroni et al. 1999, 7 = Magro and Pennisi 1991, 8 = Marty et al. 1994, 9 = Tedesco and Nagao 1996, 10 = Nakai et al. 1997, 11 = Tedesco et al. 1998, 12 = Nagao et al. 1981.

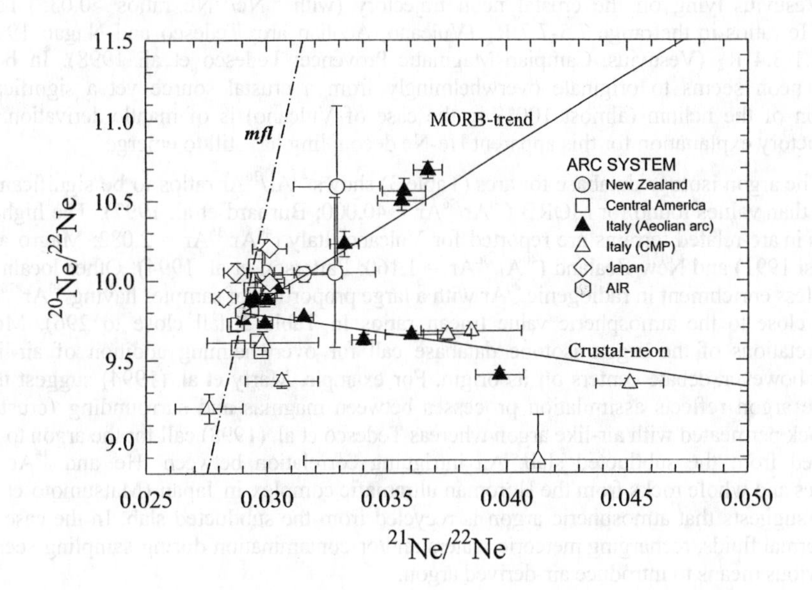

Figure 2. Neon isotope systematics of arc-related volcanism. Arc data from Table 2, MORB-trend from Sarda et al. (1988) and crustal-neon trend from Kennedy et al. (1990). mfl = mass fractionation line.

In Figure 2 we plot arc-related neon isotope ratios on the traditional 3-isotope neon plot using only those samples (n = 34) with reported $^{20}Ne/^{22}Ne$ and $^{21}Ne/^{22}Ne$ ratios that are distinct from the air value (at the 1σ level). The data appear to follow three trends:

- Samples from all five arc regions plotted have values which lie along the so-called MORB-trend (Sarda et al. 1988; Graham 2002, this volume). This is most convincingly seen for samples from the Aeolian arc (Etna region; Nakai et al. 1997) and, to a lesser extent, New Zealand (Patterson et al. 1994).
- Fumarole gases from Vulcano (Aeolian arc) and Vesuvius (Campanian Magmatic Provence) have nucleogenic neon (i.e., neon enriched in ^{21}Ne and, to a lesser extent, ^{22}Ne) and follow the trajectory labeled 'Crustal-neon.' To date, this observation is unique to Italian volcanism and has been attributed to the influence of nucleogenic neon produced in (subducted) crustal material (Tedesco and Nagao 1996; Tedesco et al. 1998).
- All arc regions have samples with neon isotope ratios overlapping with the mass fractionation line (mfl). No petrogenetic significance is attached to these observations as they are probably related to degassing phenomena.

Samples lying along the MORB-trend in 3-isotope neon space can be interpreted as tapping a mantle source (the mantle wedge). For the most part, the $^{3}He/^{4}He$ ratios of the same samples are MORB-like, or close to MORB, indicating a common origin for both light noble gases. In this respect, the helium and neon isotope systematics are coupled (within error) as postulated by the solar neon hypothesis (Honda et al. 1993a, see also Graham 2002, this volume). In contrast, samples that lie along the crustal-neon trajectory (Fig. 2) present an extreme example of helium-neon decoupling. According to the solar neon hypothesis, samples with 'crustal' neon should also be characterized by radiogenic helium (<0.1 R_A) given that nucleogenic ^{21}Ne and radiogenic ^{4}He are produced at a constant ratio ($^{21}Ne/^{4}He = 4.5\times10^{-8}$; Yatsevich and Honda 1997). However, samples from Vulcano and Vesuvius lying on the crustal neon trajectory (with $^{21}Ne/^{22}Ne$ ratios >0.035) have $^{3}He/^{4}He$ ratios in the range 7.5-7.7 R_A (Vulcano, Aeolian arc; Tedesco and Nagao 1996) and 2.1-3.4 R_A (Vesuvius, Campian Magmatic Provence; Tedesco et al. 1998). In both cases, neon seems to originate overwhelmingly from a crustal source yet a significant fraction of the helium (almost 100% in the case of Vulcano) is of mantle derivation. A satisfactory explanation for this apparent He-Ne decoupling has still to emerge.

The argon isotope database for arcs (Table 3) shows $^{40}Ar/^{36}Ar$ ratios to be significantly lower than values found for MORB ($^{40}Ar/^{36}Ar$ > 40,000; Burnard et al. 1997). The highest values in arc-related terrains are reported for Vulcano, Italy ($^{40}Ar/^{36}Ar$ = 2,082; Magro and Pennisi 1991) and New Zealand ($^{40}Ar/^{36}Ar$ = 1,160; Patterson et al. 1994). Other localities show less enrichment in radiogenic ^{40}Ar with a large proportion of samples having $^{40}Ar/^{36}Ar$ ratios close to the atmospheric value (mean ratios in Table 3 fall close to 296). Most interpretations of the argon isotope database call for overwhelming addition of air-like argon: however, debate centers on its origin. For example, Marty et al. (1994) suggest that air-like argon reflects assimilation processes between magmas and surrounding (crustal) wallrock permeated with air-like argon whereas Tedesco et al. (1998) call for the argon to be recycled from the subducted slab. An intriguing correlation between ^{3}He and ^{36}Ar in olivines and whole rocks from the Horoman ultramafic complex in Japan (Matsumoto et al. 2001) suggests that atmospheric argon is recycled from the subducted slab. In the case of geothermal fluids, recharging meteoric waters and/or contamination during sampling seems an obvious means to introduce air-derived argon.

The origin of the radiogenic argon is more problematic as both mantle and crustal reservoirs are characterized by high $^{40}Ar/^{36}Ar$ ratios. Kennedy et al. (1991) attributed radiogenic ^{40}Ar in Honduran geothermal fluids to a crustal source based upon measured

^{3}He/^{40}Ar* ratios (where ^{40}Ar* represents 'radiogenic' ^{40}Ar added to air-like argon to increase the ^{40}Ar/^{36}Ar ratio) which were typically 20 times less than MORB values. Other authors (e.g., Patterson et al. 1994) have attempted to scale the radiogenic argon to the contribution of radiogenic ^{4}He (from the crust). Based on the fact that most arc-related ^{3}He/^{4}He ratios are close to the MORB value, i.e., the contribution of crustal ^{4}He is relatively small (see Table 1), most of the radiogenic ^{40}Ar is then attributed to the mantle wedge.

To date, there is no convincing evidence in the arc-related database that ^{38}Ar/^{36}Ar ratios deviate significantly from the air value except in cases of mass fractionation (Nagao et al. 1981).

Krypton and xenon isotope systematics of arc-related volcanism

There are no reports of krypton isotope anomalies in arc-related terrains—the small number of krypton isotopic analyses (e.g., Patterson et al. 1994) reveal only atmospheric-like ratios. For the most part, the situation is similar for xenon—atmospheric-like ratios dominate the few analyses reported. An exception is found in the work of Nakai et al. (1997) which reports two samples from the vicinity of Mt. Etna with enrichments in both ^{129}Xe and ^{136}Xe relative to air. The anomalies are correlated such that they appear to reflect mixing between air and an enriched source with a xenon isotope signature similar to MORB.

Relative noble gas abundance systematics of arc-related volcanism

Patterns of relative noble gas abundances are usually reported in the F_m notation where $F_m = (^mX/^{36}Ar)/(^mX/^{36}Ar)_{atm}$ where ^{m}X represents a noble gas element X of mass m. Data are available from the following arc segments: Japan (Matsubayashi et al. 1978; Nagao et al. 1981), New Zealand (Marty and Giggenbach 1990; Patterson et al. 1994), Central America (Kennedy et al. 1991) and the Aeolian arc (Nakai et al. 1997; Tedesco and Nagao 1996). Irrespective of sampling medium (phenocryst or geothermal fluids) there appear to be a number of common features in the relative noble gas abundance patterns of arc-related samples:

1. Extreme enrichments in helium (both ^{3}He and ^{4}He) compared to air. For example, Patterson et al. (1994) report a maximum F_3 value of ~22,000 for New Zealand phenocrysts while Kennedy et al. (1991) found a maximum F_4 of 107 for Honduras geothermal fluids. Based on the ^{3}He/^{4}He ratio of the helium, the enrichment is usually attributed to the input of magmatic gas originating from the mantle wedge.
2. F_{22} and F_{84} values which resemble air or the noble gas abundances of air-saturated water (asw) i.e., $F_{22} < 1$ and $F_{84} > 1$. In the case of geothermal fluids, slight differences from the actual air or asw values have been ascribed to either uncertainties in the recharge temperature of the fluids or to fractionation processes associated with either vapor-phase separation or bubble formation (e.g., Kennedy et al. 1991). In the case of phenocrysts (e.g., from New Zealand), Patterson et al. (1994) explained the variations in F_{22} and F_{84} about the atmospheric values by solubility-controlled elemental fractionation associated with mass transfer between basaltic melt and a distinct vapor phase.
3. Significant enrichments in F_{132}, with values as high as 7.3 reported for both phenocrysts (Patterson et al. 1994) and geothermal fluids (Kennedy et al. 1991). This is greater than values anticipated by either vapor-magma models or by fractionating asw respectively. It has been suggested that adsorption-desorption processes, possibly associated with host rocks, could create such a xenon-rich component (see Marty and Giggenbach 1990).

NOBLE GASES IN BACK-ARC BASINS

The majority of back-arc basins are formed by extension and seafloor spreading behind, or within, an island arc: their formation is, therefore, associated with contemporaneous subduction-zone activity (Saunders and Tarney 1991). From the earliest geochemical studies of back-arc basin basalts (BABB) (e.g., Hart et al. 1972; Hawkins 1976; Gill 1976; Tarney et al. 1977) it has been recognized that although similar to MORB—in mineralogy and major chemistry—BABB were enriched in large ion lithophile (LIL) trace elements such as Ba compared to high field strength (HFS) elements like Ti and Zr. Differences in trace element geochemistry extend also to volatile compositions: for the most part, BABB have higher volatile contents—particularly water. In part, the higher water (and other volatile) contents of some BABB result from the propensity of back-arc basins to erupt more evolved magma types and the fact that water acts incompatibly during fractionation. However, it is apparent when comparing BABB with MORB at equivalent TiO_2 or MgO contents that BABB still have higher absolute volatile contents. This observation has been taken as evidence that back-arc basin lavas contain a volatile component originating from the associated subducting slab. For this reason, they provide insights complementary to those from arc volcanism on the topic of noble gases in subduction zones.

Helium isotopes in back-arc basins

In Table 4 we summarize the available helium isotope data for back-arc basins (see Fig. 1 for locations). For the most part, the database has been obtained using vacuum crushing techniques on glass rinds of pillow lavas—therefore, the helium is representative of the vesicle phase of such samples. A notable exception is the study by Ishibashi et al. (1994) who measured $^3He/^4He$ ratios in hydrothermal fluids from the North Fiji Basin. Comparisons of $^3He/^4He$ ratios between oceanic glasses and hydrothermal fluids (e.g., at Loihi Seamount; Hilton et al. 1998) indicate that both sampling media provide valuable means of obtaining He isotope information on underlying magma sources.

Helium isotope ratios at the four back-arc basins studied to date cover almost the complete range observed in the terrestrial environment (excluding samples containing cosmogenic ^3He; see Niedermann 2002, this volume). $^3He/^4He$ ratios > MORB (i.e., higher than 8±1 R_A) are observed at 3 of the back-arc basins: the Manus Basin (15.2 R_A), the Lau Basin (22.1 R_A) and the North Fiji Basin (10.3 R_A). Such high values are normally considered diagnostic of a deep-seated mantle plume component (Craig and Lupton 1976), and the respective authors in the three cases above have adhered to this interpretation. For example, Poreda (1985) and Poreda and Craig (1992) suggested that channeling of plume material from the nearby Samoa hotspot could explain the $^3He/^4He$ ratios at Rochambeau Bank towards the northern end of the Lau Basin (see Turner and Hawkesworth (1998) for further discussion on mantle flow between Samoa and the Lau Basin). Similarly, Macpherson et al. (1998) explained high $^3He/^4He$ ratios in the Manus Basin as reflecting a mantle plume originating at the underlying core-mantle boundary. While noting the presence of a "^{3}He-rich, hotspot-like component" in the North Fiji Basin, Ishibashi et al. (1994) did not comment on its origin. In all three cases, however, there was no attempt to relate the high $^3He/^4He$ ratios to shallow processes associated with the back-arc, so it appears safe to assume that the high values observed in these regions bear little or no information on the subduction process.

Whereas the majority of the Manus Basin samples (particularly from the Manus Spreading Centre and the Extensional Transform Zone) are characterized by $^3He/^4He$ ratios greater than those of MORB, this is not the case for lavas from the Lau Basin or the North Fiji Basin. Indeed, a significant number of samples from both basins have MORB-like $^3He/^4He$ values (see references in Table 4). Specifically, lavas with $^3He/^4He$

Table 4. Summary of helium isotope variations in back arc-related volcanics and hydrothermal fluids.

Location (segment)	*R/R_A (max)*	*R/R_A (min)*	*R/R_A*(mean) (± 1 S.D.)*	$N^{\$}$	*References*[1]
Lau Basin					
Rochambeau Bank	22.1	11.0	16.3 ± 5.3	5	[1-2]
Peggy Ridge	10.9	1.2	8.3 ± 3.1	8	[1-2]
Northern Basin	8.1	5.9	7.2 ± 0.7	9	[2-4]
Central Spreading Center	11.2	7.5	8.6 ± 0.8	14	[2-4]
East Spreading Center	8.9			1	[4]
Valu Fa Ridge	7.8	1.2	5.7 ± 2.6	10	{3, 5]
Total (Lau Basin)	22.1	1.2	8.5 ± 3.8	47	[1-5]
Mariana Trough					
North	8.6	5.7	7.8 ± 0.8	15	[6-7]
Central	8.5	0.85	6.8 ± 2.8	8	[1, 7]]
South	8.8	7.9	8.3 ± 0.3	15	[8]
Total (Mariana Trough)	8.8	0.85	7.8 ± 1.4	38	[1, 6-8]
North Fiji Basin					
160°N segment	8.8	8.6	8.7 ± 0.1	3	[9]
Triple Junction Area	10.0	7.9	9.1 ± 0.7	16	[9-10]
15 °N segment	9.0			1	[9]
N-S segment	10.3	3.8	8.4 ± 1.5	13	[9-11]
174 ° E segment	8.8	8.1	8.4 ± 0.9	6	[9]
Total (North Fiji Basin)	10.3	3.8	8.7 ± 1.1	36	[9-11]
Manus Basin					
Seamount	15.2			1	[12]
Extensional Transform Zone	13.5	0.92	10.1 ± 4.3	7	[12]
Manus Spreading Center	12.7	0.67	10.1 ± 4.2	17	[8, 12]
Southern Rifts	8.8	3.4		2	[12]
East Manus Rift	6.4	0.61	3.7 ± 2.4	8	[8, 12-13]
Total (Manus Basin)	15.1	0.61	8.5 ± 4.7	35	[8, 12-13]
Summary	22.1	0.61	8.4 ± 3.2	156	[1-13]

Footnotes to Table 4

* Includes only air- or ASW-corrected $^3He/^4He$ data except where no He/Ne ratios are given.

\$ N = number of individual results incorporated into mean. In the case of duplicate analysis, the higher number is selected.

1. Helium isotope data sources: 1= Poreda (1985), 2 = Poreda and Craig (1992), 3 = Hilton et al. (1993b), 4 = Honda et al. (1993b), 5 = Bach and Niedermann (1998), 6 = Ikeda et al. (1998), 7 = Sano et al. (1998), 8 = Macpherson et al. (2000), 9 = Nishio et al. (1998), 10 = Ishibashi et al. (1994), 11 = Marty and Zimmermann (1999), 12 = Macpherson et al. (1998), 13 = Marty et al. (2001).

values of 8±1 R_A are erupted in the central Lau Basin as well as along all spreading centers of the North Fiji Basin. Therefore, in addition to a contribution from a deep mantle plume, the other principal source of helium in active back-arc basins is the same as that supplying mid-ocean ridges. This conclusion reinforces that derived from the majority of island arc studies—namely, that the mantle is the predominant source of helium in subduction zone environments with little or no contribution from the subducted slab.

There are, however, a number of samples in each of the four back-arc basins that have $^3He/^4He$ values significantly lower than the MORB range. Oceanic glasses with $^3He/^4He$ ratios ≤1 R_A (implying that ≥80% of the helium is radiogenic in origin; Andrews 1985) are observed along the Peggy Ridge and Valu Fa Ridge (Lau Basin), the central Mariana Trough and throughout the Manus Basin (Table 4). Assuming that, in all cases, the glasses used in these studies were young enough to rule out post-eruptive in-growth of ^4He, then any radiogenic helium must have been inherited prior to eruption of the lavas onto the sea floor. Three explanations have been advanced to account for the occurrence of these low $^3He/^4He$ values:

1. The mantle wedge is depleted in helium so that it cannot mask/dilute radiogenic helium derived from the subducted slab (subducted sediments or underlying oceanic basement).
2. Magmas are contaminated by radiogenic helium trapped in crust through which the magmas are erupted. Contamination is accentuated by pre-eruptive degassing of magmas which lowers absolute concentrations thereby making magmas more susceptible to contamination.
3. The mantle wedge is enriched in U- and Th-series isotopes derived from the subducted slab so that the wedge itself is characterized by $^3He/^4He$ ratios < MORB.

The first explanation was suggested by Poreda (1985) to explain a small number of low $^3He/^4He$ ratios (<< MORB) at various localities in the Mariana Trough. By invoking a low concentration of mantle wedge helium, presumably caused by extensive degassing, it should be possible to observe the effects of the addition of slab-derived radiogenic helium on resultant $^3He/^4He$ ratios. In this way, Poreda (1985) reconciled the observation of low $^3He/^4He$ ratios with the high (and isotopically heavy: $\delta D \sim -40‰$) water and alkali contents that frequently characterize BABB lavas. Hilton et al. (1993b) re-interpreted the Mariana Trough data in terms of degassing of mantle-derived melts prior to assimilation of hydrated minerals and/or other alteration products in older oceanic crust (arc basement) immediately prior to eruption (second explanation above). This scenario is consistent with both the anticipated D/H and $^3He/^4He$ signature of upper crustal materials. This interpretation follows from observations of low $^3He/^4He$ ratios (and low helium contents) of differentiated lavas erupted along the Valu Fa Ridge (Lau Basin). The low helium contents make such samples susceptible to any assimilation and/or contamination effects. Bach and Niedermann (1998) challenged this (degassing-contamination) interpretation suggesting that the mantle wedge (in the Valu Fa region) is both depleted in mantle helium and metasomatized by slab-derived fluids introducing U and Th (third explanation). Therefore, over time, it would have evolved to a $^3He/^4He$ ratio slightly lower than that of MORB. Clearly, observations of back-arc basin $^3He/^4He$ ratios less than those found in MORB represent a contentious issue in noble gas isotope geochemistry with far-reaching implications for issues such as whether helium can be subducted, and how the volatile content of the mantle wedge has evolved. As more BABB helium data are produced, debate is likely to continue regarding the relative merits of these three possibilities.

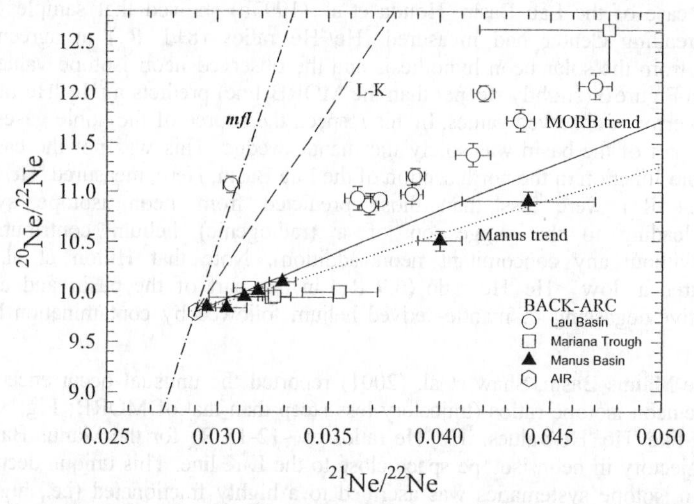

Figure 3. Neon isotope systematics of back-arc volcanism. Back-arc data from: Honda et al. 1993b (Lau Basin); Ikeda et al. 1998 and Sano et al. 1998a (Mariana Trough); and Shaw et al., 2001 (Manus Basin). Loihi-Kilauea (L-K) trend from Honda et al. (1991), Manus trend from Shaw et al. (2001). mfl = mass fractionation line.

Neon isotopes in back-arc basins

In contrast (again) to the helium isotope database, neon data in BABB are sparse. To date, neon isotope results are available for the Lau Basin (Honda et al. 1993b), the Mariana Trough (Ikeda et al. 1998; Sano et al. 1998a) and the Manus Basin (Shaw et al. 2001). In Figure 3, we plot the complete database in the traditional 3-isotope neon plot. There are three points to note:

1. Most samples from the Mariana Trough (with the exception of one outlier) plot close to the MORB correlation line.
2. The Lau Basin data lie on a linear trajectory slightly steeper than the MORB-line but nowhere near the Loihi-Kilauea (L-K) correlation lines. One sample from the Mariana Trough (which has the most extreme enrichments in $^{20}Ne/^{22}Ne$ and $^{21}Ne/^{22}Ne$ of the entire database) also appears to lie along the same trajectory.
3. The Manus Basin samples lie along a trajectory with a slope less than the MORB correlation line (the so-called 'Manus trend').

As all samples appear to follow linear trajectories in 3-isotope neon space, they must represent binary mixtures between two end-members with distinct neon isotope systematics. One end-member is clearly air reflecting the inevitable contamination of all terrestrial samples with atmospheric neon. The other end-member consists predominantly of 'mantle neon' which is itself composed of a mixture of solar neon (high $^{20}Ne/^{22}Ne \sim 13.8$) and nucleogenic neon (high $^{21}Ne/^{22}Ne$)—in various proportions.

According to the solar-neon hypothesis (Honda et al. 1993a), He and Ne isotopes are coupled and reflect addition of radiogenic ^{4}He and nucleogenic ^{21}Ne, continuously produced throughout Earth history, to primordial He and Ne captured at the time of planetary accretion. Therefore, the predicted $^{3}He/^{4}He$ ratio of those samples lying along the MORB neon correlation line is 8 R_A. The fact that the Mariana Trough samples all fall within error of the MORB $^{3}He/^{4}He$ ratio substantiates the solar-neon hypothesis and indicates negligible input of helium and neon from any reservoir other than the mantle wedge.

In the case of the Lau Basin, Honda et al. (1993b) showed that samples from the Central Spreading Center had measured $^3He/^4He$ ratios (8±1 R_A) in agreement with predictions from the solar neon hypothesis and the observed neon isotope values i.e., the trajectory in Figure 3 (slightly steeper than the MORB line) predicts a $^3He/^4He$ ratio of 8.6 R_A—within error of observed values. In this respect, the source of the noble gases (He and Ne) in this part of the basin was solely the mantle wedge. This was not the case for the King's Triple Junction in the north section of the Lau Basin. Here, measured $^3He/^4He$ ratios (5.9 to 7.4 R_A) were less than those predicted from neon isotope systematics (8.6 R_A) leading to the suggestion of a (radiogenic) helium contribution from the slab (without any concomitant neon addition). Note that Hilton et al. (1993b) also measured a 'low' $^3He/^4He$ ratio (6.9 R_A) in this part of the basin and ascribed it to pre-eruptive degassing of mantle-derived helium followed by contamination by crustal helium.

For the Manus Basin, Shaw et al. (2001) reported the unusual occurrence of highly nucleogenic neon isotope ratios (trajectory less steep than that of MORB; Fig. 3) together with plume-like $^3He/^4He$ values. $^3He/^4He$ ratios of ~12-15 R_A for the Manus Basin would predict a trajectory in neon isotope space close to the L-K line. This unique decoupling of helium-neon isotope systematics was ascribed to a highly fractionated (i.e., high He/Ne) source region resulting from either prior plume degassing—followed by in-growth, or to mantle heterogeneity inherited from the accretion process (Shaw et al. 2001). Significantly, no evidence was found to support the subducting slab, either the present-day Solomon Sea plate or ancient recycled slab, as a potential source of the nucleogenic neon (Shaw et al. 2001).

In summary, therefore, there is little or no firm evidence from neon isotopes, either taken alone or when considered along with He isotopes, for a significant input of slab-derived volatiles to back-arc basin lavas.

Argon, krypton and xenon isotopes in back-arc basins

Non-atmospheric $^{40}Ar/^{36}Ar$ ratios are found in all four back-arc basins considered above (Table 5). The highest ratio (5300) is reported for the northern section of the Lau Basin (Honda et al. 1993b) but the other three basins (Mariana Trough, North Fiji Basin and the Manus Basin) all have $^{40}Ar/^{36}Ar$ values >1000. There appears consensus on the origin of the high ratios—namely, the mantle wedge, even though the absolute ratios fall far short of MORB-type values (up to 40,000; Burnard et al. 1997). This relatively narrow range in $^{40}Ar/^{36}Ar$ ratios for BABB (296 to 5000) reflects the ubiquitous presence of air-derived argon in all samples, which acts to lower mantle $^{40}Ar/^{36}Ar$ values. Debate on the origin of the low $^{40}Ar/^{36}Ar$ ratios in back-arc basins (subduction-related or upper crustal contamination) mirrors that for arc localities.

As all samples in Table 5 are oceanic glasses, and therefore erupted in a submarine setting, the most pressing concern is seawater contamination. Honda et al. (1993b) noted a correlation between absolute abundances of krypton and xenon (which are relatively enriched in seawater) with total water content, and suggested interaction with seawater as one possible explanation for the atmospheric-like heavy noble gas systematics. They also pointed to a correlation between Kr and Xe and Ba/Nb ratio (a key slab fluid tracer), opening up the possibility that seawater interaction may have occurred in the down-going slab and not necessarily close to the site of eruption. In this way, the heavy noble gases (and by implication the low $^{40}Ar/^{36}Ar$ ratios) would be indicative of volatiles recycled through the subduction zone. Bach and Niedermann (1998) have also taken up this point as they observed the same type of correlations between absolute abundances of noble gases and both water contents and Ba/Nb ratios for samples from the Valu Fa Ridge (Lau Basin). The alternative explanation, that low $^{40}Ar/^{36}Ar$ ratios in back-arc basins trace shallow-level contamination of arc-rifted crust, was advanced by Hilton et al. (1993b) who argued that such crust has the necessary attributes of radiogenic He, low $^{4}He/^{40}Ar^*$ values (where $^{40}Ar^*$ is radiogenic argon) and low $^{40}Ar/^{36}Ar$ ratios (~ 300) resulting from ageing and modification by hydrothermal alteration. Interaction of previously-degassed primary magmas with such crust would result in the superimposition of a slab-like signature (high water, high Ba/Nb) onto crustal noble gases characteristics, and result in geochemical features such as seen along the Valu Fa Ridge. As in the case of argon isotope variations in arc-related volcanism, considerable controversy still surrounds assigning provenance to potential end-member contributors.

Table 5. Summary of $^{40}Ar/^{36}Ar$ isotope variations in back arc-related volcanics.

Location segment	$^{40}Ar/^{36}Ar$ (max)	$^{40}Ar/^{36}Ar$ (min)	$^{40}Ar/^{36}Ar$ (mean) (± 1 S.D.)	N[$]	Refs[1]
Lau Basin					
Northern Basin	5300	444	2758 ± 2337	7	[1-2]
Central Spreading Center	4900	1206	2429 ± 1140	10	[1-2]
East Spreading Center	461			1	[2]
Valu Fa Ridge	488	293	337 ± 66	11	[1, 3]
Total (Lau Basin)	5300	293	1647 ± 1686	29	[1-3]
Mariana Trough					
North	4480	295	1647 ± 1686	12	[4-5]
Central	4400	338		2	[5]]
South	2734	296	613 ± 636	15	[6]
Total (Mariana Trough)	4480	295	927 ± 1139	29	[4-6]
North Fiji Basin					
N-S segment	3339	532	1512 ± 1583	3	[7]
Total (North Fiji Basin)	3339	532	1512 ± 1583	3	[7]
Manus Basin					
Seamount	1310			1	[8]
Extensional Transform Zone	2680	296	1203 ± 1046	6	[8]
Manus Spreading Center	1764	295	846 ± 563	9	[8]
Southern Rifts	274			1	[8]
East Manus Rift	301	278	293 ± 8	7	[8-9]
Total (Manus Basin)	2680	274	770 ± 701	24	[8-9]
Summary	5300	274	1148 ± 1312	85	[1-9]

* Includes only air- or ASW-corrected $^{3}He/^{4}He$ data except where no He/Ne ratios are given.

$ N = number of individual results incorporated into mean. In the case of duplicate analysis, the higher number is selected.

Helium isotope data sources: 1 = Hilton et al. (1993b), 2 = Honda et al. (1993b), 3 = Bach and Niedermann (1998), 4 = Ikeda et al. (1998), 5 = Sano et al. (1998), 6 = Hilton et al. (unpubl), 7 = Marty and Zimmermann (1999), 8 = Shaw et al. (unpubl), 9 = Marty et al. (2001).

Finally, we note that deviations from atmospheric-like krypton isotope ratios have not been found for any back-arc basin: however, there are two reports of anomalous xenon isotope variations for the Mariana Trough. Ikeda et al. (1998) found coupled $^{134}Xe/^{130}Xe$ and $^{129}Xe/^{130}Xe$ deviations from air—similar to those found in MORB, whereas Sano et al. (1998a) reported only ^{129}Xe excesses (relative to ^{132}Xe). These signatures reinforce the idea that the mantle wedge is the principal source of volatiles in the Mariana Trough particularly where circumstances limit the amount of atmosphere-derived contributions. At all other localities, only atmospheric-like xenon isotope ratios have been found.

RECYCLING OF VOLATILES AT SUBDUCTION ZONES: A MASS BALANCE APPROACH

In the previous two sections, we considered the noble gas systematics of arc and back-arc regions. We now turn our attention to the exploitation of the noble gases in understanding volatile mass balance at subduction zones. Our focus is on the ability of noble gases to both quantify mantle-degassing rates and identify volatiles of differing provenance. Specifically, we discuss the utility of noble gases in determining the volatile output via subduction zones from various sources—both the intrinsic output of noble gases themselves as well as the other (major) volatile phases that comprise the terrestrial atmosphere.

In addition, we adopt a new approach to calculate volatile outputs for individual arc segments worldwide. This allows us to assess the volatile mass balance for a number of convergent margins worldwide. In this way, we can determine more realistically how the range in volatile input parameters affect volatile output. Summing the input versus output parameters globally, we consider the long-term recycling efficiency of both the major volatiles and the noble gases, and discuss implications for Earth evolution.

The global volatile output at arc volcanoes: The ^3He approach

There have been numerous attempts at estimating volatile fluxes associated with arc-related volcanism. In this respect, the noble gas isotope that has received most attention is ^3He as its primordial origin makes it an unambiguous tracer of mantle-derived volatiles. Therefore, if the arc flux of ^3He can be established, it would lead to the derivation of other volatile fluxes by simple measurement of the ratio $x_i/^3He$, where x_i is any volatile species discharging from volcanoes (CO_2, SO_2, H_2S, HCl, N_2, etc.). Two distinct approaches have been taken to estimate the ^3He flux from arc volcanoes:

1. Using the relatively well constrained figure of mid-ocean ridge degassing flux (~1000 mol^{3}He/yr; Craig et al. 1975), together with the assumption that the magma production rate of arcs is 20% that of MOR (Crisp 1984), an arc ^3He flux of ~200±40 mol/yr is derived (Torgersen 1989). This method further assumes that the ^3He content of magma in the mantle wedge is the same as that beneath spreading ridges.

2. Using total ^3He fluxes calculated for subaerial volcanism and estimating the fraction contributed by arc volcanism. For example, Allard (1992) derived an estimate for the total flux of ^3He into the atmosphere by subaerial volcanism to be 240-310 moles/yr (based upon integrating the CO_2 flux from 23 individual volcanoes worldwide and coupling this flux with measurements of the $CO_2/^3He$ ratios). Of the total subaerial ^3He flux, he suggested that approximately 70 mol/yr was arc-related. Adopting a similar approach, Marty and LeCloarec (1992) used polonium-210 (^{210}Po) as the flux indicator (along with the $^{210}Po/^3He$ ratio) to estimate a total subaerial volcanic ^3He flux of 150 mol/yr—of which over half (>75 mol/yr) was due to arc volcanism.

It should be noted that neither approach attempts a direct measurement of the arc ^3He flux. Whereas scaling to estimated magma production rates is used in the first instance, the second methodology relies on knowledge of an absolute flux of some chemical species from volcanoes together with a measurement of the ratio of that species to ^3He. The most widely used species to derive absolute chemical fluxes from subaerial volcanoes is SO_2 using the correlation spectrometer technique (COSPEC) (Stoiber et al. 1983). Carbon dioxide (see Brantley and Koepenick 1995) as well as ^{210}Po (Marty and LeCloarec 1992) have also been used to calibrate absolute fluxes of other species albeit to a much lesser extent.

In Table 6, we compile various estimates of the volatile flux from arcs for a number of noble gas isotopes (^3He, ^4He and ^{36}Ar) and major volatile phases (N_2, CO_2 and H_2O). The reader is referred to the literature given in the footnote for details of the methodology adopted to derive each estimate and the inherent assumptions involved. For comparison, we also compile estimates of the total subaerial volcanic fluxes. Note that all flux estimates in Table 6 are made on a global basis, i.e., the fluxes are assumed to represent the integrated output from all arcs worldwide. Although such fluxes are essential in addressing large-scale geochemical questions (e.g., mass transfer of volatiles through the upper mantle; Porcelli and Wasserburg 1995), they are of limited use in assessing the state of mass balance (input via the subducting slab versus output via the arc and back-arc) at individual arcs worldwide, and the (localized) effect of variations in the type and amount of subducted sediment on volatile output. For these questions, an estimate of volatile fluxes at individual arc segments is required. In the next section, we adopt an integrated flux approach (using SO_2) to produce volatile flux estimates but limit ourselves to distinct arc segments. A test of the success of this approach will be whether the sum of the individual arc fluxes equals that derived on a global basis (Table 6).

Volatile output at individual arcs—SO_2 and the power law distribution

In this section, we provide estimates of the ^3He (and other volatile) fluxes from individual arc segments worldwide. To achieve this objective, we use time-averaged SO_2 flux measurements from 43 passively degassing arc volcanoes (Andres and Kasgnoc 1998) coupled with our own compilation of approximately 700 volcanic and hydrothermal gas compositions. It must be borne in mind that the accuracy of individual COSPEC measurements may vary from 10-40% (Stoiber et al. 1983) and this error will propagate through to the final estimates. The time-averaged volcanic SO_2 emission rates were compiled from approximately 20,000 individual measurements dating from the early 1970s to 1997 (Andres and Kasgnoc 1998). Based on this extensive data set, Andres and Kasgnoc (1998) found that SO_2 emissions during eruptions contributed only 1% to the total volcanic SO_2 flux. This conclusion supports earlier work of Berresheim and Jaeschke (1983) who showed that 90% of the SO_2 is emitted during non-eruptive events. Although high SO_2 flux rates undoubtedly occur during eruptive periods, most eruptions are short-lived and extrapolation of the flux rates to longer time intervals are probably questionable. Therefore, in our treatment of SO_2 fluxes in this section, we feel justified in using fluxes which were derived for passively degassing volcanoes only. Furthermore, we recognize that it is impossible to measure the emissions from every single volcano; therefore, this necessitates adoption of an extrapolation procedure to account for the (small) SO_2 flux that is not measured directly. We have followed the methodology of Brantley and Koepenick (1995) who found that if the distribution of volcanic emissions follows a power law, then there is an empirical relationship between the cumulative number of volcanoes (N) having or exceeding a given emission rate (f) and the emission rate itself. The relationship takes the form:

$$N = a_* (f)^{-c} \quad (1)$$

where N is the number of volcanoes having an emission rate ≥f and a and c are constants (the value of a is determined by the choice of units of f). If the constant c < 1, the total volcanic flux (f_{tot}) can be calculated using the following approximation:

$$f_{tot} = f_1 + f_2 + \ldots\ldots f_N \; [\, (c/\{1-c\})*(N+1)*(N/\{N+1\})^{1/c}] \quad (2)$$

Table 6. Compilation of volatile fluxes (mol/yr) of arc-related and global subaerial volcanism.

Species	*Arc Flux*	*Reference*	*Global Subaerial Flux*	*Reference*
^{3}He	200 ± 40	1	275 ± 35	1
	70 ± 25	2	240-310	2
	>75	3	150	3
			3-150	4
	92	5		
^{4}He	2.0 ± 0.13 (×10^7)	1	3.75 ± 1.35 (×10^8)	1
	1.23×10^7	5		
^{40}Ar	1.9×10^8	6		
^{36}Ar	4.7×10^5	6		
CO_2	3.1×10^{12}	7	0.77 ± 0.58 (×10^{12})	10
	0.3 ± 0.2 (×10^{12})	4	1.8×10^{12}	11
	0.5 ± 0.4 (×10^{12})	10	1.5×10^{12}	2
	0.7×10^{12}	2	3.3×10^{12}	3
	1.5×10^{12}	8	2.5 ± 0.5 (×10^{12})	12
	2.5×10^{12}	9	1.1×10^{12}	13
			5.5×10^{12}	9
	1.6×10^{12}	14		
N_2	6.4×10^8	15	2.8×10^9	15
	2.0×10^{10}	16		
H_2O	8.0×10^{12}	17	55-550×10^{12}	13

Reference	Method
1. Torgersen (1989)	Scaling to MOR ^{3}He flux
2. Allard (1992)	^{3}He/CO_2 ratios plus CO_2 flux
3. Marty and Le Cloarec (1992)	^{210}Po/SO_2 – CO_2/SO_2 and CO_2/^{3}He ratios
4. Marty et al.(1989)	CO_2/^{3}He and CO_2/SO_2 ratios plus SO_2 flux
5. This work (Table 9)	Summation of individual arcs (arc = 5.4R_A)
6. This work (Table 9)	Summation of individual arcs (arc ^{40}Ar/^{36}Ar = 401)
7. Sano and Williams (1996)	C/^{3}He plus ^{3}He flux
8. Varekamp et al. (1992)	C/^{3}He plus mass flux
9. Marty B. and Tolstikhin I. N. (1998)	CO_2/^{3}He ratios plus magma emplacement
10. Williams et al. (1992)	C/S plus S flux
11. Gerlach (1991)	Flux measurement
12. Brantley SL and Koepenick K. W. (1995)	Flux measurement
13. LeGuern F. (1982)	Flux measurement
14. This work (Table 9)	Summation of individual arcs
15. Sano et al. (2001)	N/^{3}He plus ^{3}He flux
16. This work (Table 9)	Summation of individual arcs
17. This work	H_2O/CO_2 = 50 and CO_2 flux from ref. 14

where f_1 is the largest flux, f_2 is the second largest flux, and f_N refers to the Nth largest flux. Brantley and Koepenick (1995) found that the global emission of SO_2 is governed by a power law distribution (c = 0.8) so that an estimate of the total SO_2 flux can be made using a relatively small number of volcanoes. They cautioned, however, that only larger volcanoes ($f \geq 320$ Mmol/yr) followed a power law distribution.

In spite of this caveat, we have attempted to use the power law distribution to estimate the SO_2 fluxes for individual arcs worldwide. Andres and Kasgnoc (1998) have tabulated the SO_2 flux for 49 volcanoes representing 11 distinct arc systems, and we have used these estimates to derive the total fluxes for each of these arcs. We adopt two different approaches:

1. We have assumed that the global value of c (= 0.8; Brantley and Koepenick 1995) is applicable to each individual arc segment. Therefore, we can calculate f_{tot} from Equation (2) above and the tabulated fluxes for volcanoes from each arc.
2. We calculate a different c-value for use with each individual arc segment. In the case of Central America, for example, we plot log (N) versus log (f_{SO2}) for 11 volcanoes, and assuming a linear correlation (power law distribution), derive c from the gradient (Fig. 4). This value of c is used in Equation (2) to derive the total flux for that arc.

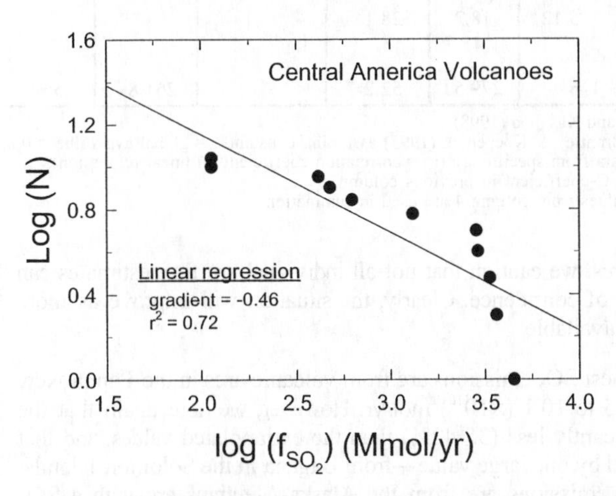

Figure 4. Log-log plot of cumulative frequency of number of Central American volcanoes (N) having SO_2 flux equal or greater than f (following methodology of Brantley and Koepenick 1995). Data from Andres and Kasgnoc (1998).

In Table 7, we show the results for 11 arc segments using these two approaches. Summing the individual arc fluxes, we obtain SO_2 flux estimates (in mol/yr) of 2.80×10^{11} (c = 0.8) and 2.62×10^{11} (various c-values). Both values compare well with previous estimates using variants of the same (power law) approach—$2\text{-}3 \times 10^{11}$ mol/yr (Brantley and Koepenick 1995); 2.92×10^{11} mol/yr (Stoiber et al. 1987) and 2.4×10^{11} mol/yr (Berresheim and Jaeschke 1983). Although the agreement is encouraging, one point of concern is that there appears a large discrepancy between summing the individual fluxes at some arcs and the total (extrapolated) flux. For example, the sum of the measured fluxes for the Andes lies between 43% and 60% of the total extrapolated flux (depending upon the chosen c-value). For comparison, Andes and Koasgnoc (1998) found that—on a global basis—the 'additional' extrapolated flux amounted to 19% of the total i.e., the sum of the measured fluxes was 81% of the total flux. It is noteworthy that calculating c-values for individual arcs seems to decrease the discrepancy between summed and extrapolated fluxes. Additionally, it seems that the large extrapolation appears to affect mostly arcs with small numbers of volcanic SO_2 fluxes (e.g., Andes, Antilles, New Zealand, Philippines; see Table 7). Therefore, while accepting the flux extrapolations give a useful first-order estimate

Table 7. Summary of SO_2 fluxes ($\times10^9$ mol/yr)—measured (meas) and extrapolated (extrap) using power-law distribution.

Arc Segment	*# volcanoes SO_2 flux*[1]	*Σ meas fluxes*	*Extrap flux*[2]	*% meas extrap*	*C-coefficient*[3] *(r^2-value)*	*Extrap flux*[4]	*% meas extrap*
Andes	3	28.2	65.9	42.8	-0.70 (0.77)	47.3	59.6
Alaska-Aleutians	3	0.41	0.56	73.3	-0.34 (0.84)	(0.34)	-
Antilles	1	1.31	4.41	29.7	-	-	-
Central America	11	20.3	25.1	80.9	-0.46 (0.72)	21.3	95.5
Indonesia	4	1.64	2.76	59.2	-0.55 (0.80)	1.88	86.9
Italy	3	27.2	29.8	91.4	-0.23 (0.92)	27.0	-
Japan	8	20.0	29.6	67.6	-0.60 (0.98)	23.3	85.9
Kamchatka-Kuriles	3	1.64	2.51	65.5	-0.41 (0.92)	1.67	98.2
New Zealand	2	4.67	15.3	30.5	>	-	-
Papua New Guinea	5	35.8	85.4	42.0	-0.84 (0.92)	100.9	35.5
Philippines	2	5.12	18.2	28.1	-	-	-
GRAND TOTAL	45	146	279.5	52.2		261.8[5]	55.8

1. SO_2 flux measurements from Andres and Kasgnoc (1998)
2. Extrapolated flux using methods of Brantley & Koepenick (1995) assuming constant c = global avg value = 0.8.
3. Recalculated c-constant using only data from specific arc (r^2 = correlation coefficient on linear regression).
4. Recalculated extrapolated flux using C-coefficient in previous column.
5. Where missing (or in parenthesis), values from column 4 are used in summation.

of the SO_2 flux for individual arcs, we caution that not all individual arc flux estimates can be treated with the same degree of confidence. Clearly, the situation will improve as more SO_2 flux measurements become available.

It is interesting that the highest SO_2 emissions are from volcanic arcs in the Papua New Guinea -SW Pacific region—8.5 to 10.1 ($\times10^{10}$) mol/yr. However, we note again that the sum of individual flux is significantly less (35-42%) than the extrapolated values, and that the flux extrapolation is weighted by one large value—from Bagana in the Solomon Islands. At the other extreme, the lowest emissions are from the Alaska-Aleutians arc with a SO_2 flux of only 5.6×10^8 mol/yr, or less than 1% of the largest value.

In Table 8 we report a compilation of volcanic gas chemistries (as molar ratios) using over 700 individual analyses of arc-related volcanic and hydrothermal fluids. In an attempt to obtain a more representative picture of the chemical variability, we present the database as median ratios which gives less weight to extreme values. In this compilation, S_t is the total S, i.e., $SO_2 + H_2S$, and $N_{2,(ex)}$ is the amount of "excess" N_2 after correction for N_2 contributed from meteoric sources in the volcanic edifice. The "excess" N_2 was calculated using:

$$[N_2]_{ex} = [N_2]_{measured} - (40 \times [Ar]_{measured}) \quad (3)$$

where square brackets designate concentration, and the numeric value 40 implies that the measured argon is derived solely from air-saturated water in equilibrium with the atmosphere, i.e., $N_2/Ar = 40$ (see Fischer et al. 1998). With reference to Table 8, there are two outliers in the CO_2/S_t column—the Philippines arc (104.5) and New Zealand (28.7)—which are both significantly higher than typical 'arc-like' values of ~5 (Giggenbach 1996). We suggest that the available database in these two cases may be biased to mainly low-temperature hydrothermal fumaroles and geothermal fluids that would act to preferentially remove sulfur species (Giggenbach 1996). Therefore, in our calculations of volatile fluxes (below), we assume a value of 5 for the CO_2/S_t ratio of both these arcs.

Table 8. Compilation of gas chemistries (median molar ratios) for 10 arc segments worldwide.

Arc Segment	*# Analyses*	*# Volcanic centers*	*CO_2/S_t*	*CO_2/N_2*	*$CO_2/N_2(ex)$*	*CO_2/He ($\times 10^5$)*	*CO_2/Ar ($\times 10^3$)*	*Refs.*
Andes	70	5	5.7	61.1	73.2	2.92	10.7	[1-4]
Alaska-Aleutians	15	6	6.1	33.8	72.8	2.00	4.61	[5-7]
Lesser Antilles	44	2	5.0	64.6	-	1.26	n.a.	[8-10]
Central America	94	10	2.7	34.1	195.4	0.81	8.96	[11-18]
Indonesia (Sunda)	42	10	4.3	60.9	95.9	1.47	7.28	[19-25]
Italy	70	13	8.5	95.9	62.3	1.45	72.0	[12, 26-31]
Japan	83	9	6.5	18.9	24.1	0.76	4.39	[15, 32-46]
Kamchatka-Kuriles	37	16	1.7	14.1	23.8	0.95	2.82	[47-59]
New Zealand	214	84	28.7	53.0	102.3	1.05	4.03	[60-62]
Philippines	48	9	104.5	115.4	140.0	0.90	13.6	[12, 63-65]
GRAND TOTAL	721	165						

Footnotes for Table 8

Gas compositional data sources:

1 = Fischer et al. (1997), 2 = Lewicki et al. (2000), 3 = Sturchio et al. (1993), 4 = Giggenbach et al. (1986), 5= Symonds et al. (1990), 6 = Sheppard et al. (1992), 7 = Motyka et al. (1993), 8 = Chiodini et al. (1996), 9 = Hammouya et al. (1998), 10 = Brombach et al. (2000), 11 = Menyailov et al. (1986a), 12 = Giggenbach (1992), 13 = Taran (1992), 14 = Rowe et al. (1992), 15 = Goff and McMurtry, (2000), 16 = Giggenbach and Corrales (1992), 17 = Janik et al. (1992), 18 = Taran et al. (1998), 19 = Allard (1983), 20 = Giggenbach et al. (2001), 21 = LeGuern (1982), 22 = Poorter et al. (1993), 23 = Fischer unpubl., 24 = Delmelle et al., (2000), 25 = Sriwana et al. (2000), 26 = Chiodini et al. (1995), 27 = Cioni and D'Amore (1984), 28 = Goff unpublished, 29 = Rogie et al. (2000), 30 = Minissale et al. (1997), 31 = Chiodini (1994), 32 = Mizutani and Sugiura (1982), 33 = Symonds et al. (1994), 34 = Giggenbach et al. (1986), 35 = Nemoto (1957), 36 = Mitzutani (1962), 37 = Mitzutani and Sugiura (1982), 38 = Symonds et al. (1996), 39 = Mitzutani (1966), 40 = Giggenbach and Matsuo (1992), 41 = Mizutani et al. (1986), 42 = Matsuo et al. (1974), 43 = Shinohara et al. (1993), 44 = Ohba et al. (1994), 45 = Ohba et al. (2000), 46 = Kiyosu and Kurahashi (1984), 47 = Fischer et al. (1998), 48 = Taran et al. (1991), 49 = Menyailov and Nikintina, (1980), 50 = Kirsanova et al. (1983), 51 = Taran et al. (1987), 52 = Taran (1985), 53 = Taran and Korbalev (1995), 54 = Taran et al. (1992), 55 = Menyailov et al. (1988), 56 = Menyailov et al. (1986b), 57 = Taran (1992), 58 = Taran et al. (1997), 59 = Taran et al. (1995), 60 = Giggenbach (1995), 61 = Christensen (2000), 62 = Giggenbach et al. (1993), 63 = Reyes et al. (1993), 64 = Delmelle et al. (1998), 65 = Giggenbach and Poreda (1993).

Table 9. Compilation of gas fluxes (mol/yr) for 11 arc segments worldwide.

Arc Segment	SO_2[1] (× 10^9)	CO_2 (× 10^9)	$N_{2\ total}$ (× 10^8)	$N_{2\ (ex)}$[2] (× 10^8)	He (× 10^4)	^3He[3]	Ar (× 10^6)
Andes	47.3	268.8	44.0	36.7	91.9	6.91	25.1
Alaska-Aleutians	0.56	3.43	1.01	0.47	1.71	0.13	0.74
Lesser Antilles	4.41	21.8	3.38	-	17.4	1.31	-
Central America	21.3	57.5	16.9	2.94	71.3	5.36	6.42
Indonesia (Sunda)	1.88	8.13	1.33	0.85	5.53	0.42	1.12
Italy	27.0	230.0	24.0	-	158.8	11.9	3.20
Japan	23.3	150.8	79.6	62.6	198.5	14.9	34.3
Kamchatka-Kuriles	1.67	2.77	1.97	1.16	2.90	0.22	0.98
New Zealand**	15.3	76.6	14.4	7.48	73.2	5.50	19.0
Papua New Guinea/ SW Pacific*	100.9	435.8	71.5	45.4	296.4	22.3	59.9
Philippines**	18.2	91.3	7.9	6.52	101.5	7.63	6.73
TOTAL (11 arcs)	261.8	1346.9	266.0	164.1	1019.1	76.6	157.5
TOTAL (Global)&	315.7	1624.2	320.8	197.9	1228.9	92.4	189.9

1. From Table 7 (column 7)
2. Excess N_2 calculated using equation 3 (see text).
3. Calculated using average arc $^3He/^4He$ ratio = 5.37 R_A (Table 1)

* Flux calculations assumes Indonesia chemistry (from Table 8)

** Assumes CO_2/SO_2 molar ratio = 5

& Assumes additional 20.6% to all fluxes (see text)

In Table 9 we couple our estimates of the SO_2 flux at various arcs (Table 7) with our compilation of the gas chemistry (Table 8) to produce an estimate of the fluxes of a number of volatile species (CO_2, N_2, $N_{2,ex}$, ^4He, ^3He and Ar) for 11 arcs worldwide. The largest emitter of volatiles (in an absolute sense) is the arc systems of Papua New Guinea and the South-west Pacific although it must be cautioned that these fluxes are derived using data from the Indonesia arcs—no gas chemistry is yet available for these regions. The northern Pacific arcs (Alaska-Aleutians and Kamchatka-Kuriles) are relatively poor emitters of volatiles, and this presumably reflects the virtual absence of carbonate sediment in this region (see Table 10).

The flux estimates in Table 9 are based on measured SO_2 emissions of 45 individual arc volcanoes from 11 regions. However, a significant number of arcs have only little or no SO_2 flux data—these include South Sandwich, Mexico, Cascadia, Izu-Bonin, Ryuku, Marianas, Makran, Andanan, East Sunda, Vanuatu and Tonga. This paucity of data is somewhat surprising as there are a total of 106 historically active volcanoes in these regions (Simkin and Siebert 1994). If we make a rather crude analogy to Central America, where approximately one-third of the 32 historically active volcanoes emit measurable SO_2, we would expect ~35 of these 106 volcanoes to contribute to the global SO_2 flux. These 35 volcanoes would be expected to contribute a total of ~0.54×10^{11} mol/yr if the median flux of the measured volcanoes (1.54×10^9 mol/yr) is an appropriate figure for their individual SO_2 fluxes. This figure can be added to the sum of SO_2 fluxes for the eleven measured arcs (2.62×10^{11} mol/yr) to yield an estimate of 3.2×10^{11} mol/yr for the total flux of SO_2 from all arc volcanoes worldwide.

Table 10. Volatile input at subduction zones.

Trench	Tonga	Kerm	Vanuatu	E.Sunda	Java	Sumatra	Andaman	Makran	Philip	Ryuku	Nankai	Marianas	Izu-Bon	Japan
Subd rate (mm/yr)	170	70	103	67	67	50	30	35	90	60	30	47.5	50	105
Trench Length (km)	1350	1400	1800	1000	2010	1000	1500	950	1550	1350	800	1400	1050	800
sediment subducted (g/y)	2.11E+13	2.74E+13	1.93E+14	5.74E+13	6.69E+13	1.36E+14	3.22E+14	2.86E+14	2.20E+13	1.81E+13	1.85E+13	5.89E+13	5.85E+13	4.82E+13
Oceanic crust subducted(g/yr)	4.64E+15	1.98E+15	3.75E+15	1.36E+15	2.72E+15	1.01E+15	9.10E+14	6.73E+14	2.82E+15	1.64E+15	4.86E+14	1.35E+15	1.06E+15	1.70E+15
FLUXES (Mmol/yr)														
H_2O_{sed}	3.68E+04	5.70E+04	3.28E+05	1.77E+05	1.81E+05	4.13E+05	1.24E+06	6.69E+05	2.99E+04	4.17E+04	5.14E+04	1.71E+05	1.22E+05	1.64E+05
H_2O_{Crust}	3.98E+04	5.18E+04	3.64E+05	1.08E+05	1.26E+05	2.58E+05	6.08E+05	5.39E+05	4.15E+04	3.43E+04	3.49E+04	1.11E+05	1.11E+05	9.10E+04
$H_2O_{Available}$	1.41E+04	1.83E+04	1.29E+05	3.83E+04	4.46E+04	9.09E+04	2.15E+05	1.90E+05	1.47E+04	1.21E+04	1.23E+04	3.93E+04	3.90E+04	3.21E+04
$CO_{2\ sed}$	0.00E+00	0.00E+00	1.06E+05	5.98E+04	9.68E+03	3.35E+03	3.01E+03	2.63E+05	1.61E+04	0.00E+00	0.00E+00	2.69E+04	3.85E+04	0.00E+00
CO_{2org}	4.79E+03	6.24E+03	4.38E+04	1.30E+04	1.52E+04	3.10E+04	7.32E+04	6.49E+04	4.99E+03	4.12E+03	4.20E+03	1.34E+04	1.33E+04	1.09E+04
$CO_{2\ crust}$	2.26E+05	9.64E+04	1.82E+05	6.59E+04	1.33E+05	4.92E+04	4.43E+04	3.27E+04	1.37E+05	7.97E+04	2.36E+04	6.54E+04	5.17E+04	8.26E+04
$N_{2\ sed}$	1.51E+02	1.96E+02	1.38E+03	4.10E+02	4.78E+02	9.74E+02	2.30E+03	2.04E+03	1.57E+02	1.30E+02	1.32E+02	4.21E+02	4.18E+02	3.44E+02
N_{2crust}	3.32E+03	1.42E+03	2.68E+03	9.68E+02	1.95E+03	7.23E+02	6.50E+02	4.80E+02	2.02E+03	1.17E+03	3.47E+02	9.61E+02	7.59E+02	1.21E+03
He_{sed}	9.28E-04	1.21E-03	8.48E-03	2.53E-03	2.94E-03	6.00E-03	1.42E-02	1.26E-02	9.67E-04	7.98E-04	8.13E-04	2.59E-03	2.58E-03	2.12E-03
He_{crust}	2.04E-01	8.72E-02	1.65E-01	5.96E-02	1.20E-01	4.45E-02	4.01E-02	2.96E-02	1.24E-01	7.21E-02	2.14E-02	5.92E-02	4.67E-02	7.48E-02
$^{36}Ar_{sed}$	2.53E-05	3.29E-05	2.31E-04	6.89E-05	8.02E-05	1.64E-04	3.87E-04	3.43E-04	2.64E-05	2.18E-05	2.22E-05	7.07E-05	7.02E-05	5.78E-05
$^{36}Ar_{crust}$	6.50E-04	2.78E-04	5.25E-04	1.90E-04	3.81E-04	1.42E-04	1.27E-04	9.42E-05	3.95E-04	2.29E-04	6.80E-05	1.88E-04	1.49E-04	2.38E-04

Footnotes for Table 10

Subduction Rate: from Plank and Langmuir (1998) and references therein.
Trench Length: from Plank and Langmuir (1998) and references therein.
Sediment subducted: calculated using subduction rate, trench length, sediment thickness and density for each arc (Plank and Langmuir 1998).
Oceanic crust subducted: calculated using subduction rate, trench length (Plank and Langmuir, 1998) and assuming a thickness of the oceanic crust of 7 km and a density of 2.89 g/cm^3.

Table 10, continued. Volatile input at subduction zones.

Trench	Kurile	Kamchat	Aleut	Alaska	Cascadia	Mexico	Centam	Colomb	Peru	SSand	N.Ant	S. Ant	TOTAL (mol/a)
Subd rate (mm/yr)	90	90	62	70	35	52	77	70	100	20	24	24	
Trench Length (km)	1650	550	1900	800	1300	1450	1450	1050	1500	800	400	400	
sediment subducted (g/y)	8.51E+13	2.95E+13	6.76E+13	7.87E+13	1.28E+14	1.75E+13	7.70E+13	3.25E+13	2.56E+13	4.77E+12	3.76E+12	3.16E+13	
Oceanic crust subducted(g/yr)	3.00E+15	1.00E+15	2.38E+15	1.13E+15	9.20E+14	1.53E+15	2.26E+15	1.49E+15	3.03E+15	3.24E+14	1.94E+14	1.94E+14	
FLUXES (Mmol/yr)													
H_2O_{sed}	2.91E+05	1.19E+05	1.76E+05	1.86E+05	4.58E+05	3.99E+04	1.01E+05	7.31E+04	4.06E+04	9.44E+03	1.66E+04	1.14E+05	5.30E+12
H_2O_{Crust}	1.61E+05	5.57E+04	1.28E+05	1.49E+05	2.42E+05	3.31E+04	1.45E+05	6.14E+04	4.84E+04	9.01E+03	7.09E+03	5.97E+04	3.62E+12
$H_2O_{Available}$	5.68E+04	1.97E+04	4.51E+04	5.25E+04	8.55E+04	1.17E+04	5.13E+04	2.17E+04	1.71E+04	3.18E+03	2.50E+03	2.11E+04	1.28E+12
$CO_{2\ sed}$	0.00E+00	0.00E+00	0.00E+00	0.00E+00	0.00E+00	0.00E+00	2.38E+05	9.62E+04	4.41E+04	0.00E+00	3.18E+02	4.26E+02	9.06E+11
CO_{2org}	1.93E+04	6.71E+03	1.54E+04	1.79E+04	2.92E+04	3.98E+03	1.75E+04	7.39E+03	5.82E+03	1.08E+03	8.53E+02	7.18E+03	4.35E+11
$CO_{2\ crust}$	1.46E+05	4.87E+04	1.16E+05	5.51E+04	4.48E+04	7.42E+04	1.10E+05	7.23E+04	1.48E+05	1.57E+04	9.45E+03	9.45E+03	2.12E+12
$N_{2\ sed}$	6.08E+02	2.11E+02	4.83E+02	5.62E+02	9.17E+02	1.25E+02	5.50E+02	2.32E+02	1.83E+02	3.41E+01	2.68E+01	2.26E+02	1.37E+10
N_{2crust}	2.15E+03	7.15E+02	1.70E+03	8.09E+02	6.57E+02	1.09E+03	1.61E+03	1.06E+03	2.17E+03	2.31E+02	1.39E+02	1.39E+02	3.11E+10
He_{sed}	3.75E-03	1.30E-03	2.97E-03	3.46E-03	5.65E-03	7.70E-04	3.39E-03	1.43E-03	1.13E-03	2.10E-04	1.65E-04	1.39E-03	8.43E+04
He_{crust}	1.32E-01	4.41E-02	1.05E-01	4.98E-02	4.05E-02	6.71E-02	9.94E-02	6.54E-02	1.34E-01	1.42E-02	8.55E-03	8.55E-03	1.92E+06
$^{36}Ar_{sed}$	1.02E-04	3.54E-05	8.11E-05	9.44E-05	1.54E-04	2.10E-05	9.24E-05	3.90E-05	3.07E-05	5.72E-06	4.51E-06	3.79E-05	2.30E+03
$^{36}Ar_{crust}$	4.21E-04	1.40E-04	3.34E-04	1.59E-04	1.29E-04	2.14E-04	3.16E-04	2.08E-04	4.25E-04	4.53E-05	2.72E-05	2.72E-05	6.10E+03

Fluxes: Calculated using amount of sediments subducted or amount of oceanic crust subducted and concentrations of volatile species in the sediments or oceanic crust listed below.

H_2O sed: water in the sediments, excluding pore waters. Calculated using the amount of water stored in the sediments for each individual arc (Plank and Langmuir, 1998)

H_2O crust: amount of water stored in the oceanic crust - 3.4 wt% (Schmidt and Poli 1998).

H_2O available: amount of water available (1.2 wt%) in the zone of arc magma generation (Schmidt and Poli 1998)

CO_2 sed: amount of (carbonate) CO_2 in sediments of each individual arc (Plank and Langmuir 1998)

CO_2 org: amount of reduced organic CO_2 stored in sediments. An average concentrations of 1 wt% is used (Bebout 1995).

CO_2 crust: amount of CO_2 stored in the oceanic crust (carbonate CO_2). A value of 0.214 wt% is used (Alt and Teagle 1999).

N_2 sed: amount of N_2 in oceanic sediments - 0.01 wt% (Bebout 1995).

N_2 crust: amount of N_2 in oceanic crust - 0.001 wt% (Bebout 1995).

He sed: amount of ^4He in oceanic sediments. An average value of 1.0×10^{-6} cm^3 STP/g is used (Staudacher and Allègre 1988).

He crust: amount of ^4He in oceanic crust. An average value of 1.0×10^{-6} cm^3 STP/g is used (Staudacher and Allègre 1988).

^{36}Ar sed: amount of ^{36}Ar in oceanic sediments. An average value of 2.7×10^{-8} cm^3 STP/g is used (Staudacher and Allègre 1988).

^{36}Ar crust: amount of ^{36}Ar in oceanic crust. An average value of 3.2×10^{-9} cm^3 STP/g is used (Staudacher and Allègre 1988).

To estimate the global flux of the volatiles other than SO_2 we need to make the assumption that the additional SO_2 added (~20%) scales proportionally to the other volatile species. In effect, the volatile ratios (e.g., CO_2/S_t etc) of the unmeasured volcanoes must equal the average values of the eleven arcs where data is available. Under this assumption, global estimates of various volatile fluxes are given in the last row of Table 9.

With respect to ^4He (Table 7), we note that the total global arc flux is 1.2×10^7 mol/yr, which gives a total ^3He flux of approximately 92 mol/yr—for a mean arc $^3He/^4He$ ratio of 5.4 R_A (from Table 1). This estimate of the ^3He flux from arcs falls within a factor of 2 of that by Torgersen (1989) but agrees within error with two other estimates—Allard (1992) estimated 70 mol/yr and Marty and LeCloarec (1992) 75 mol/yr (Table 6). This consistency bodes well for our approach of assuming a power law distribution to volatile fluxes at individual arcs, and for our admittedly crude method of estimating fluxes from non-represented arcs. Similarly, our estimate for the total arc flux of CO_2 (1.6×10^{12} mol/yr) shows remarkable agreement with other published values (Table 6). As we discuss in the next section, our estimate of the global arc N_2 flux is higher than that of Sano et al. (2001) due the availability of a significantly more extensive (and presumably more representative) database of gas chemistries.

Using helium to resolve volatile provenance

Volatile flux estimates derived in the previous section, both for arcs individually as well as arc-related volcanism globally, make no distinction as to the source or provenance of the volatiles. However, in order to assess the chemical mass balance between output at arcs and input associated with the subducting slab, the total arc output flux must be resolved into its component structures. In this way, the fraction of the total output that is derived from the subducted slab can be quantified and compared with estimates of the input parameter. As we show in this section, helium has proven remarkably sensitive in discerning volatile provenance. We use CO_2 and N_2 to illustrate the case.

CO_2 provenance at arcs. The $CO_2/^3He$ ratio is significantly higher in arc-related terrains compared to mid-ocean ridge (MOR) spreading centers. MOR spreading centers have $CO_2/^3He$ ratios $\sim2\times10^9$ (Marty and Jambon 1987; Marty and Tolstikhin 1998) whereas island arcs have $CO_2/^3He$ values $\geq 10^{10}$ (Marty et al. 1989; Varekamp et al. 1992, Sano and Marty 1995, Sano and Williams 1996; Van Soest et al. 1998). Such high values have been used to argue for addition of slab carbon to the source region of arc volcanism. However, in addition to the slab (both the sedimentary veneer and underlying oceanic basement), there are other potential contributors to the total carbon output—the mantle wedge and/or the arc crust through which magmas traverse en route to the surface. Distinguishing between these various sources is possible by considering carbon and helium together (both isotopic variations and relative abundances).

In the first instance, a number of workers (Varekamp et al. 1992; Sano and Marty 1995; Sano and Williams 1996) approximated the C-output at arcs using a three end-member model consisting of MORB mantle (M), and slab-derived marine carbonate/limestone (L) and (organic) sedimentary components (S). Sano and Marty (1995) used the following mass balance equations:

$$(^{13}C/^{12}C)_o = f_M(^{13}C/^{12}C)_M + f_L(^{13}C/^{12}C)_L + f_S(^{13}C/^{12}C)_S \quad (4)$$

$$1/(^{12}C/^3He)_o = f_M/(^{12}C/^3He)_M + f_L/(^{12}C/^3He)_L + f_S/(^{12}C/^3He)_S \quad (5)$$

$$f_M + f_S + f_L = 1 \quad (6)$$

where O = observed and f is the fraction contributed by L, S and M to the total carbon output. It should be noted that application of these equations first involves conversion of carbon isotope ratios (in the δ-notation) to absolute $^{13}C/^{12}C$ values (see Sano and Marty 1995). Then it is possible to determine the relative proportions of M-, L- and S-derived carbon in individual samples of arc-related geothermal fluids. Appropriate end-member compositions must be selected, and both Sano and Marty (1995) and Sano and Williams (1996) suggest $\delta^{13}C$ values of -6.5‰, 0‰, and -30‰ (relative to PDB) with corresponding $CO_2/^3He$ ratios of 1.5×10^9; 1×10^{13} and 1×10^{13} for M, L, and S respectively. The fractions derived for M-, L- and S-derived carbon are particularly sensitive to the choice of the mantle $CO_2/^3He$ ratio as well as the sedimentary $\delta^{13}C$ value.

Based on the analysis of arc-related geothermal samples from 30 volcanic centers worldwide and utilizing high, medium and low temperature fumaroles, Sano and Williams (1996) estimated that between 10 and 15% of the arc-wide global CO_2 flux is derived from the mantle wedge—the remaining 85-90% coming from decarbonation reactions involving subducted marine limestone, slab carbonate and pelagic sediment. Subducted marine limestone and slab carbonate supply the bulk of the non-mantle carbon—approximately 70-80% of the total carbon—the remaining ~10-15% is contributed from subducted organic (sedimentary) carbon. With this approach, therefore, it is possible to attempt more realistic volatile mass balances at arcs. It is noteworthy, however, that most studies ignore the arc crust as a potential source of carbon. Although this omission may not be significant in intra-oceanic settings, this is unlikely to be the case at all localities (see discussion regarding the Lesser Antilles; see next section).

N_2 provenance at arcs. Using an approach analogous to that for carbon, Sano and co-workers (Sano et al. 1998b, 2001) have recently directed attention at understanding the nitrogen cycle at subduction zones. Again, the problem is to identify and quantify the various contributory sources to the volcanic output: however, a major concern in this case is atmospheric nitrogen. There are three major sources of nitrogen at subduction zones: the mantle (M), atmosphere (A) and subducted sediments (S), and each has a diagnostic $\delta^{15}N$ value and $N_2/^{36}Ar$ ratio. Therefore, observed (O) variations in these two parameters for individual samples can be resolved into their component structures using the following equations (Sano et al. 1998b):

$$(\delta^{15}N)_o = f_M(\delta^{15}N)_M + f_A(\delta^{15}N)_A + f_S(\delta^{15}N)_S \tag{7}$$

$$1/(N_2/^{36}Ar)_o = f_M/(N_2/^{36}Ar)_M + f_A/(N_2/^{36}Ar)_A + f_S/(N_2/^{36}Ar)_S \tag{8}$$

$$f_M + f_S + f_A = 1 \tag{9}$$

where f_M is the fractional contribution of mantle-derived nitrogen, etc. Note that the noble gas isotope ^{36}Ar is used in this case since degassing is not expected to fractionate the $N_2/^{36}Ar$ ratio due to the similar solubilities of nitrogen and argon in basaltic magma—this is not the case for helium and nitrogen, and degassing corrections may be necessary if the $N_2/^3He$ ratio is used (Sano et al. 2001). In the above scheme, end-member compositions are generally well constrained: the mantle and sedimentary end members both have $N_2/^{36}Ar$ ratios of 6×10^6 (air is 1.8×10^4) but their $\delta^{15}N$ values are distinct. The upper mantle has a $\delta^{15}N$ value of -5±2‰ (Marty and Humbert 1997, Sano et al. 1998b) whereas sedimentary nitrogen is assumed to be +7±4‰ (Bebout 1995, Peters et al. 1978) (air has $\delta^{15}N$ = 0 ‰). The wide difference in $\delta^{15}N$ between the potential end-members makes this approach a sensitive tracer of N_2 provenance.

Gas discharges from island arc volcanoes and associated hydrothermal systems have $N_2/^{36}Ar$ ratios that reach a maximum of 9.7×10^4, with $\delta^{15}N$ values up to +4.6‰ (Sano et al. 2001). This would indicate that a significant proportion (up to 70%) of the N_2 could be derived from a subducted sedimentary or crustal source. The situation is reversed in the case of BABB glasses, which have significantly lower $\delta^{15}N$ values (-2.7 to +1.9 ‰): this implies that up to 70% of the nitrogen could be mantle-derived. A first-order conclusion from this observation is that N_2 is efficiently recycled from the subducting slab to the atmosphere and hydrosphere through arc and back-arc volcanism, with the flux probably weighted towards the arc flux. After correction for the atmospheric contribution, Sano et al. (1998b) estimate the total flux of N_2 through subduction zones (arc and back-arc regions) worldwide at 6×10^8 mol/yr (by normalizing to the arc ^3He flux). Interestingly, this value is almost 40-times lower than our estimate of N_{ex} output (2×10^{10} mol/yr; Table 9) based on integrating fluxes from different arcs. This is a consequence of Sano et al. (2001) adopting a $N_2/^3He$ ratio of 5.6×10^6 for arc volcanics based on a relatively small database of 11 geothermal samples. Our database has >700 geothermal and hydrothermal fluid analyses with a median $N_2/^3He$ ratio of 2.6×10^8 —approximately a factor of 40 higher.

Problematic issues regarding estimates of volatile output at arcs

There are two major issues of concern with the approach of using $CO_2/^3He$ and $N_2/^{36}Ar$ (or $N_2/^3He$) ratios in combination with $\delta^{13}C$ and $\delta^{15}N$ values to constrain the sources of volatiles at arcs. The first issue is the selection of end-member isotopic and relative elemental abundances—this factor has a profound effect on the deduced provenance of the volatile of interest. The second is the assumption that various elemental (and isotopic) ratios observed in the volcanic products are representative of the magma source. Both have the potential to compromise the accuracy of the output flux estimates.

In the case of CO_2, the methodology of Sano and co-workers assumes that subducted marine carbonate and sedimentary organic matter can be distinguished as potential input parameters based solely on their perceived C-isotopic compositions prior to subduction (0‰ versus -30‰ respectively). However, this approach ignores the anticipated evolution of organic-derived CO_2 to higher $\delta^{13}C$ as a function of diagenetic and/or catagenetic changes experienced during subduction (Ohmoto 1986). In the southern Lesser Antilles, for example, Van Soest et al. (1998) calculated that >50% of the total carbon would be assigned to an organic, sedimentary origin if an end-member S-value of -10‰ were chosen, as opposed to <20 % for an adopted end-member value of -30‰. In this scenario, it was suggested that the large sedimentary input implied by adopting a heavier $\delta^{13}C$ sedimentary end member could be accommodated by loss of CO_2 from the arc crust that is particularly thick in this portion of the arc. This example illustrates the point that a realistic mass balance at arcs is impossible without taking into account (a) the effect of subduction on the evolution of the C-isotopic signature of the sedimentary input, and (b) the possibility of an additional input from the arc crust. Note, however, that not all arcs require a crustal input of volatiles. For example, Fischer et al. (1998) showed that the volatiles discharged from the Kurile Islands arc (75 tons/day per volcano) could be supplied from subducted oceanic crust and mantle wedge alone.

The second potential complication is the possible fractionation of CO_2, N_2, and He during subduction and/or subsequent magma degassing. Little has been reported on elemental fractionation during the subduction process but recent studies at Loihi Seamount have shown that magma degassing can exert a strong control on resultant $CO_2/^3He$ ratios as sampled in hydrothermal fluid discharges. Hilton et al. (1998) reported large variations in $CO_2/^3He$ ratios at Loihi Seamount that were correlated with the composition of the magma undergoing degassing. For example, as helium is more soluble in tholeiitic basalt than CO_2 (i.e., $S_{He}/S_{CO2} > 1$ where S = solubility), the $CO_2/^3He$ ratio in the melt phase will evolve to lower values as a function of fractionation style (Rayleigh or Batch) and extent of degassing. Measured $CO_2/^3He$ ratios in fluids during periods of tholeiitic volcanism were low (~ 5×10^8). In contrast, $CO_2/^3He$ ratios ~10^{10} were recorded in other active periods, which is consistent with degassing of alkalic magmas (in this case, $S_{He}/S_{CO2} < 1$). In order to obtain a meaningful estimate of the carbon budget in arcs, therefore, the initial (pre-degassing) $CO_2/^3He$ ratio is of prime importance, and it is often assumed that the measured $CO_2/^3He$ ratio equates to the initial magmatic value—as shown above for Loihi Seamount, this may not necessarily be the case. The same issues of degassing-induced changes to elemental ratios apply also to the N_2-He-Ar systematics used to resolve the provenance of nitrogen in arcs (Sano et al. 2001).

A related concern is that of isotopic fractionation of carbon (or nitrogen) during subduction and/or magma degassing. Sano and Marty (1995) have concluded that arc-related high-temperature fluids are likely to preserve the $\delta^{13}C$ values of the (magmatic) source based on comparisons of $\delta^{13}C$ values in fluids and phenocrysts. Furthermore, they cite evidence of overlapping $\delta^{13}C$ values between high- and medium-to-low-enthalpy hydrothermal fluids, leading to the general conclusion that any fractionation induced by degassing and/or interactions within the hydrothermal system must be minimal. On the other hand, Snyder et al. (2001) have argued that geothermal fluids in Central America have experienced 1 to 2‰ shifts in $\delta^{13}C$ resulting from removal of bicarbonate during slab dewatering and/or by precipitation of calcite in the hydrothermal system. It should be noted, however, that even if observed values of $\delta^{13}C$ in arc-fluids are fractionated, the magnitude of the isotopic shifts proposed by Snyder et al. (2001) will make an minor difference only to calculations involving the source of carbon. To date, there is no evidence of nitrogen isotopic fractionation during magmatic degassing (Marty and Humbert 1997).

Volatile output at the back-arc and fore-arc regions

In addition to supplying volatiles that are lost via arc-related volcanism, the subducting slab may also contribute volatiles to both the back-arc and fore-arc regions. To complete a realistic mass balance for subduction zones, therefore, it is essential to quantify volatile fluxes at the back-arc and fore-arc. As we discuss below, both fluxes are severely under-constrained at present.

Volatile data, encompassing both major volatile phases (CO_2, N_2) and noble gases (He and Ar) are available for a number of back-arc basins including the North Fiji Basin, Mariana Trough and Manus Basin (Ishibashi et al. 1994, Nishio et al. 1998, Sano et al. 1998a, Marty and Zimmermann 1999, Sano et al. 2001). A wide range in $CO_2/^3He$ ratios has been reported—from 2.5×10^8 (North Fiji Basin) to 2.2×10^{11} (Manus Basin). There are considerable differences in the treatment of this database. On the one hand, Nishio et al. (1998) reporting data for the North Fiji Basin argue that the value of the mantle $CO_2/^3He$ ratio in this region is 2×10^8: in this way, they calculate that samples with higher $CO_2/^3He$ ratios (up to 9.2×10^9) contain up to 90% slab-derived carbon. On the other hand, Sano et al. (1998a) recognize that measured $CO_2/^3He$ ratios in the Mariana Trough could be fractionated during degassing. After taking account of changes in the ratio due to degassing, the measured $CO_2/^3He$ ratios correct to a value close to that of MORB (~2×10^9) implying that the carbon is wholly of mantle derivation. The different interpretations have profound effects on the estimated output flux from back-arc basins. If we scale the mid-ocean ridge ^3He flux (1000 mol/yr for ~ 6×10^4 km of ridges) to the length of back-arc ridges (~2.4×10^4 km; Uyeda and Kanamori 1979) and use the extreme $CO_2/^3He$ ratio of the North Fiji Basin (9.2×10^9) as representative of a back-arc basin source region (with a slab contribution), then the CO_2 flux from back-arc basins (globally) is 3.7×10^{12} mol/yr—of which up to 90% (3.3×10^{12} mol/yr) could be derived from the slab. This estimate would make the back-arc CO_2 flux comparable with that for MOR ($2.2\pm0.9\times10^{12}$ mol/yr; Marty and Tolstikhin 1998), and significantly higher than most estimates for arc volcanism (Table 6). Alternatively, if we accept the arguments of Sano et al. (1998a), the output of slab-derived CO_2 via back-arc volcanism is essentially zero.

The same uncertainty in back-arc flux rates applies also to N_2. Limited data on BABB glasses from the Mariana Trough, the Manus and North Fiji Basin show $N_2/^3He$ ratios ranging from 1.16×10^6 to 1.37×10^7 and $\delta^{15}N$ values varying between -2.7 and +1.9‰ (Sano et al. 2001). Therefore, depending upon the chosen end-member composition, the sediment contribution to the back-arc N_2 flux can vary between zero and close to 40 %. In the latter case, a significant fraction (one-third) of the available slab-derived nitrogen is recycled to the surface via back-arc volcanism as opposed to two-thirds via the arc (Sano et al. 2001).

Although volatile flux estimates from back-arc regions vary greatly, the situation is even more poorly constrained for fore-arc regions. We are unaware of any flux estimates for either major volatiles or noble gases. This is in spite of abundant circumstantial evidence for fluid venting in fore-arc regions e.g., serpentinite diapirism in the Marianas (Fryer et al. 1985) and mud volcanism at various accretionary prisms worldwide (Brown 1990). Indeed, there are reports of $^3He/^4He$ ratios > crustal production rates (~0.05 R_A; Andrews 1985) in fore-arc regions such as the Nankai Trough (1.9 R_A; Kastner et al. 1993), the Cook Inlet of Alaska (0.8 R_A; Poreda et al. 1988) and Kavachi volcano in the Solomon Islands (6.9 R_A; Trull et al. 1990)—this would indicate a direct output of mantle-derived volatiles. However, there are no estimates of the scale of this phenomenon.

The alternative approach to estimating fore-arc losses of volatiles is somewhat circular and based upon mass balance. For example, the estimated carbon load of incoming material at trenches worldwide (sedimentary carbonate, sedimentary organic carbon and oceanic meta-basalts) is 1.2, 0.8, and 3.4 ($\times10^{12}$ mol/yr) respectively (Bebout 1995, Alt and Teagle 1999, Plank and Langmuir 1998). This total (5.4×10^{12} mol/yr) far exceeds that of the carbon released via arc magmatism (~3×10^{12} mol/yr; Table 6), implying that a substantial amount of carbon is either released at the fore-arc or by-passes the zone of magma generation and is recycled into the mantle. Whereas there is considerable leeway in all these estimates, the uncertainty is compounded by debate on whether the zone of magma generation acts as a barrier to further subduction of volatiles (Staudacher and Allègre 1988) or if various C-bearing phases are stable at the likely P-T conditions below the arc so that deep(er) subduction can occur (cf. Kerrick and Connolly 2001). In this latter scenario, if the input of carbon to the mantle via subduction matches output via MOR (i.e., steady-state upper mantle and no losses to the back-arc) then approximately 2×10^{12} mol/yr (the MOR output; Marty and Tolstikhin 1998) by-passes the zone of magma generation leaving a residual 0.4×10^{12} mol/yr potentially available for loss via the fore-arc. There is considerable latitude in these figures (and assumptions inherent in the approach): for these reasons, therefore, we caution that constraining volatile losses at the fore-arc by this mass balance approach be treated with skepticism.

The volatile input via the trench

In this section, we provide estimates of the volatile flux input—of both noble gases (He, and Ar) and major volatiles (CO_2, N_2 and H_2O)—via the trench. We consider both the subducted sediment load as well as the underlying crustal basement (of thickness 7 km). Note that the volatile input parameter can be calculated for an extensive listing of subduction zones: however, the output parameter via arcs (Table 9) is well constrained by flux data for a limited number of arcs only. For these arcs, at least, we have a direct comparison between input and output for specific volatile species. However, by extrapolating the (limited) output data to arcs worldwide, we can address the question of volatile recycling between the mantle and crust, hydrosphere and atmosphere from a global perspective.

In Table 10, we compute the volatile fluxes into the subduction zone for individual arcs using (a) available noble gas, H_2O, CO_2 and N_2 concentration data for oceanic sediments and oceanic (crustal) basement, and (b) estimates of sediment and crust fluxes at each arc segment. In this compilation, it is important to note that the fluxes refer to volatiles potentially entering the subduction zone at the site of the trench—the effects of sediment off-scraping and/or underplating (Von Huene and Scholl 1991) on the volatile fluxes are not considered. The following points are relevant in the computation of the volatile flux data (Table 10):

1. There is limited noble gas data available for oceanic crust. However, Staudacher and Allègre (1988) report He and Ar data for 2 fine-grained basalt samples from the Indian Ocean. We note that although we adopt these data, the samples are of Jurassic age (~108 Ma) so their noble gas contents may not necessarily be appropriate for all types of subducting oceanic crust. Similarly, there is limited noble gas data available for sedimentary material, and we again use concentrations from Staudacher and Allègre (1988) obtained for sediments from the South West Indian Ridge. With the exception of helium, noble gas contents of oceanic sediments are generally higher than oceanic basalts.
2. There are 3 separate values for the water content of the subducting slab. The sediment H_2O contents are given on an arc-to-arc basis by Plank and Langmuir (1998) and represent bound (not porewater) contents of various drill-core lithologies proximal to the arc in question. A value of 3.4 wt % is used for water bound in oceanic crust assuming a 50:50 mix of basalt and gabbro (Schmidt and Poli 1998). Finally, a value of 1.2 wt % H_2O and termed 'available water', is considered 'available' for the generation of arc mamas following water loss from dehydration and diagenetic processes associated with the early stages of subduction (Schmidt and Poli 1998).
3. In an analogous fashion to water, the CO_2 concentrations in sediments come from an arc-by-arc compilation of carbonate contents of various drill core lithologies (Plank and Langmuir 1998). The amount of reduced (organic) carbon in sediments is taken as 1% based on average contents of metamorphosed sediments of the Catalina Schist (Bebout 1995). A value of 0.214 wt % is used for the amount of CO_2 stored in altered oceanic crust (Alt and Teagle 1999).
4. The N_2 concentrations are average values of oceanic sediments (0.01 wt %) and crust (0.001 wt %) (from Bebout 1995).

In their compilation of the characteristics of oceanic sediments, Plank and Langmuir (1998) has shown that both the amount and chemical composition of sediments being subducted are highly variable: we emphasize that the same observation is true also for the major volatile components, H_2O and CO_2. However, due to the paucity of data, it is currently not possible to evaluate if there is any regional variability in the noble gas contents of oceanic sediments and their crystalline basements.

The input fluxes of various volatile species (Table 10) can now be compared to various output fluxes through arc volcanism (Table 9) to assess the extent of volatile mass balance. At this stage, we ignore possible volatile losses at the back-arc given the large uncertainty in actual values. Also, we note that in the case of subducted carbon, it is important to distinguish between reduced sedimentary carbon, sedimentary carbonate and carbonate of the altered oceanic crust.

Table 11. Fractional contributions of carbonate- (*L*), Sedimentary/organic- (*S*) and MORB- (*M*) derived CO_2 to total arc CO_2 discharges*.

Arc Segment	*L*	*S*	*M*
Andes	0.56	0.33	0.09
Alaska-Aleutians	Nd	Nd	Nd
Lesser Antilles	0.71	0.17	0.03
Central America	0.59	0.31	0.06
Indonesia	0.74	0.13	0.12
Italy	Nd	Nd	Nd
Japan	0.67	0.13	0.18
Kamchatka-Kuriles	0.51	0.27	0.22
New Zealand	0.85	0.10	0.06
Papua New Guinea	Nd	Nd	Nd
Philippines	Nd	Nd	Nd

Nd = no data available (for Alaska-Aleautians, Italy, PNG and the Philippines). * Endmember compositions for M, L and S used in the computations are $\delta^{13}C$ (-6.5‰, 0‰ and -20‰) with corresponding $CO_2/^3He$ ratios of 1.5×10^9; 1×10^{13}; 1×10^{13}, respectively.

Volatile mass balance at subduction zones

Before comparing the output to input fluxes for the various volatile species, it is necessary first to resolve the carbon output via the arc into its components parts reflecting carbon provenance from the mantle wedge, and slab-derived carbonates and organic carbon. In Table 11, we report % contributions to the CO_2 output using the mass balance equations of Sano and Marty (1995) but adopting a $\delta^{13}C$ value for the sedimentary end member (S) of -20‰. Bebout (1995) shows convincing evidence of a progressive increase in $\delta^{13}C$ of carbonaceous material with increasing metamorphic grade in the Catalina Schist, and Van Soest et al. (1998) discuss the effect of adopting a higher $\delta^{13}C$ value for the S-end member on the calculation of CO_2 provenance. Resolution of the CO_2 output into M-, L- and S-components is shown in Figure 5 for arcs where sufficient data is available. In each case, we report the median values for M, L and S. The limestone-derived component dominates at all localities.

In Tables 12 (major volatiles) and 13 (He and Ar) we compute ratios of output to input fluxes for the various individual arcs. The output fluxes are given in Table 9—modified in the case of carbon for provenance (L, S, and M; see Table 11). The input fluxes are given in Table 10. It is noteworthy that output/input ratios vary significantly among individual arcs. We point out the following features of interest in Tables 12 and 13.

1. The sedimentary output of CO_2 (S) significantly exceeds the input of organic sedimentary carbon for 4 of the 9 arcs where data are available (Andes, Japan, New Zealand and the Philippines). For these localities, an additional source of organic carbon is required: the uppermost arc crust seems the only likely possibility. At the other arcs, the input of organic carbon exceeds the output via the arc, and suggests that either sediment accretion and/or off-scraping of subducting sediments could prohibit the transport of organic carbon to the zones of magma generation. For Central America, the ratio is very close to unity suggesting that the output from the arc is nearly balanced by the input of organic sedimentary carbon.

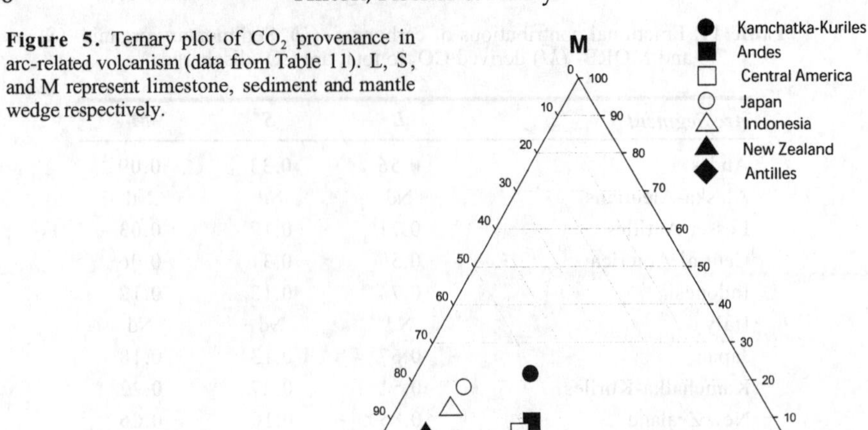

Figure 5. Ternary plot of CO_2 provenance in arc-related volcanism (data from Table 11). L, S, and M represent limestone, sediment and mantle wedge respectively.

Table 12. Ratio of output (via arc) to input (via trench) for major volatile species (CO_2, N_2 and H_2O) for selected arc segments.

	CO_2[1]			N_2 (excess)[2]	H_2O[3]	
Arc Segment	S OUT/ Org IN	L OUT/ Car[4] IN	L OUT/ (Car+ crst[5]) IN	OUT/ (Sed + crst) IN	OUT/ IN total[6]	OUT/ IN avl.[7]
Andes[8]	6.71	1.07	0.42	1.00	60.1	346.4
Alaska-Aleutians[9]	0.028	∞ [$]	0.010	0.013	0.27	1.76
Lesser Antilles[10]	0.46	20.8	0.79	-	5.52	46.2
Central America	1.02	0.14	0.097	0.14	11.7	56.0
Indonesia (Sunda)[11]	0.023	0.462	0.031	0.021	0.42	3.0
Italy						
Japan	1.80	∞	1.22	4.03	29.6	234.9
Kamchatka-Kuriles[12]	0.029	∞	0.0073	0.031	0.22	1.81
New Zealand[13]	1.29	∞	0.68	0.46	35.2	209.3
Papua New Guinea/ SW Pacific						
Philippines	2.38	4.20	0.44	0.30	63.9	310.5

1. CO_2 output resolve into L-, S-, and M-components (from Table 11).
2. N_2 excess = non-atmospheric N_2 (assumes all argon is air-saturated water derived – equation 3 in text).
3. H_2O output assumes arc H_2O/CO_2 ratio = 50 (Symonds et al. 1994).
4. Car = sedimentary carbonate input (CO_2 sed - from Table 10)
5. Crst = oceanic crust carbonate input (CO_2 crust – from Table 10).
6. Total water = sediment- and basement-hosted water only (excludes pore water).
7. Water available at zone of magma generation (1.2 wt.% - Schmidt and Poli 1998).
8. Input from Columbia and Peru arcs (Table 10).
9. Includes input from Aleutians and Alaska arc (Table 10).
10. Includes input from both northern and southern Lesser Antilles (Table 10).
11. Includes input from Java and Sumatra (Table 10).
12. Includes input from Kuriles and Kamchatka arcs (Table 10).
13. Input taken for Kermadec arc (Table 10).

$ Infinity (zero carbonate input).

Table 13. Ratio of output (via arc) to input (via trench) for noble gas isotopes (^{3}He, ^{4}He, ^{36}Ar and ^{40}Ar) for selected arc segments.

	^{3}He	^{4}He	^{36}Ar	^{40}Ar[3]
Arc Segment[1]	OUT/ IN	OUT/ IN[2]	OUT/ IN	OUT/ IN
Andes	488.8	4.55	88.9	103.3
Alaska-Aleutians	11.5	0.11	2.75	3.20
Lesser Antilles	1002	9.32		
Central America	744.9	6.94	39.1	45.5
Indonesia (Sunda)	34.6	0.32	3.63	4.22
Italy				
Japan	2767	25.8	288.5	335.3
Kamchatka-Kuriles	17.4	0.16	3.49	4.06
New Zealand	888.8	8.28	152.0	176.7
Papua New Guinea/ SW Pacific				
Philippines	879	8.19	39.7	46.2

1. See Table 12 for arcs used in input computation.
2. Input ^{3}He/^{4}He ratio = 0.05 R_A (radiogenic helium)
3. Output ^{40}Ar/^{36}Ar ratio = 401 (Table 4); input ^{40}Ar/^{36}Ar ratio = 345 (average of sediment and oceanic crust—from Staudacher and Allègre 1988).

2. The output of carbonate-derived CO_2 (L) cannot be balanced with the amount of sedimentary carbonate CO_2 being subducted except for the Central American and Sunda (Indonesia) volcanic arcs. This would suggest that, in both these localities, sediment-derived carbonate may be efficiently transported to the zones of magma generation. When carbonate CO_2 from the subducting altered oceanic crust is taken into account, the output can be supplied solely from the slab: indeed, at all localities, except Japan, only a fraction of the input CO_2 is necessary to supply the output. These figures reinforce the notion that subduction zones act as conduits for the transfer of carbon into the mantle (Kerrick and Connolly 2001).
3. In general, the amount of excess (non-atmospheric) N_2 being emitted from arc volcanoes is less than the amount of N_2 being subducted (with the exception of the Andes and Japan). This would imply that a significant amount of subducted N_2 may not reach the zones of arc magma generation and/or it is retained and lost to the (deeper) mantle. The observation of large increases in the C/N ratio of Catalina Schists as a function of increasing metamorphic grade (Bebout 1995)—interpreted as enhanced devolatilization (loss) of N_2—would seem to favor the former possibility.
4. The flux of H_2O from each arc segment greatly exceeds the amount of water available at the zone of arc magma generation (1.2 wt %; Schmidt and Poli 1998). Even when the total amount of water potentially subducted is taken into account (IN total; Table 12), for the most part (excepting Alaska-Aleutians, Indonesia and Kamchatka-Kuriles), output via the arc exceeds input via the trench. Water from the mantle wedge and/or the arc crust must contribute to the output in these cases. Alternatively, the 'excess' water may be meteoric in origin although it must first become incorporated by magma (and contribute to the high 'magmatic' H_2O/CO_2 ratio of 50).
5. The ^{3}He output flux exceeds the input parameter in all cases. This is true even in cases

(Alaska-Aleutians, Indonesia and Kamchatka-Kuriles) where the potential input of ^{4}He exceeds the output via the arc. The observation that at each locality the ^{3}He/^{4}He ratio of the output flux approaches the MORB value indicates that little, if any, of the slab-derived helium reaches the zones of magma generation.

6. A similar pattern is observed for ^{36}Ar; output via the arc greatly exceeds potential input via the trench. The output volatile flux appears dominated by atmosphere-derived ^{36}Ar (thereby validating its use to calculate 'excess' N_2). Likewise, output of ^{40}Ar far exceeds potential input. In the case that all the ^{36}Ar is atmospheric, the excess ^{40}Ar (to raise the average arc ^{40}Ar/^{36}Ar value to 401; Table 4) can be either crustal (Kennedy et al. 1991) or mantle (Patterson et al. 1994) in origin.

Global implications of volatile recycling at subduction zones

In this final section, we extend our considerations of subduction zone processing of volatiles (input versus output fluxes; see previous section) to address the issue of long-term volatile exchange between the surface of Earth and the mantle. We follow the approach of Marty and Dauphas (2002) by relating the ratio of the volatile flux through arc volcanism and the amount of volatiles carried by subducting plates (F_{arc}/F_{sub}) to the Mean Degassing Duration (MDD) of a particular volatile species. The MDD (in years) is the ratio of the total surface inventory of a volatile element (ocean + crust + atmosphere) to the present-day mantle flux, approximated by the MOR flux (assuming that plume degassing is minor compared to ridges). As discussed in general terms in the previous section, a F_{arc}/F_{sub} ratio close to unity implies little or no recycling of subducted volatiles into the mantle beyond the zone of magma generation, a ratio of <<1 implies possible (deep) recycling, and a ratio of >>1 implies that additional contributions to the volatile flux (other than from subducting sediments and oceanic crust) are required. A MDD value less than the age of Earth implies rapid recycling of volatiles between the mantle and surface reservoirs. A MDD value greater than 4.55 Ga suggests either a decreasing degassing rate with time or the occurrence of a volatile component at the Earth's surface not derived from the mantle, or both. Global averages of F_{arc}/F_{sub} ratios together with MDD values are given in Table 14 and plotted against each other in Figure 6.

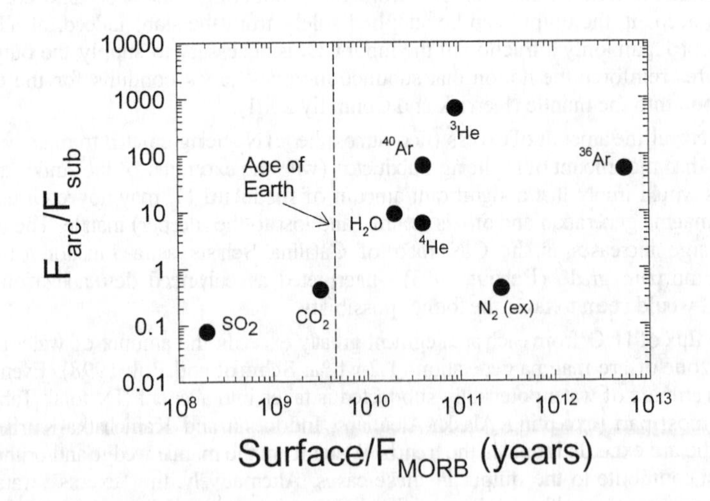

Figure 6. Volatile flux ratio (F_{arc}/F_{sub}) vs. mean degassing duration (MDD) for four major (S, C, H_2O, N_2) and two noble-gas (He, Ar) volatile species, following the approach of Marty and Dauphas (2002).

Table 14. Average global volatile fluxes out of volcanic arcs (F_{arc}) and mid-ocean ridges (F_{MORB}), and into subduction zones (F_{sub}), together with total surface inventories (atmosphere + oceans + crust).

Volatile	F_{arc} mol/yr	F_{sub} mol/yr	F_{arc}/F_{sub}	Surface Invent. (mol)	F_{MORB} mol/yr	Surface/F_{MORB} yr
H_2O	8.12E+13	8.92E+12	9.10	5.80E+23	2.90E+13	2.0E+10
SO_2	3.16E+11	(4.11E+12)	0.077	5.33E+20	(2.67E+12)	2.0E+08
CO_2 (L+S)	1.45E+12	3.46E+12	0.42	7.00E+21	2.20E+12	3.2E+09
N_2 (S + Crst)	1.98E+10	4.48E+10	0.44	2.74E+20	1.00E+09	1.8E+11
^3He	9.24E+01	1.40E-01	660	9.14E+13	1.00E+03	9.1E+10
^4He	1.23E+07	2.00E+06	6.2	3.61E+18	8.93E+07	4.0E+10
^{40}Ar	1.90E+08	2.90E+06	65.5	1.64E+18	4.00E+07	4.1E+10
^{36}Ar	4.74E+05	8.40E+03	56.4	5.55E+15	1.00E+03	5.55E+12

Footnotes to Table 14

F_{arc}: total global flux of volatile species from subduction zones - calculated by summing individual arc volatile fluxes and adding additional 20.6% (see Table 9). The water flux was calculated using the CO_2 flux (Table 9) and assuming a H_2O/CO_2 ratio of 50. The CO_2 (L + S) – limestone- + sediment-derived CO_2 is the total arc CO_2 flux (Table 9) multiplied by 0.89 (the average L+S value for 7 arcs worldwide – Table 11). ^{36}Ar flux is calculated using the total Ar flux (Table 9) and an average $^{40}Ar/^{36}Ar$ ratio of 401 (Table 4).

F_{sub}: total global subduction of volatile species. Calculated by summing flux values of individual arcs (Table 10).

Surface Invent.: global inventory of volatile species currently at the Earth's surface.

- **H_2O**: from Marty and Dauphas (2002)
- **SO_2**: from Marty and Dauphas (2002)
- **CO_2**: from Zhang and Zindler (1993)
- N_2: from Pepin (1991)
- **^3He**: from Ozima and Podosek (1983) – assumes ^{20}Ne inventory of atmosphere and planetary $^3He/^{20}Ne$ ratio of 0.0315. Present atmosphere abundance is lower because of escape to space.
- **^4He**: from Ozima and Podosek (1983) – radiogenic ^4He released into atmosphere taken as 2.2 times atmosphere ^{40}Ar abundance. Present abundance is much lower because of escape to space.
- **^{36}Ar & ^{40}Ar**: from Ozima and Podosek (1983)

F_{MORB}: flux of volatiles from the mantle, approximated by MORB flux.

- **H_2O**: from Jambon and Zimmermann (1990)
- **SO_2**: from Alt et al. (1985)
- **CO_2** :from Marty and Tolstikhin (1998)
- N_2: using ^3He flux (1000 mol/yr) from Craig et al. (1975) and MORB $N_2/^3He$ ratio of 1×10^6 (Sano et al. 2001).
- **^3He**: from Craig et al. (1975)
- **^{36}Ar**: calculated from the MORB ^3He flux and MORB $^{36}Ar/^3He$ ratio = 1 (Tolstikhin and Marty 1998).

In Figure 6, we observe a generally positive correlation between F_{arc}/F_{sub} and MDD indicative of decreasing efficiency in atmosphere-mantle exchange in the order SO_2, CO_2, H_2O, $^3He \approx {}^4He \approx {}^{40}Ar$, N_2 (ex), ^{36}Ar. Sulfur has a F_{arc}/F_{sub} ratio <<1 and MDD value of <<4.5 Ga, implying that subduction leads to a rapid and efficient exchange between the atmosphere and mantle. Carbon also has a F_{arc}/F_{sub} ratio < 1 and a MDD value of <4.5 Ga—again consistent with efficient recycling between the mantle and atmosphere. Water has a F_{arc}/F_{sub} ratio > 1 and an MDD value greater than the age of Earth. In terms of arc magmatism, this implies that an additional source of water is required other than the amount released from the subducting slab. This additional H_2O may be of surficial (meteoric) or crustal origin. The MDD value greater than the age of the Earth would be consistent with the notion that the recycling of H_2O into the deeper mantle is less efficient than for CO_2, i.e., only a small proportion of the subducted water bypasses the zone of arc magma generation to the deeper mantle for later re-emergence at MOR. Helium-4 has a F_{arc}/F_{sub} ratio similar to H_2O and a MDD value also greater than the age of the Earth implying an additional source of He in arcs—most likely from the mantle wedge. This is even more pronounced for ^{3}He and ^{40}Ar, which have the highest F_{arc}/F_{sub} ratios, consistent with the idea that He and Ar are not subducted—either to the deeper mantle nor to the zone of magma generation at arcs. Nitrogen and ^{36}Ar lie off the trend described by SO_2, CO_2, H_2O, He, and ^{40}Ar in Figure 6. Nitrogen has a similar F_{arc}/F_{sub} ratio as CO_2 (<1), indicating that N_2 is either partially recycled into the mantle or released in the fore-arc. The large MDD value suggests that recycling into the deep mantle is not very efficient and/or that the mantle degassing rate for N_2 has probably decreased over time. This is consistent with the model of Tolstikhin and Marty (1998) which suggests a vigorous N_2 degassing of the upper mantle early after accretion and possible addition of N_2 from late impacting bodies (e.g., comets). Argon-36 has a F_{arc}/F_{sub} ratio of >1 and the greatest MDD value of all volatiles shown. This implies that Ar is not recycled into the deep mantle, and that the amount discharging from arc volcanoes is mostly of shallow crustal or atmospheric origin.

In terms of the isotopic composition of volatiles in different Earth reservoirs, efficient recycling between the surface and mantle would tend to minimize isotopic differences between reservoirs. This is the case for sulfur and carbon where isotopic differences between the mantle and surface reservoirs are at the level of <10‰. Isotopic differences for water between the mantle and surface (e.g., ocean water) are larger (~70 to 100‰; Poreda et al. 1986). However, there is a huge difference in ^{40}Ar/^{36}Ar ratios between the (upper) mantle (>40,000) and the atmosphere (296). This is consistent with very inefficient recycling of ^{36}Ar into the deeper mantle, and supports the notion of a subduction barrier for argon (Staudacher and Allègre 1988).

The case of nitrogen is more enigmatic as both the uppermost region of the mantle, supplying mid-ocean ridges, and the mantle source of diamonds seem to be characterized by the same nitrogen isotopic composition (δ^{15}N about -5‰). Because some of the diamonds might have been stored in the sub-continental lithosphere for time periods of 1 Ga or more (e.g., Richardson et al. 1984), the homogeneity of the mantle sampled by MORB and diamonds suggests little or no contribution of recycled nitrogen that is likely to be heavy (^{15}N-rich)—certainly over the last few Ga. The initial δ^{15}N value of mantle nitrogen could have been much lighter than presently seen in mantle-derived rock, as low as -30‰ if one assumes an enstatite chondrite composition (Javoy et al. 1986); therefore, the -5‰ composition of MORB and diamonds could be the result of recycling of heavy nitrogen from the surface. In this case, nitrogen recycling would have proceeded early in Earth history. The recent observation by Dauphas and Marty (1999) of heavy nitrogen in rocks and minerals from the Kola magmatic province (Russia), in samples characterized by plume-like helium and neon (Marty et al. 1998), raises the interesting possibility that recycled volatiles are stored in the deep mantle source of plumes. The timescale for residence time of recycled radiogenic tracers in the mantle is of the order of 1 Ga or more (Hofmann 1997), suggesting by analogy that recycling of nitrogen in the deep mantle took place essentially in the Proterozoic. If correct, this possibility has far reaching implications for the geodynamics of the mantle, particularly the issue of the duration of plate tectonics. More volatile data on plume-like material is clearly needed to provide a complementary viewpoint to that given by subduction zone volatiles.

SUMMARY AND FUTURE WORK

This review has focused on the noble gas systematics of subduction zones. Although a considerable database has been accumulated on the noble gas (and major volatile) characteristics of both arc and back-arc regions, and these are summarized above, there are numerous topics which warrant further and more detailed attention. We suggest that the following list contains realistic objectives where advances can be made:

1. Although the helium database is significant for arcs and back-arcs, the same cannot be said for neon and argon—much less so for krypton and xenon. This is due to a combination of obtaining suitable samples (not overwhelmed by air contamination) and analytical limitations. We envisage that future refinements and developments in sample processing (e.g., laser ablation) and/or measurement capabilities will lead to improved and more complete noble gas datasets—which ideally will have complementary major volatile data.
2. It is encouraging that techniques are being developed to treat and test noble gas datasets for processes such as magmatic degassing and/or crustal assimilation which may disturb both isotopic and abundance characteristics of magmatic systems. In the case of oceanic basalts, for example, Marty and Zimmermann (1999) and Burnard (2001) provide approaches to recognize and correct for such extraneous effects. In a subaerial environment, Hilton et al. (1995) considers crustal contamination of phenocrysts, whereas Goff et al. (2000) consider the effects of magmatic degassing on the chemistry of fumarolic gases. More consideration should be given to this topic for all types of samples and localities.
3. There remain large uncertainties in estimating volatile fluxes—both for arcs individually as well as for arcs globally. Although progress has been made in the determination of SO_2 (and CO_2) fluxes for a number of arcs, the fluxes of related volatiles (H_2O, N_2 and the noble gases) is not as well known as a result of sparse datasets (particularly for the noble gases other than helium). Additional measurements using both ground- and satellite-based remote sensing techniques have the potential to improve the current database of the magnitude and speciation of volcanic emissions. These techniques need to be applied to both passively degassing volcanoes as well as those actively erupting. Moreover, monitoring studies of single volcanoes are needed to better evaluate flux and compositional variations over long time periods (e.g., see studies of Galeras Volcano, Colombia (Zapata et al. 1997; Fischer et al. 1997) and Vulcano, Italy (Nuccio et al. 1999)). Surprisingly, there are a significant number of arcs with virtually no volatile emission or gas compositional data. A high priority of future work should be the study of these arcs to provide a complementary dataset to well-studied arcs such as Central America.
4. The volatile output and composition at fore-arcs and back arcs remains poorly constrained. Currently, we are unaware of any representative volatile flux measurements for fore-arcs. These measurements are critical to arrive at an accurate volatile mass balance of subduction zones, and to evaluate if volatiles are transferred beyond this region to the zone of magma generation and/or to the deeper mantle.
5. Better constraints are required for the input parameters (particularly for the noble gases and nitrogen). Additional information on the volatile composition of both oceanic sediments and crustal basement is needed to improve estimates of mass balance at subduction zones.
6. More data are required on possible elemental (and isotopic) fractionation of the major and noble gases particularly during both metamorphic devolatilization reactions but also during magma storage, crystallization and degassing. This will provide more realistic end-member compositions to be assigned to slab- and mantle-

derived fluids in modeling studies. Progress in this area may come from experimental work on vapor-melt partitioning of volatiles allied with improved datasets on different arc systems with contrasting forcing functions (angle of slab dip, rate of subduction, etc.).

ACKNOWLEDGMENTS

Collectively, we gratefully acknowledge funding from a number of national funding agencies for the opportunity to work on subduction zones: USA (National Science Foundation and NASA), Germany (Deutsche Forschungsgemeinschaft), The Netherlands (Nederlandse Organisatie voor Wetenschappelijk Onderzoek) and France (Programme Intérieur de la Terre, Institut National des Sciences de l'Univers). This presentation benefited from comments by D. Porcelli, G. Bebout, Y. Sano and one anonymous reviewer. We also thank A. Shaw for critical remarks.

REFERENCES

Allard P (1983) The origin of hydrogen, carbon, sulphur, nitrogen and rare gases in volcanic exhalations; evidence from isotope geochemistry. *In* Forecasting Volcanic Events. Vol. 1. Tazieff H, Sabroux J (eds) Elsevier, New York, p 337-386

Allard P (1992) Global emissions of helium-3 by subaerial volcanism. Geophys Res Lett 19:1479-1481

Allard P, Jean-Baptiste P, D'Alessandro W, Parello F, Parisi B, Flehoc C (1997) Mantle derived helium and carbon in groundwaters and gases of Mount Etna, Italy. Earth Planet Sci Lett 148:501-516

Alt JC, Teagle DAH (1999) The uptake of CO_2 during alteration of the oceanic crust. Geochim Cosmochim Acta 63:1527-1535

Alt J, Satzman ES, Price D (1985) Anhydrite in hydrothermally altered basalts: DSDP hole 504B. *In* Anderson RN, Honnorez J, Baker K, et al. (eds) Initial Reports DSDP 83:283-288. U S Govt Printing Office, Washington, DC

Anderson, DL (1993) Helium-3 from the mantle: Primordial signal or cosmic dust? Science 261:170-176

Andres RJ, Kasgnoc AD (1998) A time-averaged inventory of subaerial volcanic sulfur emissions. J Geophys Res 103:25251-25261

Andrews JN (1985) The isotopic composition of radiogenic helium and its use to study groundwater movement in confined aquifers. Chem Geol 49:339-351

Bach W, Niedermann S (1998) Atmospheric noble gases in volcanic glasses from the southern Lau Basin: Origin from the subducting slab? Earth Planet Sci Lett 160:297-309

Ballentine C, O'Nions RK, Coleman ML (1996) A magnus opus: helium, neon, and argon isotopes in a North Sea Oil field. Geochim Cosmochim Acta 60:831-849

Bebout GE (1995) The impact of subduction-zone metamorphism on mantle-ocean chemical cycling. Chem Geol 126:191-218

Berresheim H, Jaeschke W (1983) The contribution of volcanoes to the global atmospheric sulfur budget. J Geophys Res 88:3732-3740

Brantley SL, Koepenick KW (1995) Measured carbon dioxide emissions from Oldoinyo Lengai and the skewed disrtibution of passive volcanic fluxes. Geology 23:933-936

Brombach T, Marini L, Hunziker JC (2000) Geochemistry of the thermal springs and fumaroles of Basse-Terre Island, Guadeloupe, Lesser Antilles. Bull Volcanol 61:477-490

Brown KM (1990) The nature and hydrogeologic significance of mud diapirs and diatremes for accretionary systems. J Geophys Res 95:8969-8962

Burnard P (2001) Correction for volatile fractionation in ascending magmas: noble gas abundances in primary mantle melts. Geochim Cosmochim Acta 65:2605-2614

Burnard PG, Stuart F, Turner G (1994) C-He-Ar variations within a dunite nodule as a function of fluid inclusion morphology. Earth Planet Sci Lett 128:243-258

Burnard P, Graham D, Turner G (1997) Vesicle-specific noble gas analyses of "popping rock"; implications for primordial noble gases in Earth. Science 276:568-571

Carroll MR, Webster JD (1994) Solubilities of sulfur, noble gases, nitrogen, chlorine, and fluorine in magmas. Rev. Mineral. 30:231-279

Chiodini G (1994) Temperature, pressure and redox conditions governing the composition of the cold CO_2 gases discharged in north Latium, Italy. Appl Geochem 9:287-295

Chiodini G, Cioni R, Marini L, Panichi C (1995) Origin of the fumarolic fluids of Vulcano Island, Italy and implications for volcanic surveillance. Bull Volcanol 57:99-110

Chiodini G, Cioni R, Frullani A, Guidi M, Marini L, Prati F, Raco B (1996) Fluid geochemistry of Montserrat Island, West Indies. Bull Volcanol 58:380-392

Christensen BW (2000) Geochemistry of fluids associated with the 1995-1996 eruption of Mt Ruhapehu, New Zealand: Signatures and processes in the magmatic-hydrothermal system. J Volcanol Geotherm Res 97:1-30

Cioni R, D'Amore F (1984) A genetic model for the crater fumaroles of Vulcano Island (Sicily, Italy). Geothermics 13:375-384

Craig H (1953) The geochemistry of stable carbon isotopes. Geochim Cosmochim Acta 3:53-92

Craig H, Horibe Y (1994) ^{3}He and methane in Sakurajima Caldera, Kagoshima Bay, Japan. Earth Planet Sci Lett 123:221-226

Craig H, Lupton JE (1976) Primordial neon, helium, and hydrogen in oceanic basalts. Earth Planet Sci Lett 31:369-385

Craig H, Clarke WB, Beg MA (1975) Excess ^{3}He in deep water on the East Pacific Rise. Earth Planet Sci Lett 26:125-132

Craig H, Lupton JE, Horibe Y (1978) A mantle helium component in Circum-Pacific volcanic gases: Hakone, the Marianas, and Mt. Lassen. *In* Terrestrial rare gases. Alexander Jr EC, Ozima M (eds) Japan Scientific Societies Press, Toyko, p 3-16

Crisp JA (1984) Rates of magma emplacement and volcanic output. J Volcanol Geotherm Res 20:177-211

Dauphas N, Marty B (1999) Heavy nitrogen in carbonatites of the Kola Peninsula: A possible signature of the deep mantle. Science 286:2488-2490

Delmelle P, Kusakabe M, Bernard A, Fischer TP, de Brouwer S, Del Mundo E (1998) Geochemical and isotopic evidence for seawater contamination of the hydrothermal system of Taal Volcano, Luzon, the Philippines. Bull Volcanol 59:562-576

Delmelle P, Bernard A, Kusakabe M, Fischer TP, Takano B (2000) The crater lake system of Kawah Ijen volcano, Indonesia: Insights from chemical and isotopic compositions obtained between 1990 and 1996. J Volcanol Geotherm Res 97:31-53

Dodson A, Brandon AD (1999) Radiogenic helium in xenoliths from Simcoe, Washington, USA: Implications for metasomatic processes in the mantle wedge above subduction zones. Chem Geol 160:371-385

Farley KA, Neroda E (1998) Noble gases in the Earth's mantle. Annu Rev Earth Planet Sci 26:189-218

Fischer TP, Sturchio NC, Stix J, Arehart GB, Counce D, Williams SN (1997) The chemical and isotopic composition of fumarolic gases and spring discharges from Galeras Volcano, Colombia. J Volcanol Geotherm Res 77:229-254

Fischer TP, Giggenbach WF, Sano Y, Williams SN (1998) Fluxes and sources of volatiles discharged from Kudryavy, a subduction zone volcano Kurile Islands. Earth Planet Sci Lett 160:81-96

Fischer JM, Fischer TP, Roggensack K, Williams SN (1999a) Magmatic volatiles in the Kamtchatka arc. EOS Trans, Am Geophys Union 80:954

Fischer TP, Roggensack K, Shuster DL, Kennedy BM (1999b) Noble gas isotopic composition of Central American magmas. EOS Trans, Am Geophys Union 80:1202-1203.

Fryer P, Ambos EL, Hussong DM (1985) Origin and emplacement of Mariana forearc seamounts. Geology 13:774-777

Gasparon M, Hilton DR, Varne R (1994) Crustal contamination processes traced by helium isotopes—Examples from the Sunda arc, Indonesia. Earth Planet Sci Lett 126:15-22

Gerlach TM (1991) Present-day CO_2 emissions from volcanoes. EOS Trans, Am Geophys Union 72: 249-255.

Giggenbach WF (1992) Magma degassing and mineral deposition in hydrothermal systems along convergent plate boundaries. Econ Geol 87:1927-1944

Giggenbach WF (1995) Variations in the chemical and isotopic composition of fluids discharged from the Taupo Volcanic Zone, New Zealand. J Volcanol Geotherm Res 68:89-116

Giggenbach WF (1996) Chemical composition of volcanic gases. *In* Monitoring and mitigation of volcano hazards. R Scarpa, R Tilling (eds) Springer-Verlag, Berlin, Heidelberg, p 221-256.

Giggenbach WF, Corrales SR (1992) Isotopic and chemical composition of water and steam discharges from volcanic-magmatic-hydrothermal systems of the Guanacaste Geothermal Province, Costa Rica. Appl Geochem 7:309-332

Giggenbach WF, Goguel RL (1989) Methods for the collection and analysis of geothermal and volcanic water and gas samples. Department of Scientific and Industrial Research, Chemistry Division. Petone, New Zealand

Giggenbach WF, Matsuo S (1991) Evaluation of results from second and third IAVCEI field workshop on volcanic gases, Mt. Usu, Japan and White Island, New Zealand. Appl Geochem 3:125-141

Giggenbach WF, Poreda RJ (1993) Helium isotopic and chemical composition of gases from volcanic-hydrothermal systems in the Philippines. Geothermics 22:369-380

Giggenbach WF, Martini M, Corazza E (1986) The effects of hydrothermal processes on the chemistry of some recent volcanic gas discharges. Per Mineral 55:15-28

Giggenbach WF, Sano Y, Wakita H (1993) Isotopic composition of helium, and CO_2 and CH_4 contents in gases produced along the New-Zealand part of a convergent plate boundary. Geochim Cosmochim Acta 57:3427-3455

Giggenbach WF, Tedesco D, Sulisiyo Y, Caprai A, Cioni R, Favara R, Fischer TP, Hirabayashi J, Korzhinsky M, Martini M, Menyailov I, Shinohara H (2001) Evaluation of results from Forth and Fifth IAVCEI Field Workshop on Volcanic Gases, Volcano Island, Italy and Java, Indonesia. J Vol Geotherm. Res 108:283-302

Gill JB (1976) Composition and age of Lau Basin and Ridge volcanic rocks: Implications for evolution of an interarc basin and remnant arc. Geol Soc Am Bull 87:1384-1395

Gill JB (1981) Orogenic andesites and plate tectonics. Springer-Verlag, Berlin

Goff F, McMurtry GM (2000) Tritium and stable isotopes of magmatic waters. J Volcanol Geotherm Res 97:347-396

Goff F, McMurtry GM, Counce D, Simac JA, Roldan-Manzo AR, Hilton DR (2000) Contrasting hydrothermal activity at Sierra Negra and Alcedo volcanoes, Galapagos Archipelago, Ecuador. Bull Volcanol 62:34-52

Graham DW (2002) Noble gas isotope geochemistry of mid-ocean ridge and ocean island basalts: Characterizatioon of mantle sourse reservoirs. Rev Mineral Geochem 47:247-318

Graham DW, Allard P, Kilburn CRJ, Spera EJ, Lupton JE (1993) Helium isotopes in some historical lavas from Mount Vesuvius. J Volcanol Geotherm Res 58:359-366

Hammouya G, Allard P, Jean-Baptiste P, Parello F, Semet MP, Young SR (1998) Pre- and syn-eruptive geochemistry of volcanic gases from Soufriere Hills of Montserrat, West Indies. Geophys Res Lett 25:3685-3688

Hart SR, Glassley WE, Karig DE (1972) Basalts and seafloor spreading behind the Mariana Island arc. Earth Planet Sci Lett 15:12-18

Hawkins JW (1976) Petrology and geochemistry of basaltic rocks of the Lau Basin. Earth Planet Sci Lett 28:283-297

Hildreth W, Moorbath S (1988) Crustal contributions to arc magmatism in the Andes of Central Chile. Contrib Mineral Petrol 98:455-489

Hilton DR (1996) The helium and carbon isotope systematics of a continental geothermal system - results from monitoring studies at Long Valley caldera (California, USA). Chem Geol 127:269-295

Hilton DR, Craig H (1989) A helium isotope transect along the Indonesian archipelago. Nature 342: 906-908

Hilton DR, Hoogewerff JA, van Bergen MJ, Hammerschmidt K (1992) Mapping magma sources in the east Sunda-Banda arcs, Indonesia: constraints from helium isotopes. Geochim Cosmochim Acta 56:851-859

Hilton DR, Hammerschmidt K, Teufel S, Friedrichsen H (1993a) Helium isotope characteristics of Andean geothermal fluids and lavas. Earth Planet Sci Lett 120:265-282

Hilton DR Hammerschmidt K, Loock G, Friedrichsen H (1993b) Helium and argon isotope systematics of the central Lau Basin and Valu Fa ridge: Evidence of crust-mantle interactions in a back-arc basin. Geochim Cosmochim Acta 57:2819-2841

Hilton DR, Barling J, Wheller GE (1995) Effect of shallow-level contamination on the helium isotope systematics of ocean-island lavas. Nature 373:330-333

Hilton DR, McMurtry GM, Goff F (1998) Large variations in vent fluid $CO_2/^3He$ ratios signal rapid changes in magma chemistry at Loihi seamount, Hawaii. Nature 396:359-362

Hilton DR, Gronvold K, Macpherson CG, Castillo PR (1999) Extreme $^3He/^4He$ ratios in northwest Iceland: constraining the common component in mantle plumes. Earth Planet Sci Lett 173:53-60

Hiyagon H (1994) Rententivity of solar He and Ne in IDPs in deep sea sediment. Science 263:1257-1259

Hiyagon H, Kennedy BM (1992) Noble gases in CH_4-rich gas fields, Alberta, Canada. Geochim Cosmochim Acta 56:1569-1589

Hofmann AW (1997) Mantle geochemistry: The message from oceanic volcanism. Nature 385:219-229

Honda M, McDougall I, Patterson DB, Doulgeris A, Clague DA (1991) Possible solar noble-gas component in Hawaiian basalts. Nature 349:149-151

Honda M, McDougall I, Patterson D. (1993a) Solar noble gases in the Earth: The systematics of helium-neon isotopes in mantle derived samples. Lithos 30:257-265

Honda M, Patterson DB, McDougall I, Falloon T (1993b) Noble gases in submarine pillow basalt glasses from the Lau Basin—detection of a solar component in backarc basin basalts. Earth Planet Sci Lett 120:135-148

Hulston JR, Lupton JE (1989) Helium isotope studies in the Taupo Volcanic Zone, New Zealand. Proc Conf Water-Rock Interaction 6:317-320, Balkema, Rotterdam

Hulston JR, Lupton JE (1996) Helium isotope studies of geothermal fields in the Taupo Volcanic Zone, New Zealand. J Volcanol Geotherm Res 74:297-321

Hulston JR, Lupton J, Rosenberg N (1986) Variations of $^3He/^4He$ isotope ratios within the Broadlands Geothermal Field, New Zealand. Proc N Z Geothermal Workshop 8:13-16

Igarashi G, Ozima M, Ishibashi J, Gamo T, Sakai H, Nojiri Y, Kawai T (1992) Mantle helium flux from the bottom of Lake Mashu, Japan. Earth Planet Sci Lett 108:11-18

Ikeda Y, Nagao K, Stern RJ, Yuasa M, Newman S (1998) Noble gases in pillow basalt glasses from the northern Mariana Trough back arc-basin. The Island Arc 7:471-478

Ikeda Y, Nagao K, Kagami H (2001) Effects of recycled materials involved in a mantle source beneath the southwest Japan arc region: Evidence from noble gas, Sr, and Nd isotopic systematics. Chem Geol 175:509-522

Ishibashi JI, Wakita H, Nojiri Y, Grimaud D, Jean-Baptiste P, Gamo T, Auzende JM, Urabe T (1994) Helium and carbon geochemistry of hydrothermal fluids from the North Fiji Basin spreading ridge (southwest Pacific). Earth Planet Sci Lett 128:183-197

Jambon A, Zimmermann JL (1990) Water in oceanic basalts: Evidence for dehydration of recycled crust. Earth Planet Sci Lett 101:323-331

Jambon A, Weber HW, Braun O (1986) Solubility of He, Ne, Ar, Kr and Xe in a basalt melt in the range 1250-1600°C. Geochemical implications. Geochim Cosmochim Acta 50:255-267

Janik CJ, Goff F, Fahlquist L, Adams AI, Roldan MA, Chipera SJ, Trujillo PE, Counce D (1992) Hydrogeochemical exploration of geothermal prospects in the Tecuamburro volcano region, Guatemala. Geothermics 21:447-481

Jarrard RD (1986) Relations among subduction parameters. Rev Geophys 24:217-284

Javoy M, Pineau F, Delorme H (1986) Carbon and nitrogen isotopes in the mantle. Chem Geol 57:41-62

Kastner M, Elderfield H, Jenkins WJ, Gieskes JM, Gamo T (1993) Geochemical and isotopic evidence for fluid flow in the western Nankai subduction zone, Japan. Proc ODP Sci Res 131:397-413

Kennedy BM, Lynch MA, Reynolds JH, Smith SP (1985) Intensive sampling of noble gases in fluids at Yellowstone: I Early overview of the data; regional patterns. Geochim Cosmochim Acta 49:1251-1261

Kennedy BM, Reynolds JH, Smith SP (1988) Noble gas geochemistry in thermal springs. Geochim Cosmochim Acta 52:1919-1928

Kennedy BM, Hiyagon H, Reynolds JH (1990) Crustal neon: A striking uniformity. Earth Planet Sci Lett 98:277-286

Kennedy BM, Hiyagon H, Reynolds JH (1991) Noble gases from Honduras geothermal sites. J Volcanol Geotherm Res 45:29-39

Kerrick DM, Connolly JAD (2001) Metamorphic devolatilization of subducted marine sediments and the transport of volatiles into the Earth's mantle. Nature 411:293-296

Kirsanova TP, Yurova LM, Vergasova LP, Taran YA (1983) Fumarolic activity of Shiveluch and Kizimen volcanoes in 1979-1980. Volcanol Seismol 3:33-43.

Kita I, Nitta K, Nagao K, Taguchi S, Koga A (1993) Difference in N_2/Ar ratio of magmatic gases from northeast and southwest Japan: New evidence for different states of plate subduction. Geology 21: 391-394

Kiyosu Y, Kurahashi M (1984) Isotopic geochemistry of acid thermal waters and volcanic gases from Zao Volcano in Japan. J Volcanol Geotherm Res 21:313-331

LeGuern F (1982) Discharges of volcanic CO_2 and SO_2 in the atmosphere. Bull Volcanol 45:197-202 (in French).

Lewicki JL, Fischer T, Williams SN (2000) Chemical and isotopic compositions of fluids at Cumbal Volcano, Colombia: Evidence for magmatic contribution. Bull Volcanol 62:347-361

Macpherson CG, Hilton DR, Sinton JM, Poreda RJ, Craig H (1998) High $^3He/^4He$ ratios in the Manus backarc basin: Implications for mantle mixing and the origin of plumes in the western Pacific Ocean. Geology 26:1007-1010

Macpherson CG, Hilton DR, Mattey DP, Sinton JM (2000) Evidence for an ^{18}O-depleted mantle plume from contrasting $^{18}O/^{16}O$ ratios of back-arc lavas from the Manus Basin and Mariana Trough. Earth Planet Sci Lett 176:171-183

Magro G, Pennisi M (1991) Noble gases and nitrogen: mixing and temporal evolution in the fumarolic fluids of Volcano, Italy. J Volcanol Geotherm Res 47:237-247

Mamyrin BA, Tolstikhin IN, Anufriev GS, Kamensky IL (1969) Anomalous isotopic composition of helium in volcanic gases. Dokl Akad Nauk SSSR 184:1197-1199 (in Russian)

Marty B, Dauphas N (2002) Formation and early evolution of the atmosphere. J Geol Soc London Sp Publ (Early history of the Earth) In Press.

Marty B, Giggenbach WF (1990) Major and rare gases at White Island volcano, New Zealand: origin and flux of volatiles. Geophys Res Lett 17:247-250

Marty B, Humbert F (1997) Nitrogen and argon isotopes in oceanic basalts. Earth Planet Sci Lett 152: 101-112

Marty B, Jambon A (1987) C/^{3}He in volatile fluxes from the solid Earth: Implication for carbon geodynamics. Earth Planet Sci Lett 83:16-26

Marty B, LeCloarec MF (1992) Helium-3 and CO_2 fluxes from subaerial volcanoes estimated from Polonium-210 emissions. J Volcanol Geotherm Res 53:67-72

Marty B, Lussiez P (1993) Constraints on rare gas partition coefficients from analysis of olivine glass from a picritic mid-ocean ridge basalt. Chem Geol 106:1-7

Marty B, Tolstikhin IN (1998) CO_2 fluxes from mid-ocean ridges, arcs and plumes. Chem Geol 145: 233-248

Marty B, Zimmermann L (1999) Volatiles (He, C, N, Ar) in mid-ocean ridge basalts: Assesment of shallow-level fractionation and characterization of source composition. Geochim Cosmochim Acta 63:3619-3633

Marty B, Jambon A, Sano Y (1989) Helium isotopes and CO_2 in volcanic gases of Japan. Chem Geol 76:25-40

Marty B, Trull T, Lussiez P, Basile I, Tanguy JC (1994) He, Ar, O, Sr and Nd isotope constraints on the origin and evolution of Mount Etna magmatism. Earth Planet Sci Lett 126:23-39

Marty B, Tolstikhin I, Kamensky IL, Nivin V, Balaganskaya E, Zimmermann JL (1998) Plume-derived rare gases in 380 Ma carbonatites from the Kola region (Russia) and the argon isotopic composition in the deep mantle. Earth Planet Sci Lett 164:179-192

Marty B, Sano Y, France-Lanord C (2001) Water-saturated oceanic lavas from the Manus Basin: volatile behaviour during assimilation-fractional crystallisation-degassing (AFCD). J Volcanol Geotherm Res 108:1-10

Matsubayashi O, Matsuo S, Kaneoka I, Ozima M (1978) Rare gas abundance pattern of fumarolic gases in Japanese volcanic areas. *In* Terrestrial rare gases. Alexander Jr EC, Ozima M (eds). Japan Scientific Societies Press, Toyko p 27-33

Matsumoto T, Chen Y, Matsuda J (2001) Concomitant occurrence of primordial and recycled noble gases in the Earth's mantle. Earth Planet Sci Lett 185:35-47

Matsuo S, Suzuoki T, Kusakabe M, Wada H, Suzuki M (1974) Isotopic and chemical composition of volcanic gases from Satsuma-Iwojima, Japan. Geochem J 8:165-173

Matsuo S Suzuki M Mizutani Y (1978) Nitrogen to argon ratio in volcanic gases. *In* Terrestrial rare gases. Alexander Jr EC, Ozima M (eds) Japan Scientific Societies Press, Toyko p 17-25.

Mazor E, Wasserburg GJ (1965) Helium, neon, argon, krypton, and xenon in gas emanations from Yellowstone and Lassen Volcanic National Parks. Geochim Cosmochim Acta 29:443-454

Menyailov IA, Nikitina LP (1980) Chemistry and metal contents of magmatic gases: The new Tolbachik volcanoes (Kamchatka). Bull Volcanol 43:197-207

Menyailov IA, Nikitina LP, Shapar VN, Pilipenko VP (1986a) Temperature increase and chemical change of fumarolic gases at Momotombo Volcano, Nicaragua, in 1982-1985; are these indicators of a possible eruption? J Geophys Res 91:12199-12214

Menyailov IA, Nikitina LP, Shapar VN, Rozhkov AM, Miklishansky AZ (1986b) Chemical composition and metal content in gas discharges from the 1981 eruption of Alaid volcano, Kuril Islands. Volcanol Seismol 1:26-31

Minissale A, Evans WC, Magro G, Vaselli O (1997) Multiple source components in gas manifestations from north-central Italy. Chem Geol 142:175-192

Mitzutani Y (1962) Origin of lower temperature fumarolic gases at Showashinzan. J Earth Sci Nagoya Univ 10:135-148

Mitzutani Y (1966) A geochemical study of the fumarolic bithmutite from Showashinzan. Bull Chem Soc Japan 39:511-516

Mizutani Y, Sugiura T (1982) Variations in chemical and isotopic compositions of fumarolic gases from Showashinzan Volcano, Hokkaido, Japan. Geochem J 16:63-71

Mizutani Y, Hayashi S, Sugiura T (1986) Chemical and isotopic compositions of fumarolic gases from Kuju-Iwoyama, Kyushu, Japan. Geochem J 20:273-285

Motyka RJ, Liss SA, Nye CJ, Moorman MA (1993) Geothermal resources of the Aleutian arc. Professional Report. Alaska Division of Geological & Geophysical Surveys, p 1-17

Nagao K, Takaoka N, Matsubayashi O (1981) Rare gas isotopic compositions in natural gases of Japan. Earth Planet Sci Lett 53:175-188

Nakai S, Wakita H, Nuccio MP, Italiano F (1997) MORB-type neon in an enriched mantle beneath Etna, Sicily. Earth Planet Sci Lett 153:57-66

Nemoto T (1957) Report on the geological, geophysical and geochemical studies of Usu volcano (Showa Shinzan). Geol Surv Japan 170:1-24

Nishio Y, Sasaki S, Gamo T, Hiyagon H, Sano Y (1998) Carbon and helium isotope systematics of North Fiji Basin basalt glasses: Carbon geochemical cycle in the subduction zone. Earth Planet Sci Lett 154:127-138

Norton FJ (1957) Permeation of gases through solids. J Appl Phys 28:34-39

Nuccio PM, Paonita A, Sortino F (1999) Geochemical modeling of mixing between magmatic and hydrothermal gases: The case of Vulcano Island, Italy. Earth Planet Sci Lett 167:321-333

Ohba T, Hirabayashi J, Yoshida M (1994) Equilibrium temperatures and redox state of volcanic gas at Unzen volcano, Japan. J Volcanol Geotherm Res 60:263-272

Ohba T, Hirabayashi J, Nogami K (2000) D/H and $^{18}O/^{16}O$ ratios of water in the crater lake at Kusatsu-Shirane volcano, Japan. J Volcanol Geotherm Res 97:329-346

Ohmoto H. (1986) Stable isotope geochemistry of ore deposits. Rev Mineral 16:491-559

Ozima M, Podosek FA (1983) Noble Gas Geochemistry. Cambridge University Press

Patterson DB, Honda M, McDougall I (1994) Noble gases in mafic phenocrysts and xenoliths from New Zealand. Geochim Cosmochim Acta 58:4411-4427

Patterson DB, Farley KA, McInnes BIA (1997) Helium isotopic composition of the Tabar-Lihir-Tanga-Feni island arc, Papua New Guinea. Geochim Cosmochim Acta 61:2485-2496

Pedroni A, Hammerschmidt K, Friedrichsen H (1999) He, Ne, Ar, and C isotope systematics of geothermal emanations in the Lesser Antilles Islands Arc. Geochim Cosmochim Acta 63:515-532

Pepin RO (1991) On the origin and early evolution of terrestrial planet atmospheres and meteoritic values. Icarus 92:1-79

Peters KE, Sweeney RE, Kaplan IR (1978) Correlation of carbon and nitrogen stable isotope ratios in sedimentary organic matter. Limnol Ocean 23:598-604

Plank T, Langmuir CH (1998) The chemical composition of subducting sediment and its consequences for the crust and mantle. Chem Geol 145:325-394

Poorter RPE, Varekamp JC, Poreda RJ, Van Bergen MJ, Kruelen R (1993) Chemical and isotopic compositions of volcanic gases from the East Sunda and Banda Arcs, Indonesia. Geochim Cosmochim Acta 55:3795-3807

Porcelli D, Ballentine CJ (2002) Models for the distribution of terrestrial noble gases and the evolution of the atmosphere. Rev Mineral Geochem 47:411-480

Porcelli D, Wasserburg GJ (1995) Mass transfer of helium, neon, argon, and xenon through a steady-state upper mantle. Geochim Cosmochim Acta 59:4921-4937

Porcelli, DR, O'Nions RK, Galer SJG, Cohen AS, Mattey DP (1992) Isotopic relationships of volatile and lithophile trace elements in continental ultramafic xenoliths. Contrib Mineral Petrol 110:528-538

Poreda RJ (1985) Helium-3 and deuterium in back-arc basalts: Lau Basin and the Mariana Trough. Earth Planet Sci Lett 73:244-254

Poreda R, Craig H (1989) Helium isotope ratios in circum-Pacific volcanic arcs. Nature 338:473-478

Poreda RJ, Craig H (1992) He and Sr isotopes in the Lau Basin mantle—depleted and primitive mantle components. Earth Planet Sci Lett 113:487-493

Poreda RJ, Schilling JG, Craig H (1986) Helium and hydrogen isotopes in ocean ridge basalts north and south of Iceland. Earth Planet Sci Lett 78:1-17

Poreda RJ, Jeffrey AWA, Kaplan IR, Craig H (1988) Magmatic helium in subduction-zone natural gases. Chem Geol 71:199-210

Reyes AG, Giggenbach WF, Saleras JRM, Salonga ND, Vergara MC (1993) Petrology and geochemistry of Alto Peak, a vapor-cored hydrothermal system, Leyte Provnce, Philippines. Geothermics 22: 479-519

Richardson SH, Gurney JJ, Erlank AJ, Harris JW (1984) Origin of diamonds in old enriched mantle. Nature 310:198-202

Rogie JD, Kerrick DM, Chiodini G, Frondini F (2000) Flux measurements of non-volcanic CO_2 emission from some vents in central Italy. J Geophys Res 105:8435-8445

Rowe Jr GL, Brantley SL, Fernandez M, Fernandez JF, Borgia A, Barquero J (1992) Fluid-volcano interaction in an active stratovolcano: The crater lake system of Poás volcano, Costa Rica. J Volcanol Geotherm Res 49:23-52

Sano Y, Marty B (1995) Origin of carbon in fumarolic gas from island arcs. Chem Geol 119:265-274

Sano Y, Wakita, H (1985) Geographical distribution of $^3He/^4He$ ratios in Japan: Implications for arc tectonics and incipient magmatism. J Geophys Res 90:8729-8741

Sano Y, Williams SN (1996) Fluxes of mantle and subducted carbon along convergent plate boundaries. Geophys Res Lett 23:2749-2752

Sano Y, Nakamura, Wakita H, Urabe A, Tominaga T. (1984) Helium-3 emission related to volcanic activity. Science 224:150-151

Sano Y, Nakamura Y, Wakita H, Notsu K, Kobayashi Y (1986) $^3He/^4He$ ratio anomalies associated with the 1984 Western Nagano earthquake: Possibly induced by a diapiric magma. J Geophys Res 91:12291-12295

Sano Y, Wakita H, Giggenbach WF (1987) Island arc tectonics of New Zealand manifested in helium isotope ratios. Geochim Cosmochim Acta 51:1855-1860

Sano Y, Nakamura Y, Notsu K, Wakita H (1988) Influence of volcanic eruptions on helium isotope ratios in hydrothermal systems induced by volcanic eruptions. Geochim Cosmochim Acta 52:1305-1308

Sano Y, Wakita H, Italiano F, Nuccio MP (1989) Helium isotopes and tectonics in southern Italy. Geophys Res Lett 16:511-514

Sano Y, Notsu K, Ishibashi J, Igarashi G, Wakita H (1991) Secular variations in helium isotope ratios in an active volcano: Eruption and plug hypothesis. Earth Planet Sci Lett 107:95-100

Sano Y, Hirabayashi J, Oba T, Gamo T (1994) Carbon and helium isotopic ratios at Kusatsu-Shirane volcano, Japan. Appl Geochem 9:371-377

Sano Y, Gamo T, Williams SN (1997) Secular variations of helium and carbon isotopes at Galeras volcano, Colombia. J Volcanol Geotherm Res 77:255-265

Sano Y, Nishio Y, Gamo T, Jambon A, Marty B (1998a) Noble gas and carbon isotopes in Mariana Trough basalt glasses. Appl Geochem 13:441-449

Sano Y, Takahata N, Nishio Y, Marty B (1998b) Nitrogen recycling in subduction zones. Geophys Res Lett 25:2289-2292

Sano Y, Takahata N, Nishio Y, Fischer TP, Williams SN (2001) Volcanic flux of nitrogen from the Earth. Chem Geol 171:263-271

Sarda P, Staudacher T, Allègre C.J. (1988) Neon isotopes in submarine basalts. Earth Planet Sci Lett 91:73-88

Saunders A, Tarney J (1991) Back-arc basins. *In* Oceanic Basalts, Floyd PA (ed). Blackie, Glasgow, p 219-263

Scarsi P (2000) Fractional extraction of helium by crushing of olivine and clinopyroxene phenocrysts: Effects on the $^3He/^4He$ measured ratio. Geochim Cosmochim Acta 64:3751-3762

Schiano P, Clocchiatti R, Ottolini L, Busa T (2001) Transition of Mount Etna lavas from a mantle-plume to an island-arc magmatic source. Nature 412:900-904

Schmidt MW, Poli S (1998) Experimentally based water budgets for dehydrating slabs and consequences for arc magma generation. Earth Planet Sci Lett 163:361-379

Shaw AM, Hilton DR, Macpherson CG, Sinton JM (2001) Nucleogenic neon in high $^3He/^4He$ lavas from the Manus back-arc basin: a new perspective on He-Ne decoupling. Earth Planet Sci Lett 194:53-66

Sheppard DS, Janik CJ, Keith TEC (1992) A comparison of gas chemistry of fumaroles on the 1912 ash flow sheet on active stratovolcanoes, Katmai National Park, Alaska. J Volcanol Geotherm Res 53: 185-197

Shinohara H, Giggenbach WF, Kazahaya K, Hedenquist JW (1993) Geochemistry of volcanic gases and hot springs of Satsuma-Iwojima, Japan. Geochem J 4/5:271-286.

Simkin T, Siebert L (1994) Volcanoes of the world. Geoscience Press, Tucson, Arizona

Snyder G, Poreda R, Hunt A, Fehn U (2001) Regional variations in volatile composition: Isotopic evidence for carbonate recycling in the Central American volcanic arc. Geochem Geophys Geosystems 2:U1-U32

Sriwana T, van Bergen ML, Varekamp JC, Sumarti S, Takano B, van Os BJH, Leng MJ (2000) Geochemistry of the acid Kawah Puti lake, Patuha Volcano, west Java, Indonesia. J Volcanol Geotherm Res 97:77-104

Staudacher T, Allègre CJ (1988) Recycling of oceanic crust and sediments: The noble gas subduction barrier. Earth Planet Sci Lett 89:173-183

Stoiber RE, Malinconico LLJ, Williams SN (1983) Use of the correlation spectrometer at volcanoes. *In* Forecasting volcanic events. Tazieff H, Sabroux J (eds) Elsevier, New York 1:425-444.

Stoiber RE, Williams SN, Hubert BJ (1987) Annual contribution of sulfur dioxide to the atmosphere by volcanoes. J Volcanol Geotherm Res 33:1-8

Sturchio NC, Williams SN, Sano Y (1993) The hydrothermal system of Volcan Purace, Colombia. Bull Volcanol 55:289-296

Sumino H, Nakai S, Nagao K, Notsu K (2000) High $^3He/^4He$ ratio in xenoliths from Takashima: Evidence for plume type volcanism in southwestern Japan. Geophys Res Lett 27:1211-1214

Symonds RB, Rose WI, Gerlach TM, Briggs PH, Harmon RS (1990) Evaluation of gases, condensates, and SO_2 emissions from Augustine Volcano, Alaska; the degassing of a Cl-rich volcanic system. Bull Volcanol 52:355-374

Symonds RB, Rose WI, Bluth GJS, Gerlach TM (1994) Volcanic-gas studies: methods, results, and applications. Rev Mineral 30:1-66

Symonds RB, Mitzutani Y, Briggs PH (1996) Long-term geochemical surveillance of fumaroles at Showa-Shinzan dome, Usu volcano, Japan. J Volcanol Geotherm Res 73:177-211

Taran YA (1985) Fumarolic activity of the Koryak volcano Kamchatka in 1983. Volcanol Seismol 5:82-85

Taran YA (1992) Chemical and isotopic composition of fumarolic gases from Kamchatka and Kuriles volcanoes. Geol Survey Japan 279:183-186

Taran YA, Korablev AN (1995) Chemical and isotopic composition of fumarolic gases from South Kurils volcanoes in 1992-1994. *In* Volcano-atmosphere interactions. Intl Chem Cong Pacific Basin Soceities (Pacifichem '95). Honolulu, Hawaii

Taran YA, Kirsanova TP, Esikov AD, Vakin EA (1987) Isotopic composition of water in fumarolic gases of some Kamchatkan volcanoes. Izv Acad Nauk USSR Geology 9:124-127 (in Russian)

Taran YA, Rozhkov AM, Serafimova EK, Esikov AD (1991) Chemical and isotopic composition of magmatic gases from the 1988 eruption of Klyuchevskoi volcano, Kamchatka. J Volcanol Geotherm Res 46:255-263

Taran YA, Pilipenko VP, Rozhkov AM, Vakin EA (1992) A geochemical model for fumaroles of the Mutnovsky volcano, Kamchatka, USSR. J Volcanol Geotherm Res 49:269-283

Taran YA, Hedenquist JW, Korzhinsky MA, Tkachenko SI, Shmulovich KI (1995) Geochemistry of magmatic gases from Kudriavy volcano, Iturup, Kuril Islands. Geochim Cosmochim Acta 59: 1749-1761

Taran YA, Connor CB, Shapar VN, Ovsyannikov AA, Bilichenko AA (1997) Fumarolic activity of Avachinsky and Koryaksky volcanoes, Kamchatka, from 1993 to 1994. Bull Volcanol 58:441-448

Taran Y, Fischer TP, Porovsky B, Sano Y, Armienta MA, Macias JL (1998) Geochemistry of the volcano-hydrothermal system of El Chichón volcano, Chiapas, Mexico. Bull Volcanol 59:436-449

Tarney J, Saunders A, Weaver SD (1977) Geochemistry of volcanic rocks from the island arcs of the Scotia Arc region. *In* Island arcs, deep sea trenches, and back-arc basins. Talwani M, Pitan III WC (eds) Am Geophys Union, Washington DC p 367-378

Tedesco D (1997) Systematic variations in the $^3He/^4He$ ratio and carbon of fumarolic fluids from active areas in Italy: Evidence for radiogenic ^4He and crustal carbon addition by the subducting African Plate? Earth Planet Sci Lett 151:255-269

Tedesco D, Nagao K (1996) Radiogenic ^4He, ^{21}Ne and ^{40}Ar in fumarolic gases on Vulcano: Implication for the presence of continental crust beneath the island. Earth Planet Sci Lett 144:517-528

Tedesco D, Allard P, Sano Y, Wakita H, Pece R (1990) Helium-3 in subaerial and submarine fumaroles of Campi Flegrei caldera, Italy. Geochim Cosmochim Acta 54:1105-1116

Tedesco D, Miele G, Sano Y, Toutain JP (1995) Helium isotopic ratio in Vulcano island fumaroles: Temporal variations in shallow level mixing and deep magmatic supply. J Volcanol Geotherm Res 64:117-128

Tedesco D, Nagao K, Scarsi P (1998) Noble gas isotopic ratios from historical lavas and fumaroles at Mount Vesuvius (southern Italy): constraints for current and future volcanic activity. Earth Planet Sci Lett 164:61-78

Tolstikhin IN, Marty B (1998) The evolution of terrestrial volatiles: a view from helium, neon, argon, and nitrogen isotope modeling. Chem Geol 147:27-52

Torgersen T (1989) Terrestrial helium degassing fluxes and the atmospheric helium budget: Implications with respect to the degassing processes of continental crust. Chem Geol 79:1-14

Torgersen T, Kennedy BM (1999) Air-Xe enrichments in Elk Hills oil field gases: Role of water in migration and storage. Earth Planet Sci Lett 167:239-253

Torgersen T, Lupton JE, Sheppard DS, Giggenbach WF (1982) Helium isotope variations in the thermal areas of New Zealand. J Volcanol Geotherm Res 12:283-298

Trull T (1994) Influx and age constraints on the recycled cosmic dust explanation for high $^3He/^4He$ ratios at hotspot volcanoes. *In* Noble gas geochemistry and cosmochemistry. Matsuda, J (ed) Terra, Tokyo, p 77-88

Trull TW, Kurz MD (1993) Experimental measurements of ^3He and ^4He mobility in olivine and clinopyroxene at magmatic temperatures. Geochim Cosmochim Acta 57:1313-1324

Trull TW, Perfit MR, Kurz MD (1990) He and Sr isotopic constraints on subduction contributions to Woodlark Basin volcanism, Solomon Islands. Geochim Cosmochim Acta 54:441-453

Turner S, Hawkesworth C (1998) Using geochemistry to map mantle flow beneath the Lau Basin. Geology 26:1019-1022

Uyeda S, Kanamori H (1979) Back-arc opening and the mode of subduction. J Geophys Res 84:1049-1061

Van Soest MC, Hilton DR, Kreulen R (1998) Tracing crustal and slab contributions to arc magmatism in the Lesser Antilles island arc using helium and carbon relationships in geothermal fluids. Geochim Cosmochim Acta 62:3323-3335

Van Soest MC, Hilton DR, Macpherson CG, Mattey DP (2002) Resolving sediment subduction and crustal contamination in the Lesser Antilles island arc: a combined He-O-Sr isotope approach. J Petrol 43: 143-170

Varekamp JC Kreulen R Poorter RPE Van Bergen MJ (1992) Carbon sources in arc volcanism, with implications for the carbon cycle. Terra Nova 4:363-373

Von Huene R, Scholl DW (1991) Observations at convergent margins concerning sediment subduction, subduction erosion, and the growth of continental crust. Rev Geophys 29:279-316

Weiss RF (1968) Piggyback sampler for dissolved gas studies on sealed water samples. Deep-Sea Res 15:695-699

Williams SN, Sano Y, Wakita H (1987) Helium-3 emission from Nevado del Ruiz volcano, Columbia. Geophys Res Lett 14:1035-1038

Williams SN, Schaefer SJ, Calvache MLV, Lopez D (1992) Global carbon dioxide emission to the atmosphere by volcanoes. Geochim Cosmochim Acta 56:1765-1770

Yatsevich I, Honda M (1997) Production of nucleogenic neon in the Earth from natural radioactive decay. J Geophys Res 102:10291-10298

Zapata G, Calvache ML, Cortés GP, Fischer TP, Garzon V, Gómez MD, Narváez ML, Ordoñez VM, Ortega EA, Stix, J, Torres CR, Williams SN (1997) SO_2 fluxes from Galeras volcano, Colombia 1989-1995: Progressive degassing and conduit obstruction of a Decade Volcano. J Volcanol Geotherm Res 77:195-208

Zhang YX, Zindler A (1993) Distribution and evolution of carbon and nitrogen in Earth. Earth Planet Sci Lett 117:331-345

Zindler A, Hart S (1986) Helium: problematic primordial signals. Earth Planet Sci Lett 79:1-8

10 Storage and Transport of Noble Gases in the Subcontinental Lithosphere

Tibor J. Dunai

Isotope Geochemistry, Faculty of Earth Sciences
Vrije Universiteit, de Boelelaan 1085
1091 RT Amsterdam, The Netherlands
dunt@geo.vu.nl

Donald Porcelli

Institute for Isotope Geology and Mineral Resources
ETH-Zürich, Sonneggstrasse 5
8092 Zürich, Switzerland
Now at: *Dept. of Earth Sciences, University of Oxford*
Parks Road, Oxford, OX1 3PR, United Kingdom

INTRODUCTION

Characteristics of the upper mantle are typically deduced from mid-ocean and ocean island volcanics that sample the convecting mantle. However, a substantial fraction of the upper mantle is isolated from this convecting reservoir as part of the subcontinental lithosphere. This is a potentially significant reservoir; not only might there be substantial abundances of noble gases and other trace element constituents, but also re-entrainment of this material into the deeper mantle may ultimately impart unique isotopic signatures to mantle domains sampled at ocean islands (McKenzie and O'Nions 1983). The characteristics of the lithosphere can be deduced from magmas derived by melting of this region and from xenolithic materials entrained by rapidly rising magmas. Unfortunately, magmas reaching the surface subaerially are generally strongly degassed and do not retain measurable noble gas concentrations, although some data can be obtained from phenocrysts. In contrast, noble gases as preserved in ultramafic xenoliths from continental volcanic provinces have provided substantial information regarding lithospheric processes and insights into the evolution of the subcontinental mantle. Diamonds also have received considerable attention, with the potential for preserving noble gases from greater depths and considerably longer time periods. Lithospheric mantle samples often show complex histories of metamorphism, deformation, and fluid interaction, and noble gases can provide constraints on the sources of trace element bearing fluids and the evolution of parent-daughter ratios.

Noble gases, in particular He, provide distinctive signatures for volatile sources. While the $^3He/^4He$ ratio of radiogenic He is very low (~0.01 R_A, where $R_A = 1.4\times 10^{-6}$, the atmospheric ratio), He found in mid-ocean ridge basalts (MORB) sampling the asthenospheric upper mantle are much higher (~8 R_A). This is due to mixing of radiogenic He with He that was trapped within the mantle since Earth formation with ≥120 R_A (see Porcelli and Ballentine 2002, this volume). The MORB range, representing a large volume of the accessible mantle, provides a benchmark for comparison for all other measured mantle compositions. Ratios higher than those in MORB have been found in ocean island (hotspot) basalts, indicating that there are mantle domains with even higher $^3He/U$ ratios. Regardless of the origin of these compositions, He isotope variations provide a method for identifying contributions to the continental lithosphere from different mantle domains, as well as from radiogenic contributions from environments with lower He/U ratios. Much of the discussions about xenolith noble gases concentrate

1529-6466/00/0047-0010$05.00

on He isotopes both for these reasons and the relative ease of making He isotope measurements, although there is some data regarding the other noble gases available.

In any sample, the trapped noble gases can be of asthenospheric, lithospheric and crustal origin, depending upon the pre-eruptive history. The main focuses of this paper are the asthenospheric and lithospheric contributions. Crustal noble gases can be important locally as contaminants of ascending mantle melts, introducing radiogenic and nucleogenic components (see Porcelli et al. 2002) to the original mantle signature, and will be treated accordingly. The data for noble gases in the underlying mantle and at subduction zones are reviewed elsewhere in this volume (Graham 2002; Hilton et al. 2002, respectively).

Some of the main questions that might be sought in the xenolith noble gas record are

- How much of the terrestrial noble gas inventory is within the continental lithosphere (how big is the reservoir)?
- What are the characteristics of this noble gas reservoir? How does this reservoir relate to other mantle domains and to what extent has it evolved independently?
- What are the sources of the noble gases? Does this provide information about the origin of continental lithosphere or trace element enrichment of the xenoliths?
- Is there a record of the characteristics of these sources in the past? For example, can we determine the past isotopic compositions of the mantle?
- What are the characteristics of mantle noble gases that might be seen in the overlying continental crust?

Unfortunately, many of these questions cannot be answered clearly. The available data will be reviewed here with the aim of identifying to what extent these issues can be addressed. This chapter will not address the data for xenoliths found in ocean island basalts (see e.g., Vance et al. 1989; Poreda and Farley 1992; Rocholl et al. 1996), which have rare gases that are typically associated with the asthenospheric source of the host magma.

SUBCONTINENTAL MANTLE AS A GEOCHEMICAL RESERVOIR

The size, composition, and age of the continental lithosphere have been the subject of numerous investigations. Following are some general considerations aimed at providing a context for viewing noble gas data. More detailed discussions are provided in the subsequent evaluations of specific noble gas studies.

Thickness of the lithosphere

The term 'lithosphere' strictly applies to the part of the mantle and crust that deforms elastically and reversibly to sustained loads. This is in contrast to the 'asthenosphere', which creeps under an applied load. Lithospheric thicknesses can be determined from observed or reconstructed flexural responses following, e.g., glacial or erosional unloading. Lithospheric thicknesses are usually <50 km for the oceanic lithosphere but are much greater under old and stable cratons. For the definition of a geochemical reservoir, the mechanical definition of the lithosphere is not entirely satisfactory, since whether or not a part of the mantle is attached and thus related to the overlying crust is not exclusively determined by its mechanical strength. Long term buoyancy (a function of temperature and composition) and (horizontal) stress levels may be of similar importance (Anderson 1990; Shapiro et al. 1999). The thermal boundary layer (TBL), separating convecting mantle with convective heat transport from overlying mantle with conductive heat transport, may be a more appropriate delineation of a geochemical reservoir. This may extend to greater depths than the mechanically defined lithosphere.

Therefore, the term lithosphere is applied here to this entire region as a chemically distinct mantle reservoir. The lithosphere has been imaged by seismic tomography (e.g., Blundell et al. 1992; Sobolev et al. 1997). The thickness of the lithosphere varies and reaches the greatest depths under Archean cratons, with estimates of ~250-400 km based on seismic data (Jordan 1975; Grand 1994; Nolet et al. 1994; Su et al. 1994; Polet and Anderson 1995; Ricard et al. 1996; Simons et al. 1999) and based on xenolith thermobarometry data (O'Reilly and Griffin 1996; Rudnick and Nyblade 1999; also Finnerty and Boyd 1987). Xenolith samples of this region are found in kimberlites and related rocks. To date, little work has been done on mantle noble gases in major silicate phases from xenoliths in these volcanics (Kaneoka et al. 1977). However, fluid inclusions within diamonds may provide important noble gas information (see *Diamonds* section), and so are the only sources of noble gas data for the thickest and oldest lithosphere. Younger continental areas such as western and central Europe typically have lithospheric thicknesses of ~100 km (Blundell et al. 1992), with regions of 50-70 km (Blundell et al. 1992; Sobolev et al. 1997). The lithosphere can be locally much thinner, especially in rifted areas that also contain the majority of volcanic centers that have yielded xenoliths containing mantle noble gases (e.g., O'Reilly and Griffin 1984). These recently erupted xenoliths have made the underlying mantle in these regions more amenable to study and are the subject of the bulk of the noble gas studies reviewed below. It is worth noting that if the lithosphere has an average thickness of 150 km and underlies 30% of the surface, then it constitutes about 7% of the upper mantle, and thus may be a significant mantle reservoir for trace constituents.

Composition of the lithosphere

Evidence for the bulk composition of the continental lithosphere has been reviewed by McDonough (1990), Rudnick et al. (1998), and Griffin et al. (1999). The main process that controls the major element geochemistry of the subcontinental lithospheric mantle is variable depletion by removal of partial melts prior to, or during, its stabilization from the underlying convecting system (see e.g., Boyd 1989; Richardson 1990; Griffin et al. 1999; Downes 2001). This depletion is also seen in Os isotopic systematics (e.g., Carlson and Irving 1994; Pearson 1999a). However, there has been considerable debate about the environment in which this takes place and the mantle domains involved, and is the subject of numerous studies. The generation of continental lithosphere may be the result of depletion of rising plume material under cratons (e.g., Jordan 1988; Stein and Hofmann 1994; Griffin et al. 1999); accreting mantle from convergent margins, which includes oceanic lithosphere, oceanic plateaus, and the mantle wedge overlying subducting slabs (Jordan 1988; Boyd 1989; Story et al. 1991; Stein and Goldstein 1996; Herzberg 1999); and underplating of asthenospheric mantle (Jochum et al. 1989). Therefore, there is the possibility of incorporating material, and noble gases, from a variety of mantle domains. There also appears to be a secular variation in the bulk composition of the lithosphere as well (Griffin et al. 1999).

In contrast to much of the major element data, many trace element and isotopic characteristics are not consistent with just depletion of mantle material, but rather show evidence of subsequent modification and both ancient and recent enrichment (e.g., Frey and Green 1974; Wass et al. 1980; Richardson et al. 1984; Menzies et al. 1985; Hawkesworth et al. 1984, 1990; Carlson and Irving 1994). This is due to infiltration by H_2O-rich fluids, silicate melts, or carbonatitic melts that re-enriches depleted rocks in LREE and other trace elements (e.g., Hawkesworth et al. 1990; Pearson 1999a; Downes 2001; Wilson and Downes 1991). These enriching fluids were likely introduced from a variety of sources with different characteristics, including fluids above subduction zones, fluids akin to the basaltic and kimberlitic volcanics that reach the surface, and highly mobile carbonate melts (see e.g., Menzies and Chazot 1995). The fluid source regions may involve different mantle domains, such as those seen in oceanic and island arc volcanism, or lithospheric material that has undergone heating or uplift. Also, multiple transits of fluid through the lithosphere can result in the generation of distinctive lithospheric compositions. The resulting enrichments evolve isotopically over different times, so that while some xenolith compositions clearly show the influence of subducted components or fluids related to the host magmas, a great diversity of compositions is observed that reflect a more complex lithospheric history (see Carlson 1995). Such modified lithosphere can then contribute to the next generation of fluids and melts (see Harry and Leeman 1995). It should be noted, however, that it is not clear to what extent such samples that are enriched in trace elements are representative of the bulk lithosphere (McDonough 1990; Rudnick et al. 1998). For example, Rudnick and Nyblade (1999) found that the geotherm below the Kalahari craton recorded in xenoliths was most compatible with no heat production within the lithosphere, implying minimal concentrations of U and Th. In contrast, O'Reilly and Griffin (2000) argue that a considerable amount of U and Th is stored in apatite and other secondary phases within the Phanerozoic lithospheric mantle, and this generates a significant amount of heat as well as radiogenic noble gases.

The details of lithospheric composition are best considered for individual locations. Here it is only worth emphasizing that the chemical composition of xenoliths reflect both major element depletion events and subsequent enrichment processes, and so have had complex open-system histories. The sources and characteristics of noble gases must consider this environment, as discussed further below.

Age of the lithosphere

The age of depletion and subsequent enrichment of the subcontinental lithosphere is related to the age of the overlying crust and the timing of tectonic processes that lead to its consolidation. Archean cratons may have lithospheric mantle roots that were depleted by up to 3.3 Ga ago, as most clearly seen in diamond inclusion Sm-Nd isochron and model age data (e.g., Richardson et al. 1984; Richardson and Harris 1997), Re-Os model depletion ages of mantle xenoliths (e.g., Walker et al. 1989; Pearson 1999a, b), and Pb isotope data for lithosphere-derived lavas and xenoliths (e.g., Rogers et al. 1994). The lithosphere underlying younger, more tectonically active continental regions such as Wyoming, southeast Australia, and western Europe may have been depleted $\leq$2 Ga ago (e.g., Carlson and Irving 1994; Handler et al. 1997; Burnham et al. 1998). While these ages may pertain to the initial stabilization of the lithosphere, subsequent enrichment events and so open system chemical behavior can extend up to even the rifting and deformation related to the eruption of the host magma. For example, subsequent enrichment can occur by subduction-related magmatism and metasomatism during orogenies or the latest magmatic activity; xenoliths from across Europe exhibit such evidence for a variety of modifications during different tectonic and magmatic events (Wilson and Downes 1991; Downes 2001). The age of the lithosphere clearly defines the time allowed for the development of noble gas isotopic compositions that are divergent from the initial noble gas source regions. Note that multiple episodic enrichment events may generate complex isotopic evolution patterns.

Possible xenolith noble gas components

A variety of components may be present in a mantle xenolith sample. As discussed in subsequent sections, considerable effort is often required to detect and confirm the presence each particular component.

Lithospheric mantle noble gases.

1. *Initially trapped mantle noble gases.* When the region of the mantle sampled by the xenolith was incorporated into the lithosphere and isolated from the rest of the mantle, noble gases were undoubtedly retained. As discussed above, these sources could include plumes, MORB source mantle (incorporated directly or through formation of oceanic lithosphere), or mantle wedge material from subduction zones. The noble gases would reflect the source isotopically, but would have had noble gas concentrations that depended upon the mechanisms of lithosphere formation. If it involved melt depletion, the noble gases would have been highly depleted, and perhaps also fractionated according to any differences in partition coefficients during melting, although partition coefficients are not known with enough precision to quantify possible fractionation effects (Caroll and Draper 1994). In such cases, these depleted gases could be easily overwhelmed by other components. Even where higher concentrations were retained, subsequent complex additions of other noble gases (see below) makes it difficult to identify initially trapped gases. Diamonds have been found to have very early formation ages, and so have been extensively studied with the hope that noble gases isolated within their structure have retained the signature of the mantle soon after lithosphere formation has been retained (see *Diamonds* section).
2. *Invading asthenospheric melts.* Volatile-rich silicate and carbonatitic melts from underlying regions may rise into the lithosphere, carrying noble gases and other trace elements. This can include melts from plumes that rise in the mantle and are responsible for hotspots, from upwelling asthenospheric mantle under rifting regions, and from subducting slabs and the overlying enriched mantle wedge. These melts carry noble gases isotopically representative of their source, but may be elementally fractionated due to processes governing their transfer into the transporting fluids. These noble gases are generally also represented at the surface in oceanic volcanism, away from possible alteration by the continental lithosphere, and so have been characterized by studies of MORB and OIB (see Graham 2002, this volume). Exceptions to this are noble gases from subducting crustal material that are released at subduction zones, which can contain dominantly radiogenic or atmospheric gases.
3. *Radiogenic components.* Radiogenic production of noble gases causes continual isotopic evolution of the lithosphere after formation, and modifies the isotopic ratios of gases introduced from the underlying mantle. However, how these compositions deviate from that of the convecting upper mantle depends upon the relative parent-daughter ratios. The noble gas isotope evolution of the lithosphere will diverge from that of the convecting mantle based upon several factors:
   - Melting during major element depletion may have preferentially removed the highly incompatible noble gases relative to parent elements, leaving a residue that becomes progressively more radiogenic than the underlying mantle.
   - It has been suggested that U is more incompatible than He during melting (Graham et al. 1990; Anderson 1998; Helffrich and Wood 2001). Similarly, K may be more incompatible than Ar. In this case, depleted mantle regions in the lithosphere may evolve with higher $^3He/^4He$ (and $^{40}Ar/^{36}Ar$) ratios than the convecting upper mantle.
   - It has been suggested that in the oceanic lithosphere, He-rich fluids can be stored in U-depleted dunites and harzburgites, and so retain their original isotopic

signature (Anderson 1998). The same might apply to regions of the continental lithosphere.

- Even without fractionation of parent elements and noble gases, the lithospheric isotopic evolution may diverge from the underlying mantle reservoirs. For example, the upper mantle evolves as an open system, with losses to the atmosphere and possible inputs from other mantle reservoirs (see Porcelli and Ballentine 2002, this volume). Therefore, lithospheric regions that are isolated from the relatively well-mixed upper mantle reservoir and are evolving as closed systems may have different isotopic evolution paths. Clearly, this will be more pronounced over greater time periods.

4. *Redistributed lithospheric volatiles.* Noble gases that are initially trapped or produced in the lithosphere can be redistributed in mobile phases, such as CO_2-rich fluids and silicate melts. Xenoliths record multiple events of fluid infiltration and trace element enrichment. Noble gases added during these events can include components that have resided elsewhere in the lithosphere and have been collected either during formation or passage of these migrating fluids. Note that the preferential movement of noble gases relative to parent elements may produce areas with high He/U ratios that therefore are resistant to change isotopically.

Post-entrainment noble gases.

1. *Transporting magma noble gases.* During transport to the surface, xenoliths are exposed to volatiles from the surrounding host magma. Transport times are rapid (Spera 1980), and so diffusion is not effective for imparting noble gases to the xenoliths. However, if a gas phase is present (i.e., CO_2), this could be incorporated into the xenoliths along cracks and grain boundaries. Note that the origin of these rare gases depends upon the provenance of the magmas, and so could also be from the lithosphere. Regardless of the source of magma volatiles, the volatiles are not derived from the source region of the xenoliths.
2. *Cosmogenic inputs.* Long residence at the surface exposes samples to production of nuclides by interaction with cosmic rays (see Niedermann 2002, this volume). This is most often seen for ^{3}He, because of the typically low abundance of this isotope. The effect is more pronounced at higher altitudes and latitudes, although it was found that ultramafic xenoliths anywhere that have been exposed for substantial times at the surface and depleted in He due to eruption or weathering can have extremely high ^{3}He/^{4}He ratios due to such inputs (Porcelli et al. 1987). It has been shown that such inputs also can be responsible for smaller shifts (Cerling and Craig 1994). Cosmogenic Ne contributions have also been measured (Staudacher and Allègre 1991, 1993). Samples studied more recently are generally collected from where exposure was limited due either to shielding by burial or recent eruption.
3. *Post-eruption grow-in.* Due to He losses during eruption, the U/He ratio of the xenoliths will increase substantially after eruption, and increasing the possibility of lowering the ^{3}He/^{4}He ratio through production of ^{4}He. Weathering at the surface will further release He, and can add U. These effects can be minimized by selecting very young unweathered samples. Corrections can also be made for older samples, although care must be taken to not lose He by sample processing, thereby altering the U/He ratio prior to measurement.

In general, in order to characterize the noble gas state of the continental lithosphere, it is necessary to measure or clearly deduce the pre-entrainment noble gas isotope inventories of xenolith samples. This first requires subtracting the effects of post-eruption cosmogenic and radiogenic additions either by careful sample selection or analytical techniques such as step heating and crushing that are designed to separate these components. A more difficult problem involves subtracting the addition of noble gases from either the host magma or from magmatic activity immediately before xenolith entrainment. It is worth noting that xenoliths are typically brought to the surface in volatile-rich alkaline, kimberlitic, or carbonatitic magmas that can potentially contaminate the volatile budget of xenoliths. The remaining noble gases are essentially asthenospheric mantle rare gases (and possibly subducted crust components), modified by radiogenic contributions during residence in the lithosphere.

NOBLE GASES IN ULTRAMAFIC XENOLITHS AND PHENOCRYSTS

Many noble gas studies have concentrated on ultramafic xenoliths and phenocrysts that are found in alkali basalts. Relatively recent eruptions from a wide range of locations provide suites of samples that are relatively unweathered, and so have retained mantle noble gases. Most of the available data is for He isotopes, and other data are discussed separately in the section *The heavier noble gases*.

Fluid inclusions as hosts for noble gases

It was recognized very early in noble gas studies that volatiles found in xenoliths are preferentially located in fluid inclusions. The noble gases are not readily incorporated into crystal lattices, and strongly partition into the CO_2-rich fluids that typically occupy mantle fluid inclusions. This was demonstrated for trapped ^{40}Ar in Hawaiian xenoliths using in vacuo crushing techniques designed to preferentially release such gases (Funkhauser and Naughton 1968), and this continues to be used to separate trapped gases from radiogenic or cosmogenic nuclides created and trapped within crystals. Bernatowicz (1981) argued that the heavier noble gases were also concentrated in fluid inclusions in a spinel lherzolite based on disaggregation experiments, although the gases were isotopically indistinguishable from air and may have been dominated by contamination. Subsequent work has documented how trapped He can be released by crushing and separated from lattice-hosted noble gases (Hilton et al. 1993b; Kurz et al. 1990; Scarsi 2000), further indicating that fluid inclusions are the predominant host of mantle noble gases in xenoliths and phenocrysts. Also, a broad correlation between fluid inclusion abundances and noble gas concentrations in xenolith suites has been observed (Dunai and Baur 1995). Note that fluids and noble gases may also reside along grain boundaries at depth. However, these are likely to be completely lost by movement along the boundaries during transport, incipient decompression melting, and sample preparation. Therefore, the quantities stored in grain boundaries at depth cannot be constrained.

Within the mantle, He in fluid inclusions is more likely to be preserved in place than is He in the mineral lattice. He diffusion coefficients in clinopyroxenes and amphibole (two major U-bearing minerals) are ~10^{-5} cm^2/sec at 1300°C (e.g., Lippolt and Weigel 1988) and so can be readily lost from minerals to grain boundaries. However, prior to migration across the lattice He in fluid inclusions must first partition into the mineral lattice, which is highly unfavored (Trull and Kurz 1993). In this way, fluid inclusion He may be retained and also remain separate from radiogenic He produced in mineral lattices. At the surface, the situation is reversed, with low temperature diffusion in the lattice sufficiently low to prevent escape, but loss readily occurring from fluid inclusions due to mineral fracturing. These considerations also hold for the other noble gases, although diffusion coefficients will be lower.

Inter-mineral differences. Systematic differences between He concentrations in coexisting mineral phases have been observed (Fig. 1). Olivines in samples from E Africa and SE Australia were generally found to contain less He than coexisting clinopyroxenes

(Porcelli et al. 1986), and a fluid inclusion study of the SE Australia suite found that pyroxenes had the greatest density of inclusions (Andersen et al. 1984). Fluid inclusion works for elsewhere report similar observations (Ertan and Leeman 1999; Andersen and Neumann 2001). More extensive He data from European locations also found a similar pattern (Dunai and Baur 1995), as well as somewhat higher He concentrations (by an average of 1.8 times) in clinopyroxenes than in orthopyroxenes. These distributions likely relate either to the ease in which fluids can invade the different crystal structures (and so preferentially adding volatiles to weaker phases), or the ability of the crystals to withstand the internal pressures of the inclusions during decompression (and so preferentially preserving volatiles in stronger phases).

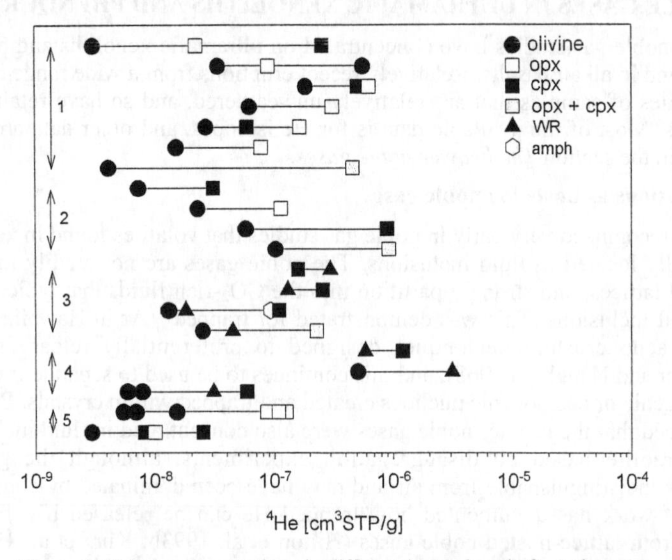

Figure 1. Helium concentrations in different host minerals from the same xenolith sample. Note that olivines tend to have the lowest concentrations, while pyroxenes have the highest. References: 1 Massif Central (Dunai and Baur 1995); 2 Eifel (Dunai and Baur 1995); Bullenmerri (Porcelli et al. 1986); 4 Tanzania (Porcelli et al. 1986); Cameroon (Barfod et al. 1999).

Trapping pressures. A common observation in fluid inclusion studies is that inclusions record maximum trapping pressures equivalent to depths that are much shallower than the source region of the xenoliths (Roedder 1984; Andersen and Neumann 2001). Such lower pressures can reflect either trapping of a CO_2 phase present in the transporting magma at the recorded pressures, or the decrepitation of inclusions that were trapped at greater depths (i.e., prior to entrainment) to form of secondary, lower pressure inclusions. The latter process is likely controlled by the strength of the host mineral. It has been argued that in some cases, a clustering around a low-pressure value reflects extended residence at that depth, for example in a magma chamber (Frezzotti et al. 1991). Distinguishing between these possibilities is important to determine whether the noble gases are from the lithospheric source region of the xenoliths or from the source region of the host magma, and requires detailed examination of crystal microstructures, such as evidence for the coincident introduction of host melt inclusions along healed fracture planes, or evidence that fluid-bearing inclusions are secondary (see Andersen and Neumann 2001). There are few locations from which samples have been studied for both fluid inclusion and He data, and are discussed in the *Regional studies* section below.

There are several observations that can be considered in evaluating the circumstances in which the CO_2-rich fluids in xenoliths were introduced and whether the host magma was the source of volatiles.

1. Melt inclusions seem to be common feature of mantle xenoliths worldwide. Detailed chemical analysis indicates that these are of different compositions to those of the host magma (Schiano and Clocchiatti 1994), and are the result of small degrees of partial melting. Upon emplacement, these melt inclusions were oversaturated in CO_2, and appear to be spatially related to CO_2 inclusions. In these cases, the noble gases in the fluid inclusions are likely to be derived from the melts invading the xenolith source region and were partitioned into fluid inclusions within the lithosphere prior to entrainment in the host magma.
2. If He in xenoliths were found to have an isotopic composition different from that in the host magma (or corresponding phenocrysts) a direct relationship would be precluded. However, there are few localities where data from both xenoliths and phenocrysts are available. In these cases, He compositions are not distinguishable (e.g., Vance et al. 1989; Barfod et al. 1999), although the possibility still remains that the source of CO_2-rich fluids in the xenolith simply originates from the same mantle region as the host magma. More data is needed before He isotopes can be used to determine whether trapped xenolith He is related to the He in surrounding magmas.
3. In order to introduce inclusions, sufficient shear stress must be applied to the xenoliths within the magma to create fractures. It is not clear under what circumstances this can be done without destroying the xenolith. CO_2 bubbles within the magma also must be available on the xenolith surfaces to enter propagating cracks. It is more likely that in many cases the fracturing and fluid introduction occur prior to xenolith entrainment, either immediately before xenolith entrainment, or by an earlier sequence of fluid invasion of the xenolith source region.

Overall, many studies assume that the rare gases measured in xenoliths were trapped in the xenolith source region and this is likely to be reasonable for many sample suites, although incorporation of fluid inclusions and noble gases at a later stage in the host magma is at least possible (Dunai and Baur 1995). The source of the noble gases trapped in xenoliths must be demonstrated in each case using fluid inclusion observations and noble gas data for the same samples. In contrast to xenoliths, phenocrysts may readily incorporate exsolved volatiles from the host magma during growth. If the He isotopic composition in the magma changes, for example due to incorporation of crustal material, this will be reflected in the compositions retained in later-growing phenocryst phases (e.g., Marty et al. 1994). It should be emphasized that it must always be clearly established using petrologic evidence that separate minerals are truly phenocrysts rather than xenocrysts derived from disrupted xenoliths.

Lithospheric He concentrations. A range of He concentrations has been found for xenoliths from each region. In an early survey of xenoliths, Porcelli et al. (1986) found (0.02-2) $\times10^{-6}$ cm^3 STP He/g in whole rock measurements, which can be compared to (4-15) $\times10^{-6}$ cm^3 STP He/g for the MORB-source mantle (see Porcelli and Ballentine 2002, this volume). While the xenolith concentrations therefore are considerably lower than that of the convecting upper mantle, the significance of this observation remains unclear. As discussed above, these xenolith concentrations reflect variable addition of volatiles into the crystals, as well as variable losses during decompression. A qualitative correlation has been observed between the abundance of fluid inclusions (filled and vacated) and He concentration (Dunai 1993), reflecting the effects of variable trapping. However, there is no data available to constrain the range of initial concentrations, nor

However, there is no data available to constrain the range of initial concentrations, nor the He concentration in inclusion fluids. Further, there is no evidence that the samples are representative of a substantial region of the lithosphere. Therefore, it is clear that noble gas data for ultramafic xenoliths cannot be easily used to quantify the noble gas abundances of the lithospheric mantle reservoir.

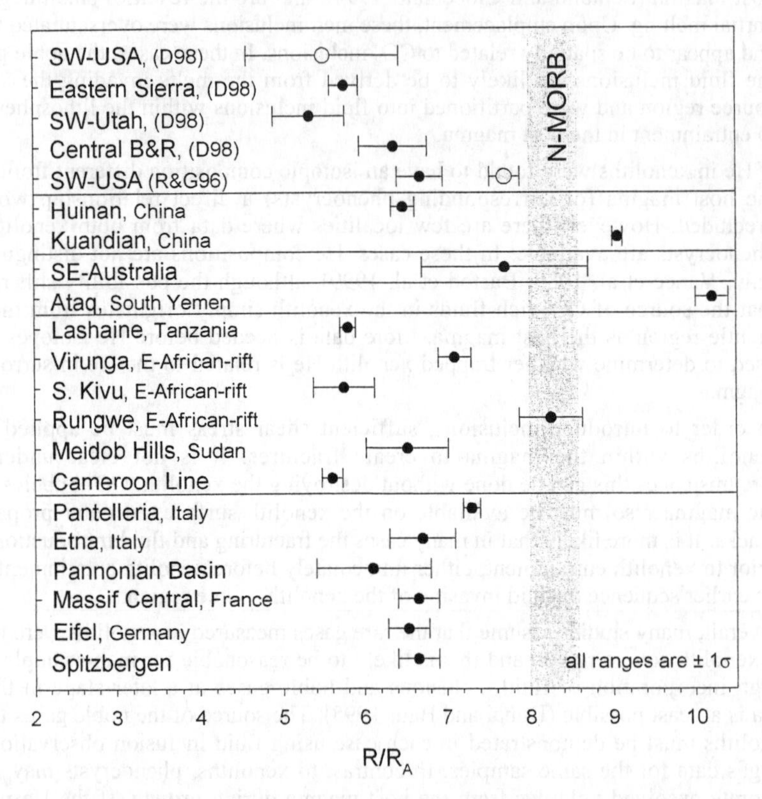

Figure 2. Published $^3He/^4He$-ratios (normalized to the atmospheric ratio $R_A = 1.39 \times 10^{-6}$) for phenocrysts and ultramafic xenoliths of various continental regions (exception: Pantelleria, modeled end member from volcanic fluids). Where weighted means were calculated, the larger, external error ($\pm 1\sigma$) is plotted. Sources are: SW-USA (D98) (Dodson et al. 1998) mean all data; Eastern Sierra, SW-Utah, Central Basin and Range (Dodson et al. 1998) weighted mean; SW-USA (R&G96) (Reid and Graham 1996) weighted mean, gray circle least enriched sample with $\varepsilon_{Nd} \sim 10$; Huinan and Kuandian (Xu et al. 1998) weighted mean, each with one outlier excluded (heat-extracted samples, radiogenic and cosmogenic contributions likely); SE-Australia (Matsumoto et al. 2000) geometric mean of three xenolith populations, error as indicated by (Matsumoto et al. 2000); Ataq, South Yemen & Lashaine, Tanzania (Porcelli et al. 1987); Virunga, South Kivu and Rungwe (Graham et al. 1995); Meidob Hills (Franz et al. 1999) weighted mean; Cameroon Line (Barfod et al. 1999) weighted mean; Pantelleria (Parello et al. 2000); Etna (Marty et al. 1994); Pannonian Basin, Massif Central and Eifel (Dunai and Baur 1995), Eifel and Massif Central values are indistinguishable from modeled corresponding end member values from crater lakes (Aeschbach-Hertig et al. 1999; Aeschbach-Hertig et al. 1996); the (depleted) N-MORB-range is that of (Kurz 1991) and (Reid and Graham 1996) who give values of 8.2±0.2 and 8.3±0.3 R_A, respectively.

Helium isotopic variations

Soon after the discovery that He with $^3He/^4He$ ratios greater than that of the atmosphere was emanating from the mantle, Tolstikhin et al. (1974) reported values of up to 10 R_A (where the ratio of the atmosphere is $R_A = 1.39 \times 10^{-6}$) in ultramafic xenoliths from alkali basalts. In a survey of mantle-derived materials, Kyser and Rison (1982) reported values of 6 and 8 R_A for two young xenoliths, and much more radiogenic values in South African samples with long surface residence times. These data indicated that He with significant fractions of primordial ^3He, rather than just radiogenic He, was present in xenoliths and was related to He found in MORB. While these and other studies (Kaneoka et al. 1978; Poreda and Basu 1984) reported a range of $^3He/^4He$ ratios between that of the atmosphere and MORB, Porcelli et al. (1986) found a much more restricted range of 6.2 to 10.6 R_A for xenoliths from a range of locations, and suggested that some of the samples studied earlier had been altered by post-eruption radiogenic production of ^4He. Subsequent studies have provided considerable data for samples that have had limited residence at the surface.

In Figure 2, the ranges of He isotopic compositions in phenocyrsts and xenoliths from locations worldwide are compiled. The first-order observation (Porcelli et al. 1986) is that the data fall into a relatively narrow range that is close to that of typical or 'normal' mid-ocean ridge basalts (N-MORB), and is distinctive from radiogenic He that characterizes crustal rocks (~0.05 R_A) as well as from basalts from major ocean islands (OIB) such as Loihi and Iceland that have ratios of up to 38 R_A (see Graham 2002 for review of oceanic data). Therefore, it appears that He in the lithosphere sampled by xenoliths is similar to that in the underlying convecting mantle, and does not contain large fractions of radiogenic He. In detail, however, the samples do not correspond completely with MORB values. A common feature of some regions is that the $^3He/^4He$-ratios are below, i.e., more radiogenic, than that of the depleted N-MORB reservoir (Dunai and Baur 1995).

Several different reasons can be considered for He isotope displacements between the xenoliths and the range of MORB values that now serve as a reference:

1. *Inclusion of post-eruption ^4He.* This may be produced in the lattice as well as in melt inclusions that can contain a significant fraction of the bulk U and Th. Samples typically lose He during eruption, thereby greatly lowering the He/U ratio of the bulk sample and so making samples more sensitive to isotopic changes due to radiogenic grow-in. Many studies have used sample-crushing extraction techniques to preferentially release trapped He in fluid inclusions and minimize contributions from the lattice or melt inclusions. In some studies of samples with young eruption ages that have used heating extraction (Porcelli et al. 1986; Patterson et al. 1994; Dunai and Baur 1995; Xu et al. 1998; Matsumoto et al. 2000), the $^3He/^4He$ ratios of sample suites generally converged on well-defined values at high He concentrations. Also, data for olivines, which are poor in U, are less likely to be affected over the age of many of the vents studied. Studies have also used direct comparison of data obtained by different methods (Matsumoto et al. 2000) to show that there are no systematic He isotope differences due to post-eruption radiogenic contributions. These values obtained by heat extraction and crushing therefore typically can be taken as reliable indicators of the $^3He/^4He$ ratios unaffected by post-eruptive radiogenic ^4He in-growth, so long as consistent results are obtained over a range of He concentrations. However, it is possible that some of the reported lower $^3He/^4He$ ratios have been affected by minor additions of post-eruption radiogenic He production.

2. *Inclusion of* in situ *produced cosmogenic ^3He.* There are some xenoliths that have $^3He/^4He$ ratios that are greater than the MORB field. While the involvement of a

minor fraction of mantle He with high $^3He/^4He$ ratios, as seen in OIB, would explain this, it is likely that some cosmogenic ^3He has been added to some samples. A correction can be made if it is assumed that the same concentration of cosmogenic ^3He has been added to each mineral, and the resulting isotopic compositions vary due to different concentrations of trapped He with a uniform isotopic composition (Cerling and Craig 1994; Dunai and Baur 1995). For example, the highest values for Lashaine, Tanzania were found in xenoliths with low He contents (Porcelli et al. 1986) during total sample fusion, and a linear correlation between the inverse of He concentration and $^3He/^4He$ ratio is consistent with addition of a cosmogenic ^3He component due to ~12,000 years of exposure (Cerling and Craig 1994). The trapped composition was found to have 5.9 R_A, consistent with other localities in the region. For the more extensive data set of Dunai and Baur (1995), corrections were made on many samples also by assuming that intermineral isotopic differences were due to cosmogenic addition of ^3He, and this resulted in a narrow range of compositions for each locality. It is not always clear whether the greater spread in values at other localities is due to post-eruption modification, or a feature of the trapped He (see discussion below). Nevertheless, this highlights the care that must be taken in interpreting minor measured differences.

3. *Addition of crustal ^4He.* The He in crustal rocks is dominated by radiogenic He and can be present in high concentrations. These rocks can include subducted materials that can add He to the source region of mantle xenoliths. Alternatively, He in a host magma that supplies He to hosted xenoliths can contain contributions from crustal materials that either were subducted and so contributed to the melt source region, or were assimilated during magma transit to the surface. Note that where such contributions are added at mantle depths, the resulting lower $^3He/^4He$ ratios are representative of a lithospheric component. However, such contributions are likely only near subduction zones.
4. *Distinctive mantle source melt source regions.* While many OIB have He characterized by $^3He/^4He$ ratios that are higher than those in MORB, some have been found to have lower ratios (see Graham 2002). As discussed in the *Regional studies* section, the presence of such a component has implications for the causes of host magma volcanism as well as the sources of lithosphere He.
5. *Distinctive lithospheric mantle He.* As discussed in the *Possible xenolith noble gas components* section, radiogenic production within the lithosphere can lower the $^3He/^4He$ ratio of He derived from the convecting mantle. Therefore, the He represents a distinctive lithospheric component. This may be created in the xenolith source region, or may involve He that is remobilized in the lithosphere. As discussed below, regional studies often contend with distinguishing between distinctive asthenospheric and lithospheric sources for such He.

Overall, samples that have been erupted recently and have had minimal surface exposure to cosmic rays therefore can be evaluated for subsurface He components. Outside of areas where subducted crustal material may be important, the sources of He are generally limited by the narrow range in xenolithic $^3He/^4He$ ratios to be the convecting upper mantle as sampled by MORB, in some cases with contributions from radiogenic production within the lithosphere or from a distinctive OIB component. The significance of these components is addressed in the more detailed discussion in the *Regional studies* section.

He-Sr relationships

The isotopic composition of He can be compared with that of other tracers of mantle sources. Available data for measured $^3He/^4He$ ratios are compared with those of $^{87}Sr/^{86}Sr$ in Figure 3. A range of $^{87}Sr/^{86}Sr$ ratios have been obtained for these samples, from values within the range of MORB to the highest value measured in a mantle sample of 0.8360 for a Tanzanian garnet lherzolite (Cohen et al. 1984). These values represent a range of mantle sources for the lithophile elements as well as examples of very long isolation from the asthenospheric mantle in a lithospheric region with high Rb/Sr ratios. Since xenoliths that are enriched in trace elements are not particularly depleted in U (e.g., Cohen et al. 1984), radiogenic ^{87}Sr is expected to be accompanied by ^4He, even though it is not known how Rb/Sr relates to U/He ratios. However, in contrast to the wide spread in Sr isotope compositions, He isotopes fall within a very restricted range. The lack of correlation between these ratios indicates that there generally has been complete decoupling of the two isotopic systems (Porcelli et al. 1986). Similar features have also been observed for xenoliths from ocean islands (Vance et al. 1989). Also, a similar decoupling is observed when He isotopes are compared to those of Nd (Stone et al. 1990). There are two obvious possibilities to explain this general observation:

1. The processes that fractionate Rb from Sr do not greatly fractionate U from He, and so the He/U ratio of the lithosphere has reached low values. However, while it might be considered reasonable that the formation of migrating fluids preferentially concentrated both U and He, there is likely to be strong fractionations within the invaded mantle as the U is introduced into the mineral lattices or newly created phases and the He remains in the volatile phase. It would be expected that the result is large fractionations between U and He between xenoliths, and so the development of a much wider range of $^3He/^4He$ ratios than observed.

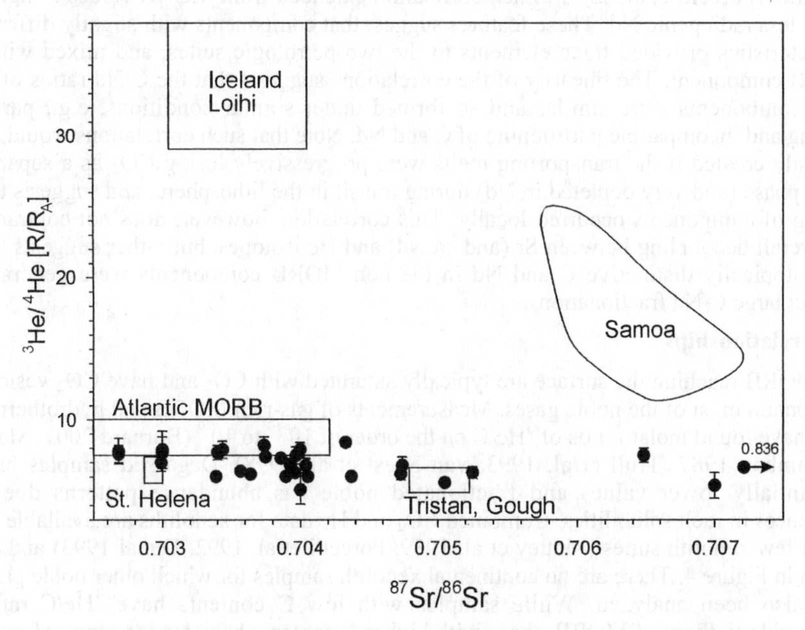

Figure 3. He-Sr-isotope relationships in xenoliths. For comparison a selection of MORB and OIB compositions are given that span the range found in present day mantle domains. A narrow range of $^3He/^4He$ ratios are observed in the xenoliths, despite the very large range in $^{87}Sr/^{86}Sr$. Data sources for xenoliths: Porcelli et al. (1986); Williams et al. (1992); Dunai and Baur (1995). Other data sources: MORB, Kurz et al. (1982); Samoa, Farley et al. (1992); St. Helena, Tristan and Gough, Kurz et al. (1982).

2. While a dominant fraction of the He has been introduced into the samples recently, the dominant fraction of Sr has been added earlier and so has had time to isotopically evolve to more radiogenic compositions. It has been shown that Nd, and certainly many other trace elements, does not preferentially enter CO_2-rich fluids (Meen et al. 1989). Where these elements are enriched, the transporting agent must have been a silicate or carbonatitic melt. Therefore, the introduction of the volatiles found in the fluid inclusions may have been the result of fluid migration that occurred much more recently than the events responsible for the enrichment of other trace elements.

This evidence suggests that CO_2-rich fluids invade the source regions of the xenoliths not long before their entrainment, or in some cases that the noble gases were trapped from the host magma degassing during ascent. In the former case the generation of the CO_2-rich fluids is likely related to upwelling of mantle material, resulting initially in very small degrees of melting. Also, these fluids must efficiently invade the xenolith source region, completely overwhelming any radiogenic He that might have been expected to accompany the radiogenic Sr and Nd. It is possible that while He/U fractionation occurs in the lithosphere, this may occur on a very restricted scale, so that the regional average is not unusually radiogenic. In this case, homogenization of lithospheric He by invading fluids may be sufficient to erase more radiogenic compositions.

An interesting observation that has some bearing on the coupling of volatiles and lithophile trace elements is that separate linear correlations were found between C and Nd isotopes in spinel lherzolites and wehrlites/pyroxenites in xenoliths from Bullenmerri, SE Australia (Porcelli et al. 1992). Each correlation extended from MORB values to lighter C and less radiogenic Nd. These features suggest that components with slightly different characteristics provided trace elements to the two petrologic suites, and mixed with a MORB component. The linearity of the correlations suggests that the C/Nd ratios of all three components were similar and so formed under similar conditions, e.g., partial melting and incompatible partitioning of C and Nd. Note that such correlations would not be easily created if the transporting melts were progressively losing CO_2 as a separate vapor phase (and very depleted in Nd) during transit in the lithosphere, and suggests that mixing of components occurred locally. This correlation, however, does not contradict the overall decoupling between Sr (and so Nd) and He isotopes, but rather suggests that the isotopically distinctive C and Nd in the non-MORB components were generated without large C/Nd fractionations.

He-C relationships

MORB reaching the surface are typically saturated with CO_2 and have CO_2 vesicles that contain most of the noble gases. Measurements of gas-rich MORB and hydrothermal vents have found molar ratios of $^3He/C$ on the order of 10^{-10} to 10^{-9} (Burnard 2001; Marty and Jambon 1987; Trull et al. 1993; van Soest et al. 1998). Degassed samples have substantially lower values and fractionated noble gas abundance patterns due to differences in melt solubilities. Combined CO_2 and He data for xenoliths are available for only a few xenolith suites (Mattey et al. 1989; Porcelli et al. 1992; Dunai 1993) and are shown in Figure 4. There are no continental xenolith samples for which other noble gases have also been analyzed. While samples with low C contents have $^3He/C$ ratios comparable to those of MORB, those with higher C contents have a wide range of much lower ratios. The samples from the Eifel and Massif Central (Dunai 1993) generally exhibit the lowest $^3He/C$ ratios, while the remaining samples are closer to the MORB range and although these come from a diverse range of locations (E. Africa, SE Australia, and Japan), there is no pattern related to location. There are several considerations that bear on the cause of this:

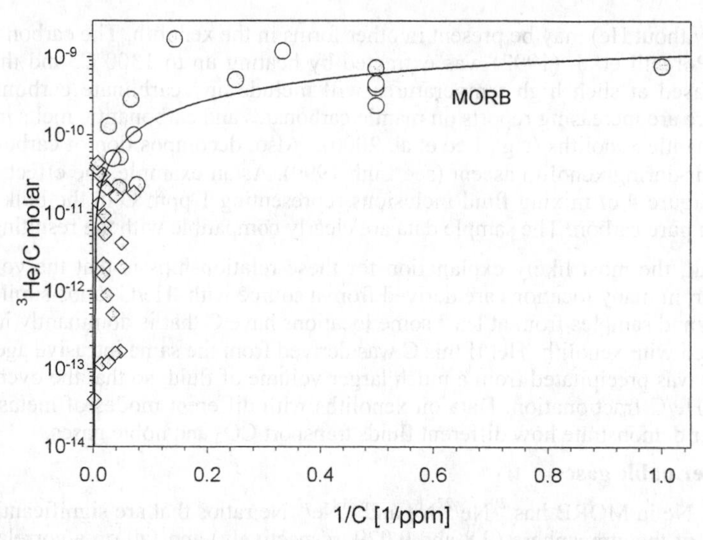

Figure 4. ^{3}He/C vs. 1/C ratios in subcontinental mantle xenoliths from locations worldwide. Circles are data from Porcelli et al. (1992) and diamonds are from Dunai (1993). The white circles indicate data that gave minimum ^{3}He/C ratios in Porcelli et al. (1992). The MORB-range is from Trull et al. (1993). The curve is a simple exponential fit to all data with two end members: pure carbon (i.e., ^{3}He-free carbonate) and ~1 ppm CO_2 inclusions with MORB ^{3}He/C-ratios.

- Silicate melts carrying volatiles from the asthenosphere will collect both ^{3}He and C during melting. If the source has similar characteristics as those of MORB, and these elements behave very incompatibly (so that the degree of melting will not affect their ratio in the melt), then the ^{3}He/C ratio will be like that of MORB (see e.g., Burnard 2001; Carroll and Draper 1994; Marty and Jambon 1987; Trull et al. 1993). This may be the main reason for similarity between MORB ratios and those of many xenoliths.
- Once a melt is saturated in CO_2, progressive loss of CO_2-rich fluids will change the ^{3}He/C ratio of the melt. This change will depend upon the difference between the behaviors of C, which is strongly dependent upon melt composition (see Brooker et al. 2001), and He, which is more weakly dependent (Carroll and Draper 1994). In alkali basalts and other low silica melts rich in depolymerizing ions (e.g., K, Na), the ^{3}He/C ratio of the melt will progressively decrease. In contrast, in basalts and more acidic melts, the ^{3}He/C ratio will increase. Therefore, a range of ratios can be generated during ascent of magmas.
- Different asthenospheric sources may not have the same ^{3}He/CO_2 ratios. There is no data available for many OIB types, and that of high ^{3}He/^{4}He hotspots is not well known (see Hilton et al. 1998). Therefore, this ratio cannot be used to distinguish possible asthenospheric sources.
- If carbonatitic melt is the carrier of volatiles into the lithosphere, the source may not have a typical ^{3}He/C ratio, and separation of a vapor phase in the lithosphere is likely to be further fractionated. ^{3}He will likely be enriched in the initial melt relative to CO_2, and then again in the initially separating vapor phase, and so decoupling from carbonatitic melts and/or solids might be easily achieved (Carroll and Draper 1994). While evidence for such processes has not been found in xenoliths, data is still very limited.

- C (without He) may be present in other forms in the xenolith. The carbon analyzed by Porcelli et al. (1992) was extracted by heating up to 1300°C, and the carbon released at such high temperatures will include any carbonate carbon present. There are increasing reports on mantle carbonates and carbonatitic melts inclusions in mantle xenoliths (e.g., Lee et al. 2000). Also, decomposition of carbonate may occur during xenolith ascent (see Luth 1999). As an example, the effect is shown in Figure 4 of mixing fluid inclusions representing 1 ppm C in the bulk xenolith with pure carbon. The sample data are clearly compatible with the resulting curve.

Overall, the most likely explanation for these relationships is that the volatiles in xenoliths from many locations are derived from a source with ^{3}He/C ratios similar to that of MORB, and samples from at least some locations have C that is dominantly in a phase unassociated with xenolith ^{3}He. If this C was derived from the same intrusive agent as the He, then it was precipitated from a much larger volume of fluid, so that the overall result is strong ^{3}He/C fractionation. Data on xenoliths with different modes of metasomatism may further demonstrate how different fluids transport CO_2 and noble gases.

The heavier noble gases

Neon. Ne in MORB has ^{20}Ne/^{22}Ne and ^{21}Ne/^{22}Ne ratios that are significantly higher than those of the atmosphere (9.8 and 0.029, respectively) and fall on a correlation line that is widely interpreted as due to mixing between atmospheric contamination and an upper mantle component. This component in turn is a mixture of 'solar' Ne with a high ^{20}Ne/^{22}Ne ratio and radiogenic ^{21}Ne. Samples from high ^{3}He/^{4}He OIB have similar ^{20}Ne/^{22}Ne ratios but lower ^{21}Ne/^{22}Ne ratios, indicating that there is a lower contribution from nucleogenic ^{21}Ne, in concert with less radiogenic He. Kyser and Rison (1982) found ^{20}Ne/^{22}Ne ratios up to 10.78±0.44 in xenoliths, but in the absence at that time of higher measured mantle values, did not discount mass fractionation of mantle Ne that has atmospheric values. Subsequent analyses (Dunai and Baur 1995; Barfod et al. 1999; Matsumoto et al. 1998, 2000) found that most xenolith Ne compositions that deviate from that of the atmosphere fall on the MORB correlation line (Fig. 5). This is in concert with the xenolith He isotope data. A few xenolith analyses have been found with Ne isotope compositions that do not fall upon the MORB correlation line, but rather have values that are closer to those of high ^{3}He/^{4}He hotspots (Dunai and Baur 1995; Matsumoto et al. 1997). This has been interpreted as an indication that such an OIB domain has contributed noble gases to the continental lithosphere, and will be discussed further in the section on the *Victorian Newer Volcanics*. Matsumoto et al. (1998) found measured ^{4}He/^{21}Ne ratios that were equal to, or lower than, the values for the convecting mantle, suggesting that recent loss of He occurred in some cases. Such losses could occur from the host mineral or from a fluid that introduced the rare gases into the xenolith, e.g., a fluid that has previously lost some CO_2 in which He preferentially entered relative to Ne.

Step heating experiments are commonly used to preferentially release and eliminate atmospheric contamination at lower temperatures. While gases in fluid inclusions are released at elevated temperatures, the highest temperatures can also release components remaining in the crystal lattice. In a suite of Australian xenoliths, Ne with a greater proportion of nucleogenic Ne relative to trapped mantle Ne was found in such high temperature steps (see Fig. 5; Matsumoto 1998, 2000). In apatites from this suite, additional nucleogenic ^{22}Ne from the reaction $^{19}F(\alpha,n)^{22}Na(\beta^+)^{22}Ne$ was found due to high apatite F concentrations (Matsumoto et al. 1997, 1998). This Ne was produced over times much greater than the eruption age of the xenoliths, and so was largely produced in the mantle. The expected accompanying excess radiogenic ^{4}He was absent, and likely had diffused away into grain boundaries and was lost, while Ne, with a lower diffusion coefficient, remained.

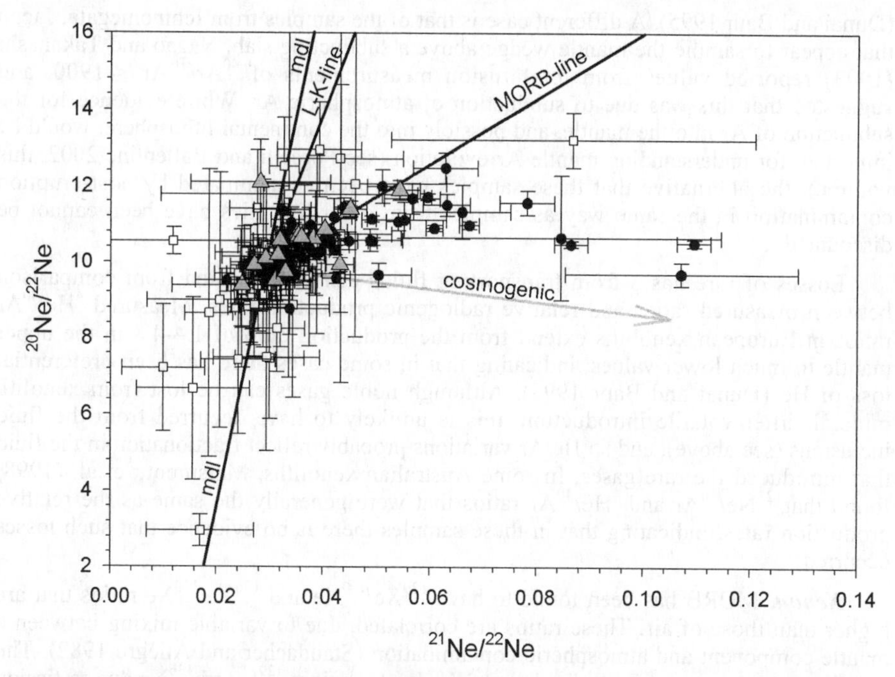

Figure 5. Neon isotopes in mantle xenoliths from locations worldwide. Many samples show values with enrichments in ^{20}Ne and ^{21}Ne and fall on the MORB correlation line. Some plot to further to the right, due to nucleogenic or cosmogenic inputs of ^{21}Ne. A few samples fall to the left, toward the 'hotspot' correlation line. Gray triangles, Cameroon Line (Barfod et al. 1999); black circles, SE-Australia (Matsumoto et al. 1998 2000); white squares, Europe (Dunai and Baur 1995). MORB and mass discrimination line (mdl) are from Sarda et al. (1988). Loihi-Kilauea Line (L-K) from Honda et al. (1991). The gray arrow indicates the direction of possible shifts if samples were exposed to cosmic rays.

Argon. A range of $^{40}Ar/^{36}Ar$ values has been measured in MORB. The highest values of (2.8-4) $\times 10^4$ (Staudacher et al. 1989; Burnard et al. 1997) are generally accepted as those of the MORB source, while lower values are interpreted to be due to ample contamination by air Ar (with $^{40}Ar/^{36}Ar = 296$). Kaneoka (1974) found excess ^{40}Ar (i.e., not attributable to in situ production) in xenoliths from the Kola Peninsula, and values for trapped Ar of up to $^{40}Ar/^{36}Ar = (3.5\pm8) \times 10^4$, before such high values had been established for the MORB source. Kyser and Rison (1982) found values ranging from close to the atmosphere to 9500 in a Massif Central sample, but did not attempt to separate contamination by step heating. Dunai and Baur (1995) also found a range of values using in vacuo crushing, with a maximum value of $(1.7\pm0.1) \times 10^4$. Barfod et al. (1999) found that while in vacuo crushing yielded a maximum value of 4.9×10^3, values of $^{40}Ar/^{36}Ar = (1.6\pm1) \times 10^4$ were obtained by single grain laser heating, demonstrating that lower values can be generated by atmospheric contamination that is not avoided simply by crushing. Matsumoto et al. (1998) found a maximum of 1.1×10^4 in Australian xenoliths and demonstrated that crushing yields higher values than step heating due to preferential release of fluid inclusion Ar. It is possible that ubiquitous atmospheric contamination is the reason there is such a range of measured values, with none as high as that of the MORB source. The other possible source of Ar at many locations is radiogenic production in the lithosphere, although this would cause even more radiogenic ratios, and would be accompanied by radiogenic ^4He and so much lower $^3He/^4He$ ratios

(Dunai and Baur 1995). A different case is that of the samples from Ichinomegata, Japan that appear to sample the mantle wedge above a subducting slab. Nagao and Takahashi (1993) reported values from total fusion measurements of $^{40}Ar/^{36}Ar \leq 1900$, and suggested that this was due to subduction of atmospheric Ar. While evidence for the subduction of Ar into the mantle, and possibly into the continental lithosphere, would be important for understanding mantle Ar evolution (see Porcelli and Ballentine 2002, this volume), the alternative that these samples have been contaminated by post-eruption contamination in the same way as samples from other localities have been cannot be discounted.

Losses of rare gases from transporting fluids can be evaluated from comparison between measured ratios and relative radiogenic production rates. Measured $^{4}He/^{40}Ar$ ratios in European xenoliths extend from the production ratio of 1.4-4.8 in the upper mantle to much lower values, indicating that in some cases there has been preferential loss of He (Dunai and Baur 1995). Although noble gases can be lost from xenolith minerals after volatile introduction, this is unlikely to have occurred from the fluid inclusions (see above), and so He/Ar variations probably reflect fractionation in the fluid that introduced the rare gases. In some Australian xenoliths, Matsumoto et al. (1998) found that $^{21}Ne/^{40}Ar$ and $^{4}He/^{40}Ar$ ratios that were generally the same as the relative production rates, indicating that in these samples there is no evidence that such losses occurred.

Xenon. MORB has been found to have $^{129}Xe/^{130}Xe$ and $^{136}Xe/^{130}Xe$ ratios that are higher than those of air. These ratios are correlated, due to variable mixing between a mantle component and atmospheric contamination (Staudacher and Allègre 1982). The mantle component contains radiogenic ^{129}Xe due to extinct ^{129}I, and ^{136}Xe due to fission of ^{238}U and extinct ^{244}Pu. Unfortunately, due to very low Xe xenolith concentrations and the difficulties of separating Xe atmospheric contamination, there have been few studies focusing on Xe in continental xenoliths. Kyser and Rison (1982) reported a few isotopic values, but all are within error of air ratios. Matsumoto et al. (1998, 2000) found samples that yielded Xe with $^{129}Xe/^{130}Xe$ and $^{136}Xe/^{130}Xe$ ratios that were greater than those of the atmosphere and coincided with the MORB trend. This indicates that xenolith Xe is derived from the upper mantle and has not been in a lithospheric environment with high U/Xe for long periods, which would add ^{136}Xe by fission decay of ^{238}U, without adding ^{129}Xe (which is the product of the extinct nuclide ^{129}I). It is interesting to note that in some cases values as high as those found in MORB were obtained, indicating that contamination lowering the mantle-derived ratios can be of the same level as that affecting MORB glasses. Unfortunately, there is little data available for ocean island basalts to determine if other mantle regions have distinctly different compositions and so can be excluded as sources of the xenolith Xe.

Regional studies of the sources of xenolith mantle rare gases

Various regional studies have sought to define the isotopic composition of He introduced into the xenoliths within the lithosphere, and so identify the regional source of mantle volatiles. The results for some of the more extensively studied locations follow.

Continental Europe. There are various Cenozoic volcanic centers located across central and Western Europe, some of which have yielded young xenoliths. An extensive study of the Eifel, Germany, and Massif Central, France (Dunai and Baur 1995) found a very restricted range of $^{3}He/^{4}He$ ratios. The samples had sufficiently high He that corrections for post-eruption additions of radiogenic ^{4}He and cosmogenic ^{3}He, using intermineral data, were small. Data for 11 vents in the Massif Central had a mean $^{3}He/^{4}He$ ratio of 6.53±0.25 R_A, while two from the Eifel had 6.03±0.14 R_A. Two other Eifel vents have somewhat lower values of 4.4-5.3 R_A. The samples include a range of lithologies, including pyroxenites, dunites, and lherzolites, and often contained hydrous phases. Lithophile isotopes and trace elements also provided evidence for both depletion and subsequent trace element enrichment events (e.g., Stosch et al. 1980). The general uniformity of the values over many vents and regardless of the lithology is remarkable (Dunai and Baur 1995). It appears that, in general, the He in these samples either was introduced shortly before eruption or was present in high concentrations prior to eruption; otherwise, it is likely that a range of $^3He/^4He$ ratios would have been generated. However, the somewhat lower values at the two Eifel vents may reflect some radiogenic production in the source (Dunai and Baur 1995). Similar He isotopic compositions were also found in olivine phenocrysts from Etna that contain He prior to modification of the host magma by shallow contamination of sediments (Marty et al. 1994).

A review of Nd, Sr, and Pb data indicates that the volcanics at each of a number of localities across Europe, including the Eifel and Massif Central, fall on correlation lines that extend from a common component that has the characteristics of HIMU mantle (Wilson and Downes 1991). The correlations have been interpreted as mixing between HIMU mantle plume material with small variable contributions at each location from a local lithospheric source. Trace element and oxygen isotopic data preclude significant crustal contamination of the magmas (Wilson and Downes 1991). Seismic imaging suggests that small plume upwellings supply volcanic centers throughout continental Europe, and the geochemical similarities suggest a common deep plume source (Granet et al. 1995; Hoernle et al. 1995). Recent seismic data suggests that broader upwelling occurs from depths of up to 2000 km, and this single source may feed smaller plumes that are responsible for the surface volcanics (Goes et al. 1999). This would account for the common isotopic characteristics over such a large region. In such a case, the overlap of He isotopic compositions in the xenoliths with that of HIMU basalts (6.8±0.9 R_A; Hanyu and Kaneoka 1997) suggests that the xenolith He is dominated by this asthenospheric mantle component, and the local lithospheric components that add other trace elements are an insignificant source of He. It appears that at many locations across Europe, He in the xenoliths was likely supplied by melts from the same source as the host magmas that invaded the lithospheric mantle in advance of xenolith entrainment, and provided exsolved CO_2 and other volatiles. At some locations, the host magma might also have provided such volatiles (Dunai and Baur 1995). A range of $^4He/^{40}Ar$ ratios below that of the production ratio indicate that the noble gas content of these invading melts was fractionated by progressive loss of CO_2 (Dunai 1993).

Victorian Newer Volcanics, SE Australia. A number of noble gas studies have focused on xenoliths from the Pliocene to Recent alkali basalts of the Newer Volcanics in SE Australia (Porcelli et al. 1986, 1987, 1989, 1992; Matsumoto et al. 1997, 1998, 2000). These basalts generally appear to be associated with plume-related volcanism (Wellman and McDougall 1974; Zhang et al. 2001), and xenoliths from a range of depths have been found in a number of very young vents. The xenoliths represent a range of petrologic types (O'Reilly and Griffin 1984) and have undergone differing degrees of metasomatic addition of trace elements, with the addition in some samples of phases such as amphibole and apatite (O'Reilly and Griffin 1988; Griffin et al. 1988). There are signs of the presence of abundant fluids, including abundant fluid inclusions and vugs up to 2 cm across, with euhedral crystal faces indicating high-pressure recrystallization of host minerals (Andersen et al. 1984). A range of values for $^3He/^4He$ ratios of 7.1-9.8 R_A has been found for anhydrous spinel lherzolites, with no systematic difference between vents (Matsumoto et al. 1998). These values clearly overlap the MORB range, and are accompanied by clearly defined Ne excesses that also overlap the MORB range. Analyses of peridotites containing amphibole and apatite, as well as garnet pyroxenites, found values of 6-10 R_A (Porcelli et al. 1986,1992; Matsumoto et al. 2000), excluding those that have higher ratios clearly due to cosmogenic inputs (Porcelli et al. 1987), although values below 7.2 R_A are generally due to radiogenic inputs that are more evident in low temperature step-heating extraction steps. Therefore, He in all the xenoliths could have been derived from the MORB source region. There is no obvious correlation between petrologic types.

Some evidence of radiogenic Ne was found in high temperature extraction steps, with shifts to the right of the MORB correlation line (Fig. 5) in pyroxenites and apatite peridotites due to nucleogenic inputs of ^{21}Ne (Matsumoto et al. 2000). Interestingly, this is not accompanied by corresponding radiogenic ^{4}He, indicating that He migrated from the lattice to grain boundaries and was lost. However, this ^{4}He was then available for re-incorporation in minerals in the presence of volatile-rich invading fluids, although it is not clear how important this effect is regionally. Note that this does not require loss of He trapped in fluid inclusions, and so cannot necessarily be related to the timing of the trapping of CO_2 fluids. Nucleogenic Ne was also found in high temperature steps of apatites, with additions of ^{22}Ne as well due to reactions involving F (Matsumoto et al. 1997, 2000). These observations require the maintenance of isotopic disequilibrium between Ne in the lattices and Ne in fluid inclusions that is released at low temperature steps.

The petrology and rare gas characteristics of the xenoliths clearly require a series of events well after formation of the lithosphere. The garnet pyroxenites appear to be cumulates from intruding melts that crosscut the peridotites (Griffin et al. 1988). Lithophile isotope indicates that multiple metasomatic episodes occurred in the source (McDonough and McCulloch 1987; Griffin et al. 1988). The amphiboles, derived from volatile-rich fluids, were formed about 300-500 Ma, based on Sr-Nd systematics (Griffin et al. 1988). The noble gases derived from a source with MORB characteristics, presumably the convecting upper mantle, were introduced either very recently or in such high concentrations that isotopic compositions were not substantially altered by radiogenic growth of ^{4}He in the lithosphere.

Matsumoto et al. (1997) reported that one sample had Ne isotope compositions that are above the MORB correlation line and fall on the line for high ^{3}He/^{4}He hotspots defined by Loihi data. This represents the first indication that Loihi-type rare gases may be present in the lithosphere. These values were found in apatite mineral separates that had low (i.e., MORB) ^{3}He/^{4}He ratios, while coexisting amphiboles had MORB Ne as well as He. It was suggested that two fluids had invaded the xenolith source region, one with MORB characteristics that precipitated amphibole, and the other with plume features that precipitated apatite. Although both phases appear to have He in fluid inclusions, the separated CO_2-rich fluid of the second invading melt must have been excluded from the phases precipitated earlier. However, there are several observations that complicate this picture. The Sr and Nd isotope compositions of the amphibole and apatite from this sample are essentially the same, suggesting a similar source, and while these compositions resemble Loihi compositions, the Pb isotopes do not (Porcelli et al. 1992). Decoupling between He and Ne is also required, since only MORB He isotope ratios are observed in the apatites. Also, Matsumoto et al. (2000) subsequently reported similar Ne characteristics in a garnet pyroxenite with MORB ^{3}He/^{4}He ratios. Since all the Ne data do overlap with MORB Ne when two sigma errors are considered (rather than the one sigma errors routinely reported for noble gas analyses), the possibility that there is hotspot Ne in the xenoliths idea requires confirmation.

Africa. There are several volcanic provinces in Africa, representing different tectonic environments. The Cameroon Line is an intraplate alkaline province extending from the continental interior to the island of São Tomé, and so across both continental and oceanic crust. Although there is no age progression across the line, with the oldest vents in the center (Lee et al. 1994), it has been identified as a hotspot. The volcanics have the characteristics of HIMU ocean islands, with Pb isotopes reflecting high time-integrated U/Pb ratios (Halliday et al. 1988). Noble gases have been studied in xenoliths from the continental volcanics and lavas by Barfod et al. (1999). Ratios of $^3He/^4He$ = 4-6.7 R_A were found in the xenoliths, with the majority having >5 R_A. Isotopic homogeneity between different phases having different He concentrations in individual samples generally ruled out the effects of post-eruption cosmogenic and radiogenic additions, although minor differences were seen between duplicates. These measured compositions overlap those of lavas from São Tomé, which have 5.9-7.1 R_A, as well as those of HIMU lavas in other ocean islands (Hanyu and Kaneoka 1997; Hilton et al. 2000). Therefore, it appears that He from an asthenospheric HIMU source can completely account for the He in the xenoliths, and there are no significant contributions from the lithosphere. The range in compositions can accommodate some MORB-type He from the surrounding convecting mantle or trapped earlier in the lithosphere. While Ne isotope compositions in the xenoliths were found to overlap MORB compositions, those in other HIMU basalts are not known and may also resemble MORB values, and so this may not have any diagnostic virtues.

The Meidob Hills are part of the Darfur Dome volcanic province in western Sudan and are another example of an isolated intracontinental volcanic field. It appears to be unrelated to the rifting to the east. A study of the lavas (Franz et al. 1999) found that Sr and Nd isotope data suggest that another HIMU asthenospheric plume has supplied the igneous activity here. Measured $^3He/^4He$ ratios of 6.6-9.2 R_A were measured for olivine phenocrysts, and there is no data available for xenoliths. It appears that this represents another example of HIMU melts traversing the lithosphere, with He isotopic compositions that extend from those of MORB to somewhat more radiogenic values, and with no evidence for anomalously radiogenic contributions supplied from the lithosphere.

Extensive volcanism can be found in eastern Africa associated with the East African Rift system. Helium isotope ratios measured in basalts from the Ethiopian Rift Valley and Afar, near the triple junction of Red Sea, Aden, and Ethiopian rift systems are 6 to 17 R_A (Marty et al 1993; Scarsi and Craig 1996), and so including MORB ratios as well as indications of both contributions from radiogenic ^4He and high $^3He/^4He$ hotspots. To the south, the rift system divides into eastern and western branches. Xenoliths from Tanzanian vents in the eastern branch have $^3He/^4He$ = 5.8-7.3 R_A (Porcelli et al. 1986, 1987). The olivines in these samples have ~7 R_A, and so coincide with MORB values, so lower values in these studies may be due to post-eruption radiogenic production in the U-rich phases. In contrast, Os isotope data of xenoliths from the nearby Labait suggests the involvement of plume material (Chelsey et al. 1999), and so complex involvement of different mantle sources may be involved. Kivu and Virunga volcanics of the western branch of the East African rift show highly variable trace element and isotopic compositions that are indicative of a lithospheric mantle source that is heterogeneous on a small scale (probably <1 km; Furman and Graham 1999). It is therefore not surprising that the $^3He/^4He$ ratios of those centers are significantly different from each other (Graham et al. 1995), despite their geographic proximity, and this appears to reflect mixing of lithospheric and asthenospheric sources. It appears that here is a source of lower $^3He/^4He$ ratios that presumably has developed within the lithosphere itself. To date, the data from the western branch of the East African rift may hold the most convincing evidence for mixing of asthenospheric and lithospheric sources of helium (Graham et al. 1995; Furman and Graham 1999).

Southwestern USA. The relative importance of lithospheric and asthenospheric contributions to the helium in basalts in the Southwestern USA was investigated in studies by Reid and Graham (1996) and Dodson et al. (1998). Reid and Graham (1996) found that the most radiogenic helium ratios (5.5 R_A) were associated with basalts that exhibit Th and Pb isotope evidence for derivation from an enriched source in the lithospheric mantle that has remained unmodified since ~1.7 Ga. Furthermore, a strong positive correlation between $^3He/^4He$ ratios and ε_{Nd} was found, suggesting an affinity between light rare earth enrichment and time-integrated $(U+Th)/^3He$ ratios. The range in observed $^3He/^4He$ ratios (5.5-7.7 R_A) most likely represents a mixture between asthenospheric and lithospheric melts. Reid and Graham (1996) concluded that the Proterozoic lithospheric mantle in the southwestern US is not a highly degassed reservoir contaminated by He derived from the underlying asthenosphere, but rather a reservoir with only slightly elevated $(U+Th)/^3He$ ratios (and so slightly lower $^3He/^4He$ ratios) compared to the depleted upper mantle MORB source. Another conclusion was that processes that enrich the lithospheric mantle in large-ion lithophile elements (LILE) also add He, so that there is no significant decoupling between LILE and helium in the processes modifying the Proterozoic lithospheric mantle beneath the SW USA. Dodson et al. (1998) significantly expanded the database of Reid and Graham (1996) and found a much larger range in $^3He/^4He$ ratios extending to lower, i.e., more radiogenic, $^3He/^4He$ ratios (2.8-7.8 R_A). This was interpreted as reflecting variations in the age of the lithospheric mantle and the degree of degassing.

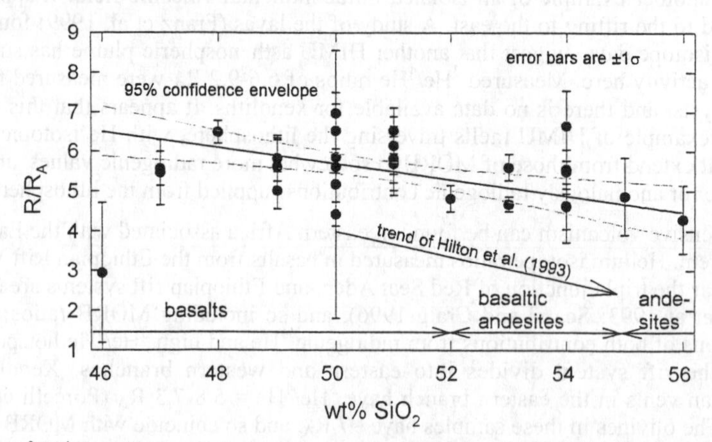

Figure 6. $^3He/^4He$-ratios (normalized to the atmospheric ratio) of SW USA phenocryst data of Dodson et al. (1998) plotted against the SiO_2-content of the host magmas. The format is similar to that used by Hilton et al. (1993a). The regression line and 95% error envelope were calculated using robust linear regression (with error weighting) of the Table2D® software. The indicated trend of Hilton et al. is based on the observation in ocean island basalts that magmas with 55% SiO_2 had helium isotopic compositions that were lower by 2 R_A than those with ~50% SiO_2. This evolution of the helium isotopic signature during magmatic differentiation (assimilation) was originally observed in lavas erupting through oceanic crust (Hilton et al. 1993a), and similar processes in the continental crust may be even more effective at changing the He isotope composition due to higher U and Th concentrations. It is clear that the evolution trend is not necessarily linear, although the considerable uncertainties of most of the data do not allow a more sophisticated approach.

One feature of the data from the SW USA that merits some attention is that the noble gas concentrations obtained by crushing are relatively low and so are potentially vulnerable to modification. In addition to changes from post-eruption addition of cosmogenic ^3He or radiogenic ^4He (Graham et al. 1998; Hilton et al. 1993b), this can occur during magmatic differentiation, as has been shown for back-arc and OIB samples (Hilton et al. 1995; Hilton et al. 1993a). In Figure 6, the $^3He/^4He$ phenocryst data of

Dodson et al. (1998) are plotted against the host magma SiO_2 contents in a format similar to that used by Hilton et al. (1993a). The host rocks are differentiated and range from basaltic to andesitic compositions, and exhibit a trend towards lower $^3He/^4He$ ratios for higher SiO_2 concentrations. This relationship is indistinguishable from that observed by (Hilton et al. 1993a) in basalts erupting through oceanic crust. The trend therefore could be explained by degassing and radiogenic in-growth during magmatic differentiation alone (Hilton et al. 1995; Hilton et al. 1993a). It therefore might well be that the $^3He/^4He$ ratios in the samples do not mirror the sources of the melt. Crustal contamination during magmatic differentiation will only aggravate this effect (Gasparon and Hilton 1994; Parello et al. 2000). The fit of the data to mixing trajectories between MORB-like asthenospheric mantle and a hypothetical lithospheric mantle end member and excluding crustal assimilation (Dodson et al. 1998), is similar to the fit that would be obtained if a slightly radiogenic asthenospheric source, with a $^3He/^4He$ ratio of ~6.5 R_A, mixed with crustal material. There is no reported SiO_2 data in the Reid and Graham (1996) study to make a similar evaluation of those samples. Overall, the question of whether the very low $^3He/^4He$ ratios represent a source feature or are due to processes occurring at shallower depths can only be settled by studying more samples from each location, especially the locations that are closest to end member compositions. A convergence of $^3He/^4He$ ratios in multiple samples from a single locality and with high He concentrations is needed to conclusively confirm source region characteristics.

Xenoliths from subduction zones. Xenoliths from various island arcs have also been studied. These are discussed in the context of island arc volcanism in Hilton et al. (2002), and are only described briefly here, since the mantle above subduction zones may include lithospheric regions directly affected by subduction zone processes or include mantle regions that are later incorporated into the continental lithosphere. In particular, this is a setting where the injection into the mantle of atmospheric or strongly radiogenic noble gases might occur. The most studied area is Japan. Samples from Ichinomegata in NE Japan have been found to have values of $^3He/^4He$ = 6.8-9.9 R_A, and so generally overlap the MORB range (Porcelli et al. 1992; Nagao and Takahashi 1993). Samples from other NE Japan vents are also within this range (Ikeda et al. 2001). There is evidence in the Ichinomegata xenoliths for additions of subducted components to other trace elements, e.g., $\delta^{13}C$ values of −3‰ to −17.5‰ that suggest additions of both light C (typical of organics) and heavy C (from carbonates) to MORB C with −5‰ (Porcelli et al. 1992). Also, fluid inclusions have been found to contain H_2O, which is unusual for mantle xenoliths (Roedder 1965; Trial et al. 1984). A lack of correlation between He and Sr isotopes indicates that while subducted Sr may be in the xenoliths, it was not accompanied by a significant He component (Ikeda et al. 2001). Since there is no evidence for a significant contribution of radiogenic ^4He from subducted components, the He found in fluid inclusions appears to be largely mobilized from the surrounding convecting mantle. However, samples with somewhat lower values may have limited contributions of radiogenic ^4He either produced in situ or from the source of invading fluids. Samples from SW Japan span a greater range in $^3He/^4He$ values. Although many overlap values for MORB, there are values as low as 0.7 R_A (Nagao and Takahashi 1993; Ikeda et al. 2001). However, the ^4He concentrations are extremely low (10^{-10}-10^{-9} cm^3 STP/g, which is 10^2 times lower than those of Ichinomegata (Porcelli et al. 1992), and some of the vents are several Ma old, so that post-eruption radiogenic contributions cannot be excluded. Nonetheless, Ikeda et al. (2001) suggested that some radiogenic ^4He from subducted components might be involved. A value of $^{40}Ar/^{36}Ar = 1.0\times10^4$ was reported that approaches that of the upper mantle of $\geq3\times10^4$, although most other values are much lower and may be due to either atmospheric contamination or Ar subduction (Ikeda et al. 2001). In contrast, Sumino et al. (2000) reported a value of $^3He/^4He$ =

16.6±1.4 R_A in a suite of samples from Takashima that otherwise exhibit MORB values, and suggested that He from a high $^3He/^4He$ plume was present. While a comparison of crushing and heating experiments did not find evidence of cosmogenic ^3He additions, a second crushing experiment produced a value of only 6.7. Clearly, this intriguing high $^3He/^4He$ ratio deserves further attention.

Xenoliths from Pliocene volcanism related to subduction below the NW US have been studied from Simcoe, Washington (Dodson and Brandon 1999). These samples have been metasomatized by a hydrous fluid that added phlogopite and trace elements. Samples measured by crushing have $^3He/^4He$ of about 7 R_A, except for the cores of two large (~8 cm) xenoliths, which have ~4.8 R_A. Post-eruption radiogenic ^4He was avoided by analyzing only olivines by crushing, and although the samples are gas-poor (10^{-10} to 10^{-8} cm^3 STP He/g) the lowest values did not correspond to the lowest concentrations (but actually included one amongst the highest concentrations). The uniformity of $^3He/^4He$ ratios in most other samples, and the consistent relationship between He composition and sample size (although there are only two large samples) further support this. It was argued that the xenoliths contained two components, with the higher $^3He/^4He$ ratios supplied by the host magma and the lower values presenting a composition generated in the mantle. It was further argued that more radiogenic He was not likely to be supplied directly from the subducted slab, but rather was generated above the slab in a region that was enriched in U and then later remobilized into the xenoliths. This is consistent with Pb isotope data for the xenoliths (Brandon et al. 1999). While it is curious that the volatile-rich fluid supplies He nonuninformly to the xenoliths in a process generally unseen elsewhere, the data still appear to provide evidence for He more radiogenic than MORB developed in the mantle wedge.

Other xenolith samples with rare gases that are unambiguously free of either post-eruption modification or contamination include those from Papua New Guinea from lavas with a strong influence by slab-derived material (Patterson et al. 1997). Here, $^3He/^4He$ ratios are either within the MORB range, or only slightly more radiogenic.

The mantle sources of xenolith He

There appears to be a general consensus that noble gases found in xenoliths have been largely introduced relatively recently and have not been trapped and stored since the stabilization of that region of the lithosphere. The similarity between the isotopic compositions of He found in continental xenoliths and that found in MORB indicates that a dominant fraction of the xenolith He originated within the convecting upper mantle. However, the shifts to somewhat lower values at some locations indicate that another component is present. With reference now to the detailed regional studies, there are several possible sources that have been discussed for the source of the mantle He trapped in the xenoliths:

MORB mantle. The convecting upper mantle sampled by MORB has $^3He/^4He$ = (8±1) R_A away from hotspots (see Graham 2002, this volume). In some areas, such as southeast Australia, xenolith He appears to have MORB He isotope compositions. This is not surprising, considering that much of the mantle underlying the continents must have this composition. It is also not incompatible with the involvement of diapiric mantle hotspots in the local volcanism. The He isotope composition of many hotspots have not been clearly characterized, and while hotspots are often assumed to have very high $^3He/^4He$ ratios as seen in Iceland and Hawaii, it has not been established that other hotspots, with very different trace element characteristics, all have such ratios. Further, the presence of such material does not preclude the involvement of MORB mantle as well within a particular lithospheric region.

HIMU mantle. The HIMU mantle component (with Pb isotopes that indicate high time-integrated U/Pb ratios or μ values) was identified in an evaluation of ocean island basalt data that defined several mixing components that appeared to control the observed lithophile element isotope compositions (Zindler and Hart 1986). Ocean islands with HIMU characteristics have also been found to have $^3He/^4He$ ratios that are somewhat lower than those of MORB and similar to the range observed for some continental xenoliths (Graham et al. 1992). Other mantle components represented in ocean island basalts have also been found to have somewhat low $^3He/^4He$ ratios. Values of 4.7-6.5 RA found in basalts from Jan Mayen, Tristan da Cunha, and Gough were attributed to contributions in the source from subducted components (Kurz et al. 1982). However, it has been argued that samples such as those with very low He concentrations may have suffered additions of crustal ^4He to the magma (Hilton et al. 1995). Other islands that have Sr, Nd, and Pb isotopic signatures for subducted material (the EM islands; Zindler and Hart 1986) have been found to have $^3He/^4He$ ratios similar to those of MORB (Hanyu and Kaneoka 1997). However, studies of HIMU islands have continued to find ratios somewhat lower than MORB. This includes data from St. Helena (4.3-5.9 R_A; Graham et al. 1992), the Cameroon Line (5.0-6.7 R_A; Barfod et al. 1999), and Mangaia, Rurutu, and Tubuai in the Cook-Austral Archipelago (6.8 R_A; Hanyu and Kaneoka 1997), although values found at La Palma extend up to those in MORB (6-9.6 R_A; Hilton et al. 2000). While the effects of magma contamination and radiogenic production cannot be discounted completely from some of these samples, the general pattern seems consistent with a mantle region of lower $^3He/^4He$ ratios. As discussed above, it appears that these mantle domains directly provide the He seen in the continental lithosphere. In western Europe, the Cameroon Line, Meidob Hills and Kivu, data from other isotopic tracers in the host basalts are consistent with such a source, which also then provides a reasonable explanation for the xenolith He. However, such domains are sampled at only a small proportion of oceanic vents (see Graham 2002, this volume), and so it is unlikely that such a source has contributed to many of the other continental locations.

Subducted components. He with MORB $^3He/^4He$ ratios can be lowered by the addition of more radiogenic components. Subducted ocean crust and sediments are depleted in He and have highly radiogenic He. Fluids generated from this material may carry this He into the overlying mantle wedge. Although it might be expected that this would cause He ratios to extend over much of the range from radiogenic (~0.02 R_A) to MORB ratios, island arc He somewhat surprisingly tends to be dominated by MORB He (See Hilton et al. 2002, this volume), and extremely radiogenic He also has not been found in xenoliths from these regions. It is interesting to note that the average $^3He/^4He$ ratio measured in island arcs is 5.9 R_A, (Hilton et al. 2002) so that the xenolith ratios range from the average value of island arcs to the average value of MORB. However, while there are some island arcs that have more radiogenic He, this has not been found in the limited data from subduction zone xenoliths. Xenoliths from Ichinomegata crater that appear to represent the underlying mantle wedge have $^3He/^4He$ = 6.8-8.4 R_A (Porcelli et al. 1992; Nagao and Takahashi 1993), similar to those of MORB, indicating that at least in this region the mantle He overlying subducting material has not been modified from that of the convecting upper mantle. It is possible, however, that U added to this region will result in the subsequent modification of the $^3He/^4He$ ratios. If this region is then accreted onto the continental lithosphere, it may then represent a low $^3He/^4He$ lithospheric domain.

Lithosphere metasomatism. It has been suggested that metasomatic fluids can transport into magma source regions trace elements that have been highly fractionated and so can generate the unique isotopic signatures seen in HIMU basalts (e.g., Halliday et al. 1995). While this process has been suggested for the oceanic asthenosphere, low xenolith $^3He/^4He$ ratios like those of HIMU ocean islands might also be generated by mantle metasomatism within the mantle lithosphere (e.g., Porcelli et al. 1986; Reid and Graham 1996). This is compatible with evidence that enrichment of U is associated with the introduction of metasomatic fluids in xenoliths (e.g., Carignan et al. 1996). However, it is not known to what extent He and U are fractionated. If both are highly incompatible during formation of the fluids and are introduced together into the lithosphere, the He isotopic composition will not diverge from that of the source region for a considerable time. However, if a CO_2 phase separates from a metasomatic melt, preferentially carrying He, and migrates separately from the parent melt, regions subsequently metasomatized by these fluids will have fractionated He/U ratios. Note that if CO_2 does not travel significant distances after separation from the melt phase, then this fractionation does not occur on significant scales. However, this might be seen on the scale of a hand sample. Some fractionation also can occur if the melt loses U through interaction with the host rock or in precipitated phases. In this way, lithospheric He isotope components can be generated, as seen in some locations. Nevertheless, the narrow range of He isotope compositions can be used to specify that any fluids that carried He into the mantle either were introduced recently or did not dramatically fractionate U from He.

High $^3He/^4He$ hotspots. It is worth noting that although not seen in xenoliths, He from high buoyancy hotspots with high $^3He/^4He$ ratios, such as Iceland, Hawaii, and Reunion (Sleep 1990) have sometimes found their surface expression in continental areas (Basu et al. 1993, 1995; Dodson et al. 1997; Graham et al. 1998; Kirstein and Timmerman 2000; Marty et al. 1996; Stuart et al. 2000) and consequently must have influenced the affected continental lithosphere. This includes 250 Ma old Siberian flood basalts (with 13 R_A; Basu et al. 1995), 42-Ma-old proto-Iceland plume basalts in Ireland (Kirstein and Timmermann 2000), 380-Ma-old Kola Peninsula carbonatites (with 19 R_A; Marty et al. 1998), and Archean komatiites (Richard et al. 1996). However, there are few indications of $^3He/^4He$ ratios greater than MORB (or hotspot Ne; see *Victorian Newer Volcanics* section) in the young xenoliths shown in Figure 2. It might be noted that the products of the earliest stages of the plume-related volcanism in a given continental area seem to have the largest chance of picking up relatively unaltered asthenospheric high $^3He/^4He$ ratios, and the isotopic expression of high $^3He/^4He$ hotspots wanes relatively quickly as the continental plate moves over the hotspot. Basalts erupting subsequently in the same area may then fail to pick up any high $^3He/^4He$ ratios (Basu et al. 1993, 1995; Dodson et al. 1997). Therefore, input of He with high $^3He/^4He$ ratios related to the high buoyancy hotspots may be a transient feature in the lithospheric mantle. This may provide at least some explanation why, while there is a record of continuing transit through the continental lithosphere of volcanics with high $^3He/^4He$ ratios, this has not left an enduring imprint on lithospheric xenoliths that have been subsequently brought to the surface.

Introducing He into the lithosphere

The introduction of He seems to occur close to eruption, and so it appears that CO_2-rich fluids that are either silicate or carbonatitic generally preceded the volcanism that hosts the xenoliths. As discussed by Matsumoto et al. (1998), such melts can more effectively traverse the lithosphere than CO_2 fluids (Brenan and Watson 1988), and so the fluids found in xenolith fluid inclusions were exsolved not far from these melts. These melts were also responsible for the introduction of many of the trace elements that are found to be enriched in metasomatized xenoliths. However, it is possible that in some circumstances, volatiles exsolved from the host magma might enter the xenoliths, either through mineral fractures during entrainment or even immediately preceding entrainment by following fracturing in advance of the host magma. The carrier melts are presumably due to mantle upwelling or regional thermal anomalies. More radiogenic Sr and Nd isotopic compositions found in the xenoliths have been attributed to evolution in the lithosphere and often are the result of multiple histories, with the He replaced by invading fluids with high He/Nd and He/Sr ratios. While several generations of such fluids may have invaded the xenolith source regions, the He isotopic signatures may be typically overwhelmed by the last introduction of CO_2-rich fluid that now bear the He. Unfortunately, the timing of the He introduction cannot be constrained because the source region U/He ratio cannot be reconstructed, and can only be constrained to precede the melting to produce the host magma. Where the source of the He can be clearly tied to a hotspot with low $^3He/^4He$ ratios or recent subduction zones, the fluids may be derived directly from the mantle underlying the lithosphere. In other circumstances, what has been introduced in the xenolith may have been at least partially remobilized from underlying lithosphere. This may occur in the vicinity of the xenoliths. Radiogenic He may even be provided within the xenolith itself, as evidenced by the possible remains of radiogenic components (e.g., Dodson and Brandon 1999), although it is not clear if such He can equilibrate with CO_2 fluids during final emplacement rather than contaminating migrating melts.

It should be emphasized that if the He budget of a xenolith is generally dominated by fluids that closely precede the volcanism that brings them to the surface, then no unambiguous information is carried about the general noble gas state of the lithospheric mantle. It might at least be argued that there is at least no evidence for a large lithospheric reservoir of He that is isotopically substantially different (i.e., either much more radiogenic or preserving very high $^3He/^4He$ ratios). Nevertheless, some of the more intriguing questions about the abundance and residence time of He in the lithosphere cannot be clearly answered from the xenolith data.

DIAMONDS

Noble gases in diamonds have received considerable attention, both for clues to the conditions of diamond formation and preservation in the mantle, and because of the possibility that diamonds have retained ambient noble gases from the lower lithosphere and from early in Earth history. There have been many reviews on diamond formation in general; more recent reviews include Haggerty (1999) and Navon (1999). Since Takeoka and Ozima (1978) reported that diamonds have sufficient He that isotopic compositions could be obtained, many studies have examined the variations recorded in diamonds of different locations and characteristics.

There are several characteristics to consider in using diamonds as noble gas samples:

- Diamonds are found in kimberlitic rocks in Archean cratons, and so are in xenoliths from beneath the oldest regions of the crust. These regions are thus likely the oldest regions of the lithosphere, and may have different characteristics from the lithosphere underlying the regions of young alkalic volcanism that have supplied the xenoliths used in other rare gas studies.
- Diamonds are stable at depths greater than ~150 km (e.g., Nickel and Green 1985). These therefore sample some of the thickest regions of the lithosphere and are from deeper than xenoliths from young alkalic volcanism.
- Inclusion assemblages have been found suggesting that at least some diamonds have formed beneath the lithosphere, at depths of up to 100 km into the lower mantle (see Harris et al. 1997; Harte et al. 1999; Davies et al. 1999). This raises the possibility that deep mantle rare gases can be found in at least some diamonds.
- The ages of diamonds have been determined from data for radiogenic isotopes

found in silicate inclusions. Kramers (1979) obtained Pb model ages of >2 Ga, while Richardson et al. (1984, 1990) found Sr and Nd model ages of 1.2-3.3 Ga. Although there has been some discussion of whether more recent events could produce old model ages (e.g., Pearson et al. 1995), Richardson and Harris (1997) produced a less ambiguous 2.0-Ga Sm-Nd isochron age. While rare gases have not been measured on dated diamonds, the data suggest that these old ages are general features, and so there is the possibility of obtaining past records of mantle gases.

- The host kimberlitic rocks are enriched in trace elements, including U and Th. Therefore, there is potential for considerable production of radiogenic nuclides both in the kimberlite source region, and in the matrix surrounding diamonds after eruption.
- Considerable quantities of rare gases have been extracted from diamonds (e.g., Takeoka and Ozima 1978; Ozima et al. 1983; Kurz et al. 1987; Honda et al. 1987), making them amenable for study. In particular, some diamonds are coated with an impure layer of varying thickness that is often rich in fluid inclusions and volatiles (Navon et al. 1988), including rare gases (Burgess et al. 1998).
- Diamonds are obtained from alluvial deposits as well as mines, and so may have had variable post-eruption histories that must be considered in deciphering surface contributions of radiogenic and nucleogenic nuclides.

The issues that will be discussed here are the possibilities of characterizing ancient He and Xe trapped in diamonds, and whether there are rare gas characteristics that can be used to identify specific sources of diamond volatiles. Ozima (1989) provides an early review of diamond data, and Lal (1994) and Begemann (1994) provide later appraisals.

Ancient He

Early measurements of He in diamonds found exceptionally high $^3He/^4He$ ratios of up to 230 R_A (Ozima and Zashu 1983; Ozima et al. 1983). These values are much higher than any found in the Earth. The only reasonable source that has been identified for producing ^3He in the mantle is $^6Li(n, \alpha)^3H(\beta^-)^3He$, and there does not appear to be sufficient Li in the diamonds to produce the excess ^3He. Therefore, it was proposed that these values represent He trapped by the diamonds very early in Earth history before significant production of ^4He occurred (Ozima et al. 1983) and even before the lithosphere was stabilized. In addition, since the highest values exceeded that of the early solar system of ~120 R_A, this required that the Earth incorporated He that had been subjected to D burning in the Sun that produced the higher value presently seen in the solar wind (see Wieler 2002, this volume). A difficulty with these samples is that their source was unknown, and could have been from alluvial deposits rather than mined at depth. It was suggested that the $^3He/^4He$ ratios were generated by cosmic ray production during exposure at the surface in such alluvial environments (Lal et al. 1987; Lal 1989; McConville and Reynolds 1989). Whether this mechanism can account for all the excess ^3He has been debated (Ozima 1990) and so analyses of samples known to have been mined at depth were required. Zadnik et al. (1987) found $^3He/^4He$ ratios of up to ~1000 R_A in a mined South African diamond, which far exceeds any plausible trapped composition and so must be explained by nucleogenic effects. Others have also argued that production of ^3He by the Li reaction can account for diamond ^3He (e.g., Kurz et al. 1987). While it still is not straightforward to produce such high values, and appears to require a close association of Li and U with the diamonds (see e.g., Begemann 1994), it no longer appears necessary to ascribe diamond He with high $^3He/^4He$ ratios to a component trapped early in Earth history.

Variations in $^3He/^4He$ ratios in diamonds have also been created by implantation of ^4He from decay of U and Th outside the diamonds (Lal 1989). It has been shown that ^4He concentrations are highest near diamond surfaces (Kurz et al. 1987; Verchovsky et al. 1993; Shelkov et al. 1998). The amounts found require a close association between diamonds and U-bearing phases.

Another issue that has been discussed is whether diamonds can retain He for several Ga at mantle temperatures and without isotopic exchange with surrounding, more radiogenic He. Since rare gases appear to be stored in fluid inclusions (Kurz et al. 1987), this involves knowing the partitioning into the diamond structure and diffusion rates in the diamond lattice. While Wiens et al. (1994) reported a diffusion coefficient of $2\times10^{-16} cm^2/sec$ at 1200°C, suggesting that He cannot be retained for periods of >1 Ga, reported values of 10^{-19} (Ozima 1989) and 10^{-21} (Shelkov et al. 1998) will allow long retention times. Also, any He that is trapped in fluid inclusions will readily partition into the mineral lattice, and so diffusive losses of this He across the diamond will be reduced. While this issue has not been completely resolved (see e.g., Begemann 1994; Wiens et al. 1994; Shelkov et al. 1998), the problem of storing ancient He will only become relevant again if unambiguous criteria are found for excluding production of ^3He in individual samples.

It must be concluded that the extensive research on rare gases in diamond has provided considerable information regarding isotopic effects in the mantle, perhaps related to environments unusually enriched in trace elements. However, it appears that no constraints regarding mantle rare gas components in the early Earth have been obtained.

Ancient Xe

Xe isotope variations have been measured in various diamond samples (Takeoka and Ozima 1978; Honda et al. 1987; McConville and Reynolds 1989; Ozima and Zashu 1991; Wada and Matsuda 1998). Excesses in both ^{129}Xe and ^{136}Xe have been found relative to the atmospheric Xe composition, with $^{129}Xe/^{130}Xe$ and $^{136}Xe/^{130}Xe$ ratios corresponding to values found for MORB (Staudacher and Allègre 1982). This indicates that the Xe has not resided in an environment with an usually high Xe/U ratio. Also, Ozima and Zashu (1991) noted that if this Xe was derived from the mantle ~3 Ga ago, this constrains the source of the excess ^{136}Xe in the upper mantle. While ^{129}Xe clearly comes from an extinct parent (^{129}I) and so excesses were generated early in Earth history, the ^{136}Xe excesses can come from either ^{238}U or extinct ^{244}Pu. If the former is the parent, a change in the ratio of ^{136}Xe to ^{129}Xe excesses will evolve with time in the mantle. Since the diamond Xe has the same ratio as the present upper mantle, this suggests that the parent of MORB Xe is ^{244}Pu. However, it has been argued that the coatings on diamonds where much of the trapped rare gases reside has been formed in association with the eruption event (Boyd et al. 1987) and so contain gases that were trapped relatively recently. Also, since large excesses in ^{136}Xe are not found in diamonds relative to the upper mantle, it appears that this Xe was not scavenged from the surrounding environment that must be rich in U to account for some measured ^4He concentrations (see above).

Mantle sources of diamond rare gases

As discussed above, there are many effects that can alter He isotope compositions. While these are clearly operative where unusual ratios are found, more subtle differences cannot be unambiguously interpreted. Therefore, while many diamonds have He with MORB $^3He/^4He$ ratios, those with somewhat higher values cannot necessarily be associated with another mantle domain. Diamond Ne isotope compositions have been reported by Ozima and Zashu (1988, 1991) and Honda et al. (1987). All measured values fall on the MORB correlation line, with the highest values of $^{20}Ne/^{22}Ne$ = 13.5±0.5

(Ozima and Zashu 1988) corresponding to the highest measured in mantle materials (see Graham 2002, this volume). Lower diamond values are likely due to the incorporation of atmospheric Ne. Ratios of $^{40}Ar/^{36}Ar$ = 300 to 15,000 have been measured, with most <1000 (Ozima and Zashu 1988, 1991; Burgess et al. 1998). These values are lower than MORB (which have values of up to 40,000; Burnard et al. 1997), but could be due to incorporation of atmospheric Ar in the analysis. Therefore, different Ar isotope ratios reflecting mantle heterogeneities or ancient upper mantle values stored in the lithosphere cannot be distinguished. The data for Xe is limited, and as discussed above, falls upon the MORB correlation for $^{129}Xe/^{130}Xe$ and $^{136}Xe/^{130}Xe$ ratios. This does not provide much information regarding noble gas sources, since compositions differing from MORB have not been identified in either oceanic basalts (see Graham 2002) or in xenoliths from the continental lithosphere (see above).

MANTLE VOLATILES IN THE CONTINENTAL CRUST

In a multitude of studies, He isotopes in groundwaters, lakes or CO_2 wells have been used as tracers for mantle contributions to crustal fluids and the rates of mantle degassing through continental areas (Oxburgh and O'Nions 1987; Oxburgh and O'Nions 1988; Aeschbach-Hertig et al. 1996, 1999; Ballentine et al. 1991; Griesshaber et al. 1992; Kipfer et al. 1994; Polyak et al. 2000; Torgersen 1993; Torgersen et al. 1995; Weinlich et al. 1999). Helium isotopes also have been used to trace the origin of metalliferous hydrothermal fluids (Simmons et al. 1988; Stuart and Turner 1992; Stuart et al. 1995). In crustal fluids and volatiles from areas close to recent active volcanism, $^3He/^4He$ ratios often are encountered that are close or identical to the values determined for some subcontinental mantle samples but somewhat lower than those of MORB. Therefore, the correct mantle value may not be the MORB ratio of ~8 R_A, but rather a slightly more radiogenic lithosphere value. In cases where values of over ~6 R_A are found at the surface, the precise mantle value is critical for determining whether there are any crustal He additions at all. Ideally, the local mantle value should be determined directly using xenoliths from the region, although this is not always possible. Of course, in locations where atmospheric or crustal components dominate crustal fluids, so that only small shifts toward mantle ratios are observed, the precise ratio for the mantle component is not important (see further discussion in Ballentine and Marty 2002, this volume).

CONCLUSIONS

The major conclusions that can be drawn from xenoliths studies of noble gases are:

1. He in ultramafic xenoliths and diamonds is located dominantly in fluid inclusions, and was likely introduced along with the CO_2-rich fluids. The distribution of rare gases in the xenoliths is controlled by the introduction and survival of these inclusions.

2. There are relatively uniform $^3He/^4He$ ratios in ultramafic xenoliths, which is remarkable considering the isotopic variations in other elements such as Sr and Nd. This decoupling of He from other isotopic signatures is likely due to relatively recent addition of He to the xenolith in fluids with high concentrations of He relative to other trace elements, either within the mantle or during transport. Any radiogenic He that resided in the lithosphere prior to this addition was overwhelmed.

3. The $^3He/^4He$ ratios in ultramafic xenoliths often are similar to those of MORB or low $^3He/^4He$ OIB and indicate that the He in the xenoliths is dominantly derived from the underlying convecting mantle. However, it is not clear whether the He came directly from the underlying mantle, or has had some residence time in the lithosphere prior to incorporation in the xenoliths.

4. In some regions, radiogenic input in the lithosphere of ^{4}He may have shifted the He isotopic compositions found in the xenoliths. However, xenolith He remains dominated by that from that of the convecting upper mantle.
5. There is no unequivocal evidence for ^{3}He/^{4}He ratios in ultramafic xenoliths that are higher than those of MORB, suggesting that high ^{3}He/^{4}He mantle sources, such as found in Iceland and Hawaii, are not important for the noble gas signature of the continental lithospheric mantle.
6. It is not possible to determine average concentration of noble gases in the lithosphere away from volcanic centers, although there is no evidence that it is more gas-rich than MORB.
7. Noble gases in xenoliths generally appear to have been introduced recently, and therefore do not contain clear records of the noble gases present prior to the most recent regional volcanic activity. While this suggests that the overall noble gas state of the lithosphere cannot be readily characterized from these samples, it does attest to the open system noble gas evolution of the lithosphere, with repeated influxes of noble gases from the underlying mantle.
8. Very high ^{3}He/^{4}He ratios found in diamonds appear to be due to production of ^{3}He by nuclear processes at the surface or in the mantle that produce values distinctive to the diamonds, rather than reflecting mantle values preserved since the time of diamond formation. Despite considerable efforts, no unambiguous evidence has been found for the noble gas isotopic composition of the Archean lithosphere.
9. Ne isotopes in ultramafic xenoliths and diamonds are not distinguishable from those of MORB.
10. There are no constraints on the heavier rare gases in ultramafic xenoliths, although Xe isotope compositions in diamonds overlap those of MORB. If this reflects Xe in the mantle 1-3 Ga ago, then this puts constraints on the origin of fissiogenic ^{136}Xe. However, it is possible that this Xe may have been added just prior to eruption.

Further progress on the origin and nature of rare gases in the lithospheric mantle may be obtained from studies that combine evidence from fluid inclusions, textures, lithophile trace element isotope systematics, and major element variations to understand how noble gas concentrations and isotopic compositions are introduced and modified within the lithosphere.

ACKNOWLEDGMENTS

Reviews by D. Graham, S. O'Reilly, and R. Wieler are greatly appreciated. DP was supported by ETH and the Swiss National Science Foundation. This is NSG publication #20020303.

REFERENCES

Aeschbach-Hertig W, Kipfer R, Hofer M, Imboden DM, Wieler R, Signer P (1996) Quantification of gas fluxes from the subcontinental mantle: the example of Laacher See, a maar lake in Germany. Geochim Cosmochim Acta 60:31-41

Aeschbach-Hertig W, Hofer M, Kipfer R, Imboden R, Wieler R (1999) Accumulation of mantle gases in a permanently stratified volcanic lake (Lac Pavin, France). Geochim Cosmochim Acta 63:3357-3372

Andersen T, Neumann E-R (2001) Fluid inclusions in mantle xenoliths. Lithos 55:301-320

Andersen T, O'Reilly SY, Griffin WL (1984) The trapped fluid phases in upper mantle xenoliths from Victoria, Australia: implications for mantle metasomatism. Contrib Mineral Petrol 88:72-85

Anderson DL (1990) Geophysics of the continental mantle: an historical perspective. *In* The Continental Mantle. Menzies MA (ed) Clarendon Press, p 1-30

Anderson DL (1998) The helium paradoxes. Proc Natl Acad Sci USA 95:4822-4827
Ballentine CJ, Burgess R, Marty B (2002) Tracing fluid origin, transport, and interaction in the crust. Rev Mineral Geochem 47:539-614
Ballentine CJ, O'Nions RK, Oxburgh ER, Horváth F, Deák J (1991) Rare gas constraints on hydrocarbon accumulation, crustal degassing and groundwater flow in the Pannonian Basin. Earth Planet Sci Lett 105:229-246
Barfod DN, Ballentine CJ, Halliday AN, Fitton JG (1999) Noble gases in the Cameroon line and the He, Ne, and Ar isotopic compositions of high μ (HIMU) mantle. J Geophys Res 104:29509-29527
Basu AR, Renne P, Das Gupta DK, Teichmann F, Poreda RJ (1993) Early and late alkali igneous pulses and high ^{3}He plume origin for the Deccan flood basalts. Science 261:902-906
Basu AR, Poreda RJ, Renne PR, Teichmann F, Vasiliev YR, Sobolev NV, Turrin BD (1995) High ^{3}He plume origin and temporal-spatial evolution of the Siberian flood basalts. Science 269:822-825
Begemann F (1994) Indigenous and extraneous noble gases in terrestrial diamonds. *In* Noble gas geochemistry and cosmochemistry. Matsuda J (ed) Terra Scientific Publ Co, Tokyo, p 217-227
Bernatowicz TJ (1981) Noble gases in ultramafic xenoliths from San Carlos, Arizona. Contrib Mineral Petrol 76:84-91
Blundell D, Freeman R, Müller S (1992) A Continent Revealed: the European Geotraverse. Cambridge University Press, Cambridge
Boyd FR (1989) Composition and distinction between oceanic and cratonic lithosphere. Earth Planet Sci Lett 96:15-26
Boyd FR, Mattey DP, Pillinger CT, Milledge HJ, Mendelssohn M, Seal M (1987) Multiple growth events during diamond genesis: an integrated study of carbon and nitrogen isotopes and nitrogen aggregation state in coated stones. Earth Planet Sci Lett 86:341-353
Brandon AD, Becker H, Carlson RW, Shirey SB (1999) Isotopic constraints on time scales and mechanisms of slab material transport in the mantle wedge: evidence from the Simcoe mantle xenoliths, Washington, USA. Chem Geol 160:387-407
Brenan JM, Watson EB (1988) Fluids in the lithosphere, 2. Experimental constraints on CO_2 transport in dunite and quartzite at elevated P-T conditions with implications for mantle and crustal carbonation processes. Earth Planet Sci Lett 91:141-158
Brooker RA, Kohn SC, Holloway JR, McMillan PF (2001) Structural controls on the solubility of CO_2 in silicate melts Part I: bulk solubility data. Chem Geol 174:225-239
Burgess R, Johnson LH, Mattey DP, Harris JW, Turner G (1998) He, Ar, and C isotopes in coated and polycrystalline diamonds. Chem Geol 146:205-217
Burnard P (2001) Correction for volatile fractionation in ascending magmas: Noble gas abundances in primary mantle melts. Geochim Cosmochim Acta 65:2605-2614
Burnard PG, Graham D, Turner G (1997) Vesicle specific noble gas analyses of popping rock: implications for primordial noble gases in Earth. Science 276:568-571
Burnham OM, Rogers NW, Person DG, van Calsteren PW, Hawkesworth CJ (1998) The petrogenesis of the eastern Pyrenean peridotites: an integrated study of their whole rock geochemistry and Re-Os isotope composition. Geochim Cosmochim Acta 62:2293-2310
Carignan J, Ludden J, Francis D (1996) On the recent enrichment of subcontinental lithosphere: a detailed U-Pb study of spinel lherzolite xenoliths, Yukon, Canada. Geochim Cosmochim Acta 60:4241-4252
Carlson RW (1995) Isotopic inferences on the chemical structure of the mantle. J Geodynamics 20:365-386
Carlson RW, Irving AJ (1994) Depletion and enrichment history of subcontinental lithospheric mantle- an Os, Sr, Nd, and Pb isotopic study of ultramafic xenoliths from the northwestern Wyoming craton. Earth Planet Sci Lett 126:457-472
Carroll MR, Draper DS (1994) Noble gases as trace elements in magmatic processes. Chem Geol 117: 37-56
Cerling TE, Craig H (1994) Geomorphology and in-situ cosmogenic isotopes. Ann Rev Earth Planet Sci 22: 273-317
Chauvel C, Hofmann AW, Vidal P (1992) HIMU-EM: the French Polynesian connection. Earth Planet Sci Lett 110:99-119
Chesley JT, Rudnick RL, Lee C-T (1999) Re-Os systematics of mantle xenoliths from the East African Rift: age, structure, and history of the Tanzanian craton. Geochim Cosmochim Acta 63:1203-1217
Cohen RS, O'Nions RK, Dawson JB (1984) Isotope geochemistry of xenoliths from East Africa: implications for development of mantle reservoirs and their interaction. Earth Planet Sci Lett 68: 209-220
Davies RM, Griffin WL, Pearson NL, Andrew AS, Doyle BJ, O'Reilly SY (1999) Diamonds from the deep: pipe DO-27, Slave Craton, Canada. Proc 7th Intl Kimberlite Conf, Cape Town, South Africa. Vol 1. Red Roof Design, Cape Town, p 148-155

Dodson A, Brandon AD (1999) Radiogenic helium in xenoliths from Simcoe, Washington, USA: implications for metasomatic processes in the mantle wedge above subduction zones. Chem Geol 160:371-385

Dodson A, Kennedy BM, DePaolo DJ (1997) Helium and neon isotopes in the Imnaha Basalt, Columbia River Basalt Group: evidence for a Yellowstone plume source. Earth Planet Sci Lett 150:443-451

Dodson A, DePaolo DJ, Kennedy BM (1998) Helium isotopes in lithospheric mantle: Evidence from Tertiary basalts of the western USA. Geochim Cosmochim Acta 62:3775-3787.

Downes H (2001) Formation and modification of the shallow sub-continental lithospheric mantle: a review of geochemical evidence from ultramafic xenolith suites and tectonically emplaced ultramafic massifs of western and central Europe. J Petrol 42:233-250

Dunai TJ (1993) Noble gases in the subcontinental mantle and the lower crust. PhD dissertation, Swiss Federal Institute of Technology (ETH), Zürich, Switzerland

Dunai TJ, Baur H (1995) Helium, neon and argon systematics of the European subcontinental mantle: Implications for its geochemical evolution. Geochim Cosmochim Acta 59:2767-2783

Ertan IE, Leeman WP (1999) Fluid inclusions in mantle and lower crustal xenoliths from the Simcoe volcanic field, Washington. Chem Geol 154:83-95

Farley KA, Natland JH, Craig H (1992) Binary mixing of enriched and undegassed (primitive?) mantle components (He, Sr, Nd, Pb) in Samoan lavas. Earth Planet Sci Lett 111:183-199

Finnerty AA, Boyd FR (1987) Thermobarometry for garnet peridotites: basis for the determination of thermal and compositional structure of the upper mantle. *In* Mantle xenoliths. Nixon PH (ed) John Wiley & Sons, p 381-402

Franz G, Steiner G, Volker F, Pudlo D, Hammerschmidt K (1999) Plume related alkaline magmatism in central Africa - the Meidob hills (W Sudan). Chem Geol 157:27-47

Frey FA, Green DH (1974) The mineralogy, geochemistry, and origin of lherzolite inclusions in Victorian basanites. Geochim Cosmochim Acta 38:1023-1059

Frezzotti ML, de Vivo B, Clocchiatti R (1991) Melt-mineral-fluid interactions in ultramafic nodules from alkaline lavas of Mount Etna (Sicily, Italy): melt and fluid inclusion evidence. J Volcanol Geotherm Res 47:209-219

Funkhauser JG, Naughton JJ (1968) Radiogenic helium and argon in ultramafic inclusions from Hawaii. J Geophys Res 73:4601-4608

Furman T, Graham DW (1999) Erosion of lithospheric mantle beneath the East African Rift system: geochemical evidence from the Kivu volcanic province. Lithos 48:237-262

Gasparon M, Hilton DR (1994) Crustal contamination processes traced by helium isotopes: examples from the Sunda Arc, Indonesia. Earth Planet Sci Lett 126:15-22

Goes S, Spakman W., Bijwaard H (1999) A lower mantle source for central European volcanism. Science 286:1928-1931

Graham DW (2002) Noble gas isotope geochemistry of mid-ocean ridge and ocean island basalts: characterization of mantle source reservoirs. Rev Mineral Geochem 47:247-318

Graham DW, Lupton F, Albarède F, Condomines M (1990) Extreme temporal homogeneity of helium isotopes at Piton de la Fournaise, Réunion Island. Nature 347:545-548

Graham DW, Humphris SE, Jenkins WJ, Kurz MD (1992) Helium isotope geochemistry of some volcanic rocks from St. Helena. Earth Planet Sci Lett 110:121-131

Graham DW, Furman TH, Ebinger CJ, Rogers NW, Lupton JE (1995) Helium, lead, strontium and neodymium isotope variations in mafic volcanic rocks from the western branch of the East African Rift. EOS Trans, Am Geophys Union 76:F686

Graham DW, Larsen LM, Hanan BB, Storey M, Pedersen AK, Lupton JE (1998) Helium isotope composition of the early Iceland plume inferred from the Tertiary picrites of West Greenland. Earth Planet Sci Lett 160:241-255

Grand SP (1994) Mantle shear structure beneath the Americas and surrounding oceans. J Geophys Res 99:11591-11621

Granet M, Wilson M, Achauer U (1995) Imaging a mantle plume beneath the French Massif Central. Earth Planet Sci Lett 136:281-296

Griesshaber E, O'Nions RK, Oxburgh ER (1992) Helium and carbon isotope systematics in crustal fluids from the Eifel, the Rhine Graben and Black Forest, F.R.G. Chem Geol 99:213-235

Griffin WL, O'Reilly SY, Stabel A (1988) Mantle metasomatism beneath western Victoria, Australia: II. Isotopic geochemistry of Cr-diopside lherzolites and Al-augite pyroxenites. Geochim Cosmochim Acta 52:449-459

Griffin WL, O'Reilly SY, Ryan CG (1999) The composition and origin of sub-continental lithospheric mantle. *In* Mantle petrology: field observations and high-pressure experimentation: A tribute to Francis R. (Joe) Boyd. Fei Y, Bertka CM, Mysen BO (eds) The Geochemical Society Spec Publ 6:13-45

Haggerty SE (1999) Diamond formation and kimberlite-clan magmatism in cratonic settings. *In* Mantle petrology: field observations and high-pressure experimentation: a tribute to Francis R. (Joe) Boyd. Fei Y, Bertka M, Mysen BO (eds) The Geochemical Society Spec Publ 6:105-123

Halliday AN, Lee D-C, Tommasini S, Davies GR, Paslick C, Fitton JG, James DE (1995) Incompatible trace elements in OIB and MORB and source enrichment in the sub-oceanic mantle. Earth Planet Sci Lett 133:379-395

Handler MR, Bennett VC, Esat TM (1997) The persistence of off-cratonic lithospheric mantle: Os isotopic systematics of variably metasomatised southeast Australian xenoliths. Earth Planet Sci Lett 151:61-75

Hanyu T, Kaneoka I (1997) The uniform and low $^3He/^4He$ ratios of HIMU basalts as evidence for their origin as recycled materials. Nature 273:273-276

Harris JW, Hutchison MT, Hursthouse M, Light M, Harte B (1997) A new tetragonal silicate mineral occurring as inclusions in lower-mantle diamonds. Nature 387:486-488

Harry DL, Leeman WP (1995) Partial melting of melt-metasomatized subcontinental mantle and the magma source potential of the lower lithosphere. J Geophys Res 100:10255-10269

Harte B, Harris JW, Hutchison MT, Watt GR, Wilding MC (1999) Lower mantle mineral associations in diamonds from São Luiz, Brazil. *In* Mantle petrology: field observations and high-pressure experimentation: a tribute to Francis R. (Joe) Boyd. Fei Y, Bertka M, Mysen BO (eds) The Geochemical Society Spec Publ 6:125-153

Hawkesworth CJ, Rogers NW, van Calsteren PWC, Menzies MA (1984) mantle enrichment processes. Nature 311:331-334

Hawkesworth CJ, Kempton PD, Rogers NW, Ellam RM, van Calsteren PWC (1990) Continental mantle lithosphere, and shallow level enrichment processes in the Earth's mantle. Earth Planet Sci Lett 96:256-168

Helffrich GR, Wood BJ (2001) The Earth's mantle. Nature 412:501-507

Herzberg C (1999) Phase equilibrium constraints on the formation of cratonic mantle. *In* Mantle petrology: field observations and high-pressure experimentation: a tribute to Francis R. (Joe) Boyd. Fei Y, Bertka CM, Mysen BO (eds) The Geochemical Society Spec Publ 6:241-257

Hilton DR, Hammerschmidt K, Loock G, Friedrichsen H (1993a) Helium and argon isotope systematics of the central Lau Basin and Valu Fa Ridge: evidence of crust/mantle interactions in a back-arc basin. Geochim Cosmochim Acta 57:2819-2841

Hilton DR, Hammerschmidt K, Teufel S, Friedrichsen H (1993b) Helium isotope characteristics of Andean geothermal fluids and lavas. Earth Planet Sci Lett 120:265-282

Hilton DR, Barling J, Wheller GE (1995) Effect of shallow-level contamination on the helium isotope systematics of ocean-island lavas. Nature 373:330-333

Hilton DR, McMurtry GM, Goff F (1998) Large variations in vent fluid $CO_2/^3He$ ratios signal rapid changes in magma chemistry at Loihi seamounts, Hawaii. Nature 396:359-362

Hilton DR, MacPherson CG, Elliott TR (2000) Helium isotope ratios in mafic phenocrysts and geothermal fluids from La Palma, the Canary Islands (Spain): implications for HIMU mantle sources. Geochim Cosmochim Acta 64:2119-2132

Hilton DR, Fischer T, Marty B (2002) Noble gases in subduction zones and volatile recycling. Rev Mineral Geochem 47:319-370

Hoernle K, Zhang Y-S, Graham D (1995) Seismic and geochemical evidence for large-scale mantle upwelling beneath the eastern Atlantic and western and central Europe. Nature 374:34-39

Hofmann AW, White WM (1982) Mantle plumes from ancient oceanic crust. Earth Planet Sci Lett 57: 421-436

Honda M, Reynolds JH, Roedder E, Epstein S (1987) Noble gases in diamonds: occurrences of solar-like helium and neon. J Geophys Res 92:12507-12521

Honda M, McDougall I, Patterson DB, Doulgeris A, Clague DA (1991) Possible solar noble gas component in Hawaiian basalts. Nature 349:149-151

Ikeda Y, Nagao K, Kagami H (2001) Effects of recycled materials involved in a mantle source beneath the southwest Japan arc region: evidence from noble gas, Sr, and Nd isotopic systematics. Chem Geol 175:509-522

Jochum KP, McDonough WF, Palme H, Spettel B (1989) Compositional constraints on the continental lithospheric mantle from trace elements in spinel peridotite xenoliths. Nature 340:548-550

Jordan TH (1975) The continental tectosphere. Rev Geophys Space Phys 13:1-12

Jordan TH (1988) Structure and formation of the continental tectosphere. J Petrol Special Lithosphere Issue, p 11-37

Kaneoka I (1974) Investigation of excess argon in ultramafic rocks from the Kola Peninsula by the $^{40}Ar/^{39}Ar$ method. Earth Planet Sci Lett 22:145-156

Kaneoka I, Takeoka N, Aoki K-I (1977) Rare gases in a phlogopite bearing peridotite in South African kimberlites. Earth Planet Sci Lett 36:181-186

Kaneoka I, Takeoka N, Aoki K-I (1978) Rare gases in mantle-derived rocks and minerals. *In* Terrestrial rare gases. Alexander Jr EC, Ozima M (eds) Japan Scientific Society, Tokyo, p 71-83

Kipfer R, Aeschbach-Hertig W, Baur H, Hofer M, Imboden DM, Signer P (1994) Injection of mantle type helium into Lake Van (Turkey): the clue for quantifying deep water renewal. Earth Planet Sci Lett 125:357-370

Kirstein LA, Timmerman MJ (2000) Evidence of the proto-Iceland plume in northwestern Ireland at 42 Ma from helium isotopes. J Geol Soc 157:923-927

Kramers JD (1979) Lead, uranium, strontium, potassium and rubidium in inclusion-bearing diamonds and mantle-derived xenoliths from southern Africa. Earth Planet Sci Lett 41:58-70

Kurz MD, Jenkins WJ, Hart SR (1982a) Helium isotope systematics of ocean islands and mantle heterogeneity. Nature 297:43-47

Kurz MD, Jenkins WJ, Schilling JG, Hart SR (1982b) Helium isotope variations in the mantle beneath the central North Atlantic Ocean. Earth Planet Sci Lett 58:1-14

Kurz MD, Gurney JJ, Jenkins WJ, Lott DE (1987) Helium isotopic variability within single diamonds from the Orapa kimberlite pipe. Earth Planet Sci Lett 86:57-68

Kurz MD, Colodner D, Trull TW, Moore RB, O'Brien K (1990) Cosmic ray exposure dating with in situ produced cosmogenic ^{3}He: results from young Hawaiian lava flows. Earth Planet Sci Lett 97:177-189

Kyser TK, Rison W (1982) Systematics of rare gas isotopes in basic lavas and ultramafic xenoliths. J Geophys Res 87:5611-5630

Lal D (1989) An important source of ^{4}He (and ^{3}He) in diamonds. Earth Planet Sci Lett 96:1-7

Lal D (1994) Helium isotopic information from diamonds: critical data available and needed. *In* Noble gas geochemistry and cosmochemistry. Matsuda J (ed) Terra Scientific Publishing, Tokyo, p 217-227

Lal D, Nishiizumi K, Klein J, Middleton R, Craig H (1987) Cosmogenic ^{10}Be in Zaire alluvial diamonds: implications for ^{3}He contents of diamonds. Nature 328:139-141

Lee C-T, Rudnick RL, McDonough WF, Ingo H (2000) Petrologic and geochemical investigations of carbonates in peridotite xenoliths from northeastern Tanzania. Contrib Mineral Petrol 139:470-484

Lippolt HJ, Weigel E (1988) ^{4}He diffusion in Ar-retentive minerals. Geochim Cosmochim Acta 52: 1449-1458

Luth RW (1999) Carbon and carbonates in the mantle. *In* Mantle petrology: field observations and high-pressure experimentation: a tribute to Francis R. (Joe) Boyd. Fei Y, Bertka M, Mysen BO (eds) The Geochemical Society Spec Publ 6:297-316

Marty B, Jambon A (1987) C/^{3}He in volatile fluxes from the solid Earth: implications for carbon geodynamics. Earth Planet Sci Lett 83:16-26

Marty B, Appora I, Barrat J-A, Deniel C, Vellutini P, Vidal P (1993) He, Ar, Sr, Nd, and Pb isotopes in volcanic rocks from Afar: evidence for a primitive mantle component and constraints on magmatic sources. Geochem J 27:219-228

Marty M, Pik R, Gerzahegn Y (1996) Helium isotopic variations in Ethiopian plume lavas: nature of magmatic sources and limit on lower mantle contribution. Earth Planet Sci Lett 144:223-237

Marty B, Tolstikhin I, Kamensky IL, Nivin V, Balaganskaya E, Zimmermann JL (1998) Plume-derived rare gases in 380Ma carbonatites from the Kola region (Russia) and the argon isotopic composition of the deep mantle. Earth Planet Sci Lett 164:179-192

Marty B, Trull T, Luissiez P, Basile B, Tanguy J (1994) He, Ar, O, Sr and Nd isotope constraints on the origin and evolution of Mount Etna. Earth Planet Sci Lett 126:23-39

Matsumoto T, Honda M, McDougall I, Yatsevich I, O'Reilly SY (1997) Plume-like neon in a metasomatic apatite from Australian lithospheric mantle. Nature 388:162-164

Matsumoto T, Honda M, McDougall I, O'Reilly SY (1998) Noble gases in anhydrous lherzolites from the Newer Volcanics, southeastern Australia: a MORB-like reservoir in the subcontinental mantle. Geochim Cosmochim Acta 62:2521-2533

Matsumoto T, Honda M, McDougall I, O'Reilly SY, Norman M, Yaxley G (2000) Noble gases in pyroxenites and metasomatized peridotites from the Newer Volcanics, southeastern Australia: implications for mantle metasomatism. Chem Geol 168:49-73

Mattey DP, Exley RA, Pillinger CT, Menzies MA, Porcelli D, Galer S, O'Nions RK (1989) Rleationships between C, He, Sr, and Nd isotopes in mantle diopsides. *In* Proc 4th Intl Kimberlite Conf, Geological Society of Australia Spec Publ 14:913-920

McConville P, Reynolds JH (1989) Cosmogenic helium and volatile-rich fluid in Sierra Leone alluvial diamonds. Geochim Cosmochim Acta 53:2365-2375

McDonough WF (1990) Constraints on the composition of the continental lithospheric mantle. Earth Planet Sci Lett 101:1-18

McDonough WF, McCulloch MT (1987) The southeast Australian lithospheric mantle: isotopic and geochemical constraints on its growth and evolution. Earth Planet Sci Lett 86:327-340

McKenzie D, O'Nions RK (1983) Mantle reservoirs and ocean island basalts. Nature 301:229-231

McKenzie D, O'Nions RK (1995) The source regions of ocean island basalts. J Petrol 36:133-159
Meen JK, Eggler DH, Ayers JC ((1989) Experimental evidence for very low solubility of rare earth elements in CO_2-rich fluids at mantle conditions. Nature 340:301-303
Menzies M, Chazot G (1995) Fluid processes in diamond to spinel facies shallow mantle. J Geodynamics 20:387-415
Menzies MA, Kempton PD, Dungan M (1985) Interaction of continental lithosphere and asthenospheric melts below the Geronimo Volcanic Field, Arizona, USA. J Petrol 26:663-693
Moreira M, Kurz MD (2001) Subducted oceanic lithosphere and the origin of the 'high μ' basalt helium isotopic signature. Earth Planet Sci Lett 189:49-57
Nagao K, Takahashi E (1993) Noble gases in the mantle wedge and lower crust- an inference from the isotopic analyses of xenoliths from Oki-Dogo and Ichinomegata, Japan. Geochem J 27:229-240
Navon O (1999) Diamond formation in the Earth's mantle. Proc 7th Intl Kimberlite Conf, Cape Town, South Africa Vol 2. Red Roof Design, Cape Town, p 584-594
Navon O, Hutcheon ID, Rossman GR, Wasserburg GJ (1988) Mantle-derived fluids in diamond micro-inclusions. Nature 335:784-789
Nickel KG and Green DH (1985) Empirical geothermobarometry for garnet peridotites and implications for the nature of the lithosphere, kimberlites, and diamonds. Earth Planet Sci Lett 73:158-170
Niedermann S (2002) Cosmic-ray-produced noble gases in terrestrial rocks as a dating tool for surface processes. Rev Mineral Geochem 47:731-784
Nolet G, Grand SP, Kennet BLN (1994) Seismic heterogeneity in the upper mantle. J Geophys Res 99:23753-23776
O'Nions RK, Oxburgh ER (1988) Helium, volatile fluxes and the development of continental crust. Earth Planet Sci Lett 90:331-347
O'Reilly SY, Griffin WL (1984) A xenolith-derived geotherm for southeastern Australia and its geophysical implications. Tectonophysics 111:41-63
O'Reilly SY, Griffin WL (1988) Mantle metasomatism beneath western Victoria, Australia I. Metasomatic processes in Cr-diopside lherzolites. Geochim Cosmochim Acta 52:433-447
O'Reilly SY, Griffin WL (1996) 4-D lithosphere mapping; methodology and examples. Tectonophysics 262:1-18
O'Reilly SY, Griffin WL (2000) Apatite in the mantle: implications for metasomatic processes and high heat production in Phanerozoic mantle. Lithos 53:217-232
Oxburgh ER, O'Nions RK (1987) Helium loss, tectonics, and the terrestrial heat budget. Science 237: 1583-1588
Ozima M (1989) Gases in diamonds. Ann Rev Earth Planet Sci 17:361-384
Ozima M (1990) Comment on "An important source of ^4He (and ^3He) in diamonds" by D. Lal. Earth Planet Sci Lett 101:107-109
Ozima M, Zashu S (1983) Primitive He in diamonds. Science 219:1067-1068
Ozima M, Zashu S (1988) Solar-type Ne in Zaire cubic diamonds. Geochim Cosmochim Acta 52:19-25
Ozima M, Zashu S (1991) Noble gas state of the ancient mantle as deduced from noble gases in coated diamonds. Earth Planet Sci Lett 105:13-27
Ozima M, Zashu S, Nitoh O (1983) $^3He/^4He$ ratio, noble gas abundance and K-Ar dating of diamonds- an attempt to search for the records of early terrestrial history. Geochim Cosmochim Acta 47:2217-2224
Ozima M, Zashu S, Mattey DP, Pillinger CT (1985) Helium, argon, and carbon isotopic compositions in diamonds and their implications in mantle evolution. Geochem J 19:127-134
Parello F, Allard P, D'Alessandro W, Federico C, Jean-Baptiste P, Catani O (2000) Isotope geochemistry of Pantelleria volcanic fluids, Sicily Channel rift: a mantle volatile end-member for volcanism in southern Europe. Earth Planet Sci Lett 180:325-339
Patterson DB, Honda M, McDougall I (1994) Noble gases in mafic phenocrysts and xenoliths from New Zealand. Geochim Cosmochim Acta 58:4411-4427
Patterson DB, Farley KA, McInnes BIA (1997) Helium isotopic composition of the Tabar-Lihir-Tanga-Feni island arc, Papua New Guinea. Geochim Cosmochim Acta 61:2485-2496
Pearson DG (1999a) Evolution of cratonic lithospheric mantle: an isotopic perspective. *In* Mantle petrology: field observations and high-pressure experimentation: A tribute to Francis R. (Joe) Boyd. Fei Y, Bertka CM, Mysen BO (eds) The Geochemical Society Spec Publ 6:57-78
Pearson DG (1999b) The age of continental roots. Lithos 48:171-194
Pearson DG, Shirey SB, Carlson RW, Boyd FR, Pokhilenko NP, Shimizu N (1995) Re-Os, Sm-Nd, and Rb-Sr isotope evidence for thick Archaean lithospheric mantle beneath the Siberian craton modified by multistage metasomatism. Geochim Cosmochim Acta 59:959-977
Polyak BG, Tolstikhin IN, Kamensky IL, Yakolev LE, Marty B, Cheshko AL (2000) Helium isotopes, tectonics and heat flow in the Northern Caucasus. Geochim Cosmochim Acta 64:1925-1944

Polet J, Anderson DL (1995) Depth extension of cratons as inferred from tomographic studies. Geology 23:205-208
Porcelli D, Ballentine CJ (2002) Models for the distribution of terrestrial noble gases and the evolution of the atmosphere. Rev Mineral Geochem 47:411-480
Porcelli D and Wasserburg GJ (1995) Mass transfer of helium, neon, argon and xenon through a steady state upper mantle. Geochim Cosmochim Acta 59:4921-4937
Porcelli D, O'Nions RK, O'Reilly SY (1986) Helium and strontium isotopes in ultramafic xenoliths. Chem Geol 54:237-249
Porcelli D, Stone JOH, O'Nions RK (1987) Enhanced $^3He/^4He$ ratios and cosmogenic helium in ultramafic xenoliths. Chem Geol 64:25-33
Porcelli D, O'Nions RK, Galer SJG, Cohen AS, Mattey DP (1992) Isotopic relationships of volatile and lithophile trace elements in continental ultramafic xenoliths. Contrib Mineral Petrol 110:528-538
Porcelli D, Ballentine CJ, Wieler R (2002) An introduction to noble gas geochemistry and cosmochemistry. Rev Mineral Geochem 47:1-18
Poreda RJ, Farley K (1992) Rare gases in Samoan xenoliths. Earth Planet Sci Lett 113:129-144
Reid MR, Graham DW (1996) Resolving lithopheric and sub-lithospheric contributions to helium isotope variations in basalts from the southwestern US. Earth Planet Sci Lett 144:213-222
Ricard Y, Nataf H-C, Montagner J-P (1996) The three-dimensional seismological model a priori constrained: confrontation with seismic data. J Geophys Res 101:8457-8472
Richard D, Marty B, Chaussidon M, Arndt N (1996) Helium isotopic evidence for a lower mantle component in depleted Archean komatiite. Science 273:93-95
Richardson SH (1990) Age and early evolution of the continental mantle. *In* The Continental Mantle. Menzies MA (ed) Clarendon Press, p 55-65
Richardson SH, Harris JW (1997) Antiquity of peridotitic diamonds from the Siberian craton. Earth Planet Sci Lett 151:271-277
Richardson SH, Gurney JJ, Erlank AJ, Harris JW (1984) Origin of diamonds in old enriched mantle. Nature 310:198-202
Richardson SH, Erlank AJ, Harris JW, Hart SR (1990) Eclogitic diamonds of Proterozoic age from Cretaceous kimberlites. Nature 346:54-56
Rocholl A, Heusser E, Kirsten T, Oehm J, Richter H (1996) A noble gas profile across a Hawaiian mantle xenolith: coexisting accidental and cognate noble gases derived from the lithospheric and asthenospheric mantle beneath Oahu. Geochim Cosmochim Acta 60:4773-4783
Rodgers NW, De Mulder M, Hawkesworth CH (1992) An enriched mantle source for potassic basanites: evidence from Karisimbi volcano, Virunga volcanic province, Rwanda. Contrib Mineral Petrol 111:543-556
Roedder E (1965) Liquid CO_2 inclusions in olivine-bearing nodules and phenocrysts from basalts. Am Mineral 50:1746-1782
Roedder E (1984) Fluid Inclusions. Rev Mineral 12
Rudnick RL, Nyblade AA (1999) The thickness and heat production of Archean lithosphere: constraints from xenolith thermobarometry and surface heat flow. *In* Mantle petrology, field observations and high-pressure experimentation: A tribute to Francis R. (Joe) Boyd. Fei Y, Bertka CM, Mysen BO (eds) Geochemical Society Spec Publ 6, p 3-12
Rudnick FL, McDonough WF, O'Connell RJ (1998) Thermal structure, thickness and composition of continental lithosphere. Chem Geol 145:399-415
Sarda P, Staudacher T, Allègre CH (1988) Neon isotopes in submarine basalts. Earth Planet Sci Lett 91: 73-88
Scarsi P (2000) Fractional extraction of helium by crushing of olivine and clinopyroxene phenocrysts: effects on the $^3He/^4He$ measured ratio. Geochim Cosmochim Acta 64:3751-3762
Scarsi P, Craig H (1996) Helium isotope ratios in Ethiopian Rift basalts. Earth Planet Sci Lett 144:505-516
Schiano P, Clocchiatti R (1994) Worldwide occurrence of silica-rich melts trapped in sub-continental and sub-oceanic mantle minerals. Nature 368:621-624
Shapiro SS, Hager BH, Jordan TH (1999) Stability and dynamics of the continental tectosphere. Lithos 48:115-133
Shelkov DA, Verchovsky AB, Milledge HJ, Pillinger CT (1998) The radial distribution of implanted and trapped ^4He in single diamond crystals and implications for the origin of carbonado. Chem Geol 149:109-116
Simmons SF, Gemmel JB, Sawkins FJ (1988) The Santo Nino silver-lead-zinc vein, Fresnillo District, Zacatecas, Mexico: Part II. Physical and chemical nature of ore-forming fluids. Econ Geol 83: 1619-1641
Simons FJ, Zielhuis A, van der Hilst RD (1999) The deep structure of the Australian continent from surface wave tomography. Lithos 48:17-43

Sobolev SV, Zeyen H, Granet M, Achauer U, Bauer C, Werling F, Altherr R, Fuchs K. (1997) Upper mantle temperatures and lithosphere-asthenosphere system beneath the French Massif Central constrained by seismic, gravity, petrologic and thermal observations. Tectonophysics 275:143-164

Sleep NH (1990) Hotspots and mantle plumes: some phenomenology. J Geophys Res 95:6715-6736

Spera FJ (1980) Aspects of magma transport. *In* Physics of magmatic processes. Hargraves RB (ed), Princeton University Press, Princeton, New Jersey, p 265-323

Staudacher T, Allègre CJ (1982) Terrestrial xenology. Earth Planet Sci Lett 60:389-406

Staudacher T, Allègre CJ (1991) Cosmogenic neon in ultramafic nodules from Asia and in quartzite from Antarctica. Earth Planet Sci Lett 91:73-88

Staudacher T, Allègre CJ (1993) The cosmic ray produced $^{3}He/^{21}Ne$ ratio in ultramafic rocks. Geophys Res Lett 20:1075-1078

Staudacher T, Sarda P, Allègre CJ (1989) Noble gases in basalt glasses from a Mid-Atlantic Ridge topographic high at 14°N: geodynamic consequences. Earth Planet Sci Lett 96:119-133

Stein M, Goldstein SL (1996) From plume head to continental lithosphere in the Arabian-Nubian shield. Nature 382:773-778

Stein M, Hofmann AW (1994) Mantle plumes and episodic crustal growth. Nature 372:63-68

Stone JOH, Porcelli D, Vance D, Galer S, O'Nions RK (1990) Volcanic traces. Nature 346:228

Storey M, Mahoney JJ, Kroenke LW, Saunders AD (1991) Are oceanic plateaus sites of komatiite formation? Geology 19:375-379

Stosch HG, Carlson RW, Lugmair GW (1980) Episodic mantle differentiation: Nd and Sr isotopic evidence. Earth Planet Sci Lett 47:263-271

Stuart FM, Turner G (1992) The abundance and isotopic composition of noble gases in ancient fluids. Chem Geol (Isot Geosci Sect) 101:97-109

Stuart FM, Burnard PG, Taylor RP, Turner G (1995) Resolving mantle and crustal contributions to ancient hydrothermal fluids: He-Ar isotopes in fluid inclusions from Dea Hwa W-Mo mineralisation, South Korea. Geochim Cosmochim Acta 59:4663-4773

Stuart FM, Ellam RM, Harrop PJ, Godfrey FJ, Bell BR (2000) Constraints on mantle plumes from the helium isotopic composition of basalts from the British Tertiary igneous province. Earth Planet Sci Lett 177:273-285

Su W-J, Woodward RL, Dziewonski AM (1994) Degree 12 model of shear velocity heterogeneity in the mantle. J Geophys Res 99:6945-6980

Sumino H, Nakai S, Nagao K, Notsu K (2000) High $^{3}He/^{4}He$ ratio in xenoliths from Takashima: evidence for plume type volcanism in southwestern Japan. Geophys Res Lett 27:1211-1214

Takaoka N, Ozima M (1987) Rare gas isotopic composition in diamonds. Nature 271:45-46

Tolstikhin IN, Mamyrin BA, Khabarin LB, Erlikh EN (1974) Isotope composition of helium in ultrabasic xenoliths from volcanic rocks of Kamchatka. Earth Planet Sci Lett 22:73-84

Torgersen T (1993) Defining the role of magmatism in extensional tectonics: helium 3 fluxes in extensional basins. J Geophys Res 98:16257-16269

Torgersen T, Drenkard S, Stute M, Schlosser P, Shapiro A (1995) Mantle helium in ground waters of eastern North America: time and space constraints on sources. Geology 23:675-678

Trial AF, Rudnick RL, Ashwal LW, Henry DJ, Bergman SC (1984) Fluid inclusions in mantle xenoliths from Ichinomegata, Japan. EOS Trans Am Geophys Union 65:306

Trull TW, Kurz MD (1993) Experimental measurements of ^{3}He and ^{4}He mobility in olivine and clinopyroxene at magmatic temperatures. Geochim Cosmochim Acta 57:1313-1324

Trull T, Nadeau S, Pineau F, Polve M, Javoy M (1993) C-He systematics in hotspot xenoliths: Implications for mantle carbon contents and carbon recycling. Earth Planet Sci Lett 118:43-64

van Soest MC, Hilton DR, Kreulen R (1998) Tracing crustal and slab contributions to arc magmatism in the Lesser Antilles island arc using helium and carbon relationships in geothermal fluids. Geochim Cosmochim Acta 62:3323-3335

Vance D, Stone JOH, O'Nions RK (1989) He, Sr and Nd isotopes in xenoliths from Hawaii and other oceanic islands. Earth Planet Sci Lett 96:147-160

Verchovsky AB, Ott U, Begemann F (1993) Implanted radiogenic and other noble gases in crustal diamonds from northern Kazakhstan. Earth Planet Sci Lett 120:87-102

Wada N, Matsuda JI (1998) A noble gas study of cubic diamonds from Zaire: constraints on their mantle source. Geochim Cosmochim Acta 62:2335-2345

Walker RJ, Carlson RW, Shirey SB, Boyd FR (1989) Os, Sr, Nd, and Pb isotope systematics of southern African peridotite xenoliths: implications for the chemical evolution of subcontinental mantle. Geochim Cosmochim Acta 53:1583-1595

Wass SY, Henderson P, Elliot CJ (1980) Chemical heterogeneity and metasomatism in the upper mantle: evidence from rare earth and other elements in apatite-rich xenoliths in basaltic rocks from eastern Australia. Phil Trans Roy Soc Lond A297:333-346

Weinlich FH, Bräuer K, Kämpf H, Strauch G, Tesar J, Weise SM (1999) An active subcontinental mantle volatile system in the western Eiger rift, Central Europe: Gas flux, isotopic (He, C, and N) and compositional fingerprints. Geochim Cosmochim Acta 63:3653-3671

Wellman P, McDougall I (1974) Cainozoic igneous activity in eastern Australia. Tectonophysics 23:49-65

Wiens RC, Lal D, Rison W, Wacker JF (1994) Helium isotope diffusion in natural diamonds. Geochim Cosmochim Acta 58:1747-1757

Williams W, Poths J, Anthony E, Olinger CT, Whitelaw M, Geissman J (1992) Magmatic ^{3}He/^{4}He signatures, ^{3}He surface exposure dating and paleomagnetism of quaternary volcanics in the Rio Grande Rift, New Mexico. EOS Trans Am Geophys Union 73:610

Wilson M, Downes H (1991) Tertiary-Quaternary extension-related alkaline magmatism in Western and Central Europe. J Petrol 32:811-849

Wilson M, Downes H (1992) Mafic alkaline magmatism associated with the European Cenozoic rift system. Tectonophysics 208:173-182

Xu S, Nakai S, Wakita H, Wang X (1995) Mantle-derived noble gases in natural gases from Songliao Basin, China. Geochim Cosmochim Acta 59:4675-4683

Xu S, Nagao K, Uto K, Wakita H, Nakai S, Liu C (1998) He, Sr and Nd isotopes of mantle-derived xenoliths in volcanic rocks of NE-China. J Asian Earth Sci 16:547-556

Zadnik MG, Smith CB, Ott U, Begemann F (1987) Crushing of a terrestrial diamond: ^{3}He/^{4}He higher than solar. Meteoritics 22:540-541

Zeyen H, Novak O, Landes M, Prodehl C, Driad L, Hirn A (1997) Refraction-seismic investigations of the northern Massif Central (France). Tectonophysics 275:99-117

Zhang M, Stephenson J, O'Reilly SY, McCulloch MT and Norman M (2001) Petrogenesis of late Cenozoic basalts in North Queensland and its geodynamic implications: trace element and Sr-Nd-Pb isotope evidence. J Petrol 42:685-719

Zindler A, Hart SR (1986) Chemical geodynamics. Ann Rev Earth Planet Sci 14:493-571

11 Models for the Distribution of Terrestrial Noble Gases and Evolution of the Atmosphere

Donald Porcelli[1] and Chris J. Ballentine[2]

Institute for Isotope Geology and Mineral Resources
ETH Zürich, Sonneggstrasse 5
8092 Zürich, Switzerland

[1] Now at: *Dept. Earth Sciences, University of Oxford,*
Parks Road, Oxford, OX1 3PR, United Kingdom
don.porcelli@earth.ox.ac.uk

[2] Now at: *Dept. Earth Sciences, University of Manchester,*
Oxford Road, Manchester, M13 9PL, United Kingdom

INTRODUCTION

Noble gases provide unique clues to the structure of the Earth and the degassing of volatiles into the atmosphere. Since the noble gases are highly depleted in the Earth, their isotopic compositions are prone to substantial changes due to radiogenic additions, even from scarce parent elements and low-yield nuclear processes. On a global scale, noble gas isotopic variations reflect planetary differentiation processes that generate fractionations between these volatiles and parent elements.

It has long been recognized that the atmosphere is not primary (i.e., solar) and so is not simply a remnant of the solar nebula. From similarities with volcanic emanations, a secondary origin of the atmosphere was proposed, that is, atmospheric volatiles are derived from degassing of the solid Earth (Brown 1949; Suess 1949; Rubey 1951). The characteristics of the atmosphere therefore reflect the acquisition of volatiles by the solid Earth during formation (see Pepin and Porcelli 2002, this volume), as well as the history of degassing from the mantle. The precise connection between volatiles now emanating from the Earth and the long-term evolution of the atmosphere are key subjects of modeling efforts.

Major advances in understanding the behavior of terrestrial volatiles have been made based upon observations on the characteristics of noble gases that remain within the Earth. Various models seek to define different components and reservoirs in the planetary interior, how materials are exchanged between them, and how the noble gases are progressively transferred to the atmosphere. These models are evaluated here after a review of the available constraints on noble gas distributions and fluxes. Note that the crust is largely ignored here, because it is not a major noble gas reservoir because of reworking and thus extensive degassing (Ballentine and Burnard 2002, this volume).

In this review, emphasis is placed on the specific features that are central to each model. Even 'unsuccessful' models that are inconsistent with available data are described, because these are valuable in discounting particular processes as dominant controls. Also, it is clear from the absence of a single, simple model that explains all the data that a combination of processes contributes to the noble gas variations observed. Indeed, new constraints have generally served to expand the complexity of models rather than conform to earlier descriptions of noble gas distribution and behavior. It is hoped that a broad review of modeling efforts will serve to highlight where inconsistencies remain.

1529-6466/00/0047-0011$10.00

Relevant decay constants and other basic elemental characteristics are found in Porcelli et al. (2002, this volume) and Ozima and Podosek (2001). For earlier reviews, see Lupton (1983), Mamyrin and Tolstikhin (1984), Ozima and Podosek (1983, 2001), O'Nions (1987), Turner (1989), Ozima (1994), Zhang (1997), Farley and Neroda (1998), and McDougall and Honda (1998).

RADIOGENIC COMPONENTS IN THE BULK EARTH AND ATMOSPHERE

The total quantities of rare gases within the Earth and the fraction of the terrestial budget of each nuclide that presently resides in the atmosphere are important model constraints. This cannot be determined easily for primordial (i.e., non-radiogenic) noble gas isotopes, because there are no good limits available on the total terrestrial budget of these nuclides; indeed, it is often simply assumed that the atmosphere is their dominant reservoir. However, the total production of nucleogenic isotopes can be constrained from parent element bulk silicate Earth (BSE) concentrations, and then compared to the abundances of these nuclides in the atmosphere to obtain a general measure of the extent of solid Earth degassing. The bulk Earth and atmosphere nucleogenic budgets are summarized in Table 1. Throughout the text, an asterisk in an isotope superscript denotes such components produced in the Earth (e.g., ^{4*}He, ^{129*}Xe) in contrast to trapped 'primordial' components.

Table 1. Noble gas daughter-nuclide budgets.

Parent nuclides	*BSE[1] concentration*	*Daughter nuclide*	*Total BSE[1] production (atoms)*	*Fraction of total BSE production in atmosphere*
^{238}U, ^{235}U,	U = 21 ppb;	^{4}He	3.9×10^{43}	(lost from atm)
^{232}Th [2]	Th/U = 3.8			
		^{21}Ne	1.81×10^{35}	?
^{40}K	K = 270 ppm	^{40}Ar	2.4×10^{42}	40%
^{129}I	I = 13 ppb	^{129}Xe	7.7×10^{35} [4]	20%
^{244}Pu	0.29 ppb [3]	^{136}Xe	1.0×10^{35} [4]	40%
^{238}U	21 ppb	^{136}Xe	7.5×10^{33}	?

Notes: 1. BSE is bulk silicate Earth, with 4.02×10^{27} g.
2. Production due to entire decay series, assuming secular equilibrium.
3. At 4.55 Ga.
4. Assuming an 82 Ma closure age for the atmosphere.

Radiogenic He and nucleogenic Ne

U, Th. The decay of U and Th series nuclides is the dominant source of ^{4}He and ^{21}Ne. There is little debate regarding the concentrations of U and Th, which are obtained from those in carbonaceous chondrites and by assuming that refractory elements (*e.g.*, Ca, U, Th) are unfractionated in the bulk Earth (e.g., O'Nions et al. 1981). In this case, a BSE concentration of 21 ppb U is obtained (Rocholl and Jochum 1993). Pb isotope systematics indicate that Th/U = 3.8 (*e.g.*, Doe and Zartman 1979).

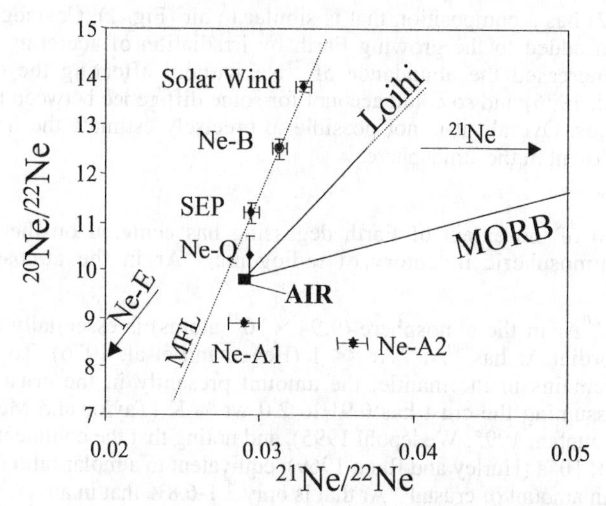

Figure 1. The Ne isotopic composition of the atmosphere is compared here with different extraterrestrial sources (from Busemann et al. 2000). While the atmospheric composition is close to that of Ne-Q, it can also be derived from simple mass fractionation (MFL = mass fractionation line) of solar Ne, along with addition of a small amount of additional radiogenic ^{21*}Ne. Data for MORB fall on a correlation line extending from air values, and reflect mixing of a mantle component composed of a trapped component with high Ne isotope ratios and air (Sarda et al. 1988; Moreira et al. 1998). OIB with high ^{3}He/^{4}He ratios generally fall close to the Loihi line, and reflect mixing between air and a different mantle component that has a lower ^{21}Ne/^{22}Ne ratio (Honda et al. 1991). While solar Ne can provide the trapped Ne seen in MORB and OIB, it has been suggested that Ne-B, a mixture of SEP and Solar Ne found in many meteorites, may be present instead (Trieloff et al. 2000).

^{4}He. The production rate of ^{4}He is well constrained, with 3.94×10^{42} atoms produced in the BSE over 4.55 Ga, or 9.8×10^{14} atoms/g in any undegassed mantle. However, the atmosphere is severely depleted from losses to space at long-term rates that cannot be easily quantified, so the fraction of ^{4}He that has degassed is unknown.

^{21}Ne. Production of ^{21}Ne is from the reaction ^{18}O(α,n)^{21}Ne, and to a much lesser extent from ^{24}Mg(n,α)^{21}Ne (which is driven by neutrons produced by (α,n) reactions), and so is directly proportional to the production of ^{4}He (α particles). There has been some uncertainty regarding the ^{21*}Ne/^{4*}He production ratio (Tolstikhin 1978; Rison 1980). Using the most recently published value of ^{21*}Ne/^{4*}He = 4.5×10^{-8} (Yatsevich and Honda 1997; Leya and Wieler 1999), a total of 1.81×10^{35} atoms of ^{21*}Ne have been produced in the BSE, which corresponds to 3.5% of the atmospheric ^{21}Ne inventory (5.19×10^{36} atoms). In contrast, production of ^{20}Ne and ^{22}Ne by similar nuclear reactions account for <<1% of the atmospheric inventory (see Ballentine and Burnard 2002). Quantification of the abundance of nucleogenic ^{21*}Ne in the atmosphere has been difficult because there is no solar system analogue that clearly defines the composition of primordial atmospheric Ne, and so can constrain the initial amount of ^{21}Ne. Atmospheric Ne can be produced by fractionation of solar Ne. Mass fractionation by Rayleigh distillation (Fig. 1) produces a primordial composition that requires that 4.4% of the present atmospheric ^{21}Ne was subsequently added as nucleogenic ^{21*}Ne, which is greater than that produced within the Earth using recently obtained production rates. Fractionation by hydrodynamic escape requires 1.6-3.8% nucleogenic ^{21*}Ne (Pepin 1991). Alternatively, it is possible that the atmosphere contains other solar system Ne components. The Ne-Q component that is widespread in meteorites (Busemann et al.

1990; Ott 2002) has a composition that is similar to air (Fig. 1). Cosmogenic Ne could also have been added to the growing Earth by irradiation of accreting materials and significantly increased the abundance of ^{21}Ne without affecting the other isotopes (Heymann et al. 1976) and so could account for some difference between primordial and present air ratios. Overall, it is not possible to precisely estimate the fraction of BSE ^{21*}Ne that is present in the atmosphere.

Radiogenic Ar

Discussion of the extent of Earth degassing has centered on the K-Ar system, because the atmospheric inventory of radiogenic ^{40}Ar in the atmosphere is well constrained.

^{40}Ar. The ^{40}Ar in the atmosphere (9.94×10^{41} atoms) is essentially all radiogenic, because primordial Ar has $^{40}Ar/^{36}Ar << 1$ (Begemann et al. 1976). To determine the amount that remains in the mantle, the amount presently in the crust also must be considered. Assuming the crust has 0.91 to 2.0 wt % K (Taylor and McLennan 1985; Rudnick and Fountain 1995; Wedepohl 1995), and noting that the continents have a mean K-Ar age of 1×10^9 a (Hurley and Rand 1969), equivalent to a molar ratio of $^{40}Ar/K = 9.1 \times 10^{-6}$, yields an amount of crustal ^{40}Ar that is only 3.1-6.8% that in air.

Potassium. K is a moderately volatile element and is depleted by a factor of ~8 in the bulk silicate Earth compared to CI chondrites (Wasserburg et al. 1964), but a precise and unambiguous concentration is difficult to obtain. If it is assumed that the MORB source value of $K/U = 1.27 \times 10^4$ (Jochum et al. 1983) is the same as that of the bulk Earth, then the BSE has 270 ppm K that has produced 2.4×10^{42} atoms ^{40}Ar. Therefore, 41% of the ^{40}Ar that has been produced is now in the atmosphere (Allègre et al. 1986, 1996; Turcotte and Schubert 1988; Table 1). However, the depleted MORB-source mantle may not have a bulk Earth K/U ratio, because a significant fraction of either element may have been added by subduction. The bulk of the K and U originally in the upper mantle is now in the continental crust, which therefore may have a BSE K/U ratio, but the K/U ratio of the crust is not sufficiently well constrained to provide a bulk Earth value. Furthermore, it has been argued (Albarède 1998; Davies 1999) that the K/U ratio and K content of the Earth are much lower, so that a much greater fraction of the ^{40}Ar may have been degassed. Cosmochemical evidence also may be considered to evaluate the K content of the bulk Earth. It has been argued that the relative proportions of moderately volatile elements in the Earth should lie on compositional trends defined by chondritic meteorite classes (Allègre et al. 1995b; Halliday and Porcelli 2001). Trends in meteoritic Rb/Sr vs. K/U are compatible with a terrestrial value of $K/U = 1.27 \times 10^4$. Only a modest reduction in the terrestrial K content would still be compatible with these relationships. Note that recent partitioning experiments indicate that the core contains <1 ppm K and so is not a significant repository of either K or ^{40}Ar (Chabot and Drake 1999).

Past atmospheric Ar characteristics. Efforts have been made to determine past atmospheric $^{40}Ar/^{36}Ar$ ratios and so constrain the past atmospheric abundances of ^{40}Ar. Cadogan (1977) reported a value of $^{40}Ar/^{36}Ar = 291.0 \pm 1.5$ from the 380-Ma old Rhynie chert, while Hanes et al. (1985) reported 258±3 for an old pyroxenite sill sample from the Abitibi Greenstone Belt that contains amphibole apparently produced during deuteric alteration at 2.70 Ga. Both samples are presumed to have atmospheric Ar trapped at these times. Using these data, the maximum possible changes in the atmospheric ^{40}Ar abundance is obtained by assuming that the ^{36}Ar abundance has been constant, so that the different $^{40}Ar/^{36}Ar$ ratios at these times reflect only lower ^{40}Ar abundances. In this case, the atmospheric ^{40}Ar abundance was 1.5% lower 380 Ma ago. At this time, there was 1.9% less ^{40}Ar in the Earth, and so it appears that the same fraction of terrestrial ^{40}Ar was in the atmosphere at that time. At 2.7 Ga ago, the calculated atmospheric ^{40}Ar abundance is 12.7% lower than at present. However, there was 66% less ^{40}Ar 2.7 Ga ago, and assuming a BSE concentration of 270 ppm K, even complete degassing of the entire Earth at that time (compared to 40% today) would not provide sufficient ^{40}Ar for the atmosphere. The alternative that the atmospheric ^{36}Ar abundance has doubled since 2.7 Ga ago is also unlikely if current arguments for early degassing of nonradiogenic nuclides are valid (see *Early Earth history* section). However, before pursuing further speculation, the possibility that the Abitibi sample contains either excess ^{40}Ar trapped during formation or subsequently produced radiogenic ^{40}Ar must be discounted. Overall, there is insufficient data on past atmospheric Ar characteristics to provide constraints on degassing models.

Radiogenic and fissiogenic Xe

The Xe budget provides strong evidence for early losses of noble gases from the atmosphere (see below), and isolation of atmospheric Xe from mantle Xe (see *Xe isotopes and a nonresidual upper mantle* section).

Plutonium. Pu is a highly refractory element represented by a single isotope, ^{244}Pu ($t_{1/2}$ = 80 Ma), which produces heavy Xe isotopes. All ^{244}Pu has decayed, and so the original Pu concentration in the Earth must be estimated from those of geochemically similar elements. Meteorite data suggests that at 4.57 Ga, $(^{244}Pu/^{238}U)_o = 6.8 \times 10^{-3}$ (Hudson et al. 1989). Other work on meteorites (Hagee et al. 1990) found that Pu appears to be correlated more closely with Nd than with U. Then using a solar Nd/U ratio, corresponding bulk meteorite Pu/U values of $(4\text{-}7) \times 10^{-3}$ were calculated, and the higher number (which is consistent with the Pu/U value found earlier) was considered more likely to represent the solar value. Therefore, the silicate Earth, or Earth-forming materials, initially had 0.29 ppb Pu, which produced 2.0×10^{35} atoms of $^{136*}Xe$. The amount produced by ^{238}U over the age of the Earth (7.5×10^{33} atoms ^{136}Xe) is much less, and so bulk Earth (and atmospheric) $^{136*}Xe$ is dominantly Pu-derived.

Iodine. Iodine is a volatile element and is depleted in the Earth relative to chondrites. The short-lived isotope ^{129}I ($t_{1/2}$ = 16 Ma) produces $^{129*}Xe$. The BSE I concentration is difficult to constrain, and data are limited. Analyses of crustal rocks were used to obtain a bulk crust abundance of 8.6×10^{18} g of I (Muramatsu and Wedepohl 1998). Analyses of MORB have yielded an upper mantle concentration of 0.8 ppb (Déruelle et al. 1992). Assuming that the crust was derived from 25% of the mantle that is now represented by the MORB source, a BSE value of 9.4 ppb is obtained. Assuming a greater volume of mantle was depleted to form the crust would result in a lower value. Based on unpublished data, Déruelle et al. (1992) quote a concentration for the crust that is 4.3 times higher, which would correspond to a BSE value of 38 ppb as calculated above, but favor a BSE value of 10 ppb based on extraction of crustal I from most of the mantle. Wänke et al. (1984) estimated a BSE concentration of 13 ppb based on the analysis of one xenolith judged to represent the undepleted mantle. Considering the poor constraints, it is remarkable that these estimates converge. At 4.57 Ga ago, $(^{129}I/^{127}I)_o = 1.1 \times 10^{-4}$ based on meteorite data (Hohenberg et al. 1967; Brazzle et al. 1999). For a BSE value of 13 ppb I, 2.7×10^{37} atoms of $^{129*}Xe$ were produced in the Earth or Earth-forming materials.

Radiogenic and fissiogenic Xe in the atmosphere. The greatest difficulty in constraining the global Xe budget has been in calculating the abundances of radiogenic and fissiogenic Xe in the atmosphere. The nonradiogenic Xe composition of the atmosphere first must be defined, but there is no suitable widespread solar system Xe component to use as a reference composition (see Ott 2002). The light isotopes of atmospheric Xe are related to both bulk chondritic and solar Xe by linear fractionation of ~4.2% per amu (Krummenacher et al. 1962), with a clear radiogenic excess of ~5% in

^{129}Xe. However, the contribution of fissiogenic heavy isotopes are more difficult to calculate; fractionated chondritic and solar Xe have greater proportions of ^{134}Xe and ^{136}Xe than is actually seen in the atmosphere and so cannot serve as the primordial terrestrial composition (see Pepin 2000). While no other suitable common solar system compositions have been found that provide the nonradiogenic heavy isotope composition, multi-dimensional isotopic correlations of chondrite data have been used to define a composition, U-Xe, that when mass-fractionated yields the light-isotope ratios of terrestrial Xe and differs from atmospheric Xe by a heavy isotope component that has the composition of ^{244}Pu-derived fission Xe (Pepin 2000). The fractionated U-Xe ratios of ^{129}Xe/^{130}Xe = 6.053 and ^{136}Xe/^{130}Xe = 2.075 are the present best estimates of the isotopic composition of nonradiogenic terrestrial Xe (see further discussion in Pepin and Porcelli 2002). Therefore, 6.8±0.30% of atmospheric ^{129}Xe ($^{129*}Xe_{atm} = 1.7\times10^{35}$ atoms) and 4.65±0.5% of atmospheric ^{136}Xe ($^{136*}Xe_{atm} = 3.81\times10^{34}$ atoms) are radiogenic. The $^{136*}Xe_{atm}$ is 20% of the total ^{136}Xe produced by ^{244}Pu in the BSE. However, the $^{129*}Xe_{atm}$ is only 0.8% of the total ^{129}Xe produced since 4.57 Ga; such a low value cannot be accounted for by incomplete degassing of the mantle nor from any uncertainties in the estimated amount of ^{129*}Xe, and requires losses to space. Note that an alternative composition for atmospheric nonradiogenic Xe, obtained by Igarashi (1995), requires that 2.8±1.3% of the atmospheric ^{136}Xe is added later. However, the composition of this component has relative proportions of the other heavy Xe isotopes that do not match the spectrum of either ^{244}Pu or ^{238}U, and so cannot be used to determine the fissiogenic Xe abundance of the atmosphere.

Xe closure ages. The depletion of radiogenic Xe in the atmosphere due to losses from the Earth to space must have occurred during early Earth history, when such heavy species could have been lost either from protoplanetary materials or from the growing Earth. Full accretion of the Earth is believed to have occurred over ~100 Ma (Wetherill 1975). Over this time, almost all of the ^{129}I, and a considerable fraction of the ^{244}Pu, had decayed to daughter Xe isotopes that could have been lost to space (Fig. 2a,b). Wetherill (1975) proposed that a 'closure age' of the Earth could be calculated by assuming that essentially complete loss of ^{129*}Xe occurred initially, followed by complete closure against further loss. The time when this closure commenced can be calculated by

$$t = \frac{-1}{\lambda_{129}} \ln\left[\left(\frac{^{129*}Xe_{atm}}{^{127}I}\right)\left(\frac{^{127}I}{^{129}I}\right)_o\right] \quad (1)$$

Wetherill obtained a closure age of ~10^8 a; using an updated value for $^{129*}Xe_{atm}$, and assuming 40% of the BSE (with ^{127}I = 13 ppb) degassed to the atmosphere, 97 Ma is obtained. The 'closure age' also can be calculated by combining the ^{129}I-^{129}Xe and ^{244}Pu-^{136}Xe systems (Pepin and Phinney 1976);

$$t = \frac{1}{\lambda_{244} - \lambda_{129}} \ln\left[\left(\frac{^{129*}Xe_{atm}}{^{136*}Xe_{atm}}\right)\left(\frac{^{238}U}{^{127}I}\right)_o\left(\frac{^{244}Pu}{^{238}U}\right)_o\left(\frac{^{127}I}{^{129}I}\right)_o {}^{136}Y_{244}\right] \quad (2)$$

Using the parent element values for the BSE ^{238}U and ^{127}I (see above), and the yield of $^{136}Y_{244} = 6.5\times10^{-3}$ for production of ^{136}Xe, a similar closure age of 82 Ma is obtained (see Fig. 2c). If atmospheric Xe loss occurred during a massive Moon-forming impact, then the closure period corresponds to the time after an instantaneous catastrophic loss event. Somewhat earlier ages of 50-70 Ma are obtained with further assumptions about the amount of Xe loss (Ozima and Podosek 1999; Porcelli et al. 2001). If radiogenic ^{136}Xe was lost over about one half-life of ^{244}Pu (80 Ma), then ~40% of the ^{136}Xe remaining in the Earth is in the atmosphere, compatible with the fraction of ^{40}Ar in the atmosphere (using a BSE value of 270 ppm K). Note that in the first 100 Ma, only 6% of the ^{40}Ar now present was produced, and losses over this time would not have significantly changed the terrestrial ^{40}Ar budget (Davies 1999).

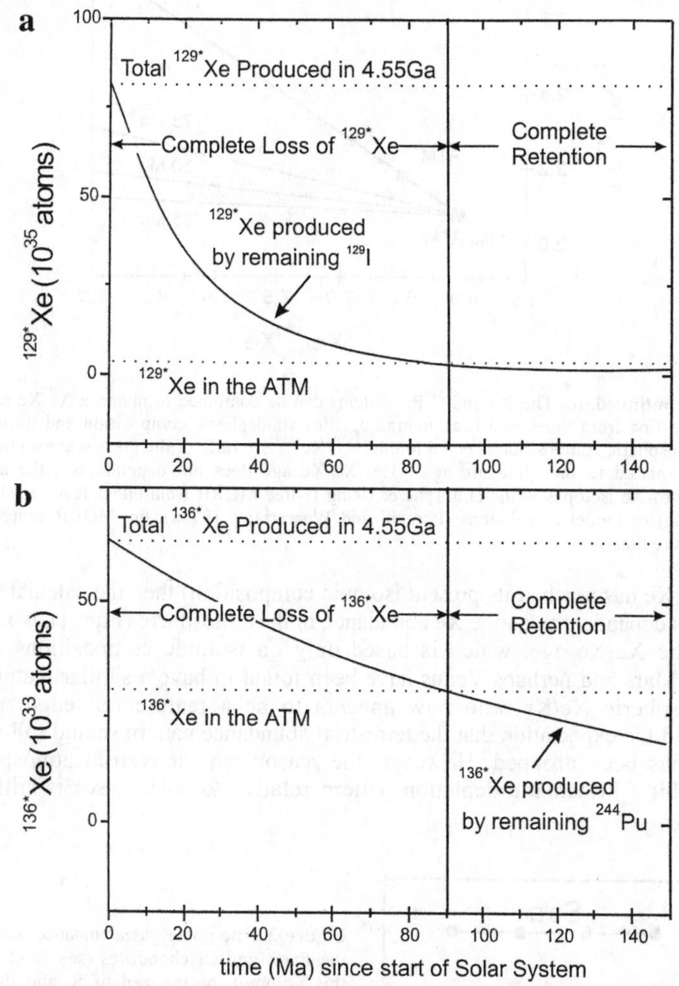

Figure 2. The abundances of $^{129*}Xe$ and $^{136*}Xe$ in the atmosphere are much lower than produced by parent nuclides in the Earth and Earth-forming materials. This has been interpreted as due to losses to space of daughter nuclides until some closure time, after which daughter Xe is retained by the Earth. (a) Abundance of $^{129*}Xe$ produced from ^{129}I after the time of closure. The amount of $^{129*}Xe$ in the atmosphere is consistent with complete loss of Xe from the Earth for the first ~80 Ma. (b) The $^{136*}Xe$ in the atmosphere from ^{244}Pu similarly reflects early Xe losses.

'Missing Xe.' Chondritic and atmosphere noble gas abundance patterns (Fig. 3) are remarkably similar, except that the atmospheric Xe/Kr ratio is much lower. This was initially thought to be due to the sequestration of ~90% of the 'missing Xe' in the crust. Models for Xe degassing would need to take this additional Xe into account. However, investigations of possible reservoirs of Xe, such as shales or glacial ice, failed to find this 'missing Xe' (Bernatowicz et al. 1984, 1985; Wacker and Anders 1984; Matsuda and Matsubara 1989; Tolstikhin and O'Nions 1994). Note also that if substantial quantities of radiogenic Xe were removed from the atmosphere long after the first 100 Ma (once the

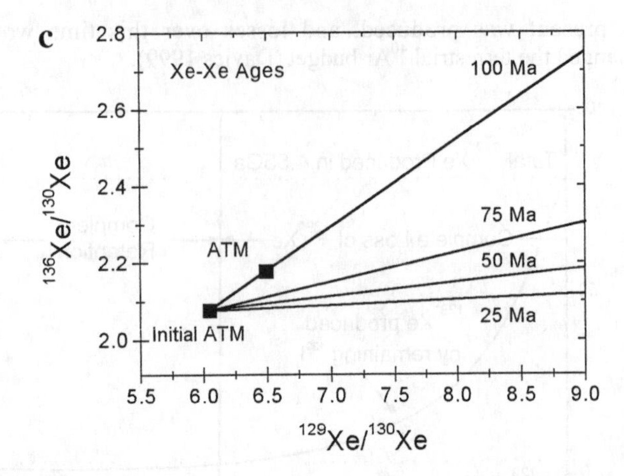

Figure 2, continued. (c) The ^{129}I and ^{244}Pu systems can be combined to obtain a Xe-Xe age. The slope of the line from the initial (i.e., nonradiogenic) atmospheric composition and through the present atmospheric composition is equal to the ^{136*Pu}Xe/^{129*}Xe ratio of the present atmosphere, and is directly related to the 'closure age.' The Xe-Xe age does not depend upon the absolute abundances of Xe isotopes in the atmosphere. Using U-free MORB from either Kunz et al (2000) (upper value) or model calculations (Porcelli and Wasserburg 1995a). the MORB source has a similar closure time.

atmospheric Xe has reached its present isotopic composition) then the calculated I-^{129}Xe age, which is dependent upon the Xe abundance in the atmosphere (Eqn. 1), would not be similar to the Xe-Xe age, which is based only on isotopic compositions (Eqn. 2). In addition, Mars and perhaps Venus have been found to have a similar feature, so the lower atmospheric Xe/Kr ratio now appears to be a feature related to planetary formation and the expectation that the terrestrial abundance pattern should follow that of chondrites has been dropped. However, the reason why terrestrial atmospheres do not all exhibit a systematic depletion pattern relative to solar gases is still without explanation.

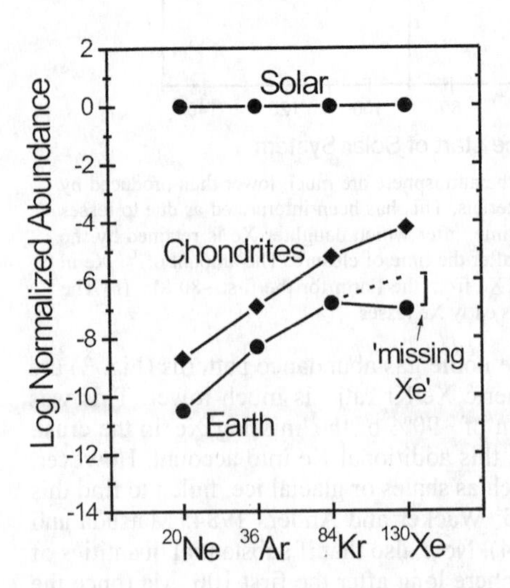

Figure 3. The noble gas abundance patterns of the Earth and CI chondrites (see Wieler 2002, this volume), normalized to Si and the solar composition. The Earth shows greater depletion in the light noble gases, similarly to CI chondrites. However, the terrestrial Xe/Kr ratio is lower than chondrites. The difference between the terrestrial pattern and the dotted line, which parallels the CI Chondrite pattern, was once interpeted to reflect 'missing Xe' sequestered some where on Earth, although the lower terrestrial Xe/Kr ratio now appears to be a planetary feature.

MANTLE NOBLE GAS CHARACTERISTICS

Following is a review of available constraints on noble gas isotope distributions and fluxes. There are no expected or observed relevant variations in Kr. For further discussion of the data, see Graham (2002) and Hilton et al. (2002) in this volume.

Helium isotopic compositions in the mantle

The ratio of $^3He/^4He$ = 120 R_A (where R_A is the atmospheric ratio of $^3He/^4He$ = 1.39×10^{-6}) for the Jupiter atmosphere (Mahaffy et al. 1998; see Wieler 2002) provides the best estimate for the solar nebula ratio and is often taken as the initial value for the Earth. The solar wind value of 330 R_A (Benkert et al. 1993) was established after D burning in the Sun, and is the correct initial Earth value if terrestrial He was derived from solar wind implantation on accreting materials (Podosek et al. 2000). Mantle He is a mixture of this 'primordial He' and radiogenic He with ~0.01 R_A (essentially all $^{4*}He$; Morrison and Pine 1955); due to the prevalence of radiogenic $^{4*}He$, primordial $^3He/^4He$ ratios are not expected to be preserved in the mantle. MORB $^3He/^4He$ ratios contain >90% radiogenic He and fall in a narrow range (Fig. 4) around 8 R_A at locations away from ocean islands (Lupton and Craig 1975; Kurz and Jenkins 1981; Graham 2002). This indicates that the MORB source region is a relatively well-mixed reservoir, although fine-scale structure has been recorded (e.g., Graham et al. 2001; Graham 2002).

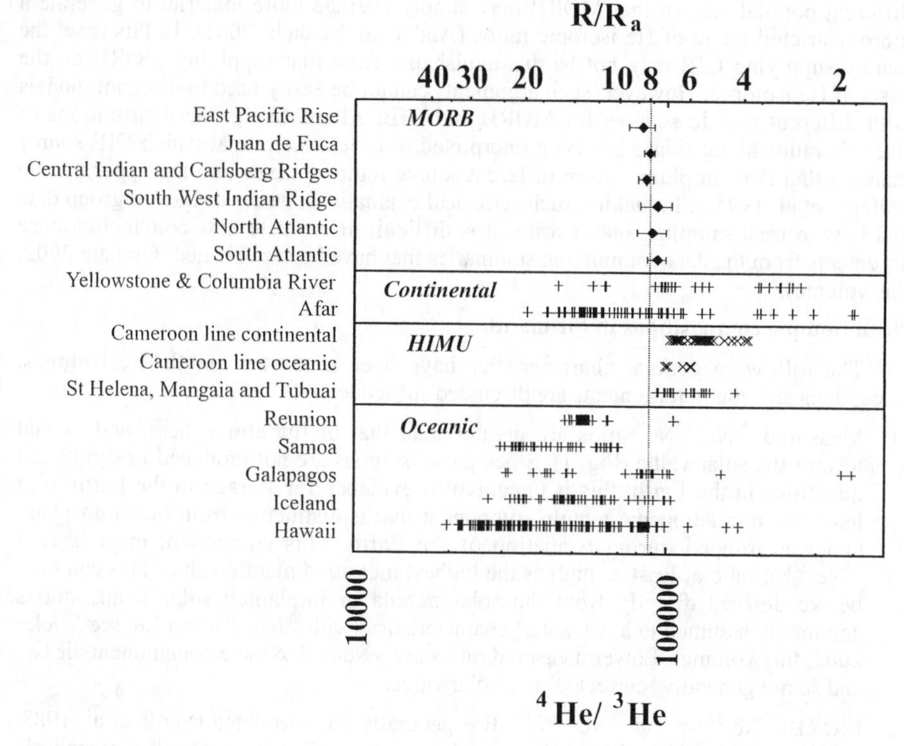

Figure 4. Compilation of He isotope data from selected mid ocean ridges (MORB), continental hotspots, ocean island volcanism (OIB), and oceanic volcanism with sources having high U/Pb ratios (HIMU) (after Barfod et al. 1999). The MORB source mantle is isotopically homogeneous compared with that supplying OIBs which, apart from HIMU, has an excess of ^3He relative to ^4He. The creation and preservation of mantle supplying high $^3He/^4He$ and its interaction with the MORB-source mantle remains the focus of both experimental work and modeling.

The $^3He/^4He$ ratios measured in OIB are more variable, with values both below and above that of MORB (Fig. 4). $^3He/^4He$ values below MORB are often associated with basalts containing Pb isotopes more radiogenic than MORB, and probably include recycled components (Kurz et al. 1982; Hanyu and Kaneoka 1998; Barfod et al. 1999). Intraplate volcanic systems with $^3He/^4He$ lower than MORB represent less than 10% of the OIB flux (e.g., Sleep 1990). It is therefore typically assumed that these OIB do not represent large-scale reservoirs that must be incorporated into global noble gas models.

Mantle models focus on the evidence for a significant long-term noble gas reservoir distinct from that of MORB with $^3He/^4He$ ratios >~10 R_A. These OIB require at least one component with a time-integrated $^3He/(U+Th)$ ratio greater than that of the MORB-source mantle (Kurz et al. 1982; Allègre et al. 1983) to produce the highest $^3He/^4He$ ratios of 32-37 R_A measured in Loihi Seamount, the youngest volcano in the Hawaiian Island chain (Kurz et al. 1982; Rison and Craig 1983; Honda et al. 1993; Valbracht et al. 1997), and Iceland (Hilton et al. 1998, 1999). Data from other hotspots has been used to suggest that there are other reservoirs with $^3He/^4He$ ratios between MORB and Loihi, although this data may reflect mixing of the source seen at Loihi with more radiogenic recycled materials (Hanyu et al. 2001).

It should be noted that a statistical evaluation of the available $^3He/^4He$ data has been presented to argue that OIB and MORB samples have not been necessarily drawn from different populations, so that MORB may simply average more material to generate a more restricted range of He isotopic ratios (Anderson 2000a,b; 2001). In this case, the mantle supplying OIB may not be distinguishable from that supplying MORB on the basis of He isotopes. However, such arguments cannot be easily used to discount models with different mantle sources for MORB and OIB. Alternatively, the distributions of $^3He/^4He$ ratios along ridges has been interpreted to reflect the rate at which OIB source material that rises in plumes from different source regions mixes into the upper mantle (Allègre et al. 1995a). In making such statistical evaluations, issues of how to group data and how to treat sampling biases arise; it is difficult to evaluate how convincing these arguments from the data compilation summaries that have been published (Graham 2002, this volume).

Neon isotopic compositions in the mantle

The following general characteristics have been noted for mantle Ne isotopes. Exceptions and counterarguments are discussed subsequently.

1. Measured $^{20}Ne/^{22}Ne$ ratios are greater than that of the atmosphere, and extend toward the solar value (Fig. 1). Since these isotopes are not produced in significant quantities in the Earth, this is unequivocal evidence for storage in the Earth of at least one nonradiogenic mantle component that is distinctive from the atmosphere and was trapped during formation of the Earth. This component must have a $^{20}Ne/^{22}Ne$ ratio at least as high as the highest measured mantle value. This can only be Ne derived directly from the solar nebula or implanted solar wind, and is commonly assumed to have 'solar' characteristics (with $^{20}Ne/^{22}Ne = 13.8$; see Wieler 2002, this volume). Lower measured ratios are assumed to have contaminant air Ne, and so are generally 'corrected' to a solar value.

2. MORB $^{20}Ne/^{22}Ne$ and $^{21}Ne/^{22}Ne$ ratios generally are correlated (Sarda et al. 1988; Moreira et al. 1998), likely due to mixing of variable amounts of atmospheric contamination with relatively uniform mantle Ne (Fig. 1). The 'corrected' or mantle $^{21}Ne/^{22}Ne$ ratio of each sample can be obtained by projecting away from the atmosphere to a solar $^{20}Ne/^{22}Ne$ ratio. These $^{21}Ne/^{22}Ne$ ratios of ~0.074 are higher than the solar value (0.033) due to additions of nucleogenic ^{21}Ne.

3. The MORB He and Ne isotopic compositions can be used to calculate the $^3He/^{22}Ne$ ratio of the source region prior to any recent fractionations created during transport and eruption. Because ^4He and ^{21}Ne production rates are directly coupled, $^3He/^4He$ and $^{21}Ne/^{22}Ne$ isotope variations should be correlated. For a source that has been closed or has lost He and Ne without fractionation,

$$\frac{\left(^{21}Ne/^{22}Ne\right)_{Source} - \left(^{21}Ne/^{22}Ne\right)_{Initial}}{\left(^{4}He/^{3}He\right)_{Source} - \left(^{4}He/^{3}He\right)_{initial}} = \left(^{3}He/^{22}Ne\right)\left(^{21*}Ne/^{4*}He\right) \quad (3)$$

   Using a corrected MORB value of $^{21}Ne/^{22}Ne = 0.074$ and $^{21*}Ne/^{4*}He = 4.5 \times 10^{-8}$, then $^3He/^{22}Ne = 11$. This is greater than the most recent estimate of 1.9 for the solar nebula (see discussion in Porcelli and Pepin 2000).

4. Measured ratios for OIB with high $^3He/^4He$ ratios span a similar range in $^{20}Ne/^{22}Ne$ ratios, but with lower corresponding $^{21}Ne/^{22}Ne$ ratios (Sarda et al. 1988; Honda et al. 1991, 1993). These OIB therefore have He and Ne isotopes that reflect higher time-integrated He/(U+Th) and Ne/(U+Th) ratios (Honda and MacDougall 1993). Using Equation (3), these yield $^3He/^{22}Ne$ source ratios that are similar to those of MORB (Poreda and Farley 1992; Honda et al. 1991, 1993).

Is solar Ne the correct trapped composition? Recently, it has been suggested that OIB and MORB data indicate that the isotopically light component in the mantle is implanted solar Ne, or Ne-B (Trieloff et al. 2000; Ozima and Igarashi 2000), which is a mixture of solar wind and solar energetic particle (SEP) Ne. It has $^{20}Ne/^{22}Ne = 12.5$, and so is fractionated relative to Ne in the sun (with $^{20}Ne/^{22}Ne = 13.8$). The few measured mantle values above this (Sarda et al. 1988; Hiyagon et al. 1992) were insufficiently precise to firmly establish the presence of Ne with a higher value. This raises the possibility that mantle Ne was supplied by irradiated material either during planetary accretion (see Pepin and Porcelli 2002), or by later subduction of irradiated meteoritic material (see *Noble Gas Mantle Models* section). However, recent analyses for Icelandic OIB samples found $^{20}Ne/^{22}Ne = 13.75\pm0.32$ (Harrison et al. 1999), demanding the presence of solar Ne in the mantle. While Trieloff et al. (2000) found a lower value for the same sample, this was using a different method that may not have fully removed the almost ubiquitous air contamination found in the basalt glass matrix (see Ballentine et al. 2001; Trieloff et al. 2001). It should be noted that it is possible that isotopic fractionation processes have modified Ne during transport and degassing, but there is no information available to rigorously evaluate this.

Statistical arguments also have been presented for the clustering of data at the lower, Ne-B value (Trieloff et al. 2000; Ozima and Igarashi 2000). Such arguments might indicate the formation of a preferential composition in the mantle, although there is insufficient data to make such a conclusion compelling. However, this does not obviate the need for a trapped mantle component with $^{20}Ne/^{22}Ne$ ratio at least as high as the highest measured value.

Is the $^3He/^{22}Ne$ ratio uniform in the mantle? A review of available data (Honda and McDougall 1998) found $^3He/^{22}Ne = 6.0\pm1.4$ for Hawaii and 10.2 ± 1.6 for MORB, with a mantle average of 7.7. This suggests that there may be a systematic difference between the MORB and OIB sources. In contrast, it has been argued that Loihi and gas-rich MORB samples have similar $^3He/^{22}Ne$ ratios (Moreira and Allègre 1998), and so the issue remains open.

Recently, Ne in Iceland samples has been found to have $^{21}Ne/^{22}Ne$ ratios that are close to that of solar Ne (Dixon et al. 2000; Moreira et al. 2001). This implies that the Iceland source has a higher Ne/(U+Th) ratio than previously calculated and so a lower

$^3He/^{22}Ne$ ratio. Two interpretations have been advanced. The source may have an unusually low ratio of $^3He/^{22}Ne \sim 4$ that is distinct from that of the source of Loihi, and so requires significant mantle heterogeneity in this ratio (Dixon et al. 2000; Moreira et al. 2001). Alternatively, magma from the MORB source may mix either directly with that of a gas-rich source (with solar Ne and a high $^3He/^4He$ ratio) to produce Loihi compositions, or may be subject to preferential loss of Ne through degassing before mixing with the gas-rich source to produce Icelandic basalts (Moreira et al. 2001; Dixon 2002).

Recent work in the Manus Basin in the SW Pacific (Shaw et al. 2001) found $^3He/^4He$ ratios of up to $12R_A$, suggesting the involvement of OIB source mantle, associated with Ne isotope ratios that lie near the MORB correlation line (even somewhat more radiogenic, with $^{21}Ne/^{22}Ne = 0.086$ when corrected to $^{20}Ne/^{22}Ne = 13.8$; Shaw et al. 2001). These ratios correspond to a source ratio of $^3He/^{22}Ne = 23.4$, which is much higher than the calculated value for MORB. While this could represent a separate trapped component in the mantle, there seems to be no reason why such an unusual component is observed only at this location. Alternatively, the source elemental ratios may have been previously fractionated due to degassing of OIB source material not long before generation of the Manus basalts (Shaw et al. 2001). In contrast, MORB-like Ne isotopes were found associated with low $^3He/^4He$ ratios of ~6 R_A in Cameroon Line OIB (Barfod et al. 1999), requiring an unusually low He/Ne ratio. Diffusive fractionation of He and Ne in the source was suggested.

Note that a compilation of MORB and OIB data has been used to indicate that measured $^{21}Ne/^{22}Ne$ and $^4He/^3He$ ratios are not correlated (Ozima and Igarashi 2000). However, this used Ne data that had not been corrected for the variable effects of atmospheric contamination.

Are there other trapped components? The only other solar system composition with $^{20}Ne/^{22}Ne$ ratios greater than the atmosphere is Q Ne ($^{20}Ne/^{22}Ne = 10.7$) (Fig. 1). Recently, Ozima and Igarashi (2000) argued that there is a preferential grouping of both MORB and OIB data at the intermediate value of $^{20}Ne/^{22}Ne = 10.8$, and suggested that this was due to the presence of Q-Ne in the mantle. If true, this would require the incorporation of a second Ne component during planetary formation and separate preservation throughout Earth history. However, Q gases have $^3He/^{22}Ne \sim 0.14$ (Busemann et al. 2000); such a large difference from the MORB value, and the absence of consequently very radiogenic $^3He/^4He$ ratios in MORB and OIB (Eqn. 3), appear to discount this as a source of mantle Ne. In general, the limited range of calculated $^3He/^{22}Ne$ ratios for the MORB source makes it difficult to accommodate a second source of trapped primordial Ne with a different origin. Also, the argument for a preferred intermediate Ne composition remains weak due to limited data.

Does the $^{20}Ne/^{22}Ne$ ratio vary in the mantle? It has been suggested (Niedermann et al. 1997) that while the maximum $^{20}Ne/^{22}Ne$ value for OIB may be solar, the MORB value is lower, ~12.5. Also, a correlation in Ne isotope data at the Shona anomaly on the mid-Atlantic ridge (Moreira et al. 1995) suggests that there is mixing between an OIB component and a MORB source with $^{20}Ne/^{22}Ne \sim 11.5$. Niedermann et al. (1997) suggested that lower values in the MORB source are due to nucleogenic production of ^{22}Ne. This requires a $^{21*}Ne/^{22*}Ne$ production ratio of ~0.4, about 150 times lower than that calculated for the mantle (Leya and Wieler 1999). Production could be increased if F is preferentially sited near U, although because the F reaction $^{19}F(\alpha,n)^{22}Na(\gamma)^{22}Ne$ only accounts for about one-third of the calculated ^{22}Ne production, production by this route must be increased by approximately 500 times. Such an increase in production appears unlikely (I. Leya, pers. comm.). Alternatively, a lower $^{20}Ne/^{22}Ne$ value might be achieved in the MORB source mantle by subduction of atmospheric Ne. While it is generally assumed that Ne is quantitatively removed from subducting material at subduction zones, the persistence of atmospheric Ne contamination, even at high temperature laboratory extraction steps, attests to the difficulty of removing Ne thermally. In order to lower the $^{20}Ne/^{22}Ne$ ratio from 13.8 to ~11.5, about 30% of the upper mantle ^{22}Ne inventory must be derived from subduction. However, because the $^{22}Ne/^{36}Ar$ ratio of the mantle is about 3 times that of the atmosphere (see *Mantle noble gas relative abundances* section), subduction of atmospheric Ne and Ar without fractionation would account for all of the ^{36}Ar in the mantle, and it is unlikely that Ne is subducted preferentially to Ar.

Regarding mantle models, Ne isotopes provide an important constraint on the relationship between mantle and atmospheric gases in models for the mantle distribution and degassing of Ne. The high mantle $^{20}Ne/^{22}Ne$ ratio indicates that simple degassing of the mantle cannot account for a dominant component of the atmosphere (see Marty and Allé 1994). This indicates that either (a) the atmosphere is derived from a different reservoir (which is no longer represented by Ne in the mantle); or (b) Ne in the atmosphere has been altered. The latter could have been caused by hydrodynamic escape to space, with preferential loss of ^{20}Ne over ^{22}Ne. Because the conditions necessary for this process could have been present only early in Earth history (Pepin 1991), it therefore requires early degassing of Ne.

Summary. Overall, the Ne data require that a separate reservoir of trapped solar (nebular or implanted solar wind) gases have been stored and largely separated from the atmosphere over Earth history. The issues of variable mantle He/Ne ratios and intermediate Ne isotope compositions remain open, and if firmly established would place demands upon the creation and preservation of such variations.

Argon isotopic compositions in the mantle

A large range in $^{40}Ar/^{36}Ar$ ratios have been measured in MORB that is generally interpreted as due to mixing of variable proportions of atmospheric Ar (with $^{40}Ar/^{36}Ar$ = 296) with a single, much more radiogenic, mantle composition. The minimum value for this mantle composition is then represented by the highest measured values of 2.8×10^4 (Staudacher et al. 1989) to 4×10^4 (Burnard et al. 1997). From correlations between $^{20}Ne/^{22}Ne$ and $^{40}Ar/^{36}Ar$ in step-heating results of a gas-rich MORB sample, a maximum value of $^{40}Ar/^{36}Ar = 4.4 \times 10^4$ was obtained (Moreira et al. 1998).

Variations in upper mantle Ar? In contradiction to the idea of a single MORB-source composition, it has recently been suggested that there are variations in the $^{40}Ar/^{36}Ar$ ratio of the MORB source, based on a correlation between Pb isotopes and $^{40}Ar/^{36}Ar$ ratios for samples with $^3He/^4He \leq 9.5$ R_A (Sarda et al. 1999). This was ascribed to subduction of a component with relatively radiogenic Pb and low $^{40}Ar/^{36}Ar$, so that there is substantial Ar subduction into the upper mantle. However, Burnard (1999) pointed out that Pb isotope compositions are known to be more radiogenic in shallow eruptive environments, and that the $^{40}Ar/^{36}Ar$ in this study correlate equally well with the depth of eruption. Therefore, the shallower samples may have lower $^{40}Ar/^{36}Ar$ ratios only due to greater gas depletion and so proportionally more air contamination. Also, Ballentine and Barfod (2000) showed that the amount of air contamination is related to basalt glass vesicularity, which in turn is related to eruption depth and volatile content, further explaining correlations of atmosphere-derived noble gases with radiogenic Pb (Sarda et al. 1999), as well as with water content or other trace element ratios (Bach and Niedermann 1998). In general, it has not been possible to separate true mantle $^{40}Ar/^{36}Ar$ variations from contamination effects either from within the magma chamber (Burnard et al. 1994; Farley and Craig 1994), through assimilation of crustal material during melt transit to the surface (Hilton et al. 1993), on equilibration with seawater during eruption (Patterson et al. 1990), or related to vesicularity and associated contamination (Ballentine and Barfod 2000).

OIB Ar. Measurements of $^{40}Ar/^{36}Ar$ in OIB with $^{3}He/^{4}He$ > 10 R_A are consistently lower than MORB values. Early values for Loihi glasses of $^{40}Ar/^{36}Ar < 10^3$ appear to reflect substantial contamination with atmospheric Ar (Fisher 1985; Patterson et al. 1990) rather than a mantle composition with a $^{40}Ar/^{36}Ar$ ratio similar to that of air (Allègre et al. 1983; Kaneoka et al. 1986; Staudacher et al. 1986). A study of basalts from Juan Fernandez (Farley et al. 1994) found atmospheric contamination contained within phenocrysts and is inferred to have been introduced into the magma chamber, thus providing an explanation for the prevalence of air contamination of OIB. Recent measurements of Loihi samples found $^{40}Ar/^{36}Ar$ values of 2600-2800 associated with high $^{3}He/^{4}He$ ratios (Hiyagon et al. 1992; Valbracht et al. 1997), while Trieloff et al. (2000) found values up to 8000 on samples with $^{3}He/^{4}He = 24$ (and so midway between MORB and the highest OIB). Poreda and Farley (1992) found values of $^{40}Ar/^{36}Ar \leq 1.2 \times 10^4$ in Samoan xenoliths that have intermediate $^{3}He/^{4}He$ ratios ((9-20) R_A). Other attempts to remove the effects of air contamination have used associated Ne isotopes and the debatable assumption that the contaminant Ne/Ar ratio is constant, and have also found low $^{40}Ar/^{36}Ar$ values (Sarda et al. 2000). It appears that $^{40}Ar/^{36}Ar$ ratios in the high $^{3}He/^{4}He$ OIB source are >3000 but lower than that of the MORB source (see also Matsuda and Marty 1995).

Nonradiogenic Ar isotopes. The solar $^{36}Ar/^{38}Ar$ ratio is ~5% lower than that of the atmosphere (see Wieler 2002, this volume). Since solar Ne has been found in the mantle, solar Ar might be expected there as well. The presence of solar Ar would constrain the relationship between atmospheric and mantle Ar. Alternatively, atmospheric Ar ratios in the mantle might reflect subduction of Ar. Measurements of nonradiogenic MORB and Loihi $^{38}Ar/^{36}Ar$ ratios are typically atmospheric within error, but have been of low precision due to the low abundance of these isotopes. Also, because most $^{40}Ar/^{36}Ar$ ratios are interpreted as having been lowered due to a substantial fraction of atmospheric contamination, the ^{36}Ar (and ^{38}Ar) of these samples are dominated by contamination. Two recent analyses of MORB and OIB samples have found $^{38}Ar/^{36}Ar$ ratios lower than that of the atmosphere (Valbracht et al. 1997; Niedermann et al. 1997). Pepin (1998) argued that these values reflect a mixture of a solar upper mantle composition with atmospheric Ar contamination. However, new high precision data for MORB with $^{40}Ar/^{36}Ar$~28,000 failed to find $^{38}Ar/^{36}Ar$ ratios that deviate from that of air (Kunz 1999). Similar $^{38}Ar/^{36}Ar$ ratios were also found in high precision OIB analyses (Trieloff et al. 2000, 2002). More recently, statistical arguments based on a compilation of published data were used to argue that the upper mantle $^{38}Ar/^{36}Ar$ ratio is atmospheric (Ozima and Igarashi 2000). However, samples with values as low as $^{40}Ar/^{36}Ar$~1000 were used and so this compilation includes samples with up to 97% contamination of ^{36}Ar and ^{38}Ar. Clearly, arguments based on this data are invalid, but careful study of individual samples with very high $^{40}Ar/^{36}Ar$ ratios have so far failed to find nonatmospheric $^{38}Ar/^{36}Ar$ ratios.

Xenon isotopic compositions in the mantle

MORB $^{129}Xe/^{130}Xe$ and $^{136}Xe/^{130}Xe$ ratios (Fig. 5) lie on a correlation extending from atmospheric ratios to higher values (Staudacher and Allègre 1982; Kunz et al. 1998), and likely reflect mixing of variable proportions of air contaminant Xe with an upper mantle component having more radiogenic $^{129}Xe/^{130}Xe$ and $^{136}Xe/^{130}Xe$ ratios. The highest measured values thus provide lower limits for the MORB source.

Origin of $^{136*}Xe$. Contributions to $^{136*}Xe$ enrichments in MORB by decay of ^{238}U or

^{244}Pu in theory can be distinguished based on the spectrum of contributions to other Xe isotopes, although analyses have typically not been sufficiently precise to identify the parent nuclide. Precise measurements of Xe in CO_2 well gases found ^{129}Xe and ^{136}Xe enrichments similar to those found in MORB, and so it is likely that this Xe is derived

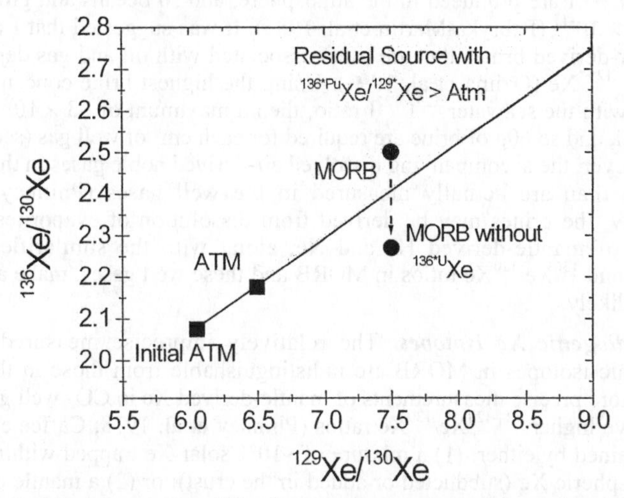

Figure 5. Xe isotope compositions for the atmosphere and upper mantle. A line through the initial (i.e., nonradiogenic) atmospheric composition and the present values has a slope of ^{136*Pu}Xe/^{129*}Xe. If this Xe was degassed from the upper mantle before complete decay of ^{244}Pu and ^{129}I, then the residual Xe in this reservoir should have a lower ratio, due to the shorter half-life of ^{129}I than ^{244}Pu. Therefore, the upper mantle composition (with contributions from ^{238}U removed) should lie above the line. The evidence indicates that corrected MORB data falls below this line, and so the upper mantle Xe cannot be residual from atmosphere degassing (Ozima et al. 1985).

from the MORB source (Staudacher 1987). These indicate that ^{244}Pu has contributed <10-20% of the ^{136}Xe that is in excess of the atmospheric composition (Phinney et al. 1978; Caffee et al. 1999). An error-weighted best fit to recent precise MORB data (Kunz et al. 1998) yielded a value of 32±10% for the fraction of ^{136}Xe excesses relative to atmosphere that are ^{244}Pu-derived, although there is considerable scatter in the data. It has been pointed out that: (a) this value was obtained by comparison with atmospheric Xe, which in turn contains Pu-derived Xe, rather than the terrestrial initial composition; and (b) if all of the data are considered rather than the filtered subset used by Kunz et al., then the parent element is not clearly discernible (Marti and Mathew 1998). Clearly, further work is warranted on the proportion of Pu-derived heavy Xe in the mantle, although it appears that the fissiogenic Xe is dominantly derived from U.

It has been argued (Ozima et al. 1993) that upper mantle ^{136*}Xe must be ^{244}Pu-derived, both because a closed system could not be dominated by ^{238}U-derived fission, and because diamonds that may be ≥1 Ga old have similar ^{129*}Xe/^{136*}Xe ratios to those of MORB, a ratio which might be expected to change due to continuing ^{136*}Xe production. However, there are models that can explain the association of ^{129*}Xe with dominantly U-derived ^{136*}Xe (see *Noble Gas Mantle Models* section). Further, not only is a fractional change in ^{136*}Xe production over 1 Ga not discounted by the data given current error limits on both diamond and MORB data, but also it is not clear when diamonds acquired their noble gases (see Dunai and Porcelli 2002, this volume).

It has been proposed that the radiogenic Xe in CO_2 well gases is derived from crustal production and does not represent the mantle (Ozima et al. 1993). This was based on the expectation that mantle radiogenic ^{129}I must be associated with ^{244}Pu-derived ^{136}Xe. While globally significant quantities of ^{129}I were only present early in Earth history, small quantities of ^{129}I are produced in the atmosphere, and so oceans and groundwater have $^{129}I/^{127}I \approx 7 \times 10^{-13}$ (Fabryka-Martin et al. 1985). It was suggested that I concentrated in the seawater-derived brines that are often associated with oil and gas deposits provided the well gas ^{129*}Xe (Ozima et al. 1993). Using the highest brine concentrations of 100 ppbI along with the seawater $^{129}I/^{127}I$ ratio, then a maximum of ~3×10^9 atoms ^{129*}Xe/g are produced, and so 60g of brine are required for each cm^3 of well gas (see Phinney et al. 1978). However, the accompanying dissolved air-derived noble gases in this brine are 10^4 times more than are actually measured in the well gases (Phinney et al. 1978). Alternatively, the brines may be derived from dissolution of evaporites, although the occurrence of mantle-derived He and Ne, along with the similarities between the $^{129}Xe/^{130}Xe$ and $^{136}Xe/^{130}Xe$ ratios in MORB and these well gases, make a crustal source of ^{129*}Xe unlikely.

Nonradiogenic Xe isotopes. The relatively imprecise measured ratios of the nonradiogenic isotopes in MORB are indistinguishable from those in the atmosphere. However, more precise measurements of mantle-derived Xe in CO_2 well gases have been found to have higher $^{124\text{-}128}Xe/^{130}Xe$ ratios (Phinney et al. 1978; Caffee et al. 1999) that can be explained by either: (1) a mixture of ~10% solar Xe trapped within the Earth and ~90% atmospheric Xe (subducted or added in the crust); or (2) a mantle component that has not been fractionated relative to solar Xe to the same extent as air Xe. In the former case, the radiogenic composition of the nonatmospheric component can be calculated (Jacobsen and Harper 1996); if 90% of the well gas Xe is derived from the atmosphere (with $^{129}Xe/^{130}Xe = 6.496$), the well gas value of $^{129}Xe/^{130}Xe = 7.2$ contains 10% of a solar component with $^{129}Xe/^{130}Xe \approx 13.5$. Note that fractionation effects have been found for Ar in other natural gas samples (Althaus et al. 2000), raising the possibility that similar effects have altered well gas Xe. While this possibility has not been rigorously evaluated, it appears that the fission component of these gases is not fractionated.

More recently, statistical arguments based on a compilation of published data were used to argue that the upper mantle $^{128}Xe/^{130}Xe$ ratio is atmospheric (Ozima and Igarashi 2000). However, most of the data used has suffered substantial atmospheric contamination, and even those data with the highest $^{129}Xe/^{130}Xe$ ratios do not have sufficiently high precision to distinguish the $^{128}Xe/^{130}Xe$ shifts seen in the well gases. Statistical arguments based on this data therefore are not valid, and confirmation whether there is atmospheric nonradiogenic Xe in the mantle will be obtained only by obtaining high precision data for relatively uncontaminated samples.

OIB Xe. Xe with atmospheric isotopic ratios in high $^3He/^4He$ OIB samples (e.g., Allègre et al. 1983) appears to be dominated by air contamination (Patterson et al. 1990; Harrison et al. 1999) rather than represent mantle Xe with an air composition. Although Samoan samples with intermediate ((9-20)R_A) He isotope ratios have been found with Xe isotopic ratios distinct from those of the atmosphere (Poreda and Farley 1992), the Xe in these samples may have been derived largely from the MORB source. Recently, Harrison et al. (1999) found slight ^{129}Xe excesses in Icelandic samples with $^{129*}Xe/^3He$ ratios that are compatible with the ratio in gas-rich MORB, but due to the uncertainties in the data it cannot be determined whether there are indeed differences between the MORB and OIB sources. Trieloff et al. (2000) reported Xe isotope compositions in Loihi dunites and Icelandic glasses that were on the MORB correlation line and had values up to $^{129}Xe/^{130}Xe = 6.9$. These were accompanied by $^3He/^4He$ ratios up to 24 R_A, and so may contain noble gases from both MORB (~8 R_A) and the highest $^3He/^4He$ ratio (37 R_A) OIB source. From these data it appears that the OIB source may have Xe that is similar to that in MORB, although some differences may not be resolvable.

Xe isotopes and a nonresidual upper mantle

The high $^{129}Xe/^{130}Xe$ ratios in MORB relative to the atmosphere have been used to model Xe degassing from the Earth (see Thomsen 1980; Staudacher and Allègre 1982). However, with better constraints on the characteristics of fissiogenic $^{136*}Xe$ in the atmosphere and mantle, there are several other constraints on the degassing of Xe:

1. Since mantle Xe has a lower $^{136*Pu}Xe/^{129*}Xe$ ratio than that of air Xe (Fig. 5), it cannot be the residue that remains after degassing of the atmosphere (Ozima et al. 1985). This can be most easily envisaged by considering that degassing occurred early and in a single event. At the time of Xe loss from the mantle to the atmosphere, the proportion of undecayed ^{244}Pu will be greater than that of ^{129}I (which has a much shorter half-life). Therefore, the remaining ^{244}Pu and ^{129}I will produce Xe with a higher $^{136*Pu}Xe/^{129*}Xe$ ratio than is in the atmosphere. This is true regardless of whether or not degassing actually occurred as a single event. Note that to compare the present upper mantle $^{136*}Xe/^{129*}Xe$ ratio to that of the atmosphere, the proportion of mantle $^{136*}Xe$ that is Pu-derived must be determined. However, as seen in Figure 5, unless all of $^{136*}Xe$ in MORB is Pu-derived, the upper mantle Xe cannot be the residue left after atmospheric Xe degassing. The data for CO_2 well gases and MORB indicate that a large fraction of the $^{136*}Xe$ is in fact from U.
2. In a closed system reservoir, Pu-derived $^{136*}Xe$ will dominate over U-derived $^{136*}Xe$ (see *Radiogenic Components* Section). Therefore, Xe from such a reservoir that has ^{129}Xe excesses will have accompanying $^{136*}Xe$ that is dominantly Pu-derived. MORB and CO_2 well gas data indicate that U-derived $^{136*}Xe$ dominates upper mantle $^{136*}Xe$. This requires that, after decay of ^{244}Pu, there is a large decrease in Xe/U in the source region, or transfer of Xe to a low Xe/U reservoir.
3. Approximately 40% of the Pu-derived ^{136}Xe produced in the Earth (after a closure age of about 1 half-life, 80 Ma) is now in the atmosphere (Table 1).
4. If isotopic fractionation of nonradiogenic Xe occurred during losses from the atmosphere to space, then degassing of nonradiogenic isotopes occurred early while this was still possible (see Pepin 1991). The early degassing of Xe corresponds to the similar requirement for Ne to account for the differences between mantle and atmospheric $^{20}Ne/^{22}Ne$. In contrast, atmospheric fissiogenic Xe does not appear to be fractionated.
5. As discussed above, nonradiogenic Xe isotopes in the mantle are fractionated with respect to atmospheric Xe, with, e.g., $^{128}Xe/^{130}Xe$ ratios that are intermediate between the solar and atmospheric ratio. This requires retention in the mantle of a component that was either only somewhat fractionated relative to solar Xe, and therefore is less fractionated than atmospheric Xe, or that is unfractionated solar Xe but has been mixed with subducted Xe.

These factors suggest the following history:

1. Nonradiogenic noble gases were degassed to the atmosphere early. The solid reservoir volume from which these gases were lost is unconstrained. This atmospheric Xe was fractionated by losses to space, so that the atmosphere became enriched in the heavy isotopes.
2. Radiogenic Xe was then degassed quantitatively from at least 40% of the mantle. This occurred either by degassing of Xe directly from the mantle, or by transport of parent Pu and I to the crust first, followed by crustal degassing.

3. Xe contained in another, deep mantle reservoir and which is not residual from atmosphere removal was subsequently added to the upper mantle. The U/Xe ratio in the upper mantle was greater than that of the deeper reservoir, so that radiogenic ^{136*}Xe in the upper mantle became dominated by U-derived Xe.

The size of the deep mantle reservoir cannot be easily constrained from Xe isotopes. Since the fraction of Pu-derived ^{136}Xe in the atmosphere is similar to that of ^{40}Ar, it might be concluded that 60% of both are retained together in a deep reservoir. However, it is possible that a significant fraction of degassed fissiogenic Xe in the atmosphere was lost during late bombardment of the Earth. This would have occurred prior to the formation of most of the ^{40}Ar, and so would not have altered the ^{40}Ar budget. Also, as discussed in the *Radiogenic Components* section, the ^{40}Ar budget of the Earth is somewhat uncertain due to debate regarding the K content of the Earth. However, if a different volume has been degassed, then the similarity between the apparent ^{40}Ar and ^{136*Pu}Xe budgets must be considered coincidental. Note that although a considerable proportion of the undegassed ^{40}Ar could be associated with K that was subducted and has subsequently decayed, this cannot be the case ^{136*Pu}Xe.

Table 2. Elemental abundance pattern of mantle reservoirs.

	Corr. † ^{20}Ne/^{22}Ne	^{4*}He/^{40*}Ar ‡	^{3}He/^{22}Ne †	^{36}Ar/^{22}Ne †	^{130}Xe/^{22}Ne †
Popping rock	12.5	1.32 to 1.99	4.7	10	0.0088
(MORB)[1]	13.8		7	5	0.0122
Iceland[2]	12.5	1.67 to 2.27	1.3	11	0.0039
	13.8		2	6.3	0.0054
Iceland[3]	12.5	2.23 to 4.09	>2.1	<8	>0.006
	13.8				
Loihi[3]	12.5	0.27 to 0.95	>1.4	<6	>0.0075
	13.8				
Air [4]			4.37×10^{-6}	18.72	0.0021
Seawater [4]			3.26×10^{-6}	63	0.027
Ocean Sediment [5]			-	192	4.03
Solar [4]			3.8	0.428	1.00×10^{-6}

[1] Moreira et al. 1998
[2] Harrison et al. 1999
[3] Trieloff et al. 2000
[4] Ozima and Podosek 1983
[5] Podosek et al. 1980
nm=not measured

† Extrapolation of measured elemental ratio to either ^{20}Ne/^{22}Ne=12.5 or ^{20}Ne/^{22}Ne = 13.8 are used to correct for air contamination. Where correlation coefficients <0.5 we have used the measured values to define the upper or lower limit on the elemental ratio.

‡ ^{4*}He/^{40*}Ar ratios are used to assess the degree of sample suite elemental fractionation (see text).

Mantle noble gas relative abundances

Abundance patterns. The measured noble gas abundance patterns of MORB and OIB scatter greatly. This is due to alteration as well as fractionation during noble gas partitioning between basaltic melts and a vapor phase that may be then preferentially gained or lost by the sample. However, MORB Ne/Ar and Xe/Ar ratios that are greater than the air values are common (Table 2). This pattern was found in a gas-rich MORB sample with high $^{40}Ar/^{36}Ar$ and $^{129}Xe/^{130}Xe$ ratios (and so relatively less air contamination) and $^4He/^{40}Ar$ ratios that are near the expected production ratio of the upper mantle (and so not fractionated) (Staudacher et al. 1989; Moreira et al. 1998).

A pattern that is consistent with radiogenic production can be calculated by assuming that the noble gases have been degassed without substantial elemental fractionation; that is, the radiogenic nuclides of the different elements are present in the mantle at relative abundances that are equal to their production ratios. In this case,

$$\frac{\left({}^{40}Ar/{}^{36}Ar\right)_{MORB} - \left({}^{40}Ar/{}^{36}Ar\right)_{Initial}}{\left({}^{4}He/{}^{3}He\right)_{MORB} - \left({}^{4}He/{}^{3}He\right)_{initial}} = \frac{{}^{3}He/{}^{36}Ar}{{}^{4*}He/{}^{40*}Ar} \quad (4)$$

and

$$\frac{\left({}^{40}Ar/{}^{36}Ar\right)_{MORB} - \left({}^{40}Ar/{}^{36}Ar\right)_{Initial}}{\left({}^{136}Xe/{}^{130}Xe\right)_{MORB} - \left({}^{136}Xe/{}^{130}Xe\right)_{initial}} = \frac{{}^{130}Xe/{}^{36}Ar}{{}^{136*}Xe/{}^{40*}Ar} \quad (5)$$

Assuming for illustration that the mantle has $K/U = 1.27 \times 10^4$ and the residence time of noble gases in the MORB source is 2 Ga, then the production ratios are $^{4*}He/^{40*}Ar = 3.5$ and $^{136*U}Xe/^{40*}Ar = 7 \times 10^{-9}$. For values of $^{40}Ar/^{36}Ar_{MORB} = 40{,}000$ and $^{136}Xe/^{130}Xe_{MORB} = 2.5$ (with $^{136*}Xe$ U-derived), along with $^{40}Ar/^{36}Ar_{Initial} = 0$ and $^{136}Xe/^{130}Xe_{Initial} = 2.08$, then $^3He/^{36}Ar_{MORB} = 1.7$ (and so $^{22}Ne/^{36}Ar_{MORB} = 0.15$) and $^{130}Xe/^{36}Ar_{MORB} = 3.3 \times 10^{-4}$. This is consistent with the pattern discussed above, with $^{22}Ne/^{36}Ar$ and $^{130}Xe/^{36}Ar$ ratios greater than those of the atmosphere (with 0.05 and 1.1×10^{-4}, respectively). Note that these arguments change qualitatively only if some noble gases are removed from the mantle with much shorter time constants, thereby maintaining ratios between radiogenic isotopes that are much different from the production ratios.

Decoupling of light and heavy noble gases? It has been suggested that MORB light and heavy noble gases are decoupled. There are several issues to consider.

1. Kaneoka (1998) showed that published mantle sample $^{20}Ne/^{22}Ne$ ratios do not correlate with $^{129}Xe/^{130}Xe$ and argued that many data could not be explained by mixing between a mantle component with a single Ne/Xe ratio and air contamination. However, large uncertainties in Xe isotope ratios, uncertainty in the mantle $^{129}Xe/^{130}Xe$ isotope ratio, and uncertainty in the range of Ne/Xe ratios of contaminants, makes this argument weak.
2. Ozima and Igarashi (2000) used a broad data compilation to argue that there are correlations between ^{36}Ar, ^{84}Kr and ^{132}Xe in MORB and OIB, but not between these nuclides and ^3He and ^{20}Ne (except for with ^{20}Ne in OIB). However, the data set included samples with heavy noble gases that were dominated by atmospheric contamination. The conclusions of this study cannot supercede those of detailed studies focused on less contaminated suites that display coherent interelement variations (Moreira et al. 1998).
3. Measured $^4He/^{21}Ne$ ratios often are greater than the production ratio in MORB, but lower in OIB (Honda and Patterson 1999; Ozima and Igarashi 2000). The wide range of ratios is in contrast to the restricted range of both $^3He/^4He$ ratios and $^3He/^{22}Ne$ ratios calculated for the mantle from isotopic compositions (see above). This indicates that the fractionation cannot be either a long-term source region feature or due to melt generation processes that would leave a fractionated mantle residue. It is possible that crustal processes increase melt $^4He/^{21}Ne$ ratios, such as by input of He from the crystallized oceanic crust (Honda and Patterson 1999), although the reason for the systematic difference between MORB and OIB is not clear. Matsuda and Marty (1995) suggest that He may preferentially diffuse into MORB magma

chambers from the surrounding mantle, thereby enriching these basalts. However, this would imply that He is removed from the mantle preferentially to Ne, leaving a range of ^{3}He/^{22}Ne ratios. This is contrary to what is calculated from He and Ne isotopes.

4. It is generally assumed that contaminant noble gases follow a regular pattern. However, many sediments appear to have Ne/Ar and Xe/Ar ratios greater than those of the atmosphere (Ozima and Alexander 1976; Podosek et al. 1980). It is not clear what causes this. While the dominant contamination source is probably unfractionated air (Ballentine and Barfod 2000), it is clear that other sources of air contamination exist and do not follow simple patterns (Harrison et al. 1999; Burnard et al., submitted; Harrison et al., submitted). This hampers efforts to correct for contamination effects.

Overall, He and Ne isotope ratios, which are least subject to uncertainties due to atmospheric contamination and measurement precision, appear to be uniform in the source region, and so He and Ne are not likely to be removed from the mantle with strong and variable fractionation. Mantle Ar, Kr, and Xe have isotope compositions that are not well constrained, but at present do not provide compelling evidence for decoupling from the light noble gases. It should be noted, however, that decoupling cannot be discounted and must be considered further if demanded by a plausible mechanism or convincing data. However, such decoupling likely would be due to addition of the heavy noble gases, e.g., from subduction, rather than preferential diffusive mobility of light noble gases (which could fractionate He and Ne).

Noble gas fluxes and mantle concentrations

The mantle fluxes that can be most easily identified are those of ^{3}He. Also, concentrations in the mantle source regions from which ^{3}He is lost through melting can be obtained where the source volume involved can be quantified. Estimated fluxes are given in Table 3.

Table 3. He flux summary.

Source	*^{3}He mol/yr*	*^{4}He mol/yr ×10^8*	*Derivation (for details, see text)*
Mid-ocean ridges	1060±250	1.16±0.27	^{3}He flux into oceans
Ocean islands	38-670	0.04-0.73*	Highest source ^{3}He concentration estimate and range in magma generation rates
Subduction zones	53-212	0.06-0.23*	Assuming MORB source ^{3}He concentration estimate and range in magma generation rates
Continental extension	8.4-84	0.009-0.092	Magmatic component only
Stable continent	2.7	3.6	Assuming steady state production and loss (Ballentine & Burnard 2002, this volume)

Mid-ocean ridge flux. The largest and most clearly defined mantle volatile flux is from mid-ocean ridges (Clarke et al. 1969; Craig et al. 1975) and is obtained by combining seawater ^{3}He concentrations in excess of dissolved air He with seawater mixing and ocean-atmosphere exchange models (see Schlosser et al. 2002). Although the original estimate of 1060±250 mol/yr ^{3}He (Lupton and Craig 1975) was revised down to 400 mol/year ^{3}He (Jean-Baptiste 1992), more recent ocean circulation models concur with the earlier estimate (Farley et al. 1995). This value represents an average over the last 1000 years. Time integrated records of this flux over longer periods do not exist and so there is substantial uncertainty in using this value as representative of mantle degassing rates over even recent geological time.

Upper mantle concentrations. Concentrations of ^{3}He in mantle-derived MORB samples have clearly been subject to modifications and so are difficult to relate to the source. However, the concentration of ^{3}He in the mantle can be determined by dividing the flux of ^{3}He into the oceans by the rate of production of melt that is responsible for carrying this ^{3}He from the mantle, which is equivalent to the rate of ocean crust production of 20 km^{3}/yr (Parsons 1981). Mid-ocean ridge basalts that degas quantitatively to produce a ^{3}He flux of 1060 mol/yr must have an average ^{3}He content of 1.96×10^{-14} mol/g or 4.4×10^{-10} cc ^{3}He(STP)/g. This ^{3}He concentration is within a factor of 2 of that obtained for the 2ΠD43 popping rock, the most gas rich basalt glass yet found, of $\sim 10.0 \times 10^{-10}$ cc ^{3}He(STP)/g (Sarda et al. 1988; Moreira et al. 1998). Although addition of magmatic gas to this sample during residence in the magma chamber cannot be ruled out, the sample vesicle size distribution is consistent with closed system formation during magma ascent (Sarda and Graham 1990) and individual inclusions in 2ΠD43 preserve He/Ar, C/He and C/N ratio variations consistent with closed system formation during ascent (Javoy and Pineau 1991). An independent estimate of the carbon content of undegassed MORB of 900-1800 ppm C (Holloway 1998), combined with a MORB-source mantle $CO_2/^{3}He$ ratio of 2×10^{9} (Marty and Jambon 1987) gives a ^{3}He concentration in undegassed MORB of $(8.4\text{-}17.0) \times 10^{-10}$ cc ^{3}He(STP)/g. The relatively close agreement between these independent calculations provides some degree of confidence in the values obtained. Assuming 10% partial melting and quantitative extraction of He from the solid phase into the melt, these undegassed melt values give a MORB-source mantle concentration of $(0.44\text{-}1.7) \times 10^{-10}$ cc ^{3}He(STP)/g, or $(1.2\text{-}4.6) \times 10^{9}$ atoms ^{3}He/g.

The mantle concentration of He cannot be usefully compared with the He inventory of the atmosphere (from where He is lost to space), but a comparison can be made with Ar. Using a time-integrated (over 2 Ga) production ratio of $^{4*}He/^{40*}Ar = 3.5$ and an upper mantle value of $^{40}Ar/^{36}Ar = 4 \times 10^{4}$ (Burnard et al. 1997) then the upper mantle has $(0.8\text{-}2.9) \times 10^{9}$ atoms ^{36}Ar/g. This is highly depleted compared to the benchmark value of 3×10^{12} atoms ^{36}Ar/g obtained by dividing the atmospheric inventory by the mass of the upper mantle.

OIB fluxes and source concentrations. The flux of ^{3}He from intraplate volcanic systems is dominantly subaerial and so it is not possible to directly obtain the time-integrated value for even a short geological period. While the Loihi hotspot in the Pacific is submarine, calculation of ^{3}He fluxes into the ocean using ocean circulation models have not required a large flux from this location that is comparable to the ^{3}He plumes seen over ridges (Gamo et al. 1987; Farley et al. 1995), although recent data has seen an extensive ^{3}He plume from Loihi (Lupton 1996). Also, major ^{3}He degassing at such oceanic hotspots may be episodic and not be typical at present. Note that the atmospheric He budget is not sufficiently well constrained to determine whether there is a large OIB flux (see e.g., Torgersen 1989).

Helium fluxes from OIB could be calculated if the rates of magmatism are known, along with the He concentrations of the source regions or concentrations of undegassed magmas. Estimates of the rate of intraplate magma production vary considerably: 0.6×10^{15} g/yr (Reymer and Schubert 1984), 4.4×10^{15} g/yr (Schilling et al. 1978), $(5\text{-}7) \times 10^{15}$ g/yr including seamounts (Batiza 1982), and $(0.4\text{-}5) \times 10^{15}$ g/yr when considering magma emplacement estimates for the last 180 Ma (Crisp 1984). These values correspond to 1-12% of the MORB production rate.

It has been difficult to determine the He concentrations of undegassed basalts. OIB samples typically have substantially lower He concentrations than MORB samples. This is paradoxical if high $^3He/^4He$ OIB ratios are derived from gas-rich source material. The very extensive degassing during subaerial eruption can account for the strong depletion of He in many samples. Therefore, it is often assumed that OIB basalt undegassed ^3He concentrations are high, but have been highly altered during eruption (see Marty and Tolstikhin 1998). However, samples from Loihi have been erupted from a range of submarine depths. There may also be other factors promoting more extensive degassing. OIB may be more volatile-rich than MORB and therefore may degas more effectively (Dixon and Stolper 1995). Also, high water contents of basalts lower CO_2 solubility, and appear to have lowered He contents in lavas with high $^3He/^4He$ ratios along the Reykjanes Ridge (Hilton et al. 2000).

Current estimates of ^3He concentrations in OIB sources rely on three approaches.

1. The ^3He flux can be estimated from volcanic CO_2 fluxes and magma generation rates. Although there are estimates of subaerial CO_2 fluxes from non-arc volcanoes (Allard 1992; Williams et al. 1992), the proportion of CO_2 from intraplate volcanism overall is poorly defined (Brantley and Koepenick 1995). The summit CO_2 flux of Kilauea volcano, Hawaii, can be combined with observed $CO_2/^3He$ values and estimated magma production rates to obtain an undegassed plume magma concentration of 4.7×10^{-10} cc ^3He(STP)/g (Hilton et al. 1997). Assuming magma production by 7% melting, this corresponds to a source concentration of 3.3×10^{-11} cc ^3He(STP)/g. However, estimates from a number of other sites are required before a global average can be reasonably deduced.
2. Direct measurements of basalt concentrations and noble gas fractionation patterns can be used to reconstruct pre-eruption concentrations. Measurements of ^3He concentrations in Hawaiian basalts (e.g., Kurz et al. 1983) of $(3.3\text{-}10.0) \times 10^{-11}$ cc ^{3}(STP)/g, are low compared to those of MORB (see Marty and Tolstikhin 1998). Moreira and Sarda (2000) assessed the extent of prior basalt degassing from the magnitude of air-corrected He, Ne and Ar fractionations in filtered MORB and OIB data sets, and using an open system Rayleigh degassing model calculated a minimum OIB source concentration of 1.10×10^{-10} cc ^3He(STP)/g, which is comparable to that of the MORB source. It was suggested that measured sample OIB He concentrations can also be subject to variable post-eruption gas loss that occurs with no elemental fractionation (e.g., by vesicle rupture during sample handling), and so the source may have had even higher concentrations. More recent developments in understanding the effects of non-ideality and variable H_2O and CO_2 contents on noble gas solubility in degassing melts (Nuccio and Paonita 2000) have yet to be applied to these modeling approaches.
3. Source He concentrations can be derived from isotopic mixing patterns at some locations. Where plumes have strong interactions with mid-ocean ridges, such as in the Reykjanes Ridge south of Iceland, basalt characteristics can be ascribed to mixing between a MORB source and a plume component. Hilton et al. (2000) found that $^3He/^4He$ and $^{206}Pb/^{204}Pb$ are linearly correlated ($r^2 = 0.87$ and $n = 18$) along the Reykjanes Ridge and may reflect mixing between components with similar He/Pb ratios. Eiler et al. (1998) found a similar relationship in Hawaiian lavas. In the limiting case where the Pb concentration of the plume is the same as that of MORB-source mantle, the ^3He concentration in the plume source is 4 times higher than that of the MORB-source mantle, taking into account end member $^3He/^4He$ isotope differences. Higher plume Pb concentrations would increase this value. For example,

adopting Pb = 0.05 ppm for the MORB-source mantle and 0.185 ppm for the plume source (the BSE value), Hilton et al. (2000) obtained a plume source component ^{3}He concentration of 15 times that of MORB-source mantle. Further, Hilton et al. (2000) suggest that the data may support a mixing line involving a higher plume source He concentration. Pre-degassing of the plume source prior to mixing with MORB-source mantle also remains a possibility, thus allowing a higher earlier plume component He concentration.

It should be emphasized that in many locations, the high ^{3}He/^{4}He source component comprises a small fraction of the sample source. A very gas-rich He source need only contribute a very small mass fraction of the source region to dominate the He budget. For example, gas-rich mantle with ^{3}He/^{4}He = 40 R_A mixed with MORB-source mantle could generate a source that has ^{3}He/^{4}He = 25 R_A and a He concentration that is at most double that of the MORB source. Therefore, in many cases (Fig. 4), source concentrations may be closer to that of MORB. In this case, the overall ^{3}He flux relative to that from MORB is proportional to the relative melt production rate, or 1-12% that of MORB. It is not clear to what extent more gas-rich OIB augment this flux.

Overall, the He content of the high ^{3}He/^{4}He source remains rather unconstrained. Estimates range from less than in the MORB source, to greater by an order of magnitude or more.

Continental areas of extension. Continental settings provide a small but significant flux from the mantle. Regional groundwater systems provide time-integrated records of this flux over large areas, but are based on short time scales and dependent on the hydrogeological model used. In the Pannonian basin (4,000 km^2 in Hungary), for example, the flux of ^{3}He has been estimated to be 8×10^4 atoms ^{3}He m^{-2} s^{-1} (Martel et al. 1989) and 0.8 to 5×10^4 atoms ^{3}He m^{-2} s^{-1} (Stute et al. 1992). Taking an area of 2×10^{14} m^2 for continents and assuming 10% is under extension, this yields a total ^{3}He flux of 8.4 to 84 mol ^{3}He/yr. This value compares with <3 mol mantle-^{3}He/yr through the stable continental crust (O'Nions and Oxburgh 1988).

Noble gas fluxes at subduction zones. The flux of ^{3}He at subduction zones from the upper mantle can be estimated from the volume of convergent zone volcanics. The estimate of Reymer and Schubert (1984) is 5% that at mid-ocean ridges. If this is largely generated by similar degrees of melting of a similar source as MORB, then the ^{3}He flux is only 5% that at the mid-ocean ridges, or 50 mol ^{3}He/year (see also Hilton et al. 2002, this volume). Torgersen (1989) estimated a higher flux of ~20% that at ridges. An estimated CO_2 flux at convergent margins of 3×10^{11} mol/year (Sano and Williams 1996) can also be used to estimate the ^{3}He flux. Using a mid-ocean ridge ratio of $CO_2/^3He$ = 2×10^9 (Marty and Jambon 1987) gives 150 mol ^{3}He/year, although the convergent margin $CO_2/^3He$ ratio may be much greater due to the presence of recycled C.

It is possible that atmospheric noble gases incorporated in subducting materials are carried into the mantle and mixed with primordial noble gas constituents. Holocrystalline mid-ocean ridge basalts and oceanic sediments contain atmospheric noble gases that are greatly enriched in the heavier noble gases. However, because concentrations vary by several orders of magnitude, an average value cannot be easily determined. Porcelli and Wasserburg (1995a) reviewed the sediment budget. Pelagic sediments have a wide range of measured concentrations of (0.05-7) $\times 10^{10}$ atoms ^{130}Xe/g (Podosek et al. 1980; Matsuda and Nagao 1986; Staudacher and Allègre 1988) with a geometric mean of 6×10^9 atoms/g. The amount of sediment subducted has been estimated to be 3×10^{15} g/yr (von Huene and Scholl 1991). Holocrystalline basalts have been found to have (4-42) $\times 10^7$ atoms ^{130}Xe/g (Dymond and Hogan 1973, Staudacher and Allègre 1988), with a geometric mean of 2×10^8 atoms ^{130}Xe/g. In the absence of any available information regarding the depth over which addition of atmospheric gases are added by alteration, it might be assumed that this occurs over the same depth as low temperature enrichment of alkaline elements, which has been estimated to be ~600 m (Hart and Staudigal 1982). An estimated 7×10^{15} g/yr of this material is subducted. In total, these numbers result in 2×10^{25} atoms ^{130}Xe/yr (33 mol/yr) reaching subduction zones in sediments and altered basalt. This can be compared to the estimate of Staudacher and Allègre (1988) determined by assuming that 40-80% of the subducting flux of oceanic crust (6.3×10^{16} g/yr) is altered and contains atmosphere-derived noble gases, and that up to 18% of this mass is ocean sediment. In this case, a flux of 13.8-39 mol ^{130}Xe/yr was obtained, similar to the value quoted above. The fluxes of the other noble gases were estimated to be 1.9-2.5 $\times10^6$ mol ^{4}He/yr, 1.9-5.9 $\times10^3$ mol ^{20}Ne/yr, 5.8-21.9 $\times10^3$ mol ^{36}Ar/yr, and 3.0-12.3 $\times10^3$ mol ^{84}Kr/yr. Subduction of noble gases at these rates over 10^9 yr would have resulted in 1%, 90%, 110% and 170% of the respective inventories of ^{20}Ne, ^{36}Ar, ^{84}Kr and ^{130}Xe to be in the upper mantle (Allègre et al. 1986). Note that the concentrations used are for dry sediments, and water-saturated values are expected to give significantly higher flux rates. It should be emphasized that these numbers are highly uncertain and can only be used for order-of-magnitude comparisons.

Subduction zone processing and volcanism may return much of the noble gases in the slab to the atmosphere. However, it is possible that the total amounts of noble gases reaching subduction zones are sufficiently high that subduction into the deeper mantle (i.e., beyond the zone of magma generation) of only a small fraction may have a considerable impact upon the composition of Ar and Xe in the upper mantle (Porcelli and Wasserburg 1995a,b). Staudacher and Allègre (1988) argued that subducting Ar and Xe must be almost completely lost to the atmosphere during subduction zone magmatism, or the high ^{129}Xe/^{130}Xe and ^{136}Xe/^{130}Xe in the upper mantle would not have been preserved throughout Earth history. However, the contrary view is that subducted noble gases are mixed with nonrecycled, mantle-derived Xe to produce the upper mantle composition (Porcelli and Wasserburg 1995a,b). Note that, as discussed above, the ^{128}Xe/^{130}Xe ratio measured in mantle-derived Xe trapped in CO_2 well gases may be interpreted as a mixture of ~90% subducted Xe with 10% trapped solar Xe. Further, direct input of the subducted slab into a gas-rich deeper reservoir that has ^{40}Ar/^{36}Ar values that are significantly lower than in the MORB source mantle (e.g., Trieloff et al. 2000) is also possible.

Undepleted mantle

He with high ^{3}He/^{4}He ratios is often assumed to be stored in a mantle reservoir that has evolved approximately as a closed system for noble gases and has BSE parent nuclide concentrations. Since this reservoir must be isolated from degassing at mid-ocean ridges and subduction zones, it is often placed within the deeper, or lower mantle. Assigning the highest OIB ^{3}He/^{4}He ratios to this reservoir, a comparison between the total production of ^{4*}He and the shift in ^{3}He/^{4}He from the initial terrestrial value to the present value provides an estimate of the ^{3}He concentration in this reservoir;

$$({}^4\mathrm{He}/{}^3\mathrm{He})_{\mathrm{present}} - ({}^4\mathrm{He}/{}^3\mathrm{He})_{\mathrm{initial}} = {}^{4*}\mathrm{He}/{}^3\mathrm{He} \qquad (6)$$

For 1.02×10^{15} atoms ^{4*}He/g (see *Radiogenic He* section), an Iceland value of ^{3}He/^{4}He = 37 R_A and an initial value of ^{3}He/^{4}He = 120 R_A (see *Mantle Noble Gas Characteristics* Section), then the reservoir has 7.6×10^{10} atoms ^{3}He/g. The concentration of another noble gas is required for comparison of lower mantle noble gas abundances with the atmosphere. Using ^{3}He/^{22}Ne = 11, a concentration in a closed system lower mantle of 7×10^9 atoms ^{22}Ne/g is obtained. A benchmark for comparison is the atmospheric ^{22}Ne abundance divided by the mass of the upper mantle (1×10^{27} g) of 1.8×10^{11} atoms ^{22}Ne/g, which is much higher, and might be taken to indicate that the atmosphere source reservoir was more gas-rich than any deep isolated reservoir.

Nonradiogenic Ar and Xe isotope concentrations of such a lower mantle reservoir cannot be directly calculated without assuming either lower mantle Ar/Ne and Xe/Ne ratios or Ar and Xe isotopic compositions. For example, a closed system lower mantle with $^{40}Ar/^{36}Ar \geq 3000$ and 270 ppm K has $^{40}Ar = 5.7 \times 10^{14}$ atoms/g and so $^{36}Ar \leq 1.9 \times 10^{11}$ atoms/g. The reservoir then has a ratio of $^{22}Ne/^{36}Ar \geq 0.9$, which is much greater than the air value of 0.05. Note that some calculations have assumed that the lower mantle concentration is equal to the atmospheric inventory divided by the mass of the upper mantle (e.g., Hart et al. 1979), which is based on the idea of an initially uniform distribution of Ar (see *Noble Gas Mantle Models* section). This is compatible with a lower mantle ratio of $^{40}Ar/^{36}Ar = 300$. However, it is possible that there are high $^{40}Ar/^{36}Ar$ ratios in the lower mantle that are due not to degassing, but rather to a lower initial trapped ^{36}Ar concentration (Porcelli and Wasserburg 1995b).

Note that these closed system considerations do not require assumptions regarding the size of the reservoir, and can involve a portion of a stratified mantle or material that is distributed as heterogeneities within another mantle reservoir. It should be emphasized that while an undepleted, undegassed mantle reservoir is a component in some models, there is no direct evidence that such a reservoir does exist.

Coupled degassing of noble gases

There have been various ideas regarding the relative behaviors of the noble gases.

1. It is often assumed that all the noble gases are highly incompatible during basalt genesis and so are efficiently extracted from the mantle without elemental fractionation. Experimental data of partitioning between basaltic melts and olivine are consistent with this for He but not for the heavier noble gases, which have been found in higher concentrations in olivine than expected (Hiyagon and Ozima 1986; Broadhurst et al. 1992). However, these results may be due to experimental difficulties (Farley and Neroda 1998). Chamorro-Perez et al. (1998) recently reported that the solubility of Ar in silicate melt is dramatically lower at pressures above ~60 kbar, presumably due to a structural change in the melt. This suggested that there may be a mechanism for preferential retention of heavy noble gases at depth in the mantle. However, more recent data (Chamorro-Perez et al. 2001) indicate that the Ar clinopyroxene/silicate melt partition coefficient is relatively constant and equal to $\sim 4 \times 10^{-4}$ at pressures up to at least 80 kbar, and so indicates that there is no structural change in the melt over that pressure range. Therefore, it appears that the noble gases are all highly incompatible in the mantle, and there is no elemental fractionation between the melt and the mantle source region.
2. While noble gases transported from the mantle may not be elementally fractionated, some fractionation may occur in the highly depleted melt residue due to small differences in partition coefficients. This remains a possibility because reliable partition coefficients for most of the noble gases are unavailable. Fractionation would be reflected in the isotopic compositions resulting from subsequent long-term radiogenic and nucleogenic additions (see *Mantle Noble Gas Relative Abundances* Section). While there is no evidence for substantial fractionations between He and Ne (see *Neon isotope compositions in the Mantle*), there is insufficient data to determine if this is the case for Ar and Xe.
3. Differences in solubility in silicate melts may create noble gas elemental fractionations. Since the solubility of noble gases in silicate melts decreases with increasing atomic radius, partitioning into exsolved CO_2-rich vapor at shallower depths will preferentially deplete the heavy noble gases in the melt. If the vapor subsequently reaches the atmosphere while the melt crystallizes and returns to the

mantle, the net effect will be preferential degassing of the lighter noble gases (Zhang and Zindler 1989). Such a process can also operate in shallow magma chambers where the CO_2 solubility is exceeded and escaping vapors can leave a fractionated melt, as well as during early degassing of a partially molten Earth (Tolstikhin and O'Nions 1994).

4. A return flux of atmospheric noble gases to the mantle may occur at subduction zones (see *Noble Gas Fluxes at Subduction Zones*). This would favor the heavy noble gases, which are more readily adsorbed and incorporated into altered or sedimentary materials, and have the net effect of changing the upper mantle relative elemental abundances.

How these processes affect the degassing of mantle reservoirs is discussed in *Noble Gas Mantle Models*.

TRACE ELEMENT ISOTOPE CONSTRAINTS ON MANTLE RESERVOIRS

Relationship between noble gases and radiogenic isotopes

Isotopic variations in the isotopes of the lithophile elements Sr, Nd, Hf, Os, and Pb have been used to define a range of mantle components. Correlations of these isotopic variations with those of noble gases could in principle define noble gas mantle distributions in these components (see Graham 2002 for discussion). Plots of $^3He/^4He$ vs lithophile isotopes (e.g., $^{206}Pb/^{204}Pb$, $^{87}Sr/^{86}Sr$, $^{187}Os/^{188}Os$, and $^{143}Nd/^{144}Nd$) in intraplate OIB show elongate arrays that are interpreted as mixing between two components (Fig. 6). The high $^3He/^4He$ component in OIB has been called FOZO (the focus zone; Hart et al. 1992) or PHEM (primitive He mantle; Farley et al. 1992). A similar high $^3He/^4He$ component in some MORB has also been resolved, and called C (common component; Hanan and Graham 1996). Although not well defined, the high $^3He/^4He$ component appears to have common features in most OIB and MORB and may represent a common mantle reservoir. This component has lithophile isotope compositions reflecting depletion along with $^3He/^4He \sim 45\pm5$Ra (Hauri et al. 1994; Hanan and Graham 1996; Hilton et al. 1999; van Keken et al. 2002). In contrast, the low $^3He/^4He$ component in the OIB elongate 'mixing' arrays varies between end-member compositions predicted for the depleted upper mantle (DMM), modern sediments (EMI), ancient sediments or depleted lithosphere (EMII) and recycled oceanic crust (HIMU) (Hart et al. 1992). Overall, there is no evidence for mantle with high $^3He/^4He$ ratios that has unmodified BSE lithophile element ratios (as reflected in isotope compositions). However, if such material exists with high ^3He concentrations, it may contribute a high $^3He/^4He$ signature to OIB sources without clearly dominating the other tracers.

Mass and character of the depleted mantle

Mass balance calculations of lithophile elements have been used to determine the fraction of an initially uniform BSE that has segregated to form the crust and depleted MORB-source mantle (e.g., Hofmann 1997). This generally assumes that the remainder of the mantle remains unmodified. Initial estimates placed the depleted volume to be between 25-50% of the mantle (Jacobsen and Wasserburg 1979; O'Nions et al. 1979; De Paolo 1980). Although later analysis of the uncertainties associated with the original input parameters placed this volume at anywhere between 30 and 90% (e.g., Zindler and Hart 1986), more recent estimates appear to converge, giving values from 45±10% (McCulloch and Bennett 1998) to between 50-65% (Turcotte et al. 2001). The upper mantle is often associated with the volume above the 670-km discontinuity, which represents only about 30% of the mantle. Therefore, a significant portion of the lower mantle also must be depleted. These figures are also consistent with the ^{40}Ar mass

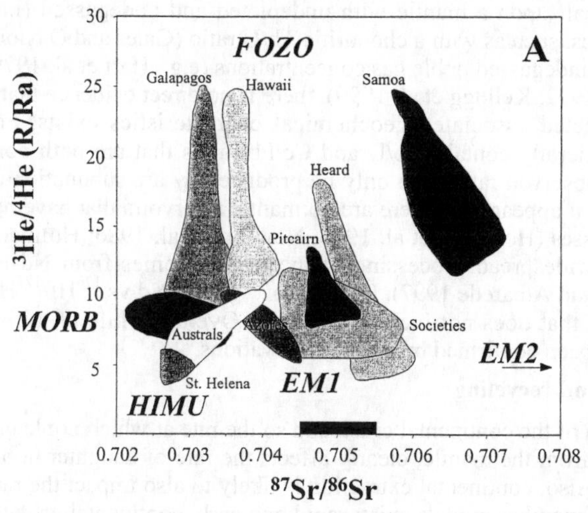

balance and the K budget discussed in the *Radiogenic components* section. However, it is possible to interpret the lower mantle as composed of components other than depleted and pristine material, such as subducted materials, so that more than 45-65% of the Earth is processed (e.g., Zindler and Hart 1986; Phipps Morgan and Morgan 1999; Helffrich and Wood 2001).

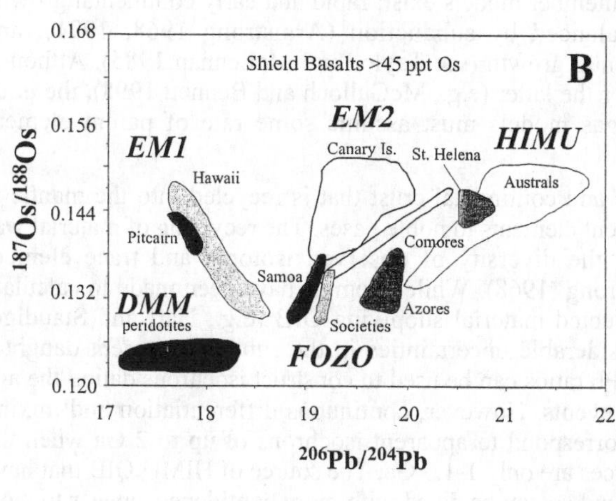

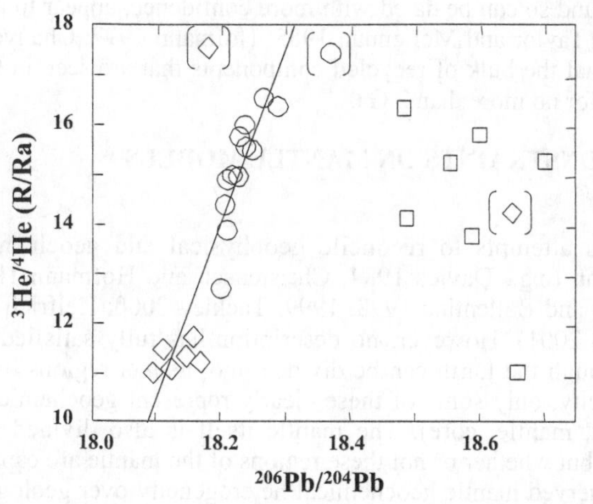

Figure 6. (A) Elongate arrays for oceanic basalts in a plot of $^3He/^4He$ against $^{87}Sr/^{86}Sr$ appear to show a common origin for the high $^3He/^4He$ (e.g. Hart et al. 1992; Hauri et al. 1994; Farley et al. 1992). (B) Data arrays for $^{187}Os/^{188}Os$ against $^{206}Pb/^{204}Pb$ again converge on a high $^3He/^4He$ component. This high-$^3He/^4He$ component has an elevated $^{187}Os/^{188}Os$ ratio indicative of recycled mafic material (after van Keken et al. 2002). (C) Plot of $^3He/^4He$ vs. $^{206}Pb/^{204}Pb$ from basalts taken from the Reykjanes Ridge (Hilton et al. 2000). Diamonds are samples south of 58.26°N, circles between 58.26°N and 61.3°N, and squares north of 61.3°N. Bracketed samples are excluded from mixing considerations, as are all samples north of 61.3°N, which have shallow eruption depth and high water content. The mixing line has a mixing constant of 1 ($r^2 = 0.87$, n = 18) and requires the ^3He concentration in the Iceland plume source to be ≥4× higher than that of the MORB-source.

Although many models call upon a mantle with undepleted and undegassed (i.e., primitive) mantle characteristics, such as with a chondritic Th/U ratio (Galer and O'Nions 1985; Turcotte et al. 2001) or undegassed noble gas concentrations (e.g., Hart et al. 1979; Kurz et al. 1982; Farley et al. 1992; Kellogg et al. 1999), there is no direct evidence that a component with all the expected associated geochemical characteristics exists. For example, OIB and MORB generally contain Nb/U and Ce/Pb ratios that are both non-chondritic and uniform. The observed ratios can only be produced by arc subduction or crust formation processes, and it appears that there are no mantle reservoirs that have not been impacted by these processes (Hofmann et al. 1986; Newsom et al. 1986; Hofmann 1997). Further evidence for widespread processing of the mantle comes from Nd-Hf isotope studies (Blichert-Toft and Albarède 1997). In a plot of $^{143}Nd/^{144}Nd$ vs $^{176}Hf/^{177}Hf$, mantle values lie on an array that does not include the BSE. Overall, while primitive mantle may exist, this has not been confirmed by direct observations.

Timing of mantle depletion and recycling

The formation and growth of the continental crust, and so the rate at which noble gas parent elements are extracted from the mantle, clearly affects the rate of daughter noble gas production in the mantle. Also, continental extraction is likely to also impact the rate of mantle degassing. Two end member models exist: rapid and early continental growth, with subsequent additions balanced by subduction (Armstrong 1968, 1991), and continuous or episodic continental growth (see Taylor and McLennan 1985). Although recent publications tend to favor the latter (e.g., McCulloch and Bennett 1998), the issue remains open, and so noble gas models must assume some rate of parent element extraction from the mantle.

Material from the oceanic and continental crust that is recycled into the mantle is likely to have high ratios of parent elements to noble gases. The recycling of material was first proposed to account for the diversity of observed isotopic and trace element compositions in OIB by Armstrong (1968). While attempts have been made to calculate the storage time of such subducted material supplying OIB (e.g., Hart and Staudigel 1989), although there are considerable uncertainties in the subducted parent daughter ratios. $^{207}Pb/^{206}Pb$ and $^{206}Pb/^{204}Pb$ ratios can be used to construct isochrons dating the age of single U-Pb differentiation events. However, continuous differentiation and mixing can produce data arrays that correspond to apparent isochrons of up to 2 Ga when the mean age of the individual sources are only 1-1.3 Ga. The source of HIMU OIB that have the highest $^{206}Pb/^{204}Pb$ values, and so can be dated with more confidence, appear to have been subducted at about 2 Ga (Taylor and McLennan 1985; Hofmann 1997; Chauvel et al. 1992). Overall, it appears that the bulk of recycled components that are seen in OIB have been stored in the mantle for no more than 2 Ga.

PHYSICAL CONSTRAINTS ON MANTLE MODELS

Mantle reservoirs

There have been various attempts to reconcile geophysical and geochemical constraints on mantle evolution (e.g., Davies 1984; Christensen and Hofmann 1994; Anderson 1998a,b; van Keken and Ballentine 1998, 1999; Tackley 2000; Helfrich and Wood 2001; van Keken et al. 2001). However, no description has fully satisfied the geochemical constraints. Although the Earth can be divided into distinct regions based upon changes in seismic velocity, only some of these clearly represent geochemically distinct reservoirs (e.g., crust, mantle, core). The mantle itself is also divided into geophysically defined regions, but whether or not these regions of the mantle are capable of preserving and sourcing observed mantle geochemical heterogeneity over geological time remains a critical aspect of current work (see *Geophysical evidence for the scale of mantle convection* section).

The principle mantle regions that may preserve geochemically distinct mantle reservoirs are defined by several geophysical characteristics. The boundary between the upper and lower mantles traditionally has been set at the 670-km seismic discontinuity. Within the upper mantle the main constituent is olivine, which undergoes an exothermic phase change to a spinel structure at 390-450 km with a Clapeyron slope of +3 MPa/K (e.g., Tackley et al. 1994). As a result of this phase change there is a ~8% increase in density below 390-450 km. Although early comparisons of seismic Earth models with mineral physics data suggested a compositionally different mantle beneath the 670-km discontinuity that is enriched in Fe or Si (Jeanloz 1989; Bina and Silver 1990; Stixrude et al. 1992), later experimental results (Chopelas and Boehler 1992; Wang et al. 1994) were interpreted to indicate that the discontinuity was produced by a combination of at least two phase changes to a post-spinel structure such as perovskite or magnesiowüstite, resulting in a greater density increase (Ito and Takahashi 1989). The sum of these phase transitions at 670 km is endothermic with a Clapeyron slope of −3 MPa/K (Ito and Takahashi 1989; Tackley et al. 1994). This would have the effect of retarding mass transfer, with earlier work indicating that this could create layering in the mantle (Christensen and Yuen 1985; Tackley et al. 1993).

It has also been suggested that the relatively high lower mantle viscosity may further promote the preservation of geochemically distinct domains. Mantle viscosity has been extensively studied using a variety of techniques. Viscosity models have been obtained from plate reconstructions and geoid inversions (Ricard and Wuming 1991; Ivins et al. 1993), postglacial rebound (Mitrovica and Forte 1995; Mitrovica 1996; Kaufmann and Lambeck 2000), plate reconstructions using lower mantle seismic tomography (Ricard et al. 1993; Lithgow-Bertelloni and Richards 1998), and geoid topography above subduction zones (Hager 1984; Hager and Richards 1989). In general these approaches all show that the lower mantle is more viscous than the upper mantle, on average by one to two orders of magnitude. Viscosity profiles show a stepwise increase at the 670-km discontinuity and a gradual increase in the lower mantle, except for a sharp decrease towards the core mantle boundary. This profile is compatible with a viscosity law based on a formulation employed by van Keken et al. (1994) that assumes that the viscosity depends on the lower mantle melting temperature of perovskite. This can be used to assess the possibility, using numerical models (see *Convection models* section) of preserving chemical heterogeneities against homogenization in the convecting mantle. The D″ region above the core mantle boundary also shows anomalous seismic velocity gradients. It is defined by an abrupt velocity increase some 200-300km above the core-mantle boundary (CMB). This discontinuity can occur over a 50-75-km transition, and shows considerable topography. The D″ region has been described as either a chemically distinct layer, a phase change, or the result of anisotropy (see reviews by Lay et al. 1998; Garnero 2000). At the base of the D″ there are the ultra-low velocity zones (ULVZ) that vary in thickness from ~5 to 50km, and which are characterized by P and S velocities that are depressed by 10% and greater. At present, studies of the ULVZ cover only 44% of the CMB surface, with 27% of the mapped region showing a clearly resolvable ULVZ. The ULVZ may originate from a combination of partial melting and chemical variations due to core-mantle interactions (see review by Garnero 2000). The geochemical implications of this reservoir remain only speculative.

Geophysical evidence for the scale of mantle convection

Evidence from high resolution seismic tomography. A critical issue for geochemical mantle models is the extent of mantle mixing by convection and the potential for maintaining separate reservoirs. The most direct information on the scale of mantle convection comes from seismic tomography imaging of subducting slabs and mantle plumes. Oceanic crust is produced at mid ocean ridges (and therefore removed at subduction zones) at a rate of $\sim 6.3 \times 10^{16}$ g/yr. This rate appears to have been approximately constant for the last 65 Ma (Parsons 1981). The entire oceanic lithosphere extends to a depth of ~100 km, and so with a subduction rate of 3 km^2/yr, is subducted at mass rate of $\sim 10^{18}$ g/yr (after Davies 1998; also see Allègre 1997). This is equivalent to processing the entire mantle in ~4 Ga. Recent high-resolution techniques have confirmed earlier seismic tomographic evidence that subducting slabs penetrated the 670-km discontinuity (Creager and Jordan 1986; Grand 1987, 1994; van der Hilst et al. 1997). Detailed seismic imaging however shows a variety of interactions between the slabs and the transition zone above 670 km. Some slabs pass into the lower mantle without any apparent obstruction, while others show deflection and eventual descent into the lower mantle. At some locations, strong deformation and accumulation occurs above 670 km, probably due to a combination of greater viscosity in the lower mantle, the dynamic effects of the endothermic phase change and, possibly, a small increase in intrinsic density in the lower mantle due to compositional differences (see review Kennett and van der Hilst 1998). Laboratory experiments (Kinkaid and Olson 1987; Griffiths et al. 1995) and numerical models simulating the fate of slabs at the 670-km discontinuity suggest that once a critical mass of relatively cold material has accumulated, there is enough negative buoyancy to sink across the phase-change boundary and 'avalanche' into the lower mantle (Machatel and Weber 1991; Tackley et al. 1993; Zhong and Gurnis 1995). Slab remnants can be identified in the lower mantle up to 150 Ma after subduction (van der Voo et al. 1999a,b).

Plumes may arise from boundary layer instabilities within the mantle or at the core-mantle boundary. An individual plume flux can be calculated from the amount of the thermally induced buoyancy required to produce the swell at an intraplate volcanic setting. For all major plumes, this totals 0.35×10^{18} g/yr (Davies 1999). The main assumption in calculating this flux estimate is that there is a thin low-viscosity channel beneath the surface swells. In the mantle beneath Iceland, clear tomographic images of the plume to a depth of 400 km appear to validate this assumption (Shen et al. 1998). In using these flux estimates for transfer between different reservoirs, the depth of origin of the plumes must be considered. There is growing evidence that plumes such as those that form the Hawaiian and Icelandic hotspots have their origin at the core-mantle boundary. The CMB ultra high seismic velocity zones beneath these islands have been interpreted as plume-induced phenomena (Helmberger et al. 1998; Russell et al. 1998). Recent images of the Iceland plume appear to show a distortion of the transition zone beneath this region that is compatible with a deep origin for this plume (Shen et al. 1998), and a mantle structure extending to the CMB (Bijwaard and Spakman 1999). A plume extending to a depth of at least 2000 km also has been imaged beneath central Europe (Goes et al. 1999). Most recently, hotspots in Hawaii, Iceland, South Pacific and East Africa have been shown to be located above slow anomalies in the lower mantle that extend down to the CMB (Zhao 2001).

Some volcanic centers, however, are better explained by either local melting anomalies (Hofmann 1997), comparatively broad upwellings associated with plate-scale flow (e.g., Darwin and African Rises; Sleep 1990), or a plume of shallow origin (e.g., Tahiti; Steinberger 2000). In order to determine the plume flux from the deep mantle, the proportion of the ocean island hotspots that are derived from the CMB, as well as the number of plumes beneath continental regions (e.g., Ritter et al. 2001), must be better constrained.

In defining the total flux of material from the deep mantle into or through the upper mantle and vice versa, Davies (1998) points out that plumes and subducting slabs are independent active features. As such, the upwellings that complement the slab flow and downwellings that complement plumes are passive and independent of each other. Neglecting the Darwin and African Rise fluxes as produced by slab-related processes, Davies (1998) estimates a total observed exchange between the upper and lower mantles of 1.35×10^{18} g/yr. At this rate, the mass of the upper mantle is exchanged in 0.8 Ga.

Convection models. Numerical mantle convection models are essential in understanding the rate and extent to which material is transfered between different regions of the mantle in the past, something that tomographic images of the present mantle are unable to resolve. Numerical models have over the last decade seen a dramatic change in the level of resolution and sophistication due to the availability of ever increasing computer resources. The earliest models were typically 2D-Cartesian boxes, bottom-heated with arbitrary boundary conditions to investigate the effect of whole box vs. layered box convection with some Earth-like values for model parameters (e.g., McKenzie and Richter 1981; Gurnis and Davies 1986). Because of limited computer resources, early models did not use realistic values for geophysical parameters, and so the results required significant extrapolation to draw conclusions for more Earth-like values. The validity of this scaling is uncertain and the effect of imposed conditions in these early models difficult to assess. Reviews by Silver et al. (1988), Kellogg (1992) and Schubert (1992) detail most of this earlier work.

More recently the emphasis in model development has focused on creating models with improved model geometries, greater resolution, more realistic formulation of geophysical parameters, and better-defined geophysical input values. These parameters include a suitably scaled combination of bottom heating and internal heating, attempts to model the rheology as a function of pressure and temperature, the addition of phase changes with appropriate Clapyron slopes, the choice of convection parameters to reproduce observed heat flow, an investigation of density contrasts, and an attempt to provide 'self-consistent' plate behavior.

The development of curvilinear geometry in models has allowed the effect of asymmetry on mixing between the upper and lower thermal boundary layers to be included in the fluid dynamical regime in both 2-D and 3-D models. For example, the resulting 2-D-cylindrical models have an upper boundary that has a larger surface area than the lower, and avoids distortion at the edge of the Cartesian box (Jarvis 1993). Schmalzl et al. (1996) investigated the differences between 2-D and 3-D mixing in isoviscous mantle convection models at comparable convective vigor. They argued that there were significant differences in mixing behavior and in particular that mixing in 3D was less efficient than in 2D. It should be noted that these experiments were limited to poloidal flow only. Ferrachat and Ricard (1998) and van Keken and Zhong (1999) show that 3-D mixing is in fact more efficient due to the effect of the torroidal component (Gable et al. 1991). Both Ferrachat and Ricard (1998) and van Keken and Zhong (1999) note that within some areas, regular 'islands' of laminar stretching persist in which unmixed material can survive. However, with temporally evolving surface boundary conditions constrained by observed plate motions, these 'islands' will be rapidly mixed back into the larger convective system (van Keken and Zhong 1999). While the incorporation of plates in numerical models clearly provides a critical element with respect to their mixing behavior, numerical models do not yet have the capability to generate self consistent 'Earth-like' plate behavior (Tackley 1998; Trompert and Hansen 1998) and are limited to proscribed behavior. Additional avenues of investigations using numerical models include assessing the survivability of highly viscous blobs in the mantle (Manga 1996; Becker et al. 1999). This particular aspect is discussed in more detail in the section *Heterogeneities preserved within the convecting mantle*.

Comparison of numerical models with geophysical observables has provided a particularly important step in understanding the mode of mantle convection. For example, Glatzmaier and Schubert (1993) use 3-D calculations to compare model results with tomographic images of temperature variations in the mantle. They conclude that the style of convection in the mantle is more like that of the whole-mantle models. Similarly, Kido et al. (1998) and Puster and Jordan (1997) compare numerical models with tomographic images, mineral physics/seismic-derived density variations, and the oceanic geoid and also concluded that most forms of layering between the upper and lower mantle, as well as more extreme versions of overturn or 'avalanching' between layers can be excluded.

A significant problem in designing numerical models has been the determination of the Rayleigh number, which in turn controls the convection rates. Van Keken and Ballentine (1998) investigated the effect of a high viscosity lower mantle on whole mantle convection by considering only models that matched the present day surface heat flow. The independently derived model surface velocities also have values that are similar to present-day plate velocities, and this gives a measure of confidence to this approach. This work found that the high viscosity of the lower mantle alone was not able to prevent large scale mixing both above and below the 670-km discontinuity and resulted in the destruction of large-scale noble gas isotopic heterogeneities. Small-scale heterogeneities do survive over time periods of several hundreds of millions of years in this model. More recent studies have reached similar conclusions (Ferrachat and Ricard 2001; Hunt and Kellogg 2001). A model for describing mantle mixing behavior during whole mantle convection that included the effects of phase transitions and depth dependent expansivity and diffusion was investigated in a similar way by van Keken and Ballentine (1999). Although the mixing rate in the lower mantle is slow compared with the upper mantle, the combined effects of phase changes and a high viscosity lower mantle did not prevent large scale mixing of the model system when the model convection is driven to reproduce realistic surface heat flow and plate motion.

Heat and helium. The present and past thermal state of the mantle provides lines of evidence that support some form of convective isolation (van Keken et al. 2001, 2002).

1. *The radioelement mass balance.* The present day BSE concentrations of U, Th and K produce 19.2 TW of heat (Van Schmus 1995). Based on radioelement concentrations, present day heat production in the continental crust accounts for between 25-50% of this total heat flux (Taylor and McLennan 1985; Rudnick and Fountain 1995), or 4.8-9.6 TW. The MORB-source mantle is estimated to be depleted in radioelements by a factor of ~2.6 relative to BSE concentrations (Jochum et al. 1983). At these concentrations the entire mantle could only produce 7.2 TW, leaving a shortfall of some 2.4-5.8 TW. The radioelements required to produce this additional heat must be stored in a region of the mantle beneath, and separate from, the continental crust and MORB-source mantle.
2. *The heat-helium imbalance* (O'Nions and Oxburgh 1983). The Earth's global heat loss amounts to 44 TW (Pollack et al. 1993). Subtracting the heat production from the continental crust (4.8-9.6 TW), and the core (3-7 TW; Buffett et al. 1996) leaves 9.6-14.4 TW to be accounted for by present day radiogenic heating and 17.8-21.8 TW as a result of secular cooling. O'Nions and Oxburgh (1983) pointed out that the present day ^{4}He mantle flux was produced along with only 2.4 TW of heat, a factor of 4-6 times lower than the total radiogenic mantle heat flux associated with present radiogenic production. If a portion of the heat from secular cooling is from past radiogenic production, this must also be associated with ^{4}He production, and so the

imbalance is even greater. O'Nions and Oxburgh (1983) suggested that this could be achieved by a boundary layer in the mantle through which heat could pass, but behind which helium was trapped. The possibility that this is due to the different mechanisms that extract heat and helium from the mantle was investigated with a secular cooling model of the Earth (van Keken et al. 2001). The rates of release of heat and helium over a 4-Ga model run varied substantially, but the ratio of the surface ^{4}He flux to heat flow equaled that of the present Earth only during infrequent periods of very short duration (Fig. 7). It is unlikely that the present day Earth happens to correspond to such a period. It has been suggested that heat and He are transported to the surface at hotspots, where the bulk of the He is lost, while the heat is lost subsequently at ridges (Morgan 1998). However, evidence for such a hotspot He flux at present or in the past, and formulation of a mantle noble gas model incorporating this suggestion, are unavailable. A boundary layer within the mantle remains a viable solution to the heat-helium imbalance.

Buoyancy fluxes. It has been argued that seafloor topography is a strong constraint on the form of present mantle convection (Davies 1988a,b, 1998; Sleep 1990; Davies and Richards 1992). The observed plume buoyancy flux only accounts for about 6% of the heat being transported out of the mantle. Because neither the upper mantle radioelement concentration (Jochum et al. 1983), nor cooling of the upper mantle can account for the amount of heat arriving at the surface, this heat must be arriving from deeper in the Earth. If this amount of heat were conducted through a boundary layer at 670 km, it would generate buoyant upwellings that would generate seafloor topography with the same magnitude as the mid-ocean ridges (which are caused by cooling and thermal contraction). As there is no such observed topography, yet 90% of the mantle heat budget must pass through the 670-km discontinuity, the heat must be transported by advection (i.e., mass flow) rather than conduction through a boundary layer (Davies 1998). In a similar study, Phipps Morgan and Shearer (1993) assessed the topography of the transition-zone seismic discontinuity to evaluate the deep buoyancy forces that drive large-scale mantle flow. This information, coupled with radial viscosity models consistent with the observed geoid, implies that significant mass must pass through the 670-km phase boundary. It should be noted that this view contrasts with that of Wen and Anderson (1997), who place a chemical boundary at 1000 km.

The combined evidence from high-resolution seismic tomography, numerical models with reasonable Earth-like characteristics, and geoid data provide a coherent and compelling case for a mantle that today is convecting a significant mass through the 670-km discontinuity. Without the additional constraints of chemically induced density or rheology contrasts, this results inefficient mixing of material both above and below the 670-km seismic discontinuity. The challenge comprehensive models face is to reconcile this evidence with both thermal constraints that provide a strong case for some degree of convective isolation, and geochemical evidence for the long term preservation of a range of mantle components.

Early Earth history

Models for the evolution of terrestrial noble gases must necessarily consider appropriate starting conditions. The initial incorporation of noble gases and the establishment of terrestrial characteristics are considered in detail in Pepin and Porcelli (2002, this volume). However, several issues are directly relevant to the models discussed here:

1. Early degassing is likely to be very vigorous due to high accretional impact energies during Earth formation over 10^8 years. Loss of major volatiles from impacting materials, and noble gases, may occur after only 10% of the Earth has been assembled (Ahrens et al. 1989).

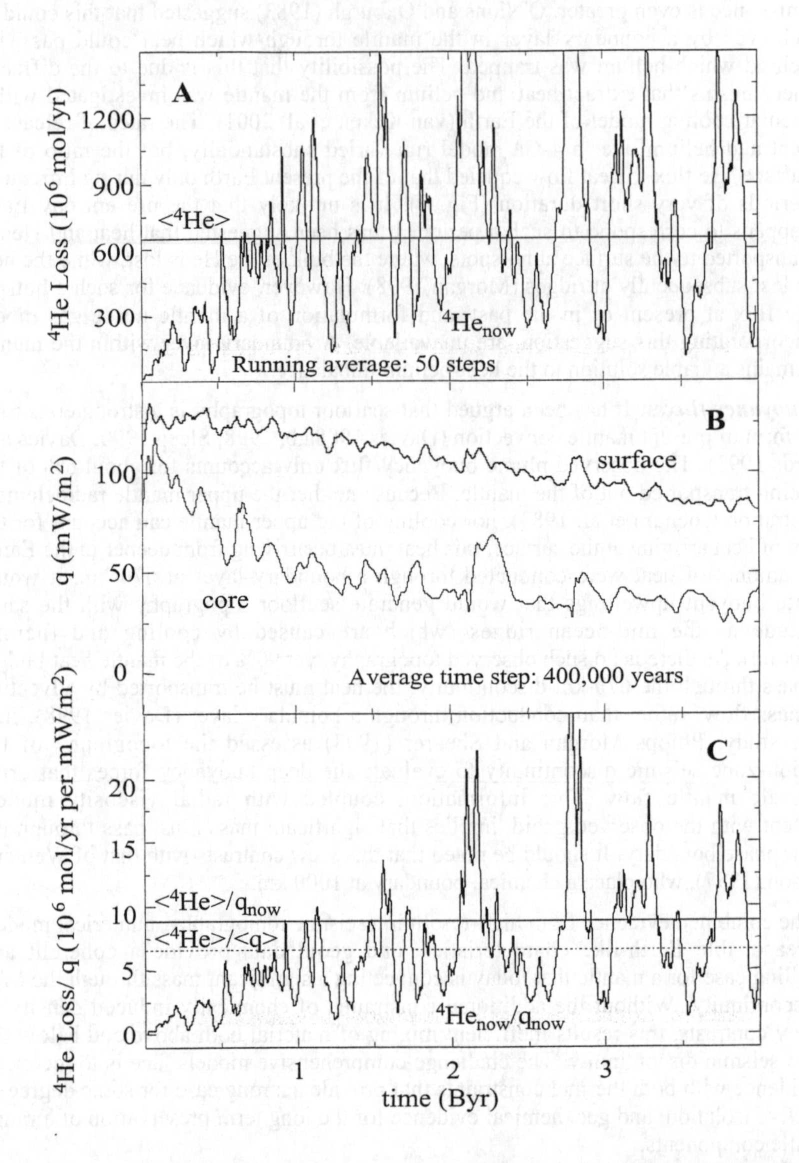

Figure 7. Numerical models can be used to assess the variance in natural heat and helium fluxes without assuming steady state conditions, as well as to consider the effect of secular cooling on the convection rate. In similar models to those used to investigate the mixing and degassing of ^{3}He and ^{4}He (van Keken and Ballentine 1998, van Keken and Ballentine 1999; see Fig. 8). van Keken, Ballentine et al. (2001) show that a model of whole mantle convection produces a large variance in the ^{4}He flux at mid ocean ridge analogues (a), but with much smaller variance in the heat flux extracted across the entire model surface (b). The variance in the combined ^{4}He/heat ratio is dominated by the variance in the He flux, and matches the very low present day ^{4}He/heat ratio ($^4He_{now}/q_{now}$) only for very short periods of time (c). It was concluded that although this approach cannot completely rule out a dynamically controlled observed low ^{4}He/q, the most reasonable explanation remains a boundary layer within the mantle system that allows heat to escape but efficiently retains ^{4}He (e.g., O'Nions and Oxburgh 1983).

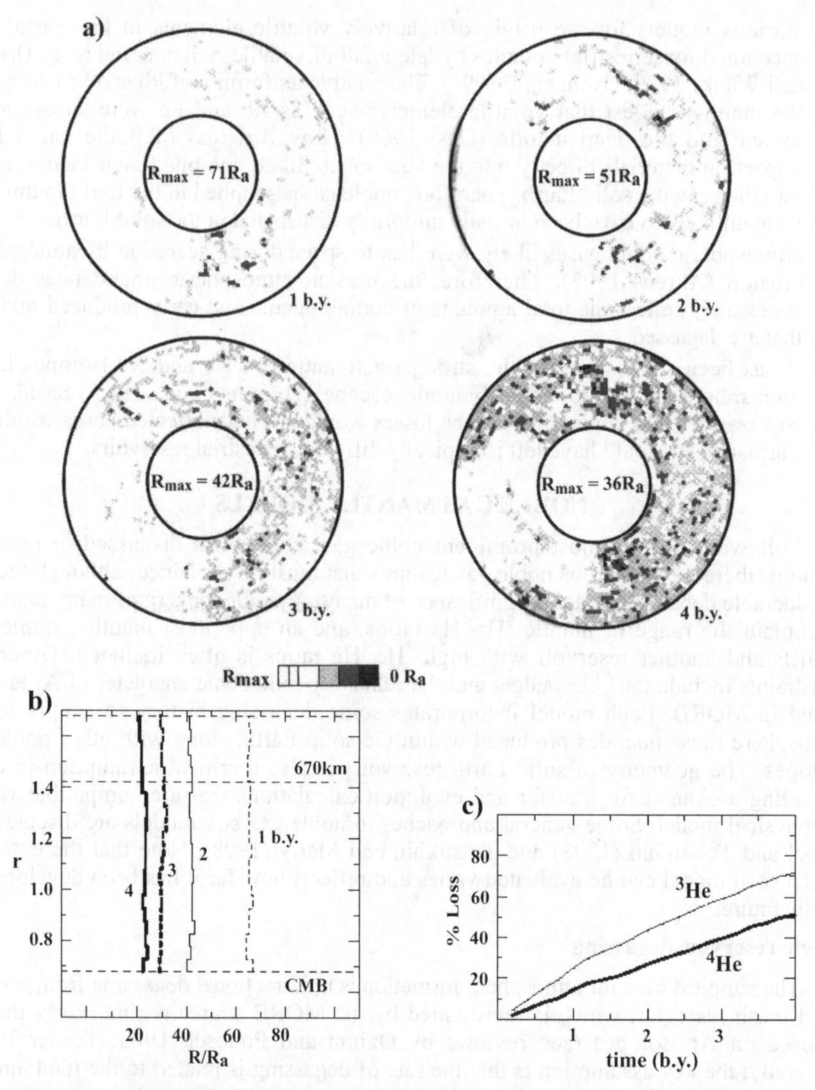

Figure 8. Numerical models that incorporate phase changes and a high viscosity lower mantle are 'driven' to reproduce present day heat flow and reproduce reasonable surface plate velocities. The incorporation of tracers that follow noble gas ingrowth and U, Th and K decay, as well as include mid ocean ridge degassing analogues allow the effect of different modes of mantle convection on ^{3}He/^{4}He isotope distribution to be followed. (a) The ^{3}He/^{4}He ratio distribution in the models after 1, 2, 3 and 4 billion years of model convection are shown as a gray scale. White represents a block in which all tracers still preserve a full primitive ^{3}He content while black shows a block where all tracers have been degassed. This model run contains the most reasonable estimates of the mantle viscosity profile and values of the Clapeyron slopes associated with the phase changes at 440km and 660km. (b) The radial average of ^{3}He/^{4}He as a function of depth. (c) The percentage ^{3}He and ^{4}He lost as a function of time. Approximately 40% of ^{4}He is lost after 4Ga, comparable with the fraction of radiogenic ^{40}Ar in the atmosphere. Phase changes combined with a high viscosity mantle alone does not preserve a lower mantle with a higher ^{3}He/^{4}He than the upper mantle (van Keken and Ballentine 1999). These results emphasize the need to provide an effective mechanism for preserving high ^{3}He/^{4}He domains.

2. Various models for the origin of relatively volatile elements in the Earth have accounted for terrestrial volatiles by late infall of volatile-rich material (e.g., Dreibus and Wänke 1989; Owen et al. 1992). The relative uniformity of Pb and Sr isotopes in the mantle suggest that volatile elements such as Rb and Pb were subsequently mixed into the deep mantle (Gast 1960). However, loss of noble gases from impacting materials directly into the atmosphere likely inhibited their incorporation into the growing solid Earth. Therefore, noble gases supplied to the Earth in this way were unlikely to have been initially uniformly distributed in the solid Earth.
3. Atmospheric noble gases likely were lost to space during accretion by atmospheric erosion (Ahrens 1993). Therefore, the present atmospheric abundances do not necessarily reflect the total amounts of nonradiogenic and early-produced nuclides that are degassed.
4. It has been suggested that the strong fractionation of Ne and Xe isotopes in the atmosphere is due to hydrodynamic escape (Hunten et al. 1987; Sasaki and Nakazawa 1988; Pepin 1991). Such losses would not have affected gases within the Earth, and so would have left isotopically different terrestrial reservoirs.

NOBLE GAS MANTLE MODELS

Following are the most prominent noble gas models. As discussed in previous sections, there are numerous noble gas features that must be explained, although there is considerable debate around the significance of many. A major concern of many models is to explain the range of mantle $^3He/^4He$ ratios, and an uppermost mantle sampled by MORB and another reservoir with high $^3He/^4He$ ratios is often included. Other key constraints include the ^{40}Ar budget and the relatively radiogenic character of Ar and Xe found in MORB. Each model incorporates some degassing history to supply to the atmosphere those nuclides produced within the solid Earth, along with other noble gas isotopes. The geometry of solid Earth reservoirs is also a critical assumption in each, providing a context for transfer and evolution calculations that are compatible with a geophysical model. Some general approaches to noble gas box models are discussed in Azbel and Tolstikhin (1993) and Tolstikhin and Marty (1998). Note that the extent to which each model can be evaluated varies and reflects how far it has been developed in the literature.

Single reservoir degassing

The simplest case for atmosphere formation is unidirectional degassing from a single solid Earth reservoir, which is represented by the MORB source region. Early models focused on Ar isotopes (see reviews by Ozima and Podosek 1983; Turner 1989). Generally, the key assumption is that the rate of degassing is related to the total amount of Ar present in the mantle. Also, there is no return flux from the atmosphere by subduction. Then:

$$\frac{d^{36}Ar_m}{dt} = -\alpha(t)^{36}Ar_m \tag{7}$$

and

$$\frac{d^{40}Ar_m}{dt} = -\alpha(t)^{40}Ar_m + \lambda_{40}y^{40}K_m \tag{8}$$

where $\alpha(t)$ is the time-dependent degassing function, $\lambda_{40} = 5.543\times10^{-10}\ a^{-1}$ is the total decay rate of ^{40}K, y = 0.1048 is the yield for ^{40}Ar, and $^{36}Ar_m$, $^{40}Ar_m$, and $^{40}K_m$ are the total Earth abundances. In the simplest case, $\alpha(t)$ is a constant (Turekian 1959; Ozima and Kudo 1972; Fisher 1978). This is reasonable if the mantle is well mixed and has been melted and degassed at a constant rate at mid-ocean ridges. Assuming there was no Ar initially in the atmosphere, combining Equations (7) and (8) then yields the atmospheric isotope composition at time t:

$$\left(\frac{^{40}\mathrm{Ar}}{^{36}\mathrm{Ar}}\right)_{\mathrm{atm}} = \left(\frac{\alpha}{\alpha-\lambda}\left(1-e^{-\lambda t}\right) - \frac{\lambda}{\alpha-\lambda}\left(1-e^{-\alpha t}\right)\right)\left(\frac{y^{40}\mathrm{K}_m e^{\lambda t}}{^{36}\mathrm{Ar}_{\mathrm{atm}}}\right) \tag{9}$$

Using a BSE value of 270 ppm K, a value of $\alpha = 1.82\times10^{-10}$ is obtained. The equation for the average mantle Ar isotope composition at time t is

$$\left(\frac{^{40}\mathrm{Ar}}{^{36}\mathrm{Ar}}\right)_{m} = \frac{\lambda_{40}}{\alpha-\lambda_{40}}\left(e^{(\alpha-\lambda_{40})t}-1\right)\frac{y^{40}\mathrm{K}_m e^{\lambda_{40}t}}{^{36}\mathrm{Ar}_{m0}} \tag{10}$$

where

$$^{36}\mathrm{Ar}_{\mathrm{atm}} = {}^{36}\mathrm{Ar}_{m0}\left(1-e^{-\alpha t}\right) \tag{11}$$

Model results. A value of $(^{40}\mathrm{Ar}/^{36}\mathrm{Ar})_m = 520$ is calculated for the mantle. Higher ratios can be obtained if an early catastrophic degassing event occurred, removing a fraction f of the ^{36}Ar from the mantle into the atmosphere (Ozima 1973). In this case, the term $(1\text{-}f)^{36}\mathrm{Ar}_{m0}$ can be substituted for $^{36}\mathrm{Ar}_{m0}$ in Equations (10) and (11) (see Ozima and Podosek 1983). This was shown to be compatible with much higher ^{40}Ar/^{36}Ar ratios; for a mantle with ^{40}Ar/^{36}Ar = 40,000, 98.6% of ^{36}Ar was degassed initially. Alternatively, a more complicated degassing function that is steeply diminishing with time can be used to match the present isotope compositions (Sarda et al. 1985; Turekian 1990), and so also involves early degassing of the bulk of the atmospheric ^{36}Ar. In general, as greater ^{40}Ar/^{36}Ar ratios were found for the mantle, degassing has been calculated to be even earlier. Regardless of the formulation used, such early degassing is required by the high measured ^{40}Ar/^{36}Ar ratios and may reflect extensive devolitilization of impacting material during accretion or a greater rate of mantle melting very early in Earth history due to higher heat flow.

Model variations. A further complexity that has been added to the basic model of a single mantle reservoir is the transfer of K from the upper mantle into the continental crust. In this case, the continental crust must be added as another reservoir, and loss of ^{40}Ar from this reservoir also must be considered. The continents may be modeled as either attaining their complete mass very early or more gradually using some growth function (see e.g., Ozima 1975; Hamano and Ozima 1978; Sarda et al. 1985). Note that there is a range of published solutions, with differences in the detailed results due to differences in the formulations of $\alpha(t)$, the rate of K transfer to the continental crust, the mass of the mantle involved, and the mantle ^{40}Ar/^{36}Ar ratio. However, all model formulations qualitatively agree that ^{36}Ar degassing dominantly occurred very early in Earth history.

Similar considerations have been applied to Xe (Staudacher and Allègre 1982; Turner 1989). In these studies, there is greater resolution of early degassing due to the short half-lives of the parent nuclides ^{129}I and ^{244}Pu. However, the scheme involving degassing of atmospheric Xe from the upper mantle, leaving a residue identified with Xe presently found in the mantle, is directly in violation of the constraints discussed in the section *Xe isotopes and a nonresidual Upper Mantle*. Models of degassing from a single mantle reservoir have also have been applied to He (Turekian 1959; Tolstikhin 1975), but have been superceded by arguments for multiple, interacting reservoirs.

A different perspective on ^{40}Ar degassing has been provided by Schwartzman (1973), who argued that K is likely to be transferred 'coherently' out of the mantle with ^{40}Ar; i.e., any ^{40}Ar that has been produced by K will be degassed when the K is transferred to the crust. While the K/^{40}Ar ratio of magmas leaving the mantle and that of the mantle source region are thus assumed to be approximately equal, the highly depleted residue could still be fractionated, so that very radiogenic upper mantle ^{40}Ar/^{36}Ar ratios could develop. It has been pointed out, using the budgets of K and ^{40}Ar, that this limits the amount of K that has been recycled to 30% of what is now in the continents (Coltice et al. 2000).

Model shortcomings. The major shortcoming of the single reservoir degassing models is that mantle heterogeneities, especially for ^{3}He/^{4}He ratios, cannot be explained. Therefore, these models have been superceded by those involving a greater number of reservoirs.

Limited interaction box models

The second generation of mantle degassing models developed with growing evidence for mantle ^{3}He/^{4}He heterogeneities (Hart et al. 1979; Kurz et al. 1982; Allègre et al. 1983, 1986). These models also incorporate the degassing of a single mantle reservoir to the atmosphere. However, to explain the high OIB ^{3}He/^{4}He ratios, there is an additional underlying gas-rich reservoir that is isolated from the degassing upper mantle. Therefore, these layered mantle models can be considered to incorporate two separate systems; the upper mantle-atmosphere, and the lower mantle (Fig. 9A). There is no interaction between these two systems, and the lower mantle is completely isolated except for a minor flux to OIB that marks its existence. It is further assumed that the mantle was initially uniform in noble gas and parent isotope concentrations, so that both systems had the same starting conditions. Note that various modifications to this basic scheme have been proposed, and are discussed below.

Model reservoirs and assumptions.

1) The atmosphere reservoir.
   a) The initial atmospheric abundances were zero, and all noble gases were once contained within the upper mantle.
   b) The atmosphere has progressively degassed from the upper mantle.
2) The upper mantle reservoir that is sampled by MORB.
   a) The volume is specified to be 25% of the bulk mantle (i.e., the volume above 670 km) (Hart et al. 1979).
   b) Initial nonradiogenic isotope concentrations are obtained by dividing the atmospheric inventory into the upper mantle volume.
   c) Initial parent element concentrations are those of the bulk silicate Earth.
   d) The evolution of noble gases in the upper mantle is a function of radiogenic production of ^{21*}Ne, ^{40*}Ar, ^{129*}Xe, and ^{136*}Xe within the mantle and gas loss to the atmosphere. The evolution follows that discussed in the *Single reservoir degassing* section, although the mantle reservoir volume is smaller.
   e) Parent elements are now depleted. For the evolution of long-lived nuclide products (e.g., ^{40}Ar), a depletion history of the parent by extraction to the continental crust is required. Sarda et al. (1985) used an exponentially diminishing depletion function, but the results are not strongly dependent upon the choice of exponent.
   f) No subduction of atmospheric noble gases back into the mantle occurs.
   g) The present MORB ^{136*}Xe excesses are interpreted as due to the decay of ^{238}U and the possible role of ^{244}Pu has not been considered in the context of this model (Allègre et al. 1983, 1986).

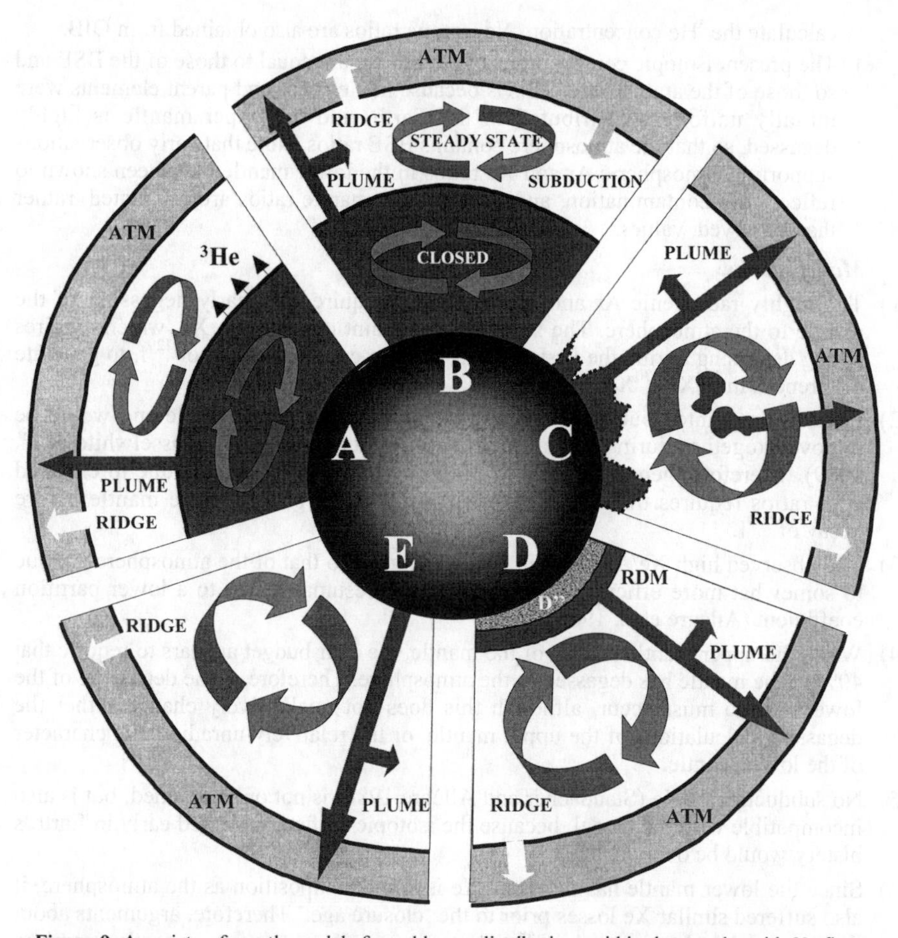

Figure 9. A variety of mantle models for noble gas distributions within the mantle, with He fluxes shown as arrows. Dark mantle regions provide He with high ^{3}He/^{4}He ratios. In the limited interaction models (A), this region comprises the lower mantle, convectively isolated by the 670-km boundary. Upper mantle characteristics reflect progressive depletion to form the atmosphere. In the steady state models (B), there is also layered convection, but the upper mantle composition is due to mixing of lower mantle, subducted, and radiogenic components. Mantle convection across the 670-km discontinuity is accomodated in the remaining models, where the high ^{3}He/^{4}He ratios reside in heterogeneities or deeper layers (C) a lower boundary layer of residual depleted mantle (RDM) of subducted oceanic lithosphere (D), or in the core (E). See text for references and detailed discussion.

3) The lower mantle reservoir that supplies OIB.
   a) The initial concentrations of nonradiogenic noble gases are the same as those for the upper mantle.
   b) The initial concentrations of the parent elements are equal to those of the bulk silicate Earth.
   c) The lower mantle has evolved essentially as a closed system (as discussed in the *Undepleted mantle* section). The fluxes to OIB are sufficiently low that the closed system approximation remains valid.
   d) The present ^{3}He/^{4}He ratio is equal to the highest OIB value, and is used to

calculate the ^{3}He concentration. Ne isotope ratios are also obtained from OIB.

e) The present isotopic compositions of Ar and Xe are equal to those of the BSE and so those of the atmosphere. This is because noble gases and parent elements were initially uniformly distributed in the Earth, and the upper mantle is highly degassed, so that the atmosphere contains BSE ratios. Note that early observations supporting atmospheric Ar and Xe ratios in the lower mantle have been shown to reflect only contamination, and so the lower mantle ratios are calculated, rather than observed, values.

Model results.

1) The highly radiogenic Ar and Xe in MORB require very early degassing of the mantle to the atmosphere. The strongest constraint comes from Xe, which requires strong degassing during the first 100 Ma, prior to complete decay of ^{129}I, to generate differences in ^{129}Xe/^{130}Xe.
2) It has been pointed out that Xe and I are both highly incompatible and would be removed together during partial melting of the upper mantle (Musselwhite et al. 1989). Therefore, generating high ^{129}Xe/^{130}Xe ratios in the mantle due to elevated I/Xe ratios requires that I was efficiently subducted back into the mantle before decay of ^{129}I.
3) The observed high Xe/Ar ratios in MORB relative to that of the atmosphere are due to somewhat more efficient degassing of Ar, presumably due to a lower partition coefficient (Allègre et al. 1986).
4) While the upper mantle is 25% of the mantle, the ^{40}Ar budget appears to require that 40% of the mantle has degassed to the atmosphere. Therefore, some degassing of the lower mantle must occur, although this does not qualitatively change either the degassing calculations of the upper mantle, or the relatively unradiogenic character of the lower mantle.
5) No subduction of Xe (Staudacher and Allègre 1988) is not only assumed, but is also incompatible with the model, because the isotopic shifts established early in Earth's history would be overwhelmed.
6) Since the lower mantle has the same Xe isotopic composition as the atmosphere, it also suffered similar Xe losses prior to the 'closure age.' Therefore, arguments about the late model closure ages of Xe isotopes (see *Xe closure ages* section) and so extensive early losses (Porcelli et al. 2001) apply to the lower mantle as well.

Model shortcomings.

1) The global ^{4}He flux at mid-ocean ridges is equal to production in the upper mantle, suggesting that the upper mantle ^{4}He concentration is in steady state and so requiring transfer of ^{3}He from the lower mantle (O'Nions and Oxburgh 1983).
2) This transfer most plausibly occurs by advection of mantle material (Kellogg and Wasserburg 1990) and so includes all the noble gases. The possibility that transfer of ^{3}He from the lower mantle occurs by diffusion (Allègre et al. 1986), so that the principles of the residual mantle model can still be applied to the other noble gases, has not been shown to be reasonable for maintaining the present global flux. Also, this would result in substantial mantle He/Ne fractionations, which are not observed (see *Neon isotopic compositions in the mantle* section).
3) The model predicts that the lower mantle is enriched in He by ~10^2 relative to the upper mantle. OIB sample suites do not provide unambiguous evidence for such a gas-rich reservoir, although the issue remains unresolved (see *OIB fluxes* section).
4) The difference between MORB and air ^{20}Ne/^{22}Ne ratios indicates either that Ne

isotopes were not initially uniformly distributed in the Earth or that Ne in the atmosphere has been modified by losses to space after degassing from the mantle. Regardless of the reason, the assumption that the atmosphere and upper mantle together form a closed system does not hold. Also, the higher $^{20}Ne/^{22}Ne$ and Ne/Ar ratios of the upper mantle limit the contribution that the presently degassing volatiles can have made to the atmosphere (Marty and Allé 1994).

5) Distinctive radiogenic Xe isotope ratios in the upper mantle, established by degassing early in Earth history, would be obliterated by contamination from plumes rising from the gas-rich lower mantle (Porcelli et al. 1986). The requirement that all rising gas-rich material is completely degassed at hotspots is probably not reasonable, and the mixing of even a small fraction of incompletely degassed material into the surrounding mantle will have an impact.
6) The ratio of ^{244}Pu-derived ^{136}Xe to radiogenic ^{129}Xe within a residual upper mantle reservoir must be greater than that of the atmosphere extracted from that reservoir due to the greater half-life of ^{244}Pu (see *Xe isotopes and a nonresidual upper mantle* section). This does not appear to be the case. Therefore, atmospheric and MORB Xe are not complementary.
7) A plausible scenario for establishing an initially uniform distribution of noble gases with the characteristics of the present atmosphere has not been advanced. For the upper mantle, accretion is likely to have prevented sequestering of volatiles and may have caused direct loss of gases in accreting material to the atmosphere (see *Early Earth history* section). However, it can be argued that this is the cause of the calculated strong early degassing, and only the physical conception of how gases were transferred then needs modification. The retention of so much gas in the lower mantle is more difficult, because many noble gas acquisition models suggest that volatiles were not necessarily added uniformly during accretion (see Pepin and Porcelli 2002).
8) Other geochemical evidence has been interpreted to indicate that much of the mantle has been processed and so largely degassed (e.g., Davies 1984; Hofmann et al. 1986; Blichert-Toft and Albarède 1997), although this objection is at odds with the general conclusions from the K-^{40}Ar budget and heat-He balance for a large undegassed reservoir. It is not clear how these issues can be resolved along with those of the noble gases.
9) The geophysical arguments against layered mantle convection have been most often used to discard this model. This difficulty might be overcome if the same principles of this model are applied to reservoirs of different volumes or different configurations.

While some objections can be overcome with model modifications, it must be concluded that the fundamental mechanism of this model, generation of atmospheric and MORB noble gas characteristics by progressive extraction of the atmosphere from the upper mantle as presently sampled by MORB, cannot account for many of the available constraints, especially those from Ne and Xe isotopes.

Model variations. A variation of the residual upper mantle model by Zhang and Zindler 1989) considers degassing of MORB by partitioning of noble gases into CO_2 vapor for transport to the surface, with the remaining noble gases in MORB returned to the mantle. Due to different solubilities, there is fractionation between noble gases in the residual MORB and those lost to the atmosphere. The net result of this partial gas loss from MORB is fractionation during mantle degassing. The model predicts that the $^{3}He/^{22}Ne$ ratio of the upper mantle is twice that of the lower mantle; and this may be possible (see *Neon isotopic compositions in the mantle* section).

There are difficulties with this model process as well. The objections raised above regarding residual mantle models also pertain to this variant. An additional difficulty, acknowledged in the original presentation, is that the model also predicts that the upper mantle Xe/Ar ratio is lower than that of air because the lower solubility of Xe in silicate melt leads to more rapid degassing of Xe than Ar. However, data from MORB indicate that the upper mantle ratio is in fact greater. Nevertheless, this degassing mechanism may have been important if there was an early magma ocean present in the first stages of terrestrial history (Ozima and Zahnle 1993), and also may account for the preferential outgassing of Xe relative to lighter noble gases that is required in models of atmospheric isotopic evolution by hydrodynamic escape (Tolstikhin and O'Nions 1994; Pepin 1997).

Another variation has been presented by Kamijo et al. (1998) and Seta et al. (2001). Here two additional factors are considered: depletion of the lower mantle reservoir by plume activity, and the subduction of parent elements into both the upper and lower mantle reservoirs. As with the other models, an initially uniform distribution of noble gases and parent elements in the mantle was assumed. Using mass fluxes between reservoirs and a linear continental growth function, the evolution of He, Ne, and Ar were considered. It was concluded that the lower mantle could contain as little as 13% of its original noble gas inventory, and have $^{40}Ar/^{36}Ar$ ratios up to 3×10^4, much higher than that of the atmosphere, due to depletion. Also, much more Ne was degassed than is now found in the atmosphere, and so allows early fractionating losses to space that accounts for the difference in atmosphere and mantle $^{20}Ne/^{22}Ne$ ratios. Such fractionating losses likely occurred early in Earth history (e.g., Pepin 1997), and in some model solutions the bulk of the Ne may be degassed sufficiently early. Again, other objections raised above regarding the other residual mantle models also pertain to this variant.

Steady-state box models

Steady-state box models (Fig. 9B) are distinguished from the limited interaction box models in being based upon the open interaction between the upper mantle and both the lower mantle and atmosphere (O'Nions and Oxburgh 1983). The steady state model was first described to explain He isotope and heat fluxes (O'Nions and Oxburgh 1983), and then applied to the U-Pb system (Galer and O'Nions 1985), and finally to the other noble gases (Kellogg and Wasserburg 1990; O'Nions and Tolstikhin 1994; Porcelli and Wasserburg 1995a,b). Although different aspects of this model have been discussed by different research groups, the studies are discussed together here as a single model. The central focus of the model is not degassing of the upper mantle to form the atmosphere, but rather mixing in the upper mantle, which has noble gas inventories that are presently not continually depleting, but rather are the result of inputs from surrounding reservoirs. In addition to the upper mantle inputs by radiogenic production from decay of U, Th, and K, atmospheric Ar and Xe are subducted into the upper mantle, and lower mantle noble gases are transported into the upper mantle within mass fluxes of upwelling material. Noble gases from these sources comprise the outflows at mid-ocean ridges. The isotopic systematics of the different noble gases are linked by the assumption that transfer of noble gases from the upper mantle to the atmosphere by volcanism, as well as the transfer from the lower into the upper mantle by bulk mass flow, occurs without elemental fractionation. It is assumed that upper mantle concentrations are in steady state, so that the inflows and outflows are equal; therefore, there are no time-dependent functions (and so additional parameters) determining upper mantle concentrations. The addition of any such free parameters would only create under-constraint in the calculated model solutions, and therefore the introduction of non-steady-state conditions are justified only when new constraints are found. However, the key upper mantle characteristic is not the time-dependence of noble gas concentrations, but the interaction with other reservoirs. Tolstikhin and Marty (1998) have explored the evolution of the mantle prior to the establishment of steady state concentrations, and some effects of the time dependence of noble gas transfers into the upper mantle.

Model reservoirs and assumptions.

1) The atmosphere reservoir.

a) The atmosphere generally has no a priori connection with other reservoirs. It simply serves as a source for subducted gases, and no assumptions are made about its origin. Since no assumption is made about the initial distribution of noble gases in the Earth, the atmospheric abundances cannot be used to derive lower mantle concentrations.

b) The daughter isotopes that are presently found in the atmosphere clearly were originally degassed from the upper mantle, possibly along with nonradiogenic nuclides. However, this occurred before the present character of the upper mantle was established. Therefore, the upper mantle no longer contains information regarding atmosphere formation.

2) The lower mantle that supplies OIB.

a) The volume is ~75% of the bulk mantle (i.e., below 670 km).

b) The initial concentrations of the parent elements are equal to those of the bulk silicate Earth, while those of the nonradiogenic noble gas isotopes are unknowns.

c) This reservoir evolves as an approximately closed system (see *Undepleted mantle* section), and so the present daughter nuclide concentrations can be calculated. Note that the amounts of radiogenic ^{129*}Xe and ^{136*}Xe depend upon the extent of their early losses (i.e., the closure age). For computational purposes, concentrations and isotopic compositions are assumed to have been approximately constant the last residence time of the upper mantle.

d) There is a bulk mass flux upwards in plumes. However, this is assumed to be small, so that the reservoir volume and concentrations are essentially unaffected, and a return flux can be ignored.

e) The high ^{3}He/^{4}He and ^{21}Ne/^{22}Ne ratios found in Loihi are used to calculate He and Ne isotope concentrations (see *Undepleted mantle* section).

f) The lower mantle Ar and Xe isotope compositions cannot be well constrained by available data and are taken as unknowns. Therefore, the isotopic compositions, and so nonradiogenic isotope abundances, are unknowns. However, these are calculated from the balance of fluxes into the upper mantle (see below).

3) The upper mantle reservoir that is sampled by MORB.

a) The noble gas influxes to this reservoir are from the lower mantle (by bulk mass transfer), radiogenic production within the upper mantle, and subduction (of the heavy noble gases). Noble gases leave the upper mantle by melting at ridges.

b) All concentrations and isotopic compositions of noble gases and parent elements are assumed to be in steady state (and uniform) over the last residence time. Note that a crustal reservoir is not included in the model, because neither the long-term depletion of parent element mantle concentrations, nor the long-term flux of daughter nuclides to the atmosphere, are considered.

c) The upper mantle is degassed to the atmosphere according to a rate constant determined by the rate of melting at ridges. Since this is the main mechanism for removing noble gases from the upper mantle, this fixes their upper mantle residence times.

d) Fractionation between noble gases during melting, or during exchange between mantle reservoirs, does not occur.

e) Present production of radiogenic noble gas nuclides (^{4*}He, ^{21*}Ne, ^{40*}Ar, and ^{136*}Xe) are known from U, Th, and K concentrations of the MORB source.

f) The MORB ^{3}He/^{4}He ratio is a result of mixing between lower mantle He and production of ^{4}He in the upper mantle (O'Nions and Oxburgh 1983). Since the flux of the latter is fixed, the rate of He transfer from the lower mantle can be calculated from (Kellogg and Wasserburg 1990):

$$\left(\frac{^4He}{^3He}\right)_{UM} - \left(\frac{^4He}{^3He}\right)_{LM} = \frac{^4P^U C_{UM} M_{UM}}{^{3He}C \dot{M}_{Plume}(1-r)} \quad (12)$$

The numerator on the right is the production rate of ^{4*}He in the upper mantle, with $^4P = 2.1 \times 10^{-9}$ atoms ^{4}He/atoms ^{238}U-g-yr (for Th/U = 2.5), $^U C_{UM} = 8 \times 10^{12}$ atoms/g (3 ppb), and an upper mantle mass of $M_{UM} = 1 \times 10^{27}$ g. Lower and upper mantle ratios are 37 R_A and 8 R_A, respectively, and the closed system lower mantle has $^{3He}C = 8.3 \times 10^{10}$ atoms ^{3}He/g. The denominator on the right is the flux of ^{3}He from the lower mantle. The mass flux rising in plumes is $\dot{M}_{Plume}$, of which some fraction r directly degasses to the atmosphere, while the remaining fraction, (1-r), is mixed into the upper mantle. Then $\dot{M}_{Plume}(1-r) = 3 \times 10^{15}$ g/yr, is the mass flux from the lower mantle that carries He into the upper mantle.

g) The fluxes of other daughter nuclides from the lower mantle are obtained from their lower mantle concentrations and the lower mantle mass flux into the upper mantle that is calculated from the He isotopes.

h) The MORB ^{21}Ne/^{22}Ne ratio (when corrected for air contamination) is the result of a mixture of lower mantle Ne and radiogenic ^{21*}Ne produced in the upper mantle (Porcelli and Wasserburg 1995b). There is no subduction of Ne, so that the ^{3}He/^{20}Ne ratios of the upper and lower mantles must be the same.

i) The MORB ^{40}Ar/^{36}Ar ratio (when corrected for atmospheric contamination of basalts) is a mixture of lower mantle Ar, radiogenic ^{40*}Ar produced in the upper mantle and subducted air Ar. While the flux of ^{40}Ar from the lower mantle is known (from the lower mantle calculated ^{40*}Ar/^{4*}He ratio), that of ^{36}Ar is not. The mixing relationship can be written as (Porcelli and Wasserburg 1995b):

$$\left(\frac{^{40}Ar}{^{36}Ar}\right)_{LM} \Big/ \left(\frac{^{40}Ar}{^{36}Ar}\right)_{UM} = \frac{1+\delta^{36}_{SUB}}{1+\delta^{40}_{PR}} \quad (13)$$

where δ^{36}_{SUB} is the ratio of subducted ^{36}Ar to lower mantle ^{36}Ar in the upper mantle, which is a free parameter. The term δ^{40}_{PR} (= 2.0) is the ratio of ^{40}Ar produced in the upper mantle to that derived from the lower mantle. Subducted ^{40}Ar is neglected. For $\delta^{36}_{SUB} \geq 0$, $(^{40}Ar/^{36}Ar)_{LM} \geq 9700$.

j) While the lower mantle ^{40}Ar concentration is fixed, the ^{36}Ar concentration is inversely proportional to the ^{40}Ar/^{36}Ar ratio, and cannot be further constrained.

k) The MORB ^{136}Xe/^{130}Xe ratio (when corrected for air contamination) is the result of mixing between lower mantle Xe, fissiogenic ^{136*U}Xe produced in the upper mantle, and subducted air Xe (Porcelli and Wasserburg 1995a). This is analogous to the Ar system. The MORB ^{129}Xe/^{130}Xe ratio (when corrected for air contamination) has no radiogenic contributions from present production, and so is simply due to mixing between lower mantle and subducted Xe. Since the MORB ^{129}Xe/^{130}Xe value is greater than that of the atmosphere, the lower mantle ratio must be equal to that of MORB (if there is no Xe subduction) or higher. Like ^{36}Ar, the ^{130}Xe flux from the lower mantle cannot be determined, but is inversely proportional to the lower mantle ^{129}Xe/^{130}Xe and ^{136}Xe/^{130}Xe ratios (Porcelli and Wasserburg 1995a).

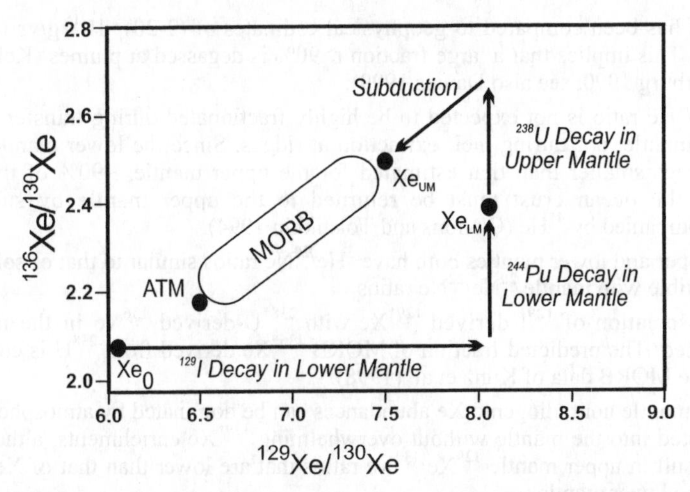

Figure 10. The systematics of Xe isotopes in a steady state mantle (Porcelli and Wasserburg 1995a). The lower mantle is characterized by increases in $^{129}Xe/^{130}Xe$ and $^{136}Xe/^{130}Xe$ due to decay of ^{129}I and ^{244}Pu, respectively. This Xe is transferred to the upper mantle (at a rate compatible with that of ^4He), where it is mixed with $^{136*}Xe$ produced by ^{238}U (raising the $^{136}Xe/^{130}Xe$ ratio) and possibly by mixing with subducted Xe, which moves the composition toward that of the atmosphere. The result is the most radiogenic composition found in MORB. Note that the amount of subduction is unconstrained. Greater proportions of subducted Xe in the upper mantle correspond to higher $^{129}Xe/^{130}Xe$ and $^{136}Xe/^{130}Xe$ ratios in the lower mantle input, corresponding to lower ^{130}Xe concentrations in the lower mantle.

Mechanisms operating in the model are more clearly seen by examining the systematics of Xe isotopes. This is shown in Figure 10. The minimum upper mantle $^{136}Xe/^{130}Xe$ and $^{129}Xe/^{130}Xe$ ratios in the upper mantle are constrained to be equal to, or greater than, the highest available MORB value. The lower mantle ratios are established early in Earth history by decay of ^{129}I and ^{244}Pu; while the $^{129}Xe/^{130}Xe$ ratio is constrained to be at least as great as that in the upper mantle, higher values are possible. Once lower mantle Xe is transported into the upper mantle, it is augmented by U-derived $^{136*U}Xe$. Subduction of atmospheric Xe is also possible, lowering the $^{136}Xe/^{130}Xe$ and $^{129}Xe/^{130}Xe$ ratios. There is therefore a trade-off between two presently unknown quantities; the lower mantle ^{130}Xe concentration (i.e., how high the lower mantle $^{136}Xe/^{130}Xe$ and $^{129}Xe/^{130}Xe$ ratios are) and the flux of subducted Xe (how much is mixed with lower mantle Xe to lower these ratios). If a specific Xe/He ratio for the lower mantle is chosen (e.g., the solar ratio), then a Xe concentration can be calculated from the He concentration, along with the amount of subducted Xe. In the model, approximately 50% of the radiogenic $^{136*}Xe$ in the upper mantle is derived from the lower mantle, where it was produced largely by ^{244}Pu.

Model results.

1) The calculated residence time of noble gases in the upper mantle is short (~1.4 Ga) and requires the long-term storage of noble gases that were trapped early in Earth history (including solar He and Ne) in the lower mantle.
2) The mass flux from the lower mantle into the upper mantle is ~50 times less than the rate of ocean crust subduction, greatly limiting the fraction of subducted material that could cross into the lower mantle (O'Nions and Tolstikhin 1996).
3) The calculated mass flux of gas-rich material from the lower mantle into the upper

mantle has been compared to geophysical estimates of (2-20) $\times10^{16}$ g/yr for plume fluxes. This implies that a large fraction r>90% is degassed at plumes (Kellogg and Wasserburg 1990; see also Davies 1999).

4) The U/^{4}He ratio is not expected to be highly fractionated during transfer from the lower mantle, nor during melt extraction at ridges. Since the lower mantle ratio is ~20 times smaller than that estimated for the upper mantle, ~90% of the U that enters the ocean crust must be returned to the upper mantle by subduction unaccompanied by ^{4*}He (O'Nions and Tolstikhin 1994).
5) The upper and lower mantles both have ^{3}He/^{20}Ne ratios similar to that of solar gases, compatible with mantle ^{20}Ne/^{22}Ne ratios.
6) The association of ^{129}I-derived ^{129*}Xe with ^{238*}U-derived ^{136}Xe in the mantle is explained. The predicted fraction of MORB ^{136*}Xe derived from ^{238}U is compatible with the MORB data of Kunz et al. (1998).
7) Upper mantle nonradiogenic Xe abundances can be dominated by atmospheric gases subducted into the mantle without overwhelming ^{129*}Xe enrichments, although this will result in upper mantle ^{129}Xe/^{130}Xe ratios that are lower than that of Xe derived from the lower mantle.
8) The model is consistent with the isotopic evidence that upper mantle Xe does not have a simple direct relationship to atmospheric Xe (see *Xe isotopes and a nonresidual upper mantle*). The radiogenic Xe presently seen in the atmosphere was degassed from the upper portion of the solid Earth prior to the establishment of the present upper mantle steady state Xe isotope compositions and concentrations.
9) The lower mantle elemental abundance pattern is distinctive from that of the atmosphere, but cannot be confined to a narrow range. In the case of no subduction of noble gases, ^{20}Ne/^{36}Ar = 3.2 and ^{130}Xe/^{36}Ar = 2×10^{-3} (close to the meteoritic 'planetary' value).
10) The lower mantle has a ^{40}Ar/^{36}Ar ratio that is much higher than the atmospheric value. This implies the ^{36}Ar/K ratio of the lower mantle is much lower than that of the reservoir that supplied the atmosphere. Assuming K was initially uniformly distributed throughout the BSE, this implies very heterogeneous incorporation of ^{36}Ar. This contrasts with the typical a priori assumption of an initially uniform distribution of ^{36}Ar.
11) The lower mantle is the storage reservoir of ^{129}I-derived ^{129*}Xe and ^{244}Pu-derived ^{136*}Xe (^{238}U-derived ^{136*}Xe is relatively negligible). Enrichments in these nuclides in MORB are inherited from the deep mantle.
12) The possible lower mantle Xe isotope compositions correspond to closure times that are similar to that of the atmosphere, indicating that early losses occurred from the deep mantle as well (Porcelli et al. 2001). These losses are considered to be prior to the assumed closed system evolution.

Model shortcomings.

1) The principal objection has been based on geophysical arguments advanced for greater mass exchange with the lower mantle below 670 km, and so the difficulty of maintaining a distinctive deep mantle reservoir. It might be possible to reformulate the model for a larger upper mantle, or greater mass fluxes between reservoirs, although in some cases nonsteady state upper mantle concentrations may be required.
2) As discussed for the limited interaction models, inclusion of a closed system lower mantle demands a greater contrast in mantle He reservoir concentrations than seen at between MORB and OIB (see *OIB fluxes and source concentrations* section). This

might be resolved with better constraints on the He concentrations of OIB source regions.

3) As discussed for the limited interaction models, other geochemical evidence has been interpreted to indicate that much of the mantle has been processed and so largely degassed, although this objection is at odds with the general conclusions from the K-^{40}Ar budget and heat-He balance for a large undegassed reservoir.
4) Albarède (1998) argued that the model mantle cannot be in steady state due to the progressive decay of the parent elements in the upper mantle, and so is self-contradictory. However, the notion of steady state upper mantle conditions clearly is an approximation. While a strict steady state does not hold, the deviation from this due to decay of the very long-lived parent nuclides over the residence time of upper mantle noble gases is minor.
5) The model predicts that the ^{3}He/^{22}Ne ratio of the two mantle reservoirs is identical. It has been argued that the upper mantle ratio may be ~70% greater (see *Neon isotopic compositions in the mantle* section). This is an important test of the model, although it is possible that other processes (e.g., subduction of Ne) might be incorporated into a revised model to explain any differences.

Overall, the objection that may be most difficult to overcome is in the lack of geophysical evidence for layered mantle convection. However, the intra-reservoir relationships involved in this model may be adaptable to other configurations of noble gas reservoirs that might be found to be compatible with geophysical data.

Break-up of a previously layered mantle

Although geophysical observations point to the present day mantle convecting as a single layer, the parameters that govern mantle convection are close to the critical point at which layered convection will occur (Tackley et al. 1993; Davies 1995). For example, based upon the temperature dependence of mantle viscosity, there may have been a higher mantle Rayleigh number in the past. This may have resulted in more complex or layered convection patterns; Steinbach et al. (1993) shows that two-layered convection results when the Rayleigh number is increased beyond a critical value in a 2-D simulation. Allègre (1997) has argued that models requiring long-term mantle layering can be reconciled with geophysical observations for present-day whole mantle convection if the mode of mantle convection changed less than 1 Ga ago from layered to whole mantle convection. A difficulty with this hypothesis is the expected consequences for the thermal history of the upper mantle. The principle process driving changes in the mode of mantle convection is the development of thermal instabilities in the lower layer that eventually result in either massive or episodic mantle overturn (Tackley et al. 1993; Davies 1995). Reviewing the consequences of models that reproduce mantle layering, Silver et al. (1988) noted that a >1000 K temperature difference develops between the two portions of the mantle, and so overturns will cause large variations in upper mantle temperatures. However, the available geological record suggests that there has been a relatively uniform mantle cooling rate of ~50-57 K/Ga (Abbott et al. 1993; Galer and Mezger 1998). Further, numerical models simulating ^{3}He/^{4}He variations in a mantle that has undergone this transition do not reproduce the observed ^{3}He/^{4}He distributions seen today (van Keken and Ballentine 1999).

The lower boundary layer reservoir

The boundary layer at the core-mantle boundary (CMB) has been explored as a reservoir for high ^{3}He/^{4}He ratio He, in the context of whole mantle convection. It has been suggested that subducted oceanic crust could accumulate there and form a distinct chemical boundary layer, accounting for the properties of the D″ layer (Christensen and

Hofmann 1994). Altered ocean crust is strongly depleted in ^{3}He and may be enriched in U, and so is likely to have relatively radiogenic He. However, a complementary harzburgitic lithosphere that is depleted in both U and He also develops during crust formation. Coltice and Ricard (1999) examined a model in which separate reservoirs of altered ocean crust and depleted harzburgitic lithosphere are maintained below the bulk convecting mantle. In this model (Fig. 9C), there are mass fluxes of altered ocean crust to a D″ reservoir, and of subducted depleted harzburgitic lithosphere to a residual depleted mantle (RDM) located immediately above D″. In order to account for the depletion of the bulk mantle, extraction of the continental crust and degassing to the atmosphere are also considered. Transfer of trace elements occurs by bulk mass fluxes that have fixed chemical fractionation factors relative to the source reservoir, so that chemical fluxes are determined by the mass fluxes. The U-Pb, Rb-Sr, and U,Th-He systems are all tracked, and the isotopic composition of all the reservoirs are fixed by observations, with D″ material being identified with the source of HIMU hotspots and RDM with that of Loihi. The model was not extended to the other noble gases. Since the model was fully and clearly developed, it is possible to examine its implications in some detail here.

Model reservoirs and assumptions.

1) The bulk mantle has a constant volume (a free parameter; depends on D″ and RDM volumes). The starting U concentration is that of the bulk silicate Earth and that of He is calculated for a closed system with a Loihi ^{3}He/^{4}He ratio. The main He outflow is to the atmosphere by degassing a constant volume with time (i.e., first order degassing). U is largely transferred to the continental crust at a set first-order removal rate. Smaller fluxes of He and U to D″ and RDM (via the altered ocean crust and depleted oceanic lithosphere, respectively) also occur. Small inputs occur from subduction of continental crust and from D″ and RDM. Present isotopic compositions are those seen in MORB.
2) The model continental crust grows linearly over 4.5 Ga to the present volume. It only serves to deplete the bulk mantle of U. Return to the bulk mantle by subduction occurs at a rate set by those of continental growth and mantle outflow.
3) The atmosphere grows progressively and serves only to deplete the bulk mantle of He. Since He is lost to space, the atmospheric He abundance is not a constraint.
4) The D″ layer grows linearly, with the present volume taken as a free parameter. There is a constant mass inflow from subduction, with a U concentration related to the bulk mantle U concentration by a large enrichment factor operating during formation of ocean crust. There is a mass flux out (dependent upon the growth rate of the layer). Present isotopic compositions are those seen in HIMU basalts.
5) The residual depleted mantle (RDM) has a constant volume, which is a free parameter. Mass outflow is specified from plume buoyancy fluxes. The mass inflow contains He and U, with concentrations related to those of the bulk mantle by specified small enrichment factors operating during formation of oceanic lithosphere. The present ^{3}He/^{4}He ratio is the highest seen at Loihi, while the starting value is solar.

Model results. The primary conclusion of the model is that Loihi and HIMU characteristics (for Sr, Pb, and He) can be generated from subducted components sequestered in two deep reservoirs. A nonlinear inversion was used to maximize the fit to the observations of continent concentrations and reservoir isotope ratios within the bounds of assigned uncertainties in the fluxes and fractionation coefficients. This was done using the U-Pb and Rb-Sr systems as well. In the best fit results, the D″ layer is 250-km thick (5% of the total mantle) and the RDM is 500-km thick.

The main issue here is whether this model provides a plausible mechanism for maintaining a reservoir with high $^3He/^4He$. The characteristics of He in the RDM Loihi source are determined by two components. The He initially incorporated into this RDM reservoir had a solar $^3He/^4He$ ratio and a concentration equal to that for a closed system reservoir that would presently have high $^3He/^4He$ OIB ratios. Therefore, without other fluxes, this would simply be a 'primitive' reservoir. The subducted depleted lithosphere has a He concentration fixed at 1% that of the bulk mantle. It may have contained a small concentration of He when it is formed prior to extensive degassing of the bulk mantle, and this will have had high $^3He/^4He$ ratios. Later it essentially only dilutes the concentration of He in the RDM. Helium is lost according to a 1.8 Ga half-life (for the best-fit results), and so the initial He has been reduced by 0.18. The U concentration decreased from 20 ppb to 3.5 ppb, and so was reduced by a similar factor. Therefore, the initial U/He ratio was not changed significantly and the reservoir evolved isotopically like a closed system, although concentrations were reduced. Changes in the U/He ratio of the bulk mantle have little effect, because the transfer from the bulk mantle to the RDM is small.

Model shortcomings. The difficulty with the model as an explanation for mantle He variations lies in the nature of the RDM reservoir. Subducted depleted material is too depleted to supply OIB alone, so that a fixed reservoir is necessary to sequester high initial He concentrations. However, there is no simple geophysical explanation for initially maintaining a fixed volume reservoir and so for intimately associating subducted and primitive material. The issue then becomes the same as that for preserving a relatively undegassed reservoir in the Earth: how to convectively isolate such a region. Whether or not this region is diluted by depleted material is then secondary. Therefore, while such depleted material may explain other characteristics such as Sr and Pb isotopes (see Coltice and Ricard 1999), the issue of how to maintain a He reservoir with a high $^3He/^4He$ ratio remains. Also, the model does not address the evolution of other noble gases. While the ^{40}Ar budget might be accommodated, additional reservoirs would need to be added to store solar Ne and radiogenic Xe separate from the atmosphere, and the evolution of the bulk mantle then would require modification. However, these changes may be possible.

The upper boundary layer reservoir

As an alternative to assigning high OIB $^3He/^4He$ ratios to an isolated lower mantle enriched in He, Anderson (1998a) argued that these ratios represent He incorporated within lithosphere that is highly depleted in U and so maintains the isotopic composition at the time of trapping. This may include harzburgitic residues that have been invaded by CO_2 and accompanying noble gases. Such a reservoir would be protected from destruction by convective mixing, and would then only require a single underlying mantle He composition. While this may plausibly occur, there are difficulties with accounting for most OIB signatures in this way. For example, in the case of Loihi, the underlying crust is 80-100-Ma old. If the He was derived from the MORB-source mantle in the past, then the upper mantle had a $^3He/^4He$ ratio of 37 R_A at 100 Ma ago. Any reasonable extrapolation of upper mantle $^3He/^4He$ ratios to the past would require Earth values far in excess of the initial, solar nebula value at 4.5 Ga. It should be noted that lithospheric xenoliths generally do not show high $^3He/^4He$ ratios, and while this may reflect imprinting by fluids not long before eruption (see Dunai and Porcelli 2002), a pre-existing higher $^3He/^4He$ ratio would only be conjectural. Overall, a compelling case for long-term lithospheric storage of He with high $^3He/^4He$ ratios has not been made.

Deeper or 'abyssal' layering

Kellogg et al. (1999) have developed and numerically tested a model in which mantle below ~1700 km has a composition, and so density, that is sufficiently different from that of the shallower mantle to largely avoid being entrained and homogenized in the overlying convecting mantle. This model has generated a great deal of interest because of its ability to preserve a region in the mantle behind which the radioelements and primitive noble gases can be preserved, while accommodating many geophysical observations. For example, it is argued that the depth of the compositional change varies, allowing slab penetration to the core mantle boundary in some locations while providing a barrier into the lower mantle elsewhere. Supporting tomographic evidence for a significant number of slabs being disrupted at 1700 km is given by van der Hilst and Kàrason (1999). More recently, this supporting evidence has been questioned because of the loss of tomographic resolution in this portion of the mantle (Kàrason et al. 2001). Although Kellogg et al. (1999) suggest that this layer will be hard to detect seismically because of its neutral thermal buoyancy, irregular shape and small density contrast, Vidale et al. (2001) argue that seismic scattering nevertheless should be resolvable, and yet is not observed. Also, if the overlying mantle has the composition of the MORB source, then the abyssal layer must contain a large proportion of the heat-producing elements, and must efficiently remove heat from the core. Therefore, a minimum 12.6-21.4 TW of mantle heat must cross from the layer into the convecting mantle, neglecting any component of secular cooling (see *Heat and helium* section). It is not yet clear what this effect would have on the thermal stability of the layer or temperature contrast with the overlying mantle.

Overall, if such a feature were found to be viable, it must be incorporated into a comprehensive noble gas model. This could include some of the features described above of other layered mantle models.

Heterogeneities preserved within the convecting mantle

Early conceptual models for preserving geochemical heterogeneities in a regime of whole mantle convection discussed the possibility of convecting material passing through and sampling compositionally different regions. Two end member models have been postulated, one in which 'blobs' or 'plums' of enriched material are passively entrained in the convecting mantle to provide OIB–source material (Davies 1984). The other has been called 'penetrative convection' (Silver et al. 1988), in which downgoing cold material drops into a compositionally different lower mantle layer. The slabs, on heating at the core mantle boundary, regain positive buoyancy and on return to the surface entrain a small portion of the deeper reservoir. In this way, lower mantle material is provided to either OIB-source or is mixed into the MORB-source mantle. Regarding the first case, numerical models have shown that more viscous 'blobs' can be preserved in a convective regime if they are at least 10 to 100 times more viscous than the surrounding mantle (Manga 1996). In these simulations, the high viscosity blobs tend to aggregate, leading to the formation of large-scale heterogeneities from smaller ones. Becker et al. (1999) have investigated the dynamical, rheological and thermal consequences of such blobs containing high radioelement and noble gas concentrations to account for the bulk Earth U budget and the high $^3He/^4He$ ratios seen in OIB. The higher heat production and resulting thermal buoyancy within these blobs must be offset by a combination of increased density (~1%) and small size (<160-km diameter) to avoid both seismic detection and thermal buoyancy that will result in the blobs rising into the upper mantle. These high viscosity zones must fill 30-65% of the mantle to satisfy geochemical mass balance constraints, and must be surrounded by low viscosity mantle to transport the resulting heat away. The thermal gradient generated through the blobs will result in a high viscosity shell around a lower viscosity core that controls the dynamical mixing behavior. For a viscosity contrast between blobs and host mantle of 100, the average viscosity of the lower mantle is predicted to be greater, by a factor of 5, than that of the upper mantle. The mechanism creating the more dense 'blob' material is uncertain, but the higher density may be explained by a different perovskite/magnesiowüstite ratio than the surrounding mantle (Becker et al. 1999). However, the higher viscosity of the 'blobs' was assumed in these models, and the mineral physics controlling the viscosity contrast remains unclear. Manga (1996) noted that clumping of 'blobs' is likely to occur and form large-scale heterogeneities, and this process is expected to work against the careful size balance required to avoid increased thermal-driven buoyancy and/or seismic detection. Note that if such a scheme is viable, the evolution of mantle noble gases, reflecting degassing and interaction between reservoirs, may still follow those calculated in layered mantle models.

Phipps Morgan and Morgan (1999) have proposed a variation on the concept of dispersed heterogeneities. In this model, a mantle has evolved in which a mixture of depleted mantle (after melt extraction), coexists with distinct veins or 'plums' of enriched components sourced from different recycled components (MORB, OIB and continental crust), and a small amount of primitive mantle with high $^3He/^4He$ ratios. The model invokes a two stage melting process. The first stage of melting occurs beneath the lithospheric mantle as a result of deeper hotter mantle material being added to this shallow mantle region. The product of this first stage of melting is dominated by material from the enriched plums or veins, including high $^3He/^4He$ components, and forms OIB. The residue is depleted in high $^3He/^4He$ ratio components, and so will have the isotopic composition of MORB. Because the residue of this process is thermally buoyant, this depleted material forms a shallow layer at the top of the asthenospheric mantle. Melting at mid-ocean ridges of this shallow asthenospheric layer provides the source for MORB.

The obvious attraction of this variant of the distributed heterogeneity model is the ability to accommodate whole mantle convection, but without the need to provide the conditions necessary to keep plums or blobs from rising to the top of the mantle and being sampled at mid-ocean ridges. Because the bulk of the mantle heterogeneity is in thin veins or small plums, no special viscosity contrast is required to maintain large-scale domains within the convecting mantle. Also, as the enriched veins or blobs are small this results in the radioelements being broadly distributed throughout the whole mantle. This then avoids the problems of overheating and thermal buoyancy faced by concentrating these elements in either large blobs or contiguous deep reservoirs.

The generation of He isotope compositions is only quantitatively described in the model. However, several difficulties with the description of He can be identified:

1) The authors recognize that the generation of the observed uniform MORB $^3He/^4He$ from the residue of heterogeneous OIB sources is a potential problem. They propose that rapid mixing of the shallow asthenosphere may homogenize this shallow reservoir. Numerical modeling may provide a future test of this.
2) Since noble gases are highly incompatible, the first melts from rising plume material that form OIB will have several orders of magnitude greater concentrations of noble gases than MORB melts subsequently produced from the same source region. The model requires that the source of all MORB melts previously supplied OIB, and so the ^3He flux from OIB must be several orders of magnitude greater than seen at MORB. This is not observed (see Table 3).
3) Because the mantle supplying mid-ocean ridges in this model is the residuum of a gas loss process, the effect of fractionation of noble gas elemental ratios in this reservoir must be considered. Mid-ocean ridge 'popping rock' basalt glass has noble gas radiogenic nuclide ratios that are almost unfractionated (see *Mantle noble gas relative abundances* section), and the $^3He/^{20}Ne$ ratio of MORB and OIB are similar (see *Neon isotopic compositions in the mantle* section). Therefore, melting of the

mantle to produce OIB must leave a relatively unfractionated source. This requires that the noble gas solid/melt phase partition coefficient for He, Ne and Ar must be very similar. At present, these partition coefficients remain poorly defined, and better definition may provide a test for this aspect of the model.

The two-stage melting model has yet to be quantitatively developed with respect to the noble gases, making it difficult to fully assess the viability of this approach.

Depleted, high He/U mantle

It is generally assumed that noble gases are the most highly incompatible elements, so that any mantle reservoir that undergoes melting would be preferentially depleted in noble gases relative to the parent elements U, Th, and K. Therefore, melt residues are expected to have low He/U ratios and so develop relatively radiogenic He isotope compositions. Alternatively, it has been suggested (Graham et al. 1990; Anderson 1998a; Helffrich and Wood 2001) that U may be more incompatible than He, so that melting can leave a residue with a higher He/U ratio than the original source material. In this case, high $^3He/^4He$ ratios seen in OIB can be generated in mantle domains that have been previously depleted by melting and can survive for some time. The central issue is whether this can explain the highest ratios in OIB by creating domains imbedded in the convecting MORB-source mantle, so that there is no requirement for convective isolation of mantle with high $^3He/^4He$ ratios. Unfortunately, this idea has not been developed into a coherent mantle model that examines the evolution of the MORB- and OIB-source mantle domains, so that it cannot be evaluated rigorously. However, several issues that must be incorporated in such a model can be considered:

1) The high $^3He/^4He$ domains must be preserved. This issue is discussed above (see *Heterogeneities preserved within the convecting mantle* section), although here the domains need not be preserved throughout Earth history. In this case, any progressive destruction of these domains by mixing into the surrounding mantle can be balanced by creation, subject to isotopic constrains discussed below.
2) The source of the domains must have high $^3He/^4He$ ratios at the time of depletion. The $^3He/^4He$ evolution of the MORB-source mantle has not been calculated for whole mantle convection, and so it is not clear when ratios as high as now observed in OIB were present. As a reference, it can be noted that in a layered mantle model, the upper mantle did not have ratios >30 R_A for the last ~2 Ga (Seta et al. 2001). If high ratios have not been present since that time, then the domains with high $^3He/^4He$ ratios must have been preserved since then.
3) The source must have started out with sufficiently high ^3He to be able to supply OIB after depletion. For example, 10% melting of a mantle source, assuming a He silicate/melt partition coefficient of 8×10^{-3} (Marty and Lussiez 1993), will leave only 8% of the He in the residue. Thus, melting of highly depleted MORB-source mantle will leave a component that will not readily impart a distinctive isotopic signature to OIB. This is a particular problem when it is considered that such high $^3He/^4He$ components must dominate the He signature in OIB source regions where there is clear evidence for recycled oceanic crust that likely contains high ^4He concen-trations, such as in Hawaii (Hauri et al. 1994). Therefore, the initial mantle source must be more gas-rich.
4) There must be a mechanism for the preferential involvement of this previously melted material at OIB.
5) The constraints provided by Xe isotopes indicate that the noble gases in the mantle cannot have evolved in a single depleting mantle reservoir (see *Xe isotopes and a nonresidual upper mantle* section). Therefore, other reservoirs must be invoked to

complement these domains.

At present, it appears that this mechanism might explain some of the He isotope variations seen in mantle-derived materials, although it is not clear if it can explain the highest $^3He/^4He$ ratios. However, a comprehensive model is necessary that explores the total volume of the high $^3He/^4He$ domains and the rates at which they are generated and destroyed, the isotopic evolution of the surrounding mantle, and the evolution of other rare gases, to fully evaluate this idea. Nonetheless, it is clear that this mechanism cannot be simply appended to a geophysical or lithophile element model for whole mantle convection.

Storage of noble gases in the core

The core has often been cited as a possible source of He with high $^3He/^4He$ ratios, in response to the difficulties of storing this He in the mantle. Any He in the core would likely have been trapped during core formation and unaccompanied by U and Th, and so would now have a primordial $^3He/^4He$ ratio. It has been argued that there is insufficient ^3He in the core to supply the mantle, because measured silicate-metal noble gas partition coefficients are very low (Matsuda et al. 1993). However, calculation of the abundances of noble gases in the core requires not only the relevant partition coefficients during core formation, but also what quantity of noble gases were available during partitioning. These parameters depend upon whether the mantle was solid or molten as the core segregated. Whether the core now can supply the noble gases seen in the mantle must also consider how reintroduction into the mantle occurs. These issues are considered below, following Porcelli and Halliday (2001).

Jephcoat (1998) observed that Xe forms a high-density solid at deep mantle pressures. Sinking of these solids could transport Xe, and possibly some fraction of other noble gases incorporated into this structure, to the deep mantle or core. However, it has not been demonstrated either that such a scarce element (~20 fg/g) is sufficiently insoluble in other high pressure phases to form a separate phase, or that it can form separate grains in the mantle sufficiently large to sink. This mechanism for transport to the core therefore cannot be evaluated further.

Partitioning of noble gases. Matsuda et al. (1993) reported molten metal/ silicate melt partition coefficients of $D(He)_{Fe/LS} \sim 0.04$ at 5 kbar and $D(He)_{Fe/LS} \sim 0.01$ at 20 kbar (both at 1600°C). Differences between noble gases were not found, although because analytical errors were large, the possibility that noble gases are fractionated during partitioning cannot be discounted. Much lower values for all the noble gases of $\sim 10^{-3}$ were reported for 60 kbar (at 1600°C) and of $\sim 3 \times 10^{-4}$ for 100 kbar (≈ 300 km depth). Comparable values were obtained using a lherzolite silicate melt for pressures between 5 and 60 kbar (Sudo et al. 1994). There are no data available for the deeper mantle, although it has been shown experimentally that Xe does not alloy with iron at core-mantle boundary pressures (Caldwell et al. 1997). A partition coefficient for He in the mantle of $D(He)_{Fe/LS} = 1 \times 10^{-4}$ is taken here for further discussion.

There are no data available for the partitioning of noble gases between solid silicates and liquid metal. However, there are some measurements for olivine and silicate melts at surface pressures. Marty and Lussiez (1993) found solid silicate/ silicate melt partitioning values of $D(He)_{SS/LS} \leq 8 \times 10^{-3}$ for coexisting olivines and glasses. Noting that the molten metal/solid silicate partition coefficient is $D(He)_{Fe/SS} = D(He)_{Fe/LS} / D(He)_{SS/LS}$; then if $D(He)_{Fe/LS} = 1 \times 10^{-4}$ and $D(He)_{SS/LS} = 8 \times 10^{-3}$, then the partitioning between liquid metal and solid silicate is $D(He)_{Fe/SS} = 1 \times 10^{-2}$.

Availability of noble gases during core segregation. The light noble gases in the mantle appear to be solar, and the most likely mechanisms for capture of such gases are by solar irradiation of accreting material and gravitational capture and dissolution of nebula gases (see Pepin and Porcelli 2002). If the mantle were solid during core segregation, then the noble gases present could only have been derived from irradiated solid material. While the irradiation models require further development, it may be possible to obtain a sufficient quantity to account for the atmospheric inventory (Sasaki 1991; Podosek et al. 2000). Porcelli and Halliday (2001) used a reference value of 6×10^{10} atoms ^3He/g for this process.

If the mantle were molten at the time of core formation, noble gases provided by accreting solid materials would have been lost to the atmosphere. The model for providing deep mantle noble gases by dissolution of a gravitationally captured atmosphere then must be invoked to provide noble gases within the Earth (Mizuno et al. 1980; Harper and Jacobsen 1996; Porcelli et al. 2001). In this scenario, the Earth reaches much of its present mass prior to dissipation of the solar nebula, and so gravitationally captures a dense atmosphere. Due to accretional energy, temperatures sufficiently high to melt a large portion of the Earth are reached. Noble gases of solar composition are dissolved from the atmosphere and convected into the deep mantle. This mechanism may be able to provide $\sim7\times10^{12}$ atoms ^3He/g in the mantle (Mizuno et al. 1980; Woolum et al. 2000; Porcelli and Halliday 2001) or possibly more.

Concentrations in the core. If He is partitioned from solid silicates to liquid metal percolating through the deep mantle, initial concentrations must have been provided by irradiated accreting material. The core represents ~32% of the bulk mass of the Earth; if this was removed from the mantle by a single batch process, then

$$^{He}C_{Fe} = {^{Hei}C_{SS}} / [0.32+(1-0.32)/D(He)_{Fe/LS}] \tag{14}$$

If the concentration of the solid mantle was initially $^{Hei}C_{M0} = 6\times10^{10}$ atoms ^3He/g and $D(He)_{Fe/SS} = 1\times10^{-2}$, then the resulting core concentration is $^{He}C_{Fe} = 1\times10^9$ atoms ^3He/g.

If He was provided by equilibration between liquid metal and a magma ocean at lower pressures, the initial mantle concentration was provided by a hot dense atmosphere. In this case, the relevant partition coefficient of $D(He)_{Fe/LS}$ is lower than that for equilibration with solid silicate, but there may have been greater concentrations of noble gases available from the overlying atmosphere. Then,

$$^{He}C_{Fe} = D(He)_{Fe/LS}\ {^{He}C_{LS}} \tag{15}$$

If $D(He)_{Fe/LS} = 10^{-4}$ and $^{HeO}C_{M0} = 7\times10^{12}$ atoms ^3He/g, then $^{He}C_{Fe} = 7\times10^8$ atoms ^3He/g, similar to the value calculated above.

Transport back into the mantle. Various mechanisms can be considered for the transport of noble gases back into the mantle.

1) Expulsion of noble gases from the outer core as the solubility limits are exceeded due to core crystallization. However, the saturation limits of both He and Ne, which would be present in the core at a ratio comparable to the solar ratio of $^4He/^{20}Ne$ = 820 (Pepin et al. 1999), are unlikely to be reached together. Therefore, He and Ne that would be provided from the core this way are likely to be highly fractionated. The evidence that mantle He/Ne ratios are not dramatically fractionated from the solar ratio appears to discount this as a viable mechanism.
2) Diffusion across the core-mantle boundary. However, there are no appropriate experimental diffusion coefficient data available to calculate the flux of He across the mantle boundary layer. Large fractionations between the noble gases also may be expected, although there is no data available to quantify this. Therefore, this possibility cannot be evaluated.

3) Partitioning into overlying partial melts. The ultra-low velocity zone at the core-mantle boundary may reflect the presence of mantle melt (see Garnero 2000; Ohtani and Maeda 2001), and partitioning from the core back into the overlying mantle may occur if conditions are favorable. However, a flux cannot be easily calculated without constraints on partition coefficients and the volume and residence time of melts at the core-mantle boundary.

4) Bulk transfer of core material into the mantle (e.g., Macpherson et al. 1998). This provides the most straightforward method for transporting He. Mass transfer may occur into the mantle by exsolution of oxides from the core (Walker 2000), although the noble gas flux cannot be calculated without knowledge of the relevant partitioning behavior during exsolution. For simple entrainment of material from the core, the mass flux is $\dot{M}_C = {}^{He}F/{}^{He}C_{Fe}$. Using the OIB ^{3}He flux of $^{He}F = 5 \times 10^{24}$ ^{3}He/yr and a core He concentration of $^{He}C_{Fe} = (0.7\text{-}1) \times 10^9$ atoms ^{3}He/g, then $\dot{M}_C$ = (5-7) $\times 10^{15}$ g/yr is obtained. This is equivalent to (2-3) $\times 10^{25}$g over 4.5 Ga, or ~0.6-0.8% of the upper mantle.

Is the core a plausible reservoir? These calculations indicate that, using presently available constraints, the core remains a plausible source of the ^{3}He found in OIB. Measured high ^{3}He/^{4}He ratios would then be the result of mixtures of He from the core that has a solar nebula ^{3}He/^{4}He ratio with radiogenic ^{4}He from the mantle. However, the implications for other geochemical parameters also must be considered. Other elements that would be affected by bulk transfer of core material into the mantle include the Pt-group elements and volatiles that may be relatively abundant in the core, such as H and C. While some limits on the amount of transferred core material are provided by these (Porcelli and Halliday 2001), it should be noted that the amount needed to sustain the ^{3}He flux is still very uncertain, and it is possible that upward revisions in the amount of noble gases within the early Earth, or in the metal/silicate partition coefficients, could reduce the required flux of core material by an order of magnitude or more.

Note that storage of He in the core remains only one component of a noble gas model that can describe the range of noble gas observations. The core has only been evaluated as a possible storage of ^{3}He. The incorporation in the core of other noble gases, and their relative fractionations, cannot be clearly evaluated without more data. Also, the distribution of radiogenic nuclides such as ^{40}Ar, ^{129}Xe, and ^{136}Xe that are produced within the mantle must be explained with a model that fully describes the mantle reservoirs. While these issues may be tractable, a comprehensive model that incorporates a core reservoir remains to be formulated. It should be emphasized that the core does not completely explain the distribution of He isotopes, because the issue of the ^{4}He-heat imbalance is not addressed at all by this model.

Subduction of meteoritic He

Anderson (1993) has proposed that subduction of interplanetary dust material in seafloor sediments is the source of mantle ^{3}He. Such material has high concentrations of implanted solar wind He due to its small grain size and so high relative surface area. However, the amount of material being deposited in pelagic sediments is presently about 10^4 times too low to account for the flux of ^{3}He released at mid-ocean ridges. While the flux of ^{3}He from OIB may be up to 10^2 times lower, only a small fraction of the subducted sediments can be assumed to be supplying OIB. Therefore, it was suggested that the ^{3}He supply to the Earth was much higher in the past. However, it has been argued that this supply has been insufficient throughout Earth history (Trull 1994), although this may still be open to debate. It should be noted that any higher past fluxes must have been composed of material that is sufficiently small to have avoided heating and degassing during atmospheric entry and surface impact (see Flynn 2001). Further, laboratory measurements indicate that He is lost from dust particles by diffusion at low temperatures (Hiyagon 1994) and so will be lost from downgoing sediments during subduction. Allègre et al. (1993) pointed out that the He/Ne ratio of meteoritic dust is much lower than that of the mantle, and so while the Ne in the mantle could be supplied by this mechanism, the ^{3}He cannot. However, the flux of extraterrestrial material to the Earth is also too low to supply the necessary Ne (Stuart 1994). Thus, in the absence of either direct evidence or plausible arguments for sufficiently high past fluxes of material of the required composition, this mechanism is not considered to be a possible dominant source of mantle noble gases (Farley and Neroda 1998).

CONCLUSIONS

A fundamental issue that all mantle models address is how to maintain separate reservoirs to account for observed mantle noble gas isotopic variations. None have been able to accommodate all noble gas constraints as well as geophysical observations. However, considerable progress has been made in understanding the processes that may be involved in the distribution, transport, and evolution of mantle noble gases.

Overall assessment of the models

Models with a boundary to convection at 670 km.

1. The limited interaction mantle models that involve extraction of the atmosphere from the upper mantle do not appear to be viable due to the lack of a simple relationship between atmospheric and MORB Ne and Xe. The models also are opposed to geophysical arguments for whole mantle convection.
2. The steady state upper mantle is compatible with much of the presently available noble gas constraints. However, the present formulation, with a mantle boundary at a depth of 670 km, stands contrary to geophysical evidence for substantial mass fluxes across this depth. The mechanism for interactions between mantle reservoirs is promising, however, and might be adapted to other configurations, although not necessarily without time-dependent distributions.
3. The idea of a break up of previously layered mantle reconciles the geochemical characteristics that can be explained by long-term evolution in a layered mantle (according to one of the layered mantle models) with geophysical observations regarding recent mantle behavior. However, numerical modeling of the thermal state of the mantle does not support this.

Models with noble gas reservoirs in other mantle layers.

4. The lower boundary layer model avoids some of the geophysical problems of larger convective layers, but must resolve how such a layer is initially stabilized with gas-rich material. Also, it must be combined with other mechanisms to account for the mantle evolution of all the noble gases. However, the model illustrates how a deep boundary may be important.
5. The upper mantle boundary layer model is not supported by any compelling observations, and its consideration emphasizes the need to store He with high ^{3}He/^{4}He ratios, as well as Xe isotopes, in the deep mantle.
6. A deeper (>670 km) mantle layer provides a potential basis for a comprehensive noble gas model. However, there are geophysical difficulties with the thermal stability of such a layer and the lack of tomographic evidence for such layering.

Models with heterogeneities within a whole mantle convection regime.

7. Dispersed domains of mantle heterogeneities, comprising a large proportion of the mantle but preferentially sampled at OIB, could provide a basis for a comprehensive noble gas model. However, this also needs further geophysical investigation into their long-term survival, particularly if they contain high concentrations of heat-producing elements. The mechanisms discussed for interaction between reservoirs in layered mantle models could be applied here.
8. The production of domains that are depleted but have high He/U ratios may also explain some He isotopic variations, although these domains do not appear to have enough ^{3}He to impart the necessary isotopic signatures to OIB. This model also requires further development to include the other rare gases.

Models with extra-mantle noble gas reservoirs.

9. The core remains a possible source of isotopically distinctive He. However, further work is required, especially on noble gas partition coefficients into core material, both to support this and to formulate the evolution of other noble gases within the mantle.

Overall, the main obstacle to a comprehensive description of mantle noble gas evolution is in finding a configuration for distinct mantle domains that is compatible with geophysical observations. In this regard, a major difficulty is the heat-helium imbalance. Models involving mantle stratification can account for this, but have been discounted on other geophysical grounds. Overall, no other adequate explanation for this imbalance has been proposed. More widespread identification of this as a major unanswered question for the distribution of U and the transport of both He and heat may provoke more geophysical investigation into possible solutions.

Another problem requiring resolution is the nature of the high ^{3}He/^{4}He OIB source region. Most models equate this with undepleted, undegassed mantle, although some models invoke depletion mechanisms. However, none of these have matched the end member components seen in OIB lithophile isotope correlations. It remains to be demonstrated that a primitive component is present and so can dominate the He and Ne isotope signatures in OIB.

Some persistent misconceptions

As understanding of terrestrial noble gas geochemistry has evolved, various erroneous conclusions have persisted both within the field and in related fields.

1) *The very radiogenic Ar and Xe isotope ratios of the upper mantle demand early degassing of the mantle.* This is a model-dependent conclusion based on the assumption that upper mantle noble gases are residual from atmosphere degassing. However, Xe isotope systematics precludes such a relationship (Ozima et al. 1985). Since other models, while not necessarily correct, can account for the observed Xe isotope variations, it is clear that the isotopic evidence can be interpreted in various ways. Nonetheless, early transfer of volatiles to the atmosphere probably did occur and was caused by impact degassing.
2) *Subduction of heavy noble gases must be very limited.* This has been based on models of the isolation of the upper mantle or arguments about preservation of nonatmospheric ^{129}Xe/^{130}Xe ratios in the mantle. In fact, upper mantle nonradiogenic Xe isotopes could be dominated by subducted Xe and admixed with very radiogenic Xe, and some models explicitly incorporate subducted Xe fluxes. Until more is conclusively known about Ar and Xe isotopic variations in the mantle, subduction must be considered a potentially important process.

3) *The processes that have dominantly formed the atmosphere continue at present.* Although primordial noble gases continue to degas, their isotopic compositions do not match those of the atmosphere and limit their contribution to a small fraction of the present atmospheric inventory. Volatile species continue to be added to the atmosphere, but the dominant inputs occurred earlier. The formation of the atmosphere cannot be easily extrapolated from the present, but must be understood by quantitative modeling of past processes.

Important parameters that are still unknown

There are, of course, many questions regarding terrestrial noble gases that remain to be explored. Some of the issues that are critical to making advances in global models of noble gas behavior are:

1) Subduction. In order to quantify the subduction flux of noble gases, including Ne, further constraints are required on the incorporation of noble gases in subducting materials, and their behavior during melting. Are there phases that can carry these elements to greater depths? Are noble gases added to the mantle wedge in metasomatizing fluids or melts?
2) Partitioning into the core. If He with high $^3He/^4He$ ratios can be stored in the core, one of the reasons for requiring a separate reservoir in the mantle is removed. More partition coefficient data at different pressures is required, along with further understanding of the conditions of core formation. Progress in geophysics, mineral physics, and petrology is necessary to determine the mechanisms for reintroduction of noble gases to the mantle.
3) Silicate partition coefficients. While noble gases are clearly highly incompatible, quantitative partitioning data is required to consider the evolution of melt source regions, particularly the ratios of noble gases to parent K, U and Th.
4) Heavy noble gas characteristics in OIB. Variations in Ar and Xe isotopic compositions are not yet clearly documented. Constraining the compositions associated with high $^3He/^4He$ ratios is clearly critical for mantle models.
5) Mantle Ar and Xe nonradiogenic isotope compositions. It is still an open question whether there are solar or fractionated heavy noble gases in the mantle.
6) The origin of noble gases. The initial sources and trapping mechanisms of noble gases affected the initial character and distribution of noble gases, and are still poorly understood. These provide the starting conditions for mantle evolution models.
7) Heat and He. A major hurdle in reconciling mantle convection patterns with noble gas distributions is uncovering the mechanism responsible for separating heat and He.
8) The connections between the distributions and evolution of noble gases and major volatiles have not been fully explored. This requires determining how initial distributions may have been related. Subduction rates also likely play a large role in major volatiles.
9) Mantle convection. Geophysical debates regarding the feasibility of maintaining chemical distinct domains or layers in the mantle need resolution.
10) Noble gas concentrations in different mantle reservoirs. Determining the noble gas concentrations of different mantle source regions has not been possible through direct measurement of volcanic materials. Understanding and correcting for magmatic degassing and secondary gas loss processes remains a challenge.

While many aspects of mantle noble gas geochemistry and atmospheric evolution remain to be explored, further progress will also be made as interpretations respond to advances in related fields such as geophysics, early solar system evolution, and trace element geochemistry. Conversely, noble gases will continue to provide strong constraints on planetary evolution as more encompassing theories are assembled.

ACKNOWLEDGMENTS

Reviews by Don Anderson, Geoff Davies, Darryl Harrison, David Hilton, Manuel Moriera, Mario Trieloff, and Rainer Wieler, and a particularly thorough review by Eleanor Dixon, have greatly improved this paper. This work was supported by the ETH and the Swiss National Science Foundation.

REFERENCES

Abbott DA, Burgess L, Longhi J, Smith WHF (1993) An empirical thermal history of the Earth's upper mantle. J Geophys Res 99:13835-13850

Ahrens TJ (1993) Impact erosion of terrestrial planetary atmospheres. Ann Rev Earth Planet Sci 21: 525-555

Ahrens TJ, O'Keefe JD, Lange MA (1989). Formation of atmospheres during accretion of the terrestrial planets. *In* Origin and evolution of planetary and satellite atmospheres. *In* Atreya SK, Pollack JB, Matthews MS (eds) University of Arizona Press, Tucson, p 328-385

Albarède (1998) Time-dependent models of U-Th-He and K-Ar evolution and the layering of mantle convection. Chem Geol 145:413-429

Allard P (1992) Global emissions of helium by subaerial volcanism. Geophys Res Lett 19:1479-1481

Allègre CJ (1997) Limitation on the mass exchange between the upper and lower mantle: the evolving convection regime of the Earth. Earth Planet Sci Lett 150:1-6

Allègre CJ, Staudacher T, Sarda P, Kurz M (1983) Constraints on evolution of Earth's mantle from rare gas systematics. Nature 303:762-766

Allègre CJ, Staudacher T, Sarda P (1986) Rare gas systematics: formation of the atmosphere, evolution and structure of the Earth's mantle. Earth Planet Sci Lett 87:127-150

Allègre CJ, Sarda P, Staudacher T (1993) Speculations about the cosmic origin of He and Ne in the interior of the Earth. Earth Planet Sci Lett 117:229-233

Allègre CJ, Moreira M, Staudacher T (1995a) $^4He/^3He$ dispersion and mantle convection. Geophys Res Lett 22:2325-2328

Allègre CJ, Poirier J-P, Humler E, Hofmann AW (1995b) The chemical composition of the Earth. Earth Planet Sci Lett 134:515-526

Allègre CJ, Hofmann AW, O'Nions RK (1996) The argon constraints on mantle structures. Geophys Res Lett 23:3555-3557

Althaus T, Niedermann S, Erzinger J (2000) Noble gas studies of fluids and gas exhalations in the East Carpathians, Romania. Chem Erde (Geochem) 60:189-207

Anders E, Grevesse N (1989) Abundances of the elements: meteoritic and solar. Geochim Cosmochim Acta 53:197-214

Anderson DL (1993) Helium-3 from the mantle—primordial signal or cosmic dust? Science 261:170-176

Anderson DL (1998a) The helium paradoxes. Proc Natl Acad Sci USA 95:4822-4827

Anderson DL (1998b) A model to explain the various paradoxes associated with mantle noble gas geochemistry. Proc Natl Acad Sci USA 95:9087-9092

Anderson DL (2000a) The statistics of helium isotopes along the global spreading ridge system and the central limit system. Geophys Res Lett 27:2401-2404

Anderson DL (2000b) The statistics and distribution of helium in the mantle. Intl Geol Rev 42:289-311

Anderson DL (2001) A statistical test of the two reservoir model for helium isotopes. Earth Planet Sci Lett 193:77-82

Armstrong RL (1968) A model for the evolution of strontium and lead isotopes in a dynamic Earth. Rev Geophys 6:175-199

Armstrong RL (1991) The persistent myth of crustal growth. Austral J Earth Sci 38:613-630

Azbel YA, Tolstikhin IN (1993) Accretion and early degassing of the earth—Constraints from Pu-U-I-Xe isotopic systematics. Meteoritics 28:609-621.

Bach W, Niedermann S (1998) Atmospheric noble gases in volcanic glasses from the southern Lau Basin: origin from the subducting slab? Earth Planet Sci Lett 160:297-309

Ballentine CJ, Barfod DN (2000) The origin of air-like noble gases in MORB and OIB. Earth Planet Sci Lett 180:39-48

Ballentine CJ, Burnard PG (2002) Production and release of noble gases in the continental crust. Rev Mineral Geochem 47:481-538
Ballentine CJ, Lee D-C, Halliday AN (1997) Hafnium isotopic studies of the Cameroon line and new HIMU paradoxes. Chem Geol 139:111-124
Ballentine CJ, Porcelli D, Wieler R (2001) Technical comment on 'Noble gases in mantle plumes' by Trieloff et al. (2000). Science 291:2269a
Barfod DN, Ballentine CJ, Halliday AN, Fitton JG (1999) Noble gases in the Cameroon line and the He, Ne, and Ar isotopic compositions of high μ (HIMU) mantle. J Geophys Res 104:29509-29527
Batiza R (1982) Abundances, distribution and sizes of volcanoes in the Pacific Ocean and implications for the origin of non-hotspot volcanoes. Earth Planet Sci Lett 60:195-206
Becker TW, Kellogg JB, O'Connell RJ (1999) Thermal constraints on the survival of primitive blobs in the lower mantle. Earth Planet Sci Lett 171:351-365
Begemann R, Weber HW, Hintenberger H (1976) On the primordial abundance of argon-40. Astrophys J 203:L155-L157
Benkert J-P, Baur H, Signer P, Wieler R (1993) He, Ne, and Ar from solar wind and solar energetic particles in lunar ilmenites and pyroxenes. J Geophys Res 98: 13147-13162
Bernatowicz TJ, Podosek FA, Honda M, Kramer FE (1984) The atmospheric inventory of xenon and noble gases in shales: the plastic bag experiment. J Geophys Res 89:4597-4611
Bernatowicz TJ, Kennedy BM, Podosek FA (1985) Xe in glacial ice and the atmospheric inventory of noble gases. Geochim Cosmochim Acta 49:2561-2564.
Bijwaard H, Spakman W (1999) Tomographic evidence for a narrow whole mantle plume below Iceland. Earth Planet Sci Lett 166:121-126
Bina CR, Silver PG (1990) Constraints on lower mantle composition and temperature from density and bulk sound velocity profiles. Geophys Res Lett 17:1153-1156
Blichert-Toft J, Albarède F (1997) The Lu-Hf isotope geochemistry of chondrites and the evolution of the mantle-crust system. Earth Planet Sci Lett 148:243-258
Brantley SL, Koepenick KW (1995) Measured carbon dioxide emissions from Oldoinyo Lengai and the skewed distribution of passive volcanic fluxes. Geology 23:933-936
Brazzle RH, Pravdivtseva OV, Meshik AP, Hohenberg CM (1999) Verification and interpretation of the I-Xe chronometer. Geochem Cosmochim Acta 63:739-760
Broadhurst CL, Drake MJ, Hagee BE, Bernatowicz TJ (1992) Solubility and partitioning of Ne, Ar, Kr, and Xe in minerals and synthetic basaltic melts. Geochim Cosmochim Acta 56:709-723
Brown H (1949) Rare gases and the formation of the Earth's atmosphere. *In* The atmospheres of the Earth and planets. Kuiper GP (ed) University of Chicago Press, Chicago, p 258-266
Buffett BA, Huppert HE, Lister JR, Woods AW (1996) On the thermal evolution of the Earth's core. J Geophys Res 101:7989-8006
Burnard PG (1999) Origin of argon-lead isotopic correlation in basalts. Science 286:871a
Burnard PG, Stuart FM, Turner G, Oskarsson N (1994) Air contamination of basaltic magmas: implications for high $^3He/^4He$ mantle Ar isotopic composition. J Geophys Res 99:17709-17715
Burnard PG, Graham D, Turner G (1997) Vesicle specific noble gas analyses of popping rock: implications for primordial noble gases in Earth. Science 276:568-571
Burnard PG, Harrison D, Turner G, and Nesbitt R (2002) The degassing and contamination of noble gases in mid-Atlantic Ridge basalts. Geochem Geophys Geosystems (in press)
Busemann H, Baur H, Wieler R (2000) Primordial noble gases in "Phase Q" in carbonaceous and ordinary chondrites studied by closed system stepped etching. Meteoritics Planet Sci 35:949-973
Cadogan PH (1977) Paleaoatmospheric argon in Rhynie chert. Nature 268:38-41
Caffee MW, Hudson GU, Velsko C, Huss GR, Alexander Jr EC, Chivas AR (1999) Primordial noble cases from Earth's mantle: identification of a primitive volatile component. Science 285:2115-2118
Caldwell WA, Nguyen JH, Pfrommer BG, Mauri F, Louie SG, Jeanloz R (1997) Structure, bonding, and geochemistry of xenon at high pressures. Science 227:930-933
Carroll MR, Draper DS (1994) Noble gases as trace elements in magmatic processes. Chem Geol 117: 37-56
Chabot NL, Drake, MJ (1999) Potassium solubility in metal: the effects of composition at 15 kbar and 1900 degrees C on partitioning between iron alloys and silicate melts. Earth Planet Sci Lett 172:323-335
Chamorro-Perez EM, Gillet P, Jambon A, Badro J, McMillan P (1998) Low argon solubility in silicate melts at high pressure. Nature 393:352-355
Chamorro-Perez EM, Brooker RA, Wartho J-A, Wood BJ, Kelley SP, Blundy JD (2001) Ar and K partitioning between clinopyroxene and silicate melt to 8 GPa. Geochim Cosmochim Acta in press
Chauvel C, Hofmann AW, Vidal P (1992) HIMU-EM: the French Polynesian connection. Earth Planet Sci Lett 110:99-119
Chopelas A, Boehler R (1992) Thermal expansivity in the lower mantle. Geophys Res Lett 19:1347-1350

Christensen UR, Hofmann AW (1994) Segregation of subducted oceanic crust in the convecting mantle. J Geophys Res 99:19867-19884
Christensen UR, Yuen DA (1985) Layered convection induced by phase transitions. J Geophys Res 90:10291-10300
Clarke WB, Beg MA, Craig H (1969) Excess ^{3}He in the sea: evidence for terrestrial primordial helium. Earth Planet Sci Lett 6:213-220
Coltice N, Ricard Y (1999) Geochemical observations and one layer mantle convection. Earth Planet Sci Lett 174:125-137
Coltice N, Albarède F, Gillet P (2000) K-40–Ar-40 constraints on recycling continental crust into the mantle. Science 288:845-847
Craig H, Clarke WB, Beg MA (1975) Excess ^{3}He in deep water on the east Pacific rise. Earth Planet Sci Lett 26:125-132
Creager KC, Jordan TH (1986) Slab penetration into the lower mantle beneath the Marianas and other island arcs of the northwest Pacific. J Geophys. Res 91:3573-3589
Crisp JA (1984) Rates of magma emplacment and volcanic output. J Volcanol Geotherm Res 89: 3031-3049
Davies GF (1984) Geophysical and isotopic constraints on mantle convection: an interim synthesis. J Geophys Res 89:6016-6040
Davies GF (1988a) Ocean bathymetry and mantle convection, 1. Large-scale flow and hotspots. J Geophys Res 93:10467-10480
Davies GF (1988b) Ocean bathymetry and mantle convection, 2. Small-scale flow. J Geophys Res 93:10481-10488
Davies GF (1995) Punctuated tectonic evolution of the earth. Earth Planet Sci Lett 136:363-379
Davies GF (1998) Topography: a robust constraint on mantle fluxes. Chem Geol 145:479-489
Davies GF (1999) Geophysically constrained mantle mass flows and the Ar-40 budget: a degassed lower mantle? Earth Planet Sci Lett 166:149-162
Davies GF, Richards MA (1992) Mantle convection. J Geol 100:151-206
De Paolo DJ (1980) Crustal growth and mantle evolution: inferences from models of element transport and Nd and Sr isotopes. Geochim Cosmochim Acta 44:1185-1196
Déruelle B, Dreibus G, Jambon A (1992) Iodine abundances in oceanic basalts: implications for Earth dynamics. Earth Planet Sci Lett 108:217-227
Dixon ET (2002) Interpretation of helium and neon isotopic heterogeneity in Icelandic basalts. Earth Planet Sci Lett (submitted)
Dixon JE, Stolper EM (1995) An experimental study of water and carbon dioxide solubilities in mid-ocean ridge basaltic liquids. 2. Applications to degassing. J Petrol 36:1633-1646
Dixon ET, Honda M, McDougall I, Campbell IH, Sigurdsson I (2000) Preservation of near-solar neon isotopic ratios in Icelandic basalts. Earth Planet Sci Lett 180:309-324
Doe BR, Zartman RE (1979) Plumbotectonics I. the Phanerozoic. *In* Geochemistry of hydrothermal ore deposits. Barnes HL (ed) Wiley, New York, p 22-70
Dreibus G, Wänke H (1989) Supply and loss of volatile constituents during accretion of terrestrial planets. *In* Origin and evolution of planetary and satellite atmospheres. Atreya SK, Pollack JB, Matthews MS (eds) University of Arizona Press, Tucson, p 268-288
Dunai T, Porcelli D (2002) The storage and transport of noble gases in the subcontinental lithosphere. Rev Mineral Geochem 47:371-409
Dymond J, Hogan L (1973) Noble gas abundance patterns in deep sea basalts—primordial gases from the mantle. Earth Planet Sci Lett 20:131-139
Eiler JM, Farley KA, Stolper EM (1998) Correlated helium and lead isotope variations in Hawaiian lavas. Geochim Cosmochim Acta 62:1977-1984
Fabryka-Martin J, Bentley H, Elmore D, Airey PL (1985) Natural iodine-129 as an environmental tracer. Geochim Cosmochim Acta 49:337-347
Farley KA, Craig H (1994) Atmospheric argon contamination of ocean island basalt olivine phenocrysts. Geochim Cosmochim Acta 58:2509-2517
Farley KA, Neroda E (1998) Noble gases in the earth's mantle. Ann Rev Earth Planet Sci 26:189-218
Farley KA, Natland JH, Craig H (1992) Binary mixing of enriched and undegassed (primitive?) mantle components (He, Sr, Nd, Pb) in Samoan lavas. Earth Planet Sci Lett 111:183-199
Farley KA, Maier-Reimer E, Schlosser P, Broecker WS (1995) Constraints on mantle He-3 fluxes and deep-sea circulation from an oceanic general circulation model. J Geophys Res 100:3829-3839
Ferrachat S, Ricard Y (1998) Regular vs. chaotic mantle mixing. Earth Planet Sci Lett 155: 75-86
Ferrachat S, Ricard Y (2001) Mixing properties in the Earth's mantle: effects of the viscosity stratification and of ocean crust segregation. Geochem Geophys Geosys 2:U1-U17

Fisher DE (1978) Terrestrial potassium abundances as limits to models of atmospheric evolution. *In* Terrestrial rare gases. Ozima M, Alexander Jr EC (eds) Japan Scientific Societies Press, Tokyo, p 173-183
Fisher DE (1985) Noble gases from oceanic island basalts do not require an undepleted mantle source. Nature 316:716-718
Flynn GJ (2001) Atmospheric entry heating of interplanetary dust. *In* Accretion of extraterrestrial matter throughout Earth's history. Peucker-Ehrenbrink B, Schmitz (eds) Kluwer, New York, p 107-127
Galer SJG, Mezger K (1998) Metamorphism, denudation and sea level in the Archean and cooling of the Earth. Precamb Res 92:389-412
Galer SJG, O'Nions RK (1985) Residence time of thorium, uranium and lead in the mantle with implications for mantle convection. Nature 316:778-782
Gamo T, Ishibashi J-I, Sakai H, Tilbrook B (1987) Methane anomalies in seawater above the Loihi seamount summit area, Hawaii. Geochim Cosmochim Acta 51:2857-2864
Garnero EJ (2000) Heterogeneity in the lowermost mantle. Ann Rev Earth Planet Sci 28:509-537
Gast PW (1960) Limitations on the composition of the upper mantle. J Geophys Res 65:1287-1297
Glatzmaier GA, Schubert G (1993) 3-dimensional spherical-models of layered and whole mantle convection. J Geophys Res 98:21969-21976
Goes S, Spakman W, Bijwaard H (1999) A lower mantle source for central European volcanism. Science 286:1928-1931
Graham DW (2002) Noble gases in MORB and OIB: observational constraints for the characterization of mantle source reservoirs. Rev Mineral Geochem 47:247-318
Graham DW, Lupton F, Albarède F, Condomines M (1990) Extreme temporal homogeneity of helium isotopes at Piton de la Fournaise, Réunion Island. Nature 347:545-548
Graham DW, Lupton JE, Spera FJ, Christie DM (2001) Upper-mantle dynamics revealed by helium isotope variations along the southeast Indian ridge. Nature 409:701-703
Grand SP (1987) Tomographic inversion for shear velocity beneath the North American plate. J Geophys Res 92:14065-14090
Grand SP (1994) Mantle shear structure beneath the Americas and surrounding oceans. J Geophys Res 99:66-78
Griffiths RW, Hackney RI, van der Hilst RD (1995) A laboratory investigation of effects of trench migration on the descent of subducting slabs. Earth Planet Sci Lett 133:1-17
Gurnis M, Davies GF (1986) The effect of depth dependent viscosity on convecting mixing in the mantle and the possible survival of primitive mantle. Geophys Res Lett 13: 541-544
Hagee B, Bernatowicz TJ, Podosek FA, Johnson ML, Burnett DS, Tatsumoto M (1990) Actinide abundances in ordinary chondrites. Geochim Cosmochim Acta 54:2847-2858
Hager BH (1984) Subducted slabs and the geoid; constraints on mantle rheology and flow. J Geophys Res 89:6003-6016
Hager BH, Richards MA (1989) Long-wavelength variations in the Earth's geoid: physical models and dynamic implications. Phil Trans R Soc Lond A328:309-327
Halliday AN, Porcelli D (2001) In search of lost planets—the paleocosmochemistry of the inner solar system. Earth and Planet Sci Lett 192:545-559
Hamano Y, Ozima M (1978) Earth-atmosphere evolution model based on Ar isotopic data. *In* Terrestrial rare gases. Ozima M, Alexander Jr. EC (eds) Japan Scientific Societies Press, Tokyo, p 155-171
Hanan BB, Graham DW (1996) Lead and helium isotope evidence from oceanic basalts for a common deep source of mantle plumes. Science 272:991-995
Hanes JA, York D, Hall CM (1985) An $^{40}Ar/^{39}Ar$ geochronological and electron microprobe investigation of an Archaean pyroxenite and its bearing on ancient atmospheric compositions. Can J Earth Sci 22:947-958
Hanyu T, Kaneoka I (1998) Open system behavior of helium in case of the HIMU source area. Geophys Res Lett 25:687-690
Hanyu T, Dunai TJ, Davies GR, Kaneoka I, Nohda S, Uto K (2001) Noble gas study of the Reunion hotspot: evidence for distinct less-degassed mantle sources. Earth Planet Sci Lett 193:83-98
Harper CL, Jacobsen SB (1996) Noble gases and Earth's accretion. Science 273:1814-1818
Harrison D, Burnard P, Turner G (1999) Noble gas behavior and composition in the mantle: constraints from the Iceland plume. Earth Planet Sci Lett 171:199-207
Harrison D, Burnard PG, Trieloff M, Turner G (2002) Resolving atmospheric contaminants in mantle noble gas analysis. Geochem Geophys Geosystems (in press)
Hart R, Dymond J, Hogan L (1979) Preferential formation of the atmosphere-sialic crust system from the upper mantle. Nature 278:156-159
Hart SR, Staudigel H (1982) The control of alkalies and uranium in seawater by ocean crust alteration. Earth Planet Sci Lett 58:202-212

Hart SR, Staudigel H (1989) Isotopic characterization and identification of recycled components. *In* Crust/mantle recycling at convergence zones. Hart SR, Gülen L (eds) Klewer Academic Publishers, Dordrecht, p 15-28
Hart SR, Hauri EH, Oschmann LA, Whitehead JA (1992) Mantle plumes and entrainment: isotopic evidence. Science 256:517-520
Hauri EH, Whitehead JA, Hart SR (1994) Fluid dynamic and geochemical aspects of entrainment in mantle plumes. J Geophys Res 99:24275-24300
Helffrich GR, Wood BJ (2001) The Earth's mantle. Nature 412:501-507
Helmberger DV, Wen L, Ding X (1998) Seismic evidence that the source of the Iceland hotspot lies at the core-mantle boundary. Nature 396:251-255
Heymann D, Dziczkaniec, Palma R. (1976) Limits for the accretion time of the Earth from cosmogenic ^{21}Ne produced in planetesimals. Proc Lunar Sci Conf 7:3411-3419
Hilton DR, Hammerschmidt K, Loock G, Friedrichsen H (1993) Helium and argon isotope systematics of the central Lau Basin and Valu Fa ridge: Evidence of crust-mantle interactions in a back-arc basin. Geochim Cosmochim Acta 57:2819-2841
Hilton DR, Barling J, Wheller OE (1995) Effect of shallow-level contamination on the helium isotope systematics of ocean-island lavas. Nature 373:330-333
Hilton DR, McMurty GM, Kreulen R (1997) Evidence for extensive degassing of the Hawaiian mantle plume from helium-carbon relationships at Kilauea volcano. Geophys Res Lett 24:3065-3068
Hilton DR, Grönvold K, Sveinbjornsdottir AE, Hammerschmidt K (1998) Helium isotope evidence for off-axis degassing of the Icelandic hotspot. Chem Geol 149:173-187
Hilton DR, Grönvold K, MacPherson CG, Castillo PR (1999) Extreme He-3/He-4 ratios in northwest Iceland: constraining the common component in mantle plumes. Earth Planet Sci Lett 173:53-60
Hilton DR, Thirlwall MF, Taylor RN, Murton BJ, Nichols A (2000) Controls on magmatic degassing along the Reykjanes Ridge with implications for the helium paradox. Earth Planet Sci Lett 183:43-50
Hilton DR, Fischer TP, Marty B (2002) Noble gases in subduction zones and volatile recycling. Rev Mineral Geochem 47:319-370
Hiyagon H (1994) Retention of helium in subducted interplanetary dust particles—Reply. Science 265:1893
Hiyagon H, Ozima M (1986) Partition of gases between olivine and basalt melt. Geochim Cosmochim Acta 50:2045-2057
Hiyagon H, Ozima M, Marty B, Zashu S, Sakai H (1992) Noble gases in submarine glasses from mid-oceanic ridges and Loihi Seamount—constraints on the early history of the Earth. Geochim Cosmochim Acta 56:1301-1316
Hofmann AW (1997) Mantle geochemistry: the message from oceanic volcanism. Nature 385:219-229
Hofmann AW, Jochum KP, Seufert M, White WM (1986) Nb and Pb in oceanic basalts: new constraints on mantle evolution. Earth Planet Sci Lett 79:33-45
Hohenberg CM, Podosek FA, Reynolds JH (1967) Xenon-iodine dating: sharp isochronism in chondrites. Science 156:233-236
Holloway JR (1998) Graphite-melt equilibria during mantle melting: constraints on CO_2 in MORB magmas and the carbon content of the mantle. Chem Geol 147:89-97
Honda M, McDougall I (1993) Solar noble gases in the Earth—the systematics of helium-neon isotopes in mantle-derived samples. Lithos 30:257-265
Honda M, McDougall I (1998) Primordial helium and neon in the Earth—a speculation on early degassing. Geophys Res Lett 25:1951-1954
Honda M, Patterson DB (1999) Systematic elemental fractionation of mantle-derived helium, neon, and argon in mid-oceanic ridge glasses. Geochim Cosmochim Acta 63:2863-2874
Honda M, McDougall I, Patterson DB, Doulgeris A, Clague DA (1991) Possible solar noble-gas component in Hawaiian basalts. Nature 349:149-151
Honda M, McDougall I, Patterson DB, Doulgeris A, Clague DA (1993) Noble gases in submarine pillow basalt glasses from Loihi and Kilauea, Hawaii—a solar component in the Earth. Geochim Cosmochim Acta 57:859-874
Hudson GB, Kennedy BM, Podosek FA, Hohenberg CM (1989). The early solar system abundance of ^{244}Pu as inferred from the St. Severin chondrite. Proc 19th Lunar Planet Sci Conf, p 547-557
Hunt DL, Kellogg LH (2001) Quantifying mixing and age variations of heterogeneities in models of mantle convection: role of depth-dependent viscosity. J Geophys Res 106:6747-6759
Hunten DM, Pepin RO, Walker JCG (1987) Mass fractionation in hydrodynamic escape. Icarus 69:532-549
Hurley PM, Rand JR (1969) Pre-drift continental nuclei. Science 164:1229-1242
Igarashi G (1995) Primitive Xe in the Earth. *In* Volatiles in the Earth and solar system. Farley KA (ed) Am Inst Phys Conf Proc 341:70-80

Ito E, Takahashi E (1989) Post-spinel transformations in the system Mg_2SiO_4 and some geophysical implications. J Geophys Res 94:10637-10646

Ivins ER, Sammis CG, Yoder CF (1993) Deep mantle viscosity structure with prior estimate and satellite constraint. J Geophys Res 98:4579-4609

Jacobsen S, Harper CJ (1996) Accretion and early differentiation history of the Earth based on extinct radionuclides. *In* Earth processes: reading the isotopic code. Basu A, Hart SR (eds) American Geophysical Union, Washington, D.C., Geophys Monograph 95:47-74

Jacobsen S, Wasserburg GJ (1979) The mean age of the mantle and crustal reservoirs. J Geophys Res 84:218-234

Jarvis GT (1993) Effects of curvature on 2-dimensional models of mantle convection-cylindrical polar coordinates. J Geophys Res 98: 4477-4485

Javoy M, Pineau F (1991) The volatiles record of a popping rock from the Mid-Atlantic Ridge at 14-degrees-N—Chemical and isotopic composition of gas trapped in the vesicles. Earth Planet Sci Lett 107:598-611

Jean-Baptiste P (1992) Helium-3 distribution in the deep world ocean; its relation to hydrothermal ^{3}He fluxes and to the terrestrial heat budget. *In* Isotopes of noble gases as tracers in environmental studies. Loosli H, Mazor E (eds) Intl Atomic Energy Agency, Vienna, Austria, p 219-240.

Jeanloz RE (1989) Density and composition of the lower mantle. Phil Trans Roy Soc Lond 328:377-389

Jephcoat AP (1998) Rare-gas solids in the Earth's deep interior. Nature 393:355-358

Jochum KP, Hofmann AW, Ito E, Seufert HM, White WM (1983) K, U, and Th in mid-ocean ridge basalt glasses and heat production, K/U and K/Rb in the mantle. Nature 306:431-436

Kamijo K, Hashizume K, Matsuda J-I (1998) Noble gas constraints on the evolution of the atmosphere-mantle system. Geochim Cosmochim Acta 62:2311-2321

Kaneoka I (1998) Noble gas signatures in the Earth's interior-coupled or decoupled behavior among each isotope systematics and problems related to their implication. Chem Geol 147:61-76

Kaneoka I, Takaoka N, Upton BGJ (1986) Noble gas systematics in basalts and a dunite nodule from Reunion and Grand Comore Islands, Indian Ocean. Chem Geol 59:35-42

Kàrason H, van der Hilst RD (2001) Tomographic imaging of the lowermost mantle with differential times of refracted and diffracted core phases (PKP, P-diff). J Geophys Res 106:6569-6587

Kaufmann G, Lambeck K (2000) Mantle dynamics, postglacial rebound and the radial viscosity profile. Phys Earth Planet Intl 121:301-324

Kellogg LH (1992) Mixing in the Mantle. Ann Rev Earth Planet Sci 20: 365-388

Kellogg LH, Wasserburg GJ (1990) The role of plumes in mantle helium fluxes. Earth Planet Sci Lett 99:276-289

Kellogg LH, Hager BH, van der Hilst RD (1999) Compositional stratification in the deep mantle. Science 283:1881-1884

Kennett BLN, van der Hilst RD (1998) Seismic structure of the mantle: from subduction zone to craton. *In* The Earth's mantle: composition, structure and evolution. Jackson I (ed) Cambridge University Press, Cambridge, p 381-404

Kido M, Yuen DA, Cadek O, Nakakuki T (1998) Mantle viscosity derived by genetic algorithm using oceanic geoid and seismic tomography for whole-mantle versus blocked-flow situations. Phys Earth Planet Intl 107:107-326

Kinkaid C, Olson P (1987) An experimental study of subduction and slab migration. J Geophys Res 92:13832-13840

Krummenacher D, Merrihue CM, Pepin RO, Reynolds JH (1962) Meteoritic krypton and barium versus the general isotopic anomalies in meteoritic xenon. Geochim Cosmochim Acta 26:231-249

Kunz J (1999) Is there solar argon in the Earth's mantle? Nature 399:649-650

Kunz J, Staudacher T, Allègre CJ (1998) Plutonium-fission xenon found in Earth's mantle. Science 280:877-880

Kurz MD, Jenkins WJ (1981) The distribution of helium in oceanic basalt glasses. Earth Planet Sci Lett 53:41-54

Kurz MD, Jenkins WJ, Hart SR (1982) Helium isotopic systematics of oceanic islands and mantle heterogeneity. Nature 297:43-46

Kurz MD, Jenkins WJ, Hart SR, Clague D (1983) Helium isotopic variations in volcanic rocks from Loihi Seamount and the island of Hawaii. Earth Planet Sci Lett 66:388-406

Lay T, Williams Q, Garnero EJ (1998) The core mantle boundary layer and deep Earth dynamics. Nature 392:461-468

Leya I, Wieler R (1999) Nucleogenic production of Ne isotopes in Earth's crust and upper mantle induced by alpha particles from the decay of U and Th. J Geophys Res 104:15439-15450

Lithgow-Bertelloni C, Richards MA (1998) The dynamics of Cenozoic and Mesozoic plate motions. Rev Geophys 36:27-78

Lupton JE (1983) Terrestrial inert gases: isotope tracer studies and clues to primordial components in the mantle. Ann Rev Earth Planet Sci 11:371-414

Lupton JE (1996) A far-field hydrothermal plume from Loihi Seamount. Science 272:976-979

Lupton JE, Craig H (1975) Excess ^{3}He in oceanic basalts, evidence for terrestrial primordial helium. Earth Planet Sci Lett 26:133-139

Machatel P, Weber P (1991) Intermittent layered convection in a model with an endothermic phase change at 670 km. Nature 350:55-57

MacPherson CG, Hilton DR, Sinton JM, Poreda RJ, Craig H (1998) High ^{3}He/^{4}He ratios in the Manus backarc basin: implications for mantle mixing and the origin of plumes in the western Pacific Ocean. Geology 26:1007-1010

Mahaffy PR, Donahue TM, Atreya SK, Owen TC, Niemann HB (1998) Galileo probe measurements of D/H and ^{3}He/^{4}He in Jupiter's atmosphere. Space Sci Rev 84:251-263

Mamyrin BA, Tolstikhin IN (1984) Helium Isotopes in Nature. Elsevier, Amsterdam

Manga M (1996) Mixing of heterogeneities in the mantle: effect of viscosity differences. Geophys Res Lett 23:403-406

Martel DJ, Deák J, Dövenyi P, Horváth F, O'Nions RK, Oxburgh ER, Stegna L, Stute M (1989) Leakage of helium from the Pannonian Basin. Nature 432:908-912

Marti K, Mathew KJ (1998) Noble-gas components in planetary atmospheres and interiors in relation to solar wind and meteorites. Proc Indian Acad Sci (Earth Planet Sci) 107:425-431

Marty B, Allé P (1994) Neon and argon isotopic constraints on Earth-atmosphere evolution. *In* Noble gas geochemistry and cosmochemistry. Matsuda J-I (ed) Terra Scientific Publishing Co., Tokyo, p 191-204.

Marty B, Jambon A (1987) C/^{3}He in volatile fluxes from the solid Earth: implications for carbon geodynamics. Earth Planet Sci Lett 83:16-26

Marty B, Lussiez P (1993) Constraints on rare gas partition coefficients from analysis of olivine-glass from a picritic mid-ocean ridge basalt. Chem Geol 106:1-7

Matsuda J-I, Marty B (1995) The ^{40}Ar/^{36}Ar ratio of the undepleted mantle; a re-evaluation. Geophys Res Lett 22:1937-1940

Matsuda J, Matsubara K (1989) Noble gases in silica and their implication for the terrestrial "missing" Xe. Geophys Res Lett 16:81-84

Matsuda J-I, Nagao K (1986) Noble gas abundances in a deep-sea core from eastern equatorial Pacific. Geochem J 20:71-80

Matsuda J, Sudo M, Ozima M, Ito K, Ohtaka O, Ito E (1993) Noble gas partitioning between metal and silicate under high pressures. Science 259:788-790

McCulloch MT, Bennett VC (1998) Early differentiation of the Earth: an isotopic perspective *In* The Earth's mantle: composition structure and evolution. Jackson I (ed) Cambridge University Press, Cambridge, p 127-158

McDougall I, Honda M (1998) Primordial solar noble gas component in the Earth: consequences for the origin and evolution of the Earth and its atmosphere. *In* The Earth's mantle; composition, structure, and evolution. Jackson I (ed) Cambridge University Press, Cambridge, p 159-187

McKenzie D, Richter FM (1981) Parameterized thermal convection in a layered region and the thermal history of the Earth. J Geophys Res 86:11667-11680

Mitrovica JX (1996) Haskell [1935] revisited. J Geophys Res 101:555-569

Mitrovica JX, Forte AM (1995) Pleistocene glaciation and the Earth's precession constant. Geophys J Intl 121:21-32

Mizuno H, Nakazawa K, Hayashi C (1980) Dissolution of the primordial rare gases into the molten Earth's material. Earth Planet Sci Lett 50:202-210

Moreira M, Allègre CJ (1998) Helium-neon systematics and the structure of the mantle. Chem Geol 147:53-59

Moreira M, Sarda P (2000) Noble gas constraints on degassing processes. Earth Planet Sci Lett 176: 375-386

Moreira M, Staudacher T, Sarda P, Schilling JG, Allègre CJ (1995) A primitive plume neon component In MORB—the Shona Ridge anomaly, South Atlantic (51-52-degrees-S). Earth Planet Sci Lett 133: 367-377

Moreira M, Kunz J, Allègre CJ (1998) Rare gas systematics in popping rock: isotopic and elemental compositions in the upper mantle. Science 279:1178-1181

Moreira M, Breddam K, Curtice J, Kurz MD (2001) Solar neon in the Icelandic mantle: new evidence for an undegassed lower mantle. Earth Planet Sci Lett 185:15-23

Morrison P, Pine J (1955) Radiogenic origin of the helium isotopes in rock. Ann NY Acad Sci 62:69-92

Muramatsu YW, Wedepohl KH (1998) The distribution of iodine in the earth's crust. Chem Geol 147: 201-216

Musselwhite DS, Drake MJ, Swindle TD (1989) Early outgassing of the Earth's mantle: implications of mineral/melt partitioning of I. Lunar Planet Sci XX:748-749
Newsom HE, White WM, Jochum KP, Hofmann AW (1986) Siderophile and chalcophile element abundances in oceanic basalts, Pb isotope evolution and growth of the Earth's core. Earth Planet Sci Lett 80:299-313
Niedermann S, Bach W, Erzinger J (1997) Noble gas evidence for a lower mantle component in MORBs from the southern East Pacific Rise: decoupling of helium and neon isotope systematics. Geochim Cosmochim Acta 61:2697-2715
Nuccio PM, Paonita A (2000) Investigation of the noble gas solubility in H_2O-CO_2 bearing silicate liquids at moderate pressure II. the extended ionic porosity (EIP) model. Earth Planet Sci Lett 183:499-512
Ohtani E, Maeda M (2001) Density of basaltic melt at high pressure and stability of the melt at the base of the lower mantle. Earth Planet Sci Lett 193:69-75
O'Nions RK (1987) Relationships between chemical and convective layering in the earth. J Geol Soc Lond 144:259-274
O'Nions RK, Oxburgh ER (1983) Heat and helium in the Earth. Nature 306:429-431
O'Nions RK, Oxburgh ER (1988) Helium, volatile fluxes and the development of continental crust. Earth Planet Sci Lett 90:331-347
O'Nions RK, Tolstikhin IN (1994) Behaviour and residence times of lithophile and rare gas tracers in the upper mantle. Earth Planet Sci Lett 124:131-138
O'Nions RK, Tolstikhin IN (1996) Limits on the mass flux between lower and upper mantle and stability of layering. Earth Planet Sci Lett 139:213-222
O'Nions RK, Evenson NM, Hamilton PJ (1979) Geochemical modeling of mantle differentiation and crustal growth. J Geophys Res 84:6091-6101
O'Nions RK, Carter, SR, Evensen, NM, Hamilton, PJ (1981) Upper mantle geochemistry. *In* The sea. Vol 7. Emiliani C (ed) Wiley, New York, p 49-71
Ott U (2002) Noble gases in meteorites—trapped components. Rev Mineral Geochem 47:71-100
Owen T, Bar Nun A, Kleinfeld I (1992) Possible cometary origin of heavy noble gases in the atmospheres of Venus, Earth and Mars. Nature 358:43-46
Ozima M (1973) Was the evolution of the atmosphere continuous or catastrophic? Nature Phys Sci 246: 41-42
Ozima M (1975) Ar isotopes and Earth-atmosphere evolution models. Geochim Cosmochim Acta 39: 1127-1134
Ozima M (1994) Noble gas state in the mantle. Rev Geophys 32:405-426
Ozima M, Alexander Jr EC (1976) Rare gas fractionation patterns in terrestrial samples and the Earth-atmosphere evolution model. Rev Geophys Space Phys 14:385-390
Ozima M, Igarashi G (2000) The primordial noble gases in the Earth: a key constraint on Earth evolution models. Earth Planet Sci Lett 176:219-232
Ozima M, Kudo K (1972) Excess argon in submarine basalts and an earth-atmosphere evolution model. Nature Phys Sci 239:23-24
Ozima M, Podosek FA (1983) Noble Gas Geochemistry. Cambridge University Press, Cambridge
Ozima M, Podosek FA (1999) Formation age of Earth from I-129/I-127 and Pu-244/U-238 systematics and the missing Xe. J Geophys Res 104:25493-25499
Ozima M, Podosek FA (2001) Noble Gas Geochemistry 2nd ed. Cambridge, Cambridge University Press, Cambridge
Ozima M, Zahnle K (1993) Mantle degassing and atmospheric evolution—noble gas view. Geochem J 27:185-200
Ozima M, Podozek FA, Igarashi G (1985) Terrestrial xenon isotope constraints on the early history of the Earth. Nature 315:471-474
Ozima M, Azuma S-I, Zashu S, Hiyagon H (1993) ^{244}Pu fission Xe in the mantle and mantle degassing chronology. *In* Primitive solar nebula and origin of planets. Oya H (ed) Terra Scientific Publishing Co, Tokyo, p 503-517
Parsons B (1981) The rates of plate creation and consumption. Geophys J Roy Astron Soc 67:437-448
Patterson DB, Honda M, McDougall I (1990) Atmospheric contamination: a possible source for heavy noble gases basalts from Loihi seamount, Hawaii. Geophys Res Lett 17:705-708
Pepin RO (1991) On the origin and early evolution of terrestrial planet atmospheres and meteoritic volatiles. Icarus 92:1-79
Pepin RO (1997) Evolution of Earth's noble gases: consequences of assuming hydrodynamic loss driven by giant impact. Icarus 126:148-156
Pepin RO (1998) Isotopic evidence for a solar argon component in the Earths mantle. Nature 394:664-667
Pepin RO (2000) On the isotopic composition of primordial xenon in terrestrial planet atmospheres. Space Sci Rev 92:371-395

Pepin RO, Phinney D (1976) The formation interval of the Earth. Lunar Sci VII:682-684

Pepin RO, Porcelli D (2002) Origin of noble gases in the terrestrial planets. Rev Mineral Geochem 47:191-246

Pepin RO, Becker RH, Schlutter DJ (1999) Irradiation records in regolith materials, I. isotopic compositions of solar-wind neon and argon in single lunar mineral grains. Geochim Cosmochim Acta 63:2145-2162

Phinney D, Tennyson J, Frick U (1978) Xenon in CO_2 well gas revisited. J Geophys Res 83:2313-2319

Phipps Morgan J, Morgan WJ (1999) Two-stage melting and the geochemical evolution of the mantle: a recipe for mantle plum pudding. Earth Planet Sci Lett 170:215-239

Phipps Morgan J, Shearer PM (1993) Seismic constraints on mantle flow and topography of the 660-km discontinuity—evidence for whole-mantle convection. Nature 365:506-511

Podosek FA, Honda M, Ozima M (1980) Sedimentary noble gases. Geochim Cosmochim Acta 44: 1875-1884

Podosek FA, Woolum DS, Cassen P, Nichols RH (2000) Solar gases in the Earth by solar wind irradiation? 10th Annual Goldschmidt Conf, Oxford

Pollack HN, Hurter SJ, Johnson JR (1993) Heat flow from the Earth's interior: analysis of the global data set. Rev Geophys 31:267-280

Porcelli D, Halliday AN (2001) The core as a possible source of mantle helium. Earth Planet Sci Lett 192:45-56

Porcelli D, Pepin RO (2000) Rare gas constraints on early Earth history. *In* Origin of the Earth and Moon. Canup RM, Righter K (eds) University of Arizona Press, Tucson, p 435-458

Porcelli D, Wasserburg GJ (1995a) Mass transfer of xenon through a steady-state upper mantle. Geochim Cosmochim Acta 59:1991-2007

Porcelli D, Wasserburg GJ (1995b) Mass transfer of helium, neon, argon, and xenon through a steady-state upper mantle. Geochim Cosmochim Acta 59:4921-4937

Porcelli D, Stone JOH, O'Nions RK (1986) Rare gas reservoirs and Earth degassing. Lunar Planet Sci XVII:674-675

Porcelli DR, Woolum D, Cassen P (2001) Deep Earth rare gases: initial inventories, capture from the solar nebula, and losses during Moon formation. Earth Planet Sci Lett 193:237-251

Porcelli D, Ballentine CJ, Wieler R (2002) An introduction to noble gas geochemistry and cosmochemistry. Rev Mineral Geochem 47:1-18

Poreda RJ, Farley KA (1992) Rare gases in Samoan xenoliths. Earth Planet Sci Lett 113:129-144

Puster P, Jordan TH (1997) How stratified is mantle convection? J Geophys Res 102:7625-7646

Reymer A, Schubert G (1984) Phanerozoic addition rates to the continental crust. Tectonics 3:63-77

Ricard Y, Wuming B (1991) Inferring the viscosity and 3-D density structure of the mantle from geoid, topography and plate velocities. Geophys J Intl 105:561-571

Ricard Y, Richards MA, Lithgow-Bertelloni C, Le Stunff Y (1993) A geodynamic model of mantle density heterogeneity. J Geophys Res 98:21895-21909

Rison W (1980) Isotopic studies of rare gases in igneous rocks: implications for the mantle and atmosphere. PhD Dissertation, University of California, Berkeley

Rison W, Craig H (1983) Helium isotopes and mantle volatiles in Loihi Seamount and Hawaiian Island basalts and xenoliths. Earth Planet Sci Lett 66:407-426

Ritter JRR, Jordan M, Christensen UR, Achauer U (2001) A mantle plume below the Eifel volcanic fields, Germany. Earth Planet Sci Lett 186:7-14

Rocholl A, Jochum, KP (1993) Th, U and other trace elements in carbonaceous chondrites—implications for the terrestrial and solar system Th/U ratios. Earth Planet Sci Lett 117:265-278

Rubey WW (1951) Geological history of seawater. Bull Geol Soc Am 62:1111-1148

Rudnick R, Fountain DM (1995) Nature and composition of the continental crust: a lower crustal perspective. Rev Geophys 33:267-309

Russell SA, Lay T, Garnero EJ (1998) Seismic evidence for small-scale dynamics in the lowermost mantle at the root of the Hawaiian hotspot. Nature 369:225-258

Sano Y, Williams S (1996) Fluxes of mantle and subducted carbon along convergent plate boundaries. Geophys Res Lett 23:2746-2752

Sarda P, Graham DW (1990) Mid-ocean ridge popping rocks: implications for degassing at ridge crests. Earth Planet Sci Lett 97:268-289

Sarda P, Staudacher T, Allègre CJ (1985) $^{40}Ar/^{36}Ar$ in MORB glasses: constraints on atmosphere and mantle evolution. Earth Planet Sci Lett 72:357-375

Sarda P, Staudacher T, Allègre CJ (1988) Neon isotopes in submarine basalts. Earth Planet Sci Lett 91: 73-88

Sarda P, Moreira M, Staudacher T (1999) Argon-lead isotopic correlation in Mid-Atlantic ridge basalts. Science 283:666-668

Sarda P, Moreira M, Staudacher T, Schilling JG, Allègre CJ (2000) Rare gas systematics on the southernmost Mid-Atlantic Ridge: constraints on the lower mantle and the Dupal source. J Geophys Res 105:5973-5996

Sasaki S (1991) Off-disk penetration of ancient solar wind. Icarus 91:29-38

Sasaki S, Nakazawa K (1988) Origin and isotopic fractionation of terrestrial Xe: hydrodynamic fractionation during escape of the primordial H_2-He atmosphere. Earth Planet Sci Lett 89:323-334

Schilling J-G, Unni CK, Bender ML (1978) Origin of chlorine and bromine in the oceans. Nature 273: 631-636

Schlosser P, Winckler G (2002) Noble gases in the ocean and ocean floor. Rev Mineral Geochem 47: 701-730

Schmalzl J, Houseman GA, Hansen U (1996) Mixing in vigorous, time-dependent three-dimensional convection and application to Earth's mantle. J Geophys Res 101: 21847-21858

Schubert G (1992) Numerical-models of mantle convection. Ann Rev Fluid Mechanics 24: 359-394

Schwartzman DW (1973) Argon degassing and the origin of the sialic crust. Geochim Cosmochim Acta 37:2479-2495

Seta A, Matsumoto T, Matsuda J-I (2001) Concurrent evolution of $^3He/^4He$ ratio in the Earth's mantle reservoirs for the first 2 Ga. Earth Planet Sci Lett 188:211-219

Shaw AM, Hilton DR, MacPherson CG, Sinton JM (2001) Nucleogenic neon in high $^3He/^4He$ lavas from the Manus Back-Arc Basin: a new perspective on He-Ne decoupling. Earth Planet Sci Lett 194:53-66

Shen Y, Solomon SC, Bjarnason IT, Wolfe CJ (1998) Seismic evidence for a lower-mantle origin of the Iceland plume. Nature 395:62-65

Silver PG, Carlson RW, Olson P (1988) Deep slabs, geochemical heterogeneity, and the large-scale structure of mantle convection—investigation of an enduring paradox. Ann Rev Earth Planet Sci 16:477-541

Sleep NH (1990) Hotspots and mantle plumes: some phenomenology. J Geophys Res 95:6715-6736

Staudacher T (1987) Upper mantle origin for Harding County well gases. Nature 325:605-607

Staudacher T, Allègre CJ (1982) Terrestrial xenology. Earth Planet Sci Lett 60:389-406

Staudacher T, Allègre CJ (1988) Recycling of oceanic crust and sediments: the noble gas subduction barrier. Earth Planet Sci Lett 89:173-183

Staudacher T, Kurz MD, Allègre CJ (1986) New noble-gas data on glass samples from Loihi seamount and Hualalai and on dunite samples from Loihi and Reunion Island. Chem Geol 56:193-205

Staudacher T, Sarda P, Allègre CJ (1989) Noble gases in basalt glasses from a Mid-Atlantic Ridge topographic high at 14°N: geodynamic consequences. Earth Planet Sci Lett 96:119-133

Steinbach V, Yuen D, Zhao W (1993) Instabilities from phase transition and the time scales of mantle thermal evolution. J Geophys Res 20:1119-1122

Steinberger B (2000) Plumes in a convecting mantle: models and observations for individual hotspots. J Geophys Res 105:11127-11152

Stixrude L, Hemley RJ, Fei Y, Mao HK (1992) Thermoelasticity of silicate perovskite and magnesiowüstite and stratification of the Earth's mantle. Science 257:1099-1101

Stuart FM (1994) Speculations about the cosmic origin of He and Ne in the interior of the Earth—comment. Earth Planet Sci Lett 122:245-247

Stute M, Sonntag C, Deak J, Schlosser P (1992) Helium in deep circulating groundwater in the Great Hungarian Plain—glow dynamics and crustal and mantle helium fluxes. Geochim Cosmochim Acta 56:2051-2067

Sudo M, Ohtaka O, Matsuda J-I (1994) Noble gas partitioning between metal and silicate under high pressures: the case of iron and peridotite. *In* Noble gas geochemistry and cosmochemistry. Matsuda J-I (ed) Terra Scientific Publishing Co., Tokyo, p 355-372

Suess HE (1949) The abundance of noble gases in the Earth and the cosmos. J Geol 57:600-607 (in German)

Tackley PJ (1998) Self-consistent generation of tectonic plates in three-dimensional mantle convection. Earth Planet Sci Lett 157: 9-22

Tackley PJ (2000) Mantle convection and plate tectonics: toward an integrated physical and chemical theory. Science 288:2002-2007

Tackley PJ, Stevenson DJ, Glatzmaier GA (1993) Effects of an endothermic phase transition at 670-km depth on spherical mantle convection. Nature 361:699-704

Tackley PJ, Stevenson DJ, Glatzmaier GA, Schubert G (1994) Effects of multiple phase transitions in a three dimensional model of convection in the Earth's mantle. J Geophys Res 99:15877-15902

Taylor SR, McLennan SM (1985) The Continental Crust: Its Composition and Evolution. Blackwell Scientific, Oxford

Thomsen L (1980) ^{129}Xe on the outgassing of the atmosphere. J Geophys Res 85:4374-4378

Tolstikhin IN (1975) Helium isotopes in the Earth's interior and in the atmosphere: a degassing model of the Earth. Earth Planet Sci Lett 26:88-96
Tolstikhin IN (1978) A review: some recent advances in isotope geochemistry. *In* Terrestrial rare gases. Alexander Jr EC, Ozima M (eds) Japan Scientific Society Press, Tokyo, p33-62
Tolstikhin IN, Marty B (1998) The evolution of terrestrial volatiles: a view from helium, neon, argon and nitrogen isotope modeling. Chem Geol 147:27-52
Tolstikhin IN, O'Nions RK (1994) The Earths missing xenon—a combination of early degassing and of rare gas loss from the atmosphere. Chem Geol 115:1-6
Torgersen T (1989) Terrestrial helium degassing fluxes and the atmospheric helium budget: implications with respect to the degassing processes of continental crust. Chem Geol 79:1-14
Trieloff M, Kunz J, Clague DA, Harrison D, Allègre CJ (2000) The nature of pristine noble gases in mantle plumes. Science 288:1036-1038
Trieloff M, Kunz J, Clague DA, Harrison D, Allègre CJ (2001) Reply to comment on noble gases in mantle plumes. Science 291:2269a
Trieloff M, Kunz J, Allègre CJ (2002) Noble gas systematics of the Réunion mantle plume source and the origin of primordial noble gases in Earth's mantle. Earth Planet Sci Lett 200:297-313
Trompert R and Hansen U (1998) Mantle convection simulations with rheologies that generate plate-like behaviour. Nature 395: 686-689
Trull T (1994) Influx and age constraints on the recycled cosmic dust explanation for high $^3He/^4He$ ratios at hotspot volcanoes. *In* Noble gas geochemistry and cosmochemistry. Matsuda J (ed) Terra Scientific Publishing Co., Tokyo, p 77-88
Turcotte DL, Schubert G (1988) Tectonic implications of radiogenic noble gases in planetary atmospheres. Icarus 74:36-46
Turcotte DL, Paul D, White WM (2001) Thorium-uranium systematics require layered mantle convection. J Geophys Res 106:4265-4276
Turekian KK (1959) The terrestrial economy of helium and argon. Geochim Cosmochim Acta 17:37-43
Turekian KK (1990) The parameters controlling planetary degassing based on ^{40}Ar systematics. *In* From mantle to meteorites. Gopolan K, Gaur VK, Somayajulu BLK, MacDougall JD (eds) Indian Acad Sci, Bangalore, p 147-152
Turner G (1989) The outgassing history of the Earth's atmosphere. J Geol Soc Lond 146:147-154
Valbracht PJ, Staudacher T, Malahoff A, Allègre CJ (1997) Noble gas systematics of deep rift zone glasses from Loihi Seamount, Hawaii. Earth Planet Sci Lett 150:399-411
van der Hilst RD, Kàrason H (1999) Compositional heterogeneity in the bottom 1000 kilometers of Earth's mantle: toward a hybrid convection model. Science 283:1885-1888
van der Hilst RD, Widiyantoro S, Engdahl ER (1997) Evidence for deep mantle circulation from global tomography. Nature 386:578-584
van der Voo R, Spakman W, Bijwaard H (1999a) Mesozoic subducted slabs under Siberia. Nature 397: 246-249
van der Voo R, Spakman W, Bijwaard H (1999b) Tethyan subducted slabs under India. Earth Planet Sci Lett 171:7-20
van Keken P, Zhong SJ (1999) Mixing in a 3D spherical model of present-day mantle convection. Earth Planet Sci Lett 171: 533-547
van Keken PE, Yuen DA, van den Berg AP (1994) Implications for mantle dynamics from the high melting temperature of perovskite. Science 264:1437-1440
van Keken PE, Ballentine CJ (1998) Whole-mantle versus layered mantle convection and the role of a high viscosity lower mantle in terrestrial volatile evolution. Earth Planet Sci Lett 156:19-32
van Keken PE, Ballentine CJ (1999) Dynamical models of mantle volatile evolution and the role of phase transitions and temperature-dependent rheology. J Geophys Res 104:7137-7151
van Keken PE, Ballentine CJ, Porcelli D (2001) A dynamical investigation of the heat and helium imbalance. Earth Planet Sci Lett 188:421-443.
van Keken PE, Hauri EH, Ballentine CJ (2002) Mixing in the mantle and the creation, preservation and destruction of mantle heterogeneity. Ann Rev Earth Planet Sci in press
Van Schmus WR (1995) Natural radioactivity of the crust and mantle. *In* Global earth physics: a handbook of physical constants. Ahrens TJ (ed) American Geophysical Union, Washington, DC, p 283-291
Vidale JE, Schubert G, Earle PS (2001) Unsuccessful initial search for a mid-mantle chemical boundary with seismic arrays. Geophys Res Lett 28:859-862
von Huene R, Scholl DW (1991) Observations at convergent margins concerning sediment subduction, subduction erosion, and the growth of continental crust. Rev Geophys 29:279-312
Wacker JF, Anders E (1984) Trapping of xenon in ice: implications for the origin of the Earth's noble gases. Geochim Cosmochim Acta 48:2373-2380

Walker D (2000) Core participation in mantle geochemistry: Geochemical Society Ingerson Lecture, GSA Denver, October 1999. Geochim Cosmochim Acta 64:2897-2911

Wang Y, Weidner DJ, Liebermann RC, Zhao Y (1994) P-V-T equation of state of (Mg,Fe)SiO_3 perovskite: constraints on the composition of the lower mantle. Phys Earth Planet Inter 83:13-40

Wänke H, Dreibus G, Jagoutz E (1984) Mantle chemistry and accretion history of the Earth. *In* Archaean geochemistry. Kröner A, Hanson GN, Goodwin AM (eds) Springer-Verlag, Berlin, p 1-24

Wasserburg GJ, MacDonald GJF, Hoyle F, Fowler WA (1964) Relative contributions of uranium, thorium, and potassium to heat production in the Earth. Science 143:465-467

Wedepohl KH (1995) The composition of the continental crust. Geochim Cosmochim Acta 59:1217-1232

Wen LX, Anderson DL (1997) Layered mantle convection: A model for geoid and topography. Earth Planet Sci Lett 146:367-377

Wetherill G (1975) Radiometric chronology of the early solar system. Ann Rev Nuclear Sci 25:283-328

Wieler R (2002) Noble gases in the solar system. Rev Mineral Geochem 47:21-70

Williams SN, Schaefer SJ, Calvache MlV, Lopez D (1992) Global carbon dioxide emission to the atmosphere by volcanoes. Geochim Cosmochim Acta 56:1765-1770

Woolum DS, Cassen P, Porcelli D, Wasserburg GJ (2000) Incorporation of solar noble gases from a nebula-derived atmosphere during magma ocean cooling. Lunar Planet. Sci. XXX

Yatsevich I, Honda M (1997) Production of nucleogenic neon in the Earth from natural radioactive decay. J Geophys Res 102:10291-10298

Yi W, Halliday AN, Lee D-C, Rehkämper M. (1995) Indium and tin in basalts, sulfide and the mantle. Geochim Cosmochim Acta 59:5081-5090

Zhao D (2001) Seismic structure and origin of hotspots and mantle plumes. Earth Planet Sci Lett 192: 251-265

Zhang Y (1997) Mantle degassing and origin of the atmosphere. Proc. 30th Geol Congr (Beijing) 1:61-78

Zhang Y, Zindler A (1989) Noble gas constraints on the evolution of the Earth's atmosphere. J Geophys Res 94:13719-13737

Zhong S, Gurnis K (1995) Mantle convection with plates and mobile faulted plate margins. Science 267:838-843

Zindler A, Hart SR (1986) Chemical geodynamics. Ann Rev Earth Planet Sci 14:493-571

12

Production, Release and Transport of Noble Gases in the Continental Crust

Chris J. Ballentine

Department of Earth Sciences
The University of Manchester
Manchester, M13 9PL, United Kingdom
cballentine@fs1.ge.man.ac.uk

Pete G. Burnard

Division of Geological and Planetary Sciences
MS 100-23, California Institute of Technology
Pasadena, California 91125

INTRODUCTION

Noble gases within the crust originate from three main sources: the atmosphere, introduced into the crust dissolved in groundwater; the mantle, in regions of magmatic activity; and those produced in the crust by the result of radioactive decay processes. The continental crust contains approximately 40% of the terrestrial radioelements (Rudnick and Fountain 1995) that produce noble gases and, after the mantle and the atmosphere, forms the third major terrestrial noble gas reservoir (neglecting the core). In addition to these sources, contributions from interplanetary dust particles (IDP), cosmic ray interaction with the crustal surface and anthropogenic noble gases can in some cases be a significant source of noble gases in crustal materials. The use of noble gases to understand the role of fluids in different geological settings relies on their low natural abundance and chemical inertness. The low abundance of noble gases in crustal systems and their distinct isotopic character means that contributions from these different sources can often be resolved and quantified. With this, information is gained about the source of associated fluids, the environment from which they originated, the physical manner in which they have been transported to the sampling site and the different phases that may have interacted within the crustal fluid system. This is only possible, however, with a detailed understanding of the processes that control the concentration and isotopic composition of the noble gases in different crustal environments.

The first part of this chapter deals with the three different mechanisms of noble gas production within the crust—radiogenic, nucleogenic, fissiogenic (Fig. 1). We show how production ratios are affected not only by the source region radioelement concentration, but in the case of nucleogenic reactions, also by the spatial distribution and concentration of the target elements. For completeness we consider cosmogenic noble gas production rates and Interplanetary Dust Particle (IDP) accumulation. We consider the total crustal budget and quantify how much of these differently sourced noble gases have, or can, contribute to the crustal fluid system. We then discuss how release from the minerals in which they are produced occurs by recoil, diffusion, fracturing and mineral alteration. The character of the mineral in which they are produced, the release process, the thermal regime and the ability of the surrounding fluid regime to transport the released noble gases, all play a role in fractionating the crustal noble gas elemental pattern and in determining their respective flux from the deeper crust into shallow systems. We detail how and where magmatic noble gases are introduced into the crust and discuss the relationship of ^{4}He and ^{3}He/^{4}He with heat flow.

1529-6466/00/0047-0012$05.00

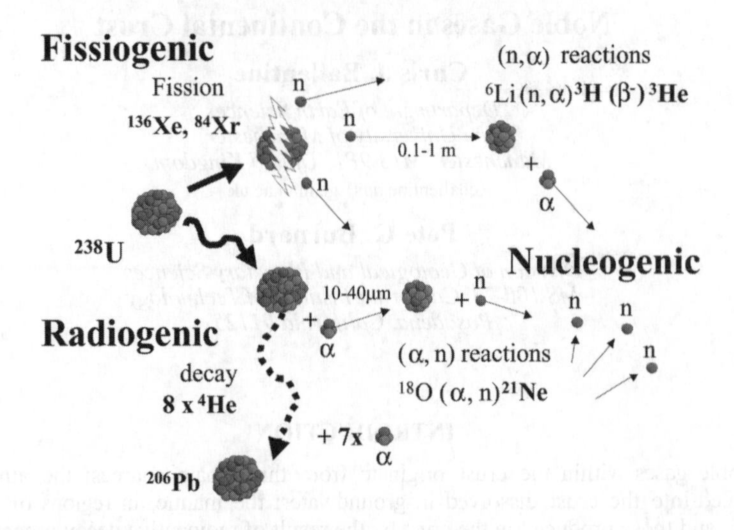

Figure 1. Schematic diagram of ^{238}U decay to ^{206}Pb, showing fission and α decay. Fission results in the direct formation of heavy Kr and Xe isotopes and neutrons. Most α particles stabilise to become ^{4}He, although a small portion react directly with local (0-40 μm) light nuclei in (α,n) reactions to create nucleogenic noble gases such as ^{21}Ne in the reaction $^{18}O(\alpha,n)^{21}Ne$. Less than 20% of the subsurface neutron flux is produced directly by fission (e.g., Yatsevich et al. 1997). The remainder is dominated by production from the (α,n) reactions. The neutrons in turn also react with nuclei on a 0.1-1 m length scale. Typical reactions include the (n,α) route producing, for example, ^{3}He in the reaction $^{6}Li(n,\alpha)^{3}H\ (\beta^{-}) \rightarrow {}^{3}He$. The nuclei available for reaction in the 0-40 μm range of an α-emitter determine the production rate of (α,n)-derived noble gases. The greater length-scale of the neutron penetration distance means that noble gases produced by the (n,α) route are not sensitive to mineral-scale elemental heterogeneity.

RADIOGENIC, NUCLEOGENIC, AND FISSIOGENIC NOBLE GASES

The subsurface neutron flux and reaction probability

Neutron interaction with atomic nuclei is directly linked to the production of several noble gas isotope species in the crust. Before considering the production of crustal noble gases it is essential to first understand the factors controlling the subsurface neutron flux. There are three main types of reaction that produce neutrons in the crust: Cosmic ray interactions; spontaneous fission; and alpha particle interaction with light nuclei. Cosmic ray interaction is only important within the top few meters of the crust (Niedermann 2002, this volume) and we neglect this source of neutrons here.

Neutrons are produced by (α,n) reactions within 0-40 μm of $^{235,238}U$ and ^{232}Th, together with a small contribution from the spontaneous fission of ^{238}U to give:

$$n = (\alpha,n)U + (\alpha,n)Th + (sf,n)U \tag{1}$$

where n is the total number of neutrons and (α,n)U, (α,n)Th, (sf,n)U are the neutrons produced by $^{235,238}U$, ^{232}Th (α,n) reactions, and spontaneous fission of ^{238}U respectively. Spontaneous fission neutron contributions from ^{232}Th and ^{235}U are negligible (Andrews and Kay 1982; Morrison and Pine 1955). The spontaneous fission neutron yield of ^{238}U is 2.2±0.3 neutrons per fission (Morrison and Pine 1955), and the decay constant for spontaneous ^{238}U fission is 8.6×10^{-17} yr^{-1} (Eikenberg et al. 1993). This compares with the ^{232}Th neutron yield of 2.5 neutrons per fission with a fission decay constant estimated to be $<2\times10^{-21}$ yr^{-1} (Wieler and Eikenberg 1999). The spontaneous fission neutron production rate is calculated by:

$$(sf,n) = (N_A/A_U) \times \lambda_f \times \text{yield} \times [U] \times 10^{-6} \quad (2)$$

where N_A = Avogadro's number (6.023×10^{23}), A_U = molar mass of uranium, λ_f = fission decay constant and [U] = the concentration of U in ppm to give

$$(sf,n)U = 0.4788\ [U]\ (g^{-1}\ yr^{-1}) \quad (3)$$

for a natural uranium composition of ^{235}U = 0.72% and ^{238}U = 99.28%. It should be noted that ^{238}U has a very high resonance capture energy at ~10eV, and in the case where these neutrons are produced within accessory phase minerals containing high concentrations of U, such as uraninite and monazite, the number of neutrons available for reaction is only between 0.2 and 0.4 times the amount produced (Gorshkov et al. 1966).

The neutron yield from (α,n) reactions is a function of the energy dependant (α,n) reaction cross section (Feige et al. 1968) and the rate of energy loss of the α-particles and reaction channels with the excited states of residual nuclei (Yatsevich and Honda 1997). The two important parameters determining the reaction cross section are the reaction energy, Q, and the height of the Coulomb barrier, B (Feige et al. 1968). B is defined by:

$$B = 2.8\ Z / (1.4A^{1/3} + 1.2)\ \text{MeV} \quad (4)$$

where Z and A are the atomic number and weight of the target nucleus respectively. The highest energy of a naturally emitted α-particle is 8.78 MeV and Coulomb interactions with the medium electrons will slow the α-particles before reaction takes place. Isotopes for which B > 8 MeV (elements with Z > 16) have a Coulomb barrier above the natural upper limit and are not therefore a significant source of subsurface neutrons. For an $m_1(\alpha,n)m_2$ reaction

$$Q = [(m_1 + m_\alpha) - (m_2 + m_n)]\ c^2 \quad (5)$$

where m_1, m_α, m_2 and m_n are the masses of m_1, the α-particle, m_2 and neutron respectively and c the speed of light. The reaction threshold for an endothermic reaction, E_{th}, is given by:

$$E_{th} = -[(m_1 + m_\alpha) / m_1]\ Q \quad (6)$$

and is the minimum kinetic energy an α-particle must have in order for the reaction to be energetically favorable. Even though the Coulomb barrier is reached, ^{16}O, ^{28}Si, which constitute some 75% of the Earth's crust by weight, as well as ^{40}Ca (also excluded by the Coulomb barrier) and ^{24}Mg do not therefore significantly participate in natural (α,n) reactions.

As demonstrated above, the reaction cross section is dependant on the energy of the reacting α-particle. To calculate the neutron production of a compound, it is necessary to know the neutron yield and the mass stopping ability of each element, the latter of which is also dependant on the α energy. Experimentally derived neutron yields exist for the range of naturally occurring α-particle energies (e.g., Feige et al. 1968; Jacobs and Liskien 1983; West and Sherwood 1982) and can be fitted to a polynomial to allow calculation of the yield at any alpha energy. From the neutron yield of the constituent elements, Y_i, the neutron yield of a particular compound, Y_c, can be calculated from

$$Y_c = \Sigma\ W_i\ S_i^m\ Y_I \quad (7)$$

where W_i and S_i^m are the mass fraction and mass stopping energy for alpha particles of element i. With the data supplied by Feige (1968), and the assumption that neutron contributions from O and Ca can be neglected an empirical U and Th derived neutron flux for an homogenous media can be calculated following:

$$(a,n)U = 0.01\,[U]\,\{13.8[Na] + 5.4[Mg] + 5.0[Al] + 1.31[Si] + 2.0[C]\}\ g^{-1}\,yr^{-1} \quad (8)$$

$$(a,n)Th = 0.01\,[Th]\,\{6.0[Na] + 2.45[Mg] + 2.55[Al] + 0.56[Si] + 0.83[C]\}\ g^{-1}\,yr^{-1} \quad (9)$$

where [U] and [Th] are the respective concentration of U and Th in ppm whereas the concentration terms for Na, Mg, Al, Si and C are their percentage concentrations in the rock. The total subsurface neutron density, N, can be derived by combining Equations (1), (3), (8) and (9) to give

$$N(\text{neutrons } g^{-1}\,yr^{-1}) = 0.01[U]\,\{13.8[Na] + 5.4[Mg] + 5.0[Al] + 1.31[Si] + 2.0[C]\}$$
$$+ 0.01[Th]\,\{6.0[Na] + 2.45[Mg] + 2.55[Al] + 0.56[Si] + 0.83[C]\} + 0.4788\,[U] \quad (10)$$

This is similar to the equation presented by Andrews (1985) but differs in neglecting Ca (see earlier discussion) instead of C as a significant neutron source.

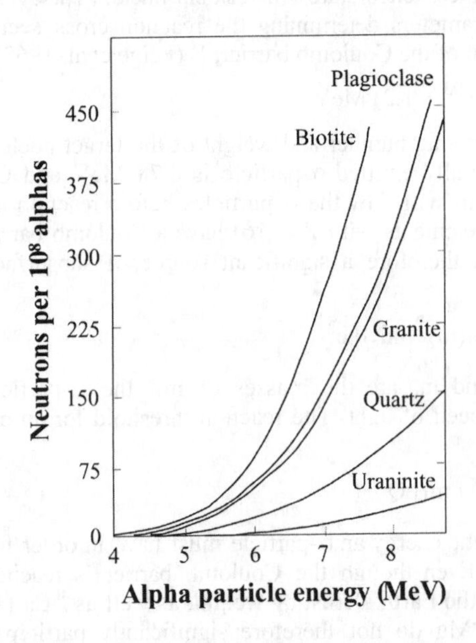

Figure 2. Martel et al. (1990) show the energy dependence of neutron production from (α,n) reactions for selected mineral and rock compositions using the data of Gorshkov et al. (1962). Neutron fluxes calculated assuming element homogeneity, would significantly overestimate the flux if the radioelements are concentrated in accessory phases such as uraninite. [Used by permission of Elsevier, from Martel et al (1990), *Chemical Geology*, Vol. 88, Fig. 5, p. 215.]

Using the light element neutron yield from Feige (1968) and the stopping energy distribution from Ziegler (1977), Martel et al. (1990) have calculated the α–energy-dependent neutron yield for selected mineral and rock compositions (Fig. 2). The results show the dependence of U and Th siting on the neutron flux produced in a rock, also showing that neutron production probability falls to zero at energies of <4.0 MeV. The effective neutron production range is therefore only 0-30 μm in major minerals and reduced to 0-20 μm in accessory phase minerals such as monazite and uraninite. Martel et al. (1990) conclude that estimates of neutron yields based on a homogenous element distribution may significantly overestimate subsurface neutron flux if the bulk of the radioelements are within the accessory phases. The size distribution of the accessory phase is also critical. Figure 3 shows the calculated neutron production rate for spherical accessory phases in a biotite matrix as a function of their radii. As the accessory mineral radius approaches zero the production rate approaches that of biotite. As the mineral radius approaches the limit of the α-particle penetration distance of 20 μm, the neutron flux becomes dominated by the spontaneous fission contribution (Martel et al. 1990).

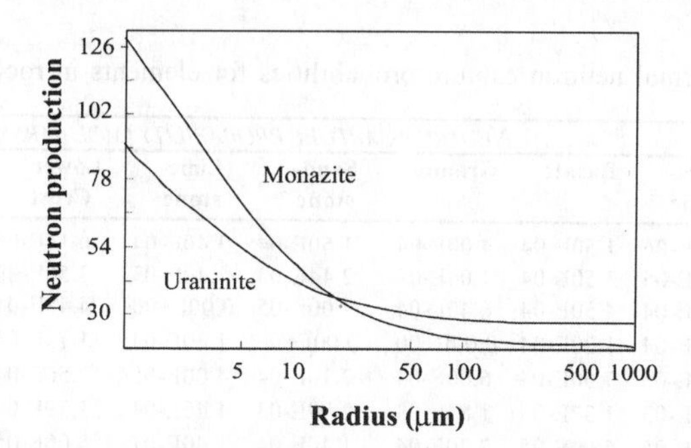

Figure 3. Martel et al. (1990, show the calculated neutron production rate as a function of grain size for spherical uraninite and monazite grains in a biotite matrix. The values are relative to an arbitrary value of 100 assigned to the yield from Al bombarded by mono-energetic 5.3MeV alpha particles. The crossover in the curves at low neutron production rates is the result of the high U/Th ratio in uraninite relative to monazite and thus a larger contribution from spontaneous fission. [Used by permission of Elsevier, from Martel et al (1990), *Chemical Geology*, Vol. 88, Fig. 6, p. 217.]

Comparison of measured *in situ* and calculated neutron fluxes were originally made by Gorshkov et al. (1966) who showed a 20% agreement between measured and predicted values. More recently, Andrews (1986) and Chazal et al. (1998) have compared measured vs. predicted values for the Stripa granite, Sweden, and in the Modane underground laboratory, Switzerland respectively. The calculated flux in the Stripa granite is 15% higher than the measured value, again showing that for this case the assumption of element homogeneity is reasonable. In contrast, Chazal et al. (1998) use a Monte Carlo approach to calculate the neutron production rate, but measure a flux four times greater than calculated values. They ascribe this difference to the difficulty in determining the water content of the rock. The moderating capacity of water (its ability to efficiently absorb neutrons) was neglected in the calculations of Andrews (1986) and may have been overestimated by Chazal et al. (1998).

Neutrons produced by (α,n) reactions have energies typically on the order of several MeV. Through elastic and inelastic scattering the neutron energy is reduced and, after many collisions, thermalizes to energies similar to gas molecules at the ambient temperature (e.g., ~0.03 eV at 25°C). Reaction probabilities for fast neutrons are generally much smaller than those for epithermal or thermal neutron reactions, particularly for light elements. Nevertheless, fast neutron reactions mean that only a portion (p_{th}) of the neutrons reach thermal energies. Morrison and Pine (1955) used an estimate for effective resonance cross section in a chain reacting nuclear pile to obtain $0.6 < p_{th} < 0.9$, with a best estimate of $p_{th} = 0.8$. Other workers argue that the high abundance of light elements (O, Si) in a typical rock matrix means that this may be an underestimate for the rock medium and, given other uncertainties, can be considered to be unity (Andrews 1985). Noble gas production rates involving reactions with thermal neutrons, such as the

$^6Li(n,\alpha)^3H(\beta^-)^3He$ reaction, are relatively simple to calculate for a given neutron flux. The fraction of neutrons in a rock matrix captured by element x is given by the equation:

$$F_x = \sigma_x N_x / \sum_i^n \sigma_i N_i \tag{11}$$

where σ and N are the neutron capture cross sections and abundances in moles, respectively, of all the elements present in the rock. The value of this summation is determined to within 1% by 17 dominant elements for all rock types (Andrews and Kay 1982) (Table 1).

Table 1. Thermal neutron capture probabilities for elements in rock matrices.

		NEUTRON CAPTURE PROBABILITY (MOL BARNS)							
Cross-section	***(barns)***	**Ultra-mafic**	**Basalt**	**Granite**	**Sand-stone**	**Lime-stone**	**Lower* Crust**	**Middle* Crust**	**Upper* Crust**
Li	71	5.00E-06	1.50E-04	4.00E-04	1.50E-04	3.40E-04	6.14E-05	7.16E-05	2.05E-04
B	755	7.00E-05	3.50E-04	1.04E-03	2.44E-03	1.40E-03	3.50E-04	3.50E-04	1.04E-03
Na	0.534	1.30E-04	4.50E-04	6.40E-04	7.00E-05	0.00E+00	4.48E-04	5.51E-04	6.72E-04
Mg	0.064	6.80E-04	1.20E-04	0.00E+00	0.00E+00	1.20E-04	1.13E-04	5.40E-05	3.49E-05
Al	0.232	4.00E-05	7.50E-04	6.60E-04	2.10E-04	4.00E-05	7.56E-04	7.06E-04	6.93E-04
Si	0.16	1.08E-03	1.37E-03	1.84E-03	2.09E-03	1.40E-04	1.39E-03	1.61E-03	1.76E-03
Cl	32.6	5.00E-05	5.00E-05	2.20E-04	1.40E-04	1.40E-04	5.00E-05	5.00E-05	2.20E-04
K	2.07	2.00E-05	4.40E-04	1.76E-03	5.60E-04	1.40E-04	2.64E-04	8.83E-04	1.49E-03
Ca	0.44	8.00E-05	7.40E-04	1.70E-04	4.20E-04	3.32E-03	7.38E-04	4.00E-04	3.30E-04
Ti	6.1	4.00E-05	1.15E-03	2.90E-04	1.90E-04	5.00E-05	6.11E-04	5.35E-04	3.82E-04
Cr	3.1	1.20E-04	1.00E-05	0.00E+00	0.00E+00	0.00E+00	1.28E-05	4.95E-06	2.09E-06
Mn	13.3	3.60E-04	4.80E-04	1.40E-04	1.00E-05	2.70E-04	1.87E-04	1.87E-04	1.50E-04
Fe	2.56	4.50E-04	3.90E-04	1.23E-03	4.40E-04	1.70E-04	2.69E-03	2.05E-03	1.44E-03
Co	37.5	1.30E-04	3.00E-05	0.00E+00	0.00E+00	0.00E+00	2.42E-05	1.59E-05	6.36E-06
Ni	4.54	1.50E-04	1.00E-05	0.00E+00	0.00E+00	0.00E+00	6.81E-06	2.55E-06	1.55E-06
Sm	5820	2.00E-05	1.90E-04	3.40E-04	3.80E-04	5.00E-05	1.08E-04	1.70E-04	1.74E-04
Gd	49000	1.60E-04	1.56E-03	2.80E-03	3.11E-03	4.00E-04	9.66E-04	1.25E-03	1.18E-03
Total capture probability		0.00359	0.00824	0.0115	0.0102	0.00658	0.00878	0.00889	0.00979

Table 1 is after Andrews and Kay (1982).
Elements that do not contribute >1% of the total capture probability for at least one rock type have not been tabulated.
Rock composition used is from Parker (1967).
*Average crust values from Rudnick and Fountain (1995); http://earthref.org/GERM/

The calculation of noble gas production rates involving fast neutron interactions is more complex, and requires a detailed calculation of the energy spectrum of the produced neutrons, and the rate and probability with which they thermalize. For example, the reactions $^{24}Mg(n,\alpha)^{21}Ne$ and $^{25}Mg(n,\alpha)^{22}Ne$ both have reaction energy thresholds in the range 2-4 MeV. Rison (1980) in calculating the Ne isotope yields from these reactions, estimated the neutron spectrum from the element (α,n) reactions as a δ function at the energy (Eα + Q), where Eα is the energy of the impinging alpha particle and Q is the reaction energy release. The number of neutrons was estimated from experimentally determined yields of (α,n) reactions. Yatsevich and Honda (1997) note that this approach does not take into account α-particle energy loss before interaction and ignored reaction channels with the excited states of residual nuclei. Consequently the calculated Ne isotope yields are significantly exaggerated. Yatsevich and Honda (1997) rectify this by estimating the energy spectrum of neutrons from the thick-target angle-integrated spectra of neutrons emitted in (α,n) reactions on light elements measured by Jacobs and Liskien (1983). This data is for the energy interval 4.0-5.5 MeV, and has been extended by fitting the data to a constant form function and incremental extrapolation to 9.0MeV to enable interpolation over the different natural α-energy intervals. Combining this approach with a neutron transport model incorporating both elastic and inelastic scattering plus residual nuclei scattering results in a neutron spectrum with a significantly lower average energy than the earlier work of Rison (1980) (Fig. 4). This in turn results in ^{21}Ne and ^{22}Ne yields by the ^{24}Mg and ^{25}Mg reactions calculated by Yatsevich and Honda (1997) being approximately five times lower than the earlier values of Rison (1980).

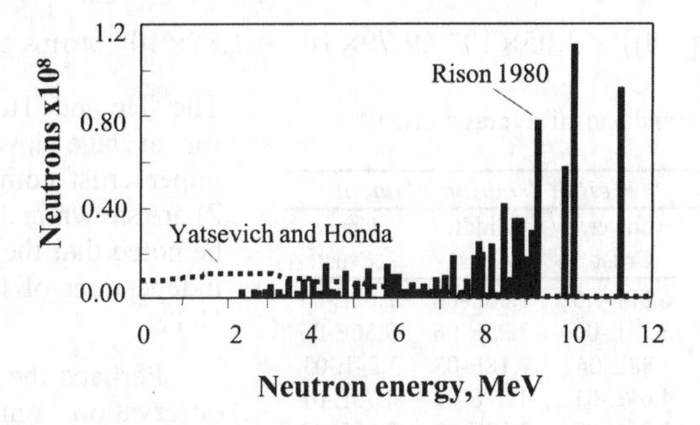

Figure 4. Yatsevich and Honda (1997) calculate the energy spectrum of neutrons produced from U and Th (α,n) reactions in an homogenous mantle over 4.5 Ga. They calculate a lower energy spread than the simplified calculation of Rison (1980) because of the inclusion of the effect of elastic and, to a lesser extent, inelastic scattering of the neutrons. Calculated yields for fast neutron reactions, such as with ^{24,25}Mg that have a 2-4 MeV threshold, are therefore much lower with the revised neutron energy spectrum. [Used by permission of the American Geophysical Union, from Yatsevich and Honda (1997), *Journal of Geophysical Research*, Vol. 104, Fig. 2, p. 10294.]

Helium

The present day ^{4}He production in the crust is dominated by the α-decay of the ^{235,238}U and ^{232}Th decay chains, and is therefore directly proportional to the concentration of these radioelements in the crust. Although some α-particles are produced by a variety of crustal nuclear reactions, these sources of ^{4}He production are many orders of magnitude smaller and can be neglected; similarly α-particles consumed by (α,n) reactions within the crust are a very small proportion of the α-particles produced. The ^{4}He produced from each radio-isotope, R, expressed as a function of present day concentration in the rock is given by

$$^4\text{He atoms g}^{-1}\text{ yr}^{-1} = X_r\,[R]\,(N_A/A_r) \times 10^{-6}\,(e^{\lambda t} - 1) \times \text{yield}_r \tag{12}$$

where X_r = fractional natural abundance of isotope R, N_A = Avogadro's number (6.023×10^{23}), A_r = molar mass of R (g), λ_r = decay constant of R (yr^{-1}), yield = number of α particles emitted in the complete decay chain, [R] = the concentration of R in ppm and t = age (yr). For ^{238}U, ^{235}U and ^{232}Th, $\lambda_{238} = 1.55 \times 10^{-10}$, $\lambda_{235} = 9.85 \times 10^{-10}$, $\lambda_{232} = 4.95 \times 10^{-11}$ (Steiger and Jäger 1977), with an α yield in each decay chain of 8, 7 and 6 and a natural abundance of $X_{238} = 0.9928$, $X_{235} = 0.0072$ and $X_{232} = 1.000$ respectively, the number of atoms of ^{4}He produced in 1 gram of rock per year becomes

$$^4\text{He atoms g}^{-1}\text{ yr}^{-1} = (3.115 \times 10^6 + 1.272 \times 10^5)\,[U] + 7.710 \times 10^5\,[Th] \tag{13}$$

^{3}He production within the crust is dominated by thermal neutron capture by ^{6}Li in the reaction ^{6}Li(n,α)^{3}H (β$^-$)^{3}He. Other reactions are detailed in Mamyrin and Tolstikhin (1984, p. 101), but for average crustal compositions, these have a ^{3}He yield at least four orders of magnitude lower than the ^{6}Li route and are neglected here. The rate of ^{3}He production therefore, is directly proportional to the subsurface thermal neutron density (Eqn. 10), the number of neutrons that reach thermal energy level (P_{th} = 0.8) and the relative capture cross section of Li (Eqn. 11). Taking the composition of a homogenous average upper crust (Table 2) as a working example, we obtain:

$$^3\text{He} = 0.8 \times \{0.01[2.8]\,(13.8[2.89] + 5.4[1.33] + 5[8.08] + 1.31[30.9] + 2[0.324]) + 0.01[10.7]\,(6.0[2.89] + 2.45[1.33] + 2.55[8.08] + 0.56[30.9] + 0.83[0.324]) + 0.4788\,[2.8]\} \times 2.05\times 10^{-4} / 9.79\times 10^{-3} = 1.87\times 10^{-1}\ \text{atoms g}^{-1}\,\text{yr}^{-1} \qquad (14)$$

Table 2. Composition of average crust.*

		Weight fraction element		
Element	Atomic Mass	Lower Crust	Middle Crust	Upper Crust
Li	6.94	6.00E-06	7.00E-06	2.00E-05
B	10.81	3.71E-06	3.22E-06	9.56E-06
C	12	5.88E-04	7.18E-03	3.24E-03
O	16	4.69E-01	4.87E-01	4.75E-01
Na	22.99	1.93E-02	2.37E-02	2.89E-02
Mg	24.31	4.28E-02	2.05E-02	1.33E-02
Al	26.92	8.78E-02	8.19E-02	8.04E-02
Si	28.09	2.44E-01	2.83E-01	3.09E-01
Cl	35.45	5.28E-05	5.28E-05	2.32E-04
K	39.1	4.98E-03	1.67E-02	2.82E-02
Ca	40.078	6.72E-02	3.64E-02	3.00E-02
Ti	47.87	4.79E-03	4.20E-03	3.00E-03
Cr	52	2.15E-04	8.30E-05	3.50E-05
Mn	54.94	7.74E-04	7.74E-04	6.20E-04
Fe	55.845	5.88E-02	4.48E-02	3.15E-02
Co	58.93	3.80E-05	2.50E-05	1.00E-05
Ni	58.69	8.80E-05	3.30E-05	2.00E-05
Sm	150.36	2.80E-06	4.40E-06	4.50E-06
Gd	157.25	3.10E-06	4.00E-06	3.80E-06
U	238.03	2.00E-07	1.60E-06	2.80E-06
Th	232.04	1.20E-06	6.10E-06	1.07E-05

*Rudnick and Fountain (1995); Parker (1967)
Wederpohl 1995; Gao et al. 1997; http://earthref.org/GERM/

Table 3. Neutron production, Li capture probability, and ^{3}He/^{4}He in average crust.

	Lower Crust	***Middle Crust***	***Upper Crust***
P_{th}	0.8	0.8	0.8
U(α,n) neutrons/g/yr	2.51E-01	1.95E+00	3.58E+00
Th(α,n) neutrons/g/yr	6.98E-01	3.42E+00	6.25E+00
Uf neutrons/g/yr	9.58E-02	7.66E-01	1.34E+00
F_{6Li}	6.99E-03	8.05E-03	2.09E-02
^{3}He atoms/g/yr	5.84E-03	3.95E-02	1.87E-01
^{4}He atoms/g/yr	1.57E+06	9.89E+06	1.73E+07
^{3}He/^{4}He	3.71E-09	3.99E-09	1.08E-08

The ^{3}He and ^{4}He production rate for average lower, middle and upper crust compositions (Table 2) are shown in Table 3. It should be noted that the ^{3}He/^{4}He ratio is independent of [U] for constant U/Th.

Perhaps the most interesting observation from these calculations is the low value of ^{3}He/^{4}He, even for the Li enriched upper crust. When we consider that the effect of radioelement heterogeneity (the preferential siting of U and Th in accessory phase minerals) would be to significantly reduce the neutron flux producing the ^{3}He (Figs. 2 and 3), these ^{3}He/^{4}He values must be considered to be upper limits for radiogenic production in average crust. Comparison with measured ^{3}He/^{4}He in whole rock samples, mineral separates and associated fluids is non-trivial. This is due to several factors:

(i) calibration and measurement of ^{3}He/^{4}He ratios $<10^{-8}$ is subject to significant error; (ii) sources of non-crustal ^{3}He, such as mantle or cosmogenic ^{3}He, can have a significant impact on the ^{3}He/^{4}He of a system; (iii) the siting and formation of ^{3}He and ^{4}He is different; and (iv) preferential release from different minerals or sites may lead to fractionation of the two isotopes.

A summary of measured and calculated values, both on whole

rocks and mineral separates is detailed in Mamyrin and Tolstikhin (1984) and more recently in Tolstikhin et al. (1996). These data compilations clearly show that a mantle ^3He signature (e.g., Graham 2002) can perturb the crustal $^3He/^4He$ even in ancient rocks (>2.8 Ga) by orders of magnitude. In samples where mantle and cosmogenic ^3He contributions to sedimentary rocks are discounted, variations from predicted ratios by a factor of three are ascribed to fractional release, with one exception, anhydrite, apparently requiring a fractionation factor of 20 (Tolstikhin et al. 1996).

Table 4. Selected measured and calculated $^3He/^4He$ in granites.

Sample ID	*Location*	*Concentration (ppm)* Li	U	Th	*$^3He/^4He$ Measured*	*$^3He/^4He$ Calc**	*$^3He/^4He$ Calc***	Ref
WR	Rapakiwi	36	6.7	35	1.60E-08	1.60E-08		1
Orthoclase	Rapakiwi	6	2.5	2.7	4.00E-07			1
Biotite	Rapakiwi	620	7.2	4.8	1.20E-07			1
Plagioclase	Rapakiwi	10	5.9	5.3	5.00E-08			1
Quartz	Rapakiwi	12	5.1	3.0	1.50E-07			1
Amphibole	Rapakiwi	37	3.5	12.7	4.50E-08			1
Zircon	Rapakiwi		990	150	2.00E-08			1
1-WR	Ukraine	37	4.8	44	1.60E-08	1.60E-08		2
2-WR	Ukraine	38	9.5	42	1.90E-08	1.70E-08		2
4-WR	Ukraine	26	11	85.4	3.00E-09	1.40E-08		2
43-WR	Tuva	10	1.6	10	8.00E-09	7.00E-09		2
27-WR	Caucasus	17	4.5	14	6.00E-08	9.00E-09		2
ss1-WR	Carnmenellis, UK	353	11.8	15.1	2.10E-09	8.00E-08	2.00E-08	3
ss1-fldspar	Carnmenellis, UK				2.38E-09			3
ss1-mica	Carnmenellis, UK				4.62E-09			3
ss1-quartz	Carnmenellis, UK				8.40E-10			3
rh11g-wr	Carnmenellis, UK	386	16.4	8.4	2.10E-09	6.40E-08	1.80E-08	3
rh12d-wr1	Carnmenellis, UK	300	10.2	6.6	1.54E-09	6.70E-08	1.80E-08	3
rh12dwr2	Carnmenellis, UK				4.90E-09	6.70E-08	1.80E-08	3
urananite1	Carnmenellis, UK				9.80E-10			3
Water	Carnmenellis, UK				3.50E-08			3,4
WR	Stripa, Sweden	11	44.1	33		5.94E-09		5,6
Water	Stripa, Sweden				5.70E-09			5

* Assuming element homogeneity WR = Whole rock

** Assuming radioelements concentrated in uraninite grains with ϕ>100μm

1) Gerling et al. (1971) (in Mamyrin and Tolstikhin 1984)
2) Tolstikhin and Drubetskoy (1977) (in Mamyrin and Tolstikhin 1984)
3) Martel et al. (1990)
4) Hilton et al. (1985)
5) Andrews et al. (1989a)
6) Andrews et al.

Granite systems provide an example where mantle influences appear to be minimal, and cosmogenic effects can be safely ruled out. This allows the effect of heterogeneity and release effects to be assessed. In Table 4 we have compiled selected $^3He/^4He$ data from different granite systems that also have theoretical values for comparison. With the exception of sample 27-WR, all measured whole rock values are either within error of calculated $^3He/^4He$ values or significantly lower. Two important observations can be made: i) more than one granite whole rock sample matches both measured and calculated $^3He/^4He$; and ii) the Stripa granite, which is the only system where the calculated neutron flux is confirmed with a measured neutron flux (Andrews et al. 1986), contains groundwater with $^3He/^4He$ indistinguishable from predicted production (Andrews et al. 1989a;b). Although a small data set, an important inference is that many systems display no resolvable fractionation from theoretical values.

Where fractionation is observed, the 'nugget' effect must be first ruled out. This is when a whole rock sample contains an accessory mineral concentration in excess of the average rock. Accessory phases, such as uraninite, contain high concentrations of He with low $^{3}He/^{4}He$ (Table 4) and will result in a $^{3}He/^{4}He$ lower than the average bulk rock. In the case of the Carnmenellis granite (Martel et al. 1990) this cause can be ruled out by multiple whole rock analyses giving the same $^{3}He/^{4}He$ value. If the measured uraninite size of ~50-μm radius is representative of the radioelement distribution, this rock would be expected to produce $^{3}He/^{4}He = 1\times10^{-8}$. Nevertheless, measured whole rock and mineral separate analyses are a factor of three too low to be accounted for by radioelement heterogeneity reduced flux alone (Table 4). An additional factor of three to five is required and is probably release related (Martel et al. 1990). A complementary ^{3}He-enriched fluid would be expected to be observed. This is indeed seen in the associated groundwater (Table 4). It would appear that the order of magnitude difference between the whole rock and groundwater $^{3}He/^{4}He$ in the Carnmenellis granite can be ascribed to a combination of radioelement heterogeneity and preferential release of ^{3}He into the surrounding fluid system. If however, the average uraninite grain size is an order of magnitude smaller, all of the difference between whole rock and groundwater $^{3}He/^{4}He$ has to be ascribed to preferential ^{3}He release into the surrounding groundwater.

In summary, the effect of radioelement heterogeneity on $^{3}He/^{4}He$ in most systems is probably small. There is some evidence from the Carnmenellis system that fractionation of $^{3}He/^{4}He$ due to preferential ^{3}He release could be as high as a factor of 10, but taking into account the evidence for radioelement element heterogeneity this is more reasonably, at most, a factor of 3 (e.g., Tolstikhin et al. 1996). In large crustal fluid systems reasonably sampling 'average' crust, $^{3}He/^{4}He$ ratios in excess of $1\text{-}3\times10^{-8}$ (i.e., more than three times the upper crust value) are due to a resolvable ^{3}He excess from sources external to the crust (e.g., Marty et al 1993).

Neon

The production of Ne isotopes in the crust is entirely due to nucleogenic routes. Recognized by (Wetherill 1954), the only significant production routes are $^{17,18}O(\alpha,n)^{20,21}Ne$, $^{19}F(\alpha,n)^{22}Na(\beta^{+})^{22}Ne$, $^{24,25}Mg(n,\alpha)^{21,22}Ne$, $^{23}Na(n,\alpha)^{20}Ne$, $^{19}F(\alpha,p)^{22}Ne$ (Yatsevich and Honda 1997). Their rate of production is therefore related to radioelement and target-element concentrations as well as the distribution of the target element with respect to any radioelement heterogeneity.

The first investigations of Ne production rates in the crust include the work by Sharif-Zade et al. (1972), Shukolyukov et al. (1973) and Verkhovskiy and Shukolyukov (1976a,b). In addition to (α,n) production, Rison (1980) also investigated (n,α) production rates. This work however, used oxygen (α,n) yields from Feige (1968) which results in a $^{20,21}Ne$ yield twice as high as more recent measurements of the (α,n) yields by West and Sherwood (1982). More recently it has also been shown by Yatsevich and Honda (1997) that the neutron yields from the (n,α) routes calculated by Rison (1980) are also too high by a factor of five because of an overestimation of the neutron energy spectrum (Fig. 4). Using the revised (α,n) yields from West and Sherwood (1982), Hünemohr (1989) and Yatsevich and Honda (1997) have calculated the Ne (α,n) yields for average crust and mantle materials. Using only the reaction cross sections Leya and Wieler (1999) have also estimated Ne production in average crust. These production rates are summarized in Table 5. Yatsevich and Honda (1997) show that with the updated Mg (n,α) yields, production by this route accounts for <0.13% of the ^{21}Ne and ^{22}Ne produced in the crust. In the mantle this production mechanism is more important, with some 3.57% and 65.35% of ^{21}Ne and ^{22}Ne produced by Mg (n,α) reactions, respectively.

Table 5. Comparison of nucleogenic Ne production calculations.

Isotope	Reference	Production over 4.5 Ga cm^3 STP/g		Present-day production rate cm^3 STP/g year	
		Mantle	Crust	Mantle	Crust
^{20}Ne	Leya and Wieler, 1999	1.10E-13	1.16E-11	1.42E-23	1.49E-21
	Yatsevich and Honda, 1997	1.35E-13	1.45E-11	1.74E-23	1.86E-21
	Hunemohr, 1989	1.45E-13	1.54E-11	1.87E-23	1.98E-21
	Kyser and Rison, 1982	2.50E-13	2.70E-11	3.40E-23	3.70E-21
^{21}Ne	Leya and Wieler, 1999	1.15E-12	1.22E-10	1.48E-22	1.56E-20
	Yatsevich and Honda, 1997	1.45E-12	1.50E-10	1.86E-22	1.92E-20
	Hunemohr, 1989	1.50E-12	1.59E-10	1.92E-22	2.04E-20
	Kyser and Rison, 1982	3.20E-12	3.00E-10	4.20E-22	4.00E-20
^{22}Ne	Leya and Wieler, 1999	1.13E-14	4.05E-11	1.48E-24	5.23E-21
	Yatsevich and Honda, 1997	3.02E-14	4.10E-11	3.98E-24	5.23E-21
	Hunemohr, 1989	1.17E-14	4.58E-11	1.50E-24	5.87E-21
	Kyser and Rison, 1982	1.00E-13	7.50E-13	1.30E-23	1.20E-22
^{4}He	Equation (13)			4.13E-15	4.13E-13

Although Hünemohr (1989) and Leya and Wieler (1999) do not consider neutron production routes, because these reactions are insignificant in the crust, this enables a direct assessment of the different Ne production rate estimates. The values given by Hünemohr are ~7% higher than those by Yatsevich and Honda, while the values of Leya and Wieler are ~25% lower. The marked difference between the approach of Leya and Wieler (1999) and the other workers was in the adoption of cross section data rather than thick target yields and is independent of limiting assumptions such as stopping ratios. Nevertheless, this comparison provides the level of uncertainty that should be used when applying these theoretical production rates.

Empirical production rates for variable U, Th and rock composition can be derived from the absolute rates for the different reactions in Yatsevich and Honda (1997) and the proportion ascribed to U relative to Th in Hünemohr (1989, p105) to give:

$$^{20}\text{Ne} = \frac{1.74\times10^{-23}}{0.02}\times\left(\frac{4.48}{6.1}[\text{U}]+\frac{1.62}{6.1}\frac{[\text{Th}]}{3}\right)\left(0.9978\frac{[\text{O}]}{44}+0.0022\frac{[\text{Na}]}{1}\right)$$

$$= (6.39[\text{U}] + 0.770[\text{Th}])(0.0226[\text{O}] + 0.0022[\text{Na}]) \times 10^{-22} \qquad (15)$$

$$^{21}\text{Ne} = \frac{1.86\times10^{-22}}{0.02}\times\left(\begin{array}{l} 0.9646\dfrac{[\text{O}]}{44}\left(\dfrac{45}{62}[\text{U}]+\dfrac{17}{62}\dfrac{[\text{Th}]}{3}\right)+ \\ 0.0357\dfrac{[\text{Mg}]}{21}\left(\dfrac{11.7}{17.7}[\text{U}]+\dfrac{6.0}{17.7}\dfrac{[\text{Th}]}{3}\right)\end{array}\right)$$

$$= \{(1.48[\text{U}] + 0.186[\text{Th}])[\text{O}] + (0.105[\text{U}] + 0.0179[\text{Th}])[\text{Mg}]\} \times 10^{-22} \qquad (16)$$

$$^{22}\mathrm{Ne} = \frac{3.98\times10^{-24}}{0.02}\times\left(\begin{array}{l}0.3465\dfrac{[\mathrm{F}]}{16}\left(\dfrac{0.340}{0.479}[\mathrm{U}]+\dfrac{0.139}{0.479}\dfrac{[\mathrm{Th}]}{3}\right)\\ +0.6535\dfrac{[\mathrm{Mg}]}{21}\left(\dfrac{2.75}{4.05}[\mathrm{U}]+\dfrac{1.30}{4.05}\dfrac{[\mathrm{Th}]}{3}\right)\end{array}\right)$$

$$= \{(3.06[\mathrm{U}] + 0.417[\mathrm{Th}])[\mathrm{F}] + (4.20[\mathrm{U}] + 0.663[\mathrm{Th}])[\mathrm{Mg}]\} \times 10^{-24} \quad (17)$$

Production rates in cm^3 STP g^{-1} yr^{-1}; U, Th and F concentrations in ppm, O, Mg and Na in wt %. Initial constants have been reduced as little as possible for the purpose of transparency with the original references. It should be noted that these equations neglect the effect of variable stopping power and should be treated with caution when applied to systems with an average atomic mass significantly different from average mantle or crust. Comparison of relative Ne production rates will be less sensitive to density variables. In addition, Leya and Wieler (1999) have shown that if you lower the O-concentration then you typically increase the stopping power and increase the production within this mineral. The net result is that the $^4He/^{21}Ne$ production ratios show no strong dependence on the target chemistry because both effects typically cancel out.

Hünemohr (1989), and later Eikenberg et al. (1993), have measured both the elemental composition of various accessory phase minerals and the nucleogenic ^{21}Ne and ^{22}Ne concentration trapped within the minerals (Table 6). Because of the small contribution from Mg in these minerals the predicted $^{21}Ne/^{22}Ne$ is shown to be roughly proportional to the O/F elemental ratio. A comparison of $^{21}Ne/^{22}Ne$ calculated from the measured O and F composition (Table 6) and using Equations (16) and (17) essentially reproduces the observed $^{21}Ne/^{22}Ne$ in these minerals (Fig. 5). Clearly on a mineral scale the relative rates of ^{21}Ne, ^{22}Ne and therefore ^{20}Ne production are well constrained theoretically and confirmed experimentally.

Ne isotopes in crustal fluids. The Ne isotope data recorded in natural gases and brines, by mass balance, reasonably preserve a record of regional crustal systems. The first studies of Ne isotopes in natural gases observed $^{21}Ne/^{22}Ne$ ratios below those found in local radioactive minerals. This was interpreted to be due to preferential thermal release of ^{22}Ne from the producing mineral (Shukolyukov et al. 1973). More recently Kennedy et al. (1990) showed that Ne isotopes in natural gases defined a crustal source with $^{21}Ne/^{22}Ne = 0.469\pm0.006$, consistent with an O/F ratio of 113 (Fig. 6). This is much lower than average crustal values of O/F = 752 or crystalline shield values of O/F = 1140, that would produce crustal $^{21}Ne/^{22}Ne$ values of 3.51 and 5.18 respectively. After assessing the potential of either regionally low O/F ratios or an underestimation of ^{22}Ne production rates, Kennedy et al. (1990) concluded that these values in fact reflect the average O/F ratio of the mineral environment in which the U and Th are sited, rather than average crustal values. The uniformity of the $^{21}Ne/^{22}Ne$ production ratio from a variety of locations lead these workers to further conclude that the source of the 'unusual' O/F ratios necessitated a common and widely distributed mineral suite such as micas and amphiboles. It was argued that other U-Th-F-rich minerals such as apatite and fluorite are more commonly found in vein deposits, and thus more likely to be inhomogeneously distributed in the crust.

A data inversion based on He and Ne isotopic mixing between crust and magmatic end-members was used to resolve an independent crustal $^{21}Ne/^{22}Ne$ production ratio of 0.57 ± 0.1 in a geographically more diverse sample suite (Ballentine 1997). These natural gases, from a variety of extensional environments, are assumed to mix with a sub-continental lithospheric mantle component with $^3He/^4He$ of 6.0 Ra (where Ra is the atmospheric $^3He/^4He$ ratio). The use of isotopes alone removes the possibility of elemental fractionation effects once the mantle fluids have mixed with the crustal gases. The resolved crustal $^{21}Ne/^{22}Ne$ production ratio lends further support to the uniformity of the mineral scale relationship between O, F and U+Th in the crust established by Kennedy et al.(1990).

Table 6. Comparison of measured and calculated $^{21}Ne/^{22}Ne$ production in accesory phase minerals.

Sample	*U ppm*	*Th ppm*	*U/Th*	*O %*	*Mg %*	*F ppm*
Bastnäsite	150	2600	0.0577	26.6	0	73500
Thorite	600	283000	0.0021	29.8	0	18600
Fregusonite	150	600	0.2500	40.4	0.09	14200
Xenotime	1000	16000	0.0625	46.5	1.58	2980
Priorite	72000	9500	7.58	27.5	0	1360
Euxenite	58000	16000	3.63	31.2	0	1100
Euxenite	41500	12000	3.46	31.3	0	910
Thorite	10000	174000	0.0575	48.9	0.11	540
Monazite	1400	32000	0.0438	27.7	0	340
Gololinite	900	2300	0.3913	35.5	0.013	210
Samarskite	63000	7400	8.51	23.4	0	180
Monazite	800	40000	0.0200	29.2	0.01	90

Sample	*$^{21}Ne/^{22}Ne$ Calc*	*$^{21}Ne/^{22}Ne$ Meas*	*error*	*Meas/Calc*
Bastnäsite	0.0171	0.0091	0.0025	0.53
Thorite	0.0758	0.1039	0.0266	1.37
Fregusonite	0.1346	0.1013	0.0265	0.75
Xenotime	0.7395	0.81	0.23	1.10
Priorite	0.9566	0.88	0.23	0.92
Euxenite	1.3418	0.82	0.21	0.61
Euxenite	1.6272	1.00	0.26	0.61
Thorite	4.2835	6.56	1.99	1.53
Monazite	3.8542	5.05	1.33	1.31
Gololinite	7.9969	5.2	1.4	0.65
Samarskite	6.1501	13.0	3.4	2.11
Monazite	15.347	7.7	2.3	0.50
			Average	1.00 ±0.1433

Composition and $^{21}Ne/^{22}Ne$ analysis of U- and Th-rich minerals by Hühemohr (1989).
Calculated $^{21}Ne/^{22}Ne$ using Equations (16) and (17).

Because of the incompatibility of F, the siting of this element in the crust has been assumed to be in accessory phase minerals. The close relationship between F and U+Th suggests that a significant portion of the crust U+Th budget must also be sited in the same accessory phase minerals. Neither Kennedy et al. (1990) or Ballentine (1997) considered preferential ^{22}Ne release. However, $^{4}He/^{40}Ar$ ratios are sensitive to thermal fractionation and release effects (e.g., Ballentine et al. 1994). Natural gases that show no fractionation from predicted crustal $^{4}He/^{40}Ar$ ratios and relatively uniform crustal $^{4}He/^{21}Ne$ still show crustal $^{21}Ne/^{22}Ne$ production rates below those predicted for average crust (see Fig. 7 for sample data sets). $^{21}Ne/^{22}Ne$ production below average crust is probably not due to excess ^{22}Ne from release fractionation effects, but is due to radioelement siting and target element heterogeneity in the crust.

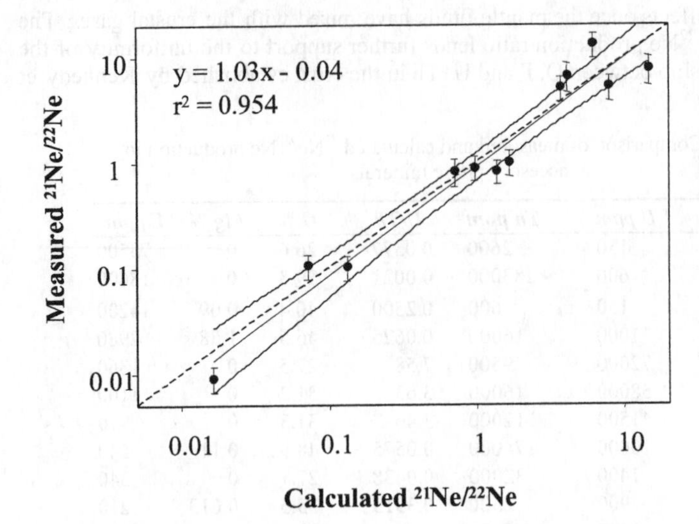

Figure 5. Comparison of values measured by Hünemohr (1989) in U- and Th-rich minerals and those calculated using the Equations (16) and (17). The outer lines show the 95% confidence limit of the fit. The dashed line is the ideal fit with a gradient of 1.

Figure 6. $^{20}Ne/^{22}Ne$ vs $^{21}Ne/^{22}Ne$ in crustal fluids after Kennedy et al. (1990) showing significant deviation from the crustal $^{21}Ne/^{22}Ne$ mixing line predicted for an average crustal O/F ratio of 752. Data from the Alberta Basin and New Mexico (Kennedy et al. 1990), Brines from the Canadian shield (Bottomley et al. 1984) and helium-rich gases from the SW USA (Emerson et al. 1966). The two arrows show vectors predicted for mass fractionation (mfl) and mantle fluid admixtures (mantle) that probably account for the scatter observed.

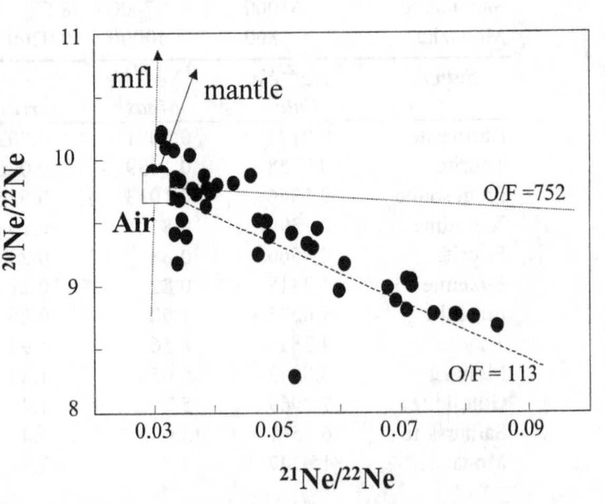

Several workers have noted the possible underestimate of ^{22}Ne production in the mantle because of similar accessory phase associations of U+Th with F (Barfod 1999; Niedermann et al. 1997). Taking the average of the two crustal studies discussed above and an error that encompasses both values in a T-test, we recommend for average crustal calculations a $^{21}Ne/^{22}Ne$ crustal production ratio of 0.52±0.04. ^{20}Ne production is dominated, like ^{21}Ne, by reactions with oxygen. The $^{21}Ne/^{20}Ne$ production ratios will not therefore affected by element heterogeneity.

Observed ^{4}***He/***^{21}***Ne.*** An important question is whether the observed production rate of Ne isotopes, relative to ^{4}He, is also in good agreement with the theoretical calculations. Since ^{21}Ne is largely dependent on the oxygen concentration, which is ubiquitous in the crust, the $^{4}He/^{21}Ne$ ratio should be the least affected by element heterogeneity. Hünemohr (1989) measured the $^{4}He/^{21}Ne$ ratio produced in the same mineral suite as the Ne-isotope study discussed above. Results from these measurements for individual minerals ranged from predicted production values to $^{4}He/^{21}Ne$ values an order of magnitude lower. This is attributed to variable but preferential ^{4}He loss from some samples.

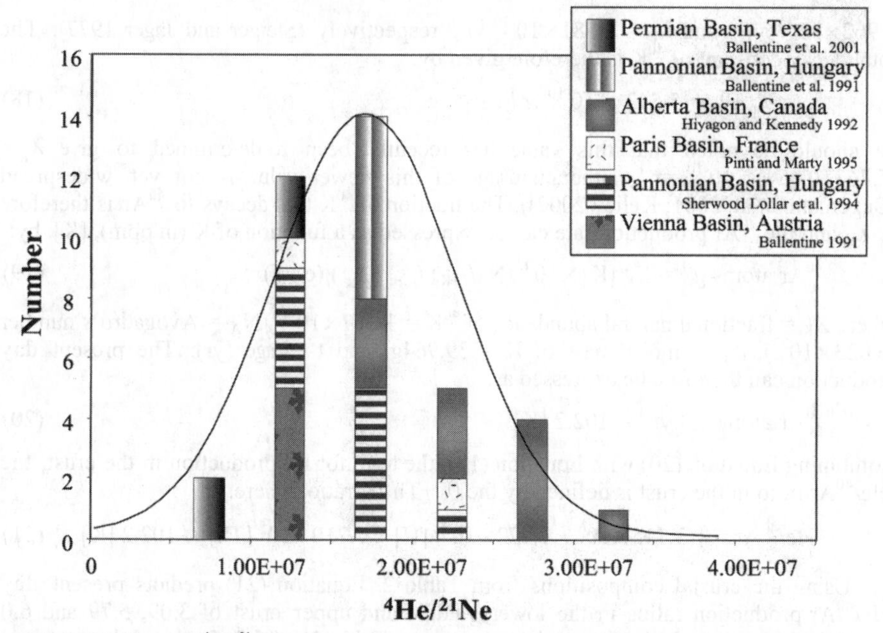

Figure 7. Observed $^4He/^{21}Ne^*$ production ratio in the crust. We have taken all natural gas data available with $^4He/^{40}Ar^*$ ratios in the range 3-6 and with $^3He/^4He < 0.6Ra$ to remove bias due to elemental fractionation or mantle components. The mean from 38 samples is $1.71 \pm 0.09 \times 10^7$ (1 standard error). This compares with theoretical estimates of between $2.02\text{-}2.64 \times 10^7$ (Table 5).

Another approach is to investigate the noble gas composition of natural gases and other crustal fluids that have collected noble gases from large crustal volumes. Many samples however, show significant elemental fractionation (e.g., Ballentine et al. 1991). Fractionation may occur by preferential release of noble gases from the producing minerals, transport effects such as diffusion, or water-oil-gas phase equilibrium fractionation. Although these processes have little effect on isotopic ratios, they provide the limiting factor in a comparison of elemental ratios and are not considered in earlier estimates of measured $^4He/^{21}Ne$ (Verchovskiy et al. 1977; Tolstikhin 1978; Kennedy et al. 1990).

Other noble gases however, provide a sensitive indicator of fractionation. Making the assumption that the present day average crustal production $^4He/^{40}Ar$ ratio is ~5 (next section), we have compiled in Figure 7 all the available data from regional crustal fluid systems for which crustal $^4He/^{40}Ar = 5.0 \pm 1.0$. We have also excluded samples with magmatic He contributions greater than ~10% ($^3He/^4He > 0.6Ra$). Data without full He, Ne and Ar abundance and isotopic information are not included. No correlation is observed between $^4He/^{40}Ar$ and $^4He/^{21}Ne$ in this filtered data set. The data distribution approximates a Gaussian distribution with a mean observed crustal $^4He/^{21}Ne$ production rate of $1.71 \pm 0.09 \times 10^7$. This compares with the theoretical estimates for average crust of between $2.02\text{-}2.64 \times 10^7$ (Table 5). At present it is unclear why the crustal record preserves a ^{21}Ne production rate 15-35% greater than that the theoretical calculations would predict.

Argon

^{40}Ar production in the crust is dominated by the decay of ^{40}K, and is therefore directly proportional to the K concentration. ^{40}K has a branched decay mode, producing ^{40}Ca by beta decay and ^{40}Ar by electron capture, with decay constants of $\lambda_\beta =$

4.962×10^{-10} yr^{-1} and $\lambda_e = 0.581\times10^{-10}$ yr^{-1} respectively (Steiger and Jäger 1977). The total decay constant of ^{40}K is therefore given by:

$$\lambda_K = \lambda_\beta + \lambda_e = 5.543 \times 10^{-10}\ \text{yr}^{-1} \tag{18}$$

(It should be noted that this value has recently been re-determined to give $\lambda_K = (5.463\pm0.054)\times10^{-10}$ yr^{-1}, although usage of this newer value is not yet widespread (Begemann et al. 2001; Kelley 2002)). The fraction of ^{40}K that decays to ^{40}Ar is therefore λ_e/λ_K and the ^{40}Ar production rate can be expressed as a function of K (in ppm), [K], by:

$$^{40}\text{Ar atoms g}^{-1} = X_K\,[K]\times10^{-6}\,(N_A/A_K)\,(\lambda_e/\lambda_K)\,(e^{\lambda t}-1) \tag{19}$$

where X_K = fractional natural abundance of ^{40}K = 1.167×10^{-4}, N_A = Avogadro's number (6.023×10^{23}), A_K = molar mass of K = 39.964g, and t = age (yr). The present day production can therefore be expressed as:

$$^{40}\text{Ar atoms g}^{-1}\ \text{yr}^{-1} = 102.2\ [K] \tag{20}$$

Combining Equation (20) with Equation (13), the term for He production in the crust, the ^{4}He/^{40}Ar ratio in the crust is defined by the (U+Th)/K ratio, where:

$$^4\text{He}/^{40}\text{Ar} = \{(3.115\times10^6 + 1.272\times10^5)\,[U] + 7.710\times10^5\,[Th]\}\,/\,102.2\,[K] \tag{21}$$

Using the crustal compositions from Table 2, Equation (21) predicts present day ^{4}He/^{40}Ar production ratios in the lower, middle and upper crust of 3.09, 5.79 and 6.0 respectively, with a production weighted average (Table 2) of 5.7. In principle measured crustal ^{4}He/^{40}Ar ratios can provide a test of these K/U ratios (Dymond and Hogan 1973), an issue that has recently provided significant controversy (e.g., Albarède 1998). In practice ^{4}He/^{40}Ar ratios are sensitive to fractionation during thermal release from their respective mineral sites (Ballentine et al. 1994; Mamyrin and Tolstikhin 1984), as well as subsequent transport and phase related fractionation processes (Ballentine et al. 1991). We have adopted in the Ne section above the assumption that ^{4}He/^{40}Ar ratios within the range 4 to 6 are relatively unaffected by these process to filter out fractionated samples, but with the implicit assumption that the K/U ratio of the crust is well defined. We discuss later the information about release, transport and phase fractionation available by considering deviations from predicted crustal values. Also see Ballentine et al. (2002).

^{36}Ar production in the crust is small compared to the ambient background of atmosphere-derived ^{36}Ar introduced into the crust dissolved in groundwater and is usually neglected. Although a small amount of muon-induced ^{36}Ar occurs close to the surface, the principle route of production is the β-decay of ^{36}Cl (Fontes et al. 1991; Hünemohr 1989). ^{36}Cl has a half-life of 3.01×10^5 yr and decays to ^{36}Ar with a branching ratio, R, of 0.95. ^{36}Cl is only produced in the crust by the thermal neutron reaction ^{35}Cl(n,γ)^{36}Cl (Bentley et al. 1986; Fontes et al. 1991). The fraction of thermal neutrons captured by any one element, F_i, is given by Equation (11). It should be noted that the reaction cross section given for Cl in Table 1 is the combined probability for ^{35}Cl and ^{37}Cl with thermal neutron capture cross sections and relative abundances of 43 and 0.43 barns and 75.77% and 24.23% respectively. These values give F_{35Cl} = 0.02206 for average upper crust. For a system in steady state the number of ^{36}Cl atoms present is given by:

$$^{36}\text{Cl atoms g}^{-1} = P_{th}\,N\,F_{35Cl}\,/\,\lambda_{36} \tag{22}$$

where P_{th} is the probability of a neutron reaching thermal energy, N is the neutron density given by Equation (10) and λ_{36} the decay constant of ^{36}Cl. Once this equilibrium has been attained the production of ^{36}Cl is equal to its rate of decay to ^{36}Ar, giving the following ^{36}Ar production rate:

$$^{36}\text{Ar atoms g}^{-1}\text{ yr}^{-1} = P_{th}\,N\,R\,F_{35Cl} \tag{23}$$

For average upper crust (Table 2), N = 10.6 neutrons g^{-1} yr^{-1} (Table 3), giving a ^{36}Ar production rate of 0.19 atoms g^{-1} yr^{-1}. This compares with ^{40}Ar production in the upper crust of 2.93×10^6 atoms g^{-1} yr^{-1}, to give a crustal $^{40}Ar/^{36}Ar$ production ratio of 1.54×10^7. Fontes et al (1991) discuss how ^{36}Ar rates can be a significant factor in specific Cl-U-Th-rich environments and further discuss cosmogenic rates of ^{36}Ar production.

^{38}Ar excess relative to ^{36}Ar observed in early studies of U- and Th-rich minerals led workers to investigate the $^{35}Cl(\alpha,p)^{38}Ar$ and $^{41}K(n,\alpha)^{38}Ar$ production routes of ^{38}Ar (Fleming 1953; Wetherill 1954). Hünemohr (1989) and Eikenberg et al. (1993) in addition investigated the production via $^{37}Cl(n,\gamma)^{38}Cl(\beta\text{-})^{38}Ar$. $^{41}K(n,\alpha)^{38}Ar$ is energetically unfavorable and can be discounted. Although energetically favorable, the thermal neutron interaction with ^{37}Cl has a reaction cross-section of only 0.43 barns and a natural abundance of 24.23%. The ^{38}Ar production rate from thermal neutrons is given by:

$$^{38}\text{Ar atoms g}^{-1}\text{ yr}^{-1} = P_{th}\,N\,F_{37Cl} \tag{24}$$

For average upper crust (Table 2) $F_{37Cl} = 6.7 \times 10^{-5}$ to give a ^{38}Ar production rate of 6×10^{-4} atoms g^{-1} yr^{-1}. No modeling of the $^{35}Cl(\alpha,p)^{38}Ar$ production rates have been made, although the ratio of the nuclear cross sections for $^{35}Cl(\alpha,p)^{38}Ar$ and $^{19}F(\alpha,n)^{22}Ne$ is predicted from Woosley et al. (1975) to be ~0.2 for α-energies between 4 and 8 MeV (Eikenberg et al. 1993). Both Hünemohr (1989) and Eikenberg et al. (1993) show that $^{38}Ar/^{22}Ne$ excesses are correlated with $^{35}Cl/^{19}F$ ratios in (U+Th)-rich minerals with a gradient consistent with this estimate. The relationship between relative F and Cl α-reaction cross sections enables an empirical equation for ^{38}Ar production rates to be derived from Equation (17):

$$^{38}\text{Ar} = \frac{3.98\times10^{-24}}{0.02} \times 0.3465 \times 0.2423 \times 0.2 \times \frac{19}{37} \times \frac{[\text{Cl}]}{16}\left(\frac{0.340}{0.479}[\text{U}] + \frac{0.139}{0.479}\frac{[\text{Th}]}{3}\right)$$

$$= \{[\text{Cl}](0.76[\text{U}] + 0.104[\text{Th}]\} \times 10^{-25} \tag{25}$$

[Cl], [U] and [Th] are concentrations in ppm and the production rate of ^{38}Ar in cm^3 STP g^{-1} yr^{-1}. For average upper crust (Table 2) this gives a production rate of 0.002 atoms g^{-1} yr^{-1}. Inclusion of $^{37}Cl(n,\gamma)$ derived ^{38}Ar results in a total 0.0026 atoms g^{-1} yr^{-1}, to give a $^{38}Ar/^{36}Ar$ crustal production ratio of ~0.014. This value is calculated assuming an homogenous distribution of elements in the upper crust. Unlike ^{36}Ar, the ^{38}Ar production is dominated by α-particle reactions and the production rate of ^{38}Ar will be sensitive to element heterogeneity and the siting of Cl relative to U+Th in the crust. This is directly analogous to ^{22}Ne production rates and is illustrated in Cl and (U+Th)-rich minerals by observed $^{38}Ar/^{36}Ar$ production ratios in excess of 14.7 (Eikenberg et al. 1993), some 10^3 times higher than ratios predicted for an elementally homogenous crust. Although some natural gases show $^{38}Ar/^{36}Ar$ ratios in excess of air ratios (Ballentine 1991), pointing to average crustal production ratios in excess of calculated average values, no systematic assessment of the ^{38}Ar production rate in the crust exists. Estimates of nucleogenic contributions to the atmosphere based on average crustal production rates will underestimate the nucleogenic ^{38}Ar contribution, while the effect of element heterogeneity on $^{38}Ar/^{36}Ar$ ratios in the mantle has yet to be assessed.

Small amounts of the unstable gases ^{37}Ar and ^{39}Ar are produced in the crust through the thermal neutron reactions $^{39}K(n,p)^{39}Ar$ and $^{40}Ca(n,\alpha)^{37}Ar$ (Lehmann et al. 1993; Pearson Jr. et al. 1991). ^{37}Ar and ^{39}Ar have half-lives of 34.95±0.08 days (Renne and Norman 2001) and 269 years respectively. Applications of these tracers are reviewed by Kipfer et al. (2002, this volume).

Krypton and xenon

Fission of ^{238}U provides the dominant mechanism for the production of $^{83,84,86}Kr$ and $^{129,131,132,134,136}Xe$ in the crust today, while production of the shielded isotopes $^{80,82}Kr$ and $^{124,126,128,130}Xe$ can be neglected. Contributions from the spontaneous fission of ^{232}Th and thermal or fast neutron induced fission of ^{235}U, ^{238}U and ^{232}Th, must also be taken into consideration. Although there has been considerable past interest in the determination of the various fission spectra and yields, recent investigation of the U-Xe-Kr dating tool pioneered by Shukolyukov et al. (1974), has resulted in a re-determination of many of these values (Eikenberg et al. 1993; Ragettli et al. 1994; Wieler and Eikenberg 1999) (Table 7). Fission products of ^{244}Pu and the decay products of ^{129}I, both now extinct, contribute important Kr and Xe isotopic components to both the terrestrial mantle and atmosphere (Porcelli and Ballentine 2002). Kr and Xe derived from these extinct radionuclei are not produced in the crust hence do not contribute to the crustal system except where carried in as components of magmatic or atmosphere-derived fluids.

It is convenient to consider the production rate of ^{136}Xe by the various routes, and scale the yield of the other Kr and Xe fissiogenic isotopes to this product (Table 7). ^{136}Xe production by spontaneous fission in the crust is directly proportional to the ^{238}U concentration. ^{238}U has a branched decay mode, producing ^{234}Th by α-decay with $\lambda_\alpha = 1.55 \times 10^{-10}$ yr^{-1} (Steiger and Jäger 1977), and spontaneous fission producing amongst other isotopes $^{136}Xe_{sf}$ with $\lambda_{sf} \sim 9 \times 10^{-17}$ yr^{-1}. Because $\lambda_{sf} << \lambda_\alpha$, $\lambda_{238U} \approx \lambda_\alpha$, and the production rate of $^{136}Xe_{sf}$ is given by:

$$^{136}Xe_{sf} \text{ atoms g}^{-1} \text{ yr}^{-1} = (N_A/A_{238U}) \times 10^{-6}\, X_{238}\, [U]\, (^{136}Y\, \lambda_{sf} / \lambda_\alpha)\, (e^{\lambda\alpha t}-1) \qquad (26)$$

where [U] is the concentration of U in ppm, ^{136}Y is the yield (~0.063), or fraction of the fission product resulting in ^{136}Xe production, X_{238} is the fractional natural abundance of ^{238}U (0.992745), N_A is Avogadro's number (6.023×10^{23}), A_{U238} is the molar mass of ^{238}U (238.051g), and t = age (yr). Although the values for ^{136}Y and λ_{sf} have been independently determined, it is more convenient and precise to determine their product, $^{136}Y\,\lambda_{sf}$. $^{136}Y\,\lambda_{sf}$ has been determined in a variety of minerals and shows a bimodal distribution of about 20% variation between U-oxides and U-rich accessory phase minerals (Eikenberg et al. 1993; Ragettli et al. 1994). Ragettli et al. (1994) argue that the lower values in U-oxides can be attributed to several possible causes, including changes in oxidation state, recoil loss and diffusion. In contrast, through a careful study Ragettli et al. (1994) show that the higher values recorded in the accessory phase minerals cannot be attributed to either underestimation of mineral formation ages, overestimation of recoil effects, excess parentless Xe or fissiogenic ^{136}Xe from other sources within the minerals. Ragettli et al. (1994) give a best estimate for $^{136}Y\,\lambda_{sf} = 6.83 \pm 0.18 \times 10^{-18}$ yr^{-1}. For the average upper crust U concentrations (Table 2), this gives a $^{136}Xe_{sf}$ production rate of 0.0480 atoms g^{-1} yr^{-1}.

Estimates of λ_{sf} for ^{232}Th are at least two orders of magnitude smaller than λ_{sf} for ^{238}U, between 5.4×10^{-19} to $<2.1 \times 10^{-21}$ yr^{-1} (Segrè 1952; Wieler and Eikenberg 1999), and this source of ^{136}Xe production can be neglected. Experimental work shows that in pitchblendes and uraninites, no fast neutron induced fission component from ^{238}U can be resolved, and that all Kr and Xe isotope data can be accounted for with only three components: air; ^{238}U spontaneous fission products, and thermal neutron induced fission of ^{235}U (Fig. 8) (Eikenberg et al. 1993). The rate of thermal neutron fission production of $^{136}Xe_{nf}$ can be calculated using the following:

$$^{136}Xe_{nf} \text{ atoms g}^{-1} \text{ yr}^{-1} = P_{th}\, N\, F_{235U}\, {}^{136}Y_{nf} \qquad (27)$$

Table 7. Xe and Kr Fission Spectra and Yields.

	$^{83}Kr/^{86}Kr$	$^{84}Kr/^{86}Kr$	$^{129}Xe/^{136}Xe$	$^{131}Xe/^{136}Xe$	$^{132}Xe/^{136}Xe$	$^{134}Xe/^{136}Xe$	$^{136}Xe/^{86}Kr$	^{136}Xe Yield (%)	Ref.
^{238}U sf	0.048 ± 0.016	0.159 ± 0.023	0.002 ± 0.002	0.076 ± 0.012	0.595 ± 0.008	0.832 ± 0.007	8.0 ± 3.7		Wetherill (1953)
	0.034 ± 0.003	0.128 ± 0.011	<0.001	0.087 ± 0.012	0.605 ± 0.008	0.857 ± 0.007	6.6 ± 0.6	6.3	Young and Thode (1960)
			0.002 ± 0.002	0.080 ± 0.008	0.595 ± 0.012	0.835 ± 0.007			Shukolyukov and Ashkinadze (1968)
	0.033 ± 0.004	0.150 ± 0.012	0.002 ± 0.001	0.082 ± 0.002	0.579 ± 0.006	0.825 ± 0.003			Shukolyukov (1970)
	0.064 ± 0.014	0.218 ± 0.011	0.005 ± 0.003	0.108 ± 0.002	0.580 ± 0.003	0.832 ± 0.004	5.5 ± 1.7		Sabu (1971)
	0.046 ± 0.013	0.110 ± 0.020	0.006 ± 0.003	0.088 ± 0.002	0.568 ±0.003	0.828 ± 0.004	6.0 ± 0.4		Hebeda (1987)
	0.049 +0.001 -0.01	0.160 +0.003 -0.015	<0.002	0.093 +0.001 -0.006	0.602+0.004 -0.006	0.843+0.007 -0.012			Hünemohr (1989)
	0.042 ± 0.001	0.0152 ± 0.001	<0.001	0.083 ± 0.002	0.570 ± 0.003	0.818 ± 0.003	6.1 ± 0.1		Eikenberg et al. (1993)
				0.087 ± 0.007	0.578 ± 0.013	0.827 ± 0.008	6.5 ± 0.5	**	Ragettli (1993); Ragettli et al., (1994)
^{238}Unf(fast)	0.309	0.634	0.14	0.454	0.73	1.12	4.28	5.9	Maeck et al. (1975)
	0.309	0.637	0.399	0.4877	0.778	1.132	5.1		Crouch (1977)
^{235}Unf(th)	0.27	0.5	0.15	0.453	0.677	1.25	3.17	6.47	Farrer and Tomlinson (1962)
	0.272	0.508	0.0912	0.4448	0.685	1.232	3.2		Crouch (1977)
^{232}Thnf (fast)	0.331	0.608		0.287	0.508	0.952	0.942	5.65	Kennet and Thode (1957)
^{244}Pu sf			<0.026 ± 0.042	0.251 ± 0.022	0.876 ± 0.031	0.921± 0.027			Alexander et al., (1971)

** $^{136}Y.\lambda sf = 6.83 \pm 0.18 \times 10^{-18}$

where P_{th} is the probability of neutrons being thermalized (0.8), N, the crustal neutron production rate (for upper crust = 11.2 neutrons g^{-1} yr^{-1}, Table 3, Eqn. 10), F_{235}, the probability that ^{235}U will capture a thermal neutron that induces fission, and $^{136}Y_{nf}$, the fraction of the resulting fission product that will form ^{136}Xe (0.0647, Table 7). The thermal neutron-capture cross-section for a fission induced reaction on ^{235}U is 586 barns (Lide 1994). Average upper crust, at 2.8 ppm U with a ^{235}U natural abundance of 0.720%, therefore gives a neutron capture probability of 5.03×10^{-8} mol barns. Total capture probability for average upper crust is 0.00979 mol barns (Table 1) to give F_{235U} = 5.13×10^{-6}, and $^{136}Xe_{nf}$ = 2.96×10^{-6} atoms g^{-1} yr^{-1}. Therefore the upper crust has $^{136}Xe_{sf}/^{136}Xe_{nf} = 1.62 \times 10^4$. Given such a high production ratio, it is indeed surprising to find resolvable $^{136}Xe_{nf}$ at all. Indeed, there are no reported $^{136}Xe_{nf}$ components in any accessory phase minerals (e.g., Ragettli et al. 1994; Wieler and Eikenberg 1999). The observation of $^{136}Xe_{sf}/^{136}Xe_{nf}$ values approaching unity in massive U-oxides such as uraninite and pitchblende (Fig. 8) (Eikenberg et al. 1993) may reflect one of the few examples of elemental heterogeneity affecting the production rate of thermal neutron induced reactions: Nevertheless, while the density of uranium oxides are larger than the average crust, the elemental neutron stopping power is lower and it is not clear that the neutrons are more efficiently stopped in UO_2 than in average crust pointing therefore, not to mineral scale heterogeneity, but perhaps high concentrations of the minerals on the meter scale. In such a case, many neutrons produced within the U oxide-rich 'layer' or deposit will be thermalized before escaping the U-rich environment, enhancing the rate of production of fission products from ^{235}U compared to average crust. There is no evidence that this is a significant process on a crustal scale.

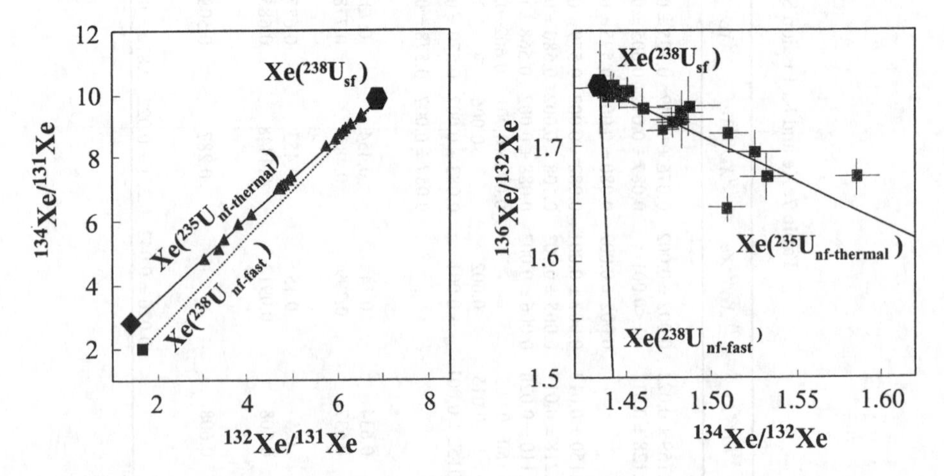

Figure 8. Xenon isotopes corrected for atmosphere contributions measured in uraninite and pitchblendes plotted after Eikenberg et al. (1993). Xe isotopes are dominated by production from the spontaneous fission of ^{238}U (Xe($^{238}U_{sf}$)). In these massive samples only, Xe from thermal neutron induced fission of ^{235}U (Xe($^{238}U_{nf\text{-}thermal}$)) is also resolvable. Xe fission products from fast neutron interactions with ^{238}U (Xe($^{238}U_{nf\text{-}fast}$)) are negligible.

CFF-Xe. Isotopic Xe anomalies, characterized by enriched ^{129}Xe, ^{131}Xe and ^{132}Xe relative to fission ^{136}Xe, were identified within minerals from the natural nuclear reactor no.2 in the Oklo uranium mine, Gabon (Shukolyukov et al. 1976). Although early theories focused on a nuclear process to account for this observation, it later became apparent that this was a chemical fractionation effect (Meshik and Shukolyokov 1986; Meshik 1988). CFF-Xe, or Chemically Fractionated Fission-Xenon is created when fast diffusion of radioactive Xe β^--precursors occurs in fine-grained minerals at high temperature. These fission fragments (Xe, Sn, Sb, Te, I, Kr, Se and Br) migrate from the parent U-bearing mineral into adjacent phases where they subsequently decay to stable Xe isotopes. The longer the lifetime of the Xe precursor (e.g., Sb, Te and I), the larger the isotopic anomaly in the stable Xe daughter-product (Meshik 1988; Meshik et al. 1996).

This process of chemical separation of fission Xe-precursors accounts for Xe and Kr anomalies from the Alamogordo nuclear test site (Pravdivtseva et al. 1986), U-bearing minerals (Gerstenberger 1980; Meshik 1988) and has also been used to explain the isotopic structure of Xe-H in meteoritic nano-diamonds (Ott 1995; 1996). Takaoka (1972) and Igarashi (1995) have compared an assumed fractionated meteorite component (Q or 'planetary') with the atmosphere Xe isotopic composition to resolve the additional atmosphere Xe fission components. The inferred fission spectrum is more enriched in ^{131}Xe and ^{132}Xe than either ^{244}Pu or ^{238}U spectra. Meshik et al. (2000) show that CFF-Xe from Okelobondo (the southern extension of Oklo) matches the theoretical fission composition calculated by Igarashi (1995). While Meshik et al. (2000) argue that the atmosphere Xe isotopic composition can therefore be accounted for with terrestrial CFF-Xe, this appears difficult to accommodate on mass balance grounds: The total fissiogenic ^{136}Xe in the atmosphere is 2.8±1.3% of the atmosphere ^{136}Xe budget (Igarashi 1995). Only 0.47% of the atmosphere ^{136}Xe can be accounted for by ^{238}U fission in the crust (Table 8). From analogy with ^{40}Ar (see the section: *Production rates today and over the history of the Earth*), only an additional ~20% can be sourced from ^{238}U decay from the mantle to account for, at most, 0.56% of the atmosphere ^{136}Xe budget. Even if this were all CFF-Xe, it could not account for the fissiogenic Xe component in the atmosphere. Furthermore, the total terrestrial fissiogenic Xe produced is unchanged. This scenario would require preferential release of CFF-Xe and efficient retention of the complementary enriched ^{136}Xe component, further degrading the mass balance. There is no reason to believe this occurs. Terrestrial CFF-Xe, while important on a mineral scale, probably has little impact on the global scale. This does not rule out an accretionary CFF-Xe contribution to the Earth's isotopic Xe composition (e.g., Shukolyukov et al. 1994).

Cosmogenic noble gas production

Noble gases are produced by spallation reactions when high energy particles, produced by interactions between cosmic rays and the Earth's atmosphere, interact with nuclei at the Earth's surface (Niedermann 2002, this volume). The purpose of this section is to evaluate how production and preservation of cosmogenic noble gases affect the crustal noble gas budget.

Of the cosmogenic noble gas isotopes that are produced at the Earth's surface, ^{3}He is produced most rapidly, for example $(^3He/^{21}Ne)_{COS}$ = 4.4 (Schäfer et al. 1999) and $(^3He/^{38}Ar)_{COS} \approx 1.5$ for [Ca] = 3% and [K] = 2.8% (Renne et al. 2001; Lal 1991). However, the effect of cosmogenic ^{3}He on the crustal ^{3}He budget is difficult to assess because ^{3}He is also lost more rapidly than the heavier noble gases; for example, most common crustal minerals (e.g., quartz, feldspar, clay minerals, carbonates) will not preserve a significant fraction of the ^{3}He produced. Only olivine, clinopyroxene and certain garnets have been conclusively shown to preserve cosmogenic ^{3}He over geologically significant periods (Dunai and Roselieb 1996; Kurz et al. 1988; Trull et al. 1991).

The column-integrated ^{3}He production at the surface at the present day due to reactions induced by cosmic-ray particles is 23×10^3 atoms ^{3}He cm^{-2} y^{-1} based on 110 atoms ^{3}He g^{-1} y^{-1} at the surface and a mean attenuation length of 160 g cm^{-2} (Niedermann 2002). Given the uncertainties indicated by Niedermann, this value is subject to an error of ~20%. The vast majority of this is not preserved because it is produced in minerals that have high He diffusivities at surface conditions; even those minerals that do preserve

Table 8. Current production rate and total noble gas contribution from the crust over the age of the earth.

	Present day Production rates cm^3 STP/yr							*Total Produced in the Crust Over 4.5Ga cm^3 STP*							
Crust	Oceanic	Lower	Middle	Upper	Total	Cosmo-	IDP	Oceanic	Lower	Middle	Upper	Cosmo-	**Total**	% of	Atmosph.
Mass (g)	5.0E+24	8.33E+24	9.63E+24	6.13E+24	2.9E+25	genic		5.00E+24	8.33E+24	9.63E+24	6.13E+24	genic		Atmosp.	(O&P'83)
^{3}He	-	3.30E+03	1.42E+04	4.26E+04	6.00E+04	1.29E+03	7.50E+02	-	2.28E+13	1.05E+14	3.16E+14	5.80E+12	4.49E+14	1549%	2.90E+13
^{4}He	4.55E+10	4.88E+11	3.54E+12	3.95E+12	7.98E+12	1.29E+03	-	3.58E+20	3.37E+21	2.63E+22	2.93E+22	5.80E+12	5.93E+22	285484%	2.08E+19
^{20}Ne	-	1.96E+03	1.60E+04	1.74E+04	3.53E+04	2.93E+02	1.50E+04	-	1.35E+13	1.18E+14	1.29E+14	1.32E+12	2.62E+14	0.0004%	6.52E+19
^{21}Ne	2.28E+03	2.04E+04	1.65E+05	1.79E+05	3.65E+05	2.93E+02	-	1.79E+13	1.41E+14	1.22E+15	1.33E+15	1.32E+12	2.71E+15	1.4051%	1.93E+17
^{22}Ne	-	4.09E+04	3.30E+05	3.58E+05	7.29E+05	2.93E+02	1.20E+03	-	2.82E+14	2.45E+15	2.66E+15	1.32E+12	5.39E+15	0.0810%	6.65E+18
^{36}Ar	-	2.64E+03	9.70E+03	4.49E+04	5.73E+04	7.31E+02	-	-	1.82E+13	7.19E+13	3.33E+14	3.29E+12	4.26E+14	0.0003%	1.24E+20
^{38}Ar	-	1.47E+03	1.19E+04	1.29E+04	2.63E+04	7.31E+02	-	-	1.02E+13	8.81E+13	9.56E+13	3.29E+12	1.97E+14	0.0008%	2.34E+19
^{40}Ar	1.17E+10	1.61E+11	6.24E+11	6.72E+11	1.46E+12	7.31E+02	-	2.34E+20	3.23E+21	1.25E+22	1.35E+22	-	2.95E+22	**79.94%**	3.69E+22
^{83}Kr	9.69E-01	6.87E+00	6.35E+01	7.08E+01	1.41E+02	-	-	6.30E+09	4.47E+10	4.13E+11	4.61E+11	-	9.25E+11	0.0002%	5.18E+17
^{84}Kr	3.51E+00	2.49E+01	2.30E+02	2.56E+02	5.11E+02	-	-	2.28E+10	1.62E+11	1.50E+12	1.67E+12	-	3.35E+12	0.0001%	2.57E+18
^{86}Kr	2.31E+01	1.64E+02	1.51E+03	1.68E+03	3.36E+03	-	-	1.50E+11	1.06E+12	9.84E+12	1.10E+13	-	2.20E+13	0.0028%	7.86E+17
^{131}Xe	1.30E+01	9.25E+01	8.55E+02	9.53E+02	1.90E+03	-	-	8.49E+10	6.02E+11	5.57E+12	6.20E+12	-	1.25E+13	0.0170%	7.31E+16
^{132}Xe	8.67E+01	6.14E+02	5.68E+03	6.33E+03	1.26E+04	-	-	5.64E+11	4.00E+12	3.70E+13	4.12E+13	-	8.27E+13	0.0892%	9.27E+16
^{134}Xe	1.24E+02	8.79E+02	8.13E+03	9.06E+03	1.81E+04	-	-	8.07E+11	5.72E+12	5.29E+13	5.89E+13	-	1.18E+14	0.3297%	3.59E+16
^{136}Xe	1.50E+02	1.06E+03	9.83E+03	1.10E+04	2.18E+04	-	-	9.76E+11	6.92E+12	6.40E+13	7.13E+13	-	1.43E+14	0.4690%	3.05E+16
neutrons/g/yr		1.05	6.19	11.21	-										

Notes
Surface area and mass of continental crust is 1.5E+8 km^2 and 2.41E+25 g respectively
Relative mass of crustal reservoirs calculated using Table 9 in Rudnick and Fountain (1995).
Mass of Oceanic crust calculated assuming surface area of 3E+8km^2, density=2.8g/cm^3 and average depth=6km
Oceanic crust taken to be U=0.047ppm; Th=0.120ppm; K=600ppm (Sun and McDonough, 1989).
^{21}Ne production calculated using Eqn. 15 and compositions in Table 2. ^{21}Ne production will be ~20% higher if observed ^{4}He/^{21}Ne relationship is used (Fig. 7)
^{21}Ne/^{22}Ne crust=0.5 ^{38}Ar crust=0.2*^{22}Ne*Cl/F (Cl/F=0.18) Cosmogenic ^{21}Ne=^{20}Ne=^{22}Ne Cosmogenic ^{3}He=^{4}He IDP ^{20}Ne/^{22}Ne=12.5

cosmogenic ^3He (olivine, clinopyroxene and garnet) release their cosmogenic ^3He at comparatively low temperatures. Consequently, it is highly unlikely that accumulation of cosmogenic ^3He is a major crustal component.

All three isotopes of Ne are produced at almost equal rates by cosmogenic particles at the Earth's surface. As ^{21}Ne is the least abundant isotope, cosmogenic production of ^{21}Ne is more likely to be a significant portion of the crustal ^{21}Ne budget than for ^{20}Ne or ^{22}Ne. Similarly, ^{38}Ar is the least abundant isotope of Ar as well as the most rapidly produced (for typical crustal compositions), therefore the effects of cosmogenic ^{38}Ar accumulation on the crustal noble gas budget will be examined.

The rate of cosmogenic ^{21}Ne production is comparatively well constrained and the column-integrated ^{21}Ne production at the Earth's surface is $\approx 5\times10^3$ atoms cm^{-2} y^{-1} (($^3He/^{21}Ne)_{COS}$ = 4.4 (Schäfer et al. 1999)). Production of cosmogenic ^{38}Ar at the Earth's surface is less well known: Lal (1991) estimated the rate of ^{38}Ar production from ^{40}Ca to be ~200 atoms g^{-1} Ca y^{-1} and analyses of cosmogenic noble gases in meteorites suggest that ^{38}Ar production from ^{39}K will be about ten times higher than that off ^{40}Ca (Freundel et al. 1986); other sources of cosmogenic ^{38}Ar are unlikely to be important relative to Ca and K production for typical upper crustal compositions. Therefore the $(^3He/^{38}Ar)_{COS}$ at the surface $\approx$ 1.5 although this is uncertain to more than a factor of 2. Typical upper crustal compositions (Ca = 3%; K = 2.8%), will result in a column-integrated ^{38}Ar production rate of 13×10^3 atoms cm^{-2} y^{-1}.

Given a land surface area of 150×10^6 km^2, the global annual production of ^{21}Ne and ^{38}Ar is of the order 300 and 700 cm^3 STP respectively (Table 8). Under the assumption of constant cosmogenic production rates and constant land surface area, the total cosmogenic ^{21}Ne and ^{38}Ar production for the Earth is of the order 1 and 3×10^{12} cc STP respectively, equivalent to <5% of the nucleogenic production of these isotopes. Given that both the flux of cosmogenic neutrons at the Earth's surface and the land surface area vary with time, the uncertainties on these estimates likely exceed one order of magnitude. However, even given these uncertainties, it is clear that cosmogenic Ne and Ar do not constitute a major contribution to the crustal system.

Interplanetary dust accumulation

While not strictly produced in the crust, it is convenient to consider noble gases from this source here. Oceanic sediments have $^3He/^4He$ ratios that are higher than expected for *in situ* radiogenic/nucleogenic production and higher than can be plausibly attributed to detrital mantle – derived material. Accretion of interplanetary dust particles (IDPs) that have high $^3He/^4He$ ratios and high He concentrations likely results in high $^3He/^4He$ ratios in slow accumulating oceanic sediments. Again, this section is not intended to review the IDP literature, but rather assess the possible impact of IDPs on the composition of the crust as a reservoir for noble gases.

It is known that IDPs can preserve their extra-terrestrial He over long time periods, certainly up to 70 Ma (Farley 1995) and possibly as long as 450 Ma (Patterson et al 1998). As a consequence of their ability to preserve ^3He over long periods of time, accumulated IDPs could represent a significant fraction of the ^3He inventory of the crust. The average IDP flux over the last 70 Ma is equivalent to 0.5×10^{-12} cm^3 STP ^3He cm^{-2} ky^{-1} (Farley 1995). If all the IDPs falling on continents were preserved, this corresponds to a ^3He IDP flux to the continental crust of the order 750 cm^3 STP y^{-1}, or about 1% of the ^3He produced in the crust by Li(n,α) reactions.

Few data exist on the heavy noble gas IDP flux to the Earth. Measurements by Matsuda (1990) suggest the average $^3He/^{20}Ne$ of IDPs is in the range 0.05, implying a continental ^{20}Ne flux of ~15,000 cm^3 STP y^{-1}, or about 30% of the ^{20}Ne produced in the crust. IDPs have high $^{20}Ne/^{21}Ne$ and $^{20}Ne/^{22}Ne$ ratios and therefore are unlikely to impact the crustal budget for ^{21}Ne or ^{22}Ne.

Uncertainties in the IDP flux over the age of the Earth preclude any realistic estimate of the total IDP noble gas inventory. It should also be noted that there is considerable uncertainty in the fate of IDPs falling onto continents. Noble gases, particularly ^{3}He, are preserved only in IDPs ≤50-μm diameter due to frictional heating during entry (Farley 1995). Small particles will be preferentially transported off the continents into oceans, considerably reducing their contributions to the continental noble gas reservoir. The noble gas carrier phase or phases in IDPs is also unknown, and their preservation during crustal processes such as diagenesis and metamorphism is speculative at this time.

Production rates in continental crust today and over the history of the Earth

Having established the factors governing the rates of noble gas production in geological materials, it is a fairly straightforward step to use current estimates of the continental crust chemical composition to calculate the present day radiogenic and nucleogenic production rates in the crust (Table 8). We have used the estimates of the continental crust chemical composition and structure following Rudnick and Fountain (1995) (Table 2). These workers have divided the crust into three portions, lower, middle and upper crust, based on observations from seismic studies summarized by Holbrook et al. (1992). The upper crust (depths from 0 to 10-15 km) is dominated by evolved rock types and contains some 66% of the heat producing elements, U, Th and K, relevant to noble gas production. The middle crust (between depths of 10-15 and 20-25 km) is composed of rocks in the amphibolite facies and is intermediate in composition between the upper and lower crust. The latter is dominated by metamorphic rocks in the granulite facies (between 20- to 25- and 40-km depth) and contains some 6% of the heat producing elements. The decrease in heat producing element concentration with depth is a result of depletion in felsic rocks caused by granulite facies metamorphism and an increase in the proportion of mafic rocks. In total the crust accounts for 42 and 32% of the U and K in the bulk silicate Earth (McDonough and Sun 1995).

Although not explicit in their estimate of the total mass of these crustal subdivisions, the average crust concentration of elements and the individual subdivision concentrations (Table 9 in Rudnick and Fountain 1995) enable us to estimate the relative proportion of these reservoirs. Taking a crustal mass of 2.41×10^{25} g (e.g., Mason and Moore 1980), this gives an upper, middle and lower crustal mass of 8.33×10^{24} g, 9.63×10^{24} g and 6.13×10^{24} g respectively. ^{4}He and ^{40}Ar production rates are directly proportional to the reservoir (U+Th) and K concentration respectively. ^{136}Xe, ^{86}Kr and related fission product production rates are directly proportional to the U concentration in each reservoir. The products of thermal neutron reactions only, ^{3}He and ^{36}Ar, are a function of the neutron production rate determined by the U+Th concentration and average chemical composition following Equation (10). The production rate of ^{20}Ne and ^{21}Ne by (α,n) and fast neutron reactions is a function of the U, Th, O, Mg and Na concentration, and we have used the empirical relationship developed in Equations (15) and (16) to estimate their production rates. It should be noted that this value is based on the theoretical production rates, the use of the observed crustal $^{4}He/^{21}Ne$ production ratio (Fig. 7) will result in a ~20% increase in the rate of ^{21}Ne production. We have not used Equation (17) (which is valid for mineral scale compositions) to estimate ^{22}Ne production because of the impact of mineral scale fluorine heterogeneity on the crustal system ^{22}Ne production rate. We have assumed for all crustal reservoirs a $^{21}Ne/^{22}Ne$ production rate of 0.52, and based ^{22}Ne production estimates on the ^{21}Ne production rate. Similarly, ^{38}Ar production is dominated by (α,n) reactions on ^{35}Cl, and is likely to be affected by mineral scale heterogeneity. We therefore neglect Equation (25) (which is valid for mineral scale compositions). We have assumed that the average crustal Cl/F ratio (0.18) is preserved on the mineral scale to give a ^{38}Ar production linked to the ^{22}Ne rate where: ^{38}Ar = 0.2 × 0.18 × ^{22}Ne.

To place some of these production rates into perspective we can consider the air-derived concentration of selected noble gases dissolved in groundwater. 1 g of water contains approximately 4.6×10^{-6} cm^3 STP ^{4}He, 6×10^{-10} cm^3 STP ^{21}Ne, 2×10^{-8} cm^3 STP ^{22}Ne, 2.5×10^{-7} cm^3 STP ^{38}Ar, 3.9×10^{-4} cm^3 STP ^{40}Ar, 1.6×10^{-8} cm^3 STP ^{86}Kr, 1.2×10^{-9} cm^3 STP ^{136}Xe. 1 g of rock at 1% porosity can contain ~3.5×10^{-3} g of groundwater. Based on present day upper crust production rates a 10% crustal contribution to these isotopes will occur in 3.9×10^3 yr (^{4}He), 7.2 Ma (^{21}Ne), 120 Ma (^{22}Ne), 42 Ga (^{38}Ar), 1.2 Ma (^{40}Ar), 20 Ga (^{86}Kr) and 234 Ma (^{136}Xe). Clearly ^{4}He is the most sensitive indicator of crustal fluid addition to near surface reservoirs, and can be used as a groundwater dating tool (Kipfer et al. 2002, this volume). Over longer, but similar, timescales ^{21}Ne and ^{40}Ar contributions will be observed, while systems that show significant crustal ^{22}Ne contributions should also contain resolvable fissiogenic ^{136}Xe. Only crustal systems with no groundwater or air contamination can hope to show ^{38}Ar or ^{86}Kr variations caused by nuclear processes in the crust.

Although the rate of crustal growth remains controversial (e.g., Armtrong 1991; Taylor and McLennan 1995), the total amount of radiogenic noble gases produced by the radioelements now in the crust provide an important cornerstone in understanding both the extent to which the silicate Earth has degassed, and the radiogenic-nucleogenic noble gas contribution to the noble gas budget of the atmosphere. While, the compatibility of noble gases relative to the heat producing elements is uncertain in some melting regimes (e.g., Chamorro-Perez et al. 1998; Barfod et al. 2002), both are highly incompatible during melting at mid-ocean ridges (e.g., Marty and Lussiez 1993), and noble gases are insoluble in silicate melts (Carroll and Draper 1994). The end result will be that noble gases produced within the mantle by radioelements or their daughter products which are now incorporated in the continental crust, will have been degassed into the atmosphere (e.g., Coltice et al. 2000). The efficiency of release of noble gases produced in the crust is discussed in some detail later in this contribution, but the concept of 'average crustal K-^{40}Ar age' provides an indication of total crust degassing efficiency. This is estimated to be 1×10^9 yr (Hurley and Rand 1969). The K in the continental crust over 4.5 Ga will have produced some 2.92×10^{22} cm^3 STP ^{40}Ar (Table 8). In the last 1×10^9 yr the crust will have produced 1.95×10^{21} cm^3 STP ^{40}Ar or 6.7% of the total continental crustal production. The 'average crustal K-^{40}Ar age' of 1×10^9 yr therefore suggests that the crust retains only 6.7% of its ^{40}Ar.

Of the 7.60×10^{22} cm^3 STP ^{40}Ar produced by K decay in the bulk silicate Earth, 2.95×10^{22} cm^3 STP ^{40}Ar has been produced in the continental and oceanic crust (Table 8). Deducting 1.95×10^{21} cm^3 STP ^{40}Ar for ^{40}Ar still trapped, crustal production accounts for 75% of the 3.69×10^{22} cm^3 STP ^{40}Ar now in the atmosphere. 1.24×10^{20} cm^3 STP ^{40}Ar in the atmosphere is from primordial sources, assuming primordial ^{40}Ar/^{36}Ar = 1 (Ozima and Podosek 1983). Some 9.23×10^{21} cm^3 STP ^{40}Ar in the atmosphere must therefore be derived from decay of K still in the mantle, or crustal K that has been recycled (e.g., Coltice et al. 2000). 3.72×10^{22} cm^3 STP ^{40}Ar must still reside in the mantle. Taking the mass of the mantle to be 4.02×10^{27} g, this would require an average mantle concentration of 9.27×10^{-6} cm^3 STP ^{40}Ar/g, some 4.2 times higher than the value of ~2.2×10^{-6} cm^3 STP ^{40}Ar/g estimated from MORB-source ^{4}He concentrations and a ^{4}He/^{40}Ar* ratio of 2 (where ^{40}Ar* is the non-atmospheric ^{40}Ar contribution assuming that all ^{36}Ar is atmospheric in origin; see Porcelli and Ballentine 2002). This forms one of the key arguments for a region of the mantle that is less degassed than the mid ocean ridge source (e.g., Allègre et al. 1996). Calculations that investigate radiogenic contributions to the

Earth's atmosphere must take the mantle degassing efficiency into consideration. For example while the crust contributes ~0.47% of the ^{136}Xe in the atmosphere, assuming the same degassing efficiency for Xe as Ar, the contribution from U still within the silicate mantle can only be 0.11% of the atmosphere ^{136}Xe budget.

RELEASE OF NOBLE GASES FROM MINERALS IN THE CRUST

The 8.0×10^{12} cm^3 STP/yr of ^{4}He produced in the crust (Table 8) compares with the ^{4}He flux from the mantle at mid ocean ridges estimated to be 2.2×10^{12} cm^3 STP ^{4}He/yr (e.g., Porcelli and Ballentine 2002, this volume). The top 10-15 km of the crust produces some 66% of the crustal radiogenic-nucleogenic budget, 28% in the middle crust (>10-15 but <20- to 25-km depth), with only 6% of the production being accounted for from deeper regions of the crust (>20- to 25-km depth) (Rudnick and Fountain 1995; Tables 2 and 8). Melting and volatile loss at mid ocean ridges efficiently degasses the material that reach this region. The crustal degassing process is very different and controlled by two stages: (1) release from the mineral in which the noble gas was produced/trapped; and (2) transport from the site of production to the surface. It is important to distinguish the relative rates of the underlying processes as very different mechanisms control mineral- and crustal-scale transport of noble gases. Noble gases produced in crustal minerals by U, Th and K decay, or by nuclear reactions with α–particles or neutrons can be released by three main mech-anisms: recoil, diffusion or chemical transformation (Fig. 9). We discuss these in turn.

Recoil loss

He and α-recoil. The high kinetic energies of alpha particles produced by U-series decay results in ejection of the alpha particles some distance from their parent nucleus. The distance traveled by the alpha particle—the stopping distance—is determined primarily by the density of the host mineral, although chemistry can be significant. Typical stopping distances are given in Figure 10 from which it is clear that α-particle stopping distances are on the same scale as grain sizes in fine-grained crustal rocks. It is important at this point to distinguish between grain size and effective grain size. This occurs when a single mineral consists of domains, or smaller regions, separated by fast diffusion pathways through the lattice, such as defects in the crystal lattice (see reviews by Farley 2002 and Kelley 2002, both this volume). When the shortest dimension of the grain hosting the parent nuclide approaches the ejection distance, a significant proportion of the He produced will be lost from that grain. Although the He atom may come to rest within any void space between grains, the stopping distance is determined by the density of the medium. Any space between grains will be occupied by low-density fluids. There is a high probability therefore that the He will more likely be buried in the adjacent grain (Rama and Moore 1984). The He atom in this case will come to rest at the end of a damage track that has one end that is "open" to porosity or fast diffusion pathways (as opposed to damage tracks that start and finish within a single grain). He diffusivity is higher along damage tracks (Hurley 1950). He will readily escape from these minerals provided the tracks do not anneal before the He atom can diffuse out of the grain. Annealing temperatures are generally slightly higher than the closure temperature for He diffusivity (e.g., apatite annealing occurs at about 110°C (Laslett et al 1987) whereas He closure in apatite occurs at around 75°C for typical grain sizes and geothermal gradients (Wolf 1996; Farley 2002)). However, fission track annealing data only exist for U-rich phases (mostly titanite, zircon and apatite) and the annealing behavior of most other mineral groups is poorly constrained. If the track does anneal, then the mineral diffusivity controls He escape, otherwise, He is rapidly released from the grains into the void space via the α-damage tracks. Water-filled porosity will be more efficient at stopping α

a) recoil

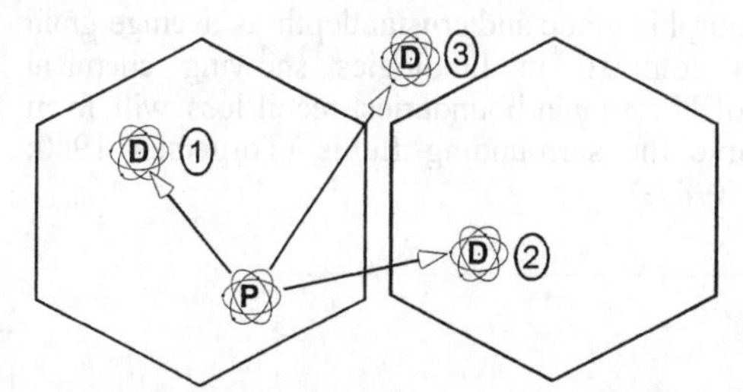

Parent, P, decays to daughter, D, with a characteristic recoil distance in the mineral, Xr. Depending on the grainsize and location of the parent, the recoiled daughter will either come to rest in the same grain (1), or in an adjacent grain (2), or in the porosity surrounding the grains (3). Note that 1 and 2 are not equivalent as the damage track for 2 has a termination in the porosity. Also, 2 is more probable than 3 because the stopping distance in the fluid between grains will be much greater than that within grains.

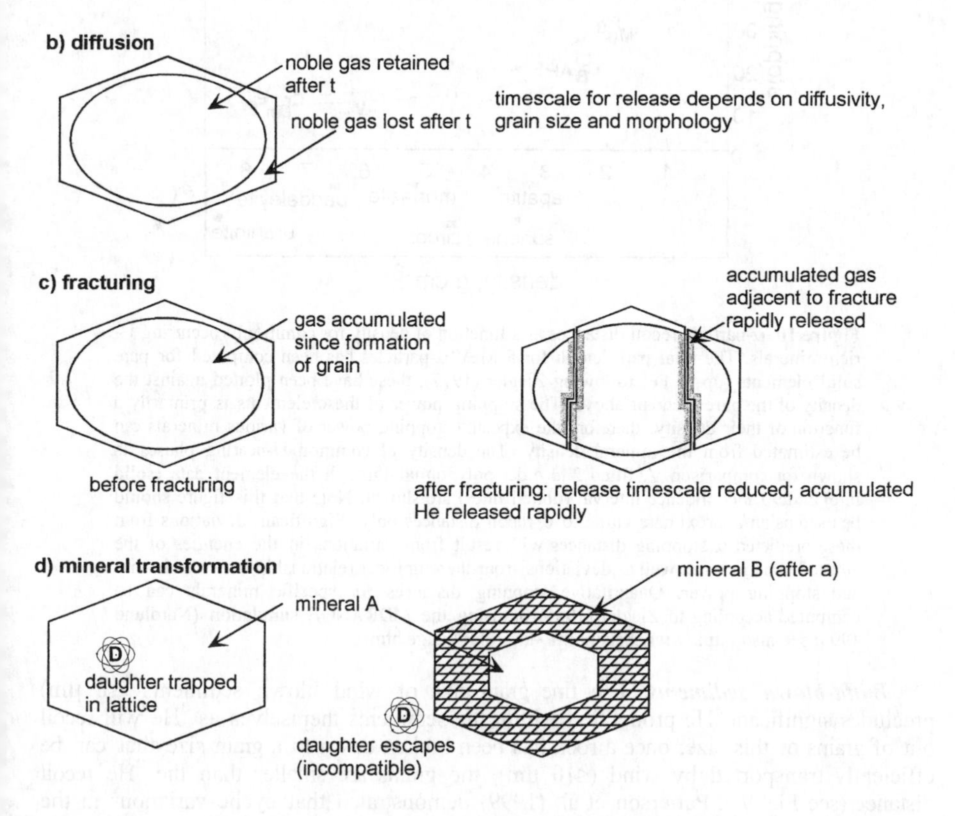

Figure 9. Mechanisms of noble gas release from crustal minerals.

particles than the equivalent dry rock, increasing the fraction of particles that are retained in the pore-space.

The fraction of ^{4}He lost by recoil is a strong function of effective grain size, or more strictly its the surface/volume ratio (Farley et al. 1996; Martel et al. 1990). Low surface area/volume ratios of <0.03 μm^{-1} result in >90% retention of ^{4}He in all but the most planar or acicular crystal morphologies. As a consequence, recoil loss of ^{4}He can be expected to decrease with increasing metamorphic grade and crustal depth as average grain sizes generally increase with depth. In contrast, in lithologies showing chemical breakdown and significant re-distribution of U on grain boundaries, recoil loss will form the dominant mechanism of ^{4}He input into the surrounding fluids (Torgersen 1980; Torgersen and Clarke 1985; Andrews et al. 1989a).

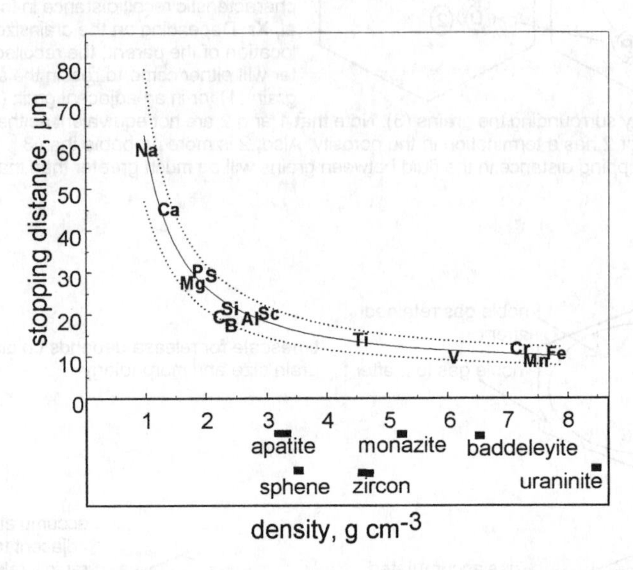

Figure 10. α-particle recoil distances as a function of density for commonly occurring U-rich minerals. The total path length for 5 MeV α-particles has been computed for pure solid elements (up to Fe) following Ziegler (1977); these have been plotted against the density of the pure element above. The stopping power of these elements is primarily a function of their density; therefore the expected stopping power of various minerals can be estimated from the mineral density. The density of common U-bearing phases is shown for comparison. A fitted 2nd order polynomial through the element data (solid line) and 90% confidence interval (dotted lines) are shown. Note that this figure should be used as an approximate guide to α recoil distances only. Significant deviations from these predicted α stopping distances will result from variations in the energies of the emitted α particles as well as deviations from the empirical relationship between density and stopping power. Quantitative stopping distances for specific minerals can be computed according to Ziegler (1985) or using the MDRANGE simulation (Nordlund 1995; see also http://www.helsinki.fi/~knordlun/range.html).

Wind-blown sediments. The fine grain size of wind blown sediments (≤8 μm) precludes significant ^{4}He production within the sediments themselves as ^{4}He will recoil out of grains of this size; once a rock has been broken down to a grain size that can be efficiently transported by wind (≤10 μm), the grains are smaller than the ^{4}He recoil distance (see Fig. 9). Patterson et al. (1999) demonstrated that cyclic variations in the ^{4}He concentration (5 to 15×10^{-9} cc STP g^{-1}) of oceanic sediments may be used to trace the wind–blown continental contribution to the deep ocean (loess). Hence, the ^{4}He in oceanic wind-blown sediments must be due to accumulation of radiogenic He in the continental regions from which the dust was derived. The radiogenic ^{4}He in the sediments is likely to be located in zircon grains; zircons are known to be He retentive (Farley 2002, this volume), and the remaining minerals (quartz, phyllosilicates) are either U-poor or not He-retentive (or both). Based on measurements of Zr concentrations in their samples,

Patterson et al. estimated that ^{4}He concentrations in zircons were ~8×10^{-6} cc STP g^{-1}, well within the range observed in zircons in general. While having potential as a tracer, volumetrically this source is very small.

Xe and Rn. Minor loss of fissiogenic ^{136}Xe, with a recoil distance ≈ 5 μm (Ziegler 1980; Ragettli et al. 1994) can result in Xe loss in fine-grained U-bearing minerals. Recoil of ^{222}Rn from mineral surfaces into groundwaters can be used to estimate groundwater flow rates; while the recoil distance of ^{222}Rn is small (0.036 μm), the short half-life of ^{222}Rn (3.8 days) relative to diffusive timescales ensures that the only realistic source of ^{222}Rn is that which is recoiled/immediately released from the minerals in which it was produced. The energy released during production of the remaining radiogenic or nucleogenic noble gas isotopes is sufficiently low to ensure that recoil is small in relation to typical effective grain sizes (Krishnaswami and Seidemann 1988), consequently these nuclides will only be released by diffusion, fracturing or chemical alteration.

Diffusive loss from minerals

Diffusion of noble gases out of minerals plays a direct role in the loss of gas at a mineral scale and is often the underlying process of other release mechanisms such as fracturing or weathering. There are relatively good estimates of the diffusivities of noble gases through different minerals, particularly for systems that have applications to geochronology. A full discussion of diffusion mechanisms and the controls on diffusion can be found in Farley (2002), Kelley (2002) and McDougall and Harrison (1988).

The efficiency of noble gas release from minerals is dependent on D/a^2 (D = diffusivity, a = radius of effective diffusion domain assuming spherical geometry). Increasing the diffusivity for example, by raising the temperature, or decreasing the effective diffusion domain size will increase the rate of noble gas loss through diffusion. Typical rates of He and Ar diffusive loss in crustal minerals are given in Figure 11 and Table 9.

The fraction of gas retained by the mineral phase can be approximated by:

$$F_R = 6/\pi^2 \exp[-\pi^2 Dt/a^2] \tag{28}$$

(after Mussett 1969), where F_R is the fraction of gas retained, D, t and a are diffusivity, time and effective grain size respectively (spherical approximation). At upper crustal conditions (<150°C) He diffusivities in fine-grained (0.1 mm) crustal minerals can range between 10^{-18} cm^2 s^{-1} to 10^{-22} cm^2 s^{-1} (Lippolt and Weigel 1988; Trull et al. 1991). At 10^{-18} cm^2 s^{-1} He is lost from the minerals more rapidly than it can accumulate. Within 10^5 years the rate of production of ^{4}He equals the rate of loss and the mineral is in 'steady-state' with respect to ^{4}He production and loss. However, for Ar with diffusivities <10^{-25} cm^2 s^{-1} (McDougall and Harrison 1988), virtually all Ar is retained in the grains (see Fig. 11). Diffusive release of noble gases from minerals at temperatures where He, but not Ar, is effectively released from the grains will result in high He/Ar ratios in any associated fluid relative to that produced in the mineral (Elliot et al. 1993; O'Nions and Ballentine 1993; Ballentine et al. 1994) (Fig. 12).

The mineral-fluid system and diffusive gradient. For diffusion to occur there must also be a chemical gradient. For noble gases in a mineral-fluid system this occurs when the noble gas equilibrium concentration in the mineral phase (Cm) is higher than the equilibrium concentration in the fluid phase (Cf), where the mineral-fluid partition coefficient (Kd) is defined as

$$K_d = C_m/C_f \tag{29}$$

Values of K_d for He and Ar are not well constrained, although are likely to be ≤10^{-3} (Brooker et al. 1998; Kelley 2002). Ignoring equilibrium partition coefficients and only

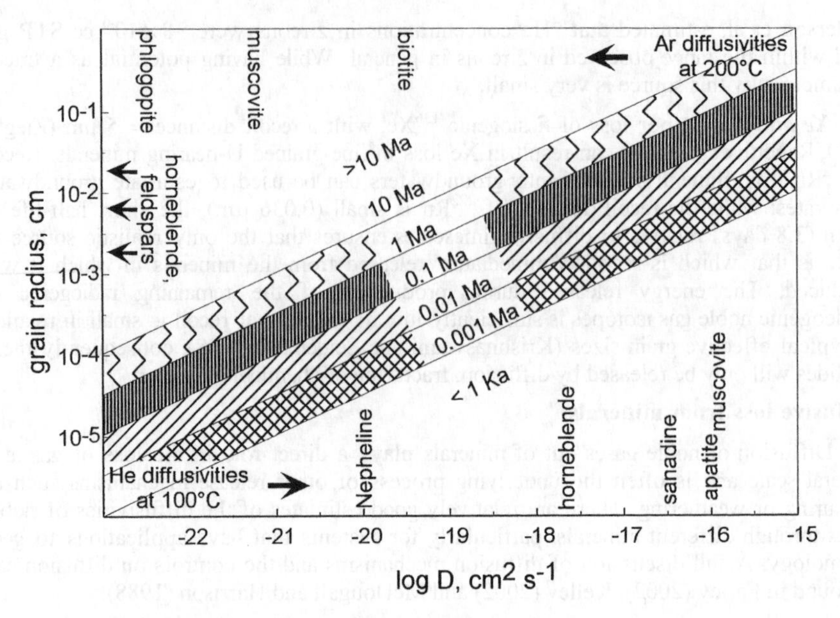

Figure 11. Mineral He and Ar retention is shown as a function of grain size and accumulation time. The time required for a grain to lose 90% of its He or Ar is contoured as a function of mineral diffusivity and grain size, using Equation (30) assuming spherical grains after Mussett (1969); the diffusivities of He at 100°C and for Ar at 300°C are shown for a few representative minerals (data from Lippolt and Weigel 1988; McDougal and Harrison 1988; Wolf et al. 1996). For example, He in a muscovite grain with an effective radius of 100 μm (10^{-2} cm) at 100°C will lose 90% of its He in a few hundred thousand years, while Ar in the same grain is still effectively retained at 300°C over >10 Ma.

considering relative mineral-water concentrations, Andrews et al (1989b) erroneously calculated that typical crustal minerals (~5 wt % K) require in excess of 650 Ma to generate sufficient radiogenic ^{40}Ar to set up a positive diffusion gradient with air-saturated water ($[^{40}Ar] = 2$ to 5×10^{-4} cm^3 STP ^{40}Ar g^{-1}; Ozima and Podosek 1983). However, when partition coefficients are considered a diffusive gradient between a mineral with 5% K and air saturated water is established within 0.65 Ma (assuming $K_d = 10^{-3}$). He concentrations in air saturated water are much lower than Ar, causing a chemical gradient to be established very rapidly.

In addition to Kd, the diffusion gradient will also be dependent on the mass ratio of solid to fluid, which together govern fluid's ability to provide a suitable sink for the noble gas released. Baxter et al (2001) for example, show that for diffusion gradients to be established:

$$K_d\,(M_R/M_F) << 1 \tag{30}$$

where M_R and M_F are the masses of rock and fluid within the characteristic distance of the production site. Because this characteristic distance is relatively small, $M_R/M_F \approx$ 1/porosity and therefore if K_d is smaller than the porosity then there will be a diffusion gradient from mineral to fluid. In environments where the porosity is low and the equilibrium concentration in the available porespace is reached, the system will then be controlled by the rate of bulk diffusive loss through the rock porespace. The observation that 'parentless' or 'excess' ^{40}Ar is found in some low porosity systems demonstrates

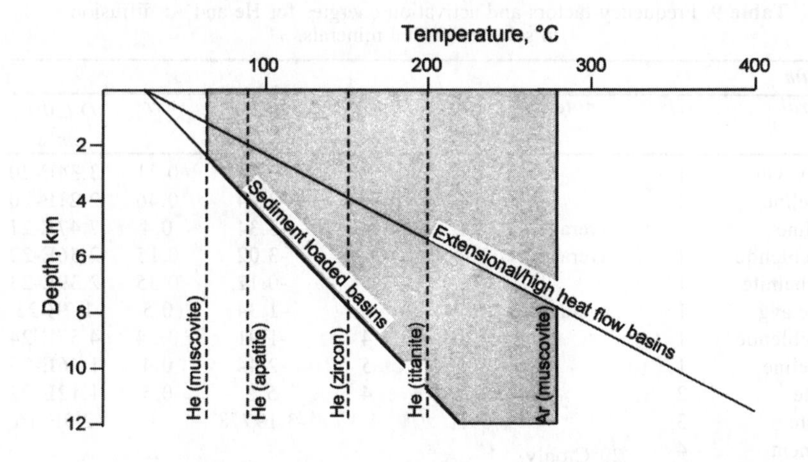

Figure 12. ^{4}He closure temperatures of common U-rich minerals compared with closure of Ar in muscovite (which has the lowest closure temperature of common K-bearing crustal phases) provides an end-member for considering relative ^{4}He vs ^{40}Ar release in different thermal regimes (O'Nions and Ballentine 1993; Ballentine et al. 1994). For typical crustal grain sizes and mineralogies, the He closure temperature is lower than that of Ar by over 100°C. If crustal minerals are held at a temperature between the He and Ar closure temperatures then the majority of the He in the rock will be released compared with only a small fraction of the Ar, resulting in elevated He/Ar ratios. Below the He closure temperature, no noble gases are released while above the Ar closure temperature, the ^{4}He/^{40}Ar* ratio will be close to production. This is illustrated above where the likely range of He and Ar closure temperatures are given by the vertical dotted lines while the solid lines show the geothermal gradients of typical loading basins (e.g., Po basin, Italy) and extensional basins (e.g., the Pannonian basin, Hungary). As expected, ^{4}He/^{40}Ar* in the Po basin (up to 124, average = 33) is considerably higher than the production ratio (≈ 5) while ^{4}He/^{40}Ar* of the Pannonian Basin fluids (up to 25, average = 7.7) approach the production value. Similar patterns are found elsewhere: see text for references.

that there are situations where Ar diffusional loss through the rock is slower than Ar production in the mineral. For example, Kelley and Wartho (2000) found ^{40}Ar in kimberlitic phlogopites that required Ar preservation in the phlogopite grains at ~700°C. This is 350°C above the phlogopite closure temperature. The phlogopite itself cannot retain Ar at these temperatures in an open system, therefore Kelley and Wartho (2000) concluded that the surrounding rock system must have retained a significant proportion of the ^{40}Ar produced at high temperatures. In this situation, diffusion out of the mineral grains does not occur due to the lack of a diffusion gradient. Instead, the ^{40}Ar will be distributed through the different minerals in the rock according to their respective K_ds, resulting in the 'parentless' ^{40}Ar observed in the phlogopite. While there have not been any equivalent studies on the distribution of He, the fact that K_d values are of the same magnitude for He as for Ar (Kelley 2002) suggests that 'parentless' He will also be trapped in minerals above the He closure temperature in low porosity environments due to the lack of a diffusion gradient. While these observations have implications for thermochronology, it is likely that the fraction of He or Ar retained in the crust, particularly the upper crust, above their respective closure temperatures is low, and will not represent a significant crustal reservoir for the noble gases.

Fracturing. Dilatant fracturing—the inelastic failure of brittle rocks that occurs during compressional loading—results from microfracturing prior to macroscopic fracturing. By compressing granites and basalts under vacuum, Honda et al (1982) showed

Table 9. Frequency factors and activation energies for He and Ar diffusion in selected crustal minerals.

Mineral	ref	note	Ea KJ mol^{-1}	err	$log(D_0)$ $cm^2 s^{-1}$	err	D (20°C) $cm^2 s^{-1}$
Helium							
muscvovite	1		87	10	-3.94	0.71	3.34E-20
nepheline	1		88	6	-3.94	0.46	2.21E-20
sanidine	1	average	94	4	-3.34	0.4	7.47E-21
horneblende	1	average	104	2	-3.02	0.15	2.56E-22
Langbeinite	1		126	5	-0.12	0.35	2.39E-23
augite avg	1	average	120	8	-1.34	0.5	1.7E-23
horneblende	1		120	4	-1.91	0.34	4.57E-24
nepheline	1		131	5	-2.44	0.4	1.46E-26
apatite	2		152.04	4	5.2	0.5	1.12E-22
granite	3		69.1	4.3	-3.19778		2.9E-16
sediment	4	20°C only					
Argon							
phlogopite	5		243.2	10.9	-0.12494		2.8E-44
biotite	6		197.4		-1.11351		4.31E-37
horneblende	6		269.22		-1.61979		2.01E-50
muscovite	6		168		-6.22185		5.97E-37

References:
1: Lippolt and Weigel (1988); 2: Wolf et al (1996); 3: Hussain (1996);
4: Ohsumi and Horibe (1984); 5: Giletti and Tullis (1974); 6: McDougall and Harrison (1988)

that dilatant fracturing of rocks releases noble gases from the minerals in which they were produced. The fraction of noble gas liberated from the rocks during compression increased linearly with the dilatant volume change (dilatancy). More He than Ar was lost at any given dilatancy, suggesting that diffusion into and along microfractures created by compression was the dominant release mechanism.

The effect of rock fracturing was examined numerically by Torgersen and O'Donnell (1991), who considered a one-dimensional slab of infinite length (Fig. 13). The gas flux out of the rock increases as a function of the fracture spacing of the slabs before (= L) and after (= l) a fracture event (for a given diffusivity). The increase in gas flux reaches a maximum of 10^4 times greater flux than the steady-state flux for L/l = 100. The timescale of release depends on D/l^2 but is likely to be short (Ka rather than Ma) for geologically reasonable conditions. For example, a 10m fracture spacing will result in loss of all the He that had accumulated prior to fracturing in ~1500 yr ($D = 10^{-7}$ $cm^2 s^{-1}$). However, the dependence of the gas release on D means that He will be released much more rapidly than Ar. Thus, the first fluids released from the slab will have high He/Ar ratios whereas later fluids lost from the slab will have low He/Ar ratios. Fracture release of noble gases can fractionate He from Ar and will likely result in episodic gas release related to tectonic movement in the brittle upper sections of the crust (Torgersen et al 1992b).

Mineral breakdown/diagenesis/metamorphism/alteration

Mineral transformations during diagenesis (e.g., illite ⇒ smectite transition), metamorphism (e.g., recrystallization of clay minerals to biotites, amphiboles, etc) or alteration (e.g., serpentinization of mafic minerals) are likely to release the radiogenic noble gases that were produced within their lattices. This assumes that:

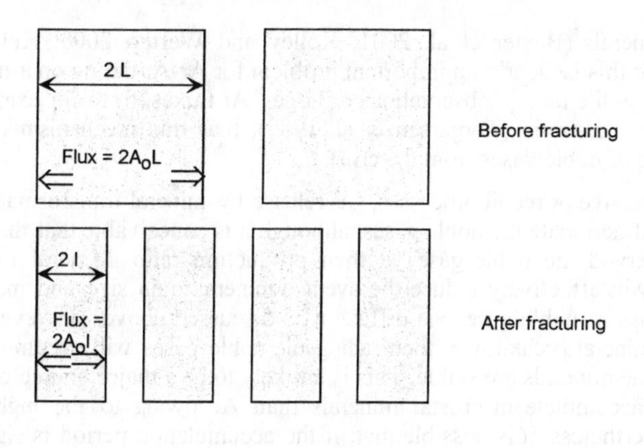

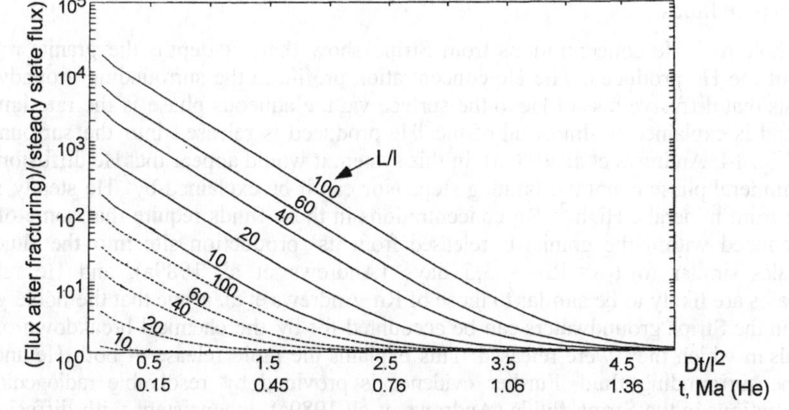

Figure 13. Predicted release of He and Ar from recently fractured rocks (after Torgersen and O' Donnell 1991) TOP: Schematic illustration of the concept behind the numerical model of Torgersen and O'Donnell. Infinite length slabs of half-width L are fractured to a new infinitely long slab of halfwidth l; the medium between slabs is assumed to be an infinite noble gas sink, therefore the steady-state flux out of the slab is a function only of the noble gas production rate (A_o) and slab thickness. BOTTOM: Computed increases in He or Ar flux (relative to the expected steady-state flux) as a function of time after fracturing. Flux computed for different unfractured/fractured ratios (L/l) for steady-state (solid lines) and open system (dashed lines) initial conditions. Time is given as the non-dimensional parameter Dt/l^2 and as likely He release times: these are calculated for l = 10 m and D_{He} = 1×10^{-7} cm^2 s^{-1} m^{2}a^{-1}. The flux of Ar from the fractured slab follows the same curves as those for He except that the timescale for release scales with D_{He}/D_{Ar}, which is likely of the order $10^4 - 10^5$, i.e., $Dt/l^2 = 1$ corresponds to 0.3 Ma for He but 3000 Ma for Ar ($D_{Ar} \approx 1 \times 10^{-12}$ cm^2 s^{-1}). Therefore, fracturing is likely an efficient mechanism for releasing He from the crust and for producing fluids with high He/Ar, but is unlikely to affect the outgassing of crustal Ar, unless fracturing occurred at relatively high temperatures (and therefore high D_{Ar}).

(a) noble gases are incompatible elements during these processes; and

(b) a mechanism exists to remove noble gases from these sites.

While there is robust evidence for the incompatible behavior of all the noble gases (Brooker et al. 1998), there may not always be a suitable pathway for the noble gases to escape the site of recrystallization. This potentially leads to parentless or 'excess' Ar trapped in minerals (Baxter et al. 2001; Kelley and Wartho 2000; Kelley 2002, this volume). While this is clearly an important problem for Ar-Ar dating on a mineral scale, it seems unlikely in the face of observations of large ^{40}Ar fluxes, from for example the Great Artesian Basin, Australia (Torgersen et al. 1989), that this mechanism can affect large scale degassing of noble gases from the crust.

Unlike diffusive or recoil processes, gas release by mineral transformation is unlikely to elementally fractionate the noble gases, although it is conceivable that the minerals may not have preserved the noble gases at their production ratio. Mineral breakdown at a constant rate will effectively reduce the average mineral grain size and make them more amenable to loss of noble gases via diffusion as discussed above. However, if there is a hiatus in the mineral breakdown, then radiogenic noble gases will accumulate during the period where the minerals are stable. This is unlikely to be a major source of ^{4}He as He is less likely to accumulate in crustal minerals than Ar owing to the higher rate of He diffusion. Nevertheless, it is possible that, if the accumulation period is significant, noble gas release from mineral breakdown can form the dominant source of radiogenic noble gases in pore fluids.

Whole rock He concentrations from Stripa show that, at depth, the granite retains ≤10% of the He produced. The He concentration profile in the surrounding groundwater suggests that diffusive loss of He to the surface via the aqueous phase is the rate limiting step, and is explained if almost all of the ^{4}He produced is released into the surrounding fluid (Fig. 14; Andrews et al. 1989a). In this system it would appear that He diffusion out of the mineral phase is not the limiting step. Nor can it be explained by ^{4}He steady state release from minerals. High ^{222}Rn concentrations in these fluids require that some of the Rn produced within the granite is released from its' production site into the fluid on timescales similar to $t_{1/2}{}^{222}$Rn = 3.5 days (Andrews et al. 1989a), and He release timescales are likely to be similar to those of Rn. Andrews et al. argue that the noble gases found in the Stripa groundwaters can be accounted for by the chemical breakdown of the minerals in which they were released. This explains the rapid release of both He and Rn into the surrounding fluid. Further evidence is provided by resolvable radiogenic Ar concentrations in the Stripa fluids (Andrews et al. 1989a), inconsistent with diffusion as the dominant mechanism for releasing noble gases from crustal minerals (e.g., Elliot et al. 1993). Similar conclusions have been made for the Carnmenellis granite, England (Andrews et al. 1989a; Martel et al. 1990).

Additional arguments for local metamorphic reactions as a significant mechanism for releasing noble gases from minerals comes from Ar-Ar dating studies that show there are lithological controls on 'parentless' ^{40}Ar; for example, 'parentless' ^{40}Ar is correlated with K content in some charnockites (Foland 1979) and eclogites (Scaillet 1996). Because the equilibrium mineral-fluid partition coefficient for noble gases, K_d, is very low, if a fluid is present it is unlikely that the released noble gases will be re-incorporated in the newly formed minerals. However, in the absence of a significant fluid volume (Eqn. 30) (e.g., in the lower crust), the released gases must re-distribute themselves depending on their solubilities in the respective new mineral phases (Baxter et al. 2001).

TRANSPORT OF NOBLE GASES FROM THE DEEP CRUST TO SHALLOW LEVEL SYSTEMS

Release of noble gases from minerals by recoil, diffusion, fracturing or chemical change defines the amount of crustal radiogenic-nucleogenic noble gas and their relative abundance released into the proximal fluid system. The transport of the noble gases from the proximity of their production to near surface systems and, ultimately, the atmosphere, is dependant on the driving force. This must be in the form of either a concentration gradient driving diffusion, or a pressure gradient resulting in advective fluid flow. With geologically reasonable timescales and parent element concentrations, noble

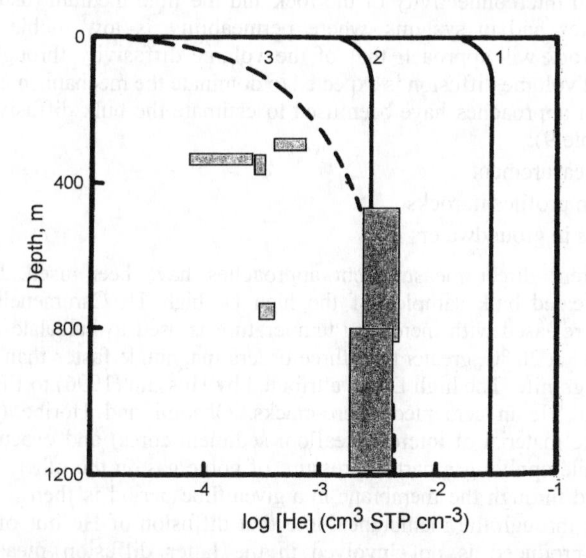

Figure 14. He concentration profiles through the shallow crust (after Andrews et al. 1989a). The He concentration profiles expected in groundwaters for He diffusion through bulk granite (solid lines) with diffusion coefficients of 1.6×10^{-10} cm^2 s^{-1} (5×10^{-7} m^2 a^{-1}) assuming that either all the He produced is available for diffusion (profile 1) or that 10% of the He produced is available for diffusion (profile 2). The dashed line shows the profile expected if all the He produced diffuses from the granite into the groundwater, followed by diffusive loss from the groundwater to the atmosphere. The boxes show the compositions measured in Stripa groundwaters (from Andrews et al. 1982), which approximately match the profile expected if all the He produced is released into the groundwater, followed by diffusive loss from the groundwater itself (D^{He}_{water} = 1×10^{-5} cm^2 s^{-1} (0.03 m^2 a^{-1})), suggesting that He loss from the minerals in which they were produced is not a significant rate-limiting step. Note that diffusion profiles for Ar through the solid rock would be even more extreme with virtually no change in Ar concentration until within a few cm of the surface.

gases are not produced in sufficient quantities to form a continuous fluid capable of advective flow alone. Crustal noble gases will remain in trace quantities in the local pore fluids. The transport of the noble gases to near surface systems is dependant therefore on the behavior of the fluid in the pore space. In a system that has no advective fluid flow, transport will occur by bulk diffusion, which in turn is controlled by the host rock character such as permeability, tortuoisity, and the fluid occupying the pore space. When movement of fluid in the pore space occurs, these fluids will remove or carry away the crustal noble gases away from the site of release. These 'carrier' fluids may be produced within the rock formation by chemical change (e.g., hydrocarbon formation or metamorphic devolatization), or be sourced externally to the system (e.g., an aquifer or magmatic fluid). In the former case, the crustal gases provide information about the source of the fluid, in the latter, information about the system through which they have passed (Ballentine et al. 2002, this volume).

Diffusion: A viable transport mechanism to degas the continental Crust?

Estimating the diffusivity of noble gases through crustal rocks—as opposed to minerals—is non-trivial because of its strong dependence on both the rock type and temperature. At low temperatures volume diffusion through minerals or along grain boundaries is slow and the dominant diffusive route is expected to depend on the permeability and interconnectivity of the rock and the fluid medium (usually water). At high temperatures and in systems where permeability is low, noble gas diffusivity through whole rock will approach that of the volume diffusivity through the dominant mineralogy, and volume diffusion is expected to dominate the mechanism of gas transport. Several different approaches have been used to estimate the bulk diffusivity in a variety of systems (Table 9):

1. Direct measurement
2. Diffusion profiles in rocks
3. He fluxes in groundwaters

Two different direct measurement approaches have been used. Hussain (1996) incrementally heated bulk samples of the high U, high Th Carnmenellis granite. The increase in ^4He released with increasing temperature is used to calculate a diffusivity of $\sim 1 \times 10^{-15}$ cm^2 s^{-1} at 20 °C, greater than three orders magnitude faster than the constituent minerals of the granite. The high D_{He} is attributed by Hussain (1996) to the availability of recoil-implanted ^4He in cemented micro-cracks. Ohsumi and Horibe (1984) create a membrane of the material of interest (seafloor sediment cores) and evacuate one side of the material while applying a partial pressure of noble gas on the other. The amount of gas that diffused through the membrane in a given time period is then a function of the bulk diffusivity through the membrane. Note that diffusion of He out of the mineral in which it was produced is not involved in the latter diffusion measurements. The diffusivities of various sediments obtained by this method ranged from 13 to 30×10^{-6} cm^2 s^{-1} for He and 3 to 6×10^{-6} cm^2 s^{-1} for Ar measured at ambient temperature. These diffusion rates are only slightly lower than those obtained for He and Ar in water of 42×10^{-6} and 9×10^{-6} cm^2 s^{-1} respectively (Jähne et al. 1987), suggesting that water-filled porosity may be the dominant diffusion medium in these seafloor sediments.

The pattern of 'excess' ^{40}Ar trapped in minerals can be used to estimate the bulk diffusivity through the rock. In theory, the concentration of ^{40}Ar in a mineral above its closure temperature should be close to zero. However, if the Ar is produced in the rock faster than it can diffuse out, then the mineral is likely to retain some 'excess' ^{40}Ar. The build-up of ^{40}Ar in a mineral above its closure temperature is a function of the relative rates of Ar production and diffusion out of the rock, not the mineral. Utilizing outcrop-scale gradients in ^{40}Ar concentrations in an amphibolite from the Simplon pass, Switzerland, Baxter et al. (2001) calculate a bulk Ar diffusivity of the order 10^{-11} cm^2 s^{-1} at a temperature of ~500°C (the conditions of their study). These results probably define the upper limit on the bulk Ar diffusivity in the upper crust (at least, for the immediate lithology of amphibolite) as diffusion rates are likely to decrease at lower temperatures. Studies by Foland (1979) and Scailliet (1996) also correlate 'excess' ^{40}Ar buildup with limited bulk diffusivity.

Torgersen (1989) formulates the steady-state He flux (F) as a function of the production rate (G) and diffusion geometry of the system. In an homogeneous slab

$$F/G = 1 - 8/\pi^{-2} f\{Dt/l^2\} \quad (31)$$

where D, l and t are the diffusivity, depth of crust and time respectively. The degassing of He into the Great Artesian Basin aquifer system, Australia, was shown to be close to whole crust production (Torgersen and Clarke 1985) and similar conclusions have been made for other sedimentary basins (e.g., Takahata and Sano 2000). From the solution of Dt/l^2, ($t = 4.5 \times 10^9$ yr; l = 40 km) and conservatively estimating F/G ≥ 0.5, Torgersen

(1989) estimate that the diffusivity of He through the crust would have to be $\geq 2 \times 10^{-5}$ $cm^2 s^{-1}$. Torgersen (1989) scale this further assuming that 75% of the radioelements are in the upper 10 km of crust to obtain a $D_{He} \geq 1 \times 10^{-6}$ $cm^2 s^{-1}$. The effective bulk diffusivity of He through upper crustal rocks is unlikely to be this high, therefore transport of He by movement of the fluid through the crust is implied.

These last calculations provide an important perspective on the importance of diffusion as a transport mechanism over long length scales in the crust, as it is clear that the crust cannot be described as a static water filled porous medium over 4.5 Ga! Taking the diffusivity of He in ocean sediments at 20°C as an example, He can diffuse about 100 m per Ma. At the other extreme the limited evidence from the granite study suggests that diffusion rates through rock with limited porespace connectivity are many orders of magnitude lower and is not therefore a viable mechanism to efficiently transport noble gases from the deep crust to the surface.

Two very different mechanical regimes exist in the crust: (1) a shallow cool portion in which the rock is strong enough to allow a hydrostatic pressure gradient to be maintained but is likely episodically fractured following lithostatic over-pressuring (see below), creating a system of interconnected pores and pathways; and (2) a deeper hotter region where the rock has insufficient mechanical strength, except locally, to hold open the pores against the lithostatic pressure exerted by the overburden (e.g., Wood and Walther 1986). The transition from one domain to another depends on the rock composition and temperature gradient. In quartz-rich sediments and with a temperature gradient of 30°C/km, this transition occurs between 3- to 6-km depth (Rubey and Hubbert 1959; Fyfe et al. 1978) and is clearly seen in many sedimentary basins where fluid pressure becomes greater than hydrostatic at depths < 3 km (Mann and Mackenzie 1990). In crystalline rocks, this transition may occur over a broad temperature range, but typically occurs between 300-450°C (Manning and Ingebritsen 1999; Simpson 1999). For typical crustal gradients of 30°C/km, transition from the brittle to the ductile region occurs between 10- to 15-km depth.

Within the shallow crust, the diffusive length-scale of He relative to the rate of groundwater flow can be a significant factor in modeling noble gas loss from aquifer systems to the atmosphere. For example, loss by diffusion to the atmosphere can be modeled assuming a zero concentration boundary at the surface and a zero concentration initial condition (Andrews et al. 1989a):

$$C(z,t) = Gt_f[1-\exp(-2z/\sqrt{(\pi Dt)}] \quad (32)$$

where C(z,t) is the concentration at a given depth (z) and time (t), G is the production rate and t_f is the formation age. Some illustrative curves for diffusive loss of noble gases from the crust are shown in Figure 14, where the He concentrations in the water are perturbed on a ~100-m length scale as the boundary layer is approached. The length scale of the water affected by Ar diffusion will only be on the order of centimeters.

Differential release and transport of helium and argon

Low He/Ar ratios in fluids will result from release of noble gases (for example, by fracturing, chemical alteration or by raising the temperature) that have accumulated in minerals within an environment of preferential ^{4}He release. This general pattern has been observed in the KTB drill-hole, Germany, where rocks from the shallow portion of the drill-hole (<3000 m) have lower He/Ar ratios than those at the bottom of the drill-hole (3000 to 7000 m) (Bach et al. 1999; Drescher et al. 1998). Curiously, the variations in He/Ar in the KTB rocks result largely from high Ar concentrations near the top of the drill-hole; without isotopic information, it is not possible to determine if this is due to variations in atmospheric contamination as opposed to preferential noble gas release from the crust. A more robust observation of this phenomenon are high $^{4}He/^{40}Ar^*$ ratios in shallow fluids from the Po basin, Italy. These high $^{4}He/^{40}Ar^*$, up to 120 compared to predicted ratios of ~5, cannot result from extreme variations in U+Th/K but must be due to preferential release of ^{4}He in the shallow crust (Elliot et al. 1993). Conversely, many fluids from the Great Artesian Basin (GAB), Australia (Torgersen et al. 1989), Permian Basin, Texas natural gases (Ballentine et al. 2001), The Trias aquifer within the Paris basin (Pinti and Marty 1995) and the Pannonian and Vienna Basins (Ballentine 1991; Ballentine et al. 1991) all have $^{4}He/^{40}Ar^*$ ratios close to the production value, consistent with their origin being from the deeper and hotter crust where diffusivities of Ar are sufficiently high to permit quantitative release of both He and Ar (Fig. 12).

In principle, the $^{4}He/^{40}Ar$ ratio in crustal fluids thus provides a tracer of the thermal conditions, or depth, from which the fluid was derived (Mamyrin and Tolstikhin 1984). In practice the regional release of ^{4}He and ^{40}Ar, and therefore the ($^{4}He/^{40}Ar$) ratio in the fluid will be dependent on the type, abundance and size distribution of the different minerals in which the respective noble gases are produced. For example, U in high heat producing rocks such as the Carnmenellis and Stripa granites is likely concentrated in fine grained (<40 μm) interstitial uraninite, facilitating ^{4}He release (Andrews et al. 1989a; Martel et al. 1990). One study from fluids trapped within the Swiss Alpine Palfris Marl, has put a limit on where the regional thermal transition from normal to high $^{4}He/^{40}Ar$ ratios occurs. Ballentine et al. (1994) use a stable isotopes and a constant $^{40}Ar^*$ concentration in CH_4-rich fluid inclusions and CH_4 from associated borehole gases to argue that these fluids have a common origin. The trapped fluids are within carbonate vein

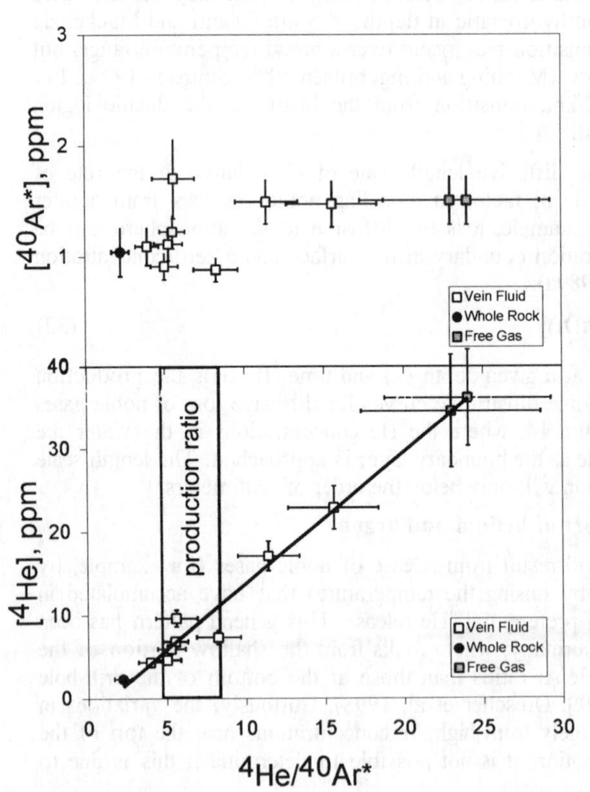

Figure 15. He and Ar concentrations are shown as a function of $^{4}He/^{40}Ar^*$ in the Palfris Marl (after Ballentine et al. 1994). Near-constant $^{40}Ar^*$ concentrations (all variation in [$^{40}Ar^*$] is within 2σ of the mean: dashed line, top figure) are found in fluids trapped in vugs within the Palfris marl and in free borehole fluids from the same horizon. Stable isotopes and a constant $^{40}Ar^*$ show that these differently sited fluids are derived from the same source. In contrast, ^{4}He concentrations are 10× higher in the free fluid than that trapped in the inclusions (lower figure) are likely due to addition of 'excess' ^{4}He to the free fluid. This is explained by ^{4}He, but not ^{40}Ar, diffusive loss into the fracture gas phase from the host marl. The transition to high $^{4}He/^{40}Ar$ ratios has occurred subsequent to the fluid formation and entrapment in the inclusions, at below 250°C and 6- to 8-km depth.

inclusions. Because of the low U+Th concentrations in the carbonate these are expected to have preserved the original fluid composition, with 6 out of 7 of the carbonates trapping fluids with $^{4}He/^{40}Ar^{*}$ ratios within error of the crustal production value (Fig. 15). In contrast the free gas phase shows a ^{4}He excess. This is explained by ^{4}He, but not ^{40}Ar, diffusive loss into the fracture gas phase from the host marl. The transition to high $^{4}He/^{40}Ar$ ratios has occurred subsequent to the fluid formation and entrapment in the inclusions. From stable isotopes, vitrinite maturity indicators and fluid inclusion formation pressures and temperatures, this occurred at below 250°C and 6- to 8-km depth.

Castro et al. (1998a,b) have modeled advective Ar and advective-diffusive He transport in a numerical simulation of different Paris Basin aquifer horizons and shown how these different transport mechanisms could account for the observed varying $^{4}He/^{40}Ar^{*}$ ratios. This interpretation is at variance with Pinti and Marty (1995) who have argued that impermeable layers act as efficient barriers for both He and Ar diffusion, and that the $^{4}He/^{40}Ar^{*}$ variation in the Paris basin results from advective mixing. Although diffusion may play a role in transporting He (but probably not the heavier noble gases) across boundary layers from one fluid system to another, in the absence of connected porosity (e.g., the crust below 3- to 15-km depth), diffusion as a significant mechanism of noble gas transport from the deep crust to shallow level systems can be neglected.

Accumulation and release: Fluid flow in the Crust

The steady-state model. There are many examples of crustal fluid systems that have concentrations of noble gases orders of magnitude higher than can be explained by local production, release and accumulation. One of the earliest examples is the series of studies on the noble gas concentration in groundwater from the Great Artesian Basin, Australia (Torgersen and Clarke 1985, 1987; Torgersen and Ivey 1985; Torgersen et al. 1989; Torgersen et al. 1992b; also see Bethke et al. 1999). An increase in concentration of noble gases with increasing age of the groundwater led to the development of the steady-state crustal degassing model. These workers argued that the flux of ^{4}He and $^{40}Ar^{*}$ into the base of the aquifer could be accounted for only if the entire crustal volume was releasing its radiogenic noble gases into the aquifer at the same rate as they were being produced. Similar rates have been inferred in other regional groundwater systems such as the Saijo Basin, Japan (Takahata and Sano 2000) and from natural gases reservoirs (Sano et al. 1986). Estimates of ^{4}He flux into lakes show similar high values (summarized by Kipfer et al. 2002, this volume), although varying over an order of magnitude in value. Hydrodynamic modeling by Castro et al. (1998a) implies a ^{4}He flux from the Paris basin equivalent to whole basin production, similar to the Great Artesian Basin. However, the model estimates are inconsistent with more direct estimates of He fluxes that use cyclic variations in the aquifer stable isotope record ($\delta^{18}O$ and δD) to estimate aquifer residence times (Marty et al 1993; Pinti and Marty 1995). The latter studies are used to indicate that the flux of ^{4}He from the basin center is likely to be lower by a factor of 10 than that produced within the aquifer.

Tectonic control of release. Both higher and lower fluxes are observed elsewhere: Within the Pannonian basin, a single gas field contains the volume of ^{4}He and $^{40}Ar^{*}$ produced in the entire crust beneath the sub-basin since the basin formation 15 Ma ago (Ballentine et al. 1991). This particular gas field shows evidence that the gases have been transported to the trapped site dissolved in an active groundwater system. A conceptual model in which basin extension releases deep stored noble gases, followed by groundwater focusing into the gas trap was proposed. Winckler et al (1997) measured exceptionally high ^{4}He and ^{40}Ar concentrations in brines from the Urania Basin of the Mediterranean with ^4He concentrations in excess of 1×10^{-3} cm^3 STP ^4He g^{-1} H_2O. These He concentrations could not be generated through any steady-state model.

A diffusive mechanism cannot account for significant noble gas mass transport from the deep crust. Nor, given the permeability of the crust, is it reasonable that there is continuous and widespread fluid loss from the deep crust. An average crust K age of 1 Ga (Hurley and Rand 1969) clearly indicates that although portions of the crust are extensively outgassed, regions of the crust must exist in which there are substantial accumulations of radiogenic and nucleogenic noble gases. Although at some point accumulations may reach a stage where discontinuous loss (both temporally and spatially) approximates the production rate, the rate of loss might be expected to be more strongly affected by large-scale tectonic events. For example, the regional strong thermal pulse caused by the formation of the Pannonian basin extensional system resulted in regional metamorphism, and the release of accumulated radiogenic noble gases at a rate far higher than they are produced in the crust (Ballentine et al. 1991), and furthermore shows clear evidence of magmatic involvement in the form of $^3He/^4He$ ratios well above crustal production values. Regions in which flux estimates are closer to theoretical steady-state values such as the Great Artesian Basin and Taiwan natural gases (Torgersen and Clarke 1987, 1987; Sano 1986) also show evidence of elevated $^3He/^4He$ ratios indicative of deep magmatic activity.

Similarly, orogenesis is also expected to result in regional metamorphism, fluid mobilization and regional degassing. Ar-Ar dating of these events clearly shows the resetting of mineral noble gas concentrations due to these processes (e.g., Vance et al. 1998), while a mass balance assessment of noble gases in the mineral and associated fluid phases in low grade Alpine metamorphism demonstrate extensive fluid loss from the system (Ballentine et al. 1994).

The concentration of radiogenic noble gases in fluids generated by regional fluid mobilization, and therefore their flux, will in turn be controlled by this history. In this respect, noble gas concentrations in the crust will also depend on the regional tectonic history. In summary:

- There is no a-priori reason, nor significant evidence, that regions of the crust do or should exhibit a uniform steady-state degassing flux.
- Crustal degassing is controlled by regional tectonic events resulting in metamorphism, such as orogeny and crustal extension.

The character of 'deep' crustal noble gases. $^4He/^{21}Ne/^{40}Ar$ ratios close to the predicted average crustal value are observed in many samples associated with either high radiogenic noble gas flux or magmatic fluids. These include natural gas samples from the Pannonian and associated Vienna basins (Ballentine 1991; Ballentine et al. 1991; Ballentine and O'Nions 1994), Red Sea brines (Winckler et al. 1997), magmatic CO_2 in the Val-Verde basin, Texas (Ballentine et al. 1991), Paris basin (Pinti and Marty 1995; Castro et al. 1998); and the Great Artesian Basin (Torgersen et al. 1989). Care must be taken within these studies to distinguish samples that have been subject to secondary fractionation process, such as phase fractionation. This can be established by identifying coherent fractionation in non-radiogenic gases such as atmosphere-derived noble gases introduced via the groundwater system (Ballentine et al. 1991). Similar studies that do not preserve predicted ratios can often be ascribed to shallow and fractionated radiogenic sources (e.g., Ballentine et al. 1994; Ballentine and Sherwood Lollar 2001). Nevertheless, the general inference is that samples ascribed to deep sources are accumulated, released and transported to near surface systems with minimal fractionation. This can occur only if:

1. Storage in the deep crust is efficient for all noble gases irrespective of site;
2. Release of different noble gases to the transporting media is not rate limited by diffusion;
3. Diffusion as a major mass transport mechanism is insignificant; and
4. Transport to the shallow systems is single phase.

HEAT AND HELIUM

The relationship between ^{4}He and heat

Radioactive decay of U-series elements accounts for >75% of the Earth's heat (O'Nions and Oxburgh 1983; Turcotte and Schubert 1982). U-series decay produces ^{4}He and Heat with a characteristic ratio of 3.7×10^{-8} cm^3 STP ^{4}He J^{-1}. However, processes in the mantle can efficiently fractionate He from heat, as the surface flux of ^{4}He from the mantle is only 5% of the equivalent Heat flux (see Graham 2002). The heat flow from the crust can be subdivided into that produced in the crust and that transported from the mantle through the crust (O'Nions and Oxburgh 1983). From estimates of the U + Th content of the continental crust, it would appear that only about half the continental heat flux was produced within the crust (O'Nions and Oxburgh 1983); this can be even lower in regions of crustal extension or tectonic deformation (Clark and Phillips 2000; Torgersen et al. 1992b). Heat and ^{4}He are produced in the mantle with the same ratio as in the crust. However, the convecting mantle has a near constant ^{3}He/^{4}He ratio ($\approx 1.2 \times 10^{-5}$), therefore the mantle also has a diagnostic ^{3}He/heat ratio that can be estimated from measurements of the He and Heat fluxes in oceanic basins. This can be used to tease apart fractionation of He and heat during transport across the mantle-crust boundary and fractionation of He and Heat that were produced in the crust.

Transport of mantle heat and helium through the crust

Given that heat transport through the crust is dominated by conduction, whereas He is transported into and through the crust primarily by advection (Bickle and McKenzie 1987), fractionation of He from heat is expected. Nevertheless, there does appear to be a global relation between the mantle heat flux and ^{3}He/^{4}He of the local crustal fluids (Polyak and Tolstikhin 1985): stable regions of crust have comparatively low conductive heat flow and radiogenic He (e.g., Russian Shield, Canadian Shield etc) whereas recently tectonized regions have high mantle heat flow associated with high ^{3}He/^{4}He (e.g., Pannonian Basin, Massif Central, Rhine, Yellowstone, Eastern Great Artesian Basin, East African Rift, Larderelo, Italy).

Polyak and Tolstikhin (1985) approached the problem of regional He isotope patterns by considering that He isotope compositions resulting from differences in radioelement or target nuclei concentrations will be averaged to some degree when entrained in crustal fluids. This enables a comparison between mantle heat flux through geotectonic units and He isotope composition. Averaging the ^{3}He/^{4}He ratios of fluids from particular geotectonic environments—primarily in Northern Europe and Asia, Polyak and Tolstikhin showed that there is a distinct temporal evolution to the ^{3}He/^{4}He ratio of groundwaters, with high ^{3}He/^{4}He ($>1 \times 10^{-6}$) limited to magmatic/tectonically 'young' regions typically <50 Ma (Fig. 16). Groundwaters from tectonically 'old' regions, such as shields, are characterized by low ^{3}He/^{4}He ratios indistinguishable from radiogenic production values (ca 10^{-8}). In a plot of log (^{3}He/^{4}He) vs. log (basement age) a smooth curve showing reducing ^{3}He/^{4}He with time is obtained (Fig. 17). Polyak and Tolstikhin observed that a plot of conductive heat flow vs. log(tectonic age) has a similar form, implying there is a relation between conductive heat flow and ^{3}He/^{4}He, i.e., mantle derived heat in the crust is accompanied by mantle derived He. This is shown in Figure 17

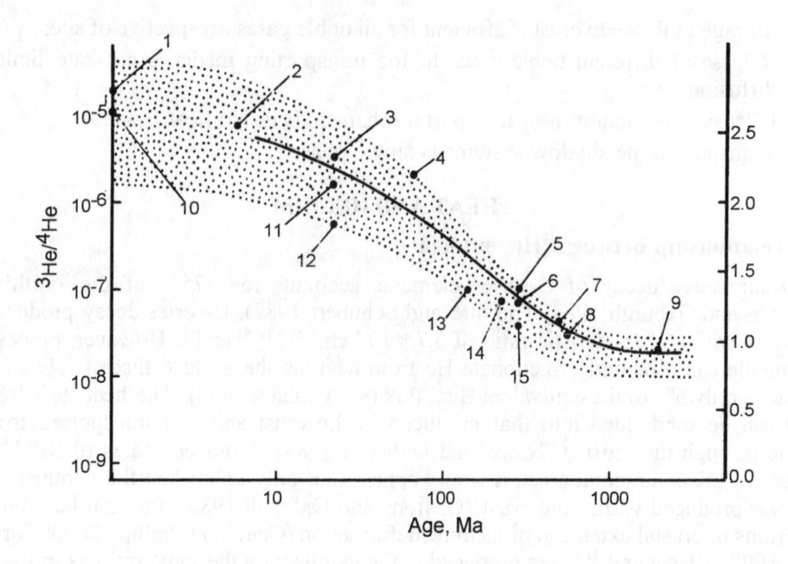

Figure 16. $^3He/^4He$ ratio as a function of age (Polyak and Tolstikhin 1985). The $^3He/^4He$ ratio of European groundwaters are plotted against the 'tectonic' age of the surrounding crust (stippled region; solid circles represent selected data). Adding more recent and/or global data increases the size of the envelope, particularly for young (<100 Ma) tectonostratigraphic units, but the relation between $^3He/^4He$ and age does not change significantly. The solid line is the conductive heat flux for the same tectono-stratigraphic units. The selected data are: 1: Iceland; 2: Eastern and Central Kamchatka; 3: Lesser Caucasus; 4: Western Kamchatka; 5: Scythian and Turanian Plates; 6: Mediterranean Foredeeps; 7, 15: Siberian Platform; 8: Baikalides; 9, 14: East European Platform; 10: Kuril Island Arc; 11: Greater Caucasus; 12: Trans-Caucasus Depression; 13: Vilyuy Syneclise (From Polyak and Tolstikhin 1985, and references therein). The groundwater compositions better represent the compositions of the geotectonic units themselves as the groundwaters naturally average out the variability of the rocks. [Used by permission of Elsevier, from Polyak and Tolstikhin (1985), *Chemical Geology*, vol. 52, Fig. 8, p. 21.]

and takes the form:

$$^3He/^4He = (^3He/^4He)_r^{\,(mQ-c)} \tag{33}$$

where $(^3He/^4He)_r$ is the radiogenic $^3He/^4He$ production ratio and m and c are constants; Q is the conductive heat flow. This relationship is only valid for mantle derived heat, with the lower intercept at the crustal $^3He/^4He = 10^{-8}$, where heat-flow is due to either crustal production or mantle heat unaccompanied by mantle He. The upper limit of the correlation assumes the mantle $^3He/^4He = 1.2 \times 10^{-5}$ and is due to the He/heat ratio of the mantle. The latter is unlikely to be exceeded during conductive heat flow because the effective diffusivity of heat through the crust is considerably greater than that of He. This value can, however, be exceeded if convective heat transfer occurs.

The relationship in some cases is less clear cut: for example, mantle He can be stored in the crust apparently without accompanying heat (Mirror Lake, Torgersen et al. 1994; Texas CO_2 well gases, Ballentine et al. 2001) and mantle heat can be transferred into the crust without affecting the $^3He/^4He$ composition of local groundwaters (North Atlantic, Oxburgh et al. 1986). Therefore, imperfect correlations are to be expected, but the considerable differences in He isotopic composition between the different potential He and heat sources means that useful information can be obtained from this approach.

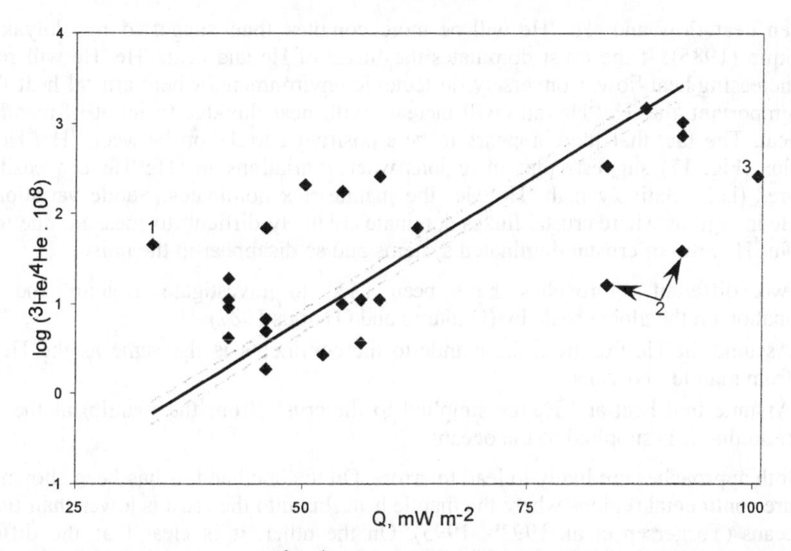

Figure 17. The relationship between ^{3}He/^{4}He and heat flux after Polyak and Tolstikhin (1985). The slope of the equation given by Polyak and Tolstikhin (1985) is shown by the solid line with error on the intercept indicated by the dashed line (slope = 6; intercept = -5.3±0.2). The poor correlation between log(^{3}He/^{4}He) and the heat flux shown here is consistent with mantle heat transported into the crust along with mantle He although with significant variability in the ^{3}He/Heat ratio supplied to the crust. It is noticeable that the outliers are limited to mobile tectonic belts where high geotherms likely result from vertical displacement: highlighted outliers are: 1. Eastern Black Sea; 2. Caucasus; 3. Baikal Rift. However, this correlation has been called into question by (Oxburgh and O'Nions 1987); further data show that high heat flow is also commonly found in regions that have low ^{3}He/^{4}He ratios. It would apear that while high ^{3}He/^{4}He ratios are usually accompanied by high heat flow, high heat flow is not always associated with high ^{3}He/^{4}He.

Oxburgh and O'Nions (1987) observed that the helium isotopic compositions of some stable cratonic areas are low and uniform, but these can have variable—and quite high—heat fluxes (e.g., U.K. mainland and shelf has ^{3}He/^{4}He $\leq 1 \times 10^{-7}$ but heat fluxes as high as 125 mW m^{-2}). Griesshaber and co-workers (Griesshaber et al. 1992, 1988) examined He-Heat relationships on a local scale and showed that there was no clear relationship between ^{3}He/^{4}He and heat flux in the extensional Rhine Graben. These observations contradict the findings of Polyak and Tolstikhin (1985), in part due to the differences in spatial scale of the two studies. A further complication is the variable amount of ^{4}He that may be stored regionally and mixed with the local fluids. A well-defined relationship between ^{3}He/^{4}He and heat over large geological provinces would seem surprising considering He and heat can readily be fractionated during relatively shallow processes.

Helium and Heat can be considered to have two components, one from the mantle and one produced in shallow crust:

$$Q_T = Q_M + Q_H \tag{34}$$

$$F_T = F_M + F_H \tag{35}$$

where Q and F are heat and helium fluxes respectively, subscripts T, M, and H are total, mantle, and crustal respectively fluxes respectively (strictly speaking, Q_H and F_H are the fluxes produced from a layer with a characteristic thickness 'H' and Q_M and F_M are the fluxes from below this layer). Given that F_M/Q_M should be constant, the relationship between heat flow and $^3He/^4He$ will be more complex than suggested by Polyak and Tolstikhin (1985). If the crust dominates the fluxes of He and heat, $^3He/^4He$ will reduce with increasing heat flow. Conversely, in tectonic environments where crustal heat fluxes are unimportant, the $^3He/^4He$ ratio will increase with heat flux due to input of mantle He and Heat. The fact that there appears to be a positive correlation between $^3He/^4He$ and heat flux (Fig. 17) suggests that in regions where variations in $^3He/^4He$ can easily be measured (i.e., relatively high $^3He/^4He$) the mantle flux dominates. Subtle variations in $^3He/^4He$ in regions where crustal fluxes dominate are likely difficult to measure due to the low $^3He/^4He$ ratio of crustal dominated systems and so disappear in the noise.

Two different approaches have been used to investigate Helium and heat fractionation on the global scale by (Oxburgh and O'Nions 1987):

1. Assume the He flux from the mantle to the continents is the same as the He flux from mantle to oceans.
2. Assume that heat and He are supplied to the crust (from the mantle) at the same ratio that it is supplied to the oceans.

Both approaches are likely to lead to errors. On the one hand, it has been shown that there are continental regions where the mantle heat flux into the crust is lower than that of the oceans (Torgersen et al. 1992b, 1995). On the other, it is clear that the different transport properties of helium and heat through both mantle and crust will result in transfer of volatiles and heat to the crust with ^3He/heat ratios different from those of the mantle.

Fractionation of He from heat has also been observed in active and ancient hydrothermal systems from continental regions, confirming that the relationship between $^3He/^4He$ and mantle heat flow is little more than an association between heat and volcanism as opposed to a well constrained causal relationship that could be used (for example) to estimate heat fluxes from $^3He/^4He$ ratios. Extreme variations in ^3He/Heat (0.01 to 10×10^{-12} cm^3 STP ^3He J^{-1}) are observed in Icelandic hydrothermal fluids (Poreda and Arnorsson 1992) in contrast to submarine hydrothermal systems that have more constant ^3He/Heat ratios of between 0.1 and 1×10^{-12} cm^3 STP ^3He J^{-1}; (Jean-Baptiste et al. 1998). Modeling by Poreda and Arnasson (1992) show that fractionation by phase separation alone, which would result in a high-^3He, low-heat vapor phase, cannot reproduce the observations. Instead, they suggest that combined fractional cooling and fractional degassing of the intrusive body is required to explain the data. From their models, He is preferentially lost from the magma such that (for each cooling-degassing increment): $(^3He/Q)_{FLUID}/(^3He/Q)_{MAGMA} \approx 20$, i.e., the magmatic body degasses twenty times faster than it cools. In the Icelandic case, convection is transferring mantle-derived volatiles into the crust more rapidly than mantle derived heat. This likely reflects the incompatible behavior of He during magmatic cooling rather than the intrinsic He and Heat transport properties.

Using the relationship derived by Turner and Stuart (1992) in order to estimate $^3He/Q_M$ (where Q_M is mantle derived heat) from measurements of noble gases in fluid inclusions, Burnard et al (1999) were able to estimate the $^3He/Q_M$ ratios of fluids related to Eocene gold deposition along the Ailaoshan Fault, Yunnan Province, China. Systematic variations along the fault showed that high ^3He/heat fluids were associated with the highest-grade gold ores and that the efficiency to extract fluids, and ^3He, from the parent intrusives decreased along with gold tonnages northward along the fault.

Transport of crustal heat and ^4He

The diffusion of heat through the crust (10^{-2} cm^2 s^{-1}) is considerably faster than that of He (10^{-5} cm^2 s^{-1}) (Torgersen et al. 1992a,b), therefore, crustal heat should be dissipated more rapidly than crustal He. Torgersen et al (1992b) calculated typical time scales for heat and mass fluxes through a 10-km-thick crustal section to be ~0.7 and 80 to 800 Ma respectively. The disparity between heat and mass flux time scales will increase for thicker crustal sections.

Table 10. Helium and heat budgets from the GAB (Torgersen et al. 1992b) and the Pannonian Basin (Martel et al. 1989).

Province	*Total Heat flux* $mW\ m^{-2}$	*Crustal Heat Flux* $mW\ m^{-2}$	*^4He flux* $cm^3\ STP\ m^{-2}$	*Heat flux from ^4He** $mW\ m^{-2}$	*Heat balance*
Central Shield	82.5	55.8	1.1×10^{-9}	30	Additional 26 $mW\ m^{-2}$ heat required
Eastern Australia	71.7	14.6	1.1×10^{-9}	30	15 $mW\ m^{-2}$ heat "lost"
Pannonian Basin	80	25♣	3×10^{-9}	81	51 $mW\ m^{-2}$ heat "lost"

* U-series decay produces 3.7×10^{-8} cm^3 $^4He\ J^{-1}$.
♣. error of ± 50%

Despite this prediction, the flux of ^4He from the GAB, Australia, is roughly correlated with the flux of heat resulting from U-series decay within the basin (Torgersen et al. 1992b). The appropriate helium and heat budgets for the GAB and the Pannonian Basin are given in Table 10. Torgersen and Clarke (1985) demonstrated that the crustal He flux (F_H) from the GAB is ~1.1×10^9 cm^3 STP $^4He\ m^{-2}\ s^{-1}$ (3×10^{10} atoms $m^{-2}\ s^{-1}$). These results show that there is a discrepancy of about 50% in the He and heat fluxes, with a heat excess in the Central shield (relative to the heat flux accounted for by the ^4He flux), while less heat (relative to the heat flux accounted for by the ^4He flux) is coming out the Eastern shield. Minor (≤25%) additional heat flux can result from K decay depending on the K/U ratio. However, considering the errors involved in these estimates, there is comparatively good agreement between He and heat fluxes for the GAB and Pannonian basins, which is not expected given their different transport properties. This may be explained if the loss of Heat and He is approaching steady state. Alternatively this may be explained if He transport out of the crust is faster than that predicted by diffusion. This is consistent with the observations of whole crust degassing, the inability of diffusion to achieve this over reasonable time scales, and the need for advective fluid transport of He out of the crust.

MAGMATIC NOBLE GASES IN THE CRUST

Tectonic control on magmatic fluid location

The general observation of Polyak and Tolstikhin (1985), that high $^3He/^4He$ is found in regions of high heat flow, restates the fact that mantle derived He in crustal fluids is found in regions of extension (Oxburgh et al. 1986; Oxburgh and O'Nions 1987). This is summarized in Table 11. In general, it can be seen that, with rare exceptions such as Mirror Lake, USA (Torgersen et al. 1995) and Val Verde Basin, Texas (Ballentine et al. 2001), fluids that have high $^3He/^4He$ ratios are from tectonically active regions and that the highest $^3He/^4He$ ratios are from regions undergoing active extension such as the Rhine Graben, Massif Central, East African Rift, or Great basin of the western United States. Active subduction also results in high $^3He/^4He$ groundwaters and geothermal fluids, e.g., Western Cordillera, Canada (Clark and Phillips 2000), Andes (Hoke et al. 1994) and the Western Pacific (Poreda and Craig 1989). Regions of crustal thickening such as the Alps do not seem to result in high $^3He/^4He$ ratios (Marty et al. 1992): while there are regions of the Alps (and also of Tibet) where $^3He/^4He$ ratios are greater than likely production

Table 11. Helium isotopes and concentrations in groundwaters, geothermal fluids, hydrocarbon and carbon dioxide wells.

Basin	*Tectonics setting*	*mean* $[^4He]$ cm^3 *STP* g^{-1}	$^3He/^4He_{mean}$ $x10^{-6}$	$^3He/^4He_{min}$ $x10^{-6}$	$^3He/^4He_{max}$ $x10^{-6}$	*Analyses*	*References*
Europe							
Pannonian	extensional	7.56	0.76	0.15	1.97	27	Martel et al. (1989); Ballentine et al. (1991); Sherwood Lollar et al (1994)
Rhine	extensional	2.32	1.03	0.21	2.14	11	Hooker et al. (1985a)
Massif Central	extensional	0.30	2.36	0.42	3.92	6	Matthews (1987)
North Sea	extensional	8480 (oil)	0.46	0.28	0.54	18	Ballentine et al. (1996)
Lardarello, Italy	convergent	0.57	1.54	0.01	3.04	16	Hooker et al. (1985b)
Paris Basin	sedimentary	533.57	0.09	0.07	0.21	20	Marty et al. (1993)
Po	sedimentary	9.04	0.04	0.03	0.05	18	Elliott et al. (1993)
Mediterannean Brine Basin	sedimentary	641.56	0.08	0.08	0.09	8	Winckler et al. (1997)
Balkan Peninsula	convergent	4015.40	0.46	0.15	1.10	19	Piperov et al. (1994)
Alps	convergent	32.59	1.12	0.03	5.10	20	Marty et al. (1992)
Stripa, Sweden	shield	1744.00	0.01	0.01	0.01	3	Andrews et al. (1989a)
Iceland	extensional	78.26	21.71	7.70	40.32	47	Poreda et al. (1992)
Switzerland	convergent	1590.19	0.12	0.07	0.20	8	Tolstikhin et al. (1996)
Busko Spa, Poland	sedimentary/foredeep	179.60	0.04	0.05	0.03	9	Zuber et al. (1997)
Carpathians, Poland	extensional	26.00	0.53	0.49	0.57	2	Lesniak et al. (1997)
Vienna Basin	extensional	126.86	0.54	0.10	3.17	20	Ballentine and O'Nions (1992)
North Amerca							
Sacramento	forearc	51.44	1.43	0.15	3.86	19	Poreda et al. (1986)
Great Basin, USA	extensional	1891.57	4.03	0.31	8.26	7	Welhan et al. (1988a;b)
West Texas	sedimentary	143.27	0.52	0.34	0.74	10	Ballentine et al. (2001); Zaikowski et al. (1987)
San Joaquin	forearc	61.05	1.31	0.15	3.86	24	Jenden et al. (1988)
Alberta Basin	sedimentary/forearc	152.42	0.20	0.01	0.69	20	Hiyagon and Kennedy (1992)

Canadian Shield	shield	15563.92	0.01	0.01	0.03	12	Bottomley et al. (1984)
Canadian Cordillara	sedimentary/forearc	6.30	3.44	9.16	0.08	7	Clarke and Phillips (2000)
Rio Grande Rift	extensional	79.95	5.84	5.41	6.65	4	Smith and Kennedy (1985)
Long Valley	extensional	1582.73	5.66	0.63	7.70	11	Hilton 1996; Welhan et al. (1988a,b)
Australia, New Zealand							
GAB, Australia		49.53	0.27	0.04	1.13	34	Torgersen et al. (1985)
Taupo Volcanic Zone	convergent	18.89	7.35	4.93	8.48	10	Hulston and Lupton (1996)
Central America							
Macuspana, Mexico	extensional?	14.9	0.34	0.06	0.70	18	Battani (1999)
South America							
Central Andes	convergent	14.14	3.01	0.11	9.06	17	Hilton et al. (1993)
Africa							
East African rift	extensional	103.65	6.43	1.26	11.20	32	Darling et al. (1995)
Egypt	extensional	270.00	0.36	n/a	n/a	1	Sturchio et al. (1996)
Asia							
Tibet	convergent	1565.41	0.08	0.02	0.31	8	Hoke et al. (2000)
Songliao, China	extensional	380.39	4.69	0.95	6.99	6	Xu et al. (1995)
Wudalianchi China	extensional	304.60	3.80	1.67	4.72	6	Dai et al. (1996)
Indus Basin	Foreland	404.44	0.04	0.01	0.08	8	Battani et al. (2000)
Indian Craton	shield	23669.58	0.23	0.74	0.01	11	Minisalle et al. (2000)
Japan	Forearc	413.16	7.91	3.96	11.40	13	Nagao et al. (1981)

Note: Data compilations that do not include He and Ne concentrations have been excluded as it is not possible to assess atmospheric contamination. Direct samples of volcanic gases have been excluded.

ratios, these do not appear to coincide with the main area of crustal thickening but rather with grabens or normal faulting at the periphery of collision (Marty et al. 1992; Hoke et al. 2000). Although transfer of other mantle-derived volatiles (such as Ne, Ar or CO_2) into crustal fluids also occurs in tectonically active regions, competing processes (such as radiogenic production in the crust) and high concentrations of these volatiles in the crust means their mantle–derived signatures are frequently obscured or complicated (e.g., Ballentine 1997): the isotopes of He remain the tool of choice for detecting mantle input into the crust.

The transfer of volatiles from the mantle into the crust is most efficient following melting of the mantle and intrusion of the melt into the crust: Watson and Brenan (1987) and Brenan and Watson (1988) have shown that fluid transport through the mantle is likely to be minimal in the absence of melt. This is also obvious in active volcanic regions where the mantle-derived ^{3}He flux decreases rapidly away from the volcanic centers (Sano and Wakita 1985). However, uncertainties in the crustal ^{4}He flux and the He concentration of the magmatic end-member make it difficult to estimate the mantle mass flux to the crust from He isotope ratios of crustal fluids. Torgersen (1993) showed that a flux as low as 10^3 atoms m^{-2} s^{-1} of ^{3}He would create a detectable change in the He isotope composition of the fluid, but depending on the efficiency of crustal ^{4}He outgassing, the range in ^{3}He/^{4}He observed in the fluid could span an order of magnitude (Fig. 18).

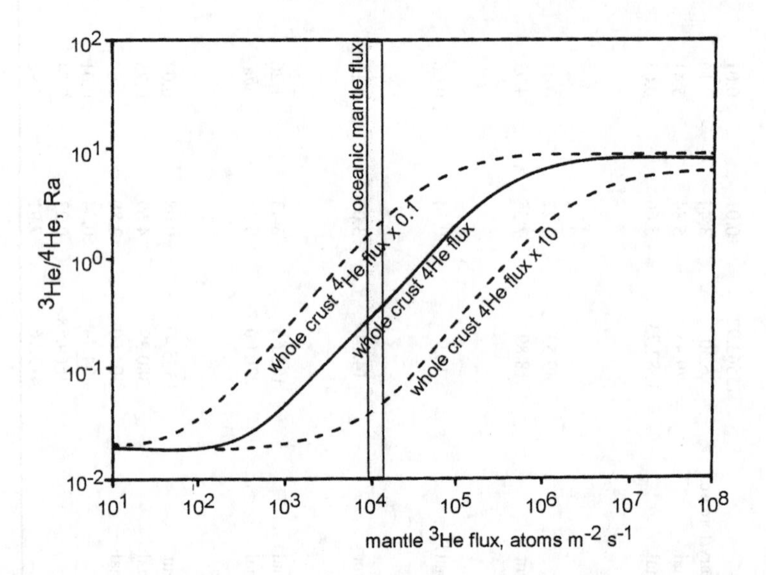

Figure 18. ^{3}He/^{4}He of groundwaters (after Torgersen 1993). The flux of ^{3}He from the mantle and ^{4}He from U series decay in the crust are the major contributions to the ^{3}He/^{4}He ratio of continental groundwaters. Because U concentration in the crust is relatively constant, the ^{3}He/^{4}He ratio of groundwaters is primarily a function of the ^{3}He flux from the subcontinental mantle. This figure illustrates the ^{3}He/^{4}He ratios expected in groundwaters for various mantle ^{3}He fluxes and crustal ^{4}He fluxes. The flux of ^{3}He from the mantle into the oceans is comparatively well constrained (gray bar): if the flux of ^{3}He from the sub-continental mantle was the same as that from the oceanic mantle, then all continental groundwaters would have ^{3}He/^{4}He ratios higher than the radiogenic production ratio, even if the flux of ^{4}He was 10 times greater than the production from U and Th decay. However, numerous groundwaters have He isotopic compositions that are indistinguishable from the radiogenic production ratio (Table 11), therefore the flux of ^{3}He under the continents is clearly lower than that under the oceans.

Fossil magmatic gases

It is clear that mantle–derived noble gases are stored in the crust as well as transported through it, as residual magmatic gases have been found trapped in fluid inclusions in ancient magmatic products [e.g., komatiites in Zimbabwe and Australia (Richard et al 1996); Greenland (Graham et al. 1998); Kola (Dauphas and Marty 1999; Marty et al. 1998), hydrothermal minerals (Burnard et al. 1999; Stuart et al. 1995; Stuart and Turner 1992), charnockites (Dunai and Touret 1993; Kamensky et al. 1990) and ultra-basic layered intrusions (Tolstikhin et al. 1992)]. While the concentrations of magmatic gases in these minerals can be surprisingly high (^{3}He concentrations can be similar to those trapped in basaltic glasses), these sites represent a small fraction of the crust and the trapped mantle noble gases are generally present in lower concentrations than radiogenic noble gases produced in the crust.

Magmatic fluids can also be preserved. For example, fossil mantle gases are preserved in the gas fields of western Texas (Ballentine et al. 2001). $^3He/CO_2$ ratio in natural gas wells in the Val Verde basin increase toward the Marathon Thrust Belt that defines the southern edge of the basin. These ratios are within the range predicted for magmatic gases, and the variation has been ascribed to Raleigh fractionation during magma degassing. Ballentine et al (2001) combine this with a simple filling model to define the direction of the magma source. Combined with the observed spatial relationship between CH_4 and CO_2 this suggest that the magmatic gases pre-dated hydrocarbon generation in the basin. Hydrocarbon generation is known to have initiated at around 280 Ma, requiring that the mantle derived CO_2 and helium has been stored in crustal traps for a period of ~300 Ma. Similarly, Marty et al. (1993) detected a ^{3}He mantle component in deep waters in the Paris basin and attributed this to Hercynian magmatism intruding the basement at >250 Ma. Young groundwaters of the eastern United States also have $^3He/^4He$ ratios (up to 1.2 Ra) that require a fossil magmatic helium source (Torgersen et al. 1994; Torgersen et al. 1995). No volcanism has occurred in the eastern United States for >95 Ma requiring crustal storage of the He for at least this length of time. Modeling shows that, given the U concentration of the crust, high gas concentrations are required in order to preserve the high $^3He/^4He$ ratios. Although extant fluids from the mantle or those released from magmatic rocks can perturb local noble gas isotopic signatures, this source of magmatic fluids does not represent a significant mass within the crustal system.

ACKNOWLEDGMENTS

Numerous discussion with, and comments from, Ingo Leya have been invaluable. We are very grateful for reviews by Tom Torgersen, Bernard Marty and Igor Tolstikhin as well as detailed comments by Sasha Verchovsky, Rainer Wieler and Ethan Baxter, which considerably improved this contribution. Help from Zhou Zheng with data compilation is very much appreciated.

REFERENCES

Albarède F (1998) Time-dependent models of U-Th-He and K-Ar evolution and the layering of mantle convection. Chem Geol 145:413-429

Alexander EC Jr, Lewis RS, Reynolds JH, Michel MC (1971) Plutonium-244: Confirmation as an extinct radioactivity. Science 172:837-840

Allègre CJ, Hofmann AW, O'Nions RK (1996) The argon constraints on mantle structure. Geophys Res Lett 23:3555-3557

Andrews JN (1985) The isotopic composition of radiogenic helium and its use to study groundwater movements in confined aquifers. Chem Geol 49:339-351

Andrews JN (1986) *In situ* neutron flux, 36-Cl production and groundwater evolution in crystalline rocks at Stripa, Sweden. Earth Planet Sci Lett 77:49-58

Andrews JN, Kay RLF (1982) Natural production of tritium in permeable rocks. Nature 298:361-363

Andrews JN, Giles IS, Kay RLF, Lee DJ, Osmond JK, Cowart JB, Fritz P, Barker JF, Gale J (1982), Radioelements, radiogenic helium and age relationships for groundwaters from the granites at Stripa, Sweden. Geochim Cosmochim Acta 46:1533-1543

Andrews JN, Hussain N, Youngman MJ (1989a) Atmospheric and radiogenic gases in groundwaters from the Stripa granite. Geochim Cosmochim Acta 53:1831-1841

Andrews JN, Davis SN, Fabryka-Martin J, Fontes J-C, Lehmann BE, Loosli HH, Michelot J-L, Moser H, Smith B, Wolf M (1989b) The *in situ* production of radioisotopes in rock matrices with particular reference to the Stripa granite. Geochim Cosmochim Acta 53:1803-1815

Armstrong RL (1991) The persistent myth of crustal growth. Austral J Earth Sci 38:613-630

Bach W, Naumann D, Erzinger J (1999) A helium, argon, and nitrogen record of the upper continental crust (KTB drill holes, Oberpfalz, Germany): implications for crustal degassing. Chem Geol 160: 81-101

Ballentine CJ (1991) He, Ne and Ar isotopes as tracers in crustal fluids. PhD dissertation, Dept Earth Sciences, University of Cambridge, Cambridge, UK

Ballentine CJ (1997) Resolving the mantle He/Ne and crustal Ne-21/Ne-22 in well gases. Earth Planet Sci Lett 152:233-249

Ballentine CJ, Burgess R, Marty B (2002) Tracing fluid origin, transport and interaction in the crust. Rev Mineral Geochem 47:319-370

Ballentine CJ, O'Nions RK (1994) The use of He, Ne and Ar isotopes to study hydrocarbon related fluid provenance, migration and mass balance in sedimentary basins. *In* Geofluids: Origin, migration and evolution of fluids in sedimentary basins. Parnell J (ed) Geol Soc Spec Publ 78:347-361

Ballentine CJ, O'Nions RK, Oxburgh ER, Horvath F, Deak J (1991) Rare gas constraints on hydrocarbon accumulation, crustal degassing and groundwater flow in the Pannonian Basin. Earth Planet Sci Lett 105:229-246

Ballentine CJ, Mazurek M, Gautschi A (1994) Thermal constraints on crustal rare gas release and migration—evidence from alpine fluid inclusions. Geochim Cosmochim Acta 58:4333-4348

Ballentine CJ, O'Nions RK, Coleman ML (1996) A Magnus opus: Helium, neon, and argon isotopes in a North Sea oilfield. Geochim Cosmochim Acta 60:831-849

Ballentine CJ, Schoell M, Coleman D, Cain BA (2001) 300-Myr-old magmatic CO_2 in natural gas reservoirs of the west Texas Permian basin. Nature 409:327-331

Ballentine CJ, Sherwood Lollar B (2002) Regional groundwater focussing of nitrogen and noble gases into the Hugoton-Panhandle giant gas field, USA. Geochim Cosmochim Acta 66:2483-2497

Barfod DN (1999) Noble gas geochemistry of the Cameroon line volcanic chain. PhD dissertation, Dept Geological Sciences, University of Michigan, Ann Arbor

Barfod DN, Lee D-C, Ballentine CJ, Halliday AN, Hall C (2002) Fractionation of helium and incompatible lithophile elements during partial melting under the Cameroon line. Earth Planet Sci Lett (submitted)

Battani A (1999) Utilisation des gaz rares He, Ne et Ar pour l'exploration pétrolière et gazière. Science de la terre, Univ Paris XI Orsay, Paris

Battani A, Sarda P, Prinzhofer A (2000) Basin scale natural gas source, migration and trapping traced by noble gases and major elements; the Pakistan Indus Basin. Earth Planet Sci Lett 181:229-249

Baxter EF, DePaolo DJ, Renne PR (2002) Spatially correlated anomalous $^{40}Ar/^{39}Ar$ "age" variations in biotites about a lithologic contact near Simplon Pass, Switzerland: A mechanistic explanation for excess Ar. Geochim Cosmochim Acta (submitted)

Bentley HW, Phillips FM, David SN (1986) Chlorine-36 in the terrestrial environment. *In* Handbook of environmental isotope geochemistry. Fritz P, Fontes J-C (eds) Elsevier, Amsterdam, p 422-480

Begemann F, Ludwig KR, Lugmair GW, Min K, Nyquist LE, Patchett PJ, Renne PR, Shih C-Y, Villa IM, Walker R. (2001) Call for an improved set of decay constants for geochronological use. Geochim Cosmochim Acta 65:111-122

Bethke CM, Zhao X, Torgersen T (1999) Groundwater flow and the ^4He distribution in the Great Artesian Basin of Australia. J Geophys Res 104:12999-13011

Bickle MJ, McKenzie D (1987) The transport of heat and matter by fluids during metamorphism. Contrib Mineral Petrol 95:384-392

Bottomley DJ, Ross JD, Clarke WB (1984) Helium and neon isotope geochemistry of some groundwaters from the Canadian Precambrian Shield. Geochim Cosmochim Acta 48:1973-1985

Brenan JM, Watson EB (1988) Fluids in the lithosphere; 2, Experimental constraints on CO_2 transport in dunite and quartzite at elevated P-T conditions with implications for mantle and crustal decarbonation processes. Earth Planet Sci Lett 91:141-158

Brooker RA, Wartho JA, Carroll MR, Kelley SP, Draper DS (1998) Preliminary UVLAMP determinations of argon partition coefficients for olivine and clinopyroxene grown from silicate melts. *In* The Degassing of the Earth. Carroll MR, Kohn SC, Wood BJ (eds) Elsevier, Amsterdam, p 185-200

Burnard PG, Hu R, Turner G, Bi X (1999) Mantle, crustal and atmospheric noble gases in Ailaoshan gold deposits, Yunnan Province, China. Geochim Cosmochim Acta 63:1595-1604
Carroll MR, Draper DS (1994) Noble gases as trace elements in magmatic processes. Chem Geol 117: 37-56
Castro MC, Jambon A, de Marsily G, Schlosser P (1998a) Noble gases as natural tracers of water circulation in the Paris Basin 1. Measurements and discussion of their origin and mechanisms of vertical transport in the basin. Water Resour Res 34:2443-2466
Castro MC, Goblet P, Ledoux E, Violette S, de Marsily G (1998b) Noble gases as natural tracers of water circulation in the Paris Basin 2. Calibration of a groundwater flow model using noble gas isotope data. Water Resour Res 34:2467-2483
Cerling TE, Craig H (1994) Geomorphology and *in situ* cosmogenic isotopes. Ann Rev Earth Planet Sci 22:273-317
Chamorro-Perez E, Gillet P, Jambon A, Badro J, McMillan P (1998) Low argon solubility in silicate melts at high pressure. Nature 393:352-355
Chazal V, Brissot R, Cavaignac JF, Chambon B, de Jesus M, Drain D, Giraud-Heraud Y, Pastor C, Stutz A, Vagneron L (1998) Neutron background measurements in the underground laboratory of Modane. Astroparticle Phys 9:163-172
Clark ID, Phillips RJ (2000) Geochemical and $^3He/^4He$ evidence for mantle and crustal contributions to geothermal fluids in the western Canadian continental margin. J Volcan Geotherm Res 104:261-276
Coltice N, Albarède F, Gillet P (2000) K-40–Ar-40 constraints on recycling continental crust into the mantle. Science 288:845-847
Crouch EAC (1977) Fission-product yields from neutron-induced fission. Academic Press, London
Cserepes L, Lenkey L (1999) Modeling of helium transport in groundwater along a section in the Pannonian basin. J Hydrol 225:185-195
Dai J, S Yan, C Dai, and D Wang (1986) Geochemistry and accumulation of carbon dioxide gases in China. AAPG Bull 80:1615-1626
Dauphas N, Marty B (1999) Heavy nitrogen in carbonatites of the Kola Peninsula: A possible signature of the deep mantle. Science 286:2488-2490
Darling WG, Griesshaber E, Andrews JN, Armannsson H, O'Nions RK (1995) The origin of hydrothermal and other gases in the Kenya Rift Valley. Geochim Cosmochim Acta 59:2501-2512
Drescher J, Kirsten T, Schafer K (1998) The rare gas inventory of the continental crust, recovered by the KTB Continental Deep Drilling Project. Earth Planet Sci Lett 154:247-263
Dunai TJ, Roselieb K (1996) Sorption and diffusion of helium in garnet: Implications for volatile tracing and dating. Earth Planet Sci Lett 139:411-421
Dunai TJ, Touret JLR (1993) A noble gas study of a granulite sample from the Nilgiri Hills, Southern India—Implications for granulite formation. Earth Planet Sci Lett 119:271-281
Dymond J, Hogan L (1973) Noble gas abundance patterns in deep-sea basalts—primordial gases from the mantle. Earth Planet Sci Lett 20:131-139
Eikenberg J, Signer P, Wieler R (1993) U-Xe, U-Kr, and U-Pb systematics for dating uranium minerals and investigations of the production of nucleogenic neon and argon. Geochim Cosmochim Acta 57: 1053-1069
Elliot T, Ballentine CJ, O'Nions RK, Ricchiuto T (1993) Carbon, helium, neon and argon isotopes in a Po Basin (Northern Italy) natural-gas field. Chem Geol 106:429-440
Farrer H, Tomlinson RH (1962) Cumulative yields of the heavy fragments in ^{235}U thermal neutron fission. Nucl Phys 34:367-381
Farley KA (1995) Cenozoic variations in the flux of interplanetary dust recorded by He-3 in a deep-sea sediment. Nature 376:153-156
Farley KA (2002) (U-Th)/He dating: Techniques, calibrations, and applications. Rev Mineral Geochem 47: 819-845
Farley KA, Wolf RA, Silver LT (1996) The effects of long alpha stopping distances on (U Th)/He ages. Geochim Cosmochim Acta 60:4223-4229
Feige Y, Oltman BG, Kastner J (1968) Production rates of neutrons in soils due to natural radioactivity. J Geophys Res 73:3135-3142
Fleming Wh (1953) Neutron and spontaneous fission in uranium ores. Phys Rev 92:378-382
Foland KA (1979) Limited mobility of argon in a metamorphic terrain. Geochim Cosmochim Acta 43: 793-801
Fontes J-C, Andrews JN, Walgenwitz F (1991) Evaluation de la production naturelle *in situ* d'argon-36 via le chlore-36: implication geochimiques et geochronologiques. C R Acad Sci Paris II 313:649-654
Fowler CMR (1990) The Solid Earth. Cambridge University Press, Cambridge
Freundel M, Schultz L, Reedy RC. (1986) Terrestrial ^{81}Kr ages of Antarctic meteorites. Geochim Cosmochim Acta 50:2663-2673

Fyfe WS, Price NJ, Thompson AB. (1978) Fluids in the Earth's Crust. Elsevier, Amsterdam
Gao S, Luo TC, Zhang BR, Zhang HF, Han YW, Zhao ZD, Hu YK (1997) Chemical composition of the continental crust as revealed by studies in East China. Geochim Cosmochim Acta 62:1959-1975
Gerling EK, Shukolyukov YuA (1960) The composition and content of xenon isotopes in uranium minerals. Radiochemistry 1:106-118
Gerling EK, Mamyrin BA, Tolstikhin IN, Yakovleva SS (1971) Isotope composition of helium in rocks. Geokhimiya 5:608-617
Gerstenberger H (1980) A xenon component of anomalous isotopic composition in neutron irradiated terrestrial materials (in German). Isotopenpraxis 16:88-91
Giletti BJ, Tullis J (1977) Studies in diffusion; IV, Pressure dependence of Ar diffusion in phlogopite mica. Earth Planet Sci Lett 35:180-183
Graham DW (2002) Noble gas isotope geochemistry of mid-ocean ridge and ocean island basalts: Characterization of mantle source reservoirs. Rev Mineral Geochem 47:247-318
Gorshkov GV, Zyabkin VA, Lyatkovskaya NM, Zvetkov YuS. (1966) Natural neutron background of the atmosphere and the Earth's crust (in Russian). Atomizdat, Mosow
Graham DW, Larsen LM, Hanan BB, Storey M, Pedersen AK, Lupton JE (1998) Helium isotope composition of the early Iceland mantle plume inferred from the tertiary picrites of West Greenland. Earth Planet Sci Lett 160:241-255
Griesshaber E, O'Nions RK, Oxburgh ER (1988) Helium and carbon isotope systematics in crustal fluids from the Eifel, Oberfalz and Schwarzwald. Chem Geol 20:37-37
Griesshaber E, O'Nions RK, Oxburgh ER (1992) Helium and carbon isotope systematics in crustal fluids from the Eifel, the Rhine Graben and Black Forest, F.R.G. Chem Geol 99:213-235
Hebeda EH, Schultz L, Freundel M (1987) Radiogenic, fissiogenic and nucleogenic noble gases in zircons. Earth Planet Sci Lett 85:79-90
Hilton DR (1996) The helium and carbon isotope systematics of a continental geothermal system; results from monitoring studies at Long Valley Caldera (California, USA). Chem Geol 127:269-295
Hilton DR, Oxbrugh ER, O'Nions RK (1985) Fluid flow through a high heat flow granite: constraints imposed by He and Ra data High Heat Production (HHP) Granites. *In* Hydrothermal Circulation and Ore Genesis. Inst Mineralogy Metallurgy, London
Hilton DR, Hammerschmidt K, Teufel S, Friedrichsen H (1993) Helium isotope characteristics of Andean geothermal fluids and lavas. Earth Planet Sci Lett 120:265-282
Hiyagon H, Kennedy BM (1992) Noble gases in CH_4-rich gas fields, Alberta, Canada. Geochim Cosmochim Acta 56:1569-1589
Hoke L, Hilton DR, Lamb SH, Hammerschmidt K, Friedrichsen H (1994) He-3 evidence for a wide zone of active mantle melting beneath the Central Andes. Earth Planet Sci Lett 128:341-355
Hoke L, Lamb S, Hilton DR, Poreda RJ (2000) Southern limit of mantle-derived geothermal helium emissions in Tibet: Implications for lithospheric structure. Earth Planet Sci Lett 180:297-308
Holbrook WS, Mooney WD, Christensen NI (1992) The seismic velocity structure of the deep continetnal crust. *In* The Continental Lower Crust. Fountain DM, Arculus R (eds) Elsevier, New York, p 1-44
Honda M, Kurita K, Hamano Y, Ozima M (1982) Experimental studies of He and Ar degassing during rock fracturing. Earth Planet Sci Lett 59:429-436
Hooker PJ, O'Nions RK, Oxburgh ER (1985a) Helium isotopes in North Sea gas fields and the Rhine. Nature 318:273-275
Hooker PJ, Bertrami R, Lombardi S, O'Nions RK, Oxburgh ER (1985b) Helium-3 anomalies and crust mantle interaction in Italy. Geochim Cosmochim Acta 49:2505-2513
Hulston JR, Lupton JE (1996) Helium isotope studies of geothermal fields in the Taupo volcanic zone, New Zealand. J Volc Geotherm Res 73:297-321
Hünemohr H. (1989) Edelgase in U- und Th-reichen Mineralen und die Bistimmung der ^{21}Ne-Dicktarget der ^{18}O(a,n)^{21}Ne-Kernreaktion im Bereich 4.0-8.8 MeV. PhD dissertation, Fachbereich Physik, Johannes-Gutenberg-Univ, Mainz
Hurley PM (1950) Distribution of radioactivity in granites and possible relation to helium age measurement. Bull Geol Soc Am 61:1-8
Hurley PM, Rand JR (1969) Pre-drift continental nuclei. Science 164:1229-1242
Hussain N (1996) Flux of 4He from Carnmenellis Granite: modeling of an HDR geothermal reservoir. Appl Geochem 12:1-8
Igarashi G. (1995) Primitive xenon in the Earth. *In* Proc of Volatiles in the Earth and Solar System. Farley K (ed) AIP press, New York 341:70-80
Jacobs GJ, Liskien H (1983) Energy spectra of neutrons produced by α-particles in thick targets of light elements. Ann Nucl Energy 10:541-552
Jähne B, Heinz G, Dietrich W (1987) Measurement of the diffusion coefficient of sparingly soluble gases in water. J Geophys Res 92:10,767-`10,776

Jean-Baptiste P, Bougault H, Vangriesheim A, Charlou JL, Radford-Knoery J, Fouquet Y, Needham D, German C (1998) Mantle ^{3}He in hydrothermal vents and plume of the Lucky Strike Site (MAR 37 degrees 17'N) and associated geothermal heat flux. Earth Planet Sci Lett 157:69-77

Jenden PD, Kaplan IR, Poreda RJ, Craig H (1988) Origin of nitrogen-rich gases in the Californian Great Valley: Evidence from helium, carbon and nitrogen isotope ratios. Geochim Cosmochim Acta 52: 851-861

Kamensky IL, Tolstikhin IN, Vetrin VR (1990) Juvenile helium in ancient rocks I: ^{3}He excess in amphiboles from 2.8-Ga charnockite series, crust-mantle fluid in intracrustal magmatic processes. Geochim Cosmochim Acta 54:3115-3122

Kelley SP, Wartho JA (2000) Rapid kimberlite ascent and the significance of Ar-Ar ages in xenolith phlogopites. Science 289:609-611

Kelley SP (2002) K-Ar and Ar-Ar dating. Rev Mineral Geochem 47:785-818

Kennedy BM, Hiyagon H, Reynolds JH (1990) Crustal neon: A striking uniformity. Earth Planet Sci Lett 98:227-286

Kennet TJ, Thode HG (1957) The cumulative yield of the krypton and xenon isotopes produced in the fast neutron fission of ^{232}Th. Can J Physics 35:969-979

Kipfer R, Aeschbach-Hertig W, Peeters F, Stute M (2002) Noble gases in lakes and ground waters. Rev Mineral Geochem 47:615-700

Krishnaswami S, Seidemann DE (1988) Comparative study of ^{222}Rn, ^{40}Ar, ^{39}Ar and ^{37}Ar leakage from rocks and minerals; implications for the role of nanopores in gas transport through natural silicates. Geochim Cosmochim Acta 52:655

Kunz J, Staudacher T, Allègre CJ (1998) Plutonium-fission xenon found in Earth's mantle. Science 280:877-880

Kurz MD, Trull TW, Colodner D, Denton G (1988) Exposure-age dating with cosmogenic ^{3}He. Chem Geol 70:39-39

Kyser KT, Rison W (1982) Systematics of rare gas isotopes in basic lavas and ultramafic xenolithes. J Geophys Res 87:5611-5630

Lehmann BE, Davis SN, Fabrykamartin JT (1993) Atmosphere and subsurface sources of stable and radioactive nuclides used for groundwater dating. Water Resour Res 29:2027-2040

Lal D (1991) Cosmic ray labeling of erosion surfaces: *In situ* nuclide production rates and erosion models. Earth Planet Sci Lett 104:424-439

Laslett GM, Green PF, Duddy IR, Gleadow AJW (1987) Thermal annealing of fission tracks in apatite—2: A quantitative analysis. Chem Geol 65:1-13

Leya I, Wieler R (1999) Nucleogenic production of Ne isotopes in Earth's crust and upper mantle induced by alpha particles from the decay of U and Th. J Geophys Res 104:15439-15450

Lesniak PM, Sakai H, Ishibashi J-I, Wakita H (1997) Mantle helium signal in the West Carpathians, Poland. Geochem J 31:383-394

Lide DR (1994) Handbook of Chemistry and Physics, 75th Edition. CRC press, London

Lippolt HJ, Weigel E (1988) ^{4}He diffusion in ^{40}Ar retentive minerals. Geochim Cosmochim Acta 52: 1449-1458

Maeck WJ, Spraktes FW, Tromp RL, Keller JH (1975) Analytical results, recommended nuclear constants and suggested correlations for the evaluation of Oklo fission product data Oklo Phenomenon. IAEA, Vienna, p 319-339

Mamyrin BA, Tolstikhin IN. (1984) Helium isotopes in nature. Elsevier, Amsterdam

Mann DM, Mackenzie AS (1990) Prediction of pore fluid pressures in sedimentary basins. Marine Petrol Geol 7:55-65

Manning CE, Ingebritsen SE (1999) Permeability of the continental crust: Implications of geothermal data and metamorphic systems. Rev Geophys 37:127-150

Martel DJ, Deak J, Dovenyi P, Horvath F, O'Nions RK, Oxburgh ER, Stegna L, Stute M (1989) Leakage of helium from the Pannonian Basin. Nature 432:908-912

Martel DJ, O'Nions RK, Hilton DR, Oxburgh ER (1990) The role of element distribution in production and release of radiogenic helium: The Carnmenellis Granite, southwest England. Chem Geol 88: 207-221

Marty B, Lussiez P (1993) Constraints on rare gas partition coefficients from analysis of olivine-glass from a picritic mid-ocean ridge basalt. Chem Geol 106:1-7

Marty B, O'Nions RK, Oxburgh ER, Martel D, Lombardi S (1992) Helium isotopes in Alpine regions. Tectonophysics 206:71-78

Marty B, Torgersen T, Meynier V, O'Nions RK, De Marsily G (1993) Helium isotope fluxes and groundwater ages in the Dogger aquifer, Paris Basin. Water Resour Res 29:1025-1035

Marty B, Tolstikhin I, Kamensky IL, Nivin V, Balaganskaya E, Zimmermann JL (1998) Plume-derived rare gases in 380 Ma carbonatites from the Kola region (Russia) and the argon isotopic composition in the deep mantle. Earth Planet Sci Lett 164:179-192
Mason B, Moore CB. (1982) Principles of Geochemistry. John Wiley and Sons, New York
Matthews A, Fouillac C, Hill R, O'Nions RK, Oxburgh ER (1987) Mantle-derived volatiles in continental crust; the Massif Central of France. Earth Planet Sci Lett 85:117-128
Matsuda J-I, Murota M, Nagao K (1990) He and Ne isotopic studies on the extraterrestrial material in deep-sea sediments. J Geophys Res 95:7111-7117
Meshik AP (1988) Xe and Kr isotopes in geochronology of uranium oxides (in Russian). PhD dissertation, Vernadsky Institute, Moscow
Meshik AP and Shukolyukov YuA (1986) Isotopic anomalies of Xe and Kr in natural nuclear reactor (uranium deposit Oklo, Gabon) as a result of radioactive precursor migration (in Russian). Proc XI All-Union Symp Stable Isotopes in Geochemistry, Moscow, p 237-238
Meshik AP, Jessberger EK, Pravdivtseva OV, and Shukolyukov YuA (1996) CFF-Xe: An alternative approach to terrestrial xenology. Proc V.M. Goldschmidt Conf, Geochem Soc, p 400
Meshik AP, Kehm K, and Hohenberg CM (2000) Anomalous xenon in zone 13 Okelobondo. Geochim Cosmochim Acta 64:1651-1661
McDonough WF, Sun S-s (1995) The composition of the Earth. Chem Geol 120:223-253
McDougall I, Harrison TM (1988) Geochronology and thermochronology by the $^{40}Ar/^{39}Ar$ method. Oxford University Press, Oxford
Minissale A, Vaselli O, Chandrasekharam D, Magro G, Tassi F, Casiglia A (2000) Origin and evolution of "intracratonic" thermal fluids from central-western peninsular India. Earth Planet Sci Lett 181: 377-394
Morrison P, Pine J (1955) Radiogenic origin of the helium isotopes in rock. Ann New York Acad Sci 62:71-92
Mussett AE (1969) Diffusion measurements and the Potassium-Argon method of dating. Geophys J Roy Astron Soc 18:257-303
Nagao K, Takaoka N, Matsubayaski O (1981) Rare gas isotopic composition in natural gases of Japan. Earth Planet Sci Lett 53:175-188
Niedermann S (2002) Cosmic-ray-produced noble gases in terrestrial rocks as a dating tool for surface processes. Rev Mineral Geochem 47:731-784
Niedermann S, Bach W, Erzinger J (1997) Noble gas evidence for a lower mantle component in MORBs from the Southern East Pacific Rise: Decoupling helium and neon isotope systematics. Geochim Cosmochim Acta 61:2697-2715
Nordlund K (1995) Molecular dynamic simulations of ion ranges in the post-keV energy region. Computation Mater Sci 3:448
Ohsumi T, Horibe Y (1984) Diffusivity of He and Ar in deep-sea sediments. Earth Planet Sci Lett 70: 61-68
O'Nions RK, Ballentine CJ (1993) Rare gas studies of basin scale fluid movement. Phil Trans Roy Soc London 344:144-156
O'Nions RK, Oxburgh ER (1983) Heat and helium in the Earth. Nature 306:429-431
O'Nions RK, Oxburgh ER (1988) Helium, volatile fluxes and the development of continental crust. Earth Planet Sci Lett 90:331-347
Ott U (1995) A new approach to the origin of Xenon-HL. Meteoritics 30:559-560
Ott U (1996) interstellar diamond xenon and timescale of supernova ejecta. Astrophys J 463:344-348
Oxburgh ER, O'Nions RK (1987) Helium loss, tectonics, and the terrestrial heat budget. Science 237: 1583-1587
Oxburgh ER, O'Nions RK, Hill RI (1986) Helium isotopes in sedimentary basins. Nature 324:632-635
Ozima M, Podosek FA (1983) Noble Gas Geochemistry. Cambridge University Press, Cambridge
Patterson DB, Farley KA, Schmitz B (1998) Preservation of extraterrestrial ^{3}He in 480-Ma-old marine limestones. Earth Planet Sci Lett 163:315-325
Patterson DB, Farley KA, Norman MD (1999) He-4 as a tracer of continental dust: A 1.9 million year record of aeolian flux to the west equatorial Pacific Ocean. Geochim Cosmochim Acta 63:615-625
Pearson Jr. FJ, Balderer W, Loosli HH, Lehman BE, Matter A, Peters T, Schmassmann H, Gautschi A (1991) Applied isotope hydrogeology—A case study in Northern Switzerland. Elsevier, Amsterdam
Pepin RO (1991) On the origin and early evolution of terrestrial planet atmospheres and meteoritic volatiles. Icarus 92:1-79
Pinti DL, Marty B (1995) Noble gases in crude oils from the Paris Basin, France—Implications for the origin of fluids and constraints on oil-water-gas interactions. Geochim Cosmochim Acta 59:3389-3404
Piperov NB, Kamensky IL, Tolstikhin IN (1994) Isotopes of the light noble gases in mineral waters in the eastern part of the Balkan Peninsula, Bulgaria. Geochim Cosmochim Acta 58:1889-1898

Polyak B. Tolstikhin IN (1985) Isotopic composition of the Earth's helium and the problem of the motive forces of tectogenesis. Chem Geol 52:9-33
Porcelli D, Ballentine CJ (2002) Models for the distribution of Terrestrial noble gases and the evolution of the atmosphere. Rev Mineral Geochem 47:411-480
Poreda R, Craig H (1989) Helium isotope ratios in circum-Pacific volcanic arcs. Nature 338:473-478
Poreda RJ, Arnorsson S (1992) Helium isotopes in Icelandic geothermal systems II: Helium-heat relationships. Geochim Cosmochim Acta 56:4229-4235
Poreda R, Jenden PD, Kaplan IR, Craig H (1986) Mantle helium in Sacramento Basin natural gas wells. Geochim Cosmochim Acta 50:9-33
Pravdivtseva OV, Meshik AP, Shukolyukov YA (1986) Isotopic composition of Xe and Kr in the samples from epicentre of the first nuclear test in Alamogordo (NM, USA) (in Russian). Proc IX All Union Symp Stable Isotopes in Geochemistry, Moscow, p 289-290
Ragettli R (1993) Vergleichende U-Xe and U-Pb Datierung an Zirkon und Monazit. PhD dissertation, ETH Zürich, Nr. 10183
Ragettli RA, Hebeda EH, Signer P, Wieler R (1994) Uranium xenon chronology: Precise determination of lambda(sf*) y-136(sf) for spontaneous fission of U-238. Earth Planet Sci Lett 128:653-670
Rama M, Moore WS (1984) Mechanism of transport of U-Th series radioisotopes from solids into ground water. Geochim Cosmochim Acta 48:395-399
Renne PR, Norman EB (2001) Determination of the half-life of Ar-37 by mass spectrometry—Article no. 047302. Phys Rev C 6304
Renne PR, Farley KA, Becker TA, Sharp WD (2001) Terrestrial cosmogenic argon. Earth Planet Sci Lett 188:435-440
Richard D, Marty B, Chaussidon M, Arndt N (1996) Helium isotopic evidence for a lower mantle component in depleted Archean komatiite. Science 273:93-95
Rison W (1980) Isotopic studies of the rare gases in igneous rocks: Implications for the mantle and atmosphere. PhD dissertation, Dept Physics, University of California, Berkeley
Rubey WW, Hubbert MK (1959) Role of fluid pressure in mechanics of overthrusting faulting I. Mechanics of fluid-filled porous solids and its application to overthrusting faulting. Geol Soc Am Bull 70: 115-166
Rudnick R, Fountain DM (1995) Nature and composition of the continental crust: A lower crustal perspective. Rev Geophys 33:267-309
Sabu DD (1971) On mass-yield of xenon and krypton isotopes in the spontaneous fission uranium. J Inorg Nucl Chem 33:1509-113
Sano Y, Wakita H (1985) Geographical distribution of $^3He/^4He$ ratios in Japan: Implications for arc tectonics and incipient magmatism. J Geophys Res 90:8729-8741
Sano Y, Wakita H, Huang CW (1986) Helium flux in a continental land area from $^3He/^4He$ ratio in northern Taiwan. Nature 323:55-57
Scaillet S (1996) Excess Ar-40 transport scale and mechanism in high-pressure phengites—A case study from an eclogitized metabasite of the Dora-Maira Nappe, Western Alps. Geochim Cosmochim Acta 60:1075-1090
Schäfer JM, Ivy Ochs S, Wieler R, Leya J, Baur H, Denton GH, Schlüchter C (1999) Cosmogenic noble gas studies in the oldest landscape on Earth: Surface exposure ages of the Dry Valleys, Antarctica. Earth Planet Sci Lett 167:215-226
Segrè E (1952) Spontaneous fission. Phys Rev 86:21-28
Sharif-Zade VB, Shukolyukov YuA, Gerling EK, Ashkinadze GSh (1972) Neon isotopes in radioactive minerals. Geochem Intl 10:199-207
Sherwood Lollar B, O'Nions RK, Ballentine CJ (1994) Helium and neon isotope systematics in carbon dioxide-rich and hydrocarbon-rich gas reservoirs. Geochim Cosmochim Acta 58:5279-5290
Simpson F (1999) Stress and seismicity in the lower continental crust: A challenge to simple ductility and implications for electrical conductivity mechanisms. Surv Geophys 20:201-227
Shukolyukov YuA. (1970) The decay of uranium nuclei in nature (in Russian). Atomizdat Publisky, Moscow.
Shukolyukov YuA, Ashkinadze GS (1968) Determining the rate constant of spontaneous fission from the accumulation of xenon isotopes in uranium minerals (in Russian). Atomnaya Energiya 25:428-429
Shukolyukov YuA, Kirsten T, Jessberger EK (1974) The Xe-Xe spectrum technique, a new dating method. Earth Planet Sci Lett 24:271-281
Shukolyukov YuA, Ashkinadze GS, Verkhovskiy AB (1976) Anomalous isotopic composition of xenon and krypton in minerals of the natural nuclear reactor. Sov Atom Energy 41:663-666
Shukolyukov YuA, Jessberger EK, Meshik AP, Minh DV, Jordan JL (1994) Chemically Fractionated Fission-Xenon in meteorites and on the Earth. Geochim Cosmochim Acta 58:3075-3092

Smith SP, Kennedy BM (1985) Noble gas evidence for two fluids in the Baca (Valles Caldera) geothermal reservoir. Geochim Cosmochim Acta 49:893-902

Stuart FM, Turner G (1992) The abundance and isotopic composition of the noble gases in ancient fluids. Chem Geol 101:97-109

Stuart FM, Burnard PG, Taylor RP, Turner G (1995) Resolving mantle and crustal contributions to ancient hydrothermal fluids: He-Ar isotopes in fluid inclusions from Dae Hwa W-Mo mineralisation, South Korea. Geochim Cosmochim Acta 59:4663-4673

Sturchio NC, Arehart GB, Sultan M, Sano Y, AboKamar Y, Sayed M (1996) Composition and origin of thermal waters in the Gulf of Suez area, Egypt. Applied Geochem 11:471-479

Stute M, Sonntag C, Deak J, Schlosser P (1992) Helium in deep circulating groundwater in the Great Hungarian Plain—Flow dynamics and crustal and mantle helium fluxes. Geochim Cosmochim Acta 56:2051-2067

Steiger RH, Jäger E (1977) Subcommission on geochronology: Convention on the use of decay constants in geo-and cosmochronology. Earth Planet Sci Lett 36:359-362

Sun S, McDonough WF (1989) Chemical and isotopic systematics of oceanic basalts: Implications for mantle composition and processes. *In* Magmatism in the Ocean Basins. Saunders AD, Norry MJ (eds) Geol Soc Spec Publ 42:313-345

Takaoka N (1972) An interpretation of general anomalies of xenon and the isotopic composition of primitive xenon. Mass Spectros 20:287-302

Takahata N, Sano Y (2000) Helium flux from a sedimentary basin. Appl Radiat Isotopes 52:985-992

Taylor SR, McLennan SM. (1985) The Continental Crust: Its Composition and Evolution. Blackwell Scientific, Oxford

Taylor SR, Mclennan SM (1995) The geochemical evolution of the continental crust. Rev Geophys 33:241-265.

Tolstikhin IN (1978) A review: Some recent advances in isotope geochemistry of light rare gases. Adv Earth Planet Sci 3:33-62

Tolstikhin IN, Drubetskoy ER (1977) The helium isotopes in rocks and minerals of the Earth's crust. *In* Problemy datirovaniya dolembriyskikh obrazovaniy. Shukolyukov YuA (ed) Nauka, Lenningrad

Tolstikhin IN, Dokuchaeva VS, Kamensky IL, Amelin YV (1992) Juvenile helium in ancient rocks II: U-He, K-Ar, Sm-Nd, and Rb-Sr systematics in the Monche Pluton; $^{3}He/^{4}He$ ratios frozen in uranium-free ultramafic rocks. *In* Origin and evolution of planetary crusts. McLennan SM, Rudnick RL (eds) Pergamon, Oxford, p 987-999

Tolstikhin I, Lehmann BE, Loosli HH, Gautschi A (1996) Helium and argon isotopes in rocks, minerals, and related groundwaters: A case study in northern Switzerland. Geochim Cosmochim Acta 60: 1497-1514

Torgersen T (1980) Controls on porefluid concentration of ^{4}He and ^{222}Rn and the calculations of $^{4}He/^{222}Rn$. J Geochem Explor 13:57-75

Torgersen T (1989) Terrestrial helium degassing fluxes and the atmospheric helium budget: Implications with respect to the degassing processes of continental crust. Chem Geol 79:1-14

Torgersen T (1993) Defining the role of magmatism in extensional tectonics; helium-3 fluxes in extensional basins. J Geophys Res 98:16,257-216,269

Torgersen T, Clarke WB (1985) Helium accumulation in groundwater I: An evaluation of sources and the continental flux of ^{4}He in the Great Artesian Basin, Australia. Geochim Cosmochim Acta 49: 1211-1218

Torgersen T, Clarke WB (1987) Helium accumulation in groundwater III: Limits on helium transfer across the mantle-crust boundary beneath Australia and the magnitude of mantle degassing. Earth Planet Sci Lett 84:345-355

Torgersen T, Ivey GN (1985) Helium accumulation in groundwater II: A model for the accumulation of the crustal ^{4}He degassing flux. Geochim Cosmochim Acta 49:2445-2452

Torgersen T, O'Donnell J (1991) The degassing flux from the solid Earth—release by fracturing. Geophys Res Lett 18:951-954

Torgersen T, Kennedy BM, Hiyagon H, Chiou KY, Reynolds JH, Clarke WB (1989) Argon accumulation and the crustal degassing flux of ^{40}Ar in the Great Artesian Basin, Australia. Earth Planet Sci Lett 92:43-56

Torgersen T, Clarke WB, Zoback MD, Lachenbruch AH (1992a) Geochemical constraints on formation fluid ages, hydrothermal heat flux, and crustal mass transport mechanisms at Cajon Pass. J Geophys Res 97:5031-5038

Torgersen T, Habermehl MA, Clarke WB (1992b) Crustal helium fluxes and heat flow in the Great Artesian Basin, Australia. Chem Geol 102:139-153

Torgersen T, Drenkard S, Farley K, Schlosser P, Shapiro A (1994) Mantle helium in groundwater of the Mirror Lake Basin, New Hampshire, USA. *In* Noble gas geochemistry and cosmochemistry. Matsuda J (ed) Terra, Tokyo, p 279-292

Torgersen T, Drenkard S, Stute M, Schlosser P, Shapiro AM (1995) Mantle helium in ground waters of eastern North America; time and space constraints on sources. Geology 23:675-678

Trull TW, Kurz MD, Jenkins WJ (1991) Diffusion of cosmogenic ^{3}He in olivine and quartz: Implications for surface exposure dating. Earth Planet Sci Lett 103:241-256

Turcotte DL, Schubert G. (1982) Geodynamics: Applications of continuum physics to geological problems. Wiley, New York

Turner G, Stuart F (1992) Helium/heat ratios and deposition temperatures of sulphides from the ocean floor. Nature 357:581-583

van Keken PE, Ballentine CJ, Porcelli D (2001) A dynamical investigation of the heat and helium imbalance. Earth Planet Sci Lett 188:421-434

Vance D, Ayres M, Kelley SP, Harris N (1998) The thermal response of a metamorphic belt to extension: Constraints from laser Ar data on metamorphic micas. Earth Planet Sci Lett 162:153-164

Verchovsky AB, Shukolyukov YuA (1976a) Isotopic composition of neon in the crustal rocks and the origin of Ne^{21}_{rad} in natural gases. Geochem Intl 13:95-98

Verchovsky AB, Shukolyukov YuA (1976b) Neon isotopes in minerals with excess of helium and argon (in Russian). Geokhimiya 3:315-322.

Verchovsky AB, Shukolyukov YuA, Ashkinadze GS (1977) Radiogenic helium, neon and argon in natural gases (in Russian). *In* The Problems of Dating Precambrain Formations. Nauka, Leningrad, p 152-170

Watson EB, Brenan JM (1987) Fluids in the lithosphere 1: Experimentally-determined wetting characteristics of CO_2-H_2O fluids and their implications for fluid transport, host-rock physical properties, and fluid inclusion formation. Earth Planet Sci Lett 85:497-515

Wedepohl KH (1995) The composition of the continental crust. Geochim Cosmochim Acta 59:1217-1232

Welhan JA, Poreda RJ, Rison W, Craig H (1988a) Helium isotopes in geothermal and volcanic gases of the western united states, I. Regional variability and magmatic origin. J Volc Geotherm Res 34: 185-199

Welhan JA, Poreda RJ, Rison W, Craig H (1988b) Helium isotopes in geothermal and volcanic gases of the western united states, II. Long Valley Caldera. J Volc Geotherm Res 34:201-209

West D, Sherwood AC (1982) Measurements of thick target (α,n) yield from light elements. Ann Nuc Energy 9:551-577

Wetherill GW (1953) Spontaneous fission yields from uranium and thorium. Phys Rev 92:907-912

Wetherill GW (1954) Variations in the isotopic abundances of neon and argon extracted from radioactive minerals. Phys Rev 96:679-683

Wieler R, Eikenberg J (1999) An upper limit on the spontaneous fission decay constant of Th-232 derived from xenon in monazites with extremely high Th/U ratios. Geophys Res Lett 26:107-110

Winckler G, Suess E, Wallmann K, deLange GJ, Westbrook GK, Bayer R (1997) Excess helium and argon of radiogenic origin in Mediterranean brine basins. Earth Planet Sci Lett 15:225-231

Wolf RA, Farley KA, Silver LT (1996) Helium diffusion and low-temperature thermochronometry of apatite. Geochim Cosmochim Acta 60:4231-4240

Wood BJ, Walther JV (1986) Fluid flow during metamorphism and its implications for fluid-rock ratios. *In* Fluid-rock Interactions During Metamorphism. Walther JV, Wood BJ (eds) Springer-Verlag, New York, p 89-108

Woosley SE, Fowler WA, Holmes JA, Zimmermann BA (1975) Tables of thermonuclear reaction data for intermediate mass nuclei. California Institute of Technology, Pasadena

Xu S, Nakai S, Wakita H, Wang X (1995) Mantle-derived noble gases in natural gases from Songliao Basin, China. Geochim Cosmochim Acta 59:4675-4683

Yatsevich I, Honda M (1997) Production of nucleogenic neon in the Earth from natural radioactive decay. J Geophys Res102:10291-10298

Young BG, Thode HG (1960) Absolute yields of xenon and krypton isotopes in ^{238}U spontaneous fission. Can J Phys 38:1-10

Zaikowski A, Kosanke BJ, Hubbard N (1987) Noble gas composition of deep brines from the Palo Duro Basin, Texas. Geochim Cosmochim Acta 51:73-84

Ziegler JF (1977) The Stopping and Ranges of Ions in Matter. Pergamon Press, New York

Ziegler JF (1980) Handbook of Stopping Cross-sections for Energetic ions in All Elements. Pergamon Press, New York

Ziegler JF, Biersack JP, Littmark U (1985) The Stopping and Range of Ions in Solids. Pergamon Press, New York

Zuber A, Weise SM, Osenbrueck K, Matenko T (1997) Origin and age of saline waters in Busko Spa (southern Poland) determined by isotope, noble gas and hydrochemical methods: Evidence of interglacial and pre-Quaternary warm climate recharges. Appl Geochem 12:643-660

13 Tracing Fluid Origin, Transport and Interaction in the Crust

Chris J. Ballentine, Ray Burgess

Department of Earth Sciences,
The University of Manchester,
Manchester, M13 9PL, United Kingdom
cballentine@fs1.ge.man.ac.uk

Bernard Marty

Centre de Recherches Pétrographiques et Géochimique
15 Rue Notre-Dame des Pauvres, B.P. 20
and
Ecole Nationale Supérieure de Géologie
Rue du Doyen Roubault
54501 Vandoeuvre les Nancy Cedex, France

INTRODUCTION

We detail here the general concepts behind using noble gases as a tracer of crustal fluid processes and illustrate these concepts with examples applied to oil-gas-groundwater systems, mineralizing fluids, hydrothermal systems and ancient ground-waters. Many of the concepts and processes discussed here are also directly applicable to the study of young ground and surface-water systems (Kipfer et al. 2002, this volume).

Noble gases in the Earth are broadly derived from two sources; noble gases trapped during the accretionary process (often called 'primitive', 'juvenile' or 'primordial' noble gases), and those generated by radioactive processes (e.g., Ballentine and Burnard 2002, this volume). Differentiation of the Earth into mantle and continental crust, degassing and early processes of atmosphere loss has resulted in the formation of reservoirs in which the abundance pattern and isotopic compositions of primitive noble gases have been variably altered. Combined with their different radioelement concentrations (U, Th, K) producing radiogenic noble gases, the mantle, crust and atmosphere are now distinct in both their noble gas isotopic composition and relative elemental abundance pattern.

Fluids that originate from these different sources will contain noble gases that are therefore isotopically distinct and resolvable (Fig. 1). Because the noble gases are chemically inert even if these fluids are lost through reaction or masked by addition of similar species from different sources, a conservative record of their presence and origin is preserved by the noble gases. Once resolved, the noble gas abundance patterns from the respective sources are particularly important, as these are sensitive to physical processes of fractionation. For example, from the distinct fractionation patterns it is possible to distinguish between for example, diffusive or advective transport processes. Similarly the abundance patterns enable the interaction of different phases to be identified and quantified. In a system that has fluids sourced from multiple terrestrial reservoirs, a fractionation pattern preserved in one component but not another (or indeed the same pattern preserved in both) gives information both about the processes and relative timing of processes operating on the differently sourced fluids either before or after mixing.

Before it is possible to exploit the unique character of the noble gases it is essential to understand how the noble gases behave in the subsurface and how the isotopic systems can be used to unambiguously resolve the different noble gas components in any crustal fluid. To this end we review first the physical chemistry of the noble gases, the various fractionation mechanisms as well as the isotopic techniques and approaches used to resolve the differently sourced noble gas components.

1529-6466/00/0047-0013$10.00

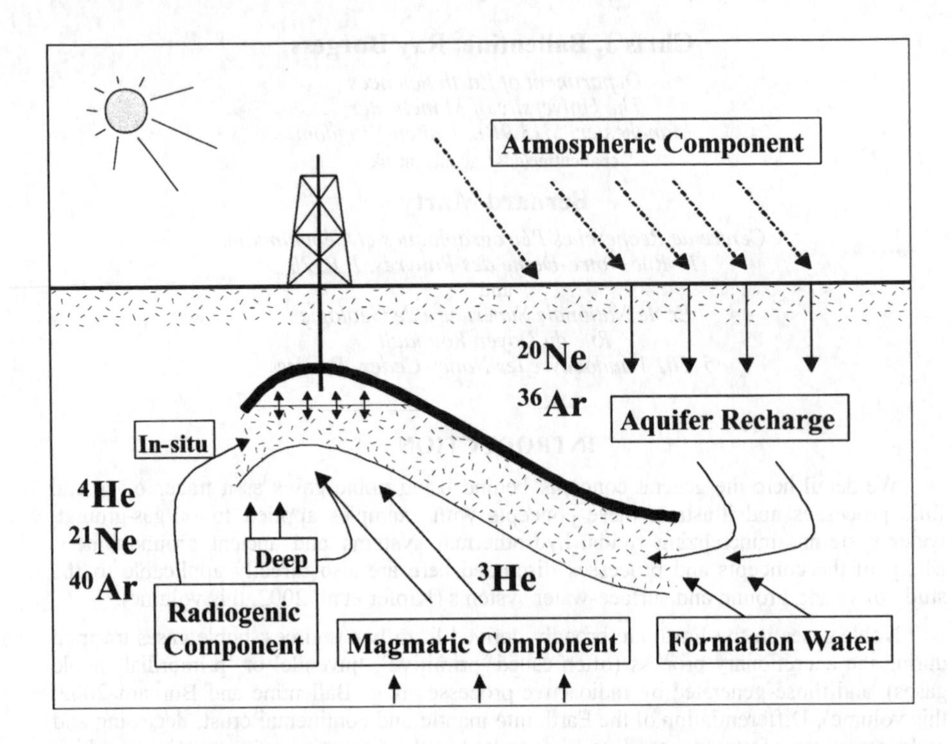

Figure 1. Schematic diagram of a gas reservoir, illustrating the different noble gas components which may occur in crustal fluids. Atmosphere-derived noble gases (e.g., ^{20}Ne and ^{36}Ar) are input into the gas phase on equilibration with the groundwater system containing dissolved atmosphere-derived noble gases. Radiogenic noble gases (e.g., ^{4}He, ^{21}Ne and ^{40}Ar) are produced by the natural decay of the radioelements U, Th and K in the crust, and are also incorporated into crustal fluids. Within areas of continental extension or magmatic activity, noble gases derived from the mantle (e.g., ^{3}He) may also be present in crustal fluids. The distinct isotopic and elemental composition of these different noble gas components allows the extent of their contribution to any crustal fluid to be quantitatively resolved and information about volumes, source and transport process of associated fluids to be identified.

PHYSICAL CHEMISTRY OF NOBLE GASES IN CRUSTAL FLUIDS

Henry's law and the assumption of ideality

It is impractical to investigate experimentally the noble gas solubilities for the range of temperatures, pressures and chemically complex systems found in the geological environment. The investigation of a few representative systems, combined with thermodynamic analysis, is the only viable approach to predict the noble gas behavior in complex natural systems. Henry's law governs the solubility of noble gases in solution. It is convenient at this stage to consider the assumptions that have to be made when applying the available data to Henry's law. A full derivation can be found in most standard texts on thermodynamics and molecular theory of gases and liquids (e.g., Atkins 1978; Nordstrom and Munoz 1985; Denbigh 1986).

Assuming ideality in both liquid and gas phase, Henry's law is

$$p_i = K_i x_i \tag{1}$$

where p_i is the partial pressure of gas i in equilibrium with a fluid containing x_i mole fraction of i in solution and K_i is the Henry's constant. More completely, the non-ideality of species i in both gas and liquid phases needs to be considered, giving

$$\Phi_i p_i = \gamma_i K_i x_i \quad (2)$$

where Φ_i and γ_i are the gas phase fugacity coefficient and liquid phase activity coefficient respectively.

Non-ideality in the gas phase

The real molar volume of the gas can be calculated from empirically derived coefficients (Dymond and Smith 1980; Table 1) for the virial equation of state V_m, where

$$PV_m/RT = 1 + B(T)/V_m + C(T)/V_m^2 \quad (3)$$

P is the total pressure, R the gas constant, T the temperature, and B(T) and C(T) the temperature dependent first and second order virial coefficients. V_m can be found by rearranging Equation (3) to a third-order polynomial and solving using Newton's method of approximation to 10 iterations. The real molar volume is used in turn to find the fugacity coefficient, where

$$\Phi\,(P,T) = \exp[\,B(T)/V_m + (C(T) + B(T)^2)/2V_m^2] \quad (4)$$

The first and second virial coefficients for pressures and temperatures corresponding to a hydrostatic pressure increase with depth and a temperature gradient of 0.03 K/m are tabulated in Table 1, together with the calculated gas phase fugacity coefficient. The fugacity coefficients for the pure noble gases as well for CH_4 and CO_2 are plotted as a function of depth in Figure 2. Non-ideality of He and Ne increases almost linearly with depth, showing up to +18% non-ideality in Ne at depths of 4500 m. In contrast, Ar, Kr and Xe show a maximum deviation from ideality at between 1000- to 1500-m depth, varying from -8% for Ar to -20% for Xe. Maximum non-ideal behavior for these gases occurs at high pressure and low temperature. For example, an increase to lithostatic pressure gradients has the effect of reducing the depth of maximum deviation from ideality in Ar, Kr and Xe by a factor of three, and increases the non-ideal behavior of Ar at this depth to -20%. The behavior of all species is coherent with the exception of Xe, which bisects the Kr fugacity at about 1500 m. This does not appear to be due to error in the set of virial coefficients used, as these are in good agreement with other sets (Dymond and Smith 1980), but may reflect the failure of the virial expansion to only the third order for Xe at these higher pressures.

Only the fugacity change with respect to pressure and temperature variation has been considered for the pure gases. For a mixed gas system, interactions between the different gas molecules and atoms must also be taken into account. The second virial coefficient, $B_m(T)$, for a binary mixture between molecules 1 and 2 can be expressed as

$$B_m(T) = B_{11}(T)x_1^2 + 2\,B_{12}(T)x_1x_2 + B_{22}(T)x_2^2 \quad (5)$$

where x_1 and x_2 are the fractions of gas 1 and 2, B_{11} and B_{22} the second order virial coefficients of the pure species, and B_{12} the interaction coefficient. For an n component mixture Equation (5) can be expressed in the more general form

$$B_m(T) = \sum_{a=1}^{n}\sum_{b=1}^{n} B_{ab}(T)x_a x_b \quad (6)$$

In principle if all the second order virial coefficients of the pure components and the interaction coefficients of all the pairs of the molecules are known, the second order virial coefficient can be calculated. For the third order virial coefficient, 112 and 122 interactions

Table 1. Noble gas and $CH_4 + CO_2$ virial and fugacity coefficients as a function of depth.

			Helium			Neon			Argon		
Depth(m)	*P(atm)*	*T(K)*	*B(T)*	*C(T)*	*Φ(P,T)*	*B(T)*	*C(T)*	*Φ(P,T)*	*B(T)*	*C(T)*	*Φ(P,T)*
300	29	298	11.74	75.1	1.01	11.42	221	1.01	-15.5	991	0.982
1200	116	323	11.58	72.3	1.05	11.86	224	1.05	-11.2	1230	0.964
2000	194	348	11.43	94.8	1.08	12.21	224	1.09	-7.14	959	0.975
2800	271	373	11.35	90.5	1.1	12.52	224	1.12	-3.84	918	1
3700	358	398	11.24	93.8	1.13	12.86	105	1.15	-1.08	877	1.03
4500	436	423	11.07	109.6	1.15	13.1	197	1.18	1.42	833	1.07

			Krypton			Xenon		
Depth(m)	*P(atm)*	*T(K)*	*B(T)*	*C(T)*	*Φ(P,T)*	*B(T)*	*C(T)*	*Φ(P,T)*
300	29	298	-52.36	2612	0.94	-130	6069	0.85
1200	116	323	-42.78	2260	0.846	-110	5306	0.804
2000	194	348	-35.21	1076	0.839	-94.5	4635	0.883
2800	271	373	-28.86	1942	0.861	-81.2	4115	0.888
3700	358	398	-23.47	1842	0.894	-70.1	3739	0.891
4500	436	423	-18.82	1759	0.93	-60.7	3469	0.886

			CH_4			CO_2		
Depth(m)	*P(atm)*	*T(K)*	*B(T)*	*C(T)*	*Φ(P,T)*	*B(T)*	*C(T)*	*Φ(P,T)*
300	29	298	-43.3	2620	0.951	-123	4931	0.856
1200	116	323	-34.6	2370	0.88	-103	4928	0.753
2000	194	348	-27.7	2335	0.887			
2800	271	373	-21.6	2144	0.915	-73	4154	0.833
3700	358	398	-16.4	1999	0.954			
4500	436	423	-11.6	1767	0.992	-52	3046	0.856

The second and third order virial coefficients B(T) and C(T) respectively, are from the compilation by Dymond and Smith (1980). These allow the real molar volume, V_{real}, to be calculated by solving $(P/RT)V_{real}^3 - V_{real}^2 - B(T)V_{real} - C(T) = 0$ (Eqns. 3 and 4), by using Newtons method of approximation, which converges at ~ 10 iterations. The fugacity coefficient $\Phi(PT) = \exp[B(T)/V_{real} + (C(T) + B(T)^2)/2V_{real}^2]$ and is calculated from B(T), C(T) and V_{real} as a function of depth assuming a temperature gradient of 0.03K/m and hydrostatic pressure to assess the deviation from ideality of the pure gas components up to 4500m depth (Fig. 2).

must be considered for a binary mix. In practice, the data set required to calculate the virial coefficients is limited and not of practical use. If however, the assumption is made that the interactions between unlike molecules is insignificant, then a model can be based on the ideal mixing of non-ideal, or real, gases, where for any species the Lewis-Randall rule can be applied where

$$f_i = f_i^\theta (P,T)\, x_i \tag{7}$$

f_i is the gas fugacity, f_i^θ is the fugacity of pure i at P and T, and x_i the molar fraction. Whilst this assumption is reasonable for a species at high concentration, for a trace gas the dominant interactions will not be with like molecules or atoms. The potential effect of this is illustrated in Figure 3. The activity of CO_2 and H_2O at a constant temperature is plotted for a binary CO_2/H_2O mixture against the molar fraction of CO_2 for pressures of 1 to 30 kbar (after Nordstrom and Munoz 1985). As the molar fraction of either H_2O or CO_2 approaches unity, the activity of that species approaches the activity predicted by the Lewis-Randall rule. However, the activity of the minor component can be significantly higher than that predicted by ideal mixing of real gases. Maximum deviation from ideal mixing occurs at low molar fraction, low temperature and high pressure. There is no data available to assess the magnitude of this deviation for the trace noble gas components in $CH_4/CO_2/H_2O$ gases.

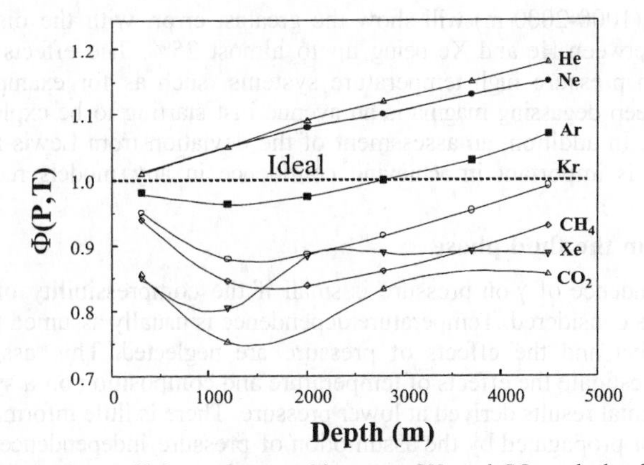

Figure 2. Fugacity coefficients of pure noble gases, CH_4 and CO_2, calculated from second and third order virial coefficients (Table 1) as a function of depth, taking a temperature gradient of 0.03 K/m and hydrostatic pressure.

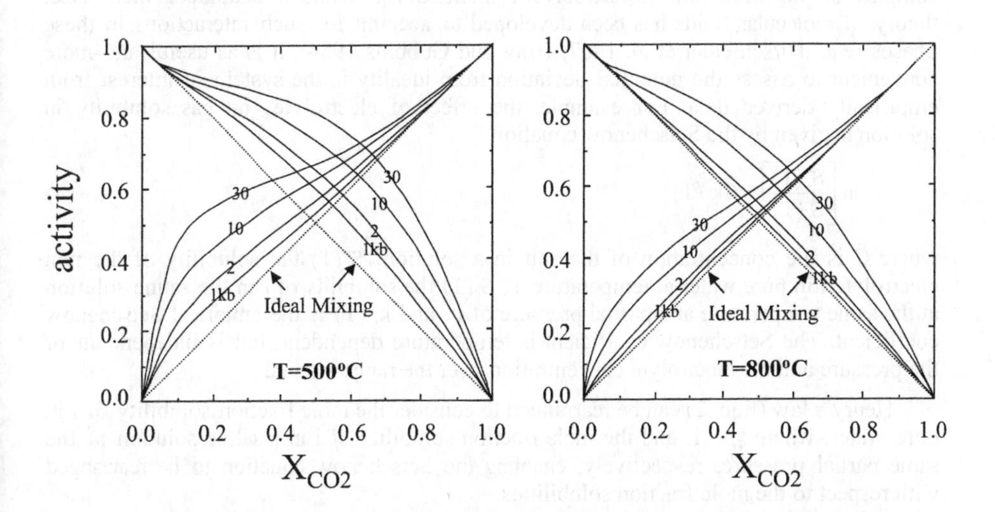

Figure 3. For a binary CO_2/H_2O mixture the activity of each species at a constant temperature is plotted against the molar fraction of CO_2 for pressures of 1 to 30 kbar (after Nordstrom and Munoz 1985). As the molar fraction of either H_2O or CO_2 approaches unity, the activity of that species approaches the activity predicted by the Lewis-Randall rule. However, the activity of the minor component can be significantly higher than that predicted by ideal mixing or real gases. Maximum deviation from ideality occurs at low concentration, low temperature and high pressure.

All solubility calculations to date assume ideal gas behavior for the noble gases irrespective of geological environment. At near surface conditions, Lewis-Randall mixing is a reasonable assumption. It is clear from Figure 2, that the assumption of ideal behavior for He, Ne, and Ar at near surface conditions is also reasonable. More care must be taken with Kr and Xe, where deviation from ideality at 300-m depth is -6% and -15% respectively. Taking a linear extrapolation from 300-m depth to the surface (and ideal conditions) for example, would suggest that at 30-m depth Kr and Xe show -0.6% and -1.5% deviation from ideality. Solubility calculations for moderate depth hydrostatic fluid systems (1000-2000 m) will show the greatest error, with the discrepancy due to non-ideality between He and Xe being up to almost 25%. The effects of non-ideal gas phases in high pressure high temperature systems, such as for example the effect on solubility in deep degassing magma is an avenue just starting to be explored (Nuccio and Paonita 2000). In addition, an assessment of the deviation from Lewis-Randall mixing in deep systems is important in achieving confidence in any models requiring solubility calculations.

Non-ideality in the fluid phase

The dependence of γ on pressure is small if the compressibility of liquids is small over the ranges considered. Temperature dependence is usually assumed to have the most significant effect and the effects of pressure are neglected. This assumption enables workers to investigate the effects of temperature and composition on a system of interest from experimental results derived at lower pressure. There is little information available to assess the error propagated by the assumption of pressure independence in high-pressure systems.

Unlike gases at low to moderate pressures, the virial series cannot be applied accurately to dissolved gases in a fluid due to short length scale ordering and resulting complex solute/solute and solute/solvent interactions. While a statistical mechanical theory of molecular fluids has been developed to account for such interactions in these phases (e.g., Hirschfelder et al. 1967; Gray and Gubbins 1984), it is as useful and more convenient to assess the potential deviation from ideality in the system of interest from empirically derived data. For example, the effect of electrolytes on gas solubility in solution is given by the Setschenow equation

$$\ln\left[\frac{S_i^o(T)}{S_i(T)}\right] = Ck_i(T) \tag{8}$$

where C is the concentration of the salt in a solution, $S_i^o(T)$ the solubility of the non electrolyte i in pure water at temperature T, $S_i(T)$ the solubility of i in the saline solution at the same temperature and partial pressure of i, and k_i (T) is the empirical Setschenow coefficient. The Setschenow coefficient is temperature dependent, but is independent of the pressure and the electrolyte concentration over the range studied.

Henry's law (Eqn. 2) can be rearranged to consider the mole fraction solubility of i in pure water, where $\gamma_i \sim 1$, and the mole fraction solubility of i in a saline solution at the same partial pressure, respectively, enabling the Setschenow equation to be rearranged with respect to the mole fraction solubilities

$$\ln\left[\frac{\varphi_i P_i / K_i(T)}{\varphi_i P_i / \gamma_i K_i(T)}\right] = Ck_i(T) \tag{9}$$

giving

$$\gamma_i = \exp[Ck_i(T)] \tag{10}$$

This satisfies the condition that as $C \to 0$, $\gamma_i \to 1$. If k_i (T) is negative this results in $\gamma_i < 1$, increasing the solubility of i. If k_i (T) is positive, this results in $\gamma_i > 1$, reducing the solubility of i. Although these effects are commonly called 'salting in' and 'salting out' of the non electrolyte, from Equation (10) it can be seen that these are no more than empirically derived changes in the activity of the dissolved gas.

Measurements of noble gas solubilty in water and NaCl solutions have focused on low temperatures (0-40°C) and salinities up to that of seawater (Weiss 1970, 1971; Clever 1979a,b). This data is used in noble gas paleotemperature investigations and is

Table 2: Noble gas Setchenow coefficients, $k_i(T)$

Species	G_1	G_2	G_3
He	-10.081	15.1068	4.8127
Ne	-11.9556	18.4062	5.5464
Ar	-10.6951	16.7513	4.9551
Kr	- 9.9787	15.7619	4.6181
Xe	-14.5524	22.5255	6.7513

Constants from Smith and Kennedy (1982) to fit $k_i(T) = G_1 + G_2/(0.01T) + G_3\ln(0.01T)$, where T is temperature in Kelvin and salinity is in units of mol^{-1} (see Eqn. 8).

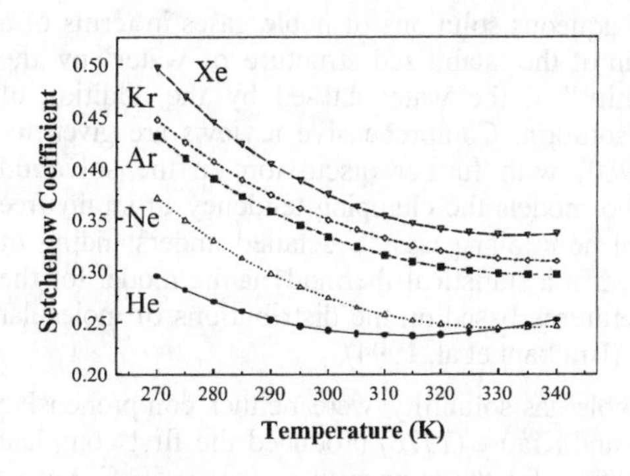

Figure 4. Setchenow coefficients calculated from Table 2 as a function of T. Only He reaches a minimum in the experimental range, limiting the ability of this data set to be extrapolated to higher temperatures.

tabulated in Kipfer et al. (2002, this volume). To expand the dataset available, specifically to investigate the range of NaCl concentrations and temperatures found in natural brines, Smith and Kennedy (1983) measured the solubilities of the noble gases in 0 to 5.2 molar NaCl solutions between 0 and 65°C. The coefficients and equation used to fit the experimental data are shown in Table 2. The variation of the Setchenow coefficients k_i, with temperature T is shown in Figure 4. Over the experimental temperature range, only k_{He} reaches a minimum. Although the data for the heavier noble gases are fit to the same functional form, it is not possible to assess the error when extrapolating to temperatures higher than 65°C for any gas except He.

Although the work of Smith and Kennedy only investigates the effect of NaCl on noble gas solubility, they note that the contribution by individual ions should be additive and in dilute brines it should be possible to estimate the salt effect of multi-electrolyte solutions. While no data exists for Mg^{++} and Ca^{++} ions, data for KI solutions show that k_{Ar} is independent of the electrolyte species (Ben-Naim and Egel-Thal 1965), suggesting that an NaCl 'equivalent' concentration provides a reasonable value from which to calculate the Setchenow coefficient. This relationship has been used in multi-ion mixtures such as seawater and for more concentrated solutions such as the Dead Sea brines (Weiss 1970; Weiss and Price 1989).

The effect of non-electrolytes such as other dissolved gases in solution is more difficult to assess due to the lack of empirically derived data. Under near surface conditions, most non-electrolytes are relatively insoluble and will have negligible effect. However, at greater pressures significant amounts of major gas species may be in solution. For example at 190atm and 70°C (hydrostatic pressure and temperature at 2km depth), about $3cm^3$(STP) of CH_4 saturate 1 cm^3 of pure water to give a 0.13M CH_4 solution (Price 1981). The assumption must be made therefore, that interactions between gaseous non-electrolytes must be small and do not significantly affect the activity coefficient of the noble gases. However, it has been shown that, for example, the CH_4 saturation concentration in water does significantly decrease when small quantities of CO_2 are present (Price 1981). While the data is not available to assess the magnitude of effect these different major species may have on noble gas solubility, this illustrates the potential shortcoming in assuming that no significant non-electrolyte interactions occur in gas-saturated solutions.

Noble gas solubility in water and oil

Water. The solubility of noble gases in water has received considerable attention from physical chemists investigating molecular models of solution in liquid water. Eley (1939) first considered the process of noble gas solution as consisting of a two step mechanism involving the creation of a cavity in the fluid. Ben-Naim and Egel-Thal (1965) described the thermodynamic behavior of aqueous solutions of noble gases in terms of a two-structure model and discuss the origin of the 'stabilized structure of water' by the noble gases, and the 'degree of crystallinity' of the water caused by the addition of electrolytes and non electrolytes to the solution. Comprehensive reviews are given by Wilhelm et al. (1977) and Ben-Naim (1980) with further discussion on the solvation structure of water by Stillinger (1980) who models the clumping tendency of strain free polyhedra cages formed on the solution of non polar gases. A detailed understanding of the noble gas behavior in water has resulted in a statistical thermodynamic model for the solubility of noble gases at varying temperatures based on the distributions of molecular populations among different energy levels (Braibant et al. 1994).

Early laboratory determinations of noble gas solubility were neither comprehensive nor over large temperature ranges. Benson and Krause (1976) produced the first complete data set for noble gas solubilities in pure water for the temperature range 0-50°C, but as only helium reaches a minimum in this range no extrapolation from this data is possible to higher temperatures. Potter and Clyne (1978) increased the data set by investigating solubilities up to the critical point of water. However, this work was subject to some error, as shown by the subsequent work of Crovetto et al. (1982) and confirmed by Smith (1985) both of whom have fitted their solubility data to curves with a third order power series between 298K and the critical temperature of water. The fit from Crovetto et al. (1982) has been taken here for Ne, Ar, Kr and Xe while the solubility of He relative to Ar has taken from Smith (1985) to calculate the Henry's constants for high temperature aqueous systems (Table 3, Fig. 5).

Table 3. Henry's constants for noble gases in water

Species	A_o	A_1	A_2	A_3
He	-0.00953	0.107722	0.001969	-0.043825
Ne	-7.259	6.95	-1.3826	0.0538
Ar	-9.52	8.83	-1.8959	0.0698
Kr	-6.292	5.612	-0.8881	-0.0458
Xe	-3.902	2.439	0.3863	-0.221

Coefficients for Ne, Ar, Kr and Xe from Crovetto et al. (1981) to fit the equation: $\ln(K_i) = A_o + A_1/(0.001T) + A_2/(0.001T)^2 + A_3/(0.001T)^3$ where K_i is Henry's constant in GPa. Coefficients for He are from Smith (1985) to fit the equation: $\ln(F_{He}) = A_o + A_1/(0.001T) + A_2/(0.001T)^2 + A_3/(0.001T)^3$ where $F_{He} = (X_{He}/X_{Ar})_{liquid}/(X_{He}/X_{Ar})_{gas}$. X is the mol fraction. T is temperature in Kelvin. Valid temperature range is from 273K to the critical point of water. 1GPa = 9870 atm. For water, K_i(atm) = 55.6 K_i^m(atm Kg/mol), where K_i^m is Henry's coefficient expressed in terms of molality.

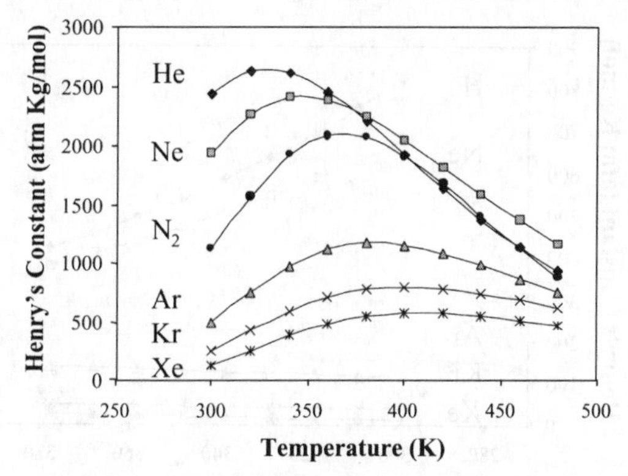

Figure 5. Henry's coefficients for noble gases in water, calculated from Table 3 following the molality convection, plotted as a function of temperature. The valid temperature range of this data set is 273K to the critical point of water.

Although it is usual to express the Henry's constant in units of pressure (Eqn. 1), to enable comparison between oil (in which the mole fraction is difficult to calculate) and water systems, we use here the molality convention where:

$$K^m_i = \Phi_i P_i / C_i \tag{11}$$

Φ_i is the gas fugacity coefficient, P_i the partial pressure of i in atm, and C_i the number of moles of i in 1000 g of the liquid phase (water).

Oil. Solubility studies of the noble gases in crude oil have mostly been limited to empirical approximations as a function of oil density and temperature (e.g., Zanker 1977; ASTM 1985). The most comprehensive study to date remains that of Kharaka and Specht (1988), who have taken two crude oils of different density and experimentally determined the solubility of He, Ne, Ar, Kr and Xe over the temperature range 278 to 373 K. The respective solubilities are fitted to a linear equation in the form $\mathrm{Log}(K^m_i) = A + BT$, where A and B are the experimentally determined coefficients, T the temperature in °C and K^m_i is the Henry's constant following the molality convention (Eqn. 11). The coefficients are given in Table 4 and plotted as a function of temperature in Figure 6.

Table 4. Henry's constants for noble gases in oil

	Heavy Oil (API=25)		Light Oil (API=34)	
Species	*A*	*B*	*A*	*B*
He	3.25	-0.0054	3.008	-0.0037
Ne	3.322	-0.0063	2.912	-0.0032
Ar	2.121	-0.0003	2.03	0.001
Kr	1.607	0.0019	1.537	0.0014
Xe	1.096	0.0035	0.848	0.0052

The solubility constant of the noble gases in oil is dependent on oil density. Coefficients taken from Kharaka and Specht (1987) for two oils fit the equation $\mathrm{Log}(K_i^m) = A + BT$ where the solubility constant K_i^m follows the molality convention and is in units of atm Kg mol^{-1}. T is temperature in °C. The valid temperature range for the determination is 25-100 °C.

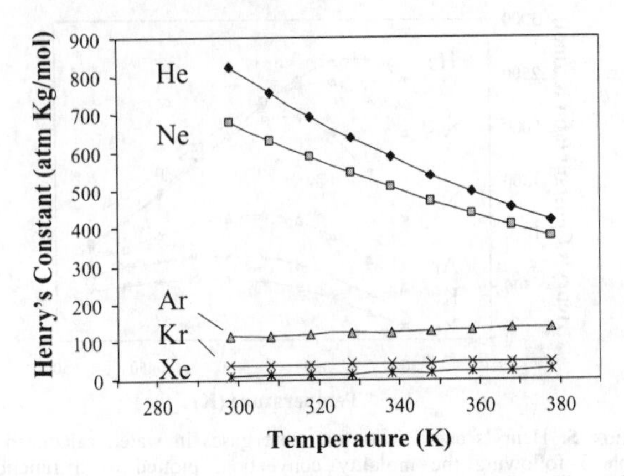

Figure 6. Henry's coefficients for noble gases in 'light' oil (API gravity = 34°), calculated from Table 4 following the molality convention, plotted as a function of temperature. Noble gas solubility constants in more dense oil are higher (Table 4). The valid temperature range of the experimental determination is 298-373 K, although with no apparent minima, extrapolation to higher temperatures is probably reasonable.

The molality convention is followed because crude oil consists of various mixtures of different molecules that make the determination of the gas mole fraction almost impossible to determine. Unlike the studies of noble gas solubilities in water, there is no comprehensive study that has investigated the noble gas solubility in oil as a function of oil density and temperature. While a linear relationship between solubility based on the two oils measured can be assumed (e.g., Ballentine et al. 1996) and to a first order is supported by the empirical approximations (Zanker 1977; ASTM 1985), this remains a limiting factor in the application of noble gas solubility studies involving an oil phase.

PHASE EQUILIBRIUM AND FRACTIONATION OF NOBLE GASES

Liquid-gas phase partitioning of noble gases

Noble gas partitioning and solubility fractionation between equilibrated subsurface phases was initially studied by Goryunov and Kozlov (1940) and further studied (Zartman et al. 1961; Bosch and Mazor 1988; Zaikowski and Spangler 1990; Ballentine et al. 1991; Ballentine et al. 1996; Hiyagon and Kennedy 1992; Pinti and Marty 1995; Torgersen and Kennedy 1999; Battani et al. 2000). Recent reviews are by O'Nions and Ballentine (1993), Ballentine and O'Nions (1994) and Pinti and Marty (2000).

Under equilibrium conditions, the distribution of noble gases between gas and liquid phases is given by Henry's Law (Eqn. 1). Following Goryunov and Kozlov (1940) and Zartman et al. (1961) and assuming ideal behavior in the gas phase, another form of this equation is derived when the concentration i in the gas phase C_g^i is related to the concentration in the liquid phase phase C_l^i by

$$C_g^i = K_i^d \, C_l^i \qquad (12)$$

Henry's constant in this form, K_i^d, is dimensionless. For a fixed volume of gas and liquid, the number of moles in the gas phase $[i]_g$ is related to the total number of moles present $[i]_T$ and the liquid to gas volume ratio, V_l/V_g, by

$$[i]_g = [i]_T \, (V_l/V_g \, K_i^d + 1)^{-1} \qquad (13)$$

Taking the limits as $V_l/V_g \rightarrow 0$, then $[i]_g \rightarrow [i]_T$, and as $V_l/V_g \rightarrow$, then $[i]_g \rightarrow 0$.

It can be seen that Henry's constants can be dimensionless (K_i^d, Eqn. 12) expressed in units of pressure, (K_i, Eqn. 2) or take the molaltity format, K_i^m (Tables 3 and 4). A similar form of Equation (13) for K_i or K_i^m can be simply derived taking due account of units. For example we consider Henry's constant, K_i, expressed in units of atm and water as the liquid phase. In this case the mole fraction, x_i, can be related to the concentrations in terms of the water density, ρ_{H2O} (g/cm^3) and the water volume, V_{H2O} (cm^3) where

$$x_i = 18\,([i]_T - [i]_g])\,(\rho_{H2O}\,V_{H2O})^{-1} \tag{14}$$

The partial pressure of i, p_i, in the gas volume V_g can be expressed in terms of the concentration of i in the gas phase at temperature T (K) assuming that at STP (1 atm, 273 K), 1 mole of gas occupies 22400 cm^3 (ideal gas behavior) by

$$p_i = [i]_g\,(22400\,T)\,(273\,V_g)^{-1} \tag{15}$$

Substitution of Equations (14) and (15) into Equation (2) and rearranging gives

$$[i]_g = [i]_T \left(\frac{22400\,T\,\rho_{H2O}\,V_{H2O}}{18 \times 273\,\frac{\gamma_i}{\phi_i} K_i\,V_g} + 1 \right)^{-1} \tag{16}$$

Equation (13), but with Henry's constant in units of molality, K_i^m (Kg atm/mol), can be similarly derived to give

$$[i]_g = [i]_T \left(\frac{22400\,T\,\rho_l\,V_l}{1000 \times 273\,\frac{\gamma_i}{\phi_i} K_i^m\,V_g} + 1 \right)^{-1} \tag{17}$$

where ρ_l is the density of liquid l (g/cm^3) at the system pressure and temperature, T. This form enables the partitioning of species i between gas and liquid phases to be calculated for any water/gas equilibrium, taking due account of any non-ideal behavior in species i in either the gas or liquid phases. In the simplest case it is possible for example to calculate the volume of gas with which groundwater has equilibrated from one noble gas concentration determination in the water phase, and an estimate of the conditions under which equilibration took place. Similarly, the volume of water with which a gas phase has equilibrated can be quantified from the determination of the concentration of one noble gas in the gas phase. This is discussed in more detail in the following sections.

Liquid-liquid phase partitioning of noble gases

The partition coefficient D_i between two phases for any species i is defined as the equilibrium concentration of i in one phase relative to the other. For two separate liquid phases (we consider here oil and water) the relationship between D_i and the Henry's constants is given as

$$D_i = C_i^{oil}/C_i^{H2O} = K_i^m{}_{(H2O)}/K_i^m{}_{(oil)} \tag{18}$$

where C_i^{oil}, C_i^{H2O}, $K_i^m{}_{(H2O)}$ and $K_i^m{}_{(oil)}$ are the number of moles of i in 1000 g of oil, 1000 g of water and the Henry's constants (atm Kg/mol) of i in water and oil respectively. Substituting $C_i^{oil} = [i]_{oil}/1000 x V_{oil}\rho_{oil}$, $C_i^{H2O} = [i]_{H2O}/1000 \times V_{H2O}\rho_{H2O}$, and $[i]_T = [i]_{oil}+[i]_{H2O}$ into Equation (18) (where ρ_{oil}, ρ_{H2O}, $[i]_T$, $[i]_{oil}$ and $[i]_{H2O}$ are the density of oil and water (g/cm^3), total number of moles of i, the number of moles of i in the oil phase and the number of moles of i in the gas phase respectively), and rearranging gives

$$[i]_{oil} = [i]_T \left(\frac{V_{H2O}\,\rho_{H2O}}{V_{oil}\,\rho_{oil}} \frac{K^m_{i\,(oil)}}{K^m_{i\,(H2O)}} + 1 \right)^{-1} \tag{19}$$

This equation form enables the partitioning of species i between any two liquid phases to be calculated as a function of liquid density and relative volumes. Similar to the gas-liquid system, in the simplest case, determination of the concentration of one groundwater-derived noble gas concentration in either the water or the oil phase enables quantification of a system oil/water volume ratio. This is discussed in detail in the following sections.

Relative fractionation

Although the absolute concentration and distribution of noble gases between different phases is often useful if the original noble gas concentration in one fluid phase is known (e.g., atmosphere-derived noble gases in groundwater), in many systems the concentration may not be so well determined, but a reasonable estimate of the initial relative concentrations of noble gases in a fluid may be available (e.g., crustal-radiogenic or diluted air/groundwater-derived noble gases). In this case it is often convenient to investigate the relative change in noble gas ratios, or fractionation, from the predicted ratio. By convention fractionation is usually assessed relative to ^{36}Ar.

Gas-liquid. From Equation (13), the relative fractionation between, for example, species i and Ar in the gas phase, $([i]/[Ar])_g$, can be related to the original ratio in the system $([i]/[Ar])_T$, after Bosch and Mazor (1988), by

$$\left(\frac{[i]}{[Ar]}\right)_g = \left(\frac{[i]}{[Ar]}\right)_T \frac{\left(\frac{V_g}{V_l} + \frac{1}{K^d_{Ar}}\right)}{\left(\frac{V_g}{V_l} + \frac{1}{K^d_i}\right)} \tag{20}$$

Taking the limits as $V_g/V_l \rightarrow \infty$ then $([i]/[Ar])_g \rightarrow ([i]/[Ar])_T$, and as $V_g/V_l \rightarrow 0$ then $([i]/[Ar])_g \rightarrow ([i]/[Ar])_T(K^d_i/K^d_{Ar})$. Therefore as V_g/V_l becomes small, the noble gases are fractionated proportionally to their relative solubilities in the liquid phase, or more precisely including the effects of non-ideality on the solubility:

$$\text{As } V_g/V_l \rightarrow 0, \quad \frac{\left(\frac{[i]}{[Ar]}\right)_g}{\left(\frac{[i]}{[Ar]}\right)_T} \rightarrow \frac{\frac{\gamma_i}{\Phi_i}K^d_i}{\frac{\gamma_{Ar}}{\Phi_{Ar}}K^d_{Ar}} = F_{gas} \tag{21}$$

where F_{gas} is [i]/[Ar] ratio in the gas phase normalized to the original system ratio to give a fractionation factor. An F_{gas} of 1 indicates that no fractionation from the original system value has occurred.

Liquid-liquid. In a similar fashion the relative change in the [i]/[Ar] ratio in a liquid-liquid system can be assessed. For example, the high solubility of noble gases in oil relative to water can potentially result in significant and distinct fractionation of the noble gases where equilibrium has occurred between these two phases. Following Bosch and Mazor (1988), the [i]/[Ar] ratio in the oil phase, $([i]/[Ar])_{oil}$, is related to the original [i]/[Ar] ratio, $([i]/[Ar])_T$, the oil/water volume ratio V_{oil}/V_{H2O} and the solubility of the noble gas species in each phase, where

$$\left(\frac{[i]}{[Ar]}\right)_{oil} = \left(\frac{[i]}{[Ar]}\right)_T \frac{\left(\frac{V_{oil}}{V_{H2O}} + \frac{K^d_{Ar(oil)}}{K^d_{Ar(H2O)}}\right)}{\left(\frac{V_{oil}}{V_{H2O}} + \frac{K^d_{i(oil)}}{K^d_{i(H2O)}}\right)} \tag{22}$$

Maximum fractionation occurs in the oil phase when

$$\text{as } V_{oil}/V_{H20} \rightarrow 0, \qquad \frac{\left(\dfrac{[i]}{[Ar]}\right)_{oil}}{\left(\dfrac{[i]}{[Ar]}\right)_{T}} \rightarrow \frac{\left(\dfrac{K^d_{Ar\,(oil)}}{K^d_{Ar\,(H2O)}}\right)}{\left(\dfrac{K^d_{i\,(oil)}}{K^d_{i\,(H2O)}}\right)} = F_{oil} \tag{23}$$

The magnitude of the fractionation seen in either phase is therefore proportional to the ratio between the noble gas relative solubilities in the two liquid phases.

Equilibrium fractionation of the Ne/Ar ratio: An example. From Equations (20) and (21) and the solubility data presented in Tables 2 and 3, the fractionation of the noble gases between gas and liquid phases can be calculated for a range of V_g/V_l ratios and at temperatures and salinities appropriate to those in an active sedimentary basin. Figure 7a shows the maximum Ne/Ar fractionation in water and gas phases for pure water and a 5M NaCl brine. For pure water, maximum fractionation in the gas phase is 3.4 at 290 K and decreases with increasing temperature, salinity and V_g/V_l. For example, at 290 K and in equilibrium with a 5M NaCl brine, Ne/Ar fractionation in the gas phase has a maximum value of 2.5.

The effect of phase equilibrium with an oil phase and the resulting magnitude of fractionation that can occur is illustrated in Figures 7b and 7c. The relative solubilities of the noble gases in an oil phase have a greater range than in water. This range increases with increasing oil density. The effect of this increased difference in solubility between the noble gases is an increase in the magnitude of fractionation that can occur in an oil/gas/water system. For example a gas phase in equilibrium with 'Heavy' crude oil (API = 25) at 330 K as V_g/V_l aproaches zero will have a Ne/Ar fractionation factor of 7.1. This decreases to a maximum of 4.3 for a light crude oil (API = 34) at the same temperature (Fig. 7b). Fractionation in any gas phase associated with oil decreases with increasing temperature, V_g/V_l and decreasing oil density.

Equilibration between water and an oil phase causes maximum fractionation in the oil phase as the salinity of the water phase approaches saturation and with increasing oil density when V_{oil}/V_{water} approaches zero (Fig. 7c). Similarly, maximum fractionation in the water fractionation in the water phase as V_{oil}/V_{water} approaches infinity, and as the salinity of the water phase approaches saturation. Unlike liquid/gas phase fractionation, which increases with decreasing temperature, water/oil fractionation reaches a maximum at moderately low temperatures. This occurs, for example, in a pure water/ 'light' oil (API = 34) system at 310 K, with a maximum Ne/Ar fractionation of 0.51 and 1.96 in the oil and water phases respectively. This can be compared with a pure water/ 'heavy' oil (API = 25) system where at 286 K a maximum Ne/Ar fractionation of 0.27 and 3.69 is obtained in the oil and water phases respectively.

Rayleigh fractionation

A simple dynamic model. The maximum magnitude of noble gas fractionation that can occur when two phases have been equilibrated is summarized for Ne/Ar in Figure 7. Although the 'phase equilibrium' model demonstrates the effect of the physical conditions in a system on the limits of noble gas fractionation, the phase equilibrium model represents only one end-member of the processes that may be occurring in a dynamic subsurface fluid environment. To convey some sense of the relevance of the phase equilibrium model in a dynamic system it is useful to consider the extent to which noble gases partition and fractionate between phases when a gas bubble passes through a column of liquid (Ballentine 1991; Fig. 8).

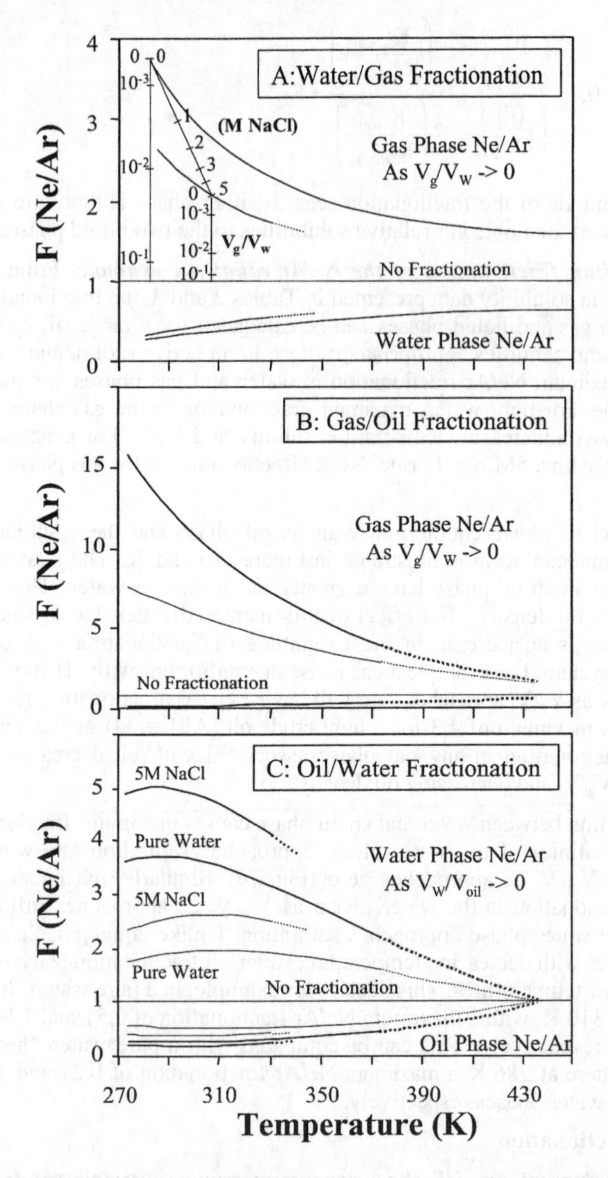

Figure 7. (A) The dependence of Ne/Ar fractionation between water and a gas phase is shown as a function of temperature, salinity and the gas/water volume ratio (Eqns. 20, 21) modified from Ballentine et al. (1991). The maximum equilibrium fractionation in the gas phase occurs when the water phase salinity and temperature is low and as V_g/V_w approaches 0. A graduated scale between pure water and 5 M NaCl brine is shown to illustrate the effect of changing salinity. Graduated scales are also shown to illustrate the effect of changing the V_g/V_w ratio for both the pure water and a 5M NaCl brine. Fractionation of the Ne/Ar ratio in the water phase is the inverse of that in the gas phase, with maximum fractionation occuring at low temperature and salinity as V_g/V_w approaches infinity. (B) The dependence of Ne/Ar fractionation in an oil/gas phase system is shown as a function of temperature (faint line is 'light' oil, API = 34; dark line 'heavy' oil, API = 25). Maximum fractionation in the gas phase occurs as the V_g/V_{oil} ratio approaches zero. Maximum fractionation occurs in an oil/gas system at low temperature and as the oil density increases.

Figure 7 caption, continued.
Fractionation in the oil phase is the inverse of that occuring in the gas phase, with maximum fractionation occuring as the V_g/V_{oil} ratio approaches infinity. (C) The dependence of Ne/Ar fractionation in an oil/water phase system ('light' oil, API = 34) is shown as a function of temperature. Solid lines represent the limit of experminetal data, dashed lines an extrapolation. A fractionation maximum occurs at low temperature, 310 K for 'light' oil, and high salinity. In the water phase maximum fractionation occurs at high salinity, high oil density and as the V_{water}/V_{oil} ratio approaches zero. Maximum fractionation in the oil phase occurs at high salinity, high oil density and as the V_{water}/V_{oil} ratio approaches infinity.

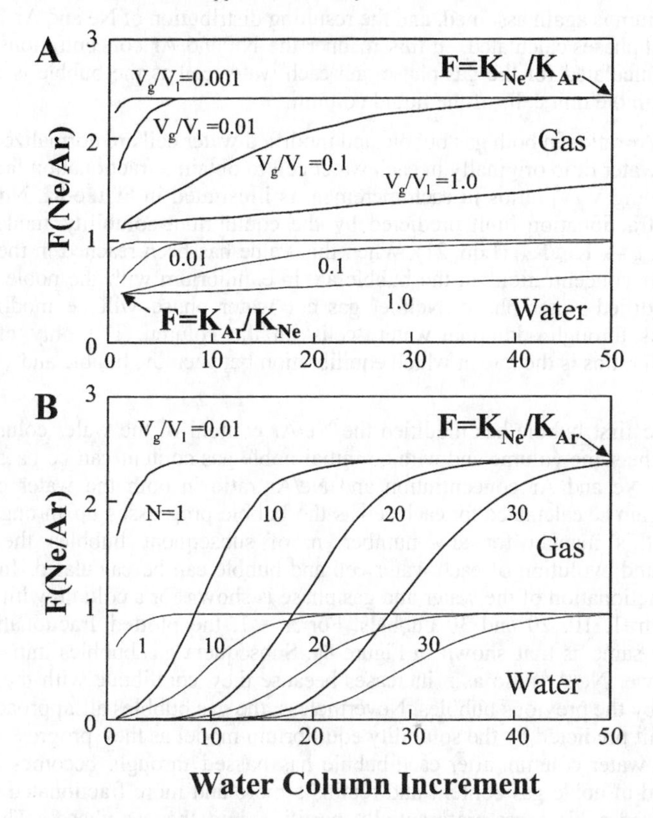

Figure 8. (A) A water column is divided into fifty equal unit cells and it is assumed there is no liquid or dissolved gas between cells. Each cell originally has the noble gas content of air-equilibrated water and all calculated Ne/Ar ratios are normalized to this value to obtain a fractionation factor F. The column temperature is taken to be 325 K, which for pure water gives K_{Ne} = 133245 atm and K_{Ar} = 55389 atm. A gas bubble of constant volume is passed sequentially through the column, equilibrium assumed to occur in each water cell and the Ne and Ar partitioned into the respective gas and water phases (Eqn. 16). The evolution of the Ne/Ar ratio in the gas bubble (bold) and each water phase increment (Faint) is shown for different gas/water volume ratios, V_g/V_l. The gas bubble Ne/Ar ratio approaches the maximum fractionation value predicted for a gas/water phase equilibrium where as $V_g/V_l \rightarrow 0$, $F \rightarrow K_{Ne}/K_{Ar}$. The cell V_g/V_l ratio only determines the rate at which this limit is approached. (B) The same water column with a fixed cell V_g/V_l ratio of 0.01. n subsequent bubbles are passed through the column and the He/Ne distribution between phases calculated at each stage. The gas bubble Ne/Ar ratio evolution for n = 1, 10, 20 and 30 is shown in bold, together with the residual Ne/Ar in the water column cells (faint lines). All gas bubbles approach the limit imposed by the phase equilibrium model. The water phase is fractioned in the opposite sense and is fractionated in proportion to the magnitude of gas loss following the Rayleigh fractionation law (Eqn. 24).

As a starting point, the liquid can be taken to be water that has equilibrated with air to obtain its noble gas content. Furthermore, it is assumed that the liquid is saturated with respect to the dominant gas species forming the bubble/gas phase. The column is divided into cells and it is assumed that there is no transport of dissolved gases or fluid between the cells. When a gas bubble, initially with no noble gas content, is introduced into the first cell the distribution of both Ne and Ar can be calculated from Equation (16) assuming complete equilibration between the gas and fluid in that cell only. The volume of the bubble is assumed to be constant and, now with a noble gas content, is moved to the next cell. Equilibrium is again assumed, and the resulting distribution of Ne and Ar between the gas and liquid phases calculated. In this manner the Ne and Ar concentrations and Ne/Ar ratio can be calculated for the gas phase and each water cell as the bubble is sequentially passed through the unit cells of the liquid column.

The Ne/Ar ratios in both gas bubble and modified water cell are normalized to the air-equilibrated water ratio originally in each water cell to obtain a fractionation factor, F. The effect of varying V_g/V_l ratios in each increment is illustrated in Figure 8a. No gas bubble exceeds the fractionation limit predicted by the equilibrium solubility model where as $V_g/V_l \rightarrow 0$, $F_{gas} \rightarrow K_{Ne}/K_{Ar}$ (Eqn. 21). When this value has been reached in the gas phase, the Ne and Ar concentrations in the bubble are in equilibrium with the noble gas content of the unmodified water phase. Neither gas nor water phase will be modified as this bubble passes through additional water 'cells' in the column. The only effect of the varying V_g/V_l ratios is the rate at which equilibration between the bubble and water cells is reached.

After the first bubble has modified the Ne/Ar content of the water column, another bubble with the same volume and with no initial noble gas content can be passed into the column. The Ne and Ar concentration and Ne/Ar ratio in both the water cell and gas bubble can again be calculated for each cell as the bubble progresses up through the water column. In this manner for any number, n, of subsequent bubbles the noble gas distribution and evolution of each water cell and bubble can be calculated. In Figure 8b, the Ne/Ar fractionation of the water and gas phase is shown for a column with a cell V_g/V_l = 0.001 for n=1, 10, 20 and 30 bubbles. For n = 1, the plotted fractionation in both phases is the same as that shown in Figure 8a. Subsequent gas bubbles initially inherit a lower and lower Ne/Ar ratio as n increases because they equilibrate with the water cells fractionated by the previous bubbles. Nevertheless, the gas bubbles all approach the same F(Ne/Ar) limit predicted by the solubility equilibrium model as they progress through the column. The water column, after each bubble has passed through, becomes increasingly more depleted in noble gas content and retains a more and more fractionated Ne/Ar ratio as the dissolved noble gases preferentially partition into the gas phases. The noble gas ratio in the water phase exceeds the fractionation predicted by the equilibrium model, and is related to the fraction of gas remaining by the Rayleigh fractionation law

$$\left(\frac{[i]}{[Ar]}\right)_{water} = \left(\frac{[i]}{[Ar]}\right)_{o} P^{(\alpha-1)} \tag{24}$$

P is the fraction of Ar remaining in the liquid (water) phase, $([i]/[Ar])_o$, the original liquid phase i/Ar ratio and α is the fractionation coefficient given for a gas/liquid system where:

$$\alpha = (K_i^{liquid}/K_{Ar}^{liquid}) \tag{25}$$

Similarly, the Rayleigh fractionation coefficient used to determine the magnitude of fractionation in a water phase that has interacted with an oil phase (instead of gas) is given by

$$\alpha = (K_i^{water}\,K_{Ar}^{oil})/(K_i^{oil}\,K_{Ar}^{water}) \tag{26}$$

The K variables are the solubilities of the noble gas i and Ar in oil and water. In summary:

- The phase equilibrium model limits the maximum noble gas fractionation that will occur in a gas phase migrating though groundwater.
- By direct analogy, noble gas fractionation in an oil phase migrating through groundwater will be similarly limited by the phase equilibrium model.

As the phase equilibrium value is approached in either the gas or the oil phase, quantitative information from the magnitude of fractionation about the volume of water that has equilibrated with the non-water phase will be lost (although minimum volumes can be inferred).

The path length and time required for either the gas or oil phase to achieve the phase equilibrium limit will depend on the relative availability of the groundwater for equilibration, which in turn is controlled by factors such as porosity, tortuosity and interconnectivity of the rock matrix, as well as the residence time of both the gas or oil and groundwater phases. It is also important to consider scale. For example, the 'bubble', can be considered to be as large as a gas or oil field, and its movement only relative to the water. In this case an oil or gas field that equilibrates with an active groundwater system cannot be distinguished from groundwater equilibration during oil or gas phase migration from source rock to trap.

In contrast to a migrating gas or oil phase, the residual groundwater phase will be fractionated following Rayleigh fractionation (Eqn. 24; Fig. 9).

- Although extensive fractionation in the residual groundwater phase can occur in the opposite sense to that of the migrating phase, the absolute concentration of the fractionated noble gases in the water phase is very much reduced.
- Strongly fractionated noble gas ratios can be transferred to the gas or oil phase from previously 'stripped' groundwater, but any such oil or gas can only contain low water-derived noble gas concentrations.

Re-solution and effervescence

A mechanism proposed to increase the magnitude of fractionation in a gas phase over that predicted by the phase equilibrium model limit is that of a multi-stage process of re-solution and effervescence (Zartman et al. 1961). In the simplest case a gas phase, containing a significant concentration of fractionated noble gases in equilibrium with air-equilibrated groundwater, could be re-dissolved by a change in physical conditions such as an increase in pressure, a decrease in temperature and salinity, or mixing with unsaturated (with respect to the major gas phase) water. This would create a local increase in both the groundwater noble gas concentration and the magnitude of noble gas fractionation in solution. Subsequent formation of a gas phase in equilibrium with the modified groundwater would show a fractionation relative to air-equilibrated water in excess of that predicted by the single stage equilibrium solubility model.

We can consider the volumes and concentrations required for this process through a worked example. In Figure 10, water and gas volumes have been chosen to produce significant fractionation in the final gas phase by a two-stage process of re-solution and effervescence, with a V_g/V_l ratio of 0.001 at each stage. The water is taken to be seawater, which has equilibrated with air at 20°C, now at conditions typical of 700-m depth: a temperature of 310K, hydrostatic pressure (68atm) and with a high salinity (5M NaCl) ($K_{Ne} = 4.72 \times 10^5$ atm and $K_{Ar} = 2.33 \times 10^5$ atm). The water is equilibrated with a gas phase and produces a $^{20}Ne/^{36}Ar$ fractionation value of 1.77 in the gas phase, which can be compared to the phase equilibrium limit of K_{Ne}/K_{Ar}=2.03 as $V_g/V_l \rightarrow 0$. It is assumed that this volume of gas is then re-dissolved in a small volume of unaltered air-equilibrated seawater. The $^{20}Ne/^{36}Ar$ ratio of the fluid is now 1.72 times greater than the unaltered water and, when equilibrated with a gas phase under the same conditions as the first stage, produces a $^{20}Ne/^{36}Ar$ fractionation in the gas phase of F = 3.00. This is ~50% greater than the maximum fractionation for a single stage process. Zartman et al. (1961) suggest that this process may be repeated several times to produce highly fractionated noble gas patterns.

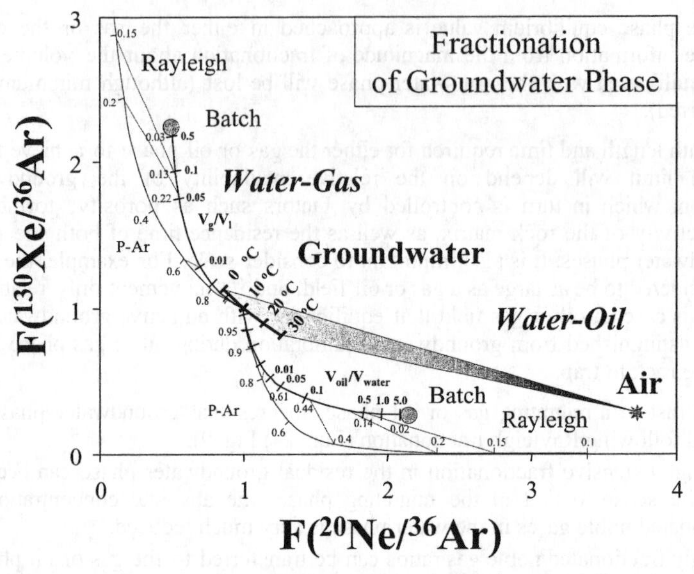

Figure 9. Fractionation of the Ne/Ar and Xe/Ar ratios in groundwater is shown by process and interacting phase. Conditions are taken to be 330 K and 2 M NaCl salinity. ***Water-Gas*** batch equilibration is shown as a function of Gas/Water volume ratio (V_g/V_l), as well as fraction of Ar remaining in the water phase, P-Ar. Batch noble gas distribution between water and gas phases is calculated following Equation (17). Rayleigh fractionation (faint line) of the water phase by gas is calculated following Equations (24) and (25). The graduated scale shows the fraction of Ar remaining in the groundwater phase. ***Water-Oil*** batch equilibration is also calculated for a light crude oil (API = 41, density = 0.82g/cm^3, Kharaka and Specht 1988) following Equation (19). The change in Ne/Ar and Xe/Ar ratios is shown as a function of Oil/Water volume ratio (V_{oil}/V_{water}), as well as fraction of Ar remaining in the water phase, P-Ar. Rayleigh fractionation (faint line) of the water phase by gas is calculated following Equations (24) and (26). The graduated scale shows the fraction of Ar remaining in the groundwater phase. Tie lines between batch and Rayleigh fractionation resulting in the same P-Ar are also shown for gas-water and oil-water systems. On the same figure we have taken fresh water equilibrated with air at 1atm pressure at 293 K as our reference composition. The figure shows the effect on the fractionation values of variable recharge temperature and air addition. The effect of either a gas phase or an oil phase passing through groundwater will be distict and in principle enables quantification of the respective oil/water or oil/gas ratios.

The first point for consideration in this two-stage model is the re-solution of the gas phase. We use CH_4 as an example. 65 m^3 CH_4 (STP) occupies 1 m^3 at a depth of 700 m (68 atm and 310 K). From Price (1981), 14.6 m^3 CH_4 (STP) saturates 10 m^3 of pure water under these. ~40 m^3 of pure water, unsaturated with respect to CH_4, are therefore required to re-dissolve this gas (more if this were a brine). If this volume is air-equilibrated water, the additional (and unfractionated) noble gas content would have the effect of lowering the final $^{20}Ne/^{36}Ar$ fractionation value to F = 1.57, less than the maximum fractionation value predicted for a single stage equilibrium. Notwithstanding more complex processes, such as the addition of a phase with no initial noble gas content, it would seem that the process of re-solution of the major gas phase provides a major limiting factor when advocating a re-solution and effervescence model.

If the fluid in equilibrium with the final gas phase is air-equilibrated water, the concentration of fractionated noble gases in solution must be significantly higher than

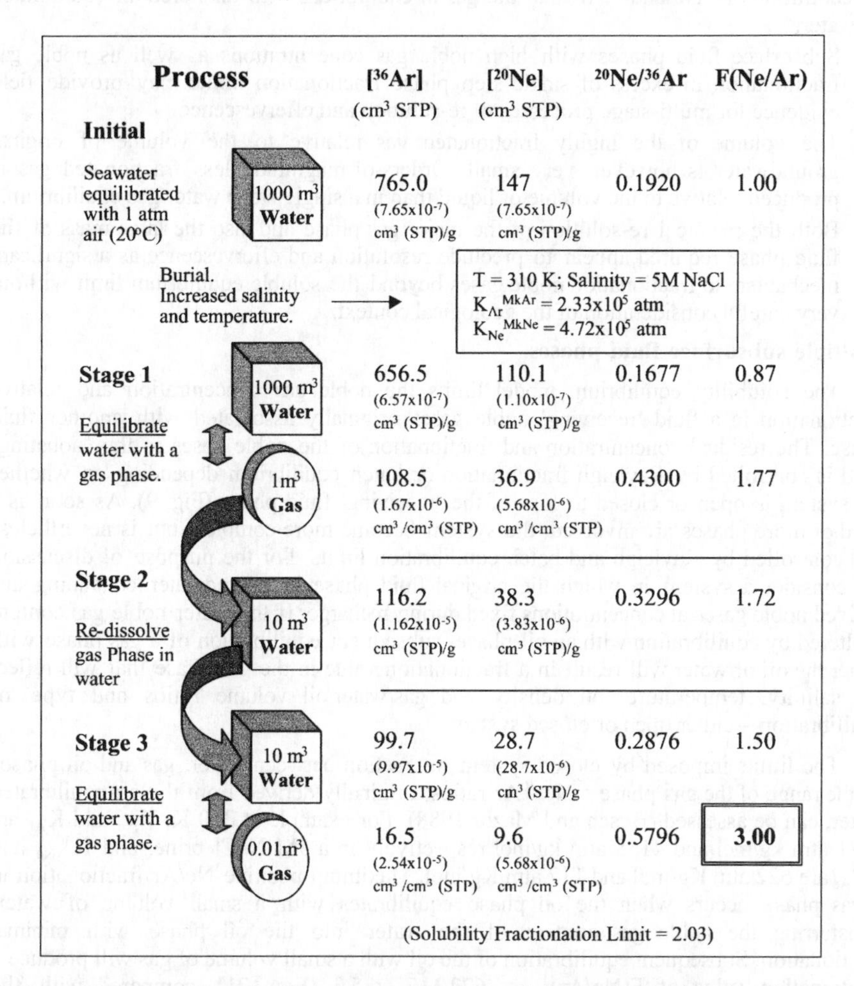

Process
[36Ar] (cm3 STP)
[20Ne] (cm3 STP)
20Ne/36Ar
F(Ne/Ar)
Initial
Seawater equilibrated with 1 atm air (20°C)
1000 m3 Water
765.0 (7.65x10-7) cm3 (STP)/g
147 (7.65x10-7) cm3 (STP)/g
0.1920
1.00
Burial. Increased salinity and temperature.
T = 310 K; Salinity = 5M NaCl
K_Ar^MkAr = 2.33x10^5 atm
K_Ne^MkNe = 4.72x10^5 atm
Stage 1
1000 m3 Water
656.5 (6.57x10-7) cm3 (STP)/g
110.1 (1.10x10-7) cm3 (STP)/g
0.1677
0.87
Equilibrate water with a gas phase.
1m3 Gas
108.5 (1.67x10-6) cm3 /cm3 (STP)
36.9 (5.68x10-6) cm3 /cm3 (STP)
0.4300
1.77
Stage 2
10 m3 Water
116.2 (1.162x10-5) cm3 (STP)/g
38.3 (3.83x10-6) cm3 (STP)/g
0.3296
1.72
Re-dissolve gas Phase in water
Stage 3
10 m3 Water
99.7 (9.97x10-5) cm3 (STP)/g
28.7 (28.7x10-6) cm3 (STP)/g
0.2876
1.50
Equilibrate water with a gas phase.
0.01m3 Gas
16.5 (2.54x10-5) cm3 /cm3 (STP)
9.6 (5.68x10-6) cm3 /cm3 (STP)
0.5796
3.00
(Solubility Fractionation Limit = 2.03)

Figure 10. Re-solution and effervescence—a 3-stage worked example. **Initial:** A volume of seawater is equilibrated with air at 20°C and 1 atm, and then buried. The temperature and salinity and temperature is increased and the salinity modified Henry's constants, K_i^{Mki}, calculated. **Stage 1:** The water is equilibrated with a gas phase and the noble gas distribution between the phase calculated (Eqn. 16). **Stage 2:** The gas phase is re-dissolved in a volume of seawater that has an unmodified noble gas composition. **Stage 3:** The volume of water containing the re-dissolved noble gases is equilibrated with a gas phase containing no noble gases and the noble gas distribution between the phases re-calculated. The fractionation achieved in the gas phase is approximately 50% higher than that predicted by the phase fractionation model limit, but requires huge quantities of water. The physical process of re-solution of gases is also problematic (see text). When invoking a multi-stage process of re-solution and effervescence to account for highly fractionated noble gas ratios, the geological implications of mass balance and mechanism of re-solution must be very carefully considered.

typical meteoric groundwaters. This would also require a proportionally high concentration of noble gases in the gas phase (although this could be masked by subsequent dilution by air-noble gas free gas addition). This is demonstrated in the worked example (Fig. 10), where the final gas phase has an order of magnitude greater concentration of ^{20}Ne and ^{36}Ar than the gas in equilibrium with unaltered air-equilibrated seawater.

- Subsurface fluid phases with high noble gas concentrations as well as noble gas fractionation in excess of single step phase fractionation limits may provide field evidence for multi-stage processes of re-solution and effervescence.
- The volume of the highly fractionated gas relative to the volume of original groundwater is however, very small. Orders-of-magnitude less fractionated gas is produced relative to the volume of liquid than in a single-stage water/gas equilibrium.
- Both the physical re-solution of the major gas phase and also the large mass of the fluid phase required appear to preclude resolution and effervescence as a significant mechanism to fractionation noble gases beyond the soluble equilibrium limit without very careful consideration of the geological context.

Multiple subsurface fluid phases

The solubility equilibrium model limits the noble gas concentration and relative fractionation in a fluid 'receiving' noble gases originally associated with another fluid phase. The residual concentration and fractionation of the noble gases in the 'donating' fluid is controlled by Rayleigh fractionation or batch equilibrium depending on whether the system is open or closed to loss of the 'receiving' fluid phase (Fig. 9). As soon as a third or more phases are involved, the system become more complex, but is nevertheless still controlled by Rayleigh and batch equilibration limits. For the purpose of discussion we consider a system in which the original fluid phase is groundwater containing air-derived noble gases at concentrations fixed during recharge. If the water noble gas content is altered by equilibration with an oil phase, subsequent equilibration of a gas phase with either the oil or water will result in a fractionation value in the gas phase that will reflect the salinity, temperature, oil density and gas/water/oil volume ratios and type of equilibration—either open or closed system

The limits imposed by closed system interaction between water, gas and oil phases on the range of the gas phase $^{20}Ne/^{36}Ar$ ratios, originally derived from the air-equilibrated water, can be assessed (Bosch and Mazor 1988). For example at 310 K, K_{Ne} and K_{Ar} are 8471 atm kg/mol and 4155 atm kg/mol respectively in a 5M NaCl brine, and $K^{oil}{}_{Ne}$ and $K^{oil}{}_{Ar}$ are 622 atm Kg/mol and 117 atm Kg/mol. Maximum positive Ne/Ar fractionation in a gas phase occurs when the oil phase equilibrates with a small volume of water, transferring the noble gas content of the water into the oil phase with minimal fractionation. Subsequent equilibration of the oil with a small volume of gas will produce a fractionation value of $F(Ne/Ar)_{gas}$ = 622/117 = 5.3 (Eqn. 21), compared with the maximum fractionation of $F(Ne/Ar)_{gas}$ = 8471/4155 = 2.03 predicted for a water/gas system under the same conditions (Fig. 7a). Maximum gas phase Ne/Ar fractionation, in the opposite direction, occurs when the oil phase equilibrates with a large volume of water to produce $F(Ne/Ar)_{oil}$ = (117/4155)/(622/8471) = 0.31 (Eqn. 23).

The addition of one extra phase, crude oil, to a gas/water system more than doubles the range of fractionation that can occur in any associated gas phase, from between $F(Ne/Ar)_{gas}$ = 1.0 to 2.03 to between $F(Ne/Ar)_{gas}$ = 0.31 to 5.3. In the case of an open system, fractionation in the residual phase will be even more extreme, and reflected in much lower concentrations (Battani et al. 2000). While it is possible to envisage a myriad of different interactions between water, gas and oil phases, depending on the order of interaction, open or closed system behavior and the relative fluid volumes, it is not a useful exercise at this point to consider all of the possibilities, as other factors may also play a role in constraining the physical model development. For example, stable isotope information from the hydrocarbon gases may rule out the involvement or association with an oil phase (Schoell 1983), and systems must be considered on a case-by-case basis.

Diffusion or kinetic fractionation

In a gas phase. Gaseous diffusion processes can generate both elemental and isotopic fractionation in natural gases. Marty (1984) reviews the processes that can affect noble gases after Present (1958) and distinguishes among:

a) free-molecule diffusion;
b) mutual diffusion; and
c) thermal diffusion.

(a) Free-molecule diffusion takes place when a gas is traveling through a conduit in which gas-wall collisions are more frequent than gas-gas collisions. For this to occur the conduit diameter must be smaller than the mean free path of the gas atoms. Because the mean speed of particles is proportional to $m^{-1/2}$, where m is the mass of the gas atom, in the case of a binary mixture the lightest component will be enriched at the outlet of a conduit. It is shown that the fractionation coefficient, α, between two elements of masses m_1 and m_2 is approximated by:

$$\alpha = [(m_2/m_1)^{1/2}] \tag{27}$$

For a system depleted by a free-molecule diffusive process the Rayleigh fractionation law can again be applied. Taking m_1 as the mass of gas i and m_2 as the mass of Ar:

$$\left(\frac{[i]}{[Ar]}\right)_{gas} = \left(\frac{[i]}{[Ar]}\right)_{o} P^{(\alpha-1)} \tag{28}$$

where P is the fraction of Ar remaining in the gas reservoir, $([i]/[Ar])_o$, the original gas phase i/Ar ratio and α is the fractionation coefficient in Equation (27).

It should be noted that for natural gases the dimension of the conduit needs to be very small (diameter < 10^{-8} m). Because the mean free path of the gas atoms is proportional to temperature and inversely proportional to pressure, this will decrease with depth. Fractionation through free molecular diffusion therefore, will only be significant in special circumstances and when the pressure of the system is low (approaching atmospheric pressure).

(b) Mutual diffusion describes the diffusion of two or more gas species when the dominant interactions are gas-gas collisions. When a gas i with mass m_1 diffuses through a gas with an average molecular mass m_g, the diffusion coefficient of i is proportional to

$$[m_1 \times m_g)/(m_1 + m_g)]^{-1/2} = (m_1^*)^{-1/2} \tag{29}$$

where m_1^* is the 'reduced mass.' The mean velocity of gas i is proportional to its diffusion coefficient. Therefore, for a second gas j with a mass of m_2 the relative velocity between i and j provides the fractionation coefficient α that can be used in Equation (28), where

$$\alpha = (m_2^*/m_1^*)^{1/2} = (m_2/m_1)^{1/2} \times [(m_1 + m_g)/(m_2 + m_g)]^{1/2} \tag{30}$$

These equations are only strictly relevant in the subsurface to a single phase gas system undergoing diffusive loss and remains to be applied in any noble gas study.

(c) Thermal diffusion occurs when a gas mixture is in a non-equilibrium state because of a tendency for lighter molecules to be concentrated at the high temperature boundary.

It is shown that the fractionation factor, f, in a steady state is:

$$f = (T'/T)^{\beta} \quad (31)$$

where T' and T are the highest and lowest temperature and β is the element or isotope pair specific thermal diffusion parameter. Isotope enrichment can only be obtained by preferential extraction of either the cold or hot gas (e.g., Clusius and Dickel 1938).

Marty (1984) considers various scenarios in the subsurface where isotopic fractionation may occur in light of the results of Nagao et al (1979, 1981). From both theoretical considerations (see above) and observed isotopic anomalies, this author concluded that the most likely cause of rare gas isotopic fractionation in natural gases is mutual diffusion taking place between atmospheric gases in soil pores and volcanic/geothermal gases ascending through the upper level of the crust and sediments.

In water. The two principle controlling factors for gas diffusion processes in liquids are the gas mass and its activation energy for diffusion. The diffusion activation energy in turn is controlled by the extent of interaction of the gas molecule or atom with the liquid phase. For noble gases, because they are monatomic and have a stable electron shell, there is little interaction with water and the rate of diffusion is almost entirely controlled by their respective masses. This is in contrast with species such as CO_2 and CH_4 where interaction occurs with water molecules through induced dipole-dipole moments. Because this is in addition to mass, these species diffuse significantly more slowly in water than noble gases of similar mass (Table 5, Fig. 11).

Table 5. Diffusion coefficients* in water (Jähne et al. 1987).

Gas	*Medium*	E_a Kj/Mol	*1σ error* %	*A* 10^{-5} cm²/s	*1σ error* %
He	water	11.70	5	818	2.1
	seawater	12.02	5	886	1.8
Ne	water	14.84	8	1608	3.5
Ar **	water	17.30	10	3141	5
Kr	water	20.20	3	6393	1.6
Xe	water	21.61	5	9007	3.5
Rn	water	23.26	11	15877	11
H_2	water	16.06	3	3338	1.6
	seawater	14.93	9	1981	4.3
CH_4	water	18.36	4	3047	2.7
CO_2	water	19.51	2	5019	1.3

* To fit equation $D = A \exp(-E_a/RT)$ where D is the diffusion coeficient, T the temperature in Kelvin, and R the gas constant

** Extrapolated, see text

The most complete study of noble gas diffusion rates in water remains the experimental determination by Jähne et al. (1987). In this work the diffusion coefficient in water was determined for systems between 0 and 35°C and the results expressed in terms of the diffusion constant, A (cm²/s), and diffusion activation energy, E_a (Kj/Mol) (Table 5), to provide a temperature dependent expression for the determination of the gas diffusion coefficient, D (cm²/s), at variable temperature following

$$D = A\, e^{-Ea/RT} \quad (32)$$

R and T are the gas constant and temperature in Kelvin respectively. For any one temperature it can be shown that the diffusion coefficient for the noble gases are well correlated with the square roots of their masses (Fig. 11). Although Ar has not been experimentally determined in this study, this clear relationship enables the values of A and E_a for Ar to be readily interpolated from the other noble gas values. The interpolated values for Ar are also shown in Table 5. The correlation of diffusion coefficient with the square root of their masses also allows the relative mass fractionation of noble gases to be calculated using Equations (27) and (28).

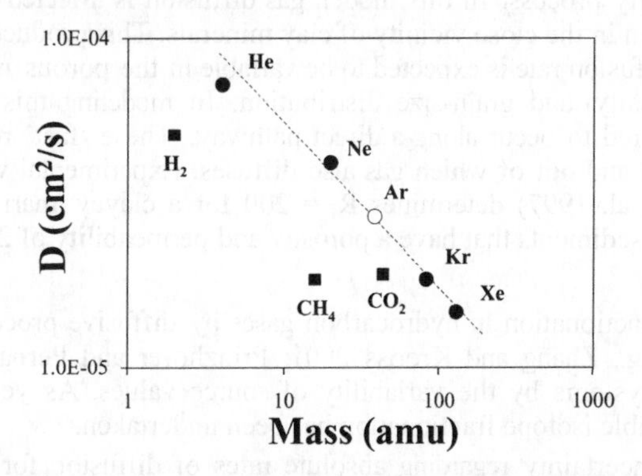

Figure 11. Diffusion coefficients for noble and selected active gases measured in water at 25°C, shown as a function of mass (after Jähne et al 1987).

Jähne et al. (1987) also investigated isotopic specific diffusion coefficients for ^{3}He and the change in $\delta C^{13}(CO_2)$ during diffusive gas loss from water. The increase in He diffusivity for the ^{3}He compared to ^{4}He was in agreement with the ratio of the square-root of their masses. This result provides further supporting evidence that the diffusion coefficients for individual isotopic noble gas species can reasonably be determined as a function of mass from Table 5 for variable temperatures. This is in contrast with the results for the study of $\delta C^{13}(CO_2)$, which showed a fractionation factor far lower than the value predicted from the square root of the reduced mass. This discrepancy indicates that in the case of active gases the difference is not just an effect of mass but of the isotope specific interaction energy with the water molecules.

The effect of salinity on gas diffusion rates is not quantitatively determined except for He and H_2 in seawater (Table 5), showing a reduction in diffusion rate with an increase in salinity.

In a water-filled porous medium. As soon as diffusion in a water-filled porous medium is considered, the effects of porosity, tortuosity, permeability as well as any interaction with the porous medium must also be considered. This enables us to define the first order 'apparent' diffusion coefficient, D_a (cm^2/s), where:

$$D_a = D/(R \times R_f) \qquad (33)$$

R_f is the retardation factor caused by the porous medium geometry and R the retardation factor due to physical or chemical interaction between the gas and porous medium. For the case of noble gases these latter interactions are usually insignificant and R = 1. R_f, determined for different rock varies over orders of magnitude but has not been correlated with physical rock properties to enable an assessment of all rock types. For example, R_f values of between 1.5 and 2.5 have been estimated for He and Ar in deep-sea sediments with a porosity of 70-80% (Ohsumi and Horibe 1984). Krooss and Leythaeuser (1988) have measured diffusion coefficients for light hydrocarbons in sedimentary rocks and calculated R_f values of between 20-50 for argillaceous sediments. Bourke et al. (1989) using both iodine and tritiated water diffusion in London clays calculates a similar value ($R_f \sim 30$). This is in contrast with Neretnieks (1982) who reports R_f values ranging from 100 to 1000 in compacted bentonite for hydrogen and methane diffusion.

Rebour et al. (1997) review the literature describing gas diffusion in a porous medium as a 'double' porosity process. In this model, gas diffusion is affected by the increase in water viscosity when in the close vicinity of clay minerals. This produces an environment in which the gas diffusion rate is expected to be variable in the porous network depending on the local tortuosity and grain-size distribution. In modeling this type of system, diffusion is considered to occur along a direct pathway. These 'fast' routes interconnect 'slow' regions, into and out of which gas also diffuses. Experimental work by the same authors (Rebour et al. 1997) determines $R_f = 200$ for a clayey marl from Paris basin Callovo-Oxfordian sediments that have a porosity and permeability of 23% and 10^{-22} m^{-2}, respectively.

The isotopic fractionation in hydrocarbon gases by diffusive processes is a topic of current research (e.g., Zhang and Krooss 2001; Prinzhofer and Pernaton 1997), but is masked in natural systems by the variability of source values. As yet no work linking noble gases with stable isotope fractionation has been undertaken.

- Despite the uncertainty regarding absolute rates of diffusion for noble gases in a water-filled medium, the relative rates remain a direct function of mass. In principle, for example, the extent of diffusive gas loss for any reservoir can be determined by the magnitude of fractionation of known noble gas elemental ratios using Equation (24) and the appropriate mass fractionation coefficient (Eqn. 27).

RESOLVING DIFFERENT NOBLE GAS COMPONENTS IN CRUSTAL FLUIDS

Terrestrial noble gases are dominated by three reservoirs: the atmosphere, crust and mantle. The isotopic compositions of noble gases produced by radioactive decay in the crust are distinct from noble gases derived from the mantle, which in turn are distinct from those in the atmosphere (Fig. 1).

- Notably, the isotopes of ^{20}Ne, ^{36}Ar, ^{82}Kr and ^{130}Xe are not produced in significant quantities by radioactive processes in the crust (Ballentine and Burnard 2002, this volume) and, in the absence of a magmatic contribution, are almost entirely dominated by atmosphere-derived sources.

Two-component mixing. The atmosphere however, does contain a significant amount of noble gases, such as ^{21}Ne, ^{40}Ar and ^{136}Xe that are also derived from crustal/radiogenic sources. To distinguish which of these species are derived from the immediate crustal system from those that are derived from atmosphere sources, the isotopic ratios can be compared with the atmospheric ratio to identify the crustal 'excess.' In a two-component crust/air mixture

$$[^{21}\text{Ne}]_{crust} = [^{21}\text{Ne}]_{tot} \times [1-(^{21}\text{Ne}/^{20}\text{Ne})_{air}/(^{21}\text{Ne}/^{20}\text{Ne})_s] \quad (34)$$

$$[^{40}\text{Ar}]_{crust} = [^{40}\text{Ar}]_{tot} \times [1-(^{40}\text{Ar}/^{36}\text{Ar})_{air}/(^{40}\text{Ar}/^{36}\text{Ar})_s] \quad (35)$$

$$[^{136}\text{Xe}]_{crust} = [^{136}\text{Xe}]_{tot} \times [1-(^{136}\text{Xe}/^{130}\text{Xe})_{air}/(^{136}\text{Xe}/^{130}\text{Xe})_s] \quad (36)$$

The subscripts crust and tot refer to the crustal and total concentrations while the subscripts air and sample refer to the isotopic composition of the atmosphere (Ozima and Podosek 1983) and sample respectively.

U and Th decay in both the mantle and crust to produce ^{4}He. However, the Earth's mantle has also preserved a significant quantity of 'primitive' ^{3}He during accretion (e.g., Porcelli and Ballentine 2002, this volume). Because ^{3}He is not produced in significant quantities by radioactive decay processes, mantle-derived He has a far higher ^{3}He/^{4}He ratio than crustal sources and even small magmatic additions to crustal fluid systems are readily resolvable (Poreda et al. 1986; Oxburgh et al. 1986). In contrast to the heavier noble gases, He, because of thermal escape from the atmosphere, has only a low abundance in the atmosphere. It is nevertheless necessary to correct any measured He isotopic composition for air-derived contributions by using the observed air-derived ^{20}Ne concentration following (Craig et al. 1978) where

$$({}^3\mathrm{He}/{}^4\mathrm{He})_c = \frac{({}^3\mathrm{He}/{}^4\mathrm{He})_s \times ({}^4\mathrm{He}/{}^{20}\mathrm{Ne})_s / ({}^4\mathrm{He}/{}^{20}\mathrm{Ne})_{air} - ({}^3\mathrm{He}/{}^4\mathrm{He})_{air}}{({}^4\mathrm{He}/{}^{20}\mathrm{Ne})_s / ({}^4\mathrm{He}/{}^{20}\mathrm{Ne})_{air} - 1} \tag{37}$$

Subscripts c, s and air refer to the corrected, measured and air-derived ratios, respectively. The $(^4\mathrm{He}/^{20}\mathrm{Ne})_{air}$ elemental ratio, unlike isotopic ratios, is subject to elemental fractionation. In applications where the air correction is large and/or critical the $(^4\mathrm{He}/^{20}\mathrm{Ne})_{air}$ ratio has to be determined with care. For example, in a groundwater that has not undergone phase fractionation this value can be determined from the recharge temperature and estimate of air in excess of recharge equilibrium or 'excess air' (e.g., Kipfer et al. 2002, this volume). In less critical applications where the air-derived component reasonably has a groundwater origin the measured $(^4\mathrm{He}/^{20}\mathrm{Ne})_{air}$ = 0.288 in 10°C air-equilibrated water is often used (e.g., Craig et al. 1978). In many old groundwaters and hydrocarbon fluids, $(^4\mathrm{He}/^{20}\mathrm{Ne})_s$ is large and the correction is negligible. In this case $(^3\mathrm{He}/^4\mathrm{He})_c \approx (^3\mathrm{He}/^4\mathrm{He})_s$.

Once corrected for atmosphere-derived He, the ^{3}He/^{4}He ratio represents the sum of only two components; the crust and the mantle. The contribution of crustal ^{4}He is then given by

$$[{}^4\mathrm{He}]_{crust} = \frac{[{}^4\mathrm{He}]_{tot} \times [({}^3\mathrm{He}/{}^4\mathrm{He})_{mantle} - ({}^3\mathrm{He}/{}^4\mathrm{He})_c]}{[({}^3\mathrm{He}/{}^4\mathrm{He})_{mantle} - ({}^3\mathrm{He}/{}^4\mathrm{He})_{crust}]} \tag{38}$$

Subscripts mantle, crust and c refer to the mantle, crust and air-corrected values. Although $(^3\mathrm{He}/^4\mathrm{He})_{crust}$ is well defined ($\sim 1\times10^{-8}$, Ballentine and Burnard 2002, this volume), the choice of $(^3\mathrm{He}/^4\mathrm{He})_{mantle}$ has to be made with care: ^{3}He/^{4}He for local subcontinental lithospheric mantle recorded in mantle xenoliths ranges between 8.54×10^{-6} to 6.53×10^{-6} (Dunai and Baur 1995; Dunai and Porcelli 2002, this volume), input from the convecting mantle typical of that supplying mid ocean ridges would have higher values at ^{3}He/^{4}He = 1.12×10^{-5} (e.g., Graham 2002, this volume), while mantle regions beneath the crust influenced by a high ^{3}He/^{4}He plume, such as Yellowstone USA (Kennedy et al. 1985), may have yet higher ^{3}He/^{4}He values. Quoted errors need to reflect the degree of end-member uncertainty.

Three-component mixing. When a significant magmatic component is present in a crustal fluid, in addition to the ubiquitous air-derived noble gases, there will also be a significant contribution from purely crustal radiogenic sources. In principle a similar approach to the resolution of the three-component He mixture can be taken (Eqns. 37, 38), by first correcting for atmosphere-derived contributions by reference to an unambiguously air-derived isotope. As above, the accuracy of this correction is entirely dependent on how well the elemental ratio of the air-derived pair is known. In many systems of interest elemental fractionation may have occurred and this approach for a single sample is then no longer appropriate.

In the case of Ar, negligible ^{36}Ar contributions to either crust or mantle components enable the atmosphere ^{40}Ar to be corrected following Equation (35), although in this case the ^{40}Ar excess is the sum of crustal and mantle contributions. Negligible or unresolvable differences among mantle, crust and air ratios of $^{38}Ar/^{36}Ar$ make it impossible for this isotope pair to be effectively used in resolving the mantle and crustal ^{40}Ar components.

In the case of Ne, the $^{21}Ne/^{22}Ne$ and $^{20}Ne/^{22}Ne$ ratios of all three components are significantly different, and given three isotopes and three components the contribution from each source to each isotope can be calculated (e.g., Ballentine and O'Nions 1992) where

$$[20]_{air} = \tag{39}$$

$$[20]_{Total} \frac{\left\{\left(\frac{21}{22}\right)_{mntl} - \left(\frac{21}{22}\right)_{rad} + \left(\frac{22}{20}\right)_{meas} \times \left(\left(\frac{21}{22}\right)_{rad} \times \left(\frac{20}{22}\right)_{mntl} - \left(\frac{21}{22}\right)_{mntl} \times \left(\frac{20}{22}\right)_{rad}\right) + \left(\frac{21}{20}\right)_{meas} \times \left(\left(\frac{20}{22}\right)_{rad} - \left(\frac{20}{22}\right)_{mntl}\right)\right\}}{\left\{\left(\frac{21}{22}\right)_{mntl} - \left(\frac{21}{22}\right)_{rad} + \left(\frac{22}{20}\right)_{air} \times \left(\left(\frac{21}{22}\right)_{rad} \times \left(\frac{20}{22}\right)_{mntl} - \left(\frac{21}{22}\right)_{mntl} \times \left(\frac{20}{22}\right)_{rad}\right) + \left(\frac{21}{20}\right)_{air} \times \left(\left(\frac{20}{22}\right)_{rad} - \left(\frac{20}{22}\right)_{mntl}\right)\right\}}$$

$$[21]_{mntl} = \tag{40}$$

$$[21]_{Total} \frac{\left\{\left(\frac{20}{22}\right)_{air} - \left(\frac{20}{22}\right)_{rad} + \left(\frac{22}{21}\right)_{meas} \times \left(\left(\frac{21}{22}\right)_{air} \times \left(\frac{20}{22}\right)_{rad} - \left(\frac{20}{22}\right)_{air} \times \left(\frac{21}{22}\right)_{rad}\right) + \left(\frac{20}{21}\right)_{meas} \times \left(\left(\frac{21}{22}\right)_{rad} - \left(\frac{21}{22}\right)_{air}\right)\right\}}{\left\{\left(\frac{20}{22}\right)_{air} - \left(\frac{20}{22}\right)_{rad} + \left(\frac{22}{21}\right)_{mntl} \times \left(\left(\frac{21}{22}\right)_{air} \times \left(\frac{20}{22}\right)_{rad} - \left(\frac{20}{22}\right)_{air} \times \left(\frac{21}{22}\right)_{rad}\right) + \left(\frac{20}{21}\right)_{mntl} \times \left(\left(\frac{21}{22}\right)_{rad} - \left(\frac{21}{22}\right)_{air}\right)\right\}}$$

$$[22]_{rad} = \tag{41}$$

$$[22]_{Total} \frac{\left\{\left(\frac{20}{22}\right)_{air} \times \left(\frac{21}{22}\right)_{mntl} - \left(\frac{20}{22}\right)_{mntl} \times \left(\frac{21}{22}\right)_{air} + \left(\frac{21}{22}\right)_{meas} \times \left(\left(\frac{20}{22}\right)_{mntl} - \left(\frac{20}{22}\right)_{air}\right) + \left(\frac{20}{22}\right)_{meas} \times \left(\left(\frac{21}{22}\right)_{air} - \left(\frac{21}{22}\right)_{mntl}\right)\right\}}{\left\{\left(\frac{20}{22}\right)_{air} \times \left(\frac{21}{22}\right)_{mntl} - \left(\frac{20}{22}\right)_{mntl} \times \left(\frac{21}{22}\right)_{air} + \left(\frac{21}{22}\right)_{rad} \times \left(\left(\frac{20}{22}\right)_{mntl} - \left(\frac{20}{22}\right)_{air}\right) + \left(\frac{20}{22}\right)_{rad} \times \left(\left(\frac{21}{22}\right)_{air} - \left(\frac{21}{22}\right)_{mntl}\right)\right\}}$$

The parentheses subscripts crust, air, mntl and meas refer to the isotopic composition of the crust, air mantle and sample respectively. The square bracket subscripts crust, mntl, air, and Total, refer to the concentration of the crust, mantle and air component relative to the total isotopic contribution, respectively. Xe is another noble gas isotopic system where the crust, mantle and air end-members are significantly different. Substitution of the appropriate Xe isotopic end-member compositions into Equations (39)-(41) enables the end-member contributions to be derived for the three component Xe isotopic system.

Element ratio mixing lines. Although it is not possible to resolve mantle and crustal ^{40}Ar contributions in a single sample, this is possible with multiple samples from environments in which the elemental ratios from the respective end-member sources are constant and unaffected by subsequent fractionation. This is achieved by extrapolation, of an element-ratio/isotope-ratio mixing line to the isotope-defined end-members. For example, a plot of $^{3}He/^{4}He$ vs $^{40}Ar^{*}/^{4}He$ (where there are negligible air contributions to He and $^{40}Ar^{*}$ is the ^{40}Ar corrected for air-derived ^{40}Ar, Eqn. 35), represents an isotope ratio and elemental pair whose component parts have only two sources – the crust and the mantle. For a system in which the mantle and crustal components have constant $^{40}Ar/^{4}He$, a mixing line will be defined. Extrapolation to mantle and crustal $^{3}He/^{4}He$ end-member isotope compositions enables resolution of the respective $^{40}Ar/^{4}He$ component ratios (e.g., Stuart et al. 1995) (Fig. 12).

- This technique is applicable to all noble gas systems that allow reduction to two component element-ratio/isotope-ratio mixing lines.

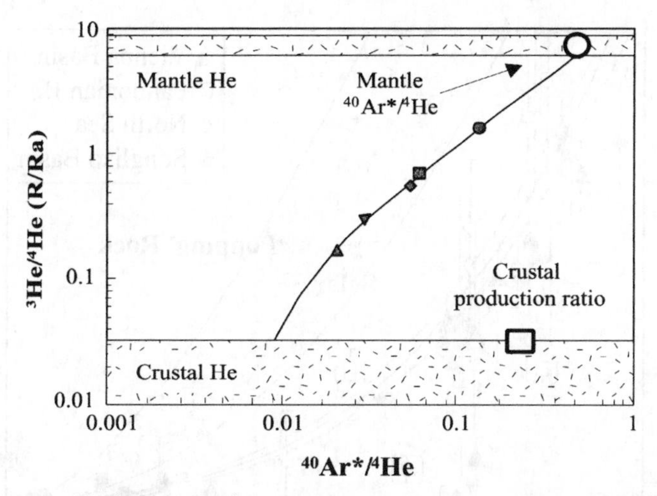

Figure 12. $^{40}Ar^*/^4He$ vs $^3He/^4He$ measured in Dae Hwa (S. Korea) W-Mo deposit fluid inclusions after Stuart et al. (1995). Extrapolation of the mixing line defined by the samples from different mineralization zones to known end-member $^3He/^4He$ isotopic compositions enables end-member $^4He/^{40}Ar$ ratios to be determined. In this example the mantle $^{40}Ar/^4He = 0.69 \pm 0.06$ and is typical of unfractionated samples from the mantle (e.g., Graham 2002, this volume), in contrast the crustal $^{40}Ar/^4He = 0.007$. The latter value is far higher than crustal production ratio of ~0.2 and typical of a fluid derived from shallow cool regions of the crust (e.g., Ballentine and Burnard 2002, this volume).

Isotope ratio mixing lines. In cases where the elemental ratios may have been variably altered, by for example phase fractionation, the element-ratio mixing lines discussed above will not be preserved. Isotopic ratios are unaffected by this form of fractionation and isotope-ratio only mixing lines can be constructed. Ballentine (1997) for example, use this approach to identify the mantle He/Ne ratio of magmatic fluids in natural gases (Fig. 13). In this study a data inversion was used to identify the best fit mixing line and crustal $^{21}Ne/^{22}Ne$ production ratio for natural gases from gas fields around the world. The Ne isotopic composition, corrected for the atmospheric contribution using Equations (39)-(41), then represents a two component mix of mantle and crustal-derived Ne. Similarly, air contributions to the $^3He/^4He$ are negligible and this ratio represents a mix of mantle and crustal-derived He. The air-corrected Ne, $(^{21}Ne/^{22}Ne)_c$, plotted against $^3He/^4He$ falls on a single mixing line for almost all samples and defines the mixing constant, r, where:

$$r = (^4He/^{21}Ne)_{crust}/(^4He/^{21}Ne)_{mntl} \quad (42)$$

$(^4He/^{21}Ne)_{crust}$ and $(^4He/^{21}Ne)_{mntl}$ are the $^4He/^{21}Ne$ ratio of the crust and mantle components before mixing respectively. Further examples of isotope mixing relationships and three-dimensional approaches when resolving three components from the isotope systematics are detailed in the section *Description and analysis of multi-component noble gas mixtures in ore fluids.*

NOBLE GASES IN HYDROCARBON GAS AND OIL RESERVOIRS

In the context of playing a role in hydrocarbon exploration or field development noble gas studies are in their infancy. In principle noble gas fractionation in groundwater can provide a sensitive and quantitative tool with which to identify both natural gas

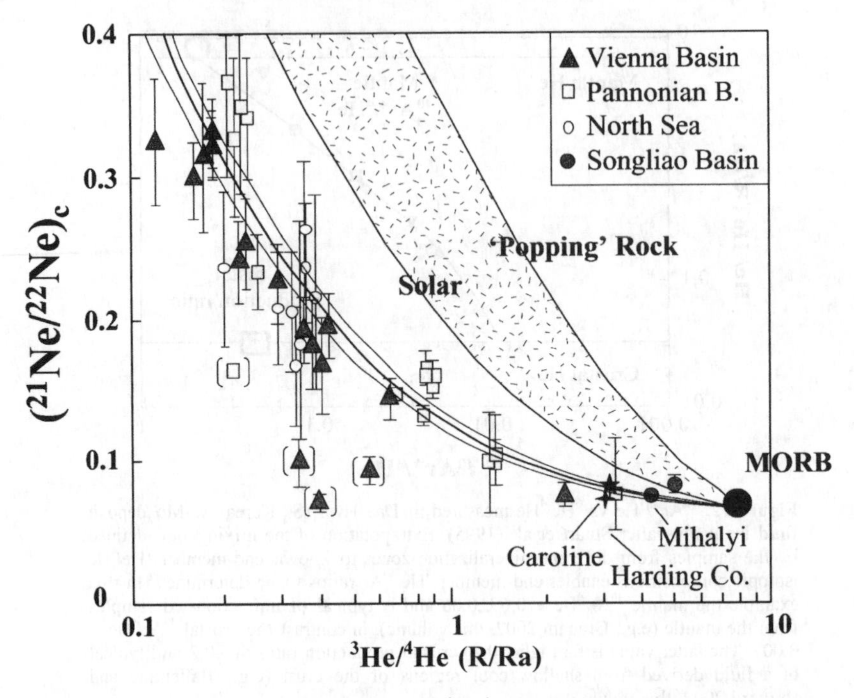

Figure 13. Plot of $(^{21}Ne/^{22}Ne)_c$ vs. measured $^3He/^4He$ (R/Ra) for natural gases containing a significant mantle-derived component (Ballentine 1997). All data falls within 2σ error of a remarkably well defined mixing line between crustal and mantle derived end-member values. The hyperbolic mixing constant, D, determined from the mixing line is defined by D = $(He/Ne)_{crust}/(He/Ne)_{mantle}$. As $(He/Ne)_{crust}$ is reasonably well defined (e.g., Ballentine and Burnard 2002, this volume) the mixing line enables the magmatic He/Ne ratio in these crustal systems to be calculated. This is quite distinct from the mixing line involving an unfractionated mantle source bounded by a mantle He/Ne ratio with 'Solar' or 'Popping rock' values (hatched region). The difference has been ascribed to only partial degassing and solubility related fractionation of the noble gases in the magmatic source supplying volatiles to the crustal system. This appears to be a common feature in many extensional systems (see also Fig. 18).

(Zaikowski and Spangler 1990) and oil migration pathways—in the absence of either the gas or oil phase (Fig. 9). In practice, the acquisition of groundwater samples that are uncontaminated by drill fluids from boreholes is difficult for commercial reasons. The extension of this type of study to groundwater trapped in fluid inclusions is at the limit of current experimental techniques and remains unexplored. Most studies to date have focused on the information available from noble gases that are readily determined in natural gas and oil samples from actively producing fields. Of particular interest is the quantification of the amount of water that has equilibrated with an oil or a gas phase. This provides an understanding of the role that groundwater has played in both the secondary migration of oil and gas (where primary migration is expulsion from the source rock, and secondary migration is transport to the reservoir) and diagenetic processes associated with groundwater movement during hydrocarbon migration or emplacement. An equally important use of noble gases has been in identifying the origin of non-hydrocarbon gases such as CO_2 and N_2 by relating these species to mantle-derived and crustal-radiogenic gases, respectively.

Identifying and quantifying groundwater/gas/oil interaction

The origin of atmosphere-derived noble gases in the subsurface. The introduction of atmosphere-derived noble gases into the subsurface is dominated by noble gases dissolved in groundwater. In the case of meteoric waters this occurs at recharge. The physical processes controlling the concentrations of noble gases in the meteoric groundwater phase are well constrained and include temperature, altitude, and salinity at recharge in addition to a small portion of air in excess of recharge equilibrium values (Kipfer et al. 2002, this volume). Figure 9 shows the typical range of Ne/Ar and Xe/Ar ratios in fresh water normalized to air-equilibrated water at 10°C.

Water associated with the sedimentary burial process ('formation' water) or density driven sinking plumes from highly saline lakes have both also equilibrated with the atmosphere, contain dissolved air noble gases, and contribute to the subsurface inventory (Zaikowski et al. 1987). These latter sources remain poorly constrained.

Podosek et al. (1980, 1981) have shown that atmosphere-derived Kr and Xe can be preferentially trapped in shales. Recently, Torgersen and Kennedy (1999) have found correlated Kr and Xe enrichments in oil-associated natural gases that are consistent with this trapped sedimentary origin of atmosphere-derived Kr and Xe. The range, type and condition of rock sequences that preferentially trap Xe and Kr, and the conditions of their release, remain poorly constrained (Table 6).

Phase fractionation. In a simple two-phase system, the recipient phase equilibrating with the groundwater will be the sampled oil or gas. The magnitude of fractionation in the oil or gas from the original groundwater values will therefore be controlled by the equilibrium solubility law (Eqns. 20, and 22) and reflect the subsurface conditions of temperature, water salinity and gas/water or oil/water volume ratio. The most uncertainty

Table 6. Fractionated atmosphere-derived Ne, Kr and Xe measured in carbon-rich crusta rocks (after Torgersen and Kennedy, 1999).

Reference	$F(^{22}Ne)$	$F(^{84}Kr)$	$F(^{132}Xe)$	$[^{36}Ar]$ $\times10^{-8}$ cm^3 STP/g
Bogard et al. 1965	4.6-1010	0.7-1145	1600-45000	0.15-45
Frick and Chang 1977	7.6-15.6	220-261	2955-4345	n.a.
Podosek et al 1980	0.1-11.3	0.8-20.5	4-1300	0.1-1.0

in making this calculation is attached to the estimation of the elemental ratios before fractionation occurred. For example in the case of a meteoric water-gas system, even though a reasonable estimate of recharge temperature may sometimes be made, the amount of 'excess air' can be highly variable (Fig. 9). In principle, given four observables (Ne, Ar, Kr, Xe) three unknowns can be resolved using a data inversion technique similar to that developed to investigate paleotemperature calculations from noble gas concentrations in groundwater (Ballentine and Hall 1999; Kipfer et al 2002, this volume). If reasonable estimates of the subsurface conditions can be made the variables may include for example, temperature of groundwater recharge (derived from the gas phase albeit with large errors!) excess air and the gas/water volume ratio. One of the principle advantages of such an approach is the rigorous propagation of errors through to the derived values, including errors due to measurement as well as errors associated with an incomplete model formulation. Development of this technique is the focus of current research (Ballentine et al. 1999). Without a full data inversion workers to date have typically relied on elemental ratio pairs and an assumed range for their original value. In this respect the non atmosphere-derived noble gases can play a role in testing the robustness of a calculation because the magnitude of fractionation of non atmosphere-derived noble gases in the groundwater will be subject to the same fractionation process and show coherent fractionation with the atmosphere-derived gases (Fig. 14). The underlying assumptions in using this robustness check are two-fold: (1) the radiogenic noble gas composition has not itself been subject to fractionation (e.g., Ballentine and Burnard 2002, this volume); and (2) the radiogenic noble gases were indeed in the fluid system prior to the fractionating process.

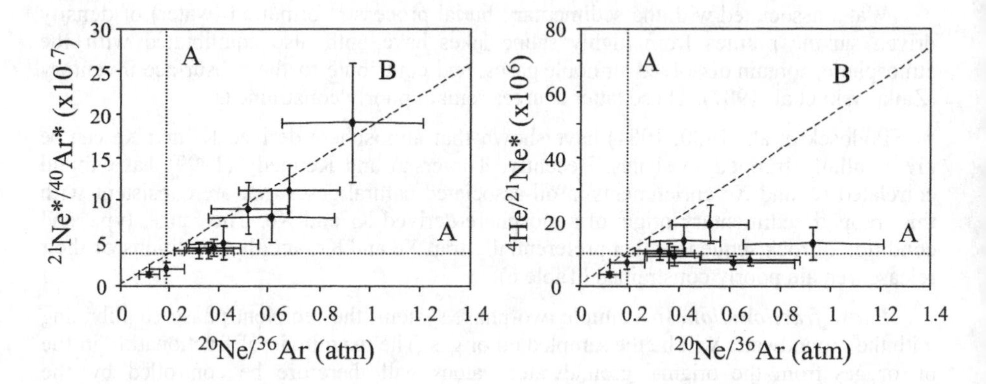

Figure 14. Plot showing radiogenic $^{21}Ne^*/^{40}Ar^*$ vs $^{20}Ne/^{36}Ar$ and $^4He/^{21}Ne^*$ vs. $^{20}Ne/^{36}Ar$ in the Hajduszoboszlo gas field, Hungary, after Ballentine et al. (1991). Left hand side figure: The crustal derived $^{21}Ne^*/^{40}Ar^*$ fractionates coherently with the groundwater-derived $^{20}Ne/^{36}Ar$ ratio (Line B) showing that these two differently sourced noble gas pairs were mixed before the fractionating processes. Right hand side figure: In contrast the crustal-derived $^4He/^{21}Ne^*$ shows little fractionation from the predicted source ratio (Lines A). This is explained by a solubility based fractionation process, because although Ne and Ar have very different solubilities in groundwater the solubilities of He and Ne in water under subsurface conditions are similar. This will result in negligible fractionation of the He/Ne ratio.

Identifying the phase of transport. The hydrogeologic system can play an important role in the transport of gas from source to trap and identification of this process can influence exploration strategies (Toth 1980; Toth and Corbett 1986). Identification of whether or not groundwater has played a significant role in natural gas transport is possible by considering the concentration of noble gases derived from the groundwater that are now in the natural gas. For example, if we consider a groundwater that has not undergone phase separation or equilibration, its concentration of atmosphere-derived noble gases, as discussed above, fall in a relatively tight range and remain constant. The saturation limit for a major gas species (we consider here CH_4) is a function of pressure and temperature (depth) as well as water salinity. The point at which CH_4 saturation is reached will result in CH_4 gas phase formation. The noble gases will partition between the gas and water phase. Because the volume of the gas phase at this stage is small relative to the water the relative abundance of the noble gases will be fractionated following Equation (20). The solubility of Ar however is similar to that of CH_4 under typical subsurface conditions. The gas phase will therefore preserve an unfractionated $CH_4/^{36}Ar$ ratio of the groundwater at the point (depth) at which saturation occurred (Ballentine et al. 1991; Fig. 15).

Gas reservoirs that preserve the 'saturation' $CH_4/^{36}Ar$ ratio consistent with their formation depth may have been transported to the trapping site dissolved in the groundwater.

Case studies

We review here selected case studies to illustrate some of the main applications of noble gases to natural gas and oil-bearing systems.

Groundwater and natural gas transport, the Pannonian basin, Hungary. The Pannonian basin forms the Great Hungarian Plain and is an extensional basin formed in the Middle Miocene that has developed into a series of deep basins separated by shallower basement blocks. Oil and gas fields are found throughout the basin, with most to-date being found on the basin margins or above the uplifted basement blocks. Several noble gas studies have focused on the basin aquifer systems (Oxburgh et al. 1986; Martel et al. 1989; Stute and Deak 1989; Stute et al. 1992). Noble gas studies of the natural gas reservoirs are by Ballentine and O'Nions (1992), Sherwood Lollar et al. (1994, 1997). We focus here on the Hajduszoboszlo gas field study by Ballentine et al. (1991). This field is a stacked gas field producing at various intervals between 700- to 1300-m depth and occupies a portion of the sub-basin high to the north west of the Derecske sub-basin. $^3He/^4He$, $^{20}Ne/^{22}Ne$, $^{21}Ne/^{22}Ne$ and $^{40}Ar/^{36}Ar$ were determined as well as He, Ne and Ar abundance. The $^3He/^4He$ ratios of between 0.18 to 0.46Ra show that between 2 to 5% of the He is mantle derived. $^{21}Ne/^{22}Ne$ and $^{40}Ar/^{36}Ar$ ratios range between 0.299-0.46 and 340-1680. These are all in excess of atmosphere derived values (0.0290 and 295.5) due to the addition of crustal radiogenic ^{40}Ar and ^{21}Ne ($^{40}Ar^*$ and $^{21}Ne^*$) contributing 3-15% and 32-82% of the ^{21}Ne and ^{40}Ar respectively. The greatest contribution of crustal-derived gases is clearly correlated with the deepest samples.

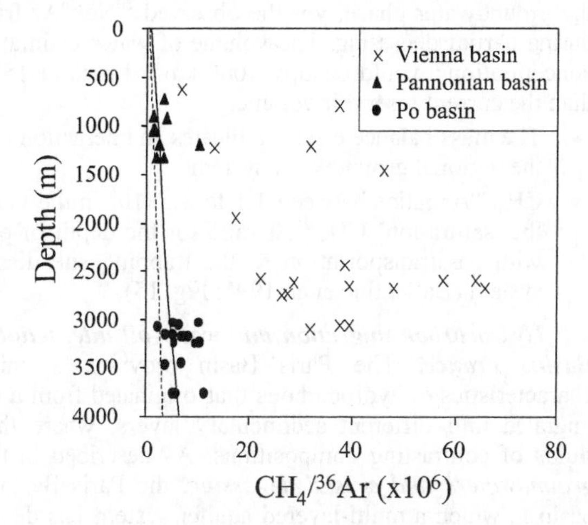

Figure 15. $CH_4/^{36}Ar$ plotted as a function of depth for gas fields in the Vienna basin, Austria (Ballentine 1991), Pannonian basin, Hungary (Ballentine et al. 1991) and the Po Basin, Italy (Elliot et al. 1993). These values are compared with the 'saturation' $CH_4/^{36}Ar$ value calculated for seawater containing 7.5×10^{-7} cm^3 (STP) ^{36}Ar, a salinity of 0.23M NaCl equivalent, a temperature gradient of 0.03 K/km and at hydrostatic pressure (Solid line). The dashed line is the saturation value if the salinity increases to 3 M NaCl equivalent. The gases in the Pannonian and Po basin studies lie on the saturation line for their depth, are closely linked to the groundwater system, and may have exsolved from solution. These contrast with the Vienna basin gases that have had far less contact with the groundwater system.

Coherent fractionation is observed in a plot of $^{21}Ne^*/^{40}Ar^*$ (crustal component) against $^{20}Ne/^{36}Ar$ (atmosphere-derived) pointing to a fractionation process operating on the system after mixing of these differently sourced species (Fig. 14). Plotting $^4He/^{21}Ne^*$ (crustal component) against $^{20}Ne/^{36}Ar$ no coherent fractionation is seen (Fig. 14). This is the classic pattern for a solubility equilibrium fractionation process. Negligible fractionation between $^4He/^{21}Ne^*$ is observed because their solubilities in water under reservoir conditions are very similar. There is no resolvable mass fractionation, and diffusion as a significant transport process in this system can therefore be neglected.

Making the assumption that the natural gas originally contained no ^{36}Ar, the ^{36}Ar now in the natural gas requires a minimum volume of groundwater to have interacted with the gas phase. This is a minimum estimate because this assumes quantitative degassing of the groundwater phase, yet the observed $^{20}Ne/^{36}Ar$ fractionation discussed above occurs during partial degassing. The volume of water estimated from typical groundwater ^{36}Ar concentrations would occupy 1000 km^3 of rock at 15% porosity, some 670 times larger than the current reservoir volume.

- The mass balance clearly indicates an interaction between the gas now in the field and the regional groundwater system.
- $CH_4/^{36}Ar$ ratios between 1.1 to 3.5×10^6 in the natural gas are indistinguishable from the 'saturation' $CH_4/^{36}Ar$ ratio for the depth of production and therefore consistent with gas transportation to the trapping site dissolved in the regional groundwater system (Ballentine et al. 1991; Fig. 15).

Hydrocarbon migration and water-oil interaction in a quiescent basin: The Paris Basin, France. The Paris Basin provides a unique opportunity to study the characteristics of hydrocarbons that originated from a common source-rock lithology and migrated into different sedimentary layers, where they subsequently interacted with fluids of contrasting compositions. As described in the section *Noble gases in ancient groundwaters and crustal degassing*, the Paris Basin is a post-Variscan intra-cratonic basin in which a multi-layered aquifer system has developed. Of interest in this context are the Upper Triassic (Keuper) Chaunoy fluvial sandstones and the Middle Jurassic oolitic limestone reservoirs. The Middle Jurassic aquifer is separated from the Triassic aquifer by 400-700 m of low-permeability Lower Jurassic (Lias) mudrocks and shales, which are the source-rocks of oils in the Paris Basin. The Middle Jurassic contains low-enthalpy geothermal waters (from 50 to 80°C) and oil accumulations. The Triassic aquifer contains waters at temperatures up to 120°C associated with oil accumulations. Oil primary migration took place from the Liassic shales upward to the Jurassic limestone and laterally to the Triassic sandstone, during the Paleocene-Oligocene time (Espitalié et al. 1988). Vertical faults, affecting the Mesozoic cover of the Paris Basin and reactivated by tectonic post-Alpine stresses, have played an important role in oil secondary migration. These vertical faults constitute the preferential pathways for oil flow through the Lias. An important hydrodynamic component flowing in the Jurassic and the Triassic aquifers and contemporary with oil migration seems to have affected both the distribution of oil pools in the Paris Basin (Poulet et Espitalié 1987) and the loss of ~90% of the hydrocarbon (Espitalié et al. 1988). Cross-formational fluid flow in the Paris Basin is also apparent in the common source of salinity (halite deposited in the eastern part of the Triassic aquifer) for both Jurassic and Triassic groundwaters (Worden & Matray 1995).

Both groundwater and oil accumulations have been studied for noble gases (Marty et al. 1993; Pinti and Marty 1995, 1998; Pinti et al. 1997). The main noble gas feature in the Paris Basin fluids is the presence of a resolvable mantle-derived noble gas component, which is weaker than the one measured in the younger and tectonically active Pannonian Basin (Ballentine et al. 1991). An important *in situ* contribution of radiogenic noble gases is masking progressively the traces of a large-scale fluid flow, which affected the Paris Basin probably in early Tertiary time. It is likely that this episodic fluid flow introduced deep-seated mantle-derived and radiogenic noble gases into the basin, and possibly triggered the hydrocarbon primary and secondary migration within the basin (Pinti and Marty 1998). The relationship between the helium $^3He/^4He$ ratios and the total amount of ^4He in the basement, Trias and Middle Jurassic groundwaters, suggests that there are at least three sources of helium occurring in the Paris Basin. The first source is fluids circulating in the southern crystalline basement and characterized by high $^3He/^4He$ ratios (0.12-0.14 Ra) due to addition of mantle-derived ^3He. The second source is Triassic groundwaters located at the center of the basin, which is characterized by $^3He/^4He$ ratios intermediate between the basement and the Middle Jurassic fluids ($^3He/^4He$ = 0.08 Ra). The third source is water located east of the Middle Jurassic aquifer, with low $^3He/^4He$ isotopic ratios ($^3He/^4He$ = 0.02 Ra) resulting from the production of helium in the local reservoir's rocks (Bathonian-Callovian limestones).

The transport of radiogenic helium, argon and neon (and associated mantle-derived helium) from the Trias to the Middle Jurassic aquifer is apparent in the distribution of the radiogenic $^4He/^{40}Ar^*$ and $^{21}Ne^*/^{40}Ar^*$ isotope ratios among Triassic and Middle Jurassic oil-field brines (see section *Noble gases in ancient groundwaters and crustal degassing* and Fig. 28, below). The $^4He/^{40}Ar^*$ and $^{21}Ne^*/^{40}Ar^*$ isotope ratios clearly show a correlation and indicate mixing between the Trias and the Middle Jurassic groundwaters. The variation of the radiogenic noble gas isotope ratios can be attributed to the initial ratio of the parent elements $^{238,235}U$, ^{232}Th and ^{40}K in minerals and rocks, which varies for different lithologies, or to preferential diffusion of ^4He and $^{21}Ne^*$ relative to $^{40}Ar^*$ from the mineral to the fluid phase. This in turn depends on the thermal and tectonic regime of the basin. The Trias groundwaters show $^4He/^{40}Ar^*$ ratios of 4-7 and $^{21}Ne^*/^{40}Ar^*$ ratios of $2.5\text{-}4\times10^{-7}$. These ratios could correspond to a source having a K/U ratio of about 35,000 and which releases He, Ne and Ar in water close to their production ratio. This source could be the Triassic sandstones, the crystalline basement, or both. The second source of radiogenic noble gases has high $^4He/^{40}Ar^*$ ratios of 40 and $^{21}Ne^*/^{40}Ar^*$ ratios of 65×10^{-7} and could correspond to the carbonate, which is characterized by very low K/U ratios.

In the Middle Jurassic oils, the elemental fractionation of atmosphere-derived noble gases was found to be consistent with oil/water phase equilibrium partitioning (Pinti and Marty, 1995). In the Triassic oils, the noble gas fractionation trends indicate a more complicated history, notably involving degassing of hydrocarbons previously equilibrated with groundwaters. Pinti and Marty (1995) have interpreted this degassing episode as the result of processes of gas stripping due to oil washing (Lafargue and Barker, 1988). Calculations indicated that both Middle Jurassic and Triassic oils have seen much larger quantities of waters with oil/water ratios possibly ranging between 0.2 and 0.01, whereas the present-day oil/water ratios in the Middle Jurassic and Triassic oil fields average ~ 1. Assuming a mean groundwater residence time of few Ma in the center of the Paris Basin (Marty et al., 1993), where most of the oil accumulations reside, and an integrated mean oil/water ratio in oil reservoirs lower by one order of magnitude than those presently observed, then the residence time of oils in their reservoirs should also be an order of magnitude higher than those of flowing waters and could be of the order of ~20-40 Ma. Such a figure is in qualitative agreement with current estimates for the timing of oil migration in the Basin (Paleocene-Oligocene, Poulet et Espitalié 1987; Espitalié et al. 1988).

Groundwater and diagenesis, the Magnus oil field, North Sea. The Magnus oilfield is located in the East Shetland Basin, northern North Sea. The field consists of a single oil phase with no associated gas cap and contained an estimated in-place oil reserve of 2.65×10^8 m^3 oil (STP). Hydrocarbon accumulation occurs in Middle Jurassic sandstones located on the dipping flank of a tilted Jurassic fault block. Petrographic and isotopic evidence from diagenetic minerals show that minerals in the crest of the reservoir grew in pore water containing significantly more meteoric water than those down dip, which are dominated by seawater (Emery et al. 1993; Macaulay et al. 1992). Cementation of the Magnus sandstone appears to have occurred concurrently with reservoir filling at ~72-62 Ma (Emery et al. 1993). Models addressing the role of groundwater in effecting regionally observed cementation of oil-bearing systems appeal to either local dissolution and re-precipitation or require the regional flow of groundwater (e.g., Bethke et al. 1988; Gluyas and Coleman 1992; Aplin et al. 1993). Because the Magnus system filling occurred at the same time as the cementation it is reasonable to assume that the noble gases in the oil phase preserve a record of the groundwater volumes during the quartz precipitation.

Depressurization during production results in both gas and oil phases being present at the surface. A pilot study determined that He, Ne and Ar are almost quantitatively partitioned into the gas phase under separator conditions. Given the flow rate of both oil and gas, the analyses of the gas phase alone enables an accurate reconstruction of the subsurface and single oil phase noble gas composition. Gas samples across the Magnus field were taken and the He, Ne and Ar isotopic ratios and abundances in the oil were determined (Ballentine et al. 1996). Both the He and the Ne isotope systematics require a contribution from a mantle source. If the mantle end-member is modeled using mid-ocean ridge values (Graham 2002, this volume), 2.3 to 4.5% of the ^{4}He and 4.3 to 6.2% of the ^{21}Ne in the Magnus oil is mantle-derived. The remainder of the ^{4}He and 9.0 to 12.0% of the ^{21}Ne is crustal-radiogenic, and the remaining ^{21}Ne is atmosphere-derived. The quantity of radiogenic noble gas associated with the Magnus oil/groundwater system can only be accounted for by production predominantly from outside the volume of the Magnus Sandstone aquifer/reservoir drainage area and the associated Kimmeridge Clay source rock formation and together with the mantle-derived noble gases, provides strong evidence for cross formational communication with deeper regions of the crust. This is not the case for the groundwater-derived noble gases.

The ^{20}Ne and ^{36}Ar have been input into the oil phase by interaction with an air-equilibrated groundwater. Because the Magnus oil field has no gas cap, this is a simple two-phase system. Similarly, because the groundwater, with the exception of a small amount of meteoric water incursion at the crest of the system, is dominated by seawater there is no variable excess air component to consider. In principle, knowing the original noble gas concentration in the seawater and the temperature and salinity of water on equilibration with the Magnus oil, together with either the ^{20}Ne *or* ^{36}Ar concentration and their respective solubilities in the oil enable Equation (19) to be used to determine the system oil/water volume ratio. In practice the noble gas solubility database for different oils is limited, and $K^{m}_{Ar(oil)}$ and $K^{m}_{Ne(oil)}$ are not known for the Magnus oil. Nevertheless noble gas solubility data is available for two oils of different density (Kharaka and Specht 1988; Table 4). If it is assumed that at any one temperature and over a small density range the relative change in solubility of both Ne and Ar is proportional, an equation can be developed that links the $K^{m}_{Ar(oil)}$ to $K^{m}_{Ne(oil)}$ in the Magnus oil (Ballentine et al. 1996). This then leaves two unknowns, the linked solubility term and the oil/water volume ratio. With two separate equations for ^{20}Ne and ^{36}Ar derived from the general Equation (19), these can be solved.

- Noble gas partitioning between a seawater-derived groundwater and the oil phase at the average Magnus Sandstone aquifer temperature requires a subsurface seawater/oil volume ratio of 110(±40) to account for both the ^{20}Ne and ^{36}Ar concentrations in the central and southern Magnus samples.
- The volume of groundwater that has equilibrated with the Magnus oil is indistinguishable from the static volume of water estimated to be in the down-dip Magnus aquifer/reservoir drainage volume. This suggests that the Magnus oil has obtained complete equilibrium with the groundwater in the reservoir drainage volume, probably during secondary migration, and further suggests that the concurrent cementation of the Magnus sandstone aquifer has occurred with little or no large-scale movement of groundwater through the aquifer system.

Tracing the CO_2 source in the west Texas Permian basin, USA. CO_2 in natural gases can originate from a number of sources including methanogenesis and oil field biodegradation, kerogen decarboxilation, hydrocarbon oxidation, decarbonation of marine carbonates and degassing of magmatic bodies. The $\delta^{13}C(CO_2)$ signature can be used to distinguish between these different sources, with the exception of magmatic and carbonate-derived CO_2, which have overlapping $\delta^{13}C(CO_2)$ (Jenden et al. 1993; Sherwood Lollar et al. 1997). Although methanogenesis or oil field biodegradation can sometimes result in gas fields with up to 40% CO_2 by volume, kerogen decarboxilation and hydrocarbon oxidation rarely result in gas containing more than a few percent CO_2. This is in distinct contrast to decarbonation/magmatic sources, which can result in gas fields containing up to 100% by volume CO_2. Noble gases can be used to distinguish between decarbonation/magmatic sources (Fig. 16; Sherwood Lollar et al. 1997; Ballentine et al. 2001).

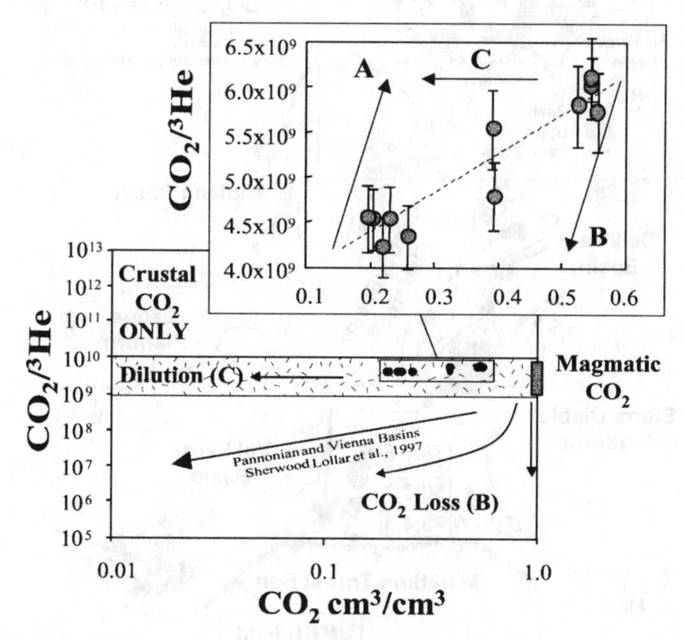

Figure 16. $CO_2/^3He$ vs. fraction of CO_2. The main figure shows the range of $CO_2/^3He$ values found in pure magmatic samples. $CO_2/^3He$ values above this range, irrespective of CO_2 content, can only be attributed to a CO_2 source containing no ^3He and provides an unambiguous identification of crustal-sourced CO_2. Values within this range or below contain a magmatic CO_2 component but have been subject to possible CO_2 loss, dilution (e.g., addition of CH_4 or N_2), and/or crustal CO_2 addition (after Sherwood Lollar et al. 1997). Inset shows the values found in CO_2 rich natural gases in the JMBB field, west Texas Permian basin, which vary within the magmatic range (after Ballentine et al. 2001). Vectors A, B and C show the effect of crustal CO_2 addition, CO_2 loss through reaction or precipitation and dilution respectively. Near constant $\delta^{13}C(CO_2)$ rules out either loss or addition of CO_2 and requires the range to be due to magmatic source variation (Fig. 18).

The west Texas Permian basin was formed as a result of the late Palaeozoic collision of South America with North America that also resulted in the Marathon-Ouachita orogenic belt as well as widespread interior continental deformation. The Val Verde basin is a foreland sub-basin of the west Texas Permian basin, and lies between the Central basin platform and the Marathon thrust belt (Fig. 17). Natural gas reservoirs show a systematic regional increase in CO_2 content towards the Marathon thrust belt, varying from an average of about 3% in the basin center to as high as 97% on the foredeep margin of the thrust belt. The main producing formation in the JM-Brown Bassett (JMBB) field is brecciated Ordovician Ellenberger dolomite. This field reflects the regional spatial trend in natural gas CO_2 content, with samples increasing from 20% to 55% CO_2 towards the Marathon thrust belt. The remaining gas is dominated by CH_4.

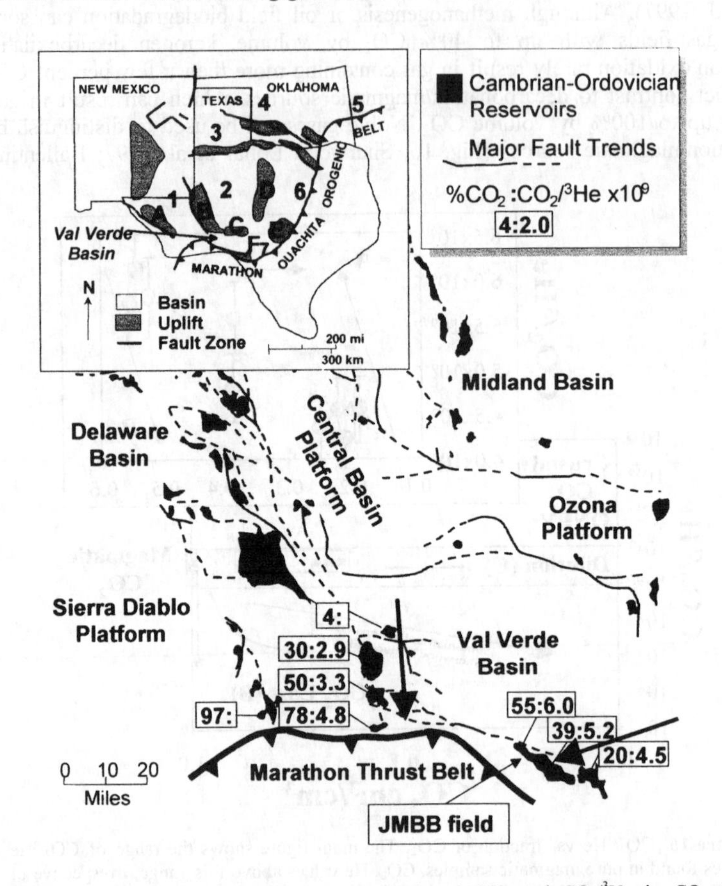

Figure 17. Showing coherent spatial variation in $\%CO_2$ and $CO_2/^3He$ in CO_2-rich natural gases in the Val Verde basin, part of the west Texas Permian basin (after Ballentine et al. 1991). Arrows show the direction of the regional increase in CO_2 content and $CO_2/^3He$ ratio towards the Marathon thrust belt. Inset shows the location of the Val Verde basin relative to the major Permian uplift and basinal features. Basins: 1, Delaware; 2, Midland; 3, Palo-Duro; 4, Anadarko; 5, Arkoma; 6, Ft Worth; 7, Kerr. Uplifts: A, Sierra Diablo; B, Central basin; C, Ozona; D, Concho arch; E, Llano; F, Devils River.

Samples from across the field were analyzed for their C, He, Ne and Ar isotopes as well as the abundance of He, Ne, Ar and major gas species. $^3He/^4He$ varies between 0.24 and 0.54 Ra. If it is assumed that the magmatic and crustal components have 8.0 and 0.02 Ra composition respectively, the measured ratios correspond to between 3.2 and 6.8% of the helium being derived from the mantle. In reality, sub-continental mantle $^3He/^4He$ ratios are believed to be slightly lower than the mantle supplying mid-ocean ridges (Dunai and Baur 1995) and the percentage mantle contributions to the JMBB are therefore a lower limit. Similarly from the $^{21}Ne/^{22}Ne$ and $^{40}Ar/^{36}Ar$ ratios the crustal components can be resolved ($^4He^*$, $^{21}Ne^*$ and $^{40}Ar^*$) showing $^4He^*/^{40}Ar^*$ and $^4He^*/^{21}Ne^*$ ratios indistinguishable from average crustal production (Ballentine and Burnard 2002, this volume). This result indicates that no fractionation process has operated on these species either during release from their source, on transport, or during their residence time in the trapping structure (eliminating for example, significant diffusive loss since trapping). It was not possible to investigate the groundwater-derived noble gases due to significant amounts of air contamination on sampling (sampling procedures used were for stable isotopes, not noble gases).

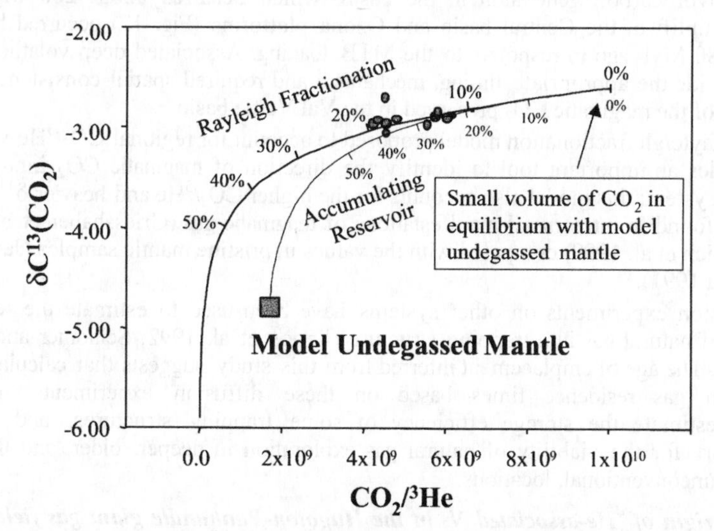

Figure 18. Evolution of $\delta^{13}C(CO_2)$ and $CO_2/^3He$ calculated for the gas phase of a degassing magma body (after Ballentine et al. 1991). Two models are shown: (i) The composition of the gas evolving from the magma by a Rayleigh fractionation process; and (ii) The composition of this gas in an accumulating reservoir. The model undegassed magma is taken to have $\delta^{13}C(CO_2)$ = -4.7‰ and $CO_2/^3He$ = 2×10^9, and is within the range estimated for the mantle source. The tick marks are the percentage loss of CO_2 from the magma body. $\delta^{13}C(CO_2)$ fractionation between a CO_2 gas phase and magma is taken to be 2‰ (Mattey 1991), and the relative solubility of He/CO_2 = 5 (Bottinga 1991). Both $CO_2/^3He$ and $\delta^{13}C(CO_2)$ of the JMBB field are consistent with partial degassing of the source magma body (Ballentine et al. 2001). In this context, samples with the highest $CO_2/^3He$ are from the earliest stages of outgassing.

The $^3He/^4He$ correlates directly with percent CO_2, showing clear two-component mixing between the hydrocarbon gas containing crustal-derived He and a CO_2 component with an elevated $^3He/^4He$. $CO_2/^3He$ for all samples are within the magmatic range but vary systematically with percent CO_2 (Fig 16). Various models were investigated to account for the $CO_2/^3He$ variation, including crustal CO_2 addition, precipitation or a combination of the latter combined with CH_4 addition/dilution. None of these models were able to satisfy both the very small variation in $\delta^{13}C(CO_2)$ and the mixing vectors shown in Figure 16. A magma-degassing model was constructed that accounted for both $\delta^{13}C(CO_2)$ and $CO_2/^3He$ (Fig. 18). In the context of a magma-degassing model (Figs. 13 and 18), samples with the highest $CO_2/^3He$ are from the earliest stages of outgassing and are located closest to the Marathon thrust belt. A simple filling model in which reservoirs closest to the magma source are filled and then diluted by subsequent outgassing (lower $CO_2/^3He$) predicts that the highest $CO_2/^3He$ ratios are furthest from the degassing magma, ruling out the Marathon thrust belt as the source of magmatic CO_2 in the Val Verde basin.

Tertiary volcanism is associated with the Basin and Range province to the west of the Val Verde basin, and some 100 km away from the JMBB study. Although this is a potential source of magmatic volatiles, this source is not consistent with the inferred direction of filling (Fig. 17). The increase in CO_2 content as the MTB is approached can be accounted for if CO_2 emplacement pre-dates CH_4 generation in the hydrocarbon `kitchens' to the north of the gas fields. Assuming simple filling, the traps closest to these `kitchens' would have the highest CH_4 content. CO_2 charging, therefore, pre-dates the onset of hydrocarbon generation in the basin, which occurred about 280 Myr ago. Maximum uplift of the Central basin and Ozona platforms (Fig. 17) occurred between 310 and 280 Myr ago in response to the MTB loading. Associated deep volatile release would provide the appropriate timing, mechanism and required spatial consistency to be the source of the magmatic CO_2 preserved in the Val Verde basin.

- The Rayleigh fractionation model proposed to account for regional $CO_2/^3He$ variation provides an important tool to identify the direction of magmatic CO_2 input into a basin system; this model also accounts for the higher $CO_2/^3He$ and heavier $\delta^{13}C(CO_2)$ often found in intracrustal manifestations of magmatic gas (Griesshaber et al. 1992; Weinlich et al. 1999) compared with the values in pristine mantle samples (Javoy and Pineau 1991).
- Diffusion experiments on other systems have been used to estimate the residence time of natural gas in a trapping structure (Kroos et al. 1992; Schlomer and Kroos 1997). The age of emplacement inferred from this study suggests that calculations of natural gas residence times based on these diffusion experiments seriously underestimate the storage efficiency of some trapping structures, and provide support for the viability of natural gas exploration in deeper, older, and therefore more unconventional, locations.

The origin of ^4He-associated N_2 in the Hugoton-Panhandle giant gas field, USA. The most abundant non-hydrocarbon gas in sedimentary basins is nitrogen. In the USA, 10%, 3.5% and 1% of natural gases contain >25%, >50% and >90% by volume nitrogen respectively (Jenden and Kaplan 1989). Nevertheless, the dominant sources and mechanisms responsible for focusing and enrichment of nitrogen within natural gas fields are poorly constrained. This is in part due to the multiple sources of nitrogen in the subsurface including atmosphere-derived nitrogen dissolved in groundwater, nitrogen released from sedimentary organic matter, nitrogen released from metasediments during metamorphism and, in areas of magmatic activity, an igneous or mantle nitrogen origin. The overlapping range of nitrogen isotopic values for the respective systems has meant that nitrogen isotopes alone cannot be used to quantify the contribution of these different sources to natural gas systems. Nitrogen gas associated with high radiogenic ^4He concentrations is particularly common (Gold and Held 1987; Jenden and Kaplan 1989; Jenden et al. 1988; Pierce et al. 1964; Poreda et al. 1986; Stilwell 1989; Hiyagon and Kennedy 1992; Hutcheon 1999). Because of the association of ^4He, and therefore other crustal noble gases such as $^{21}Ne^*$ and $^{40}Ar^*$, the noble gases are particularly appropriate for tracing the origin of He-associated nitrogen (N_2^*) (Ballentine and Sherwood Lollar 2002).

The Hugoton-Panhandle giant gas field is the case type example of a system containing N_2^* (Pierce et al. 1964). Extending 350 km across SW Kansas and the Oklahoma/Texas Panhandles this field contained more than 2.3×10^{12} m^3 (STP) of recoverable gas, and produces from Permian carbonates between 400-900m depth on the south and western margins of the Anadarko basin. The isotopic compositions of the hydrocarbon gases across the entire field are indistinguishable, are hence co-genetic and reasonably originate from the Anadarko sedimentary basin. N_2 concentrations throughout the field vary between 5-75%, averaging ~15%. The highest concentrations of nitrogen in the Texas Panhandle are found on the SSW margin of the field, on the side of the field furthest from the Anadarko basin hydrocarbon 'kitchen'. In the Oklahoma and Kansas Hugoton, the highest nitrogen content is found to the north and west of the field, again on the opposite edge of the field to the Anadarko basin. The nitrogen is locally proportional to the ^4He content, although the $^4He/N_2$ systematically increases from 0.02 in the Kansas Hugoton to 0.077 in the Texas-Panhandle (Gold and Held 1987; Jenden et al. 1988; Pierce et al. 1964).

Ballentine and Sherwood Lollar (2002) show that the nitrogen isotopic composition also changes systematically in this field, decreasing from $\delta^{15}N(N_2)$ = +9.4‰ in the Kansas Hugoton to $\delta^{15}N(N_2)$ = +2.7‰ in the Texas-Panhandle. $^3He/^4He$, $^{21}Ne/^{22}Ne$ and $^{40}Ar/^{36}Ar$ ratios enable noble gas contributions from mantle, crustal and groundwater sources to be resolved and quantified in the samples. Crustal radiogenic $^4He/^{21}Ne^*$ and $^4He/^{40}Ar^*$ ratios show a 60% excess of ^4He compared to predicted crustal production values, and are typical of noble gases released from the shallow crust (Ballentine and Burnard 2002, this volume). Although significant and resolvable mantle and groundwater-derived noble gases are present, mantle $^3He/N_2$ and groundwater $^{36}Ar/N_2$ values rule out significant mantle or atmosphere contributions to the gas field N_2, which is crustal in origin.

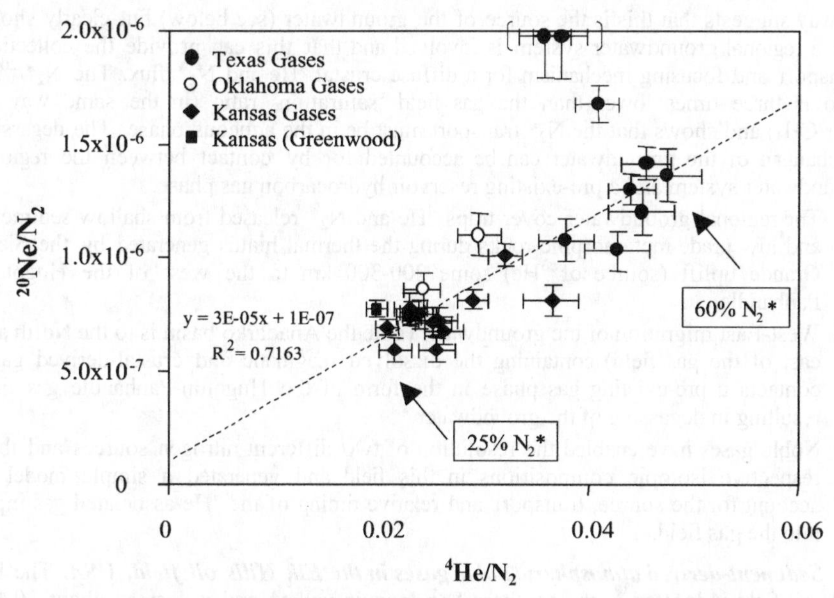

Figure 19. Plot of $^{20}Ne/N_2$ vs. $^4He/N_2$ in natural gases from the Giant Hugoton-Panhandle gas field in Texas-Oklahoma-Kansas, USA after Ballentine and Sherwood Lollar (2002). This natural gas field is the case-type system in which N_2 content is related to ^4He concentration. Most samples fall on a line indicating simple two-component mixing between one nitrogen component that is associated with both crustal ^4He and groundwater-derived ^{20}Ne (N_2^*) and another nitrogen component that has no resolvable association with any noble gases. Identifying one He/N_2 end-member ratio enables the relative contribution of these two nitrogen components to any one sample to be calculated. The nitrogen isotopes also vary systematically with He/N_2 and from the noble gas mixing relationship, the end-member nitrogen isotopic compositions can also be determined (see text).

A plot of $^4He/N_2$ vs. $^{20}Ne/N_2$ shows that almost all samples lie on a simple mixing line between two crustal nitrogen components (Fig. 19). One N_2 component is associated with the crustal ^4He and groundwater-derived ^{20}Ne (N_2*). The other nitrogen component has no resolvable association with either crustal- or groundwater-derived noble gases. An end-member $^4He/N_2^* = 0.077$ (Pierce et al. 1964) is used to define the He-associated component and enables the relative contribution of non-He associated nitrogen to each sample to be calculated (Fig. 19). In turn this enables the $\delta^{15}N(N_2)$ for each end-member to be calculated, where $\delta^{15}N(N_2)$=-3‰ and +13‰ for N_2* and the non-He associated N_2 respectively. The $\delta^{15}N(N_2^*)$ value is not compatible with a crystalline or high grade metamorphic source and, similar to the ^4He, probably originates from a shallow or low metamorphic grade source rock. ^4He mass balance nevertheless requires a regional crustal source; its association with a resolvable magmatic ^3He contribution pointing to a source to the recently active Sierra Grande uplift to the west of the gas field, in the opposite direction to the Anadarko basin hydrocarbon source which is unlikely to be a source of magmatic ^3He.

The ratio of radiogenic ^4He to groundwater-derived ^{20}Ne is almost constant throughout the entire system and clearly indicates a link between the crustal-derived ^4He (and hence the N_2*) and the groundwater system. To place perspective on the volume of groundwater constrained by the ^{20}Ne mass balance this is equivalent to the water in a 100m thick static aquifer that covers three times the area of the Anadarko basin. This in no way suggests that this is the source of the groundwater (see below) but clearly shows that a regional groundwater system is involved and that this can provide the collection, transport and focusing mechanism for a diffuse crustal ^4He and N_2* flux. The $N_2^*/^{20}Ne$ ratio is three times lower than the gas field 'saturation' ratio (in the same way as $^{36}Ar/CH_4$) and shows that the N_2* transport must be in the aqueous phase. The degassing mechanism of the groundwater can be accounted for by contact between the regional groundwater system and a pre-existing reservoir hydrocarbon gas phase.

- The regional groundwater cover traps ^4He and N_2* released from shallow sediments and low grade metamorphic rocks during the thermal hiatus generated by the Sierra Grande uplift (source of ^3He) some 200-300 km to the west of the Hugoton-Panhandle.
- West-East migration of the groundwater (note the Anadarko basin is to the North and east of the gas field) containing the dissolved magmatic and crustal-derived gases contacts a pre-existing gas phase in the form of the Hugoton-Panhandle gas field resulting in degassing of the groundwater.
- Noble gases have enabled the resolution of two different nitrogen sources and their respective isotopic compositions in this field and generated a simple model to account for the source, transport and relative timing of the ^4He-associated gas input into the gas field.

Sediment-derived atmospheric noble gases in the Elk Hills oil field, USA. The Elk Hills oil field is located in the southern San Joaquin valley, and is located about 30 km southeast of Bakersfield, California, USA. Production to date from the Elk Hills anticline has exceeded 2 billion barrels of oil from five producing intervals ranging from the 'Dry Gas Zone' (2-5 Ma) to the 'Santos Oil Zone' (30-35 Ma). Samples were collected from gas-oil separators and ^4He, ^{36}Ar, ^{22}Ne, ^{84}Kr and ^{132}Xe abundance determined (Torgersen and Kennedy 1999). These workers did not tabulate the isotopic compositions, with the exception of ^4He and $^{40}Ar^*$, but report that the Ne, Kr and Xe isotopes are consistent with an atmosphere-derived source. Torgersen and Kennedy give the concentrations normalized to ^{36}Ar and relate this to the air value to obtain a fractionation value where $F(^nN_g)= \{[^nN_g]/[^{36}Ar]_{sample}\}/\{[^nN_g]/[^{36}Ar]_{air}\}$. There is an increasing enrichment related to atomic number where $F(^{22}Ne) < F(^{84}Kr) < F(^{132}Xe)$. Although such enrichments have been previously noted in oil-related systems (Bosch and Mazor 1988; Hiyagon and Kennedy 1992; Pinti and Marty 1995), the Elk Hills study shows an average Xe enrichment factor of ~30, with the highest ratio 576 times the air ratio representing the highest $^{132}Xe/^{36}Ar$ ratio yet measured in a terrestrial fluid.

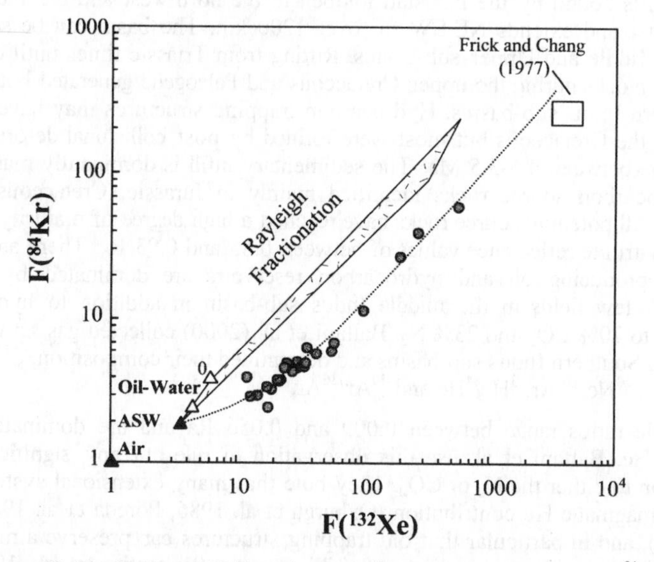

Figure 20. The Kr/Ar and Xe/Ar ratios normalized to air values to give $F(^{84}Kr)$ and $F(^{132}Xe)$ from the Elk Hills oil wells (California, USA) are plotted after Torgersen and Kennedy (1999). The filled triangles represent the composition of air and air-equilibrated water at 20°C (ASW). The open triangles represent the ratios predicted for oil-water equilibration at different oil/water volume ratios, and the shaded region labeled 'Rayleigh Fractionation' the range of values predicted after continuous gas loss from either the water or oil phase. The dashed line shows the weighted least squares fit through the Elk Hills data, forced through the ASW value. This data cannot be accounted for by solubility equilibrium or open system fractionation of groundwater-derived atmospheric noble gases. The data is explained by a trapped sedimentary source for atmospheric Kr and Xe that is released into the oil phase during oil formation and primary migration. The values measured in the Elk Hills samples can be compared with those measured in carbon rich extracts from cherts measured by Frick and Chang (1977).

$F(^{84}Kr)$ is strongly correlated with $F(^{132}Xe)$, but unlike earlier observations, such as the Paris Basin (Pinty and Marty 1995), the Elk Hills data cannot be attributed to fractionation from an oil-water-gas system (Fig. 20). Torgersen and Kennedy argue that the extreme values observed in the Elk Hills system are due to the preferential trapping of Xe > Kr by carbon rich sediments (the oil source rocks) which are subsequently released from the sediment during oil formation and primary migration. This argument is supported by observations of very high $F(^{84}Kr)$ and $F(^{132}Xe)$ values in a variety of carbon rich rocks including chert, thucolite and shale (Bogard et al. 1965; Frick and Chang 1977; Podosek et al 1980; Table 6). A model is proposed that considers mixing and dilution of the highly fractionated sediment-derived Kr and Xe with groundwater-derived species, and relates the decrease in $F^{84}Kr$ and $F^{132}Xe$ with an increase in ^{36}Ar concentration in the hydrocarbons that is proportional to the groundwater/hydrocarbon ratio. This correlation is not yet well established.

- The identification of sediment-derived atmospheric Kr and Xe in some sample types means that the extension of solubility fractionation models (that assume *a priori* a groundwater origin for all atmospheric noble gases) to include Kr and Xe has to be assessed with caution on a case-by-case basis.

$^3He/^4He$ closure and $^{20}Ne/^{36}Ar$ fractionation in the Indus basin, Pakistan. The Indus basin is bound by the Pakistan foldbelt to the northwest and the Indian shield to the southeast and extends NE-SW for over 1200 km. The basin can be separated into Southern, Middle and Upper sub-basins. Rifting from Triassic times until collision with the Afghan blocks during the upper Cretaceous and Paleogene generated both the Middle and Southern Indus sub-basins. Hydrocarbon trapping structures may have been formed as early as the Cretaceous but most were formed by post collisional deformation during the Pliocene between 4.5-3.5 Ma. The sedimentary infill is dominantly marine in origin, with hydrocarbon source rocks identified mainly in Jurassic, Cretaceous and Eocene sequences. All potential source rocks have reached a high degree of maturity in the study area, with vitrinite reflectance values of between 0.85 and 0.93 R_o. There are no trapping structures producing oil and hydrocarbon reservoirs are dominated by thermogenic methane. A few fields in the middle Indus sub-basin in addition to hydrocarbon gas contain up to 70% CO_2 and 23% N_2. Battani et al. (2000) collected gas samples from the Middle and Southern Indus sub-basins and determined their composition, $d^{13}C(CO_2$, CH_4, C_2-$C_4)$, ^4He, ^{20}Ne, ^{36}Ar, $^3He/^4He$ and $^{40}Ar/^{36}Ar$.

$^3He/^4He$ ratios range between 0.009 and 0.056 Ra and are dominated by crustal radiogenic He. Battani et al. use this observation to rule out any significant magmatic contribution to either the N_2 or CO_2. They note that many extensional systems contain a resolvable magmatic He contribution (Oxburgh et al. 1986; Poreda et al. 1986; Ballentine et al. 1991), and in particular that old trapping structures can preserve a mantle $^3He/^4He$ signature for tens if not hundreds of million years (Ballentine et al. 1996; also see Ballentine et al. 2001). Battani et al (2000) argue that the lack of a resolvable magmatic $^3He/^4He$ signature in the extensional Indus basin can only be explained if the basinal fluid system was open during rifting, and that no significant amounts of magmatic fluid have been preserved in the present day hydrocarbon traps. This is entirely consistent with the late development of trapping structures within the basin.

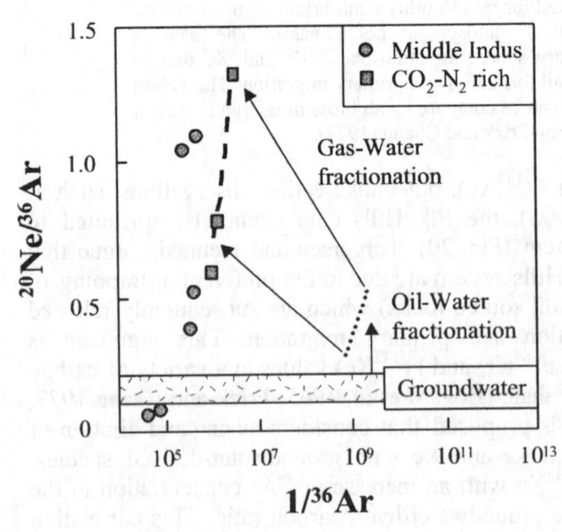

Figure 21. Plot of $^{20}Ne/^{36}Ar$ vs. $1/^{36}Ar$ for natural gases from the middle Indus sub-basin, Pakistan (after Battani et al. 2000). Mixing is observed between two components, one with high $^{20}Ne/^{36}Ar$ and low ^{36}Ar and the second with $^{20}Ne/^{36}Ar$ similar to unfractionated groundwater. Battani et al. have explored a variety of models to account for the high $^{20}Ne/^{36}Ar$ and conclude that the most viable mechanism is a two-stage process involving an oil phase. The first stage is Rayleigh fractionation of the residual noble gases in the groundwater after interacting with oil. The second stage is transfer of a small amount of the noble gases, accompanied by further fractionation, into the gas phase. Model conditions are discussed in the text.

$^{20}Ne/^{36}Ar$ ratios and ^{36}Ar concentrations provides another important perspective on this system. In a plot of $^{20}Ne/^{36}Ar$ vs. $1/^{36}Ar$, there is clear mixing between a gas component that is characterized by low ^{36}Ar and high $^{20}Ne/^{36}Ar$ values, and a second component that has higher ^{36}Ar and groundwater $^{20}Ne/^{36}Ar$ values (Fig. 21). Battani et al interpret the low $^{20}Ne/^{36}Ar$ values to have an unfractionated groundwater source. The origin of the high $^{20}Ne/^{36}Ar$ values, up to 1.3 compared with a groundwater range of 0.13 to 0.19, is investigated in more detail. Other workers have also observed high $^{20}Ne/^{36}Ar$ values in both natural gases (e.g., Ballentine et al. 1991) and waters (Castro et al. 1998). Battani et al. note that a single stage equilibrium model for fractionation between groundwater and a gas (Bosch and Mazor 1988; Ballentine et al. 1991) cannot produce $^{20}Ne/^{36}Ar$ values greater than ~0.6. This limit also applies for the gas phase during Rayleigh fractionation of a simple water-gas system. Battani et al. also investigate the possibility of a process of re-solution and effervescence. This would result in high ^{36}Ar being correlated with high $^{20}Ne/^{36}Ar$ in the gas phase and can be discounted.

Battani et al. (2000) argue that if groundwater is first equilibrated with an oil phase, the concentration of atmosphere-derived noble gases in the water phase will be reduced and the $^{20}Ne/^{36}Ar$ ratio increased due to the higher relative solubility of Ar in the oil phase. If this process is by a Rayleigh distillation process, the fractionation in the residual water can be extreme. Subsequent gas-water interaction can produce much more highly fractionated $^{20}Ne/^{36}Ar$ ratios than a single water-gas phase system, and the most highly fractionated values would correlate with the lowest ^{36}Ar concentrations. Battani et al. model this process for a variety of conditions. They consider for example, the conditions required to account for the CO_2-N_2 rich gases in the middle Indus. The oil-water distillation process is taken to have left 30% of the ^{36}Ar in the groundwater at ~120°C and at a depths of 3700m. Subsequent gas-water Rayleigh fractionation is modeled as leaving 90% of the ^{36}Ar in the groundwater phase, at a temperature of 29°C and at a depth of 630 m (Fig. 21). There are clearly many variables in this modeling process that include the density of the oil and the salinity of the groundwater in addition to the parameters that have been varied to obtain a fit to the data set. In this respect it is not straightforward to assess the uniqueness of the model parameters used. Nevertheless, the model is sensitive to the depth of the gas water equilibration (pressure has a large effect on the water/gas volume ratio) and the model depths of <630 m for the middle Indus gases are far lower than current reservoir depths of 1300-3000 m. Battani et al (2000) suggest that this discrepancy can be accounted for by burial of the trapping structures. Confirmation of sediment build-up/basin subsidence rates may provide a future test for this model.

- Radiogenic $^{3}He/^{4}He$ ratios in Indus basin natural gases rule out magmatic contributions to high CO_2-N_2 gases and are consistent with open system behavior prior to late trap formation.
- Extreme $^{20}Ne/^{36}Ar$ and low ^{36}Ar concentrations in some natural gases are explained by equilibration of an oil phase with the groundwater before gas phase formation.

NOBLE GASES IN ANCIENT GROUNDWATERS AND CRUSTAL DEGASSING

Quantifying flow paths and velocities of groundwaters in sedimentary basins and in the crystalline crust is of prime importance for the exploration and exploitation of economic resources (e.g., petroleum, water resources, geothermal energy) as well as for assessing the potential of underground storage of hydrocarbons and of chemical/nuclear wastes. Estimating groundwater circulation requires the documentation of variations in permeability and in fluid physical and chemical compositions. For deep, aged groundwaters, these parameters need to be constrained independently. The compositional evolution of groundwater can be chemically investigated but estimates of flow rates must be based upon isotopic methods. The generally long residence time of such fluids (typically 10^4 to 10^7 years) are often in excess of time intervals that can be addressed by methods based on the decays of cosmogenic radionuclides. ^{3}H ($T_{1/2}$ = 12.4 yr) dating can only be used for very young groundwaters (Kipfer et al. 2002, this volume). Groundwaters up to 25,000 yr old may be dated using ^{14}C ($T_{1/2}$ = 5730 yr), provided that correction can be made for secular variation of ^{14}C activity in air (e.g., Bard et al. 1990), dissolution of ^{14}C-free carbonate and for ^{14}C exchange between dissolved carbonates and carbonate minerals in the aquifer (e.g., Fontes and Garnier 1979). This requires detailed and somewhat model-dependent corrections and, for aged groundwaters, is in practice best applicable to the case of sandstone aquifers. Dating groundwaters with residence time exceeding the limit of the ^{14}C method is subject to large errors. U-series disequilibria does not depend uniquely on the water residence time, but also on other parameters such as the water redox state. The use of cosmogenic ^{36}Cl ($T_{1/2}$ = 3.5×10^5 yr) may be complicated by *in situ* production of ^{36}Cl in the aquifer matrix.

Noble gas isotopic ratios and abundances and stable (H, O) isotopic ratios set constraints on the origin of waters and of their dissolved species, as well as on their residence time. The isotopic ratio of helium, displays a large range of variation between the isotopic composition of radiogenic helium (<0.02 Ra, Ballentine and Burnard 2002, this volume) produced in the continental crust and the isotopic composition of mantle helium (8 Ra for the upper mantle, Graham 2002, this volume). Consequently, helium can be used as a sensitive tracer of fluid circulation in basins developing in tectonically active areas (e.g., Martel et al. 1989; Marty et al. 1993; O'Nions and Oxburgh 1988; Sano et al. 1986; Stute et al. 1992). The accumulation of radiogenic helium in basin aquifers provides a suitable geochronometer for aged groundwaters because (1) the production of radiogenic helium is significant and easily identifiable from the natural atmospheric background, and (2) the transfer of helium from rock to fluid is fast on a geological time scale (Ballentine and Burnard 2002, this volume). The use of this potential geochronometer requires the quantification of the accumulation rate of He in groundwater. However, the origin of radiogenic isotopes such as ^{4}He in basin waters is highly debated. Helium water age, computed assuming that radiogenic helium produced in the aquifer rock has been quantitatively transferred into the aquifer water, is often in excess of the ^{14}C age when available (Dewonck et al. 2001; Heaton 1984), or the hydrologic age (Torgersen and Clarke 1985). The origin of such ^{4}He "excess" is a matter of debate and several studies advocate a contribution from the deep continental crust (e.g., Andrews et al. 1985b, Castro et al. 1998a, Dewonck et al. 2001; Heaton 1984; Marty et al. 1993; Stute et al. 1992; Torgersen and Clarke 1985), and/or from intra-basinal sources such as old and stagnant waters, aquifer rocks and adjacent aquitard rocks (Pinti and Marty 1998; Tolstikhin et al. 1996), or detrital minerals rich in inherited He and altered at low temperature (Solomon et al. 1996; Torgersen 1980). These sources are probably contributing variable amounts of radiogenic He to deep aquifer waters, and their respective strengths can only be assessed by studies addressing well-documented aquifers.

Given the economic and societal importance of this area of research, in this subsection we clarify the origin of helium in deep groundwaters in cases where independent constraints are available and focus on the use of noble gases to understand groundwater circulation for aquifers where residence times are in excess of those covered by classical hydrological tracers. We address the problem of accumulation of radiogenic helium, and of other noble gas isotopes and discuss several case studies, namely the Great Artesian Basin, Australia, the Paris Basin, France, and the Pannonian basin, Hungary.

Sources of He isotopes in groundwaters

The abundance of air-derived He (Air-Saturated He, or ASW-He) is fixed by the equilibrium solution of atmospheric helium and is low due to the low He content of air (5.22 ppm vol; Oliver et al. 1984). ASW-He is around 2.10×10^{-12} mol/g, to be compared with abundances of 10^{-10} to 10^{-7} mol/g typically observed in aged basinal groundwaters. The accumulation rate of radiogenic ^{4}He in groundwater from the decay of U and Th in aquifer rocks is typically 10^{-16}-10^{-17} mol/g.yr (see below). Considering that with modern installations it is possible to measure the He content of groundwaters with a precision better than 5%, it should be possible to measure residence times as low as a few thousand years, which intersects nicely the time span covered by ^{14}C dating. Therefore, considerable effort has been made to understand the processes responsible for the accumulation of radiogenic helium in groundwater and to attempt inter-calibration of the ^{4}He dating method with the ^{14}C dating method for residence times lower then 40,000 yr.

The ^{4}He dating method, first suggested by Savchenko (1935) and by Davis and DeWeist (1966), was developed by Andrews (1983), Andrews and Lee (1979), Heaton (1981,) Heaton and Vogel (1979), Marine (1979) and Torgersen (1980). The simplest expression of the method is based on the *in situ* accumulation of radiogenic ^{4}He in groundwater residing in a porous rock and depends on the U and Th content of the rock, the porosity, and the efficiency of rock-fluid transfer. When only the *in situ* contribution of radiogenic ^{4}He is considered, the amount of ^{4}He in a groundwater, [^{4}He], is

$$[^4He] = [^4He]_{ASW} + P_4 . t \tag{43}$$

$[^4He]_{ASW}$ is ASW-He (this concentration is a function of temperature and pressure, although its variation is generally small compared to the radiogenic ^{4}He content), P_4 is the accumulation rate of ^{4}He from *in situ* production and t is the residence time of groundwater (the exponential function of radioactive decay has been linearized because the residence time of groundwaters are always small compared to the decay constants of U and Th).

The accumulation rate can be expressed after as

$$P_4 = \frac{\rho . \Lambda . J_4}{\phi} \tag{44}$$

where ρ is the density of porosity rock (the density of water is neglected here as its natural variation is minor with respect to that of [^{4}He]), Λ is a parameter defining the efficiency of transfer from the rock matrix to water, ϕ is the porosity, and J_4 is the source function of radioactive production of ^{4}He in the aquifer matrix (e.g., Torgersen 1980). Assuming He loss from matrix minerals is fast on a geological time scale (Ballentine and Burnard 2002, this volume), Λ is considered to be close to 1 and J_4 can be expressed as

$$J_4 = 1.05\times10^{-17} . [U] . \{1 + 0.123 . ([Th]/[U] - 4)\} \tag{45}$$

[^{4}He] is expressed in mol/g_{H2O}, with J_4 in mol/g_{rock}.yr and U and Th in ppm. Such accumulation rates are valid if detrital minerals in the matrix are devoid of radiogenic ^{4}He accumulated before sedimentation and not released completely since then. Torgersen (1980) has discussed the potential of inherited ^{4}He contributions to groundwater by silicate weathering and concluded that this source was small compared to other sources. However, Solomon et al. (1996) have argued that, under specific conditions and in the case of young groundwaters, diffusion of inherited ^{4}He could become the dominant contribution of non-atmospheric helium. In a detailed study of a well-documented (^{3}H-^{3}He, CFCs) site of Northern Ontario where the ^{4}He content of groundwater increases linearly with distance along the flow path, these authors concluded that the accumulation rate of ^{4}He was two orders of magnitude higher than that predicted by *in situ* production. They have argued that the horizontal fluid velocities are too great to allow upward migration of radiogenic helium from the underlying shield rocks and that another source of

^{4}He internal to the aquifer was needed. They have conducted laboratory experiments to document the diffusive properties of ^{4}He in aquifer solids, and concluded that the release rate at aquifer temperature of inherited radiogenic He was sufficient to provide the additional ^{4}He source. These authors suggested finally that this effect was widespread among young aquifers (see Kipfer et al. 2002, this volume). In the deeper and older systems discussed here, the contribution of inherited He trapped in the aquifer rocks are a minor component compared with other sources.

Deep aquifers and the crustal He flux

In a study of the Auob Sandstone of Namibia (Southwest Africa), (Heaton 1981; Heaton 1984), found *in situ* ^{4}He ages much in excess of ^{14}C ages, and advocated the contribution of a ^{4}He source external to the aquifer having a flux of 0.4-1.6×10^{-13} mol/g_{H2O} yr whereas the *in situ* accumulation rate was computed to be 1.8×10^{-16} mol/g_{H2O} yr. A similar observation was made by Torgersen and Clarke (1985) in a classical study of the J-aquifer of the Great Artesian Basin (GAB), Australia. For groundwater dated by ^{14}C to be younger than 50,000 yr, the accumulation rate is estimated at 2.0 x 10^{-16} mol/g_{H2O} yr, very comparable to the *in situ* accumulation rate of 1.8×10^{-16} mol/g_{H2O} yr, whereas for groundwaters older than 100,000 yr according to hydrological modeling, the accumulation rate was 74 times the *in situ* value (Fig. 22). Torgersen and Clarke (1985) pointed out that localized concentration of U and Th are unlikely, given the large-scale integration of the ^{4}He accumulation over the area of the basin. In the case of a ^{4}He flux external to an aquifer, F_4, entering it from below, neglecting $[^4He]_{ASW}$, the observed $[^4He]$ becomes

$$[^4He] = \{P_4 + (F_4/\phi.h)\}.t \tag{46}$$

h is the aquifer thickness. F_4 can be computed from the observed ^{4}He content, and independent estimate of residence time t, to give

$$F_4 \simeq [^4He].\ \phi.h/t \tag{47}$$

More sophisticated models, notably taking into account the transverse dispersion, have been developed subsequently (e.g., Castro et al. 1998b, Stute et al. 1992; Torgersen and Ivey 1985), but the basic information needed to estimate the magnitude of external ^{4}He fluxes into aquifers is the residence time of groundwater, either measured by an independent geochronometer, or estimated from hydrological modeling. Hence computed fluxes depend critically on the way residence times are estimated, which is currently a matter of debate as we shall see later. Torgersen and Clarke (1985) computed that the flux of ^{4}He entering the aquifer is 1.6×10^{-6} mol/m^{2}yr. Given reasonable U and Th distribution in the continental crust and assuming no He transfer at the mantle-crust boundary (which is reasonable, given the low ^{3}He/^{4}He ratios of the samples), degassing of the whole continental crust provides a ^{4}He flux of 1.4-1.6×10^{-6} mol/m^2 yr (O'Nions and Oxburgh 1983; Torgersen and Clarke 1985), very similar to the GAB flux. Furthermore, Torgersen and Clarke (1985) interpreted the data from Heaton (1981; 1984) to conclude that in the case of the Auob Sandstone of Namibia the required external flux is also well matched by the whole crustal degassing flux.

Torgersen et al. (1989) further conducted a Ar isotope study on the GAB waters and proposed that radiogenic ^{40}Ar* could not be entirely sourced from the aquifer rocks and required, as in the case of ^{4}He, a source external to the aquifer characterized by a ^{4}He/^{40}Ar* ratio close to that of the crustal production. The required ^{40}Ar* flux was estimated to be comparable to that produced by the whole crust. However, they noted that this flux is decoupled from the ^{4}He flux and has to be discontinuous in time and space. The difference between He and Ar was attributed to several factors including heterogeneous distribution of U, Th and K, differential release of noble gas isotopes from minerals, the variability of tectonic and hydraulic stress, and the rapid rate of crustal fluid transport during metamorphism, which rapidly fractionates released He from Ar.

This important finding has far-reaching consequences not only in the field of hydrology and groundwater dating, but also in that of noble gas geochemistry; for the first time large-scale crustal degassing, otherwise required by noble gas degassing models (e.g., Ozima and Podosek 1983), was apparently observed. This view nevertheless raises two conceptual questions. First, why is the large scale flux seen only for already aged groundwaters and not for those younger than 50,000 yr. Second, the increase of [^{4}He] with distance suggests that the aquifer is accumulating He, i.e., helium does not escape from the top of the aquifer. In order to address these points, Torgersen and Ivey (1985) developed a model of He invasion of the GAB taking into account the effective vertical diffusion (transverse dispersion) of He in addition to its horizontal advective transport, justified by the large thickness (600 m) of the aquifer. They treated this problem in 2D,

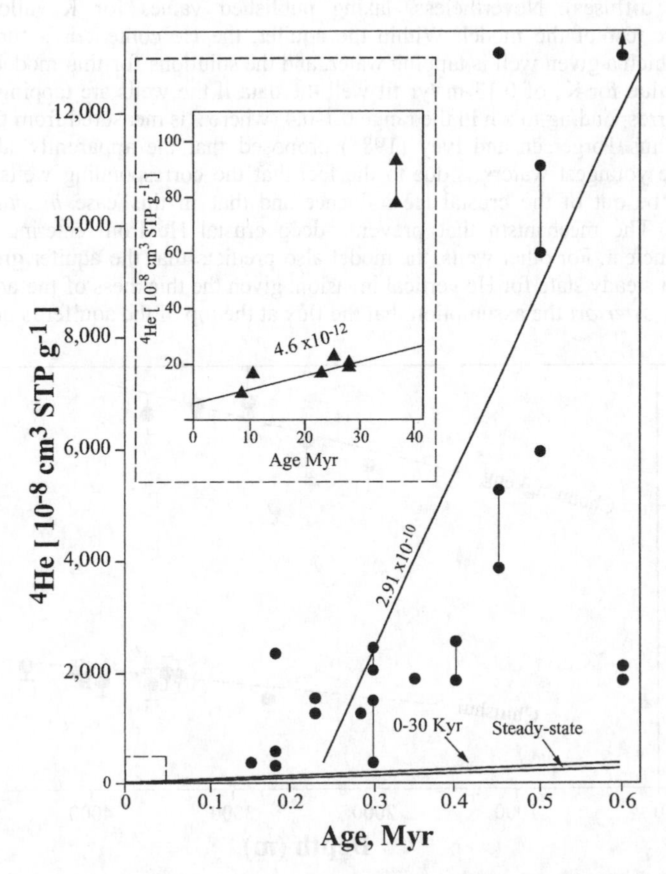

Figure 22. Helium content of groundwaters in the J1 aquifer of the Great Artesian Basin, Australia, as a function of groundwater residence time. For residence times lower than 50 kyr as estimated from ^{14}C activity measurements, the increasing He content along the flow path can be accounted for by *in situ* accumulation of radiogenic He produced in the porosity rock. For residence times older than 100 kyr, the He accumulation rate requires contribution of a radiogenic ^{4}He source external to the aquifer, the strength of which is comparable to that derived from degassing of the whole underlying continental crust (after Torgersen and Clarke, 1985).

and the general equation governing the ^{4}He content is

$$U\frac{\partial[^4He]}{\partial x} = K_a\frac{\partial^2[^4He]}{\partial^2 z} + P_4 \qquad (48)$$

Assuming that the flux out of the aquifer is null, the general solution is in the form

$$[^4He]_{x,z} = (P_4/U).x + (F_4/U).\{(x/h) + (h.U/K_a)\ .\ f(x, z)\} \qquad (49)$$

x and z represent the horizontal distance from recharge and the vertical axis is depth. U is the Darcy velocity, K_a is the effective vertical diffusion coefficient, also sometimes called the dispersivity coefficient, and f is a function of x and z. Thus the crossover point for which the left-hand term becomes dominant over the *in situ* production is a function of the dimensions of the system, particularly of the x/h ratio, and of U and K_a. It comes out immediately that the least defined parameter is K_a, which is a poorly defined function of the tortuosity of the aquifer structure (that is, the microsopic path filled with water in which He diffuses). Nevertheless, taking published values for K_a allows a semi-quantitative test of the model. Within the aquifer, the He content is a function of the depth at which a given well is tapping water, and the solutions for this model, assuming a constant value for K_a of 0.13 m^2/yr fit well the data if the wells are tapping water in an interval corresponding to z/h in the range 0.1-0.4 (where z is measured from the top of the aquifer). Thus Torgersen and Ivey (1985) proposed that the apparently absent crustal flux for the youngest waters is due to the fact that the corresponding wells are shallow enough to be out of the crustal He influence and that in this case *in situ* production dominates. The mechanism that prevents deep crustal He from entering the shallow waters is unclear. For other wells, the model also predicts that the aquifer groundwater is not yet in a steady state for He vertical invasion, given the thickness of the aquifer, which justifies *a posteriori* the assumption that the flux at the top of the aquifer is null.

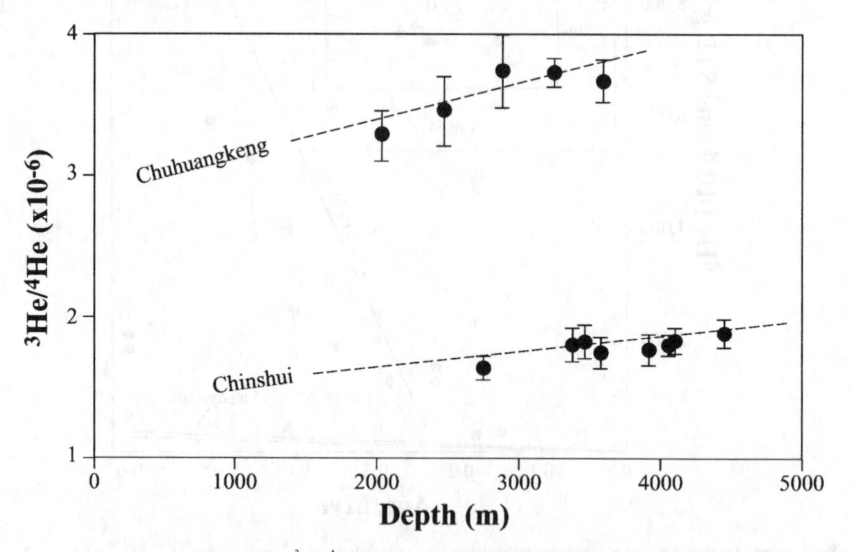

Figure 23. Variation of the ^{3}He/^{4}He ratios as a function of depth in basinal natural gas wells in Northern Taiwan. For two localities, Sano et al. (1986) found that the ^{3}He/^{4}He ratios increase with increasing depth, a trend attributed to the dilution of a partly mantle-derived He component by radiogenic ^{4}He produced in the sediments. These authors concluded that (i) the required ^{4}He flux is equivalent to that derived from the whole underlying continental crust, and (ii) the required ^{3}He flux is similar to that originating at mid-ocean ridges.

Sano et al. (1986) used a somewhat different method to estimate the flux of radiogenic ^{4}He in Northern Taiwan, based on two He isotope gradients in natural gas wells of different depths (Fig. 23). They found that the ^{3}He/^{4}He ratios increase with depth and interpreted this as the result of dilution of a He component, enriched in mantle-derived ^{3}He, by radiogenic He produced in the sedimentary pile. They concluded that this area was subject to a flux of ^{3}He (2-4×10^{-12} mol/m^{2}yr) equivalent to that of mid-ocean ridges, probably related to regional Tertiary magmatism, and a ^{4}He flux of 1.3-1.4×10^{-6} mol/m^{2}yr of continental origin, very similar to the ^{4}He flux from the whole crust discussed above.

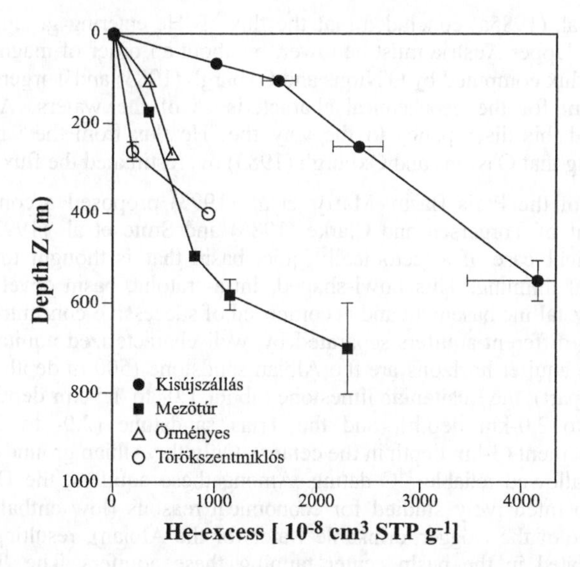

Figure 24. Variation of the He content as a function of depth in the discharge area of the PL2 aquifer, Great Hungarian Plain (after Stute et al. 1992). From the He gradient and the water velocity, these authors derived a ^{4}He crustal flux one order of magnitude lower than that due to degassing of the whole continental crust.

Stute et al. (1992) developed a model similar to that of Torgersen and Ivey (1985) for the Great Hungarian Plain (GHP) basin. In addition, these authors considered the case where aquifers do not accumulate He, that is, He is allowed to exit the aquifers at their top. Note that in this case, no flux can be computed if a steady state is reached, because [^{4}He] becomes constant and independent of time t in Equation (5). Stute et al. (1992) considered that the model of Torgersen and Ivey (1985) was valid for the PL2 aquifer of the GHP and that the measured [^{4}He] in a given well at distance x from recharge was representative of the ^{4}He content averaged over the thickness z. They used a range for F_4 values determined for different areas of the GHP to derived a mean residence time of the PL2 groundwater of 0.5-2.3×10^6 yr. Furthermore, the GHP offers the opportunity to sample helium in the discharge area at different depth, yielding good correlations between [^{4}He] and depth. These gradients enable the calculation of the He fluxes entering the studied aquifers from below (Fig. 24). The total helium flux is given as

$$J = v_z.\phi.\left\{\frac{[^4\mathrm{He}]z_2 - [^4\mathrm{He}]z_1}{1-\exp(\Delta_z.v_z/K_a)} + [^4\mathrm{He}]z_1\right\} \quad (50)$$

where z_1 and z_2 are two different well depths, and v_z is the upward vertical velocity.

Stute et al. (1992) computed a ^{4}He crustal flux in the range 0.03-0.24×10^{-6} mol/m^{2}yr, one order of magnitude lower than the whole crust degassing flux. They attributed this difference to the shielding of crustal degassing by Tertiary layers in the GHP. It is interesting to note that, working also on groundwaters in the Great Hungarian Plain, Martel et al. (1986) derived ^{4}He and ^{3}He fluxes one order of magnitude higher than those derived by Stute et al. (1992), by estimating the total volume of waters circulating in the basin multiplied by their average ^{3}He and ^{4}He contents. Stute et al. (1992) attributed this discrepancy to a biased estimate of groundwater circulation as, according to these authors, the decrease of water velocities with depth was not taken into account by Martel et al.

Andrews et al. (1985a) concluded that the flux of ^{4}He entering groundwaters in the Molasse basin of Upper Austria must be lower, by about an order of magnitude, than the whole crust ^{4}He flux computed by O'Nions and Oxburgh (1983) and Torgersen and Clarke (1985) to account for the geochemical characteristics of the waters. Andrews et al. (1985a) attributed this discrepancy to the way the ^{4}He flux from the whole crust was computed, arguing that O'Nions and Oxburgh (1983) overestimated the flux of crustal He.

In the case of the Paris Basin, Marty et al. (1993) proposed a conceptual model distinct from that of Torgersen and Clarke (1985) and Stute et al. (1992). This basin represents a typical case of a tectonically quiet basin that is thought to have evolved mainly by crustal thinning. This bowl-shaped, intra-cratonic basin developed over the Ante-Permian crystalline basement and is composed of successive concentric sedimentary deposits, hosting different aquifers separated by well characterized aquitards. From the surface, the main aquifer horizons are the Albian sandstone (500 m depth on average in the basin central part), the Lusitanian limestone (about 1.0- to 1.5-km depth), the Dogger limestone (1.5- to 2.0-km depth), and the Trias sandstone (2.0- to 2.5-km depth) overlying the basement (3-km depth in the center). Only the Albian groundwater is young enough to have allowed reliable ^{14}C dating. Among these aquifers, the Dogger and the Albian have been intensively studied for economical reasons (low enthalpy geothermal energy in the case of the Dogger, drinkable water for the Albian), resulting in numerous wells mostly located in the basin center tapping these aquifers. The ^{4}He content of groundwater in the Dogger aquifer does not show any relationship with distance from the recharge areas, strongly suggesting that a steady state has been reached for this aquifer, with a ^{4}He flux entering the aquifer from below equivalent to the ^{4}He flux going out from the aquifer (Marty et al. 1993).

The occurrence of cross formational He transfer in the basin is further attested by ^{3}He water contents in excess of those predicted by radiogenic production, which are regarded as representing a small mantle He contribution originating in the basement underlying basin sedimentary strata. Thus Marty et al. (1993) postulated that the basin aquifers are in a steady state for He and that the flux entering the Albian aquifer is equivalent to that leaving the Dogger after transition through intermediate aquifers such as the Lusitanian one, plus the radiogenic production of He isotopes in the sedimentary strata between the Dogger and the Albian aquifers. This production was found to be minor compared to the total He flux, which was computed from the ^{4}He content in the Albian groundwater and its ^{14}C-based residence time (about 10,000 yr in average, within a factor of 2) to be 2 x 10^{-7} mol/m^2 yr, one order of magnitude lower than the whole crust flux. The residence time of helium in the Dogger aquifer was then computed to be 4 Myr (within a factor of 2), which is consistent with the stable isotope data of water. Indeed, the δ^{18}O value of -4 ‰ vs. SMOW of the meteoric end-member is drastically different from the δ^{18}O value of around -8‰ for present-day precipitations and at the recharge areas (Morvan and Lorraine) and of around -10‰ that characterizes waters recharged at basin margins during glacial periods of the Quaternary (Dewonck et al. 2001) (Fig. 25).

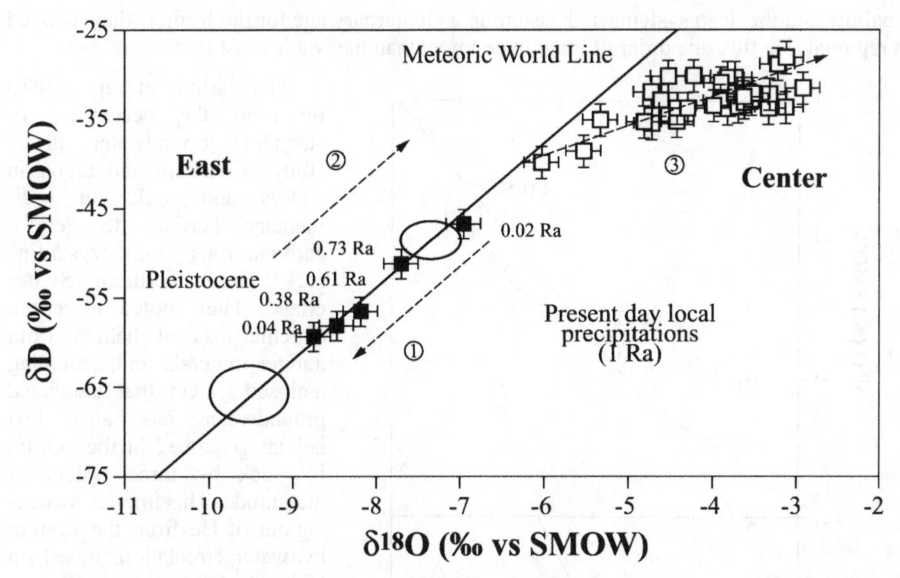

Figure 25. Stable isotope composition of groundwaters in the Trias and Dogger aquifers of the Paris Basin, France (after Dewonck 2001; and Matray et al. 1995) from the eastern recharge in the Lorraine region (black squares) towards the basin center (open squares). Starting from the present-day local precipitation composition, δD and $\delta^{18}O$ values become more negative along the meteoric world line in the recharge area (trend 1), recording the colder climate of the Quaternary, with $^3He/^4He$ ratios (expressed as fractions of Ra where Ra is the atmospheric value) decreasing by dilution with radiogenic ^4He. Groundwaters towards the basin center have more positive δD and $\delta^{18}O$ values, recording pre-Quaternary warmer conditions (trend 2), and depart from the meteoric world line due to dissolution of evaporites in the Trias aquifer and subsequent transfer in the Dogger aquifer (trend 3).

Worden and Matray (1995) have used the stable isotope ratios to provide evidence for very low groundwater circulation in the Dogger aquifer. Furthermore, Pinti et al. (1997) determined groundwater noble gas paleotemperatures (NGT) and the relationship between stable isotopes and NGT for several Dogger groundwater wells. Pinti et al. found NGT ranging between 14°C and 26°C along a presumed flow line running from north to south. Such high paleorecharge temperatures suggest strongly that waters in the Dogger aquifer were recharged before the Quaternary, possibly in the Eocene for the most ancient waters (Fig. 26). Pinti et al. (1997) concluded that deep basin aquifers can host groundwaters for geologically significant periods of time, contrary to hydrological models that predict significant water velocity. Somewhat different conclusions where derived for the same basin by Castro et al. (1998b) who developed a 2-D hydrologic model of groundwater circulation in the Paris basin calibrated using noble gas measurements, taking into account selective diffusion of noble gases between aquifers. Their model resulted in a He flux similar to that from the whole crust and by consequence Quaternary ages for groundwater in the Dogger, in contradiction with stable isotope and NGT data discussed above.

Zuber et al. (1997) studied saline waters in Busko Spa (Southern Poland) and argued from stable isotope and NGT data that the shallow water system was recharged during the Quaternary but that waters from the deep system (typically 300-m depth) were recharged in much warmer period, up to 24°C and were pre-Quaternary, implying long residence time (>5 Ma) for the latter. From the concentration gradient of ^4He between the shallow and the deep systems and assuming a Quaternary age for the former, they derived a regional ^4He flux one order of magnitude lower than the whole crust flux.

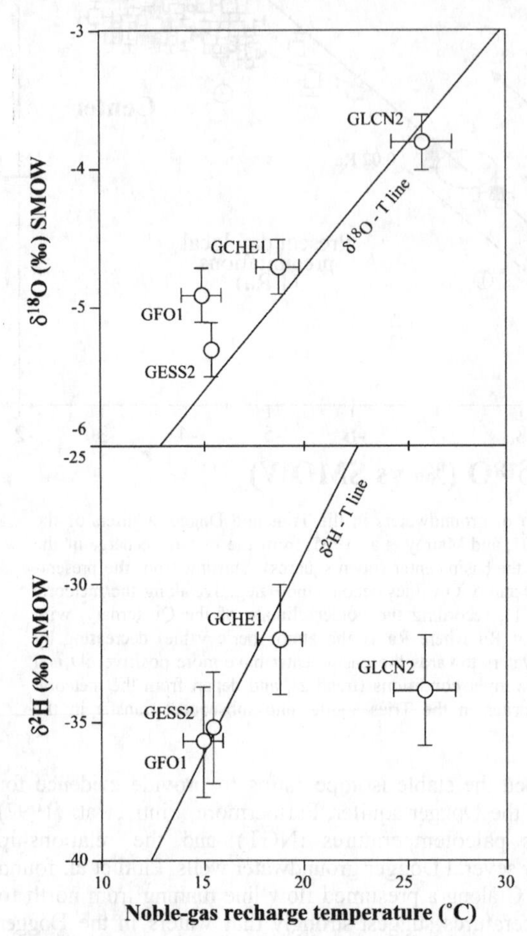

Figure 26. Noble gas recharge temperatures (NGT) versus δD and $\delta^{18}O$ values for four wells tapping groundwaters from the Dogger aquifer in the central area of the Paris Basin (after Pinti et al. 1997). Both stable isotope compositions and NGT indicate that these waters were recharged during a much warmer climate than today, most likely during the Tertiary, which demonstrates that basin aquifers can store groundwaters for long geological periods.

Tolstikhin et al. (1996) proposed the occurrence of stagnant groundwater in a study of helium and argon in waters and rocks in sedimentary Tertiary to Permo-carboniferous and crystalline rocks from Northern Switzerland. They noted a general extreme loss of helium from aquifer minerals and, assuming a closed system, that associated groundwater has also lost helium generated in the porosity rock by three orders of magnitude. This implies sweeping out of He from the system by water circulation. Based on He and Ar isotope measurements in rocks and fluids (Fig. 27), they outlined the role of intra-basinal sources filled with old, stagnant waters from where He was diffusing into freshwater circulating along more permeable layers constituting the aquifers. They also concluded that no external source was required in their case study. Pinti and Marty (1998) argued also that, if the Dogger waters of the Paris basin are very ancient, as suggested by stable isotope and NGT data, then *in situ* production and production in the underlying Liassic shales can become a significant source of radiogenic He. Such contribution could account for the elevated $^4He/^{40}Ar^*$ ratios up to 100 measured in the Dogger waters, whereas the $^4He/^{40}Ar^*$ ratios of the Triassic aquifer below the Dogger aquifer, and separated by the Liassic shales, are close to the radiogenic production ratio, providing evidence for exchange with the underlying basement.

Advective versus diffusive transfer of noble gases in basins

These studies illustrate the complex problem of the nature of noble gas transfer in basins. There is no doubt that both advection and diffusion play major roles, but the respective strengths of each process are very different depending on the scale of observation. Diffusion allows quantitative transfer of noble gas isotopes from minerals to waters and is a temperature-, and atomic radius-dependent process (see Ballentine and Burnard 2002, this volume). At larger scales such as that of basins or even that of the whole continental crust, we have seen in this section that contrasting conclusions can be drawn. High He contents in deep aquifers can be accounted for only by mixing between stagnant waters sampling aquitards and waters flowing in aquifers, or by advocating the occurrence of large-scale He fluxes from the whole continental crust. The main difficulty of this problem is the efficiency of helium transfer in the water-saturated crust and in overlying sediments. The diffusion rate of any species (e.g., He) in semi-permeable layers can be expressed as the product of the species diffusivity at local temperature by a coefficient taking into account the geometry of the transfer path, and often defined as the tortuosity of the medium. The tortuosity can be reliably estimated for permeable sedimentary rocks that have a well-characterized porosity, such as oolithic limestone, porous sandstones, etc. In this case the diffusion rate is higher and comparable to the diffusivity in water.

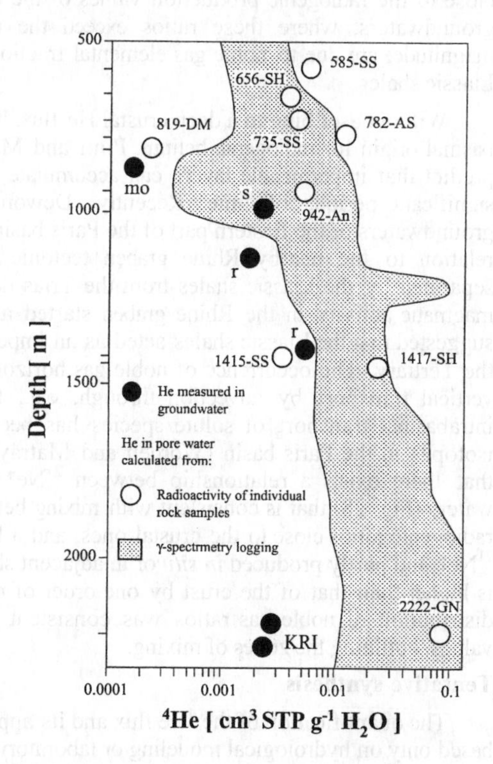

Figure 27. Measured and computed He contents of groundwaters in sedimentary Tertiary to Permo-Carboniferous sequence of Northern Switzerland after Tolstikhin et al. (1996). The measured He contents of rocks and waters are much lower than those expected for a closed system, pointing to loss of helium by groundwater circulation. These authors concluded that no He source external to the aquifers were required and that the He contents of sampled groundwaters could all be accounted for by mixing between circulating groundwaters and old, stagnant waters.

The problem is much more difficult in the case of semi-permeable or even impermeable layers such as shales or marls where the connectivity between pockets of interstitial water is uncertain and difficult to evaluate from laboratory measurements. Laboratory experiments indicated either relatively high He mobility in analogs of deep-sea sediments at rates comparable to, although slightly lower than, diffusion in water ($1.3\text{-}3.0\times10^{-9}$ mol/m^2.s, Ohsumi and Horibe 1984). However, Rebour et al. (1997) designed a new diffusion experiment for estimating diffusion of He from He-saturated water through cm-thick shale discs and found an extremely low diffusivity for a shale sample from the Callovo-Oxfordian aquitard of the Paris basin of 2×10^{-12} mol/m^{2}s. Considering an aquitard thickness of typically 300 m in the Paris basin, it would take helium about 1 Ma to diffuse through the aquitard using the Ohsumi and Horibe (1984) value, and about 1 Ga using the Rebour et al. (1997) measurement. Even taking into account the increase of temperature with depth, it can be concluded either that noble gases can diffuse quantitatively through semi permeable layers, or that shales can constitute efficient traps for noble gases over geological periods, depending on which diffusion value is adopted.

The case of high diffusivity of He in sediments supports models advocating large scale vertical diffusion of helium through basins (e.g., Andrews et al. 1985a, Castro et al. 1998b, Torgersen and Clarke 1985). Notably, Castro et al. (1998a,b) proposed that the difference of $^{21}Ne^{*}/^{40}Ar^{*}$ and $^{4}He/^{40}Ar^{*}$ ratios between the Trias, where these ratios are close to the radiogenic production values of the crust, and the overlying Dogger aquifer groundwaters, where these ratios exceed the production values up to an order of magnitude, are due to noble gas elemental fractionation by diffusion in the inter-bedded Liassic shales.

Without ascribing to a deep crustal He flux, Tolstikhin et al. (1996) suggest an intra-basinal origin for radiogenic helium. Pinti and Marty (1998) and Winckler et al. (1997) predict that impermeable layers can accumulate and store noble gases over geologically significant periods of time. Recently, Dewonck et al. (2001) noted that Triassic groundwaters of the Eastern part of the Paris basin contain a significant amount of ^{3}He, in relation to the nearby Rhine graben tectonic activity, whereas the Dogger waters separated by the Liassic shales from the Trias do not yet show such a signal. Since the magmatic activity in the Rhine graben started at least in the Miocene, Dewonck et al. suggested that the Liassic shales acted as an impermeable barrier since at least the end of the Tertiary. The occurrence of noble gas horizontal barriers in basins requires noble gas vertical transport by advection through, e.g., faults and major tectonic events. The intrabasinal transport of solute species has been demonstrated for salinity, Sr and C isotopes in the Paris basin (Worden and Matray 1995). Pinti and Marty (1995) noted that there exists a relationship between $^{21}Ne^{*}/^{40}Ar^{*}$ and $^{4}He/^{40}Ar^{*}$ ratios in Dogger waters (Fig. 28) that is consistent with mixing between a Triassic end-member presenting radiogenic ratios close to the crustal ones, and a Dogger end-member enriched in ^{4}He and $^{21}Ne^{*}$ and partly produced *in situ* or in adjacent shales (the U/K ratio of Dogger limestone is higher than that of the crust by one order of magnitude). They furthermore noted the distribution of noble gas ratios was consistent with those of salinity and Sr isotope values, outlining the zones of mixing.

Tentative synthesis

The quantification of the ^{4}He flux and its application to deep groundwater studies, if based only on hydrological modeling or laboratory analogs, would give disparate answers. Nevertheless, a large of number of case studies have accumulated pertinent observations. In the following we attempt a synthesis of these, taking into account the tectonic and structural contexts of each case study.

Concerning the estimation of deep groundwater movement using noble gases:

It does not appear possible to assume a priori a He accumulation rate in aged groundwaters. The studies presented in this section show that the ^{4}He accumulation rates are highly variable, from virtually no external ^{4}He contribution required, to ^{4}He accumulation rates apparently exceeding the ^{4}He flux from the whole continental crust. Hence the expectation that the ^{4}He flux from the crust into aquifers could be used to estimate groundwater residence time is not valid.

Noble gases studies suggest that flow rates in basins are discontinuous and variable in time and space. For example, the water velocity in the Triassic aquifer of the Paris basin decreases drastically with distance from the recharge area and therefore with depth (Dewonck et al. 2001). Stute et al. (1992) also noted a similar decrease of water movement with depth in the case of the Great Hungarian Plain. It has also been suggested that deep fluid movement is controlled by tectonic events affecting the basins (Ballentine et al. 1991, 2001; Pinti and Marty 1995).

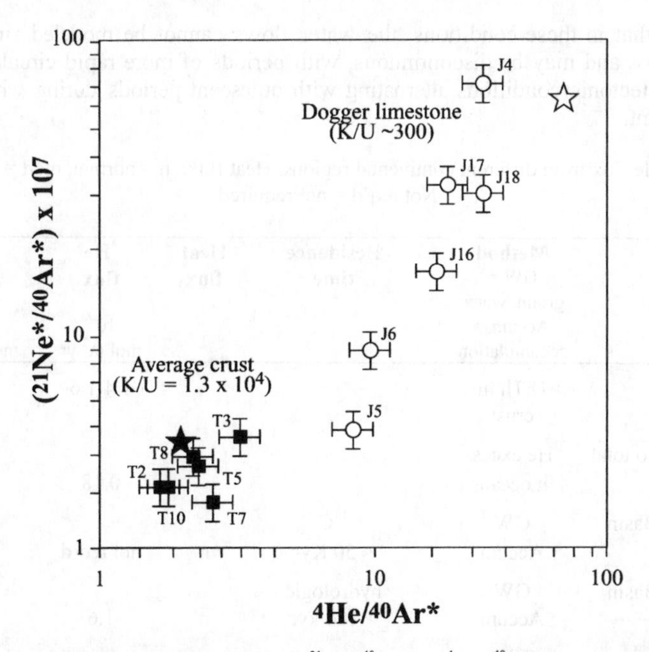

Figure 28. Relationship shown between $^{21}Ne^{*}/^{40}Ar^{*}$ and $^{4}He^{*}/^{40}Ar^{*}$ in groundwaters from the Dogger (open circles) and the Trias (black squares) aquifers, Paris Basin, France (after Pinti & Marty 1995). The average crustal production and the Dogger limestone production values are indicated by filled and open stars, respectively. The Trias values can be accounted for by contribution of crust-derived noble gases, whereas the Dogger values illustrate cross-formational fluid circulation allowing mixing between Trias and Dogger groundwaters.

It is possible in some cases to extend the geochronology of groundwaters beyond the range of ages addressed by ^{14}C using ^{4}He accumulation. At present this requires calibration by an external constraint on water residence time (e.g., stable isotopes, noble gas paleotemperatures) or a simple hydrological model that enables the assumption of continuity in the groundwater flow to be made. A more detailed understanding of the processes that control this calibration have the potential to increase the utility of this dating tool.

Unique information about groundwater residence times and cross formational flow can be derived from noble gases. The occurrence of ^{3}He excesses that cannot be accounted for by production in the crust or in basins and are attributed to a mantle origin demonstrate without ambiguity the occurrence of cross formational flows. $^{4}He/^{21}Ne^{*}/^{40}Ar^{*}$ ratios close to the production values found in shallow aquifers can hardly be established *in situ* because it would require temperatures much in excess of those prevailing in these aquifers (Ballentine et al., 1994; Ballentine and Burnard 2002, this volume). They instead provide evidence for a contribution of a fluid component coming from much deeper regions of the basins or of the crust.

Noble gas paleotemperatures coupled with stable isotope data indicate that basins can host waters for geologically significant periods of time. In two studies of Northern Poland (Zuber et al. 1997) and the Paris basin (Pinti et al. 1997), stable isotope data and NGT up to 25 °C require waters in deep aquifers to have been recharged in much warmer periods than the Quaternary, implying residence times of several Ma at least. This would translate to groundwater (Darcy) velocities on the order of a cm/yr. It is probable that in these conditions, the water flows cannot be modeled simply using Darcy's law and may be discontinuous, with periods of more rapid circulation under favorable tectonic conditions alternating with quiescent periods during which waters are stagnant.

Table 7. He flux from different continental regions. Heat flux: n = normal, mod = moderate. Not req'd = not required.

Area	**Method** GW = groundwater Accum = accumulation	**Residence time**	**Heat flux**	**^{4}He flux** 10^{-6} mol/m^2.yr	**^{3}He flux** 10^{-12} mol/m^{2}yr	**Ref.**
continents	U,Th in crust			1.4-1.6	0.03	1-3
global (relative to total Earth's surface)	^{3}He excess in oceans			0.18	2	3,4
Great Artesian Basin Australia	GW Accum	^{14}C < 50 Kyr	n	not req'd	-	2
Great Artesian Basin Australia	GW Accum	hydrologic > 100 kyr	n	1.6	0.1	2
Auob sanstone Namibia	GW Accum	^{14}C	n	0.4-1.5	-	2,5
Northern Taiwan	^{3}He/^{4}He in gas wells	-	mod	1.3-1.4	2-4	6
Great Hungarian Plain Hungary	GW Accum	^{4}He gradient discharge area	high	0.03-0.24	0.09-0.26	7
Great Hungarian Plain Hungary	GW Accum	Total water discharge X ^{4}He	high	4.1	3.8	8
Molasse basin Upper Austria	GW Accum	geochemistry	n	~ 0.1	not reported	9
Molasse Northern Switzerland	GW Accum	noble gas mass balance	n	not req'd	not req'd	10
Paris basin, centre France	GW Accum	multi-layer model ^{14}C in Albian	n	0.2	0.02	11
Paris basin, centre France	GW Accum	stable isotopes NGT	n	< 0.2	< 0.02	12,13
Paris basin, centre France	GW Accum	2-D model noble gas calib.	n	2	0.2	14
Eastern Paris basin Lorraine, France	GW Accum	^{14}C	mod	2-8	1-3	15

1) O'Nions and Oxburgh 1983; 2) Torgersen and Clarke 1985; 3) Torgersen and Clarke 1987; 4 : Craig et al. 1975; 5) Heaton 1984; 6) Sano et al, 1986; 7) Stute et al. 1992; 8) Martel et al. 1989; 9) Andrews 1985; 10) Tolstikhin et al. 1996; 11) Marty et al. 1993; 12) Pinti and Marty, 1995; 13) Pinti et al. 1997; 14) Castro et al. 1998; 15) Dewonck et al. 2001

Concerning the magnitude of noble gas fluxes in the continental crust and through basins **(Table 7):**

Estimating the flux of He isotopes using groundwater requires us to have constraints on groundwater residence times. The best of these is ^{14}C groundwater dating but is generally limited to groundwater recharge areas and enables the derivation of He fluxes only for these areas (Dewonck et al. 2001; Heaton 1984; Torgersen and Clarke 1985). Hydrologic (Darcy) ages are often not applicable because variations in hydrolic conductivity along the flow path cannot be quantitatively estimated. Independent constraints such as stable isotope signatures or noble gas paleotemperatures can help in constraining water residence time in a qualitative way (Pinti et al. 1997; Zuber et al. 1997). Volumetric estimates of groundwater discharges (Martel et al. 1989; Stute et al. 1992) or $^{3}He/^{4}He$ isotope gradients with depth (Sano et al. 1986) offer promising opportunities to derive quantitative estimates of He isotope fluxes.

The magnitude of He flux from the continental crust is related to the volcano-tectonic setting. ^{3}He and ^{4}He fluxes tend to be higher in areas having experienced Tertiary and Quaternary volcanism and/or being affected by active tectonics such as rifting and/or stretching (Rhine graben; eastern margin of the Paris Basin; Dewonck et al. (2001), Taiwan; Sano et al. (1986), Pannonian basin; Martel et al. (1989), Ballentine et al. (1991), Stute et al. (1992)). In these areas, not only does ^{3}He degassing underline the occurrence of magmatism at depth, but also enhanced degassing of crustal ^{4}He takes place at rates that are comparable to, or higher than, that representing degassing of the whole continental crust (Sano et al. 1986; Ballentine et al. 1991, Dewonck et al. 2001).

The degassing flux is not continuous as previously proposed. High ^{4}He fluxes from active regions suggest that thermal and tectonic events allow degassing of radiogenic ^{4}He stored in the crust during quiescent periods. Crustal degassing is therefore intermittent and not in a steady state. Concerning He isotope fluxes in stable continental areas, the situation is less clear. Whereas studies of the Auob sandstone, Namibia, and the Great Artesian basin, Australia, concluded that a ^{4}He flux similar to that produced by the whole continental crust was required, those of the Paris basin in its central part led to contrasting conclusions, requiring either limited ^{4}He flux (0.1 times or less than the whole crust one, (Marty et al. 1993; Pinti and Marty 1998)), or similar to the latter (Castro et al. 1998b), depending on the approach adopted. The high ^{4}He fluxes for the GAB and the Paris basin were derived mostly from estimates of hydrologic parameters combined with noble gas measurements, whereas the low flux of the Paris basin was derived from independent geochemical arguments (^{14}C dating, stable isotope and noble gas paleotemperatures). For deep groundwater in southern Poland, a flux one order of magnitude lower than the whole crust flux was derived. For aquifers in the peri-Alpine Molasse, a low flux was derived from water geochemistry constraints (Andrews et al. 1985b) or no flux was required from a noble gas mass balance approach (Tolstikhin et al. 1996). For these basins, the ^{3}He flux from the mantle was found to be two orders of magnitude lower than that estimated for global mantle degassing.

This area of research certainly deserves more investigation, given its importance in the fields of noble gas geochemistry and of groundwater research. It can be anticipated that the increasing shortage in water resources and the underground waste disposal programs will promote well-focused case studies in the not-far future.

MAGMATIC FLUIDS IN THE CRUST

Mantle degassing in the continental crust: The noble gas imprint

Noble gases in the mantle are trapped in minerals and, without magma generation and transport, could not reach the Earth's surface because diffusion, even at mantle temperatures, does not enable significant transport distances over geological time (Ballentine and Burnard 2002, this volume). It is likely that the occurrence of mantle-derived ^3He in basin fluids indicates the generation and the degassing of magmas at depth (Oxburgh et al. 1986).

In fluids present in the continental crust, the isotopic ratio of helium varies widely between the isotopic composition of radiogenic helium (~0.02 Ra, Ballentine and Burnard 2002, this volume) and that of mantle helium (8Ra for the upper mantle, Graham 2002, this volume). $^3He/^4He$ ratios higher than 0.1 Ra provide strong evidence for the presence of mantle-derived fluids. $^3He/^4He$ ratios between 0.02 Ra and 0.1 Ra are more ambiguous to interpret. He isotope fractionation may occur during helium transfer from minerals to rocks, although a detailed study of the Carnmenellis granite, S.W. England has indicated that ^4He is released into crustal fluids preferentially to ^3He (Martel et al. 1990). Because ^3He is produced by the thermal neutron activation of ^6Li which produces ^3H and then ^3He, the $^3He/^4He$ due to natural nuclear reactions in crustal rocks is a direct function of the Li content of rocks and minerals (Ballentine and Burnard 2002). In order to get $^3He/^4He$ ratios in the range 0.1 Ra in the producing rocks, several hundreds of ppm Li are required, which represent exceptional situations for continental rocks. Consequently, $^3He/^4He$ ratios higher than 0.02 Ra may also be interpreted as representing addition of mantle-derived He (Marty et al. 1993).

It has become clear in recent years that the isotopic composition of helium, dissolved in continental groundwaters or other near-surface fluids, can vary markedly from place to place. This variation appears to correlate with geological setting: stable regions are characterized by helium generated by radioactive decay processes within the continental crust while that collected in zones of active extension or young volcanism is everywhere marked by the presence of a mantle-derived component (e.g., Hooker et al. 1985b, Mamyrin and Tolstikhin 1984; Martel et al. 1989; Sano and Wakita 1985). A review of the relationship between helium isotopic ratios and tectonic settings can be found in Polyak and Tolstikhin (1985). Two recent studies have addressed the relationship of helium isotope ratios with tectonics in the continental crust. Polyak et al. (2000) have shown that there existed for the western Caucasus a correlation between the heat flow and $^3He/^4He$ ratios of fluids. Such relationship can be understood as two different consequences of a common cause: the development of rifting and mantle decompression beneath a thinned continental crust, which ultimately allows partial melting and mantle degassing. Without magma generation, it would be impossible for the mantle to lose its volatile elements since the process of even grain boundary diffusion is too slow to enable quantitative extraction of noble gases (Ballentine and Burnard 2002, this volume).

Another example of a relationship between He isotopes and tectonics is given by the western Alps and their adjacent regions (Marty et al. 1992). It is evident from Figure 29 that mantle helium degassing occurs in the crust peripheral to the Alpine orogenic belt and that the occurrence of mantle-derived helium on an European wide scale coincides approximately with the recent volcanic provinces around the Alps. Furthermore, helium presenting radiogenic isotopic ratios is often found in nitrogen-rich gases of presumably crustal origin whereas He enriched in mantle-derived ^3He is associated with CO_2 having an isotopic composition typical of mantle-derived carbon ($\delta^{13}C$ around -5‰). Thus the isotopic composition of helium in natural gases and waters enables the unambiguous determination of the areas where mantle degassing is taking place through the continental crust. The coincidence between the geographic distribution of He isotopic ratios and the recent volcanic provinces around the Alps is consistent with the view that the release of deep-seated gases occurs in regions where partial melting has taken place (Oxburgh et al.

1986). In these cases, mantle-derived helium and undoubtedly other volatiles released during melting are emplaced in the crust and reach the near surface. In contrast, the thickened crust of the Alpine block shows little or no evidence of mantle ^{3}He or indeed associated volcanism.

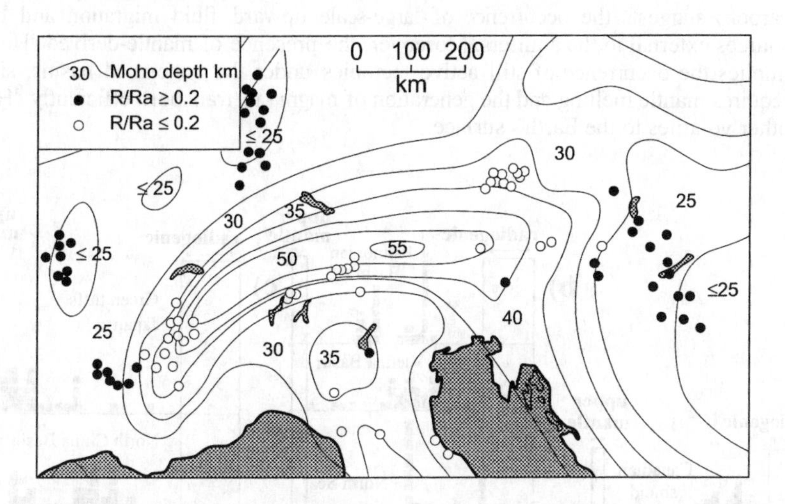

Figure 29. Distribution of $^3He/^4He$ ratios (expressed as R/Ra, where Ra is the atmospheric value) measured in fluids (mineral spings, natural gases) in the western Alps and adjacent regions (after Marty et al. 1992). Fluids sampled in the Alpine regions where the Moho depth exceeds 30 km have $^3He/^4He$ ratios dominated by crustal helium (R/Ra < 0.2) whereas fluids sampled in the peripheral regions subject to rifting, Tertiary to Quaternary volcanism, and lithospheric thinning are characterized by the contribution of mantle-derived helium rich in ^{3}He.

On a regional scale, the area over which mantle gas escapes exceeds that of surface volcanism. For example, mantle He is already apparent in the Drôme region 60 km away from the first volcanic center. Likewise, mantle He is present in the Lorraine groundwaters about 100 km west of the Rhine graben (Dewonck, 2001), which demonstrates that groundwater circulation is likely to be responsible for lateral transport of mantle-derived volatiles in the continental crust. Helium isotope data from the sedimentary basins around the Alps are consistent with the view that active extensional basins (e.g., the Pannonian basin) are associated with igneous activity and are degassing helium that has a clear mantle signature while loading basins (e.g., the Po basin) often have helium that is of crustal origin.

Because helium can be used as a sensitive tracer of the contribution of mantle-derived volatiles, its isotopic composition in basinal waters bears important information on the tectonic context having led to the formation of basins, on the origin of the heat flux through basins, and on cross formational flows within basins (e.g., Martel et al. 1989; Marty et al. 1993; O'Nions and Oxburgh 1988; Sano et al. 1986; Stute et al. 1992). Figure 30a-c represents the statistical distribution of He isotopic ratios in fluids sampled in different basins worldwide. In the case of Europe, basins developing in extensional domains such as the Pannonian Basin (Martel et al. 1989), the Vienna Basin (Ballentine 1997), or the North Sea rifting systems (Ballentine et al. 1996), show the highest mantle-^{3}He contributions, whereas a loading basin like the Po Basin in the foreground of the Alpine orogenesis contains purely radiogenic helium (Elliot et al. 1993). In many gas fields of Canada and the U.S., mantle He is found (Hiyagon and Kennedy 1992, Jenden et al. 1988, 1993; Ballentine and Sherwood Lollar 2002), and the highest mantle signals are observed in hydrocarbon fields trapped in the Green Tuffs of Northeast Japan (Sano and Wakita 1985).

- The detection of mantle-derived helium in sedimentary fluids (water and hydrocarbon) suggests the occurrence of large-scale upward fluid migration and helium sources external to the aquifers. Moreover, the presence of mantle-derived ^{3}He often implies the occurrence of still active tectonics under the concerned basins, since it requires mantle melting and the generation of magma to transport efficiently ^{3}He and other volatiles to the Earth's surface.

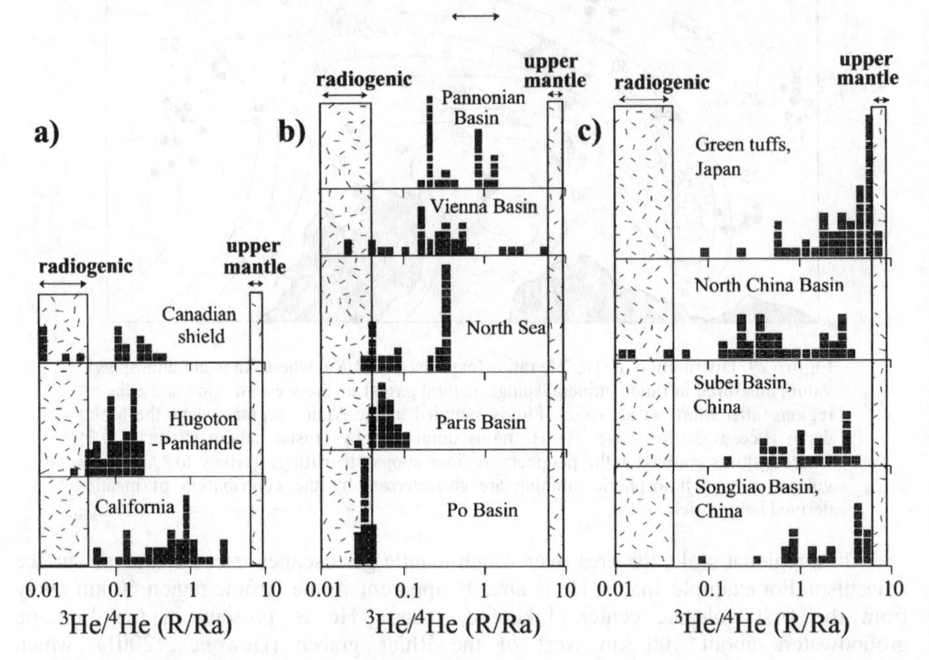

Figure 30. Histograms showing the distribution of ^{3}He/^{4}He ratios measured in fluids (hydrocarbons, groundwaters) sampled in various sedimentary basins worldwide. a - Europe; b-North America; c - China and Japan (see relevant references in the text). In most cases ^{3}He/^{4}He ratios vary between a radiogenic end-member characteristic of crustal or sedimentary production, and a mantle end-member enriched in ^{3}He, suggesting the contribution of mantle-derived fluids and possibly of mantle-derived heat in some cases.

NOBLE GASES IN MINERAL DEPOSITS AND HYDROTHERMAL FLUIDS

The conservative behavior and chemical inertness of noble gases has led to their extensive use to provide information on the fluxes, movement and interactions of contemporary crustal fluids. A major advance has been in extending these studies to characterize ancient waters in fluid inclusions (Turner et al. 1993). The principal aims of these studies are to provide data on the noble gas geochemistry of inclusion fluids and understand the observations in terms of likely sources and interactions. This information provides novel insight into a palaeofluid's evolutionary history and can assist in discriminating between different models for the formation of hydrothermal mineral deposits. Because of the large differences in end-member isotopic and elemental compositions, He and Ar are the most informative and widely studied noble gases in

Table 8. Important nuclear reactions for Ar-Ar methodology.

Reaction[a]	*Approximate conversion: element/product (mole/mole)*[b]	*Approximate detection limit of parent element (g)*[c]
$^{37}Cl(n,\gamma)^{38}Cl(\beta)^{38}Ar$	10^{-6}	10^{-10}
$^{39}K(n,p)^{39}Ar$	10^{-7}	10^{-9}
$^{40}Ca(n,\alpha)^{37}Ar$	10^{-7}	10^{-9}
$^{79}Br(n,\gamma)^{80}Br(\beta)^{80}Kr$ $^{81}Br(n,\gamma)^{82}Br(\beta)^{82}Kr$	10^{-5}	10^{-12}
$^{127}I(n,\gamma)^{128}I(\beta)^{128}Xe$	10^{-6}	10^{-12}
^{235}U(n, fission)Kr and Xe isotopes[d]	10^{-5}	10^{-12}

a) All products listed stable except ^{39}Ar ($t_{½}$ = 269 days) and ^{37}Ar ($t_{½}$ = 35.04 days)

b) Conversion factors depend upon isotopic abundance of parent isotope, fast and thermal neutron flux and neutron cross sections. Approximate values shown are for typical fast and thermal neutron fluxes of 10^{18} and 10^{19} cm^{-2} respectively.

c) Detection limits based upon 10× typical blank or background level product noble gas isotope in MS1 mass spectrometer (University of Manchester).

d) Detection limit of ^{235}U based upon production of $^{134}Xe_U$.

mineralizing fluids (Simmons et al. 1987; Stuart and Turner 1992; Stuart et al. 1995; Burnard et al. 1999). Furthermore, the use of ^{40}Ar-^{39}Ar methodology provides a means to combine noble gases with a range of geochemically important elements in the inclusion fluids including K, Ca, Cl, Br and I (Table 8). In a series of groundbreaking studies, Turner and co-workers showed how ^{40}Ar-Cl and K-^{36}Ar correlations could be used to disentangle mixtures of Ar components in order to determine the origin of the fluids, the source of salinity and the age of mineralization (Kelley et al. 1986; Turner 1988; Turner and Bannon 1992). The method was soon extended to include measurements of Br, I and noble gas isotopes of Kr and Xe (Bohlke and Irwin 1992a). Halogens (Cl, Br and I) are particularly important constituents of crustal fluids. They can be used to fingerprint individual fluid sources, but are especially useful for identifying modification processes such as seawater evaporation, halite precipitation and interaction with organic-rich or evaporite-bearing units. Moreover, halogens do not exhibit changes in elemental ratios as the result of vapor phase separation (e.g., boiling), a mechanism whereby noble gas concentrations and elemental ratios can become strongly modified.

The main difficulty with applying noble gas studies to ancient fluids has been: (1) the presence of atmospheric noble gases as a ubiquitous contaminant; and (2) assessing modifications to noble gas isotopes by post-entrapment processes including *in situ* production and diffusion in the inclusion fluids and host minerals. Lower abundance and higher mobility means that post-entrapment processes are more serious for He than Ar. A persistent and seemingly intractable problem with noble gas studies of hydrothermal fluids is the presence of modern-day atmospheric noble gases in the samples. Atmospheric gases are either adsorbed on the mineral surfaces, or sealed in voids and cracks within mineral grains. This problem can be partially alleviated by appropriate choice of noble gas extraction technique and most often *in vacuo* crushing, or laser microprobe to target groups of related inclusions, are employed to preferentially extract gases from fluid inclusions (Bohlke and Irwin 1992a, Kendrick et al. 2001a). However, the difficulty remains of resolving a mixture of atmospheric and air saturated water (ASW) noble gases.

Post-entrapment modification of He and Ar isotopes

He and Ar isotopes that are trapped within minerals can be modified in one of four ways and must be either ruled out as significant sources or a correction applied before interpretation can be made:

(1) U, Th and K may be present in solution or in daughter minerals within fluid inclusions. ^{4}He and ^{40}Ar produced from decay since mineral precipitation may therefore be released with the inclusion fluids. The production of ^{4}He and ^{40}Ar can be calculated using Equations (13) and (19) in Ballentine and Burnard (2002, this volume), respectively. Alternatively, if the proportion of fluid-derived ^{4}He and ^{40}Ar can be determined, then the excess ^{40}Ar or ^{4}He may be used to calculate a deposit age.

(2) Radioactive decay of U and Th and their radioactive daughter products within the lattice of a sample produces ^{4}He, which may diffuse or recoil (~20 μm) into fluid inclusions. The amount of He entering fluid inclusions will depend on the inclusion size distribution, the distribution of the parent elements and on the He diffusion processes within the host mineral, all of which are difficult to quantify. Shorter recoil lengths and lower diffusivity make this mechanism unimportant for ^{40}Ar. Significant volume diffusion of He in host minerals can be identified by progressively lower ^{3}He/^{4}He ratios during sequential crushing consistent with smaller fluid inclusions having higher radiogenic He contents because of their higher surface area/volume (Stuart et al. 1995). In these cases it is only the ^{3}He/^{4}He of the initial crush that provides a minimum value for the fluid ^{3}He/^{4}He at the time of trapping, as this stage releases the largest inclusions. Dense, poorly cleaved sulfides (e.g., pyrite) appear to be the best retainers of noble gases and hence noble gas isotopic and elemental signatures. The incorporation of lattice trapped ^{40}Ar would be expected to create an increase in the ^{40}Ar/^{36}Ar ratio with crush step, however, decreasing contamination of air or the rupturing of increasingly primary inclusions would also have the same effect.

(3) Nucleogenic production of ^{3}He dominantly via the reaction Li(n,α)^{3}H(β)^{3}He (Morrison and Pine 1955). The principal source of neutrons is from (α,n) reactions on light elements. The α-particles are supplied from decay of U and Th and decay products. Production of ^{3}He is controlled by the Li content (Mamyrin and Tolstihkin 1984). The concentration of these elements can be measured within each sample and appropriate correction applied if necessary.

(4) Cosmogenic He is produced in the uppermost 1 m of the Earth's surface with a ^{3}He/^{4}He of about ~0.1 and may be a very important post-entrapment mechanism. Therefore, it is essential that samples are mined at depth and have spent very little (if any) time at the Earth's surface.

The effects of these modification mechanisms will obviously increase with sample age and with lithophile element content. Previous studies have shown that the post-entrapment production of ^{3}He and ^{4}He within sulfides (pyrite, sphalerite and galena) is insignificant when compared to the measured values. Thrower (1999) studied sulfides from deposits of the Colorado mineral belt (60-15 Ma) and determined post-entrapment isotope production to be <0.6% in all cases which was well below the error for each measurement, making corrections unnecessary. However, in studies of older deposits or of samples containing a high lithophile element content, the post-entrapment production of these isotopes should always be considered.

Noble gas mixtures in mineralizing fluids

Noble gases in a hydrothermal fluid are likely to be a mixture of at least three components: (1) air present as modern and/or palaeoatmospheric gases, or dissolved in air-saturated water (meteoric, seawater and sedimentary formation waters); (2) mantle; and (3) crustal sources. Air-saturated waters associated with mineral deposit formation are generally heated to some extent during mineralization. This produces a modified ASW with noble gas contributions from the atmosphere (Ar) and from the crust (He ± Ar) by chemical leaching or diffusion. These low-temperature fluids generally represent late stage and peripheral fluids. Similarly, no pure mantle noble gas isotope signature has been isolated in ore deposit studies (except modern ocean floor sulfides; Stuart et al. 1994). Instead, fluids exhibiting a dominantly mantle signature but containing a crustal component often characterize a 'magmatic' end-member. This reflects addition of crustal volatiles to a mantle component prior to mineralization (Burnard et al. 1999). Physical processes including boiling, formation of a separate gas phase (e.g., CO_2 and H_2O) or unmixing of immiscible fluids can further modify the signatures of these end-member components. These processes will result in elemental fractionation of noble gases (and noble gas-halogen ratios) but negligible isotopic fractionation or change in halogen ratios (Br/Cl or I/Cl).

Description and analysis of multi-component noble gas mixtures in ore fluids

The procedure by which multi-component mixtures of noble gas (He and Ar) isotopes and halogens can be represented in terms of 2 and 3 dimensional linear mixing diagrams is described in this section. The treatment is not intended to be exhaustive but to illustrate how potentially useful pieces of mineral deposit information can be extracted from the data.

Magmatic and ASW components. Two-dimensional plots of He and Ar isotopes of ore minerals can be used to investigate the mixing relationships between fluid end-members. A plot of ^{3}He/^{4}He *versus* ^{40}Ar/^{36}Ar in Figure 31, shows mixing between magmatic and low-temperature modified ASW components for deposits from the Ailoshan Gold Province, China (Burnard et al. 1999). Model defined mixing curves can be constructed using the equations of Langmuir et al. (1978) and are dependent on the contribution of each of the two components present within the fluid. The curvature of the line (r) is given by

$$r = {}^4\text{He}/{}^{36}\text{Ar}_{\text{(magmatic)}} / {}^4\text{He}/{}^{36}\text{Ar}_{\text{(modified ASW)}} \quad (51)$$

Because ^{36}Ar is concentrated within the meteoric component, r values for mixing curves are large and typically ~1200. Generally, available data fall into one of two categories: (1) a constant ^{3}He/^{4}He value while exhibiting increasing ^{40}Ar/^{36}Ar ratio (Daping deposit, Fig. 31); or (2) showing increasing ^{3}He/^{4}He values with constant ^{40}Ar/^{36}Ar (Zhenyuan and Mojiang deposits, Fig. 31). The latter data show only a vertical section of the mixing curve and a constant ^{3}He/^{4}He value is not attained. In such cases one could use a typical r-value for type 1 data (1200) and model a minimum magmatic ^{3}He/^{4}He value. We show later how these binary mixtures can be resolved into ASW, crustal and mantle end-members using 3-D mixing diagrams.

Further insight into the magmatic component can be gained from the ^{40}Ar*/Cl ratio obtained from irradiated samples (Kelley et al. 1986; Turner and Bannon 1992). Data from quartz vein samples from the Globe Miami porphyry copper deposit are plotted on a ^{40}Ar/^{36}Ar *versus* Cl/^{36}Ar diagram (Fig. 32, Kendrick et al. 2001b). The correlation

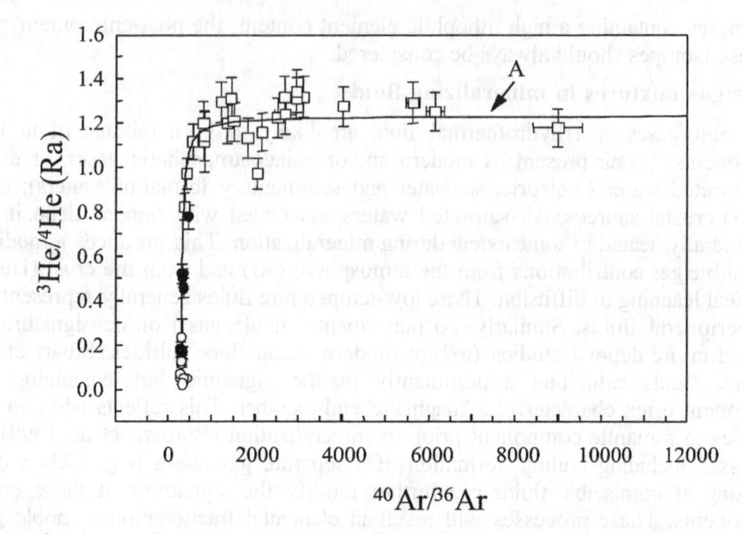

Figure 31. $^3He/^4He$ vs. $^{40}Ar/^{36}Ar$ for the Ailaoshan gold deposits modelled as a binary mixture between two fluids. The magmatic fluid has high $^3He/^4He$ (~1.23 Ra) and high $^{40}Ar/^{36}Ar$ (>8000) and is formed by mixing of mantle and crustal components prior to mineralization. The modified ASW end-member has low $^3He/^4He$ (0.01 Ra) and low $^{40}Ar/^{36}Ar$ (≥296) formed by mixing of atmosphere (Ar) and crustal He most likely during heating and interaction of meteoric fluid heated during mineral deposit formation. The mixing line A is constructed by making assumptions about the end-member He/Ar ratio and the $^{40}Ar/^{36}Ar$ value of the mantle component. The line has curvature, r = 1400 (modified from Burnard et al. 1999).

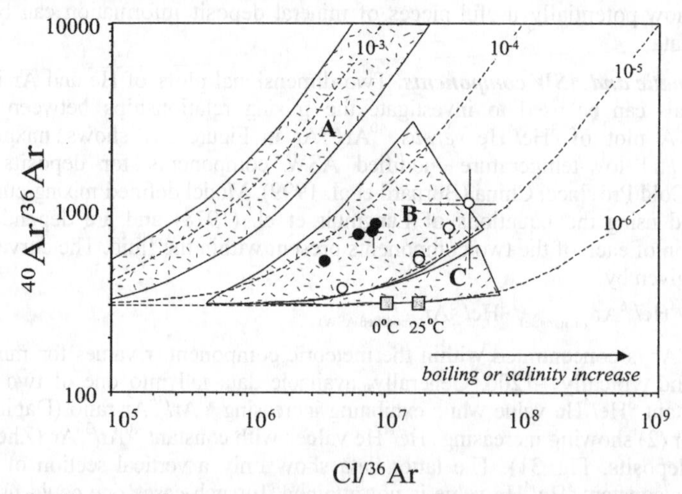

Figure 32. $^{40}Ar/^{36}Ar$ vs. $Cl/^{36}Ar$ for vein quartz samples from the Globe Miami copper prophyry deposit, Arizona, USA. Solid symbols - quartz samples from the inner zone of the deposit; open symbols - quartz samples from the outer zone of the deposit. Data from both zones are compatible with mixing of a low salinity meteoric end member (high ^{36}Ar) and a high salinity magmatic end member (high ^{40}Ar and Cl). Samples from the inner zone are characterized by having a higher magmatic component (higher $^{40}Ar/Cl$ value) compared to those from the outer zone. Data fields shown for comparison: (A) Mantle fluids in diamonds (Johnson et al. 2000; Burgess et al. 2002); (B) Copper porphyry deposits (Kendrick et al. 2001b, Irwin and Roedder 1995); (C) Crustal fluids (Kelley et al. 1986; Bohlke and Irwin 1992b,c, Turner and Bannon 1992; Burgess et al. 1992; Polya et al. 2000; Gleeson et al. 2001; Kendrick et al. 2002a).

indicates mixing between a high salinity magmatic fluid and low salinity meteoric fluid. The magmatic component in Figure 32 has elevated $^{40}Ar/^{36}Ar$ and a $^{40}Ar^{*}/Cl$ given by the slope of the correlation. Samples from interior zones of the deposit are dominated by the magmatic component with $^{40}Ar^{*}/Cl$ up to 57×10^{-6}. In contrast outer zones of the deposit are dominated by ASW and have lower $^{40}Ar^{*}/Cl$ of 19×10^{-6}. Wide variations exist in $^{40}Ar^{*}/Cl$ compositions of fluids enabling likely sources to be identified. Mantle fluids in diamond show high $^{40}Ar^{*}/Cl$ values in the range $550\text{-}1350\times10^{-6}$ (Johnson et al. 1999; Burgess et al. 2002). In contrast, typical crustal fluids exhibit $^{40}Ar^{*}/Cl$ values between 30 and 2,500 times lower ($0.6\text{-}5\times10^{-6}$; Turner and Bannon 1992) while $^{40}Ar^{*}/Cl$ of porphyry copper deposits are intermediate. Higher $^{40}Ar^{*}/Cl$ are predicted for metamorphic fluids which may acquire high concentrations of excess ^{40}Ar by thermal release from minerals without a concomitant release of Cl.

The effects of atmospheric Ar can be examined using the $Cl/^{36}Ar$ ratio. The concentration of atmospheric Ar can be calculated if absolute abundances of chlorine in the inclusions can be determined by an independent method. The main limitation to this approach is in the accuracy with which the bulk salinity of the fluid inclusion can be obtained. This is done using thermometric analyses of individual fluid inclusion types combined with a qualitative assessment of their abundance. The concentration of ^{36}Ar in inclusion fluids can be considerably above ASW values (10^{-6} cm^3/g H_2O) probably reflecting the presence of trapped modern atmosphere. Furthermore, it is difficult to resolve air and ASW mixtures because of their similar isotopic compositions. Elemental noble gas ratios are important for partially distinguishing these components, for example, the low salinity end-member ($Cl/^{36}Ar \leq 10^6$) in the Globe Miami porphyry copper deposit (Fig. 32) has $^{40}Ar/^{36}Ar \sim 296$, but $^{84}Kr/^{36}Ar$ are between the values for air (0.02) and ASW (0.04) indicating it is a mixture of meteoric water and air (see Fig. 6 in Kendrick et al. 2001b). The present data do not enable distinction between modern-day air and palaeoatmospheric Ar, and it is possible that both are present in the samples.

The effect of boiling. Evolution of a hydrothermal fluid may involve boiling and gas loss, this leads to fractionation of noble gases from Cl, but does not affect the isotopic signature. After boiling the fluid has higher salinity but is depleted in noble gases that are preferentially partitioned into the vapor phase. An interesting example is Bingham Canyon copper porphyry deposit, Utah which contains a high proportion of vapor phase inclusions and low ^{36}Ar concentration in the fluid of $0.2\text{-}0.6\times10^{-6}$ cm^3/g, below that of ASW (Irwin and Roedder 1995; Kendrick et al. 2001b). The Kr/Ar and Xe/Ar ratios of these fluids are significantly higher than ASW and $Cl/^{36}Ar$ are an order of magnitude higher than in similar porphyry copper deposits from Arizona (Kendrick et al. 2001b). These characteristics are most easily explained as the effects of boiling (Kendrick et al. 2001b), rather than being indicative of a magmatic fluid (Irwin and Roedder 1995).

Radiogenic nobles gases ($^{40}Ar^*$ and ^4He). The effects of atmosphere and ASW can be removed in order to investigate the relative contributions of the crustal and mantle components by considering only the radiogenic noble gases. The mantle beneath the crust has $^{40}Ar^*/^4He = 0.5$ and $R/R_a = 6\text{-}8$, while the crustal values are $^{40}Ar^*/^4He = 0.2$ and $R/R_a = 0.01\text{-}0.05$. Thus, mixing between crust and mantle components will form linear mixing trends on a plot of $^3He/^4He$ *versus* $^{40}Ar^*/^4He$ (Figs. 12, 33). However, fractionation of He and Ar will produce a horizontal trajectory on the figure. Extrapolation of mixing trends to known $^3He/^4He$ values for each end-member can give $^{40}Ar^{*}/^4He$ values or fractionation state of a given end-member. Available data indicate that Ar and He in both mantle and crustal end-members are fractionated. Mississippi Valley-type (MVT) deposits from the

Pennine orefield, UK, show low $^3He/^4He$ and $^{40}Ar^*/^4He$ values consistent with models of formation from basinal brines in which a mantle component is absent (Stuart and Turner 1992; Kendrick et al. 2002a,b). The $^{40}Ar^*/^4He$ values are lower than the crustal value consistent with chemical leaching and incorporation of crustal He by low temperature fluids (Fig. 33). This is supported by fluid inclusion homogenization temperatures that are mostly between 90-200°C (Kendrick et al. 2002a and references therein), below the closure temperature for Ar in most crustal minerals (McDougall and Harrison 1999). In contrast, fluids associated with porphyry copper deposit mineralization, intrusion-related hydrothermal systems where magmatic fluids are invaded and progressively dominated by meteoric water, show magmatic $^3He/^4He$ ratios, but have $^{40}Ar^*/^4He$ above the mantle production ratio (Fig. 33). The intrusion will have a high temperature thermal front associated with it, which may be capable driving ^4He from the crustal basement rocks more efficiently than ^{40}Ar; assimilation of this crust will lead to a high $^{40}Ar^*/^4He$ ratio. Other deposit types show mixing trends between the fractionated crust and mantle end-members (Fig. 33).

Combined with the Cl abundance, the $^{40}Ar^*/Cl$ ratio obtained from irradiated samples can be used to calculate the concentration of excess ^{40}Ar in the inclusion fluids. After appropriate correction for radiogenic ^{40}Ar formed by *in situ* decay of K in the inclusions, the remainder is excess ^{40}Ar carried by the brines from which the host mineral was precipitated. The concentration of excess ^{40}Ar is useful as a measure of the extent of fluid-rock interaction (Kelley et al. 1986), residence time of the fluid in an aquifer (Turner and Bannon 1992) and for evaluating fluid circulation patterns in sedimentary basins (Kendrick et al. 2002a).

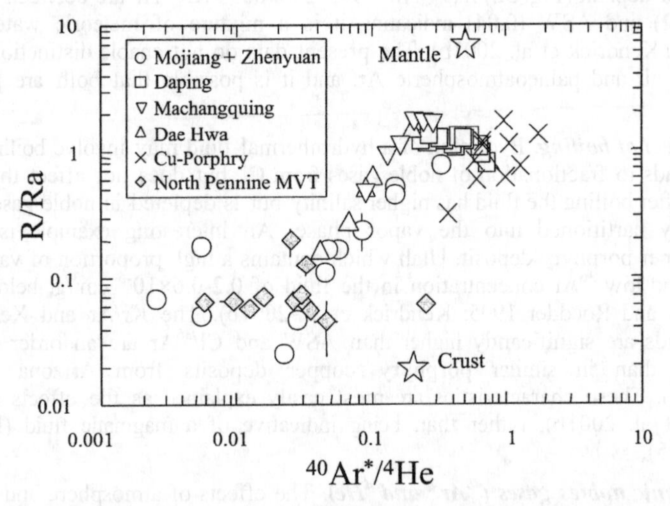

Figure 33. $^3He/^4He$ vs. $^{40}Ar^*/^4He$ for inclusion fluids in different ore deposits. Estimates of the $^{40}Ar^*/^4He$ ratios for the crust and mantle are plotted on the figure. The data in most deposits are compatible with mixing between a high $^3He/^4He$ fluid having a mantle-like $^{40}Ar^*/^4He$ and a fluid enriched in ^4He. It is assumed that this fluid is low temperature ASW capable of releasing ^4He (but not $^{40}Ar^*$) during circulation in the crust. Data sources: Mojiang, Zhenyuan and Daping (Ailaoshan Au deposits, China; Burnard et al.1999); Machangquing Cu deposit, China (Hu et al. 1998); Dae Hwa W-Mo deposit, South Korea (Stuart et al. 1995); Cu-porphyry deposits, USA (Kendrick et al. 2001b); Pennine MVT deposits (Stuart and Turner 1992; Kendrick et al. 2002a).

Age of mineralization. Radiogenic ^{40}Ar arises from *in situ* decay of K within inclusions and can, in principle, be used to calculate a K-Ar age. In practice, the most accurate ages have come from radiogenic ^{40}Ar located in small authigenic micas or K-feldspar trapped in the fluid inclusions (Kelley et al. 1986; Turner and Bannon 1992; Irwin and Roedder 1995; Qui 1996; Kendrick et al. 2001a). Dating of quartz veins that host the mineralization is advantageous as it is likely to give a more accurate mineralization age than isotopic dating of host rock alteration minerals. A successful approach to dating quartz vein samples has been used by Kendrick et al. (2001) whereby *in vacuo* crushing initially releases excess ^{40}Ar, followed by stepped heating to preferentially release radiogenic ^{40}Ar and K-derived ^{39}Ar from solid inclusions. When plotted on a ^{40}Ar/^{36}Ar-Cl/^{36}Ar-K/^{36}Ar diagram the data define a mixing plane (Fig. 34). Intersection of the mixing plane with the ^{40}Ar/^{36}Ar-K/^{36}Ar axes gives the ^{40}Ar/K ratio and hence K-Ar age of the mineralization.

Further developments

3-D characterization of He and Ar in ore-deposit systems. Ore fluids are multicomponent mixtures that can be analyzed using 3D linear mixing diagrams in which, linear mixing forms planes and trends that are constrained by data. A useful advantage of this approach is in describing the changing composition of a fluid composition during mineralization reflecting the progressive change in dominance from one fluid type to the next (e.g., from mantle to crust or ASW). Thrower (1999) has provided a method to represent and analyze the contributions and trends involving ASW, crustal and mantle components on a ^{36}Ar/^{40}Ar-^{3}He/^{4}He-^{3}He/^{40}Ar mixing diagram (Fig. 35). ^{40}Ar is used as the denominator in this diagram as it is the only isotope present in all three components. On such a three-dimensional mixing plot, each end-member is clearly distinguished (Fig. 35a). Mantle and crustal components are effectively positioned on the basal x-y plane defined by their respective ^{3}He/^{4}He and ^{3}He/^{40}Ar values. The vertical position is effectively

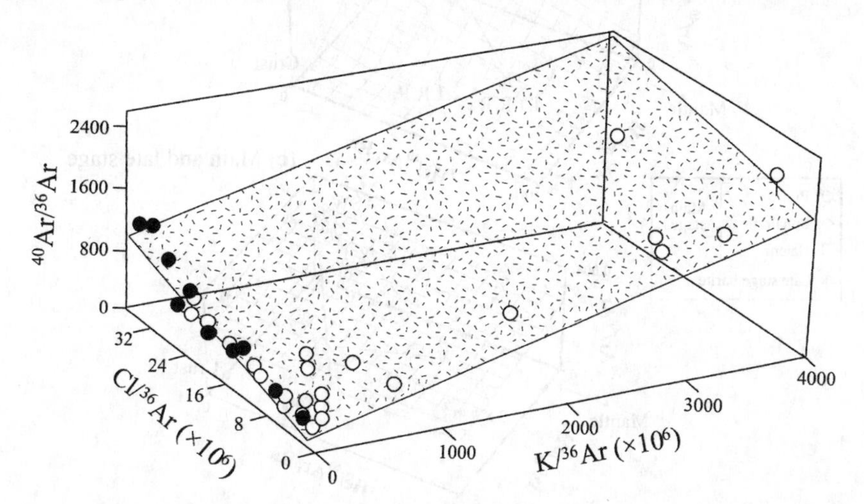

Figure 34. ^{40}Ar/^{36}Ar-Cl/^{36}Ar-K/^{36}Ar mixing diagram for combined *in vacuo* crushing and stepped heating analyses of vein quartz from the Silverbell copper porphyry deposit, Arizona, USA. Data form a plane representing a mixture of atmosphere, fluid and solid components. The orientation of the plane depends on the balance of radiogenic and excess ^{40}Ar, Cl and K. Data obtained by *in vacuo* crushing show relatively high excess ^{40}Ar and Cl contents and are aligned along the Cl/^{36}Ar axis, while data obtained by step heating has variable K/Cl, and higher K/^{36}Ar and are positioned on the plane. Projection of the plane onto the ^{40}Ar/^{36}Ar–K/^{36}Ar axes enables determination of the ^{40}Ar/K ratio and a K-Ar age of 56±2 Ma (after Kendrick et al. 2001).

controlled by ^{36}Ar content and hence indicates ASW contribution. Elemental fractionation (e.g., boiling) alters both the $^4He/^{40}Ar$ and the $^3He/^{40}Ar$ values and, since the $^{40}Ar/^{36}Ar$ value remains unchanged, results in horizontal movement from the plane.

In deposits containing contributions from all three end-members data will form a unique mixing plane for that mineralization type and also for each mineralization stage, if the fluid sources change between each stage. Figure 35 illustrates this point using data from samples representing early, main and late stage mineralization of the Black Cloud

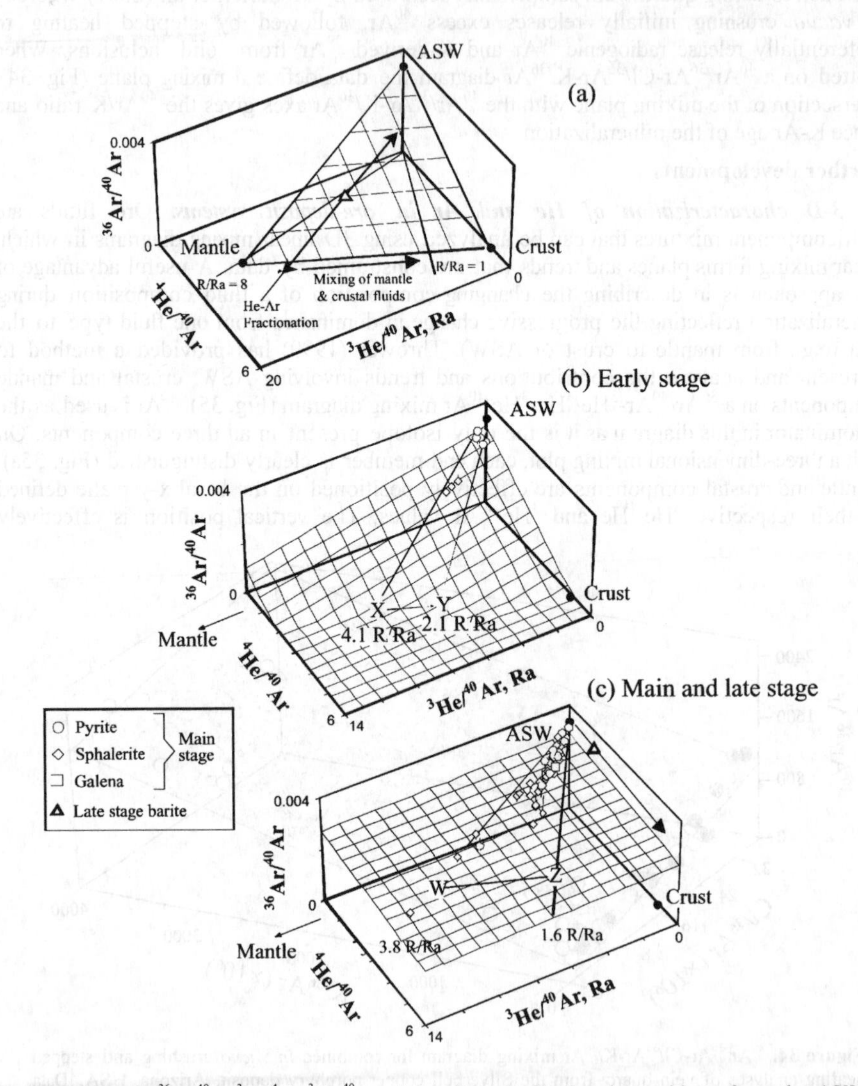

Figure 35. (a) $^{36}Ar/^{40}Ar$-$^3He/^4He$-$^3He/^{40}Ar$ mixing diagram showing the position of mantle, ASW and crustal end-members and the mixing plane between them. Data for sulfide samples are shown in (b) for early (c) for main and late stage mineralization stages of the Black Cloud replacement Pb-Zn deposit, Colorado mineral belt. End-member compositions vary between stages forming separate mixing planes. Mixing trends are evident during early (X-Y) and main (W-Z) stages (modified from Thrower 1999).

Pb-Zn deposit, Colorado mineral belt (Thrower 1999). Early stage samples formed from a magmatic fluid with $^3He/^4He$ of 4.1 Ra and ASW ($^{36}Ar/^{40}Ar$ = 0.0034). Continued addition of a crustal component during deposition further lowered the $^3He/^4He$ to 1.6 (main stage minerals) and 0.05 for the late stage barite. A more detailed explanation of the data is beyond this review but it is worth pointing out that early, main and late stage minerals lie on different mixing planes that have a progressively more horizontal aspect reflecting the increasing importance of the ASW component as mineralization proceeded.

Halogens (Br/Cl and I/Cl). In this review we have concentrated on some of the most important geological information that can be elucidated from combined noble gas-halogen studies. For separate reviews of halogens as fluid tracers see Worden (1996) and for their specific application to studies of mineral deposits see Bohlke and Irwin (1992b) and Wilkinson (2001). Our aim here is to show the complimentary nature of noble gas and halogen data using as an example, halogen data for replacement Pb-Zn deposits of the Colorado mineral belt (Fig. 36; Thrower 1999). On a plot of Br/Cl versus I/Cl (Fig. 36) data lie on a mixing trend between mantle and crustal fluid sources and provide evidence for a genetic relationship between these deposits. Furthermore, data for different stages of mineralization of the Black Cloud deposit clearly show the relative change in influence of these two fluid sources throughout ore deposition. It is noted that the latter observation is in excellent agreement with the He and Ar systematics of Black Cloud sulfide samples (see discussion above and Fig. 35). Other examples of the combined use of noble gases and halogens as fluid tracers in ore deposits include studies of Mississippi Valley Type deposits (Turner and Bannon 1992; Bohlke and Irwin 1995; Kendrick et al. 2002a,b), copper porphyry deposits (Kendrick et al. 2001b), granite-related systems (Bohlke and Irwin 1992b, Burgess et al. 1992; Turner and Bannon 1992; Polya et al. 2000; Gleeson et al. 2001), geothermal (Bolke and Irwin 1992b), and Au deposits (Bohlke and Irwin 1992b).

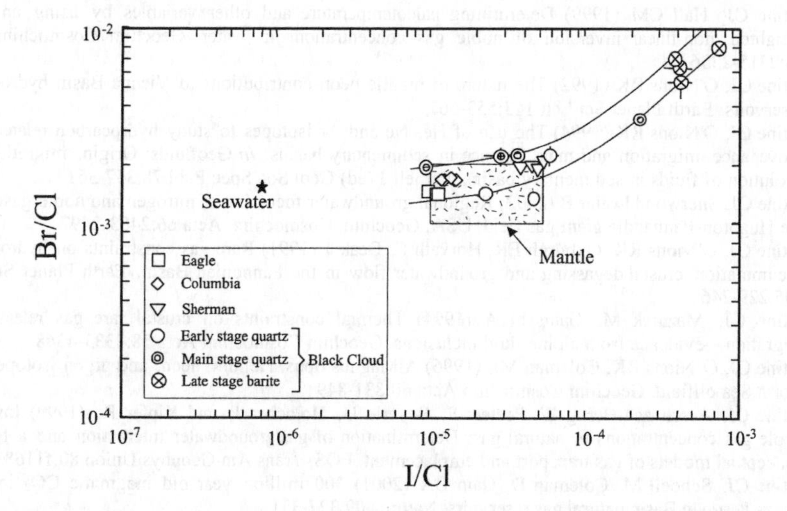

Figure 36. Br/Cl vs. I/Cl quartz and barite vein data for replacement Pb-Zn deposits of the Colorado mineral belt (Thrower 1999). The data are compatible with mixing between a mantle component similar in composition to fluids in diamond, and a crustal component with higher Br/Cl and I/Cl values. Note that Black Cloud deposit quartz and barite are co-genetic with sulfide samples shown in Figure 5. Shaded area labeled "mantle" defined by the range determined for fluids in diamonds (Burgess et al. 2002).

ACKNOWLEDGMENTS

CJB thanks Keith O'Nions, Alex Halliday, Max Coleman, Barbara Sherwood Lollar and Martin Schoell for their inspiration, support and collaboration in this field. RB thanks Clint Thrower, Mark Kendrick, Pete Burnard, Grenville Turner and Richard Pattrick for assistance and discussions. BM thanks Daniele Pinti, Max Coleman, Jean-Michel Matray, Christian Fouillac, Richard Worden and Frédérique Walgenwitz for support and collaboration over five years and more. We thank Zhou Zheng for many useful comments, Don Porcelli for his detailed review and editorial handling, and Richard Hartley for drafting many of the figures.

REFERENCES

Andrews JN (1983) Dissolved radioelements and inert gases in geothermal investigations. Geothermics 12:67-82

Andrews JN, Lee DJ (1979) Inert gases in groundwater from the Butter sandstone of England as indicators of age and trends. J Hydrol 41:233-252

Andrews JN, Goldbrunner JE, Darling WG, Hooker PJ, Wilson GB, Yougman MJ, Eichinger L, Rauert W, Stichler W (1985) A radiochemical, hydrochemical and dissolved gas study of groundwaters in the Molasse basin of Upper Austria. Earth Planet Sci Lett 73:317-332

Aplin AC, Warren EA, Grant SM, Robinson AG (1993) Mechanisms of quartz cementation in North Sea sandstone: Constraints from fluid compositions. *In* Diagenesis and Basin Development. Horbury A, Robinson AG (eds) AAPG Studies in Geology 36:7-22

ASTM (1985) Standard methods for estimating solubility of gases in petroleum liquids. *In* Annual Book of ASTM Standards, American Society of Testing and Materials, p 669-673

Atkins PW (1979) Physical Chemistry. Oxford University Press, Oxford, UK

Ballentine CJ (1991) He, Ne and Ar isotopes as tracers in crustal fluids. PhD dissertation, University of Cambridge, Cambridge, UK

Ballentine CJ (1997) Resolving the mantle He/Ne and crustal $^{21}Ne/^{22}Ne$ in well gases. Earth Planet Sci Lett 152:233-249

Ballentine CJ, Burnard PG (2002) Production and release of noble gases in the continental crust. Rev Mineral Geochem 47:481-538

Ballentine CJ, Hall CM (1999) Determining paleotemperature and other variables by using an error-weighted, non-linear inversion of noble gas concentrations in water. Geochim Cosmochim Acta 63:2315-2336

Ballentine CJ, O'Nions RK (1992) The nature of mantle neon contributions to Vienna Basin hydrocarbon reservoirs. Earth Planet Sci Lett 113:553-567

Ballentine CJ, O'Nions RK (1994) The use of He, Ne and Ar isotopes to study hydrocarbon related fluid provenance, migration and mass balance in sedimentary basins. *In* Geofluids: Origin, migration and evolution of fluids in sedimentary basins. Parnell J (ed) Geol Soc Spec Publ 78:347-361

Ballentine CJ, Sherwood Lollar B (2002) Regional groundwater focussing of nitrogen and noble gases into the Hugoton-Panhandle giant gas field, USA. Geochim. Cosmochim. Acta 66:2483-2497

Ballentine CJ, O'Nions RK, Oxburgh ER, Horvath F, Deak J (1991) Rare gas constraints on hydrocarbon accumulation, crustal degassing and groundwater flow in the Pannonian Basin. Earth Planet Sci Lett 105:229-246

Ballentine CJ, Mazurek M, Gautschi A (1994) Thermal constraints on crustal rare gas release and migration—evidence from alpine fluid inclusions. Geochim Cosmochim Acta 58:4333-4348

Ballentine CJ, O'Nions RK, Coleman ML (1996) A Magnus opus: Helium, neon, and argon isotopes in a North Sea oilfield. Geochim Cosmochim Acta 60:831-849

Ballentine CJ, Aeschbach-Hertig W, Peeters F, Beyerle U, Holocher J, and Kipfer R. (1999) Inverting noble gas concentrations in natural gas: Determination of gas/groundwater interaction and a test for conceptual models of gas transport and emplacement. EOS, Trans Am Geophys Union 80:f1168

Ballentine CJ, Schoell M, Coleman D, Cain BA (2001) 300 million year old magmatic CO_2 in West Texas Permian Basin natural gas reservoirs. Nature 409:327-331

Bard E, Hamelin B, Fairbanks RG, Zindler A (1990) Calibration of the ^{14}C timescale over the past 30,000 years using mass spectrometric U-Th ages from Barbados corals. Nature 345:405-409

Battani A, Sarda P, Prinzhofer A (2000) Basin scale natural gas source, migration and trapping traced by noble gases and major elements; the Pakistan Indus Basin. Earth Planet Sci Lett 181:229-249

Bethke CM, Harrison WJ, Upson C, Altaner ST (1988) Supercomputer analysis of sedimentary basins. Science 239:261-267

Ben-Naim A (1980) Hydrophobic Interactions. Plenum Press, New York
Ben-Naim A, Egel-Thal M (1965) Thermodynamics of aqueous solutions of noble gases I-III. J Sol Chem 69:3240-3253
Benson B, Krause DJ (1976) Empirical laws for dilute aqueous solutions of non-polar gases. J Chem Phys 64:689-709
Bogard DD, Rowe MW, Manuel OK, Kuroda PK (1965) Noble gas anomalies in the mineral thucholite. J Geophys Res 70:703-708
Bohlke JK, Irwin JJ (1992a) Laser microprobe analyses of noble gas isotopes and halogens in fluid inclusions: Analyses of microstandards and synthetic inclusions in quartz. Geochim Cosmochim Acta 56:187-201
Bohlke JK, Irwin JJ (1992b) Laserprobe analyses of Cl, Br, I, and K in fluid inclusions: Implications for the sources of salinity in some ancient hydrothermal fluids. Geochim Cosmochim Acta 56:203-225
Bohlke JK, Irwin JJ (1992c) Brine history indicated by argon, krypton, chlorine, bromine, and iodine analyses of fluid inclusions from the Mississippi Valley type lead-fluorite-barite deposits at Hansonburg, New Mexico. Earth Planet Sci Lett 110:51-66
Bosch A, Mazor E (1988) Natural gas association with water and oil depicted by atmospheric noble gases: case studies from the southern Mediterranean Coastal Plain. Earth Planet Sci Lett 87:338-346
Bottinga Y, Javoy M (1991) MORB degassing: bubble growth and ascent. Chem Geol 81:255-270
Bourke PJ, Gilling D, Jeffries NL, Lever DA, Lineham TR (1989) Laboratory experiments of mass transfer in London clay. Proc Material Res Soc Symp 127:805-812
Braibant A, Fisicaro E, Dallevalle F, Lamb JD, Oscarson JL, Rao RS (1994) Molecular thermodynamic model for the solubility of noble gases in water. J Phys Chem 98:626-634
Burgess R, Taylor RP, Fallick AE, Kelley SP (1992) ^{40}Ar-^{39}Ar laser microprobe study of the fluids in different colour zones of a hydrothermal scheelite crystal from the Dae Hwa W-Mo mine, South Korea. Chem Geol 102:259-267
Burgess R, Layzelle E, Turner G, Harris JW (2002) Constraints on the age and halogen composition of Siberian coated diamonds. Earth Planet Sci Lett (in press)
Burnard PG, Hu R, Turner G, Bi XW (1999) Mantle, crustal and atmospheric noble gases in Ailoshan Gold deposits, Yunnan Province, China. Geochim Cosmochim Acta 63:1595-1604
Castro MC, Jambon A, de Marsily G, Schlosser P (1998a) Noble gases as natural tracers of water circulation in the Paris Basin 1. Measurements and discussion of their origin and mechanisms of vertical transport in the basin. Water Resources Res 34:2443-2466
Castro MC, Goblet P, Ledoux E, Violette S, de Marsily G (1998b) Noble gases as natural tracers of water circulation in the Paris Basin 2. Calibration of a groundwater flow model using noble gas isotope data. Water Resources Res 34:2467-2483
Clever HL (1979a) Helium- and neon-gas solubilities. *In* Solubility Data Series, Vol 1. Intl Union of Pure and Applied Chemistry (IUPAC). Pergamon Press, Oxford
Clever HL (1979b) Krypton, Xenon, and Radon - gas solubilities. In Solubility data series, Vol 2. Intl Union of Pure and Applied Chemistry (IUPAC). Pergamon Press, Oxford
Clusius K, Dickel G (1938) Neues Verfehren zur Gasentmischung und Isotopentrennung. Naturwissenschaften 26:546
Craig H, Clarke WB, Beg MA (1975) Excess ^{3}He in deep waters on the East Pacific Rise. Earth Planet Sci Lett 26:125-132
Craig H, Lupton JE, Horibe Y (1978) A mantle helium component in circum Pacific volcanic gases: Hakone, the Marianas, and Mt Lassen. *In* Terrestrial rare gases. Alexander EC, Ozima M (eds), Japan Sci Societies Press, Tokyo, p 3-16
Crovetto R, Fernandez-Prini R, Japas ML (1982) Solubilities of inert gases and methane in H_2O and in D_2O in the temperature range of 300 to 600 K. J Chem Phys 76:1077-1086
Davis SN, DeWeist RJM (1966) Hydrology. Wiley, New York
Denbigh K (1986) The principles of chemical equilibrium. Cambridge University Press, Cambridge, UK
Dewonck S, Marty B, France-Lanord C (2002) Geochemistry of the Lorraine groundwaters, France: Paleoclimatic record, mantle helium flux and hydrodynamic implications. Geochim Cosmochim Acta submitted
Dunai TJ, Baur H (1995) Helium, neon and argon systematics of the European subcontinental mantle: Implications for its geochemical evolution. Geochim Cosmochim Acta 59:2767-2783
Dunai TJ, Porcelli D (2002) The storage and transport of noble gases in the subcontinental mantle. Rev Mineral Geochem 47:371-409
Dymond JH, Smith EB (1980) The virial coefficients of gases and mixtures. Clarendon Press, Oxford, UK
Eley DD (1939) Aqueous solutions of noble gases. Trans Faraday Soc 35:1281

Emery D, Smalley PC, Oxtoby NH (1993) Synchronous oil migration and cementation in sandstone reservoirs demonstrated by quantitative description of diagenesis. Phil Trans R Soc London A 344: 115-125

Espitalié J, Maxwell JR, Chenet Y, Marquis F (1988) Aspects of hydrocarbon migration in the Mesozoic of the Paris Basin as deduced from an organic geochemical survey. Organic Geochem 13:467-481

Fontes JC, Garnier JM (1979) Determination of the initial ^{14}C activity of total dissolved carbon: a review of existing models and a new approach. Water Resources Res 15:399-413

Frick U, Chang S (1977) Ancient carbon and noble gas fractionation. Proc Lunar Sci Conf 8:263-272

Gleeson SA, Wilkinson JJ, Stuart FM, Banks DA (2001) The origin and evolution of base metal mineralizing brines, South Cornwall, U.K. Geochim Cosmochim Acta 65:2067-2079

Gluyas J, Coleman M. (1992) Material flux and porosity changes during sediment diagenesis. Nature 356:52-54

Goryunov MS, Kozlov AL (1940) Voprosy geokhimie gelinosnykh gazov i usloviya nakopleniya geliya v eyemnoy kore (A study of the geochemistry of helium-bearing gases and the conditions for the accumulation of helium in the Earth's crust.) State Sci-Tech Pub Co Oil and Solid Fuel Lit, Leningrad-Moscow

Gold T, Held M (1987) Helium-Nitrogen-Methane systematics in natural gases of Texas and Kansas. J Petrol Geol 10:415-424

Graham DW (2002) Noble gas isotope geochemistry of mid-ocean ridge and ocean island basalts: Characterization of mantle source reservoirs. Rev Mineral Geochem 47:247-318

Gray CG and Gubbins KE (1984) Theory of molecular fluids. Clarendon Press, Oxford, UK

Griesshaber E, O'Nions RK, Oxburgh ER (1992) Helium and carbon isotope systematics in crustal fluids from the Eifel, the Rhine Graben and Black Forest, F.R.G. Chem Geol 99:213-235

Heaton THE (1981) Dissolved gases: some applications to groundwater research. Trans Geol Soc Afr 84:91-97

Heaton THE (1984) Rates and sources of ^{4}He accumulation in groundwater. Hydrol Sci J 29:29-47

Heaton THE, Vogel JC (1979) Gas concentrations and ages of groundwaters in the Beaufort group sediments, South Africa. Water S A 5:160-170

Hirschfelder JO, Curtiss CF, and Bird RB (1967) Molecular theory of gases and liquids. John Wiley & Sons Ltd, London, UK

Hiyagon H, Kennedy BM (1992) Noble gases in CH_4-rich gas fields, Alberta, Canada. Geochim Cosmochim Acta 56:1569-1589

Hu R, Burnard PG, Turner G, Bi X (1998) Helium and argon isotope systematics in fluid inclusions of Machangquing copper deposit in west Yunnan province, China. Chem Geol 146:55-63

Hutcheon I (1999) Controls on the distribution of non-hydrocarbon gases in the Alberta Basin. Bull Can Petrol Geol 47:573-593

Irwin JJ, Roedder E (1995) Diverse origins of fluid inclusions at Bingham (Utah, USA), Butte (Montana, USA), St. Austell (Cornwall, UK) and Ascension Island (mid-Atlantic, UK), indicated by laser microprobe analysis of Cl, K, Br, I, Ba + Te, U, Ar, Kr, and Xe. Geochim Cosmochim Acta 59: 295-312

Jähne B, Heinz G, Dietrich W (1987) Measurement of the diffusion coefficient of sparingly soluble gases in water. J Geophys Res 92:10,767-10,776

Javoy M, Pineau F (1991) The volatiles record of a popping rock from the Mid-Atlantic Ridge at 14-degrees-N—Chemical and isotopic composition of gas trapped in the vesicles. Earth Planet Sci Lett 107:598-611

Jenden PD and Kaplan IR (1989) Origin of natural-gas in Sacramento Basin, California. AAPG Bull 73:431-453

Jenden PD, Kaplan IR, Poreda RJ, Craig H. (1988) Origin of nitrogen rich gases in the Californian Great Valley: Evidence from helium, carbon and nitrogen isotope ratios. Geochim Cosmochim Acta 52: 851-861

Jenden PD, Hilton DR, Kaplan IR, Craig H (1993) Abiogenic hydrocarbons and mantle helium in oil and gas fields. *In* Howel DG (ed) The future of energy gases. U S Geol Surv Prof Paper 1570, U S Geological Survey, p 31-56

Kelley S, Turner G, Butterfield AW, Shepherd TJ (1986) The source and significance of argon isotopes in fluid inclusions from areas of mineralization. Earth Planet Sci Lett 79:303-318

Kendrick MA, Burgess R, Pattrick RAD, Turner G (2001a) Halogen and Ar-Ar age determinations of inclusions within quartz veins from porphyry copper deposits using complementary noble gas extraction techniques. Chem Geol 177:351-370

Kendrick MA, Burgess R, Pattrick RAD, Turner G (2001b), Fluid inclusion noble gas (He, Ar, Kr, Xe) and halogen (Cl, Br, I) evidence on the origin of Cu-Porphyry mineralising fluids. Geochim Cosmochim Acta 65:2641-2658

Kendrick MA, Burgess R, Pattrick RAD, Turner G (2002a) The origin of a fluorite rich MVT brine: combined noble gas (He, Ar, Kr) and halogen (Cl, Br, I) analysis of fluid inclusions from the South Pennine Orefield, U.K. Econ Geol (in press)

Kendrick MA, Burgess R, Leach D, Pattrick RAD (2002b) Hydrothermal fluid origins in Mississippi Valley-Type ore deposits: Noble gas and halogen evidence from the Illinois-Kentucky Fluorspar district, Viburnum Trend and Tri-State deposits, Mid-continent, U.S.A. Econ Geol (in press)

Kennedy BM, Lynch MA, Reynolds JH, Smith SP (1985) Intensive sampling of noble gases in fluids at Yellowstone: I. Early overview of the data; regional patterns. Geochim Cosmochim Acta 49: 1251-1261

Kharaka YK and Specht DJ (1988) The solubility of noble gases in crude oil at 25-100°C. Appl Geochem 3:137-144

Kipfer R, Aeschbach-Hertig W, Peeters F, Stute M (2002) Noble gases in lakes and ground waters. Rev Mineral Geochem 47:615-700

Krooss BM, Leythaeuser D (1988) Experimental measurements of the diffusion parameters of light hydrocarbons in water-saturated sedimentary rocks, II. Results and geochemical significance. Organic Geochem 11:193-199

Krooss BM, Leythaeuser D, Schaefer RG (1992) The quantification of diffusive hydrocarbon losses through cap rocks of natural gas reservoirs—a re-evaluation. AAPG Bull 76:403-406

Lafargue E, Barker C (1988) Effect of water washing on crude oil composition. AAPG Bull 72:263-276

Langmuir CH, Vocke RD Jr, Hanson GN, Hart SR (1978) A general mixing equation with applications to Icelandic basalts. Earth Planet Sci Lett 37:380-392

Macaulay CI, Haszeldine RS, Fallick AE (1992) Diagenetic pore waters stratified for at least 35 Million years: Magnus oil field, North Sea. AAPG Bull 76:1625-1634

Mamyrin, BA, Tolstikhin IN (1984) Helium isotopes in nature. Elsevier, Amsterdam

Marine IW (1979) The use of naturally occurring He to estimate groundwater velocities for studies of geological storage waste. Water Resources Res 15:1130-1136

Martel DJ, Deak J, Dövenyi P, Horvath F, O'Nions RK, Oxburgh ER, Stegena L, Stute M (1989) Leakage of helium from the Pannonian basin. Nature 342:908-912

Marty B (1984) On the noble gas isotopic fractionation in naturally occurring gases. Geochem J 18:157-162

Marty B, Torgersen T, Meynier V, O'Nions RK, de Marsily G (1993) Helium isotope fluxes and groundwater ages in the Dogger Aquifer, Paris Basin. Water Resources Res 29:1025-1035

Mattey D (1991) Carbon-dioxide solubility and carbon isotope fractionation in basaltic melt. Geochim Cosmochim Acta 55:3467-3473

Morrison P, Pine J (1955) Radiogenic origin of helium isotopes in rocks. Ann N Y Acad Sci 62:71-92

Nagao K, Takaoka N, Matsubayashi O, (1979) Isotopic anomalies of rare gases in the Nigorikawa geothermal area, Hokkaido, Japan. Earth Planet Sci Lett 44:82-90

Nagao K, Takaoka N, Matsubayashi O (1981) Rare gas isotopic ratios in natural gases of Japan. Earth Planet Sci Lett 53:8175-188

Neretnieks I (1982) Diffusivities of some dissolved constituents in compacted wet bentonite clay-MX80 and the impact of radionuclide migration in the buffer. SKBF/KBS Teknisk Rapport, Stockholm, p 82-87

Nordstrom DK, Munoz L (1985) Geochemical thermodynamics. Benjamin-Cumming Publ Co, London

Nuccio P, Paonita A (2000) Investigation of the noble gas solubility in H_2O-CO_2 bearing silicate liquids at moderate pressure II: the extended ionic porosity (EIP) model. Earth Planet Sci Lett 183:499-512

Ohsumi T, Horibe Y (1984) Diffusivity of He and Ar in deep-sea sediments. Earth Planet Sci Lett 70: 61-68

Oliver BM, Bradley JG, Farrar IV H (1984) Helium concentration in the Earth's lower atmosphere. Geochim Cosmochim Acta 48:1759-1767

O'Nions RK, Ballentine CJ (1993) Rare gas studies of basin scale fluid movement. Phil Trans R Soc London 344:144-156

O'Nions RK, Oxburgh ER (1983) Heat and helium in the Earth. Nature 306:429-431

O'Nions RK, Oxburgh ER (1988) Helium, volatile fluxes and the development of the continental crust. Earth Planet Sci Lett 90:331-347

Oxburgh ER, O'Nions RK, Hill RI (1986) Helium isotopes in sedimentary basins. Nature 324:632-635

Ozima M, Podosek FA (1983) Noble gas geochemistry. Cambridge University Press, Cambridge, UK

Pierce AP, Gott GB, Mytton JW (1964) Uranium and helium in the Panhandle gas field, Texas, and adjacent areas. U S Geol Surv Prof Pap 454-G:1-57

Pinti DL, Marty B (1995) Noble gases in crude oils from the Paris basin, France—Implications for the origin of fluids and constraints on oil-water interactions. Geochim Cosmochim Acta 59:3389-3404

Pinti DL, Marty B (1998) The origin of helium in deep sedimentary aquifers and the problem of dating very old groundwaters. *In* Parnell J (ed) Dating and duration of fluid flow and fluid-rock interaction. Geol Soc Spec Publ 144:53-68

Pinti DL and Marty B (2000) Noble gases in oil and gas fields: Origin and processes. *In* Kyser K (ed) Fluids and Basin Evolution. Mineral Soc Can Short Course 28, Mineralogical Society of Canada, Toronto, p 160-196

Pinti DL, Marty B, Andrews JN (1997) Atmosphere-derived noble gas evidence for the preservation of ancient waters in sedimentary basins. Geology 25:111-114

Podosek FA, Honda M, Ozima M (1980) Sedimentary noble gases. Geochim Cosmochim Acta 44: 1875-1884

Podosek FA, Bernatowcz TJ, Kramer FE (1981) Adsorption of xenon and krypton on shales. Geochim Cosmochim Acta 45:2401-2415

Polya DA, Foxford KA, Stuart F, Boyce A, Fallick AE (2000) Evolution and paragenetic context of low δD hydrothermal fluids from the Panasqueira W-Sn deposit, Portugal: new evidence from microthermometric, stable isotope, noble gas and halogen analyses of primary fluid inclusions. Geochim Cosmochim Acta 64:3357-3371

Porcelli D, Ballentine CJ (2002) Models for the distribution of Terrestrial noble gases and the evolution of the atmosphere. Rev Mineral Geochem 47:411-480

Poreda R, Jenden PD, Kaplan IR, Craig H (1986) Mantle helium in Sacramento Basin natural gas wells. Geochim Cosmomochim Acta 50:9-33

Potter RWI, Clyne MA (1978) The solubility of the noble gases He, Ne, Ar, Kr and Xe in water up to the critical point. J Solution Chem 7:837-844

Poulet M, Espitalié J (1987) Hydrocarbon migration in the Paris Basin. *In* Doligez B (ed) Migrations of hydrocarbons in sedimentary basins. Editions Technip, Paris, p 131-171

Present RD (1958) Kinetic Theory of Gases. McGraw Hill, New York

Price LC, Blount CW, MacGowan D, Wenger L (1981) Methane solubility in brines with application to the geopressured resource. *In* Proc 5th Conf Geopressured Geothermal Energy, Baton Rouge, Louisiana, p 205-214

Prinzhofer A, Pernaton E (1997) Isotopically light methane in natural gas: bacterial imprint or diffusive fractionation? Chem Geol 142:193-200

Qui H-N (1996) ^{40}Ar-^{39}Ar dating of the quartz samples from two mineral deposits in western Yunnan (SW China) by crushing in vacuum. Chem Geol 127:211-222

Rebour V, Billiotte J, Deveughele M, Jambon A, le Guen C (1997) Molecular diffusion in water saturated rocks: a new experimental method. J Cont Hydrol 28:71-93

Sano Y, Wakita H, Huang CW (1986) Helium flux in a continental land area from $^3He/^4He$ ratio in northern Taiwan. Nature 323:55-57

Savchenko VP (1935) The problems of geochemistry of helium. Natural Gases 9:53-197 (in Russian)

Sherwood Lollar B, O'Nions RK, Ballentine CJ (1994) Helium and neon isotope systematics in carbon dioxide-rich and hydrocarbon-rich gas reservoirs. Geochim Cosmochim Acta 58:5279-5290

Sherwood Lollar B, Ballentine CJ, O'Nions RK (1997) The fate of mantle-derived carbon in a continental sedimentary basin: Integration of C/He relationships and stable isotope signatures. Geochim Cosmochim Acta 61:2295-2307

Schlomer S, Kroos BM (1997) Experimental characterisation of the hydrocarbon sealing efficiency of cap rocks. Marine Petrol Geol 14:563-578

Schoell M (1983) Genetic characterisation of natural gases. AAPG Bull 67:2225-2238.

Simmons SF, Sawkins FJ, Sclutter DJ (1987) Mantle-derived helium in two Peruvian hydrothermal ore deposits. Nature 329:429-432

Smith SP (1985) Noble gas solubility in water at high temperature. EOS, Trans Am Geophys Union 66:397

Smith SP Kennedy BM (1983) The solubility of noble gases in water and in NaCl brine. Geochim Cosmochim Acta 47:503-515

Solomon DK, Hunt A, Poreda RJ (1996) Source of radiogenic helium-4 in shallow aquifers: Implications for dating young groundwater. Water Resources Res 32:1805-1813

Stillinger FH (1980) Water revisited. Science 209:451-457

Stilwell DP (1989) CO_2 resources of the Moxa Arch and the Madison Reservoir. *In* Eisert JL (ed) Gas Resources of Wyoming. Wyoming Geol Assoc 40th Field Conf, Casper, p 105-115

Stuart FM, Turner G (1992) The abundance and isotopic composition of the noble gases in ancient fluids. Chem Geol 101:97-109

Stuart FM, Turner G, Duckworth RC, Fallick AE (1994) Helium isotopes as tracers of trapped hydrothermal fluids in ocean-floor sulphides. Geology 22:823-826

Stuart FM, Burnard PG, Taylor RP, Turner G (1995) Resolving mantle and crustal contributions to ancient hydrothermal fluids: He-Ar isotopes in fluid inclusions from Dae Hwa W-Mo mineralization, South Korea. Geochim Cosmochim Acta 59:4663-4673

Stute M, Deak J (1989) Environmental isotope study (^{12}C, ^{13}C, ^{18}O, D, noble gases) on deep groundwater circulation systems in Hungary with reference to paleoclimate. Radiocarbon 31:902-918

Stute M, Sonntag C, Deak J, Schlosser P (1992) Helium in deep circulating groundwater in the Great Hungarian plain: Flow dynamics and crustal and mantle helium fluxes. Geochim Cosmochim Acta 56:2051-2067

Thrower CD (1999) The evolution of mineralizing fluids in the Colorado mineral belt, defined by combined noble gas and halogen analyses. PhD dissertation, University of Manchester, Manchester, UK

Tolstikhin IN, Lehmann BE, Loosli HH, Gautschi A (1996) Helium and argon isotopes in rocks, minerals, and related groundwaters: A case study in northern Switzerland. Geochim Cosmochim Acta 60: 1497-1514

Torgersen T (1980) Controls on pore fluid concentration of ^4He and ^{222}Rn and the calculation of $^4He/^{222}Rn$ ages. J Geochem Explor 13:57-75

Torgersen T, Clarke WB (1985) Helium accumulation in groundwater. I: an evaluation of sources and the continental flux of crustal ^4He in the Great Artesian Basin, Australia. Geochim Cosmochim Acta 49:1211-1218

Torgersen T, Clarke WB (1987) Helium accumulation in groundwater. III: Limits on helium transfer across the mantle-crust boundary beneath Australia and the magnitude of mantle degassing. Earth Planet Sci Lett 84:345-355

Torgersen T, Ivey GN (1985) Helium accumulation in groundwater. II: a model for the accumulation of crustal ^4He degassing flux. Geochim Cosmochim Acta 49:2445-2452

Torgersen T, Kennedy BM (1999) Air-Xe enrichments in Elk Hills oil field gases: role of water in migration and storage. Earth Planet Sci Lett 167:239-253

Torgersen T, Kennedy BM, Hiyagon H, Chiou KY, Reynolds JH, Clarke WB (1989) Argon accumulation and the crustal degassing flux of ^{40}Ar in the Great Artesian Basin, Australia. Earth Planet Sci Lett 92:43-56

Toth J (1980) Cross formational gravity-flow of groundwater: a mechanism of transport and accumulation of petroleum. AAPG Studies in Geology 10:121-167

Toth J, Corbett T (1986) Post-Paleocene evolution of regional groundwater flow-system and their relation to petroleum accumulations, Taber area, Southern Alberta, Canada. Bull Can Petrol Geol 34:339-363

Turner G (1988) Hydrothermal fluids and argon isotopes in quartz veins and cherts. Geochim Cosmochim Acta 52:1443-1448

Turner G, Bannon MP (1992) Argon isotope geochemistry of inclusion fluids from granite-associated mineral veins in southwest and northeast England. Geochim Cosmochim Acta 56:227-243

Weinlich FH, Braeuer K, Kampf H, Strauch G, Tesar J, Weise SM (1999) An active subcontinental mantle volatile system in the western Eger rift, Central Europe: Gas flux, isotopic (He, C, and N) and compositional fingerprints. Geochim Cosmochim Acta 63:3653-3671

Weiss RF (1970) The solubility of nitrogen, oxygen and argon in water and seawater. Deep-Sea Res 17:721-735

Weiss RF (1971) Solubility of helium and neon in water and seawater. J Chem Eng Data 16:179-188

Weiss RF, Price BA (1989) Dead Sea gas solubilities. Earth Planet Sci Lett 92:7-10

Wilhelm E, Battino R, Wilcock RJ (1977) Low pressure solubility of gases in liquid water. Chem Rev 77:219-262

Wilkinson JJ (2001) Fluid inclusions in hydrothermal ore deposits. Lithos 55:229-272

Winckler G, Suess E, Wallman K, deLange GJ, Westbrook GK, Bayer R (1997) Excess helium and argon of radiogenic origin in Mediterranean brine basins. Earth Planet Sci Lett 15:225-231

Worden RH (1996) Controls on halogen concentrations in sedimentary formation waters. Mineral Mag 60:259-274

Worden RH, Matray JM (1995) Cross formational flow in the Paris basin. Basin Res 7:53-66

Zaikowski A, Spangler RR (1990) Noble gas and methane partitioning from groundwater. An aid to natural gas exploration and reservoir evaluation. Geology 18:72-74

Zaikowski A, Kosanke BJ, and Hubbard N (1987) Noble gas composition of deep brines from the Palo Duro Basin, Texas. Geochim Cosmochim Acta 51:73-84

Zanker A (1977) Inorganic gases in petroleum. Hydrocarbon Process 56:255-256

Zartman RE, Wasserburg GJ, Reynolds JH (1961) Helium, argon and carbon in some natural gases. J Geophys Res 66:277-306

Zhang TW, Krooss BM (2001) Experimental investigation on the carbon isotope fractionation of methane during gas migration by diffusion through sedimentary rocks at elevated temperature and pressure. Geochim Cosmochim Acta 65:2723-2742

Zuber A, Weise SM, Osenbrück K, and Matenko T (1997) Origin and age of saline waters in Busko Spa (Southern Poland) determined by isotope, noble gas and hydrochemical methods: evidence of interglacial and pre-Quaternary warm climate recharges. Appl Geochem 12:643-660

14 Noble Gases in Lakes and Ground Waters

Rolf Kipfer[1,2], Werner Aeschbach-Hertig[1,3], Frank Peeters[1], and Marvin Stute[4]

[1]Department of Water Resources and Drinking Water
Swiss Federal Institute of Environmental Science and Technology, EAWAG
8600 Dübendorf, Switzerland
rolf.kipfer@eawag.ch

[2]Institute for Isotope Geology and Mineral Resources
[3]Environmental Physics
Swiss Federal Institute of Technology, ETH Z
8092 Zürich, Switzerland

[4]Barnard College, Columbia University
New York, New York

INTRODUCTION

In contrast to most other fields of noble gas geochemistry that mostly regard atmospheric noble gases as 'contamination,' air-derived noble gases make up the far largest and hence most important contribution to the noble gas abundance in meteoric waters, such as lakes and ground waters. Atmospheric noble gases enter the meteoric water cycle by gas partitioning during air / water exchange with the atmosphere.

In lakes and oceans noble gases are exchanged with the free atmosphere at the surface of the open water body. In ground waters gases partition between the water phase and the soil air of the quasi-saturated zone, the transition between the unsaturated and the saturated zone. Extensive measurements have shown that noble gas concentrations of open waters agree well with the noble gas solubility equilibrium according to (free) air / (free) water partitioning, whereby the aquatic concentration is directly proportional to the respective atmospheric noble gas abundance (Henry law, Aeschbach-Hertig et al. 1999b).

In applications in lakes and ground waters the gas specific Henry coefficient can simplifying be assumed to depend only on temperature and salinity of the water. Hence the equilibrium concentrations of noble gases implicitly convey information on the physical properties of the water during gas exchange at the air / water interface, i.e., air pressure, temperature and salinity of the exchanging water mass. The ubiquitous presence of atmospheric noble gases in the meteoric water cycle defines a natural baseline, which masks other noble gas components until their abundance is sufficiently large that these components can be separated against the natural atmospheric background. For most classical geochemical aspects this typical feature of natural waters may look at first sight as a disadvantage. In fact it turns out to be advantageous because in most cases the noble gas abundance in water can be understood as a binary mixture of two distinct noble gas components—a well-constrained atmospheric component and a residual component of non-atmospheric origin.

Only very few processes are able to fractionate atmospheric noble gases. All these processes are controlled by well-understood physical mechanisms, which in consequence constrain air-derived noble gases and any other component completely. In addition to atmospheric noble gases basically two non-atmospheric noble gas components are present in most natural waters: *radiogenic* noble gases and *terrigenic* noble gases from different geochemical compartments of the Earth.

Radiogenic noble gases are generated by all kinds of disintegrations of radioactive

1529-6466/00/0047-0014$10.00

precursors and succeeding nuclear reactions (see Ballentine and Burnard 2002 in this volume). Only $^{4}He_{rad}$, $^{3}He_{rad}$, occasionally $^{40}Ar_{rad}$, and very rarely $^{21}Ne_{rad}$ have sufficiently large production yields that these isotopes can be observed in natural waters. Whereas rocks and minerals generate all four isotopes, ^{3}He (as tritiogenic $^{3}He_{tri}$) is also produced by the decay of atmospheric tritium (^{3}H, half-life: 4500 d, Lucas and Unterweger 2000) that is bound in meteoric water molecules. On the one hand the combined analysis of tritiogenic $^{3}He_{tri}$ and ^{3}H allows the quantitative dating of young waters having residence times of up to 50 years (e.g., Tolstikhin and Kamenskiy 1969; Schlosser et al. 1988; Solomon and Cook 2000). On the other hand radiogenic $^{4}He_{rad}$ generally yields qualitative ages for old ground waters recharging on millennium time scales (e.g., Solomon 2000).

Terrigenic noble gases denote as a collective term noble gases originating from well-defined geochemical reservoirs with a distinct geochemical composition. In meteoric waters two terrigenic components can be found: noble gases of crustal and of mantle origin. Terrigenic fluids are defined by the characteristic isotopic composition of He. The particular $^{3}He/^{4}He$ ratio is a defining feature of the respective terrigenic component. The continental crust is dominated by isotopically heavy He that is produced *in situ* by nuclear reactions in crustal rocks and minerals ($^{3}He/^{4}He < 10^{-7}$, Mamyrin and Tolstikhin 1984, Ballentine and Burnard 2002 in this volume). Note that in the literature crustal He is often identified by radiogenic He. Here, we distinguish between He that is produced by nuclear processes ('radiogenic He') and He that originates from a geochemical reservoir ('terrigenic He'). The Earth mantle contains besides newly produced He relicts of isotopic light He inherited during planet formation ($^{3}He/^{4}He > 10^{-5}$, Mamyrin and Tolstikhin 1984; see Porcelli and Ballentine 2002 and Graham 2002 in this volume).

In summary, in most cases the noble gas concentrations in meteoric waters can be understood as a mixture of *atmospheric* noble gases (He, Ne, Ar, Kr, Xe) with variable amounts of *radiogenic* and / or *terrigenic* He. The noble gas concentrations in meteoric waters can be translated into

1. climatic conditions during air / water exchange (*atmospheric noble gas component*),
2. water-residence times and renewal rates (*radiogenic noble gas component, mainly He*), and
3. geochemical fingerprints and the origin of non-atmospheric fluids (*terrigenic noble gas component, mainly He*).

In lakes and young ground waters the determination of water residence times by ^{3}H-^{3}He dating in order to study transport and mixing processes has been the prevalent use of noble gas isotopes (e.g., Torgersen et al. 1977; Schlosser et al. 1988). Recent applications in lakes also used the concentrations of atmospheric noble gases to reconstruct possible lake level fluctuations in the past recorded as change of the partial pressure of the (noble) gases due to variation of the altitude of the lake surface (Craig et al. 1974; Kipfer et al. 2000).

Similarly, the dependence of noble gas solubility equilibrium on the physical conditions during gas exchange, in particular the sensitivity of the Henry coefficients on temperature, has successfully been used in ground-water studies to reconstruct the soil temperature prevailing during ground-water recharge. If an aquifer contains ground waters that recharged during different climatic conditions in the past, noble gas concentrations provide information about the past temperature evolution (Mazor 1972; Andrews and Lee 1979). This approach has been applied to reconstruct the continental temperature regime of the Pleistocene / Holocene transition in the tropics (e.g., Stute et al. 1995b; Weyhenmeyer et al. 2000) as well as in mid latitudes (Stute et al. 1995a; Beyerle et al. 1998; Stute and Schlosser 2000; Aeschbach-Hertig et al. 2002b).

Results of all these studies were hampered by the common observation that ground waters in contrast to surface waters always contain atmospheric (noble) gases in excess, i.e., the measured dissolved noble gas concentrations are usually significantly larger than the expected solubility equilibrium. Such noble gas excesses, being characteristic for ground waters, are known in the literature as 'excess air' (Heaton and Vogel 1981). Despite experimental evidence that showed that the elemental composition of the noble gas excess is 'air-like', the formation processes responsible for the observed gas excess in ground waters remain unclear. Only recently, new concepts have been developed that link the formation of excess air with the physical processes that control air / water partitioning at the transition between the unsaturated and the saturated zone in soils. In particular two models, the partial re-equilibration and closed system equilibration model, are now available, which describe the relative abundances of dissolved atmospheric noble gases as well as the possible fractionation relative to pure air (Stute 1989; Stute et al. 1995b; Aeschbach-Hertig et al. 2000).

Apart from lakes and ground water, noble gases are also useful in the study of other continental water reservoirs, such as pore waters in sediments and rocks. However, little work has been done in these areas so far. The one 'aqueous' system that has received considerable attention in recent years is ice. The polar ice sheets have proven to be excellent archives of past environmental conditions, and noble gases as conservative tracers play a role in extracting information from these archives. A short review of recent applications of noble gases in ice is therefore included at the end of this chapter.

ANALYTICAL TECHNIQUES

The determination of noble gases in water can be divided into three successive analytical steps: (1) noble gas extraction from the water, (2) purification and separation of the extracted noble gases, and (3) quantitative (mass spectrometric) analysis. For extended discussions of methods for noble gas analysis in waters (and other terrestrial fluids) the readers are referred to Clarke et al. (1976), Rudolph (1981), Bayer et al. (1989), Stute (1989), Gröning (1994), Ludin et al. (1997), and Beyerle et al. (2000a).

Following the classical approach (Rudolph 1981; Stute 1989; Gröning 1994) the dissolved noble gases are degassed from a water sample typically collected in copper tubes (10-45 cm^3 volume) by vacuum extraction. To extract the dissolved gases the copper tubes are connected to an extraction vessel mounted on an ultra-high vacuum extraction / purification system. After pumping down the head space of the extraction vessel (<10^{-4} Torr) the copper tube is opened and connected over a capillary to cold traps that are filled with adsorbent media, such as char coal or zeolite, and cooled by an acetone / dry-ice mixture or by liquid nitrogen. Vigorous shaking or stirring of the water facilitate the liberation of the volatile gases into the gas phase. The liberated gases including water vapor are transported through the capillary in the direction of the cold traps where the condensable fluids are adsorbed. The capillary limits, like a bottleneck, the amount of water being transported. The velocity of the gas stream increases to such an extent that the extracted non-condensable gases over the cold traps are prevented from back diffusion into the extraction vessel. Experiments demonstrate that a degassing procedure of about 5 min extracts more than 99.995% of the dissolved gases, but only about 0.5 g of water. As a result the extraction leads to the virtually complete separation of the gas from the water phase. If required, the degassed water can be transferred back into the original copper tube or a glass bulb and sealed off again to allow ingrowth of tritiogenic $^3He_{tri}$ for a later ^3H determination (Bayer et al. 1989; Beyerle et al. 2000a).

In the classical analytical procedure the extracted gases are dried and cleaned by passing various getter pumps. Finally a cryostatic cold trap entirely adsorbs the purified noble gases from where they are released sequentially element by element by progressive heating. Each noble gas element is expanded into an appropriate sector and/or quadrupole mass spectrometer for final determination. The mass spectrometer and especially the set up of the ion source are tuned to guarantee long term stability and reproducibility of the noble gas analysis. Commonly the most abundant noble gas isotopes are determined on Faraday cups. The rare ^{3}He isotope is counted on an electron multiplier. The ^{3}He measurement is carried out in parallel with the measurement of the abundant ^{4}He to determine a precise ^{3}He/^{4}He ratio.

Although the classical approach is analytically fairly robust and yields reliable results for the elemental abundance of noble gases in waters, there are some experimental difficulties that are delicate to deal with. The most critical point is the element specific gas release from the cryostatic cold trap. In particular the elemental selectivity of cryostatic cold traps is too coarse to completely isolate Ar from Kr. As common ion-sources are tuned to maximum sensitivity their response to different gas amounts is not only highly non-linear, but also depends on the relative presence of all components in the rest gas. Hence the Ar residual in the Kr determination may cause analytical problems, because of the overwhelming atmospheric Ar abundance.

The problem of cryostatic separation can be circumvented by a gas preparation / purification procedure that operates without any separation of the heavier noble gases, as developed by Beyerle et al. (2000a). According this analytical method the gas phase containing the heavy noble gases is diluted to such an extent that Ar does not interfere with the measurement of the other noble gases, and Ar, Kr and Xe can be analyzed in a single measuring cycle.

According to the Beyerle et al. (2000a) extraction scheme a first series of traps cooled by liquid nitrogen adsorbs all condensable gases being extracted from a water sample. The free gas phase, that contains all the He and Ne, is cleaned and expanded for the consecutive ^{3}He, He and Ne measurement. To analyze the heavy noble gases the water vapor and the rest of the condensable fluids are released by heating up the cold traps. The wet gas phase is dried by passing a zeolite trap and expanded into a large reservoir (~2000 cm^3) from which small aliquots (~1 cm^3) are taken, cleaned and prepared for the final simultaneous mass spectrometric analysis of Ar, Kr and Xe (Beyerle et al. 2000a).

Advantages of the approach 'dilution instead of separation' are that the reactive components, mainly N_2, O_2, CH_4 and CO_2, are removed easily and that repeated measurements of a particular sample are possible. Most importantly, however, the dilution step can be adjusted such, that the Ar does not perturb the Kr and Xe detection. All three heavy noble gases can be determined in one single measurement: Ar is detected on a Faraday cup, whereas Kr and Xe are counted on an electron multiplier.

As Xe concentrations depend most sensitively on the temperature during air / water exchange, an additional gas split can be prepared for a more precise Xe measurement. Note that there is no requirement of complete or even reproducible elemental separation, only the quantitative Xe trapping has to be guaranteed.

The mass spectrometric systems and the extraction procedures are regularly calibrated against air (0.4-2 cm^3 STP), a high quality noble gas standard and/or an internal water standard (Beyerle et al. 2000a). Deep water of lakes is suggested for the use as long-term water standard, because of the stable physical conditions and the virtually constant stable noble gas concentrations of such deep-water bodies. As lake waters contain reasonable tritium amounts (e.g., Torgersen et al. 1977, 1981; Aeschbach-Hertig et al. 1996a; Hohmann et al. 1998; Peeters et al. 2000a) the water standard can even be used for ^{3}H calibration.

The experimental methods for the analysis of the noble gases in meteoric water achieve an overall precision—expressed as relative error of an individual concentration—of about 1%. Selected elemental (He/Ne, Ar/Kr, Kr/Xe) and isotopic ratios (^{3}He/^{4}He, ^{20}Ne/^{22}Ne, ^{36}Ar/^{40}Ar) have errors of less than 0.6% (Beyerle et al. 2000a).

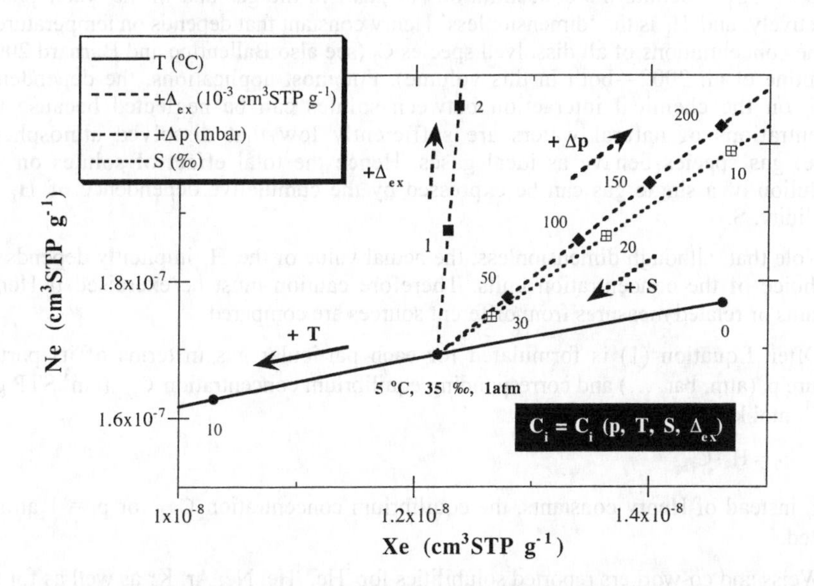

Figure 1. Effect of pressure (p), temperature (T), salinity (S), and excess air (Δ_{ex}) on gas / partitioning of Xe and Ne. Arrows indicate the increase of the noble gas concentrations in reaction to an increase of the respective property. The symbols mark the concentrations C_i[p, T, S, Δ_{ex}] being calculated for the physical conditions indicated by the numbers. Xe reacts sensitively to T changes but remains almost unaffected by injection and dissolution of air (+Δ_{ex}). Ne behaves in the opposite way. Changing either pressure (+Δp) or salinity (S) has almost the same effect on all (noble) gases. Changes of the physical conditions prevailing gas exchange are imprinted and retained the specific noble gas pattern, which in turn can be back-translated in a most direct manner into information on environmental change.

NOBLE GAS COMPONENTS IN WATER

Atmospheric noble gases 1: Solubility equilibrium

Open waters with a free interface to the atmosphere tend to dissolve atmospheric gas until the concentrations of the gas and the water phase are in a thermodynamic equilibrium. The kinetic aspects of the air / water exchange are beyond the scope of this chapter. An extended review is given by Schwarzenbach et al. (1993). Gas exchange at the lake / air interface can saturate a water layer at a rate of about 1 m per day, i.e., the controlling gas transfer velocities are on the order of 10^{-5} m s^{-1}. As gas exchange phenomena are fairly fast, the surface water of open water bodies is expected to have atmospheric noble gases in equilibrium with the atmosphere at the physical conditions prevailing (Fig. 1, Table 1). This axiomatic statement is supported by experimental evidence (Craig and Weiss 1971; Aeschbach-Hertig et al. 1999b) and holds accordingly for all atmospheric gases that have neither additional sources nor sinks and hence are bio-geochemically conservative.

Gas partitioning at the free air / water interface can be described reasonably well by Henry's law, which assumes that the concentrations in the two phases are directly proportional to each other:

$$C_i^{gas} / C_i^{water} = H_i'\left(T, C_j^{water}, ...\right) \approx H_i'(T,S) \quad (1)$$

where C_i^{gas}, C_i^{water} denote the concentrations of gas i in the gas and in the water phase, respectively, and H_i is the 'dimensionless' Henry constant that depends on temperature T and the concentrations of all dissolved species C_j (see also Ballentine and Burnard 2002; Ballentine et al. 2002—both in this volume). For most applications, the dependence of H_i on the chemical interaction between solutes can be neglected because the concentrations of natural waters are sufficiently low that dissolved atmospheric (noble) gas species behave as ideal gases. Hence the total effect of solutes on the dissolution of a single gas can be expressed by the cumulative dependence of H_i on the salinity, S.

Note that, although dimensionless, the actual value of the H_i implicitly depends on the choice of the concentration units. Therefore caution must be exercised if Henry constants or related measures from different sources are compared.

Often Equation (1) is formulated for each particular gas in terms of its partial pressure p_i (atm, bar, ...) and corresponding equilibrium concentration $C_{i,eq}$ (cm^3 STP g^{-1}, mol l^{-1}, mol kg^{-1}, mol mol^{-1}, ...)

$$p_i = H_i \cdot C_{i,eq} \quad (2)$$

Often, instead of Henry constants, the equilibrium concentration $C_{i,eq}$ for $p_i = 1$ atm is reported.

Weiss and co-workers reported solubilities for ^3He, ^4He, Ne, Ar, Kr as well as for O_2 and N_2 as a function of temperature and salinity for fresh and ocean waters (Table 1; Weiss 1970, 1971; Weiss and Kyser 1978). As this fundamental piece of work was strongly motivated by practical oceanographic research, the noble gas solubilities were expressed in the form of equilibrium concentrations with moist atmospheric air. For the atmosphere, it is justified to assume that its major elemental composition remains constant over the relevant time scales controlling gas exchange. Hence the gas partial pressure p_i can be expressed by the total atmospheric pressure p_{tot} corrected for water vapor content, $e_w(T)$, and the volume or mole fraction z_i of the gas i in dry air (Ozima and Podosek 1983).

$$p_i = z_i \cdot \left[\sum_j p_j - e_w(T)\right] = z_i \cdot \left[p_{tot} - e_w(T)\right] \quad (3)$$

The decrease in atmospheric pressure with increasing altitude h can be described by a (local) barometric altitude formula.

$$p_{tot}(h) \approx p_{tot}^{sl} \cdot e^{-h/h_{atm}} \quad (4)$$

where $p_{tot}(h)$ and p_{tot}^{sl} are the local and the sea level pressure, and h_{atm} the typical local scale height (8000-8300 m). This conversion is not unique and the pressure dependence on altitude has to be adapted to local conditions (e.g., Gill 1982).

Benson and Krause (1976), Clever (1979a,b; 1980) and others (Top et al. 1987) re-evaluated the solubilities for noble gases in pure water, whereas Smith and Kennedy (1983) expanded the noble gas solubilities to the typical salinity range of natural brines. These later works had a more chemical focus and consequently solubilities were quoted in terms of Henry constants or related measures. Comparing solubility data of different sources can cause serious pitfalls as the sets often hide principle experimental caveats. Whereas moist air equilibrium concentrations can be applied directly, other forms of solubility data have to be converted in manageable equilibrium concentrations including a salinity term (see Eqn. 11).

Table 1. Equilibrium concentrations of noble gases, nitrogen and oxygen for environmental conditions met in open water bodies.

	Equilibrium concentrations [cm^3 STP g^{-1}]						
T [°C]	**He** [10^{-8}]	**Ne** [10^{-7}]	**Ar** [10^{-4}]	**Kr** [10^{-8}]	**Xe** [10^{-8}]	**N_2** [10^{-2}]	**O_2** [10^{-3}]
	S = 0.1‰ *'Baikal' water*, p_{tot}= 1atm						
4	4.78	2.15	4.47	10.89	1.64	1.66	9.16
10	4.64	2.02	3.86	9.10	1.32	1.45	7.89
20	4.47	1.85	3.12	6.96	0.95	1.19	6.36
	S = 6‰ *'Issyk-Kul' water*, p_{tot}= 1atm						
4	4.62	2.06	4.28	10.41	1.63	1.59	8.76
10	4.49	1.94	3.70	8.71	1.31	1.37	7.56
20	4.34	1.79	3.00	6.68	0.95	1.14	6.11
	S = 12.6‰ *'Caspian Sea' water*, p_{tot}= 1atm						
4	4.45	1.98	4.07	9.89	1.63	1.51	8.35
10	4.33	1.87	3.53	8.29	1.30	1.32	7.22
20	4.19	1.72	2.87	6.38	0.94	1.09	5.85
	S = 35‰, p_{tot}=1 atm ≡ 0 masl, **Ocean**						
4	3.90	1.71	3.44	8.32	1.60	1.26	7.07
10	3.82	1.63	3.00	7.01	1.28	1.11	6.15
20	3.73	1.52	2.47	5.45	0.93	0.93	5.04
	S = 0.1‰, p_{tot}= 0.947 atm ≡ 458 masl, **Lake Baikal**						
4	4.53	2.03	4.23	10.31	1.55	1.57	8.67
10	4.39	1.91	3.65	8.60	1.25	1.37	7.47
20	4.23	1.75	2.95	6.58	0.90	1.13	6.01
	S = 6‰, p_{tot} = 0.813 atm ≡ 1608 masl, **Lake Issyk-Kul**						
4	3.76	1.68	3.48	8.45	1.33	1.29	7.11
10	3.65	1.58	3.01	7.06	1.06	1.12	6.13
20	3.53	1.45	2.44	5.40	0.77	0.92	4.94

For additional information on the gas concentrations or the water bodies refer to:

- General aspects: Aeschbach-Hertig et al. 1999b; Peeters et al. 2002.
- Lake Baikal: Hohmann et al. 1998; Aeschbach-Hertig et al. 1999b.
- Caspian Sea: Aeschbach-Hertig et al. 1999b, Peeters et al. 2000a.
- Lake Issyk-Kul: Hofer et al. 2002.

Table 1 is continued on the next page.

Table 1B. Noble gas concentrations at equilibrium may be calculated using the following parameterizations: *Equilibrium concentration of **He, Ne, Ar, Kr, (^{3}He/^{4}He)$_{eq}$ = R$_{eq}$** and **Xe**:*

$$C_{eq}^{i} = \exp\begin{pmatrix} t_1 + t_2 \cdot (100/T) + t_3 \cdot \ln(T/100) + t_4 \cdot (T/100) + \\ S \cdot \left[s_1 + s_2 \cdot (T/100) + s_3 \cdot (T/100)^2\right] \end{pmatrix} \cdot \frac{p_{tot} - e_w}{(p_{norm} - e_w) \cdot 1000}$$

$$R_{eq} = R_a / \exp\left((r_1 + r_2/T + r_3/T^2) \cdot (1 + r_4 \cdot S)\right)$$

$$X_{Xe} = \exp\left(x_1 + x_2 \cdot (100/T) + x_3 \cdot \ln(T/100)\right)$$

$$C_{eq}^{Xe} = X_{Xe} \cdot \frac{V_{Xe} \cdot z_{Xe}}{M_{H2O}} \cdot \exp\left(-NaCl \cdot [x_4 + x_5 \cdot (100/T) + x_6 \cdot \ln(T/100)]\right)$$

C^{i}_{eq}, C^{Xe}_{eq}:	Equilibrium concentration of He (= ^{4}He), Ne, Ar, Kr and Xe (cm^3 STP g^{-1})
p_{tot}, p_{norm}:	Total local atmospheric pressure (atm) and reference pressure (≡ 1 atm)
e_w:	Water vapor pressure (atm)
T, S:	Water temperature (K) and salinity (g kg^{-1})
t_i, s_i, r_i, x_i:	'Temperature', 'salinity', 'R_{eq}' and 'Xe' coefficient i (-)
R_a, R_{eq}:	Atmospheric ^{3}He/^{4}He (1.384·10^{-6}, Clarke et al. 1976) and ^{3}He/^{4}He ratio at equilibrium
X_{Xe}, V_{Xe}:	Mole fraction solubility (-) and molar volume of Xe at STP (22280.4 cm^3 STP mol^{-1})
z_{Xe}:	Atmospheric (dry air) volume fraction of Xe (8.7·10^{-8})
NaCl:	NaCl concentration (mol l^{-1}), mass of salt being interpreted as the same mass of NaCl
M_{H2O}:	Molar weight of water (~18.016 g mol^{-1})
ρ(T,S):	Density of water as function of T and S [kg m^{-3}]

coefficient	***He***	***Ne***	***Ar***	***Kr***
t_1	-167.2178	-170.6018	-178.1725	-112.684
t_2	216.3442	225.1946	251.8139	153.5817
t_3	139. 2032	140.8863	145.2337	74.4690
t_4	-22.6202	-22.6290	-22.2046	-10.0189
s_1	-0.044781	-0.127113	-0.038729	-0.011213
s_2	0.023541	0.079277	0.017171	-0.001844
s_3	-0.0034266	-0.0129095	-0.0021281	0.0011201
	R$_{eq}$		*coefficient*	***XE***
r_1	-0.0299645		x_1	-74.7398
r_2	19.8715		x_2	105.21
r_3	-1833.92		x_3	27.4664
r_4	0.000464		x_4	-14.1338
			x_5	21.8772
			x_6	6.5527

To adapt $C_{i,eq}$ to local conditions, the local atmospheric pressure has to be translated into noble gas partial pressures. This transformation has to account for the facts that the ratio of water vapor pressure to total pressure p_{tot} is variable, and that noble gas volume fractions are only known in dry air (Ozima and Podosek 1983). Hence any adaptation of solubility data to the local altitude of the surface of an open water mass has to be corrected for atmospheric water vapor pressure (see Eqn. 12).

Beyerle et al. (2000a) and Aeschbach-Hertig et al. (1999b) compared the different sets of noble gas solubilities with noble gas concentrations measured in artificially equilibrated water and in samples from lakes that can be assumed to be in equilibrium with the atmosphere (Fig. 2). Although the different solubility data sets agree on the percent level there are some deviations, especially in case of He and Ne. As a conclusion of this comparison Aeschbach-Hertig et al. (1999b) suggested to use for common application in natural meteoric waters 'Weiss' solubilities and 'Clever' solubilities being generalised by a Setchenow equation to account for salinity-dependence (see Eqn. 11).

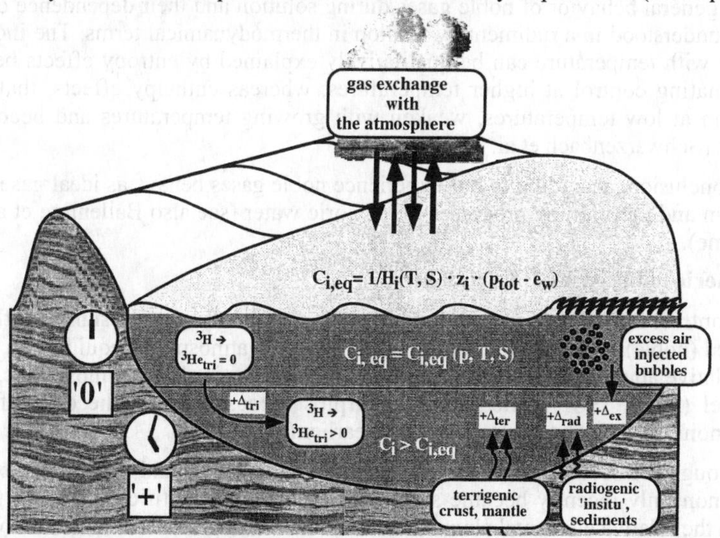

Figure 2. Origin of noble gases in lakes. Lakes have a large free surface in close contact with the atmosphere. Hence the concentrations of atmospheric (noble) gases are expected to coincide with equilibrium concentrations and to be large realtive to the radiogenic and terrigenic components. Wind-exposed lakes show small, but significant excesses of atmospheric noble gases, especially Ne is supersaturated by up to 5% ($+\Delta_{ex}$). This characteristic atmospheric excess that affects mainly light noble gas isotopes are attributed to the complete dissolution of air bubbles being submerged by wave activity. If a water parcel is transported away from the interface with the atmosphere gas exchange ceases and possible (noble) gas excesses are retained as the escape to the atmosphere is prevented. In particular non-atmospheric He ($+\Delta_{rad\ /\ tri,\ ter}$) accumulates in the deep water of lakes. $^3He_{tri}$ produced by ^3H decay in the water leads to an increase of ^3He concentration with time. The ratio $^3He_{tri}/^3H$ can be used for dating purposes. The deep water of lakes are only gently renewed during seasonal or even longer lasting stagnation periods and accumulate terrigenic and radiogenic He emanating from the sediments into the water body ($+\Delta_{rad,\ ter}$).

The noble gas equilibrium concentrations record changes in physical conditions during air / water partitioning, such as changes in (soil) temperature, altitude or pressure, and salinity. In the temperature range relevant for the environment, the noble gas solubilities in water—as for all other poorly soluble gases—generally decrease with increasing temperature (Fig. 1, Table 1). Interestingly, at temperatures higher than 60°C noble gas concentrations increase dramatically with increasing temperature (Crovetto et al. 1982). Generally noble gases become less soluble with increasing salinity (Fig. 1, Table 1, Smith and Kennedy 1983; Suckow and Sonntag 1993). The temperature and salinity dependence increases with the atomic mass of the noble gas and is most evident for Xe (Fig. 1, Table 1).

Both effects—the dependence on T and on S—can readily be explained by a conceptual model that describes poorly soluble substances in water (Schwarzenbach et al.

1993). The model conceptualises the solution of (noble) gases on the microscopic level as cavities built by water molecules that trap individual (noble) gas atoms. The attracting forces between water and host increase with the atomic radius and the dielectric constant of the (noble) gas. In consequence, the intermolecular forces increase with molecular mass. This explains why the ratios of elemental noble gas concentrations in water at atmospheric equilibrium are enriched with respect to the atmospheric abundance in favor of the heavier noble gases.

The general behavior of noble gases during solution and their dependence on T and S can be understood in a rudimentary fashion in thermodynamical terms. The increase in solubility with temperature can be qualitatively explained by entropy effects becoming the dominating control at higher temperatures, whereas enthalpy effects, that govern dissolution at low temperatures, weaken with growing temperatures and become less important (Schwarzenbach et al. 1993).

In conclusion, according to our experience noble gases behave as ideal gases during dissolution and partitioning processes in meteoric water (see also Ballentine et al. 2002, this volume).

Atmospheric noble gases 2: Excess Air

In contrast to surface waters, ground waters commonly contain significantly larger air-derived (noble) gas concentrations than the expected atmospheric equilibrium (Fig. 3). As the relative abundance of the noble gases in the excess is almost atmospheric, Heaton and Vogel (1979, 1981) called the gas surplus 'excess air'. In the case of Ar the phenomenon was described at least 20 years earlier (Oana 1957).

Although the occurrence of excess air is ubiquitous in ground waters, the excess air phenomenon only recently became subject of a detailed scientific discussion (Fig. 4). Although the relative elemental abundance of the observed gas excess of ground waters is often atmospheric, some ground waters show noble gas excesses that are not simply caused by injection and complete dissolution of pure air (Stute et al. 1995b; Aeschbach-Hertig et al. 1999b, 2000; Ballentine and Hall 1999; Weyhenmeyer et al. 2000; Holocher et al. 2001). All these waters have atmospheric noble gases in excess, but the heavier noble gases are more enriched than the light noble gases, i.e., the excess is fractionated relative to pure air (Figs. 4 and 5).

Until now, two models have been proposed to conceptualisze the fractionation of excess air in ground waters (Fig. 5). Although both models describe only the bulk behavior of atmospheric gas dissolution, the two concepts are able to describe the observed noble gas abundances, including the possible fractionation of the excess in favor of the heavy noble gases.

In the first model Stute (1989; 1995b) interpreted the fractionation in terms of a two-step concept. In a first step, ground water inherits pure excess air during ground-water recharge (unfractionated excess, UA-model). In the second step the initial pure air excess is partly lost by diffusively controlled gas exchange through the water / soil air interface which results in the partial re-equilibration of the ground water with respect to free atmospheric conditions (PR-model). Although there is no direct experimental evidence to support the PR-model, the development of a correction scheme to separate the fractionated excess component from the equilibrium concentrations allowed Stute et al. (1995b) to reconstruct successfully the temperature shift during the Pleistocene / Holocene transition in tropical Brazil.

In many cases the PR-model requires large amounts of initially dissolved excess air which can be several hundred percent of the equilibrium Ne concentration. In contrast, the observed supersaturation of ground waters due to the presence of excess air is typically in the range of 10-50%. Hence the PR-model implies that the diffusive processes expel most of the initially dissolved excess air. Except for some rare Pleistocene ground waters, being recharged possibly in close vicinity of large inland ice masses of northern Europe (Juillard-Tardent 1999), the extreme amounts of initial excess air required by the PR-model appear to be unrealistic. The dissolution of enormous amounts of air would either force degassing or would demand that hydrostatic pressure in response to ground-water recharge increase by several atmospheres to keep the excess air in solution.

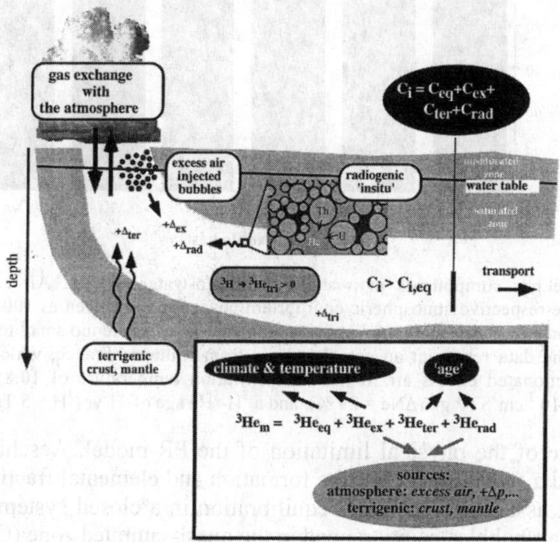

Figure 3. Origin of noble gas components in ground waters. Atmospheric gases are dissolved in ground waters near the water table by gas exchange between water and soil air. However during soil air / water partitioning ground waters dissolve atmospheric (noble) gases in significant excess ($+\Delta_{ex}$) compared to the equilibrium concentrations. This excess air component is most probably generated by re-equilibration between water with entrapped air that can cover up to 10% of the available pore space in the quasi-saturated zone. The slightly enhanced pressure prevailing in this soil zone leads to a new secular equilibrium between entrapped air and the super saturated water that is enriched relative to air in favor of heavy noble gases (note, hydrostatic overload prevents the supersaturated gases from degassing). During the water flow through the aquifer the ground water continuously accumulates non-atmospheric noble gases, in particular He of terrigenic ($+\Delta_{ter}$) and radiogenic ($+\Delta_{rad\ /\ tri}$) origin. The measured noble gas concentrations exceed commonly the expected equilibrium concentrations because the noble gas abundance represents a mixture of various components in varying proportions. Thereby air-derived noble gases can be interpreted in terms of the physical and climatic conditions prevailing ground-water recharge, radiogenic and terrigenic noble gases carry information on water residence time and on the origin of terrigenic fluids.

The conceptual problem of the PR-model can be overcome if it is assumed that theexcess gas does not enter ground water in a large single event of injection and degassing, but in a cycle of various injection and degassing steps (multi-step re-equilibration, MR-model, Fig. 5). Each step injects a much smaller amount of excess air and hence remains physically more acceptable. It can be shown that a multiple stepwise build-up of excess air leads in the limiting case to a similar fractionation pattern as single step formation (Eqn. 26). Aquifers that undergo ground-water table fluctuations in response to intermittent recharge events might conceptually experience such step-wise formation of excess air.

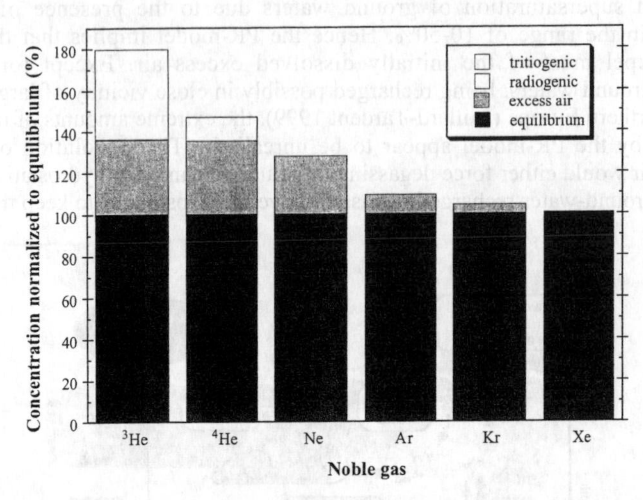

Figure 4. Noble gas components in a shallow ground-water sample. All concentrations are normalized to the respective atmospheric equilibrium concentration (given as 100 %). Note that the excess air component for Xe and the radiogenic component for ^{3}He are too small to be clearly visible on this scale. The data represent an actual sample from southern France, which was interpreted assuming unfractionated excess air. It has an infiltration temperature of 10.8°C, an excess air component of 3×10^{-3} cm^{3}STP g^{-1} (ΔNe = 29 %), and a ^{3}H-^{3}He age of 21 yr (^{3}H = 5 TU).

Being aware of the physical limitation of the PR-model, Aeschbach-Hertig et al. (2000) suggested a second model of the formation and elemental fractionation of excess air. The concept assumes (noble) gas equilibration in a closed system between ground water and small air bubbles being trapped in the quasi-saturated zone (CE-model, Fig. 5). The unsaturated and saturated zones of soils are not separated by a sharp interface. Rather the wetting conditions change over the transition of the quasi-saturated zone, where water saturation widely prevails, but where up to 10-20% of the pore space is still occupied by immobile entrapped soil air bubbles. This 'entrapped air' is a major control of ground-water recharge. It builds a rigid, but transient barrier against water infiltration into soils (Christiansen 1944; Faybishenko 1995).

Such entrapped air is conceptually thought to be a gas reservoir of restricted volume that does not exchange with the free soil air. In the quasi-saturated zone, water velocities are small (<0.5 m d^{-1}, Faybishenko 1995) and the pressure is slightly enhanced relative to the free atmospheric pressure due to the hydrostatic loading. Under these conditions it can be assumed that the (noble) gases re-partition by establishing a new equilibrium between the entrapped air and the water. As the soil gas volume is limited and the total amount of gas has to be conserved, the volume ratios of the atmospheric gases in the bubbles as well as the gas concentrations in the water have to change according to the different physical condition of the quasi-saturated zone. Because partitioning of individual gas species is controlled by the specific Henry coefficient, the more soluble gases are enriched in the water phase with respect to pure air. The relative elemental (noble) gas abundance in the gas phase is fractionated in favor of the less soluble (larger Henry coefficient) gases. In consequence the CE-model predicts the typical fractionation pattern of excess air in ground water with its characteristic enrichment of the heavier noble gases.

Although both concepts on the formation of excess air can produce the 'ground-water-like' enrichment of heavy noble gases, it has to be realised that the underlying

a) Unfractionated excess Air (UA) by complete dissolution of entrapped air

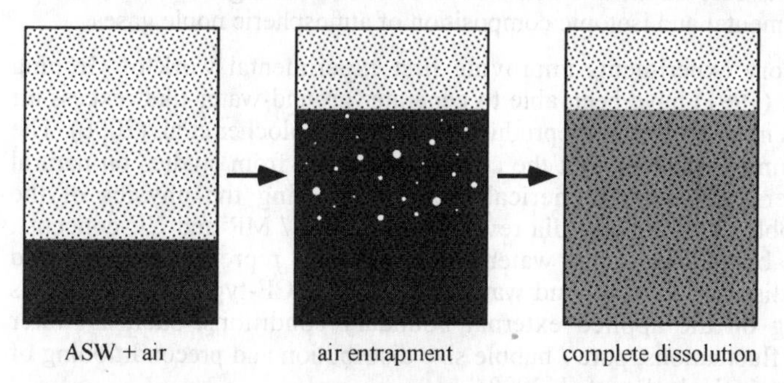

b) Partial Re-equilibration (PR) with the atmosphere

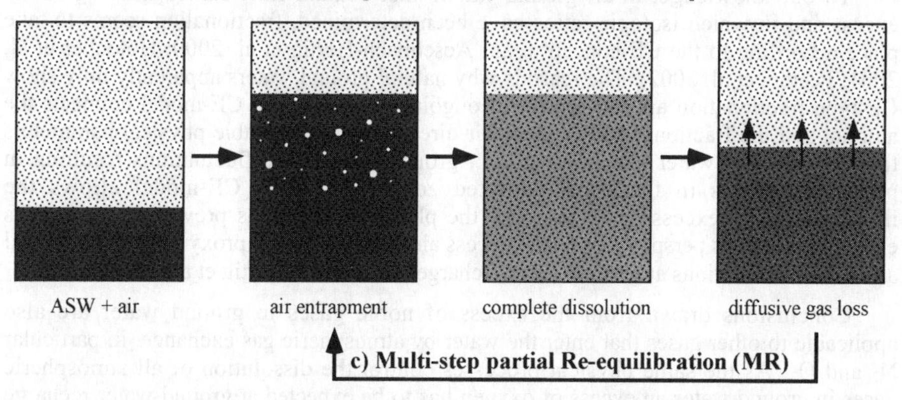

d) Closed-system Equilibration (CE) with entrapped air

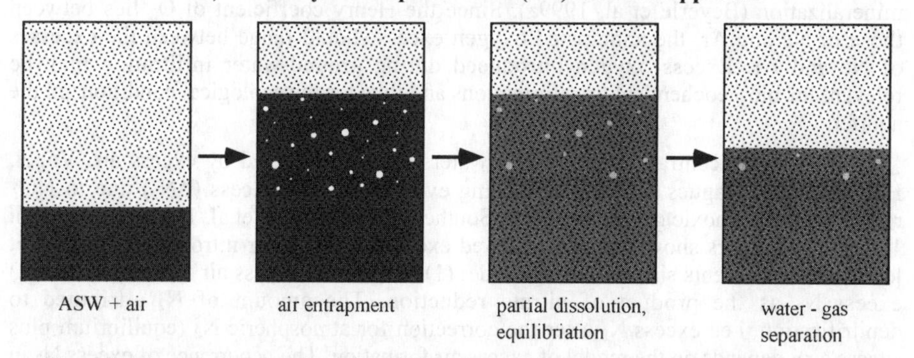

Figure 5. Schematic representation of the different model concepts for the formation of excess air in ground water. Note that different greyscales indicate changes in the noble gas composition of the respective reservoir.

physical processes are completely different. Whereas the PR / MR-model assumes that noble gas fractionation is controlled by molecular diffusion, the CE-model explains fractionation as the result of a new equilibrium between the dissolved gases in the water and the gas phase of the entrapped air. In case of moderate amounts of excess air and small fractionation both concepts lead to similar predictions of elemental (noble) gas fractionation. But in the case of intense fractionation the models significantly differ in their predictions of elemental and isotopic composition of atmospheric noble gases.

Recently, laboratory experiments improving first experimental work of Gröning (1989) and Osenbrück (1991) have been able to generate 'ground-water-like' excess air in simple soil columns in a reliable and reproducible manner (Holocher et al. (2000). The results from the column experiments put the conclusion drawn from studies on natural systems in a new perspective. A numerical model describing the physics of the dissolution of gas bubbles in porous media reveals that the PR / MR and CE-concepts, that capture only the bulk behavior of water and excess air, represent extreme and limiting cases of gas dissolution in ground water. PR / MR or CE-type fractionation is reproduced depending on the applied external boundary conditions, such as water discharge, water table fluctuations, pore / bubble size distribution and preconditioning of the entrapped air reservoir (Holocher et al. 2000).

To our knowledge, in all ground waters that contain fractionated noble gases in excess and in which isotopic ratios have been determined, fractionation seems to take place according to the CE-concept (e.g., Aeschbach-Hertig et al. 2000; Holocher et al. 2001; Peeters et al. 2002). The reasons why natural ground waters apparently only show CE-type fractionation are the subject of ongoing research. The CE-model connects the amount and the fractionation of excess air directly with observable physical parameters that control air / water partitioning, e.g., ground-water table fluctuations resulting in pressure changes in the quasi-saturated zone. Hence the CE-model allows the interpretation of excess air in terms of the physical conditions prevailing during gas exchange. It opens perspectives to use excess air as an additional proxy to reconstruct soil and climate conditions at ground-water recharge (Aeschbach-Hertig et al. 2001).

Conclusions drawn from the excess of noble gases in ground water are also applicable to other gases that enter the water by atmospheric gas exchange, in particular N_2 and O_2. As the same physical processes control the dissolution of all atmospheric gases in ground water an excess of oxygen has to be expected at ground-water recharge assuming that oxygen partial pressure of the soil air is not already reduced by mineralization (Beyerle et al. 1999a). Since the Henry coefficient of O_2 lies between those of Ne and Ar, the anticipated oxygen excess should range between the excesses of Ne and Ar. Excess oxygen introduced during ground-water infiltration may be relevant for bio-geochemical transformations and thus for the ecological conditions in the aquifer.

Nitrogen concentrations in ground waters are also affected by excess air. Vogel, Heaton and colleagues provide convincing evidence for N_2 excess (ΔN_2) due to NO_3 reduction in old anoxic ground waters in Southern Africa (Vogel et al. 1981; Heaton et al. 1983). The authors show that the observed excess of molecular nitrogen consists of at least two components similar in magnitude: (1) ΔN_2 due to excess air formation, and (2) excess N_2 as the product of nitrate reduction. The amount of N_2 attributed to denitrification, i.e., excess N_2 after the correction for atmospheric N_2 (equilibrium plus excess air), depends on the model of excess air formation. The occurrence of excess N_2 in anoxic meteoric waters does not necessarily indicate denitrification and may be related to excess air. ΔN_2 is only an indicator for denitrification, if excess air and possible fractionation processes are accounted for.

Excess air is a ubiquitous and important constituent of atmospheric (noble) gases in ground waters, whereas atmospheric (noble) gas excesses in the open waters of lakes are of far less importance. Nevertheless large wind exposed lakes show sometimes small air excess amounts ranging up to few percent in case of Ne (Fig. 2, Aeschbach-Hertig et al. 1999b; Kipfer et al. 2000; Peeters et al. 2000a).

Radiogenic He (and Ar)

Radiogenic noble gases are produced either directly by radioactive decay or indirectly by subsequent nuclear reactions triggered by the initial radioactive disintegration. ^{3}He, ^{4}He, and to a lesser extent also ^{40}Ar, are the most prominent noble gas isotopes that are produced by the decay of their radioactive precursors ^{3}H, U, Th, other α emitting isotopes, and ^{40}K. α decay induces secondary nuclear reactions in rock and minerals that produce several noble gas isotopes (Ozima and Podosek 1983; Mamyrin and Tolstikhin 1984; Andrews et al. 1989; Lehmann and Loosli 1991; Ballentine and Burnard 2002 in this volume). But only ^{3}He is generated by these secondary reactions in sufficient amounts that it can be observed routinely in natural waters. All other noble gas isotopes have such low production yields that they are hardly detectable against their atmospheric background. The only exception may be $^{21}Ne_{rad}$ in very old ground waters (e.g., Bottomley et al. 1984).

Due to its unique chemical properties that are also for noble gases exceptional—low atmospheric abundance in combination with low solubility, high diffusivity and high mobility—He tends to accumulate in fluid phases if they are not in direct contact with the atmosphere. Hence, both He isotopes of radiogenic origin convey information about the time elapsed since gas exchange with the atmosphere. This so called 'water age' is a measure of the water residence time. ^{4}He concentrations are commonly interpreted in qualitative terms as residence time of old (ground) waters that recharged on times scales of thousands of years. ^{3}He generated by the decay of atmospheric ^{3}H (tritiogenic $^{3}He_{tri}$) is successfully used in quantitative terms ...

i. ...to date young ground waters having residence times up to 50 years (e.g., Tolstikhin and Kamenskiy 1969; Schlosser et al. 1988, 1989; Solomon et al. 1993; Szabo et al. 1996; Cook and Solomon 1997; Stute et al. 1997; Aeschbach-Hertig et al. 1998; Dunkle Shapiro et al. 1998, 1999; Beyerle et al. 1999a; Holocher et al. 2001), and

ii. ...to analyze deep water exchange and determine mixing rates in lakes (Torgersen et al. 1977; Aeschbach-Hertig et al. 1996a; Hohmann et al. 1998) and oceans (Peeters et al. 2000a; Schlosser and Winckler 2002, this volume).

Terrigenic He

The term 'terrigenic noble gases' describes noble gas components originating from different geochemical reservoirs of the solid Earth, such as the Earth's mantle or rocks of the continental crust. Noble gases are among the most prominent defining features of fluids from these large geochemical regions that are characterized by their particular isotope geochemistry.

Part of the radiogenic ^{3}He production is causally linked to α decay, i.e., the production of ^{4}He (e.g., Mamyrin and Tolstikhin 1984; Ballentine and Burnard 2002 in this volume). Such type of He is defined by a uniform ^{3}He/^{4}He ratio that intrinsically characterizes the geochemical environment.

As He is the most mobile noble gas, it emanates and ascends from deeper strata towards the Earth surface. This general He degassing from the Earth is often assessed in terms of pseudo-diffusive He fluxes from different geochemical reservoirs (e.g., O'Nions and Oxburgh 1983; Oxburgh and O'Nions 1987; O'Nions and Oxburgh 1988; Torgersen 1989). Deep circulating ground waters trap the ascending He. Hence He accumulates continuously in ground waters. Independent of its geochemical classification the accumulating He is a qualitative measure for the ground-water residence time (Fig. 3). Radiogenic and / or terrigenic He concentrations can even be interpreted quantitatively if the respective (terrigenic) He fluxes and / or the (radiogenic) He production rates are known.

Similarly He from the geological basin enters a lake where He is transported passively by the general mixing processes that control water exchange within the water body (Fig. 2). If the mixing dynamics of a specific lake is known, the continental terrigenic noble gas fluxes can be determined, i.e., noble gas fluxes from the continental crust and / or the Earth's mantle can be quantified (Torgersen et al. 1981; Igarashi et al. 1992; Kipfer et al. 1994; Hohmann et al. 1998).

Radioactive noble gas isotopes

Nuclear reactions produce not only stable, but also some radioactive noble gas isotopes. Due to their radioactivity the natural background of these isotopes is small and the nuclei are easily detected by their radioactive decay. Although radioactive, these isotopes are noble gases and thus behave chemically inert. As a result, any change in concentration is controlled solely by physical processes, such as radioactive production / decay and mixing of different components. The concentrations of radioactive noble gas isotopes therefore can most directly be employed to calculate the time elapsed since the system was isotopically closed, i.e., the time since radioactive decay alone determined the change of the concentrations of the radioactive isotopes. Some radioactive noble gas isotopes have half-lives similar to renewal times of natural water resources and hence can be used to determine water residence times.

The most prominent noble gas isotopes used in hydrology are ^{222}Rn, ^{85}Kr, ^{39}Ar and ^{81}Kr with half-lives of 3.82 d, 10.76 yr, 269 yr, and 230,000 yr (Fig. 6, Loosli et al. 2000). Due to its short half-life ^{222}Rn is suitable to trace processes in the aquatic environment that have short time constants. ^{39}Ar fills the time gap between ^{14}C and tracers used to date young ground water (e.g., ^{3}H-^{3}He, ^{85}Kr, and man made trace gases). ^{81}Kr is the isotope of choice to determine water residence times in ground waters that recharge on times scales of 10^5 years (Collon et al. 2000; Loosli et al. 2000).

With the exception of ^{222}Rn that can be continuously measured by α counting, detection and quantitative measurement of the other isotopes are experimental challenges as the particular noble gas abundances / activities are very small. Whereas the atomic abundance of ^{85}Kr, i.e., the ^{85}Kr/Kr ratio, is on the order of $1:10^{11}$, which is still relatively high for radioactive noble gases, ^{39}Ar and ^{81}Kr have extremely low atomic abundances on the order of $1:10^{15}$. Detection of these radioactive noble gases requires sample sizes of 500 l of water in case of ^{39}Ar, and about 15'000 l in case of ^{81}Kr. The water is usually degassed during field operations to avoid the transport of huge amounts of water (Loosli 1983; Collon et al. 2000; Loosli et al. 2000).

Because the half-lives of ^{85}Kr and ^{39}Ar are comparatively short, the radioactive activity and hence the concentration of the nuclides can be measured by ultra-low level counting techniques. The proportional counters need to be installed in shielded underground laboratories constructed by materials with very low natural activity (Loosli 1983; Loosli et al. 2000). The ^{81}Kr analysis is even more ambitious because it requires access either to synchrotron accelerators (Collon et al. 2000) or ultra sensitive laser techniques that are able of ionising a single isotope of an element (Ludin and Lehmann 1995). Because of the extreme requirements on sampling and detection, only few groups world wide, including the Institute of Environmental Physics of the University of Bern, have the experimental experience to extract and determine noble gas radioisotopes in natural waters. Comprehensive reviews on the use of radioisotopes in water are given in Loosli (1983), Lehmann et al. (1993), and Loosli et al. (2000).

In spite of the experimental difficulties there are at least two convincing reasons why radioactive noble gases are measured and applied to study transport in aquatic environments.

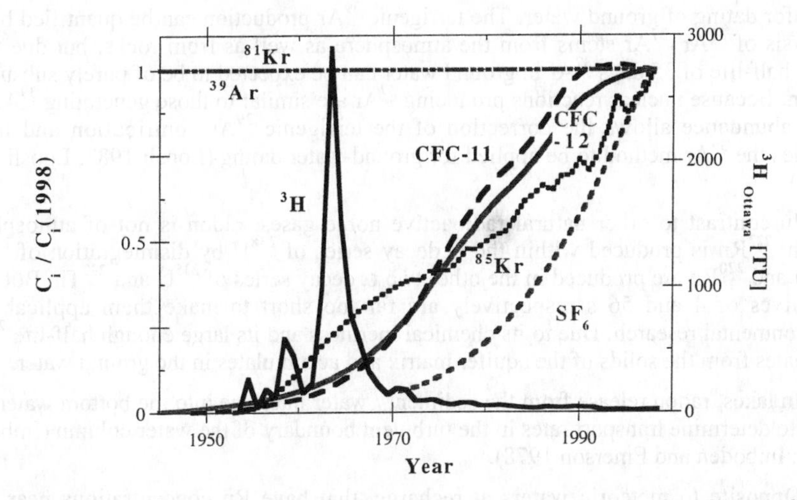

Figure 6. Atmospheric concentrations as functions of the radiogenic noble gases ^{39}Ar, ^{81}Kr, and ^{85}Kr and of the man made traces gases CFC-11, CFC-12, and SF_6. Concentrations are normalized to the respective atmospheric abundances in 1998. In addition a typical (input) curve of ^{3}H in precipitation is shown. Note, nuclear bomb tests in the atmosphere increased the global tritium inventory by up a factor of 1000. Since then ^{3}H concentrations steadily decreased, but still today tritium activity in young meteoric waters is commonly one order of magnitude larger than the natural background due to cosmic rays production.

i. The natural time scales of the dynamics of oceans and ground waters agree with the time-range assessable by radioactive noble gases and which hardly can be approached by any other tracer.
ii. The time information is recorded by the ratio of the radioactive and the stable isotope (NG_{ra}/NG_{st}) and thus does not depend on absolute concentrations.

^{39}Ar, ^{81}Kr and ^{222}Rn are of natural origin and are produced by nuclear reactions. The two lighter isotopes are generated predominantly in the atmosphere by cosmic ray / air interaction, e.g., spallation of ^{40}Ar yields ^{39}Ar, whereas ^{81}Kr originates from nuclear reactions with stable Kr isotopes (Loosli 1983; Lehmann et al. 1993; Loosli et al. 2000). As the typical mixing time of the atmosphere of only a few years is much shorter than the half-lives of the two isotopes, the atmospheric abundances of ^{39}Ar and ^{81}Kr are in first approximation at steady state and reflect the equilibrium between production due to cosmic ray interaction and radioactive decay (Fig. 6). Together with the stable isotopes of the respective element the 'atmospheric' radioactive noble gases, ^{39}Ar, ^{81}Kr (as well as ^{85}Kr) enter the meteoric water cycle via gas exchange with the atmosphere. Air / water partitioning determines the NG_{ra}/NG_{st} ratio in the water, whereby the inherited ratio coincides with the atmospheric ratio. After the water has been moved away from the air / water interface, radioactive decay remains the only process affecting the radioisotope concentrations. As a result, the ratio NG_{ra}/NG_{st} decreases with time and can be used as a measure of the time that elapsed since the water was last in contact with the atmosphere or soil air.

Besides atmospheric sources radioactive noble gases, in particular ^{37}Ar and ^{39}Ar, are produced in rocks an minerals (Loosli 1983, Loosli et al. 2000). The Ar isotopes are produced by natural nuclear reactions in the solids of the aquifer matrix, from where the generated nuclei tend to emanate and to accumulate in the surrounding ground waters. The subsurface production sets a natural limit down to which the radioisotope can be used for dating of ground water. The terrigenic ^{39}Ar production can be quantified by the analysis of ^{37}Ar. ^{37}Ar stems from the atmosphere as well as from rocks, but due to its short half-life of 35 days ^{37}Ar in ground water can be expected to be of purely subsurface origin. Because nuclear reactions producing ^{37}Ar are similar to those generating ^{39}Ar, the ^{37}Ar abundance allows the correction of the terrigenic ^{39}Ar contribution and hence enables the ^{39}Ar method to be applied for ground-water dating (Loosli 1983; Loosli et al. 2000).

In contrast to other natural radioactive noble gases, radon is not of atmospheric origin. ^{222}Rn is produced within the α decay series of ^{238}U by disintegration of ^{226}Ra. ^{219}Rn and ^{220}Rn are produced in the other two α decay series of ^{235}U and ^{232}Th. But their half-lives of 4 and 56 s respectively are far too short to make them applicable in environmental research. Due to its chemical inertness and its large enough half-life ^{222}Rn emanates from the solids of the aquifer matrix and accumulates in the ground water.

In lakes, radon release from the sediment / water interface into the bottom water was used to determine transport rates in the turbulent boundary of the water column (Imboden 1977; Imboden and Emerson 1978).

Opposite to meteoric waters at recharge, that have Rn concentrations near zero reflecting the virtual absence of Rn in the atmosphere, ground waters show Rn concentrations near the expected production equilibrium characterizing the surrounding aquifer matrix. Hence all deviation of the Rn concentration towards smaller values can be interpreted in terms of admixture of very recent recharge into 'older' ground water or in terms of the travel time of the recent recharge, for example the transgression of riverine water through the transition of hyporeic zone, i.e., the water just below the river bed, into the near by ground water (Hoehn and von Gunten 1989).

^{85}Kr is not of natural origin but man made. The isotope is released from re-processing plants during recycling of spent nuclear fuel (Lehmann et al. 1993; Lehmann and Purtschert 1997; Loosli et al. 2000). As a consequence of the increased re-processing demands the ^{85}Kr activity in the atmosphere continuously and steadily increased since the 1950s reaching an activity of 1.4 Bq m^{-3} in 2000.

As long as water is in contact with the atmosphere, the ^{85}Kr/Kr ratio in the water reflects the atmospheric input function (Fig. 6). After the water is moved away from the air / water interface, gas exchange is interrupted and the ^{85}Kr/Kr ratio decreases due to radioactive decay. In order to date a young ground water that recharged on time scales similar to the half-life of ^{85}Kr, the measured ^{85}Kr/Kr ratio is calculated back in time to compensate for radioactive decay. The back-extrapolated ^{85}Kr/Kr ratio is compared with the known atmospheric input function. It can be shown that, as long as the atmospheric ^{85}Kr activity monotonously increases in time, the two functions only have one single intersection. The intersection marks the time of the last gas exchange between the water and the atmosphere.

The application of ^{39}Ar, ^{81}Kr, and ^{85}Kr for dating purposes requires only the measurement of a single isotopic ratio NG_{ra}/NG_{st}. In contrast to any individual gas concentration, the ratio NG_{ra}/NG_{st} is hardly affected by fractionation processes. Therefore dating based on radioactive noble gases can be employed even if the water is affected by strong elemental fractionation or by phase separation, e.g., during degassing upon sampling. Hence especially the use of ^{85}Kr is expected to have a large potential to date young ground water being affected by organic contamination. Such contaminated ground waters produce large amounts of CH_4 and CO_2 during organic degradation. The generated gases may trigger water / gas partitioning that alters the absolute gas concentrations, but cannot significantly bias isotopic ratios.

Man-made pseudo-conservative trace gases

Certain man-made compounds, such as chloro-fluoro-carbons, CFCs, and sulphur-hexafluoride, SF_6, are chemically non-reactive and behave almost as inert gases. SF_6 and the CFCs, particularly CFC-11 (CCl_3F) and CFC-12 (CCl_2F_2), stem from industrial production and are released to the atmosphere. There, these trace components accumulate due to their long atmospheric residence time of up to hundreds of years (Cunnold et al. 1994).

In response to increasing industrial use, atmospheric CFC concentrations continuously and rapidly rose between the 1930s and 1995 (Fig. 6, Busenberg and Plummer 1992). Then the international environmental regulations banning production of CFCs showed first impacts. The atmospheric SF_6 concentration increases exponentially since 1970 (Fig. 6). With regard to the changing atmospheric partial pressure of the conservative anthropogenic trace gases, their particular concentrations in a water parcel can be interpreted in terms of the time elapsed since the last gas exchange. Hence CFCs and SF_6 concentrations yield information on water renewal and recharge rates in lakes (Hofer et al. 2002; Vollmer et al. 2002) and ground waters (Busenberg and Plummer 2000) and thus complement the information on water dynamics obtained from noble gas dating techniques.

The combination of dating methods restricts the estimation of the mean residence time of aquatic systems much more than the use of only one technique (e.g., Hofer et al. 2002; Vollmer et al. 2002; Zeollmann et al. 2001). Because these trace gases are mostly chemically and biologically inert, they are mainly subject of physical processes in the water cycle. Thus in principle the chemical and biological alterations in the water body can be separated from the common physical transport.

Gas exchange at the air–water interface and mixing processes have different effects on different transient tracers, e.g., CFCs and rare gases are much more influenced by gas exchange with the free atmosphere and/or soil air than tritium being bound in water molecules ($^1H^3HO$). Hence the combination of tracer techniques enables not only the dating of water masses (e.g., Busenberg et al. 1993; Busenberg and Plummer 2000, Schlosser and Winckler 2002), but also the study of the dynamics of the gas–water interaction in the unsaturated zone (Cook and Solomon 1995; Brennwald et al. 2001).

DATA EVALUATION AND INTERPRETATION

Several different noble gas (particularly He) components can be present in natural waters. The information that we hope to obtain from the study of dissolved noble gases is related to the individual components. For instance, the radiogenic and tritiogenic He components contain age information, whereas the solubility equilibrium components of all noble gases may be used to derive the equilibration temperature. For the interpretation and use of dissolved noble gases, it is therefore necessary to evaluate the data in terms of individual components. In this section, we discuss techniques to separate the components contributing to the measured total noble gas concentrations and isotope ratios.

The component separation approach is based on the assumption that each source of noble gases contributes a single, well-defined portion to the total noble gas concentrations in a sample. This assumption can only be strictly valid if the water parcels retain their unique identity, i.e., if mixing and diffusion are negligible. The effect of mixing or diffusion depends on concentration gradients, which may differ between elements and even isotopes, thus resulting in noble gas compositions that cannot be interpreted as a sum of single components. For instance, mixing of two water parcels that equilibrated with air at different temperatures creates a composition that does not exactly correspond to equilibrated water of any temperature. Of course, mixing, diffusion, and dispersion are always present in lakes and ground waters. Thus, in principle, the correct approach to interpret the data is to model the evolution of noble gases in the complete aquatic system under investigation and to use only the measured total noble gas concentrations to constrain model parameters. However, in practice this approach is often not feasible, whereas the component separation provides a direct way to derive valuable information about the studied system. In many cases, the approximation underlying the component separation appears to be reasonably well justified and to lead to reliable and useful results.

Any attempt to calculate individual noble gas components present in a sample must be based on a conceptual model, specifying which components are assumed to be present and what their composition is. In the first part of this section, we will discuss the models that up to the present have been developed. In the second part, actual techniques to perform the component separation are discussed. Finally, we introduce the basic concepts that are needed to interpret the inferred components in terms of age, temperature, or other quantities.

Conceptual models for noble gases in water

All conceptual models that have been applied to interpret dissolved noble gases in lakes and ground waters suppose that the measured concentrations are a sum of (some of) the components introduced in the previous section. In a general way, the basic model equation may be written as (compare Figs. 2, 3 and4):

$$C_{i,m} = C_{i,eq} + C_{i,ex} + C_{i,rad} + C_{i,ter} \tag{5}$$

where C_i stands for the concentration of the noble gas isotope i, and the subscripts indicate the components (m: measured (total), eq: atmospheric solubility equilibrium, ex: excess air, rad: radiogenic, ter: terrigenic). Obviously, a separation of the components in Equation (5) requires measurement of several noble gas isotopes and some knowledge of the elemental and / or isotopic composition of the individual components.

An important simplification results from the fact that the components of non-atmospheric origin (radiogenic and terrigenic) play a role only for certain isotopes. With the exception of He, all stable noble gases have at least one isotope that usually is only of atmospheric origin (e.g., ^{20}Ne, ^{36}Ar, and virtually all Kr and Xe isotopes). Therefore, it is usually sufficient to have a model for the noble gas components of atmospheric origin (equilibrium and excess air), and to calculate the non-atmospheric contributions to certain isotopes from the difference between their measured concentrations and the model predictions for their atmospheric components. Hence, we focus on models of the dissolved noble gases of atmospheric origin in ground water. We first address the atmospheric equilibrium component, and then discuss several descriptions for the excess air component.

Atmospheric solubility equilibrium. The dissolved concentrations of the noble gases in equilibrium with the atmosphere can easily be calculated from Henry's law (Eqn. 2), using the noble gas partial pressures in moist air given by Equation (3). The practical problem is to calculate the Henry coefficients $H_i(T,S)$ in appropriate units. Several different expressions for gas solubilities are common in the literature and a variety of units for the concentrations in the two phases are used. In Equation (2), the units atm for p_i and cm^3 STP g^{-1} for C_i are widely used in noble gas studies. For some calculations it is convenient to use the same concentration units (e.g., mol m^{-3}) for both phases, resulting in the 'dimensionless' Henry coefficient H_i', as in Equation (1). The conversion from H_i' in volumetric units (e.g., (mol/l$_{gas}$)(mol/l$_{water}$)$^{-1}$) to H_i in units atm (cm^3 STP g^{-1})$^{-1}$ is a simple application of the ideal gas law yielding

$$H_i = H_i' \cdot \frac{T}{T_0} \cdot P_0 \cdot \rho_w \tag{6}$$

where T_0 and P_0 are standard temperature and pressure, and ρ_w is the density of water. In these units, the coefficients H_i and H_i' have almost the same numerical values, because P_0 = 1 atm, $\rho_w \approx 1$ g cm^{-3}, and $T/T_0 \approx 1$ for temperatures not too far from T_0 = 273.15 K.

Yet, none of these units is common in the literature on gas solubilites. For the determination of gas solubilities in controlled lab experiments, often a pure gas phase at known pressure (usually 1 atm) is brought into contact with a known mass or volume of the solvent (e.g., pure water). A convenient way to express gas solubilities is then the Bunsen coefficient β_i, which is the volume of gas i at STP adsorbed per volume of solvent if the gas pressure p_i is 1 atm. Alternatively, the so-called mole fraction solubility X_i states the mole fraction of gas i in the solution at equilibrium with the pure gas at 1 atm. In field samples one usually determines the amount of dissolved gases per mass of solution.

Strictly speaking, the concentration measures 'per volume of solvent' (Bunsen), 'per number of moles in the solution' (mole fraction), and 'per mass of solution' (samples) are not linearly related, and hence Henry's law cannot simultaneously be valid for all forms. To illustrate the problem, consider the conversion from mole fraction concentrations x_i to per weight concentrations C_i:

$$x_i \left[\frac{mol_i}{mol_{solution}} \right] = \frac{n_i}{n_s + n_i} \tag{7}$$

$$C_i \left[\frac{mol_i}{g_{solution}} \right] = \frac{n_i}{M_s n_s + M_i n_i} \tag{8}$$

where n is the number of moles and M the molar mass (g mol^{-1}) and the subscripts refer to the dissolved gas i and the solvent s. The conversion from x_i to C_i is:

$$C_i = \frac{1}{M_s} \frac{1}{1/x_i - 1 + M_i/M_s} \tag{9}$$

which obviously is non-linear. If C_i were defined as 'per mass of solvent', then the term M_i/M_s in the denominator of Equation (9) would disappear. This expression is sometimes found in the literature. However, since actual measured concentrations of dissolved gases in field samples are per mass of solution, it is hardly appropriate in practice.

For low soluble gases such as noble gases and at low pressures, the contribution of the dissolved gases to the mass or number of moles of the solution is for all practical purposes negligible ($n_i << n_s$). Hence all terms with n_i can be neglected in the denominators of Equations (7) and (8), and the conversion reduces to the simple linear relationship

$$C_i = \frac{x_i}{M_s} \tag{10}$$

We recommend using this approximation, which is usually extremely good. The conversion of units for the Henry coefficients is then straightforward.

Another problem that is encountered in the calculation of atmospheric equilibrium concentrations is the dependence of the Henry coefficient on the concentration of dissolved ions, the so-called salting out effect. Difficulties are that (1) sometimes freshwater solubilities from one source have to be combined with salting-out coefficients from another source, (2) one should know the effect of each ion species, and (3) salt concentrations are also expressed in different units that may not be strictly convertible. In this problem, some approximations are inevitable. Aeschbach-Hertig et al. (1999b) gave an equation to calculate equilibrium concentrations from mole fraction solubilities X_i as given by Clever (1979a,b; 1980) and the salting coefficients K_i determined by Smith and Kennedy (1983). According to the above discussion, we rewrite this equation (Eqn. 5 of Aeschbach-Hertig et al. 1999b) as follows:

$$C_{i,eq} = \frac{X_i(T)}{M_s} \cdot \frac{(P - e_w(T))z_i}{P_0} \cdot \frac{\rho(T, S = 0)}{\rho(T, S)} \cdot V_i \cdot e^{-K_i(T)c} \tag{11}$$

where $\rho(T,S)$ is the density of water (Gill 1982), V_i is the molar volume of gas i, and c is the molar concentration of NaCl. An approximative conversion between c [mol L^{-1}] and salinity S [g kg^{-1}] can be made by equating the two measures expressed in terms of total mass of salt per volume of solution (Aeschbach-Hertig et al. 1999b).

For He, Ne, Ar, and Kr, Weiss (1970, 1971) and Weiss and Kyser (1978) gave empirical equations that directly yield the atmospheric solublity equilibrium concentrations $C_{i,eq}$ at a total pressure of 1 atm. While this form of gas solubilities is very convenient in practice, it should be noted that such equilibrium concentrations are—as shown by Equations (2) and (3)—not directly proportional to P. The correct way to calculate the equilibrium concentrations at any pressure from those at P_0 = 1 atm is:

$$C_{i,eq}(T,S,P) = C_{i,eq}(T,S,P_0)\frac{P - e_w(T)}{P_0 - e_w(T)} \tag{12}$$

Unfractionated excess air. The most straightforward explanation of the phenomenon of a gas excess above atmospheric equilibrium in ground waters is complete dissolution of small air bubbles trapped in soil pores (Fig. 5a, Heaton and Vogel 1979, 1981; Rudolph et al. 1984). This traditional—but experimentally never directly verified—concept implies that the excess has the same composition as the entrapped gas, which is assumed to be atmospheric air, hence the name 'excess air'. We refer to this concept as the 'unfractionated air' (UA) model. Because the composition of air is given by the volume fractions z_i, the UA-model needs only one parameter to describe the excess air component, namely the amount of trapped and completely dissolved air per mass of water (A_d). The excess air term in Equation (5) is written as:

$$C_{i,ex}^{UA} = A_d \cdot z_i \tag{13}$$

It is important to address the physical interpretation of the UA-model. The parameter A_d is the concentration of dissolved excess air, but because the model assumes complete dissolution, it should also reflect the concentration of entrapped air in the soil of the recharge area. Values of A_d in units of cm^3 STP air per g of water may therefore approximately be interpreted as volume ratios of entrapped air to water. Typical values are on the order of a few times 10^{-3} cm^3 STP g^{-1}. Assuming that such excesses originate from complete dissolution of entrapped air, they indicate that entrapped air initially occupies only a few per mil of the available pore space volume. This is in contrast to the literature about air entrapment in soils (Christiansen 1944; Fayer and Hillel 1986; Stonestrom and Rubin 1989; Faybishenko 1995), in which entrapped air volume ratios ranging from a few percent up to several 10s of percent in extreme cases are reported.

The driving force for dissolution of trapped air bubbles is pressure. In order to keep gases permanently in solution, the total pressure must at least equal the sum of the partial pressures corresponding to equilibrium with the dissolved gas concentrations. Because the water is assumed to have been at atmospheric equilibrium before the dissolution of the excess air, the sum of the partial pressures corresponding to the equilibrium components $C_{i,eq}$ is exactly balanced by the atmospheric pressure P. Thus, we can write the dissolution condition for the excess pressure and excess air only:

$$P_{ex} = P_{tot} - P \geq \sum_i p_{i,ex} = \sum_i H_i C_{i,ex} = A_d \sum_i H_i z_i \tag{14}$$

The sum on the right hand side of Equation (14) is dominated by the most abundant atmospheric gases nitrogen and oxygen. According to Equation (14) an excess pressure of 0.1 atm is needed (at 13°C) to keep a typical amount of excess air of 2×10^{-3} cm^3 STP g^{-1} of air permanently in solution.

There are two sources of excess pressure on entrapped air in ground water: hydrostatic head and surface tension. In addition, the noble gas partial pressures may be increased in the soil air due to oxygen consumption. Surface tension is inversely related to the size of the trapped bubbles. For spherical bubbles, it reaches the magnitude of 0.1 atm at a diameter of ~30 µm. Thus, surface tension plays an important role in very fine-grained sediments, but may not be the dominant force in productive (e.g., sandy) aquifers. In the case of hydrostatic head, an excess pressure of 0.1 atm corresponds to a column height of the water overload of about 1 m. Water table fluctuations of the order of 1 m probably occur in most recharge areas, thus hydrostatic pressure has the potential to contribute significantly to the dissolution of entrapped air. Complete removal of oxygen from the soil air would increase the partial pressures of the conservative gases by about 25%. However, the oxygen in the soil air may not be completely consumed and at least part of it is replaced by CO_2 (Frei 1999). Therefore, it is questionable whether the effect of oxygen consumption can reach the magnitude of 0.1 atm.

Despite the initial success of the UA-model, there is increasing evidence that excess air tends to be fractionated relative to atmospheric air, with an enrichment of the heavy gases compared to the light gases (e.g., Stute et al. 1995b; Clark et al. 1997; Aeschbach-Hertig et al. 2000). Thus, strictly speaking the traditional name 'excess air' is somewhat misleading.

If excess air is fractionated, it can no longer be expressed by a single parameter such as A_d. A convenient and widely used measure for the size of the gas excess, which is independent of its composition, is the relative Ne excess ΔNe:

$$\Delta Ne(\%) = \left(\frac{C_{Ne,m}}{C_{Ne,eq}} - 1 \right) \cdot 100\% \tag{15}$$

Ne is best suited to quantify excess air because it has almost only atmospheric components and a low solubility, resulting in a relatively large excess. Moreover, because the equilibrium concentration $C_{Ne,eq}$ does not strongly depend on temperature, an approximative value for ΔNe can be calculated even if the exact equilibration temperature is not known. Thus, ΔNe is practically an observable quantity. As a rule of thumb, in the case of unfractionated excess air 10% ΔNe correspond approximately to 10^{-3} cm^3 STP g^{-1} of dissolved air, requiring at least 0.05 atm overpressure.

Partial re-equillibration. The conceptual idea of the partial re-equilibration (PR) model introduced by Stute et al. (1995b) is that entrapped air bubbles are initially completely dissolved as in the UA-model, but later a part of the resulting excess is lost by molecular diffusion across the water table (Fig. 5b). Due to the different molecular diffusivities D_i, this process leads to a fractionation of the excess air with respect to atmospheric air. Stute et al. (1995b) parameterized the degree of re-equilibration by the remaining fraction of the initial Ne excess and wrote the model equation for the excess components as:

$$C_{i,ex}^{PR} = C_{i,ex}(0) \cdot \left(\frac{C_{Ne,ex}}{C_{Ne,ex}(0)} \right)^{D_i/D_{Ne}} \tag{16}$$

By using $A_d \cdot z_i$ instead of $C_{i,ex}(0)$ and introducing the re-equilibration parameter $R = -\ln[C_{Ne,ex}/C_{Ne,ex}(0)]$, the PR-model may be rewritten as (Aeschbach-Hertig et al. 2000):

$$C_{i,ex}^{PR} = A_d \cdot z_i \cdot e^{-R\frac{D_i}{D_{Ne}}} \tag{17}$$

This formulation highlights the physical interpretation of the PR-model: The exponential term describes the result of gas exchange between water and soil air. The PR-model is based on the assumption that the gas exchange rate is proportional to the molecular diffusivity. This is the case in the stagnant boundary layer model of gas exchange, in which the exchange rate r_i is given by (Schwarzenbach et al. 1993):

$$r_i = \frac{D_i}{\delta \cdot h} \tag{18}$$

where δ is the thickness of the boundary layer and $h = V/A$ is the (mean) depth of the water body (with surface area A and volume V) affected by the exchange.

The dynamic development of dissolved gas concentrations under the influence of gas exchange is given by:

$$C_i(t) = C_{i,eq} + C_{i,ex}(0) \cdot e^{-r_i t} \tag{19}$$

Comparison with the PR-model Equation (17) yields the following interpretation:

$$C_{i,ex}(0) = A_d z_i \text{ and } r_i t = R\frac{D_i}{D_{Ne}} \tag{20}$$

Thus, the parameter $R = r_{Ne}t$ describes the degree of re-equilibration for Ne. By defining

$$R_i = R\frac{D_i}{D_{Ne}} = r_i t = \frac{D_i}{\delta \cdot h} t \tag{21}$$

we can re-write the PR-model equation as:

$$C_{i,ex}^{PR} = A_d \cdot z_i \cdot e^{-R_i} \tag{22}$$

The physical quantities determining the degree of re-equlibration are the thickness of the boundary layer (δ), the depth of the exchanging water body (h), and the duration of gas exchange (t). The differences between the individual gases originate from their different molecular diffusivities (D_i).

Because the molecular diffusivity decreases with the mass of the isotopes, the PR-model predicts that the depletion of the initially unfractionated excess air is largest for the light noble gases, whereas the heavy gases are enriched relative to air in the remaining excess. In particular, He should be strongly depleted in the excess air component, because of its extraordinarily high diffusion coefficient. However, due to the presence of non-atmospheric He components, it is difficult to verify this prediction.

Another consequence of the diffusion controlled fractionation process is that significant isotopic fractionations should occur. Peeters et al. (2002) suggested the use of $^{20}Ne/^{22}Ne$ ratios to test the validity of PR-model to describe gas partitioning in ground waters. In the few aquifers where Ne isotopes have been analyzed, no significant diffusive isotopic fractionation has been found so far (Peeters et al. 2002).

Multi-step partial re-equilibration. The PR-model frequently predicts rather high values for the initial concentration of dissolved air (A_d). For instance, in the study of a Brazilian aquifer, in which the model was introduced Stute et al. (1995b) found initial Ne excesses up to about 300%, corresponding to A_d-values of up to 3×10^{-2} cm^3 STP g^{-1}. Complete dissolution of such amounts of air would require excess pressures of more than 1.5 atm, or—if hydrostatic pressure dominates—water table fluctuations of more than 15 m, which seem rather unrealistic. Moreover, if air bubbles would be trapped and dissolved at such large depths, it appears unlikely that a diffusive flux to the water table would still occur. Such findings render the physical interpretation of the PR-model concept problematic.

To circumvent the problem of high initial excess air, while retaining the concept of diffusive re-equilibration, we propose a model of multi-step partial re-equilibration (MR). The conceptual idea of this new model is that the water table periodically fluctuates, thereby repeating cycles of trapping, dissolution and diffusive loss of air (Fig. 5c). Starting as usual with equilibrated water, the excess air component after the first dissolution-degassing step is given by Equation (22). After the second step, we have

$$C_{i,ex}^{MR,2} = \left(A_d \cdot z_i \cdot e^{-R_i} + A_d \cdot z_i\right) \cdot e^{-R_i} = A_d \cdot z_i \left(e^{-R_i} + e^{-2R_i}\right) \tag{23}$$

and after n steps, the general equation is:

$$C_{i,ex}^{MR,n} = A_d \cdot z_i \cdot \sum_{k=1}^{n} e^{-kR_i} = A_d \cdot z_i \cdot \sum_{k=1}^{n} \left(e^{-R_i}\right)^k \tag{24}$$

The sum in Equation (24) is the sum of a finite geometric series, and from the respective theory we finally find the MR-model equation:

$$C_{i,ex}^{MR,n} = A_d \cdot z_i \cdot e^{-R_i} \frac{1-e^{-nR_i}}{1-e^{-R_i}} \tag{25}$$

Note that the MR-model has three parameters: the number of steps (n) and the parameters A_d and R as in the PR-model. The quantities R_i in Equation (25) are related to the universal parameter R by Equation (21).

To discuss the predictions of the MR-model, we consider four limiting cases:

(1) $R_i \to \infty$: This case of complete re-equilibration is trivial ($C_{i,ex} = 0$).

(2) $R_i = 0$: No degassing at all, the model reduces to the UA-model ($C_{i,ex} = n \cdot Ad \cdot zi$).

(3) n = 1: This case is equivalent to the PR-model ($C_{i,ex} = A_d \cdot z_i \cdot e^{-R_i}$).

(4) $n \to \infty$ This is the most interesting case. Provided that $R_i > 0$, the term e^{-nR_i} approaches zero and Equation (25) becomes (proceeding from the finite to the infinite geometric series):

$$C_{i,ex}^{MR,\infty} = A_d \cdot z_i \cdot \sum_{k=1}^{\infty} \left(e^{-R_i}\right)^k = A_d \cdot z_i \cdot \frac{e^{-R_i}}{1-e^{-R_i}} \tag{26}$$

This case differs from the PR-model only by the correction factor $\left(1-e^{-R_i}\right)^{-1}$. For $R_i \gg 1$ (strong degassing), the correction factor tends to unity and the MR-model approaches the PR-model. In reality, neither n nor R_i need to be particularly large to approach the PR-model rather closely. Thus, the MR concept may be seen as a more realistic description of excess air formation for samples with a strong diffusive fractionation pattern, that according to the PR-model would require unrealistically large initial dissolution of entrapped air.

However, as the MR-model approaches the PR-model, it also predicts diffusive isotopic fractionation, i.e., the $^{20}Ne/^{22}Ne$ should be significantly lower than the atmospheric ratio (Peeters et al. 2002). Since such a depletion of air-derived light noble gas isotopes has not been observed in field studies, a concept is needed that explains the enrichment of heavy noble gases in the excess air component without diffusive fractionation of noble gas elements *and* isotopes.

Closed-system equilibration. Such a concept is provided by the closed-system equilibration (CE) model (Aeschbach-Hertig et al. 2000). The idea of this model is that the entrapped air dissolves only partially, and a new solubility equilibrium is attained in a closed system consisting initially of air-saturated water and a finite volume of entrapped air under elevated total pressure P_{tot} (Fig. 5d). The CE-model equation is (Aeschbach-Hertig et al. 2000):

$$C_{i,ex}^{CE} = \frac{(1-F)A_e z_i}{1+FA_e z_i / C_{i,eq}} \tag{27}$$

where A_e is the initial amount of entrapped air per unit mass of water and F is a fractionation parameter. Note that A_e (entrapped air) is the same as A_d (dissolved air) only in the case of total dissolution.

The model parameters A_e and F in the CE-model have a clear physical interpretation (Aeschbach-Hertig et al. 2000, 2001). A_e given in cm^3 STP g^{-1} approximately equals the initial volume ratio between entrapped air and water, whereas F describes the reduction of the gas volume by partial dissolution and compression. Correspondingly, F can be expressed as the ratio of two parameters v and q, where v is the ratio of the entrapped gas volumes in the final (V_g) and initial state (V_g^0), and q is the ratio of the dry gas pressure in the trapped gas to that in the free atmosphere (P):

$$F = \frac{v}{q} \quad \text{with} \quad v \equiv \frac{V_g}{V_g^0} \quad \text{and} \quad q \equiv \frac{P_{tot} - e_w}{P - e_w} \tag{28}$$

Provided that $P_{tot} \geq P$, the following inequalities hold: $q \geq 1$ and $0 \leq v$, $F \leq 1$. Note that only the combined parameter $F = v/q$ is needed to define the excess air component, not v and q individually. The parameters q, v, and A_e are coupled by the physical requirement that the sum of the partial pressures of all gases in the trapped volume equals P_{tot}. Any pair of the parameters A_e, F, q, and v fully determines the amount and composition of excess air, the most intuitive choice being A_e (≈ air / water volume ratio) and q (≈ pressure exerted on the entrapped air). The CE-model thus allows a direct physical interpretation of the excess air component.

Typical values for the CE-model parameters A_e and q found in several aquifers support the notion that these parameters reflect actual physical conditions during infiltration. Aeschbach-Hertig et al. (2001) found that typical values of A_e in six aquifers ranged from 0.02 to 0.04 cm^3 STP g^{-1}, indicating that a few percent of the pore space were occupied by entrapped air during infiltration, in full agreement with expectations (Fayer and Hillel 1986; Holocher et al. 2002). Typical q-values were ~1.2 for three temperate zone aquifers and ~1.5 for three semi-arid sites. Explaining these values by hydrostatic pressure due to water table fluctuations implies typical amplitudes of 2 and 5 m, respectively. Such amplitudes of water table fluctuations are probably at the upper limit of what can be expected.

These results provide a natural solution to the apparent discrepancy between typical concentrations of entrapped air (several percent of the pore space, corresponding to A_e-values of the order of 10^{-2} cm^3 STP g^{-1}) and those of actually dissolved excess air (A_d-values of the order of 10^{-3} cm^3 STP g^{-1} in the UA-model). It seems that typically more air is entrapped than can be dissolved at the prevailing pressure. In this situation, the resulting excess air component is fractionated according to the CE-model and its size (expressed, e.g., by ΔNe) is limited not by the available air reservoir A_e, but by the pressure acting on that reservoir (q). As a result, the observable quantity ΔNe is strongly correlated with q (Aeschbach-Hertig et al. 2001) and obtains a direct physical interpretation: it is essentially a measure of the pressure on the entrapped air.

If the excess pressure is ascribed to the hydrostatic load of infiltrating water, then ΔNe should be a measure of the amplitude of water table fluctuations. A verification of a direct link between water table fluctuations and ΔNe or q under field conditions is still missing. Experiments with sand columns (Holocher et al. 2002) support the correlation of q with the amplitude of water level changes.

When comparing the different excess air models, which are summarized in Table 2, we may first note that the UA-model is contained as a special case (R or F = 0) in the more advanced models. There is growing evidence that this special case is not generally realized in natural systems (Stute et al. 1995b; Ballentine and Hall 1999; Aeschbach-Hertig et al. 2000, 2001). Thus, the problem of model choice essentially reduces to the PR / MR and CE models. Although the experience with the comparison of these models is still limited at present, it seems that in many cases all of them provide a reasonable fit to the measured concentrations of Ne, Ar, Kr, and Xe. The models differ strongly in their predictions for He and the He, Ne, and Ar isotope ratios (Aeschbach-Hertig et al. 2000; Peeters et al. 2002). Because of the non-atmospheric components, He can rarely be used to test the models, and $^{40}Ar/^{36}Ar$ may also be equivocal for the same reason. Thus, analysis of the $^{20}Ne/^{22}Ne$ ratio provides the best option to distinguish between the models and so far appears to favor the CE-model (Peeters et al. 2002). Field and laboratory studies under various conditions are needed to understand which model provides the best approximation of reality under which conditions.

Table 2. Excess air models.

Model	*Equation*	*Parameters*	*Reference*
Unfractionated excess **A**ir (UA)	$C_{i,ex}^{UA} = A_d \cdot z_i$	A_d: Conc. dissolved excess air	Heaton and Vogel (1981)
Partial **R**e-equilibration (PR)	$C_{i,ex}^{PR} = A_d \cdot z_i \cdot e^{-R \cdot D_i/D_{Ne}}$	A_d: Initial conc. of dissolved excess air R: Degree of re-equilibration	Stute et al. (1995b)
Multi-step partial **R**e-equilibration (MR)	$C_{i,ex}^{MR,n} = A_d \cdot z_i \cdot e^{-R_i} \dfrac{1-e^{-nR_i}}{1-e^{-R_i}}$	$R_i = R \cdot D_i/D_{Ne}$ A_d, R: as in PR, for each step n: Number of dissolution–degassing steps	This work
Closed-system **E**quilibration (CE)	$C_{i,ex}^{CE} = \dfrac{(1-F)A_e z_i}{1+FA_e z_i/C_{i,eq}}$	A_e: Initial conc. entrapped air F: Reduction of entrapped volume by dissolution and compression	Aeschbach-Hertig et al. (2000)

Separation of the components

Separation of He components using Ne. In applications of the He isotopes, particularly the tritiogenic $^3He_{tri}$, for dating purposes, usually only He and Ne are analyzed. The Ne concentration is used to estimate the atmospheric He components in order to calculate the non-atmospheric He components that carry the time information. This approach is based on the fact that in Equation (5) Ne usually has only two components (equilibrium and excess air), whereas ^{4}He has one additional component (terrigenic), and ^{3}He has two non-atmospheric components (terrigenic and radiogenic / tritiogenic). The equations for Ne, ^{4}He, and ^{3}He are linked by the elemental composition L_{ex} = (He/Ne)$_{ex}$ of the excess air component as well as the isotopic compositions R_{ex} = (^{3}He/^{4}He)$_{ex}$ and R_{ter} = (^{3}He/^{4}He)$_{ter}$ of both the excess air and terrigenic He components. Further introducing R_{eq} = (^{3}He/^{4}He)$_{eq}$, we can write Equation (5) explicitly as:

$$Ne_m = Ne_{eq} + Ne_{ex} \tag{29a}$$

$$^4He_m = {}^4He_{eq} + L_{ex} \cdot Ne_{ex} + {}^4He_{ter} \tag{29b}$$

$$^3He_m = R_{eq} \cdot {}^4He_{eq} + R_{ex} \cdot L_{ex} \cdot Ne_{ex} + R_{ter} \cdot {}^4He_{ter} + {}^3He_{tri} \tag{29c}$$

This equation system can easily be solved for the non-atmospheric He components. From Equation (29a) we determine Ne_{ex}, insert the result into Equation (29b) to obtain $^4He_{ter}$, and finally solve Equation (29c) for the sought-after $^3He_{tri}$:

$$\begin{aligned} ^3He_{tri} = {}& {}^4He_m \cdot (R_m - R_{ter}) - {}^4He_{eq} \cdot (R_{eq} - R_{ter}) \\ & - L_{ex} \cdot (Ne_m - Ne_{eq}) \cdot (R_{ex} - R_{ter}) \end{aligned} \tag{30}$$

Of course, the applicability of Equation (30) depends crucially on our knowledge of the quantities on the right hand side. If the infiltration conditions – in particular the recharge temperature – are known, the equilibrium components can be calculated. The elemental and isotopic compositions of the excess air component (L_{ex} and R_{ex}) are usually assumed to be atmospheric. Finally, the terrigenic He component usually originates from the crust and a typical value of 2×10^{-8} is assigned for R_{ter}. These traditional assumptions should be critically assessed in each particular case (Holocher et al. 2001). Some general features are discussed in the following.

In lakes, often even further simplifications can be made. Ne concentrations are usually close to atmospheric equilibrium, and also the terrigenic He component is usually small, although it can be important in lakes with high water residence times or high terrigenic fluxes. In the simplest case, both Ne and ^{4}He concentrations are at equilibrium at the water temperature. The concentration of tritiogenic $^3He_{tri}$ can then be calculated as follows:

$$^3He_{tri} = {}^4He_m \cdot (R_m - R_{eq}) \tag{31}$$

Note that the difference is taken between isotope ratios rather than between ^{3}He concentrations, because the ratios are usually measured with higher precision than the concentrations. The analytical precision of the ^{3}He/^{4}He ratio measurement usually ranges between 0.2 and 1%. This uncertainty essentially determines the precision and detection limit for $^3He_{tri}$ and hence for the derived ^{3}H-^{3}He age. It should also be noted that $^3He_{tri}$ and thus the age becomes zero if R_m equals the ^{3}He/^{4}He ratio at solubility equilibrium ($R_{eq} \approx 1.36\times10^{-6}$, Benson and Krause 1980), not the atmospheric ^{3}He/^{4}He ratio ($R_a = 1.384\times10^{-6}$, Clarke et al. 1976).

In many lakes, significant ^{4}He excesses are found while ΔNe is close to zero, indicating the presence of terrigenic helium. In such cases, the correction for excess air based on Ne only adds noise, and $^3He_{tri}$ may be calculated from a simplified version of Equation (30) without the Ne-term. However, a value for R_{ter} is needed. Usually the excess ^{4}He is radiogenic, originating from crustal rocks. Because typical values for R_{ter} are then much smaller than R_m and R_{eq}, the correction for terrigenic $^3He_{ter}$ is small, and the uncertainty of the assumed value of R_{ter} does not strongly increase the error of $^3He_{tri}$. However, in several lakes in volcanic areas, terrigenic He of mantle origin has been found (Sano and Wakita 1987; Igarashi et al. 1992; Kipfer et al. 1994; Aeschbach-Hertig et al. 1996b, 1999a; Clark and Hudson 2001). Because typical mantle helium isotope ratios (about 10^{-5}, e.g., Mamyrin and Tolstikhin 1984) are much larger than R_m and R_{eq}, the mantle-derived ^3He may become the dominant ^3He-component, rendering it difficult or even impossible to calculate $^3He_{tri}$.

The best way to reduce the uncertainty about the terrigenic He isotope ratio R_{ter} is to look for samples that contain large concentrations of terrigenic He from the local or regional source, from which R_{ter} can be derived. Such samples may be found in thermal or mineral springs, or in deep ground waters. For example, hydrothermal water entering the northern basin of Lake Baikal was found to have a similar He isotopic composition as nearby hot springs on land (Kipfer et al. 1996). In studies of shallow aquifers, samples of deeper, He rich ground water have proven to be very valuable (Aeschbach-Hertig et al. 1998; Dunkle Shapiro et al. 1998; Holocher et al. 2001).

The most general case, when both excess air and terrigenic He are present, is rather unusual for lakes but typical for ground waters. The decisive parameter is then the He/Ne ratio of the excess air component, L_{ex}. In view of the different excess air models discussed above, it appears questionable whether the traditional assumption that L_{ex} equals the atmospheric He/Ne ratio (L_{air} = 0.288) is valid in general. If the excess air is fractionated, L_{ex} is always lower than L_{air} (Fig. 7). The important point to note in Figure 7 is that in the case of the CE-model, L_{ex} is restricted to the range between L_{air} and the He/Ne ratio of air-saturated water (L_{eq}). Depending on temperature, L_{eq} varies between 0.22 and 0.25. L_{ex} is thus restricted to the range between 0.22 and 0.288. In contrast, if the PR-model applies, L_{ex} decreases exponentially with R. Because R is not restricted, L_{ex} can in principle approach zero. R-values of the order of 1 (corresponding to a decrease of the Ne excess by a factor of 1/e) are not unusual, implying significantly lower values for L_{ex} than in the case of the CE-model.

Based only on He and Ne data, the appropriate value of L_{ex} cannot be determined. Additional information from other noble gas concentrations or isotope ratios is needed to distinguish between the different models and to determine the model parameters (Peeters et al. 2002). However, up to now only very few ^3H-^3He studies of shallow ground waters provide such additional information (Beyerle et al. 1999a; Holocher et al. 2001). In all previous studies, only He and Ne were measured and the assumption $L_{ex} = L_{air}$ was applied. Because L_{air} is actually only the upper limit of L_{ex}, this approach tends to overestimate the atmospheric and thus to underestimate the non-atmospheric He components. A clear sign that this approach is not always appropriate is the common occurrence of negative values for $^4He_{ter}$ or even $^3He_{tri}$.

For example, in the large ^3H-^3He study of a shallow sandy aquifer on Cape Cod of Dunkle Shapiro et al. (1999), 55 out of a total of 91 samples yielded negative $^4He_{ter}$ concentrations, clearly indicating that L_{ex} was overestimated. For each sample, the upper limit for L_{ex} is defined by the physical requirement $^4He_{ter} \geq 0$. Most of these upper limits are higher than 0.25, thus well within the range predicted by the CE-model. Only two samples require values of L_{ex} that are significantly below the range of the CE-model and can only be explained by the PR-model. We conclude that for the majority of the samples the CE-model would be sufficient to remove the inconsistency of negative $^4He_{ter}$ concentrations.

Using L_{ex}-values lower than L_{air} yields higher results for $^4He_{ter}$, $^3He_{tri}$, and the ^3H-^3He ages. Yet, as long as the value of L_{ex} is restricted to the range allowed by the CE-model, the differences in the calculated quantities are usually small. In the study of Dunkle Shapiro et al. (1999) practically no changes result, because $^4He_{ter} = 0$ was assumed in the calculation of most ages. In principle, however, for samples with a large and strongly fractionated excess air component the usual atmospheric excess air correction can result in significantly underestimated ages.

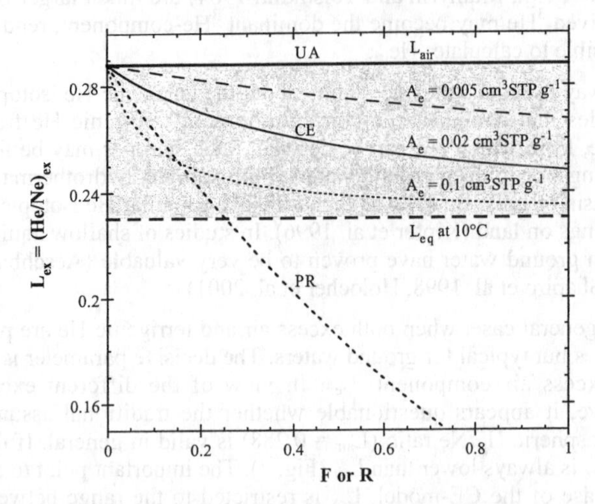

Figure 7. He/Ne ratios of the excess air component (L_{ex}) for different excess air models as functions of the fractionation parameters F or R. The atmospheric He/Ne ratio L_{air} = 0.288 defines the upper limit for L_{ex}. In the CE-model, the He/Ne ratio of air saturated water (L_{eq}) defines a lower limit for L_{ex}, which depends on both A_e and F (curves for three A_e values are shown). In the PR-model, L_{ex} is independent of A_d and decreases exponentially with R towards zero. For R > 0.25, the PR-model predicts L_{ex}-values below the range allowed by the CE-model.

Separation using all noble gases: Iterative approaches. If the equilibrium component is unknown (e.g., because of unknown equilibration temperature) and excess air is present, we have for each noble gas at least two unknown components (equilibrium and excess air) in Equation (5). However, because both components have defined (although in general unknown) elemental compositions, the equations are linked and can be solved. This is the typical situation in applications of the 'noble gas thermometer', where the equilibration temperature is to be derived from the concentrations of dissolved noble gases in ground water.

In principle, the measurement of two noble gases of purely atmospheric origin (usually all except He) and the assumption of atmospheric excess air enables the calculation of the two unknown components. The solution can be found by a graphical method, which has been used in some early studies involving the measurement of Ar and N_2 (e.g., Heaton and Vogel 1981; Heaton et al. 1986). Of course, N_2 is not a noble gas and can be affected by biogeochemical processes (e.g., denitrification, Heaton et al. 1983). A better choice of gases would be Ne and Xe, because their solubilities differ strongly (compare Fig. 1). Ne is strongly affected by excess air, whereas Xe reacts most sensitive to temperature. In a plot of Ne versus Xe concentrations (Fig. 8), the data align along straight lines representing addition of excess air to air saturated water (ASW). The intersection of these lines with the ASW-curve yields the equilibrium component (and the temperature), whereas the excess air component is given by the offset along the lines. The disadvantage of this approach is that an explicit assumption about the elemental composition of excess air (slope of the lines) has to be made in order to obtain unequivocal results.

However, if the effort is made to measure Ne and Xe, it is natural to measure Ar and

Kr as well. The traditional approach to determine recharge temperatures from full noble gas data sets involves an iterative correction for excess air (Andrews and Lee 1979; Rudolph 1981; Rudolph et al. 1984; Stute and Deák 1989; Pinti and Van Drom 1998). The measured concentrations of Ne, Ar, Kr, and Xe are corrected for excess air—assumed to be of atmospheric composition—and the equilibration temperature is calculated from the corrected concentration of each noble gas. This process is iteratively repeated with varying amounts of excess air until optimum agreement between the four temperatures is reached.

The extra information gained by analysing all noble gases can be used to estimate the uncertainty of the derived temperature from the standard deviation of the individual temperatures. Systematic trends in the temperatures derived from the individual noble gases indicate that the composition of the gas excess differs from atmospheric air. The iterative approach to correct for excess air is still applicable in such cases, as long as the composition of the gas excess can be described by one additional parameter. Stute et al. (1995b) used this approach in connection with the PR-model.

Separation using all noble gases: Inverse techniques. Although the iterative method works well in most cases, there is a more fundamental way to solve the problem, which has a number of advantages. The problem is to solve Equation (5), which can be written as an explicit equation system by inserting Equations (2) and (3) for the equilibrium component and one of the model Equations (13), (17), (25), or (27) for the excess air component. Restricting the problem to atmospheric gases, we are left with equations for the equilibrium and excess air components of Ne, Ar, Kr, and Xe. In this form, we recognize that the actual unknowns are no longer the individual components, but their defining parameters T, S, P, A_d or A_e, and R or F.

Thus, in general we have 4 measured atmospheric noble gas concentrations but 5 unknown parameters. Obviously, this general problem cannot be solved. Yet, in most practical cases, some of the 5 parameters are well constrained. We can usually set $S \approx 0$, because infiltrating ground water is fresh, and we can estimate P from the altitude of the recharge area (Eqn. 4). We are then left with 3 unknown parameters and 4 measured concentrations, thus the system is over-determined. It can be solved for the model parameters by inverse modeling techniques based on error weighted least squares fitting (Aeschbach-Hertig et al. 1999b; Ballentine and Hall 1999).

The inverse approach determines those parameter values that within the framework of each conceptual model yield the best fit to the observed concentrations. The best fit is the one that minimizes χ^2, which is the sum of the weighted squared deviations between the modeled and measured concentrations:

$$\chi^2 = \sum_i \frac{\left(C_i^{meas} - C_i^{mod}\right)^2}{\sigma_i^2} \tag{32}$$

where C_i^{mod} are the modeled concentrations, and $C_i^{meas} \pm \sigma_i$ are the measured concentrations with their experimental 1σ-errors. This solution approach is very flexible with regard to the choice of the parameters to be varied, the constraints (measured concentrations) to be used, as well as the incorporation of different conceptual models for excess air.

An essential feature of the inverse approach is the use of the experimental errors as weights in Equation (32). This assures that the influence of each individual measurement on the final parameter values is properly weighted. Moreover, the use of the experimental errors allows the derivation of objective error estimates for the obtained parameter values. In the theory of least squares fitting, uncertainties of the estimated parameters are

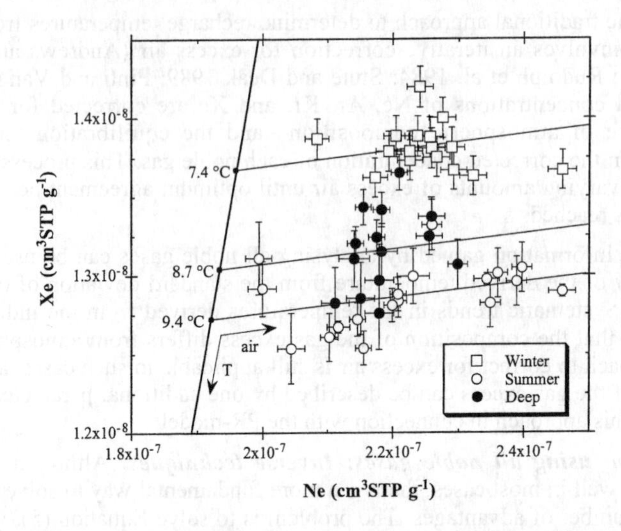

Figure 8. Ne vs. Xe concentrations from an alluvial aquifer (Beyerle et al. 1999). The data are explained by variations of the recharge temperature (solid line) and varying amounts of excess air (thin lines). The excess air lines were calculated by fitting straight lines with prescribed slope (atmospheric Xe/Ne ratio) to three groups of samples. Samples from shallow boreholes taken in winter (open squares) have the highest Xe concentrations, corresponding to the lowest temperatures (7.4°C). Shallow samples from summer and fall (open circles) indicate warmer recharge temperatures (9.4°C). The samples from deeper boreholes (full circles) lie in between (8.7°C). These mean temperatures calculated only from Ne and Xe data lie close to the mean temperatures calculated from all noble gases (7.4, 9.2, and 8.6°C, respectively).

derived from the covariance matrix. These errors correspond to a rigorous propagation of the experimental uncertainties. Besides the uncertainties of the parameters, also their mutual correlation can be obtained from the covariance matrix.

Most importantly, however, the use of the experimental errors allows an objective judgement of the agreement between model and data, i.e., the validity of the conceptual model that was adopted to describe the data. The model selection is based on the χ^2-test. The expected minimum value of χ^2 is the number of degrees of freedom $\nu = n - m$, where n is the number of data points and m is the number of free parameters. The probability p for χ^2 to be higher than a given value due to random analytical errors, although the model description is correct, can be obtained from the χ^2-distribution with ν degrees of freedom. If p is lower than some cut-off value p_c ($p_c = 0.01$ proved to be appropriate), the model is rejected.

The χ^2-test can be generalized to assess the applicability of a conceptual model to a whole data set consisting of N samples. Applying the same model to each sample of the data set may be interpreted as fitting one model with N·m free parameters to N·n data points. The χ^2-value for the whole data set is then the sum of the χ^2-values of the individual samples, and the number of degrees of freedom is N·ν. This data set χ^2-value also follows a χ^2-distribution, but with a much larger number of degrees of freedom than for each individual sample. Therefore, a conceptual model may not be consistent with a whole data set although it cannot be rejected based on any single sample.

The results of the inverse approach are not the individual components, but values for the parameters from which the equilibrium and excess air components for all species (including those that were not used in the inverse procedure, such as ^3He and ^4He) can be calculated. The result of such calculations can then be used to determine possible non-atmospheric components using Equation (5). Alternatively, the non-atmospheric components may be treated as fit parameters in an inverse modeling approach based on an ensemble of samples, as demonstrated by Peeters et al. (2002).

Interpretation

The whole effort of component separation is undertaken to provide quantities that can be interpreted in terms of meaningful parameters such as water residence times ('ages'), recharge temperatures, or other environmental conditions at recharge. Here we discuss the basic steps that lead from the individual components to these interpreted quantities.

^{3}H-^{3}He age. As ^{3}H decays, $^3He_{tri}$ is produced. However, as long as the water is in contact with the atmosphere, the resulting excess $^3He_{tri}$ can continuously escape into the air. As soon as a water parcel is isolated from the gas exchange with the atmosphere, the ^{3}H decay is matched by a corresponding increase of $^3He_{tri}$ (Fig. 9). The ^{3}H-^{3}He clock starts ticking, and measures the isolation time of the water parcel.

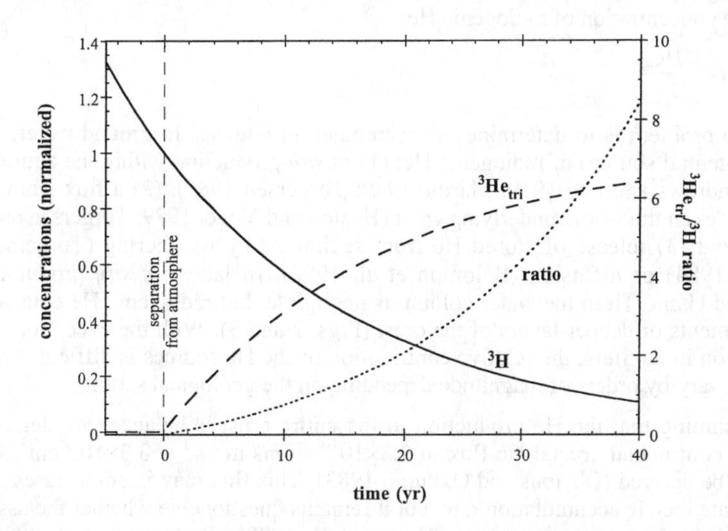

Figure 9. Temporal evolution of ^{3}H and $^3He_{tri}$ concentrations, normalized to $^3H(t = 0)$, and the $^3He_{tri}/^3H$ ratio in a water parcel. As long as the water is in contact with the atmosphere, ^{3}H decays but $^3He_{tri}$ remains zero due to gas exchange. After isolation from the atmosphere, $^3He_{tri}$ increases as ^{3}H decreases. The $^3He_{tri}/^3H$-ratio is a unique function of the time elapsed since isolation, the so-called ^{3}H-^{3}He water age.

As shown in Figure 9, the $^3He_{tri}/^3H$ ratio steadily increases with time, and thus the ^{3}H-^{3}He water age τ is a unique function of this ratio. The expression for τ is easily obtained from the law of radioactive decay (e.g., Tolstikhin and Kamenskiy 1969):

$$\tau = \frac{1}{\lambda} \cdot \ln\left(1 + \frac{^3He_{tri}}{^3H}\right) \tag{33}$$

where λ is the decay constant of tritium (half-life 4500 days or 12.32 yr, $\lambda = 0.05626\ yr^{-1}$, Lucas and Unterweger 2000). ^{3}H is usually given in TU (tritium units, 1 TU is equivalent to a $^3H/^1H$ ratio of 10^{-18}), $^3He_{tri}$ in $cm^3\ STP\ g^{-1}$. In order to evaluate Equation (33), these units have to be converted as follows (for fresh water): $1\ cm^3\ STP\ g^{-1} = 4.019 \cdot 10^{14}$ TU.

The water age τ depends in a non-linear way on the concentrations of $^3He_{tri}$ and ^3H. As a result, mixing of water parcels with different concentrations results in water ages that deviate from the true age of the mixture. For this reason, τ is sometimes referred to as the 'apparent water age'. In situations with strong mixing, e.g., in lakes, τ should be interpreted with some caution. The distortion of τ due to mixing is less important for small ages. If $^3He_{tri}/^3H << 1$ or equivalently $\tau << 1/\lambda$, Equation (33) can be linearised to give:

$$\tau = \frac{1}{\lambda} \cdot \frac{^3He_{tri}}{^3H} \tag{34}$$

In this case, τ is a linear function of the ratio $^3He_{tri}/^3H$, but not of ^3H. The age behaves only linearly if water parcels with equal ^3H concentrations are mixed. Otherwise, the apparent age of the mixture is always biased towards the component with the higher ^3H concentration (see also Schlosser and Winckler 2002, this volume).

^{4}He accumulation age. Radiogenic He produced by α-decay of U and Th series nuclides in crustal minerals accumulates both in ground water and deep water of lakes. Assuming a constant accumulation rate J_{He}, the water residence time τ follows directly from the concentration of radiogenic He:

$$\tau = \frac{^4He_{rad}}{J_{He}} \tag{35}$$

The problem is to determine the accumulation rate J_{He}. In ground water, there are three potential sources of radiogenic He: (1) *in situ* production within the aquifer matrix (e.g., Andrews and Lee 1979; Marine 1979; Torgersen 1989), (2) a flux from adjacent layers or even the whole underlying crust (Heaton and Vogel 1979; Torgersen and Clarke 1985), and (3) release of stored He from sediments by weathering (Torgersen 1980; Heaton 1984) or diffusion (Solomon et al. 1996). In lakes, *in situ* production from dissolved U and Th in the water column is negligible, but radiogenic He emanates from the sediments or deeper layers of the crust (Figs. 2 and 3). With the exception of *in situ* production in aquifers, the relative contribution of the He sources is difficult to estimate as it can vary by orders of magnitude depending on the geological setting.

Assuming that the He production in the entire crust is balanced by degassing, an average continental crustal He flux of 2.8×10^{10} atoms m^{-2} s^{-1} = 3.3×10^{-6} cm^3 STP cm^{-2} yr^{-1} can be derived (O'Nions and Oxburgh 1983). This flux may in some cases be useful to estimate the He accumulation rate, but it remains questionable whether the assumption of a uniform flux is justified (e.g., Castro et al. 1998b; Ballentine et al. 2002 in this volume; Ballentine and Burnard 2002 in this volume). Therefore, He accumulation usually provides only a qualitative timescale.

If it can be demonstrated that *in situ* production is the dominant ^{4}He source, quantitative dating is possible *in situ* ^{4}He production rate and the resulting accumulation rate J_{He} can be calculated from the U and Th concentrations of the aquifer material (Andrews and Lee 1979; Castro et al. 2000):

$$J_{He} = \Lambda_{He} \frac{\rho_r}{\rho_w} (C_U \cdot P_U + C_{Th} \cdot P_{Th}) \cdot \left(\frac{1-\Theta}{\Theta} \right) \tag{36}$$

The release factor Λ_{He} (fraction of produced ^{4}He which is released from minerals into the water) is usually taken to be 1 (e.g., Andrews and Lee 1979; Torgersen and Clarke 1985). ρ_r and ρ_w are the densities of the aquifer material and the water, respectively. The ^{4}He production rates from U and Th decay are $P_U = 1.19 \times 10^{-13}$ cm^3 STP μg_U^{-1} yr^{-1} and $P_{Th} = 2.88 \times 10^{-14}$ cm^3 STP μg_{Th}^{-1} yr^{-1} (Andrews and Lee 1979). The U and Th concentrations (C_U, C_{Th}, in μg g^{-1}) and the porosity (Θ) of the aquifer matrix have to be measured.

Recharge temperature, salinity, and altitude. From the equilibrium concentrations $C_{i,eq}$ we can potentially learn something about the water temperature T and salinity S, as well as the atmospheric pressure P. The water temperature is related to the soil and eventually to the mean annual air temperature, whereas salinity may indicate evaporative enrichment, and pressure is closely related to altitude. Although the focus has, in most cases, been directed to temperature, the other options should not be neglected.

After the separation of components, each individual parameter T, S, or P can be calculated by inversion of the solubility equations $C_{i,eq}$ (T,S,P) if the other two parameters are known. The inverse techniques discussed above offer the possibility to determine all three parameters at once. However, these parameters are rather strongly correlated, in particular in the presence of excess air (Fig. 1; Aeschbach-Hertig et al. 1999b). The reason for this problem is that the dependence of noble gas concentrations on the parameters follows some systematic trends. The effects of changes of T and S on the concentrations increase with molar mass of the gas, whereas the effect of excess air (A_d) decreases, because the solubilities strongly increase with molar mass. P has a uniform effect relative to equilibrium concentrations, but in the presence of excess air it is relatively more important for the heavy noble gases. As a result, especially the effects of P and S are very similar, and both can be approximated by a combination of T and A_d (see Fig. 1).

Parameter combinations that have very different patterns of effects on the concentrations (e.g., T-A_d, Fig. 1) are readily identifiable. In contrast, parameter pairs that have similar effects (e.g., S-P, Fig. 1) are hard to separate. Attempts to simultaneously fit two or more correlated parameters result in large uncertainties. Table 3 lists the errors obtained for T, S, and P from fitting various parameter combinations to a synthetic data set calculated with T = 10°C, S = 0‰, P = 1 atm, and $A_d = 3\times10^{-3}$ cm^3 STP g^{-1}, assuming experimental errors of ±1% on all concentrations. Each parameter on its own can be determined quite precisely. The errors increase only slightly if in addition an unknown amount of excess air (A_d) is present. Fractionated excess air enlarges the errors by a factor of two to three. Combinations of two parameters among T, S, and P yield even larger errors, and if in addition excess air is present, the uncertainties increase strongly. Finally, it is practically impossible to simultaneously fit T, S, and P.

It should however be stressed that each individual parameter of the equilibrium component can be well determined even in the presence of fractionated excess air. This fact provides the basis for the application of noble gas concentrations in ground water as indicators of paleotemperature.

Excess air parameters. Separation of the components, in particular by the inverse technique, also provides the parameters that define the excess air components such as the concentrations of dissolved or entrapped air (A_d or A_e), and the fractionation parameters R or F. Even without a perfect description of excess air, ΔNe can be quantified. Although the potential of excess air as an indicator of past recharge conditions was discussed in the first studies that identified the excess air component (Heaton and Vogel 1981; Heaton et al. 1983), little further progress has been made in the interpretation of the excess air signal.

The recent development of different models for the formation of excess air, and the methods to distinguish between these models and to determine their parameters based on

Table 3. Uncertainties of the inverse parameter estimation for different sets of free parameters, using a synthetic data set corresponding to T = 10°C, S = 0 ‰, P = 1 atm, $A_d = 3\times10^{-3}$ cm³ STP g⁻¹, with errors of 1% on all concentrations.

Parameters	*ΔT [°C]*	*ΔS [‰]*	*ΔP [atm]*
T or S or P	0.19	0.75	0.005
T, A_d	0.21	-	-
S, A_d	-	1.0	-
P, A_d	-	-	0.007
T, A_d, R	0.49	-	-
T, A_e, F	0.38	-	-
S, A_d, R	-	2.5	-
S, A_e, F	-	1.9	-
P, A_d, R	-	-	0.021
P, A_e, F	-	-	0.015
T, S	0.82	3.0	-
T, P	0.62	-	0.017
S, P	-	9.3	0.067
T, S, A_d	2.6	11	-
T, P, A_d	2.0	-	0.067
S, P, A_d	-	39	0.30
T, S, P	7.6	110	0.63

field data (inverse fitting, isotope ratios) provide the basis for further investigations of the information potentially available from the excess air (Aeschbach-Hertig et al. 2001). The models suggest that excess air is related to physical conditions in the quasi-saturated zone, where air is trapped during ground-water infiltration. Potentially important parameters are the air / water volume ratio, the pore size distribution, and the pressure in this zone. The pressure acting on the entrapped air may in turn be related to the amplitude of water table fluctuations and thus ultimately to the amount or variability of recharge.

APPLICATIONS IN LAKES

Applications of noble gases in lakes are mostly concerned with the identification of transport processes and the estimation of exchange rates between different regions within a lake. Torgersen et al. (1977) were the first to propose ^{3}H-^{3}He ages as a methodology in physical limnology and suggested that gas exchange rates and residence times can be determined from ^{3}H-^{3}He water ages. The time information provided by the combination of ^{3}H and ^{3}He is especially useful in estimating vertical water exchange in deep lakes, e.g., Lake Lucerne (Aeschbach-Hertig et al. 1996a), Lake Baikal (Peeters et al. 1997; Hohmann et al. 1998; Peeters et al. 2000b), Lake Issyk-Kul (Hofer et al. 2002; Vollmer et al. 2002), the Caspian Sea (Peeters et al. 2000a), the great lakes of North America (Torgersen et al. 1977), and in chemically stratified lakes, e.g., Green Lake (Torgersen et al. 1981), Lake Zug (Aeschbach-Hertig 1994) and Lake Lugano (Wüest et al. 1992). In addition to vertical mixing horizontal exchange between different lake basins can be studied (Zenger et al. 1990; Aeschbach-Hertig et al. 1996a). The ^{3}H-^{3}He ages may also be employed as time information required for the calculation of oxygen depletion and of the terrigenic ^{4}He flux from the sediments (Top and Clarke 1981; Mamyrin and Tolstikhin 1984; Aeschbach-Hertig 1994; Aeschbach-Hertig et al. 1996a; Hohmann et al. 1998; Peeters et al. 2000a).

The atmospheric noble gases Ne, Ar, Kr and Xe that have been measured in Lake Baikal (Hohmann et al. 1998; Aeschbach-Hertig et al. 1999b), in the Caspian Sea (Peeters et al. 2000a), and in Lake Tanganyika (Kipfer et al. 2000) carry information on the conditions at the lake surface during gas exchange. This information has been used to study lake level fluctuation in Lake Tanganyika (Craig et al. 1973; Kipfer et al. 2000) but might also be exploitable with respect to paleoclimate if the noble gases can be measured in the pore water of lake sediments. The behavior of the noble gas ^{222}Rn differs from that of the other noble gases because it is lost rapidly by radioactive decay. ^{222}Rn has been employed to study horizontal and vertical mixing in the near sediment region of lakes (Imboden and Emerson 1978; Imboden and Joller 1984; Weiss et al. 1984; Colman and Armstrong 1987) and to trace river flow in Lake Constance (Weiss et al. 1984).

Mixing and the distribution of dissolved substances in lakes

In most lakes the distribution of dissolved substances is dominated by transport due to turbulent motions. Mixing in the horizontal direction is rapid, resulting in nearly homogeneous concentrations horizontally. In the vertical direction, density stratification suppresses turbulence, thus reducing vertical exchange of dissolved substances and heat. As a consequence vertical concentration and heat gradients can build up. Because density stratification varies seasonally, vertical mixing varies substantially over the year.

During spring and summer, warming of the surface waters due to the heat flux from the atmosphere leads to a warm surface layer that is mixed by wind forcing and convection during nighttime cooling. This layer is separated from the deep-water body by a strong thermocline (May and July profiles, Fig. 10). Associated with the thermocline is a strong density stratification preventing exchange of heat and dissolved substances between the surface layer and the cold deep water. In fall and winter, surface cooling causes convection that successively erodes the thermocline from the top thus increasing the depth of the homotherm surface layer (November profile, Fig. 10). This process leads

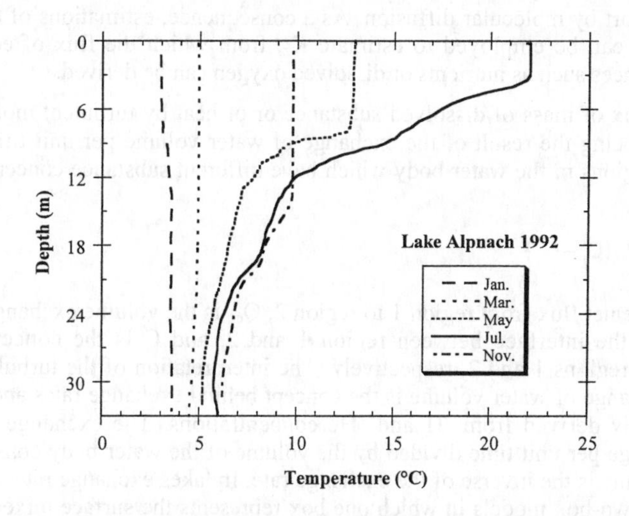

Figure 10. Seasonal variation of the vertical temperature distribution in temperate lakes exemplified by temperature profiles from Lake Alpnach (Switzerland).

to a vigorous redistribution of all dissolved substances. In quite a few lakes convective mixing by surface cooling continues until the entire water column becomes homotherm that is commonly termed 'full turnover'. Note however, that this does not necessarily imply a fully mixed water column. In freshwater lakes further winter cooling can lead to a cold surface layer and an associated density-stratification (January profile, Fig. 10), because below 4°C freshwater density decreases with decreasing temperature. Warming of the surface waters in early spring removes the cold surface layer leading again to homotherm conditions with vigorous mixing (March profile, Fig. 10). Then the seasonal mixing cycle with the development of a surface layer begins anew.

The pattern of seasonal mixing can vary substantially between lakes and depends on climatic and meteorological conditions. In addition it may be altered e.g., by density stratification due to gradients in the concentration of dissolved substances, a feature typical for lakes with anoxic deep water. In very deep lakes the deep-water body may not be affected by seasonal mixing at all (e.g., Lake Tanganyika). River inflows introduce dissolved substances to the lakes and may contribute to vertical transport if the river water is denser than the lake water and sinks to greater depths resulting in upwelling of lake water (e.g., in Lake Baikal). However, in most lakes turbulent motions are predominantly responsible for vertical transport of dissolved matter and thus central for the understanding of the lake ecosystem. Hence, the quantification of the vertical exchange is one of the central questions in lakes research to which noble gases and especially ^{3}H-^{3}He dating have been applied.

The vertical flux F of a dissolved substance resulting from turbulent motion is commonly described in analogy to transport by molecular diffusion:

$$F = -K_z \cdot \frac{\partial C}{\partial z} \tag{37}$$

where C is the concentration, z the depth (positive upwards) and K_z the vertical turbulent diffusivity. In contrast to molecular diffusivities, the turbulent diffusion coefficient is a characteristic property of the turbulent motion. Hence, K_z is independent of the specific properties of the dissolved substance and the same K_z applies to substance concentrations and heat. In most lakes the transport by turbulence is several orders of magnitude larger than the transport by molecular diffusion. As a consequence, estimations of the vertical flux of tracers can be employed to estimate K_z, from which the flux of ecologically relevant substances such as nutrients or dissolved oxygen can be derived.

The net flux of mass of dissolved substance or of heat by turbulent motion can be interpreted as being the result of the exchange of water volume per unit time between neighboring regions in the water body which have different substance concentrations or temperatures:

$$F_{12} = \frac{Q_{ex}}{A_{12}} \cdot (C_1 - C_2) \tag{38}$$

where F_{12} is the net flux from region 1 to region 2, Q_{ex} is the volume exchange rate, A_{12} is the area of the interface between region 1 and 2, and C is the concentration or temperature in regions 1 and 2, respectively. The interpretation of the turbulent flux as caused by exchange of water volume is the concept behind exchange rates and residence times commonly derived from ^{3}H and ^{3}He concentrations. The exchange rate is the volume exchange per unit time divided by the volume of the water body considered and the residence time is the inverse of the exchange rate. In lakes exchange rates are usually calculated for two-box models in which one box represents the surface mixed layer and the other the deep water layer, because the exchange of nutrient rich deep water with surface water across the thermocline is a key question to understand biological production in surface waters.

Distribution of noble gases in lakes

Atmospheric noble gases Ne, Ar, Kr and Xe. The predominant source of atmospheric noble gases in lakes is gas exchange between atmosphere and water (Fig. 2). Usually the concentrations of the atmospheric noble gases Ne, Ar, Kr and Xe measured in the deep water of lakes are close to atmospheric equilibrium at the water temperature of the sample and atmospheric pressure at the lake surface. Profiles of ^{20}Ne concentrations in Lake Baikal are shown in Figure 11b. Assuming that concentrations in the water remain unchanged after the water is transported from the lake surface to greater depths, concentrations of Ne, Ar, Kr and Xe carry information on temperature, salinity and atmospheric pressure during gas exchange which can be inferred from the data using inverse fitting techniques (Aeschbach-Hertig et al. 1999b). In Lake Baikal and in the Caspian Sea temperatures derived from noble gas concentrations agree well with temperatures measured with temperature sensors (Fig. 12; Aeschbach-Hertig et al. 1999b; Peeters et al. 2000a) In large lakes, noble gas concentrations can be raised compared to atmospheric equilibrium by unfractionated excess air which probably stems from dissolution of air bubbles injected into the water by breaking waves (Fig. 2). Excess air expressed as Ne excess above atmospheric equilibrium, ΔNe, is on average 1% ΔNe in Lake Baikal (Aeschbach-Hertig et al. 1999b) but can be up to 5% ΔNe in Lake Tanganyika (Kipfer et al. 2000) and 4% ΔNe in the Caspian Sea (Peeters et al. 2000a). The latter values are on the same order as in the ocean (Bieri 1971). The information captured by the dissolved atmospheric noble gas concentrations is potentially stored in the pore water of lake sediments and thus could be an excellent tool in paleolimnology to study temperature changes in tropical lakes or variations of the salt content in closed-basin lakes indicating lake level fluctuations.

If atmospheric noble gases at the lake surface are always in equilibrium with the atmosphere, one would expect that in lakes located in temperate regions, noble gas concentrations and in particular the temperature sensitive Xe concentration should have strong vertical gradients during summer stratification. Assuming 25°C in the surface and 4°C in the deep water, the corresponding Xe concentration should differ by about 50%. Because the Xe flux at the lake surface due to gas exchange is limited and vertical transport would cause a flux of Xe from deep to shallow waters, Xe concentrations in the surface waters can be expected to be above atmospheric equilibrium in summer. Thus Xe and other atmospheric noble gases might be applied to study gas and deep-water exchange. To our knowledge, a profile of all atmospheric noble gases has not yet been measured in temperate lakes during periods of strong temperature stratification.

^{3}He and ^{4}He. Gas exchange between the atmosphere and the water at the lake surface is an important source or sink of ^{3}He and ^{4}He. Therefore the concentrations of ^{3}He and ^{4}He near the lake surface are usually close to atmospheric equilibrium (Fig. 11a,d). However, in addition to the atmosphere, He of terrigenic / radiogenic origin enters the lake via the sediments (Fig. 2). In most cases $^{4}He_{ter}$ stems from the Earth's crust but in some cases also from the Earth's mantle. If the contribution of the mantle component is significant, $^{3}He_{ter}$ can be important. However, in most lakes $^{3}He_{ter}$ is predominantly of crustal origin and is small compared to the other components contributing to the total concentration of ^{3}He. In contrast to $^{3}He_{ter}$, tritiogenic $^{3}He_{tri}$ is in most lakes an important source of ^{3}He which depends on the distribution of ^{3}H. The sources of ^{4}He and ^{3}He below the lake surface increase the contributions of $^{4}He_{ter}$, $^{3}He_{ter}$ and $^{3}He_{tri}$ to the total concentration of ^{4}He and ^{3}He, respectively, and thus lead to an increase in the concentration of both helium isotopes at larger depths. ^{3}He and ^{4}He concentrations usually decrease towards the lake surface where the terrigenic and tritiogenic excess is lost to the atmosphere by gas exchange (see Fig. 11a,d).

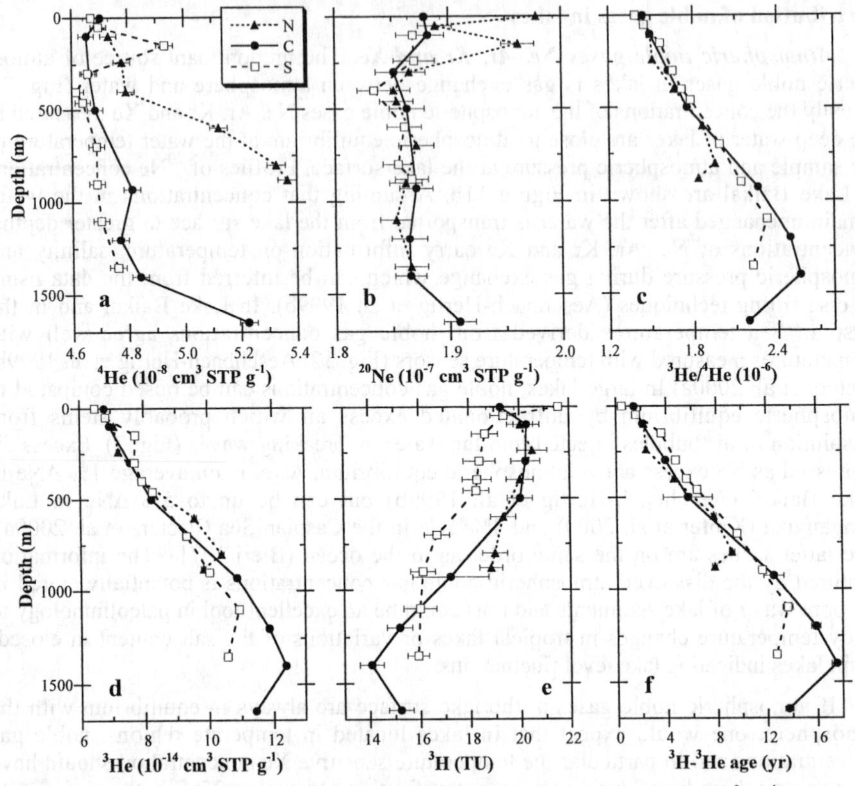

Figure 11. The vertical distribution of the ^{4}He, ^{20}Ne, ^{3}He and ^{3}H concentrations, the ^{3}He/^{4}He ratio and the ^{3}H-^{3}He water ages in the northern (N), central (C) and southern (N) basin of Lake Baikal in 1992 (data from Hohmann et al. 1998).

^{222}Rn. The noble gas isotope ^{222}Rn is of radiogenic origin and its concentration distribution is closely linked to the distribution of its mother element ^{226}Ra. In lakes, Ra is predominantly concentrated in the sediments and to a lesser extent adsorbed to particles suspended in the water. Therefore, the sediments are the most important source of ^{222}Rn and the ^{222}Rn concentrations are highest near the sediments, decreasing from there towards the lake surface. Because of the short half-life of ^{222}Rn (3.82 d) its concentration distribution has not only vertical but can also have large horizontal gradients which both depend on the location of ^{222}Rn release, i.e., on the slope of the sediments, and on the location of sampling with respect to lake morphometry. Therefore, the distribution of ^{222}Rn is affected by horizontal and vertical mixing processes. Although this enables an assessment of horizontal and vertical mixing processes from the distribution of ^{222}Rn, their quantification is non-trivial. In general, the estimation of mixing rates from the distribution of ^{222}Rn requires computation of transport in 2 or 3 dimensions (Imboden and Joller 1984).

Application of ^{3}H-^{3}He dating in lakes

Assuming that at the lake surface ^{3}He is in equilibrium with the atmosphere $^3He_{tri}$, and consequently the ^{3}H-^{3}He age, should be zero. As soon as the water is moved away from the lake surface and gas exchange stops, $^3He_{tri}$ accumulates and the ^3H-^3He age increases with time (Fig. 2). Thus the ^3H-^3He age is a measure of the 'isolation age', defined as the time elapsed since the water was last in contact with the atmosphere (Tolstikhin and Kamenskiy 1969; Torgersen et al. 1977). Note that water masses in lakes are usually mixtures and the isolation age of a sample must be understood as the volume weighted mean of the isolation age of the waters involved in the mixing. Because mixing leads to volume weighted mean concentrations of ^3H and ^3He and the ^3H-^3He age depends non-linearly on the concentrations of ^3H and ^3He, in mixed waters the ^3H-^3He age does not exactly agree with the isolation age (see Schlosser and Winckler 2002, this volume). In general, the ^3H-^3He water age of a mixture of water masses is biased towards the component with the larger ^3H concentration. However, in most lakes the simplified interpretation of the ^3H-^3He age as being equivalent to the isolation age can serve as a tool to study transport processes in a qualitative manner.

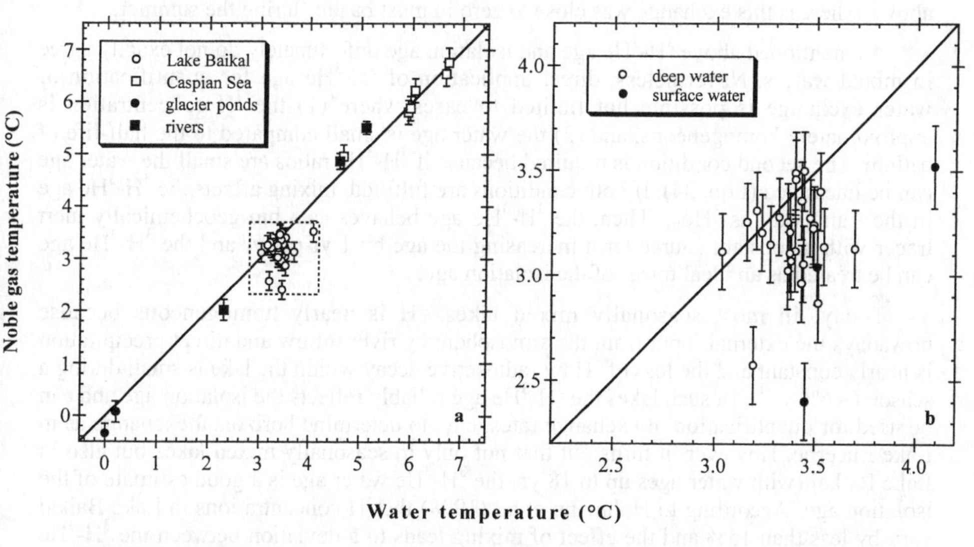

Figure 12. Comparison of temperatures derived from noble gas concentrations (NGT) with temperatures measured using thermistors (Aeschbach-Hertig et al. 1999b). Shown are several examples from surface waters including Lake Baikal and the Caspian Sea. In almost all cases the NGT agree very well with the data from temperature sensors.

For example: ^3H-^3He ages in the three basins of Lake Baikal are close to zero near the surface (Fig. 11f). With increasing depth the ^3H-^3He age increases, indicating that the deeper water was isolated from the surface for a longer period of time than the water at shallower depth. The increase of the ^3H-^3He age with depth is a measure of the intensity of vertical exchange. In the bottom 300 to 400 m, ^3H-^3He ages are about constant or even slightly decrease with increasing depth. In a one-dimensional interpretation it is impossible to explain younger water located below older water except by inflow of young water. However, inflow of young water from ground-water sources at great depths is unlikely, because typical ground-water flow is orders of magnitudes slower than vertical transport in lakes, deep ground waters generally have very large ages, and most ground-water inflow to lakes occurs near the edge of a lake, not in the middle. In Lake Baikal the young ages at largest depths originate from water of near surface regions which sinks advectively as density plumes along the basin boundaries to greatest depths where it spreads laterally (Peeters et al. 1996b; Hohmann et al. 1997). Because the ^3H-^3He age near the surface is less than at greater depth the density plumes transport comparatively young water. Several processes causing such density plumes in Lake Baikal have been identified, e.g., river inflow, inter basin exchange, wind forcing, cabbeling (horizontal mixing of waters with temperature above and below the temperature of maximum density) and inflow of hydrothermal waters (Weiss et al. 1991; Shimaraev et al. 1993; Kipfer et al. 1996; Peeters et al. 1996a; Hohmann 1997; Hohmann et al. 1997).

In Lake Lucerne, which is a complex system of several basins separated by sills leaving 5 to 100 m water depth for inter basin exchange, ^{3}H-^{3}He age inversions indicate that horizontal water exchange between basins generates density plumes ventilating the bottom regions of the different basins. Aeschbach-Hertig et al. (1996a) estimated the horizontal and vertical exchange between neighboring basins responsible for deep-water renewal by a simple calculation of mixing ratios based on ^{3}H-^{3}He ages. During the winter in all basins the water deeper than the sills was exchanged more than once by water from above, whereas this exchange was close to zero in most basins during the summer.

As mentioned above ^{3}H-^{3}He age and isolation age unfortunately do not exactly agree in mixed waters. Nevertheless, direct application of ^{3}H-^{3}He age for quantification of water exchange is possible but limited to cases where (1) the ^{3}H concentration is approximately homogeneous, and (2) the water age is small compared to the half-life of tritium. The second condition is required because if ^{3}H-^{3}He ratios are small the water age can be linearized (Eqn. 34). If both conditions are fulfilled, mixing affects the ^{3}H-^{3}He age in the same way as $^3He_{tri}$. Then, the ^{3}H-^{3}He age behaves as a bio-geochemically inert tracer with a constant source term increasing the age by 1 yr per yr and the ^{3}H-^{3}He age can be treated as an ideal tracer of the isolation age.

Today, in most seasonally mixed lakes, ^{3}H is nearly homogeneous because nowadays the external input from the atmosphere by river inflow and direct precipitation is nearly constant and the loss of ^{3}H by radioactive decay within the lake is small during a season (~ 6% yr^{-1}). In such lakes the ^{3}H-^{3}He age reliably reflects the isolation age and can be used for quantification of exchange rates, e.g., to determine horizontal exchange as in Lake Lucerne. However, it turns out that not only in seasonally mixed lakes but also in Lake Baikal (with water ages up to 18 yr) the ^{3}H-^{3}He water age is a good estimate of the isolation age. According to Hohmann et al. (1998) the ^{3}H concentrations in Lake Baikal vary by less than 15% and the effect of mixing leads to a deviation between the ^{3}H-^{3}He age and the isolation age of less than 10%, the ^{3}H-^{3}He age being generally larger than the isolation age of the mixed water. Even in the Caspian Sea, where ^{3}H-^{3}He ages reach up to 25 yrs and the ^{3}H concentration in the southern basin decreases with depth to about 50% of the surface value (Peeters et al. 2000a), the deviation between ^{3}H-^{3}He age and isolation age due to mixing is less than 10%.

Although the discussion above demonstrates that ^{3}H-^{3}He ages provide a reliable estimate of the isolation age in a wide range of lakes, this is not always the case, as has been demonstrated for Lake Lugano (Aeschbach-Hertig 1994). Therefore, it is advisable to complement the age estimates based on ^{3}H and ^{3}He concentrations by independent estimations based on other tracers such as CFC-12 or SF_6. Dating based on these tracers compares the concentration in the water sample with the historic atmospheric equilibrium concentration of the tracer at the lake surface to determine the apparent date of equilibration. The elapsed time between the sampling date and the apparent equilibration date gives the CFC-12 age or the SF_6 age, both measuring the time elapsed since the water was in last contact with the atmosphere, i.e., the isolation age. However, because the historic atmospheric concentrations of CFC-12 and SF_6 vary in a non-linear fashion, mixing of water masses also leads to a deviation between tracer age (CFC-12 age or SF_6 age) and isolation age. Because sources and transport of the transient tracers ^{3}He, ^{3}H, SF_6, and CFC-12 differ, the deviation between tracer age and isolation age caused by mixing is different between the ^{3}H-^{3}He, SF_6, and CFC-12 dating techniques (Hofer et al. 2002).

In the Caspian Sea ^{3}H-^{3}He ages closely agree with the CFC-12 ages (Fig. 13; Peeters et al. 2000a). The agreement of the ages supports the reliability of the ages derived from the tracers and their interpretation as a reliable proxy for the isolation age. In Lake Issyk-Kul, ^{3}H-^{3}He ages closely agree with the SF_6 ages but are significantly smaller than the CFC-12 ages (Hofer et al. 2002; Vollmer et al. 2002). Hofer et al. (2002) demonstrated that mixing and the change in the atmospheric CFC-12 concentration since the 1990s leads to CFC-12 ages that significantly overestimate the isolation age. In Lake Lugano ^{3}H-^{3}He ages are significantly larger than SF_6 ages (Fig. 13; Holzner 2001). This indicates that mixing significantly affects the ^{3}H-^{3}He ages and that they do not agree with the isolation age.

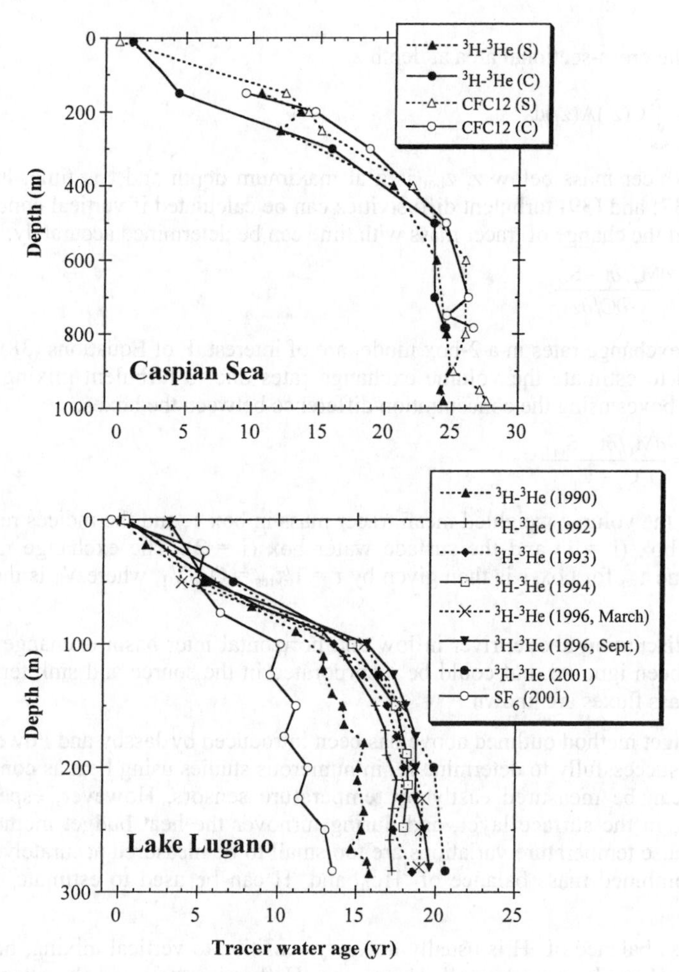

Figure 13. Comparison of ^{3}H-^{3}He ages with CFC-12 and SF_6 ages. (Upper panel) In case of the central and southern basin of the Caspian Sea (Peeters et al. 2000a), ^{3}H-^{3}He ages agree very well with CFC-12 ages indicating that the tracer water ages are a reliable proxy for the isolation age. (Lower panel) In case of Lake Lugano (Holzner 2001), ^{3}H-^{3}He ages are significantly larger than SF_6 ages suggesting that the isolation age in Lake Lugano cannot be described very reliably by ^{3}H-^{3}He ages.

Quantification of vertical exchange rates and vertical turbulent diffusivities

Budget methods. Vertical exchange rates and turbulent diffusivities K_z can be calculated from the heat balance or the mass balance of tracers for which transformation rates are known. Assuming horizontal homogeneity, the temporal change of tracer mass below a given depth z must be the sum of the net vertical mass flux through the cross-section at z and all sources and sinks of tracer mass below z. In the case of conservative tracers sources and sinks below z must be mass fluxes across the sediment-water interface. In the case of ^3H, radioactive decay is an additional sink. In the case of ^3He, tritium decay represents a source. If the increase of mass due to all sources and sinks, S_M, is known, the net mass flux can be calculated:

$$F \cdot A = -\left(\frac{dM_z}{dt} - S_M\right) \tag{39}$$

where A is the cross-sectional area at depth z,

$$M_z = \int_{z_{bot}}^{z} C(z')A(z')dz'$$

is the total tracer mass below z, z_{bot} is z at maximum depth and t is time. From F in Equations (37) and (39) turbulent diffusivities can be calculated if vertical concentration gradients and the change of tracer mass with time can be determined accurately:

$$K_z = \frac{\partial M_z/\partial t - S_M}{A \cdot \partial C/\partial z} \tag{40}$$

If only exchange rates in a 2-box model are of interest, F of Equations (38) and (39) can be used to estimate the volume exchange rates due to turbulent mixing between neighboring boxes using the concentration difference between the boxes:

$$Q_{ex} = \frac{\partial M_1/\partial t - S_{M,1}}{C_2 - C_1} \tag{41}$$

where M_1 is the volume weighted mean tracer mass in box 1, and the indices refer to the deep water box (i = 1) and the surface water box (i = 2). The exchange rate r and residence time τ_{res} for box 1 is then given by $r = 1/\tau_{res} = Q_{ex}/V_1$, where V_1 is the volume of box 1.

In the discussion above, river inflow and horizontal inter basin exchange at larger depth have been ignored, but could be incorporated in the source and sink terms if the water and mass fluxes are known.

The budget method outlined above has been introduced by Jassby and Powell (1975) and applied successfully to determine K_z in numerous studies using heat as conservative tracer that can be measured easily by temperature sensors. However, especially in deep waters, in the surface layer, and during turnover the heat budget method is not reliable because temperature variations are too small to be measured accurately. In these cases the combined mass balance of $^3He_{tri}$ and ^3H can be used to estimate exchange rates.

The mass balance of ^3H is usually not very sensitive to vertical mixing, because in many lakes ^3H gradients are small. Hence the ^3H flux is small and the change in ^3H content below a given depth is not significantly larger than the precision of the measurements. The mass balance of $^3He_{tri}$ is linearly dependent on the mass balance of ^3H, because $^3He_{tri}$ is coupled to ^3H via tritium decay. This problem could be overcome, if the estimation of exchange rates is based on the sum $^3H^* = {}^3H + {}^3He_{tri}$:

$$Q_{ex} = \frac{\partial M_{^3H^*,1}/\partial t}{^3H_2^* - ^3H_1^*} \tag{42}$$

However, the experimental errors of ^{3}H concentrations are usually significantly larger than those of ^{3}He, making the error of ^{3}H* rather large. Thus in the case of water bodies with long residence times and / or with a homogeneous ^{3}H concentration it can be advantageous to estimate volume exchange rates from ^{3}He concentrations

$$Q_{ex} = \frac{\partial M_{^3He_{tri},1}/\partial t - \lambda M_{^3H,1}}{^3He_{tri,2} - ^3He_{tri,1}} \tag{43}$$

because the error introduced by neglecting the transport and loss of ^{3}H due to decay during the observational period is small (Aeschbach-Hertig 1994).

In studies where the data are insufficient to calculate the temporal change of e.g., $^3He_{tri}$, water exchange can be estimated from the ^{3}H-^{3}He ages, provided that the ^{3}H-^{3}He ages agree with the isolation age. Then, the ^{3}H-^{3}He age can be treated as ideal tracer for water age τ which increases by 1 yr per yr. In case of steady state conditions, volume exchange between box 1 and box 2 can be calculated from the budget method (Eqn. 41):

$$Q_{ex} = \frac{V_1}{\tau_2 - \tau_1}. \tag{44}$$

Thus, if the assumptions underlying Equation (44) are fulfilled, the difference between volume weighted mean ^{3}H-^{3}He age of surface and deep water can be used as the residence time of the deep water (Peeters et al. 2000a). In Lake Baikal the exchange rates derived from this estimate of residence time can be highly inaccurate at times when ^{3}H gradients are large (e.g., close to 1963 during the atmospheric bomb peak) but agree reasonably well with exchange rates estimated from inverse modeling of several transient tracers if ^{3}H gradients are small (Peeters et al. 1997). In the case of the Caspian Sea the deep-water residence time was estimated based on the difference in ^{3}H-^{3}He ages and on the difference in CFC-12 ages between surface and deep water (Peeters et al. 2000a). Because mixing has a different effect on ^{3}H-^{3}He and on CFC-12 ages, the agreement of the residence times obtained supports the reliability of these estimates.

The two box models discussed above can be extended to a multibox model. In the limit of infinitely small boxes the multibox model corresponds to the continuous model of Jassby and Powell (1975). Using Equation (40), the vertical turbulent diffusivity as function of depth can be obtained from τ if τ is at steady state and can be treated as ideal tracer with source strength of 1yr/yr:

$$K_z = \frac{V}{A} \cdot \left(\frac{\partial \tau}{\partial z}\right)^{-1}. \tag{45}$$

Thus if the ^{3}H-^{3}He age closely agrees with the isolation age and is at steady state, K_z can be obtained directly from a vertical profile of ^{3}H-^{3}He ages. In Figure 14 measured ^{3}H-^{3}He ages in Lake Zug (Switzerland) are compared with the water age simulated by using Equation (45) and assuming that water age is at steady state and that mixing can be characterized by a diffusivity independent of depth and time. The latter assumption is unrealistic because mixing varies seasonally. Hence, the estimated diffusivity is a measure of the long-term mean effect of mixing on tracer distributions rather than the true vertical turbulent diffusivity. Model results and data agree reasonably well if a K_z betwen 6×10^{-5} m^{2}s^{-1} and 7×10^{-5} m^{2}s^{-1} is employed (Fig. 14). This range of K_z agrees well with the long-term mean K_z for Lake Zug derived from the budget of ^{3}He and from inverse numerical modeling of the temporal development of ^{3}He concentrations (Aeschbach-Hertig 1994).

Numerical modeling. In the deep water of very deep lakes such as Lake Baikal temporal changes in the concentration of ^{3}He or ^{3}H can be very small and a long time period must pass until changes can be determined with reasonable accuracy. In addition, in Lake Baikal river water sinks as density plumes from near surface regions to greatest depth and causes deep-water renewal (Hohmann et al. 1997). This process results in upwelling in the open water and thus contributes to the vertical transport of tracers. In such a case instead of the budget methods outlined above, 1-D numerical models are recommended to analyze deep-water exchange. The model must be capable of describing the vertical advective and vertical turbulent diffusive flux of the tracer at each depth and also simulate the input and loss of tracers by gas exchange, outflow and inflows at the lake surface. Because the atmospheric concentration of ^{3}He remains constant over time and the tritium concentration in precipitation can be reconstructed from the ^{3}H survey of the IAEA (IAEA/WMO 1998), it is possible to simulate ^{3}He and ^{3}H concentrations over long time periods and to predict values for ^{3}He and ^{3}H for the time of observation. By adjusting turbulent diffusivities and upwelling velocities such that data and model prediction agree, transport parameters can be determined (inverse fitting procedure). The constraint on the model parameters increases if data are available from different years and if several tracers are included. In the case of Lake Baikal the inverse fitting technique has been based on the combination of the tracers ^{3}He and ^{3}H with the gaseous tracers CFC-11 and CFC-12 (Peeters et al. 2000b). The transport rates estimated from the transient tracers can serve as the basis to estimate oxygen depletion in the deep water or the He flux from the sediments (see below).

Long term numerical modeling of the concentrations of ^{3}He and ^{3}H has also been successfully applied in smaller lakes that undergo strong seasonal changes (Aeschbach-Hertig 1994; Aeschbach-Hertig et al. 1999a, 2002a; Holzner 2001). However, in such

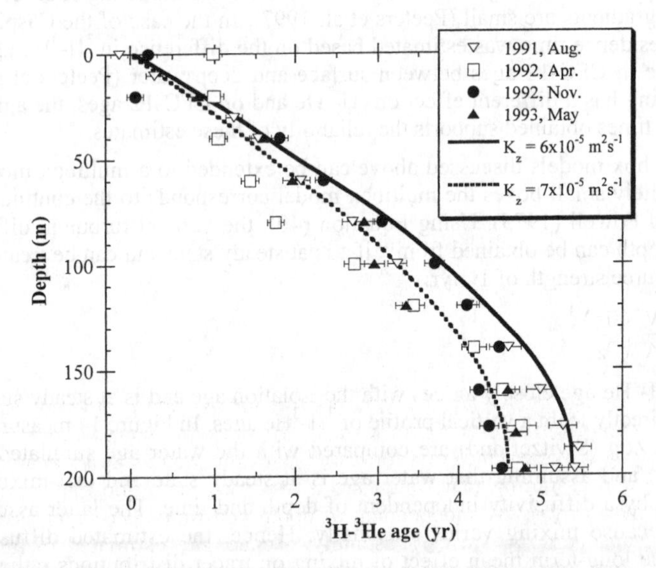

Figure 14. ^{3}H-^{3}He ages in Lake Zug determined from concentrations of noble gases and ^{3}H measured in water samples collected between 1991 to 1993 (Aeschbach-Hertig 1994) and isolation ages derived from a simplified exchange / mixing model. The model assumes that water ages are at steady state and that vertical transport can be described in analogy to Ficks law using a diffuivity of 6×10^{-5} m^{2}s^{-1} (solid line) and 7×10^{-5} m^{2}s^{-1} (dashed line) independent of depth and time.

lakes inverse fitting of the tracer data can only produce reliable results if the number of parameters describing the seasonal changes of the vertical distribution of K_z is small. Hence, a substantial amount of additional information must be available to describe the seasonally changing thickness of the surface layer and the vertical and temporal variation of K_z.

He flux from the continental crust and oxygen depletion

Estimation of the ^{4}He flux. In lakes, He emanates from the sediments and accumulates in the water column. The accumulation rate can be used to estimate the He flux from the sediments. Clarke et al. (1977) and Top and Clarke (1981) investigated the ^{4}He flux in several lakes in Labrador to localize zones of high ^{4}He fluxes that they assumed to be indicative of uranium and thorium deposits. However, Mamyrin and Tolstikhin (1984) demonstrated that the large ^{4}He flux observed in one of the lakes could also be explained by a ^{4}He flux from the deeper crust. O'Nions and Oxburgh (1988) have demonstrated that the radiogenic ^{4}He flux in lakes in Labrador is close to the steady state ^{4}He flux of 3×10^{10} atoms m^{-2} s^{-1} from the continental crust (O'Nions and Oxburgh 1983).

Assuming that the ^{4}He flux is constant in time and that the ^{4}He flux from the sediments does not vary within the typical depth range of lakes, the ^{4}He flux can be determined from the ^{4}He budget in a 2-box model:

$$F_{4He,sed} \cdot A_1 \approx \frac{dM_{^4He_1}}{dt} - Q_{ex} \cdot \left({}^4He_2 - {}^4He_1\right) \tag{46}$$

where $F_{4He,sed}$ is the ^{4}He flux per sediment area and Q_{ex} can be determined from the balance of e.g., $^3He_{tri}$. In good approximation the total sediment area below a given depth z is equal to the cross-sectional area at depth z. Equation (46) states that the mass flux of ^{4}He introduced from the sediments below depth z corresponds to the change per unit time of the total mass of ^{4}He below z minus the ^{4}He lost by exchange with the upper layer. The boundary between upper and lower layer is somewhat arbitrary and the use of volume weighted mean ^{4}He concentrations in upper and lower layer to calculate ^{4}He exchange is an approximation. In seasonally mixed lakes with time varying ^{4}He profiles in which mixing across the thermocline is very small in the summer months, the accumulation over the summer season of ^{4}He in the deep water below the thermocline can be used directly to estimate $F_{4He,sed}$ (e.g., Laacher See Aeschbach-Hertig et al. 1996b).

In lakes where the ^{4}He profile is at steady state, $F_{4He,sed} = -Q_{ex} \cdot ({}^4He_2 - {}^4He_1)/A_1$. If also the water age is at steady state, introducing the volume exchange estimated from water age differences (Eqn. 44) leads to

$$F_{4He,sed} = \frac{V_1}{A_1} \cdot \frac{{}^4He_2 - {}^4He_1}{\tau_2 - \tau_1} \tag{47}$$

where τ_1 and τ_2 are the volume weighted mean water ages in box 1 and 2.

A simplified procedure to estimate $F_{4He,sed}$ directly compares the excess of ^{4}He above atmospheric equilibrium Δ^4He with the ^{3}H-^{3}He age (Mamyrin and Tolstikhin 1984). Assuming that the ^{3}H-^{3}He age measures the water residence time it can be interpreted as the time during which ^{4}He of terrestrial origin accumulates in a water parcel. Thus the ratio of Δ^4He to ^{3}H-^{3}He age gives the ingrowth of ^{4}He per unit volume and time. The total ^{4}He flux from the sediments can be obtained by multiplication with the lake volume V_o. Division by the surface area A_o gives the ^{4}He flux per unit area:

$$F_{4He,sed} \approx h_o \frac{\Delta^4He}{\tau} \tag{48}$$

where $h_o = V_o/A_o$ is the mean depth of the lake.

In Equation (48) the ratio Δ^4He/τ from an individual sample has been taken as representative for the entire lake. If several measurements are available the volume weighted mean values of Δ^4He and τ could be applied. This would correspond to the two box model (Eqn. 47) with box 2 representing surface water in equilibrium with the atmosphere. At the lake surface ${}^4He_2 = {}^4He_{equ}$ and ${}^3He_2 = {}^3He_{equ}$, implying $\tau_2 = 0$. Because ${}^4He_{equ}$ varies only very little with temperature ${}^4He_{equ}$ is approximately constant. Then $F_{4He,sed} = V_o/A_o \cdot \Delta^4He_1/\tau_1$ with Δ^4He_1 and τ being volume weighted mean values.

As an alternative, a representative value for Δ^4He/τ in Equation (48) can be obtained from all data simultaneously by using a regression of Δ^4He versus τ (e.g., Aeschbach-Hertig 1994; Hohmann et al. 1998). This technique implicitly assumes that the ratio Δ^4He/τ is the same for all samples. Figure 15 demonstrates that in many lakes a linear correlation between Δ^4He and τ is a reasonable approximation.

All methods outlined above are simplifications because mixing is assumed to have the same effect on ${}^3H$-${}^3He$ age and ${}^4He$, and, even more severely, the ${}^4He$ flux is treated as a volume source. In reality ${}^4He$ enters from the sediments and the accumulation of ${}^4He$ per unit volume should be largest near the lake bottom where the area to volume ratio $dA/dV = 1/A \cdot dA/dz$ is largest. In addition the ${}^3H$-${}^3He$ age only approximates the true age. All these difficulties can be avoided if $F_{4He,sed}$ can be determined by inverse fitting of the ${}^4He$ concentrations using a 1-D vertical continuous model with known transport parameters. Assuming homogeneous conditions in the horizontal direction the equation describing the vertical transport of ${}^4He$ is:

$$\frac{\partial^4 He}{\partial t} = \frac{1}{A}\frac{\partial}{\partial z}\left(A \cdot K_z \frac{\partial^4 He}{\partial z}\right) + F_{4He,sed}\frac{1}{A}\frac{\partial A}{\partial z} \tag{49}$$

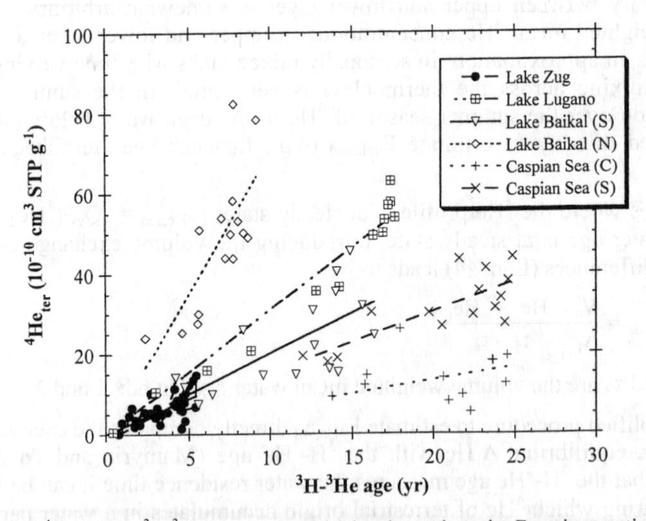

Figure 15. Δ^4He versus ${}^3H$-${}^3He$ age in several lakes located on the Eurasian continent. A linear correlation (lines) between Δ^4He and ${}^3H$-${}^3He$ age appears to be a reasonable approximation to the data from the different lakes. Data were compiled from several sources: Lake Zug and Lake Lugano (Aeschbach-Hertig 1994), central (C) and southern (S) basin of the Caspian Sea (Peeters et al. 2000a) and southern (S) and northern (N) basin of Lake Baikal (Hohmann et al. 1998).

The calculation of ^{4}He from Equation (49) requires as upper boundary condition a model describing the loss of ^{4}He by gas exchange at the lake surface. The best estimate of $F_{4He,sed}$ can be obtained by adjusting $F_{4He,sed}$ such that the squared deviation between simulated and measured ^{4}He normalised by the errors of the ^{4}He measurements becomes minimal (χ^2 minimisation). This technique has been applied to estimate $F_{4He,sed}$ in Lac Pavin (Aeschbach-Hertig et al. 1999a) and in Lake Lugano and Lake Zug (Aeschbach-Hertig 1994). Aeschbach-Hertig (1994) demonstrated for Lake Zug and Lake Lugano that $F_{4He,sed}$ obtained from inverse numerical modeling and from the Δ^4He versus ^{3}H-^{3}He age correlation technique agree reasonably well.

The flux of ^{4}He from the continental crust. The ^{4}He fluxes determined in several lakes in Switzerland and in the Caspian Sea using Equation (48) agree within a factor of 3, except for Urner basin of Lake Lucerne in which ^{4}He is possibly affected by ground-water sources (Table 4; Aeschbach-Hertig 1994). Ground-water inflow can introduce water with very high ^{4}He concentrations and thus increase the total ^{4}He flux into a lake. Vertical profiles of the ^{3}He and ^{4}He concentrations in the lakes considered suggest that the excess of ^{4}He is predominantly of crustal origin. Because inflow of ^{4}He-rich ground water leads to an increase in the total ^{4}He flux into a lake, the values given in Table 4 must be considered as an upper limit of the crustal ^{4}He flux.

The ^{4}He flux in the different basins of Lake Baikal is higher than the ^{4}He flux in the lakes mentioned above. The ^{3}He/^{4}He ratio measured in hot springs around Lake Baikal is about 2.2×10^{-7} indicating a small but significant contribution of mantle helium (Hohmann et al. 1996). In the northern basin of Lake Baikal Kipfer et al. (1996) observed substantial hydrothermal inflow accompanied by large ^{4}He concentrations. This explains that the ^{4}He flux in the northern basin is significantly larger than in the other basins of Lake Baikal (Hohmann et al. 1996). Hydrothermal inflows might also affect the ^{4}He flux in the central

Table 4. ^{4}He flux estimated in several lakes located on the Eurasian continent.

Lake	*Location*	*mean depth [m]*	*^{4}He accumulation rate [10^{-10} cm^3 STP g^{-1} yr^{-1}]*	*^{4}He flux [10^{10} atoms m^{-2} s^{-1}]*
Alpnach	Switzerland	22	5.7	1.1
Lucerne	Switzerland			
Vitznau basin		75	3.3	2.1
Gersau basin		146	1.8	2.2
Urner basin		144	12	14
Zug	Switzerland	84	2.3	1.6
Lugano	Switzerland	171	3.3	4.8
Baikal[1)]	Siberia			
Southern basin		844	1.9	14
Central basin		854	1.7	12
Northern basin		576	5.7	28
Caspian Sea[1)]	Middle Asia			
Southern basin		461	1.5	5.9
Central basin		259	0.8	1.9

[1)] Data and mean depth below 200 m.

Flux estimates were taken from Aeschbach-Hertig et al. (1996a: Lake Alpnach, Lake Lucerne, Lake Lugano, Lake Zug); Hohmann et al. (1998: Lake Baikal); Peeters et al. (2000a: Caspian Sea).

and southern basin, but direct evidence for hydrothermal activity in these basins does not exist. Additionally, the massive and up to 30 million year old sediments in Lake Baikal might release more ^{4}He than the host rock of the sediments (see hypothesis of Solomon et al. 1995). Hence, the ^{4}He flux obtained from the data of the southern and central basin of Lake Baikal can be expected to be higher than the common crustal flux and should therefore be considered as an upper limit of the ^{4}He flux from the continental crust.

The estimations of the crustal ^{4}He flux based on the experimental observations from the lakes located in Central Europe and Asia are of the same order of magnitude and range between 10^{10} and 10^{11} atoms m^{-2} s^{-1}. Based on these results one can speculate that the crustal ^{4}He flux over the Eurasian continent is approximately constant and agrees within a factor of 3 with the theoretical value of 3×10^{10} atoms m^{-2} s^{-1} (O'Nions and Oxburgh 1983). The range of the experimental values also agrees with the ^{4}He flux estimates derived from ground-water studies (Torgersen and Clarke 1985; Torgersen and Ivey 1985)

Oxygen depletion. By analogy to the estimation of the ^{4}He flux, oxygen consumption can also be estimated from a regression of concentration of dissolved oxygen versus ^{3}H-^{3}He age (Jenkins 1976). Dissolved oxygen is introduced into lakes by gas exchange at the lake surface and by biological production in the photic zone. In most lakes dissolved oxygen decreases below the photic zone with increasing depth and ^{3}H-^{3}He age, because O_2 is consumed by decomposition of organic material. In seasonally mixed lakes the rate of oxygen depletion can be estimated from the decrease with time of the concentration of dissolved oxygen during the period of stratification in summer and fall. If O_2 concentrations vary only very little with time, oxygen depletion can be estimated from the correlation between ^{3}H-^{3}He age and O_2 assuming that the decrease of oxygen per unit volume is constant in time. However, oxygen depletion comprises of two components: (1) a volume sink due to consumption in the open water and (2) an areal sink due to consumption at the sediment water interface. Especially in the deeper regions of lakes the latter may dominate the overall oxygen depletion. Nevertheless, the regression of O_2 concentration versus ^{3}H-^{3}He age provides a reasonable estimate of overall oxygen depletion which agrees with the values on oxygen depletion derived from traditional methods (Aeschbach-Hertig 1994). Especially in lakes where concentrations of dissolved oxygen remain the same over years (e.g., Lake Baikal) the regression of O_2 versus ^{3}H-^{3}He age is a comparatively simple method to estimate oxygen depletion. In the deep water of Lake Baikal oxygen depletion estimated in this manner is about 140 mg O_2 m^{-3} yr^{-1} (Hohmann et al. 1998) which is somewhat larger than the mean oxygen depletion of 80 mg O_2 m^{-3} yr^{-1} estimated from inverse numerical modeling (Peeters et al. 2000b). The inverse modeling approach has the advantage that a volume and an areal sink for oxygen can be distinguished. However, numerical modeling requires a substantial amount of information on the exchange processes and still can describe the transport of dissolved oxygen only in a very simplified manner.

Noble gases from the Earth's mantle

Lakes located in volcanically and tectonically active zones are suitable systems to obtain information on the noble gas composition in the Earth's mantle and to study mantle volatile fluxes because a lake can act as a collector of mantle fluids and gases (Fig. 2). The ^{3}He/^{4}He ratio has been employed as an indication of a mantle component (e.g., Sano et al. 1990; Collier et al. 1991; Giggenbach et al. 1991; Igarashi et al. 1992; Kipfer et al. 1994; Aeschbach-Hertig et al. 1996b, 1999a; Clark and Hudson 2001). In the lakes studied, the ^{3}He/^{4}He ratio varies between 7.4×10^{-6} and 10.3×10^{-6} (Table 5). The concentrations of ^{3}He and ^{4}He measured in several of the lakes mentioned above are shown in Figure 16. In the case of Laacher See the atomic ratios of additional gases in the mantle component were also determined: $(^{20}Ne/^{3}He)_{man} = 1.8±0.3$ and $(^{36}Ar/^{3}He)_{man} \leq 3.5±0.6$ (Aeschbach-Hertig et al. 1996b). These ratios lie between the ratios calculated from Staudacher et al. (1989) for the upper and lower mantle.

Table 5. Helium fluxes, C/^{3}He and heat/^{3}He ratios in lakes located in volcanic or tectonically active regions (extended from Aeschbach-Hertig et al. 1996b).

Lake (country)	*Area [km^2]*	*^{4}He flux ×10^{-12} [atoms m^{-2} s^{-1}]*	*^{3}He/^{4}He [10^{-6}]*	*C/^{3}He [10^9]*	*heat/^{3}He [10^{-9} J atom^{-1}]*	*Reference*
Lake Nyos (Cameroon)	1.49	300±40	7.84±0.04	30±15	0.29±0.04	Sano et al. (1990)
Crater Lake (USA)	53	0.55	9.9	40	100	Collier et al. (1991)
Lake Mashu (Japan)	20	0.92	9.43±0.17	180	170	Igarashi et al. (1992)
Lake Nemrut (Turkey)	11	≤6	10.32±0.06	37±22 [a]	12±1	Kipfer et al. (1994)
Lake Van (Turkey)	3600	0.2-0.3	10.3 [b]	-	-	Kipfer et al. (1994)
Laacher See (Germany)	3.31	10±2	7.42±0.03	8.6±1.0	1.0±0.2	Aeschbach-Hertig et al. (1996b)
Lac Pavin (France)	0.44	0.6±0.2	9.1±0.01	13	6	Aeschbach-Hertig et al. (1999a, 2002a)

[a] Average of three gas samples from the Nemrut Caldera that were not discussed by Kipfer et al. (1994).
[b] Value taken from the nearby Lake Nemrut.

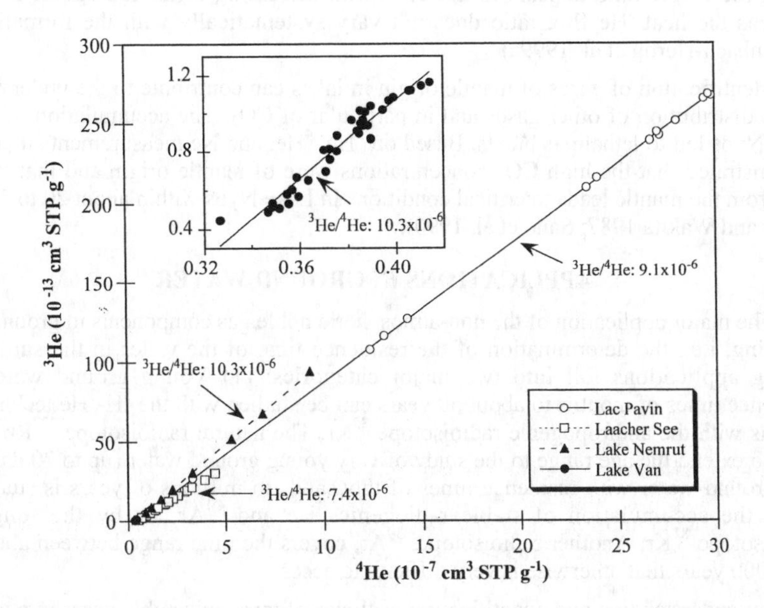

Figure 16. ^{3}He versus ^{4}He concentrations in several lakes affected by gas fluxes from the Earth's mantle. In all lakes shown, the ^{3}He/^{4}He ratios are more than two orders of magnitude higher than the ^{3}He/^{4}He ratio in the helium flux from the continental crust. Data from Kipfer et al. (1994: Lake Nemrut and Lake Van, Turkey), from Aeschbach-Hertig et al. (1996b: Laacher See, Germany; 1999a: Lac Pavin, France).

The estimation of the fluxes of the mantle component is complicated by the fact that in most cases ^3H-^3He dating cannot be employed in the presence of a significant mantle component because the separation of $^3He_{tri}$ and $^3He_{ter}$ becomes unreliable. In seasonally mixed lakes the accumulation of mantle derived gases in the deep water below the thermocline over the stratified period can be used to estimate fluxes. This technique gives a lower estimate of the mantle fluxes because it neglects the loss of gases due to transport through the thermocline. In addition, sources of mantle gases in near surface regions cannot be included. In lakes where gas concentrations are in steady state, the gas flux can be estimated from the gas exchange at the lake surface.

The local ^4He flux in volcanic lakes can be significantly higher than the ^4He flux from the continental crust. The ^3He flux in most volcanic lakes, e.g., 7.4×10^7 ^3He atoms m^{-2} s^{-1} (Laacher See, Aeschbach-Hertig et al. 1996b), is significantly larger than the mean oceanic flux of 4×10^4 ^3He atoms m^{-2} s^{-1} (Craig et al. 1975). Because of the comparatively small surface area of the lakes from tectonically active regions their overall contribution to the global ^3He flux is small (e.g., Laacher See 0.01%, Aeschbach-Hertig et al. 1996b, Lake Van 0.05%, Kipfer et al. 1994).

The flux ratio of heat to ^3He in volcanic lakes could be indicative of the type of volcanism involved. In Lac Pavin (France, Aeschbach-Hertig et al. 1999a), Laacher See (Germany, Aeschbach-Hertig et al. 1996b) and Lake Nyos (Cameroon, Sano et al. 1990) the heat/^3He flux ratio of 6×10^{-9} J atom^{-1}, 1×10^{-9} J atom^{-1} and 0.3×10^{-9} J atom^{-1}, respectively, is small compared to the heat/^3He flux ratio of 170×10^{-9} J atom^{-1} and 100×10^{-9} J atom^{-1} in Lake Mashu (Japan, Igarashi et al. 1992) and Crater Lake (USA, Collier et al. 1991), respectively. The former three lakes are maar lakes while the latter lakes are related to subduction volcanism at tectonic plate boundaries. For the three maar lakes, the C/^3He ratio appears to decrease with increasing formation age of the maar, whereas the heat/^3He flux ratio does not vary systematically with the formation age (Aeschbach-Hertig et al. 1999a)

Identification of gases of mantle origin in lakes can contribute to the understanding of the distribution of other gases and in particular of CO_2. The accumulation of CO_2 in Lake Nyos led to lethal gas bursts. Based on ^3He, ^4He, and Ne measurements it could be demonstrated that the high CO_2 concentrations were of mantle origin and that the CO_2 flux from the mantle leads to critical conditions in Lake Nyos within about 20 to 30 years (Sano and Wakita 1987; Sano et al. 1990).

APPLICATIONS IN GROUND WATER

The major application of the non-atmospheric noble gas components in ground water is dating, i.e., the determination of the residence time of the water in the subsurface. Dating applications fall into two major categories: (1) Young ground water with residence times of months to about 50 years can be studied with the ^3H-^3He technique as well as with the anthropogenic radioisotope ^{85}Kr. The natural radioisotope ^{222}Rn can be used to extend the age range to the study of very young ground water (up to 20 days). (2) Old ground water with residence times of thousands to millions of years is studied by using the accumulation of stable radiogenic ^4He and ^{40}Ar, or by the long-lived radioisotope ^{81}Kr. Another radioisotope, ^{39}Ar, covers the time range between about 100 and 1000 years that otherwise is very difficult to access.

An understanding and quantification of fluxes of terrigenic noble gases into aquifers is a prerequisite to their use for dating purposes. Conversely, the study of non-atmospheric He and Ar isotopes in ground water contributes to our understanding of the degassing of the Earth's mantle and crust. At this point, the field of noble gases in ground water is strongly linked to the noble gas geochemistry of other crustal fluids and the solid

Earth (see Ballentine et al. 2002 and Ballentine and Burnard 2002 in this volume).

The major application of the atmospheric noble gas components in ground water is paleotemperature reconstruction, i.e., determination of the temperature at which the infiltrating ground water equilibrated with the atmosphere in the past. This method offers a unique possibility to derive paleotemperatures based on simple physical principles. However, the atmospheric noble gases not only yield information about recharge temperatures, but also on the excess air component, and thus on the processes of gas-water exchange during ground-water infiltration. Some recent studies suggest that excess air may yield valuable paleoclimate information, but this is still a young field that leaves many open questions to be investigated.

Dating of young ground waters

In this section, applications of several tracer methods used for dating of ground waters with ages up to about 50 yr are presented. It is discussed how the resulting data are being used to derive hydraulic quantities such as recharge rates or flow velocity, to quantify mixing of different water components, and even to calibrate numerical ground-water models. More comprehensive reviews of these methods may be found in Cook and Solomon (1997) and Cook and Herczeg (2000).

^{3}H-^{3}He. Starting in the early 1950s, tritium (^{3}H) from anthropogenic sources (mainly atmospheric nuclear weapon tests) was added to the atmosphere in significant amounts (Fig. 6). After oxidation to ^{1}H^{3}HO, tritium participates in the global hydrological cycle. ^{3}H concentrations in precipitation have been monitored by an international network of stations beginning in the 1950s, and the delivery of ^{3}H to surface waters, ground water, and the oceans is reasonably well known (Doney et al. 1992). ^{3}H has been used extensively as a dye for better understanding the dynamics of ground-water flow systems (e.g., Kaufman and Libby 1954; Begemann and Libby 1957; Brown 1961; Münnich et al. 1967; see Schlosser 1992 for additional references on early tritium applications). However, as a consequence of ^{3}H decay and mixing, it has become increasingly difficult to quantify ground-water flow dynamics by relating measured ^{3}H concentrations to the ^{3}H input functions. The simultaneous measurement of ^{3}H and its decay product ^{3}He allows us to compensate radioactive decay and to determine ground-water ^{3}H concentrations as if ^{3}H were a stable isotope. In addition, we can determine an apparent age based on the concentration ratio of radioactive mother (^{3}H) and daughter ($^3He_{tri}$), the ^{3}H-^{3}He age τ (Eqn. 33).

This method, originally proposed by Tolstikhin and Kamenskiy (1969), has found many applications in hydrogeology during the past twenty years (e.g., Torgersen et al. 1979; Takaoka and Mizutani 1987; Weise and Moser 1987; Poreda et al. 1988; Schlosser et al. 1988, 1989; Solomon and Sudicky 1991; Solomon et al. 1992, 1993; Ekwurzel et al. 1994; Szabo et al. 1996; Beyerle et al. 1999a). While ground water percolates through the unsaturated zone, all produced ^{3}He escapes fairly rapidly into the atmosphere. However, just below the water table the loss of ^{3}He is limited considerably (Schlosser et al. 1989), and ^{3}He begins to accumulate and the ^{3}H-^{3}He age to increase. Tritiogenic ^{3}He can be fairly easily separated if the other He components are dominated by atmospheric sources with only small additions of radiogenic or mantle He (see Eqns. 29-31). Dispersion (including molecular diffusion) affects ^{3}H-^{3}He ages in a systematic way by mixing ^{3}H and ^{3}He from the bomb test peak into younger and older water and thus shifting ^{3}H-^{3}He ages somewhat towards the early 1960s (Fig. 17).

^{3}H-^{3}He ages have been used to determine ground-water flow velocities in different situations. Vertical ^{3}H-^{3}He profiles allowed determination of ground-water recharge rates (e.g., Schlosser et al. 1988; Solomon and Sudicky 1991; Solomon et al. 1993). Horizontal transects of ^3H-^3He ages with increasing distance from rivers served to quantify river infiltration (Stute et al. 1997; Beyerle et al. 1999a). ^3H-^3He data have also been used for

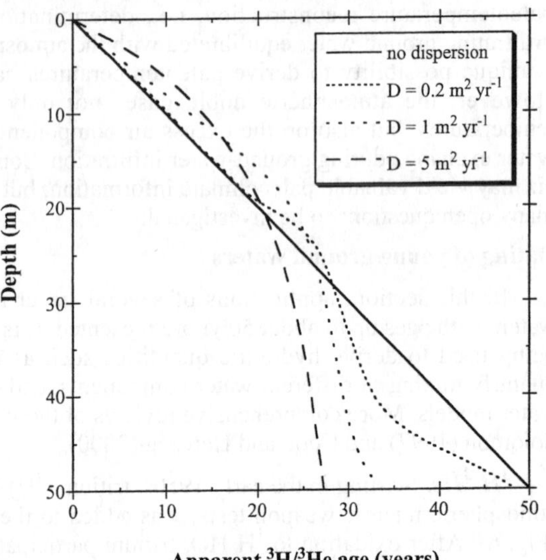

Figure 17. Effect of dispersion on the ^3H-^3He age in a system dominated by vertical water movement. A ^3H input function typical for the north-eastern United States was used and a vertical flow velocity of 1 m yr^{-1} assumed. The calculation was conducted for the year 1992. Similar one-dimensional models were published by Schlosser et al. (1989), Solomon and Sudicky (1991), Ekwurzel et al. (1994), and others.

the estimation of hydraulic conductivity, effective porosity, and dispersivity on a range of scales (Solomon et al. 1995). Another important application of the method is to provide chronologies for records of past environmental change, in particular histories of ground-water contamination (Böhlke et al. 1997; Aeschbach-Hertig et al. 1998; Johnston et al. 1998; Schlosser et al. 1998; Dunkle Shapiro et al. 1999). ^3H-^3He data can also provide constraints on mixing of different water components, e.g., in aquifers affected by river infiltration (Plummer et al. 1998b; Plummer et al. 2000; Holocher et al. 2001).

^3H-^3He dating has also been applied in fractured rock aquifers (Cook et al. 1996; Aeschbach-Hertig et al. 1998). Typical problems encountered in fractured systems are their extreme heterogeneity, double porosity causing differences between hydrodynamic and ^3H-^3He age, and geochemical complications due to the presence of other He sources.

From the beginning, other environmental tracers such as chlorofluorocarbons (CFC-11, CFC-12), ^{85}Kr, and more recently SF_6 have been used in parallel with ^3H and ^3He. Because of the different shape of their input functions, sources and sinks, the combination of these tracers provides better constraints on the ground-water flow regime than the use of a single method alone. In the following, two ^3H-^3He case studies are discussed in some more detail.

^3H-^3He, CFCs, and ^{85}Kr were studied in a sandy, unconfined aquifer on the Delmarva Peninsula in the eastern USA by Ekwurzel et al. (1994). ^3H and ^3H+^3He depth-profiles show peak-shaped curves that correspond to the time series of ^3H concentration precipitation, smoothed by dispersion (Fig. 18a). The peak occuring at a depth of about 8m below the water table therefore most likely reflects the ^3H peak in precipitation that occurred in 1963 (Fig. 6). The ^3H-^3He ages show a linear increase with depth, reaching a maximum of about 32 years. The ^3H-^3He ages are also supported by CFC-11, CFC-12, and ^{85}Kr tracer data (Fig. 18b). The latter tracers are used here as 'dyes' and their concentrations are converted into residence times by using the known history of the atmospheric concentrations and their solubility in water. From the vertical ^3H-^3He age profile at well nest 4 at the Delmarva site, the vertical flow velocity can be estimated as 0.5 m yr^{-1}. Assuming a porosity of 0.3, this is equivalent to a recharge rate of 0.15 m yr^{-1}.

An example for the use of ^3H-^3He data to determine the origin and spreading rate of pollutants is the extensive study of a sewage plume spreading in an unconfined aquifer in Cape Cod, Massachusetts (USA) by Dunkle Shapiro et al. (1999). ^3H-^3He ages at the center of the plume were found to increase linearly as a function of distance from the source (Fig. 19). Similar to the Delmarva study, the ^3H+^3He depth profiles resemble the ^3H precipitation input curves. The resulting distribution of ^3H-^3He ages was found to be consistent with the beginning of the use of detergents and their expected history in sewage. The ages also compared well with those from particle-tracking simulations.

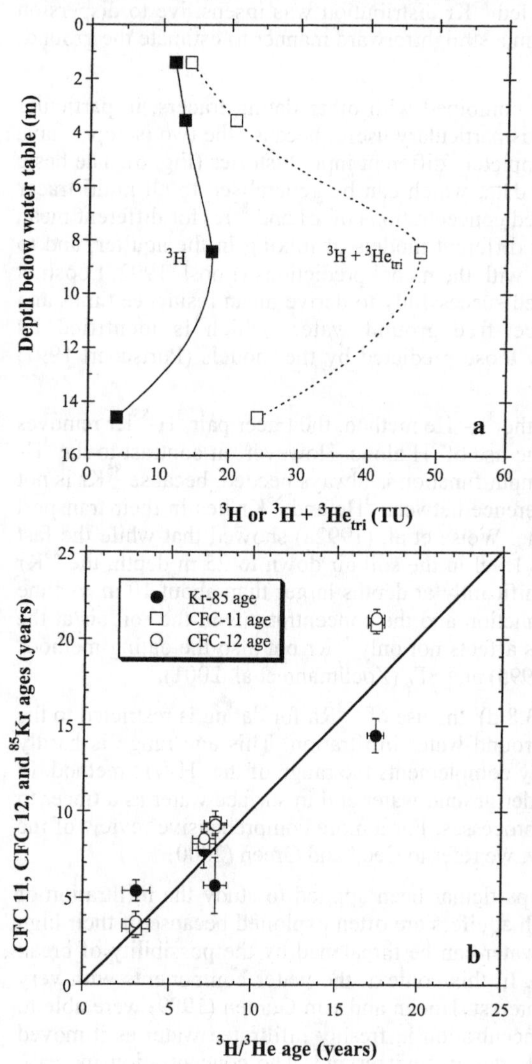

Figure 18. Results from a multi-tracer study, Delmarva Peninsula, USA. (a) ^3H and $^3H + {}^3He_{tri}$ versus depth for well nest 4. (b) Comparison of ^3H-^3He, CFC-11, CFC-12, and ^{85}Kr ages (after Ekwurzel et al. 1994).

^{85}Kr. Although ^{85}Kr is radioactive, the age information is not derived from the radioactive decay itself, but from the increase of the atmospheric concentrations since the 1950s. In this regard, the method has strong similarities with the transient tracer methods based on ^3H, CFCs, and SF_6. In comparison with these related techniques, the ^{85}Kr-method has several advantages: (1) The input function is still steadily increasing today; (2) locally enhanced atmospheric concentrations occur only near the sources (nuclear fuel reprocessing plants e.g., in Europe, Weiss et al. 1992); (3) local contamination is unlikely and subsurface production is usually small; (4) conservative behavior can be taken for granted, as usual for noble gases; and (5) the ^{85}Kr method is unaffected by excess air and recharge temperature, because only the $^{85}Kr/Kr$ ratio and not the absolute concentration is relevant.

The reason why ^{85}Kr has not been applied as widely as ^3H, ^3H-^3He, or CFCs lies in the technical difficulties of detecting its minute concentrations in natural waters. Sampling large volumes of water is time-consuming and not always feasible. Laser-based resonance ionization mass spectrometry for ^{85}Kr analysis is currently being developed (Thonnard et al. 1997), which should reduce the sample size to a few liters. If such techniques can be established, ^{85}Kr could become a major tool for dating shallow ground waters.

Up to the present, ^{85}Kr studies of shallow ground waters are rare and limited to small numbers of samples. An early assessment of the feasibility of the method with a few examples was presented by Rozanski and Florkowski (1979). In a study of ground-water flow based exclusively on ^{85}Kr conducted in the Borden aquifer in Canada a monotonic increase of the ^{85}Kr age along the ground-water flow path was found and compared with the results of a two-dimensional advection dispersion model of ^{85}Kr transport (Smethie et al. 1992). It was found that the modeled ^{85}Kr distribution was insensitive to dispersion and that therefore ^{85}Kr could be used in a straightforward manner to estimate the ground-water residence time.

More commonly, ^{85}Kr has been combined with other dating tracers, in particular with tritium. This tracer combination is particulary useful because the two isotopes have similar half-lives but experienced completely different input histories (Fig. 6). The basic idea in the interpretation of ^{3}H-^{85}Kr data, which can be generalised to all multi-tracer approaches, is to calculate the expected concentrations of ^{3}H and ^{85}Kr for different mean ground-water residence times and/or different models of mixing in the aquifer, and to compare the measured concentrations with the model predictions (Loosli 1992; Loosli et al. 2000). This approach has been used successfully to derive mean residence times and to quantify admixing of old, tracer-free ground water, which is identified by concentrations of ^{3}H and ^{85}Kr below those predicted by the models (Purtschert 1997; Mattle 1999).

Comparable in many respects to the ^{3}H-^{3}He method, the tracer pair ^{3}H-^{85}Kr removes many of the difficulties inherent to the use of ^{3}H alone. However, in contrast to the ^{3}H-^{3}He method, a knowledge of the ^{3}H-input function is always needed, because ^{85}Kr is not linked to the ^{3}H-decay. Another difference between ^{3}He and ^{85}Kr lies in their transport behavior through the unsaturated zone. Weise et al. (1992a) showed that while the fast diffusing ^{3}He was at the atmospheric level in the soil air down to 25 m depth, the ^{85}Kr activity in the soil gas decreased significantly at depths larger than about 10 m. A time lag between the atmospheric input function and the concentrations in the soil air at the water table in deep unsaturated zones affects not only ^{85}Kr but also the dating methods based on CFCs (Cook and Solomon 1995) and SF_6 (Zoellmann et al. 2001).

^{222}Rn. Due to its short half-life (3.8 d), the use of ^{222}Rn for dating is restricted to the study of the first few weeks after ground-water infiltration. This age range is hardly accessible by other methods but nicely complements the range of the ^{3}H-^{3}He method. In addition, ^{222}Rn can also be used in older ground water and in surface water as a tracer to study hydrological and geochemical processes. For a more comprehensive review of the ^{222}Rn-method in subsurface hydrology, we refer to Cecil and Green (2000).

As a dating tracer, ^{222}Rn has in particular been applied to study the infiltration of river water into alluvial aquifers. Such aquifers are often exploited because of their high yield, but the quality of the ground water can be threatened by the possibility of break through of contaminated river water. In this context, the water components with very short residence times are of central interest. Hoehn and von Gunten (1989) were able to monitor the increase of the ^{222}Rn concentration in freshly infiltrated water as it moved away from the rivers in two alluvial sites in Switzerland. The concentration increase could be converted to residence times of up to about 15 days and the ground-water flow velocity could be derived.

A difficulty of the ^{222}Rn dating method is to distinguish between changes of the ^{222}Rn concentration that are due to actual aging of a water parcel and variations that

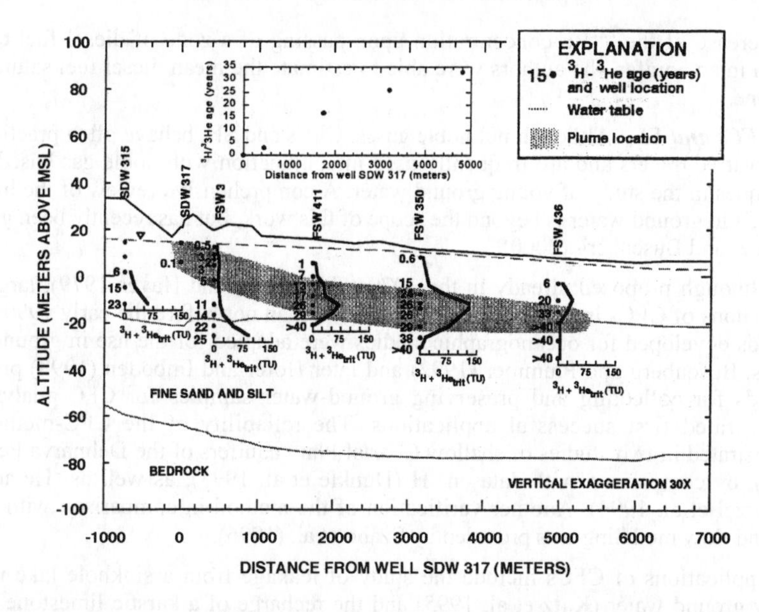

Figure 19. Distribution of ^{3}H-^{3}He ages across a sewage plume (shaded) for an unconfined aquifer in Cape Cod, Massachusetts. Depth profiles in piezometer clusters along the plume are shown. The inset shows the increase of ^{3}H-^{3}He ages along a single flowline through the core of the plume where sewage (boron, detergents) concentrations have maximum values (adapted from Dunkle Shapiro et al. 1999).

merely reflect different mixing ratios between Rn-rich ground water and virtually Rn-free surface water. To solve this problem, Bertin and Bourg (1994) used chloride as an additional tracer to quantify the mixing ratios. On the other hand, the large contrast in ^{222}Rn concentrations between ground and surface water renders this isotope useful for the study of interaction between ground water and surface water. For example, Cable et al. (1996) used ^{222}Rn in conjunction with CH_4 to quantify ground-water discharge to the coastal ocean. Ellins et al. (1990) applied ^{222}Rn to study the discharge of ground water into a river. In these cases, the ^{222}Rn concentrations were interpreted in terms of mixing between surface and ground water, rather than in terms of age.

In ground waters with residence times above about 20 days, the concentration of ^{222}Rn is at equilibrium with the production rate in the aquifer matrix. Instead of age information, ^{222}Rn then conveys information about the characteristics of the aquifer matrix, such as U content, porosity, grain size distribution, and release efficiency for ^{222}Rn (Andrews and Lee 1979). Because both radiogenic ^{4}He and ^{222}Rn originate from the U decay series, the study of ^{222}Rn may support the application of ^{4}He as a dating tool (Torgersen 1980). ^{222}Rn data from ground water in the Stripa granite in conjunction with data and production rate calculations of other noble gas radioisotopes (^{37}Ar, ^{39}Ar, ^{85}Kr) were used to draw conclusions about the distribution of U and the release and transport mechanisms for these isotopes (Loosli et al. 1989). In this case all investigated isotopes were found to be in equilibrium with their local subsurface production.

A relatively new and promising application of ^{222}Rn is its use as partitioning tracer in studies of aquifers contaminated by non-aqueous-phase liquids (NAPLs). Hunkeler et al. (1997) demonstrated in laboratory and field studies that Rn partitions strongly into the NAPL phase, resulting in a corresponding decrease in the water phase. By observation of the decrease of the ^{222}Rn concentration upon passing of a zone of diesel fuel contamination in an aquifer, the authors were able to estimate the mean diesel fuel saturation in this zone.

CFCs and SF_6. Although not noble gases, CFCs and SF_6 behave often practically as conservative tracers and are frequently used in conjunction with noble gas based dating techniques in the study of young ground water. A comprehensive review of the literature on CFCs in ground water is beyond the scope of this work, but has recently been given by Plummer and Busenberg (2000).

Although proposed already in the 1970s (Thompson and Hayes 1979), large-scale applications of CFCs in ground-water hydrology began not before the early 1990s, when methods developed for oceanographic studies were adapted for the use in ground-water studies. Busenberg and Plummer (1992) and later Hofer and Imboden (1998) presented methods for collecting and preserving ground-water samples for CFC analysis and demonstrated first successful applications. The reliability of the CFC-method was demonstrated in two studies of shallow Coastal Plain aquifers of the Delmarva Peninsula (USA), by comparison with data on ^3H (Dunkle et al. 1993), as well as ^3He and ^{85}Kr (Ekwurzel et al. 1994). Another verification of the method in comparison with ^3H-^3He ages and flow modeling was provided by Szabo et al. (1996).

Applications of CFCs include the study of leakage from a sinkhole lake into the nearby ground water (Katz et al. 1995) and the recharge of a karstic limestone aquifer from a river (Plummer et al. 1998a; Plummer et al. 1998b). In the latter case, ^3H-^3He ages were used to reconstruct the initial CFC concentration in the river water fraction. Conversely, Modica et al. (1998) used CFCs to constrain the source and residence time of ground-water seepage into a gaining stream. Detailed vertical CFC profiles and tracking of the ^3H peak were used to assess vertical recharge velocities to within ±10% in a shallow, silty sand aquifer (Cook et al. 1995). Vertical profiles of CFCs as well as ^3H and ^3He were also used to infer shallow ground-water flow in fractured rock (Cook et al. 1996). Important applications of ground-water dating techniques are related to the source and history of ground-water contaminations. For example, Böhlke and Denver (1995) and Johnston et al. (1998) used CFCs in combination with ^3H or ^3H-^3He dating to reconstruct the recharge history of nitrate in agricultural watersheds.

Complications of the CFC-method are increased atmospheric concentrations near industrial source regions, degradation of at least CFC-11 under anoxic conditions, and contamination of ground and surface waters. The 'urban air' problem was assessed by Oster et al. (1996) for the case of western Europe, whereas Ho et al. (1998) provide data from a large urban area in eastern North America. Microbial degradation of CFC-11 under anaerobic conditions has been observed in several studies (e.g., Lovley and Woodward 1992; Cook et al. 1995; Oster et al. 1996). Local contamination is a serious and widespread, but still not fully understood problem (Thompson and Hayes 1979; Busenberg and Plummer 1992; Böhlke et al. 1997; Beyerle et al. 1999a). A fundamental problem for the CFC-method is the nearly constant atmospheric mixing ratios in recent years, making alternatives such as SF_6 attractive for dating of very young ground water.

SF_6 is just emerging as a new tracer in ground-water hydrology. A detailed assessment of the method including first applications and comparisons with ages derived from CFCs, ^3H-^3He, ^{85}Kr, as well as radiogenic ^4He accumulation, has been presented by Busenberg and Plummer (2000). Zoellmann et al. (2001) used SF_6 together with ^3H to calibrate a transport model that was used to predict nitrate levels in wells for different land-use scenarios. Bauer et al. (2001) found evidence for CFC-113 retardation in comparison to SF_6, ^3H, and ^{85}Kr. Plummer et al. (2001) found SF_6 and ^3H-^3He to be the most reliable dating methods in a multi-tracer study including also CFCs, ^{35}S, and stable isotopes of water.

Multitracer studies and modeling. A straightforward interpretation of any tracer concentration in terms of ground-water age is only possible under the assumption that the sampled water parcel has not been affected by mixing along its way between the points of recharge and sampling. This assumption is often referred to as the plug flow or piston flow model. It is obviously a simplification of the real flow conditions in aquifers, where dispersion always creates some degree of mixing.

To improve the interpretation of tracer data, simple input-output models of the effect of mixing on tracer concentrations have been developed since the early years of tracer hydrology (Erikson 1958; Vogel 1967; Maloszewski and Zuber 1982; Zuber 1986a; b) and applied up to the present (Cook and Böhlke 2000; Plummer et al. 2001). These 'lumped-parameter' models view the ground-water system as a 'black box', which transfers the tracer input curve into a time-dependent output concentration that can be compared to the observations. The transfer functions that characterize the different types of models specify the fraction of the input of each year that is present in the mixed output water. The transfer function of the piston flow model is a delta function that picks the input concentration of a single year and yields the respective output without any mixing. Mixing is often described by the 'exponential' or the 'dispersion' model. The transfer function of the exponential model prescribes an exponential decrease of the fraction of each input year with increasing elapsed time, which is equivalent to modeling the ground-water reservoir as a mixed reactor. The dispersion model weighs the fraction of the different input years by a Gauss-type distribution.

The more recent development of sophisticated ground-water flow and transport modeling software and the availability of powerful computers allow a much more detailed modeling of the tracer evolution in ground-water systems. Consequently, applications of tracer data in conjunction with multi-dimensional flow and transport models began to show up in the 1990s (Reilly et al. 1994; Szabo et al. 1996; Sheets et al. 1998). Despite the undisputable power of this approach, it should be noted that in cases with few tracer data and little knowledge about the ground-water system, the simple lumped-parameter models still have their justification (Richter et al. 1993). In any case, it is clear that the full potential of the tracer methods can only be exploited by simultaneous application of several tracer methods and by interpretation of the data in the framework of a model.

The combination of several independent tracer methods has a great potential in identifying mixing or other disturbing effects in a ground-water system. Close agreement between the apparent (piston-flow) ages derived from different tracers , as found in some comparative studies (Ekwurzel et al. 1994; Szabo et al. 1996), indicate that mixing and dispersion are of minor importance. Many of the early tracer studies were aimed at verifying the methods and were therefore conducted in homogeneous, sandy aquifers with relatively simple and uniform flow fields. In more complex settings, larger deviations between the different tracer ages are to be expected and also have been observed (Cook et al. 1996; Plummer et al. 1998a).

Within the framework of the lumped-parameter models, it is relatively simple to derive information about the mixing regime from multi-tracer data. As discussed above for the ^{3}H-^{85}Kr tracer pair, the predicted concentrations for any tracer can be calculated for different lumped-parameter models with various parameter values, and the model that best fits the data can be found. This inverse modeling procedure can be illustrated by plotting the predicted output concentrations for any pair of tracers versus each other (Fig. 20). By plotting the measured concentrations in such a figure containing a series of curves that represent different models and parameter values, the curve that best fits the data can easily be found (Loosli 1992; Plummer et al. 2001). A more rigorous approach to solve the problem involves least square fitting, analogous to the inverse techniques to derive model parameters from the data on atmospheric noble gases. A first example of this approach was presented by Purtschert et al. (1999). Such an approach can go beyond classical lumped-parameter models by including effects such as binary mixing, in particular with a component of old, tracer-free water. Under favorable conditions, the combination of different tracers can even be used to decompose mixed samples into the primary components (Beyerle et al. 1998).

Of course, inverse-modeling approaches can also be used to determine the parameters of numerical ground-water transport models from fits to observed tracer data. Estimates of residence times for a number of locations in an aquifer provide a powerful calibration target for numerical ground-water transport models. If enough data are available to constrain the numerous unknowns in such models, this is presumably the most effective way to extract useful information from tracer data, in particular because the numerical models can be used to make predictions for the future development of the investigated system. Such predictions may become much more constrained and trustworthy if the model has been calibrated against tracer data.

Typically, tracer data are used to calibrate certain model parameters, such as recharge rate, hydraulic conductivity, or (effective) porosity (Reilly et al. 1994). Solomon et al. (1992) compared the spatial distribution of ^{3}H-^{3}He ages with a previously calibrated ground-water transport model of the Borden aquifer (Canada). Although the ^{3}H-^{3}He age profiles were vertically offset from the modeled travel times, the gradients in travel time and ^{3}H-^{3}He age compared very well. In a study of a sandy aquifer in the southern New Jersey coastal plain, ^{3}H-^{3}He ages were compared to two-dimensional ground-water flow models that were calibrated without using information from large-scale tracer measurements (Szabo et al. 1996). Steady-state finite difference ground-water flow models were calibrated at three sites by adjusting horizontal and vertical hydraulic conductivities to match measured hydraulic heads. Travel times were then calculated using particle tracking codes and compared to the measured ^{3}H-^{3}He (and CFC) ages. The agreement between ^{3}H-^{3}He, CFC, and particle tracking ages indicated that the influence of dispersion was very small at this site and that, with few exceptions, the ground-water flow model did not require major adjustments. Comparison of a water bound tracer such as ^{3}H with gaseous tracers such as ^{85}Kr or SF_6 provides constraints on the thickness of the unsaturated zone and its field capacity (Bauer et al. 2001; Zoellmann et al. 2001). In a study of river infiltration by means of a three-dimensional transport model, Mattle et al. (2001) found the hydraulic conductance of the riverbed to be the decisive parameter that could be calibrated by using data on tritiogenic ^{3}He.

In recent studies, attempts have been made to merge hydrogeological models and ^{3}H-^{3}He data in more complex systems. Particle tracking ages derived from independently calibrated three-dimensional flow models for two sites in a hydrogeologically complex buried-valley aquifer in Ohio (Sheets et al. 1998) compared reasonably well with ^{3}H-^{3}He ages (Dunkle Shapiro et al. 1998). The agreement decreased with depth. Selected conceptual and parameter modifications (porosity, transmissivity) to the models resulted in improved agreement between ^{3}H-^{3}He ages and simulated travel times. The first attempts to calibrate fractured rock ground-water flow models using ^{3}H-^{3}He ages demonstrate some of the difficulties encountered in these even more complex environments (e.g., Cook et al. 1996).

It is important to note that in some cases environmental tracer data not only help to constrain model parameters, but also can indicate weaknesses of the conceptual model that underlies a numerical model. Reilly et al. (1994) and Sheets et al. (1998) showed that by changing their conceptual assumptions about the distribution of recharge, the

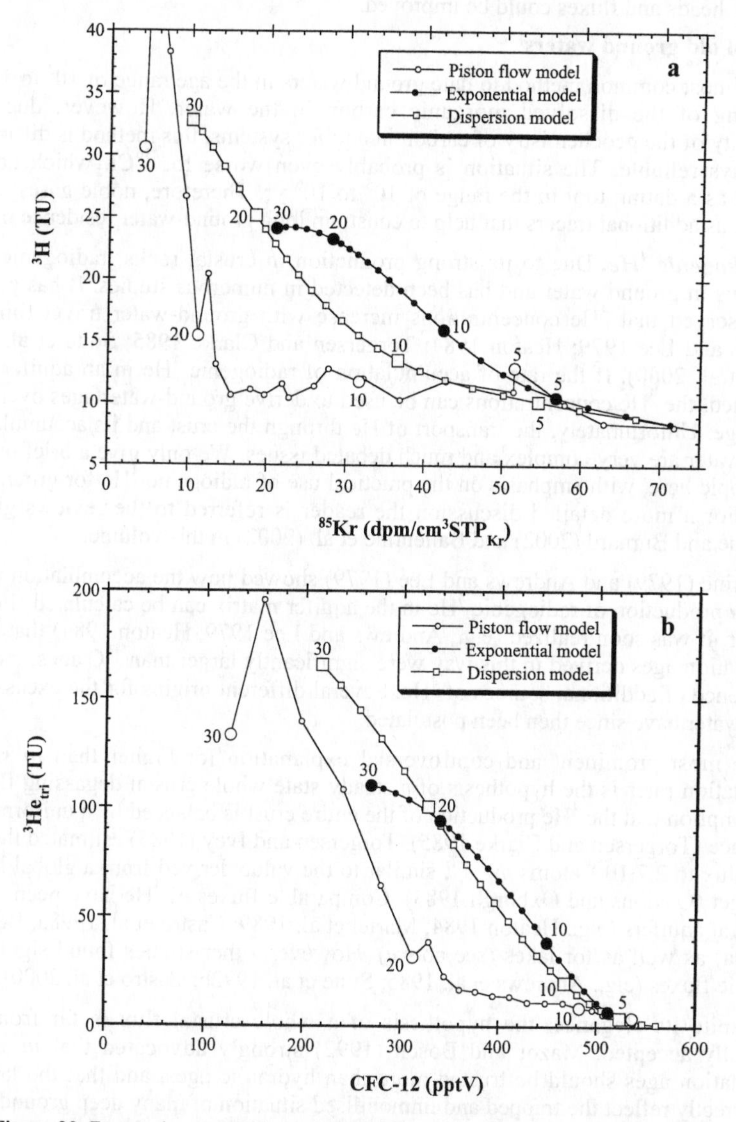

Figure 20. Expected concentrations of several tracers used for dating of young ground waters as functions of mean water residence time for different lumped-parameter models. (a) ^{3}H (in tritium units, TU) versus ^{85}Kr (in decay per minutes per cm^{3}STP of Kr); (b) ^{3}He$_{tri}$ (in TU) versus CFC-12 (in equivalent atmospheric volume fractions, pptV). All output concentrations were calculated for the year 2000 and mean residence times between 1 and 30 yr, using the lumped-parameter models (see Maloszewski and Zuber 1982) 'piston flow', 'exponential' and 'dispersion' (with a value of 0.2 for the dimensionless dispersion parameter) and input functions typical for central Europe. Selected values of the mean residence time (5, 10, 20, 30 yr) are highlighted by larger symbols and labels. By comparing measured concentrations with these diagrams, the appropriate model can be selected. Mixtures with old, tracer-free ground water lie on mixing lines toward the origin.

agreement between the modeled and measured tracer concentrations as well as the hydraulic heads and fluxes could be improved.

Dating of old ground waters

The most common method to date ground waters in the age range of 10^3 to 10^4 yr is ^{14}C-dating of the dissolved inorganic carbon in the water. However, due to the complexity of the geochemistry of carbon in aquifer systems, this method is difficult and not always reliable. The situation is probably even worse for ^{36}Cl, which has been explored as a dating tool in the range of 10^5 to 10^6 yr. Therefore, noble gases are very welcome as additional tracers that help to constrain long ground-water residence times.

Radiogenic ^{4}He. Due to its strong production in crustal rocks, radiogenic ^{4}He is ubiquitous in ground water and has been detected in numerous studies. It has generally been observed that ^{4}He concentrations increase with ground-water travel time (e.g., Andrews and Lee 1979; Heaton 1984; Torgersen and Clarke 1985; Stute et al. 1992b; Castro et al. 2000). If the rate of accumulation of radiogenic ^{4}He in an aquifer can be determined, the ^{4}He concentrations can be used to derive ground-water ages over a wide time range. Unfortunately, the transport of He through the crust and its accumulation in ground water are very complex and much debated issues. We only give a brief overview of this topic here, with emphasis on the practical use of radiogenic ^{4}He for ground-water dating. For a more detailed discussion the reader is referred to the reviews given by Ballentine and Burnard (2002) and Ballentine et al. (2002) in this volume.

Marine (1979) and Andrews and Lee (1979) showed how the accumulation rate due to *in situ* production of radiogenic ^{4}He in the aquifer matrix can be calculated (Eqn. 36). However, it was soon realized (e.g., Andrews and Lee 1979; Heaton 1984) that the ^{4}He accumulation ages derived in this way were significantly larger than ^{14}C ages, indicating the presence of additional sources of ^{4}He. Several different origins for the excess ^{4}He in ground water have since then been postulated.

The most prominent and controversial explanation for higher than *in situ* ^{4}He accumulation rates is the hypothesis of a steady state whole crustal degassing flux, i.e., the assumption that the ^{4}He production of the entire crust is balanced by a uniform flux to the surface (Torgersen and Clarke 1985). Torgersen and Ivey (1985) estimated the whole crustal flux to $2.7 \cdot 10^{10}$ atoms m^{-2} s^{-1}, similar to the value derived from a global heat and He budget (O'Nions and Oxburgh 1983). Comparable fluxes of ^{4}He have been reported for several aquifers (e.g., Heaton 1984; Martel et al. 1989; Castro et al. 1998a; Beyerle et al. 1999a) as well as for lakes (see above). However, other studies found significantly lower ^{4}He fluxes (e.g., Andrews et al. 1985; Stute et al. 1992b; Castro et al. 2000).

Despite its elegance, the hypothesis of a whole crustal flux is far from being universally accepted. Mazor and Bosch (1992) strongly advocated that *in situ* ^{4}He accumulation ages should be trusted more than hydraulic ages, and that the large ^{4}He ages correctly reflect the trapped and immobilised situation of many deep ground waters. Even more important are conceptual objections against the idea of a whole crustal flux. Andrews (1985) showed that only the uppermost part of the crust should be affected by a diffusive ^{4}He flux. Solomon (2000) as well as Ballentine and Burnard (2002) argue that the high diffusivity of ^{4}He required to support the typical whole crustal flux is unlikely to be valid for the bulk crust.

Between the extreme views of *in situ* production and whole crustal flux, there are alternative explanations for the observed ^{4}He concentrations in ground water. Andrews and Lee (1979) and Tolstikhin et al. (1996) suggested that diffusion from the adjacent confining layers constituted the major source of ^{4}He. This hypothesis is supported by the widespread correlation between ^{4}He and Cl^- in ground waters, since both species could originate from the pore water of the aquitards (Lehmann et al. 1996). Another possible He source is the release of ^{4}He stored in the minerals of the aquifer matrix over geologic time by weathering or diffusion. Heaton (1984) discussed the He release due to chemical weathering, but considered it unlikely to represent an important source in two aquifers in southern Africa. Solomon et al. (1996) explained the occurrence of significant concentrations of radiogenic ^{4}He in young, shallow ground water by diffusion of ^{4}He from sediment grains, which only relatively recently had been produced by erosive fragmentation of old, ^{4}He-rich rocks. This mechanism may be important in shallow aquifers consisting of relatively recent sediment (e.g., Beyerle et al. 1999a) or in lake sediments, but is hardly relevant for deep ground water hosted by old sediments.

From a practical point of view, with respect to the applicability of ^{4}He as a dating tool, it can be stated that *in situ* production is rarely the only source of ^{4}He in ground water, but that the origin and in particular the strength of the additional sources is uncertain. This uncertainty has practically prevented the application of ^{4}He accumulation as a quantitative dating tool. Quantitative dating using ^{4}He requires calibration with another dating tool, such as ^{14}C, as suggested by Andrews and Lee (1979). Such a calibration was used in a study in Niger to date ground water beyond the limit of the ^{14}C method (Beyerle et al. submitted). Beyerle et al. (1999b) used the age information from the four first examples of successful ^{81}Kr-dating (see below) to calibrate a ^{4}He age scale for part of the GAB. Even if calibration of the ^{4}He age is not possible, ^{4}He may still provide a useful qualitative chronology, e.g., for noble gas paleotemperature records (Clark et al. 1997).

In some aquifers, or at least in some parts of them (upstream and uppermost layers), it may be possible to rule out major contributions from He sources other than *in situ* production. Significant diffusive release of stored He from sand grains can occur at most for 50 million years after deposition of the sediments (Solomon et al. 1996). The crustal He flux may be shielded by underlying aquifers that flush the He out of the system before it can migrate across them (Torgersen and Ivey 1985; Castro et al. 2000). Such favorable conditions enabled Aeschbach-Hertig et al. (2002b) to calculate ^{4}He accumulation ages from the U- and Th-concentrations in the Aquia aquifer (USA). The resulting ages were consistent with the paleoclimate information provided by the atmospheric noble gases and made it possible to construct a paleotemperature record (Fig. 21) despite the lack of reliable ^{14}C ages due to the complex carbonate chemistry at this site.

Apart from its potential use as dating tool, the study of terrigenic He in aquifers contributes to our understanding of the degassing of the Earth's crust and mantle, as discussed by Ballentine and Burnard (2002) and Ballentine et al. (2002) in this volume. Although crustal (radiogenic) He dominates in most aquifers, mantle He has occasionally been observed in ground water. Strong and unambiguous mantle He signatures can be observed in springs in volcanic areas, such as the Eifel and Rhine Graben regions of Germany (Griesshaber et al. 1992; Aeschbach-Hertig et al. 1996b), the Massif Central in France (Matthews et al. 1987; O'Nions and Oxburgh 1988), or western Turkey (Gülec 1988). Away from point sources, the contribution of a mantle-derived component to the non-atmospheric He in ground water has to be inferred from ^{3}He/^{4}He ratios that are elevated compared to typical crustal values. A large-scale flux of mantle He may be found in zones of active extension, such as the Great Hungarian Plain, where Stute et al. (1992b) found a mantle-contribution of 6 to 9%. Such findings show that vertical transport of He through the crust can occur at least in active regions. Quite surprising in this context is the detection of a mantle He component in ground water of the Mirror Lake Basin (New Hampshire, USA), in an area that is neither tectonically nor volcanically active (Torgersen et al. 1994; Torgersen et al. 1995).

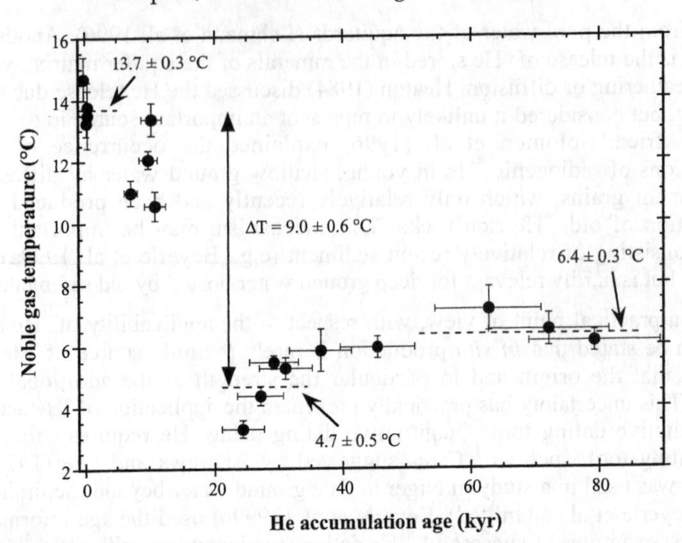

Figure 21. Noble gas temperature record from the Aquia aquifer, Maryland (USA), modified from Aeschbach-Hertig et al. (2002b). The chronology for this record could be established by ^{4}He accumulation ages based on the assumption of pure in-situ production. Only data from Aquia aquifer wells that could consistently be interpreted are shown. Mean noble gas temperatures for selected groups of samples, thought to best represent the Holocene, last glacial maximum (LGM), and preceding last glacial period are indicated, as well as the derived Holocene–LGM temperature difference.

^{40}Ar and ^{21}Ne. Although ^{40}Ar is produced from ^{40}K in similar amounts as $^{4}He_{rad}$ from U and Th in crustal rocks, the detection of radiogenic $^{40}Ar_{rad}$ is much less frequent than that of radiogenic He. The main reason is the much higher atmospheric abundance of Ar, but also the better retention of Ar in minerals may play a role. Ballentine et al. (1994) discussed the thermal constraints on the release of ^{4}He and ^{40}Ar, showing that preferential release of ^{4}He is expected in the upper part of the crust. In a large study of He and Ar isotopes in rocks and ground water, Tolstikhin et al. (1996) found that the ratio $(^{4}He/^{40}Ar)_{rad}$ in the ground water was higher than the production ratio in the aquifer matrix, whereas the opposite was true for the rock and mineral samples.

Torgersen et al. (1989) found a large variation in the $(^{4}He/^{40}Ar)_{rad}$ ratio in ground water from the GAB with a mean value close to their estimate of the whole crustal $^{4}He/^{40}Ar$ production ratio of 4.5. From these findings it was concluded that the overall degassing flux of both ^{40}Ar and ^{4}He was close to steady state with the whole crustal production. In contrast, Beyerle et al. (2000b) reported data from another part of the GAB that indicated a rather uniform $(^{4}He/^{40}Ar)_{rad}$ ratio of 50, an order of magnitude larger than the production ratio. Thus, the existence of a whole crustal degassing flux appears less likely for ^{40}Ar than for ^{4}He.

A method to date very old pore waters in impermeable rocks using radiogenic ^{4}He and ^{40}Ar has been developed by Osenbrück et al. (1998). In their case study in a cap rock above a salt dome, large $^{40}Ar_{rad}$ concentrations were found, but it appeared that they originated from the K-rich salts in the salt deposits and entered the cap rock by diffusion, rather than from *in situ* production. Therefore, ^{40}Ar dating was not possible in this case.

The only other radiogenic noble gas isotope that has some potential for ground-water dating is ^{21}Ne. Kennedy et al. (1990) noticed a very uniform radiogenic $^{21}Ne/^{4}He$ ratio of about $5{\times}10^{-8}$ in crustal fluids. Bottomley et al. (1984) were able to detect radiogenic ^{21}Ne

in granitic rocks, whereas Weise et al. (1992b) for the first time reported a ^{21}Ne excess in a sedimentary aquifer. Although the applicability of ^{21}Ne has been demonstrated by these studies, this method is likely to remain a specialty that may help to improve the interpretation of ^{4}He data in some cases.

^{39}Ar. With its half-life of 269 yr, ^{39}Ar covers a very important age range between the ranges accessible by ^{3}H-^{3}He, ^{85}Kr, CFCs and SF_6 (younger than 50 yr) and by ^{14}C (older than about 1 kyr). It also has quite ideal properties, similar to those of ^{85}Kr, with the exception that subsurface production cannot *a priori* be neglected and may in some cases lead to deviations of the evolution of the ^{39}Ar concentration from the pure radioactive decay. In crystalline rocks, dating may even be impossible, but ^{39}Ar may then be used to study water-rock interactions (Loosli et al. 1989; Loosli et al. 1992). As for ^{85}Kr, the reason why ^{39}Ar has not been applied more widely lies in the difficulty of its analysis.

The applicability of ^{39}Ar for ground-water dating was demonstrated by Andrews et al. (1984) in a multi-isotope study of the East Midlands Triassic sandstone aquifer (UK), where subsurface production was found to be unimportant and ^{39}Ar activities were consistent with those of ^{14}C, ^{3}H, and ^{85}Kr. ^{39}Ar ages from 25 up to 240 yr could be calculated for wells in or near the recharge area, whereas for some wells further downstream only a lower limit of 1300 yr could be assigned. Some further examples of ^{39}Ar dating of ground water were discussed by Loosli (1992). ^{39}Ar data played an important role in a multi-tracer study of a confined aquifer in the Glatt Valley, Switzerland (Beyerle et al. 1998; Purtschert et al. 2001). In between young waters in the recharge area (^{3}H and ^{85}Kr active) and late Pleistocene waters downstream (^{39}Ar in production equilibrium, ^{14}C low), some samples were found for which ^{39}Ar indicated significantly younger ages than ^{14}C. By using the water chemistry (mainly Cl^- concentrations) as additional constraint, these samples could be decomposed into an old and a relatively young (^{39}Ar active, but ^{3}H and ^{85}Kr dead) component.

^{81}Kr. Among the noble gas radioisotopes with the potential for application in hydrology, ^{81}Kr is certainly the most exotic. Due to the enormous effort of ^{81}Kr analysis, there are so far only two examples of actual ^{81}Kr-measurements on ground-water samples. Lehmann et al. (1991) were able to determine the ^{81}Kr/Kr ratio in one sample from the Milk River aquifer (Canada) to be (82±18)% of the modern atmospheric ratio. The measurement was done by laser resonance spectroscopy following an elaborate multi-step isotope enrichment procedure. In view of the achieved accuracy, which yielded only an upper limit of 140 kyr for the age of the sample, this first measurement was hardly more than a demonstration of the feasibility of ^{81}Kr-dating.

The first real application of ^{81}Kr was conducted recently on four samples from the GAB by using accelerator mass spectrometry on a cyclotron (Collon et al. 2000). In this study, ages between 220 and 400 kyr could be determined with reasonable uncertainties of 40 to 50 kyr. These reliable ages offered a unique opportunity to interpret ^{36}Cl, ^{4}He, and noble gas data that were analyzed in the framework of the same study on timescales of 10^5 yr (Beyerle et al. 1999b; Lehmann et al. 1999). Except for the difficulty of measurement, ^{81}Kr appears to be an almost ideal dating tracer for ground water in this age range, which is appropriate for deep ground waters in large sedimentary basins.

Noble gas recharge temperatures

The possibility to derive paleotemperature records from dissolved noble gases in ground water, on the basis of the temperature dependency of their solubilities in water, is probably the application of noble gases in subsurface hydrology that received most attention in recent years. A large number of ground-water studies over the past 40 years used this approach to reconstruct paleoclimate conditions.

Method and historic development. Water percolating through the unsaturated zone equilibrates continuously with ground air until it reaches the capillary fringe and the quasi-saturated zone where entrapped air resides and excess air is formed. Model calculations and field observations have shown that after correcting for excess air formation, noble gas temperatures closely reflect the mean ground temperature at the water table, except in cases where the water table is very close (1 to 2 m) to the surface, or the recharge rates are very high (exceeding several hundred mm yr^{-1}; Stute and Schlosser 1993). Unless the water table is very deep (>30 m), the noble gas temperatures are usually found to be within 1°C of the ground (soil) temperature at the surface. Ground temperatures are typically slightly (1±1°C) warmer than mean annual air temperatures, although larger temperature differences may occur under extreme climate conditions (Smith et al. 1964; Beyerle et al. submitted). The conversion of noble gas temperatures into air temperatures should therefore be based on local relationships between ground and air temperature or calibrated locally by analysing young ground water that infiltrated under known climate conditions.

After leaving the water table, ground water moves towards the discharge area carrying information imprinted in the recharge area in the form of atmospheric noble gas concentrations. Mixing occurring in the aquifer or induced by sampling through wells with long screens tends to result in smoothing the variability of the recorded signals (Stute and Schlosser 1993). Due to this smoothing and also the difficulties of dating old ground waters, the noble gas method cannot be used to study short-term climate fluctuations, but it is well-suited to derive quantitative estimates of mean temperatures during major climate states, and in particular the difference between the last glacial maximum (LGM, about 21 kyr BP) and the Holocene.

The earliest applications of the noble gas thermometer include a study by Oana (1957), who found that N_2 and Ar concentrations in precipitation followed the seasonal temperature cycle whereas the concentrations in ground water were fairly constant and on average slightly higher than in precipitation. Sugisaki (1961) used the seasonal cycle of N_2 and Ar in ground water recharged by a stream to determine ground-water flow velocities. Mazor (1972) found that dissolved atmospheric noble gases in thermal ground waters of the Jordan Rift Valley in Israel reflected the modern air temperature in the recharge areas. He suggested that ground water could be used as an archive of paleoclimate, because it appears to retain the dissolved gases over many thousands of years.

The first glacial–interglacial paleotemperature record was established for the Bunter Sandstone in England (Andrews and Lee 1979). Ground water with a ^{14}C age between 20,000 and 35,000 years was characterized by noble gas temperatures 4 to 7°C lower than ground water of Holocene (0-10,000 years) origin. The gap in the record between 10,000-20,000 years was explained by the absence of recharge due to permafrost conditions in England during the last glacial maximum (Fig. 22). Heaton (1981) derived paleotemperatures from N_2 and Ar concentrations in ground water in southern Africa and found a systematic temperature difference between recent and glacial ground water. Similar patterns were found in confined aquifers in Germany and northern Africa, not only in the noble gas temperatures but also in stable isotope ($\delta^{18}O$, δD) records (Rudolph et al. 1984). Since these early studies many paleoclimate studies were conducted using the concentrations of atmospheric noble gases (and N_2-Ar) to reconstruct climate conditions for the last ~30,000 years (e.g., Heaton et al. 1986; Stute et al. 1992a, 1995a,b; Clark et al. 1997, 1998; Dennis et al. 1997; Beyerle et al. 1998; Clark et al. 1998; Weyhenmeyer et al. 2000).

Besides paleoclimate reconstructions, noble gas temperatures are also useful in

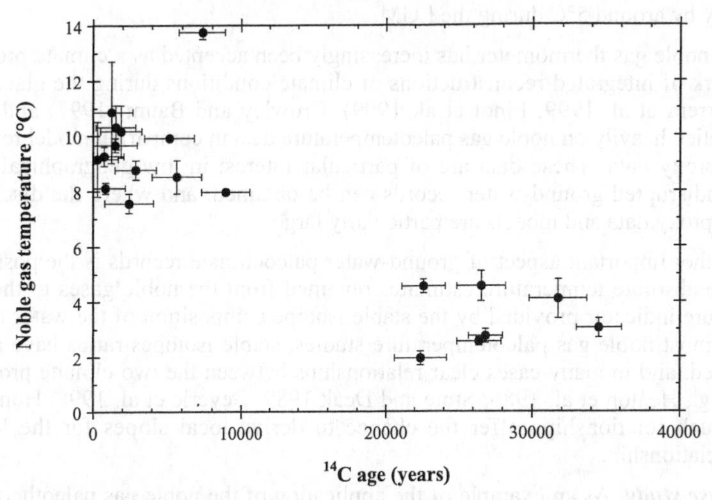

Figure 22. Noble gas temperature record from the Bunter Sandstone in the United Kingdom (modified from Andrews and Lee 1979).

helping to constrain ground-water flow regimes. For example, in cases where radiocarbon ages have large uncertainties due to complicated hydrochemical conditions, the glacial–interglacial temperature transition in the noble gas temperatures can be used as a stratigraphic marker to check the derived ages (e.g., Clark et al. 1997; Aeschbach-Hertig et al. 2002b). Commonly used dating techniques for shallow ground water (^{3}H-^{3}He, CFCs, SF_6) require knowledge of the recharge temperature and the amount of excess air, in order to separate the contributing mechanisms or to relate the concentrations of dissolved gases to atmospheric input functions. Dissolved noble gas concentrations can provide this information.

Paleoclimatic implications. The presently available noble gas paleotemperature records may be divided into two groups: those from northern temperate latitudes and those from tropical and sub-tropical areas. A major question in studies from higher latitudes is whether a complete record was obtained or whether the coldest periods, namely the LGM, may be missing because glaciation or permafrost prevented recharge during this period. Such a gap, first suggested by Andrews and Lee (1979) and later confirmed by Beyerle et al. (1998), may also be present in other European records (e.g., Rudolph et al. 1984; Andrews et al. 1985; Bertleff et al. 1993) from locations which may have cooled below the freezing point during the LGM. Therefore, estimates of 5 to 7°C glacial cooling obtained in such records may not represent the maximum cooling. Two noble gas records from northern latitudes appear to be complete and indicate a maximum LGM cooling of ~9°C (Fig. 21, Stute and Deák 1989; Aeschbach-Hertig et al. 2002b).

The first reliable tropical noble gas paleotemperature record found a LGM cooling of about 5°C (Stute et al. 1995b), a finding that received considerable attention in the ongoing debate about the magnitude of the glacial cooling in the tropics (e.g., Broecker 1996; Crowley 2000). The result of this particular study was challenged on grounds of insufficient modelling of the excess air fractionation by Ballentine and Hall (1999), but later confirmed by using the CE-model for excess air (Aeschbach-Hertig et al. 2000). Other noble gas studies from tropical sites also yielded similar coolings (Andrews et al. 1994; Edmunds et al. 1998; Weyhenmeyer et al. 2000), suggesting that the tropics and subtropics (Stute et al. 1992a; Clark et al. 1997; Stute and Talma 1998) cooled rather uniformly by around 5°C during the LGM.

The noble gas thermometer has increasingly been accepted as a climate proxy in the framework of integrated reconstructions of climate conditions during the glacial period (e.g., Farrera et al. 1999; Pinot et al. 1999). Crowley and Baum (1997) and Crowley (2000) relied heavily on noble gas paleotemperature data in comparing model results with climate proxy data. These data are of particular interest in low geographical latitudes where undisrupted ground-water records can be obtained, and where the discrepancies between proxy data and models are particularly large.

Another important aspect of ground-water paleoclimate records is the possibility to relate the absolute temperature estimates obtained from the noble gases to the relative temperature indicator provided by the stable isotope composition of the water ($\delta^{18}O$ and δ^2H). In most noble gas paleotemperature studies, stable isotopes ratios have also been determined, and in many cases clear relationships between the two climate proxys were found (e.g., Heaton et al. 1986; Stute and Deák 1989; Beyerle et al. 1998; Huneau et al. 2001). Such relationships offer the chance to derive local slopes for the long-term $\delta^{18}O$/T-relationship.

A case study. As an example of the application of the noble gas paleothermometer, we discuss a study conducted on the Stampriet artesian aquifer in Namibia. This confined sandstone aquifer is located in southeastern Namibia and has been extensively studied since the 1970s (e.g., Vogel et al. 1981; Heaton et al. 1983; Heaton 1984; Stute and Talma 1998). The noble gas temperature record was re-evaluated applying the CE-model and the inverse method developed by Aeschbach-Hertig et al. (1999b).

The average noble gas temperature of ground water less than 10,000 years is about 26°C (Fig. 23a), in close agreement with today's mean ground temperature in the region as determined from the temperature of shallow ground water (26.7°C). The average noble gas temperature of water samples 15,000 years and older, i.e., of glacial origin, is 21°C, indicating that the mean annual ground temperature in Namibia was about 5°C lower during the last glacial maximum as compared to today. This estimate of the temperature change is consistent with the the 5.5°C cooling obtained with the N_2/Ar technique at the Uitenhage aquifer (Heaton et al. 1986) and a maximum cooling of 6°C derived from a speleothem in Cango Caves in South Africa (Talma and Vogel, 1992). The glacial period is also characterized by an increased $\delta^{18}O$ (Fig. 23b), suggesting that the dominating moisture source shifted from the Atlantic to the Indian Ocean at the transition from the Pleistocene (>10,000 years) to the Holocene (<10,000 years). The noble gas record of the aquifer also shows a clear trend in the concentrations of excess air (Fig. 23c). The peak in ΔNe reaches values of up to 225% around 6,000 radiocarbon years ago, indicating a transition from a drier to a wetter climate, which probably caused the water table to rise and consequently to trap and dissolve air. Such a transition is also indicated by independent paleoclimatic evidence. The concentrations of non-atmospheric He show an increase as a function of radiocarbon age which is initially dominated by in situ production and later by the accumulation of a crustal He flux (Fig. 23d, Heaton 1984; Castro et al. 2000).

In summary, the studies of the Stampriet aquifer highlight all main features of noble gas paleoclimate studies: (1) The noble gas temperatures in conjunction with a ^{14}C timescale provide a quantitative record of the glacial/interglacial temperature change, (2) the local long-term relationship between $\delta^{18}O$ and noble gas temperature can be quantified, (3) the excess air record appears to provide additional climatic information related to precipitation, and iv) the accumulation of He provides a qualitative age scale and information on the crustal He flux.

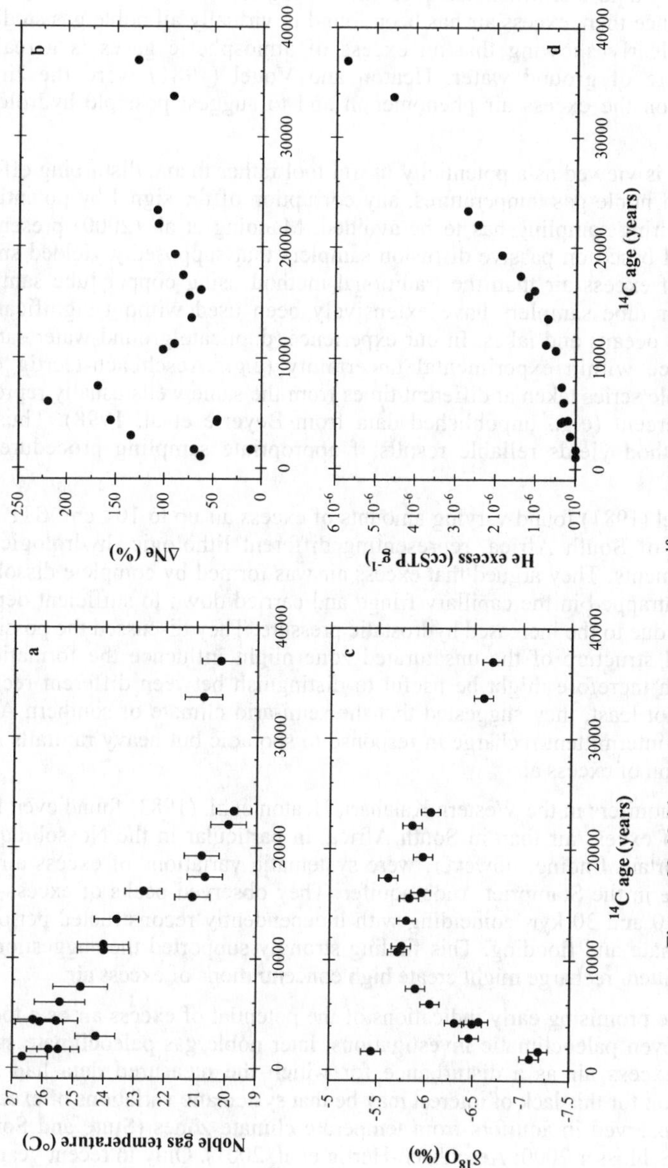

Figure 23. Noble gas temperatures, ΔNe, $\delta^{18}O$, and He excess as a function of radiocarbon ages derived from ground water in the Stampriet aquifer, Namibia (after Stute and Talma 1998). The noble gas data were re-evaluated using the CE-model for excess air (Aeschbach-Hertig et al. 2000).

Excess air

Herzberg and Mazor (1979) found significant amounts of excess air only in some karstic springs, suggesting that excess air was a characteristic feature of karstic systems. However, both Andrews and Lee (1979) and Heaton and Vogel (1981) found excess air also in sedimentary aquifers, which had different lithologies and were located in different climate zones. Since then, excess air has been found in virtually all noble gas studies of ground water, clearly showing that an excess of atmospheric gases is a real and ubiquitous feature of ground water. Heaton and Vogel (1981) were the first to focus explicitly on the excess air phenomenon and to suggest possible hydrological applications.

If excess air is viewed as a potentially useful tool rather than a disturbing effect in the calculation of noble gas temperatures, any corruption of the signal by potential air contamination during sampling has to be avoided. Manning et al. (2000) presented a sampling method based on passive diffusion samplers that supposedly yielded smaller concentrations of excess air than the traditional method using copper tube samplers. However, copper tube samplers have extensively been used without significant air contamination in oceans and lakes. In our experience, duplicate ground-water samples usually reproduce within experimental uncertainty (e.g., Aeschbach-Hertig et al. 2002b) and sample series taken at different times from the same wells usually reproduce within a few percent (e.g., unpublished data from Beyerle et al. 1998). Thus, the copper tube method yields reliable results if appropriate sampling procedures are applied.

Heaton and Vogel (1981) found varying amounts of excess air up to 10^{-2} cm^3 STP g^{-1} in several aquifers of South Africa, representing different lithologic, hydrologic, and climatic environments. They argued that excess air was formed by complete dissolution of air bubbles entrapped in the capillary fringe and carried down to sufficient depth to force dissolution due to the increased hydrostatic pressure. They discussed the possibility that the physical structure of the unsaturated zone might influence the formation of excess air, which therefore might be useful to distinguish between different recharge areas. Last but not least, they suggested that the semi-arid climate of southern Africa, characterized by intermittent recharge in response to sporadic but heavy rainfalls might favor the formation of excess air.

In confined aquifers in the Western Kalahari, Heaton et al. (1983) found even larger concentrations of excess air than in South Africa, in particular in the Nossob aquifer. Their most important finding, however, were systematic variations of excess air with ground-water age in the Stampriet Auob aquifer. They observed peaks of excess air at ages of around 10 and 30 kyr, coinciding with independently reconstructed periods of more humid climate and flooding. This finding strongly supported the suggestion that heavy but intermittent recharge might create high concentrations of excess air.

Despite these promising early indications of the potential of excess air as a tool for hydrologic and even paleoclimatic investigations, later noble gas paleoclimate studies mostly treated excess air as a disturbance for which the measured data had to be corrected. A reason for this lack of interest may be that systematic variations of excess air have not been observed in aquifers from temperate climate zones (Stute and Sonntag 1992; Stute and Schlosser 2000; Aeschbach-Hertig et al. 2001). Only in recent years, the interest in the excess air signal has been revived.

In an aquifer in a semi-arid region of tropical Brazil, Stute et al. (1995b) found large concentrations of clearly fractionated excess air. They noted that ΔNe was roughly two to three times higher in the glacial than in the Holocene waters but did not interpret this finding any further. In contrast, in the case study of the Stampriet Auob aquifer, Stute and Talma (1998) confirmed the existence of an excess air peak in this aquifer (Fig. 23c) and interpreted it in terms of a climatic transition.

Wilson and McNeill (1997) compared the Ne excess from aquifers in different lithologies and climates. Their main conclusion was that lithology had a strong influence on the Ne excess, which was found to increase from granites over sandstones to limestones. Notably, they found no excess air in ground water from granites of southwest England. Relating the Ne excess to the recharge temperature, Wilson and McNeill (1997) found no relationship for limestone aquifers of the UK, a certain decrease of the Ne excess with temperature for a sandstone aquifer in the UK, and a strong decrease for a sandstone aquifer in Niger (data from Andrews et al. 1994). The strong difference of excess air between the warmer Holocence and the cooler Pleistocene waters, the latter having unusually high Ne excesses, in semi-arid tropical Niger is fully consistent with the findings of Heaton et al. (1983), Stute et al. (1995b), and Stute and Talma (1998). Wilson and McNeill (1997) noted that not temperature itself, but rather other climatic factors such as precipitation or frequency of flooding were likely to influence the excess air content of ground water. Indeed they found an increase of the Ne excess with mean annual precipitation when comparing different aquifers, but ascribed this effect to different lithologies and inferred a decrease of the Ne excess with precipitation when comparing data from the same lithologies.

In the light of the new approaches for a physical interpretation of the excess air component provided by the CE-model, Aeschbach-Hertig et al. (2001) made another attempt to find systematical patterns in the excess air data from several aquifers. In a comparison of three aquifers from temperate, humid climates and three aquifers from tropical, semi-arid regions (one of them being the Brazilian aquifer of Stute et al. 1995b), they found systematic differences between the two groups. In agreement with previous studies, the tropical aquifers exhibited higher mean values of ΔNe and a strong negative correlation between ΔNe and the noble gas temperature (Fig. 24).

The CE-model suggests that in general not the amount of entrapped air (A_e), but the pressure acting on this entrapped reservoir (q) determines the final excess (ΔNe). This argues against a major influence of lithology, unless very fine pores cause the surface tension to become a major source of excess pressure on the trapped bubbles, or very low porosity severely limits air entrapment. Otherwise, ΔNe is expected to depend mainly on the hydrostatic pressure. Persistent bubbles of entrapped air, which exist long enough to allow complete equilibration between water and gas phase are likely bound to certain sites in the porous medium, and cannot easily be transported up- or down-wards. Increasing the hydrostatic pressure in order to dissolve more gas therefore requires an increase of the water table. As a result, ΔNe is thought to be related mainly to the amplitude of water table fluctuations, and thus to depend on the variability of precipitation rather than on its mean value. In semi-arid climates with a distinct rainy season, and strong intermittent recharge events, this variability is likely higher than in humid climates with more continuous recharge. An additional factor that favors relatively high excess air in semi-arid regions are deep unsaturated zones, offering the potential for water table increases of many meters.

However, laboratory and field studies under defined conditions are needed to firmly establish the relationships between excess air parameters and environmental conditions during infiltration. Early laboratory experiments with sand columns designed to create and study excess air used somewhat artificial constructions to increase the hydrostatic pressure in the column and facilitate the dissolution of bubbles (Gröning 1989; Osenbrück 1991). More recent laboratory studies have demonstrated that excess air can

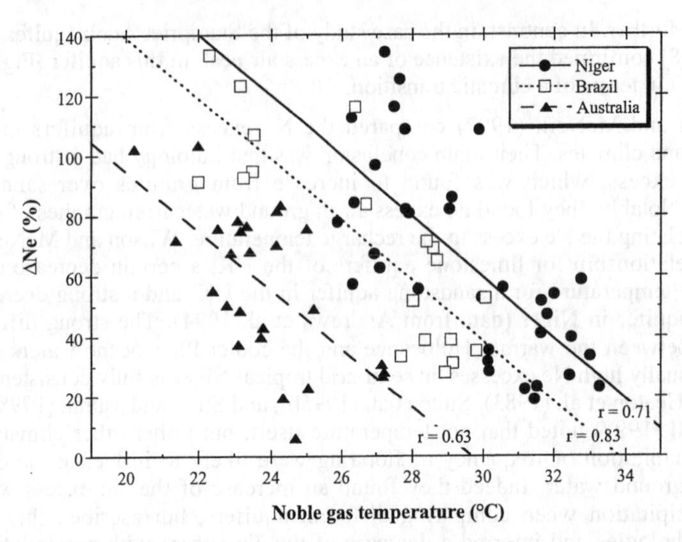

Figure 24. Excess air (expressed by the relative Ne excess ΔNe) in relation to noble gas temperature for three aquifers from tropical, semi-arid sites. Data are from (Beyerle et al. submitted) for Niger, (Stute et al. 1995b) for Brazil, and (Beyerle et al. 1999b) for Australia. Despite a large scatter, all three aquifers show strong and very similar decreasing trends of ΔNe with temperature (about -10 % per °C), which presumably reflect a change in infiltration condition (e.g. variability of recharge) between the Pleistocene and the Holocene.

be generated by simple water table fluctuations of less than 1 m amplitude, and that the size of the excess depends mainly on the hydrostatic pressure (Holocher et al. 2000, 2002). Complete dissolution of entrapped air was only achieved if the sand column was continuously flushed with air-saturated water at relatively high flow rates. Numerical modeling of the kinetics of gas exchange between entrapped gas and water enabled an understanding of the mechanisms that lead to different type of excess air fractionations depending on the experimental setup. Both the numerical model and the data showed that after a water table increase a new equilibrium between water and entrapped air was quickly achieved, although in the first phase complete dissolution of the smallest bubbles is possible.

In summary, there is little evidence for systematic variations of excess air in aquifers from temperate, humid climate zones (Wilson and McNeill 1997; Aeschbach-Hertig et al. 2001). In contrast, past periods of strongly increased excess air contents have been identified in five aquifer systems from warm, semi-arid climates: the Stampriet Auob aquifer in Namibia (Heaton et al. 1983; Stute and Talma 1998), the Serra Grande and Cabecas aquifer in Brazil (Stute et al. 1995b), the Continental Intercalaire aquifer in the Irhazer Plain in Niger (Andrews et al. 1994; Wilson and McNeill 1997); the Continental Terminal aquifers in the Iullemeden Basin in Niger (Aeschbach-Hertig et al. 2001; Beyerle et al. submitted), and the Great Artesian Basin in Australia (Beyerle et al. 1999b; Aeschbach-Hertig et al. 2001). In several of these studies, the periods of enhanced excess air could be correlated to known periods of more humid climate. These findings provide strong support for the hypothesis that excess air is a proxy for infiltration conditions, in particular the intensity and variability of recharge in semi-arid regions. The questions related to excess air are of central importance for the contemporary research on paleoclimate reconstruction based on noble gases in ground water.

NOBLE GASES IN ICE

The polar ice sheets constitute one of the most important archives of past environmental conditions. A prominent example is the reconstruction of the atmospheric composition in the past, in particular with respect to the greenhouse gases CO_2 and CH_4, based on air trapped in polar ice (e.g., Raynaud et al. 1993). No major changes of the atmospheric abundances of the noble gases are to be expected over the timescale of several 100 kyr accessible by the ice archive. Therefore, the isotopic and elemental composition of the noble gases in trapped air provides a tool to study physical processes acting during the entrapment of air in ice.

Gravitational separation

Atmospheric gases trapped in polar ice are enriched in the heavy species due to gravitational separation. This fractionation takes place in the thick firn layer (up to around 100 m) where the pore space is still connected. Below this layer, air bubbles are closed off in the ice and henceforth essentially preserved. The change in the ratio of any two gases from the initial atmospheric value f_0 to the value f in the trapped gas depends on the absolute mass difference ΔM between the two gases, the depth Z of the firn layer, and temperature T, as follows (Craig and Wiens 1996):

$$\frac{f}{f_0} = e^{\frac{g \cdot Z \cdot \Delta M}{R \cdot T}} \tag{50}$$

where g is the gravitational acceleration and R the gas constant.

Craig et al. (1988) demonstrated the occurrence of gravitational fractionation in polar firn based mainly on the isotopic composition of nitrogen and oxygen, but also on the elemental ratios O_2/N_2 and Ar/N_2. Craig and Wiens (1996) used the $^{84}Kr/^{36}Ar$ ratio in trapped air to confirm the dominance of gravitational separation over kinetic fractionation by processes such as effusion from compressed air bubbles, which depend on the relative mass difference ΔM/M rather than the absolute difference ΔM. The large ΔM between Kr and Ar allowed a clear distinction between different processes. Graf and Craig (1997) presented both Kr/Ar and Xe/Ar data that clearly confirmed the presence of gravitational separation.

However, elemental ratios can be fractionated due to gas loss from the samples through microfractures or crystal imperfections, which is governed by differences in molecular volumes (Craig et al. 1988; Bender et al. 1995). This process does not significantly affect isotope ratios. For this reason, much of the work on fractionation of trapped gases in ice has focused on high-precision isotope ratio measurements.

Because the magnitude of the gravitational fractionation depends on the thickness of the firn layer, which in turn is correlated to ambient temperature during firn accumulation, Craig and Wiens (1996) suggested the use of noble gas ratios in ice as a paleothermometer. Indeed, isotope ratios of N_2 and Ar in trapped air have recently been successfully employed to quantify past temperature changes, yet based on an additional fractionating effect, namely thermal diffusion.

Thermal diffusion

Thermal diffusion denotes an effect derived from kinetic gas theory, which predicts that a gas mixture subjected to a temperature gradient tends to unmix, producing an enrichment of the heavier species in the colder regions (Grew and Ibbs 1952; Chapman and Cowling 1970). A rapid temperature change at the surface of an ice sheet will lead to a temperature gradient in the firn column, which will cause the gases to separate by thermal diffusion. Heavier isotopes or elements will preferentially migrate towards the cooler regions. This separation will continue until it is balanced by diffusion along the concentration gradient in the opposite direction.

The fractional difference δ (in per mil) of the isotope ratio f at temperature T to the ratio f_0 at temperature T_0 is given by (Chapman and Cowling 1970):

$$\delta = \left(\frac{f}{f_0} - 1\right) \cdot 10^3 = \left(\left[\frac{T_0}{T}\right]^{\alpha} - 1\right) \cdot 10^3 \tag{51}$$

where α is the thermal diffusion factor characteristic for the investigated pair of gases.

The occurrence of thermal diffusion in nature was first observed by Severinghaus et al. (1996) for soil gas in sand dunes. The fractionation of soil air by water vapor diffusion, gravitational settling, and thermal diffusion studied by these authors may also have a small effect on dissolved noble gases in ground water, and hence the calculation of noble gas paleotemperatures. However, these comparatively minor effects have not yet been further studied in the soil–air–ground-water system.

Severinghaus et al. (1998) demonstrated that thermal fractionation of the air in polar firn layers can be detected and separated from the gravitational effect by high precision analysis of $^{15}N/^{14}N$ and $^{40}Ar/^{36}Ar$ ratios. Rapid temperature changes at the surface induce a transient temperature gradient in the upper part of the firn, which by way of thermal diffusion induces an isotopic signal that propagates downwards and is recorded in the trapped air in the ice. The resulting isotopic anomaly can directly be compared to other changes in the trapped air record. In this way, Severinghaus et al. (1998) showed that the increase of atmospheric methane at the end of the Younger Dryas cold period (about 12 kyr BP) began within 0 to 30 years after the warming event. This finding provides important constraints on the mechanisms of climate change at that time.

Moreover, by relating the isotopic anomaly in the trapped air as a marker of warming to the corresponding change of $\delta^{18}O$ in the ice, the age difference between ice and trapped air can be determined. This gas-age-ice-age difference can then be translated into a paleotemperature estimate using an empirical model of the snow densification process, which slows with decreasing temperature. For the Younger Dryas in central Greenland, Severinghaus et al. (1998) deduced 15°C lower than present temperatures. This estimate is in good agreement with values obtained from borehole temperature profiles, but significantly larger than inferred previously from the $\delta^{18}O$ paleothermometer calibrated by the modern spatial $\delta^{18}O$/temperature relationship.

Further studies have confirmed the usefulness of the thermal diffusion signal stored in polar ice cores. Severinghaus and Brook (1999) investigated the termination of the last glacial period (about 15 kyr BP), with similar results as in the previous study of the termination of the Younger Dryas. Lang et al. (1999) studied the abrupt warming at the start of a Dansgaard-Oeschger event about 70 kyr BP, whereas Leuenberger et al. (1999) focused on the largest temperature excursion during the Holocene, a cool period around 8.2 kyr BP. The latter studies were based on N_2 isotopes only and focused on the calibration of the $\delta^{18}O$ paleothermometer.

Severinghaus et al. (2001) directly observed thermal fractionation of the present-day firn air in response to seasonal temperature gradients. Elemental and isotopic ratios of N_2, O_2, Ar, Kr, and Xe analyzed in the air of the top 15 m of the firn at two Antarctic sites were found to match a model without adjustable parameters reasonably well. Relative thermal diffusion sensitivities for different gas pairs could be derived.

Analytical methods. Detection of the small thermal fractionation signals requires high precision isotope ratio mass spectrometry, which differs substantially from the methods used to analyze noble gases in water samples as discussed earlier in this chapter. The most prominent difference is the use of dynamic rather than static mass spectrometry. Severinghaus et al. (2002) discuss in detail their methods used to measure $^{40}Ar/^{36}Ar$ and $^{84}Kr/^{36}Ar$ ratios in ice samples.

Ice samples are put into pre-cooled extraction vessels for evacuation. The ice is then melted and a gas extraction and purification similar to that used for water samples is performed. Because dynamic analysis requires far more sample than the static mode, ultrapure N_2 is added to increase bulk pressure by a factor of 10. The mass spectrometric analysis follows conventional procedures of dynamic isotope ratio mass spectrometry, with modifications designed to avoid any fractionating effects, such as thermal diffusion during volume splitting steps. An external precision of the order of ±0.01‰ for $^{40}Ar/^{36}Ar$ is achieved, yielding a good resolution of signals associatied with abrupt climate change that are about 0.4‰ in Greenland ice cores. The reproducibility of $^{84}Kr/^{36}Ar$ ratios is about ±1‰ for ice core samples. The $^{84}Kr/^{36}Ar$ ratio is primarily measured to identify samples that have experienced argon leakage out of the bubbles, e.g., during storage.

Helium isotopes

Helium isotopes have also been analyzed in polar ice cores, although very few results have been published so far. The first data showed a strong depletion of He in trapped air relative to atmospheric air (Craig and Chou 1982; Jean-Baptiste et al. 1993). The $^3He/^4He$ ratio was also found to be somewhat depleted compared to air. These findings clearly indicate loss of helium, presumably due to upward diffusion through the ice in response to increasing compression of the trapped air bubbles. However, some uncertainty remained as to what extent He loss during sampling influenced the results. The high diffusivity of He in ice clearly complicates its use in this archive.

Radiogenic He. Strong signals of crustal He have been found near the base of the ice sheets. Craig and Scarsi (1997) observed an approximately constant and slightly lower than atmospheric $^3He/^4He$ ratio in the upper 2.8 km, but a clearly radiogenic signature in the lowermost 250 m of the GISP2 ice core. The profile showed sharp peaks of radiogenic ^4He, indicating a disturbed stratigraphy of the deepest section of the core.

In the Vostok ice core, Jean-Baptiste et al. (2001) found a very sharp transition both in He concentration as well as in the isotopic ratio at a depth of 3539 m, the boundary between glacier ice and accreted ice from re-freezing of water of the underlying Lake Vostok. In the glacier ice, the He concentration was found to be nearly constant although depleted relative to the atmosphere, and the isotopic ratio was close to the air value. In the accreted ice, the He content was higher by a factor of more than three, and the $^3He/^4He$ ratio was correspondingly lowered ($R/R_a = 0.25\pm0.04$). Apparently, radiogenic He that had accumulated in the lake water was incorporated into the ice during freezing. Jean-Baptiste et al. (2001) used the assumption of a whole crustal He flux to estimate the water renewal time of the lake to about 5000 yr. Given the uncertainties related to the crustal He flux, this result must be regarded with caution, although it is in reasonable agreement with geophysical inferences.

Extraterrestrial ^{3}He. Polar ice also archives the influx of extraterrestrial matter in the form of interplanetary dust particles (IDPs), similar to the ocean sediments (see Schlosser and Winckler 2002 in this volume). Ice cores may provide less ambiguous results for the IDP flux than ocean sediments, as accumulation rates and time scales are well known in this archive. By analyzing He isotopes in particles separated from polar ice, Brook et al. (2000) demonstrated the utility of the ice core record to study the IDP flux and found similar results as obtained from marine sediments.

REFERENCES

Aeschbach-Hertig W (1994) Helium und Tritium als Tracer für physikalische Prozesse in Seen. PhD dissertation, ETH Zürich, Zürich

Aeschbach-Hertig W, Beyerle U, Holocher J, Peeters F, Kipfer R (2001) Excess air in ground water as a potential indicator of past environmental changes. *In* Intl Conf on the Study of Environmental Change Using Isotope Techniques (IAEA-CN-80). IAEA, Vienna, p 34-36

Aeschbach-Hertig W, Hofer M, Kipfer R, Imboden DM, Wieler R (1999a) Accumulation of mantle gases in a permanently stratified volcanic Lake (Lac Pavin, France). Geochim Cosmochim Acta 63: 3357-3372

Aeschbach-Hertig W, Hofer M, Schmid M, Kipfer R, Imboden DM (2002a) The physical structure and dynamics of a deep, meromictic crater lake (Lac Pavin, France). Hydrobiol (in press)

Aeschbach-Hertig W, Kipfer R, Hofer M, Imboden DM, Baur H (1996a) Density-driven exchange between the basins of Lake Lucerne (Switzerland) traced with the ^3H-^3He method. Limnol Oceanogr 41: 707-721

Aeschbach-Hertig W, Kipfer R, Hofer M, Imboden DM, Wieler R, Signer P (1996b) Quantification of gas fluxes from the subcontinental mantle: The example of Laacher See, a maar lake in Germany. Geochim Cosmochim Acta 60:31-41

Aeschbach-Hertig W, Peeters F, Beyerle U, Kipfer R (1999b) Interpretation of dissolved atmospheric noble gases in natural waters. Water Resour Res 35:2779-2792

Aeschbach-Hertig W, Peeters F, Beyerle U, Kipfer R (2000) Palaeotemperature reconstruction from noble gases in ground water taking into account equilibration with entrapped air. Nature 405:1040-1044

Aeschbach-Hertig W, Schlosser P, Stute M, Simpson HJ, Ludin A, Clark JF (1998) A $^3H/^3He$ study of ground-water flow in a fractured bedrock aquifer. Ground Water 36:661-670

Aeschbach-Hertig W, Stute M, Clark J, Reuter R, Schlosser P (2002b) A paleotemperature record derived from dissolved noble gases in ground water of the Aquia Aquifer (Maryland, USA). Geochim Cosmochim Acta 66:797-817

Andrews JN (1985) The isotopic composition of radiogenic helium and its use to study ground-water movement in confined aquifers. Chem Geol 49:339-351

Andrews JN, Balderer W, Bath AH, Clausen HB, Evans GV, Florkowski T, Goldbrunner JE, Ivanovich M, Loosli H, Zojer H (1984) Environmental isotope studies in two aquifer systems: A comparison of ground-water dating methods. *In* Isotope Hydrology 1983 (IAEA-SM-270). IAEA, Vienna, p 535-577

Andrews JN, Davis SN, Fabryka-Martin J, Fontes J-C, Lehmann BE, Loosli HH, Michelot J-L, Moser H, Smith B, Wolf M (1989) The in situ production of radioisotopes in rock matrices with particular reference to the Stripa granite. Geochim Cosmochim Acta 53:1803-1815

Andrews JN, Fontes J-C, Aranyossy J-F, Dodo A, Edmunds WM, Joseph A, Travi Y (1994) The evolution of alkaline ground waters in the continental intercalaire aquifer of the Irhazer Plain, Niger. Water Resour Res 30:45-61

Andrews JN, Goldbrunner JE, Darling WG, Hooker PJ, Wilson GB, Youngman MJ, Eichinger L, Rauert W, Stichler W (1985) A radiochemical, hydrochemical and dissolved gas study of ground waters in the Molasse basin of Upper Austria. Earth Planet Sci Lett 73:317-332

Andrews JN, Lee DJ (1979) Inert gases in ground water from the Bunter Sandstone of England as indicators of age and palaeoclimatic trends. J Hydrol 41:233-252

Ballentine CJ, Burgess R, Marty B (2002) Tracing fluid origin, transport and interaction in the crust. Rev Mineral Geochem 47:539-614

Ballentine CJ, Burnard PG (2002) Production, release and transport of noble gases in the continental crust. Rev Mineral Geochem 47:481-538

Ballentine CJ, Hall CM (1999) Determining paleotemperature and other variables by using an error-weighted, nonlinear inversion of noble gas concentrations in water. Geochim Cosmochim Acta 63:2315-2336

Ballentine CJ, Mazurek M, Gautschi A (1994) Thermal constraints on crustal rare gas release and migration: Evidence from Alpine fluid inclusions. Geochim Cosmochim Acta 58:4333-4348

Bauer S, Fulda C, Schäfer W (2001) A multi-tracer study in a shallow aquifer using age dating tracers ^3H, ^{85}Kr, CFC-113 and SF_6—indication for retarded transport of CFC-113. J Hydrol 248:14-34

Bayer R, Schlosser P, Bönisch G, Rupp H, Zaucker F, Zimmek G (1989) Performance and Blank Components of a Mass Spectrometric System for routine measurement of helium isotopes and tritium by the ^{3}he ingrowth method. *In* Sitzungsberichte der Heidelberger Akademie der Wissenschaften. 5:241-279

Begemann F, Libby WF (1957) Continental water balance, ground-water inventory and storage times, surface ocean mixing rates and world-wide water circulation patterns from cosmic-ray and bomb tritium. Geochim Cosmochim Acta 12:277-296

Benson BB, Krause D (1976) Empirical laws for dilute aqueous solutions of nonpolar gases. J Chem Phys 64:689-709

Benson BB, Krause D (1980) Isotopic fractionation of helium during solution: A probe for the liquid state. J Solution Chem 9:895-909

Bender M, Sowers T, Lipenkov V (1995) On the concentrations of O_2, N_2, and Ar in trapped gases from ice cores. J Geophys Res 100:18651-18660

Bertin C, Bourg ACM (1994) Radon-222 and chloride as natural tracers of the infiltration of river water into an alluvial aquifer in which there is significant river/ground-water mixing. Environ Sci Technol 28:794-798

Bertleff B, Ellwanger D, Szenkler C, Eichinger L, Trimborn P, Wolfendale N (1993) Interpretation of hydrochemical and hydroisotopical measurements on palaeoground waters in Oberschwaben, south German alpine foreland, with focus on quanternary geology. *In* Isotope techniques in the study of past and current environmental changes in the hydrosphere and the atmosphere (IAEA-SM-329). IAEA, Vienna, p 337-357

Beyerle U, Aeschbach-Hertig W, Hofer M, Imboden DM, Baur H, Kipfer R (1999a) Infiltration of river water to a shallow aquifer investigated with $^3H/^3He$, noble gases and CFCs. J Hydrol 220:169-185

Beyerle U, Aeschbach-Hertig W, Imboden DM, Baur H, Graf T, Kipfer R (2000a) A mass spectrometric system for the analysis of noble gases and tritium from water samples. Environ Sci Technol 34: 2042-2050

Beyerle U, Aeschbach-Hertig W, Peeters F, Kipfer R (2000b) Accumulation rates of radiogenic noble gases and noble gas temperatures deduced from the Great Artesian Basin, Australia. Beyond 2000—New Frontiers in Isotope Geoscience 1:21

Beyerle U, Aeschbach-Hertig W, Peeters F, Kipfer R, Purtschert R, Lehmann B, Loosli HH, Love A (1999b) Noble gas data from the Great Artesian Basin provide a temperature record of Australia on time scales of 10^5 years. *In* Isotope techniques in water resources development and management (IAEA-SM-361, IAEA-CSP-C/2). IAEA, Vienna, p 97-103

Beyerle U, Purtschert R, Aeschbach-Hertig W, Imboden DM, Loosli HH, Wieler R, Kipfer R (1998) Climate and ground water recharge during the last glaciation in an ice-covered region. Science 282:731-734

Beyerle U, Rüedi J, Aeschbach-Hertig W, Peeters F, Leuenberger M, Dodo A, Kipfer R (2002) Evidence for periods of wetter and cooler climate in the Sahel between 6 and 40 kyr. Geophys Res Lett (submitted)

Bieri RH (1971) Dissolved Noble Gases in Marine Waters. Earth Planet Sci Lett 10:329-333

Böhlke JK, Denver JM (1995) Combined use of ground water dating, chemical, and isotopic analyzes to resolve the history and fate of nitrate contamination in two agricultural watersheds, Atlantic coastal plain, Maryland. Water Resour Res 31:2319-2339

Böhlke JK, Révész K, Busenberg E, Deak J, Deseö E, Stute M (1997) Ground water record of halocarbon transport by the Danube river. Environ Sci Technol 31:3293-3299

Bottomley DJ, Ross JD, Clarke WB (1984) Helium and neon isotope geochemistry of some ground waters from the Canadian Precambrian Shield. Geochim Cosmochim Acta 48:1973-1985

Brennwald MS, Peeters F, Beyerle U, Rüedi J, Hofer M, Kipfer R (2001) Modeling the transport of CFCs and 3H through the unsaturated zone. Geophys Res Abstr EGS HSA6

Broecker W (1996) Glacial climate in the tropics. Science 272:1902-1903

Brook EJ, Kurz MD, Curtice J, Cowburn S (2000) Accretion of interplanetary dust in polar ice. Geophys Res Lett 27:3145-3148

Brown RM (1961) Hydrology of tritium in the Ottawa Valley. Geochim Cosmochim Acta 21:199-216

Busenberg E, Plummer LN (1992) Use of chlorofluorocarbons (CCl_3F and CCl_2F_2) as hydrologic tracers and age-dating tools: The alluvium and terrace system of central Oklahoma. Water Resour Res 28:2257-2284

Busenberg E, Plummer NL (2000) Dating young ground water with sulfur hexafluoride: Natural and anthropogenic sources of sulfur hexafluoride. Water Resour Res 36:3011-3030

Busenberg E, Weeks EP, Plummer LN, Bartholomay RC (1993) Age dating ground water by use of chlorofluorocarbons (CCl_3F and CCl_2F_2), and distribution of chlorofluorocarbons in the unsaturated zone, Snake River plain aquifer, Idaho National Engineering Laboratory, Idaho. U S Geol Surv 93-4054 93:1-47

Cable JE, Bugna GC, Burnett WC, Chanton JP (1996) Application of ^{222}Rn and CH_4 for assessment of ground water discharge to the coastal ocean. Limnol Oceanogr 41:1347-1353

Castro MC, Goblet P, Ledoux E, Violette S, de Marsily G (1998a) Noble gases as natural tracers of water circulation in the Paris Basin, 2. Calibration of a ground water flow model using noble gas isotope data. Water Resour Res 34:2467-2483

Castro MC, Jambon A, de Marsily G, Schlosser P (1998b) Noble gases as natural tracers of water circulation in the Paris Basin, 1. Measurements and discussion of their origin and mechanisms of vertical transport in the basin. Water Resour Res 34:2443-2466

Castro MC, Stute M, Schlosser P (2000) Comparison of ^{4}He ages and ^{14}C ages in simple aquifer systems: implications for ground water flow and chronologies. Appl Geochem 15:1137-1167

Chapman S, Cowling TG (1970) The mathematical theory of non-uniform gases. Cambridge University Press, Cambridge, London, New York

Cecil LD, Green JR (2000) Radon-222. *In* Environmental tracers in subsurface hydrology. Cook P, Herczeg AL (eds) Kluwer Academic Publishers, Boston, p 175-194

Christiansen JE (1944) Effect of entrapped air upon the permeability of soils. Soil Sci 58:355-365

Clark JF, Davisson ML, Hudson GB, Macfarlane PA (1998) Noble gases, stable isotopes, and radiocarbon as tracers of flow in the Dakota aquifer, Colorado and Kansas. J Hydrol 211:151-167

Clark JF, Hudson GB (2001) Quantifying the flux of hydrothermal fluids into Mono Lake by use of helium isotopes. Limnol Oceanogr 46:189-196

Clark JF, Stute M, Schlosser P, Drenkard S, Bonani G (1997) A tracer study of the Floridan aquifer in southeastern Georgia: Implications for ground water flow and paleoclimate. Water Resour Res 33: 281-289

Clarke WB, Jenkins WJ, Top Z (1976) Determination of tritium by mass spectrometric measurement of ^{3}He. Intl JAppl Radiat Isotopes 27:515-522

Clarke WB, Top Z, Beavan AP, Gandhi SS (1977) Dissolved helium in lakes: Uranium prospecting in the precambrian terrain of Central Labrador. Econ Geol 72:233-242

Clever HL (ed) (1979a) Helium and neon-gas solubilities. Solubility Data Series. Vol 1. Pergamon Press, Oxford, New York, Toronto, Sydney, Paris, Frankfurt

Clever HL (ed) (1979b) Krypton, xenon and radon-gas solubilities. Solubility Data Series. Vol 2. Pergamon Press, Oxford, New York, Toronto, Sydney, Paris, Frankfurt

Clever HL (ed) (1980) Argon. Solubility Data Series. Vol 4. Pergamon Press, Oxford, New York, Toronto, Sydney, Paris, Frankfurt

Collier RW, Dymond J, McManus J (1991) Studies of hydrothermal processes in Crater Lake, OR. College of Oceanography, Oregon State University, Report #90-7:1:201

Collon P, Kutschera W, Loosli HH, Lehmann BE, Purtschert R, Love A, Sampson L, Anthony D, Cole D, Davids B, Morrissey DJ, Sherrill BM, Steiner M, Pardo RC, Paul M (2000) ^{81}Kr in the Great Artesian Basin, Australia: a new method for dating very old ground water. Earth Planet Sci Lett 182:103-113

Colman JA, Armstrong DE (1987) Vertical eddy diffusivity determined with Rn-222 in the benthic boundary-layerof ice-covered lakes. Limnol Oceanogr 32:577-590

Cook PG, Böhlke J-K (2000) Determining timescales for ground water flow and solute transport. *In* Environmental tracers in subsurface hydrology. Cook P, Herczeg AL (eds) Kluwer Academic Publishers, Boston, p 1-30

Cook PG, Herczeg AL (eds) (2000) Environmental tracers in subsurface hydrology. Kluwer Academic Publishers, Boston

Cook PG, Solomon DK (1995) Transport of atmospheric trace gases to the water table: Implications for ground water dating with chlorofluorocarbons and krypton 85. Water Resour Res 31:263-270

Cook PG, Solomon DK (1997) Recent advances in dating young ground water: chlorofluorocarbons, ^{3}H/^{3}He and ^{85}Kr. J Hydrol 191:245-265

Cook PG, Solomon DK, Plummer LN, Busenberg E, Schiff SL (1995) Chlorofluorocarbons as tracers of ground water transport processes in a shallow, silty sand aquifer. Water Resour Res 31:425-434

Cook PG, Solomon DK, Sanford WE, Busenberg E, Plummer LN, Poreda RJ (1996) Inferring shallow ground water flow in saprolite and fractured rock using environmental tracers. Water Resour Res 32:1501-1509

Craig H, Chou CC (1982) Helium isotopes and gases in Dye 3 ice cores. EOS Trans Am Geophys Union 63:298

Craig H, Clarke WB, Beg MA (1975) Excess ^{3}He in deep water on the East Pacific Rise. Earth Planet Sci Lett 26:125-132

Craig H, Dixon F, Craig VK, Edmond J, Coulter G (1974) Lake Tanganyika geochemical and hydrographic study: 1973 expedition. Scripps Institution of Oceanography Series 75-5:1-83

Craig H, Horibe Y, Sowers T (1988) Gravitational separation of gases and isotopes in polar ice caps. Science 242:1675-1678

Craig H, Scarsi P (1997) Helium isotope stratigraphy in the GISP2 ice core. EOS Trans Am Geophys Union 78:F7

Craig H, Weiss RF (1971) Dissolved gas saturation anomalies and excess helium in the ocean. Earth Planet Sci Lett 10:289-296

Craig H, Wiens RC (1996) Gravitational enrichment of $^{84}Kr/^{36}Ar$ ratios in polar ice caps: a measure of firn thickness and accumulation temperature. Science 271:1708-1710
Crovetto R, Fernandez-Prini R, Japas ML (1982) Solubilities of inert gases and methane in H_2O and in D_2O in the temperature range of 300 to 600 K. J Chem Phys 76:1077-1086
Crowley TJ (2000) CLIMAP SSTs re-revisited. Clim Dyn 16:241-255
Crowley TJ, Baum SK (1997) Effect of vegetation on an ice-age climate model simulation. J Geophys Res 102:16463-16480
Cunnold DM, Fraser PJ, Weiss RF, Prinn RG, Simmonds PG (1994) Global trends and annual releases of CCl_3F and CCl_2F_2 estimated from ALE/GAGE and other measurements from July 1978 to June 1991. J Geophys Res 99:1107-1126
Dennis F, Andrews JN, Parker A, Poole J, Wolf M (1997) Isotopic and noble gas study of Chalk ground water in the London Basin, England. Appl Geochem 12:763-773
Doney SC, Glover DM, Jenkins WJ (1992) A model function of the global bomb tritium distribution in precipitation, 1960-1986. J Geophys Res 97:5481-5492
Dunkle SA, Plummer LN, Busenberg E, Phillips PJ, Denver JM, Hamilton PA, Michel RL, Coplen TB (1993) Chlorofluorocarbons (CCl_3F and CCl_2F_2) as dating tools and hydrologic tracers in shallow ground water of the Delmarva Peninsula, Atlantic Coastal Plain, United States. Water Resour Res 29:3837-3860
Dunkle Shapiro S, LeBlanc D, Schlosser P, Ludin A (1999) Characterizing a sewage plume using the ^3H-^3He dating technique. Ground Water 37:861-878
Dunkle Shapiro S, Rowe G, Schlosser P, Ludin A, Stute M (1998) Tritium–helium-3 dating under complex conditions in hydraulically stressed areas of a buried-valley aquifer. Water Resour Res 34:1165-1180
Edmunds WM, Fellman E, Goni IB, McNeill G, Harkness DD (1998) Ground water, palaeoclimate and palaeorecharge in the southwest Chad Basin, Borno State, Nigeria. *In* Isotope techniques in the study of environmental change (IAEA-SM-349). IAEA, Vienna, p 693-707
Ekwurzel B, Schlosser P, Smethie WM, Plummer LN, Busenberg E, Michel RL, Weppernig R, Stute M (1994) Dating of shallow ground water: Comparison of the transient tracers $^3H/^3He$, chlorofluorocarbons, and ^{85}Kr. Water Resour Res 30:1693-1708
Ellins KK, Roman-Mas A, Lee R (1990) Using ^{222}Rn to examine ground water/surface discharge interaction in the Rio Grande de Manati, Puerto Rico. J Hydrol 115:319-341
Erikson E (1958) The possible use of tritium for estimating ground water storage. Tellus 10:472-478
Farrera I, Harrison SP, Prentice IC, Ramstein G, Guiot J, Bartlein PJ, Bonnefille R, Bush M, Cramer W, von Grafenstein U, Holmgren K, Hooghiemstra H, Hope G, Jolly D, Lauritzen SE, Ono Y, Pinot S, Stute M, Yu G (1999) Tropical climates at the Last Glacial Maximum: a new synthesis of terrestrial palaeoclimate data. I. Vegetation, lake levels and geochemistry. Clim Dyn 15:823-856
Faybishenko BA (1995) Hydraulic behavior of quasi-saturated soils in the presence of entrapped air: Laboratory experiments. Water Resour Res 31:2421-2435
Fayer MJ, Hillel D (1986) Air encapsulation: 1. Measurement in a field soil. Soil Sci Soc Am J 50:568-572
Frei MW (1999) Entwicklung und erster Feldeinsatz einer neuen massenspektrometrischen Methode zur Echtzeitbestimmung der Hauptgaskomponenten in Bodenluft. Diploma thesis, ETH-Zürich / EAWAG, Zürich / Dübendorf
Giggenbach WF, Sano Y, Schmincke HU (1991) CO_2-rich gases from Lakes Nyos and Monoun, Cameroon; Laacher See, Germany; Dieng, Indonesia, and Mt. Gambier, Australia—variations on a common theme. J Volcanol Geotherm Res 45:311-323
Gill AE (1982) Atmosphere-Ocean Dynamics. Academic Press, San Diego
Graf T, Craig H (1997) GISP2: Gravitational enrichment of Kr/Ar and Xe/Ar ratios. EOS Trans Am Geophys Union 78:F7
Graham DW (2002) Noble gas isotope geochemistry of mid-ocean ridge island basalts: characterization of mantle source reservoirs. Rev Mineral Geochem 47:247-318
Grew KE, Ibbs TL (1952) Thermal Diffusion in Gases. Cambridge University Press, Cambridge, London, New York
Griesshaber E, O'Nions RK, Oxburgh ER (1992) Helium and carbon isotope systematics in crustal fluids from the Eifel, the Rhine Graben and Black Forest, F.R.G. Chem Geol 99:213-235
Gröning M (1989) Entwicklung und Anwendung einer neuen Probenahmetechnik für Edelgasmessungen an Grundwasser und Untersuchungen zum Luftüberschuss im Grundwasser. Diploma thesis, Universität Heidelberg, Heidelberg
Gröning M (1994) Edelgase und Isotopentracer im Grundwasser: Paläo-Klimaänderungen und Dynamik regionaler Grundwasserfliesssysteme. PhD dissertation, Universität Heidelberg, Heidelberg
Gülec N (1988) Helium-3 distribution in Western Turkey. Mineral Res. Expl. Bull. 108:35-42
Heaton THE (1981) Dissolved gases; some applications to ground water research. Trans Geol Soc South Africa 84:91-97

Heaton THE (1984) Rates and sources of ^{4}He accumulation in ground water. Hydrol Sci J 29:29-47
Heaton THE, Talma AS, Vogel JC (1983) Origin and history of nitrate in confined ground water in the western Kalahari. J Hydrol 62:243-262
Heaton THE, Talma AS, Vogel JC (1986) Dissolved gas paleotemperatures and ^{18}O variations derived from ground water near Uitenhage, South Africa. Quat Res 25:79-88
Heaton THE, Vogel JC (1979) Gas concentrations and ages of ground waters in Beaufort Group sediments, South Africa. Water SA 5:160-170
Heaton THE, Vogel JC (1981) "Excess air" in ground water. J Hydrol 50:201-216
Herzberg O, Mazor E (1979) Hydrological applications of noble gases and temperature measurements in underground water systems: Examples from Israel. J Hydrol 41:217-231
Ho DT, Schlosser P, Smethie WM, Simpson HJ (1998) Variability in atmospheric chlorofluorocarbons (CCl_3F and CCl_2F_2) near a large urban area: Implications for ground water dating. Environ Sci Technol 32:2377-2382
Hoehn E, von Gunten HR (1989) Radon in ground water: A tool to assess infiltration from surface waters to aquifers. Water Resour Res 25:1795-1803
Hofer M, Imboden DM (1998) Simultaneous determination of CFC-11, CFC-12, N_2 and Ar in water. Analyt Chem 70:724-729
Hofer M., Peeters F, Aeschbach-Hertig W, Brennwald M, Holocher J, Livingstone DM., Romanovski V, and Kipfer R (2002) Rapid deep-water renewal in lake Issyk-Kul (Kyrgyzstan) indicated by transient tracers. Limnol Oceanogr 47:1210-1216
Hohmann R (1997) Deep-Water Renewal in Lake Baikal. PhD dissertation, ETH-Zürich, Zürich
Hohmann R, Hofer M, Kipfer R, Peeters F, Imboden DM (1998) Distribution of helium and tritium in Lake Baikal. J Geophys Res 103:12823-12838
Hohmann R, Kipfer R, Peeters F, Piepke G, Imboden DM, Shimaraev MN (1997) Processes of deep water renewal in Lake Baikal. Limnol Oceanogr 42:841-855
Holocher J, Matta V, Aeschbach-Hertig W, Beyerle U, Hofer M, Peeters F, Kipfer R (2001) Noble gas and major element constraints on the water dynamics in an alpine floodplain. Ground Water 39:841-852
Holocher J, Peeters F, Aeschbach-Hertig W, Beyerle U, Brennwald M, Hofer M, Kipfer R (2000) New insights into the mechanisms controlling the formation of excess air in ground water. EOS Trans Am Geophys Union 81:F441
Holocher J, Peeters F, Aeschbach-Hertig W, Beyerle U, Hofer M, Brennwald M, Kinzelbach W, Kipfer R (2002) Experimental investigations on the formation of excess air in quasi-saturated porous media. Geochim Cosmochim Acta (in press)
Holzner C (2001) Untersuchung der Tiefenwassererneuerung in meromiktischen Seen mittels transienter Tracer und numerischer Modellierung. Diploma thesis, ETH-Zürich / EAWAG-Zürich-Dübendorf
Huneau F, Blavoux B, Aeschbach-Hertig W, Kipfer R (2001) Palaeoground waters of the Valréas Miocene aquifer (Southeastern France) as archives of the LGM/Holocene climatic transition in the Western Mediterranean region. *In* Intl Conf Study of Environmental Change Using Isotope Techniques (IAEA-CN-80). IAEA, Vienna, p 27-28
Hunkeler D, Hoehn E, Höhener P, Zeyer J (1997) ^{222}Rn as a partitioning tracer to detect diesel fuel contamination in aquifers: laboratory study and field observations. Environ Sci Technol 31:3180-3187
IAEA/WMO (1998) Global Network of Isotopes in Precipitation. The GNIP Database www.iaea.or.at/programs/ri/gnip/gnipmain.htm. IAEA, Vienna
Igarashi G, Ozima M, Ishibashi J, Gamo T, Sakai H, Nojiri Y, Kawai T (1992) Mantle helium flux from the bottom of Lake Mashu, Japan. Earth Planet Sci Letters 108:11-18
Imboden DM (1977) Natural radon as a limnological tracer for the study of vertical and horizontal eddy diffusion. *In* Isotopes in lake studies. IAEA, Vienna p 213-218
Imboden DM, Emerson S (1978) Natural radon and phosphorus as limnologic tracers: Horizontal and vertical eddy diffusion in Greifensee. Limnol Oceanogr 23:77-90
Imboden DM, Joller T (1984) Turbulent mixing in the hypolimion of Baldeggersee (Switzerland) traced by natural radon-222. Limnol Oceanogr 29:831-844
Jassby A, Powell T (1975) Vertical patterns of eddy diffusion during stratification in Castle Lake, California. Limnol Oceanogr 20:530-543
Jean-Baptiste P, Petit JR, Lipenkov VY, Raynaud D, Barkov N (2001) Constraints on hydrothermal processes and water exchange in Lake Vostok from helium isotopes. Nature 411:460-462
Jean-Baptiste P, Raynaud D, Mantisi F, Sowers T, Barkov N (1993) Measurement of helium isotopes in Antarctic ice: preliminary results from Vostok. C R Acad Sci Paris 316:491-497
Jenkins WJ (1976) Tritium-helium dating in the Sargasso Sea: A measurement of oxygen utilization rates. Science 196:291-292
Johnston CT, Cook PG, Frape SK, Plummer LN, Busenberg E, Blackport RJ (1998) Ground water age and nitrate distribution within a glacial aquifer beneath a thick unsaturated zone. Ground Water 36:171-180

Juillard-Tardent M (1999) Gasübersättigung an Paläogrundwässern in Estland. Diploma thesis, Universität Bern, Bern

Katz BG, Lee TM, Plummer LN, Busenberg E (1995) Chemical evolution of ground water near a sinkhole lake, northern Florida -1. Flow patterns, age of ground water, and influence of lake water leakage. Water Resour Res 31:1549-1564

Kaufman S, Libby WF (1954) The natural distribution of tritium. Phys Rev 93:1337-1344

Kennedy BM, Hiyagon H, Reynolds JH (1990) Crustal neon: a striking uniformity. Earth Planet Sci Lett. 98:277-286

Kipfer R, Aeschbach-Hertig W, Baur H, Hofer M, Imboden DM, Signer P (1994) Injection of mantle type helium into Lake Van (Turkey): The clue for quantifying deep water renewal. Earth Planet. Sci Lett 125:357-370

Kipfer R, Aeschbach-Hertig W, Beyerle U, Goudsmit G, Hofer M, Peeters F, Klerkx J, Plisnier J-P, Kliembe EA, Ndhlovu R (2000) New evidence for deep water exchange in Lake Tanganyika. EOS Trans Am Geophys Union 80:OS239

Kipfer R, Aeschbach-Hertig W, Hofer M, Hohmann R, Imboden DM, Baur H, Golubev V, Klerkx J (1996) Bottomwater formation due to hydrothermal activity in Frolikha Bay, Lake Baikal, eastern Siberia. Geochim Cosmochim Acta 60:961-971

Lang C, Leuenberger M, Schwander J, Johnsen S (1999) 16°C rapid temperature variation in central Greenland 70,000 years ago. Science 286:934-937

Lehmann BE, Davis S, Fabryka-Martin J (1993) Atmospheric and subsurface sources of stable and radioactive nuclides used for ground-water dating. Water Resour Res 29:2027-2040

Lehmann BE, Loosli HH (1991) Isotopes formed by underground production. *In* Applied Isotope Hydrogeology, a Case Study in Northern Switzerland. Pearson FJ, Balderer W, Loosli HH, Lehmann BE, Matter A, Peters T, Schmassmann H, Gautschi A (eds) Elsevier, Amsterdam, p 239-296

Lehmann BE, Loosli HH, Purtschert R, Andrews JN (1996) A comparison of chloride and helium concentrations in deep ground waters. *In* Isotopes in water resources management (IAEA-SM-336). IAEA, Vienna, p 3-17

Lehmann BE, Loosli HH, Rauber D, Thonnard N, Willis RD (1991) ^{81}Kr and ^{85}Kr in ground water, Milk River aquifer, Alberta, Canada. Appl Geochem 6:419-423

Lehmann BE, Purtschert R (1997) Radioisotope dynamics—the origin and fate of nuclides in ground water. Appl Geochem 12:727-738

Lehmann BE, Purtschert R, Loosli HH, Love A, Sampson L, Collon P, Kutschera W, Beyerle U, Aeschbach-Hertig W, Kipfer R (1999) ^{81}Kr-, ^{36}Cl- and ^{4}He-dating in the Great Artesian Basin, Australia. *In* Isotope techniques in water resources development and management (IAEA-SM-361). IAEA, Vienna, p 98

Leuenberger M, Lang C, Schwander J (1999) Delta^{15}N measurements as a calibration tool for the paleothermometer and gas-ice age differences: A case study for the 8200 B.P. event on GRIP ice. J Geophys Res 104:22163-22170

Loosli HH (1983) A dating method with ^{39}Ar. Earth Planet Sci Lett 63:51-62

Loosli HH (1992) Applications of ^{37}Ar, ^{39}Ar and ^{85}Kr in hydrology, oceanography and atmospheric studies. *In* Isotopes of noble gases as tracers in environmental studies. IAEA, Vienna, p 73-85

Loosli HH, Lehmann BE, Balderer W (1989) Argon-39, argon-37 and krypton-85 isotopes in Stripa ground waters. Geochim Cosmochim Acta 53:1825-1829

Loosli HH, Lehmann BE, Smethie WM (2000) Noble gas radioisotopes: ^{37}Ar, ^{85}Kr, ^{39}Ar, ^{81}Kr. *In* Environmental tracers in subsurface hydrology. Cook P, Herczeg AL (eds) Kluwer Academic Publishers, Boston, p 379-396

Loosli HH, Lehmann BE, Thalmann C, Andrews JN, Florkowski T (1992) Argon-37 and argon-39: Measured concentrations in ground water compared with calculated concentrations in rock. *In* Isotope techniques in water resources development (IAEA-SM-319). IAEA, Vienna, p 189-201

Lovley DR, Woodward JC (1992) Consumption of freons CFC-11 and CFC-12 by anaerobic sediments and soils. Environ Sci Technol 26:925-929

Lucas LL, Unterweger MP (2000) Comprehensive review and critical evaluation of the half-life of Tritium. J Res Natl Inst Stand Technol 105:541-549

Ludin A, Wepperning R, Bönisch G, Schlosser P (1997) Mass spectrometric measurement of helium isotopes and tritium in water samples. Techn Report Lamont-Doherty Earth Observatory, Palisades, New York

Ludin AI, Lehmann BE (1995) High-resolution diode-laser spectroscopy on a fast beam of metastable atoms for detecting very rare krypton isotopes. Appl Phys B61:461-465

Maloszewski P, Zuber A (1982) Determining the turnover time of ground water systems with the aid of environmental tracers, I. Models and their applicability. J Hydrol 57:207-231

Mamyrin BA, Tolstikhin IN (1984) Helium isotopes in nature. Elsevier, Amsterdam, Oxford, New York, Tokyo
Manning AH, Sheldon AL, Solomon DK (2000) A new method of noble gas sampling that improves excess air determinations. EOS Trans Am Geophys Union 81:F449
Marine IW (1979) The use of naturally occurring helium to estimate ground water velocities for studies of geologic storage of radioactive waste. Water Resour Res 15:1130-1136
Martel DJ, Deak J, Horvath F, O'Nions RK, Oxburgh ER, Stegena L, Stute M (1989) Leakage of Helium from the Pannonian Basin. Nature 342:908-912
Matthews A, Fouillac C, Hill R, O'Nions RK, Oxburgh ER (1987) Mantle-derived volatiles in continental crust: the Massif Central of France. Earth Planet Sci Lett 85:117-128
Mattle N (1999) Interpretation von Tracermessungen mittels Boxmodellen und numerischen Strömungs-/Transportmodellen. PhD dissertation, Universität Bern, Bern
Mattle N, Kinzelbach W, Beyerle U, Huggenberger P, Loosli HH (2001) Exploring an aquifer system by integrating hydraulic, hydrogeologic and environmental tracer data in a three-dimensional hydrodynamic transport model. J Hydrol 242:183-196
Mazor E (1972) Paleotemperatures and other hydrological parameters deduced from gases dissolved in ground waters, Jordan Rift Valley, Israel. Geochem Cosmochim Acta 36:1321-1336
Mazor E, Bosch A (1992) Helium as a semi-quantitative tool for ground water dating in the range of 10^4-10^8 years. *In* Isotopes of noble gases as tracers in environmental studies. IAEA, Vienna, p 163-178
Modica E, Burton HT, Plummer LN (1998) Evaluating the source and residence times of ground water seepage to streams, New Jersey Coastal Plain. Water Resour Res 34:2797-2810
Münnich KO, Roether W, Thilo L (1967) Dating of ground water with tritium and ^{14}C. *In* Isotopes in Hydrology (IAEA-SM-83). IAEA, Vienna, p 305-320
O'Nions RK, Oxburgh ER (1983) Heat and helium in the earth. Nature 306:429-431
O'Nions RK, Oxburgh ER (1988) Helium, volatile fluxes and the development of continental crust. Earth Planet Sci Lett 90:331-347
Oana S (1957) Bestimmung von Argon in besonderem Hinblick auf gelöste Gase in natürlichen Gewässern. J Earth Sci Nagoya Univ 5:103-105
Osenbrück K (1991) Laborversuche zur Bildung des Luftüberschusses im Grundwasser. Diploma thesis, Universität Heidelberg, Heidelberg
Osenbrück K, Lippmann J, Sonntag C (1998) Dating very old pore waters in impermeable rocks by noble gas isotopes. Geochim Cosmochim Acta 62:3041-3045
Oster H, Sonntag C, Münnich KO (1996) Ground water age dating with chlorofluorocarbons. Water Resour Res 32:2989-3001
Oxburgh ER, O'Nions RK (1987) Helium loss, tectonics, and the terrestrial heat budget. Science 237: 1583-1588
Ozima M, Podosek FA (1983) Noble gas geochemistry. Cambridge Univ. Press, Cambridge, London, New York
Peeters F, Beyerle U, Aeschbach-Hertig W, Holocher J, Brennwald MS, Kipfer R (2002) Improving noble gas based paleoclimate reconstruction and ground water dating using $^{20}Ne/^{22}Ne$ ratios. Geochim Cosmochim Acta in press
Peeters F, Kipfer R, Achermann D, Hofer M, Aeschbach-Hertig W, Beyerle U, Imboden DM, Rozanski K, Fröhlich K (2000a) Analysis of deep-water exchange in the Caspian Sea based on environmental tracers. Deep-Sea Res I 47:621-654
Peeters F, Kipfer R, Hofer M, Imboden DM, Domysheva VM (2000b) Vertical turbulent diffusion and upwelling in Lake Baikal estimated by inverse modeling of transient tracers. J Geophys Res 105:14283 and 3451-3464
Peeters F, Kipfer R, Hohmann R, Hofer M, Imboden DM, Kodenev GG, Khozder T (1997) Modelling transport rates in Lake Baikal: gas exchange and deep water renewal. Environ Sci Technol 31: 2973-2982
Peeters F, Piepke G, Kipfer R, Hohmann R, Imboden DM (1996a) Description of stability and neutrally buoyant transport in freshwater lakes. Limnol Oceanogr 41:1711-1724
Peeters F, Piepke G, Kipfer R, Hohmann R, Imboder DM (1996b) Neutrally buoyant transport in deep freshwater systems: the case of Lake Baikal (Siberia). Am Soc Limnology Oceanography 73
Pinot S, Ramstein G, Harrison SP, Prentice IC, Guiot J, Stute M, Joussaume S (1999) Tropical paleoclimates at the Last Glacial Maximum: comparison of Paleoclimate Modeling Intercomparison Project (PMIP) simulations and paleodata. Clim Dyn 15:857-874
Pinti DL, Van Drom E (1998) PALEOTEMP: a MATHEMATICA program for evaluating paleotemperatures from the concentration of atmosphere-derived noble gases in ground water. Computers & Geosci 24:33-41

Plummer LN, Busenberg E (2000) Chlorofluorocarbons. *In* Environmental tracers in subsurface hydrology. Cook P, Herczeg AL (eds) Kluwer Academic Publishers, Boston, p 441-478

Plummer LN, Busenberg E, Böhlke JK, Nelms DL, Michel RL, Schlosser P (2001) Ground water residence times in Shenandoah National Park, Blue Ridge Mountains, Virginia, USA: a multi-tracer approach. Chem Geol 179:93-111

Plummer LN, Busenberg E, Drenkard S, Schlosser P, Ekwurzel B, Weppernig R, McConnell JB, Michel RL (1998a) Flow of river water into a karstic limestone aquifer—2. Dating the young fraction in ground water mixtures in the Upper Floridan aquifer near Valdosta, Georgia. Appl Geochem 13: 1017-1043

Plummer LN, Busenberg E, McConnell JB, Drenkard S, Schlosser P, Michel RL (1998b) Flow of river water into a Karstic limestone aquifer. 1. Tracing the young fraction in ground water mixtures in the Upper Floridan Aquifer near Valdosta, Georgia. Appl Geochem 13:995-1015

Plummer LN, Rupert MG, Busenberg E, Schlosser P (2000) Age of irrigation water in ground water from the eastern Snake River Plain Aquifer, South-Central Idaho. Ground Water 38:264-283

Poreda RJ, Cerling TE, Solomon DK (1988) Tritium and helium isotopes as hydrologic tracers in a shallow unconfined aquifer. J Hydrol 103:1-9

Purtschert R (1997) Multitracer-Studien in der Hydrologie, Anwendungen im Glattal, am Wellenberg und in Vals. PhD dissertation, Universität Bern, Bern

Purtschert R, Beyerle U, Aeschbach-Hertig W, Kipfer R, Loosli HH (2001) Palaeowaters from the Glatt Valley, Switzerland. *In* Palaeowaters in Coastal Europe: evolution of ground water since the late Pleistocene. Vol 189. Edmunds WM, Milne CJ (eds) Geological Soc London, London, p 155-162

Purtschert R, Loosli HH, Beyerle U, Aeschbach-Hertig W, Imboden D, Kipfer R, Wieler R (1999) Dating of young water components by combined application of $^{3}H/^{3}He$ and ^{85}Kr measurements. *In* Isotope techniques in water resources development and management (IAEA-SM-361). IAEA, Vienna, p 59-60

Reilly TE, Plummer LN, Phillips PJ, Busenberg E (1994) The use of simulation and multiple environmental tracers to quantify ground water flow in a shallow aquifer. Water Resour Res 30: 421-433

Richter J, Szymczak P, Abraham T, Jordan H (1993) Use of combinations of lumped parameter models to interpret ground water isotopic data. J Contam Hydrol 14:1-13

Raynaud D, Jouzel J, Barnola JM, Chappellaz J, Delmas RJ, Lorius C (1993) The ice record of greenhouse gases. Science 259:926-934

Rozanski K, Florkowski T (1979) Krypton-85 dating of ground water. *In* Isotope Hydrology 1978 (IAEA-SM-228/2). IAEA, Vienna, p 949-961

Rudolph J (1981) Edelgastemperaturen und Heliumalter ^{14}C-datierter Paläowässer. PhD dissertation, Universität Heidelberg, Heidelberg

Rudolph J, Rath HK, Sonntag C (1984) Noble gases and stable isotopes in ^{14}C-dated palaeowaters from central Europa and the Sahara. *In* Isotope Hydrology 1983 (IAEA-SM-270). IAEA, Vienna, p 467-477

Sano Y, Kusakabe M, Hirabayashi J, Nojiri Y, Shinohara H, Njine T, Tanyileke G (1990) Helium and carbon fluxes in Lake Nyos, Cameroon: Constraint on next gas burst. Earth Planet Sci Lett 99:303-314

Sano Y, Wakita H (1987) Helium isotope evidence for magmatic gases in Lake Nyos, Cameroon. Geophys Res Lett 14:1039-1041

Schlosser P (1992) Tritium/^{3}He dating of waters in natural systems. *In* Isotopes of noble gases as tracers in environmental studies. IAEA, Vienna, p 123-145

Schlosser P, Shapiro SD, Stute M, Aeschbach-Hertig W (1998) Tritium/^{3}He measurements in young ground water: Chronologies for environmental records. *In* Isotope techniques in the study of environmental change (IAEA-SM-349). IAEA, Vienna, p 165-189

Schlosser P, Stute M, Dörr C, Sonntag C, Münnich KO (1988) Tritium/^{3}He-dating of shallow ground water. Earth Planet Sci Lett 89:353-362

Schlosser P, Stute M, Sonntag C, Münnich KO (1989) Tritiogenic ^{3}He in shallow ground water. Earth Planet Sci Lett 94:245-256

Schlosser P, Winckler G (2002) Noble gases in the ocean and ocean floor. Rev Mineral Geochem 47: 701-730

Schwarzenbach RP, Gschwend PM, Imboden DM (1993) Environmental organic chemistry. John Wiley & Sons, New York

Severinghaus JP, Bender ML, Keeling RF, Broecker WS (1996) Fractionation of soil gases by diffusion of water vapor, gravitational settling, and thermal diffusion. Geochim Cosmochim Acta 60:1005-1018

Severinghaus JP, Brook EJ (1999) Abrupt climate change at the end of the last glacial period inferred from trapped air in polar ice. Science 286:930-934

Severinghaus JP, Grachev A, Battle M (2001) Thermal fractionation of air in polar firn by seasonal temperature gradients. Geochem Geophys Geosyst 2:2000GC000146

Severinghaus JP, Luz B, Caillon N (2002) A method for precise measurement of argon 40/36 and krypton/argon ratios in trapped air in polar ice with applications to past firn thickness and abrupt climate change. Geochim Cosmochim Acta (in press)

Severinghaus JP, Sowers T, Brook EJ, Alley RB, Bender ML (1998) Timing of abrupt climate change at the end of the Younger Dryas interval from thermally fractionated gases in polar ice. Nature 391: 141-146

Sheets RA, Bair ES, Rowe GL (1998) Use of ^{3}H/^{3}He ages to evaluate and improve ground water flow models in a complex buried-valley aquifer. Water Resour Res 34:1077-1089

Shimaraev MN, Granin NG, Zhadanov AA (1993) Deep ventilation of Lake Baikal due to spring thermal bars. Limnol Oceanogr 38:1068-1072

Smethie WMJ, Solomon DK, Schiff SL, Mathieu G (1992) Tracing ground water flow in the Borden aquifer using krypton-85. J Hydrol 130:279-297

Smith GD, Newhall F, Robinson LH, Swanson D (1964) Soil temperature regimes: Their characteristics and predictability. U S Dept Agricul , Soil Conservation Service Report SCS-TP-144

Smith SP, Kennedy BM (1983) Solubility of noble gases in water and in NaCl brine. Geochim Cosmochim Acta 47:503-515

Solomon DK (2000) ^{4}He in ground water. *In* Environmental tracers in subsurface hydrology. Cook P, Herczeg AL (eds) Kluwer Academic Publishers, Boston, p 425-439

Solomon DK, Cook PG (2000) ^{3}H and ^{3}He. *In* Environmental tracers in subsurface hydrology. Cook P, Herczeg AL (eds) Kluwer Academic Publishers, Boston, p 397-424

Solomon DK, Hunt A, Poreda RJ (1996) Source of radiogenic helium 4 in shallow aquifers: Implications for dating young ground water. Water Resour Res 32:1805-1813

Solomon DK, Poreda RJ, Cook PG, Hunt A (1995) Site characterization using ^{3}H/^{3}He ground-water ages, Cape Cod, MA. Ground Water 33:988-996

Solomon DK, Poreda RJ, Schiff SL, Cherry JA (1992) Tritium and helium 3 as ground water age tracers in the Borden aquifer. Water Resour Res 28:741-755

Solomon DK, Schiff SL, Poreda RJ, Clarke WB (1993) A validation of the ^{3}H/^{3}He-method for determining ground water recharge. Water Resour Res 29:2591-2962

Solomon DK, Sudicky EA (1991) Tritium and helium 3 isotope ratios for direct estimate of spatial variations in ground water recharge. Water Resour Res 27:2309-2319

Staudacher T, Sarda P. Richardson SH, Allègre CJ, Sagna I, Dmitriev LV (1989) Noble gases in basalt glasses from a Mid-Atlantic Ridge topographic high at 14°N: geodynamic consequences. Earth Planet Sci Lett 96:119-133

Stonestrom DA, Rubin J (1989) Water content dependence of trapped air in two soils. Water Resour Res 25:1947-1958

Stute M (1989) Edelgase im Grundwasser—Bestimmung von Paläotemperaturen und Untersuchung der Dynamik von Grundwasserfliesssystemen. PhD dissertation, Universität Heidelberg, Heidelberg

Stute M, Clark JF, Schlosser P, Broecker WS (1995a) A 30'000 yr continental paleotemperature record derived from noble gases dissolved in ground water from the San Juan Basin, New Mexico. Quatern Res 43:209-220

Stute M, Deák J (1989) Environmental isotope study (^{14}C, ^{13}C, ^{18}O, D, noble gases) on deep ground water circulation systems in Hungary with reference to paleoclimate. Radiocarbon 31:902-918

Stute M, Deák J, Révész K, Böhlke JK, Deseö É, Weppernig R, Schlosser P (1997) Tritium/^{3}He dating of river infiltration: An example from the Danube in the Szigetkös area, Hungary. Ground Water 35: 905-911

Stute M, Forster M, Frischkorn H, Serejo A, Clark JF, Schlosser P, Broecker WS, Bonani G (1995b) Cooling of tropical Brazil (5°C) during the Last Glacial Maximum. Science 269:379-383

Stute M, Schlosser P (1993) Principles and applications of the noble gas paleothermometer. *In* Climate Change in Continental Isotopic Records. Vol 78. Swart PK, Lohmann KC, McKenzie J, Savin S (eds) American Geophysical Union, Washington, DC, p 89-100

Stute M, Schlosser P (2000) Atmospheric noble gases. *In* Environmental tracers in subsurface hydrology. Cook P, Herczeg AL (eds) Kluwer Academic Publishers, Boston, p 349-377

Stute M, Schlosser P, Clark JF, Broecker WS (1992a) Paleotemperatures in the Southwestern United States derived from noble gases in ground water. Science 256:1000-1003

Stute M, Sonntag C (1992) Paleotemperatures derived from noble gases dissolved in ground water and in relation to soil temperature. *In* Isotopes of noble gases as tracers in environmental studies. IAEA, Vienna, p 111-122

Stute M, Sonntag C, Déak J, Schlosser P (1992b) Helium in deep circulating ground water in the Great Hungarian PlaIn Flow dynamics and crustal and mantle helium fluxes. Geochim Cosmochim Acta 56:2051-2067

Stute M, Talma AS (1998) Glacial temperatures and moisture transport regimes reconstructed from noble gases and delta ^{18}O, Stampriet aquifer, Namibia. *In* Isotope techniques in the study of environmental change (IAEA-SM-349). IAEA, Vienna, p 307-318

Suckow A, Sonntag C (1993) The influence of salt on the noble gas thermometer. *In* Isotope techniques in the study of past and current environmental changes in the hydrosphere and the atmosphere (IAEA-SM-329). IAEA, Vienna, p 307-318

Sugisaki R (1961) Measurement of effective flow velocity of ground water by means of dissolved gases. Am J Sci 259:144-153

Szabo Z, Rice DE, Plummer LN, Busenberg E, Drenkard S, Schlosser P (1996) Age dating of shallow ground water with chlorofluorocarbons, tritium/helium 3, and flow path analysis, southern New Jersey coastal plain. Water Resour Res 32:1023-1038

Takaoka N, Mizutani Y (1987) Tritiogenic ^{3}He in Ground water in Takaoka. Earth Planet Sci Letters 85:74-78

Talma AS, Vogel JC (1992) Late Quaternary paleotemperatures derived from a speleothem from Cango Caves, Cape Province, South Africa. Quat Res 37:203-213

Thompson GM, Hayes JM (1979) Trichloroflouromethan in ground water—a possible tracer and indicator of ground water age. Water Resour Res 15:546-554

Thonnard N, McKay LD, Cumbie DH, Joyner CF (1997) Status of laser-based krypton-85 analysis development for dating of young ground water. Geol Soc Am Abstr 29(6):A-78

Tolstikhin I, Lehmann BE, Loosli HH, Gautschi A (1996) Helium and argon isotopes in rocks, minerals, and related ground waters: A case study in northern Switzerland. Geochim Cosmochim Acta 60: 1497-1514

Tolstikhin IN, Kamenskiy IL (1969) Determination of ground-water ages by the T-^{3}He Method. Geochem Intl 6:810-811

Top Z, Clarke WB (1981) Dissolved helium isotopes and tritium in lakes: Further results for uranium prospecting in Central Labrador. Econ Geol 76:2018-2031

Top Z, Eismont WC, Clarke WB (1987) Helium isotope effect and solubility of helium and neon in distilled water and seawater. Deep-Sea Res 34:1139-1148

Torgersen T (1980) Controls on pore-fluid concentration of ^{4}He and ^{222}Rn and the calculation of ^{4}He/^{222}Rn ages. J Geochem Explor 13:57-75

Torgersen T (1989) Terrestrial degassing fluxes and the atmospheric helium budget: Implications with respect to the degassing processes of continental crust. Chem Geol 79:1-14

Torgersen T, Clarke WB (1985) Helium accumulation in ground water, I: An evaluation of sources and the continental flux of crustal ^{4}He in the Great Artesian Basin, Australia. Geochim Cosmochim Acta 49:1211-1218

Torgersen T, Clarke WB, Jenkins WJ (1979) The tritium/helium-3 method in hydrology. *In* Isotope Hydrology 1978 (IAEA-SM-228/2). IAEA, Vienna, p 917-930

Torgersen T, Drenkard S, Farley K, Schlosser P, Shapiro A (1994) Mantle helium in the ground water of the Mirror Lake Basin, New Hampshire, U.S.A. *In* Noble Gas Geochemistry and Cosmochemistry. Matsuda J (ed) Terra Scientific Publishing Company, Tokyo, p 279-292

Torgersen T, Drenkard S, Stute M, Schlosser P, Shapiro A (1995) Mantle helium in ground waters of eastern North America: Time and space constraints on sources. Geology 23:675-678

Torgersen T, Hammond DE, Clarke WB, Peng T-H (1981) Fayetteville, Green Lake, New York: ^{3}H-^{3}He water mass ages and secondary chemical structure. Limnol Oceanogr 26:110-122

Torgersen T, Ivey GN (1985) Helium accumulation in ground water, II: A model for the accumulation of the crustal ^{4}He degassing flux. Geochim Cosmochim Acta 49:2445-2452

Torgersen T, Kennedy BM, Hiyagon H, Chiou KY, Reynolds JH, Clark WB (1989) Argon accumulation and the crustal degassing flux of ^{40}Ar in the Great Artesian Basin, Australia. Earth Planet. Sci Lett 92:43-56

Torgersen T, Top Z, Clarke WB, Jenkins WJ, Broecker WS (1977) A new method for physical limnology—tritium-helium-3 ages: Results for Lakes Erie, Huron and Ontario. Limnol Oceanogr 22:181-193

Vollmer MK, Weiss RF, Schlosser P, Williams RT (2002) Deep-water renewal in Lake Issyk-Kul. Geophys Res Lett 29:124/1-124/4

Vogel JC (1967) Investigation of ground-water flow with radiocarbon. *In* Isotopes in Hydrology (IAEA-SM-83). IAEA, Vienna, p 355-369

Vogel JC, Talma AS, Heaton THE (1981) Gaseous nitrogen as evidence for denitrification in ground water. J Hydrol 50:191-200

Weise S, Eichinger L, Forster M, Salvamoser J (1992a) Helium-3 and krypton-85 dating of shallow ground waters: diffusive loss and correlated problems. *In* Isotopes of noble gases as tracers in environmental studies. IAEA, Vienna, p 147-162

Weise S, Moser H (1987) Ground water dating with helium isotopes. *In* Isotope techniques in water resources development. IAEA, Vienna, p 105-126

Weise SM, Faber P, Stute M (1992b) Neon-21—a possible tool for dating very old ground waters? *In* Isotope techniques in water resources development (IAEA-SM-319). IAEA, Vienna, p 179-188

Weiss RF (1970) The solubility of nitrogen, oxygen and argon in water and seawater. Deep-Sea Res 17:721-735

Weiss RF (1971) Solubility of helium and neon in water and seawater. J Chem Eng Data 16:235-241

Weiss RF, Carmack EC, Koropalov VM (1991) Deep-water renewal and biological production in Lake Baikal. Nature 349:665-669

Weiss RF, Kyser TK (1978) Solubility of krypton in water and seawater. J Chem Eng Data 23:69-72

Weiss W, Sartorius H, Stockburger H (1992) Global distribution of atmospheric ^{85}Kr: A database for the verification of transport and mixing models. *In* Isotopes of noble gases as tracers in environmental studies. IAEA, Vienna, p 29-62

Weiss W, Zapf T, Baitter M, Kromer B, Fischer KH, Schlosser P, Roether W, Münnich KO (1984) Subsurface horizontal water transport and vertical mixing in Lake Constance traced by radon-222, tritium, and other physical and chemical tracers. *In* Isotope Hydrology 1983 (IAEA-SM-270). IAEA, Vienna, p 43-54

Weyhenmeyer CE, Burns SJ, Waber HN, Aeschbach-Hertig W, Kipfer R, Loosli HH, Matter A (2000) Cool glacial temperatures and changes in moisture source recorded in Oman ground waters. Science 287:842-845

Wilson GB, McNeill GW (1997) Noble gas recharge temperatures and the excess air component. Appl Geochem 12:747-762

Wüest A, Aeschbach-Hertig W, Baur H, Hofer M, Kipfer R, Schurter M (1992) Density structure and tritium-helium age of deep hypolimnetic water in the northern basin of Lake Lugano. Aquat Sci 54:205-218

Zenger A, Ilmberger J, Heinz G, Schimmele M, Schlosser P, Imboden D, Münnich KO (1990) Behavior of a medium-sized basin connected to a large lake. *In* Large Lakes: Ecological Structure and Function. Tilzer MM, Serruya C (eds) Springer-Verlag, Berlin, Heidelberg, p 133-155

Zoellmann K, Kinzelbach W, Fulda C (2001) Environmental tracer transport (^{3}H and SF_6) in the saturated and unsaturated zones and its use in nitrate pollution management. J Hydrol 240:187-205

Zuber A (1986a) Mathematical models for the interpretation of environmental radioisotopes in ground water systems. *In* Handbook of Environmental Geochemistry. Vol 2. Fritz P, Fontes JC (eds) Elsevier, Amsterdam, p 1-59

Zuber A (1986b) On the interpretation of tracer data in variable flow systems. J Hydrol 86:45-57

15 Noble Gases in Ocean Waters and Sediments

Peter Schlosser[1,2] and Gisela Winckler[1]
Lamont-Doherty Earth Observatory
Columbia University
Palisades, New York 10964

[2] *Department of Earth and Environmental Sciences*
and *Department of Earth and Environmental Engineering*
Columbia University
New York, New York 10027
peters@ldeo.columbia.edu

INTRODUCTION

Noble gases are widely used in studies of the basic properties and dynamics of natural systems including the ocean. This chapter describes some of the more extensive applications of noble gases (mainly helium isotopes) to studies of oceanographic problems. They include the modern oceanic circulation, paleo-oceanography, hydrothermal and cold brine systems in the deep ocean, and ocean/atmosphere gas exchange.

Originally, oceanic noble gas studies were focused on geochemical problems such as the question of the existence of excess helium in seawater derived from radioactive decay of the elements of the uranium and thorium series in ocean sediments (Suess and Wänke 1965; Bieri et al. 1966). After the discovery of mantle (primordial) helium during the 1960s (Clarke et al. 1969; Craig and Weiss 1971; Craig et al. 1975) and ^{3}He derived from decay of tritium in the near-surface waters of the ocean (Jenkins and Clarke 1976), large programs were developed to exploit helium isotopes, frequently in combination with tritium, for studies of water mass formation, circulation and variability (e.g., Jenkins 1987; Schlosser et al. 1991).

More recent studies started to explore the use of helium isotopes to paleoceanographic objectives. These studies build upon the discovery of extraterrestrial helium in deep-sea sediments in the early 1960s (Merrihue 1964). This signal is now being systematically examined for its potential as a tool for investigations of sediment accumulation rates on very long time scales (millions of years) and their correlation with changes in paleoceanography (e.g., Farley 1995; Farley and Patterson 1995; Marcantonio et al. 1995, 1996, 2001).

Soon after the discovery of mantle helium in seawater, noble gases were applied to studies of hydrothermal systems. Originally, these studies were focused on the source of the fluids emanating from hydrothermal vents as well as the related heat flux. More recent studies use the full set of noble gases to infer processes of the formation and dynamics of hydrothermal systems at the sea floor.

Field studies of ocean/atmosphere gas exchange originally were concentrated around naturally existing isotopes such as ^{222}Rn and ^{14}C (Roether and Kromer 1978; Broecker and Peng 1982; Smethie et al. 1985). Later studies included so-called 'bomb' ^{14}C. However, since gas exchange rates derived from these methods had considerable uncertainties, new methodologies were explored. One of them includes the injection of two gas tracers with similar solubility and strong differences in their diffusion constants (e.g., Watson et al. 1991; Wanninkhof et al. 1993, Nightingale et al. 2000a,b). The most promising dual tracer pair is the combination of ^{3}He and SF_6.

The following paragraphs outline the basic methodologies behind the use of helium

1529-6466/00/0047-0015$05.00

isotopes and the heavier noble gases in the areas described above. Instead of presenting a comprehensive review of all studies that were performed in the individual fields, we focus on the basic principles and use exemplary studies to demonstrate applications and results.

TRACING OCEAN CIRCULATION USING ^{3}HE AND TRITIUM

^{3}He can be used in studies of the oceanic circulation in two ways: (1) through the simultaneous measurement of ^{3}He and its radioactive parent isotope tritium, and (2) by mapping the plumes of mantle helium injected into the intermediate depth waters of the world oceans.

Tritium/^{3}He method

Tritium geochemistry. Tritium is the radioactive isotope of hydrogen that decays by β-decay to the noble gas ^{3}He ($t_{1/2}$ = 12.43 yr; Unterweger et al. 1980; note that Kipfer et al. use a slightly different half-life in the previous chapter). Natural tritium is produced by interaction of cosmic rays with nitrogen and oxygen mainly in the upper atmosphere. After oxidation to HTO, tritium takes part in the hydrological cycle. The potential of tritium for applications to studies of the natural water cycle has been realized shortly after its detection in the environment (von Faltings and Harteck 1950; Grosse et al. 1951; Kaufman and Libby 1954; Begemann and Libby 1957; Giletti et al. 1958).

Natural tritium concentrations in ocean surface waters and in young groundwater are of the order of 1 tritium atom per 10^{18} hydrogen atoms. This is the reason that tritium concentrations are reported as TU (Tritium Units). One TU stands for a tritium to hydrogen ratio [^{3}H]/[H] of 10^{-18}. The activity of a water sample with a tritium concentration of 1 TU is equivalent to 3.2 pCi or 0.12 Bq per liter of H_2O. The production rate of natural tritium is about 0.5±0.3 ^{3}H atoms cm^{-2} s^{-1} (Craig and Lal 1961) leading to natural tritium values in ocean surface waters of about 0.2 TU (Dreisigacker and Roether 1978).

Before natural tritium could be fully exploited for studies of natural water systems, tritium from anthropogenic sources (mainly nuclear weapon tests) was added to the atmosphere in considerable amounts. By the mid 1960s the natural background of tritium in precipitation was practically masked by so-called bomb tritium (e.g., Weiss et al. 1979; Fig. 1). For the past 4 to 5 decades, bomb tritium severely limited the use of natural tritium as a tracer because only few uncontaminated tritium data are available from the pre-bomb era. However, bomb tritium offered a new tool for studies of water movement in natural system. It is equivalent to a 'dye' that was introduced into the environment on a global scale at a relatively well-known rate. Most of the bomb tritium was added to the environment in three pulses during 1954, 1958-1959 and, predominantly, 1963.

During the following three decades the application of bomb tritium has been developed to a routine tool in studies of natural water systems (e.g., Münnich and Roether 1967; Roether et al. 1970; Atakan et al. 1974; Östlund 1982; Broecker et al. 1986).

The surface waters of the ocean are the largest sink for tritium. Transfer into the surface ocean occurs by water vapor exchange, precipitation and continental run-off (e.g., Weiss et al. 1979; Weiss and Roether 1980). Most of the nuclear weapons tests were performed in the northern hemisphere, leading to a strong asymmetry in the global north-south distribution of bomb tritium in the ocean surface waters. Concentrations in the northern hemisphere are relatively high compared to those in the southern hemisphere (Weiss and Roether 1980; Fig. 1).

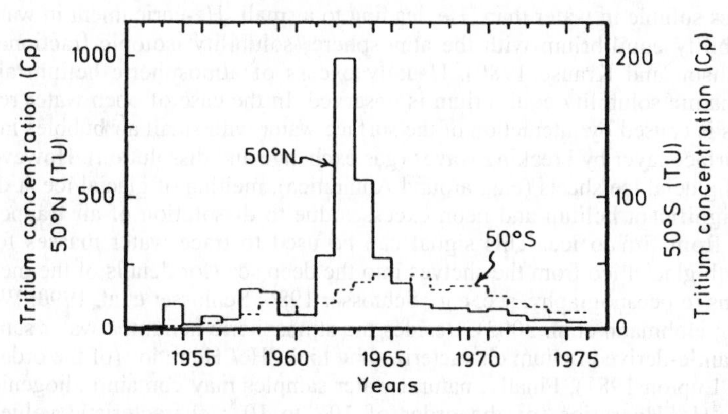

Figure 1. Tritium concentration in marine precipitation at 50°S and 50°N as a function of time (after Weiss and Roether 1980). Note the different scales for the tritium concentrations in the Northern and Southern hemisphere.

Presently, maximum tritium concentrations in the surface waters of the oceans are approximately 1.5 to 2 TU in the northern hemisphere and 0.20 to 0.75 TU in the southern hemisphere. Regions that receive high concentrations of runoff such as the Arctic Ocean or locations surrounded by continents such the Mediterranean and Red seas exhibit elevated tritium concentrations.

Tritiogenic ^{3}He and tritium/^{3}He age. Limitations in using tritium alone as oceanographic tracer arise from the shape of its 'input function', i.e., the evolution of the transfer rate of tritium to the surface waters of the ocean (Fig. 1). The gradient in the surface waters decreased significantly during the 1970s and 1980s, and presently it is too small to be of much use for determination of relative age fields. Additionally, it is difficult to detect the bomb peak in the interior of the ocean, typically preventing the use of tritium as an absolute age marker.

Simultaneous measurement of tritium and its radioactive decay product, ^{3}He, helps to at least partially eliminate these problems. Combined measurement of tritium and ^{3}He allows us to calculate the tritium/^{3}He age. This tracer age is a measure for the time that elapsed since a water parcel has been isolated from exchange with the atmosphere. During this process tritium is taken up and ^{3}He is lost to the atmosphere by gas exchange, i.e., the 'tritium/^{3}He 'clock' is set to zero. Jenkins and Clarke (1976) introduced this method to oceanography. The tritium/^{3}He age is calculated as follows (see also Kipfer et al. 2002, this volume):

$$\tau = \frac{T_{1/2}}{\ln 2} \cdot \ln\left(1 + \frac{[^3He_{tri}]}{[^3H]}\right) \tag{1}$$

where $T_{1/2}$ is the half-life of tritium, $[^3He_{trit}]$ the tritiogenic ^{3}He concentration of the water sample, and $[^3H]$ the tritium concentration of the water at the time of sample collection. Solution of this equation requires the calculation of the tritiogenic ^{3}He concentration, i.e., the fraction of the measured ^{3}He concentration that has been added to the water by radioactive decay of tritium.

Helium observed in natural water samples has several sources that may be distinguished by their $^3He/^4He$ ratio. Generally, atmospheric helium with a $^3He/^4He$ ratio of 1.384×10^{-6} (Clarke et al. 1976) is the major component present in natural waters. Its solubility is a function of the temperature and salinity of the water (Weiss 1971). ^{3}He is slightly less soluble in water than ^{4}He, leading to a small ^{4}He-enrichment in water which is in solubility equilibrium with the atmosphere (solubility isotopic fractionation $\alpha \approx$ 0.983, Benson and Krause 1980). Usually excess of atmospheric helium above the thermodynamic solubility equilibrium is observed. In the case of open water reservoirs, this excess is caused by interaction of the surface water with small air bubbles introduced into the surface layer by breaking waves (gas exchange and dissolution). However, in the vicinity of glacial ice sheets (e.g., around Antarctica), melting of glacial ice at depth can produce significant helium and neon excesses due to dissolution of air trapped during transition from firn to ice. This signal can be used to trace water masses formed in contact with glacial ice from the shelves into the deep sea (for details of the method and applications to oceanography, see, e.g., Schlosser 1986; Schlosser et al. 1990; Weppernig et al. 1996; Hohmann et al. 2002). Besides the atmospheric helium, a water sample may contain mantle-derived helium characterized by high ^{3}He/^{4}He ratios (of the order of 10^{-5}, Craig and Lupton 1981). Finally, natural water samples may contain radiogenic helium with low ^{3}He/^{4}He ratios (of the order of 10^{-7} to 10^{-8}; characteristic value: 2×10^{-8}, Mamyrin and Tolstikhin 1984). Radiogenic helium is produced by the α-decay of the natural radioactive decay series and contains a ^{3}He component originating mainly from the ^{6}Li(n,α)^{3}H reaction. The ^{3}He balance, including tritiogenic ^{3}He, is given in Equation (2):

$$^3\mathrm{He_m} = {}^3\mathrm{He_{eq}} + {}^3\mathrm{He_{ex}} + {}^3\mathrm{He_{ter}} + {}^3\mathrm{He_{tri}} \quad (2)$$

with: $^3\mathrm{He_m}$ = measured ^{3}He concentration of the water sample, $^3\mathrm{He_{tri}}$ = tritiogenic ^{3}He, $^3\mathrm{He_{eq}}$ = ^{3}He concentration in solubility equilibrium with the atmosphere, $^3\mathrm{He_{ex}}$ = ^{3}He originating from excess air, and $^3\mathrm{He_{ter}}$ = terrigenic ^{3}He (crustal and mantle helium).

To separate tritiogenic ^{3}He from the total ^{3}He of the water sample (Eqn.3), both ^{4}He and neon are used as indicators of atmospheric and terrigenic helium.

$$^3\mathrm{He_{tri}} = {}^4\mathrm{He_m} \cdot \mathrm{R_m} - ({}^4\mathrm{He_m} - {}^4\mathrm{He_{ter}}) \cdot \mathrm{R_a} + {}^4\mathrm{He_{eq}} \cdot \mathrm{R_a} \cdot (1-\alpha) - {}^4\mathrm{He_{ter}} \cdot \mathrm{R_{ter}} \quad (3)$$

where $\mathrm{R_m}$ and $^4\mathrm{He_m}$ are the measured ^{3}He/^{4}He ratio and ^{4}He concentration of the water sample, respectively, $\mathrm{R_{ter}}$ and $^4\mathrm{He_{ter}}$ are the ^{3}He/^{4}He ratio and ^{4}He concentration of the terrigenic helium component, respectively, and $\mathrm{R_a}$ is the ^{3}He/^{4}He ratio of atmospheric helium. If both mantle helium and radiogenic helium are present in a water sample in addition to atmospheric helium, the individual helium components cannot be separated in a straightforward way. If only mantle helium or radiogenic helium is contained in the water, it is easier to separate the tritiogenic ^{3}He because in this case reasonable estimates for the ^{3}He/^{4}He ratios of the added helium components are available. In the Pacific and Indian oceans, the mixing of mantle helium from intermediate waters with low tritium concentrations into the near-surface waters for which most of the oceanic tritium/^{3}He age dates are calculated, complicates tritium/^{3}He dating. In this case ^{3}He/silica correlations can be used to separate the mantle-derived ^{3}He from the tritiogenic ^{3}He (see Jenkins 1996).

Details on the separation of $^3\mathrm{He_{tri}}$ are given by Schlosser (1992) and Well et al. (2001), among others. In most ocean environments, the contribution of the radiogenic ^{4}He signal is small and typically can be neglected.

^{3}He results of oceanographic water samples are usually reported in the δ notation where δ^3He means the percent deviation of the measured ^{3}He/^{4}He ratio of a water sample ($\mathrm{R_m}$) from that of an air standard ($\mathrm{R_a}$)

$$\delta^3\mathrm{He} = \left(\frac{\mathrm{R_m}}{\mathrm{R_a}} - 1\right) \cdot 100\% \quad (4)$$

In many cases it is more convenient to use the TU notation for ^{3}He concentrations. ^{3}He concentrations in TU are obtained from Equation (5):

$$^3He_{trit} = 4.021 \times 10^{14} \left[\frac{^4He_m \cdot (R_m - R_a) + {^4He_{eq}} \cdot R_a \cdot (1-\alpha)}{1 - \frac{S}{1000}} \right] \quad (5)$$

where S is the salinity of the water sample in psu. If all the tritium of a water sample (S = 0) with a tritium concentration of 1 TU decays, it results in a tritiogenic ^{3}He concentration of 2.49×10^{-15} cm^3 STP. This leads to an increase in δ^3He of about 3.86% based on a salinity of 0 and a temperature of 9°C (^{4}He solubility equilibrium concentration: 4.66×10^{-8} cm^3 g^{-1}; Weiss 1971).

The tritium/^{3}He age does not depend on the initial tritium concentration of the water parcel. This means that the initial tritium concentration of a water parcel that is determined by the tritium delivery does not have to be known for calculating a tracer age. Typically, the delivery rate of tritium to natural water bodies is not known exactly and in certain regions estimates of the delivery rate can be very difficult and inaccurate. The tritium/^{3}He age is an apparent age and can only be taken as the true age of the water if ^{3}He sources other than tritium decay (^{3}He derived from mantle helium or nucleogenic ^{3}He found in radiogenic helium) can be excluded or corrected for, and mixing with water of different tritium and ^{3}He concentrations is negligible or can be quantified. The tritium/^{3}He age is non-linear with respect to mixing (Jenkins and Clarke 1976). For illustration of this problem, two mixing lines are drawn in a plot of the isochrones calculated from Equation (10) (A-B and A′-B′ in Fig. 2). In both cases the tritium/^{3}He age of the mixture is weighted in favor of the water with the higher tritium concentration (Fig. 2). The logarithmic function of the tritium/^{3}He age causes non-linearities even in cases of equal tritium concentrations. If water with no tritium and no ^{3}He is added to a certain water parcel, its tritium/^{3}He age is not affected, since the mixing is along the isochrones.

The age resolution of the tritium/^{3}He method is determined mainly by the measurement precision of the ^{3}He/^{4}He ratio (e.g., Schlosser 1992). On the basis of a

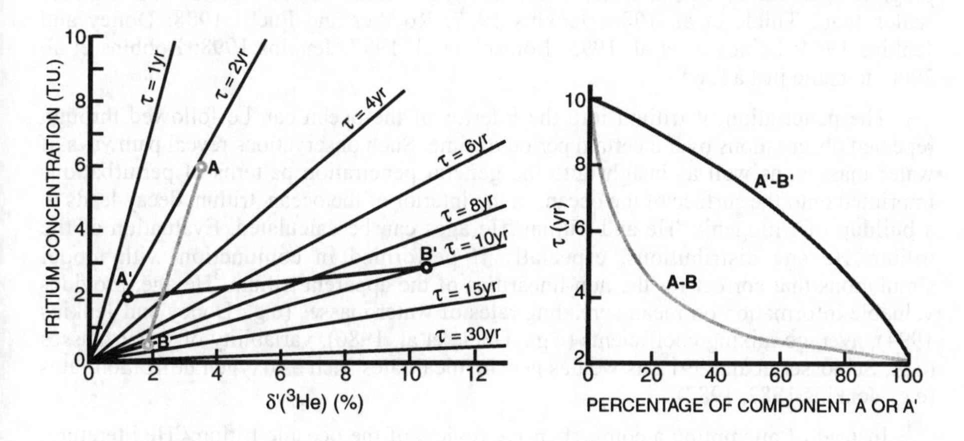

Figure 2. Tritium/^{3}He isochrones in years (left) and non-linearity of the tritium/^{3}He age with respect to mixing along the lines A-B and A'-B' (right), after Jenkins (1974). In both cases, the tritium/^{3}He age of the mixture is weighted in favor of the water mass with the higher tritium concentration.

$^3He/^4He$ precision of ±0.2%, the age resolution is calculated and plotted in Figure 3. For waters with tritium concentrations of about 50 TU (shallow groundwater) an age resolution of about one week can be achieved. For northern hemisphere ocean surface waters with tritium concentrations of about 2 TU the time resolution is of the order of 6 to 12 months, and for Southern Hemisphere surface waters (≈ 0.5 TU) the time resolution is no better than about 3 years if it is possible to separate the tritiogenic ^3He from the total ^3He measured in the water. The age resolution given here is that of the apparent tritium/^3He age and does not take into account systematic errors such as mixing.

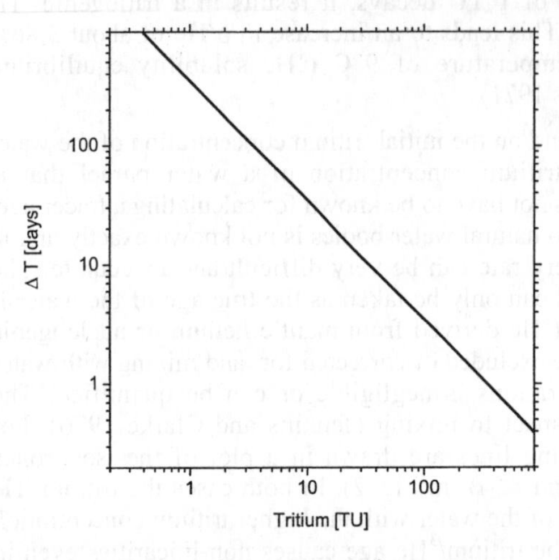

Figure 3. Time-resolution of the tritium/^3He method as a function of the tritium concentration of a water sample calculated on the basis of a measurement precision of ±0.2% for δ^3He. After Schlosser (1992).

Exemplary results from tritium/^{3}He studies

Applications of the tritium/^3He method to problems of the oceanic circulation started in the early 1970s when, as part of the GEOSECS (Geochemical Ocean Sections) program, Jenkins and Clarke (1976) conducted a systematic study of the distribution of these isotopes in the Atlantic Ocean. This study was followed by numerous tritium/^3He programs that addressed ocean circulation or oxygen utilization on regional and/or global scales (e.g., Thiele et al. 1986; Jenkins 1987; Roether and Fuchs 1988; Doney and Jenkins 1994; Schlosser et al. 1995; Bönisch et al. 1997; Jenkins 1998; Robbins et al. 2000, to name just a few).

The penetration of tritium into the interior of the ocean can be followed through repeated observations over a certain period of time. Such observations reveal pathways of water masses, as well as insight into the general penetration patterns of perturbations imprinted onto the surface of the ocean. In the interior of the ocean, tritium decay leads to a buildup of tritiogenic ^3He and tritium/^3He ages can be calculated. Evaluation of the tritium/^3He age distributions, especially if performed in conjunction with model simulations that correct for the non-linearities of the apparent tritium/^3He age, provides valuable information on mean spreading rates of water masses (e.g., Doney and Jenkins 1994), average mixing coefficients (e.g., Thiele et al. 1986), variability of water masses (e.g., Schlosser et al. 1991), as well as geochemical rates such as oxygen utilization rates (e.g., Jenkins 1982, 1987).

Instead of attempting a comprehensive review of the oceanic tritium/^3He literature, we present examples of two types of applications: dye-type penetration of tritium into the surface layers of the ocean and variability in deep water formation rates.

Ventilation of the oceanic thermocline. The spatial resolution of the WOCE (World

Ocean Circulation Experiment) tritium/helium isotope survey exceeds that achieved during the GEOSECS program by a factor of ~10. The data density obtained from this survey is sufficiently high to allow 3-dimensional visualizations of the tritium plumes in the ocean. In this way, the WOCE tritium/^{3}He sections (for an example, see Fig. 4)

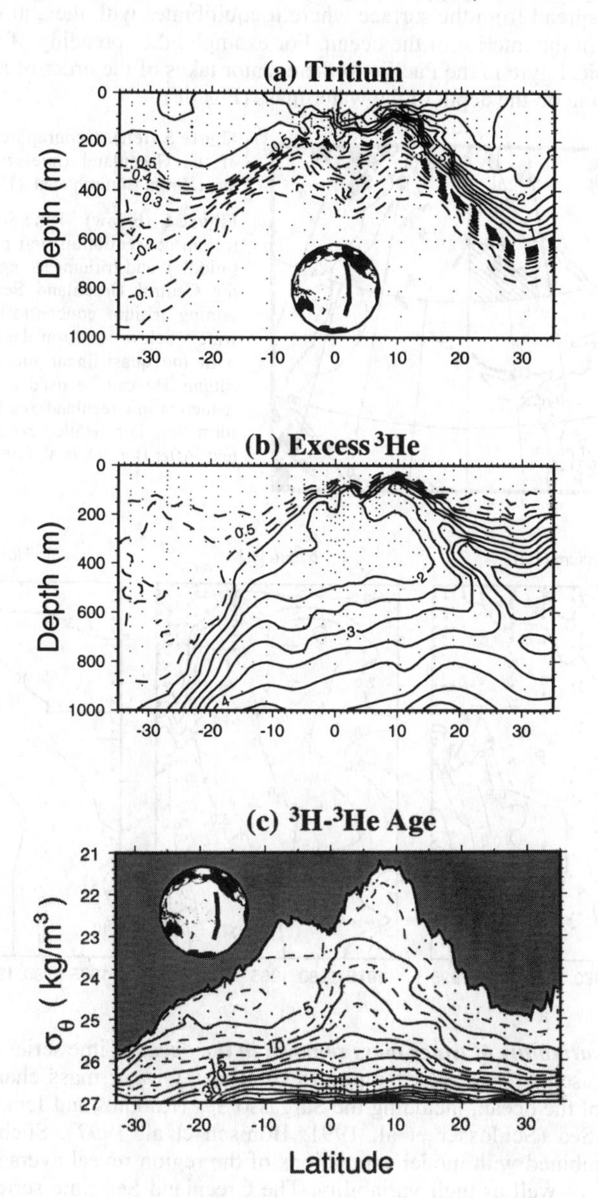

Figure 4. Distribution of tritium (a), ^{3}He (b), and tritium/^{3}He age (c) along 135°W (WOCE section P17C) occupied in 1991. The data delineate the penetration pattern of bomb tritium from the surface into the interior of the Pacific Ocean. The apparent tritium/^{3}He ages can be used to derive information on the mean residence times of the waters in the thermocline of the Pacific Ocean. After Schlosser et al. (2001).

provide valuable information on the spreading of water masses from the surface into the interior of the ocean (e.g., Jenkins 1998). If we combine the tritium and the tritiogenic ^{3}He distributions, we can derive information on mean residence times ('ages') of specific water masses (Fig. 4c). These apparent ages provide a good measure for the time it takes a water mass to spread from the surface where it equilibrates with the atmosphere to the point of interest in the interior of the ocean. For example, the spreading of water masses from the subtropical gyre in the Pacific to the equator takes of the order of a few years to a decade depending on the depth of the water mass (Fig. 4c).

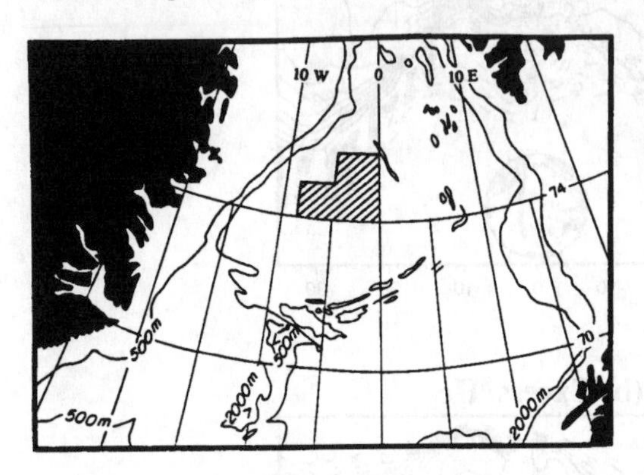

Figure 5 (left). Geographical location of the Greenland time-series experiment. From Bönisch et al. (1997).

Figure 6 (below). Time series of the tritium concentration (left panel), δ^3He (middle), and tritium/^{3}He age (right) in the Central Greenland Sea. The declining tritium concentrations in the waters below ~2000-m depth, together with the quasi-linear increase in the tritium/^{3}He can be used to derive the reduction in Greenland Sea Deep Water formation. For detailed explanation, see text. After Bönisch et al. (1997).

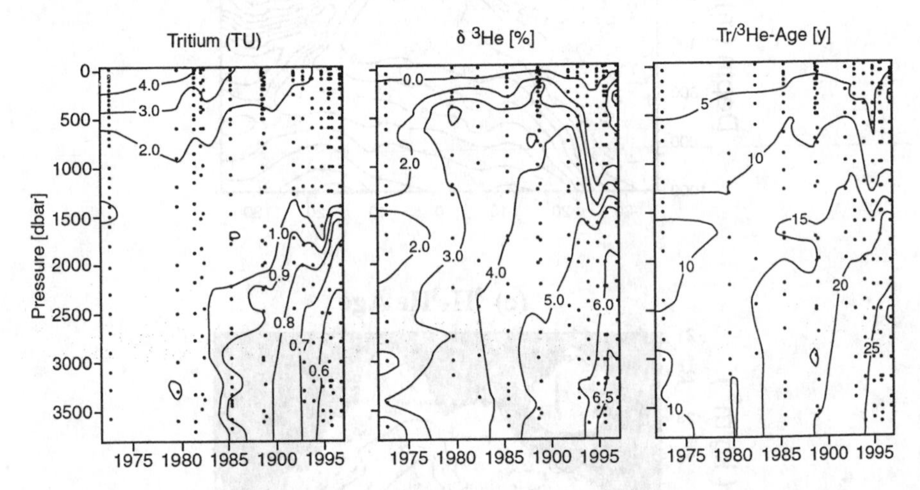

Studies of variability of water mass renewal in the ocean. Time series of tritium and ^{3}He have been used to study the temporal evolution of water mass characteristics in certain regions of the ocean, including the Sargasso Sea (Robbins and Jenkins 1998) and the Greenland Sea (Schlosser et al. 1991; Bönisch et al. 1997). Such time series, especially if combined with model simulations of the region reveal average water mass formation rates, as well as their variability. The Greenland Sea time series (Fig. 6; for geographical position of the stations, see Fig. 5) indicates a steady decrease of tritium throughout the water column, whereas the concentrations of ^{3}He and the tritium/^{3}He age are increasing in the deep water. The decreasing near-surface tritium concentrations reflect the decrease in the tritium delivery from the atmosphere to the surface ocean. The increase in ^{3}He and the tritium/^{3}He age in the deep water (depths below 1500 m), on the other hand, are evidence for a change in the formation rate of deep water in the Greenland Sea at around 1980. The tritium/^{3}He data, together with CFC data and a simple time-dependent mass balance (box model) were used to quantify the reduction in the rate of deep-water formation that occurred at this time in the center of the Greenland Sea. The reduction rate was estimated to be about 80% (from 0.5 Sv to 0.1 Sv, Schlosser et al. 1991; Bönisch and Schlosser 1995; 1 Sv = 10^6 m^3 sec^{-1}). Without tracer data the reduction in deep water renewal could not have been quantified with the same accuracy.

This change is significant and persisted between 1980 and 2000, the time of the last tracer observations in the central Greenland Sea. It led to a significant change in the characteristics of the deep waters in the Greenland Sea. For example, the temperature and salinity properties of Greenland Sea Deep Water observed in the mid- and late 1990s fall outside the classical definitions of this water mass (Bönisch et al. 1997). The deep water that in the past was formed by deep convection from the surface is slowly replaced by warmer and saltier waters originating in the Arctic Ocean and the Norwegian Sea. Whereas we were able to document the change in deep water formation in the central Greenland Sea by measuring a variety of variables and to quantify its reduction using transient tracers including tritium/^{3}He, we do not yet have a solid explanation for the cause of the reduction in deep water formation. The high northern latitude oceans, especially the Arctic Ocean, are presently undergoing significant changes (e.g., Morison et al. 1998, 2000). There is a distinct possibility that part of these changes are going beyond the natural variability and represent early signs of anthropogenically induced trends.

MANTLE ^{3}He

General background

Although the early search for helium in the oceans was driven by the hypothesis that the ocean should show a signature of excess ^{4}He derived from α -decay of elements of the uranium and thorium decay series in the sediments (Suess and Wänke 1965), the attention turned to mantle-derived ^{3}He as soon as the first helium isotope profiles had been measured in the mid/late 1960s in the South Pacific and as part of the preparations for the GEOSECS program in the North Pacific (Clarke et al. 1969; Clarke et al. 1970). These measurements revealed elevated ^{3}He/^{4}He ratios in the deep waters of the Pacific Ocean (up to several ten percent). The excess ^{3}He estimated from the helium isotope measurements were attributed to so-called primordial helium, i.e., helium that was trapped in the interior of the Earth during its formation (Clarke et al. 1969). Measurements of helium isotopes in thermal fluids (Mamyrin et al. 1969) and deep sea basalts confirmed this hypothesis and the ^{3}He/^{4}He ratio of the mantle-derived helium was determined to be approximately 8 times that of the atmosphere, i.e., roughly 1×10^{-5} (Krylov et al. 1974; Lupton and Craig 1975). These findings provided unambiguous evidence not only that the earth contains a primordial volatile component from its accretion but also that degassing of primordial volatiles is still occurring at mid-ocean ridges and hot spots.

Geochemical background

Hydrothermal activity is a common phenomenon found in the deep sea on a global scale, mainly along the mid-ocean ridges. Hydrothermal vents discharge fluids that contain a variety of trace elements, including noble gases. Helium emanated by deep-sea hydrothermal vents is strongly enriched in the light isotope ^{3}He. Frequently, helium isotope ratios in fluids from deep-sea hydrothermal vents are close to those of Mid Ocean Ridge Basalt (MORB, R/R_a = 8-9, Craig and Lupton 1981). However, due to the large variation in helium isotope ratios in fluids emanating from all possible deep-sea sources, including seamounts, the range of observed helium isotope ratios is large (ca. 3 to 30 times the atmospheric ratio). Such a variety in helium isotope ratios can be used to track different sources of helium observed in the intermediate waters of the ocean (Lupton 1996, 1998; see below). Along with the elevated helium isotope ratios, the absolute ^{3}He concentrations in hydrothermal fluids (ca. 0.2 to 2.5×10^{-9} cm^3 STP g^{-1}) are elevated by several orders of magnitude above background ^{3}He concentrations in water that has been equilibrated with the atmosphere (ca. 6×10^{-14} cm^3 STP g^{-1}, e.g., Butterfield et al. 1990). Helium injected in the deep sea follows the circulation of the water masses to which it is added. Eventually, it is released from the ocean to the atmosphere by air/sea gas exchange. Due to different source strength (spreading rates of the mid-ocean ridges; Lupton 1998) and mean residence times in the major ocean basins, the mean ^{3}He excess related to injection of mantle helium into the intermediate depth waters of the ocean differs from basin to basin. The highest values are found in the Pacific (ca. 20%; Lupton 1998), the lowest in the Atlantic (1 to 2%; e.g., Rüth et al. 2000). The Indian Ocean excess has been estimated on the basis of model simulations to be about 10% (Farley et al. 1995; see also Östlund et al. 1987 for data tables and sections).

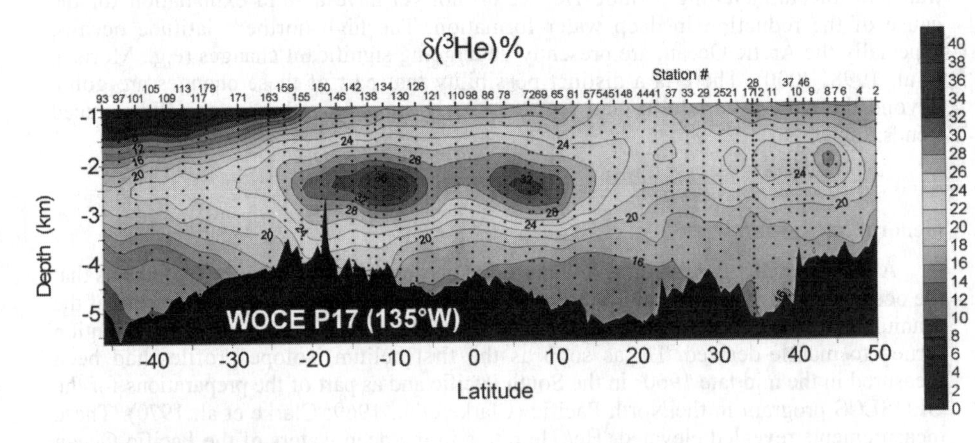

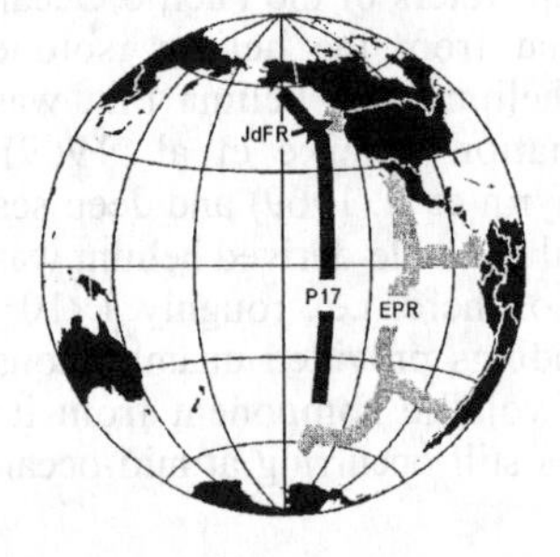

Figure 7. δ^{3}He distribution along 135°W (WOCE section P17C) from 1000-m depth to the bottom (contour interval is 2%). From Lupton (1998).

^{3}He plumes in the ocean

After initial systematic surveys could be completed during the 1970s, it became apparent that the ^{3}He emanating from the mid ocean ridges formed large-scale plumes (e.g., Lupton et al. 1980; Lupton and Craig 1981) that can be used to study the oceanic circulation on mid-depth strata. Further investigations during the past two decades, including sections from the World Ocean Circulation Experiment, led to systematic mapping of large-scale ^{3}He plumes in the major ocean basins. This work has been progressed the farthest for the Pacific Ocean (Lupton 1996, 1998; Schlosser et al. 2001).

In addition to the large plume emanating from the East Pacific Rise at 10°N and 15°S (EPR, Lupton et al. 1980) discovered early on, at least two other major ^{3}He sources have been detected and described in the Pacific Ocean, e.g., Lupton, 1996, 1998). One is emanating from the Juan de Fuca Ridge in the northeastern Pacific at 48°N and one from Loihi Seamount near Hawaii at 20°N. All three plumes can be recognized in a N/S section across the central Pacific at about 135W (WOCE line P17, Fig. 7) and can be separated by depth (Lupton 1998). The EPR plume is situated deep in the water column (~2500 m). The Juan de Fuca Ridge plume is found somewhat higher in the water column (~2000-m depth) and the Loihi seamount plume is rather shallow (~1100 m). The Loihi plume is different from the others in that it is fed from a hot spot source and not from mid-ocean ridge vents. The spreading pattern of the plumes can be derived from horizontal ^{3}He maps in the Pacific as demonstrated by Lupton (1996, 1998). Whereas the EPR and Juan de Fuca Ridge plumes indicate westward spreading (the EPR plume spreads across the entire Pacific Basin), the Loihi seamount plume has an eastward movement (Fig. 8). Work similar to that in the Pacific Ocean was performed in the South Atlantic by Rüth et al. (2000). Here, the mantle ^{3}He signals are much smaller due to the slower spreading rates of the mid ocean ridge and the ^{3}He signatures and patterns are more difficult to translate into information on the oceanic circulation. The Indian Ocean ^{3}He data from WOCE are presently being evaluated and first unpublished results look very promising, indicating ^{3}He plumes from at least three different sources (mid ocean ridge, Indonesian Throughflow, and Gulf of Aden).

Most of the evaluation of the ^{3}He plumes in the ocean has been qualitative or semi-quantitative in nature. However, there are increasing attempts to incorporate ^{3}He into Ocean General Circulation Models (OGCMs) in order to understand how such models ventilate the mid-depth and deep-water masses. An early attempt of ^{3}He simulations in an OGCM has been documented by Farley et al. (1995). The results of the model simulations indicate that the model can roughly close the helium balance and reproduces most of the major observed features of the ^{3}He field in the ocean. At the same time, the simulations show many shortcomings of the model and future work will contribute to both a better understanding of the ^{3}He plumes and the quality of ocean circulation models.

THE FLUX OF ^{3}HE AND ^{4}HE FROM THE SEAFLOOR

^{4}He

On the basis of mass balance considerations, Suess and Wänke (1965) postulated that the flux of ^{4}He produced by U and Th decay in the deep-sea sediments should be detectable as a ^{4}He excess in the abyssal waters of the ocean. Early measurements of helium dissolved in seawater were difficult and interpretation of the results was challenging. For example König et al. (1964) interpreted their results as consistent with solubility data, whereas Suess and Wänke (1965) used the same measurements to support their hypothesis that there should be a measurable helium excess related to a sediment source. Bieri et al. (1966) reported noble gas measurements from Pacific waters and concluded that there has to be a terrrestrial helium source at the sediment interface that supplies the excess helium observed in the water column. Whereas these early studies focused on radiogenic helium as the source of the observed excess helium, the introduction of helium isotope measurements revealed that there are two sources of terrigenic helium that contribute to the observed helium excess. In addition to the crustal (radiogenic) ^{4}He there is a contribution of mantle-derived helium that is enriched in ^{3}He by roughly a factor of 8. The presence of helium from two terrigenic sources and the added complication of a helium excess due to the (partial) dissolution of air at the sea surface and close to floating ice shelves (e.g., Schlosser 1986; Schlosser et al. 1990)

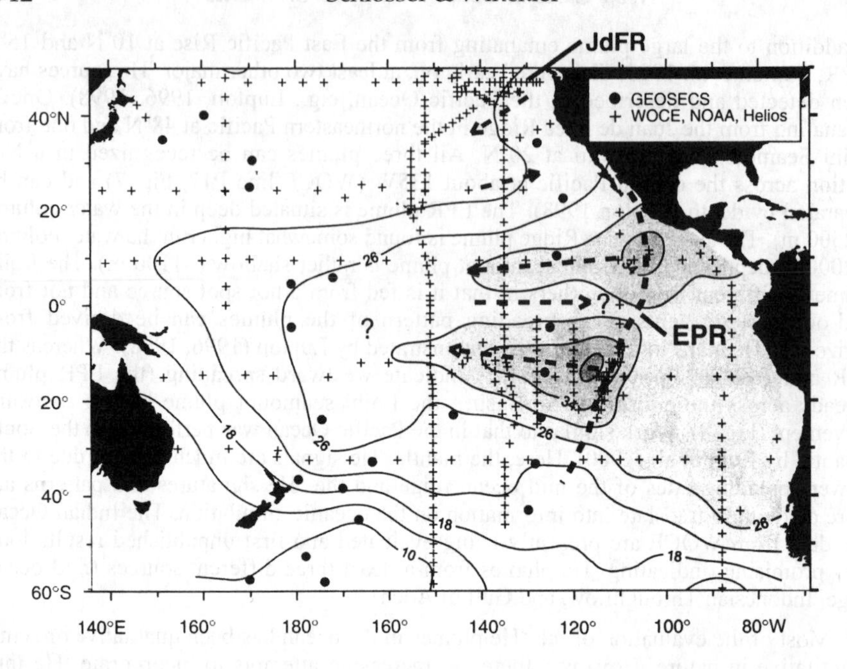

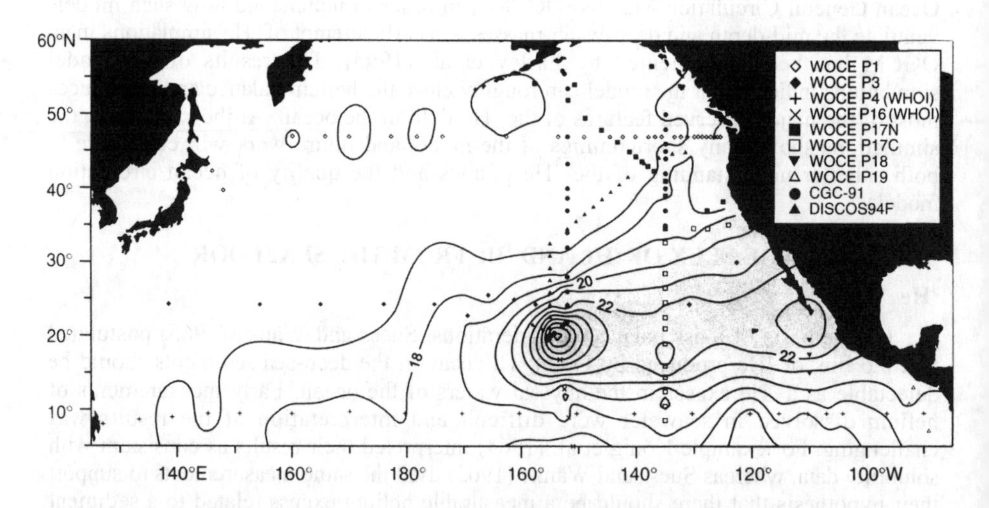

Figure 8. Map of δ^3He (%) contoured on a surface at 2500 m depth (upper panel, contour interval is 4%) and on a surface of 1100 m depth (lower panel, contour interval is 1%) in the Pacific Ocean. Large ^{3}He-rich plumes emanate from the East Pacific Rise (EPR) at 15°S and 10°N and Juan de Fuca Ridge (JdFR) systems and spread westward, as indicated by the dashed arrows. Question marks denote regions where flow patterns are not obvious from the δ^3He distribution. The helium signal emanating from Loihi Seamount, on the southeastern flank of Hawaii, is transported eastward as a continuous plume to the coast of Mexico. Plot produced by John Lupton using data from Lupton (1998) and unpublished data produced by W. J. Jenkins. [Used by permission of Academic Press, from Schlosser et al. (2001), *International Geophysics Series*, Vol. 77, Fig. 5.8.14 and 5.8.16, p. 442 and 443]

makes it fairly difficult to accurately quantify the ^{4}He flux from the sea floor. In typical oceanographic water masses, ^{4}He from the individual helium sources is difficult to separate and the amounts of excess ^{4}He in typical water samples is relatively small if compared to the ^{3}He excesses related to the addition of mantle helium.

Based on the observed ^{4}He excess and turnover time of the Pacific Ocean Craig et al. (1975) estimated the oceanic ^{4}He flux to be $3\pm1\times10^5$ atoms cm^{-2} s^{-1} if referred to the area of the entire oceanic crust. This estimate includes both the hydrothermal ^{4}He flux emanating from mid-ocean ridges as well as the crustal ^{4}He flux by degassing of the oceanic crust. The oceanic ^{4}He flux is small compared to the total ^{4}He flux of ca. 1.3×10^6 atoms cm^{-2} s^{-1} (normalized to the entire earth surface).

There are specific regions in the ocean where strong ^{4}He signals can be observed. For example, ^{4}He excesses in the Black Sea (Top and Clarke 1983) and the Eastern Mediterranean (Roether et al. 1998) were converted into ^{4}He fluxes and yielded values of $1.3\pm0.5\times10^6$ and $3.1\pm1.2\times10^6$ atoms cm^{-2} s^{-1}, respectively. These values are comparable to the fluxes estimated for the continental crust ($2.7\pm1\times10^6$ atoms cm^{-2} s^{-1} normalized to the continental area; e.g., Torgersen 1989). This shows that on a local/regional basis there are significant radiogenic ^{4}He fluxes from the ocean sediment into the water column. Such features are not unexpected because at certain locations the seafloor is closer to the composition of continental crust than oceanic crust. However, averaged over the entire ocean basins, the flux of crustal ^{4}He is small compared to that from the continental crust.

Well et al. (2001) evaluated high-quality data from the Pacific Ocean that became available during the World Ocean Circulation Experiment (WOCE). Separating the crustal ^{4}He component from the mantle component, they revealed widespread occurrence of a crustal radiogenic ^{4}He flux out of deep-sea sediments and the oceanic crust into the water column (Fig. 9). Although the flux of ^{4}He is small in these regions, Well et al. (2001) were able to estimate a flux of roughly $1\pm0.4\times10^5$ atoms cm^{-2} s^{-1}. Further evaluation of the WOCE helium isotope and neon data will hopefully improve our knowledge on the distribution of helium sources and their relative flux values in the global ocean.

^{3}He

After the discovery of mantle-derived helium in the oceans (Clarke et al. 1969), the first estimate of the ^{3}He flux yielded a value of about 2 atoms cm^{-2} s^{-1} (Clarke et al. 1969). This estimate was based on simple advection/diffusion or straight advection models that utilize the mean concentration of excess ^{3}He in the deep waters of the ocean (Δ^3He) and mean residence time of the deep waters. The flux of this mantle-derived helium is concentrated in areas of active mid-ocean ridge spreading or ocean islands (e.g., Craig and Lupton 1981). Later estimates by Craig et al. (1975) were based on a larger, although still sparse, data set. In their new assessment of the ^{3}He flux from the ocean, these authors use two methods: one based on the average mean upwelling rate and the mean ^{3}He excess in the deep ocean, the other based on the ^{4}He flux and the ^{3}He/^{4}He ratio of the excess helium calculated from profiles located on the East Pacific Rise. They obtain values of 3.3 atoms cm^{-2} s^{-1} and 4.8 atoms cm^{-2} s^{-1}, respectively (average value given by Craig et al. 1975: 4 ± 1 atoms cm^{-2} s^{-1}.

Since these estimates of ^{3}He fluxes in the mid-1970s, little work has been done that would have resulted in significant refinements of these fluxes. Simulations of the mantle helium distribution using a global general ocean circulation model (Farley et al. 1995) used a simple parameterization of the ^{3}He flux from mid-ocean ridges and determined the

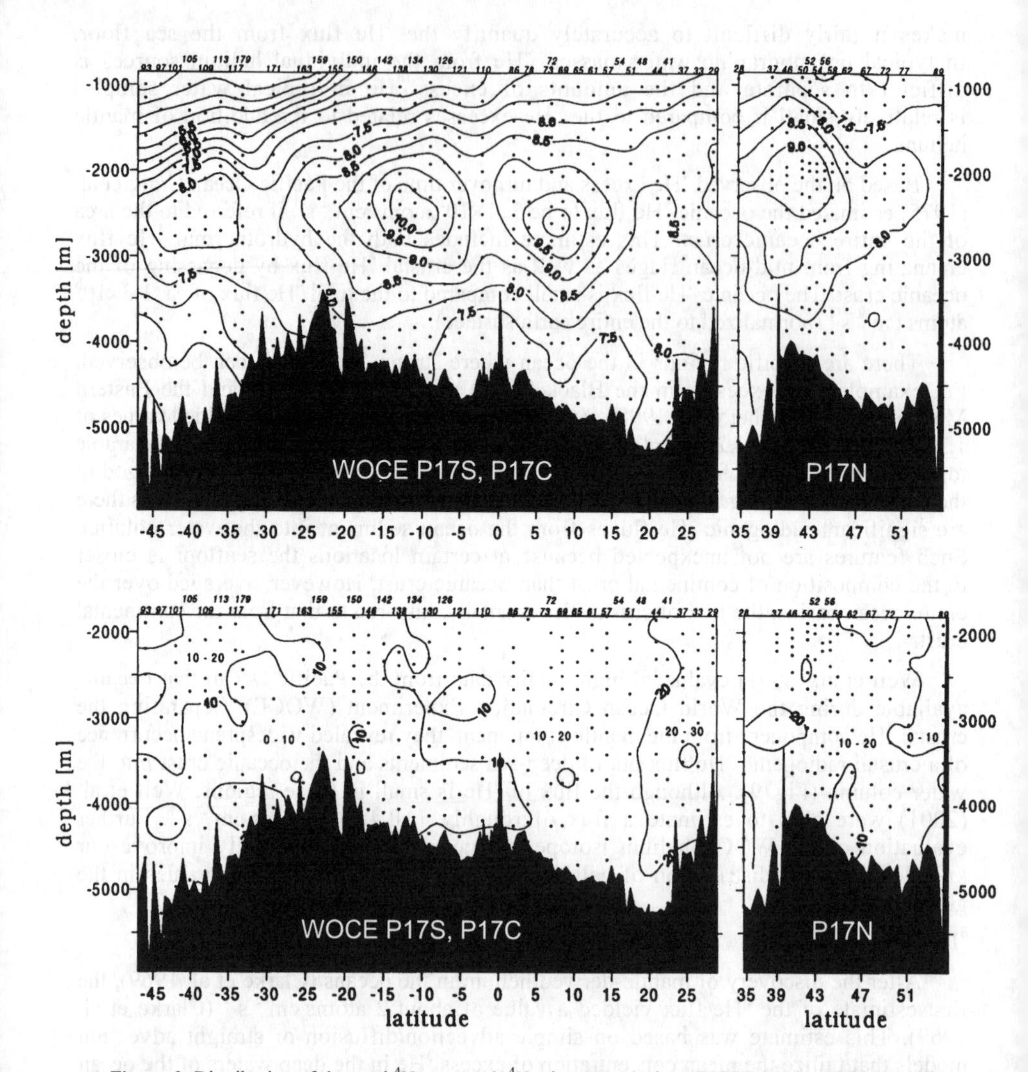

Figure 9. Distribution of the total ^{4}He excess (Δ^4He, in percent) along WOCE section P17 at 135°W in the Pacific Ocean (upper panels). The data can be used, together with measurements of neon and the ^{3}He/^{4}He ratio, to determine the fraction of radiogenic ^{4}He (in pmol kg^{-1}) contained in the deep waters below ~2000 m (lower panels). The data indicate the presence of a small radiogenic ^{4}He component in the deep and bottom waters of the Pacific Ocean. From Well et al. (2001).

distribution of the ^{3}He emanated from these sources throughout the world ocean. These simulations showed that with the specific source parameterization used in their study, Farley et al. (1995) could reproduce the global oceanic balance of ^{3}He reasonably well. However, at the same time the simulations clearly revealed that there is still much work to be done to understand the details of the simulated fields. Comparison with high-density surveys of ^{3}He in the ocean that are now becoming available (e.g., Lupton 1998, Rüth et al 2000) will enable us to reach a deeper level of understanding of the ^{3}He patterns observed in the ocean. Such progress should make new efforts to refine the estimate of the oceanic ^{3}He flux a worthwhile exercise.

EXTRATERRESTRIAL ^{3}He IN DEEP-SEA SEDIMENTS

Delivery of extraterrestrial ^{3}He to the ocean sediments

Four decades ago, ^{3}He of apparent extraterrestrial origin was identified in ocean sediments. Merrihue (1964) observed ^{3}He/^{4}He ratios in marine sediments roughly 2 orders of magnitude higher than those observed in atmospheric helium and attributed it to the presence of cosmic material.

Systematic studies of noble gases in marine sediments were started in the early 1980s by Ozima et al. (1984). Shortly thereafter, Takayanagi and Ozima (1987) recognized the potential of ^{3}He as a proxy of sediment accumulation rates. Most studies of noble gases in marine sediments (including this chapter) focus on helium isotopes, however, neon (Nier and Schlutter 1990, 1993) and argon isotopes (Tilles 1966, 1967; Amari and Ozima 1988) have also been studied. In the early 1990s, extraterrestrial ^{3}He in ocean sediments received considerable attention when Anderson (1993) questioned the understanding of mantle geochemistry and proposed that the mantle ^{3}He might not represent volatiles trapped in the earth in early earth history but rather is derived from subducted interplanetary dust particles (IDPs). However, this hypothesis could be refuted based on mass flux considerations (Allègre et al. 1993) and experimental evidence showing that helium would not be retained during subduction due to sufficiently high diffusion coefficients (Hiyagon 1994).

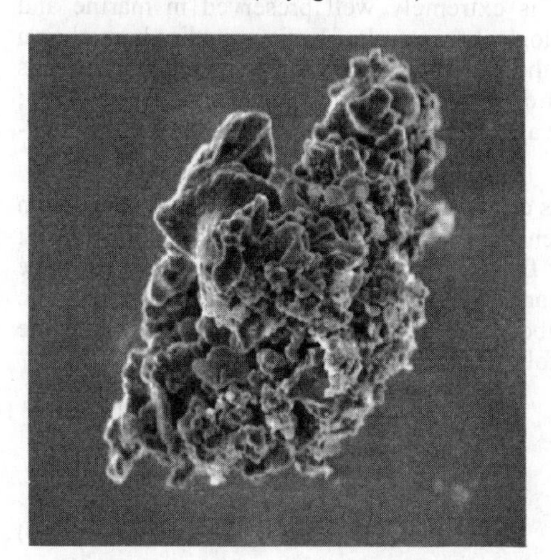

Figure 10. Electron microscope picture of an interplanetary dust particle. The size of the particles carrying the extraterrestrial ^{3}He that are accumulated in marine sediments typically is between 3 and 35 μm (e.g., Farley et al. 1997) [photograph courtesy of Scott Messenger, http://stardust.wustl.edu].

Extraterrestrial ^{3}He is delivered to the earth surface by interplanetary dust particles (IDP, Fig. 10). IDPs are derived from asteroid collisions as well as cometary debris (Dohnanyi 1976) and are thought to acquire their characteristic helium signature from implantation of solar wind and solar flare gases (e.g., Nier and Schlutter 1990). Approximately 40,000 tons of IDPs are deposited annually on the earth's surface (Love and Brownlee 1993). However, the major fraction of the IDPs is heated to temperatures >800°C during entry into the earth's atmosphere and looses its helium signal (Farley et al. 1997). Only IDPs with diameters ≤ 35 microns, corresponding to 0.5% of the total IDP mass flux, transit the atmosphere at temperatures of 500-800°C or below (Fraundorf et al. 1982) and retain their extraterrestrial helium signature. After being removed from the troposphere mainly by wet deposition and rapid settling through the oceanic water column, the IDPs continuously accumulate in marine sediments (e.g., Takayanagi and

Ozima 1987) or on ice shields (Brook et al. 2000).

The helium isotope characteristics of IDPs are well constrained from analysis of individual particles collected from the stratosphere. They have a $^3He/^4He$ ratio of 2.4×10^{-4}, as well as a fairly constant ^3He concentration of 1.9×10^{-5} cm^3 STP g^{-1} (Nier and Schlutter 1992). As the cosmic dust is enriched in ^3He by ~8 orders of magnitude compared to terrigenous matter it can be readily detected. However, in spite of various noble gas studies of single IDP grains, both from the stratosphere and ocean sediments (e.g., Fukumoto et al. 1986; Nier et al. 1990; Nier and Schlutter 1992) the carrier phase that actually hosts the extraterrestrial helium has not been unambiguously identified. Amari and Ozima (1985) found that the helium resides mainly in the magnetic fraction of sediments and therefore identified magnetite to be the main carrier of the extraterrestrial ^3He signal. The magnetite is thought to be produced by heating of the IDPs during atmospheric entry (Amari and Ozima 1985). In a subsequent study, Fukumoto et al. (1986) found significant helium contributions from a non-magnetic fraction and suggested extraterrestrial silicates to be another likely host mineral. The presence of a non-magnetic carrier phase was later confirmed by Patterson et al. (1998) and Farley (2001). Future work is needed to identify the relative importance of different carrier phases and whether they are associated to magnetic and/or non-magnetic sediment fractions.

The extraterrestrial helium signal is extremely well preserved in marine and terrestrial sedimentary archives over geological time scales. Various studies have shown that the extraterrestrial helium carried to the seafloor by IDPs can be retained for periods of at least 65 million years (Ma) against diffusive and/or diagenetic loss (Farley 1995; Farley et al. 1998). Recently, Patterson et al. (1998) identified extraterrestrial ^3He in 480-Ma old sedimentary rocks.

Helium contained in ocean sediments can be interpreted as a mixture of helium from two sources, extraterrestrial and terrigeneous helium (Takayanagi and Ozima 1987; Marcantonio et al. 1995). Contributions from an atmospheric helium component have been shown to be negligible (e.g., Farley and Patterson 1995). Assuming that the isotopic compositions of both mixing end-members are known, one can easily calculate the amount of extraterrestrial ^3He, using the following equation:

$$\frac{^3He_{ET}}{^3He_m} = \frac{1 - \dfrac{\left(\dfrac{^3He}{^4He}\right)_{ter}}{\left(\dfrac{^3He}{^4He}\right)_m}}{1 - \dfrac{\left(\dfrac{^3He}{^4He}\right)_{ter}}{\left(\dfrac{^3He}{^4He}\right)_{ET}}} \tag{6}$$

where m denotes the measured, ET the extraterrestrial, and ter the terrigeneous component. Whereas the extraterrestrial IDP component is well constrained ($^3He/^4He = 2.4\times10^{-4}$, $[^3He] = 1.9\times10^{-5}$ cm^3 STP g^{-1}, Nier and Schlutter 1992), the helium isotope composition of the terrigeneous end-member is less well constrained and may vary regionally depending on the age and composition of the source material (Marcantonio et al. 1998; Farley 2001). A series of mixing lines that have the extraterrestrial component as one end-member and fan out towards different terrigeneous end-members can be drawn (Fig. 11). In regions of low terrigeneous supply (i.e., far away from ocean-margin sites and volcanic input) such as the central and eastern equatorial Pacific where most of the work on extraterrestrial ^3He in deep-sea sediments has been conducted, $^3He/^4He$ ratios range from 10^{-5} to 10^{-4} (Marcantonio et al. 1995, 1996; Patterson and Farley 1998; Winckler et al. 2001b) and virtually all ^3He (>99.5%) is of extraterrestrial origin. Here, the choice of the $^3He/^4He$ ratio of the sedimentary end-member is not very critical and does not significantly change the mixing lines. Typical IDP concentrations in sediments that are not diluted by a terrigeneous component are of the order of 0.1 to 1 ppm; in sediments dominated by terrigeneous components they can be lower than 1 ppb.

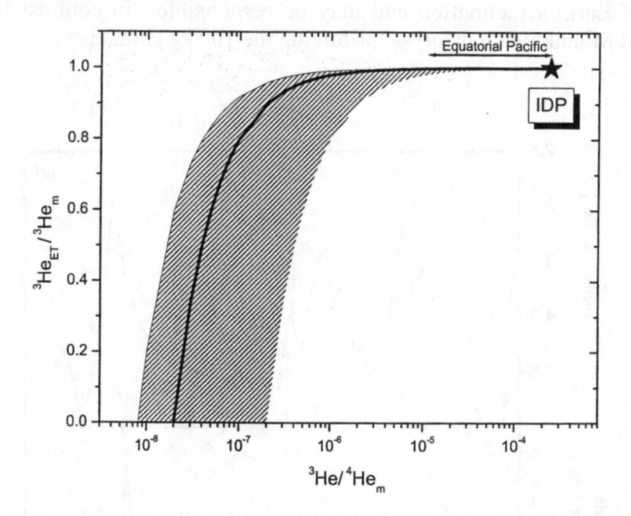

Figure 11. Relative fraction of extraterrestrial ^3He in the total observed ^3He concentration of a sediment sample plotted versus the $^3He/^4He$ ratio (modified from Farley 2001). The heavy black line represents mixing of an IDP component and an average terrigeneous component ($^3He/^4He = 2\times10^{-8}$, e.g., Farley and Patterson 1995). The envelope represents mixing with the two extreme cases of terrigeneous helium signatures analyzed to date: sediments derived from old continental crust (e.g., North Atlantic ice-rafted debris with $^3He/^4He = 6\times10^{-9}$, Marcantonio et al. 1998) and sediments from the Amazon River fan ($^3He/^4He = 2\times10^{-7}$, Marcantonio et al. 1998).

Applications of IDP-derived ^3He

Following the pioneering work by Takayanagi and Ozima (1987) several recent studies have investigated extraterrestrial ^3He in sediments over times scales of thousands to tens of millions of years, both to constrain sedimentation rates in paleoceanographic studies and variations of the IDP accretion to obtain astrogeophysical information.

The Cenozoic record. Farley and coworkers used the extraterrestrial ^3He signal in marine sediments to study astrogeophysical events (for a review, see Farley 2001). In a 70 Ma record from the central North Pacific, they identified ^3He flux variations and attributed them to asteroidal breakup events or the passage of comets through the inner solar system (Farley 1995). In a more detailed study focusing on the Late Eocene they found evidence for large comet shower impact events in the Late Eocene by tracing the IDP fluxes associated with them (Farley et al. 1998).

The Quaternary record. Studies of Pleistocene sediments from the North Atlantic and the equatorial Pacific revealed that the accumulation of extraterrestrial ^3He exhibits a strong cyclicity with a period of ~100 ka (Farley and Patterson 1995; Marcantonio et al.

1995, 1996, 2001; Patterson and Farley 1998; for an example see Fig 12). However, the interpretation of those sediment records is still controversial. Interestingly, both interpretations link the ^{3}He signal to climate change:

- Farley and Patterson (1995) and Patterson and Farley (1998) interpreted the variability of the ^{3}He accumulation in deep-sea sediments to directly reflect the cyclicity of the accretion rate of IDPs, i.e. changes in delivery of IDPs from space. This interpretation supports a recent hypothesis by Muller and MacDonald (1995, 1997a,b,c) that the accretion rate of cosmic dust to the earth varies with the 100 ka cycle of Earth's inclination and may be responsible—in contrast to the widely accepted Milankovitch theory—for forcing the 100 ka climate cycle.

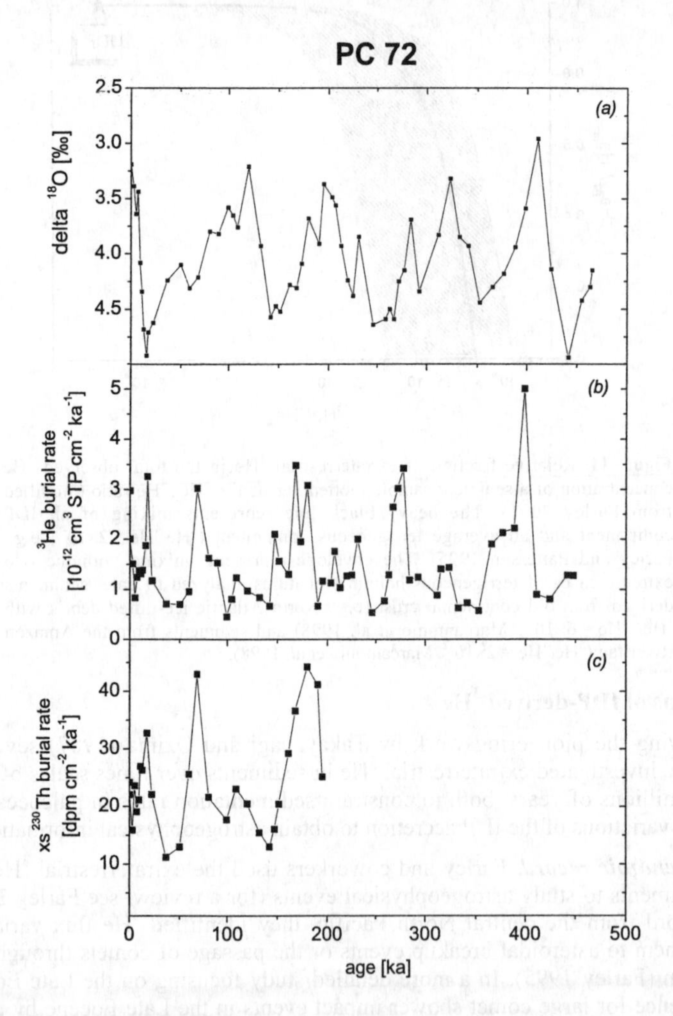

Figure 12. Multi-proxy record for core PC72 from the equatorial Pacific Ocean (0.1°N, 140°W) after Marcantonio et al. (1996): $\delta^{18}O$ (a), accumulation ("burial") rates of ^{3}He (b), and ^{230}Th (c) are plotted versus age. The ^{3}He and ^{230}Th accumulation rates are derived by multiplying the concentrations with the ^{18}O-derived bulk sedimentation rates and show synchronous 100 kyr variability.

- Marcantonio et al. (1995, 1996, 2001) reached a different conclusion. These authors compared the ^{3}He record to that of $^{230}Th_{ex}$, which is well established as a constant flux tracer due to its homogeneous production in the water column and its high particle reactivity (e.g., François et al. 1990). Both records show a strong correlation for the last 210 ka. Given that both tracers are derived from a different and decoupled source, the small variability in the $^{3}He/^{230}Th_{ex}$ ratio is a strong indication that the flux of IDPs and consequently extraterrestrial ^{3}He to the seafloor has been constant during that period. Marcantonio et al. (1995, 1996, 2001) postulate that the variability of the ^{3}He accumulation is caused by variable sediment redistribution by deep-sea currents, so-called sediment focusing. The focusing patterns are correlated with the global climate record such that periods of maximum focusing correspond to interglacials leading to the hypothesis that the temporal pattern of sediment accumulation reflects climate-related reorganization of deep-sea currents (Fig.12).

Potential of ^{3}He as a constant flux tracer. Marcantonio et al.'s (1995, 1996, 2001) findings and interpretation open up the potential of IDP-derived extraterrestrial ^{3}He as a constant flux proxy. In various recent studies, the ^{3}He flux has been determined independently. The estimate of Marcantonio et al. (1996, 2001) for the $^{3}He_{ET}$ flux in the equatorial Pacific ($8.0 \pm 2.4\times10^{-13}$ cm^{3} STP cm^{-2} ka^{-1}) is in agreement with the global average Holocene $^{3}He_{ET}$ flux ($7.7 \pm 2\times10^{-13}$ cm^{3} STP cm^{-2} ka^{-1}) reported by Higgins (2001) and the $^{3}He_{ET}$ flux determined from ice core studies ($6.2\pm2.7\times10^{-13}$ cm^{3} STP cm^{-2} ka^{-1} for Greenland/GISP2 and $7.7\pm2.5\times10^{-13}$ cm^{3} STP cm^{-2} ka^{-1} for Vostok, Brook et al. 2000). Further support for this approach on even longer timescales is derived from comparative studies of ^{3}He and ^{10}Be in the Pacific Ocean confirming a constant ^{3}He delivery to the earth over 7 Ma (Higgins 2001). However, there is clearly a need for more calibration work to unambiguously confirm the constancy of the $^{3}He_{ET}$ flux over these time scales and thus to finally establish the use of ^{3}He as constant flux tracer.

In the paleo-oceanographic context, constant flux tracers are valuable tools because they enable the reconstruction of particulate fluxes. One of the advantages of this approach is that it allows us to establish mass accumulation rates independently from single-point age models (e.g., ^{18}O). These models frequently are biased and sensitive to sediment redistribution effects. For example, this holds true for paleoceanographic studies in the Quaternary, where many interpretations of the sediment record rely on potentially erroneous sediment accumulation rates derived from δ^{18}O stratigraphy.

Under the assumption of a constant ^{3}He flux ($F_{^3He}$) the instantaneous mass sediment accumulation rate (MAR) is inversely proportional to the ^{3}He concentration

$$MAR = \frac{F_{^3He}}{^3He} \tag{7}$$

Accordingly, the particle fluxes (AR) of other sedimentary components i are

$$AR_i = MAR \cdot [i] = \frac{F_{^3He}}{^3He} \cdot [i] \tag{8}$$

Unlike the Thorium method, which is limited to the past 250 ka due to radioactive decay, ^{3}He, a stable isotope, should be useful as a constant flux proxy on longer time scales.

Whereas there is an ongoing controversy about application of the constant flux proxy approach for Pleistocene records (see above), there is agreement (Farley et al. 2001) about the recently developed application of the method to determine the duration of 'special' extreme events in the geological past. Mukhopadhyay et al. (2001) used the constant flux characteristics of ^{3}He to determine the time interval over which the deposition of the K/T boundary clay occurred to be 11±2 ka. Eltgroth and Farley (2001) used a similar approach to evaluate the duration of the Late Paleocene Thermal Maximum.

NOBLE GASES IN DEEP-SEA BRINES

Deep-sea brines are hypersaline water bodies that have been observed in various deep-sea environments, such as the Red Sea, the eastern Mediterranean and the Gulf of Mexico. Their high salt concentrations (up to 15 times the concentration of seawater) are attributed to leaching of underlying evaporites. Due to their high density, they are typically found in depressions on the seafloor, so-called brine basins. Noble gases have proven to be valuable tools to decode the complex geochemical processes underlying the origin and formation of these brines.

The Red Sea

In 1966, Miller et al. (1966) discovered the first deep-sea brine, the Atlantis II brine, in the central rift zone of the Red Sea, which contains more than 20 morphological depressions filled by highly saline brines (Hartmann et al. 1998). The brines are part of the geological setting of the Red Sea: whereas the southern part is a young spreading center, the northern part is tectonically inactive. Noble gas studies have significantly contributed to understanding the origin and evolution of these deep-sea brine systems. Lupton et al. (1977) detected MORB-derived helium ($^3He/^4He = 1.2\times10^{-5}$) and demonstrated that the Atlantis II brine is part of an active hydrothermal system below the Red Sea. In a more detailed study in the late 90s, elevated $^{40}Ar/^{36}Ar$ ratios up to 305 (compared to 295.5 for atmospheric $^{40}Ar/^{36}Ar$ ratios) were identified in the Atlantis II brine confirming the transport of mantle-derived argon in hydrothermal fluids along with mantle-derived helium at active seafloor hydrothermal systems (Winckler et al. 2001a). The hydrothermal activity is thought to accompany the recent start of seafloor spreading in the transition zone between the Southern and Northern Red Sea (Bonatti 1985).

Systematic studies of atmospheric noble gases were used to reconstruct the dynamic processes underlying the formation of the Atlantis II brine. The atmospheric noble gases (Ne, Ar_{atm}, Kr, Xe) in the deepest brine layer of the Atlantis II brine were found to be depleted by 20-30% compared to the initial concentrations in ambient Red Sea Deep Water. No systematic fractionation between the different noble gases was observed, which suggests sub sea floor boiling and subsequent phase separation to be responsible for the observed depletion pattern (Winckler et al. 2000). Together with the MORB characteristics of the helium and argon isotopes, the geochemical evolution of the brine before injection into the deep water could be reconstructed. After having circulated through evaporites, where it became enriched in salt, and through young oceanic crust, where it became enriched in MORB derived helium and argon, the ascending fluid boils, and consequently the residual liquid is depleted in atmospheric noble gases. The depleted fluid rises to the sediment surface and feeds the Atlantis II brine system. The noble gas study of Winckler et al. (2000) provides evidence that the Red Sea brines are oceanic hydrothermal systems, similar to vent systems at sediment-free ridges (e.g., the Juan de Fuca Ridge), where boiling and phase separation control the hydrothermal chemistry.

Extending the study of the Red Sea brines to the inactive northern part, Winckler et al. (2001a) observed significant differences of the helium and argon isotope characteristics along the Red Sea rift system. The Kebrit brine, located in the northern Red Sea, has a low helium isotope signature ($^3He/^4He = 1\times10^{-6}$) indicating a predominantly crustal origin and no signs of active hydrothermal input. The "radiogenic" signature is interpreted as the result of a helium flux from the continental crust that accumulates in the brine. Again, this is consistent with the geological setting of the northern part of the Red Sea where the continental crust is stretched and thinned representing pre-seafloor spreading conditions. The striking agreement of the geochemical fingerprints and the tectonic situation of the Red Sea, i.e. the northward progression of the seafloor spreading, confirms the unique potential of noble gases to contribute to the study of complex geochemical processes (Fig 13).

The Eastern Mediterranean

The Eastern Mediterranean is the site of 5 deep-sea basins that were discovered in the 1980s (e.g., Jongsma et al. 1983) and 1990s (MEDRIFF-Consortium 1996). Detailed noble gas studies were completed in three of them, the Urania, Atalante and Discovery brines, in order to investigate their origin and fluid kinematics.

In the Urania Basin, extraordinary helium and argon signatures were observed. The ^{4}He concentration of this brine is enriched by 4 to 5 orders of magnitude compared to ambient seawater. Low ^{3}He/^{4}He ratios of 1×10^{-7} and high ^{40}Ar/^{36}Ar ratios of up to 470 indicate a radiogenic source. On the basis of these data, Winckler et al. (1997) concluded that the Urania Basin is fed by advective transport from a deep fluid reservoir below the evaporite sequence that efficiently seals the crustal flux from lower sedimentary strata. The unique enrichment in ^{4}He is explained by a helium accumulation process over several million years, likely since the formation of the evaporites during the Messinian salinity crisis 6 Ma ago. Due to its extremely high ^{4}He concentration, injection of radiogenic helium from local features like the Urania brine may contribute significantly to the helium inventory and may be responsible for the high radiogenic helium excess (~6% compared to ambient sea water) in the deep (>1200 m) eastern Mediterranean (Roether et al. 1992, 1994, 1996). The case of the Urania Brine emphasizes the potential importance of brines as interface between local geochemical and oceanographic features on a regional scale.

AIR/SEA GAS EXCHANGE STUDIED BY DUAL TRACER RELEASE EXPERIMENTS

Background

Knowledge of air/sea gas exchange rates is central to understanding the balances and fluxes of a significant number of gases with impact on climate or biogeochemical cycles. They include dissolved O_2, CO_2, dimethyl sulphide (DMS), nitrous oxide, halocarbons, or methane (Bange et al. 1996; Bates et al. 1996; Blake et al. 1997; Liss 1999). These compounds may influence atmospheric chemistry and/or affect atmospheric ozone levels. They also might contribute to climate change (Hahn and Crutzen 1982; Charlson and Rodhe 1982; Charlson et al. 1987; Solomon et al. 1994). Due to the great difficulty in direct measurement of the gas transfer rate between the ocean and the atmosphere, considerable uncertainty still exists in the balances and air/sea fluxes for these gases. Early measurements of the air/sea gas exchange rates were focused on the ^{14}C and ^{222}Rn methods (Roether and Kromer 1978; Broecker and Peng 1982; Smethie et al. 1985). Fundamental methodological problems with these methods led to the development and application of dual gas tracer experiments. In these experiments, two inert gases with similar solubility and different diffusion coefficients are injected into a near-surface water body and the change in the ratio of the gases due to gas exchange across the molecular boundary layer is converted into gas transfer coefficients. Measurement of the ratio of two gases avoids some of the difficulties with inhomogeneities in the tracer distributions in the water that affected the ^{222}Rn method and the large scale averaging that was used for the ^{14}C method. The most widely used tracers are ^{3}He in combination with SF_6.

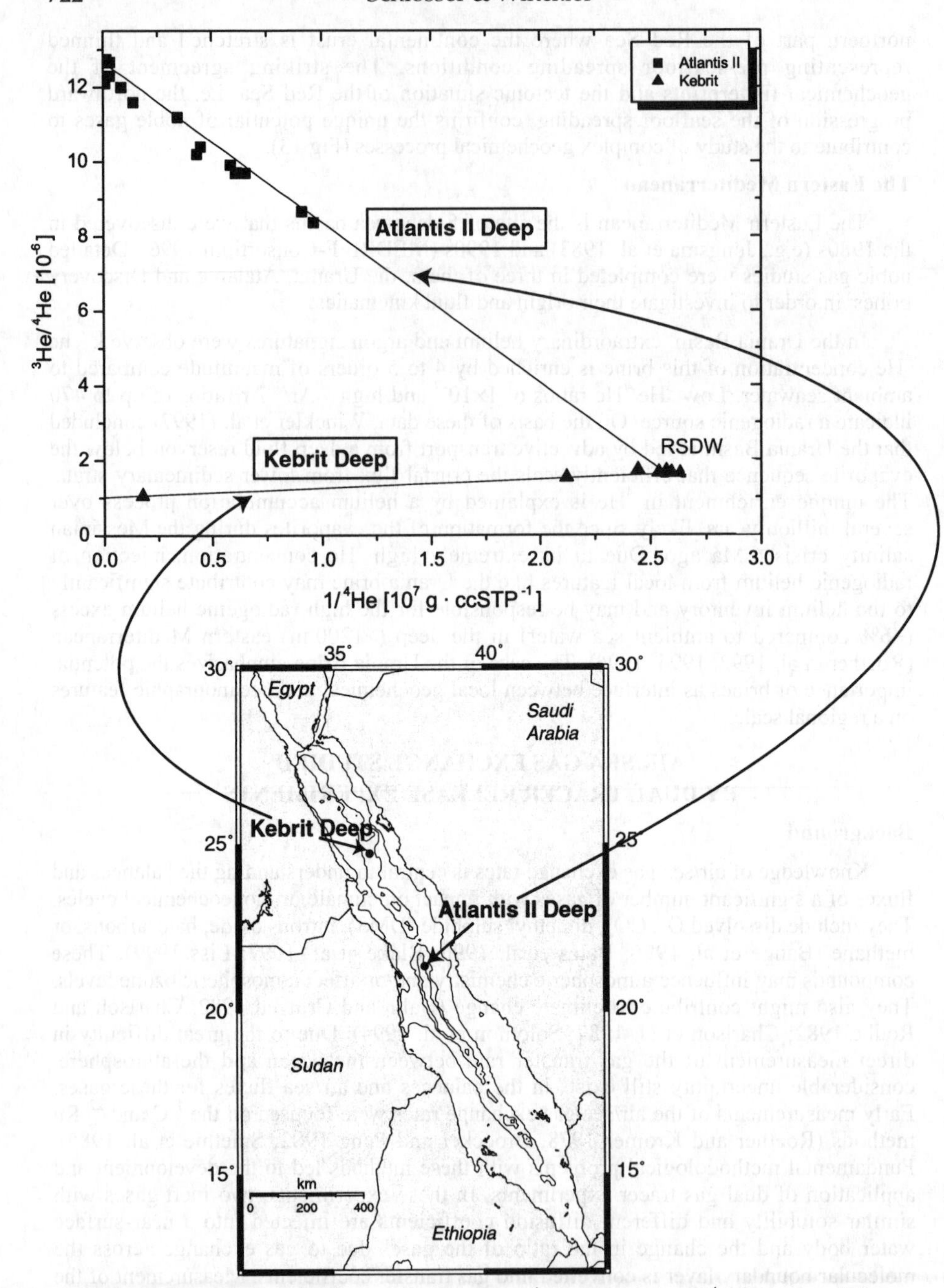

Figure 13. Correlation of $^3He/^4He$ with reciprocal of ^4He concentration ($1/^4He$) from two brine basins in the Red Sea after Winckler et al. (2001a). Samples from the Atlantis brine define a mixing between RSDW (Red Sea Deep Water) and a MORB-derived component ($^3He/^4He = 1.27\times10^{-5} = 9.2\ R_a$), samples from Kebrit brine lie on a mixing line between RSDW and a ^4He-enriched 'crustal' end-member ($^3He/^4He = 1\times10^{-6}$). The different helium isotopic finger prints of the two brines reflect the geological setting, i.e., the northward progression of seafloor spreading in the Red Sea.

$SF_6/^3He$ Method

Dispersion and gas exchange rates in unconfined natural water systems can be determined using two volatile gaseous tracers with different gas exchange rates if the ratio of the gas exchange rates is known. The gas exchange rate of a gas is proportional to the Schmidt number (defined as the kinematic viscosity of water divided by the molecular diffusivity of the gas in water) of the gas raised to an exponent *n* (Jähne et al. 1987). For wavy, unbroken water surfaces without bubble entrainment, *n* has been shown to be -0.5 in both laboratory experiments (Jähne et al. 1984; Ledwell 1984; Asher et al. 1992) and field measurements (Watson et al. 1991). Under these conditions, the ratio of *k* for SF_6 and ^3He can be expressed as:

$$\frac{k_{SF_6}}{k_{^3He}} = \left(\frac{Sc_{SF_6}}{Sc_{He}}\right)^{-0.5} \tag{9}$$

where k_{SF6} and k_{3He} are the gas exchange rates of SF_6 and ^3He, Sc_{SF6} and Sc_{3He} are the Schmidt numbers for SF_6 and ^3He, respectively. The gas exchange rate of ^3He can be determined by combining the advection-diffusion equation for ^3He and SF_6 in water and incorporating Equation (9):

$$k_{^3He} = h\frac{d}{dt}\left(\frac{\ln(^3He/SF_6)}{1-\left(Sc_{SF_6}/Sc_{^3He}\right)^{-0.5}}\right) \tag{10}$$

where *h is* the average water depth, -0.5 is the Schmidt number exponent, and ^3He and SF_6 are the excess of ^3He and SF_6 in the mixed layer. Conversely, k_{SF6} can be determined by substituting SF_6 for ^3He and vice versa in Equation (10). After separating the gas exchange component, the dispersion can be accounted for with a method described by Clark et al. (1996).

Results from oceanic dual gas tracer releases

The dual gas tracer method (typically ^3He and SF_6) has been used to study gas exchange in rivers/estuaries (e.g., Clark et al. 1994), the coastal ocean (e.g., Watson et al. 1991; Wanninkhof et al. 1993; Wanninkhof et al. 1997; Nightingale et al. 2000a), and the open ocean (Nightingale et al. 2000b). For such experiments, mixtures of ^3He and SF_6 are injected into the surface waters and the decrease of the $^3He/SF_6$ ratio is monitored as a function of time and wind speed. Results from an experiment on Georges Bank indicate that the method can provide valuable data that yield gas exchange rates averaged over periods of several days. During the course of the experiment the $^3He/SF_6$ ratio decreased by a factor of about 3 over a period of 8 days due to gas loss to the atmosphere (Fig. 14); ^3He escapes much faster than SF_6 due to its higher diffusion coefficient in the molecular boundary layer. The measured $^3He/SF_6$ ratios were translated into gas transfer coefficients using Equation (10). The resulting gas transfer coefficients provided valuable information on the relationship between gas transfer velocities and wind speed (Fig. 15). For a long time, this relationship was very poorly defined and even today there are still large gaps in our understanding of air/sea gas exchange, as well as in our capability to measure the variables that would define it better.

PERSPECTIVES

Noble gases have been firmly established as routine tools in studies of a variety of oceanographic fields. Since the first pioneering measurements of noble gases in the ocean and its sedimentary environment, rich data sets, some of them with global coverage at high spatial resolution, were collected and evaluated. In many cases such as the application of tritium/^3He and mantle ^3He to ocean circulation, detailed descriptions of

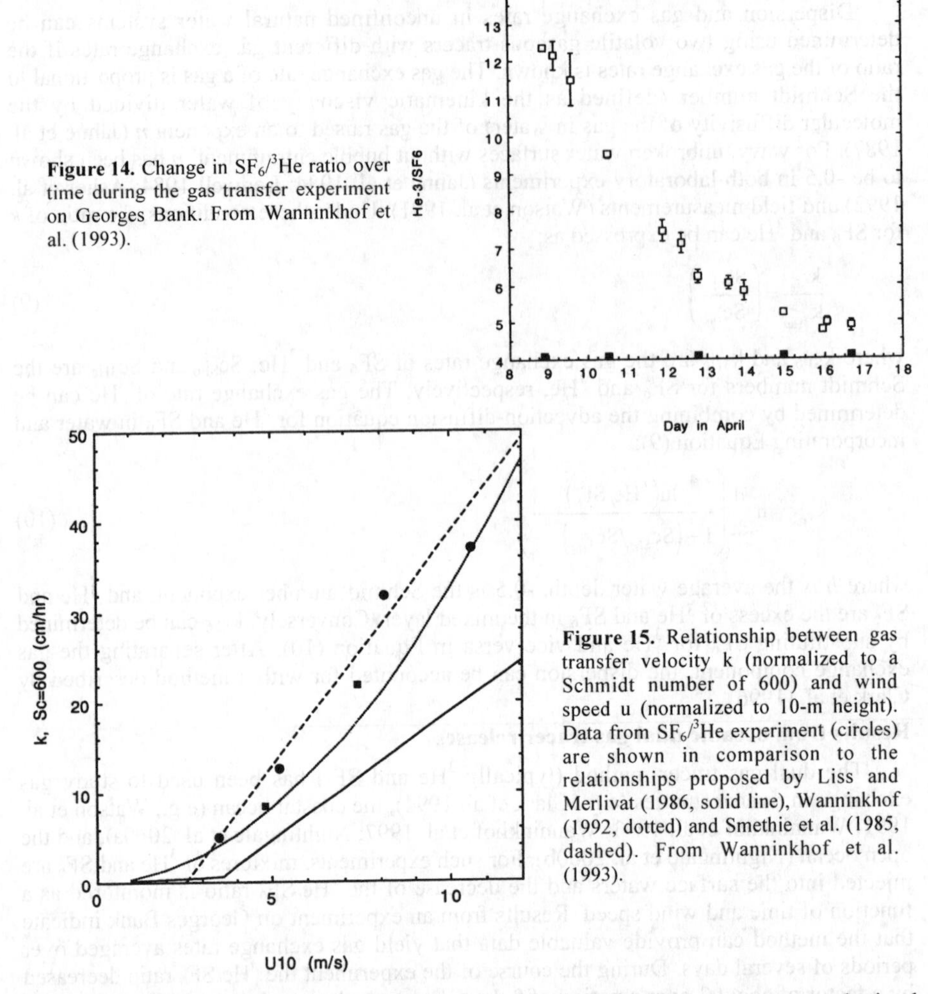

Figure 14. Change in $SF_6/^3He$ ratio over time during the gas transfer experiment on Georges Bank. From Wanninkhof et al. (1993).

Figure 15. Relationship between gas transfer velocity k (normalized to a Schmidt number of 600) and wind speed u (normalized to 10-m height). Data from $SF_6/^3He$ experiment (circles) are shown in comparison to the relationships proposed by Liss and Merlivat (1986, solid line), Wanninkhof (1992, dotted) and Smethie et al. (1985, dashed). From Wanninkhof et al. (1993).

the tracer fields consisting of several tens of thousands of measurements were obtained mainly during the 1990s in the framework of the WOCE. Synthesis and analysis of these data is underway and will contribute to our understanding of the formation, circulation and variability of oceanic water masses. They are increasingly used in model calibration efforts and they guide future sampling campaigns. Although the tritium signal is decreasing in the oceans due to the spike-type delivery of tritium with the main peak having occurred in the early 1960s, as well as the radioactive decay of tritium, there is still a sufficiently large signal left to obtain a signal to noise ratio of ~100 in the northern hemisphere. It is anticipated that tritium/^{3}He will yield valuable information on ocean circulation that is complementary to that obtained from other methods, including transient tracers such as the CFCs (chlorofluorocarbons) CFC 11, CFC 12, and CFC 113. Other applications of noble gases to oceanographic studies are still in the stage of establishing and firming up the methodology. For example, extraterrestrial ^{3}He in deep-sea sediments has shown potential as tracer for estimating sediment accumulation rates and first results have been obtained that show its value for paleoceanographic studies. The next steps in the development of this tracer have to be systematic surveys of deep-sea sediment cores and careful examination of the constancy of the ^{3}He flux to the sediment over long time scales. Similarly, the studies of deep-sea brines have shown the value of noble gases for such studies. Further results will reveal the full strength of these tracers in investigations of the origin and dynamics of such brines. The few coastal and open ocean ^{3}He/SF$_6$ release experiments showed that this method yields valuable results. Future applications should be integrated with direct flux measurements of the air/sea exchange that allow us to resolve short time scales (the ^{3}He/SF$_6$ experiments provide gas exchange rates averaged over longer time scales and larger space scales).

Future developments in Accelerator Mass Spectrometry might enable development of ^{39}Ar to a routine tool in oceanography. If the technical difficulties in measuring this noble gas isotope on small water samples can be mastered, it would offer a unique tool for studies of ocean circulation on time scales ranging from decades to the order of one thousand years, a good match for mean residence times of intermediate and deep waters (e.g., Loosli 1989).

Overall, during the past roughly four decades noble gases have significantly contributed to the field of oceanography. We anticipate that future studies will provide further extensive data sets that will form the foundation for improvement in our knowledge of a wide variety of topics. These topics reach from modern circulation of the ocean to understanding processes that shaped the oceanic environment over many millions of years.

ACKNOWLEDGMENTS

This chapter was written with support from the U.S. National Science Foundation (Grant # OCE 9820130) and the Deutsche Akademie der Naturforscher Leopoldina (Leopoldina Fellowship to G. Winckler). Comments by W. Aeschbach-Hertig and an anonymous reviewer considerably improved the manuscript. This is L-DEO contribution #6353.

REFERENCES

Allègre CJ, Sarda P, Staudacher T (1993) Speculations about the cosmic origin of He and Ne in the interior of the Earth. Earth Planet Sci Lett 117:229-233

Amari S, Ozima M (1985) Search for the origin of exotic helium in deep-sea sediments. Nature 317: 520-522

Amari S, Ozima M (1988) Extraterrestrial noble gases in deep-sea sediments. Geochim Cosmochim Acta 52:1087-1095

Anderson DL (1993) Helium-3 from the mantle: primordial signal or cosmic dust? Science 261:170-176

Asher WE, Farley PJ, Wanninkhof R, Monahan EC, Bates TS (1992) Laboratory and field measurements concerning the correlation of fractional area foam coverage with air/sea gas transport. *In* Precipitation Scavenging and Atmosphere-Surface Exchange, Vol. 2. SE Schwartz, WGN Slinn (eds) Hemisphere, p 815-828

Atakan Y, Roether W, Münnich KO, Matthess G (1974) The Sandhausen shallow-groundwater tritium experiment. *In* Isotope Techniques in Groundwater Hydrology (Proc Symp Vienna), Intl Atom Energy Agency, Vienna, p 21-43

Bange HW, Rapsomanikis S, Andreae MO (1996) Nitrous oxide emissions from the Arabian Sea. Geophys Res Lett 23:3175-3178

Bates TS, Kelly KC, Johnson JE, Gammon RH (1996) A re-evaluation of the open ocean source of methane to the atmosphere. J Geophys Res 101(D3):6953-6961

Begemann F, Libby WF (1957) Continental water balance, ground water inventory and storage times, surface ocean mixing rates and worldwide water circulation patterns from cosmic-ray and bomb tritium. Geochim Cosmochim Acta 12:277-296

Benson BB, Krause D (1980) Isotopic fractionation of helium during solution: A probe for the liquid state. J Solution Chem 9:895-909

Bieri RH, Koide M, Goldberg ED (1966) Noble gas contents of Pacific seawaters. J Geophys Res 71: 5243-5246

Blake NJ, Blake DR, Chen TY, Collins JE, Sachse GW, Anderson BE, Rowland FS (1997) Distribution and seasonality of selected hydrocarbons and halocarbons over the western Pacific basin during PEM-West A and PEM-West B. J Geophys Res 102(D23):28315-28331
Bönisch G, Schlosser P (1995) Deep water formation and exchange rates in the Greenland/Norwegian seas and the Eurasian Basin of the Arctic Ocean derived from tracer balances. Prog Oceanogr 35:29-52
Bönisch G, Blindheim J, Bullister JL, Schlosser P, Wallace DWR (1997) Long-term trends of temperature, and transient tracers in the central Greenland Sea. J Geophys Res 102:18553-18571
Bonatti E (1985) Punctiform initiation of seafloor spreading in the Red Sea during transition from a continental to an oceanic rift. Nature 316: 33-37
Broecker WS, Peng T-H (1982) Tracers in the Sea. Eldigio Press, Palisades, NY
Broecker WS, Peng T-H, Östlund HG (1986) The distribution of bomb tritium in the ocean. J Geophys Res 91:14331-14344
Brook EJ, Kurz MD, Curtice J, Cowburn S (2000) Accretion of interplanetary dust in polar ice. Geophys Res Lett 27:3145-3148
Butterfield DA, Massoth GJ, McDuff RE, Lupton JE, Lilley MD (1990) Geochemistry of hydrothermal fluids from Ashes Vent Field, Axial Seamount, Juan de Fuca Ridge: Sub-seafloor boiling and subsequent fluid-rock interaction. J Geophys Res 95(B8):12895-12921
Charlson RJ, Rodhe H (1982) Factors controlling the acidity of natural rainwater. Nature 295:683-685
Charlson RJ, Lovelock JE, Andreae MO, Warren SG (1987) Oceanic phytoplankton, atmospheric sulfur, cloud albedo and climate. Nature 326:655-661
Clark JF, Wannikhof R, Schlosser P, Simpson HJ (1994) Gas-exchange rates in the tidal Hudson River using a dual tracer technique. Tellus B 46:274-285
Clark JF, Schlosser P, Stute M, Simpson HJ (1996) SF_6-^3He tracer release experiment: a new method of determining longitudinal dispersion coefficients in large rivers. Environ Sci Techn 30:1527-1532
Clarke WB, Beg MA, Craig H (1969) Excess ^3He in the sea: Evidence for terrestrial primordial helium. Earth Planet Sci Lett 6:213-220
Clarke WB, Beg MA, Craig H (1970) Excess He-3 in East Pacific. EOS Trans Am Geophys Union 51: 325-326
Clarke WB, Jenkins WJ, Top Z (1976) Determination of tritium by mass spectrometric measurement of ^3He. Intl J Appl Rad Isotopes 27:515-522
Craig H, Lal D (1961) The production rate of natural tritium. Tellus 13:85-105
Craig H, Lupton JE (1981) Helium-3 and mantle volatiles in the ocean and the oceanic crust. *In* The Sea, Vol. 7. C Emiliani (ed) John Wiley & Sons, New York, p 391-428
Craig H, Weiss RF (1971) Dissolved gas saturation anomalies and excess helium in the ocean. Earth Planet Sci Lett 10:289-296
Craig H, Clarke WB, Beg MA (1975) Excess ^3He in deep water on the East Pacific Rise. Earth Planet Sci Lett 26:125-132
Dohnanyi JS (1976) Sources of interplanetary dust: Asteroids. *In* Interplanetary dust and zodiacal light 48:187-206. H Elsasser, H Fechtig (eds) Springer-Verlag, Berlin
Doney SC, Jenkins JW (1994) Ventilation of the deep western boundary current and abyssal western North Atlantic - Estimates from tritium and He-3 distributions. J Phys Oceanogr 24:638-659
Dreisigacker E, Roether W (1978) Tritium and ^{90}Sr in north Atlantic surface water. Earth Planet Sci Lett 38:301-312
Eltgroth SF, Farley KA (2001) High resolution timing of the Late Paleocene Thermal Maximum by extraterrestrial ^3He implied sedimentation rates at ODP site 690B. EOS Trans Am Geophys Union, Fall Mtg Suppl 82(:F1140
Farley KA (1995) Cenozoic variations in the flux of interplanetary dust recorded by ^3He in a deep-sea sediment. Nature 376:153-156
Farley KA (2001) Extraterrestrial helium in seafloor sediments: Identification, characteristics, and accretion rate over geological time. *In* Accretion of extraterrestrial matter throughout Earth's history. B Peucker-Ehrenbrink, B Schmitz (eds) Kluwer Academic/Plenum Publishers, p 179-204
Farley KA, Patterson DB (1995) A 100-kyr periodicity in the flux of extraterrestrial ^3He to the sea floor. Nature 378:600-603
Farley KA, Maier-Reimer E, Schlosser P, Broecker WS (1995) Constraints on mantle ^3He fluxes and deep-sea circulation from an oceanic general circulation model. J Geophys Res 100(B3):3829-3839
Farley KA, Love SG, Patterson DB (1997) Atmospheric entry heating and helium retentivity of interplanetary dust particles. Geochim Cosmochim Acta 61:2309-2316
Farley KA, Montanari A, Shoemaker EM, Shoemaker CS (1998) Geochemical evidence for a comet shower in the Late Eocene. Science 280:1250-1253
Farley KA, Mukhopadhyay S, Eltgroth S (2001) Assessment of the distribution of time in seafloor sediments using extraterrestrial ^3He. EOS Trans Am Geophys Union Fall Mtg Suppl 82:F1140

François R, Bacon MP, Suman DO (1990) ^{230}Th profiling in deep-sea sediments: high-resolution records of flux and dissolution of carbonate in the equatorial Atlantic during the last 24,000 y. Palaeoceanography 5:761-787
Fraundorf P, Lyons T, Schubert P (1982) The survival of solar flare tracks in interplanetary dust silicates on deceleration in the Earth's atmosphere. J Geophys Res 87:409-412
Fukumoto H, Nagao K, Matsuda J (1986) Noble gas studies on the host phase of high ^{3}He/^{4}He ratios in deep-sea sediments. Geochim Cosmochim Acta 50:2245-2253
Giletti BJ, Bazan F, Kulp JL (1958) The geochemistry of tritium. Trans Am. Geophys. Union 39:807-818
Grosse AV, Johnston WM, Wolfgang RL, Libby WF (1951) Tritium in nature. Science 113:1-2
Hahn J, Crutzen PJ (1982) The role of fixed nitrogen in atmospheric photochemistry. Phil Trans R Soc London, Ser B 296:521-541
Hartmann M, Scholten JC, Stoffers P, Wehner F (1998) Hydrographic structure of brine-filled deeps in the Red Sea: new results from the Shaban, Kebrit, Atlantis II and Discovery Deep. Mar Geol 144:311-330
Higgins S. (2001) Extraterrestrial tracer in the sea: evaluation and application of ^{3}He in interplanetary dust particles as a 'constant flux' tracer in marine sediments. PhD dissertation, Columbia University, New York, New York
Hiyagon H (1994) Retention of solar helium and neon in IDPs in deep sea sediments. Science 263: 1257-1259
Hohmann R, Schlosser P, Ludin A, Weppernig R (2002) Excess helium and neon in the Southeast Pacific: tracers for glacial meltwater. J Geophys Res (in press)
Jähne B, Huber W, Dutzi A, Wais T, Ilmberger J (1984) Wind/Wave-tunnel experiment on the Schmidt number and wave field dependence of air/water gas exchange. *In* Gas Transfer at Water Surfaces W Brutsaert, GH Jirka (eds) Reidel, Norwell, Massachusetts, p 303-309
Jähne B, Münnich KO, Bösinger R, Dutzi A, Huber W, Libner P (1987) On parameters influencing air-water gas exchange. J Geophys Res 92:1937-1949
Jenkins WJ (1974) Helium Isotope and Rare Gas Oceanology. PhD dissertation, McMaster University, Toronto, 170 p
Jenkins WJ (1982) Oxygen utilization rates in North-Atlantic sub-tropical gyre and primary production in oligotrophic systems. Nature 300:246-248
Jenkins WJ (1987) H-3 and He-3 in the Beta-Triangle: Observations of gyre ventilation and oxygen utilization rates. J Phys Oceanogr 17:763-783
Jenkins WJ (1996) Tritium and ^{3}He in the WOCE Pacific Programme. Intl WOCE Newslett 23:6-8.
Jenkins WJ (1998) Studying subtropical thermocline ventilation and circulation using tritium and He-3. J Geophys Res 103(C8):15817-15831
Jenkins WJ, Clarke WB (1976) The distribution of ^{3}He in the Western Atlantic Ocean. Deep-Sea Res. 23:481-494
Jongsma D, Fortuin AR, Huson W, Troelstra SR, Klaver GT, Peters JM, Van Harten D, De Lange GJ, Ten Haven L (1983) Discovery of an anoxic basin within the Strabo Trench, eastern Mediterranean. Nature 305:795-797
Kaufman S, Libby WF (1954) The natural distribution of tritium. Phys Rev 93:1337-1344
Kipfer R, Aeschbach-Hertig W, Peeters F, Stute M (2002) Noble gases in lakes and groundwaters. Rev Mineral Geochem 47:615-700
König H, Wänke H, Bien GS, Rakestraw NW, Suess HE (1964) Helium, neon and argon in the oceans. Deep-Sea Res. 11:243-247
Krylov AY, Mamyrin BA, Khabarin LA, Mazina TI, Silin YI (1974) Helium isotopes in ocean-floor bedrock, Geochem Int 11:839-844
Ledwell JR (1984) The variation of the gas transfer coefficient with molecular diffusivity. *In* Gas Transfer at Water Surfaces. W Brutsaert, GH Jirka (eds) Reidel, Norwell, Massachusetts, p 299-302
Liss PS (1999) Biogeochemistry - Take the shuttle-from marine algae to atmospheric chemistry. Science 285:1217-1218
Liss PS, Merlivat L (1986) Air-sea gas exchange rates: Introduction and synthesis. *In* The Role of Air-Sea Exchange in Geochemical Cycling. P Buat-Menard (ed) Reidel, Norwell, Massachusetts, p 113-129
Loosli HH (1989) Argon-39: A tool to investigate ocean water circulation and mixing. *In* Handbook of Environmental Isotope Geochemistry. P Fritz, J Ch Fontes (eds) Elsevier, Amsterdam, p 385-392.
Love SG, Brownlee DE (1993) A direct measurement of the terrestrial mass accretion rate of cosmic dust. Science 262:250-253
Lupton JE (1996) A far-field hydrothemal plume from Loihi Seamount. Science 272:976-979
Lupton JE (1998) Hydrothermal helium plumes in the Pacific Ocean. J Geophys Res 103:15853-15868
Lupton JE, Craig H (1975) Excess ^{3}He in oceanic basalts: evidence for terrestrial primordial helium. Earth Planet Sci Lett 26:133-139
Lupton JE, Craig H (1981) A major ^{3}He source at 15°S on the East Pacific Rise. Science 214:13-18

Lupton JE, Weiss RF, Craig H (1977) Mantle helium in the Red Sea brines. Nature 266:244-246
Lupton JE, Klinkhammer GP, Normark WR, Haymon R, Macdonald KC, Weiss RF, Craig H (1980) Helium-3 and manganese at the 21°N East Pacific Rise hydrothermal site. Earth Planet Sci Lett 50:115-127
Mamyrin BA, Tolstikhin IN (1984) Helium isotopes in nature. Elsevier, Amsterdam
Mamyrin BA, Tolstikhin IN, Anufriyev GS, Kamenskii IL (1969) Isotopic analysis of terrestrial helium on a magnetic resonance mass spectrometer. Geochem Intl 6:517-524
Marcantonio F, Kumar N, Stute M, Anderson RF, Seidl MA, Schlosser P, Mix A (1995) A comparative study of accumulation rates derived by He and Th isotope analysis of marine sediments. Earth Planet Sci Lett 133:549-555
Marcantonio F, Anderson RF, Stute M, Kumar N, Schlosser P, Mix A (1996) Extraterrestrial ^{3}He as a tracer of marine sediment transport and accumulation. Nature 383:705-707
Marcantonio F, Higgins S, Anderson RF, Stute M, Schlosser P, Rasbury ET (1998) Terrigeneous helium in deep-sea sediments. Geochim Cosmochim Acta 62:1535-1543
Marcantonio F, Anderson RF, Higgins S, Stute M, Schlosser P, Kubik P (2001) Sediment focusing in the central equatorial Pazific Ocean. Paleoceanography 16:260-267
MEDRIFF-Consortium (1996) Three brine lakes discovered in the seafloor of the eastern Mediterranean. EOS Trans Am Geophys Union 76:313
Merrihue C (1964) Rare gas evidence for cosmic dust in modern Pacific red clay. Ann. NY Acad. Sci. 119:351-367
Miller AR, Densmore CD, Degens ET, Hathaway JC, Manheim FT, McFarlin PF, Pocklington R, Jokela A (1966) Hot brines and recent iron deposits in deeps of the Red Sea. Geochim Cosmochim Acta 30: 341-359
Morison J, Steele M, Andersen R (1998) Hydrography of the upper Arctic Ocean measured from the nuclear submarine USS Pargo. Deep-Sea Res Part I 45:15-38
Morison J, Aagaard K, Steele M (2000) Recent environmental changes in the Arctic: A review. Arctic 53:359-371
Mukhopadhyay S, Farley KA, Montanari A (2001) A short duration of the Cretaceous-Tertiary boundary event: Evidence from extraterrestrial helium-3. Science 291:1952-1955
Muller RA, MacDonald GJ (1995) Glacial cycles and orbital inclination. Nature 377:107-108
Muller RA, MacDonald GJ (1997a) Glacial cycling and astronomical forcing. Science 277:215-218
Muller RA, MacDonald GJ (1997b) Spectrum of 100kyr glacial cycle: Orbital inclination, not eccentricity. Proc Nat Acad Sci USA 94:8329-8334
Muller RA, MacDonald GJ (1997c) Simultaneous presence of orbital inclination and eccentricity in proxy climate records from Ocean Drilling Program Site 806. Geology 25:3-6
Münnich KO, Roether W (1967) Transfer of bomb ^{14}C and tritium from the atmosphere to the ocean: Internal mixing of the ocean on the basis of tritium and ^{14}C profiles. *In* Radioactive dating and methods of Low Level Counting (Proc Symp Monaco), Intl Atom Energy Agency, Vienna, p 93-104
Nier AO, Schlutter DJ (1990) Helium and neon in stratospheric particles. Meteoritics 25:263-267
Nier AO, Schlutter DJ (1992) Extraction of helium from individual interplanetary dust particles by step-heating. Meteoritics 27:166-173
Nier AO, Schlutter DJ (1993) The thermal history of interplanetary dust particles collected in earth's stratosphere. Meteoritics 28:675-681
Nier AO, Schlutter DJ, Brownlee DE (1990) Helium and neon isotopes in deep Pacific Ocean sediments. Geochim Cosmochim Acta 54:173-182
Nightingale PD, Malin G, Law CS, Watson AJ, Liss PS, Liddicoat MI, Boutin J, Upstill-Goddard RC (2000a) *In situ* evaluation of air-sea gas exchange parameterizations using novel conservative and volatile tracers. Global Biogeochem Cycl 14:373-387
Nightingale PD, Liss PS, Schlosser P (2000b) Measurement of air-sea gas transfer during an open ocean algal bloom. Geophys Res Lett 27:2117-2120
Östlund HG (1982) The residence time of the freshwater component in the Arctic Ocean. J Geophys Res 87:2035-2043
Östlund HG, Craig H, Broecker WS, Spencer D (1987) GEOSECS Atlantic, Pacific, and Indian Ocean Expeditions. Vol 7, Shore-based Data and Graphics. National Science Foundation, Washington DC
Ozima M, Takayanagi M, Zashu S, Amari S (1984) High ^{3}He/^{4}He ratios in ocean sediments. Nature 311:449-451
Patterson DB, Farley KA (1998) Extraterrestrial ^{3}He in seafloor sediments: Evidence for correlated 100kyr periodicity in the accretion rate of interplanetary dust, orbital parameters, and Quarternary climate. Geochim Cosmochim Acta 62:3669-3682
Patterson DB, Farley KA, Schmitz B (1998) Preservation of extraterrestrial ^{3}He in 480-Ma-old marine limestones. Earth Planet Sci Lett 163:315-325

Robbins PE, Jenkins WJ (1998) Observations of temporal changes of tritium-He-3 age in the eastern North Atlantic thermocline: Evidence for changes in ventilation? J Mar Res 56:1125-1161

Robbins PE, Price JF, Owens WB, Jenkins WJ (2000) The importance of lateral diffusion for the ventilation of the lower thermocline in the subtropical North Atlantic. J Phys Oceanogr 30:67-89

Roether W, Fuchs G (1988) Water mass transport and ventilation in the Northeast Atlantic derived from tracer data. Philos Trans R Soc London, Ser. A 325:63-69

Roether W, Kromer B (1978) Field determination of air-sea gas-exchange by continuous measurement of Rn-222. Pure Appl Geophysics 116:476-485

Roether W, Münnich KO, Östlund HG (1970) Tritium profile at the North Pacific (1969) Geosecs Intercallibration Station. J Geophys Res 75:7672-7675

Roether W, Schlosser P, Kuntz R, Weiss W (1992) Transient-tracer studies of the thermohaline circulation of the Mediterranean. *In* Winds and Currents of the Mediterranean Basin: Reports in Metereology and Oceanography 41, Vol. 2. H Charnock (ed) p 291-317

Roether W, Roussenov VM, Well R (1994) A tracer study of the thermohaline circulation of the eastern Mediterranean. *In* Ocean processes in climate dynamics: Global and Mediterranean examples. Malanotte-Rizzoli P, Robbinson AR (eds) Kluwer Academic, Norwell, Massachusetts, p 371-394

Roether W, Manca BB, Klein B, Bregant D, Georgopoulos D, Beitzel V, Kovacevic V, Luchetta A (1996) Recent changes in eastern Mediterranean deep waters. Science 271:333-335

Roether W, Well R, Putzka A, Rüth C (1998) Component separation of oceanic helium. J Geophys Res 103(C12):27931-27946

Rüth C, Well R, Roether W (2000) Primordial He-3 in South Atlantic deep waters from sources on the Mid-Atlantic Ridge. Deep-Sea Res Part I 147:1059-1075

Schlosser P (1986) Helium: a new tracer in Antarctic oceanography. Nature 321:233-235

Schlosser P (1992) Tritium/^{3}He dating of waters in natural systems. *In* Isotopes of noble gases as tracers in environmental studies, Intl Atom Energy Agency, Vienna, p 123-145

Schlosser P, Bayer R, Foldvik A, Gammelsrod T, Rohardt G, Münnich KO (1990) O-18 and helium as tracers of ice shelf water and water/ice interaction in the Weddel Sea. J Geophys Res 95(C3):3253-3263

Schlosser P, Bönisch G, Rhein M, Bayer R (1991) Reduction of deep water formation in the Greenland Sea during the 1980s - Evidence from tracer data. Science 251:1054-1056

Schlosser P, Bönisch G, Kromer B, Loosli HH, Buehler R, Bayer R, Bonani G, Koltermann KP (1995) Mid-1980s distribution of tritium, ^{3}He, ^{14}C and ^{39}Ar in the Greenland/Norwegian Seas and the Nansen Basin of the Arctic Ocean. Prog Oceanogr 35:1-28

Schlosser P, Bullister JL, Fine R, Jenkins WJ, Key R, Lupton J, Roether W, Smethie WM (2001) Transformation and age of water masses. *In* Ocean circulation and climate. Academic Press, p 431-452

Smethie WM, Takahashi T, Chipman DW, Ledwell JR (1985) Gas exchange and CO_2 flux in the tropical Atlantic Ocean determined from ^{222}Rn and pCO_2. J Geophys Res 90:7005-7022

Solomon S, Burkholder JB, Ravishankara AR, Garcia RR (1994) Ozone depletion and global warming potential of CF3I. J Geophys Res 99(D10):20929-20935

Suess HE, Wänke H (1965) On the possibility of a helium flux through the ocean floor. Prog Oceanogr 3:347-353

Takayanagi M, Ozima M (1987) Temporal variation of ^{3}He/^{4}He ratio recorded in deep-sea sediment cores. J Geophys Res 92:12531-12538

Thiele G, Roether W, Schlosser P, Kuntz R, Siedler G, Stramma L (1986) Baroclinic flow and transient tracer fields in the Canary Cape Verde Basin. J Phys Oceanogr 16:814-826

Tilles D (1966) Atmospheric noble gases from extraterrestrial dust. Science 151:1015-1018

Tilles D (1967) Extraterrestrial excess ^{36}Ar and ^{38}Ar concentrations as possible accumulation-rate indicators for sea sediments. Icarus 7:94-100

Top Z, Clarke WB (1983) Helium, neon and tritium in the Black Sea. J Mar Res 41:1-17

Torgersen T (1989) Terrestrial helium degassing fluxes and the atmospheric helium budget: Implications with respect to the degassing processes of continental crust. Chem Geol 79:1-14

Unterweger MP, Coursey BM, Schima FJ, Mann WB (1980) Preparation and calibration of the 1978 National Bureau of Standards tritiated-water standards. Intl J Appl Rad Isotopes 31:611-614

von Faltings V, Harteck P (1950) Der Tritiumgehalt der Atmosphäre. Z Naturforsch 5a:438-439

Wanninkhof R (1992) Relationship between wind-speed and gas exchange over the ocean. J Geophys Res 97 (C5):7373-7382

Wanninkhof R, Asher W, Weppernig R, Chen H, Schlosser P, Langdon C, Sambrotto R (1993) Gas transfer experiment on Georges Bank using 2 volatile deliberate tracers. J Geophys Res 98(C11):20237-20248

Wanninkhof R, Hitchcock G, Wiseman WJ, Vargo G, Ortner PB, Asher W, Ho DT, Schlosser P, Dickson ML, Masserini R, Fanning K, Zhang JZ (1997) Gas exchange, dispersion, and biological productivity on the west Florida shelf: Results from a Lagrangian tracer study. Geophys Res Lett 24:1767-1770

Watson AJ, Upstill-Goddard RC, Liss P S (1991) Air-sea gas exchange in rough and stormy seas measured by a dual tracer technique. Nature 349:145-147

Weiss RF (1971) Solubility of helium and neon in water and seawater. J Chem Eng Data 16:235-241

Weiss W, Roether W (1980) The rates of tritium input to the world oceans. Earth Planet Sci Lett 49: 435-446

Weiss W, Bullacher W, Roether W (1979) Evidence of pulsed discharge of tritium from nuclear energy installations in Central European precipitation. *In* Behavior of Tritium in the Environment (Proc Symp, San Francisco), Intl Atom Energy Agency, Vienna, p 17-30

Well R, Lupton J, Roether W (2001) Crustal helium in deep Pacific waters. J Geophys Res 106(C7): 14165-14177

Weppernig R, Schlosser P, Khatiwala S, Fairbanks RG (1996) Isotope data from Ice Station Weddell: implications for deep water formation in the western Weddell Sea. J Geophys Res :25,723-25,739

Winckler G, Suess E, Wallmann K, De Lange GJ, Westbrook GK, Bayer R (1997) Excess helium and argon of radiogenic origin in Mediterranean brine basins. Earth Planet Sci Lett 151:225-233

Winckler G, Kipfer R, Aeschbach-Hertig W, Botz R, Schmidt M, Schuler S, Bayer R (2000) Sub-seafloor boiling of Red Sea Brines: New indication from noble gas data. Geochim Cosmochim Acta 64:1567-1575

Winckler G, Aeschbach-Hertig W, Kipfer R, Botz R, Rübel AP, Bayer R, Stoffers P (2001a) Constraints on origin and evolution of Red Sea brines from helium and argon isotopes. Earth Planet Sci Lett 184: 671-683

Winckler G, Anderson RF, Schlosser P, Stute M (2001b) Constant export productivity in the Equatorial Pacific through the Mid-Pleistocene Climate Transition - New evidence from an extraterrestrial ^{3}He record. EOS Trans Am Geophys Union, Fall Mtg Suppl 8:F636

16 Cosmic-Ray-Produced Noble Gases in Terrestrial Rocks: Dating Tools for Surface Processes

Samuel Niedermann

GeoForschungsZentrum Potsdam
Telegrafenberg
D-14473 Potsdam, Germany

nied@gfz-potsdam.de

INTRODUCTION

The production of noble gas isotopes by interactions of high-energy cosmic ray particles with rocks was first recognized half a century ago when Paneth et al. (1952) showed that the high $^{3}He/^{4}He$ ratios in iron meteorites must be due to cosmic ray interactions and provide information about their ages. Other investigations in the late 1950s and the 1960s revealed the presence of a whole spectrum of cosmic-ray-produced noble gases in meteorites, showing characteristic isotopic abundances deviating substantially from those of all other known noble gas reservoirs (e.g., Marti et al. 1966). Likewise, a number of radionuclides with half-lives of some 10^{5} to 10^{7} years were also detected, including the unstable noble gas isotope ^{81}Kr. As cosmic ray particles can penetrate a few meters at most into rock material, an obvious application of these "cosmogenic" nuclides was to determine the time during which meteorites had traveled through space as small objects, which of course required the knowledge of production rates. After the return of lunar samples by the Apollo and Luna missions, the new technique was successfully applied to date the surface exposure of lunar rocks and, thereby, determine for instance the ages of lunar craters (e.g., Eugster et al. 1977). A review of the basics, methods, and results of cosmogenic nuclide studies in extraterrestrial material is given in another chapter of this book (Wieler 2002).

On Earth, most interactions of cosmic rays with matter occur in the upper layers of the atmosphere, whose shielding depth down to sea level is 1033 g/cm^{2}, corresponding to about 3.5 m of rock. Therefore it is not surprising that interactions with terrestrial rock material are several orders of magnitude less abundant than in space. Nevertheless, the first observation of a cosmogenic nuclide in a terrestrial rock dates back almost as long as that in a meteorite: Davis and Schaeffer (1955) determined the activity of the radionuclide ^{36}Cl in a Cl-rich rock from Cripple Creek, Colorado, and already developed some of the principles of surface exposure dating. Based on a calculated saturation activity, they estimated a surface exposure age of 24,000 years for that rock. Furthermore, they discussed the potential applications of the method for dating glaciation events or volcanic eruptions and also mentioned the effects of erosion.

In the twenty years following Davis and Schaeffer's pioneering work, attention within the field of *in situ* produced cosmogenic nuclides focused on extraterrestrial matter as described above, probably both due to the fascinating attraction of the rapidly evolving field of space technology and to the generally much higher concentrations of cosmogenic nuclides better amenable to mass spectrometric detection at that time. The first report of cosmogenic noble gas isotopes in a terrestrial surface rock is that of Srinivasan (1976), who found excesses of the light Xe isotopes in barites from Southern Africa and Australia and estimated exposure ages of between 50 and 270 ka for the two rocks. Curiously, this has remained the only study of terrestrial cosmogenic Xe until today. Three years later Craig et al. (1979) attributed the presence of excess ^{21}Ne (along with ^{3}He) in native

1529-6466/00/0047-0016$10.00

metals from Greenland to cosmic ray production. Activity in cosmogenic nuclide studies exploded in the year 1986. Nishiizumi et al. (1986) and Klein et al. (1986) observed the radionuclides ^{10}Be and ^{26}Al produced *in situ* in quartz and SiO_2 glass, respectively. Phillips et al. (1986) studied another radionuclide, ^{36}Cl, in volcanic rocks. Kurz (1986a,b) and Craig and Poreda (1986) reported the presence of cosmogenic ^{3}He in olivine and pyroxene grains separated from Hawaiian basalts. And finally, early in the following year Marti and Craig (1987) announced their detection of cosmic-ray-produced ^{21}Ne along with ^{3}He in olivine and clinopyroxene phenocrysts from Maui.

These studies laid the foundations of surface exposure dating, showing the feasibility of detecting and quantitatively recording various *in situ* produced cosmogenic nuclides in terrestrial rocks and demonstrating a wealth of possible applications. What followed was a combination of methodological improvements, theoretical considerations, and application studies, which over the next decade slowly succeeded in convincing the Earth science community of the usefulness of the new technique. Several major reviews (Lal 1988; Cerling and Craig 1994a; Gosse and Phillips 2001), several workshops and conference sessions as well as special editions of journals (e.g., *Radiocarbon* 38(1), 1996; *Geomorphology* 27(1-2), 1999; *Nuclear Instruments and Methods in Physics Research B* 172(1-4), 2000) have dealt with the issue, and today cosmogenic nuclides have become widely accepted as a quantitative tool for investigating various processes acting on the surface of the Earth, which in many cases are not amenable to other methods (see *Application examples* section at the end of this chapter).

The present review will focus on the cosmogenic noble gas isotopes (mainly ^{3}He and ^{21}Ne), particularly where experimental issues are concerned. In view of the close relation and interdependence with radionuclide studies however, the latter will be included in the considerations wherever it seems appropriate, and all general information on surface exposure dating, such as the basic equations, will be given in a form applicable to radionuclides as well. For a more detailed discussion of cosmogenic radionuclides I refer to the extensive review of Gosse and Phillips (2001).

THE FUNDAMENTALS OF SURFACE EXPOSURE DATING

Production mechanisms of cosmogenic nuclides in terrestrial rocks

At the top of the Earth's atmosphere, cosmic rays are composed of ~ 87% protons, ~ 12% α particles, and minor contributions from heavier nuclei, electrons, and positrons. Typical energies range from a few MeV up to ~ 10^{20} eV, with a flux maximum at a few hundred MeV/nucleon (e.g., Simpson 1983). Upon entering the atmosphere, these particles interact with the constituents of air. Due to the high energies involved, which are far above the binding energies of atomic nuclei, the main type of nuclear reaction taking place is spallation, in which a few nucleons are sputtered off the target nucleus. In this way a cascade of secondary particles is produced as cosmic rays approach the terrestrial surface. Concurrently, the composition changes from proton-dominated to neutron-dominated, both due to the higher probability of neutrons to escape from a nucleus (because they do not have to overcome the Coulomb barrier) and to their enhanced range as compared to charged particles (because they do not lose energy by electromagnetic interactions). Weakly-interacting particles (electrons and muons) as well as photons are also present in the particle cascade. A detailed account of the propagation of cosmic rays in the atmosphere has been given by Lal and Peters (1967); a summary can be found in Desilets and Zreda (2001).

Spallation. The cosmic ray energy spectrum below ~ 12 km altitude in the atmosphere is invariant despite the roughly exponential decrease of its flux (Lal and

Peters 1967). Therefore secondary cosmic ray particles reaching the surface of the Earth still have ample energy to cause spallation to be the most important type of nuclear interaction taking place in surface rocks as well. The term "spallation" is used for a nuclear reaction of a fast incoming particle with a nucleus in which a few protons and neutrons are sputtered off, leaving behind a lighter nucleus. In Table 1, I have compiled spallation reactions important to cosmogenic nuclide studies. The probability of producing a given daughter nuclide (e.g., ^{21}Ne) from a target nuclide (e.g., ^{28}Si) depends on nuclear excitation functions (i.e., functions describing the response of a nucleus to specific nuclear interactions), which are qualitatively based on the following principles:

- Neutrons may escape more easily from a nucleus than protons because they do not have to overcome the so-called Coulomb barrier, an energy wall which only affects the charged protons. This favors the production of neutron-poor daughter nuclei, i.e., the light isotopes of an element.
- Alpha particles are particularly stable and are therefore preferably ejected. Therefore reactions such as ^{24}Mg(n,α)^{21}Ne have large cross sections (the cross section is a measure for the probability of a nuclear reaction).
- The difference in mass between the target nucleus and the product is typically a few atomic mass units. Therefore the most important contributions to the production of a given nuclide are from elements slightly higher in atomic number. An exception is He which is produced from all elements, because the nuclei of ^{4}He (i.e., alpha particles), ^{3}He, and ^{3}H (which subsequently decays to ^{3}He) are among the particles sputtered off.

Table 1. Examples of different nuclear reaction types producing nuclides which can be used in surface exposure studies.

Target element	*Spallation*	*Thermal neutron capture*	*Negative muon capture*
Li		^{6}Li(n,α)^{3}H(β$^-$)^{3}He	
O	^{16}O(n,2pn)^{14}C	^{17}O(n,α)^{14}C	^{16}O(μ$^-$,pn)^{14}C
	^{18}O(n,αn)^{14}C		^{16}O(μ$^-$,αpn)^{10}Be
	^{16}O(n,4p3n)^{10}Be		
Na	^{23}Na(n,pn)^{22}Ne		^{23}Na(μ$^-$,n)^{22}Ne
	^{23}Na(n,p2n)^{21}Ne		^{23}Na(μ$^-$,2n)^{21}Ne
	^{23}Na(n,p3n)^{20}Ne		
Mg	^{25}Mg(n,α)^{22}Ne		^{24}Mg(μ$^-$,pn)^{22}Ne
	^{24}Mg(n,α)^{21}Ne		^{24}Mg(μ$^-$,p2n)^{21}Ne
	^{24}Mg(n,αn)^{20}Ne		
Al	^{27}Al(n,3p3n)^{22}Ne		^{27}Al(μ$^-$,αn)^{22}Ne
	^{27}Al(n,3p4n)^{21}Ne		
	^{27}Al(n,2n)^{26}Al		
Si	^{29}Si(n,2α)^{22}Ne		^{28}Si(μ$^-$,2n)^{26}Al
	^{28}Si(n,2α)^{21}Ne		^{28}Si(μ$^-$,αpn)^{22}Ne
	^{28}Si(n,p2n)^{26}Al		
Cl		^{35}Cl(n,γ)^{36}Cl†	
K	^{39}K(n,pn)^{38}Ar	^{39}K(n,α)^{36}Cl†	^{39}K(μ$^-$,n)^{38}Ar
	^{41}K(n,p3n)^{36}Cl†		
Ca	^{40}Ca(n,αp)^{36}Cl†		^{40}Ca(μ$^-$,α)^{36}Cl†
	^{40}Ca(n,2pn)^{38}Ar		^{40}Ca(μ$^-$,pn)^{38}Ar
Fe	^{56}Fe(n,8p11n)^{38}Ar		^{54}Fe(μ$^-$,n)^{53}Mn
	^{54}Fe(n,pn)^{53}Mn		
	^{56}Fe(n,p3n)^{53}Mn		

† ^{36}Cl decays to ^{36}Ar with a half-life of 3×10^5 a

Thermal neutron capture. Some of the neutrons produced in the nuclear cascade are slowed down to thermal energies and can be captured by nuclei having a large cross section for such reactions. Among them are $^{6}Li(n,\alpha)^{3}H(\beta^{-})^{3}He$ and $^{35}Cl(n,\gamma)^{36}Cl$ (Table 1). Unlike spallation reactions, the product nucleus may be one atomic mass unit heavier than the target nucleus. In the case of (n,α) or (n,p) reactions the energy necessary for the disintegration of the nucleus is not provided by the incoming neutron but results from an excited state after neutron capture. Thermal neutron capture is of minor importance for production of light cosmogenic noble gases in terrestrial rocks, since Li is a very rare target element and Ne is not produced by such reactions. However, the radionuclide ^{36}Cl (half life 300,000 years) decays to ^{36}Ar, and there are also thermal neutron capture reactions producing Kr and Xe isotopes, such as $^{79}Br(n,\gamma\beta^{-})^{80}Kr$, $^{81}Br(n,\gamma\beta^{-})^{82}Kr$, $^{127}I(n,\gamma\beta^{-})^{128}Xe$.

Negative muon capture. Muons belong to the lepton particle family and can be viewed as the heavier brothers of electrons. They are the decay products of pions, very short-lived particles originating in proton-proton or proton-neutron collisions in the upper layers of the atmosphere. Due to their half life of 2.2×10^{-6} s these muons only reach the Earth's surface owing to the relativistic time dilatation prevailing in their system of reference. As weakly-interacting particles they have a longer range than nucleons and are indeed the most abundant cosmic ray particles present at sea level (e.g., Lal 1988).

Slowed down to thermal energies, negative muons can be captured by atoms and quickly cascade to the K shell (the innermost band of electron orbits), where they may decay to an electron and two neutrinos or interact with the nucleus, by reactions such as those given in Table 1. The energy available for excitation of the nucleus is smaller than in spallation reactions; therefore the mass difference between target and product nucleus is usually small (Lal 1988; Stone et al. 1998). In surface rocks at sea level, negative muon capture contributes a few percent at most to the total production of a nuclide by cosmic rays (e.g., Brown et al. 1995a; Heisinger et al. 1997; Heisinger and Nolte 2000) and is therefore often neglected in exposure age studies. However, due to the penetrating nature of muons, their relative contribution increases with depth and becomes dominant over spallation at a few meters below the surface. This is shown in Figure 1 for the production of ^{10}Be, ^{14}C, and ^{26}Al in quartz. Unfortunately, such data are not currently available for the noble gases, but it can be expected that they would look similar (E. Nolte, pers. comm. 2001). Negative muon capture may thus be an important process in all investigations dealing with shielded samples or wherever complex exposure histories or high erosion rates are involved.

Fast muon induced reactions. In addition to the capture of stopped negative muons, fast muons can generate a cascade of particles (nucleons, pions, photons) in rocks, which will react with target nuclei. These energetic reactions are comparable to spallation reactions except that they follow a different depth dependence. In surface rocks, fast muon induced reactions are about an order of magnitude less important than negative muon captures, but may surpass the latter at a few tens of meters depth (e.g., Heisinger et al. 1997; Heisinger and Nolte 2000; cf. Fig. 1).

The spatial variation of cosmogenic nuclide production

In space, the flux and energy spectrum of cosmic rays are entirely isotropic. However on approaching the Earth, modulation by the magnetic field, absorption by the atmosphere, and shading by the body of the Earth result in distinct variations of the cosmic ray intensity, flux direction and energy distribution depending on the exact location within the troposphere. The main factors controlling these variations and thus production rates of cosmogenic nuclides are as follows:

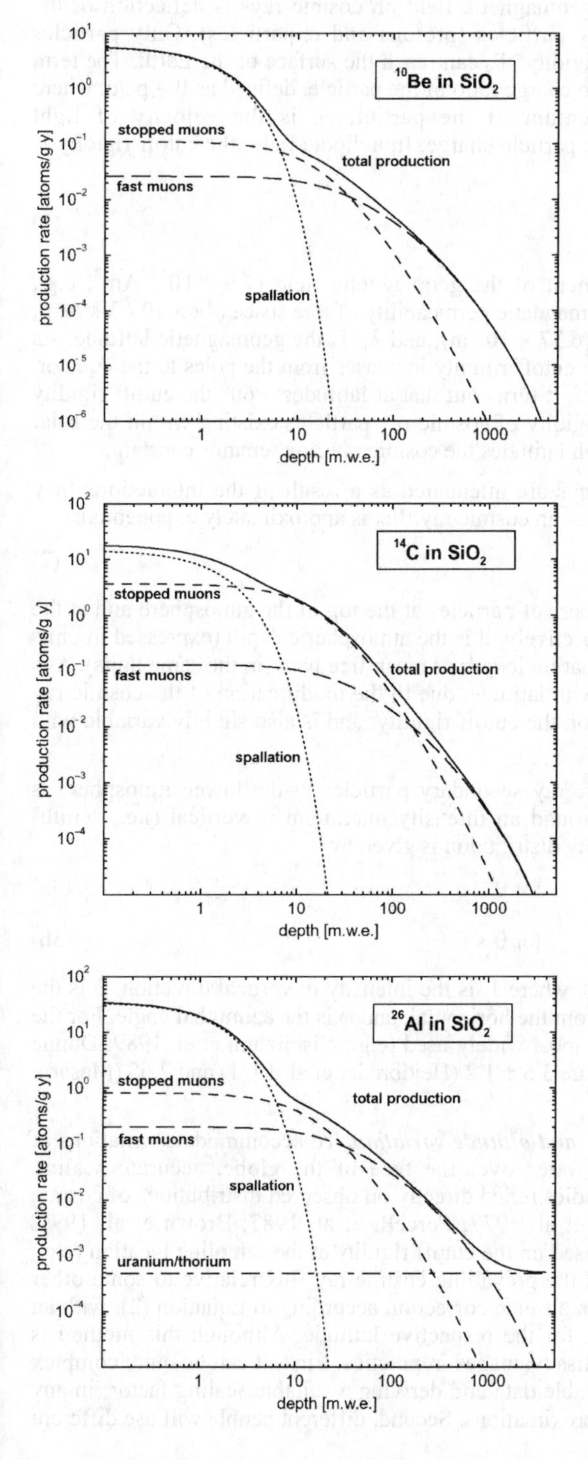

Figure 1. Depth profiles of ^{10}Be, ^{14}C, and ^{26}Al production in quartz by spallation reactions, capture of stopped negative muons, and fast muon interactions, respectively. The data are based on both experimental determinations and model calculations (Heisinger and Nolte 2000); data for the noble gas isotopes are not currently available, but would look at least qualitatively similar. Depth is given in meters of water equivalent, where 1 m.w.e. corresponds to a shielding depth of 100 g/cm^2. [Used by permission of the editor of Nucl. Instr. and Meth. in *Phys. Res. B*, from Heisinger and Nolte (2000), Fig. 1, p. 791.]

- The principle effect of the geomagnetic field on cosmic rays is deflection of the charged primary cosmic ray particles (protons and α particles). Only particles exceeding a certain "cutoff rigidity" P_c can reach the surface of the Earth. The term "rigidity" means the energy to charge ratio of the particle, defined as $P = pc/q$, where p is the relativistic momentum of the particle, c is the velocity of light (3.00×10^8 m/s), and q is the particle charge. In a dipole field, the cutoff rigidity is given by

$$P_c = \frac{M \mu_0 c}{16 \pi R^2} \cos^4 \lambda_m \quad (1)$$

where M is the dipole moment of the geomagnetic field (7.9×10^{22} Am^2; e.g., Merrill et al. 1998), μ_0 is the magnetic permeability of free space ($4\pi \times 10^{-7}$ Vs/Am), R is the radius of the Earth (6.37×10^6 m), and λ_m is the geomagnetic latitude. As according to Equation (1) the cutoff rigidity increases from the poles to the equator, the cosmic ray flux decreases. It turns out that at latitudes > 60° the cutoff rigidity drops below the minimum rigidity of cosmic ray particles existing within the solar system. Therefore at such high latitudes the cosmic ray flux remains constant.

- In the atmosphere cosmic rays are attenuated as a result of the interactions they undergo. The resulting decrease in cosmic ray flux is approximately exponential:

$$N = N_0 e^{-d/\Lambda} \quad (2)$$

Here, N_0 and N are the numbers of particles at the top of the atmosphere and at the location of the observer, respectively, d is the atmospheric depth (expressed in units of g/cm^2), and Λ is the attenuation length or mean free path (in the same units). Λ is not a constant but changes with latitude, due to the modification of the cosmic ray energy spectrum depending on the cutoff rigidity, and is also slightly variable with altitude (e.g., Lal 1991).

- The flux direction of cosmic ray secondary particles in the lower atmosphere is distributed symmetrically around an intensity maximum in vertical (i.e., zenith) direction. The angular intensity distribution is given by

$$J(\theta,\varphi) = J_0 \sin^m \theta \qquad \text{for } \theta \geq 0 \quad (3a)$$

$$J(\theta,\varphi) = 0 \qquad \text{for } \theta < 0 \quad (3b)$$

(e.g., Heidbreder et al. 1971), where J_0 is the intensity in vertical direction, θ is the inclination angle measured from the horizontal, and φ is the azimuthal angle. For the exponent m a value of 2.3 is most widely used (e.g., Nishiizumi et al. 1989; Dunne et al. 1999). Other estimates are 3.5 ± 1.2 (Heidbreder et al. 1971) and 2.65 (Masarik et al. 2000).

Scaling methods for latitude and altitude variation. To accommodate variations of cosmogenic nuclide production rates over the face of the globe, accurate scaling procedures are required. Early studies relied directly on observed distributions of cosmic ray intensities (e.g., Yokoyama et al. 1977; Porcelli et al. 1987; Brown et al. 1991; Staudacher and Allègre 1991). Based on the cutoff rigidity at the sampling location (e.g., Shea et al. 1987), they estimated the prevailing cosmic ray flux relative to some other coordinate position and applied an altitude correction according to Equation (2), with an attenuation length Λ appropriate for the respective latitude. Although this method is correct in principle, it has some disadvantages in practice. First, it can be quite complex searching the literature for applicable data and deriving a suitable scaling factor; in any case one will have to resort to approximations. Second, different people will use different data and different approximations, therefore the methods used by various researchers will not be easily comparable. The introduction of an easily treatable formalism applicable to all altitudes and latitudes by Lal (1991) was thus soon accepted as a sort of standard method to convert production rates from one location to another. Lal's method is based on third-order polynomials in altitude fitted to nuclear disintegration rates in the atmosphere (Lal and Peters 1967), which are given for geomagnetic latitudes from 0° to 60° with a 10° spacing. The polynomial coefficients are reproduced in Table 2. The production rate of any spallation-produced nuclide in a surface rock is expected to be proportional to the nuclear disintegration rate, so if the production rate at one location is known it can be scaled to any location. If P_n is the production rate of a nuclide at sea level and high latitudes (≥60°), the place usually taken for normalization, then

$$P(\lambda_m,h) = N(\lambda_m,h) \times P_n / 563.4 \tag{4}$$

where $P(\lambda_m,h)$ is the production rate at geomagnetic latitude λ_m and altitude h and $N(\lambda_m,h)$ is the nuclear disintegration rate at the same location as calculated from Lal's (1991) polynomials. Figure 2a displays the resulting altitude and latitude dependence.

Lal's (1991) scaling method has recently been challenged by Dunai (2000a), who claims that some of the approximations inherent in Lal's procedure may lead to large systematic errors, especially for high altitudes and for latitudes around 30°. Among these approximations are the description of the geomagnetic field by an axial dipole and the extrapolation of attenuation path lengths from high to low altitudes. Dunai has proposed a different scaling method, in order to avoid such critical approximations. To include the non-dipole components, Dunai uses the geomagnetic field inclination instead of the geomagnetic latitude as the parameter describing the field and then derives the dependence of the neutron flux on latitude and altitude by fitting two separate five-parameter sigmoidal functions to the observational data according to

$$N_{1030}(I) = Y + \frac{A}{[1+\exp((X-I)/B)]^C} \tag{5}$$

$$\Lambda(I) = y + \frac{a}{[1+\exp((x-I)/b)]^c} \tag{6}$$

(Dunai 2000a, 2000b). Here, $N_{1030}(I)$ is the sea level (1030 g/cm^2 atmospheric pressure) nuclear disintegration rate (or production rate) relative to I = 90°, $\Lambda(I)$ is the attenuation path length in g/cm^2, and I is the inclination in degrees. For the case of a dipole field, I can easily be calculated from the geomagnetic latitude λ_m using the relation

$$\tan I = 2 \tan \lambda_m \tag{7}$$

whereas in other cases an average inclination value must be derived from paleomagnetic records. The parameters A, B, C, X, Y and a, b, c, x, y are given in Table 3. The nuclear disintegration rate (or production rate) at altitude h is then given by

$$N(\Delta d,I) = N_{1030}(I) \times e^{\Delta d/\Lambda(I)} \tag{8}$$

with Δd being the difference in atmospheric depth between sea level and altitude h, which is proportional to the difference in atmospheric pressure Δp (in SI units: $\Delta d = \Delta p/g$, where $g = 9.80665$ m/s^2 is the standard gravitational acceleration). The conversion of altitude to atmospheric depth is described in the Appendix of Dunai (2000a). Although the calculation according to Dunai's method is somewhat more complicated, it shares with Lal's method the advantage of offering a consistent set of scaling parameters which enables direct comparison of results obtained by various workers from locations all over

Table 2. Polynomial coefficients for the calculation of the nuclear disintegration rate $N(\lambda_m,h)$ at geomagnetic latitude λ_m and altitude h in the atmosphere according to Lal (1991): $N(\lambda_m,h) = a_0 + a_1h + a_2h^2 + a_3h^3$, with h in km and N in $g^{-1}a^{-1}$.

λ_m [°]	a_0	a_1	a_2	a_3
0	330.7	255.9	98.43	20.50
10	337.9	252.1	111.0	20.73
20	382.1	272.1	132.5	24.83
30	469.3	394.6	97.76	47.20
40	525.6	505.4	142.0	58.87
50	571.1	588.1	170.9	76.12
60-90	563.4	621.8	177.3	78.91

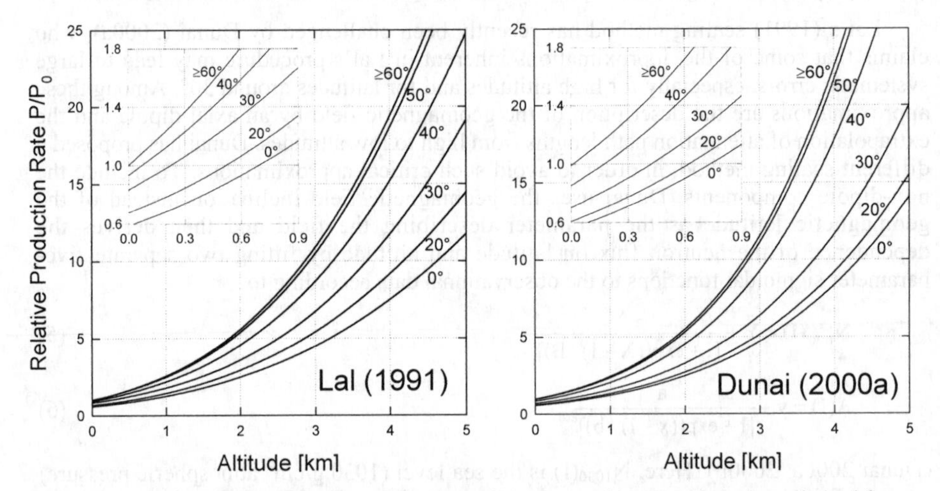

Figure 2. Dependence of cosmogenic nuclide production rates on altitude and latitude, as derived from the scaling methods of Lal (1991) and Dunai (2000a), respectively. Production rates are normalized to P_0, the value at sea level and high latitudes. Though the curves look similar at first glance, differences between the two scaling methods become increasingly evident at higher altitudes. Near sea level, the largest differences occur at latitudes 20-40° as shown in the insets.

Table 3. Numerical values of the coefficients for the calculation of the nuclear disintegration rate $N_{1030}(I)$ and the attenuation length $\Lambda(I)$ according to Equations (5) and (6), respectively (Dunai 2000a).

Coefficients for $N_{1030}(I)$		*Coefficients for $\Lambda(I)$*	
A	0.4450	a	19.85
B	4.1703	b	-5.430
C	0.3350	c	3.590
X	62.698	x	62.050
Y	0.5555	y	129.55

the world. In Figure 2b, the altitude and latitude dependence of cosmogenic nuclide production according to Dunai (2000a) is shown and can be compared to that for Lal's (1991) method. Although the curves are quite similar, distinct differences by more than 10% occur at high altitudes (> 3-4 km) and, for sea level (visible in the insets of Fig. 2), at 20-40° latitude.

Desilets et al. (2001) claim that Dunai's (2000a) scaling model is based on several false assumptions and thus does not provide an improvement over Lal's (1991) method. In a reply, Dunai (2001a) gives reasons for his choices and maintains his view regarding the correctness of the procedure. It is not the purpose of this review to evaluate the arguments of either side nor to give recommendations as to which scaling method should be preferred in future. According to Desilets and Zreda (2001), both methods have shortcomings as they do not adequately appreciate the influence of effects such as energy-dependence of the nucleon attenuation length, energy sensitivity and background correction of instruments used for cosmic ray monitoring, parameters of the real geomagnetic field, and solar activity. However, no alternative scaling model is provided by these authors. Whichever method is preferred, it is important to remember that consistent sets of production rates and scaling methods should be used. It is, for example, not advisable to take an experimentally determined production rate which had been scaled to sea level and high latitude using Lal's method and then scale it to the sampling location by Dunai's method. For this reason sea level/high latitude production rates will be quoted according to both methods in the *Production rates of cosmogenic nuclides* section.

Both the scaling procedures of Lal (1991) and Dunai (2000a) are strictly only valid for the spallation-produced component of cosmogenic nuclides. If production by muons cannot be neglected, their contribution must be scaled for altitude independently because of their higher attenuation path length. Lal (1991) offered explicit polynomial coefficients for the altitude and latitude dependence of the ^{10}Be and ^{26}Al production rates, which included a "muogenic" contribution of 15-17% at sea level (cf. Nishiizumi et al. 1989). However, according to Brown et al. (1995a) that contribution is much lower, only ~ 1-3%. Therefore it is not advisable to use the data of Lal's (1991) Table 1 to scale ^{10}Be and ^{26}Al production rates, as the scaling error is probably smaller when the muon-produced component is completely neglected. Dunai (2000a) gives a modified form of Equation (8):

$$N(\Delta d,I) = (1-\alpha)\, N_{1030}(I) \times e^{\Delta d/\Lambda(I)} + \alpha\, N_{1030}(I) \times e^{\Delta d/\Lambda_\mu} \qquad (9)$$

where α is the fraction of the nuclide produced by muons at sea level and $\Lambda_\mu = 247$ g/cm^2 (Lal 1988).

Stone (2000) has pointed out an additional difficulty in scaling production rates. Strictly, they do not depend on altitude, but on air pressure, which is usually equivalent according to their relationship in the "standard atmosphere" (cf. Appendix of Dunai 2000a). However, the mean sea level air pressure is not completely uniform over the globe, but in some regions high-pressure (e.g., Siberia) or low-pressure systems (e.g., Iceland) persist. The most prominent deviations from the standard atmosphere occur in Antarctica, with mean ground level pressures 20-40 mbar lower than expected (Radok et al. 1996). Stone (2000) provides a formalism translating Lal's (1991) scaling method to a function of pressure instead of altitude. He concludes that production rate variations are restricted to a few percent at most locations. In Antarctica, however, cosmogenic nuclide production may be 25-30% higher than implied by conventional scaling methods, which may have a significant impact on exposure histories of Antarctic rocks (see *Glacier movement and ice sheet evolution* in the *Application examples* section).

Effects of dip angle and shading. On an extended horizontal surface, cosmic ray particles are arriving from the whole upper half-space, i.e., from a solid angle of 2π. However if the surface is inclined, part of the flux is faded out, resulting in smaller cosmogenic nuclide production rates. Likewise, big objects on the horizon (such as nearby mountains) may block out a substantial portion of the cosmic ray flux. Dunne et al. (1999) have calculated scaling factors for both cases, assuming an angular distribution of the cosmic ray intensity according to Equations (3a) and (3b). For a surface of uniform slope with dip angle δ, the azimuth angle φ is defined as φ = 0 in the direction of maximum slope. The fraction of the total cosmic ray flux hitting the surface is then given by

$$f(\delta) = \frac{m+1}{2\pi} \int_{\varphi=0}^{2\pi} \int_{\theta=\gamma(\varphi,\delta)}^{\pi/2} (J/J_0) \cos\theta \, d\theta \, d\varphi \tag{10}$$

with the slope angle γ in direction φ inserted as

$$\gamma(\varphi,\delta) = \arctan(\cos\varphi \tan\delta) \tag{11}$$

and J from Equations (3a), (3b). m is the exponent of the angular distribution (Eqn. 3a). Figure 3 depicts the dependence of the fractional cosmic ray flux on the dip angle.

For the case of a horizontal surface shaded by a rectangular obstruction reaching up to an inclination angle θ_0 and extending over an azimuth angle Δφ, the remaining fraction of cosmic ray flux is calculated as

$$f(\theta_0, \Delta\varphi) = 1 - \frac{\Delta\varphi}{2\pi} \sin^{m+1}\theta_0 \tag{12}$$

(Dunne et al. 1999). The same formula can also be applied to a triangular obstruction having a baseline extension Δφ and attaining a maximum inclination θ_T, if θ_0 is replaced by

$$\theta_R = 0.62\,\theta_T - 0.00065\,\theta_T^2 \tag{13}$$

(for θ_R, θ_T in degrees and m = 2.3; see Dunne et al. 1999). The dependence of the remaining fraction of cosmic ray flux ("shielding factor") on the inclinational and azimuthal extension of a rectangular obstruction is plotted in Figure 4.

Temporal variation of cosmogenic nuclide production

Solar activity dependence. Galactic cosmic rays reaching the solar system are modulated by solar magnetic fields, which reduce the intensity of the low-energy branch (up to ~ 10 GeV/nucleon). Since these magnetic fields vary with the 11-year solar cycle, the cosmic ray flux at the top of the Earth's atmosphere is anticorrelated with indices of solar activity, such as the sunspot number. In consequence the cosmic ray neutron flux in the atmosphere and hence the production rates of cosmogenic nuclides change accordingly from solar minimum to maximum (e.g., Simpson 1983). Figure 5 shows that for a solar modulation parameter (the solar equivalent to the cutoff rigidity) between 300 and 900 MeV, production rates deviate by about ± 25% from the average at high latitudes (Masarik and Beer 1999). Near the geomagnetic equator, however, they remain almost constant because there the low-energy branch of primary cosmic ray particles (those which are most affected by solar modulation) is prevented from entering the atmosphere anyway due to the high cutoff rigidity.

Typical time spans relevant to cosmic ray exposure dating are at least tens to hundreds of solar cycles; therefore the dependence of production rates on the 11-year

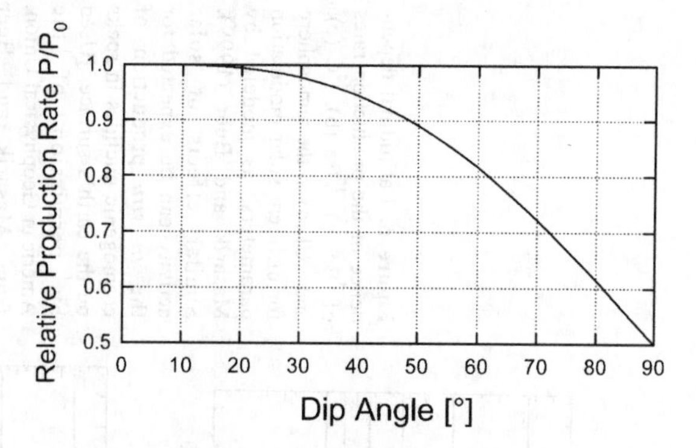

Figure 3. Dependence of the cosmogenic nuclide production rate on the dip angle of an inclined surface, as calculated from Equation (10), with m = 2.3. P_0 is the production rate on a flat horizontal surface.

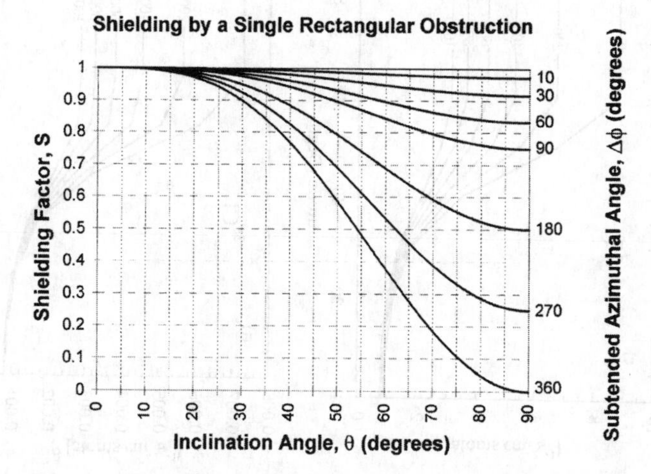

Figure 4. Shielding factor S (= f(θ,Δφ) from Eqn. 12) resulting from a cosmic-ray-blocking rectangular obstruction which extends over an azimuthal angle Δφ and up to a zenith angle θ. m=2.3 was assumed for the exponent of the angular cosmic ray distribution. [Used by permission of the editor of *Geomorphology*, from Dunne et al. (1999), Fig. 1, p. 6.]

cycle can usually be neglected. One exception is the direct determination of cosmic ray production rates by exposing a target to cosmic ray irradiation for a few years and subsequently measuring the cosmogenic nuclides produced (Yokoyama et al. 1977; Graf et al. 1996; Nishiizumi et al. 1996; Brown et al. 2000). In this case the normalized production rate P_n can be estimated from the measured production rate P_m by comparing the average neutron counting rate (per hour or day) over one or several complete solar cycles N_{av} with the average counting rate during the exposure period:

$$P_n = P_m \times N_{av} \times \frac{n}{\sum_{i=1}^{n} N_i} \tag{14}$$

where n is the number of exposure days and N_i is the counting rate on day i. Neutron counting rates should be obtained from nearby observatories. In Equation (14) the decay of a radionuclide during the exposure period has been neglected since it is only relevant for very short-lived nuclides such as ^{7}Be (cf. Nishiizumi et al. 1996).

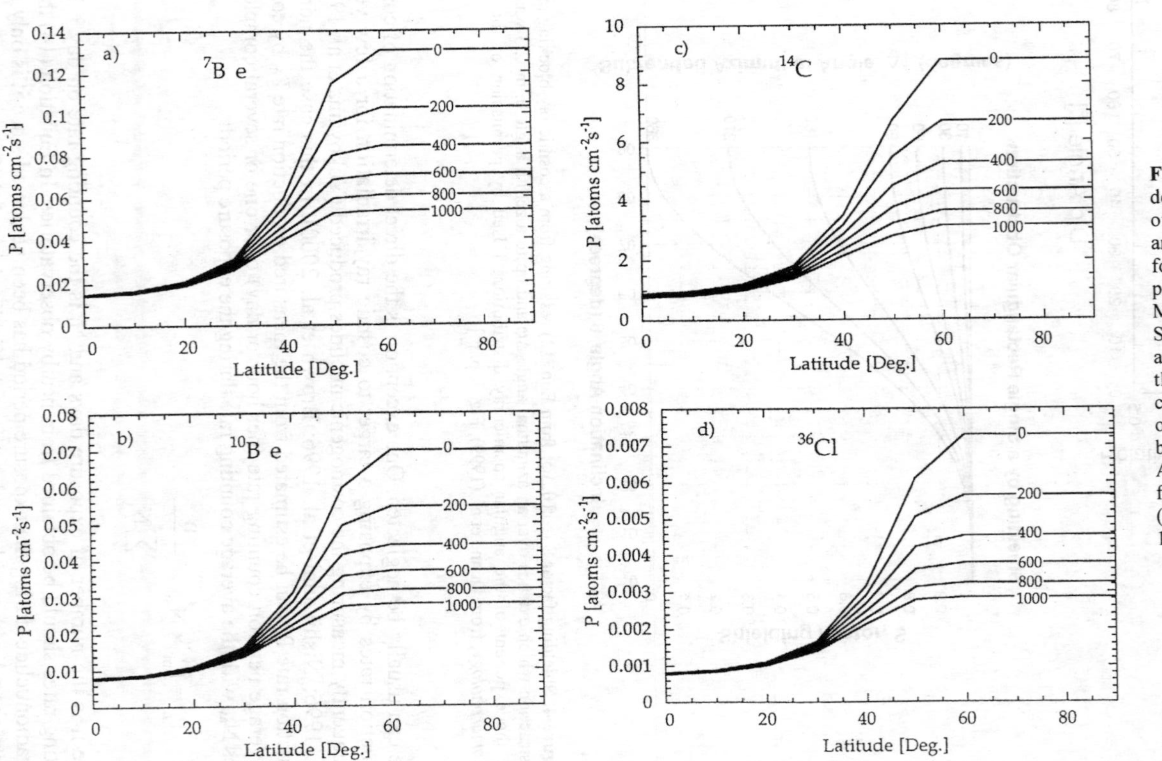

Figure 5. Latitudinal dependence of the production rates of ^{7}Be (a), ^{10}Be (b), ^{14}C (c), and ^{36}Cl (d) in the atmosphere for different solar modulation parameters, as modeled by Masarik and Beer (1999). Similar effects of solar activity can be expected for the *in situ* production of cosmogenic nuclides in rocks on the earth's surface. [Used by permission of the American Geophysical Union, from Masarik and Beer (1999), *J. Geophys. Res.*, Vol. 104, Fig. 8, p. 12108.]

The solar activity is also subject to variations on longer timescales. A prominent minimum of solar activity, during which almost no aurora and sunspots were observed, has been reported for the years 1645-1715 A.D., the so-called Maunder minimum (Eddy 1976). Other minima have been identified from ^{14}C and ^{10}Be records in tree rings and polar ice cores (e.g., Beer et al. 1988; Bard et al. 1997). To resolve such excursions from deviations caused by the geomagnetic field as discussed below, records from polar regions are best suited as the effects of solar modulation are maximal there whereas the geomagnetic field influence is minimal. Castagnoli and Lal (1980) have estimated that a solar modulation parameter of 100 MeV may describe the cosmic ray flux during the Maunder minimum and similar periods of quiet sun, corresponding to a doubling of cosmogenic nuclide production at high latitudes. Fortunately the time-integrated production rates over the whole exposure history of a rock are much less affected by such excursions.

Secular variations of the geomagnetic pole position. The geomagnetic field undergoes secular variations, i.e., changes in the direction and intensity of its dipole and non-dipole components over periods of years to several millennia (e.g., Sternberg 1992). Such changes affect production rates of cosmogenic nuclides because they control the cutoff rigidity and thus the cosmic ray flux and energy spectrum. Three parameters are required to describe the geomagnetic field at a certain location, namely intensity, inclination, and declination. The latter two give the direction of the field vector; their evolution with time is quite well known for various places on Earth (e.g., Ohno and Hamano 1992, 1993). Averaging the magnetic field vectors all over the planet is expected to average out the non-dipole components and yield the positions of the geomagnetic poles. The movement of the north geomagnetic pole over the last 10 ka is plotted in Figure 6. It is qualitatively evident that for periods of more than a few thousand years, the geographic pole is a good approximation to the average geomagnetic pole position, i.e., geographic latitude can be used instead of geomagnetic latitude when scaling production rates. Sternberg (1996) estimates that the deviation is only ~ 1.5° for exposure times longer than several centuries. Therefore geographic latitude is the better choice in the majority of cosmogenic nuclide studies. Only for exposure ages shorter than ~ 1000 a is it recommended to estimate an average geomagnetic pole position from the data of Ohno and Hamano (1992, 1993) and calculate the average geomagnetic latitude from

$$\sin\lambda_m = \cos\varepsilon \cos\lambda_g \cos\Delta\varphi_g + \sin\varepsilon \sin\lambda_g \tag{15}$$

where λ_m and λ_g are the geomagnetic and geographic latitudes of the site, ε is the geographic latitude of the average geomagnetic pole position, and $\Delta\varphi_g$ is the difference in geographic longitude of the site and the geomagnetic pole.

The scaling method of Dunai (2000a) involves also non-dipolar components of the geomagnetic field. Although these components should cancel out when averaged over the whole Earth, they may yield substantial contributions at a specific location. Therefore Dunai based the scaling of production rates on inclination rather than geomagnetic latitude. Nevertheless, for time periods of > 10-20 ka, the time-averaged geomagnetic field at any location can be approximated by a geocentric axial dipole field (e.g., Merrill et al. 1998), and in these cases the geographic latitude may also be used with Dunai's (2000a) scaling procedure, by converting inclination I to latitude λ according to Equation (7). For shorter exposures an average inclination should be estimated based on suitable paleomagnetic records from the vicinity of the sampling site.

Long-term geomagnetic field intensity changes. Not only the direction of the geomagnetic field vectors, but also their intensity may change with time. Both paleomagnetic studies (e.g., Valet et al. 1998; Guyodo and Valet 1999; Juarez and Tauxe

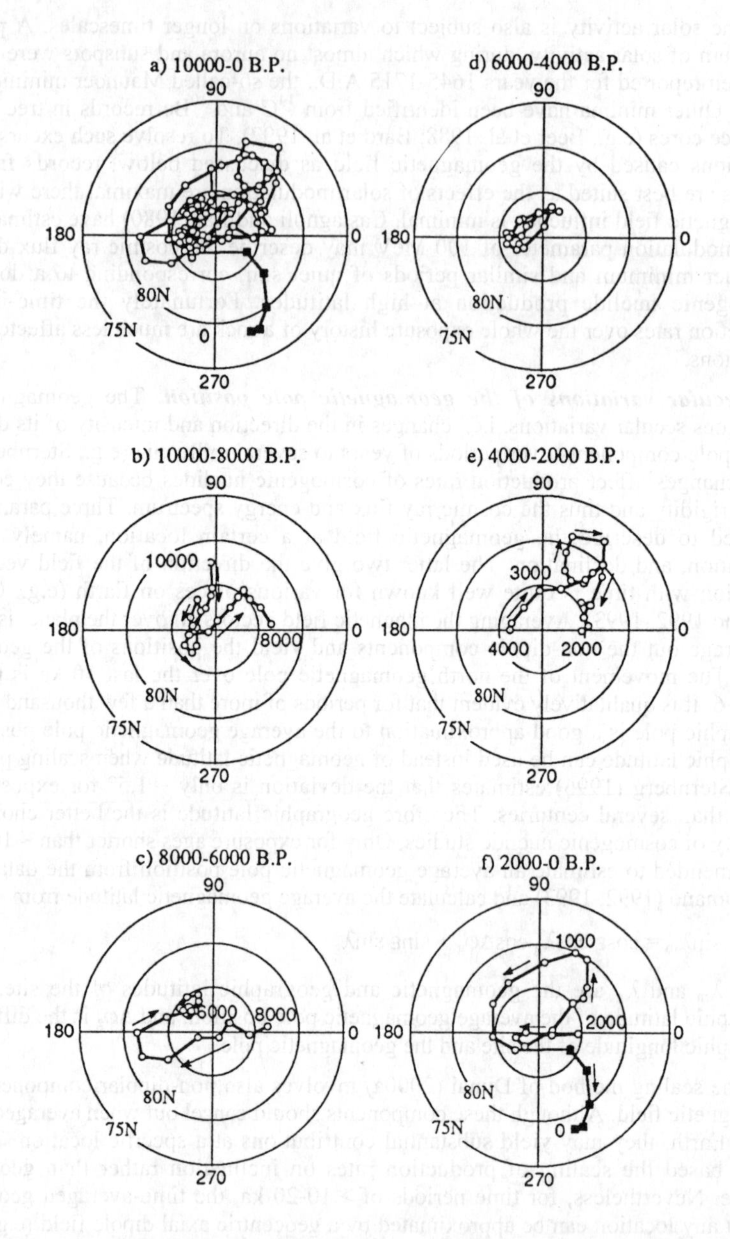

Figure 6. Location of the north geomagnetic pole from 10,000 a to present in 100 a increments, according to Ohno and Hamano (1992, 1993). Numbers denote years BP, arrows indicate the directions of the polar movement. It is qualitatively evident that for periods of the order of millennia, the geographic pole is a good approximation to the average geomagnetic pole position. [Used by permission of the American Geophysical Union, from Ohno and Hamano (1992), *Geophys. Res. Lett.*, Vol. 19, Fig. 2, p. 1717.]

2000) and investigations of cosmogenic nuclide production in earlier epochs (e.g., Frank et al. 1997; Plummer et al. 1997; Wagner et al. 2000) indicate variations of the dipole moment, which directly influence production rates. The relative paleointensities as derived from the ^{10}Be deposition rate (Frank et al. 1997) and from the paleomagnetic record (Guyodo and Valet 1999; see Figure 7) in stacked deep-sea sediments agree remarkably well. However, there are some difficulties involved in the interpretation of these data. There has been a dispute as to the level of climatic influence in such data sets (e.g., Kok 1999; Frank 2000). Moreover, the pattern of temporal variation as shown in Figure 7 is not reproduced by all paleomagnetic data: Channell et al. (1997) did not observe the sharp increase in paleointensity from ~ 40 ka to the present at ODP site 983, and Goguitchaichvili et al. (1999) derive a geomagnetic dipole moment indistinguishable from the modern value for the Pliocene field, in contrast to the results of Juarez and Tauxe (2000).

Although the magnitude of geomagnetic field intensity variations in the past may be disputed, it seems clear that variations did occur. The dependence of cosmogenic nuclide production rates on such variations has been estimated in several studies (e.g., Cerling and Craig 1994a; Shanahan and Zreda 2000). At high latitudes production rates remain constant because the cutoff rigidity is too low to influence the cosmic ray spectrum, whereas at low latitudes the marked changes in cutoff rigidity accompanying a change in the dipole moment result in corresponding modifications of production rates. Equation (1) describes the relation between cutoff rigidity, dipole strength, and geomagnetic latitude. In a modified form, it can be written as

$$P_c = 14.9\ \text{GV} \times (M_t/M_0)\cos^4\lambda_m \tag{16}$$

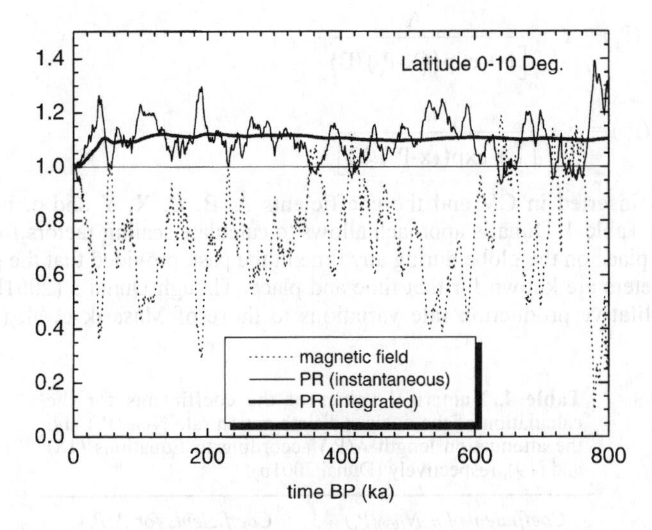

Figure 7. Relative variations of the geomagnetic field intensity from 800 ka to present (dotted line) and their influence on instantaneous (thin solid line) and time-integrated (thick solid line) cosmogenic nuclide production rates at latitudes 0-10°, according to model calculations of Masarik et al. (2001). The geomagnetic field reconstruction relies on dendrochronologically derived $\Delta^{14}C$ (Stuiver et al. 1998) for the last 10 ka and on paleomagnetic records from deep-sea sediments (Guyodo and Valet 1999) for the remainder of the period. The influence of geomagnetic field intensity on production rates decreases towards higher latitudes (cf. Fig. 8). [Used by permission of the editor of *Geochim. Cosmochim. Acta*, from Masarik et al. (2001), Fig. 1, p. 2998.]

where M_0 is the present dipole moment and M_t the dipole moment at an earlier time t. A variation of the dipole field is thus equivalent to a change in geomagnetic latitude, according to

$$\cos \lambda_m' = (M_t/M_0)^{1/4} \cos \lambda_m \tag{17}$$

where λ_m' is the geomagnetic latitude where the cutoff rigidity (and consequently the production rate) at dipole strength M_0 is equal to that for λ_m at dipole strength M_t. Production rates at M_t can therefore be determined by scaling the sea level, high latitude production rates to latitude λ_m' instead of λ_m. This method fails to work if $(M_t/M_0)^{1/4} > 1/\cos \lambda_m$, i.e., for high magnetic field intensities and low latitudes.

Two recent studies have addressed the issue in more detail. Masarik et al. (2001) use model calculations of cosmic ray particle interactions with matter to estimate the relation between changes in cutoff rigidities and cosmogenic nuclide production rates. Their calculations are based on the $\Delta^{14}C$ and paleomagnetic records of Stuiver et al. (1998) for the last 10 ka and Guyodo and Valet (1999) for the period 10-800 ka, respectively, and result in relatively modest variations of the time-integrated production rates. Figure 7 shows that at low latitudes (0-10°), the deviations from the present-day production rates are always < 1% during the last 10 ka, and less everywhere else. From ~ 5 to 40 ka, production rates increase steadily to ~ 10% above present-day values (at 0-10°) and remain ~ 9-12% higher than today back to 800 ka (Fig. 7). At latitudes of 30-40°, the deviations are always < 2.5%, and at > 40° production rates are virtually unaffected by variations of the geomagnetic field intensity.

Dunai (2001b) has fitted his scaling parameters N_{1030} and Λ to cutoff rigidity instead of inclination in order to enable a full description of cosmic ray flux variations, including non-dipole components:

$$N_{1030}(P_c) = Y + \frac{A}{\left[1 + \exp\left((X\text{-}P_c)/B\right)\right]^C} \tag{18}$$

$$\Lambda(P_c) = y + \frac{a}{\left[1 + \exp\left((x\text{-}P_c)/b\right)\right]^c} \tag{19}$$

where P_c is inserted in GV and the coefficients A, B, C, X, Y and a, b, c, x, y are compiled in Table 4. Dunai's approach allows to calculate scaling factors for production rates at any place on the globe during any time in the past, provided that the geomagnetic field parameters are known for that time and place. Though Dunai's (2001b) data yield similar qualitative production rate variations to those of Masarik et al. (2001) when

Table 4. Numerical values of the coefficients for the calculation of the nuclear disintegration rate $N_{1030}(P_c)$ and the attenuation length $\Lambda(P_c)$ according to Equations (18) and (19), respectively (Dunai 2001b).

Coefficients for $N_{1030}(P_c)$		*Coefficients for $\Lambda(P_c)$*	
A	0.5221	a	17.183
B	-1.7211	b	2.060
C	0.3345	c	5.9164
X	4.2822	x	2.2964
Y	0.4952	y	130.11

evaluated based on the Guyodo and Valet (1999) record (Fig. 8), there is an obvious quantitative discrepancy. At low latitudes, time-integrated production rates reach a maximum of 20% above present-day values at ~140 and ~200 ka, compared to 12% according to Masarik et al. (2001). At mid-latitudes of 30 and 40°, Dunai's (2001b) results imply maximum deviations of 15% and 7%, respectively, as compared to ~2.5% (Masarik et al. 2001). Furthermore, Dunai (2001b) obtains a marked altitude-dependence of production rate variations, whereas Masarik et al. (2001) claim that the corrections are virtually identical between sea level and 5000 m.

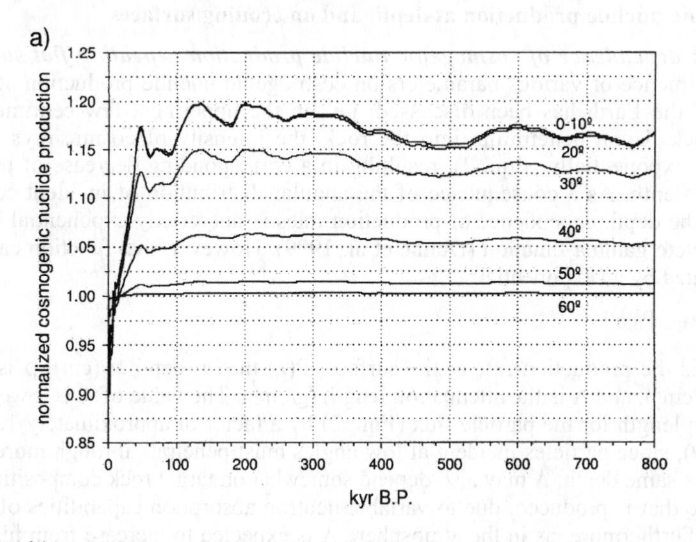

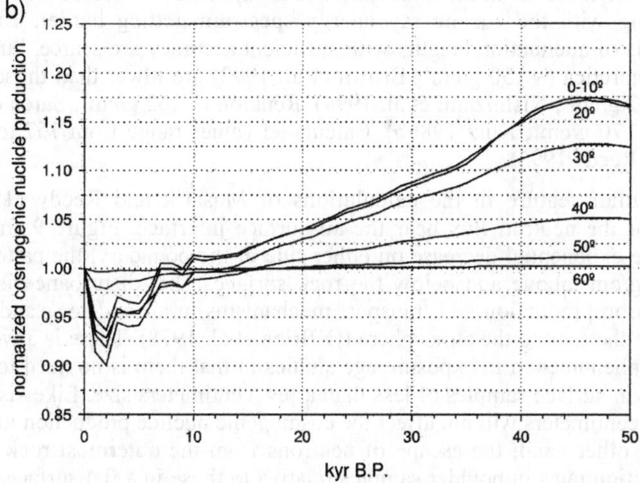

Figure 8. Time-integrated cosmogenic nuclide production rate at sea level, for various latitudes, relative to the present production rate at the same latitude according to the scaling method of Dunai (2001b). (a) shows the period 0-800 ka, (b) is an enlargement of the 0-50 ka range. While based on the same paleomagnetic record (Guyodo and Valet 1999) for the period 10-800 ka, this model predicts substantially higher production rate variations than that of Masarik et al. (2001; Fig. 7). [Used by permission of the editor of *Earth Planet. Sci. Lett.*, from Dunai (2001b), Fig. 6, p. 206.]

At present, the issue of past variations of the geomagnetic field strength and how they control cosmogenic nuclide production rates is obviously not well enough understood and needs further investigation. Fortunately most cosmogenic nuclide studies in the past, including the majority of production rate determinations (see *Experimental determinations of production rates* section), have focused on mid- and high-latitude sites, where production rate variations are minor. However in future the accuracy of absolute exposure ages, particularly for locations at low latitudes, will critically depend on reliable correction factors for geomagnetic field variation.

Cosmogenic nuclide production at depth and on eroding surfaces

Depth dependence of cosmogenic nuclide production beneath a flat surface. So far, the influence of various parameters on cosmogenic nuclide production at the very surface of the Earth has been discussed, i.e., in the uppermost few centimeters of a surface rock. Upon penetrating into the rock, the intensity of cosmic rays is further attenuated exponentially (Eqn. 2), resulting in a corresponding decrease of production rates with depth. As a consequence of the angular distribution of incident cosmic ray particles, the depth dependence of production rates is not strictly exponential but obeys an incomplete gamma function (Dunne et al. 1999). However, that function can well be approximated by an exponential:

$$P(z) = P_0\, e^{-\rho z/\Lambda} \tag{20}$$

where P_0 is the production rate at the surface, P(z) that at depth z (cm), ρ is the rock density (g/cm^3), and Λ is the attenuation length (g/cm^2). The value of Λ is lower than the attenuation length for the particle flux (Eqn. 2) by a factor of approximately 1.3 (Dunne et al. 1999), since particles incident at low angles must penetrate through more material to reach the same depth. Λ may also depend somewhat on target rock composition and on the nuclide that is produced, due to variable neutron absorption capabilities of different elements. Furthermore, as in the atmosphere Λ is expected to increase from high to low latitudes along with the cosmic ray energy spectrum getting harder. Experimental determinations of attenuation lengths with sufficient accuracy are scarce, but indeed the data from Antarctica (~ 150 g/cm^2; Brown et al. 1992) are lower than those from New Mexico (~ 162 g/cm^2; Nishiizumi et al. 1994), Réunion (~ 162 g/cm^2; Sarda et al. 1993), or Hawaii (~ 170 g/cm^2; Kurz 1986b). Calculated values range from 157 to 167 g/cm^2 (Masarik and Reedy 1995).

One peculiar feature in the calculations of Masarik and Reedy (1995) is the disturbance of the neutron flux near the air-surface interface. Figure 9 shows that in contrast to the exponential decrease on either side of the boundary, the pattern is flat to about 10-20 g/cm^2 above and below the rock surface. That disturbance is caused by different neutron production and transport mechanisms in air and rock and the loss of neutrons from the rock to the atmosphere (O'Brien et al. 1978). There is a welcome side effect to this phenomenon in exposure age studies in that there is no need to correct for self-shielding in surface samples of less than a few centimeters size. Likewise, erosional loss by a few centimeters will not affect the cosmogenic nuclide production in the surface layer. On the other hand, the escape of neutrons from the outermost rock layers may reduce production rates in boulder samples relative to those in a flat surface (Masarik et al. 2000), by as much as 12% for the surface of a hemisphere with 1m radius. It must be cautioned that these results have so far not been tested experimentally. If the flat profile down to 10 g/cm^2 is correct, production rates at depth would be some 6% higher than for an exponential decrease starting at the very surface.

The exponential decrease according to Equation (20) is only valid for the spallation-produced component. Thermal neutron capture reactions have a distinct depth

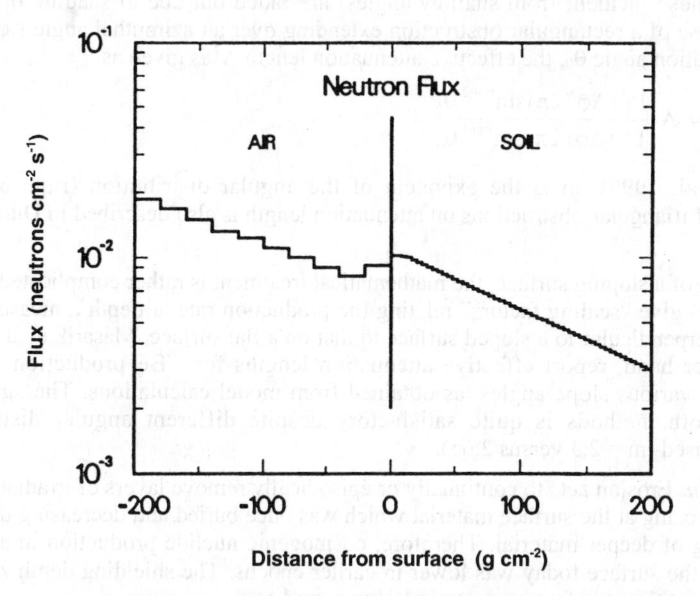

Figure 9. Total cosmic ray neutron flux on either side of the air-surface interface as calculated by Masarik and Reedy (1995). Boundary effects produce a flat pattern for ~ 10-20 g/cm^2 on both sides of the interface. [Used by permission of the editor of *Earth Planet. Sci. Lett.*, from Masarik and Reedy (1995), Fig. 1, p. 387.]

dependence showing a maximum at some depth below the surface, ~30-50 g/cm^2 (Liu et al. 1994). Since these reactions are only important for a few nuclides, in particular ^{36}Cl, but not for cosmogenic He and Ne production (cf. *Production mechanisms of cosmogenic nuclides in terrestrial rocks* section above), they are not discussed further here. Reference is made to the investigations of Liu et al. (1994) and Phillips et al. (1996).

For muon-induced cosmogenic nuclide production, a roughly exponential decrease with depth can again be assumed, with

$$P_\mu(z) = P_{\mu,0}\, e^{-\rho z/\Lambda_\mu} \tag{21}$$

where Λ_μ ~ 1300 g/cm^2 (Barbouti and Rastin 1983; Brown et al. 1995a), although this is only an approximation since the exact pattern of muon propagation and stopping rate depends much more strongly on the chemical composition of the target than in the case of neutrons, as is evident from the marked difference between the attenuation lengths in rock and air (247 g/cm^2; Lal 1988). At depths greater than ~5000 g/cm^2, Λ_μ increases further as the relative contribution of fast muon interactions increases (Bilokon et al. 1989; Heisinger and Nolte 2000). More accurate mathematical representations of the evolution of cosmogenic nuclide production rates with depth, including interactions with both stopped and fast muons, can be found in Stone et al. (1998) or Granger and Smith (2000), whereas Heisinger and Nolte (2000) have shown graphic illustrations based on their model calculations (cf. Fig. 1).

Effective attenuation length beneath inclined or partially shaded surfaces. As the attenuation length Λ for cosmogenic nuclide production by spallation (Eqn. 20) is determined by the angular cosmic ray neutron distribution, it increases when part of the neutrons (those incident from shallow angles) are faded out due to shading or surface slope. In case of a rectangular obstruction extending over an azimuthal angle $\Delta\varphi$ and up to an inclination angle Θ_0, the effective attenuation length Λ^* is given as

$$\Lambda^* = \Lambda \frac{1-(\Delta\varphi/2\pi)\sin^{m+2}\theta_0}{1-(\Delta\varphi/2\pi)\sin^{m+1}\theta_0} \tag{22}$$

(Dunne et al. 1999). m is the exponent of the angular distribution (Eqn. 3a). The influence of triangular obstructions on attenuation length is also described in Dunne et al. (1999).

In case of a sloping surface, the mathematical treatment is rather complicated. Dunne et al. (1999) give "scaling factors" relating the production rate at depth z measured in a direction perpendicular to a sloped surface to that on a flat surface. Masarik et al. (2000), on the other hand, report effective attenuation lengths for ^{10}Be production beneath surfaces of various slope angles, as obtained from model calculations. The agreement between both methods is quite satisfactory despite different angular distribution exponents used (m = 2.3 versus 2.65).

Erosion. Erosion acts to continually or episodically remove layers of irradiated rock, thereby exposing at the surface material which was once buried and decreasing the depth of shielding of deeper material. Therefore, cosmogenic nuclide production in a sample exposed at the surface today was lower in earlier epochs. The shielding depth z(-t) of a rock sample at time t before present can be expressed as

$$z(-t) = z_p + \int_{-t}^{0} \varepsilon(t')\,dt' \tag{23}$$

where $\varepsilon(t)$ is the erosion rate and z_p the present shielding depth. For the case of a constant erosion rate, one obtains

$$z(-t) = z_p + \varepsilon t \tag{24}$$

and

$$P(z,-t) = P_0\, e^{-\rho(z_p+\varepsilon t)/\Lambda} + P_{\mu,0}\, e^{-\rho(z_p+\varepsilon t)/\Lambda_\mu} \tag{25}$$

for the total production rate from spallation and negative muon capture (cf. Eqn. 20, 21). The latter term becomes important not only for deep samples, but also for surface samples where high erosion rates have prevailed for long periods of time.

Derivation of exposure ages and erosion rates

Simple exposure histories. The determination of the exposure age or erosion rate of a certain geomorphic feature is easiest in the case of a simple exposure history, i.e., if the rock studied was excavated from a completely shielded location by a brief process (such as a volcanic eruption) and then experienced cosmic ray irradiation at constant (or no) erosion without ever being moved. Temporal variations of the surface production rate due to solar activity or geomagnetic field changes should be negligible as well. In such a case, the temporal evolution of a cosmogenic nuclide concentration C in the rock is given by

$$dC(z,t)/dt = P(z,t) - \lambda C(z,t) \tag{26}$$

where λ is the decay constant of a radionuclide ($\lambda = 0$ for stable nuclides). Defining t = 0 for the time of excavation and z_0 the initial shielding depth, we obtain $z(t) = z_0 - \varepsilon t$ (cf. Eqn. 24) and P(z,t) as in Equation (25), but replacing $z_p + \varepsilon t$ by $z_0 - \varepsilon t$. Integration of (26) then yields the solution

$$C(z,t) = C(z,0)\, e^{-\lambda t} + \frac{P_0}{\lambda + \rho\varepsilon/\Lambda} e^{-\rho(z_0 - \varepsilon t)/\Lambda} \left(1 - e^{-(\lambda + \rho\varepsilon/\Lambda)t}\right) + \frac{P_{\mu,0}}{\lambda + \rho\varepsilon/\Lambda_\mu} e^{-\rho(z_0 - \varepsilon t)/\Lambda_\mu} \left(1 - e^{-(\lambda + \rho\varepsilon/\Lambda_\mu)t}\right) \tag{27}$$

There are two important limiting cases for such simple exposure histories: The case of no erosion ($\varepsilon = 0$) and that of steady-state erosion. For $\varepsilon = 0$ Equation (27) reduces to

$$C(z_0,t) = C(z_0,0)\, e^{-\lambda t} + \frac{1 - e^{-\lambda t}}{\lambda} \left(P_0\, e^{-\rho z_0/\Lambda} + P_{\mu,0}\, e^{-\rho z_0/\Lambda_\mu}\right) \tag{28}$$

If the initial concentration $C(z_0,0) = 0$, as can usually be assumed for radionuclides, and neglecting the muogenic production, the exposure age T is calculated from

$$T = -\frac{1}{\lambda} \ln\left(1 - \frac{C(z_0,T)\,\lambda}{P_0\, e^{-\rho z_0/\Lambda}}\right) \tag{29}$$

For $\lambda = 0$ (stable nuclides), the differential equation (26) reduces to $dC/dt = P(z_0)$ which has the solution

$$C(z_0,t) = C(z_0,0) + P_0\, e^{-\rho z_0/\Lambda}\, t + P_{\mu,0}\, e^{-\rho z_0/\Lambda_\mu}\, t \tag{30}$$

and therefore (for zero initial concentration or, equivalently, if only the cosmogenic component is considered)

$$T = \frac{C(z_0,T)}{P_0\, e^{-\rho z_0/\Lambda} + P_{\mu,0}\, e^{-\rho z_0/\Lambda_\mu}} \tag{31}$$

In the case of steady-state erosion (cf. Lal 1991), the cosmogenic nuclide concentration is not determined by the exposure age T but by the erosion rate: When $T \gg 1/(\lambda + \rho\varepsilon/\Lambda)$ and $T \gg 1/(\lambda + \rho\varepsilon/\Lambda_\mu)$, Equation (27) reduces to

$$C(z,t) = C(z,0)\, e^{-\lambda t} + \frac{P_0}{\lambda + \rho\varepsilon/\Lambda} e^{-\rho(z_0 - \varepsilon t)/\Lambda} + \frac{P_{\mu,0}}{\lambda + \rho\varepsilon/\Lambda_\mu} e^{-\rho(z_0 - \varepsilon t)/\Lambda_\mu} \tag{32}$$

It should be noted that due to the higher attenuation length, it takes longer for the muogenic component to reach erosion equilibrium, in particular for long-lived or stable nuclides ($\lambda \approx 0$). For a sample on the present-day surface ($z_0 - \varepsilon T = 0$) and again assuming $C(z_0,0) = 0$, we can nevertheless neglect the third term of (32) in first-order approximation since $P_0\Lambda$ is clearly higher than $P_{\mu,0}\Lambda_\mu$, though not by orders of magnitude ($P_{\mu,0}/P_0 \le 0.03$, $\Lambda_\mu/\Lambda \approx 8$). Using this approximation the steady-state erosion rate can be expressed as

$$\varepsilon = \frac{\Lambda}{\rho}\left(\frac{P_0}{C} - \lambda\right) \tag{33}$$

For a more exact solution including production by muons, a numerical approach or the use of modeled depth vs. concentration profiles (e.g., Heisinger et al. 1997; Heisinger and Nolte 2000) is required.

Neither of the two limiting cases (no erosion or steady-state erosion) may be common in practice. The assumption of no erosion may be taken if less than a few centimeters of rock have been removed during the whole time of exposure, corresponding to a change in production rate by < 10%, which is typically well within the general precision of exposure dating. Using the same criterion, steady-state erosion is fulfilled when $T > 3/(\lambda + \rho\varepsilon/\Lambda)$ or, for stable nuclides, $T > 150$ cm/ε (using $\rho = 3$ g/cm^3 and $\Lambda = 160$ g/cm^2). In all other cases the cosmogenic nuclide concentration is determined by a combination of exposure time and erosion rate, and Equations (29) or (31) and (33), respectively, give only the *minimum* exposure age and the *maximum* erosion rate compatible with the data. The temporal evolution of a cosmogenic nuclide concentration for various erosion rates is plotted in Figure 10. To derive the real exposure age (or the real erosion rate), independent information on either quantity is required. In principle, one possibility to achieve simultaneous information on both parameters is to study two (or more) cosmogenic nuclides in the same rock, yielding two equations such as (27) for the two unknowns. However, the method only works under certain conditions which have been documented in detail by Gillespie and Bierman (1995): First, the half-lives of the two nuclides must be sufficiently different. The ratio of the stable cosmogenic nuclides ^{3}He and ^{21}Ne equals that of the production rates for any combination of ε and T, so the terms containing these quantities in Equation (27) cancel out, leaving only one equation for two unknowns. Second, one of the nuclides should be close to equilibrium between production and the combined effects of erosion and radioactive decay, while the other should still be growing; otherwise the precision of the determination will be poor. This is shown in Figure 11, for the pairs ^{10}Be-^{26}Al and ^{21}Ne-^{26}Al. Such plots are a common means to estimate both exposure age and erosion rate from the determination of two cosmogenic nuclides (e.g., Lal 1991; Nishiizumi et al. 1991a; Graf et al. 1991). The area between the two curves ("steady-state erosion island"; Lal 1991) comprises all the possible combinations of ε and T for the case of simple exposure histories.

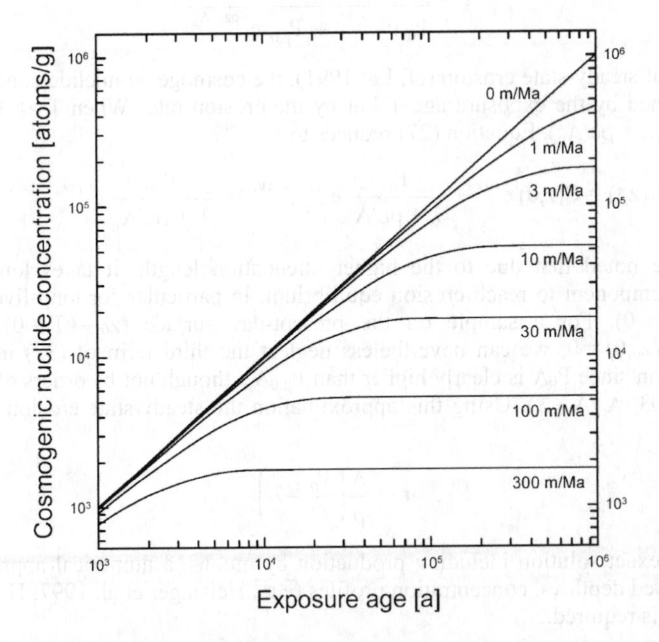

Figure 10. Temporal evolution of the concentration of a stable cosmogenic nuclide in dependence of the erosion rate, for a production rate of 1 atom g^{-1} a^{-1}. Expected concentrations in real samples can be calculated by multiplying the values from this plot with the actual production rate.

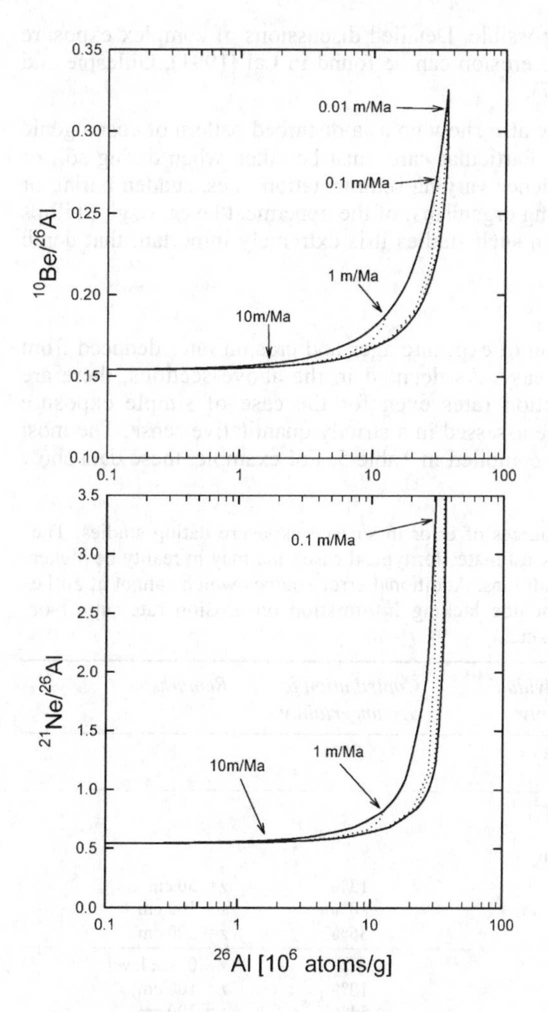

Figure 11. Plots of the ratios $^{10}Be/^{26}Al$ and $^{21}Ne/^{26}Al$, respectively, versus the ^{26}Al concentration for rocks with a simple exposure history. The lower solid lines represent the temporal evolution of these quantities in the absence of erosion, the upper solid lines in erosion equilibrium. Dotted lines indicate the temporal evolution for erosion rates of 10, 1, 0.1, and 0.01 m/Ma (1 m/Ma = 10^{-4} cm/a). The areas between the curves ("steady-state erosion island;" Lal 1991) comprise all possible combinations of exposure ages and erosion rates for simple exposure histories, data outside of these areas indicate complex exposure histories or experimental error. Curves are plotted for ^{10}Be, ^{21}Ne, and ^{26}Al production rates of 5.42, 19.0, and 35.2 atoms g^{-1} a^{-1}, respectively (modified from Kubik et al. 1998; Niedermann 2000; see Table 6).

Complex exposure histories. Complex exposure histories are those which involve, for example, several stages of cosmogenic nuclide production at various depths, variable erosion rates, episodic burial, changes in altitude of exposure, etc. Such scenarios cannot usually be resolved by cosmogenic nuclide studies alone, but require independent information from other methods. Even if such information is not available, it is most important to recognize complex exposure histories in order to avoid gross mistakes in the interpretation of exposure age data. Evidence for complex exposure histories can be taken from plots such as Figure 11. Experimental data which do not fit into the allowed area of the plot clearly indicate that the assumptions on which Equation (27) is based are not met (provided that experimental errors can be excluded and contributions from non-cosmogenic sources have been corrected for). For example, a data point left of the steady-state erosion island may result from a rock buried for some period of time, causing a change in the ratio of the two nuclides because the shorter-lived one decays more rapidly. Data points to the right of the allowed zone would imply exposure conditions under a higher production rate (e.g., at higher altitude). Ratios of short-lived to long-lived nuclide above the production ratio are not possible. Detailed discussions of complex exposure scenarios involving non-steady-state erosion can be found in Lal (1991), Gillespie and Bierman (1995), or Small et al. (1997).

Complex exposure histories may also show up as a disturbed pattern of cosmogenic nuclide concentration versus depth. Particular care must be taken when dating soil or alluvial deposits, which often experience varying sedimentation rates, sudden burial, or bioturbation (i.e., soil mixing by living organisms) of the uppermost layers (e.g., Phillips et al. 1998; Braucher et al. 2000). In such studies it is extremely important that depth profiles are taken.

Error considerations

Estimating the absolute precision of exposure ages and erosion rates deduced from cosmogenic nuclide studies is not easy. As detailed in the above sections, there are various factors controlling production rates even for the case of simple exposure histories, only some of which can be assessed in a strictly quantitative sense. The most important of these contributions are compiled in Table 5. For example, these data show

Table 5. Compilation of important sources of error in surface exposure dating studies. The quoted uncertainties (2σ level) are only estimates for typical cases and may in reality be higher or lower depending on the special conditions. Additional error sources which cannot at all be quantified in any general manner include lacking information on erosion rate, uplift or subsidence, complex exposure histories, etc.

Error source	*Typical individual uncertainty*	*Contribution to age uncertainty*	*Remarks*
Measured nuclide concentration C	5-20%	5-20%	
Attenuation length Λ	10 g/cm^2		
Shielding depth z	5 cm		
Host rock density ρ	0.2 g/cm^3		
→ Shielding correction		13%	z = 50 cm
		20%	z = 100 cm
		36%	z = 200 cm
Neglect of muogenic component	2%	2%	z = 0, sea level
		10%	z = 100 cm
		54%	z = 200 cm
		7400%	z = 500 cm
Production rate	10-20%	10-20%	at sea level, ≥ 60°
Scaling method		10%[1]	Lal (1991)
		2-10%	Dunai (2000a)
Scaling with elevation instead of air pressure	1%	6-8%	outside Antarctica
	− 2-4%	+15-30%	in Antarctica
Use of geographic latitude	1.5°	2%	for T > 1 ka
Non-dipole components of geomagnetic field	10-25%	0-30%	for T < 10-20 ka, depends on location
Long-term solar activity variations	± 400 MeV	10%?	time-integrated at high latitudes
Long-term variations of geomagnetic field	$\pm 4 \times 10^{22}$ Am^2	≤ 12%	Masarik et al. (2001)
		≤ 20%	Dunai (2001b)

[1] Dunai (2000a) argues that Lal's (1991) uncertainty estimate does not include all error sources and that 20% would be more realistic.

the increasing influence of relatively small individual errors with increasing shielding depth, due to the exponential dependence. In consequence, erosion will also have a non-trivial effect on production rate uncertainties even if the value of the erosion rate is well constrained. Moreover, it is obvious that scaling of cosmogenic nuclide production rates in space and time remains a major source of error due to our lack of knowledge about relevant parameters, such as the local geomagnetic field, and their control on production rates.

The usual way to circumvent trouble arising from such difficulties is to quote error limits which only include experimental uncertainties of the measurement but not those of the production rate determination, the scaling method, geomagnetic field variations, neglect of the muon-produced component, assumptions regarding simple or complex exposure histories, etc. This may not be a major problem as long as both authors and readers are aware of it and do not over-interpret the data. Also, it does not make much sense to include the (systematic) errors of production rate determinations or scaling when only results from one nuclide for a limited geographical area are compared. Whenever absolute ages or erosion rates are concerned however, such systematic errors must be taken into account. Bierman (1994) gives an excellent presentation on the limitations of the use of cosmogenic nuclides for geomorphic applications, although some of his error assessments may seem a little pessimistic.

Systematic errors may be introduced in various stages of a cosmogenic nuclide study. Especially important to noble gas studies is the presence of non-cosmogenic components (see *Discrimination of cosmogenic against trapped, radiogenic, and nucleogenic components* in the *Experimental issues* section), which may lead to wrong conclusions if corrections for such components are based on faulty assumptions. Difficulties in the geomorphic setting may be another important error source. For example, snow or soil cover during part of a rock's exposure is not always easy to detect but may have a substantial influence on cosmogenic nuclide production. Similar to erosion, neglect of these effects will yield too low exposure ages; it is therefore better to quote minimum exposure ages when they cannot be assessed. For studies of two cosmogenic nuclides with different half-lives (or one stable and one radioactive nuclide), I refer to the rigorous treatment of the precision of exposure age and erosion rate estimates by Gillespie and Bierman (1995).

Though uncertainties resulting from cosmogenic nuclide studies may thus seem rather high when all error sources are considered, the unique possibilities of this method must be borne in mind (see *Application examples*). In many cases other techniques rely on much weaker assumptions or provide no quantitative answers at all. Furthermore, many error sources (e.g., those connected with production rate scaling) will certainly become considerably smaller in near future.

PRODUCTION RATES OF COSMOGENIC NUCLIDES

Experimental determinations of production rates

When the early studies of cosmogenic nuclides in terrestrial surface rocks were published, production rates were quite poorly known. Davis and Schaeffer (1955) and Phillips et al. (1986) estimated the ^{36}Cl production rate from the cosmic ray neutron flux and the thermal neutron capture cross section of Cl. Srinivasan (1976) scaled the ^{126}Xe production rate from the lunar surface to the Earth's surface by considering the attenuation of cosmic rays all through the atmosphere. Kurz (1986a,b), Craig and Poreda (1986), and Porcelli et al. (1987) used the ^{3}H production rate as calculated by Yokoyama et al. (1977) to estimate the ^{3}He production rate. The problem with such estimates was a large influence of model calculations which were in part based on distant extrapolations or ill-constrained parameters, such as nuclear excitation functions for reactions involving neutrons (Lal 1988). Therefore, experimental determinations of cosmogenic nuclide production rates in rocks or minerals of various target element composition were urgently needed. Volcanic rocks are probably among the best-suited samples for such studies, since simple exposure histories are common, the influence of erosion can be judged from the preservation of surface textures, and exposure ages are identical to eruption ages, which can often be determined by conventional methods such as ^{14}C or ^{40}Ar-^{39}Ar dating. Other possibilities to determine production rates experimentally include rocks exposed by glacial scouring, giant floods or landslides and artificial targets exposed for a few years at mountain altitudes. The following summary of experimental production rate determinations will focus on the noble gas isotopes ^{3}He and ^{21}Ne but will also include the radionuclides ^{10}Be and ^{26}Al, which have often been used in combined studies along with He and Ne.

Production rates in basaltic rocks and their constituents. Volcanic rocks (mostly of basaltic composition) have provided the largest data set of experimental production rates, especially for ^{3}He. The first such determination was that of Kurz (1986b), who measured the cosmogenic ^{3}He (hereafter $^{3}He_c$) concentration in olivine phenocrysts from a radiocarbon-dated (28 ka) lava flow from Mauna Loa, Hawaii, and deduced a production rate of 97 atoms g^{-1} a^{-1} at sea level and 37°N geomagnetic latitude (Kurz 1987). That study was extended by Kurz et al. (1990) to a whole suite of Hawaiian lava flows, ranging in age from 600 to 14,000 a. The production rates obtained from that data set scatter by about a factor of 4, which can hardly be explained by real production rate variations. Dunai (2001b) has reevaluated these data and concludes that the association of samples to dated lava flows is unreliable in several cases. Cerling (1990), Poreda and Cerling (1992), and Cerling and Craig (1994b) studied olivine and pyroxene separates from various basalts in the western USA and France and derived a sea level, high latitude ^{3}He production rate of 115 ± 4 atoms g^{-1} a^{-1}. In addition, Poreda and Cerling (1992) determined the $(^{3}He/^{21}Ne)_c$ ratio in olivines as a function of their Fo content (Fig. 12), providing the first experimental value for a ^{21}Ne production rate: 45 ± 4 atoms g^{-1} a^{-1} in Fo_{81} olivine. The $(^{3}He/^{21}Ne)_c$ production ratio was also measured by Staudacher and Allègre (1991, 1993a) in various ultramafic nodules and olivine separates. They obtained values differing from each other and from that of Poreda and Cerling (1992). The scatter is probably related to difficulties in correcting for the magmatic He and Ne components, but may in part also be explained by variations in the composition of the minerals. Whereas the ^{3}He production rate is rather insensitive to such variations (e.g., Eugster 1988) because ^{3}He (and ^{3}H) nuclei are sputtered off any target nucleus, the production rates of most other nuclides depend critically on nuclear excitation functions which may vary substantially for different target nuclides (cf. *Production mechanisms of cosmogenic nuclides in terrestrial rocks* section). For example, the Ne production rate from Mg is substantially higher than that from Al or Si, because of the large cross sections of the $^{24}Mg(n,\alpha)^{21}Ne$ and $^{25}Mg(n,\alpha)^{22}Ne$ reactions (e.g., Hohenberg et al. 1978; Masarik and Reedy 1996; Leya et al. 2000).

More recently, the ^{3}He production rate has been determined in 2450-7090 year old basalt flows from Oregon by Licciardi et al. (1999). These authors applied a correction for variations of the geomagnetic pole position (Ohno and Hamano 1992, 1993) to their production rate values, the difference between corrected and uncorrected data is < 5% in all cases. They also revised the earlier production rate determinations of Kurz (1987), Kurz et al. (1990), Cerling (1990), and Cerling and Craig (1994b), applying a consistent method for the conversion of radiocarbon ages to calendar ages (Stuiver et al. 1998), and derived a mean Holocene ^{3}He production rate (normalized to sea level and high latitude)

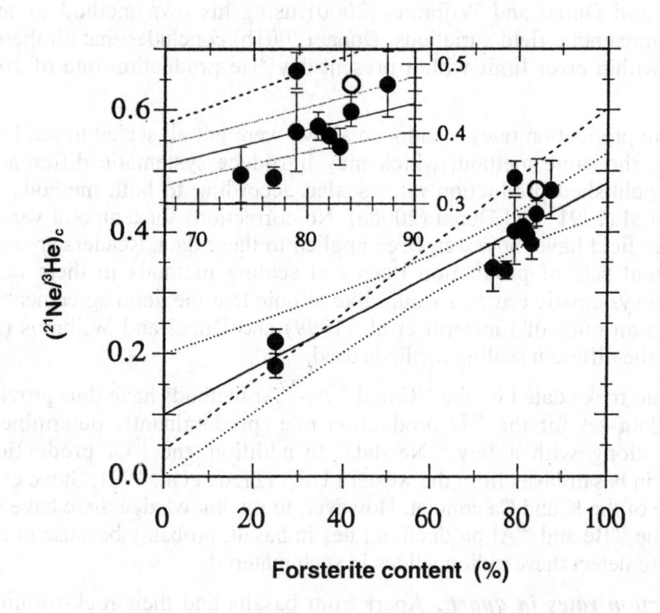

Figure 12. Compositional dependence of the $(^{21}Ne/^{3}He)_c$ ratio on the forsterite content of olivine, according to Poreda and Cerling (1992). Data around Fo_{80} are plotted enlarged in the inset. The solid line is a least squares fit through all the filled circles, with 95% confidence intervals (dotted lines). The dashed line gives the dependence expected by Lal (1991). The open circle is from Marti and Craig (1987). [Used by permission of the American Geophysical Union, from Poreda and Cerling (1992), *Geophys. Res. Lett.*, Vol. 19, Fig. 2, p. 1865.]

of 119 atoms g^{-1} a^{-1}. A minor mistake in the calculation of the weighted means, along with a new calibration of Kurz's (1987) lava flow age, led J.M. Licciardi (pers. comm. 2001) to slightly adjust that value to 116 ± 4 atoms g^{-1} a^{-1}.

Whereas all the above production rate determinations were performed on rocks < 20 ka in age, Dunai and Wijbrans (2000) studied much older lava flows from Lanzarote, Canary Islands. In olivines from two flows with ^{40}Ar-^{39}Ar ages of 152 and 281 ka, they derived a normalized ^3He production rate of 118 ± 11 atoms g^{-1} a^{-1}, whereas a 1.35 Ma old flow yielded a nominally lower value. However, as the authors point out, erosion effects on that old surface can easily explain the ^3He deficiency, and the production rate averaged over 1.35 Ma is also consistent with that derived from the younger flows. Interestingly, the ^3He production rates derived by Licciardi et al. (1999) and Dunai and Wijbrans (2000) agree remarkably well, although the geomagnetic field intensity in the Holocene is considered to have been some 50% above the average of the preceding epochs (e.g., Guyodo and Valet 1996; Valet et al. 1998; Frank 2000; cf. Fig. 7). Therefore, as the Lanzarote calibration site is at relatively low latitude (29°N), one would expect a sizable effect on the production rate (Dunai 2001b; Masarik et al. 2001; Figs. 7 and 8). The lack of a resolvable difference may indicate that the influence of geomagnetic field intensity variations is not too large (cf. *Long-term geomagnetic field intensity changes* section). On the other hand, in an evaluation of the ^3He production rate determinations of Kurz (1987), Kurz et al. (1990), Cerling and Craig (1994b), Licciardi et al. (1999), and Dunai and Wijbrans (2000) using his own method to accommodate reported geomagnetic field variations, Dunai (2001b) concludes that all these studies are consistent within error limit with a present-day ^{3}He production rate of 103 ± 4 atoms g^{-1} a^{-1}.

The ^{3}He production rates mentioned above were not all scaled to sea level and high latitudes by the same method, which may introduce systematic differences. Table 6 shows the published production rates scaled according to both methods widely used today, i.e., Lal (1991) and Dunai (2000a). No corrections for temporal variations of the geomagnetic field have, however, been applied to these data. Readers are encouraged to use consistent sets of production rates and scaling methods in their work to avoid unnecessary systematic errors. I would like to note that the good agreement between the ^{3}He production rates of Licciardi et al. (1999) and Dunai and Wijbrans (2000) is not affected by the different scaling methods used.

Volcanic rocks dated by the ^{14}C and ^{40}Ar-^{39}Ar methods have thus provided quite an extensive data set for the ^{3}He production rate (predominantly determined in olivine separates), along with a few ^{21}Ne data. In addition, the ^{36}Cl production rate was determined in basalt lavas from the western USA (Zreda et al. 1991; Stone et al. 1996), in dependence of the K and Ca content. However, to my knowledge there have not been any studies of the ^{10}Be and ^{26}Al production rates in basalt, probably because of experimental difficulties to detect these radionuclides in such material.

Production rates in quartz. Apart from basalts and their rock-forming minerals, most experimental determinations of cosmogenic nuclide production rates have dealt with quartz. Because of its simple chemical composition (SiO_2) and its abundance in a wide variety of rock types, quartz is considered a very well-suited mineral for surface exposure dating. Nishiizumi et al. (1989) used quartz separates from glacially polished granitic rocks from the Sierra Nevada, California, to derive the production rates of ^{10}Be and ^{26}Al. The surfaces sampled had been exposed by the retreating glaciers of the Tioga period of the last ice age and showed evidence that at least several meters of rock had been removed by glacial scouring. Therefore simple exposure histories could be assumed for these rocks. Assuming an age of 11,000 a, which was based on several radiocarbon dates for the deglaciation of the Sierra Nevada, Nishiizumi et al. (1989) derived ^{10}Be and ^{26}Al production rates of 61.9 and 373.6 atoms g^{-1} a^{-1}, respectively, at 3340 m altitude and 44°N geomagnetic latitude. A subset of the same samples was later used by Niedermann et al. (1994), who determined a ^{21}Ne/^{26}Al production ratio of 0.65 ± 0.11 in quartz, corresponding to a ^{21}Ne production rate of 243 atoms g^{-1} a^{-1} at the given altitude and latitude. A ^{3}He production rate was not established because ^{3}He is not quantitatively retained in quartz (cf. *Retentivity of minerals for cosmogenic He and Ne* section below). A revised deglaciation age (~ 13,000 a; Clark et al. 1995) and inadequate assumptions used for scaling to sea level and high latitudes led the authors of these studies to reconsider their production rate estimates (Nishiizumi et al. 1996; Niedermann 2000). The values given in Table 6 are based on the revised figures and, in the case of ^{10}Be and ^{26}Al, neglecting the production by muons whose contribution was overestimated in the studies of Nishiizumi et al. (1989, 1996) as shown by Brown et al. (1995a).

An independent production rate determination of ^{10}Be and ^{26}Al in quartz was carried out by Kubik et al. (1998). They sampled quartz veins from large boulders derived from the landslide deposits of Köfels, Austria, which had been dated at 9800 ± 100 a by ^{14}C. The precision to which that age is known is clearly better than for the Sierra Nevada samples. If scaled to sea level and high latitudes by the same method, the Köfels production rates are 4-19% higher than those of Nishiizumi et al. (1989), yielding a ^{26}Al/^{10}Be ratio of 6.52 versus 6.02. It is difficult to judge to what extent such differences

Table 6. Production rates of the cosmogenic noble gas isotopes ^3He and ^{21}Ne and of the radionuclides ^{10}Be and ^{26}Al as determined in rocks and minerals of well-known exposure history. Most reported data are weighted means from several individual production rate determinations, which were all separately scaled to sea level and high latitudes by both the methods of Lal (1991) and Dunai (2000a). Temporal variations of the geomagnetic field and contributions from non-dipole components of the geomagnetic field were neglected to facilitate comparison of different data sets. Several data were revised according to more recent age calibrations, details are given in the remarks.

Nuclide	*Reference*	*Target rock or mineral*	*Calibration site*	*Age range (ka)*	*Scaling method*		*Remarks*
					Lal	*Dunai*	
^3He	Kurz (1986b,1987)	olivine	Hawaii	32.9	123	138	Revised age 32.94 ± 0.94 ka (Bard et al. 1998)
	Kurz et al. (1990)	olivine	Hawaii	0.55-10.7	109	120	Revised ages (Licciardi et al. 1999); individual data scatter by factor 4
	Cerling (1990); Cerling and Craig (1994b)	olivine, pyroxene	Western USA and France	2.3-17.4	112	117	Revised ages (Licciardi et al. 1999)
	Licciardi et al. (1999)	olivine	Oregon	2.5-7.1	113	122	
	Dunai and Wijbrans (2000)	olivine	Canary Islands	152-281	99	118	
^{21}Ne	Poreda and Cerling (1992)	Fo_{81} olivine	Utah	17.3	46	48	Revised age (Licciardi et al. 1999)
	Staudacher and Allègre (1991)	ultramafic nodules, olivine	Russia and Mongolia	~7-27	~80	~87	Based on ^3He production rates of 113/122 atoms g^{-1} a^{-1}
	Staudacher and Allègre (1993a)	ultramafic nodules, olivine	Russia, Mongolia, Réunion	~14-130	~32	~35	Based on ^3He production rates of 113/122 atoms g^{-1} a^{-1}
	Niedermann et al. (1994); Niedermann (2000)	SiO_2	California	13	20.3	19.0	Revised age (Clark et al. 1995; Niedermann 2000)
^{10}Be	Nishiizumi et al. (1989)	SiO_2	California	13	5.16	4.90	Revised age (Clark et al. 1995); without muon contribution
	Nishiizumi et al. (1991a)	SiO_2	Antarctica	> 4000	5.53	5.61	Saturation assumed
	Kubik et al. (1998)	SiO_2	Austria	9.8	5.35	5.42	Without muon contribution
^{26}Al	Nishiizumi et al. (1989)	SiO_2	California	13	31.1	29.5	Revised age (Clark et al. 1995); without muon contribution
	Nishiizumi et al. (1991a)	SiO_2	Antarctica	> 4000	33.9	34.4	Saturation assumed
	Kubik et al. (1998)	SiO_2	Austria	9.8	34.8	35.2	Without muon contribution
For comparison:							
^3He	Masarik and Reedy (1995)	Fo_{81} olivine			105	105	Model calculations
		SiO_2			124	124	
^{21}Ne		Fo_{81} olivine			41.1	41.1	
		SiO_2			18.4	18.4	
^{10}Be		SiO_2			5.97	5.97	
^{26}Al		SiO_2			36.1	36.1	

are caused by the scaling method (e.g., non-dipole components of the geomagnetic field), to variations in geomagnetic field intensity or solar activity during different periods of exposure, or just to experimental error. A determination of the ^{21}Ne production rate in the Köfels samples, as planned in the Zurich lab (R. Wieler, pers. comm. 2001), will be most informative.

Nishiizumi et al. (1991a) report ^{10}Be and ^{26}Al data from an Antarctic quartz in which the concentrations of both radionuclides were at saturation, corresponding to an exposure age of > 4 Ma and negligible erosion. In such a case Equation (27) reduces to $C = P_0 / \lambda$ (for a surface sample and neglecting the muogenic component), so the production rate can readily be calculated from the concentration and the decay constant. The resulting production rates compare quite well to those of Kubik et al. (1998), but are clearly higher than those of Nishiizumi et al. (1989), especially for ^{10}Be (Table 6). Hudson et al. (1991) determined ^{21}Ne in six of Nishiizumi et al.'s (1991a) quartz samples and deduced a minimum ^{21}Ne production rate of 80 atoms g^{-1} a^{-1} at 1800m elevation and 77.6°S, converting to ~ 17 atoms g^{-1} a^{-1} at sea level. This value is consistent with that of Niedermann et al. (1994). These determinations did, however, not consider the exceptionally low air pressure in Antarctica, which increases production rates and leads to a shift of the "steady state erosion island" (cf. Fig. 11), possibly invalidating the argument of saturation concentration for ^{10}Be (Stone 2000).

Production rates in artificial targets. Cosmogenic nuclide production rates have also been determined in artificial targets exposed for a few years to cosmic ray irradiation. In an early study, Yokoyama et al. (1977) measured ^{22}Na and ^{24}Na in three metal targets (mainly Al) exposed in the Mont Blanc region, France. Of more relevance to surface exposure dating today are the experiments of Nishiizumi et al. (1996) and Graf et al. (1996), who exposed water and SiO_2 targets, respectively, at mountain altitudes in Colorado, and that of Brown et al. (2000), who exposed water tanks at different altitudes in France. Nishiizumi et al. (1996) converted their results from H_2O to SiO_2 based on neutron irradiation experiments (Reedy et al. 1994) and argued that the average production rate over 10,000 years should be ~ 15% higher than that during the last four solar cycles due to geomagnetic field changes. In this way they achieved an agreement within 2% with the value obtained by Nishiizumi et al. (1989) in Sierra Nevada quartz (Table 6). However, a 15% difference between the present "instantaneous" production rate and that integrated over 10 ka is at odds with calculations of Dunai (2001b) and Masarik et al. (2001), who expect production rate variations of less than a few percent within that period of time. On the other hand, Brown et al. (2000) found an inconsistency between the neutron irradiation results of Reedy et al. (1994) and their observation of similar ^{3}He/^{10}Be production ratios in water and quartz, rendering the conversion of the ^{10}Be production rate from H_2O to SiO_2 uncertain. In their water targets, Brown et al. (2000) also determined the ^{3}H/^{3}He production ratio from oxygen as 0.32 ± 0.08.

The ^{21}Ne production rate in a SiO_2 target during exposure on Mt. Evans (4250 m) was 410 ± 60 atoms g^{-1} a^{-1} (Graf et al. 1996). As detailed by Niedermann (2000), this converts to a sea level, high latitude value of 17.7 atoms g^{-1} a^{-1} over four solar cycles if Dunai's (2000a) scaling method is applied (or 21.6 atoms g^{-1} a^{-1} according to Lal 1991). The agreement with the production rate in Sierra Nevada quartz is well within experimental uncertainties and does not indicate a substantial production rate variation either.

Production rates by muons. Experimental determinations of cosmogenic nuclide production by muons are scarce, and for noble gases even absent. In a depth profile from lateritic soil in the Congo, Brown et al. (1995a) were able to discern the muon-produced component of ^{10}Be in quartz and constrain its contribution at the surface (300 m altitude) to between 1 and 3% of the total ^{10}Be production, i.e., ~ 0.05-0.15 atoms g^{-1} a^{-1} scaled to sea level and high latitudes. A detailed study of ^{36}Cl production in a 20 m marble profile from an Australian quarry was carried out by Stone et al. (1998). These authors derived a normalized ^{36}Cl production rate by negative muon capture in calcite of 2.1 ± 0.4 atoms g^{-1} a^{-1}, ~ 10% of the total production. The higher fraction of muon-produced ^{36}Cl as compared to ^{10}Be can be explained by a relatively high cross section for α emission after μ^- capture, favoring the reaction $^{40}Ca(\mu^-,\alpha)^{36}Cl$.

No experimental determinations of production rates by muons for other cosmogenic nuclides are available in the current literature. Therefore production models (e.g., Heisinger et al. 1997; Heisinger and Nolte 2000) should be used in all cases where production by muons might be relevant.

Future needs. The data in Table 6 show the present state of production rate determinations in natural samples. There are quite a lot of consistent data for ^{3}He in olivine, a few less consistent ones for ^{21}Ne in olivine, and just one for ^{21}Ne in SiO_2. For ^{10}Be and ^{26}Al, the three determinations in SiO_2 agree within ~ 10-20%. One possible problem is the bias of production rate determinations to locations in the western USA and Hawaii, which may introduce a systematic error caused by the scaling method. In this respect, the Canary Island data of Dunai and Wijbrans (2000) are closer to the western hemisphere data when scaled by Dunai's (2000a) method. On the other hand, the ^{10}Be and ^{26}Al production rates from California, Antarctica, and Austria scatter less if scaled by Lal's (1991) procedure. However, different ages of calibration samples may also contribute to scatter of production rates. To investigate the reasons for such scatter, more determinations at various locations all over the globe are required. Ideally, these determinations should be made using samples of similar age. In addition, to better resolve temporal variations samples of sufficiently different ages from a limited area should be studied.

It is also evident from Table 6 that cosmogenic nuclide production rates should be determined for more compositions of target elements. Though not all minerals are likewise suited for surface exposure dating with ^{3}He and ^{21}Ne (see *Experimental issues* below), there are certainly more than just olivine and quartz. And finally, a determination of the contribution to ^{3}He and ^{21}Ne production from muon-induced reactions would be important to assess exposure histories of samples from shielded locations or where erosion is high.

Production rates obtained by model calculations

In the early years of surface exposure dating production rate estimates of terrestrial cosmogenic nuclides relied mainly on models. Yokoyama et al. (1977) calculated production rates of various radionuclides in granite, basalt, and limestone, based on the flux and energy spectrum of cosmic ray particles in the atmosphere and on nuclear excitation functions, mainly for protons. Their ^{3}H production rate, when scaled to sea level and high latitude, is ~ 60 atoms g^{-1} a^{-1}, in good agreement with the ^{3}He production rate established later (Table 6) if ^{3}He is produced in equal parts directly and via ^{3}H, although the latter assumption is questioned at least for O spallation by the results of Brown et al. (2000). The ^{10}Be and ^{26}Al production rates are not directly comparable since measured values are only available for SiO_2. However, a value of < 2 atoms g^{-1} a^{-1} for ^{10}Be in granite (assumed SiO_2 content 73%) is obviously much too low, whereas ~ 40 atoms g^{-1} a^{-1} for ^{26}Al seem rather high.

Lal (1991) estimated production rates for the noble gas isotopes ^{3}He, ^{20,21,22}Ne, and ^{36,38}Ar from the target elements O, Mg, Al, Si, Ca, and Fe. His ^{3}He rates (47-83 atoms g^{-1} a^{-1} for O, Mg, Al, Si) are clearly below those determined in real samples (Table 6), and the ^{21}Ne production rate in Si of 18 atoms (g Si)$^{-1}$ a^{-1} is 2.5 times lower than that found by Niedermann et al. (1994). Lal (1991) was aware of the limited validity of his estimates which he ascribed mainly to the lack of applicable nuclear excitation functions.

In this respect the work of Masarik and Reedy (1995) was a major step forward. These authors relied on cross sections as derived from measurements in extraterrestrial samples, where at sufficient shielding depths cosmogenic nuclide production is dominated by neutrons also. Indeed their production rates agree remarkably well with experimental values (Table 6). In a follow-up paper, Masarik and Reedy (1996) reported elemental coefficients for the production rates of ^{10}Be, ^{14}C, ^{26}Al, ^{36}Cl as well as ^{3}He and ^{21}Ne, which are reproduced in Table 7. Using these coefficients it is possible to estimate production rates for any elemental composition of the target rock or mineral, i.e., also for minerals for which no experimental production rate determinations have been carried out. The good agreement of the "tested" production rates in quartz and olivine (Masarik and Reedy 1995) could suggest a high confidence in the elemental coefficients as well, at least for the major target elements. Another set of elemental production rates for the Ne isotopes (Schäfer et al. 1999), which is based on calculated neutron spectra in the centers of spherical meteorites, does however not agree very well with Masarik and Reedy's (1996) data (see Table 7). Schäfer et al. (1999) argue that their coefficients provide a better consistency with experimental evidence from pyroxenes.

EXPERIMENTAL ISSUES

Retentivity of minerals for cosmogenic He and Ne

As for other geochronological methods (cf. Kelley 2002; Farley 2002), a prerequisite to successful surface exposure investigations is the quantitative retention of cosmogenic nuclides in the minerals studied. In general, it can be assumed that minerals with typically high noble gas concentrations from other sources (e.g., Carroll and Draper 1994), such as mantle or radiogenic gases, should also retain cosmogenic noble gases. However, a caveat to that contention is the distinct siting of different components within minerals. Trapped noble gases, especially those from crustal sources, are often concentrated in fluid inclusions, whereas the cosmogenic nuclei are produced within the crystal lattice. Unless there are cracks rendering the fluid inclusions leaky, diffusion loss from the lattice may be more severe than from inclusions. Because gases are preferentially partitioned in the

Table 7. Elemental coefficients for the calculation of cosmogenic nuclide production rates (atoms g^{-1} a^{-1}) in dependence of the chemical composition of the target mineral (Masarik and Reedy 1996). For example, P(^{26}Al) is calculated as 225 [Al] + 77 [Si] + 0.15 [Fe], with element concentrations in weight fractions (g/g). For ^{21}Ne, an alternative data set from Schäfer et al. (1999) is also reported.

Nuclide	*O*	*Na*	*Mg*	*Al*	*Si*	*K*	*Ca*	*Ti*	*Fe*
^{3}He	135	-	116	107	111	-	61	-	40
^{10}Be	10.87	-	0.52	0.45	0.39	-	-	-	0.16
^{14}C	31.3	-	5.3	4.2	4.3	-	-	-	1.2
^{21}Ne	-	98	131	65	39	-	4	-	0.20
^{21}Ne [1]	-	-	196	54	45	-	-	-	-
^{26}Al	-	-	-	225	77	-	-	-	0.15
^{36}Cl	-	-	-	-	-	129	65	16	0.9

[1] Schäfer et al. (1999)

fluid phase, they will tend to remain within fluid inclusions rather than diffuse through the crystal.

Nevertheless, the minerals olivine and pyroxene, which are commonly used to investigate mantle noble gases, do retain cosmogenic He and Ne quantitatively, as implied by their low diffusion coefficients (Hart 1984; Trull et al. 1991) and demonstrated by ample data, such as identical concentrations of ^3He_c in coexisting olivine and pyroxene grains (e.g., Craig and Poreda 1986; Cerling 1990) or the consistent ^3He production rates discussed in the preceding section. Plagioclase is another mineral which has been tested for cosmogenic ^3He and ^{21}Ne retention. Cerling (1990) found virtually no ^3He_c; obviously this mineral is unsuited for ^3He exposure dating. Poreda and Cerling (1992) concluded that the retention of ^{21}Ne in volcanic plagioclase is satisfactory, based on constant ratios of $^{21}Ne_c$ in plagioclase to ^3He_c or $^{21}Ne_c$ in coexisting olivine, independent of the age. They derived a normalized ^{21}Ne production rate of 16.8 ± 1.7 atoms g^{-1} a^{-1} for Ab_{36} plagioclase, which seems, however, somewhat low. According to the formula of Masarik and Reedy (1996; cf. Table 7), 23.4 atoms g^{-1} a^{-1} ^{21}Ne are expected in Ab_{36} plagioclase, i.e., 40% more. In the case of Fo_{81} olivine, the agreement between Poreda and Cerling's (1992) production rate and that calculated from Masarik and Reedy (1996) is very good: 45 ± 4 versus 44.4 atoms g^{-1} a^{-1}. Therefore it is doubtful whether plagioclase indeed retains ^{21}Ne quantitatively. Bruno et al. (1997) report diffusion losses of ~ 50% for plagioclase from Antarctic dolerites, confirming a poor retentivity of plagioclase for Ne as well.

The extent to which ^3He_c is lost from quartz by diffusion was a matter of debate in the early nineties (e.g., Cerling 1990; Graf et al. 1991; Trull et al. 1991; Brook and Kurz 1993). It can be concluded that ^3He_c may be retained in quartz more or less quantitatively in favorable cases, e.g., in large grains and at prevailing low temperatures, but is more often lost to high percentages and can therefore not commonly be applied to exposure age studies. In contrast, $^{21}Ne_c$ in quartz is obviously well-suited, in spite of relatively low degassing temperatures (Niedermann et al. 1993; see next section). This is indicated by a general consistency of ^{10}Be, ^{26}Al, and ^{21}Ne exposure ages (e.g., Bruno et al. 1997; Schäfer et al. 1999; Hetzel et al. 2002a).

Minerals other than those mentioned above have so far not been extensively used in cosmogenic noble gas studies. Judging from its generally high noble gas content (e.g., Carroll and Draper 1994), amphibole could be another candidate for quantitative He and Ne retention among relatively widespread minerals, but this supposition would have to be checked. Dunai and Roselieb (1996) reported a very high He retentivity of garnet and therefore expected a great potential for use in exposure age studies. However, high U and Th concentrations are quite common in garnet, so radiogenic He and nucleogenic Ne may interfere with the cosmogenic components (see next section). The same problem is expected for micas although they might as well retain Ne and possibly He. Farley et al. (2001) studied cosmogenic ^3He in fluorapatite and expected good retention characteristics of this mineral, which is widely used in U/Th-He dating (cf. Farley 2002), but again radiogenic and nucleogenic components may be abundant.

Discrimination of cosmogenic against trapped, radiogenic, and nucleogenic components

To allow for an accurate determination of the concentration of a cosmogenic noble gas isotope, noble gas components from other sources must be effectively discriminated against. These components include:

- trapped components, i.e., atmospheric, mantle, or crustal gases residing in the crystal lattice or in fluid inclusions. They may originate from solution in the magma, incorporation of air or water during eruption, fluid interaction during metamorphic events, etc.

- radiogenic components produced by radioactive decay of U, Th, and ^{40}K.
- nucleogenic components produced by naturally occurring nuclear reactions, e.g., $^{6}Li(n,\alpha)^{3}H(\beta^{-})^{3}He$, $^{18}O(\alpha,n)^{21}Ne$, $^{19}F(\alpha,n)^{22}Na(\beta^{+})^{22}Ne$, $^{24,25}Mg(n,\alpha)^{21,22}Ne$. In these reactions the α particles are derived from U and Th decay and the neutrons from other (α,n) reactions or from U fission.

Sample selection and preparation. Whereas cosmic ray irradiation is often the only important source of radionuclides in a rock, the stable noble gases are always mixtures of several components. To minimize uncertainties arising from decomposition of components, samples with high ratios of cosmogenic to non-cosmogenic gases should be used. High concentrations of cosmogenic isotopes, i.e., old samples taken from the surface at high altitude and latitude and where erosion is low, are of course most favorable, but choice in this respect is usually limited when specific questions are to be solved. Low concentrations of non-cosmogenic components are therefore essential.

There is no general recipe to avoid high concentrations of trapped noble gases. Depending on their source and the mechanism of incorporation, they may occupy various sites in a rock or mineral, but often they are concentrated in fluid inclusions. Therefore a low abundance of fluid inclusions in the minerals studied may be a useful criterion for sample selection. In addition, the sample may be crushed to small grain sizes before loading, so that gases are released from part of the fluid inclusions. At least for quartz, this method proved useful in reducing the amount of trapped Ne (Niedermann et al. 1994). On the other hand, care must be taken to avoid "irreversible" adsorption of atmospheric noble gases on fresh grain surfaces (e.g., Niedermann and Eugster 1992), which may not be a problem for Ne in quartz but might affect certain other minerals.

Radiogenic and nucleogenic noble gas components are abundant in minerals with high concentrations of U, Th, and K, particularly when their age of formation is relatively high. Dating young lava flows, for which the mineral formation age and the surface exposure age are roughly identical, is therefore easier than studying landforms consisting of old rock which was exposed relatively recently. Typically low concentrations of U, Th, and K in olivine and pyroxene are another argument for studying those minerals, besides their good He and Ne retentivity. In quartz, these elements are also rare, but in practice there are often inclusions or intergrowths of U-rich minerals, such as biotites, especially in granitic host rocks. Clean quartz separates are therefore essential to avoid problems with interfering nucleogenic Ne components. Recent experience in Potsdam indicates that quartz clasts derived from very low-grade metasediments contain much less nucleogenic Ne than those from granites (Niedermann and Hermanns 1999), but on the other hand vein quartz may carry a lot of trapped gases in fluid inclusions (Hetzel et al. 2002a). Since α particles from U/Th decay have a range of 10-40 μm (e.g., Ziegler 1977), they may also be implanted from neighboring minerals into the quartz. Etching the quartz grains is therefore advisable to remove the surface layer. In addition to the quartz separation procedure of Kohl and Nishiizumi (1992), which was developed for ^{10}Be and ^{26}Al analyses and includes etching by HF and HNO_3, Niedermann et al. (1994) used a heavy liquid to obtain density fractions. The 2.66-2.69 g/cm^3 fraction was least affected by nucleogenic Ne, whereas the heavy and light fractions were probably contaminated by U/Th-bearing mineral and fluid inclusions, respectively.

Nucleogenic ^{3}He produced by the reaction $^{6}Li(n,\alpha)^{3}H(\beta^{-})^{3}He$ is only important in minerals with high Li content. The $^{3}He/^{4}He$ ratio resulting from U/Th decay and ^{6}Li neutron capture can be estimated from the chemical composition of the mineral and relevant nuclear parameters (Mamyrin and Tolstikhin 1984) and is, for common rock compositions, on the order of 10^{-8}. In Li-rich minerals, such as hornblende or micas, substantially elevated $^3He/^4He$ ratios could however be established, rendering the distinction of a cosmogenic ^3He component difficult. Therefore such minerals are probably not well suited for ^3He surface exposure dating.

Crushing versus melting. Those non-cosmogenic components which cannot be avoided by careful sample selection and preparation must be corrected for. Usually the first step in such a procedure is to calculate the excess of a noble gas isotope (e.g., ^3He, ^{21}Ne) over the composition of the trapped gas:

$$^3He_{ex} = [(^3He/^4He)_m - (^3He/^4He)_{tr}] \times {}^4He_m \quad (34a)$$

$$^{21}Ne_{ex} = [(^{21}Ne/^{20}Ne)_m - (^{21}Ne/^{20}Ne)_{tr}] \times {}^{20}Ne_m \quad (34b)$$

where indices ex, m, and tr mean excess, measured, and trapped. These equations are approximations assuming that $^4He_{ex}$ and $^{20}Ne_{ex}$ can be neglected. The isotope ratio in the trapped component must either be assumed or determined independently, if possible. Since $(^3He/^4He)_{tr}$ ratios may vary by several orders of magnitude in terrestrial samples, from $\sim 1 \times 10^{-8}$ to $\sim 5 \times 10^{-5}$, it is essential to use correct values. A reliable method to determine the $(^3He/^4He)_{tr}$ ratio in a sample is crushing in vacuo, which releases only gases stored in fluid inclusions but not those from the crystal lattice, where cosmogenic nuclides reside. This technique was already applied in the earliest studies of terrestrial cosmogenic He (Kurz 1986a,b; Craig and Poreda 1986) and yields consistent results under the prerequisite that the trapped He compositions in fluid inclusions and matrix are identical and that there is no contribution from radiogenic ^4He produced *in situ*. These conditions are met at least for olivines from young volcanic rocks. In older rocks and in U/Th-rich minerals, a contribution from radiogenic ^4He in the crystal lattice must be expected, which does not show up by crushing and requires a correction. Likewise, implantation of α particles may have a sizable effect if U/Th concentrations in neighboring minerals or the host rock are distinctly higher (cf. Dunai and Wijbrans 2000).

Trapped Ne is usually assumed to have an atmospheric isotopic composition ($^{21}Ne/^{20}Ne = 0.002959$, $^{22}Ne/^{20}Ne = 0.1020$; Eberhardt et al. 1965), but this is not always correct. In the mantle rocks studied by the Paris group (Staudacher and Allègre 1991, 1993a,b; Sarda et al. 1993), the presence of mantle Ne was demonstrated by elevated $^{20}Ne/^{22}Ne$ ratios; these authors assumed mixtures of atmospheric and MORB-type Ne for the trapped component. Niedermann et al. (1994) found evidence for two distinct trapped Ne components in quartz from a Sierra Nevada granite, being somewhat fractionated from air in both directions. Hetzel et al. (2002a) applied in vacuo crushing to quartzite from the northeastern margin of the Tibetan Plateau and reported $^{21}Ne/^{20}Ne$ ratios up to 0.0039 (30% higher than atmospheric). Unlike He, for which part of the cosmogenic component is released along with the trapped one when crushing quartz (Brook and Kurz 1993), cosmogenic Ne is obviously retained. The non-atmospheric Ne in vein quartz was most probably trapped from crustal fluids. The crushing data are indeed representative of the trapped Ne composition, as shown by consistent ^{21}Ne and ^{10}Be ages, whereas assuming atmospheric trapped Ne yields erroneously high ^{21}Ne ages (Hetzel et al. 2002a).

Stepwise heating. Stepwise heating is another method for discriminating between cosmogenic and non-cosmogenic components. Although a quantitative separation of different components is seldom possible, distinct degassing characteristics may nevertheless yield significant information. The few papers which report stepwise heating data for cosmogenic He show the major release of ^3He_c from mafic and ultramafic minerals (olivine, pyroxene) below ~ 900-1100°C (Kurz 1986a; Staudacher and Allègre 1991, 1993a; Sarda et al. 1993; Schäfer et al. 2000). Extraction temperatures for mantle He are somewhat higher in some samples, but similar in others, whereas radiogenic ^{4}He is released concurrently with ^{3}He$_c$ (Kurz 1986a; Schäfer et al. 2000). The benefit of the stepwise heating technique for studies of cosmogenic helium is thus limited, at least in the minerals mentioned.

Ne is different. In olivine and pyroxene, only minor amounts of cosmogenic Ne are degassed at temperatures ≤ 900°C, whereas the release of trapped Ne is relatively uniform over the whole temperature range (Staudacher and Allègre 1991, 1993a; Schäfer et al. 2000). A partial separation of trapped and cosmogenic Ne is therefore achievable by stepwise heating at ~ 900°C and 1600-1800°C, allowing for a more accurate determination of excess Ne due to enhanced ^{21}Ne/^{20}Ne and ^{22}Ne/^{20}Ne ratios in the steps with a reduced "background" of trapped Ne.

A good resolution of different components is even more important for quartz, because the contribution of nucleogenic Ne is often higher than in pyroxenes and olivines. Niedermann et al. (1993) showed that in quartz ALH 85-4, separated from an Antarctic sandstone, > 97% of ^{21}Ne$_c$ is degassed at temperatures below 600°C. The main release of trapped Ne occurs at somewhat higher temperatures of ~ 400-800°C. These results are confirmed by studies of other quartz samples (Hudson et al. 1991; Phillips et al. 1998; Niedermann et al. 2001a), though it seems safer to increase the limit below which complete degassing of cosmogenic Ne occurs to 800°C. Figure 13 shows the ^{21}Ne$_c$ release pattern of six quartz samples. Despite some differences in the details, which are probably caused by distinct diffusion characteristics due to variations in the effective grain size (e.g., Trull and Kurz 1993), about 90% of ^{21}Ne$_c$ or more is degassed below

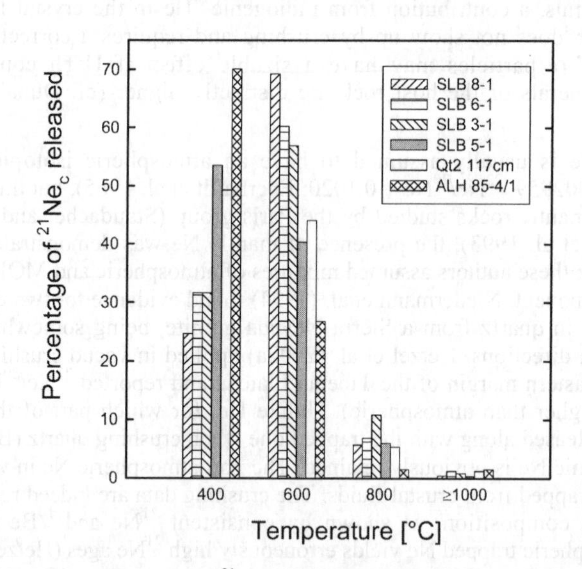

Figure 13. Release pattern of cosmogenic ^{21}Ne for six quartz samples degassed by stepwise heating at 400°C, 600°C, 800°C, and ≥ 1000°C. Samples SLB 8-1, 6-1, 3-1, and 5-1 are from metamorphic and granitic host rocks (Sierra Laguna Blanca, Argentina; Niedermann et al. 2001a), Qt2 117cm is a soil sample consisting of sand-sized grains (Pajarito Plateau, New Mexico; Phillips et al. 1998), and ALH 85-4/1 is from a sandstone boulder (Allan Hills, Antarctica; Niedermann et al. 1993). Despite some differences in the details, it is obvious that virtually no cosmogenic Ne is left in these quartz samples above 800°C.

600°C, and < 1.5% is left above 800°C in all samples. Since analytical uncertainties of Ne abundance determinations are typically ≥5%, any contributions from heating steps > 800°C can therefore safely be neglected, provided that the six samples plotted are representative for any quartz. However, in two quartz separates of Bruno et al. (1997), the fractions of excess ^{21}Ne remaining at 800°C are 43% and 23%, respectively. There are no indications for a significant nucleogenic Ne component in these samples, so the excesses observed at high temperatures must be cosmogenic as well. There are two possible reasons for the unusual degassing pattern: Bruno et al. (1997) acknowledge that sample temperatures may have been considerably below those of the crucible because of relatively short extraction times. In addition, they used rather large grain sizes of 300-500 μm. If these grains were exceptionally free of internal fractures, which determine the effective grain size for diffusion (Trull and Kurz 1993), it is possible that indeed higher temperatures were required to extract $^{21}Ne_c$. In addition to reducing the trapped component, crushing the quartz grains to ~100 μm or less will most probably ensure complete release of cosmogenic Ne at 800°C.

The completeness of $^{21}Ne_c$ extraction at 800°C is important whenever nucleogenic Ne is present. Niedermann et al. (1994) found two distinct components of nucleogenic Ne in quartz. One of them was released along with cosmogenic Ne predominantly between 100 and 600°C and consisted only of ^{21}Ne produced by the $^{18}O(\alpha,n)$ reaction. The other one, termed Ne_{HT} (for "high-temperature"), was mainly extracted above 800°C and was characterized by a $^{21}Ne/^{22}Ne$ ratio of ~ 1, close to that of cosmogenic Ne. Whereas the monoisotopic low-temperature component can easily be distinguished from cosmogenic Ne based on three-isotope systematics (see next section), the discrimination of Ne_c from Ne_{HT} depends critically on the distinct release characteristics. The two components can also be identified in granitic quartz from Argentina (Niedermann et al. 2001a). Probably the monoisotopic component is produced by $^{18}O(\alpha,n)^{21}Ne$ reactions in the quartz crystal lattice, which would explain the similar release pattern to cosmogenic Ne. Ne_{HT}, however, may be located in solid or fluid inclusions, in which the presence of F along with U and Th enables ^{22}Ne production through the reaction $^{19}F(\alpha,n)^{22}Na(\beta^+)^{22}Ne$. The higher temperatures required to extract Ne_{HT} would then be connected to the distinct retentivity of the inclusions.

Significant contributions of nucleogenic ^{22}Ne at temperatures < 600°C are reported in granitic quartz from Tibet by Schäfer et al. (2002). These authors suppose that micro-inclusions of biotite are the source of this component. It is indeed possible that reduced diffusion lengths in very small-grained biotites may decrease the temperature required for release of nucleogenic Ne. On the other hand, Schäfer et al. (2002) observed the nucleogenic signature also by in vacuo crushing, indicating that this component resides at least partly in fluid inclusions. Hence another explanation could be the presence of a crustal trapped component (see *Crushing versus melting* above) rather than *in situ* production. A better understanding of the relative importance of different components in various rock types will require more experimental experience as well as detailed petrographic observations, combined with other microanalytical techniques.

Neon three-isotope systematics. For Ne, information about different components can also be extracted from three-isotope systematics. Figure 14 schematically shows the signatures of Ne components and their mixing lines with atmospheric Ne in three-isotope space. It is customary in cosmogenic Ne studies to use the ^{20}Ne normalization, not ^{22}Ne as in most other applications, because it facilitates the identification of ^{21}Ne and ^{22}Ne excesses. Moreover, the determination of $^{21}Ne/^{20}Ne$ is sometimes more accurate than that of $^{21}Ne/^{22}Ne$ because of the isobaric interference of CO_2^{++} on m/e = 22.

As cosmogenic Ne in a certain mineral is characterized by a unique $^{21}Ne/^{22}Ne$ ratio,

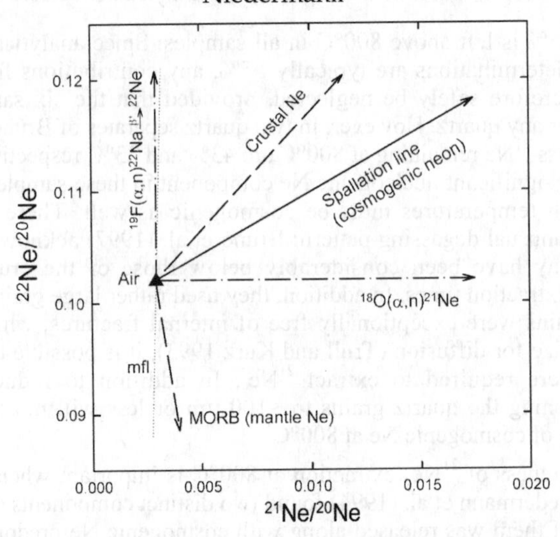

Figure 14. Neon three-isotope diagram (^{22}Ne/^{20}Ne vs. ^{21}Ne/^{20}Ne) showing the compositions and trends of various Ne components. Mixtures of atmospheric and cosmogenic Ne plot on the spallation line, the slope of which has been experimentally determined for quartz and pyroxene (Table 8). Contributions from nucleogenic ^{21}Ne and ^{22}Ne are characterized by shifts in horizontal and vertical directions, respectively. The trends for crustal Ne (Kennedy et al. 1990) and MORB-type Ne (Sarda et al. 1988) are also given. The dotted line labeled mfl is the mass fractionation line.

any two-component mixture of atmospheric and cosmogenic Ne must lie on the "spallation line" in Figure 14. Deviations from that line indicate contributions from other components, such as nucleogenic Ne or non-atmospheric trapped Ne (be it mantle, crustal, or mass-fractionated air Ne). Of course, a position on the spallation line does not necessarily imply a cosmogenic origin, as fortuitous mixtures of nucleogenic ^{21}Ne and ^{22}Ne (such as Ne_{HT}) or mantle and crustal Ne in the right proportions might also plot there. However, as long as there are not more than three components the contribution of each one can unequivocally be calculated. For example, if a three-component mixture of atmospheric Ne, cosmogenic Ne, and nucleogenic ^{21}Ne is indicated with sufficient confidence, the $^{21}Ne_c$ concentration is obtained from

$$^{21}\mathrm{Ne_c} = {}^{20}\mathrm{Ne_m} \times \frac{(^{21}\mathrm{Ne}/^{20}\mathrm{Ne})_c}{m} \frac{\left[(^{22}\mathrm{Ne}/^{20}\mathrm{Ne})_{air} - (^{22}\mathrm{Ne}/^{20}\mathrm{Ne})_m\right]}{\left[(^{21}\mathrm{Ne}/^{20}\mathrm{Ne})_{air} - (^{21}\mathrm{Ne}/^{20}\mathrm{Ne})_c\right]} \quad (35)$$

where m is the slope of the spallation line. Equation (35) corresponds to a horizontal shift of a data point in the three-isotope diagram to the spallation line and calculation of $^{21}Ne_{ex}$ for the shifted point according to (34b). Such a procedure is of course only applicable for data points lying to the right of the spallation line.

The slope of the spallation line depends on the relative production ratios of cosmogenic ^{20}Ne, ^{21}Ne, and ^{22}Ne and, therefore, on the mineral composition. It has been determined experimentally for quartz and pyroxene (Niedermann et al. 1993; Bruno et al. 1997; Phillips et al. 1998; Schäfer et al. 1999). The relevant data are compiled in Table 8. The values for $(^{21}\mathrm{Ne}/^{20}\mathrm{Ne})_c$ were assumed based on observations in well-shielded lunar samples and meteorites and model calculations (Hohenberg et al. 1978; Schäfer et al.

Table 8. Slope of the spallation line in a $^{22}Ne/^{20}Ne$ vs. $^{21}Ne/^{20}Ne$ three-isotope plot and Ne production ratios for quartz and pyroxene. The values for $^{22}Ne/^{20}Ne$ and $^{22}Ne/^{21}Ne$ are based on the experimental determination of the slope and an assumed $^{21}Ne/^{20}Ne$ ratio.

Mineral	*Slope*	$^{21}Ne/^{20}Ne$	$^{22}Ne/^{20}Ne$	$^{22}Ne/^{21}Ne$	*Reference*
Quartz	1.120±0.021	0.8±0.1	0.99±0.13	1.243±0.022	Niedermann et al. (1993)
		1.05±0.20 [1]	1.26±0.25 [1]		Bruno et al. (1997)
	1.10±0.10	0.8±0.1	0.98±0.15	1.22±0.10	Phillips et al. (1998)
	1.143±0.038	0.8±0.1	1.01±0.13	1.266±0.040	Schäfer et al. (1999)
Pyroxene	1.055±0.017	1.05±0.20	1.21±0.23	1.15±0.03	Bruno et al. (1997)
	1.069±0.035	1.10±0.20	1.27±0.24	1.159±0.040	Schäfer et al. (1999)
Model calculations:					
Quartz		0.78	1.02	1.31	Hohenberg et al. (1978) [2]
		0.83	0.94	1.14	Schäfer et al. (1999)
Pyroxene		0.91	0.79	0.86	Hohenberg et al. (1978) [2]
		1.10	1.11	1.01	Schäfer et al. (1999)

[1] Assumed $^{21}Ne/^{20}Ne$ too high, see text

[2] Lunar surface, shielding depth 500 g/cm^2

1999; Leya et al. 2000). The $(^{21}Ne/^{20}Ne)_c$ ratio of 1.05 ± 0.20 assumed by Bruno et al. (1997) is probably too high for quartz, since the production of ^{21}Ne and ^{22}Ne is only favored over that of ^{20}Ne in Mg-containing minerals (such as pyroxene) due to large cross sections of the $^{24,25}Mg(n,\alpha)^{21,22}Ne$ reactions. However the $(^{22}Ne/^{21}Ne)_c$ ratio is rather insensitive to that assumption. Likewise, the large error limits of the $(^{21}Ne/^{20}Ne)_c$ assumption have only minor influence on the precision of $(^{22}Ne/^{21}Ne)_c$. Indeed, the agreement between individual determinations of the slope and the $(^{22}Ne/^{21}Ne)_c$ ratio is excellent, with error limits of only 2-3%. Any deviations from a two-component mixture of atmospheric and cosmogenic Ne in quartz or pyroxene can therefore clearly be recognized and the due corrections be applied.

Cosmogenic noble gases as a nuisance

Even in a paper intended to boost the prospects of cosmogenic nuclides, their unpleasant sides should not remain unmentioned. Cosmogenic He and Ne components are not always welcome, because they may overprint other noble gas signatures and can even lead to erroneous conclusions when they are not recognized. A drastic example is the discovery of extremely high $^3He/^4He$ ratios up to 3.2×10^{-4} (230 R_A) in South African diamonds by Ozima et al. (1983), prior to the first reports of cosmogenic ^3He in terrestrial rocks. Based on these data the authors inferred that the initial terrestrial $^3He/^4He$ ratio was higher than that of planetary He and close to the solar value, and they discussed implications for the Earth's accretion. However, as Lal et al. (1987) and McConville and Reynolds (1989) have shown, such high $^3He/^4He$ ratios are most probably due to cosmic ray irradiation during surface residence of the diamonds, which had been purchased from commercial sources.

Today it has become clear that samples from the surface (or undocumented samples) are unsuitable for investigations of the isotopic compositions of mantle He and Ne, unless they are very young or were excavated only recently from well-shielded locations. Suboceanic rocks are of course not affected, but for continental samples, depending on the age and the altitude and latitude of the sampling location, a few meters of shielding are typically required to ensure that cosmogenic ^{3}He and ^{21}Ne will be low enough not to interfere with isotopic signatures of mantle, crustal, or nucleogenic components. Fortunately, since samples from well-shielded locations are not always easy to obtain, there is one possibility to extract information on trapped gases even from old surface samples, namely crushing in vacuo. As already mentioned, crushing only liberates gases from fluid inclusions, whereas the lattice-bound cosmogenic components are obviously retained (except for He in quartz; Brook and Kurz 1983).

In any noble gas investigation dealing with continental rocks, possible contributions of cosmogenic isotopes have thus to be assessed. Despite careful sampling, problems cannot always be avoided. Matsumoto et al. (2000) studied mantle xenoliths from the Newer Volcanics, Australia, and found ^{3}He/^{4}He ratios up to 8×10^{-5} in some low temperature steps, although the rocks were sampled in active quarries where they were buried until recently. These authors report a striking decoupling of He and Ne in the xenoliths, with MORB-like ^{3}He/^{4}He and "nucleogenic" ^{21}Ne/^{22}Ne ratios, which they interpret in terms of mantle metasomatism. While Matsumoto et al. (2000) acknowledge the presence of cosmogenic He and Ne in a few of their samples, they dismiss it for most of them. However, as judged from the isotope systematics, I believe that the observed ^{21}Ne excesses may well be of cosmogenic origin. Noble gas data on Romanian mantle xenoliths obtained in Potsdam have revealed a similar component with high ^{3}He/^{4}He and high ^{21}Ne/^{22}Ne ratios (Althaus et al. 1998), despite 5-10 m of shielding by volcanic rocks. Though we earlier interpreted the extremely high ^{3}He/^{4}He ratios of > 30 R_A as plume-derived, we now consider the possibility whether muon-induced reactions might have produced enough cosmogenic ^{3}He and ^{21}Ne at such depths since eruption (~ 0.8 Ma), assuming reasonable erosion rates.

Cosmogenic noble gases versus radionuclides

In the early years of terrestrial cosmogenic nuclide studies, the numbers of publications dealing with radionuclides and with noble gas isotopes, respectively, were more or less equal. However, later the radionuclide studies began to overwhelm, especially those with ^{10}Be and ^{26}Al. Among the reasons for such imbalance are experimental difficulties with noble gases as described above, first of all the interference of non-cosmogenic components. It is clear that radionuclides are superior in this respect, as contributions from such components are usually negligible or (as in the case of meteoric ^{10}Be) can reliably be removed (cf. Kohl and Nishiizumi 1992), but noble gases have other advantages:

- As stable nuclides, the noble gases record cosmic ray irradiation at any time in the past. This offers possibilities to investigate exposure histories which are longer than the limit of a few million years imposed by the half-life of ^{10}Be, the longest-lived radionuclide commonly used (see *Glacier movement and ice sheet evolution* below). Equivalently, very low erosion rates can only be quantified with noble gases. Even for exposure ages which are still within reach of the ^{10}Be method, results get inaccurate and depend strongly on assumptions for the production rate when saturation levels are approached (e.g., Stone 2000). Moreover, complex exposure histories involving prior surface residence followed by burial on a timescale of several million years can only be identified using cosmogenic noble gases along with radionuclides. It is even possible to study "paleo-exposures" (Libarkin et al. 2002), i.e., surface exposures which occurred many million years ago in rock which was later covered permanently.

- Typically, the sample size needed for noble gas studies is much smaller than for radionuclide analyses. Whereas 0.5-1 g of quartz is usually sufficient to determine $^{21}Ne_c$ concentrations on the 10^6 atoms/g level with < 10% uncertainty (2σ), some 20-30 g of quartz are required to achieve a similar accuracy for ^{10}Be and ^{26}Al. Therefore noble gases may be the better choice when suitable material is scarce.

- So far, the ^{10}Be and ^{26}Al methods have been successfully applied in quartz only. In other minerals, the separation of meteoric ^{10}Be has not reliably been attained, and the high concentrations of stable ^{27}Al render the detection of ^{26}Al impossible (e.g., Gosse and Phillips 2001). ^{36}Cl can be determined in more minerals, but the interpretation of results is often difficult due to various production mechanisms and uncertain production rates (Gosse and Phillips 2001), and relatively few labs have used it for surface exposure dating. The noble gases provide an alternative for mafic minerals and may work with other minerals also (***Retentivity of minerals for cosmogenic He and Ne*** section).
- The production rate of ^3He is much higher than for any other cosmogenic nuclide. In minerals where it is quantitatively retained, ^3He thus provides the chance to date exceptionally young surfaces. For example, we have derived an age of years for a sample from the 1993 lava flow of Láscar volcano (Chile) taken at 4540 m altitude a few months after eruption (Niedermann et al. 2001b), which illustrates the potential of ^3He on timescales also relevant to archeological studies. New developments in noble gas mass spectrometry, such as the "compressor ion source" which improves the mass spectrometer sensitivity for He and Ne by two orders of magnitude (Baur 1999), may further increase the precision of ^3He (as well as ^{21}Ne) determinations in the future.

For many applications a combined study of cosmogenic noble gases and radionuclides will be useful, for example in order to identify and evaluate complex exposure histories. Clearly, both techniques should be further developed without concentrating too much on just one of them.

APPLICATION EXAMPLES

In the last part of this review, I will present an overview of relevant work on terrestrial *in situ* produced cosmogenic nuclides. The intention is to show the broad range of possible applications and refer interested readers to the cited literature, without trying to provide detailed recapitulations of the various investigations nor claiming for completeness. Many of the cited studies were carried out with radionuclides; however there is no reason why similar studies should not be possible with noble gases as well.

Dating of lava flows

Lava flows were among the first surface features to be studied by cosmogenic nuclides. In many cases they are indeed ideal subjects, because the conditions for simple exposure histories are obviously met, and concentrations of radiogenic/nucleogenic isotopes are generally low since the rocks are only as old as their surface residence. The only major difficulty in dating lava flows is the influence of erosion, as was recognized even in the earliest studies (Craig and Poreda 1986; Kurz 1986a,b). Erosion effects tend to be more severe in humid climates (e.g., Hawaii) and for old lava flows. Cerling (1990) dated four lava flows at Owens Valley, California, and concluded that the ^3He ages of the younger flows (~ 13 and ~ 57 ka) were reasonable, while flows with K-Ar ages of > 2 Ma yielded ^3He ages too low by an order of magnitude. Erosion rates of 2-3 m/Ma or a part-time cover are required to account for the difference. Anthony and Poths (1992) compared their ^3He ages for lavas in the Potrillo volcanic field, New Mexico, with age estimates based on K-Ar and accumulation of calcium carbonate in soils and found consistent results, except for one set of K-Ar data. Linking their findings with compositional trends, they were able to establish a time sequence of magma evolution in individual eruptive centers and the volcanic field overall. Staudacher and Allègre (1993b) measured cosmogenic ^{3}He and ^{21}Ne in rocks from two eruption phases of Piton de la Fournaise volcano (Réunion) and derived an age of 23.8 ± 2.0 ka for the collapse of its second caldera. These examples illustrate the potential of cosmogenic nuclides for dating of lavas and volcanic eruptions, especially when other methods fail due to the young age of volcanic rocks or lack of ^{14}C-datable material.

Glacier movement and ice sheet evolution

Another important field of *in situ* cosmogenic nuclide studies is the history of glacier advance and retreat or the evolution of polar ice sheets. Such questions are directly linked to global climate variations. Many papers have therefore dealt with glacial landforms on the Antarctic continent, trying to gain information regarding their stability under changing climatic conditions. Apart from their potential significance to social or political issues, such studies are particularly suited for the method because exceptionally high exposure ages of several million years and exceptionally low erosion rates (~ 0.1-1m/Ma) promote accurate determinations of cosmogenic nuclide concentrations. This was shown in the earliest studies of Antarctic rocks (Nishiizumi et al. 1986, 1991a; Brown et al. 1991; Graf et al. 1991). One possible problem in the interpretation of Antarctic exposure histories is the low air pressure over that continent, which causes production rates substantially higher than expected as recognized only recently (Stone 2000). In this respect the stable noble gas isotopes provide more robust data than radionuclides having concentrations close to the saturation level (Stone 2000). In the following summary of relevant work it must be remembered that the reported exposure ages, which do not take account of the pressure effect, may be too high by ~ 25%.

Investigations have concentrated on the Dry Valleys in East Antarctica, for obvious reasons because the overwhelming majority of the continent is permanently covered by ice. Brook et al. (1993) studied boulders from a sequence of Taylor Glacier moraines and found significant scatter of ages within a single deposit. Moraine boulders are certainly not ideal samples for surface exposure dating because of frequent complex exposure histories (e.g., pre-exposure prior to glacier transport, past soil cover, boulders of an older deposit exposed between younger material, etc.). Nevertheless Brook et al. (1993) could place constraints on regional uplift rates and ice thickness variability within a timeframe of 2-3 Ma, and other studies also show that equivocal results can largely be avoided by careful sampling. Brook et al. (1995) dated clasts from drift sheets deposited by advances of the Ross Sea ice and again got widely scattered results. However, in this case the authors argued that the scatter may indicate several distinct drift depositions connected with episodes of ice sheet grounding, because the relatively young ages (8-106 ka for the younger drift) are less likely to be affected by erosion or soil cover and pre-exposure can in most cases be ruled out as the material is derived from the sea floor.

Ivy-Ochs et al. (1995), Bruno et al. (1997), Schäfer et al. (1999), Summerfield et al. (1999), and Van der Wateren et al. (1999) determined ^{10}Be, ^{3}He, and ^{21}Ne minimum exposure ages of erratic boulders, glacial sediments, and bedrock samples from various locations in the Dry Valleys region and the Transantarctic Mountains. These data include the highest exposure ages for terrestrial rocks reported to date, 10.08 ± 0.24 Ma for a dolerite boulder from Mount Fleming (Schäfer et al. 1999) and 11.2 ± 0.2 Ma for a bedrock sample from Daniels Range (Van der Wateren et al. 1999). Such high exposure ages can only be resolved with the stable noble gas isotopes, whereas radionuclides are at secular equilibrium between production and decay. Moreover, erosion rates have to be exceptionally low, < 6 cm/Ma in the case of the Mount Fleming dolerite. These data prove the extreme stability of the Dry Valleys landscape since at least Miocene time and exclude any significant climate variations during the Pliocene warm period. They support models of a stable East Antarctic Ice Sheet and limit uplift rates of the Transantarctic Mountains to ~ 100 m/Ma.

Schäfer et al. (2000) investigated the sublimation rate of a remnant ice body covered by a layer of till in the Dry Valleys region. Based on the exposure ages of two erratic boulders, one on top of the till and the other one sticking in the ice, they derived a sublimation rate of a few m/Ma at most and a minimum age of the remnant ice of 3 Ma. Ackert et al. (1999) measured ^{3}He and ^{36}Cl in moraine boulders deposited by a high-stand of the West Antarctic Ice Sheet and found evidence against a significant contribution to the eustatic sea level rise at 11 ka from West Antarctic Ice Sheet meltwater. Such studies have undoubtedly demonstrated the relevance of cosmogenic nuclide studies in Antarctica to global climate issues.

A few cosmogenic nuclide papers have dealt with glaciation history in temperate regions. ^{36}Cl dating of moraine boulders at Bloody Canyon, California, enabled Phillips et al. (1990) to establish the chronology of glacial sequences in the eastern Sierra Nevada and correlate them to global ice volume and the marine oxygen isotope record. In a similar study, Fernandez Mosquera et al. (2000) examined the sequence of deglaciation in the northwestern Iberian Peninsula (Spain and Portugal), based on cosmogenic ^{21}Ne in moraine boulders and bedrock. Bierman et al. (1999) sampled bedrock outcrops in Minnesota and Baffin Island, at the southern and northern paleo-margins of the Laurentide Ice Sheet, respectively. Based on ^{26}Al/^{10}Be ratios which were lower than expected for a simple exposure history, these authors proposed a model involving periods of temporary burial and concluded that the surfaces were at least 0.5 Ma old, despite individual exposure age data of only ~ 100 ka. An important conclusion from that work is the fact that the burial periods would have remained undetected if only one nuclide (or only stable nuclides) had been measured.

The extent of Pleistocene glaciations in the Himalayas and the Tibetan Plateau has been the subject of several recent studies. The paleoclimate in that region my have significant consequences for the South Asian monsoon system and even for global climate. Owen et al. (1999) and Phillips et al. (2000) dated glacial deposits in Northern Pakistan with cosmogenic radionuclides and arrived at different conclusions regarding the timing of glacial advances. Schäfer et al. (2002) applied ^{10}Be, ^{26}Al, and ^{21}Ne dating of erratic boulders to two locations in Central and Eastern Tibet. They found surprisingly old ages of 160-170 ka for moraines in the Central Plateau, excluding a major advance of glaciers in that area ever since. Extensive glaciations of the Tibetan Plateau, which could have substantially influenced the monsoon system or even triggered the onset of glacial advances in the whole northern hemisphere, are obviously not consistent with these data.

Rates of erosion and soil accumulation

Studies of erosion rates have been another important application for *in situ* produced cosmogenic nuclides, since quantitative data from other methods are scarce. The earlier papers focused mainly on exposure ages and provided erosion rate estimates only as a by-product. Craig and Poreda (1986), Kurz (1986b), and Nishiizumi et al. (1990) derived erosion rates of 8-12 m/Ma for lavas from Haleakala volcano, Maui (Hawaii), by comparing the nominal surface exposure ages with established lava ages. Sarda et al. (1993) estimated a somewhat lower erosion rate in a similar climate at Piton de la Fournaise volcano, Réunion. These authors pointed out that erosion rates derived from river loads or from neotectonic studies are often much higher, up to 1000 m/Ma, most probably because samples for cosmogenic studies are typically taken from relatively stable locations and are therefore not representative for vast areas where erosion is dominated by river incision. A similar conclusion was reached by Small et al. (1997), who assessed erosion rates of alpine bedrock summits in the western USA and inferred that the mean value of 7.6 ± 3.9 m/Ma was similar to erosion rates in other environments, excluding very arid regions.

Extremely low erosion rates were reported for the Antarctic Dry Valleys, as already mentioned in the preceding section. Values on the order of 0.1 m/Ma or less are obviously typical for the cold and arid climate conditions prevailing there (Nishiizumi et al. 1991a; Graf et al. 1991; Ivy-Ochs et al. 1995; Bruno et al. 1997; Schäfer et al. 1999; Summerfield et al. 1999). Even for samples taken from slopes inclined by 36-38°, Summerfield et al. (1999) found erosion rates ≤ 1m/Ma. In sharp contrast, locations affected by glacier incision may have suffered downcutting rates of some 1000m/Ma (Van der Wateren et al. 1999). In warmer climates under arid conditions, denudation is about an order of magnitude faster than in the extremely stable Dry Valleys surfaces. Bierman and Turner (1995) deduced a rate of ~ 0.7 m/Ma for the tops of granitic inselbergs in South Australia. Stone and Vasconcelos (2000) determined lowest rates of 1-2 m/Ma for the highest surface settings near Mt Isa (Queensland), whereas 2-5 m/Ma were typical at lower elevations. These authors also pointed out the dependence of erosion rates on rock types, being two to five times lower in silicate rocks than in limestone.

Fleming et al. (1999) and Cockburn et al. (2000) used erosion rate determinations with cosmogenic radionuclides to assess the rates of escarpment retreat at the passive continental margins of eastern South Africa and Namibia, respectively. In both cases the retreat rates were shown to be much too low to be compatible with an escarpment origin at the continental margin during the time of continent break-up.

Brown et al. (1995b) and Bierman and Steig (1996) developed a method to estimate mean erosion rates for entire river basins from cosmogenic nuclide concentrations in river sediment. Under the assumptions that production rates within the catchment area do not vary much, that cosmogenic nuclide concentrations are in erosion equilibrium, and that sediments are not stored for long times on the valley floor, the basin-wide erosion rate is given by

$$\langle\varepsilon\rangle = \frac{\langle P_0\rangle \Lambda}{\langle C\rangle \rho} \tag{36}$$

where symbols are as in Equations (26)-(33) and the brackets indicate average values. Brown et al. (1995b) demonstrated the suitability of their method by comparing the derived $\langle\varepsilon\rangle$ value of 43 m/Ma for the Icacos river basin, Puerto Rico, with erosion rates determined from ^{10}Be concentrations in samples from ridge crests and hill slopes and from river-load mass balance studies. In a subsequent investigation Brown et al. (1998a) applied the technique to assess the anthropogenic influence on denudation in another Puerto Rican watershed affected by agricultural exploitation. Schaller et al. (2001) studied the catchments of four middle European rivers with drainage areas from 3000 to 43,000 km^2 and found that erosion rates based on cosmogenic nuclides were higher by a factor of 1.5-4 than those obtained from river-load gauging. They proposed several possible causes for the difference and concluded that cosmogenic nuclides may provide better estimates of long-term erosion rates over timescales of some 10^4 years. Granger et al. (2001) assessed the dependence of erosion rates on slope gradient by studying ^{10}Be and ^{26}Al in bedrock and river sediment from granitic terrain in the northern Sierra Nevada, California. Their results demonstrate that erosion is controlled by the extent of bedrock outcropping, due to much faster weathering of soil-covered rock compared to bare rock, which may in some cases lead to higher erosion rates on valley bottoms and gentle footslopes than on steep mountain slopes.

Soil accumulation on a stable surface may, in principle, be regarded as negative erosion. However, there are some complications. One of these is the presence of cosmogenic nuclides inherited from earlier exposure on hillslopes or during transport, which will vary among different sediment clasts. As shown by Anderson et al. (1996) and Repka et al. (1997), such inheritance can be treated by amalgamating a statistically adequate number of clasts and by comparing measured depth profiles of cosmogenic nuclide concentrations with those expected for zero inheritance. There may be several types of depth profiles, depending on the rate and persistence of soil accretion (short-term deposition or steady accumulation) and the role of bioturbation within the soil. Explicit treatments of the theoretical background as well as application examples are given by Phillips et al. (1998) and Braucher et al. (2000).

Rates of tectonic uplift

Cosmogenic nuclides can be used in two ways to estimate rates of tectonic uplift. For samples which were exposed long enough, such as those in the Antarctic Dry Valleys (see *Glacier movement and ice sheet evolution* section), the mere concentration of a cosmogenic nuclide provides constraints on the uplift rate because of the dependence of production rates on altitude. A maximum uplift rate is obtained under the assumption of steady uplift from sea level to the present elevation (cf. Bruno et al. 1997). This method has been applied to limit uplift in the Transantarctic Mountains to < 170 m/Ma (Brook et al. 1993; Ivy-Ochs et al. 1995; Bruno et al. 1997), as opposed to certain studies suggesting much higher rates up to 1000 m/Ma.

The second method utilizes the relationship between river incision and uplift. Rivers in tectonically active regions compensate for uplift by downcutting. In the course of this process, river benches (straths) or fluvial terraces are created, which are later abandoned as incision continues. Therefore the time since abandonment of the strath or terrace, along with its elevation above the present valley floor, provides a measure for tectonic uplift, provided that a climatic control on terrace formation and abandonment can be ruled out. Burbank et al. (1996) and Leland et al. (1998) investigated incision of the Indus river in the northwestern Himalayas. They derived incision rates between 2 and 12 km/Ma (or mm/a) and argued for an equilibrium between incision and bedrock uplift in this region, because otherwise the longitudinal river profile would have been severely perturbed. In other tectonic settings the direct equation of incision and regional uplift may not apply as well. However, on a more local scale, the relative shifts in altitude caused by tectonic fault scarps can be assessed quite safely. Brown et al. (1998b) used ^{10}Be to date alluvial fans on the southern margin of the Tien Shan mountains (China), which were offset by tectonic faults. Combining their ages with the vertical offsets, they deduced vertical slip rates of 0.4-1.1 km/Ma. In a similar study at the northeastern margin of the Tibetan plateau, Hetzel et al. (2002b) found slip rates of 0.3-0.5 km/Ma for the Yumen thrust fault north of the Qilian Shan mountains, an order of magnitude less than assumed earlier based on the assumption of Holocene ages for the tectonically offset alluvial fan surfaces.

Earthquakes and landslides

Alluvial fans cut by tectonic faults can also yield information on the frequency of earthquakes. An offset debris flow fan in Owens Valley (California) was dated with ^{10}Be by Bierman et al. (1995). Based on boulder ages from different parts of the flow and the record of three faulting events, the authors deduced earthquake recurrence intervals of 5800-8000 years for that location. Zreda and Noller (1998) measured ^{36}Cl in a bedrock fault scarp at Hebgen Lake (Montana) and were able to extract the ages of six prehistoric earthquakes dating from 24 to 0.4 ka.

Hermanns et al. (2001) determined ^{21}Ne exposure ages of eight giant landslides and one tectonically uplifted terrace at Sierra Laguna Blanca (Argentina). Based on landslide ages of 150-430 ka, these authors argued for tectonic oversteepening of the mountain front generating repeated collapses, probably triggered by earthquakes larger than magnitude 7.5, whereas a relation to wetter climatic conditions was deemed unlikely. The cessation of landsliding after ~ 150 ka was interpreted as marking thrust-front migration towards the western piedmont, consistent with uplift of the terrace dated at ~ 85-130 ka.

Further applications

A few more examples are finally presented to illustrate the range of applications of terrestrial cosmogenic nuclide studies.

On the moon, cosmogenic nuclides have been a common means of dating meteorite impact craters (e.g., Eugster et al. 1977). A similar study on a terrestrial impact crater, namely Meteor Crater in Arizona (also known as Barringer crater or Canyon Diablo), was carried out by Nishiizumi et al. (1991b) and Phillips et al. (1991), using ^{10}Be, ^{26}Al, and ^{36}Cl as well as rock-varnish ^{14}C. They obtained consistent ages of 49.2 and 49.7 ka, respectively, in agreement with thermoluminescence studies but a factor of two higher than earlier estimates based on soil development and ^{14}C in lake sediments on the crater floor.

Wells et al. (1995) tested several competing models of desert stone pavement formation by comparing the ^{3}He exposure ages of pavement clasts with those of basalt outcrops from which they were derived. They obtained identical ages, which they judged as clear evidence that clasts had remained at the surface continuously and were not concentrated randomly by processes such as surface runoff or upward migration within soils.

Small et al. (1999) developed a method to estimate the rate of regolith production from bedrock based on cosmogenic nuclide concentrations and a mass balance model. They found regolith production rates of 7-20 m/Ma on alpine hillslopes in the Wind River Range, Wyoming, about a factor of 2 higher than bedrock erosion in similar environments, perhaps due to increased frost weathering in the water-bearing regoliths.

Cerling et al. (1999) determined ^{3}He exposure ages of boulders collected from four different debris flow fans at a tributary river mouth in Grand Canyon, Arizona. Together with radiocarbon dates obtained for the younger surfaces, they were able to establish a chronology for the episodic debris flows and draw conclusions regarding the relation between magnitudes and recurrence intervals of debris flows at the studied location.

Future prospects

Besides the various themes which have already been covered by cosmogenic nuclide studies as shown in the above sections, new fields may become amenable to the method in near future. I can only give a very limited outlook based on recent case studies.

Three investigations have tried to apply cosmogenic nuclides to issues of paleontology or archeology. Boaretto et al. (2000) made an effort to date cave sediments and flint tools from a prehistoric site in Israel. Although they could not derive the ages of the investigated layers due to experimental difficulties, their concept might be promising in the future. Ivy-Ochs et al. (2001) dated chert flakes from Egypt, which were produced by prehistoric men when making flintstone tools. Although they found reasonable ages of ~ 300 ka, the question what proportion of the exposure took place before the cobbles were worked on could not yet be answered. Farley et al. (2001) observed cosmogenic ^{3}He in fossil tooth enamel fluorapatite. Provided that He is quantitatively retained in tooth enamel, which the authors expect based on the retention of radiogenic ^{4}He, cosmogenic He could provide a new tool for dating fossil remains.

Libarkin et al. (2002) measured cosmogenic ^{21}Ne in quartz from the 28 Ma old Fish Canyon Tuff (Colorado), which is covered by later-emplaced tuff and sediment layers. Comparing the $^{21}Ne_c$ concentration with the time interval during which Fish Canyon Tuff was exposed at the surface, which is constrained by ^{40}Ar/^{39}Ar dating of the two tuff layers, these authors discuss implications for paleoerosion rates and paleoelevation in that area of the Colorado Rocky Mountains. Libarkin et al.'s (2002) study shows a novel perspective especially for applications of cosmogenic noble gas isotopes, as radionuclides do not record such ancient exposures.

Renne et al. (2001) have presented the first report of cosmogenic Ar in terrestrial rocks. Though the background of atmospheric Ar is usually higher than for Ne, a substantial production rate in Ca-rich rocks (e.g., Hohenberg et al. 1978; Lal 1991) makes its detection feasible. In fluorapatite, fluorite, sphene, and plagioclase samples from Antarctica and Namibia, they found elevated ^{38}Ar/^{36}Ar ratios up to 0.289 by total fusion and 0.364 by stepwise heating (atmosphere: 0.188). The ^{38}Ar/^{36}Ar ratios in Antarctic apatites correlate with ^3He_c concentrations, confirming the cosmogenic nature of the ^{38}Ar excesses. However, estimates for the production rates of ^{38}Ar and ^{36}Ar are not consistent for apatite and fluorite samples and also disagree with calculations of Lal (1991). The Ar production rate systematics are complicated by ^{36}Ar production through decay of the cosmogenic radionuclide ^{36}Cl (half life 3×10^5a), which is produced in part by neutron capture of ^{35}Cl (Table 1). Nevertheless, cosmogenic Ar may provide a valuable additional tool for surface exposure dating once production rates are better constrained, as it will be applicable in minerals which do not retain He and Ne quantitatively and in particular in Ca-rich rocks such as limestones.

In the future, investigations of terrestrial cosmogenic nuclides produced *in situ* in rock material will certainly become more and more widespread. As shown in the preceding sections, applications of the method span many fields of Earth sciences and beyond. Accuracy of results is expected to improve as both experimental methods and knowledge of the background, such as production rates, are refined. The use of other minerals and additional nuclides will further expand the range of treatable subjects.

ACKNOWLEDGMENTS

I thank Joe Licciardi for unpublished data and Ralf Hetzel and Rainer Wieler for comments on the manuscript. Reviews by Ken Farley, Simon Kelley, William Phillips, and Jörg Schäfer were very helpful to better achieve this article's purpose to become a survey of the current state of the art in cosmic ray exposure dating using noble gases.

REFERENCES

Ackert RP Jr, Barclay DJ, Borns HW Jr, Calkin PE, Kurz MD, Fastook JL, Steig EJ (1999) Measurements of past ice sheet elevations in interior west Antarctica. Science 286:276-280

Althaus T, Niedermann S, Erzinger J (1998) Noble gases in ultramafic mantle xenoliths of the Persani Mountains, Transylvanian Basin, Romania. Mineral Mag 62A:43-44

Anderson RS, Repka JL, Dick GS (1996) Explicit treatment of inheritance in dating depositional surfaces using *in situ* ^{10}Be and ^{26}Al. Geology 24:47-51

Anthony EY, Poths J (1992) ^{3}He surface exposure dating and its implications for magma evolution in the Potrillo volcanic field, Rio Grande Rift, New Mexico, USA. Geochim Cosmochim Acta 56:4105-4108

Barbouti AI, Rastin BC (1983) A study of the absolute intensity of muons at sea level and under various thicknesses of absorber. J Phys G: Nucl Phys 9:1577-1595

Bard E, Arnold M, Hamelin B, Tisnerat-Laborde N, Cabioch G (1998) Radiocarbon calibration by means of mass spectrometric $^{230}Th/^{234}U$ and ^{14}C ages of corals: an updated database including samples from Barbados, Mururoa and Tahiti. Radiocarbon 40:1085-1092
Bard E, Raisbeck GM, Yiou F, Jouzel J (1997) Solar modulation of cosmogenic nuclide production over the last millennium: comparison between ^{14}C and ^{10}Be records. Earth Planet Sci Lett 150:453-462
Baur H (1999) A noble-gas mass spectrometer compressor source with two orders of magnitude improvement in sensitivity. EOS Trans, Am Geophys Union 46:F1118
Beer J, Siegenthaler U, Bonani G, Finkel RC, Oeschger H, Suter M, Wölfli W (1988) Information on past solar activity and geomagnetism from ^{10}Be in the Camp Century ice core. Nature 331:675-679
Bierman P, Steig EJ (1996) Estimating rates of denudation using cosmogenic isotope abundances in sediment. Earth Surf Proc Landforms 21:125-139
Bierman P, Turner J (1995) ^{10}Be and ^{26}Al evidence for exceptionally low rates of Australian bedrock erosion and the likely existence of pre-Pleistocene landscapes. Quat Res 44:378-382
Bierman PR (1994) Using *in situ* produced cosmogenic isotopes to estimate rates of landscape evolution: A review from the geomorphic perspective. J Geophys Res 99:13885-13896
Bierman PR, Gillespie AR, Caffee MW (1995) Cosmogenic ages of earthquake recurrence intervals and debris flow fan deposition, Owens Valley, California. Science 270:447-450
Bierman PR, Marsella KA, Patterson C, Davis PT, Caffee M (1999) Mid-Pleistocene cosmogenic minimum-age limits for pre-Wisconsinan glacial surfaces in southwestern Minnesota and southern Baffin Island: a multiple nuclide approach. Geomorphology 27:25-39
Bilokon H, Cini Castagnoli G, Castellina A, D'Ettorre Piazzoli B, Mannocchi G, Meroni E, Picchi P, Vernetto S (1989) Flux of the vertical negative muons stopping at depths 0.35-1000 hg/cm^2. J Geophys Res 94:12145-12152
Boaretto E, Berkovits D, Hass M, Hui SK, Kaufman A, Paul M, Weiner S (2000) Dating of prehistoric caves sediments and flints using ^{10}Be and ^{26}Al in quartz from Tabun Cave (Israel): Progress report. Nucl Instr Meth Phys Res B172:767-771
Braucher R, Bourlès DL, Brown ET, Colin F, Muller J-P, Braun J-J, Delaune M, Edou Minko A, Lescouet C, Raisbeck GM, Yiou F (2000) Application of *in situ*-produced cosmogenic ^{10}Be and ^{26}Al to the study of lateritic soil development in tropical forest: theory and examples from Cameroon and Gabon. Chem Geol 170:95-111
Brook EJ, Kurz MD (1993) Surface-exposure chronology using *in situ* cosmogenic ^3He in Antarctic quartz sandstone boulders. Quat Res 39:1-10
Brook EJ, Kurz MD, Ackert RP Jr, Denton GH, Brown ET, Raisbeck GM, Yiou F (1993) Chronology of Taylor Glacier advances in Arena Valley, Antarctica, using *in situ* cosmogenic ^3He and ^{10}Be. Quat Res 39:11-23
Brook EJ, Kurz MD, Ackert RP, Raisbeck G, Yiou F (1995) Cosmogenic nuclide exposure ages and glacial history of late Quaternary Ross Sea drift in McMurdo Sound, Antarctica. Earth Planet Sci Lett 131: 41-56
Brown ET, Bourlès DL, Burchfiel BC, Deng Q, Li J, Molnar P, Raisbeck GM, Yiou F (1998b) Estimation of slip rates in the southern Tien Shan using cosmic ray exposure dates of abandoned alluvial fans. GSA Bulletin 110:377-386
Brown ET, Bourlès DL, Colin F, Raisbeck GM, Yiou F, Desgarceaux S (1995a) Evidence for muon-induced production of ^{10}Be in near-surface rocks from the Congo. Geophys Res Lett 22:703-706
Brown ET, Brook EJ, Raisbeck GM, Yiou F, Kurz MD (1992) Effective attenuation lengths of cosmic rays producing ^{10}Be and ^{26}Al in quartz: implications for exposure age dating. Geophys Res Lett 19:369-372
Brown ET, Edmond JM, Raisbeck GM, Yiou F, Kurz MD, Brook EJ (1991) Examination of surface exposure ages of Antarctic moraines using *in situ* produced ^{10}Be and ^{26}Al. Geochim Cosmochim Acta 55:2269-2283
Brown ET, Stallard RF, Larsen MC, Bourlès DL, Raisbeck GM, Yiou F (1998a) Determination of predevelopment denudation rates of an agricultural watershed (Cayaguás River, Puerto Rico) using *in situ*-produced ^{10}Be in river-borne quartz. Earth Planet Sci Lett 160:723-728
Brown ET, Stallard RF, Larsen MC, Raisbeck GM, Yiou F (1995b) Denudation rates determined from the accumulation of *in situ*-produced ^{10}Be in the Luquillo Experimental Forest, Puerto Rico. Earth Planet Sci Lett 129:193-202
Brown ET, Trull TW, Jean-Baptiste P, Raisbeck G, Bourlès D, Yiou F, Marty B (2000) Determination of cosmogenic production rates of ^{10}Be, ^3He and ^3H in water. Nucl Instr Meth Phys Res B 172:873-883
Bruno LA, Baur H, Graf T, Schlüchter C, Signer P, Wieler R (1997) Dating of Sirius group tillites in the Antarctic Dry Valleys with cosmogenic ^3He and ^{21}Ne. Earth Planet Sci Lett 147:37-54
Burbank DW, Leland J, Fielding E, Anderson RS, Brozovic N, Reid MR, Duncan C (1996) Bedrock incision, rock uplift and threshold hillslopes in the northwestern Himalayas. Nature 379:505-510

Carroll MR, Draper DS (1994) Noble gases as trace elements in magmatic processes. Chem Geol 117:37-56
Castagnoli G, Lal D (1980) Solar modulation effects in terrestrial production of carbon-14. Radiocarbon 22:133-158
Cerling TE (1990) Dating geomorphologic surfaces using cosmogenic ^{3}He. Quat Res 33:148-156
Cerling TE, Craig H (1994a) Geomorphology and *in situ* cosmogenic isotopes. Ann Rev Earth Planet Sci 22:273-317
Cerling TE, Craig H (1994b) Cosmogenic ^{3}He production rates from 39°N to 46°N latitude, western USA and France. Geochim Cosmochim Acta 58:249-255
Cerling TE, Webb RH, Poreda RJ, Rigby AD, Melis TS (1999) Cosmogenic ^{3}He ages and frequency of late Holocene debris flows from Prospect Canyon, Grand Canyon, USA. Geomorphology 27:93-111
Channell JET, Hodell DA, Lehman B (1997) Relative geomagnetic paleointensity and δ^{18}O at ODP Site 983 (Gardar Drift, North Atlantic) since 350 ka. Earth Planet Sci Lett 153:103-118
Clark DH, Bierman PR, Larsen P (1995) Improving *in situ* cosmogenic chronometers. Quat Res 44: 367-377
Cockburn HAP, Brown RW, Summerfield MA, Seidl MA (2000) Quantifying passive margin denudation and landscape development using a combined fission-track thermochronology and cosmogenic isotope analysis approach. Earth Planet Sci Lett 179:429-435
Craig H, Poreda R, Lupton JE, Marti K, Regnier S (1979) Rare gases and hydrogen in josephinite. EOS Trans, Am Geophys Union 60:970
Craig H, Poreda RJ (1986) Cosmogenic ^{3}He in terrestrial rocks: The summit lavas of Maui. Proc Natl Acad Sci USA 83:1970-1974
Davis R Jr, Schaeffer OA (1955) Chlorine-36 in nature. Ann NY Acad Sci 62:107-121
Desilets D, Zreda M (2001) On scaling cosmogenic nuclide production rates for altitude and latitude using cosmic ray measurements. Earth Planet Sci Lett 193:213-225
Desilets D, Zreda M, Lifton NA (2001) Comment on 'Scaling factors for production rates of *in situ* produced cosmogenic nuclides: a critical reevaluation' by Tibor J. Dunai. Earth Planet Sci Lett 188:283-287
Dunai TJ (2000a) Scaling factors for production rates of *in situ* produced cosmogenic nuclides: a critical reevaluation. Earth Planet Sci Lett 176:157-169
Dunai TJ (2000b) Erratum to "Scaling factors for production rates of *in situ* produced cosmogenic nuclides: a critical reevaluation." Earth Planet Sci Lett 178:425
Dunai TJ (2001a) Reply to comment on 'Scaling factors for production rates of *in situ* produced cosmogenic nuclides: a critical reevaluation' by Darin Desilets, Marek Zreda and Nathaniel Lifton. Earth Planet Sci Lett 188:289-298
Dunai TJ (2001b) Influence of secular variation of the geomagnetic field on production rates of *in situ* produced cosmogenic nuclides. Earth Planet Sci Lett 193:197-212
Dunai TJ, Roselieb K (1996) Sorption and diffusion of helium in garnet: implications for volatile tracing and dating. Earth Planet Sci Lett 139:411-421
Dunai TJ, Wijbrans JR (2000) Long-term cosmogenic ^{3}He production rates (152 ka-1.35 Ma) from ^{40}Ar/^{39}Ar dated basalt flows at 29°N latitude. Earth Planet Sci Lett 176:147-156
Dunne J, Elmore D, Muzikar P (1999) Scaling factors for the rates of production of cosmogenic nuclides for geometric shielding and attenuation at depth on sloped surfaces. Geomorphology 27:3-11
Eberhardt P, Eugster O, Marti K (1965) A redetermination of the isotopic composition of atmospheric neon. Z Naturforschung 20a:623-624
Eddy JA (1976) The Maunder minimum. Science 192:1189-1202
Eugster O (1988) Cosmic-ray production rates for ^{3}He, ^{21}Ne, ^{38}Ar, ^{83}Kr, and ^{126}Xe in chondrites based on ^{81}Kr-Kr exposure ages. Geochim Cosmochim Acta 52:1649-1662
Eugster O, Eberhardt P, Geiss J, Grögler N, Jungck M, Mörgeli M (1977) The cosmic-ray exposure history of Shorty Crater samples; The age of Shorty Crater. Proc Lunar Sci Conf 8th:3059-3082
Farley KA (2002) (U/Th)/He dating: Techniques, calibrations, and applications. Rev Mineral Geochem 47:819-845
Farley KA, Cerling TE, Fitzgerald PG (2001) Cosmogenic ^{3}He in igneous and fossil tooth enamel fluorapatite. Earth Planet Sci Lett 185:7-14
Fernandez Mosquera D, Marti K, Vidal Romani JR, Weigel A (2000) Late Pleistocene deglaciation chronology in the NW of the Iberian Peninsula using cosmic-ray produced ^{21}Ne in quartz. Nucl Instr Meth Phys Res B172:832-837
Fleming A, Summerfield MA, Stone JO, Fifield LK, Cresswell RG (1999) Denudation rates for the southern Drakensberg escarpment, SE Africa, derived from *in situ*-produced cosmogenic ^{36}Cl: initial results. J Geol Soc London 156:209-212

Frank M (2000) Comparison of cosmogenic radionuclide production and geomagnetic field intensity over the last 200 000 years. Phil Trans R Soc London A358:1089-1107

Frank M, Schwarz B, Baumann S, Kubik PW, Suter M, Mangini A (1997) A 200 kyr record of cosmogenic radionuclide production rate and geomagnetic field intensity from ^{10}Be in globally stacked deep-sea sediments. Earth Planet Sci Lett 149:121-129

Gillespie AR, Bierman PR (1995) Precision of terrestrial exposure ages and erosion rates estimated from analysis of cosmogenic isotopes produced *in situ*. J Geophys Res 100:24637-24649

Goguitchaichvili AT, Prévot M, Camps P (1999) No evidence for strong fields during the R3-N3 Icelandic geomagnetic reversal. Earth Planet Sci Lett 167:15-34

Gosse JC, Phillips FM (2001) Terrestrial *in situ* cosmogenic nuclides: theory and application. Quat Sci Rev 20:1475-1560

Graf T, Kohl CP, Marti K, Nishiizumi K (1991) Cosmic-ray produced neon in Antarctic rocks. Geophys Res Lett 18:203-206

Graf T, Marti K, Wiens RC (1996) The ^{21}Ne production rate in a Si target at mountain altitudes. Radiocarbon 38:155-156

Granger DE, Riebe CS, Kirchner JW, Finkel RC (2001) Modulation of erosion on steep granitic slopes by boulder armoring, as revealed by cosmogenic ^{26}Al and ^{10}Be. Earth Planet Sci Lett 186:269-281

Granger DE, Smith AL (2000) Dating buried sediments using radioactive decay and muogenic production of ^{26}Al and ^{10}Be. Nucl Instr Meth Phys Res B 172:822-826

Guyodo Y, Valet J-P (1996) Relative variations in geomagnetic intensity from sedimentary records: the past 200,000 years. Earth Planet Sci Lett 143:23-36

Guyodo Y, Valet J-P (1999) Global changes in intensity of the Earth's magnetic field during the past 800 kyr. Nature 399:249-252

Hart SR (1984) He diffusion in olivine. Earth Planet Sci Lett 70:297-302

Heidbreder E, Pinkau K, Reppin C, Schönfelder V (1971) Measurements of the distribution in energy and angle of high-energy neutrons in the lower atmosphere. J Geophys Res 76:2905-2916

Heisinger B, Niedermayer M, Hartmann FJ, Korschinek G, Nolte E, Morteani G, Neumaier S, Petitjean C, Kubik P, Synal A, Ivy-Ochs S (1997) *In situ* production of radionuclides at great depths. Nucl Instr Meth Phys Res B 123:341-346

Heisinger B, Nolte E (2000) Cosmogenic *in situ* production of radionuclides: Exposure ages and erosion rates. Nucl Instr and Meth in Phys Res B172:790-795

Hermanns RL, Niedermann S, Villanueva Garcia A, Sosa Gomez J, Strecker MR (2001) Neotectonics and catastrophic failure of mountain fronts in the southern intra-Andean Puna Plateau, Argentina. Geology 29:619-622

Hetzel R, Niedermann S, Ivy-Ochs S, Kubik PW, Tao M, Gao B (2002a) 21Ne versus 26Al exposure ages of fluvial terraces: The influence of crustal Ne in quartz. Earth Planet Sci Lett (in press)

Hetzel R, Niedermann S, Tao M, Stokes S, Kubik PW, Ivy-Ochs S, Gao B, Strecker MR (2002b) Low slip rates and long-term preservation of geomorphic features in Central Asia. Nature 417:428-432

Hohenberg CM, Marti K, Podosek FA, Reedy RC, Shirck JR (1978) Comparisons between observed and predicted cosmogenic noble gases in lunar samples. Proc Lunar Planet Sci Conf 9th:2311-2344

Hudson GB, Caffee MW, Beiriger J, Ruiz B, Kohl CP, Nishiizumi K (1991) Production rate and retention properties of cosmogenic ^{3}He and ^{21}Ne in quartz. Eso Trans, Am Geophys Union 72:575

Ivy-Ochs S, Schlüchter C, Kubik PW, Dittrich-Hannen B, Beer J (1995) Minimum ^{10}Be exposure ages of early Pliocene for the Table Mountain plateau and the Sirius Group at Mount Fleming, Dry Valleys, Antarctica. Geology 23:1007-1010

Ivy-Ochs S, Wüst R, Kubik PW, Müller-Beck H, Schlüchter C (2001) Can we use cosmogenic isotopes to date stone artifacts? Radiocarbon (in press)

Juarez MT, Tauxe L (2000) The intensity of the time-averaged geomagnetic field: the last 5 Myr. Earth Planet Sci Lett 175:169-180

Kelley S (2002) K-Ar and Ar-Ar dating. Rev Mineral Geochem 47:785-818

Kennedy BM, Hiyagon H, Reynolds JH (1990) Crustal neon: a striking uniformity. Earth Planet Sci Lett 98:277-286

Klein J, Giegengack R, Middleton R, Sharma P, Underwood JR Jr, Weeks RA (1986) Revealing histories of exposure using *in situ* produced ^{26}Al and ^{10}Be in Libyan desert glass. Radiocarbon 28:547-555

Kohl CP, Nishiizumi K (1992) Chemical isolation of quartz for measurement of *in situ*-produced cosmogenic nuclides. Geochim Cosmochim Acta 56:3583-3587

Kok YS (1999) Climatic influence in NRM and ^{10}Be-derived geomagnetic paleointensity data. Earth Planet Sci Lett 166:105-119

Kubik PW, Ivy-Ochs S, Masarik J, Frank M, Schlüchter C (1998) ^{10}Be and ^{26}Al production rates deduced from an instantaneous event within the dendro-calibration curve, the landslide of Köfels, Ötz Valley, Austria. Earth Planet Sci Lett 161:231-241

Kurz MD (1986a) Cosmogenic helium in a terrestrial igneous rock. Nature 320:435-439

Kurz MD (1986b) *In situ* production of terrestrial cosmogenic helium and some applications to geochronology. Geochim Cosmochim Acta 50:2855-2862

Kurz MD (1987) Erratum. Geochim Cosmochim Acta 51:1019

Kurz MD, Colodner D, Trull TW, Moore RB, O'Brien K (1990) Cosmic ray exposure dating with *in situ* produced cosmogenic ^{3}He: results from young Hawaiian lava flows. Earth Planet Sci Lett 97:177-189

Lal D (1988) *In situ*-produced cosmogenic isotopes in terrestrial rocks. Ann Rev Earth Planet Sci 16: 355-388

Lal D (1991) Cosmic ray labeling of erosion surfaces: *in situ* nuclide production rates and erosion models. Earth Planet Sci Lett 104:424-439

Lal D, Nishiizumi K, Klein J, Middleton R, Craig H (1987) Cosmogenic ^{10}Be in Zaire alluvial diamonds: implications for ^{3}He contents of diamonds. Nature 328:139-141

Lal D, Peters B (1967) Cosmic ray produced radioactivity on the earth. *In* Handbook of Physics, 46/2. Springer, Berlin, p 551-612

Leland J, Reid MR, Burbank DW, Finkel R, Caffee M (1998) Incision and differential bedrock uplift along the Indus River near Nanga Parbat, Pakistan Himalaya, from ^{10}Be and ^{26}Al exposure age dating of bedrock straths. Earth Planet Sci Lett 154:93-107

Leya I, Lange H-J, Neumann S, Wieler R, Michel R (2000) The production of cosmogenic nuclides in stony meteoroids by galactic cosmic-ray particles. Meteorit Planet Sci 35:259-286

Libarkin JC, Quade J, Chase CG, Poths J, McIntosh W (2002) Measurement of ancient cosmogenic ^{21}Ne in quartz from the 28 Ma Fish Canyon Tuff, Colorado. Chem Geol 186:199-213

Licciardi JM, Kurz MD, Clark PU, Brook EJ (1999) Calibration of cosmogenic ^{3}He production rates from Holocene lava flows in Oregon, USA, and effects of the Earth's magnetic field. Earth Planet Sci Lett 172:261-271

Liu B, Phillips FM, Fabryka-Martin JT, Fowler MM, Stone WD (1994) Cosmogenic ^{36}Cl accumulation in unstable landforms 1. Effects of the thermal neutron distribution. Water Resour Res 30:3115-3125

Mamyrin BA, Tolstikhin IN (1984) Helium Isotopes in Nature. Elsevier Science Publishers, Amsterdam

Marti K, Craig H (1987) Cosmic-ray-produced neon and helium in the summit lavas of Maui. Nature 325:335-337

Marti K, Eberhardt P, Geiss J (1966) Spallation, fission, and neutron capture anomalies in meteoritic krypton and xenon. Z Naturforschung 21a:398-413

Masarik J, Beer J (1999) Simulation of particle fluxes and cosmogenic nuclide production in the Earth's atmosphere. J Geophys Res 104:12099-12111

Masarik J, Frank M, Schäfer JM, Wieler R (2001) Correction of *in situ* cosmogenic nuclide production rates for geomagnetic field intensity variations during the past 800,000 years. Geochim Cosmochim Acta 65:2995-3003

Masarik J, Kollar D, Vanya S (2000) Numerical simulation of *in situ* production of cosmogenic nuclides: Effects of irradiation geometry. Nucl Instr Meth Phys Res B172:786-789

Masarik J, Reedy RC (1995) Terrestrial cosmogenic-nuclide production systematics calculated from numerical simulations. Earth Planet Sci Lett 136:381-395

Masarik J, Reedy RC (1996) Monte Carlo simulation of *in situ* produced cosmogenic nuclides. Radiocarbon 38:163-164

Matsumoto T, Honda M, McDougall I, O'Reilly SY, Norman M, Yaxley G (2000) Noble gases in pyroxenites and metasomatised peridotites from the Newer Volcanics, southeastern Australia: implications for mantle metasomatism. Chem Geol 168:49-73

McConville P, Reynolds JH (1989) Cosmogenic helium and volatile-rich fluid in Sierra Leone alluvial diamonds. Geochim Cosmochim Acta 53:2365-2375

Merrill RT, McElhinny MW, McFadden PL (1998) The Magnetic Field of the Earth. Academic Press, San Diego

Niedermann S (2000) The ^{21}Ne production rate in quartz revisited. Earth Planet Sci Lett 183:361-364

Niedermann S, Althaus T, Hahne K (2001b) A minimum age for Llullaillaco south flow from cosmogenic ^{3}He: much older than 19th century. *In* III South American Symposium on Isotope Geology, Extended Abstracts Volume (CD), Sociedad Geológica de Chile, Santiago, Chile, p 56-59

Niedermann S, Eugster O (1992) Noble gases in lunar anorthositic rocks 60018 and 65315: Acquisition of terrestrial krypton and xenon indicating an irreversible adsorption process. Geochim Cosmochim Acta 56:493-509

Niedermann S, Graf T, Kim JS, Kohl CP, Marti K, Nishiizumi K (1994) Cosmic-ray-produced ^{21}Ne in terrestrial quartz: the neon inventory of Sierra Nevada quartz separates. Earth Planet Sci Lett 125: 341-355

Niedermann S, Graf T, Marti K (1993) Mass spectrometric identification of cosmic-ray-produced neon in terrestrial rocks with multiple neon components. Earth Planet Sci Lett 118:65-73

Niedermann S, Hermanns RL (1999) The chronology of giant landslides at Sierra Laguna Blanca, Argentina, as deduced from cosmogenic ^{21}Ne. European Geophysical Society, Geophys Res Abstr 1:43
Niedermann S, Hermanns RL, Strecker MR (2001a) ^{21}Ne surface exposure dating of giant landslides at Sierra Laguna Blanca, Argentina: Evidence for tectonic control on slope oversteepening. *In* III South American Symposium on Isotope Geology, Extended Abstracts Volume (CD), Sociedad Geológica de Chile, Santiago, Chile, p 413-416
Nishiizumi K, Finkel RC, Caffee MC, Southon JR, Kohl CP, Arnold JR, Olinger CT, Poths J, Klein J (1994) Cosmogenic production of ^{10}Be and ^{26}Al on the surface of the earth and underground. Proc 8th Intl Conf Geochron, Cosmochron and Isotope Geol, U S Geol Surv Circular 1107:234
Nishiizumi K, Finkel RC, Klein J, Kohl CP (1996) Cosmogenic production of ^{7}Be and ^{10}Be in water targets. J Geophys Res 101:22225-22232
Nishiizumi K, Klein J, Middleton R, Craig H (1990) Cosmogenic ^{10}Be, ^{26}Al, and ^{3}He in olivine from Maui lavas. Earth Planet Sci Lett 98:263-266
Nishiizumi K, Kohl CP, Arnold JR, Klein J, Fink D, Middleton R (1991a) Cosmic ray produced ^{10}Be and ^{26}Al in Antarctic rocks: exposure and erosion history. Earth Planet Sci Lett 104:440-454
Nishiizumi K, Kohl CP, Shoemaker EM, Arnold JR, Klein J, Fink D, Middleton R (1991b) *In situ* ^{10}Be-^{26}Al exposure ages at Meteor Crater, Arizona. Geochim Cosmochim Acta 55:2699-2703
Nishiizumi K, Lal D, Klein J, Middleton R, Arnold JR (1986) Production of ^{10}Be and ^{26}Al by cosmic rays in terrestrial quartz *in situ* and implications for erosion rates. Nature 319:134-136
Nishiizumi K, Winterer EL, Kohl CP, Klein J, Middleton R, Lal D, Arnold JR (1989) Cosmic ray production rates of ^{10}Be and ^{26}Al in quartz from glacially polished rocks. J Geophys Res 94: 17907-17915
O'Brien K, Sandmeier HA, Hansen GE, Campbell JE (1978) Cosmic ray induced neutron background sources and fluxes for geometries of air over water, ground, iron, and aluminum. J Geophys Res 83:114-120
Ohno M, Hamano Y (1992) Geomagnetic poles over the last 10,000 years. Geophys Res Lett 19:1715-1718
Ohno M, Hamano Y (1993) Global analysis of the geomagnetic field: Time variation of the dipole moment and the geomagnetic pole in the Holocene. J Geomag Geoelectr 45:1455-1466
Owen LA, Caffee M, Finkel RC, Gualtieri L, Spencer JQ, Richards B (1999) Timing of multiple glaciations throughout the Himalayas. Geol Soc Am Abstr Progr 31:A141
Ozima M, Zashu S, Nitoh O (1983) ^{3}He/^{4}He ratio, noble gas abundance and K-Ar dating of diamonds – An attempt to search for the records of early terrestrial history. Geochim Cosmochim Acta 47:2217-2224
Paneth FA, Reasbeck P, Mayne KI (1952) Helium 3 content and age of meteorites. Geochim Cosmochim Acta 2:300-303
Phillips FM, Leavy BD, Jannik NO, Elmore D, Kubik PW (1986) The accumulation of cosmogenic chlorine-36 in rocks: a method for surface exposure dating. Science 231:41-43
Phillips FM, Zreda MG, Flinsch MR, Elmore D, Sharma P (1996) A reevaluation of cosmogenic ^{36}Cl production rates in terrestrial rocks. Geophys Res Lett 23:949-952
Phillips FM, Zreda MG, Smith SS, Elmore D, Kubik PW, Dorn RI, Roddy DJ (1991) Age and geomorphic history of Meteor Crater, Arizona, from cosmogenic ^{36}Cl and ^{14}C in rock varnish. Geochim Cosmochim Acta 55:2695-2698
Phillips FM, Zreda MG, Smith SS, Elmore D, Kubik PW, Sharma P (1990) Cosmogenic chlorine-36 chronology for glacial deposits at Bloody Canyon, eastern Sierra Nevada. Science 248:1529-1532
Phillips WM, McDonald EV, Reneau SL, Poths J (1998) Dating soils and alluvium with cosmogenic ^{21}Ne depth profiles: case studies from the Pajarito Plateau, New Mexico, USA. Earth Planet Sci Lett 160:209-223
Phillips WM, Sloan VF, Shroder JF Jr, Sharma P, Clarke ML, Rendell HM (2000) Asynchronous glaciation at Nanga Parbat, northwestern Himalaya Mountains, Pakistan. Geology 28:431-434
Plummer MA, Phillips FM, Fabryka-Martin J, Turin HJ, Wigand PE, Sharma P (1997) Chlorine-36 in fossil rat urine: An archive of cosmogenic nuclide deposition during the past 40,000 years. Science 277:538-541
Porcelli DR, Stone JOH, O'Nions RK (1987) Enhanced ^{3}He/^{4}He ratios and cosmogenic helium in ultramafic xenoliths. Chem Geol 64:25-33
Poreda RJ, Cerling TE (1992) Cosmogenic neon in recent lavas from the western United States. Geophys Res Lett 19:1863-1866
Radok U, Allison I, Wendler G (1996) Atmospheric surface pressure over the interior of Antarctica. Antarct Sci 8:209-217
Reedy RC, Nishiizumi K, Lal D, Arnold JR, Englert PAJ, Klein J, Middleton R, Jull AJT, Donahue DJ (1994) Simulations of *in situ* cosmogenic nuclide production. Nucl Instr Meth Phys Res B92:297-300
Renne PR, Farley KA, Becker TA, Sharp WD (2001) Terrestrial cosmogenic argon. Earth Planet Sci Lett 188:435-440

Repka JL, Anderson RS, Finkel RC (1997) Cosmogenic dating of fluvial terraces, Fremont River, Utah. Earth Planet Sci Lett 152:59-73

Sarda P, Staudacher T, Allègre CJ (1988) Neon isotopes in submarine basalts. Earth Planet Sci Lett 91: 73-88

Sarda P, Staudacher T, Allègre CJ, Lecomte A (1993) Cosmogenic neon and helium at Réunion: measurement of erosion rate. Earth Planet Sci Lett 119:405-417

Schäfer JM, Baur H, Denton GH, Ivy-Ochs S, Marchant DR, Schlüchter C, Wieler R (2000) The oldest ice on Earth in Beacon Valley, Antarctica: new evidence from surface exposure dating. Earth Planet Sci Lett 179:91-99

Schäfer JM, Ivy-Ochs S, Wieler R, Leya I, Baur H, Denton GH, Schlüchter C (1999) Cosmogenic noble gas studies in the oldest landscape on Earth: surface exposure ages of the Dry Valleys, Antarctica. Earth Planet Sci Lett 167:215-226

Schäfer JM, Tschudi S, Zhao Z, Wu X, Ivy-Ochs S, Wieler R, Baur H, Kubik PW, Schlüchter C (2002) The limited influence of glaciations in Tibet on global climate over the past 170000 yr. Earth Planet Sci Lett 194:287-297

Schaller M, von Blanckenburg F, Hovius N, Kubik PW (2001) Large-scale erosion rates from *in situ*-produced cosmogenic nuclides in European river sediments. Earth Planet Sci Lett 188:441-458

Shanahan TM, Zreda M (2000) Chronology of Quaternary glaciations in East Africa. Earth Planet Sci Lett 177:23-42

Shea MA, Smart DF, Gentile LC (1987) Estimating cosmic ray vertical cutoff rigidities as a function of the McIlwain L-parameter for different epochs of the geomagnetic field. Phys Earth Planet Inter 48: 200-205

Simpson JA (1983) Elemental and isotopic composition of the galactic cosmic rays. Ann Rev Nucl Part Sci 33:323-381

Small EE, Anderson RS, Hancock GS (1999) Estimates of the rate of regolith production using ^{10}Be and ^{26}Al from an alpine hillslope. Geomorphology 27:131-150

Small EE, Anderson RS, Repka JL, Finkel R (1997) Erosion rates of alpine bedrock summit surfaces deduced from *in situ* ^{10}Be and ^{26}Al. Earth Planet Sci Lett 150:413-425

Srinivasan B (1976) Barites: anomalous xenon from spallation and neutron-induced reactions. Earth Planet Sci Lett 31:129-141

Staudacher T, Allègre CJ (1991) Cosmogenic neon in ultramafic nodules from Asia and in quartzite from Antarctica. Earth Planet Sci Lett 106:87-102

Staudacher T, Allègre CJ (1993a) The cosmic ray produced ^{3}He/^{21}Ne ratio in ultramafic rocks. Geophys Res Lett 20:1075-1078

Staudacher T, Allègre CJ (1993b) Ages of the second caldera of Piton de la Fournaise volcano (Réunion) determined by cosmic ray produced ^{3}He and ^{21}Ne. Earth Planet Sci Lett 119:395-404

Sternberg RS (1992) Radiocarbon fluctuations and the geomagnetic field. *In* Taylor RE et al. (eds) Radiocarbon After Four Decades. Springer, New York, p 93-116

Sternberg R (1996) Workshop on secular variations in the rates of production of cosmogenic nuclides on Earth: Paleomagnetic averages of geomagnetic latitude. Radiocarbon 38:169-170

Stone J, Vasconcelos P (2000) Studies of geomorphic rates and processes with cosmogenic isotopes—examples from Australia. J Conf Abstr 5:961

Stone JO (2000) Air pressure and cosmogenic isotope production. J Geophys Res 105:23753-23759

Stone JO, Allan GL, Fifield LK, Cresswell RG (1996) Cosmogenic chlorine-36 from calcium spallation. Geochim Cosmochim Acta 60:679-692

Stone JO, Evans JM, Fifield LK, Allan GL, Cresswell RG (1998) Cosmogenic chlorine-36 production in calcite by muons. Geochim Cosmochim Acta 62:433-454

Stuiver M, Reimer PJ, Bard E, Beck JW, Burr GS, Hughen KA, Kromer B, McCormac G, van der Plicht J, Spurk M (1998) INTCAL98 radiocarbon age calibration, 24,000-0 cal BP. Radiocarbon 40:1041-1083

Summerfield MA, Stuart FM, Cockburn HAP, Sugden DE, Denton GH, Dunai T, Marchant DR (1999) Long-term rates of denudation in the Dry Valleys, Transantarctic Mountains, southern Victoria Land, Antarctica based on *in situ* produced cosmogenic ^{21}Ne. Geomorphology 27:113-129

Trull TW, Kurz MD (1993) Experimental measurements of ^{3}He and ^{4}He mobility in olivine and clinopyroxene at magmatic temperatures. Geochim Cosmochim Acta 57:1313-1324

Trull TW, Kurz MD, Jenkins WJ (1991) Diffusion of cosmogenic ^{3}He in olivine and quartz: implications for surface exposure dating. Earth Planet Sci Lett 103:241-256

Valet J-P, Tric E, Herrero-Bervera E, Meynadier L, Lockwood JP (1998) Absolute paleointensity from Hawaiian lavas younger than 35 ka. Earth Planet Sci Lett 161:19-32

Van der Wateren FM, Dunai TJ, Van Balen RT, Klas W, Verbers ALLM, Passchier S, Herpers U (1999) Contrasting Neogene denudation histories of different structural regions in the Transantarctic Mountains rift flank constrained by cosmogenic isotope measurements. Global Planet Change 23: 145-172

Wagner G, Masarik J, Beer J, Baumgartner S, Imboden D, Kubik PW, Synal H-A, Suter M (2000) Reconstruction of the geomagnetic field between 20 and 60 kyr BP from cosmogenic radionuclides in the GRIP ice core. Nucl Instr Meth Phys Res B172:597-604

Wells SG, McFadden LD, Poths J, Olinger CT (1995) Cosmogenic ^{3}He surface-exposure dating of stone pavements: Implications for landscape evolution in deserts. Geology 23:613-616

Wieler R (2002) Cosmic-ray-produced noble gases in meteorites. Rev Mineral Geochem 47:125-170

Yokoyama Y, Reyss J-L, Guichard F (1977) Production of radionuclides by cosmic rays at mountain altitudes. Earth Planet Sci Lett 36:44-50

Ziegler JF (1977) Helium: Stopping powers and ranges in all elemental matter. Pergamon, New York

Zreda MG, Noller JS (1998) Ages of prehistoric earthquakes revealed by cosmogenic chlorine-36 in a bedrock fault scarp at Hebgen Lake. Science 282:1097-1099

Zreda MG, Phillips FM, Elmore D, Kubik PW, Sharma P, Dorn RI (1991) Cosmogenic chlorine-36 production rates in terrestrial rocks. Earth Planet Sci Lett 105:94-109

17 K-Ar and Ar-Ar Dating

Simon Kelley
Department of Earth Sciences
The Open University
Milton Keynes MK7 6AA, United Kingdom
s.p.kelley@open.ac.uk

INTRODUCTION — A BIT OF HISTORY

The aim of this chapter is to present the K-Ar and Ar-Ar dating techniques in the context of noble gas studies, since there are already several recent texts on K-Ar and Ar-Ar dating (Dickin 1995; McDougall and Harrison 1999). The focus of this section will be aspects of argon transport and storage in the crust, which affect K-Ar and Ar-Ar dating including Ar-loss from minerals by diffusion and Ar-gain by minerals or "excess argon."

The K-Ar dating technique was one of the earliest isotope dating techniques, developed soon after the discovery of radioactive potassium, and provided an important adjunct to U-Pb and U-He dating techniques. The ease of measurement and ideal half-life (1250 million years; see Table 2 below), for dating geological events has made this the most popular of isotopic dating techniques. Aldrich and Nier (1948) first demonstrated that ^{40}Ar was the product of the decay of ^{40}K, and soon after K-Ar ages were being measured in several laboratories most often using an absolute method such as a McLeod gauge to measure argon concentrations. The first published K-Ar results by such a technique were those of Smits and Gentner (1950) who analyzed sylvite from the Buggingen Oligocene evaporite deposits, obtaining an age of 20 million years. Mass spectrometers, which simultaneously measured very small amounts of gas, and the isotope ratios necessary to make corrections for atmospheric contamination, quickly replaced manometric techniques. Crucially the use of static vacuum techniques, pioneered by John Reynolds at the University of California-Berkeley, meant that mass spectrometers were sufficiently sensitive to analyse the small amounts of gas released from common rocks and minerals. Although the earliest mass spectrometers were built 'in house', the introduction of the commercially available MS10 (Farrar et al. 1964), a small 180° metal mass spectrometer built for leak testing, made K-Ar dating generally available. Complete descriptions of early K-Ar development and techniques can be found in Schaeffer and Zähringer (1966) and Dalrymple and Lanphere (1969).

Although Thorbjorn Sigurgeirsson proposed the principles of Ar-Ar dating in an unpublished Icelandic laboratory report in 1962, he never succeeded in publishing or testing the idea. The Ar-Ar dating technique as it is practised today originated in the noble gas laboratory of John Reynolds in Berkeley, where Craig Merrihue and Grenville Turner were working on neutron irradiated meteorite samples using the I-Xe dating technique. Merrihue recognised that a ^{39}Ar signal seen in the chart recorder traces was the result of neutron irradiation and published the idea in an abstract (Merrihue 1965). The publication of Merrihue and Turner (1966) saw the birth of the Ar-Ar dating technique. Written by Turner after the untimely death of Merrihue, this paper unusually describes a fully formed isotope dating technique (compare this with the slow emergence of the full K-Ar technique), possibly because the Berkeley Laboratory had been recording the full traces of all noble gases for some time, allowing Merrihue and Turner rapid access to a considerable database of measurements. The advantage of the Ar-Ar technique is that potassium and argon are effectively measured simultaneously on the same aliquot of sample, providing greater internal precision and also the ability to analyse very small and

1529-6466/00/0047-0017$05.00

heterogeneous samples. Ar-Ar dating proved to be an ideal technique for dating meteorites because it made the best use of the extremely limited number of samples and also provided thermal histories. Indeed when lunar samples were returned from the Apollo 11 mission, Ar-Ar provided a crucial dating technique. Some samples were dated using K-Ar and yielded ages in the broad range 3 to 4 Ga, testifying to the antiquity of the lunar surface, although this much had been estimated from crater densities. In contrast, the Ar-Ar dating technique provided a wealth of precise ages and thermal histories. Using very small samples, Grenville Turner was able to unravel the crystallization histories, thermal histories during post-eruption heating and the cosmic ray exposure histories in a classic series of papers (Turner 1970b,c; 1971b, 1972). Turner applied quantitative diffusion concepts to stepwise argon release and recovered information from partially outgassed samples, establishing techniques and protocols that are still used to interpret stepwise heating Ar-Ar spectra today. This work and much of the early history of the Ar-Ar dating technique are set out in detail in McDougall and Harrison (1999).

Although this chapter describes both K-Ar and Ar-Ar techniques, it should be noted that K-Ar dating is now important in only limited situations including standardization (i.e., first principles dating of standards), dating fine grained clay samples, dating young basalts and obtaining dates in rapid turnaround times. Ar-Ar dating is now used in a very wide range of geological applications, dating samples as old as lunar basalts and primitive meteorites, and volcanic rocks erupted only 2000 years ago. Ar-Ar dating has been applied to many areas of Earth Sciences for dating igneous, metamorphic and sedimentary events. In recent years the introduction of laser techniques for single spot and laser heating analysis has widened the range of applications for Ar-Ar dating and the introduction of more sophisticated models for stepwise heating continue to provide ever more detailed thermal histories from K-feldspars.

THE K-AR AND AR-AR DATING METHODS

Introduction

Both K-Ar and Ar-Ar dating techniques are based upon the decay of a naturally occurring isotope of potassium, ^{40}K to an isotope of argon, ^{40}Ar (Fig. 1). The decay of ^{40}K is by a branching process; 10.48% of ^{40}K decays to ^{40}Ar by β^+ decay (Beckinsale and Gale 1969, also proposed gamma-less electron capture decay but this has never been verified), followed by γ decay to the ground state, and by electron capture direct to the ground state, and 89.52% decays to ^{40}Ca by β^- to the ground state (Fig. 1). ^{40}K-^{40}Ca dating using the more common branch is also possible (e.g., Marshall and DePaolo 1982), but its application is generally restricted to old potassium-rich rocks since ^{40}Ca is the most abundant naturally occurring isotope (96.94%), making the small amounts of radiogenically produced ^{40}Ca very difficult to measure. Argon, in contrast, is a rare trace element and radiogenically produced ^{40}Ar generally exceeds the levels of trapped ^{40}Ar (although this is not always the case—see later). The naturally occurring isotopes of argon are measured by mass spectrometry for K-Ar dating (^{36}Ar, ^{38}Ar and ^{40}Ar). The $^{36}Ar/^{38}Ar$ ratio is almost constant (see Table 1), although cosmogenic ^{38}Ar can be detected in some Ca-rich samples (Renne et al. 2001). Absolute argon concentrations, required for the K-Ar technique, are measured as a ratio against a known amount of ^{38}Ar tracer gas. Mass spectrometry for Ar-Ar dating requires only isotope ratios between naturally occurring isotopes and also reactor produced isotopes ^{39}Ar and ^{37}Ar which have half lives of 269 years and 34.95 days respectively. As we shall see later, the irradiation procedure produces not only the radioactive isotopes but also small amounts of stable isotopes of argon, and it is thus important to measure all argon masses precisely by mass spectrometry in order to correct for neutron-induced interferences.

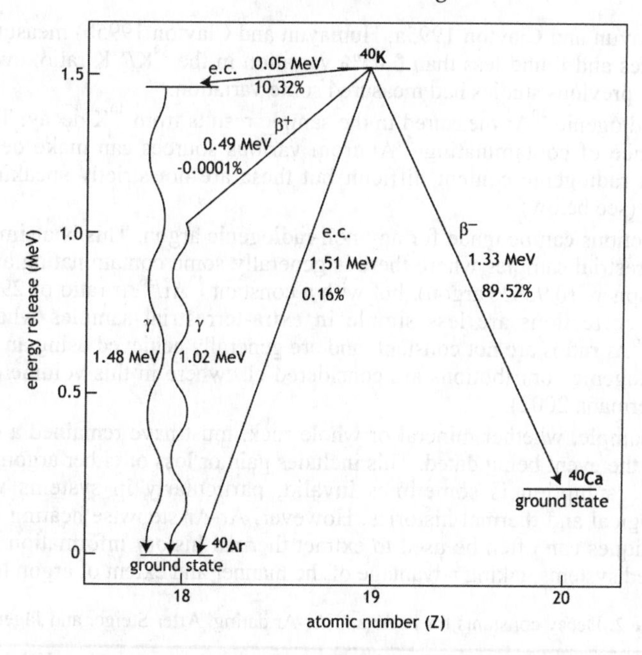

Figure 1. Branching diagram showing the decay scheme for ^{40}K, showing decay to ^{40}Ar and ^{40}Ca (after McDougall and Harrison 1999).

Table 1. Naturally occurring isotopes of argon and potassium.

Isotope	*Abundance (%)*
^{40}Ar	99.600
^{38}Ar	0.632
^{36}Ar	0.336
^{39}K	93.2581
^{40}K	0.01167
^{41}K	6.7302

After Steiger and Jäger (1977).

The essential difference between K-Ar and Ar-Ar dating techniques lies in the measurement of potassium. In K-Ar dating, potassium is measured generally using flame photometry, atomic absorption spectroscopy, or isotope dilution and Ar isotope measurements are made on a separate aliquot of the mineral or rock sample. In Ar-Ar dating, as the name suggests, potassium is measured by the transmutation of ^{39}K to ^{39}Ar by neutron bombardment and the age calculated on the basis of the ratio of argon isotopes.

Assumptions

The "date" measured by both K-Ar and Ar-Ar techniques reflects the time since radiogenic argon produced by decay of ^{40}K, became trapped in the mineral or rock. This may be the "age" of the rock or the most recent cooling event and in some samples may even reflect an integrated cooling age for a range of sub-grains. However, like all isotope-dating techniques, there is a set of assumptions that must be valid if the number measured is to be interpreted as the age of a geological event:

1. The decay of the parent nuclide, potassium must be independent of its physical state. This is the standard assumption that must be valid for any isotope dating technique.
2. The ^{40}K/K ratio must be a constant at any given time (Table 1). Most recently

(Humayun and Clayton 1995a; Humayun and Clayton 1995b) measured a range of samples and found less than 0.05% variation in the $^{39}K/^{41}K$ ratio, even in samples where previous studies had measured some variation.

3. All radiogenic ^{40}Ar measured in the sample results from ^{40}K decay. The occasional presence of contaminating ^{40}Ar from various sources can make determining the actual radiogenic content difficult but these are not strictly speaking radiogenic argon (see below).
4. Corrections can be made for any non-radiogenic argon. This is a simple procedure in terrestrial samples where there is generally some contaminating argon from the atmosphere (0.934% argon), but with a constant $^{40}Ar/^{36}Ar$ ratio of 295.5 (Table 1). Such corrections are less simple in extra-terrestrial samples where the initial $^{40}Ar/^{36}Ar$ ratios are not constant, and are generally achieved using an isochron plot. Cosmogenic contributions are considered elsewhere in this volume (Wieler 2002; Niedermann 2002).
5. The sample, whether mineral or whole rock, must have remained a closed system since the event being dated. This includes gain or loss of either argon or potassium. This assumption is sometimes invalid, particularly in systems with complex geological and thermal histories. However, Ar-Ar stepwise heating and laser spot techniques can often be used to extract thermal history information from partially opened systems, taking advantage of the manner and extent of argon loss.

Table 2. Decay constants for K-Ar and Ar-Ar dating. After Steiger and Jäger (1977).

Decay	*Decay factor*	*Value*
$^{40}K \rightarrow ^{40}Ca$ by β^-	λ_{β^-}	4.962×10^{-10} a^{-1}
$^{40}K \rightarrow ^{40}Ar$ by electron capture and γ	λ_e	0.572×10^{-10} a^{-1}
$^{40}K \rightarrow ^{40}Ar$ by electron capture	λ'_e	0.0088×10^{-10} a^{-1}
combined value	$\lambda = \lambda_{\beta^-} + \lambda_{ec} + \lambda'_{ec}$	5.543×10^{-10} a^{-1}
	present day $^{40}K/K$	0.0001167

CALCULATING K-AR AND AR-AR AGES

The age equation for the K-Ar isotope system is:

$$t = \frac{1}{\lambda} \ln\left[1 + \frac{\lambda}{\lambda_e + \lambda'_e} \frac{^{40}Ar^*}{^{40}K}\right] \tag{1}$$

where t is the time since closure, λ is the total decay of ^{40}K, and $(\lambda_e + \lambda'_e)$ is the partial decay constant for ^{40}Ar (Begemann et al. 2001) (Table 2). $^{40}Ar^*/^{40}K$ is the ratio of radiogenic daughter product (shown conventionally as $^{40}Ar^*$ to distinguish it from atmospheric ^{40}Ar) to the parent ^{40}K. Since there is no common natural fractionation of potassium isotopes (Humayun and Clayton 1995a,b), the modern ratio of $^{40}K/K$ is a constant (Table 1), and thus measurement of potassium and argon concentrations together with isotope ratios of Ar, enable an age to be calculated.

The Ar-Ar technique, first described by Merrihue and Turner (1966), is based on the same decay scheme as K-Ar, but instead of measurement on a separate aliquot, potassium is measured by creating ^{39}Ar from ^{39}K by neutron bombardment in a nuclear reactor. The reaction induced is:

$$^{39}_{19}K(n,p)^{39}_{18}Ar \tag{2}$$

The ratio of ^{39}K to ^{40}K is effectively constant (see above) and thus the critical $^{40}Ar^*/^{40}K$

ratio is proportional to the ratio of the two argon isotopes $^{40}Ar/^{39}Ar$. Although ^{39}Ar is radioactive, decaying with a half-life of 269 years, this effect is small for the period between irradiation and analysis (generally less than 6 months) and is easily corrected for.

Mitchell (1968) showed that the number of ^{39}Ar atoms formed during irradiation can be described by the equation:

$$^{39}Ar = {}^{39}K\,\Delta \int \varphi(\varepsilon)\sigma(\varepsilon)d(\varepsilon) \tag{3}$$

where ^{39}K is the number of atoms, Δ is the duration of the irradiation, φ(ε) is the neutron flux density at energy ε, and σ(ε) is the neutron capture cross section of ^{39}K for neutrons of energy ε for the neutron in/proton out reaction shown in Equation (2).

Rearranging Equation (1) in terms of $^{40}Ar^*$ yields:

$$^{40}Ar^* = {}^{40}K\frac{\lambda_e + \lambda_e'}{\lambda}\left[(e^{\lambda t}) - 1\right] \tag{4}$$

Combining Equations (3) and (4) for a sample of age t yields:

$$\frac{^{40}Ar^*}{^{39}Ar} = \frac{^{40}K}{^{39}K}\frac{\lambda_e + \lambda_e'}{\lambda}\frac{1}{\Delta T}\frac{\left[(e^{\lambda t}) - 1\right]}{\int \varphi(\varepsilon)\sigma(\varepsilon)d(\varepsilon)} \tag{5}$$

This can be simplified by defining a dimensionless irradiation-related parameter, J, as follows:

$$J = \frac{^{39}K}{^{40}K}\frac{\lambda}{\lambda_e + \lambda_e'}\Delta T \int \varphi(\varepsilon)\sigma(\varepsilon)d(\varepsilon) \tag{6}$$

The J value is determined by using mineral standards of known age to monitor the neutron flux. Substituting Equation (6) into Equation (5) and rearranging, yields the Ar-Ar age equation:

$$t = \frac{1}{\lambda}\ln\left[1 + J\frac{^{40}Ar^*}{^{39}Ar}\right] \tag{7}$$

The ratio of the two isotopes of argon, naturally produced radiogenic ^{40}Ar and reactor-produced ^{39}Ar is thus proportional to the age of the sample. For terrestrial samples, the ^{40}Ar peak measured in the mass spectrometer most often has two components (neglecting the $^{40}Ar_K$ interference reaction), radiogenic and atmospheric. The $^{40}Ar/^{36}Ar$ ratio of the atmosphere was determined by IUGS convention as 295.5 (Table 1; Steiger and Jäger 1977), though Nier determined a value of 296 (Nier 1950). When the $^{40}Ar/^{36}Ar$ ratio of contaminating argon components is >295.5, the extra argon is known as extraneous argon. The term extraneous argon includes both excess and inherited argon following the terminology of Dalrymple and Lanphere (1969) and McDougall and Harrison (1999). Excess argon is the component of argon incorporated into samples by processes other than *in situ* decay, generally from a fluid or melt at the grain boundary. Inherited argon results from the incorporation of older material in a sample, such as for example grains of sand caught up in an ignimbrite eruption. However, in the simple case, assuming that all the non-radiogenic argon is atmospheric, the daughter/parent ratio ($^{40}Ar^*/^{39}Ar$) can be determined from the equation:

$$^{40}Ar^*/^{39}Ar = [^{40}Ar^*/^{39}Ar]_m - 295.5[^{36}Ar/^{39}Ar]_m \tag{8}$$

where subscript m denotes the measured ratio. This equation is always a simplification; in most terrestrial samples surface contamination ensures that some atmospheric argon is present, though fluids at depth rarely have atmospheric ratios (see below). However, in extraterrestrial samples atmospheric argon is recognised as a modern contamination or the result of weathering. In this light, it might seem strange to assume that all contaminating argon in terrestrial samples has an atmospheric isotope ratio, given that many trap argon at depth, not in equilibrium with the atmosphere. This puzzle will be discussed in some detail in the later section on argon diffusion and solubility.

Sample irradiation for Ar-Ar dating induces not only Reaction (2) but also a series of interfering reactions caused by neutron bombardment of potassium, calcium, chlorine and argon. The complete series of interfering reactions is detailed in Table 3, but most have low production rates relative to the reaction in Equation (3) and can be ignored. The most important reactions are those involving calcium and potassium. The corrections are generally small, though they are critical for samples younger than 1 Ma when the interfering reactions producing ^{40}Ar from ^{40}K are important, and for samples with Ca/K > 10, when reactions producing ^{36}Ar and ^{39}Ar from isotopes of Ca become important. The magnitude of the interference from these reactions varies with the irradiation time and neutron flux energy spectrum. The range of measured interference factors for many of the world's reactors are listed in McDougall and Harrison (1999).

The ^{42}Ca(n,α)^{39}Ar and ^{40}Ca(n,nα)^{36}Ar production ratios do not vary a great deal, because they are caused by fast neutrons and the energy spectrum of fast neutrons in most reactors is fairly similar. The far larger variation in the interference in the ^{40}K(n,p)^{40}Ar reaction is caused by its higher sensitivity to the ratio of fast to thermal neutrons in the reactor. This ratio varies between reactors and also between different irradiation positions within a reactor. In fact samples are often shielded with cadmium foil to reduce the thermal neutron flux and lower the efficiency of the ^{40}K(n,p)^{40}Ar reaction. The precise correction factors can be determined by irradiating pure salts of Ca and K (often CaF_2, KCl and K_2SO_4). An additional correction must also be made for the decay of ^{37}Ar (produced by neutron bombardment of calcium) which has a half-life of 34.95±0.08 days (Renne and Norman 2001). The short half-life of ^{37}Ar means that all Ca-rich samples must be analyzed within about 6 months of irradiation otherwise the precision determining the original ^{37}Ar concentrations may be affected, compromising the corrections to ^{36}Ar and ^{39}Ar for Ca irradiation.

Another factor affecting the accuracy of Ar-Ar dating is ^{39}Ar recoil. This effect is crucial when studying very fine scale argon distributions or fine grained minerals such as clays, but ^{39}Ar recoil from mineral surfaces can also affect high precision dating. Turner and Cadogan (1974), calculated the likely distances of ^{39}Ar recoil during irradiation to be a mean of 0.08 μm, a study which was refined by Onstott et al. (1995) and measured directly by Villa (1997). The effects are most obviously detected in measurements of fine grained clays (e.g., Foland et al. 1983), but are commonly cited as a reason for variable ages produced from altered minerals (Lo and Onstott 1989), and basaltic rocks (e.g., Baksi 1994; Feraud and Courtillot 1994).

The Ar-Ar technique is able to achieve higher levels of internal precision than K-Ar dating since it does not depend upon separate absolute measurements but instead requires only the ratios of Ar isotopes and can achieve precision of better than 0.25%. However, external precision and accuracy are affected by the uncertainty in the age of mineral standards, as we will see in the following section. In order to achieve optimum precision in the mass spectrometric measurements, the neutron flux (which affects the magnitude of the J value) must be carefully selected. The flux must be sufficient for precise measurement of ^{39}Ar and a ^{40}Ar*/^{39}Ar ratio within the dynamic range of the mass spectrometer (generally less than 100 for good precision). Further, at higher flux levels the interfering reactions on Ca and K also become more important, degrading the precision and accuracy with which the ^{40}Ar*/^{39}Ar ratio may be determined. Therefore, for

Table 3. Interfering reactions on Ca, K, Ar and Cl.

Important interfering reactions are shown in bold type. The main ^{39}Ar-producing reaction is shown with a single border.

Argon isotope	*Ca*	*K*	*Ar*	*Cl*
^{36}Ar	**^{40}Ca(n,nα)^{36}Ar**			^{35}Cl(n,γ)^{36}Cl→β$^-$→^{36}Ar
^{37}Ar	**^{40}Ca(n,α)^{37}Ar**	^{39}K(n,nd)^{37}Ar	^{36}Ar(n,γ)^{37}Ar	
^{38}Ar	^{42}Ca(n,nα)^{38}Ar	^{39}K(n,d)^{38}Ar	^{40}Ar(n,nd)^{38}Cl→β$^-$→^{38}Ar	^{37}Cl(n,γ)^{38}Cl→β$^-$→^{38}Ar
		^{41}K(n,α)^{38}Cl→β$^-$→^{38}Ar		
^{39}Ar	**^{42}Ca(n,α)^{39}Ar**	**^{39}K(n,p)^{39}Ar**	^{38}Ar(n,γ))^{39}Ar	
	^{43}Ca(n,nα)^{39}Ar	^{40}K(n,d)^{39}Ar	^{40}Ar(n,d)^{39}Cl→β$^-$→^{39}Ar	
^{40}Ar	^{43}Ca(n,α)^{40}Ar	**^{40}K(n,p)^{40}Ar**		
	^{44}Ca(n,nα)^{40}Ar	^{41}K(n,d)^{40}Ar		

The terminology (a,b) used here refers to nuclear reactions taking place during irradiation where a is the incident particle and b is the resulting emission. The terms are n=neutron, p = proton, d = deutron, α=and alpha particle,γ = a gamma particle and β$^-$ = a positron.

each sample there is an optimum flux level and given that many samples are irradiated together, each package sent for irradiation is a compromise. Turner (1971a) calculated the fields for optimum J value, and correspondingly integrated neutron flux, which were upgraded by McDougall and Harrison (1999) in the light of higher sensitivity, higher resolution mass spectrometers (Fig. 2).

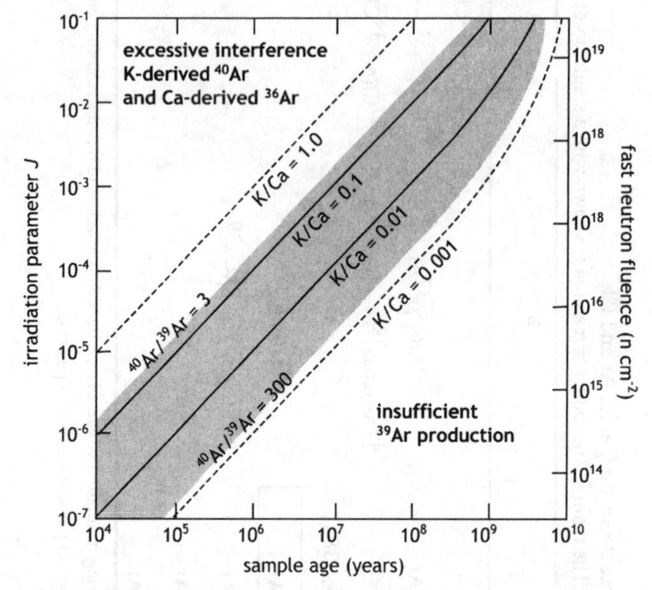

Figure 2. A Figure for optimizing irradiation parameters, taking account of age and Ca/K. The irradiation parameter is plotted against Age (Ma) and zones of optimum irradiation level are highlighted (after McDougall and Harrison 1999; Turner 1971a).

The availability of five argon isotopes provided by the Ar-Ar technique facilitates isotope correlation plots, the most common of which is the three isotope plot $^{36}Ar/^{40}Ar$ vs. $^{39}Ar/^{40}Ar$ (Fig. 3). Samples containing a mixture of radiogenic and atmospheric Ar plot along a line with negative slope between the $^{39}Ar/^{40}Ar$ ratio representing the age and the atmospheric $^{36}Ar/^{40}Ar$ ratio of 0.003384 (= 1/295.5) (Fig. 3a). The correlation plot also allows Ar-Ar ages to be calculated for samples with contamination other than modern air, since the age can equally be determined from lines passing through the $^{36}Ar/^{40}Ar$ axis at values other than the atmospheric ratio (Fig. 3b). However, a mixture of contaminating phases with more than one isotope composition in a sample results in a scatter of points not defining a line, and no age can be calculated. In many cases atmospheric 'blank' argon released from furnaces during heating is the only detected contaminating argon component. In cases where the contaminating argon is not homogeneous, physical techniques such as stepped heating, *in vacuo* crushing and laser spot dating have been used to separate components (see below).

The values of constants and estimation of errors

As the internal precision of Ar-Ar ages has improved over the years, the following have been the focus of debate:

1. The commonly accepted values for the K decay constants (Steiger and Jäger 1977)
2. The inter-laboratory and inter-standard calibration of Ar-Ar ages.

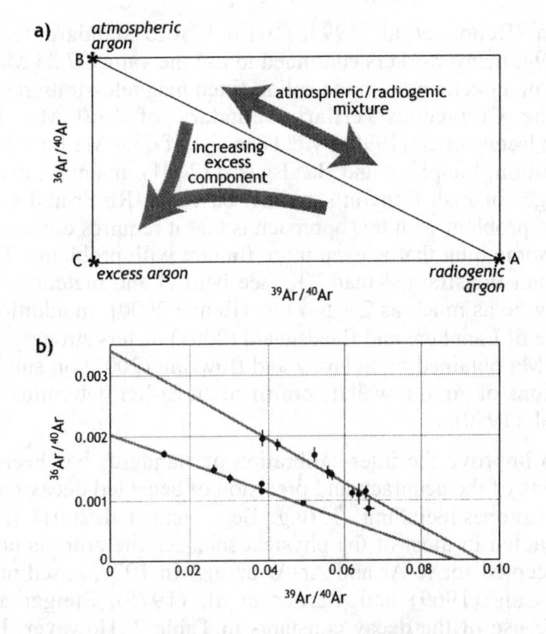

Figure 3. (a) An argon isotope correlation diagram, showing a correlation between atmospheric and radiogenic argon components which form an isochron. Any pure ^{40}Ar component would lie at the origin and thus any excess argon component tends to pull the point B towards the origin. (b) Two samples of amphibole analyzed by laser spot technique. The upper line intercepts within errors of atmospheric argon, the lower line yields a very similar age yet intercepts at a much lower ^{36}Ar/^{40}Ar ratio and contains excess argon.

K-Ar can be regarded as an absolute dating technique, dependent only on the value of the decay constant, and calibration of ^{38}Ar spike. However, all Ar-Ar ages are derived relative to the age of mineral standards, which are irradiated at the same time as the sample. The external precision of Ar-Ar ages is thus limited by the external precision of the age of the mineral standard as determined by the K-Ar method. The most widely used international standards are the hornblendes Hb3gr and MMHb1; biotites GA1550, GHC-305 and B4B, muscovite B4M, and sanidines from the Fish Canyon Tuff, Taylor Creek and Alder Creek (widely accepted ages for these standards are found in McDougall and Harrison 1999). Many other pure mineral samples are used as internal standards and several have been proposed as international standards but are not mentioned here since they are not in wide use. The advantage of using mineral standards is that they are freely available but since they are natural, series errors can be introduced if the various standards are not inter-calibrated. This has been an area of particular controversy in recent years, somewhat masking the improvements in internal precision. Fish Canyon Tuff is a prime example of the problems that are faced by those attempting to achieve accurate, and high precision Ar-Ar ages. Fish Canyon sanidine was proposed as an international standard by Cebula et al. (1986), who reported an age of 27.79 Ma but this was relative to another standard, MMHb1 with an age of 518.9 Ma (Alexander et al. 1978). When the age of the MMHb1 was revised to 520.4 Ma (Samson and Alexander 1987), the age of Fish Canyon sanidine became 27.84 Ma, though some workers used a value of 27.55 Ma, based on a different value for the age of MMHb1. In 1994, Renne et al. determined the age of Fish Canyon sanidine to be 28.03 Ma, an age later confirmed by cross calibration (Renne et al. 1998b) with biotite standard GA1550. However, subsequent to 1994, many workers continued to use the value 27.84 Ma possibly because this yielded ages in agreement with the spline-fitted magneto-stratigraphic timescale, and in particular, the Cretaceous/Tertiary boundary of 65.0 Ma. Using the value recommended by Renne et al. (1998b) yields an age of 65.4 Ma for tektites from the K/T boundary. In addition, Lanphere and Baadsgaard (2001), maintain that a value of 27.51 Ma is the best age for Fish Canyon sanidine, based on Rb/Sr and U/Pb dates on Fish Canyon Tuff. The problem with this approach is that it requires cross calibration between dating methods, something that is even more fraught with problems. The decay constant of ^{87}Rb is no better constrained than ^{40}K (see below) and meteorite cross calibrations indicate ages may be as much as 2% too low (Renne 2000). In addition, the 27.52±0.09 Ma bulk U/Pb age of Lanphere and Baadsgaard (2001) differs strongly from the U/Pb age of 28.476±0.064 Ma obtained by Schmitz and Bowring (2001) on single grain and small multi-grain fractions of zircon, which confirmed an earlier determination of 28.41±0.05 Ma by Oberli et al. (1990).

The work to improve the inter-calibration of standards has been accompanied by parallel discussions of the accuracy and precision of accepted decay constants of several important parent isotopes including ^{40}K (e.g., Begemann et al. 2001). It is notable that the decay constant quoted in most of the physical sciences literature is not the same as the one generally accepted for K-Ar and Ar-Ar dating. In 1977, based mostly on work by Beckinsale and Gale (1969) and Garner et al. (1975), Steiger and Jäger (1977) recommended the use of the decay constants in Table 2. However, Endt and Van der Leun (1973) later compiled the same data as Beckinsale and Gale (1969) to produce different results mainly by different selection criteria and statistical techniques. In a recent summary of ^{40}K decay constants, Audi et al. (1997) report a total decay constant of 5.428×10^{-10} a^{-1}, as previously cited in the nuclear physics literature, a branching ratio of 89.28% and ^{40}K/K = 1.17×10^{-4}. Min et al. (2000) showed that a better correlation could be achieved between Ar-Ar and U-Pb ages using the decay constants of Endt and Van der Leun (1973) together with modern physical constants resulting in a total decay constant of $5.463 \pm 0.054 \times 10^{-10}$ a^{-1} which corresponds to a half-life of 1269±13 Ma. Renne (2000) further demonstrated that in one of the oldest rapidly cooled meteorites, called Acapulco, nuclear physics decay constants (Audi et al. 1997) produced Ar-Ar ages within errors of U-Pb ages, in line with the petrologic interpretation of these samples as rapidly cooled, whereas the existing Ar-Ar ages indicated slow cooling. Note that the interpretation of this important result is controversial (Trieloff et al. 2001; Renne 2001) and final resolution may await a more complete characterization of the thermal history of the Acapulco meteorite. (UTh)/He studies of Acapulco phosphates (Min et al. 2000) appear to validate the rapid cooling which supports the Audi et al. (1997) constants, particularly for the β^- decay (Min et al. 2001).

Reading this section may leave those new to K-Ar and Ar-Ar dating bemused by the current controversy over decay constants and the inter-calibration of standards in such a mature isotope dating technique. It must be emphasised that this controversy has arisen recently as the precision on age determinations has improved, and attempts are made to correlate with U-Pb ages, where decay constants are better constrained. Ages in the literature other than those cited above are still calculated using the decay constants recommended by Steiger and Jäger (1977), and Renne et al. (1998b) represents the most precise inter-calibration of standards combined with a precise K-Ar date on the GA1550 biotite standard using isotope dilution for the K analysis.

Analytical errors on the Ar-Ar age have generally been calculated using the simple error propagation of Dalrymple et al. (1981):

$$\sigma_t^2 \approx \frac{J^2\sigma_R^2 + F^2\sigma_J^2}{\lambda^2(1+FJ)^2} \tag{9}$$

where σ_t is the error on the age, J is the value from Equation (6), F is the $^{40}Ar^*/^{39}Ar$ ratio, λ is the combined decay constant, σ_R is the error on the $^{40}Ar^*/^{39}Ar$ ratio and σ_J is the error on the J value. A more complete numerical error analysis is given by Scaillet (2000). Errors on the age generally include the error in determining the J value, but not the external errors on determining the K-Ar age of the standard (e.g., GA1550) used to determine the neutron flux or the decay constants. However, when comparing Ar-Ar ages with ages determined by other isotope techniques such as U-Pb, these errors must be considered, something that is particularly important in dating short lived events such as terrestrial meteorite impacts or flood basalts. Renne et al. (1998a) present discussion of this problem and equations for calculating full external errors and appropriate errors when ages for samples measured against two different standards where those standards have been inter-calibrated.

ARGON DIFFUSION AND SOLUBILITY

K-Ar and Ar-Ar dating are isotope dating techniques, not isotope tracer techniques, but the trace element chemistry of argon plays an integral role in the assumptions and interpretation of K-Ar and Ar-Ar data. The use of the Ar-Ar dating technique to investigate cooling histories and understand extraneous argon are premised on an understanding of argon diffusion rates and argon solubility in hydrous fluids, melts and minerals. Below we will see how diffusion data provide the link between K-Ar or Ar-Ar age and the thermal history of the sample being investigated. We will also see how the assumption that all radiogenic daughter product ($^{40}Ar^*$) in a mineral results from the decay of ^{40}K is commonly contravened. In order to understand how this can happen, we need to consider partition and solubility of argon in fluid, melts and minerals.

Argon diffusion (and its use to determine thermal histories)

A full exposition of diffusion theory and diffusion mechanisms is beyond the scope of this chapter but the reader is referred to McDougall and Harrison (1999). Many observations show that argon diffusion rates in natural and laboratory experiments follow an Arrhenius relationship. There are however, important departures where fast track diffusion dominates in nature (Kramar et al. 2001; Reddy et al. 2001b) and laboratory analysis of hydrous minerals (Gaber et al. 1988; Lee 1993; Lo et al. 2000). Even in natural cases which might earlier have been identified as volume diffusion effects, careful compositional control shows that phase mixing can mimic argon loss profiles (e.g., Onstott and Peacock 1987; Wartho 1995). In such cases the data can not easily be inverted to produce thermal histories.

Having considered cases in which lattice or 'volume' diffusion does not explain the data, it should be noted that lattice diffusion seems to dominate many natural systems, in argon and other noble gases (cf. Farley 2002, this volume). Lattice diffusion follows a second order diffusion mechanism (McDougall and Harrison 1999) with an Arrhenius relationship given by the equation:

$$D = D_0 e^{\left(\frac{-E}{RT}\right)} \tag{10}$$

where D is the diffusion coefficient (m^2s^{-1}), D_0 is the pre-exponential factor (the theoretical diffusion constant at infinite temperature or a measure of 'conduction' in the mineral), E is the activation energy (J), R is the gas constant ($J\ K^{-1}\ mol^{-1}$) and T is temperature (K). This logarithmic relationship between diffusion rate and temperature means that at high temperatures, argon can diffuse through a mineral lattice and escape into the grain boundary network much faster than it is produced by decay of ^{40}K. At low temperatures, diffusion rates are so slow that the daughter atoms are quantitatively retained in the mineral lattice (Fig. 4). There is also a temperature range of partial retention similar to the helium partial retention zone (HePRZ) of (U-Th)/He dating (Farley 2002, this volume), though this is generally quite narrow and represents a small length of time relative to the age of the sample. However the partial retention zone may become significant in slowly cooled metamorphic rocks (Dodson 1986; Lister and Baldwin 1996; Wheeler 1996). In early work, K-Ar ages of minerals such as biotite and muscovite were regarded as recording instantaneous closing temperatures and thus times in the cooling thermal history of the rock and closure temperatures were largely qualitative (e.g., Purdy and Jäger 1976), but Dodson (1973), provided an important advance by quantifying the closure temperature effect. By making a simplifying assumption, Dodson, showed that the closure temperature of a mineral could be described by the equation:

$$\frac{E}{RT_c} = \ln\left(\frac{A\,R\,T_c^2\,D_0/r^2}{E\,dT/dt}\right) \tag{11}$$

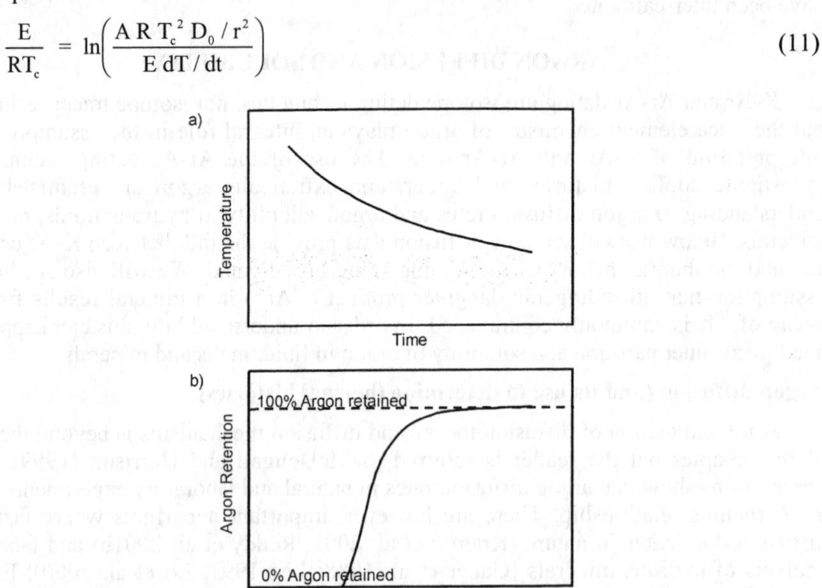

Figure 4. (a) A Temperature-time (Tt) curve for a cooling metamorphic terrain, in which the cooling rate is linear in 1/T. (b) Although the cooling rate changes slowly, argon retention within a mineral, following the Tt path changes rapidly between rapid exchange to the grain boundary and no exchange. This sudden flip from open to closed gives rise to the closure temperature concept.

where E and D_0 are the activation energy and pre-exponential factor for argon diffusion, T_c is the closure temperature, R is the gas constant, r is the effective grain radius for argon diffusion (see discussion of diffusion domains below), dT/dt is the cooling rate at closure and A is a geometrical factor describing the variation of diffusion in the mineral (A = 55 for spherical geometry or equal diffusion in all three dimensions, A = 27 for cylindrical or diffusion dominantly in only two dimensions, and A = 8.7 for planar or dominantly one dimensional diffusion). The main simplifying assumption made by Dodson (1973) in order to produce this simple equation is that the cooling path of the rocks is linear in 1/T. Although metamorphic cooling is commonly not linear in 1/T, the results are quite robust unless rocks are cooling very slowly. At cooling rates slower than around 5°C per million years, inner zones of grains close significantly earlier than the outer zones, leading to a closure temperature profile within the grain. Dodson (1986) took account of this slow cooling effect by the addition of an expression for the closure temperature as a function of position within the crystal. The function is tabulated in McDougall and Harrison (1999) and still suitable for rapid hand calculation. In more complex cases the diffusion equations can be solved numerically by finite difference, and programs such as DIFFARG (Wheeler 1996) allow the user to input any thermal history, limited only by the power of the computer. DIFFARG is suitable for calculations on single grains, but Lovera et al. (1997) have provided a technique for analysing multiple sub-grains or 'domains' by modeling ^{39}Ar release during laboratory experiments (see below).

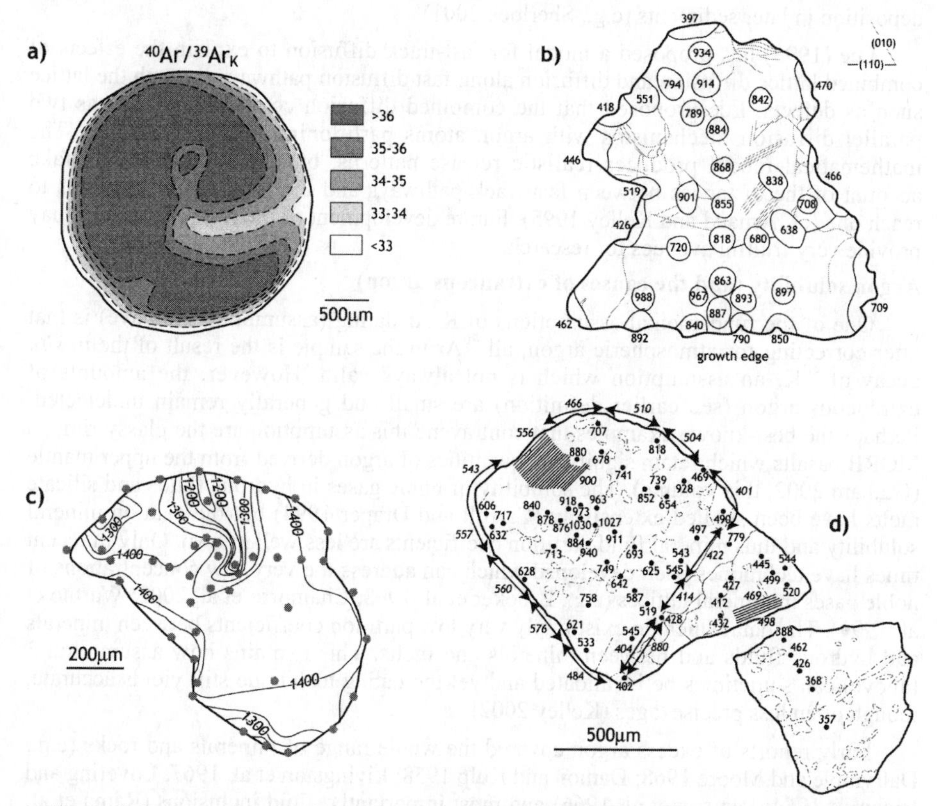

Figure 5. Laser profiles showing the effects of cracks and defects in minerals. (a) Biotite in an argon loss experiment, showing an indentation in the age contours resulting from a crack (Onstott et al. 1991). (b) Muscovite exhibiting young ages along an obvious fracture, shown left to right just below the center of the grain (Hames and Cheney 1997). (c) Biotite from a slowly cooled terrain again showing an indentation resulting from a crack (Hodges et al. 1994). (d) K-feldspar from a deformed sandstone, note that old ages are preserved in the center of this boudinaged grain but close to the boudinage and in the smaller fragment, only younger ages were measured (Reddy et al. 2001).

The discussion above considered only simple whole grain or sub-grain diffusion when lattice diffusion was the transporting mechanism. In recent years, the advent of laser analysis has demonstrated just how variable the actual within grain patterns can be (Fig. 5) (e.g., Hodges et al. 1994; Kelley and Turner 1991; Kramar et al. 2001; Lee et al.

1990; Phillips and Onstott 1988; Scaillet et al. 1990; Villa et al. 1996). Even before laser analysis had illustrated the patterns produced, bulk mineral analysis had shown that the relationship between grain size and argon loss broke down at larger grain sizes. Both natural (Layer et al. 1989; Wright et al. 1991) and laboratory (Grove and Harrison 1996) bulk heating experiments showed that the closure temperature did not increase as biotite grain size increased beyond around 150 μm, other bulk stepwise heating work determined effective radii up to 340 μm (Copeland et al. 1987), and yet age gradients have been measured in even larger grains (Hames and Cheney 1997). It seems likely that the effective diffusion radii are actually the result of variable cracking and defect density of the mica (Kramar et al. 2001; Mulch et al. 2001), particularly biotite which is the physically weaker of the two most commonly analyzed micas (Dahl 1996). Muscovite grains seem to retain metamorphic cooling age gradients even after erosion, transport and deposition in later sediments (e.g., Sherlock 2001).

Lee (1995) has proposed a model for fast-track diffusion to explain the effects of combined lattice diffusion and diffusion along fast diffusion pathways through the lattice such as defects. Lee proposed that the combined diffusion could be modeled as two parallel diffusion mechanisms with argon atoms partitioning between the two. The mathematical model produces realistic release patterns, but does not currently take account of the distances between fast track pathways and the time taken for atoms to reach one (cf. Arnaud and Kelley 1995). Future development of the fast track model may provide very fruitful avenues for research.

Argon solubility (and the causes of extraneous argon)

One of the fundamental assumptions of K-Ar dating (assumption 3, above) is that after correcting for atmospheric argon, all ^{40}Ar in the sample is the result of the *in situ* decay of ^{40}K, an assumption which is not always valid. However, the amounts of extraneous argon (see earlier definition) are small and generally remain undetected. Perhaps the best-known examples that contravene this assumption are the glassy rims to MORB basalts which retain significant quantities of argon derived from the upper mantle (Graham 2002, this volume). The solubility of noble gases in hydrous fluids and silicate melts have been studied extensively (Carroll and Draper 1994) but the data on mineral solubility and thus mineral fluid partition coefficients are less well known. Only in recent times have techniques been developed which can address the very low concentrations of noble gases in mineral lattices (e.g., Brooker et al. 1998; Chamorro et al. 2002; Wartho et al. 1999). The data which do exist imply very low partition coefficients between minerals and hydrous fluids and between minerals and melts. This explains how assumption 3 (above) can sometimes be invalidated and yet the dating technique still yields accurate, though often less precise, ages (Kelley 2002).

Early reports of excess argon covered the whole range of minerals and rocks (e.g., Dalrymple and Moore 1968; Damon and Kulp 1958; Livingston et al. 1967; Lovering and Richards 1964; Pearson et al. 1966) and most importantly, fluid inclusions (Rama et al. 1965). The step heating technique has been quite successful in discriminating against low concentrations of homogeneously distributed excess argon (e.g., Renne et al. 1997) which can be plotted on an isochron diagram (e.g., Heizler and Harrison 1988; Roddick 1978). In fact many published Ar-Ar ages contain small amounts of excess argon reflected in an initial ^{40}Ar/^{36}Ar ratio that is within a few percent of the atmospheric ratio. Such determinations yield precise results when the ratio of the contaminating component is close to that of atmospheric argon. However, as the ratio of the contaminant increases and the small ^{36}Ar peak becomes more difficult to measure, and the possibility of obtaining a precise age is quickly compromised. In extreme cases, excess argon may be undetected (e.g., Arnaud and Kelley 1995; Foland 1983; Pankhurst et al. 1973; Sherlock and Arnaud

1999). Further, the initial ratio correction only works when the isotope ratio within the samples is homogeneous, in cases of heterogeneous excess argon, a spread of data makes precise age determination impossible (e.g., Cumbest et al. 1994; Pickles et al. 1997).

The development of new analytical techniques for Ar-Ar dating has led to several advances in our understanding of excess argon. Ar-Ar stepped heating provides a physical technique to separate and analyse phases within individual samples as a result of their different breakdown temperatures (e.g., Belluso et al. 2000). Stepwise heating also produces decrepitation of fluid inclusions at low temperatures resulting in the high initial ages commonly observed in release spectra. Stepwise heating has also demonstrated excess argon diffusion into grain boundaries (Harrison and McDougall 1981). One feature of excess argon commonly associated with low potassium rocks and minerals such as plagioclase, amphibole and clinopyroxene is the saddle or 'U'-shaped Ar-Ar stepped heating release spectrum (Harrison and McDougall 1981; Lanphere and Dalrymple 1976; Wartho et al. 1996) (Fig. 6). Several explanations have been offered for this release pattern, first described by Dalrymple et al. (1975) and Lanphere and Dalrymple (1976) in samples from dykes intruding Pre-Cambrian rocks in Liberia. The initial Ar release yields old apparent ages, which decrease with progressive ^{39}Ar release, approaching the true age of the sample, and finally returning to older ages towards the end of argon release. Although arguments have been made for excess argon incorporation via special diffusion mechanisms such as anion diffusion (Harrison and McDougall 1981), recent data seem to demonstrate that the most likely candidates are fluid inclusions, released at low temperature, and melt or solid inclusions, released at high temperature during mineral breakdown (Boven et al. 2001; Esser et al. 1997) (Fig. 7).

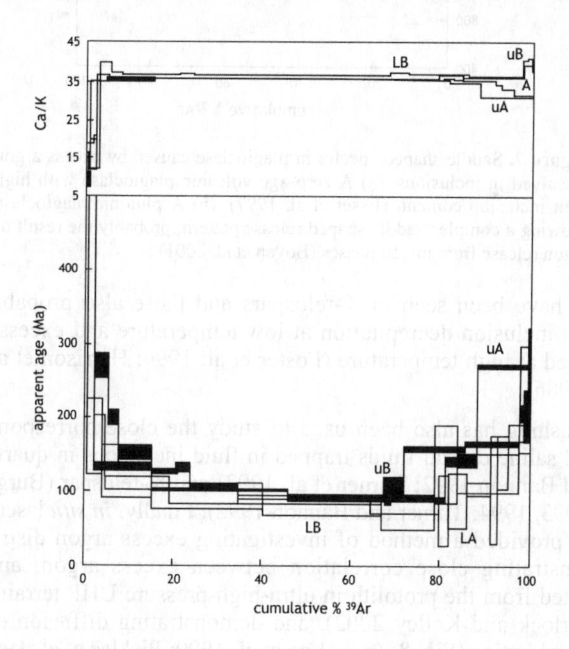

Figure 6. Excess argon in amphibole producing 'U' or 'saddle' shaped release spectrum. Dark spectra are original grains, light spectra are the same samples after acid leaching which reduced the excess argon component at both low and high ends of the release (Wartho et al. 1996).

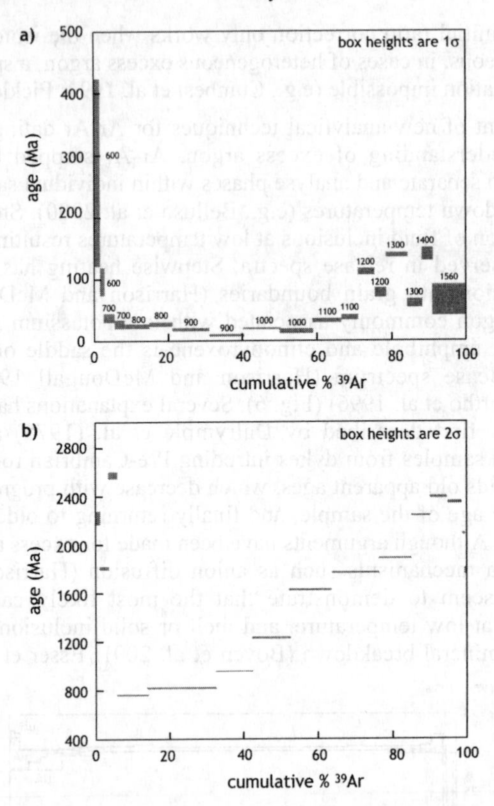

Figure 7. Saddle shaped spectra in plagioclase caused by excess argon dissolved in inclusions. (a) A zero age volcanic plagioclase with high melt inclusion content. (Esser et al. 1997). (b) A plutonic plagioclase showing a complex saddle shaped release pattern, probably the result of argon release from mixed phases (Boven et al. 2001).

Similar patterns have been seen in K-feldspars and these also probably relate to the interplay of fluid inclusion decrepitation at low temperature and excess argon in large sub-grains released at high temperature (Foster et al. 1990; Harrison et al. 1994; Zeitler and FitzGerald 1986).

In vacuo crushing has also been used to study the close correspondence between excess argon and saline crustal fluids trapped in fluid inclusions in quartz (Kelley et al. 1986; Turner and Bannon 1992; Turner et al. 1993) and K-feldspar (Burgess et al. 1992; Harrison et al. 1993, 1994; Turner and Bannon 1992). Finally, *in situ* laser spot extraction techniques have provided a method of investigating excess argon distributions within minerals, demonstrating close correlation between excess argon, and composition signatures inherited from the protolith in ultra-high-pressure UHP terrains (e.g., Giorgis et al. 2000; Sherlock and Kelley 2002), and demonstrating diffusion of excess argon through the mineral lattice (Fig. 8) (e.g., Lee et al. 1990; Pickles et al. 1997; Reddy et al. 1996).

Excess argon is most commonly found in metamorphic rocks, and is less common in volcanic systems where outgassing to the atmosphere provides a release mechanism. Excess argon is also particularly common in hydrothermal systems associated with large granite intrusions (Kelley et al. 1986; Kendrick et al. 2001a,b; Turner et al. 1993), shear zones in ancient metamorphic terrains (Allen and Stubbs 1982; Smith et al. 1994; Vance et al. 1998), contact metamorphic aureoles (Harrison and McDougall 1980) and high and ultra-high-pressure metamorphic rocks (Arnaud and Kelley 1995; Inger et al. 1996; Li et al. 1994; Ruffet et al. 1997; Scaillet 1996; Scaillet et al. 1990, 1992; Sherlock et al. 1999; Sherlock and Kelley 2002).

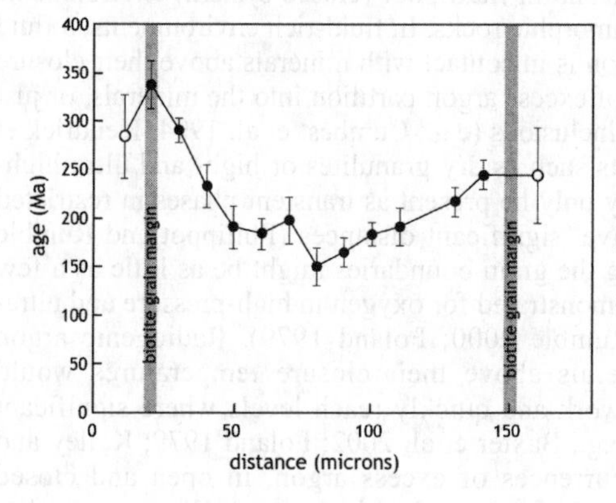

Figure 8. Excess argon diffusing into a biotite grain from the Seconda Zona Diorito Kinzigitica (IIDK) in the western Alps, the true cooling age of the biotites samples is around 40 Ma (from Pickles et al. 1997).

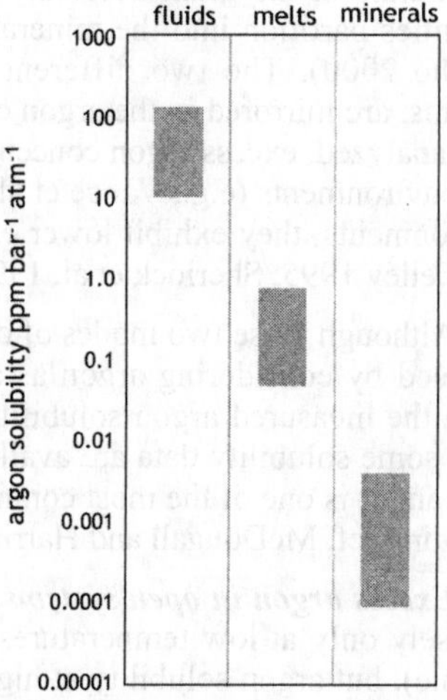

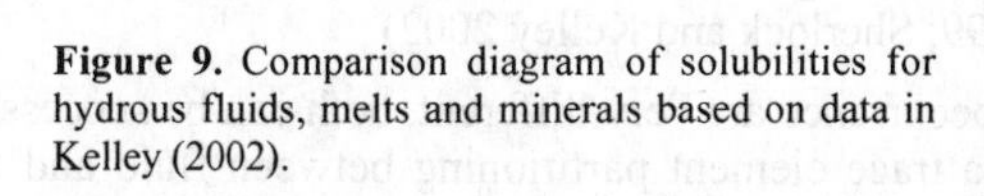

Figure 9. Comparison diagram of solubilities for hydrous fluids, melts and minerals based on data in Kelley (2002).

The importance of argon geochemistry is demonstrated clearly in Figure 9, which shows that argon is a highly incompatible trace element, strongly favoring partition from minerals into grain boundary fluids in metamorphic rocks, or from crystals into melts or melts into bubbles in magmatic systems, leaving minerals as the most highly depleted part of any system. Small amounts of excess argon may often be present but the concentrations are so low as to be masked by the larger radiogenic component. The corollary to this hypothesis is that more cases of excess argon are detected in low potassium minerals and the excess Ar is very often present in quartz (e.g., Kelley et al. 1986; Kendrick et al. 2001b; Rama et al. 1965; Turner and Bannon 1992; Vance et al. 1998). The highly incompatible nature of the argon in solid/melt and solid/fluid systems makes the fluids and melts effectively "infinite reservoirs" for radiogenic argon in these systems. More importantly however, the fact that some experimental data are available means that it is possible to model simple systems and define the limits of K-Ar and Ar-Ar dating. Although mineral data are sparse, reliable data now exist for olivine, phlogopite, clinopyroxene and K-feldspar (Brooker et al. 1998; Roselieb et al. 1999; Wartho et al.

1999). Emerging work on quartz, plagioclase, leucite and other minerals may add to this database (Roselieb et al. 1999; Wartho et al. unpublished data). In the following sections we will explore how the data for K-feldspar can be used to explore how natural systems might work.

Excess argon has been detected in both fluid-rich (open system) environments such as shear zones or hydrothermal systems and in fluid-poor (closed system) environments such as granulites or high-pressure metamorphic rocks. In fluid-rich environments, a fluid with a high concentration of excess argon is in contact with minerals above their closure temperature, and significant quantities of excess argon partition into the minerals, or just as likely, become incorporated as fluid inclusions (e.g., Cumbest et al. 1994; Kendrick et al. 2001a,b). In fluid-poor environments such as dry granulites or high- and ultra-high-pressure metamorphic rocks, fluids may only be present as transient phases in restricted zones. In such rocks, fluids do not travel significant distances (Philippot and Rumble 2000) and thus transport of argon along the grain boundaries might be as little as a few centimetres over millions of years as demonstrated for oxygen in high-pressure and ultra-high-pressure rocks (Philippot and Rumble 2000; Foland 1979). Radiogenic argon produced in potassium bearing minerals above their closure temperatures would accumulate in the grain boundary network and quickly reach levels where significant quantities partition into the minerals (e.g., Baxter et al. 2002; Foland 1979; Kelley and Wartho 2000). The two different occurrences of excess argon, in open and closed systems, are mirrored in the argon contents of quartz (fluid inclusions). Where quartz has been analyzed, excess argon concentrations signal excess argon in other minerals in fluid-rich environments (e.g., Vance et al. 1998; Kendrick et al. 2001a, b) though in fluid-poor environments, they exhibit lower excess argon contents than co-existing micas (Arnaud and Kelley 1995; Sherlock et al. 1999; Sherlock and Kelley 2002).

Although these two modes of occurrence are very different, both can be successfully modeled by considering argon as a trace element partitioning between fluid and solid using the measured argon solubility. K-feldspar is a good example for a simple model since some solubility data are available (Wartho et al. 1999; Wartho and Kelley unpubl. data) and it is one of the most common potassium bearing minerals used in K-Ar and Ar-Ar dating (cf. McDougall and Harrison 1999).

Excess argon in open systems. Argon solubility in saline waters has been measured precisely only at low temperatures (Smith and Kennedy 1983; Kipfer et al. 2002, this volume), but argon solubility at high temperatures can be estimated by extrapolating the salinity data in proportion with the high-temperature solubility data. Such extrapolation indicates that argon solubility in pure and saline grain boundary fluids up to 300°C (peak closure temperature for K-feldspar grains) lies in the range 25 to 100 ppm bar^{-1} atm^{-1}. The large range reflects uncertainty in these values but they serve to constrain the model. The solubility of argon in K-feldspar is 0.66 ppb bar^{-1} atm^{-1} (data derived from Wartho et al. 1999) thus the D_{Ar} for the K-feldspar grain boundary fluid system lies in the range 6.6×10^{-6} to 2.6×10^{-5}. Figure 10 illustrates concentrations of excess argon in K-feldspar (expressed as the increase in age they would cause in an orthoclase K-feldspar) in equilibrium with variable concentrations of argon in the corresponding grain boundary fluid. The shaded area indicates concentrations of (atmospheric) argon found in near surface ground waters and some deeper waters (Smith and Kennedy 1983; Torgersen et al. 1989; Kipfer et al. 2002, this volume). More extreme concentrations of excess argon are found in hydrothermal fluids and in fluid inclusions which range up to 21 ppm (Burgess et al. 1992; Harrison et al. 1993; Harrison et al. 1994; Kelley et al. 1986; Turner and Wang 1992). If such concentrations of excess argon were introduced into grain boundary fluids in the model, they would increase the apparent ages of K-feldspar samples by less than 0.6 Ma. Even if 100 ppm of excess argon was introduced into the model fluid, K-feldspar ages would rise by only 4 Ma. Therefore in the huge majority of cases, the level of excess argon in K-feldspar will be 1 to 2 orders of magnitude below detection limits (around 0.1 Ma in a 100-Ma K-feldspar). Even when excess argon is present in fluids at the highest levels measured in the upper crust, excess argon in K-feldspar would be a minor component in most analyzed samples. This observation corroborates the many measurements showing that excess argon is uncommon in K-feldspar, much of the excess argon which is detected in K-feldspar is confined to fluid inclusions (Burgess et al. 1992; Harrison et al. 1994).

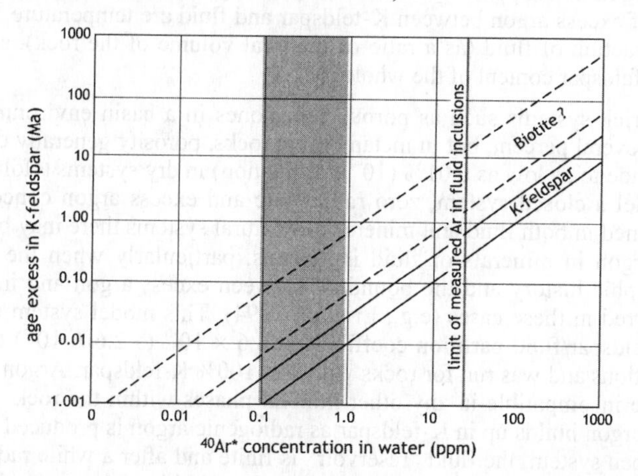

Figure 10. A plot of age excess vs ^{40}Ar concentration in the grain boundary fluid for an open system. Solid lines are those derived from the mineral/fluid partition coefficients (6.6×10^{-6} to 2.6×10^{-5}) for K-feldspar, dashed lines are those for another mineral with higher partition coefficients such as biotite. The shaded area indicates concentrations of (atmospheric) argon found in near surface ground waters and some deeper waters (Smith and Kennedy 1983; Torgersen et al. 1989; Kipfer et al. 2002, this volume). The vertical line indicates the current limit of excess argon concentration found in crustal fluids (Harrison et al. 1994).

The model confirms how robust the Ar-Ar system is when applied to K-feldspar, but minerals with higher mineral/fluid partition coefficients such as biotite will behave differently in a similar system. Note that biotite has been shown to yield ages over 100 Ma older than the expected age (e.g., Brewer 1969; Smith et al. 1994) and several studies describe excess ages in biotite but not in co-existing muscovite (e.g., Roddick et al. 1980). Roddick et al. (1980) argued that this reflected greater solubility of argon in biotite, an observation corroborated by crystal chemical (ionic porosity) arguments that argon solubility in biotite should be higher than muscovite (Dahl 1996). The dashed lines on Figure 10 illustrate the behavior of a mineral with a partition coefficient of 1×10^{-3} and similar radiogenic argon concentrations in grain boundary fluids to those seen in fluid inclusions. Such a mineral would commonly yield ages of the order of 1 Ma older than the true closure age but under extreme excess argon conditions, might yield ages more than 100 Ma older than the true closure age (compared with 4 Ma for K-feldspar in the same fluid). Where fluids are derived from basement rocks in fluid-rich regimes such as orogenic thrust belts, excess argon is unsurprisingly common (Brewer 1969; Kelley 1988; Smith et al. 1994; Reddy et al. 1996; Vance et al. 1998). The widespread influx of fluids in such regimes is illustrated by the occurrence of excess argon over broad areas (e.g., Brewer et al. 1969; Smith et al. 1994).

Excess argon in closed systems. Modeling excess argon in a closed system is less well constrained since there are no measured argon solubilities in hydrous fluids at high temperatures and even the presence of a fluid on the grain boundaries of such rocks may amount to an absorbed OH-layer or a CO_2-rich fluid. Further, any fluids which were present may have been isolated in pores or as grain edge tubules (Holness 1997) and are likely to have existed only transiently. However, the model illustrates how excess argon develops in a system without an infinite reservoir of hydrous fluid or melt. The same K-feldspar/fluid system used in the previous section can also be used to investigate the behavior of argon in fluid-poor systems. In a closed system, the controlling factors on the distribution of excess argon between K-feldspar and fluid are temperature, fluid salinity, the volume fraction of fluid (as a ratio of the total volume of the rock), and potassium content or K-feldspar content of the whole rock.

In fluid-rich systems such as porous sandstones in a basin environment, porosity might reach several percent, but in metamorphic rocks, porosity generally decreases with increasing grade to as low as 0.01% (10^{-4} vol fraction) in dry systems (Holness 1997). In order to model a closed system, zero radiogenic and excess argon concentrations are initially assumed in both fluid and minerals. In natural systems there may be pre-existing radiogenic argon in minerals or fluid inclusions, particularly when the rocks have a polymetamorphic history and the boundary between excess argon and inherited argon becomes blurred in these cases (e.g., Li et al. 1994). This model system uses the same range of K-feldspar/fluid partition coefficients (6.6×10^{-6} to 2.6×10^{-5}) to account for salinity variations and was run for rocks with 1 to 100% K-feldspar. Argon is assumed to be even more incompatible in any other mineral phases within the rock. As the model runs, excess argon builds up in K-feldspar as radiogenic argon is produced by ^{40}K decay. Unlike the open system, the fluid "reservoir" is finite and after a while radiogenic argon increases to the level where partition back into the K-feldspar becomes detectible. The fractional age excess has been calculated for fluid filled porosity of 1%, 0.1% and 0.01%. Figure 11 shows that samples with greater than 0.1% porosity do not generate significantly old ages in K-feldspar. If the porosity is 0.1%, excess argon in the same K-feldspar would yield ages as much as 1.5 Ma too old in a 100 Ma rock with 100% K-feldspar rock, but only 0.03 to 0.8 Ma in the same age rock with 5 to 30% K-feldspar (common to crustal rocks such as granite). In all probability this would still be below detection levels. Only in the most fluid-poor systems with 0.01% porosity, does the system start to exhibit detectable excess argon with up to 8 Ma excess argon in a rock with 30% 100 Ma old K-feldspar, and even in this case only the most K-feldspar-rich rocks containing very saline fluids will produce detectable excess argon. Furthermore, in this model it was assumed that the other minerals in the rock exhibited lower argon solubility but if another mineral such as biotite (with higher partition coefficient) is present, excess argon concentrations in K-feldspar quickly drop below detection levels even in the most fluid-poor terrains (e.g., Foland 1979). K-feldspar ages measured in eclogite terrains which exhibit closed system excess argon in phengite (Arnaud and Kelley 1995), sometimes reveal high-temperature excess argon, but this may also result from plagioclase or inclusions outgassing during the cycle-heating experiment (Arnaud and Kelley 1995; Boven et al. 2001).

It is not clear how rigorously this model can be applied to the most fluid-poor rocks since the grain boundary fluid phase in dry systems such as eclogites and granulites may be as little as a layer of OH- molecules at the grain boundaries. However, the model serves to illustrate how a closed system can explain phenomena observed in high-pressure and ultra-high-pressure terrains, particularly the occurrence of excess argon in phengite when it is the predominant potassium-bearing mineral in the rock. In several UHP terrains, it has been noted that excess argon is more prevalent in rocks with protracted histories or old protoliths (Li et al. 1994; Arnaud and Kelley 1995; Inger et al. 1996; Giorgis et al. 2000). Although much of this 'inherited argon' will partition into the limited fluid phase, excess argon is more likely in rocks with older protoliths. Another prediction of a closed system model is that potassium-rich rocks will contain greater concentrations of excess argon than potassium-poor rocks, even when only one potassium bearing mineral phase is present (e.g., Sherlock and Kelley 2002). Ironically, in fluid-poor high-pressure rocks it would be more appropriate to measure ages from low-K meta-basalts or meta-sandstones rather than mica-rich schists or K-feldspar-rich rocks.

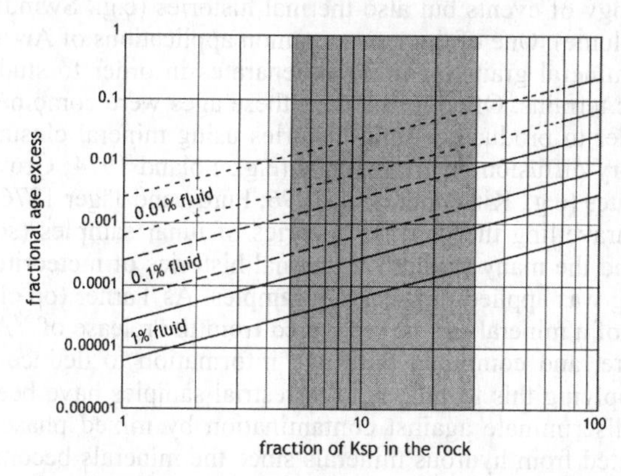

Figure 11. A plot of age excess vs. ^{40}Ar concentration in the grain boundary fluid for a closed system. Solid lines are those derived from the mineral/fluid partition coefficients (6.6×10^{-6} to 2.6×10^{-5}) for K-feldspar in a rock with 1% fluid. Dashed lines are those for the same rock with 0.1% fluid and the dashed/dotted lines are those for a rock with 0.01% fluid. The shaded area represents the range of rock compositions from which K-feldspar is commonly analyzed (5 to 50%).

In summary, the presence of excess argon in most crustal fluids means that assumption No. 3 of the K-Ar and Ar-Ar may never be strictly valid, but in the majority of cases, the concentration of excess argon in minerals is swamped by the *in situ* radiogenic component and is not a factor in determining an accurate age.

Excess argon in fluid and melt inclusions. There is considerable overlap between the studies of excess argon in minerals and crustal noble gas studies (Ballentine and Burnard 2002; Ballentine and Marty 2002, both in this volume) in studies of fluid and melt inclusions. The recent publications of Kendrick et al. (2001a, b) demonstrate just how close the two sets of studies have become. Fluid inclusions and melt inclusions are commonly important sources of excess argon in minerals analyzed for K-Ar and Ar-Ar dating, particularly in low potassium minerals such as amphibole and plagioclase. The simple reason for the importance of fluid and melt inclusions is illustrated by Figure 11. Melt inclusions incorporated from a magma containing excess argon will contain ~100 times more argon than minerals crystallising from the same melt, and fluid inclusions in equilibrium may contain as much as 10,000 times the excess argon concentration (by weight) of the mineral structure. Inclusions provide some of the most intractable analytical problems in K-Ar and Ar-Ar dating although they can also occasionally be used to advantage in preserving mineral ages (e.g., Kelley et al. 1997; van der Bogaard and Schirnick 1995). Cumbest et al. (1994) showed how the interaction of metamorphic fluids could result in different generations of excess argon in fluid inclusions and Harrison et al. (1994) showed that fluid inclusions were present in many K-feldspar samples but repeated cycles of step heating could reduce and correct for their effects.

Melt inclusions are common in mineral separates used to date volcanic and hyperbyssal rocks, but have received little attention. However, the effects of melt inclusions from a historic flow on Mount Erebus were studied by Esser et al. (1997) showing that their presence resulted in saddle-shaped age spectra during stepped ^{39}Ar release (Fig. 7a).

APPLICATIONS

The applications cited here focus upon the Ar-Ar dating technique, simply because most recent developments have utilized the Ar-Ar dating technique. The aim of this section is to highlight applications which illustrate the breadth of problems which have been addressed using Ar-Ar dating.

Thermochronology

The diffusion of radiogenic noble gas daughter products provides several routes for determining not only the chronology of events but also thermal histories (e.g., Swindle 2002; Farley 2002, both in this volume). One of the most common applications of Ar-Ar geochronology has been dating mineral grains or mineral separates in order to study cooling and uplift in metamorphic terrains. Often in the past, these ages were combined with U-Pb and Rb-Sr ages in order to produce cooling histories using mineral closure temperatures based upon laboratory diffusion determinations (e.g., Foland 1974; Grove and Harrison 1996) or field estimates (e.g., Kirschner et al. 1996; Purdy and Jäger 1976). However, after the success of unravelling the thermal histories of lunar samples (see Turner 1977 for a full review), and the many studies on thermal histories of meteorites (Turner 1969, 1970a), step heating was applied to terrestrial samples. As Turner (op cit) showed, the diffusion parameters of a mineral can be recovered from the release of ^{39}Ar during the step heating procedure, and combined with age information to deduce a thermal history. The results of applying this technique to terrestrial samples have been mixed. While step heating can discriminate against contamination by mixed phases, thermal histories are rarely extracted from hydrous minerals since the minerals become unstable long before they melt, releasing argon not only from outer layers but along cleavages parted by explosive loss of water in biotite (Gaber et al. 1988) or via breakdown reactions in the case of amphibole (Lee 1993; Lee et al. 1991; Wartho 1995; Wartho et al. 1991) and biotite (Lo et al. 2000). However this aspect has recently been exploited (Belluso et al. 2000), using the preferential breakdown of different amphibole compositions in an attempt to date the formation of separate generations.

Feldspars do not break down during short term *in vacuo* heating experiments prior to melting since they are anhydrous, and thus do not suffer the same effects as hydrous minerals. However, the complex sub-solidus reactions and exsolution in plagioclase feldspar means that Ar-Ar step heating generally reflects mixed phases, and interpretation may be difficult (e.g., Boven et al. 2001). Stepwise heating of K-feldspar on the other hand, provides an opportunity to explore thermal histories without the need to resort to other isotope systems or even other minerals. The microstructure of K-feldspar is relatively simple in volcanic rocks, but in plutonic and basement rocks, progressive episodes of reaction during cooling result in complex microtexture of intergrowths and exsolution (Parsons et al. 1999). Plutonic and basement feldspars are thus composed of many variably sized and shaped diffusion domains within single grains of K-feldspar, some as small as a micron. The bulk closure temperature of large K-feldspar grains would be around 300°C if they acted as whole grains (cf. Foland 1974) but in fact the 'effective' closure temperature is closer to 150-200°C, and this is the reason why they were regarded as "leaky" by early workers. Several workers suggested that argon loss from K-feldspars might be the result of variable sub-grain sizes, but it was not until Lovera et al. (1989, 1991) that this effect was quantified using cycled heating steps designed to preferentially outgas different diffusion domain sizes, rather than the standard monotonic temperature increase normally used in step heating. Lovera et al. (1989, 1991) also formulated a mathematical model using the deviation of the ^{39}Ar release from a simple Arrhenius relationship, to monitor argon diffusion in domains of differing sizes. The individual step Ar-Ar ages were then combined with derived diffusion information to produce a continuous thermal history over a range of as much as 200°C. This was a considerable advance over a single meaningless closure temperature. In their initial work, Lovera et al. used an age spectrum to monitor the cooling of the Chain of Ponds Pluton in northwestern Maine (Fig. 12), showing that the age spectrum could not be modeled using a single diffusion domain but that a reasonably good fit could be achieved using three domains of different sizes (i.e., diffusion radii) and specific volume fractions.

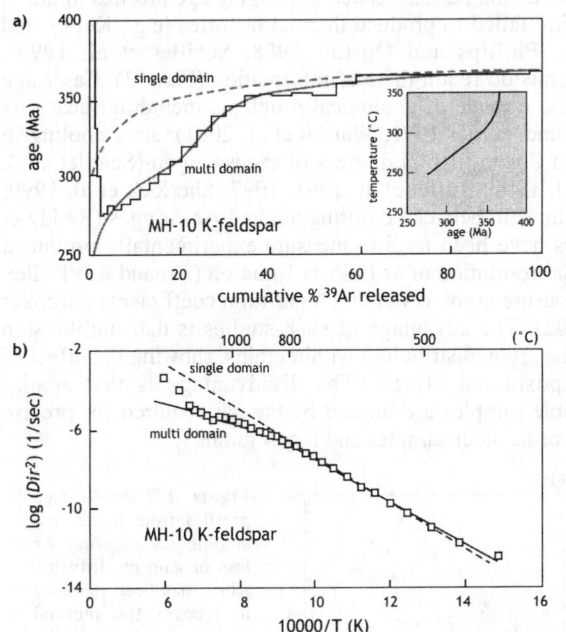

Figure 12. Cycle heating results for a K-feldspar sample from the Chain of Ponds pluton (Lovera et al. 1989), showing the fit of modeled argon release for a single diffusion domain lines and the improved fit which can be obtained by using three diffusion domains. The lower figure shows the departure of argon diffusion from a simple Arrhenius relationship.

A great deal of discussion of multi-domain modeling (MDD) has ensued mainly surrounding the diffusion mechanisms (Arnaud and Kelley 1997; Harrison et al. 1991; Lovera et al. 1989, 1991, 1993, 1997; McDougall and Harrison 1999; Villa 1994; Wartho et al. 1999) and the physical reality of the diffusion domain sizes used in the model (Arnaud and Eide 2000; Burgess et al. 1992; FitzGerald and Harrison 1993; Lovera et al. 1989, 1991, 1993; Parsons et al. 1999; Reddy et al. 2001a). The model has also evolved considerably since its inception with the use of an increasing number of discrete diffusion domain sizes (e.g., Mock et al. 1999a,b), evocation of multiple activation energies (Harrison et al. 1991) which was later discontinued, a realization that fluid inclusions played an important role in apparent excess argon seen at low temperatures (Turner and Wang 1992; Burgess et al. 1992; Harrison et al. 1993, 1994), and more recently focus on the low-temperature release to determine an activation energy used throughout (Lovera et al. 1997) since higher temperatures may be affected by pre-melting and other artifacts. Monte Carlo inversion techniques have also been used to define domain sizes and thermal histories (Warnock and Zeitler 1998; Lovera et al. 1997) and there has been discussion of the effects of recoil in very small domains upon the release patterns and their interpretation (Onstott et al. 1995; Villa 1997).

If the MDD technique had produced no useful information, the controversy and constantly evolving modeling techniques would have dissuaded workers from applying it to geological problems. However, as McDougall and Harrison (1999) document, the cycle-heating MDD model technique has resulted in reasonable thermal histories in a wide range of geological situations. Moreover, McDougall and Harrison (1999) set out a full set of testable assumptions and techniques for accepting or rejecting data. In summary, the cycle-heating technique must be undertaken with great care and awareness of the potential advantages and pitfalls.

The cycle-heating MDD modeling technique offers the chance to extract thermal information from bulk K-feldspar separates, but what of hydrous minerals such as micas and amphiboles? In fact, laser spot dating has succeeded in imaging age profiles in micas and amphiboles where step heating failed to produce thermal histories (e.g., Kelley and Turner 1991; Lee et al. 1991; Phillips and Onstott 1988; Scaillet et al. 1990), demonstrating that hydrous minerals do retain thermal information (Fig. 13). Laser age profiles have been used to address a range of geological problems including; heat loss from igneous intrusions (Kelley and Turner 1991; Wartho et al. 2001); slow cooling in metamorphic terrains (Hames and Cheney 1997); ingress of excess argon (Scaillet et al. 1990; Scaillet 1996; Pickles et al. 1997; Ruffet et al. 1991, 1997; Sherlock et al. 1999; Giorgis et al. 2000) and the relationship between deformation and Ar-Ar ages (Reddy et al. 1996, 1997, 2001). UV lasers have been used to measure experimentally produced argon diffusion profiles at a spatial resolution of as little as 1 micron (Arnaud and Kelley 1997; Wartho et al. 1999) and measure argon crystal/melt partition coefficients (Brooker et al. 1998; Chamorro et al. 2002). The advantage of such studies is that, unlike step heating, they provide images of the argon distribution within grains, showing the effect of sub-grain boundaries and compositional effects. The disadvantage is that spatial resolution and the range of suitable samples are limited by the gas required for precise isotope measurement—in other words, older samples and larger grains.

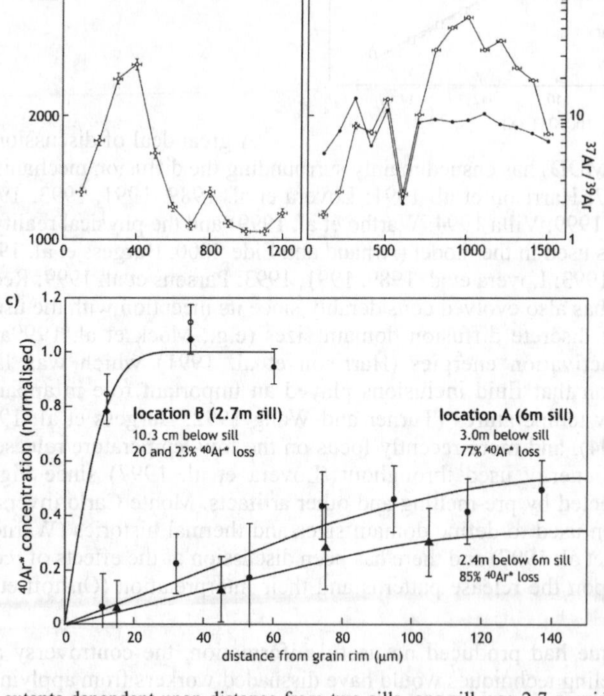

Figure 13. Ar-Ar age profiles from micas and amphiboles showing Ar loss or gain by diffusion which has been modeled to recover the thermal history of the sample. (a),(b) show two-amphibole grains from a rock close to the Duluth Gabbro. The age profiles show diffusion domains defined by the presence of biotites in the outer zone of the grains (Note the Ca/K monitored by $^{37}Ar/^{39}Ar$ on the right). Both domains yielded the same time-integrated diffusion parameters (Kelley and Turner 1991). (a) A small domain within the grain has suffered 78% argon loss. (b) A larger domain showing only 43% Ar loss. (c) Ar-loss profiles from biotite grains close to a sill on the Isle of Mull, Scotland, outgassed to varying extents dependent upon distance from two sills, one sill was 2.7 m wide, the other was 6 m wide. The variable outgassing was used to produce a magma flow history for the sill (Wartho et al. 2001).

Dating young volcanic eruptions

K-Ar and Ar-Ar have long been used to date young (Pre-historic) volcanic eruptions (cf. Curtis in Schaeffer and Zähringer 1966) but in recent years, the advent of low blank furnaces and laser extraction techniques, particularly the CO_2 laser, have pushed the boundaries. It remains unclear whether laser extraction or furnace extraction is the superior technique for very young volcanic samples. Lasers offer slightly lower blanks, though the quantities of sample can be too large for focussed laser extraction (>0.5 g in some cases), but furnaces offer better temperature control. In a study of zero age samples Esser et al. (1997) showed that anorthoclase samples suffered small amount of excess argon which were better discriminated using the furnace extraction technique. There are particular analytical challenges offered by dating young volcanic samples, including precise calibration of the mass spectrometer since many samples contain only small amounts of radiogenic argon. Contamination and control of the sample quality is also particularly difficult since sample sizes are often 0.1-1 g. Finally, melting large quantities of sample also produces large quantities of H_2O, CO_2 and other gases released from minerals, and requires larger getters to clean the gas prior to mass spectrometric measurement of isotope ratios.

The ages of the youngest volcanic rocks dated by the Ar-Ar technique have now reached the historic realm. For example, Renne et al. (1997) have measured the age of the Vesuvius eruption observed and recorded by Pliny in 79 AD. Other work has focussed upon dating human evolution (Deino et al. 1998; Ludwig and Renne 2000; Renne et al. 1997), the relationship between volcanic eruptions and recent climate change (Ton-That et al. 2001), glacial advances (Singer et al. 2000; Wilch et al. 1999), and precise calibration of young magnetic reversals (e.g., Singer and Pringle 1996).

High-precision ages on altered basalts

One of the reasons why precision of the standards and problems of the decay factors used for Ar-Ar dating came to light was the push to achieve high precision ages for huge flood basalt provinces, which have been linked by some workers to climate change and global mass extinctions. A great deal of effort has gone into analytical techniques and protocols in order to obtain high precision ages for flood basalts (e.g., Foland et al. 1993; Hofmann et al. 2000; Min et al. 2000; Renne 1995; Renne and Basu 1991; Renne et al. 1992, 1995, 1996a,b; Storey et al. 1995; Duncan et al. 1997) this work also contributed strongly to the problems concerning accuracy and precision, decay constants and international standards as a result. Precision (excluding decay and standard errors), even on samples with low potassium are commonly less than 0.5%, sometimes as low as 0.25%. In addition, analytical techniques involving acid clean-up have been developed to analyse extremely difficult samples from the sea floor (Pringle et al. 1991).

Dating low-temperature processes

Both the K-Ar and Ar-Ar techniques have been used to date low-temperature phases such as clay minerals and in particular the mineral illite. The problem for Ar-Ar dating of such phases is that recoil of ^{39}Ar during neutron irradiation can compromise the measured age. In fine grained phases such as clays, this can cause up to 30% ^{39}Ar loss by recoil and resulting ^{40}Ar*/^{39}Ar ratios, calculated without the recoiled ^{39}Ar are 30% too high. This problem has been overcome to some extent by encapsulating clay samples in evacuated silica vials, breaking the vials after irradiation, and measuring the amount of recoiled ^{39}Ar (e.g., Foland et al. 1983; Smith et al. 1993). More recently, Onstott et al. (1997) showed that the process could be miniaturized, using micro-ampoules and opening them with a UV laser. The technique has been utilized (Dong et al. 1995, 1997a,b) to investigate variations in argon retention properties of clays. Other low-temperature K-bearing minerals such as cryptomelane and alunite have been used to date surface processes (Vasconcelos 1999), including the formation of ore gangue (Itaya et al. 1996; Vasconcelos 1999) and cave deposits (Polyak et al. 1998).

Ar-Ar dating of authigenic K-feldspar overgrowths has also been investigated using dissolution (Hearn et al. 1987), stepped heating (Mahon et al. 1998) and laser extraction (e.g., Girard and Onstott 1991) with varying success. Recent work using a UV laser extraction technique provides further evidence that such overgrowths can provide meaningful ages (Hagen et al. 2001).

Another growing field of work has been dating detrital minerals in sediments as a provenance tool, but also in order to study the sedimentary systems supplying basins and uplift or thermal history of the source. Clauer (1981) and Adams and Kelley (1998) have shown that while biotite is often quickly altered in the sedimentary environment, white micas survive and retain ages even in second cycle sediments (those which were originally deposited and then re-excavated during later basin inversions; e.g., Sherlock 2001). K-feldspar also retains pre-erosion cooling ages (e.g., Copeland et al. 1990) but the lower closure temperatures mean that it has more commonly been used for thermal analysis of sedimentary basins (e.g., Harrison and Be 1983).

Unique samples

The Ar-Ar dating technique depends only upon two isotopes of argon, and thus it is possible to analyse individual particles such as lunar soil grains (Burgess and Turner 1998), terrestrial sand grains (Kelley and Bluck 1989), and glass spherules (Culler et al. 2000). Ar-Ar dating has also successfully been used to date small amounts of heterogeneous samples, such as pseudotachylytes in fault zones (Kelley et al. 1994; Magloughlin et al. 2001; Muller et al. 2001), and even terrestrial meteorite impacts (Spray et al. 1995). Finally, in its earliest days the Ar-Ar dating was called upon to extract a great deal of temporal and thermal information from unique and precious moon rocks (Turner 1977; Swindle 2002, this volume), and this is still the case as demonstrated by the recent dating of the unique ALH84001 meteorite sample from Mars (Turner et al. 1997).

ACKNOWLEDGMENTS

The author was greatly assisted in writing this chapter by discussion with Sarah Sherlock. John Taylor kindly drew the figures. Paul Renne, Nicolas Arnaud, Matt Heizler and Mike Cosca provided detailed and insightful reviews which greatly improved the text. Rainer Wieler spotted the many mistakes and typos.

REFERENCES

Adams CJ, Kelley SP (1998) Provenance of Permian-Triassic and Ordovician metagreywake terranes in New Zealand: Evidence from 40Ar/39Ar dating of detrital micas. Geol Soc Am Bull 110:422-332

Aldrich LT, Nier AO (1948) Argon 40 in potassium minerals. Phys Rev 74:876-877

Alexander ECJ, Mickelson GM, Lanphere MA (1978) MMhb-1: A new ^{40}Ar-^{39}Ar dating standard. U S Geol Surv Open-file Rept 70-701:6-8

Allen AR, Stubbs D (1982) An ^{40}Ar/^{39}Ar study of a polymetamorphic complex in the Arunta Block, central Australia. Contrib Mineral Petrol 79:319-332

Arnaud NO, Kelley SP (1995) Evidence for widespread excess argon during high-pressure metamorphism in the Dora Maira (Western Alps, Italy), using an ultra-violet laser ablation microprobe ^{40}Ar/^{39}Ar technique. Contrib Mineral Petrol 121:1-11

Arnaud NO, Kelley SP (1997) Argon behavior in gem-quality orthoclase from Madagascar: Experiments and some consequences for Ar-40/Ar-39 geochronology. Geochim Cosmochim Acta 61:3227-3255

Arnaud NO, Eide EA (2000) Brecciation-related argon redistribution in alkali feldspars: An in naturo crushing study. Geochim Cosmochim Acta 64:3201-3215

Audi G, Bersillon O, Blachot J, Wapstra AH (1997) The NUBASE evaluation of nuclear and decay properties. Nucl Phys A624:1-124
Baksi AK (1994) Geochronological Studies On Whole-Rock Basalts, Deccan Traps, India—Evaluation of the Timing of Volcanism Relative to the K-T Boundary. Earth Planet Sci Lett 121:43-56
Ballentine CJ, Burnard P (2002) Production and release of noble gases in the continental crust. Rev Mineral Geochem 47:481-538
Ballentine CJ, Marty B (2002) Tracing fluid origin, transport and interaction in the crust. Rev Mineral Geochem 47:539-614
Baxter EF, DePaulo DJ, Renne PR (2002) Spatially correlated Anomalous $^{40}Ar/^{39}Ar$ "Age" Variations in Biotites about a lithologic contact near Simplon Pass, Sqitzerland: A mechanistic explanation for "excess Ar". Geochim Cosmochim Acta 66:1067-1083
Beckinsale RD, Gale NH (1969) A reappraisal of the decay constants and branching ratio of ^{40}K. Earth Planet Sci Lett 6:289-294
Begemann F, Ludwig KR, Lugmair GW, Min K, Nyquist LE, Patchett PJ, Renne PR, Shih CY, Villa IM, Walker RJ (2001) Call for an improved set of decay constants for geochronological use. Geochim Cosmochim Acta 65:111-121
Belluso E, Ruffini R, Schaller M, Villa IM (2000) Electron-microscope and Ar isotope characterization of chemically heterogeneous amphiboles from the Palala Shear Zone, Limpopo Belt, South Africa. Eur J Mineral 12:45-62
Boven A, Pasteels P, Kelley SP, Punzalan L, Bingen B, Demaiffe D (2001) $^{40}Ar/^{39}Ar$ study of plagioclases from the Rogaland anorthosite complex (SW Norway); an attempt to understand argon ages in plutonic plagioclase. Chem Geol 176:105-135
Brewer MS (1969) Excess Radiogenic Argon in Metamorphic Micas from the Eastern Alps, Austria. Earth Planet Sci Lett 6:321-331
Brooker RA, Wartho J-A, Carroll MR, Kelley SP, Draper DS (1998) Preliminary UVLAMP determinations of argon partition coefficients for olivine and clinopyroxene grown from silicate melts. Chem Geol 147:185-200
Burgess R, Turner G (1998) Laser argon-40-argon-39 age determinations of Luna 24 mare basalts. Meteorit Planet Sci 33:921-935
Burgess R, Kelley SP, Parsons I, Walker FDL, Wordon RH (1992) ^{40}Ar-^{39}Ar analysis of perthite microtextures and fluid inclusions in alkali feldspars from the Klokken syenite, South Greenland. Earth Planet Sci Lett 109:147-167
Carroll MR, Draper DS (1994) Noble gases as trace elements in magmatic processes. Chem Geol 117: 37-56
Cebula GT, Kunk MJ, Mehnert HH, Naser CW, Obradovich JD, Sutter JF (1986) The Fish Canyon Tuff, a potential standard for the ^{40}Ar-^{39}Ar and fission track dating methods. Terra Cognita (abstr) 6:139-140
Chamorro EM, Brooker RA, Wartho J-A, Wood BJ, Kelley SP, Blundy JD (2002) Ar and K partitioning between clinopyroxene and silicate melt to 8 GPa. Geochim Cosmochim Acta 66:507-519
Clauer N (1981) Strontium and argon isotopes in naturally weathered biotites, muscovites and K-feldspars. Chem Geol 31:325-334
Copeland P, Harrison TM, Heizler MT (1990) $^{40}Ar/^{39}Ar$ single-crystal dating of detrital muscovite and K-feldspar from Leg 116, Southern Bengal Fan: Implications for the uplift and erosion of the Himalayas. Proc Ocean Drill Progr 116:93-114
Copeland P, Harrison TM, Kidd WSF, Ronghua X, Yuquan Z (1987) Rapid early Miocene acceleration of uplift in the Gandese Belt, Xizang (southern Tibet), and its bearing on accommodation mechanisms of the India-Asia collision. Earth Planet Sci Lett 86:240-252
Culler TS, Becker TA, Muller RA, Renne PR (2000) Lunar impact history from Ar-40/Ar-39 dating of glass spherules. Science 287:1785-1788
Cumbest RJ, Johnson EL, Onstott TC (1994) Argon composition of metamorphic fluids: Implications for $^{40}Ar/^{39}Ar$ geochronology. Geol Soc Am Bull 106:942-951
Dahl PS (1996) The crystal-chemical basis for Ar retention in micas: Inferences from interlayer partitioning and implications for geochronology. Contrib Mineral Petrol 123:22-39
Dalrymple GB, Lanphere MA. (1969) Potassium-Argon Dating, Principle Techniques and Applications to Geochronology. Freeman and Co., San Francisco, 258 p
Dalrymple DG, Moore JG (1968) Argon 40: Excess in submarine pillow basalts from Kilauea Volcano, Hawaii. Science 161:1132-1135
Dalrymple GB, Grommé CS, White RW (1975) Potassium-argon age and paleomagnetism of diabase dykes in Liberia: Initiation of central Atlantic rifting. Bull Geol Soc Am 86:399-411
Damon PE, Kulp JL (1958) Excess helium and argon in beryl and other minerals. Am Mineral 43:433-459
Deino AL, Renne PR, Swisher CC (1998) Ar-40/Ar-39 dating in paleoanthropology and archeology. Evol Anthropol 6:63-75

Dickin AP. (1995) Radiogenic Isotope Geology. Cambridge University Press, Cambridge, UK, 490 p
Dodson MH (1973) Closure temperature in cooling geochronological and petrological systems. Contrib Mineral Petrol 40:259-274
Dodson MH (1986) Closure profiles in cooling systems. Mater Sci Forum 7:145-154
Dong HL, Hall CM, Peacor DR, Halliday AN (1995) Mechanisms of Argon Retention in Clays Revealed By Laser Ar-40- Ar-39 Dating. Science 267:355-359
Dong HL, Hall CM, Halliday AN, Peacor DR (1997a) Laser Ar-40-Ar-39 dating of microgram-size illite samples and implications for thin section dating. Geochim Cosmochim Acta 61:3803-3808
Dong HL, Hall CM, Halliday AN, Peacor DR, Merriman RJ, Roberts B (1997b) Ar-40/Ar-39 illite dating of Late Caledonian (Acadian) metamorphism and cooling of K-bentonites and slates from the Welsh Basin, UK. Earth Planet Sci Lett 150:337-351
Duncan RA, Hooper PR, Rehacek J, Marsh JS, Duncan AR (1997) The timing and duration of the Karoo igneous event, southern Gondwana. J Geophys Res–Solid Earth 102:18127-18138
Endt PM, Van der Leun C (1973) Energy levels of A = 21-44 nuclei (V). Nucl Phys A214:1-625
Esser RP, McIntosh WC, Heizler MT, Kyle PR (1997) Excess argon in melt inclusions in zero-age anorthoclase feldspar from Mt Erebus, Antarctica, as revealed by the $^{40}Ar/^{39}Ar$ method. Geochim Cosmochim Acta 61:3789-3801
Farley KA (2002) (U-Th)/He dating: techniques, calibrations, and applications. Rev Mineral Geochem 47:819-845
Farrar E, Macintyre RM, York D, Kenyon WJ (1964) A simple mass spectrometer for the analysis of argon at ultra-high vacuum. Nature, p 531-533
Feraud G, Courtillot V (1994) Did Deccan volcanism pre-date the Cretaceous-Tertiary transition?—Comment. Earth Planet Sci Lett 122:259-262
Fitz Gerald JD, Harrison TM (1993) Argon diffusion domains in K-feldspar: I. Microstructures in MH-10. Contrib Mineral Petrol 113:367-380
Foland KA (1974) Ar^{40} diffusion in homogenous orthoclase and an interpretation of argon diffusion in K-feldspars. Geochim Cosmochim Acta 38:151-166
Foland KA (1979) Limited mobility of argon in a Metamorphic Terrain. Geochim Cosmochim Acta 43:793-801
Foland KA (1983) $^{40}Ar/^{39}Ar$ incremental heating plateaus for biotites with excess argon. Chem Geol (Isotop Geosci Sect) 1:3-21
Foland KA, Linder JS, Laskowski TE, Grant NK (1983) $^{40}Ar/^{39}Ar$ dating of glauconites; measured ^{39}Ar recoil loss from well-crystallized specimens. Chem Geol (Isotop Geosci Sect) 2:241-264
Foland KA, Fleming TH, Heimann A, Elliot DH (1993) Potassium-argon dating of fine-grained basalts with massive argon loss: Applications of the $^{40}Ar/^{39}Ar$ technique to plagioclase and glass from the Kirkpatrick Basalt, Antartica. Chem Geol (Isotop Geosci Sect) 107:173-190
Foster DA, Harrison TM, Copeland P, Heizler MT (1990) Effects of excess argon within large diffusion domains on K-feldspar age spectra. Geochim Cosmochim Acta 54:1699-1708
Gaber LJ, Foland KA, Corbató CE (1988) On the significance of argon release from biotite and amphibole during $^{40}Ar/^{39}Ar$ vacuum heating. Geochim Cosmochim Acta 52:2457-2465
Garner EL, T.J. M, J.W. G, Paulsen PJ, Barnes IL (1975) Absolute isotopic abundance ratios and the atomic weight of a reference samples of potassium. J Res Natl Bur Stand 79A:713-725
Giorgis D, Cosca MA, Li S (2000) Distribution and significance of extraneous argon in UHP eclogite (Sulu terrane, China): Insights from *in situ* $^{40}Ar/^{39}Ar$ UV-laser ablation analysis. Earth Planet Sci Lett 181:605-615
Girard J-P, Onstott TC (1991) Application of $^{40}Ar/^{39}Ar$ laser-probe and step-heating techniques to the dating of diagenetic K-feldspar overgrowths. Geochim Cosmochim Acta 55:3777-3793
Graham CM (2002) Noble gas isotope geochemistry of mid-ocean ridge and ocean island basalts: characterization of mantel source reservoirs. Rev Mineral Geochem 47:247-318
Grove M, Harrison TM (1996) $^{40}Ar^*$ diffusion in Fe-rich biotite. Am Mineral 81:940-951
Hagen E, Kelley SP, Dypvik H, Nilsen O, Kjolhamar B (2001) Direct dating of authigenic K-feldspar overgrowths from the Kilombero Rift of Tanzania. J Geol Soc 158:801-807
Hames WE, Cheney JT (1997) On the loss of Ar-40* from muscovite during polymetamorphism. Geochim Cosmochim Acta 61:3863-3872
Harrison TM, McDougall I (1980) Investigations of an intrusive contact, northwest Nelson, New Zealand: II. Diffusion of radiogenic and excess ^{40}Ar in hornblende revealed by $^{40}Ar/^{39}Ar$ age spectrum analysis. Geochim Cosmochim Acta 44:2005-2020
Harrison TM, McDougall I (1981) Excess ^{40}Ar in metamorphic rocks from Broken Hill, New South Wales: Implications for $^{40}Ar/^{39}Ar$ age spectra and the thermal history of the region. Earth Planet Sci Lett 55:123-149

Harrison TM, Be K (1983) Ar-40/Ar-39 age spectrum analysis of detrital microclines from the Southern San-Joaquin Basin, California—An approach to determining the thermal evolution of sedimentary basins. Earth Planet Sci Lett 64:244-256

Harrison TM, Lovera OM, Heizler MT (1991) $^{40}Ar/^{39}Ar$ results for alkali feldspars containing diffusion domains with differing activation energies. Geochim Cosmochim Acta 55:1435-1448

Harrison TM, Heizler MT, Lovera OM (1993) *In vacuo* crushing experiments and K-feldspar thermochronology. Earth Planet Sci Lett 117:169-180

Harrison TM, Heizler MT, Lovera OM, Chen W, Grove M (1994) A chlorine disinfectant for excess argon released from K-feldspar during step heating. Earth Planet Sci Lett 123:95-104

Hearn PP, Sutter JF, Belkin HE (1987) Evidence for Late-Paleozoic brine migration in Cambrian carbonate rocks of the central and southern Appalachians—Implications for Mississippi Valley-type sulfide mineralization. Geochim Cosmochim Acta 51:1323-1334

Heizler MT, Harrison TM (1988) Multiple trapped argon isotope components revealed by $^{40}Ar/^{39}Ar$ isochron analysis. Geochim Cosmochim Acta 52:1295-1303

Hodges KV, Hames WE, Bowring SA (1994) $^{40}Ar/^{39}Ar$ age gradients in micas from a high-temperature-low-pressure metamorphic terrain: Evidence for very slow cooling and implications for the interpretation of age spectra. Geology 22:55-58

Hofmann C, Feraud G, Courtillot V (2000) Ar-40/Ar-39 dating of mineral separates and whole rocks from the Western Ghats lava pile: further constraints on duration and age of the Deccan traps. Earth Planet Sci Lett 180:13-27

Holness MB (1997) The permeability of non-deforming rock. *In* Holness MB (ed) Deformation-Enhanced Fluid Transport in the Earth's Crust and Mantle. Chapman and Hall, London, p 9-39

Humayun M, Clayton RN (1995a) Potassium isotope geochemistry: Genetic implications of volatile element depletion. 59:2131-2148

Humayun M, Clayton RN (1995b) Precise determinations of the isotopic composition of potassium: Application to terrestrial rocks and lunar soils. Geochim Cosmochim Acta 59:2115-2130

Inger S, Ramsbottom W, Cliff RA, Rex DC (1996) Metamorphic evolution of the Sesia-Lanzo Zone, Western Alps: Time constraints from multi-system geochronology. Contrib Mineral Petrol 126: 152-168

Itaya T, Arribas A, Okada T (1996) Argon release systematics of hypogene and supergene alunite based on progressive heating experiments from 100 to 1000 degrees C. Geochim Cosmochim Acta 60: 4525-4535

Kelley SP (1988) The relationship between K-Ar mineral ages, mica grain sizes and movement on the Moine Thrust Zone, NW Highlands, Scotland. J Geol Soc London 145:1-10

Kelley SP (2002) Excess Argon in K-Ar and Ar-Ar Geochronology. Chem Geol (in press)

Kelley SP, and Bluck BJ (1989) Detrital mineral ages from the Southern Uplands using $^{40}Ar/^{39}Ar$ laser probe. J Geol Soc London, 146:401-403

Kelley SP, Turner G (1991) Laser probe ^{40}Ar-^{39}Ar measurements of loss profiles within individual hornblende grains from the Giants Range Granite, northern Minnesota, USA. Earth Planet Sci Lett 107:634-648

Kelley SP, Wartho J-A (2000) Rapid kimberlite ascent and the significance of Ar-Ar ages in xenolith phlogopites. Science 289:609-611

Kelley SP, Turner G, Butterfield AW, Shepherd TJ (1986) The source and significance of argon in fluid inclusions from areas of mineralization. Earth Planet Sci Lett 79:303-318

Kelley SP, Reddy SM, Maddock R (1994) $^{40}Ar/^{39}Ar$ laser probe investigation of a pseudotachylyte vein from the Moine Thrust Zone, Scotland. Geology 22:443-446

Kelley SP, Bartlett JM, Harris NBW (1997) Pre-metamorphic ages from biotite inclusions in garnet. Geochim Cosmochim Acta 61:3873-3878

Kendrick MA, Burgess R, Pattrick RAD, Turner G (2001a) Fluid inclusion noble gas and halogen evidence on the origin of Cu-porphyry mineralizing fluids. Geochim Cosmochim Acta 65:2651-2668

Kendrick MA, Burgess R, Pattrick RAD, Turner PG (2001b) Halogen and Ar-Ar age determinations of inclusions within quartz veins from porphyry copper deposits using complementary noble gas extraction techniques. Chem Geol 177:351-370

Kipfer R, Aeschbach-Hertig W, Peeters F, Stute M (2002) Noble gases in lakes and groundwaters. Rev Mineral Geochem 47:615-700

Kirschner DL, Cosca MA, Masson H, Hunziker JC (1996) Staircase Ar-40/Ar-39 spectra of fine-grained-white mica: Timing and duration of deformation and empirical constraints on argon diffusion. Geology 24:747-750

Kramar N, Cosca MA, Hunziker JC (2001) Heterogeneous Ar-40* distributions in naturally deformed muscovite: *In situ* UV-laser ablation evidence for micro structurally controlled intra-grain diffusion. Earth Planet Sci Lett 192:377-388

Lanphere MA, Dalrymple GB (1976) Identification of Excess ^{40}Ar by the ^{40}Ar/^{39}Ar age spectrum technique. Earth Planet Sci Lett 32:141-148
Lanphere MA, Baadsgaard H (2001) Precise K-Ar, ^{40}Ar-^{39}Ar, Rb-Sr and U/Pb mineral ages from the 27.5 Ma Fish Canyon Tuff reference standard. Chem Geol 175:653-671
Layer PW, Kroner A, McWilliams M, York D (1989) Elements of the Archean thermal history and apparent polar wander of the Eastern Kaapvaal Craton, Swaziland, from single grain dating and paleomagnetism. Earth Planet Sci Lett 93:23-34
Lee JKW (1993) The argon release mechanisms of hornblende *in vacuo*. Chem Geol (Isotop Geosci Sect) 106:133-170
Lee JKW (1995) Multipath diffusion in geochronology. Contrib Mineral Petrol 120:60-82
Lee JKW, Onstott TC, Hanes JA (1990) An ^{40}Ar/^{39}Ar investigation of the contact effects of a dyke intrusion, Kapuskasing Structural Zone, Ontario: A comparison of laser microprobe and furnace extraction techniques. Contrib Mineral Petrol 105:87-105
Lee JKW, Onstott TC, Cashman KV, Cumbest RJ, Johnson D (1991) Incremental heating of hornblende *in vacuo*: Implications for ^{40}Ar/^{39}Ar grochronology and the interpretation of thermal histories. Geology 19:872-876
Li S, Wang S, Chen Y, Liu D, Qiu J, Zhou H, Zhang Z (1994) Excess argon in phengite from eclogite: evidence from dating of eclogite minerals by Sm-Nd, Rb-Sr and ^{40}Ar/^{39}Ar methods. Chem Geol (Isot Geosci Sect) 112:343-350
Lister GS, Baldwin SL (1996) Modeling the effect of arbitrary P-T-t histories on argon diffusion in minerals using the MacArgon program for the Apple Macintosh. Tectonophysics 253:83-109
Livingston DE, Damon PE, Mauger RL, Bennet R, Laughlin AW (1967) Argon-40 in cogenetic feldspar-mica mineral assemblages. J Geophys Res 72:1361-1375
Lo C-H, Onstott TC (1989) ^{39}Ar recoil artifacts in chloritized biotite. Geochim Cosmochim Acta 53: 2697-2711
Lo CH, Lee JKW, Onstott TC (2000) Argon release mechanisms of biotite *in vacuo* and the role of short-circuit diffusion and recoil. Chem Geol 165:135-166
Lovera OM, Richter FM, Harrison TM (1989) The ^{40}Ar/^{39}Ar thermochronometry for slowly cooled samples having a distribution of diffusion domain sizes. J Geophys Res 94:917-935
Lovera OM, Richter FM, Harrison TM (1991) Diffusion domains determined by ^{39}Ar released during step heating. J Geophys Res 96:2057-2069
Lovera OM, Heizler MT, Harrison TM (1993) Argon diffusion domains in K-feldspar, II: Kinetic parameters of MH-10. Contrib Mineral Petrol 113:381-393
Lovera OM, Grove M, Harrison TM, Mahon KI (1997) Systematic analysis of K-feldspar ^{40}Ar/^{39}Ar step heating results: I. Significance of activation energy determinations. Geochim Cosmochim Acta 61:3171-3192
Lovering JF, Richards JR (1964) Potassium-argon age study of possible lower-crust and upper-mantle inclusions in deep-seated intrusions. J Geophys Res 69:4895-4901
Ludwig KR, Renne PR (2000) Geochronology on the paleoanthropological time scale. Evol Anthropol 9:101-110
Magloughlin JF, Hall CM, van der Pluijm BA (2001) Ar-40-Ar-39 geochronometry of pseudotachylytes by vacuum encapsulation: North Cascade Mountains, Washington, USA. Geology 29:51-54
Mahon KI, Harrison TM, Grove M (1998) The thermal and cementation histories of a sandstone petroleum reservoir, Elk Hills, California. Chem Geol 152:227-256
Marshall BD, DePaolo DJ (1982) Precise age determinations and petrogenetic studies using the K-Ca method. 46:2537-2545
McDougall I, Harrison TM. (1999) Geochronology and Thermochronology by the ^{40}Ar/^{39}Ar method. Oxford University Press, New York, 212 p
Merrihue CM (1965) Trace-element determinations and potassium-argon dating by mass spectroscopy of neutron irradiated samples. EOS Trans Am Geophys Union 46:125 (abstr)
Merrihue CM, Turner G (1966) Potassium-argon dating by activation with fast neutrons. J Geophys Res 71:2852-2857
Min K, Mundil R, Renne PR, Ludwig KR (2000) A test for systematic errors in ^{40}Ar-^{39}Ar geochronology through comparison with U-Pb analysis of a 1.1 Ga rhyolite. Geochim Cosmochim Acta 64:73-98
Min K, Farley KA, Renne P (2001) Single-grain (U-Th)/He Ages from apatites in Acapulco meteorite. EOS Trans Am Geophys Union Fall Mtg 2001 Suppl Abstr V22C-1059
Mitchell JG (1968) The argon-40/argon-39 method of potassium-argon age determination. Geochim Cosmochim Acta 32:781-790
Mock C, Arnaud NO, Cantagrel JM (1999a) An early unroofing in northeastern Tibet? Constraints from Ar- 40/Ar-39 thermochronology on granitoids from the eastern Kunlun range (Qianghai, NW China). Earth Planet Sci Lett 171:107-122

Mock C, Arnaud NO, Cantagrel JM, Yirgu G (1999b) Ar-40/Ar-39 thermochronology of the Ethiopian and Yemeni basements: reheating related to the Afar plume? Tectonophysics 314:351-372

Mulch A, Cosca MA, Handy MR (2002) *In situ* UV-laser $^{40}Ar/^{39}Ar$ geochronology of a micaceous mylonite: an example of defect-enhanced argon loss. Contrib Mineral Petrol 142:738-752

Muller W, Prosser G, Mancktelow NS, Villa IM, Kelley SP, Viola G, Oberli F (2001) Geochronological constraints on the evolution of the Periadriatic Fault System (Alps). Intl J Earth Sci 90:623-653

Niedermann S (2002) Cosmic-ray-produced noble gases in terrestrial rocks as a dating tool for surface processes. Rev Mineral Geochem 47:731-784

Nier AO (1950) A redetermination of the relative abundances of the isotopes of carbon, nitrogen, oxygen, argon and potassium. Phys Rev 77:789-793

Oberli F, Fischer H, Meier M (1990) High-resolution ^{238}U-^{206}Pb zircon dating of Tertiary bentonites and the Fish Canyon Tuff: a test for age "concordance" by single-crystal analysis. Geol Soc Australia 27:74 (abstr)

Onstott TC, Peacock MW (1987) Argon retentivity of hornblendes: A field experiment in a slowly cooled metamorphic terrane. Geochim Cosmochim Acta 51:2891-2903

Onstott TC, Phillips D, Pringle-Goodell L (1991) Laser Microprobe Measurement of Chlorine and Argon Zonation in Biotite. Chem Geol 90:145-168

Onstott TC, Miller ML, Ewing RC, Arnold GW, Walsh DS (1995) Recoil refinements: Implications for the $^{40}Ar/^{39}Ar$ dating technique. Geochim Cosmochim Acta 59:1821-1834

Onstott TC, Mueller C, Vrolijk PJ, Pevear DR (1997) Laser Ar-40/Ar-39 microprobe analyses of fine-grained illite. Geochim Cosmochim Acta 61:3851-3861

Pankhurst RJ, Moorbath S, Rex DC, Turner G (1973) Mineral age patterns in ca. 3700 m.y. old rocks from West Greenland. Earth Planet Sci Lett 20:157-170

Parsons I, Brown WL, Smith JV (1999) $^{40}Ar/^{39}Ar$ thermochronology using alkali feldspars: Real thermal history or mathematical mirage of microtexture? Contrib Mineral Petrol 136:92-110.

Pearson RC, Hedge CE, Thomas HH, Stearn TW (1966) Geochronology of the St Kevin Granite and neighboring Precambrian rocks, northern Sawatch Range, Colorado. Geol Soc Am Bull 77:1109-1120

Philippot P, Rumble D (2000) Fluid-rock interactions during high-pressure and ultrahigh-pressure metamorphism. Intl Geol Rev 42:312-327

Phillips D, Onstott TC (1988) Argon isotopic zoning in mantle phlogopite. Geology 16:542-546

Pickles CS, Kelley SP, Reddy SM, Wheeler J (1997) Determination of high spatial resolution argon isotope variations in metamorphic biotites. Geochim Cosmochim Acta 61:3809-3883

Polyak VJ, McIntosh WC, Guven N, Provencio P (1998) Age and origin of Carlsbad cavern and related caves from Ar-40/Ar-39 of alunite. Science 279:1919-1922

Pringle MS, Staudigel H, Gee J (1991) Jasper Seamount—7 million years of volcanism. Geology 19: 364-368

Purdy JW, Jäger E. (1976) K-Ar ages on rock-forming minerals from the Central Alps. Mem Geol Inst Univ Padova 30, 31 p

Rama SNI, Hart SR, Roedder E (1965) Excess radiogenic argon in fluid inclusions. J Geophys Res 70: 509-511

Reddy SM, Kelley SP, Wheeler J (1996) A $^{40}Ar/^{39}Ar$ laser probe study of micas from the Sesia Zone, Italian Alps: Implications for metamorphic and deformation histories. J Metamorph Geol 14:493-508

Reddy SM, Kelley SP, Magennis L (1997) A microstructural and argon laserprobe study of shear zone development at the western margin of the Nanga Parbat-Haramosh Massif, western Himalaya. Contrib Mineral Petrol 128:16-29

Reddy SM, Potts GJ, Kelley SP (2001) $^{40}Ar/^{39}Ar$ ages in deformed potassium feldspar: Evidence of microstructural control on Ar isotope systematics. Contrib Mineral Petrol 141:186-200

Renne PR (1995) Excess ^{40}Ar in biotite and hornblende from the Norilsk-1 Intrusion, Siberia—Implications for the age of the Siberian Traps. Earth Planet Sci Lett 134:225-225

Renne PR (2000) Ar-40/Ar-39 age of plagioclase from Acapulco meteorite and the problem of systematic errors in cosmochronology. Earth Planet Sci Lett 175:13-26

Renne PR (2001) Reply to Comment on "Ar-40/Ar-39 age of plagioclase from Acapulco meteorite and the problem of systematic errors in cosmochronology" by Mario Trieloff, Elmar K. Jessberger and Christine Fieni. Earth Planet Sci Lett 190:271-273

Renne PR, Basu AR (1991) Rapid eruption of the Siberian Traps flood basalts at the Permo-Triassic boundary. Science 253:176-179

Renne PR, Norman EB (2001) Determination of the half-life of Ar-37 by mass spectrometry. Phys Rev C 6304:7302

Renne PR, Ernesto M, Pacca IG, Coe RS, Glen JM, Prevot M, Perrin M (1992) The age of Parana flood volcanism, rifting of Gondwanaland, and the Jurassic-Cretaceous boundary. Science 258:975-979

Renne PR, Deino AL, Walter RC, Turrin BD, Swisher CC, Becker TA, Curtis GH, Sharp WD, Jaouni AR (1994) Intercalibration of astronomical and radioisotopic time. Geology 22:783-786

Renne PR, Zhang ZC, Richards MA, Black MT, Basu AR (1995) Synchrony and causal relations between Permian-Triassic boundary crises and Siberian flood volcanism. Science 269:1413-1416

Renne PR, Glen JM, Milner SC, Duncan AR (1996a) Age of Etendeka flood volcanism and associated intrusions in southwestern Africa. Geology 24:659-662

Renne PR, Deckart K, Ernesto M, Feraud G, Piccirillo EM (1996b) Age of the Ponta Grossa dike swarm (Brazil), and implications to Parana flood volcanism. Earth Planet Sci Lett 144:199-211

Renne PR, Sharp WD, Deino AL, Orsi G, Civetta L (1997) Ar-40/Ar-39 dating into the historical realm: Calibration against Pliny the Younger. Science 277:1279-1280

Renne PR, Karner DB, Ludwig KR (1998a) Radioisotope dating—Absolute ages aren't exactly. Science 282:1840-1841

Renne PR, Swisher CC, Deino AL, Karner DB, Owens TL, DePaolo DJ (1998b) Inter-calibration of standards, absolute ages and uncertainties in $^{40}Ar/^{39}Ar$ dating. Chem Geol 145:117-152

Renne PR, Farley KA, Becker TA, Sharp WD (2001) Terrestrial cosmogenic argon. Earth Planet Sci Lett 188:435-440

Roddick JC (1978) The application of isochron diagrams in $^{40}Ar/^{39}Ar$ dating: A discussion. Earth Planet Sci Lett 41:233-244

Roddick JC, Cliff RA, Rex DC (1980) The evolution of excess argon in Alpine biotites—A ^{40}Ar-^{39}Ar analysis. Earth Planet Sci Lett 48:185-208

Roselieb K, Wartho J-A, Buttner H, Jambon A, Kelley SP (1999) Solubility and diffusivity of noble gases in synthetic phlogopite: A UV LAMP investigation. 4:368

Ruffet G, Gruau G, Ballèvre M, Féraud G, Philipot P (1997) Rb-Sr and ^{40}Ar-^{39}Ar laser probe dating of high-pressure phengites from the Sesia Zone (Western Alps): Underscoring of excess argon and new age constraints on high-pressure metamorphism. Chem Geol 141:1-18

Samson SD, Alexander EC (1987) Calibration of the interlaboratory ^{40}Ar-^{39}Ar dating standard—MMhb-1. Chem Geol (Isotop Geosci Sect) 66:27-34

Scaillet S (1996) Excess ^{40}Ar transport scale and mechanism in high-pressure phentites: A case study from an eclogitised metabasite of the Dora-Maira nappe, western Alps. Geochim Cosmochim Acta 60: 1075-1090

Scaillet S (2000) Numerical Error Analysis in Ar-40/Ar-39 Dating. Chem Geol 162:269-298

Scaillet S, Feraud G, Lagabrielle Y, Ballevre M, Ruffet G (1990) Ar-40/Ar-39 laser-probe dating by step heating and spot fusion of phengites from the Dora-Maira nappe of the Western Alps, Italy. Geology 18:741-744

Scaillet S, Féraud G, Ballèvre M, Amouric M (1992) Mg/Fe and [(Mg,Fe)Si-Al_2] compositional control on argon behavior in high-pressure white micas: A $^{40}Ar/^{39}Ar$ continuous laser-probe study from the Dora-Maira nappe of the internal western Alps, Italy. Geochim Cosmochim Acta 56:2851-2872

Schaeffer OA, Zähringer J. (1966) Potassium Argon Dating. Springer-Verlag, New York, 234 p

Schmitz MD, Bowring SA (2001) U-Pb zircon and titanite systematics of the Fish Canyon Tuff: an assessment of high-precision U-Pb geochronology and its application to young volcanic rocks. Geochim Cosmochim Acta 65:2571-2587

Sherlock SC (2001) Two-stage erosion and deposition in a continental margin setting: a $^{40}Ar/^{39}Ar$ laserprobe study of offshore detrital white micas in the Norwegian Sea. J Geol Soc London 158: 793-800

Sherlock SC, Arnaud NO (1999) Flat plateau and impossible isochrons: Apparent ^{40}Ar-^{39}Ar geochronology in a high-pressure terrain. Geochim Cosmochim Acta 63:2835-2838

Sherlock SC and Kelley SP (2002) Excess argon evolution in HP-LT rocks: A UVLAMP study of phengite and K-free minerals, NW Turkey. Chem Geol (in press)

Sherlock SC, Kelley SP, Inger S, Harris NBW, Okay A (1999) ^{40}Ar-^{39}Ar and Rb-Sr geochronology of high-pressure metamorphism and exhumation history of the Tavsanli Zone, NW Turkey. Contrib Mineral Petrol 137:46-58

Singer B, Hildreth W, Vincze Y (2000) Ar-40/Ar-39 evidence for early deglaciation of the central Chilean Andes. Geophys Res Lett 27:1663-1666

Singer BS, Pringle MS (1996) Age and duration of the Matuyama-Brunhes geomagnetic polarity reversal from Ar-40/Ar-39 incremental heating analyses of lavas. Earth Planet Sci Lett 139:47-61

Smith PE, Evensen NM, York D (1993) First successful ^{40}Ar-^{39}Ar dating of glauconites: Argon recoil in single grains of cryptocrystalline material. Geology 21:41-44

Smith PE, York D, Easton RM, Özdemir Ö, Layer PW (1994) A laser $^{40}Ar/^{39}Ar$ study of minerals across the Grenville Front: Investigations of reproducible excess Ar patterns. Can J Earth Sci 31:808-817

Smith SP, Kennedy BM (1983) The Solubility of noble gases in water and in NaCl brine. Geochim Cosmochim Acta 47:503-515

Smits F, Gentner W (1950) Argonbestimmummungen an Kalium-Mineralien I. Bestimmungen an tertiären Kalisalzen. Geochim Cosmochim Acta 1:22-27

Spray JG, Kelley SP, Reimold WU (1995) Laser probe argon^{40}/argon^{39} dating of coesite- and stishovite-bearing pseudotachylytes and the age fo the Vredefort impact event. Meteoritics 30:335-343

Steiger RJ, Jäger E (1977) Subcommission on geochronology: Convention on the use of decay constants in geo- and cosmocchronology. Earth Planet Sci Lett 36:359-362

Storey M, Mahoney JJ, Saunders AD, Duncan RA, Kelley SP, Coffin MF (1995) Timing of hot spot-related volcanism and the breakup of Madagascar and India. Science 267:852-855ds

Swindle TD (2002) Noble gases in the moon and meteorites—Radiogenic components and early volatile chronologies. Rev Mineral Geochem 47:101-124

Ton-That T, Singer B, Paterne M (2001) Ar-40/Ar-39 dating of latest Pleistocene (41 Ka) marine tephra in the Mediterranean Sea: implications for global climate records. Earth Planet Sci Lett 184:645-658

Torgersen T, B.M. K, Hiagon H, Chiou KY, J.H. R, Clarke WB (1989) Argon accumulation and the crustal degassing flux of ^{40}Ar in the Great Atresian Basin, Australia. Earth Planet Sci Lett 92:43-56

Trieloff M, Jessberger EK, Fieni C (2001) Comment on "Ar-40/Ar-39 age of plagioclase from Acapulco meteorite and the problem of systematic errors in cosmochronology" by Paul R. Renne. Earth Planet Sci Lett 190:263-265

Turner G (1969) Thermal histories of meteorites by the 39Ar-40Ar method. *In* Millman PM (ed) Meteorite Research. D. Reidel Publishing, Dordrecht, The Netherlands, p 407-417

Turner G (1970a) Thermal histories of meteorites. *In* Runcorn SK (ed) Paleogeophysics. Academic Press, London, p 491-502

Turner G (1970b) ^{40}Ar-^{39}Ar age determination of lunar rock 12013. Earth Plant Sci Lett 9:177-180

Turner G (1970c) Argon-40/argon-39 dating of lunar rock samples. Science 167:466-468

Turner G (1971a) ^{40}Ar-^{39}Ar dating: The optimization of irradiation parameters. Earth Planet Sci Lett 10:227-234

Turner G (1971b) ^{40}Ar-^{39}Ar ages from the lunar maria. Earth Planet Sci Lett 11:169-191

Turner G (1972) ^{40}Ar-^{39}Ar age and cosmic ray irradiation history of the Apollo 15 anorthosite 15415. Earth Planet Sci Lett 14:169-175

Turner G (1977) Potassium-argon chronology of the moon. Phys Chem Earth 10:145-195

Turner G, Cadogan PH (1974) Possible effects of ^{39}Ar recoil in ^{40}Ar-^{39}Ar dating. Geochim Cosmochim Acta 5:1601-1615

Turner G, Bannon MP (1992) Argon isotope geochemistry of inclusion fluids from granite-associated mineral veins in southwest and northeast England. Geochim Cosmochim Acta 56:227-243

Turner G, Wang S (1992) Excess argon, crustal fluids and apparent isochrons from crushing K-feldspar. Earth Planet Sci Lett 110:193-211

Turner G, Burnard P, Ford JL, Gilmour JD, Lyon IC, Stuart FM (1993) Tracing fluid sources and interactions. Phil Trans Roy Soc London Ser A 344:127-140

Turner G, Knott SF, Ash RD, Gilmour JD (1997) Ar-Ar chronology of the Martian meteorite ALH84001: Evidence for the timing of the early bombardment of Mars. Geochim Cosmochim Acta 61:3835-3850

van der Bogaard P, Schirnick C (1995) ^{40}Ar/^{39}Ar laser-probe ages of Bishop Tuff quartz phenocrysts substantiate long-lived silicic magma chamber at Long Valley, United States. Geology 23:759-762

Vance D, Ayres M, Kelley SP, Harris NBW (1998) The thermal response of a metamorphic belt to extension: constraints from laser Ar data on metamorphic micas. Earth Planet Sci Lett 162:153-164

Vasconcelos PM (1999) K-Ar and Ar-40/Ar-39 geochronology of weathering processes. Ann Rev Earth Planet Sci 27:183-229

Villa IM (1994) Multipath Ar transport in K-feldspar deduced from isothermal heating experiments. Earth Planet Sci Lett 122:393-401

Villa IM (1997) Direct determination of ^{39}Ar recoil distance. Geochim Cosmochim Acta 61:689-691

Villa IM, Grobéty B, Kelley SP, Trigila R, Wieler R (1996) Assessing Ar transport paths and mechanisms in the McClure Mountains hornblende. Contrib Mineral Petrol 126:67-80

Warnock AC, Zeitler PK (1998) Ar-40/Ar-39 thermochronometry of K-feldspar from the KTB borehole, Germany. Earth Planet Sci Lett 158:67-79

Wartho JA (1995) Apparent Argon Diffusive Loss Ar-40/Ar-39 Age Spectra in Amphiboles. Earth Planet Sci Lett 134:393-407

Wartho JA, Dodson MH, Rex DC, Guise PG (1991) Mechanisms of Ar release from Himalayan metamorphic hornblende. Am Mineral 76:1446-1448

Wartho J-A, Rex DD, Guise PG (1996) Excess argon in amphiboles linked to greenschist facies alteration in Kamila amphibolite belt, Kohistan island arc system, northern Pakiston: Insights from ^{40}Ar/^{39}Ar step-heating and acid leaching experiments. Geol Mag 133:595-609

Wartho JA, Kelley SP, Brooker RA, Carroll MR, Villa IM, Lee MR (1999) Direct measurement of Ar diffusion profiles in a gem-duality Madagascar K-feldspar using the ultra-violet laser ablation microprobe (UVLAMP). Earth Planet Sci Lett 170:141-153
Wartho JA, Kelley SP, Blake S (2001) Magma flow regimes in sills deduced from Ar isotope systematics of host rocks. J Geophys Res-Solid Earth 106:4017-4035
Wieler R (2002) Cosmic-ray-produced noble gases in meteorites. Rev Mineral Geochem 47:125-170
Wheeler J (1996) A program for simulating argon diffusion profiles in minerals. Computers Geosciences 28:919-929
Wilch TI, McIntosh WC, Dunbar NW (1999) Late Quaternary volcanic activity in Marie Byrd Land: Potential Ar-40/Ar-39-dated time horizons in West Antarctic ice and marine cores. Geol Soc Am Bull 111:1563-1580
Wright N, Layer PW, York D (1991) New insights into thermal history from single grain Ar-40/Ar-39 analysis of biotite. Earth Planet Sci Lett 104:70-79
Zeitler PK, FitzGerald JD (1986) Saddle-shaped $^{40}Ar/^{39}Ar$ age spectra from young, microstructurally complex potassium feldspars. Geochim Cosmochim Acta 50:1185-1199

(U-Th)/He Dating: Techniques, Calibrations, and Applications

Kenneth A. Farley

Division of Geological and Planetary Sciences
California Institute of Technology
Pasadena, California 91125
farley@gps.caltech.edu

INTRODUCTION

The possibility of dating minerals by the accumulation of ^{4}He from U and Th decay has been recognized for many years (e.g., Strutt 1905), but in the century since the idea was first conceived, the method has rarely been applied successfully. After several investigations of (U-Th)/He dating of various minerals (e.g., Damon and Kulp 1957; Fanale and Kulp 1962; Damon and Green 1963; Turekian et al. 1970; Bender 1973; Leventhal 1975; Ferreira et al. 1975) the technique was essentially abandoned as yielding unreliable and usually low ages, presumably as a result of diffusive He loss possibly associated with radiation damage. In 1987, Zeitler and coworkers rekindled interest in the method by proposing that in the case of apatite, He ages might be meaningfully interpreted as ages of cooling through very low temperatures. Laboratory diffusion data presented by these authors indicated a closure temperature of about 100ºC, a value supported by more recent studies (Lippolt et al. 1994; Wolf et al. 1996b; Warnock et al. 1997). Consistent with this interpretation Wolf et al. (1996a) found that apatite He ages increase systematically with sample elevation in a mountain range, as expected for exhumation-induced cooling through a low closure temperature. Based on the strength of these results and additional laboratory (Farley 2000) and natural (Warnock et al. 1997; House et al. 1999; Stockli et al. 2000) constraints on He diffusivity, recent attention has focused on applications of apatite He thermochronometry. There is also renewed interest in He dating of other U- and Th-bearing minerals both for dating mineral formation and for thermochronometry. For example, Lippolt and coworkers have undertaken detailed studies of He diffusion and dating of various phases, most notably hematite formed in hydrothermal systems (Lippolt and Weigel 1988; Wernicke and Lippolt 1992; Lippolt et al. 1993; Wernicke and Lippolt 1994a,b).

Here I present an overview of recent techniques, calibrations, and applications of the (U-Th)/He dating method; Hurley (1954) provides an excellent summary of earlier work in this field. Much of this paper focuses on apatite, because the He behavior and requisite analytical techniques are better established for this phase than for other target minerals, such as zircon and titanite. Similarly, much of this paper concerns He diffusivity behavior required for thermochronometric applications, yet recent work is also considering applications to direct dating, for example, of young tephras (Farley et al. 2001).

Rationale

A wide variety of dating techniques are already well developed and widely applied. Given the methodological complexity, the limited analytical precision, and the rather small observational base presently available, it seems appropriate to consider why development and application of (U-Th)/He dating has received renewed attention. Two immediate answers come to mind. First, many materials are inappropriate for dating by existing techniques, most obviously because they are too poor in parent or radiogenic daughter isotope or too rich in non-*in situ* produced daughter. Thus a new technique with different requirements is potentially advantageous. This is indeed true for the (U-Th)/He

1529-6466/00/0047-0018$05.00

system: the sensitivity for measurement of U, Th, and He is extremely high and the background of "excess" He is low, so both young materials and those with only a trace of U and Th are potentially datable. As a result young volcanic rocks lacking sanidine for Ar/Ar dating are a potential target of the (U-Th)/He method (e.g., Graham et al. 1987; Farley et al. 2002). At most this is likely to be a niche market.

Greater interest in (U-Th)/He dating arises from the fact that He ages of various minerals can be used to delineate the cooling history of rocks through a temperature range that is only partially accessed by existing dating techniques. There is a large family of minerals potentially suitable for He thermochronometry, only a few of which have been explored. Figure 1 presents an overview of estimated He closure temperatures of several minerals (described in detail below) compared with closure temperatures of other thermochronometric methods. He dating complements existing techniques, and of particular interest is the apatite helium method, which is sensitive to temperatures substantially lower than any other method. Geologic applications which illustrate the unique uses of (U-Th)/He dating are presented at the end of this chapter.

TECHNICAL ASPECTS

He ingrowth

^{4}He nuclei (α particles) are produced by the series decay of ^{238}U, ^{235}U, and ^{232}Th and by α decay of ^{147}Sm. In essentially all minerals the overwhelming majority of radiogenic helium derives from actinide decay, so the ingrowth equation is:

$$^4\text{He} = 8\,^{238}\text{U}\left(e^{\lambda_{238}t}-1\right) + 7\left(^{238}\text{U}/137.88\right)\left(e^{\lambda_{235}t}-1\right) + 6\,^{232}\text{Th}\left(e^{\lambda_{232}t}-1\right) \qquad (1)$$

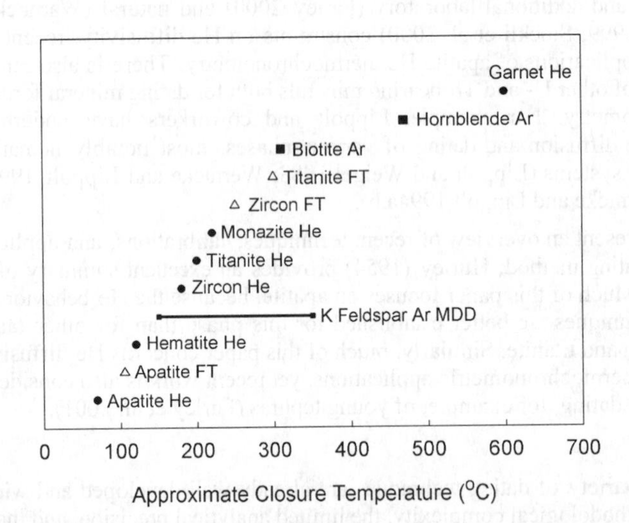

Figure 1. Nominal closure temperatures of various thermochronometers showing how He systems (●) complement existing techniques (■: Ar; Δ: fission track). Systems are simply ordered by closure temperature on the Y-axis. With the exception of apatite and titanite, the He closure temperatures are not yet well-known. Data sources: apatite He (Farley 2000); apatite fission track (Gallagher et al. 1998); hematite He (Wernicke et al. 1994); K feldspar multi-diffusion domain Ar (MDD) and other Ar methods (McDougall and Harrison 1999 and references therein); zircon fission track (Yamada et al. 1995); titanite fission track (Coyle and Wagner 1995), garnet He (Dunai and Roselieb 1996).

where ^4He, U, and Th refer to present-day amounts, t is the accumulation time or He age, and λ is the decay constant (λ_{238} = 1.551×10^{-10} yr^{-1}, λ_{235} = 9.849×10^{-10} yr^{-1}, λ_{232} = 4.948×10^{-11} yr^{-1}). The coefficients preceding the U and Th abundances account for the multiple α particles emitted within each of the decay series, and the factor of (1/137.88) is the present day $^{235}U/^{238}U$ ratio. This equation assumes secular equilibrium among all daughters in the decay chain, a condition guaranteed for crystals formed more than ~350 kyr prior to the onset of He accumulation. For most applications this assumption is valid, but in certain cases the effects of secular disequilibrium must be considered (see below).

Equation (1) assumes the absence of initial ^4He in the crystal being dated, and this is in general a good assumption. For example, while atmospheric Ar frequently accounts for a substantial fraction of the ^{40}Ar in a K/Ar or Ar/Ar analysis (Kelley 2002, this volume), the concentration of He in the atmosphere is so low (5 ppm vs. ~1% for ^{40}Ar) that trapped atmospheric He is unlikely to be important. In some cases fluid inclusions may carry crustal or mantle helium, but for U,Th-rich minerals like apatite, zircon, and titanite, the He concentration of such fluids and/or the inclusion density would have to be high to affect He ages except when the He ages are very low. Nevertheless some workers believe they have detected fluid-inclusion-hosted helium in a few apatites (Lippolt et al. 1994; Stockli et al. 2000). The presence of helium "inherited" from some prior history, for example due to incomplete degassing of a crystal stoped into a magma chamber, is unlikely given the high diffusivity of He in most solids. However, beryl and other cyclosilicates sometimes harbor very high concentrations of "excess" helium from a poorly understood source (e.g., Toyoda and Ozima 1988). This helium probably enters the cyclosilicates through the large central channels of these minerals, making them unsuitable for He dating. A final potential source of excess He is solution from surrounding fluids into grain interiors (see Kelley 2002, this volume, for a discussion of the analogous problem in Ar geochronology). At present neither laboratory data (such as He solubilities in minerals at relevant temperatures) nor sufficient age data are available to quantitatively evaluate the significance of this phenomenon.

DIFFUSION BEHAVIOR

Knowledge of the He-retention characteristics of the phase being dated is critical for correct interpretation of (U-Th)/He data. For example, it is known that He is not retained under Earth-surface conditions in quartz (Trull et al. 1991), sanidine, and muscovite (Lippolt and Weigel 1988), so these phases have little obvious potential for any type of helium dating. However, helium is thought to be retained at the Earth's surface in olivine (Trull et al. 1991), pyroxene (Lippolt and Weigel 1988), amphibole (Lippolt and Weigel 1988), garnet (Dunai and Roselieb 1996), non-metamict zircon (Hurley 1952; Damon and Kulp 1957), non-metamict titanite (Hurley 1952; Reiners and Farley 1999), apatite (Zeitler et al. 1987), allanite (Wolf 1997), magnetite (Fanale and Kulp 1962), hematite (Wernicke and Lippolt 1994a,b; Bahr et al. 1994) and submarine basaltic glass (Graham et al. 1987). Several of these phases, especially olivine and pyroxene, have been used extensively for cosmogenic ^3He studies, so He retention is robustly established (Niedermann 2002, this volume). In others the conditions for retention (temperature, grain size, degree of radiation damage) have not been explored.

While in the case of rapidly cooled rocks at the Earth's surface (e.g., volcanics) demonstration of quantitative retention at ~25°C is sufficient to successfully apply He dating, thermochronometry of slowly cooled rocks requires precise knowledge of how diffusivity scales with temperature. Typically, laboratory experiments are used to constrain the parameters of the Arrhenius relationship:

$$\frac{D}{a^2} = \frac{D_o}{a^2} e^{-E_a/RT} \quad (2)$$

where D is the diffusivity, D_o the diffusivity at infinite temperature, E_a the activation energy, R the gas constant, T the Kelvin temperature, and a the diffusion domain radius (Fechtig and Kalbitzer 1966). If this relationship is obeyed, then measurements of $\ln D/a^2$ as a function of reciprocal temperature will plot on a straight line with intercept $\ln D_o/a^2$ and slope $-E_a/R$. If the measurements do not plot on a straight line, more complex behavior such as multiple diffusion mechanisms or domain sizes may be involved.

It is important to note that such laboratory measurements may not apply under natural conditions. For example, diffusion coefficients are commonly measured at temperatures far higher than are relevant in nature, so large and potentially inaccurate extrapolations are often necessary. Similarly, some minerals undergo chemical or structural transformations and possibly defect annealing during vacuum heating; extrapolation of laboratory data from these modified phases to natural conditions may lead to erroneous predictions.

Substantial effort is required to measure the He diffusivity parameters in a given phase and to determine how those parameters vary with mineral characteristics such as grain size and shape, chemical composition, and defect and/or radiation damage density. Only apatite has been studied in detail, but limited He diffusivity information is available on other minerals as well, as discussed below.

Apatite

Zeitler et al. (1987) initiated interest in He thermochronometry by demonstrating that apatite has a closure temperature of about 100ºC (here and elsewhere closure temperatures are referenced to a cooling rate of 10ºC/Myr (Dodson 1973; Kelley 2002, this volume). More recent efforts (Lippolt et al. 1994; Wolf et al. 1996b; Warnock et al. 1997; Farley 2000) confirm this approximate closure temperature, and in addition suggest:

1) He diffusion from Durango apatite (Young et al. 1969, a common "standard" apatite), as well as a variety of other apatites, obeys an Arrhenius relationship (Eqn. 2; illustrated in Fig. 2), suggesting that He diffusion from apatite is a single-mechanism thermally activated volume diffusion process, at least at temperatures <300ºC. Above 300ºC the Arrhenius plot is curved. The origin and significance of this curvature has been explored, but remains poorly understood; it is probably not relevant for helium diffusion in the natural setting.
2) High-precision experiments indicate activation energies between ~32 and ~38 kcal/mol. It is not clear whether this spread is real or analytical in origin; there is no persuasive evidence that this quantity is correlated with observable apatite characteristics.
3) In Durango apatite, the quantity D/a^2 varies with grain size in the manner expected if the diffusion domain is the grain itself, i.e., the quantity "a" is the physical grain dimension. He diffusion from Durango apatite is crystallographically isotropic. The relevant dimension for diffusion is thus the prism radius, as this is the shortest pathway for He loss. These characteristics have not yet been demonstrated to apply to more typical apatites, though Reiners and Farley (2001b) present He age data that support such an effect.
4) Taken together the most precise observations suggest an apatite closure temperature of 70ºC in apatites of ~70-90 µm radius. Variation of the closure temperature with grain size and cooling rate based on the Durango observations is shown in Figure 3.

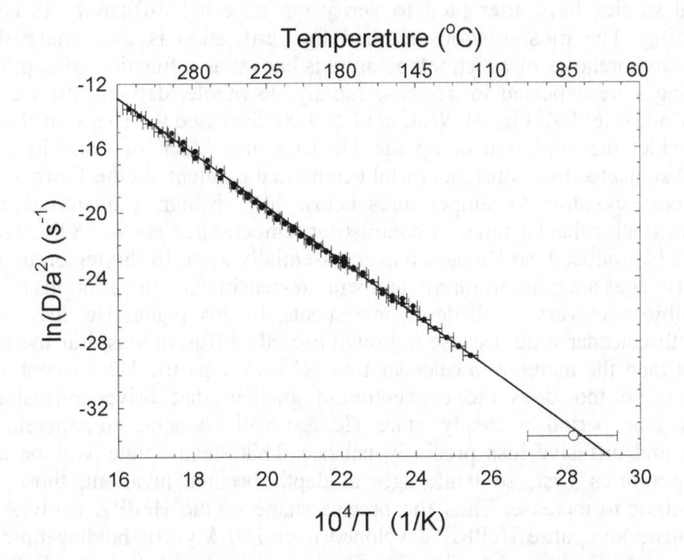

Figure 2. Helium diffusion Arrhenius plot with Otway Basin borehole constraint. The laboratory-measured He diffusion from this sample of Durango apatite defines an extremely linear array (filled symbols), consistent with simple thermally activated volume-diffusion. He age measurements from the Otway Basin provide a completely independent estimate of He diffusivity (open symbol), under natural conditions, which is in excellent agreement with the extrapolated laboratory data. Laboratory data are from Farley (2000). The Otway Basin constraint is from House et al. (1999).

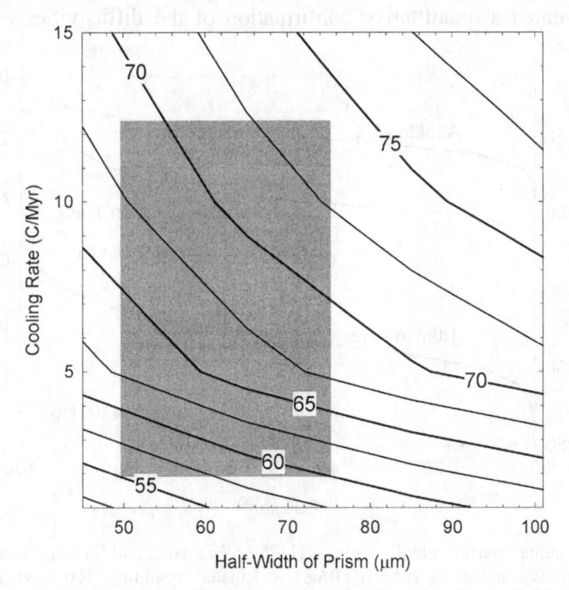

Figure 3. Helium closure temperature (T_c) as a function of grain size and cooling rate. T_c was calculated assuming an activation energy of 33 kcal/mol and D_o = 50 cm^2/sec, assuming spherical geometry and including the effects of α ejection on He diffusion (see Farley (2000) for details and justification of geometry). The shaded region indicates ranges typically observed in nature. This figure differs slightly from the one presented by Farley (2000) because slightly different diffusion parameters were adopted.

Several studies have attempted to verify the expected diffusivity behavior in the natural setting. The most obvious method for verification is to examine the He age distribution in boreholes in which temperature is known as a function of depth. In such a setting He ages are expected to decrease rapidly downhole, defining the Helium Partial Retention Zone (HePRZ; Fig. 4). Wolf et al. (1998) discussed the shape of the HePRZ in detail. Consider the evolution of apatite He ages in a block of crust instantaneously created and subjected to a constant crustal geothermal gradient. At the Earth's surface and at depths corresponding to temperatures below 40°C, helium is quantitatively retained and He ages track calendar time. In contrast, at temperatures above ~80°C, He is lost as rapidly as it is produced, so He ages remain essentially zero. In the region between these two limits He ages are extraordinarily temperature-sensitive, with variations of millions of years possible over very small depth increments. In this region He ages will initially increase with calendar time because ingrowth exceeds diffusive loss, but the age increase will be less than the increase in calendar time. However as the He concentration in the crystal rises, so too does the concentration gradient that drives diffusion. After a sufficiently long period a steady state He age will be achieved wherein radiogenic production and diffusive loss precisely balance. This steady state will be achieved at higher temperatures first, so while ages at depth become invariant, those at shallow depths continue to increase. Thus the precise shape of the HePRZ evolves with time. Figure 4 shows an apatite HePRZ developed over 100 Myr of holding time computed from laboratory diffusivity data. This profile is compared with the essentially analogous fission track partial annealing zone in the figure.

Efforts by Wolf (1997), Warnock et al. (1997) and House et al. (1999) to confirm the existence of the HePRZ were broadly successful. In three different borehole settings the apatite helium ages were found to decrease rapidly at about the predicted temperature, but problems arising from mineral inclusions, poorly known thermal histories, and other phenomena prevented a quantitative confirmation of the diffusivities extrapolated from

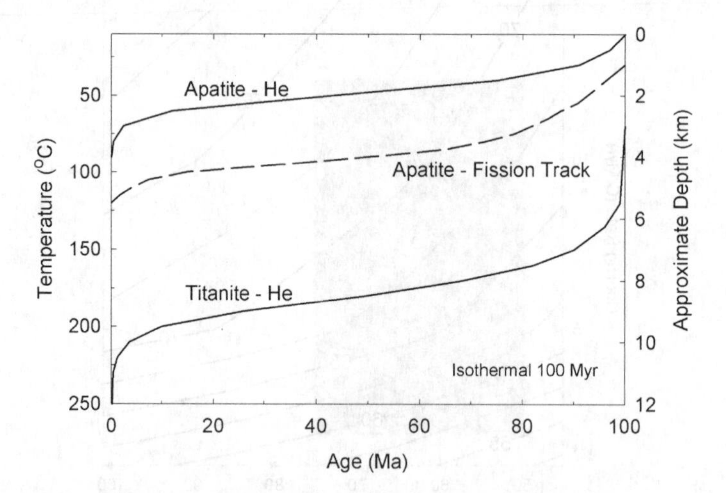

Figure 4. The helium partial retention zones (HePRZ) for apatite and for titanite, and the apatite fission track partial annealing zone (FTPAZ), calculated assuming 100 Myr of isothermal holding at the temperatures indicated on the Y-axis. The right-hand axis scales temperature into sub-crustal depth assuming a geothermal gradient of 20°C/km and a mean surface temperature of 10°C. The apatite helium curve was calculated using parameters in the caption to Figure 3, for titanite, an activation energy of 44.6 kcal/mol, D_o = 50 cm^2/sec, and r = 250 μm were assumed (Reiners and Farley 1999). The fission track profile was calculated using AFTSolve (Ketcham and Donelick 2000) assuming kinetic parameters appropriate for Durango apatite.

laboratory measurements. These data leave little doubt that the laboratory data are approximately correct, but do not constitute unassailable evidence.

House et al. (1999) recognized that problems associated with thermal history could be circumvented by taking advantage of the fact that He ages of samples held close to the closure temperature will rapidly achieve the steady state age where He production and diffusive loss are in balance (Wolf et al. 1998). Provided a sample is in steady state, the measured age and downhole temperature provide a diffusivity–temperature pair that is completely independent of the laboratory measurements. Samples thought to be in steady state in boreholes from the Otway Basin, Australia, yielded diffusivities in excellent agreement with the extrapolated laboratory data (Fig. 2). Unfortunately the downhole temperatures from these industry wells are not very well known, so some uncertainty remains.

Using a completely different approach Stockli et al (2000) also confirmed the extrapolation of laboratory data. In the White Mountains of California, a rapidly exhumed crustal block, these authors discovered a very well defined HePRZ lying above an equally well-defined apatite FTPAZ. The FTPAZ in conjunction with an unconformity defining the pre-exhumation Earth surface was used to compute the pre-exhumation geothermal gradient. The HePRZ predicted in such a geothermal gradient using extrapolated laboratory data matched observations extremely well. This study is described more fully in the section on applications.

These studies provide compelling evidence that laboratory data adequately describe He diffusion characteristics for most apatites so far investigated. Nevertheless additional work to establish the generality of these data and the possible effects of chemical substitution and defects (Farley 2000) on He diffusivity is warranted. The strong effects of even modest substitution of Cl for F on apatite fission track annealing characteristics (Green et al. 1986) demonstrate that extreme caution is necessary when attempting to generalize kinetic behavior from a specific and limited set of observations.

Hematite

Lippolt and coworkers have undertaken several studies of the He retention characteristics of hematite. In the case of specular hematite, reasonably linear He diffusion Arrhenius plots obtained at temperatures <1000°C were interpreted as evidence for volume diffusion (Bahr et al. 1994; Lippolt et al. 1993). He diffusivity scales approximately as expected if the diffusion domain is the grain itself, and a closure temperature of ~220°C was estimated for grains of radius ~500 μm. He diffusion from botryoidal hematite plots on linear Arrhenius arrays only at temperatures below 250°C, and is apparently controlled by volume diffusion from small (0.1 to 10 μm) crystals within the larger botryoidal mass (Bahr et al. 1994; Wernicke and Lippolt 1994a). A closure temperature of 120°C was estimated from the linear part of the Arrhenius plot for botryoidal hematite grains larger than ~10 μm. This value is only a general estimate; for example, a 50°C difference in closure temperature was inferred for two zones of a single botryoidal hematite specimen (Wernicke and Lippolt 1994a). Bahr et al. (1994) attributed deviations from linearity on the Arrhenius plots of the two hematite varieties to variations in the grain size distribution of the samples. The role of grain size and chemistry on He diffusion from hematite warrants further investigation.

Titanite

Based on laboratory stepped-heating experiments, Reiners and Farley (1999) concluded that He diffusion from titanite is generally consistent with a thermally activated volume diffusion process, although some subtleties in the diffusion characteristics remain poorly understood. An activation energy of about ~45 kcal/mol was established, and, like apatite and specular hematite, the diffusion domain was found to be the grain itself. Based on these observations, a closure temperature of ~200ºC was estimated for grains of ~500 μm diameter. Insufficient data are available to evaluate possible compositional controls on He diffusion from this phase. The titanite HePRZ computed from these results is expected to lie between 150 and 200ºC, substantially deeper than both the FTPAZ and the apatite HePRZ (Fig. 4).

The effects of radiation damage on He loss from radioactive minerals like zircon and titanite have long been recognized. For example, Hurley (1952) showed that He ages decrease with increasing U content in a cogenetic suite of Precambrian titanites. Hurley interpreted this observation to indicate radiation-damage induced He loss. Although he presented a method for correcting ages for this loss, it is unclear how such a correction can be applied to samples with time-varying temperature histories in which both radiation damage (and annealing) and He diffusivity vary. Based on fission track studies, radiation damage in titanite anneals over geologic timescales at temperatures above ~300ºC (Coyle and Wagner 1998), so the relevant measure of radiation damage is the α activity coupled with the time elapsed since cooling through this temperature. Using Hurley's model such He loss should not strongly affect He ages until a few hundred Myr after a titanite cools below ~300ºC, at least in titanites with equivalent U contents of less than 1000 ppm. This conclusion is supported by the observation that He ages of the quickly-cooled Mount Dromedary (99 Ma) and Fish Canyon Tuff (28 Ma) titanite standards indicate quantitative He retention (Reiners and Farley 1999; House et al. 2000).

These observations provide no insight to how He diffusion and closure temperature vary with radiation damage. Reiners and Farley (1999) investigated this phenomenon but saw no obvious indication of enhanced He diffusivity, not even in titanites which likely cooled below 300ºC several hundred million years ago. However, in step-heating analyses of titanites with large He accumulations and dark colors (both reasonable proxies of radiation damage) unexpectedly high and erratic He release was observed in some steps, presumably related to radiation damage. A mechanistic explanation for this behavior has not yet been developed, nor is it obvious how this behavior affects He retention in the natural setting. Quantitative application of titanite thermochronometry will require careful assessment of this phenomenon.

To test the measured diffusivity data in the natural setting and to assess the role of radiation damage, Reiners et al. (2000) analyzed titanite He ages in the Proterozoic Gold Butte Block of Southern Nevada, a rapidly and very deeply exhumed crustal block. He ages define a well developed HePRZ positioned approximately at the structural position expected based on extrapolation of the laboratory data. At least for the degree of radiation damage experienced in these samples, He diffusivity is not so severely compromised as to preclude meaningful thermochronometry.

Zircon

Limited data suggest that He diffusion from zircon does not obey a simple Arrhenius relationship (Reiners and Farley 2001a). Instead, in most analyzed samples a distinct increase in slope and decrease in diffusivity occurs as the experiment proceeds. This observation is consistent with annealing of radiation damage or with multiple diffusion domains of different sizes, possibly induced by radiation damage. In addition, the most heavily radiation damaged zircons yield extremely erratic Arrhenius profiles. Although these data are difficult to interpret, Reiners and Farley (2000) proposed a minimum closure temperature of about 180ºC for He release from zircon, with lower values arising from radiation damage effects. He age-paleodepth relationships in the Gold Butte Block (see titanite section) also indicate a He closure temperature in zircon of about 200°C. Overall these data are not sufficient to reliably establish the He diffusivity of zircon.

Like titanite, radiation damage is known to promote He loss from zircon. For example, Damon and Kulp (1957) showed that near-quantitative He retention occurs in Sri Lankan zircons up to a total irradiation of about 6×10^{15} α's/mg (equivalent to 1000 ppm of U decaying for ~1 Gyr), but drops dramatically at higher dosages. Based on fission track annealing studies, radiation damage accumulates in zircons only at temperatures below ~250°C (Yamada et al. 1995), so He dating of typical zircons that have been cooler than this temperature for up to a few hundred Myr should not be greatly compromised by radiation damage. However as noted by Hurley (1954), strong zonation in U and Th in zircons may cause some areas to be severely damaged before the entire crystal becomes metamict; even local metamictization is likely to perturb He diffusion.

Garnet

Dunai and Roselieb (1996) presented the results of He solution experiments suggesting a very high closure temperature for garnet, ~600°C. No data are available to assess the role of composition on He diffusion, but given the diversity of substitutions possible in garnet some compositional effects are likely. In many cases the U and Th in garnet derives from inclusions of radioactive minerals, and this is a potential pitfall for garnet thermochronometry.

The α-emission correction

For any application of (U-Th)/He dating, whether as an absolute chronometer or for thermochronometry, and regardless of the material being analyzed, careful consideration of the consequences of the MeV energies of α decay is critical. Energetic decay and associated long α-stopping distances greatly limit the materials suitable for He dating and present perhaps the greatest impediment to determination of ages with the sub-percent accuracy and precision of which isotopic dating techniques are commonly capable. Given its unique and as yet inadequately appreciated significance for He dating, this section elaborates the α-stopping phenomenon and considers ways to accommodate and/or correct for it. α particles of the U and Th series are sufficiently energetic that they travel about 20 μm through solid matter before coming to rest. As a result, α decay induces a spatial separation between parent and daughter nuclei. To the extent that the parent abundance within a rock is not uniform, e.g., because distinct crystals are present, spatial variability in daughter/parent ratio will result. This unavoidably leads to the erroneous appearance of He age heterogeneity within the rock, with regions or crystals that experience a net import of α particles being "too old" and regions with a net export being "too young." The effect can be substantial in small crystals and will lead to erroneous "ages" if not accommodated in some fashion.

Each α decay within the U and Th series has a characteristic energy and hence a characteristic and well-known (Ziegler 1977) stopping distance within a given material. As a result, an α particle will come to rest on the surface of a sphere centered on the site of the parent nucleus and with a radius equivalent to the stopping distance. As shown in Figure 5, there are three relevant outcomes of α decay in a crystal. If the parent nucleus is located more than the stopping distance away from the edge of the crystal, then that α particle will be retained within the crystal regardless of the trajectory the particle takes. For a parent nucleus lying within one stopping distance of the crystal boundary there is some probability that the α particle will be ejected. The ejection probability rises to a maximum of about 50% on the grain edge, e.g., if the edge of the grain is considered a flat surface, then in a statistical sense half of the trajectories taken by the α particle will lead to ejection, and half to retention. The increase in ejection with proximity to grain edge is shown schematically in the lower part of Figure 5. It is also important to consider that decay occurring outside of the crystal can lead to implantation into the crystal of interest.

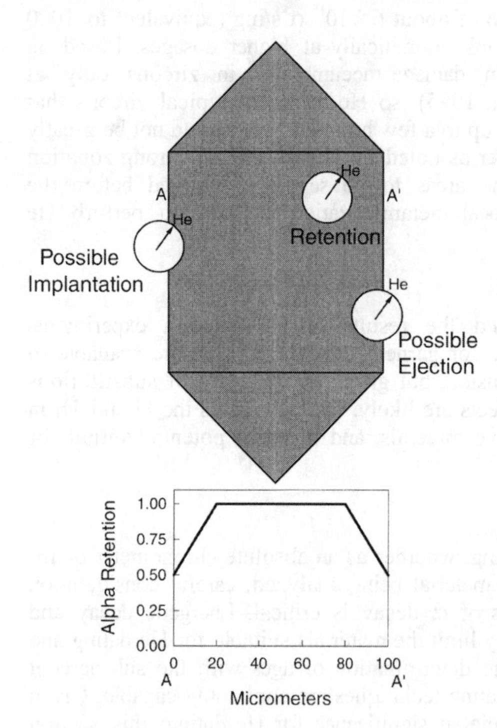

Figure 5. The effects of long α–stopping distances on He retention. The upper figure illustrates the three relevant possibilities within a schematic crystal: α retention, possible α ejection, and possible α implantation. The center of the circle denotes the site of the parent U or Th nuclide, and the edge of the white circle labeled He indicates the locus of points where the α particle may come to rest; the arrow indicates one possible trajectory. The lower plot shows schematically how α retention changes from rim to core to rim along the path A-A'; exact equations defining the shape of this curve as a function of grain size were given by Farley et al. (1996).

Because this phenomenon is restricted to the outermost ~20 μm of a crystal, chemical or mechanical removal of the grain surface will eliminate the effect. However, as discussed above, the He diffusion domain in some minerals is the grain itself (Bahr et al. 1994; Reiners and Farley 1999; Farley 2000). Because the grain edge is the site of diffusional helium loss, it will tend to have a lower He concentration than the grain interior as a result of diffusive transport. Removal of the outermost portion will bias the age of the remaining crystal toward erroneously high values. As discussed by Reiners and Farley (1999), the magnitude of the effect is a sensitive function of the crystal's thermal history. For some applications, such as dating of quickly cooled minerals (e.g., from tephras), or large crystals in which the diffusion gradient is on a longer length-scale than α-ejection/implantation, this approach may be appropriate, but in general erroneous ages will result from surface removal when the diffusion and α-ejection boundaries coincide.

As an alternative, Farley et al. (1996) developed a quantitative model for correcting He ages for the effects of long α stopping distances based on measured grain geometry and size. Several assumptions are required:

1) Implantation from the surrounding matrix is insignificant, so only α ejection need be

considered. In most minerals of interest for He dating, the parent nuclide concentration contrast between the mineral being dated and the host rock is so large that implantation is trivial compared to *in situ* produced He. In some cases, such as very U-Th poor apatites, this assumption may be violated.

2) The distribution of U and Th in the crystal being dated must be specified. Because the distribution will in general *not* be known, Farley et al. (1996) presented results assuming a homogeneous distribution and discussed the error introduced by various types of zonation.

In accord with intuition, the model shows that the two most important variables controlling the total fraction of alphas retained in a crystal (the "F_T" parameter, that is, the factor by which the measured age must be divided to obtain the "α-ejection-corrected" age) are the surface to volume ratio (β) of the crystal, and the α stopping distance. Crystals with small β have less "skin" affected by α ejection, and hence require smaller corrections. While each decay in the U and Th chains has a characteristic stopping distance, the mean F_T obtained by modeling each decay separately does not differ substantially from simply using a single mean stopping distance for each parent. However, because stopping distances vary significantly with the density and to a lesser extent the composition of the stopping medium, it is necessary to use different stopping distances for different minerals. Analytical and Monte Carlo results were presented that allow computation of F_T from measured dimensions for several simple grain geometries including a sphere, a cylinder, and a cube (Farley et al. 1996).

Table 1 expands on the work of Farley et al. (1996) by presenting a second-order polynomial relating F_T to β for both U and Th series decay for two systems of particular interest: a hexagonal prism of apatite density, and a tetragonal prism of zircon density. In both cases the prism terminations were ignored. The coefficients were obtained by Monte Carlo modeling as described previously (Farley et al. 1996).

A typical result of this modeling is shown in Figure 6 in which F_T is plotted as a function of prism width (defined here to be the distance between opposed apices) for an apatite hexagonal prism of length/width ratio of 3. For relatively large grain widths, F_T values are fairly constant, in the range 0.8 to 0.9, decreasing only slowly with decreasing width. However the curve becomes increasingly steep for widths less than 80 μm. The message from this plot is that in general the largest grains will have the least uncertainty on the correction, and that typical corrections for small accessory minerals will be in the range 0.65 (or even less) to 0.9.

Figure 6 and Table 1 show that α retentivity is slightly higher for ^{238}U than for ^{232}Th, reflecting the higher mean energy of α decays in the ^{232}Th series. This distinction is relatively subtle but can be accommodated by computing a weighted mean of the F_T values for U and Th, where the weighting factor is the fraction of He derived from each parent. Specifically,

$$^{Mean}F_T = a_{238}\,{}^{238U}F_T + \left(1 - a_{238}\right){}^{232Th}F_T \tag{3}$$

Because ^{235}U and ^{232}Th have very similar decay energies, the above expression associates the He from these two parents with a single F_T value. The fraction of He derived from ^{238}U (a_{238}) can be calculated exactly from Equation (1), or approximated from the measured Th/U ratio for integration periods of less than ~200 Myr as:

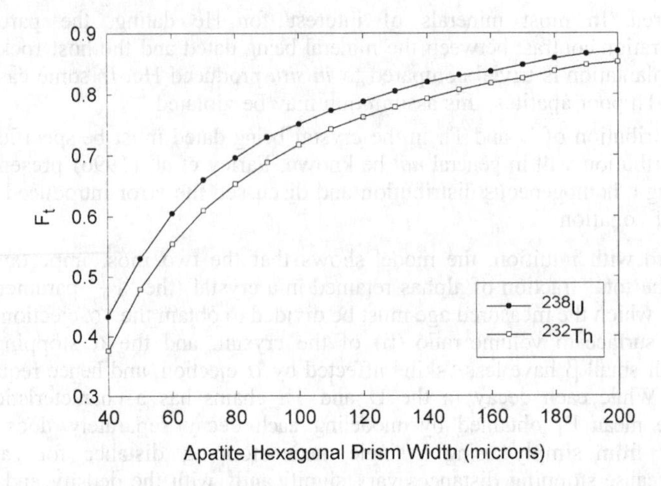

Figure 6. The effects of α-ejection on He retention in an apatite hexagonal prism. F_T is the total fraction of alphas retained within the crystal, assumed here to have a length/width ratio of 3. The ^{238}U and ^{232}Th series lie on slightly different curves because of differences in decay energy; ^{235}U would plot on top of the ^{232}Th curve. The curves were calculated using the equation and coefficients in Table 1.

$$A_{238} = (1.04 + 0.245\ (Th/U))^{-1} \tag{4}$$

This model has been implemented in many different studies using the following procedure:

1) Grains to be analyzed are selected on the basis of good crystal morphology. In many cases it is necessary to search for grains that match the ideal geometry as closely as possible; for example, we have found that in some rocks a substantial fraction of apatites have a tabular habit that is not well-approximated by a hexagonal prism model. Grains to be analyzed together as a single aliquot are selected to be of similar size.
2) Grains are measured using a reticule in a binocular microscope. In the case of apatite, the prism diameter and length are measured.

Table 1. Geometry and mineral-specific fit parameters for F_T calculation.

Geometry-Mineral		a_1	a_2
Apatite hexagonal prism	^{238}U series	-5.13	6.78
	^{232}Th series	-5.90	8.99
Zircon tetragonal prism	^{238}U series	-4.31	4.92
	^{232}Th series	-5.00	6.80

$F_T = 1 + a_1\beta + a_2\beta^2$. β is the surface to volume ratio; for a hexagonal prism given by $\beta = (2.31L+2R)/(RL)$ where R is half the distance between the opposed apices and L is the length. For a tetragonal prism $\beta = (4L+2W)/(LW)$ where L is the length and W is the width of the prism. ^{235}U parameters are essentially identical to ^{232}Th.

3) An F_T value is computed for the grain based on the grain's dimensions and geometry and the α ejection model described above. In the case of apatite, many grains break perpendicular to the c-axis during sample preparation, such that measured lengths are shorter than existed in the rock. As demonstrated by Farley et al. (1996), F_T is not very sensitive to the original length; at Caltech it is standard practice to multiply the observed length of all grains by a factor of 1.5 to account for breakage unless the grains are demonstrably unbroken.
4) The mean F_T of the entire population of grains in the aliquot is computed, weighting each grain by its mass contribution to the aliquot. The weighting is based on observed grain dimensions. This weighting implicitly assumes that grains contribute helium in proportion to their mass. If grains of very different sizes have very different U or Th contents, this weighting will be incorrect. Hence the effort in step 1 to pick grains of a common size. Indeed we observe that many apatite and zircon samples show large grain-to-grain heterogeneity in U and Th concentration, presumably reflecting the compositional evolution of magmas during crystallization. Thus this precaution is important.

This technique works well for minerals that retain their characteristic size and morphology when separated from the host rock, as both apatite and zircon commonly do. It is less successful for minerals that shatter, leaving no indication of original grain morphology or dimensions. This problem is particularly acute for titanite, which in our experience is very commonly broken during sample preparation.

Experiments at Caltech show that F_T corrections in the range of ~0.65 to ~0.85 can be reproduced by individual observers to better than a few percent. Systematic observer bias in F_T determinations has not been observed. At smaller grain sizes, where the variation in F_T with grain size becomes increasingly steep (Fig. 6), errors in F_T will become increasingly large. We avoid analyzing grains with $F_T < 0.65$.

The biggest problem with this approach to correcting α ejection effects is the assumption of a uniform parent nuclide distribution; U- and Th-zoned crystals violate this assumption. It is useful to consider how large an error zonation can induce in the final computed He age. Consider an apatite hexagonal prism of typical dimensions yielding an F_T value of 0.75. If instead of a homogenous distribution, 100% of the U and Th is located more than one stopping distance from the grain boundary, then the true fraction of alphas retained would be unity, and the model F_T-corrected age would be 33% greater than the true age. Alternatively, if all of the parent is located on the prism rim, then the fraction of alphas retained would by ~0.5. In this case the model F_T-corrected He age would be 33% younger than the true age. For all other degrees and styles of zonation, the error introduced will be between these two bounds, so the maximum error is ±33%. Farley et al. (1996) considered several different scenarios and concluded that only extreme zonation will produce large errors in F_T correction. For example, an 11-fold linear decrease in parent abundance from core to rim yielded an F_T value differing from the homogeneous case by just 10%.

The degree to which U and Th are zoned in phases relevant for He dating has not been extensively explored, but in zircons and to a lesser extent apatites zonation is generally thought to be fairly common. Of greatest concern are situations in which the parent concentration of the outermost 20 μm is dramatically different from the remainder of the grain; oscillatory zonation is no more problematic than monotonic zonation that affects the surface in a comparable fashion. In the case of apatite, rare earth elements continue to diffuse even at fairly low temperatures (Watson et al. 1985). If this observation applies to U and Th as well, then slowly cooled apatites may in general have fairly uniform parent distributions. Further documentation of the magnitude and style of zonation in phases of interest such as apatite, titanite and zircon would be useful for evaluating how large an uncertainty to place on the ejection-correction component of He age determinations.

A final point to consider is that because α ejection and diffusion occur at the same boundary—at least in specular hematite (Bahr et al. 1994), apatite (Farley 2000) and titanite (Reiners and Farley 1999), the He gradient that drives diffusion will be more rounded than diffusion alone would produce. For example, in a quickly cooled sample that has experienced α ejection, the He concentration profile would look similar to Figure 2. In contrast, without α ejection, there would be a step function right at the grain boundary. As a result, diffusive transport will be slightly reduced; for most cooling scenarios the effect is equivalent to just a few degree increase in closure temperature (Farley 2000). Note however that during laboratory diffusion experiments, this rounding may significantly bias estimates of diffusion coefficients made when the first few percent of the helium is released (Farley 2000).

Analytical procedures, accuracy, precision and mineral standards

Once grains have been selected for analysis and measured for α ejection correction, a (U-Th)/He age determination involves ^{4}He extraction and measurement, and U and Th measurement on either the same or a separate aliquot. Analytical techniques, especially for the small amounts of ^{4}He obtained from most samples, have evolved rapidly in the last five years, and so are described in some detail here.

Helium extraction. Early work by us and others involved the analysis of separate aliquots for He and for U and Th. While this approach permits straightforward vacuum fusion of the grains being analyzed for helium, it has the serious drawback that aliquots must be large enough to average out any grain-to-grain variability in U-Th-He content. Without a detailed study of a given sample it is hard to estimate how large an aliquot will ensure homogeneity but aliquots of at least a few mg are likely required. Preparation of such large aliquots can be very tedious, hence we have developed several techniques for analyzing parent and daughter on the same aliquot. The single-aliquot technique has the additional advantage that the aliquot need never be weighed: the age is computed directly from the *relative* parent and daughter abundances using Equation (1).

Initially we loaded grains for analysis into stainless steel capsules that could be retrieved from the furnace after outgassing of He at temperatures lower than the melting point of the mineral being dated. This technique, described in detail in the Appendix, is particularly well suited to apatite, which can be quantitatively outgassed of ^{4}He in 20 min at 950°C. However the technique is awkward or unsuccessful when high temperatures (above the melting point of the steel capsule) are required for complete He extraction, e.g., from zircon and titanite. Blank levels for the furnace technique average about 0.6 fmol of ^{4}He, sufficiently low for the analysis of aggregates of about ten crystals in typical apatite samples as young as a few million years. However this blank is too high to permit routine single crystal dating of apatite unless the crystal is unusually large or has a He age of more than a few tens of million years.

Laser extraction is a more elegant alternative that has recently been adapted for He dating (House et al. 2000). Even the most refractory minerals can be heated to the melting point using a laser of appropriate wavelength. In addition, laser-based methods have far lower He blanks compared with resistance heating. Unfortunately laser heating of bare mineral grains is known to cause volatilization and loss of U, and in some cases Th, from the fused sample. This presumably occurs when the most intensely heated part of the sample melts, causing it to couple even better with the laser. Local thermal runaway and boiling results. Loss of U or Th precludes analyzing parent and daughter on the same aliquot. For example, Reiners and Farley (1999) found that, even at very low power from a Nd-YAG laser, fusion of bare titanite grains to a glassy bead was associated with substantial U loss. The U is presumably vaporized from the sample and deposited somewhere in the vacuum chamber. A similar result was obtained by Stuart and Persano (1999) upon CO_2 laser heating of bare apatite.

House et al. (2000) overcame this problem by wrapping the sample in a very small (1 mm × 1 mm) Pt foil envelope which is heated by a Nd-YAG laser. The Pt is a sufficiently good heat conductor that uniform heating results, as indicated by a uniform incandescence. Using an optical pyrometer the temperature of the sample can be estimated and controlled with a feedback loop to the laser output. As expected, the laser blanks are far better than those on the resistance furnace; holding an empty Pt-foil envelope at 1350°C for 30 min yields about 0.02 fmol of ^{4}He. Because the technique is far faster (many samples can be loaded in the laser vacuum chamber and pumped simultaneously) and has a lower blank, we now routinely outgas apatites, zircons, and titanites with this technique. The blank is sufficiently low that single crystal dating of nearly all apatite samples is possible. Except for the method by which the sample is heated, the laser procedure follows the resistance heating procedure (see the Appendix).

U and Th are analyzed by isotope dilution inductively-coupled plasma mass spectrometry performed on the same aliquot analyzed for He (Appendix).

Accuracy and precision. Based on reproducibility of pure standard gases and aqueous standard solutions, the overall analytical precision of He ages determined by this procedure should be about 2% (2σ, excluding errors in α-ejection correction) when ages are well above blank levels. Most of this uncertainty arises from the He measurement. In actual practice, we obtain He ages that reproduce to about 6% (2σ), demonstrating some natural variability within grain populations.

Accuracy of He ages is as important as precision for most geologic interpretations. Obtaining accurate U and Th abundances is a routine procedure in, e.g., U-Pb geochronology laboratories. Standard solutions accurate to better than a few per mil can be obtained by dilution of commercial ICPMS standards, which routinely carry a stated accuracy of 1 per mil. Obtaining precise and accurate He aliquots is more challenging, requiring knowledge of both pipette volume and the partial pressure of He in the standard tank. At Caltech the volume of the standard He pipette was determined by manometric comparison to a primary calibrated volume (determined by weighing after filling with water or Hg). He pressure was determined by capacitance manometry corrected for thermal transpiration effects. The accuracy of a He aliquot calibrated in this fashion is probably ~1%.

The He dating method as described above is an absolute dating technique, based on fundamental measured quantities rather than by reference to independently dated mineral standards. In this regard it is different from other low temperature thermochronometers like Ar/Ar and fission track counting. Nevertheless it is useful to have mineral standards to verify both analytical calibrations and the degree to which the system faithfully acts as a chronometer. Unfortunately identifying appropriate standards is not easy, particularly for a low temperature system like He-in-apatite. The key requirement is that the mineral of interest cooled through both He closure and closure of the independent chronometer at the same time. Like apatite fission track standards, appropriate standards for He dating include rapidly cooled volcanic and very shallow plutonic samples. However at least in the case of apatite, He ages are very susceptible to post-eruption diffusive He loss that may be problematic. For example, consider the case of an apatite of 55 μm radius, erupted 30 Ma and immediately buried under 750 m of overburden. Given a mean annual temperature of 15°C and a geothermal gradient of 27°C/km, this sample will sit at 35°C.

Modeling based on He diffusivity data shown in Figure 2 indicates that such a sample would yield a He age of just 27.6 Ma, almost 10% lower than the "expected age" obtained from other techniques such as Ar/Ar and fission track counting. If such a sample were recently exposed and used as a He age standard, problems would clearly result. Thus standards for He dating, at least for apatite, must be carefully selected. No consensus yet exists on the most appropriate specimens for such standards.

He ages are available for only a few "standards." At Caltech the mean of several dozen age determinations on Durango apatite is 32.0 Ma, with a standard deviation of the population of 1 Ma (Farley 2000; House et al. 2000 and Farley, unpublished). This age is in excellent agreement with the accepted age of this apatite (31.6±1 Ma) (Jonckheere et al. 1993). Warnock et al. (1997) reported a slightly younger He age (27.5 Ma) for this apatite, while Wolf et al. (1996b) reported a slightly older value. Note that Durango apatite crystals are so large (cm) that the α ejection-affected rim can be excluded from analysis, and furthermore that the closure temperature of these large grains means that even substantial post-eruptive burial is not an issue. At present Durango apatite is perhaps the most appropriate age standard, although its unusual size and chemistry (e.g., high Th/U ratio) make it atypical of most unknowns to be dated. Reiners and Farley (1999) reported a titanite He age for Fish Canyon Tuff of 30.1±2 Ma, while House et al. (2000) reported a value of 27.9±1 Ma for aliquots of the same material. Unpublished work at Caltech reveals significantly lower He ages and large scatter from Fish Canyon apatite, possibly because of burial. Reiners and Farley (2001a) reported a zircon He age of 28.1±2.8 for Fish Canyon Tuff. These values are similar to the accepted age of 28.5±0.1 for the Fish Canyon Tuff (Schmitz and Bowring 2001).

Mineral inclusions. In our experience at Caltech the single biggest difficulty in He dating of apatite is the presence of small U-Th-rich inclusions within the dated grains. This difficulty was first noted by Lippolt et al. (1994) and was further described by House et al. (1997). The most common inclusions are zircon and monazite, but we have occasionally encountered xenotime and allanite. We have also observed inclusions of quartz, feldspar, pyrite and graphite(?), but these are unlikely to carry sufficient U and Th to be a problem.

Inclusions cause several different problems for (U-Th)/He dating. Although some effect might be expected from differing He closure temperatures of host and inclusion, for inclusions smaller than ~15 μm diameter, essentially all αs will be ejected into the host mineral, so only the He diffusion characteristics of the host are important (Farley et al. 1996). However, localization of U and Th may bias the α ejection effect and may also modify the diffusion behavior by changing the He concentration gradient. More importantly, many inclusion phases, especially zircon, survive the dissolution technique we use on apatite (see the Appendix). Hence the inclusions will contribute He to the analysis, but not U and Th. As a result, some He will be "parentless," and anomalously old ages will result. These anomalously old ages will tend to be irreproducible because the inclusions are not in equal abundance in the analyzed aliquots.

In many cases inclusions in apatite can be detected during the grain selection process. At Caltech apatites to be dated are inspected at ~120 × under a binocular microscope using transmitted light and crossed-polarizers. When the apatites are taken to extinction even tiny inclusions of phases like zircon stand out—in some cases entire grains appear to be shot full of inclusions. These grains are easily removed prior to analysis. In rare cases this technique has been found inadequate, usually because the inclusions are oriented parallel to the *c*-axis (possibly from exsolution of monazite) and are extinct at the same time as the apatite host. In these cases the re-extract test (see the Appendix) and age irreproducibility are sufficient to identify problem samples.

Unlike apatite, microscopic detection of inclusions is very difficult in dark or opaque minerals such as hematite or hornblende, and in minerals with high indices of refraction such as zircon and titanite. Insufficient evidence exists to evaluate the prevalence of problem inclusions in such phases.

INTERPRETATION OF HE AGES AND EXAMPLES

He cooling ages

The simplest way to interpret He cooling ages is to associate them with the time of cooling of the sample through a specific closure temperature, as defined by Dodson (1973). However, this approach is restricted to t-T paths involving monotonic cooling. A more robust way to interpret He ages has been described by Wolf et al. (1998), who presented a numerical forward model which yields a He age given an arbitrary t-T path and measured diffusivity parameters. This is analogous to computational models for interpreting Ar cooling ages (Kelley 2002, this volume).

A wide variety of t-T paths can yield a given He age, so a unique interpretation of a single He age is not generally possible without additional information. Age-elevation profiles can be used to more tightly constrain cooling histories (Wolf et al. 1998), as can comparison with other techniques, such as fission track methods (Stockli et al. 2000). For the latter to be successful, further efforts must be made to ensure compatibility of the thermal calibrations of the dating techniques. For example, a recent study suggests that cooling histories derived from apatite fission track-length models (Gallagher 1995) are inconsistent with apatite He ages (House et al. 1999). Such a discrepancy is not surprising given the comparative insensitivity of fission tracks to temperatures at which He ages are extremely sensitive.

There are two additional ways to obtain more detailed cooling information from a *single* sample: analysis of He ages of crystals of different sizes, and examination of the He concentration profile within an individual crystal. Consider three very different cooling histories: instantaneous cooling at 7 Ma, cooling at 10 C/Myr for the past 10 Myr, and a 100-Myr isothermal period at 65°C. All three histories yield apatite He ages of ~7 Ma in grains of 65 μm radius. How can these histories be distinguished?

1) *Age–grain size relationships*. Because the diffusion domain in apatite is the grain itself (Farley 2000), as the grain size decreases, the He diffusivity increases (i.e., “a” in D/a^2 varies as the minimum grain dimension). Grains of different sizes are sensitive to different temperatures and have different closure temperatures. As shown in Figure 7, the three histories yield distinctly different He age–grain size relationships. In the case of instantaneous cooling, all He ages are the same because cooling through the range of closure temperatures happens instantly. In the sample cooling at 10°C/Myr, the ~12°C variation in closure temperature from 35 to 105 μm radius causes He ages to vary by ~1.2 Myr. In the isothermal case, He ages achieve equilibrium between diffusive loss and production, and hence are enormously sensitive to grain size. Provided high precision ages can be obtained on grains spanning a fairly large size range, this technique can be used to narrow down allowed thermal histories. Such work also presupposes that the apatites lack fast sub-grain He diffusion pathways such as cracks; this may not always be the case (see Kelley 2002, this volume, for a discussion of fast pathways for Ar loss). In apatites from the Bighorn Mountains of Wyoming, Reiners and Farley (2001b) found a strong correlation between grain size and He age which they attributed to this effect and used to elaborate the thermal history of the mountain range.

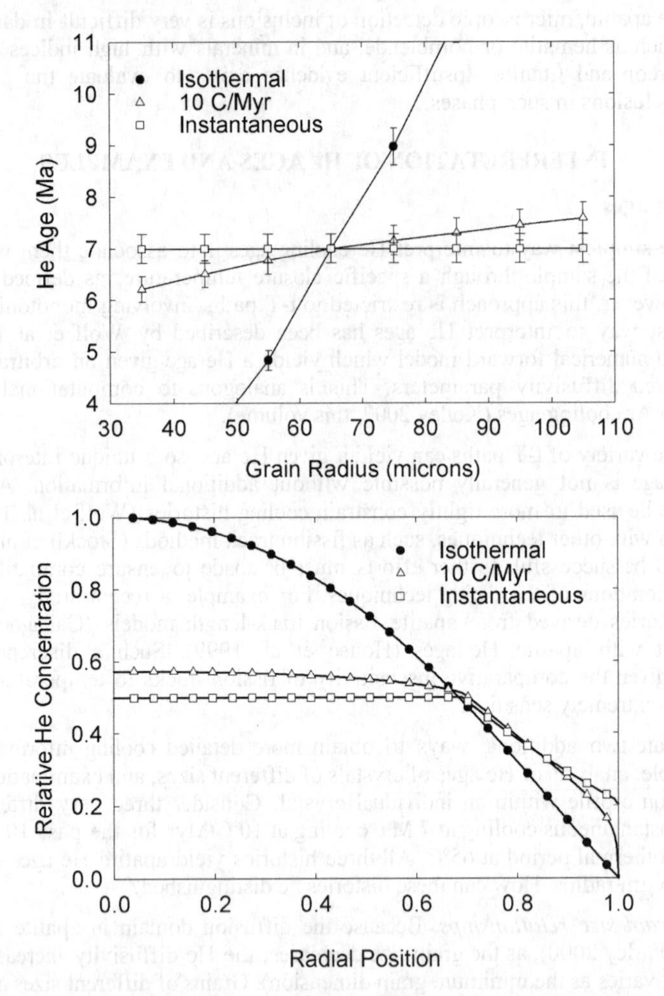

Figure 7. Additional cooling history information from grain size variation of He age (upper panel) and He concentration gradient (lower panel). Three different cooling histories—instantaneous cooling at 7 Ma, constant cooling from 10 Ma to present at 10°C/Myr, and isothermal for 100 Myr at 65ºC, yield the same helium age of ~7 Ma for apatite grains of 65 μm radius. The upper panel shows that the three histories yield distinct relationships between He age and grain size; the ability to actually distinguish among the histories depends on the range of grain sizes available for analysis, as well as analytical precision. The lower panel shows that each history also yields a characteristic He concentration profile within an apatite grain that might be distinguished using step-heating experiments. Ages in the upper panel were computed using the apatite diffusion parameters listed in Figure 3; ages were assigned an arbitrary uncertainty of 5%. The concentration profiles in the lower panel were computed for a grain of 65 μm radius. For both panels the α ejection effect was included and corrected for in the computations.

2) *He-concentration profile.* The concentration gradient within a crystal is a sensitive function of the thermal history experienced by the sample. As shown in Figure 7, the three thermal histories yield concentration profiles that are very different. In the case of instantaneous cooling, no diffusional rounding is observed at all (only the α-

ejection effect is present). The sample cooling at 10°C/Myr shows limited diffusional rounding, while the isothermal sample shows the extreme rounding characteristic of a balance between radiogenic production and diffusive loss. In particular note the differences in the profiles over the outermost 5% of the grain. If these profiles could be measured, detailed information on cooling history could be obtained. The shape of the concentration profile can to some extent be deduced from step-heating experiments (Albarede 1978), but it is not yet clear whether such an approach is practical.

Some case studies

Applications of (U-Th)/He thermochronometry have only recently come under investigation, so it is too early to predict how the technique may ultimately be used. Compared to other techniques, the two obvious attractions of the He method are sensitivity to previously inaccessible temperature ranges, and comparatively high precision. By combining He ages on various phases with other methods such as feldspar multidomain thermochronometry (Lovera et al. 1989) and fission track techniques (Gallagher 1995; Gallagher et al. 1998), more detailed and robust cooling histories may be obtained than previously available. In the case of apatite, the He closure temperature of 70ºC is lower than that of any other known technique, so apatite He ages provide unique information on the final stages of cooling. In addition, in some cases He ages provide better age precision and better temperature control than existing techniques. This is particularly true in samples that have experienced recent cooling and hence have young ages.

The following are summaries of several studies that illustrate these applications.

Sensitivity to smaller degrees of exhumation. Compared to apatite fission track techniques commonly used to assess the timing and rate of mountain range exhumation, the apatite (U-Th)/He method is sensitive to temperatures ~25ºC lower, or about 1 km shallower in the crust. As a consequence, smaller degrees of total exhumation are required to expose the thermochronometric record of the exhumation process. In the case of rapid cooling, such exhumation may reveal an exhumed HePRZ, in which the lower break-in-slope indicates the base of the HePRZ and yields the age of the onset of exhumation (see Fitzgerald and Gleadow 1990 for a discussion of the analogous and now well understood concept in fission track dating). For example, as indicated in Figure 6, in a typical geothermal gradient of 20ºC/km the base of the apatite HePRZ will be brought to the surface after ~3.5 km of exhumation. In contrast, ~5 km of exhumation are required to reveal the base of the FTPAZ.

This application has been explored by Stockli et al. (2000) in the Basin and Range province of the Western US. In the White Mountains of Eastern California, He ages decrease with structural paleodepth from 55 Ma to just a few Ma, in precisely the pattern expected for an exhumed HePRZ (Fig. 8) established between ~55 Ma and 12 Ma. Extending for ~1 km below the HePRZ the He ages are invariant at 12 Ma (Fig. 7). This invariance occurs because these initially sub-HePRZ apatites had zero age while the HePRZ developed, then were rapidly exhumed and cooled at 12 Ma. At the base of the section the He ages again decline, recording more recent cooling. Apatite fission track ages capture the 12-Ma event, but only in the structurally deepest samples; were the total exhumation only slightly less, the timing of the event would be very difficult to estimate from fission track ages alone.

Paleotopography. Apatite He ages are sensitive to temperatures found at depths of between 1 and 3 km within the Earth's crust (Fig. 4). At these depths the crustal temperature field is strongly influenced by the position of the free cooling surface, such

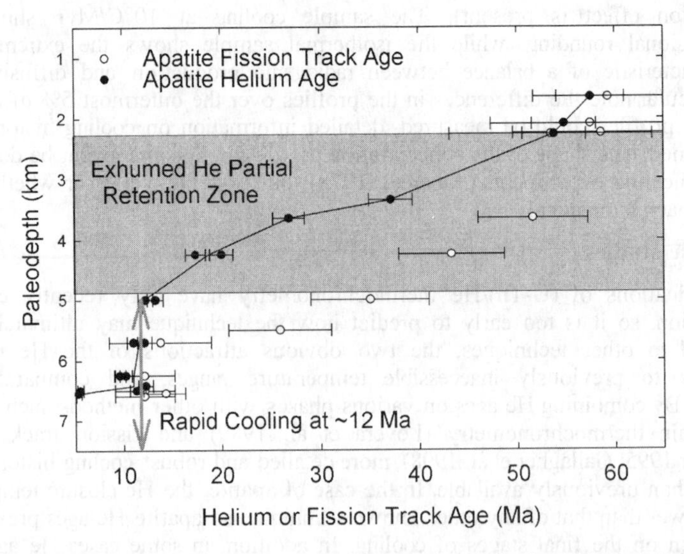

Figure 8. Vertical profile of apatite He and fission track ages in the N. White Mountains of California, after Stockli et al. (2000). He ages plot on a well-defined HePRZ consistent with ~7 km of rapid exhumation at 12 Ma. A similar pattern is seen in the fission track data, but only the deepest samples directly date the exhumation event. Unlike the He ages, the fission track ages do not constrain the total amount of exhumation at 12 Ma. The position of the HePRZ is in excellent agreement with predictions based on laboratory diffusion measurements. Younger He ages at the base of the range document renewed cooling and exhumation.

that isotherms track the overlying topography. The amplitude of relief on the isothermal surfaces depends on the wavelength of the overlying topography, decreasing rapidly with decreasing wavelength. The amplitude also decreases with increasing crustal depth and hence temperature. This phenomenon has been recognized as a complexity for interpreting low temperature cooling ages in terms of exhumation (Stuwe et al. 1994; Mancktelow and Grasemann 1997), but House et al. (1998) turned it to advantage to investigate the history of topography, in the Sierra Nevada of California. These authors proposed that if the wide and deeply incised canyons transverse to the range were an ancient feature, then He ages on an orogen-parallel transect at a constant elevation might be affected. In the limit of exhumation through steady-state topography, as the range is exhumed, rocks at a given elevation lying below deep canyons will cool before rocks lying below major ridges. As a result, He ages would be older in samples presently exposed near canyon bottoms than in samples from the same elevation between canyons.

Figure 9 shows the results obtained by House et al. (1998) on a horizontal transect at 2000-m elevation from the southern and central Sierra Nevada. Apatite He ages in the vicinity of the San Joaquin and Kings rivers are older by almost 20 Myr than samples from the intervening ridge locations. The simplest interpretation of these results is that the modern river drainages are in much the same place as when these He ages were set 60 to 80 million years ago. Modeling by House et al. (1998) suggests that the amplitude of the He age variation is consistent with mean canyon-ridge relief of about 3 km by 80 Ma.

The importance of this work is that it demonstrates that paleotopography, as distinct from exhumation, may be obtained from apatite He cooling ages. While other thermochronometers may also record this effect, because the relief on isotherms declines rapidly with increasing temperature, it is most pronounced in the apatite (U-Th)/He system. As discussed by House et al. (1998), a variety of complications must be considered in choosing a suitable study locality, in developing an appropriate sampling strategy, and in interpreting the results, but such a tool has great potential value in understanding long-term landscape evolution.

High-precision ages even on young samples. In principle (U-Th)/He ages of fairly high precision (few percent) can be obtained even on samples that are very young. This suggests two additional applications for (U-Th)/He dating: absolute dating of very young volcanic units, and obtaining precise cooling histories of recently exhumed rocks. In the case of young volcanic units such as tephras, existing techniques such as ^{14}C, U-series disequilibrium and $^{40}Ar/^{39}Ar$ have limited ability to date samples in the range ~50 kyr to ~1.5 Myr, yet this age range is of considerable interest in a variety of disciplines including hominid evolution and climate history. In U-Th rich minerals such as zircon, He ingrowth is so rapid that high precision ages are possible in samples as young as 50 kyr. For example, 1 mg of zircon with 200 ppm each of U and Th will carry more than 300 times typical helium blank levels using laser extraction in 50 kyr, and a single zircon crystal of ~10 μg mass will exceed blank by 10 times in just 175 kyr.

However, an important issue to be dealt with in tephra dating is that young crystals might acquire He prior to establishment of secular equilibrium in the ^{238}U system. In particular zircons likely crystallize with a significant $^{230}Th/^{238}U$ deficit. As a result, each decay of ^{238}U initially produces less than 8 α particles (Eqn. 1) until secular equilibrium is achieved. The magnitude of the discrepancy introduced by secular disequilibrium (relative to Eqn. 1) varies systematically with eruption age, the Th/U ratio of the crystal, and the magma residence time prior to eruption (Farley et al. 2002). For example, in the case of a typical zircon, the ^{230}Th deficit would cause He ages computed from Equation (1) to underestimate the eruption age by 50% in very young samples (10 kyr) declining to <10% by 1 Myr. Farley et al. (2002) presented equations for accommodating secular disequilibrium, and demonstrated that analysis of samples from a single eruption spanning a range of Th/U ratio can be used to identify and eliminate secular disequilibrium effects on computed eruption ages. These authors presented measurements of apatite and zircon separates of a New Zealand Tephra indicating an eruption age of 330 ± 10 kyr. Although this application is in its infancy, even with the uncertainties introduced by secular disequilibrium the technique appears promising.

In a similar fashion, cooling ages obtained from the (U-Th)/He system compare favorably in precision, especially at young ages, with alternative techniques such as apatite fission track counting. For example, Spotila et al. (2001) showed that apatite He ages of ~1.5 Ma can be obtained with a precision of 100 kyr (2σ) on a rapidly exhumed crustal sliver lying between two strands of the San Andreas fault where it passes through San Gorgonio Pass in Southern California. In a 1 km vertical transect, 8 samples yielded He ages ranging from 1.39±0.10 (2σ) to 1.64±0.11 Ma. The high precision on these ages constrains not only the onset of rapid exhumation of the sliver, but places fairly restrictive limits on the exhumation rate. Precision on both of these geologic quantities is critical for establishing the mechanism responsible for exhumation along this major strike-slip fault.

FUTURE PROSPECTS

Work over the past five years suggests a variety of techniques and applications for (U-Th)/He dating. Some of these, such as apatite He thermochronometry, have been applied in a range of settings and have yielded geologically reasonable and scientifically interesting results. Other techniques, while tantalizing, are not yet sufficiently developed

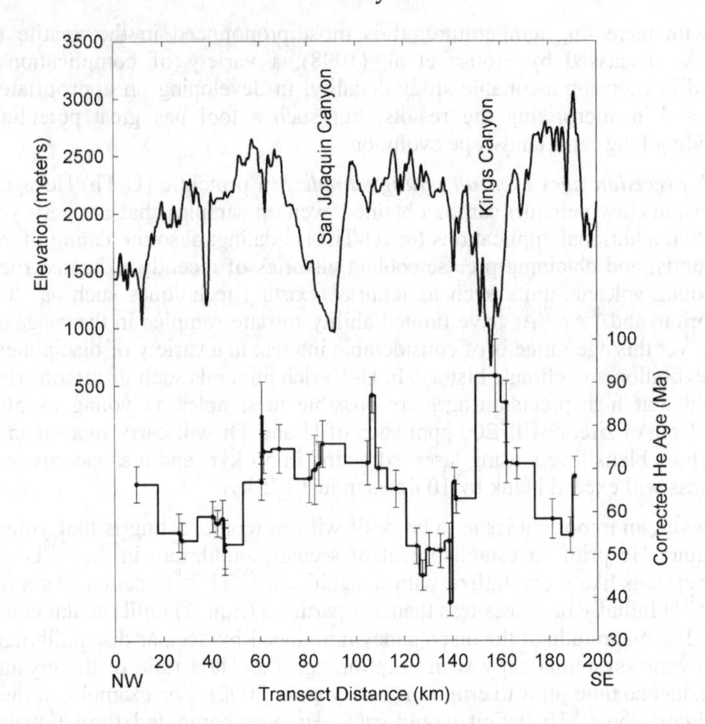

Figure 9. Apatite He ages on an orogen-parallel 2000-m elevation transect in the Sierra Nevada, compared with modern elevation profile, after House et al. (1998). The general correspondence between the location of deep and wide valleys (San Joaquin and Kings Canyons) and higher He ages suggests that He ages were influenced by perturbations in isotherms associated with paleocanyons located approximately where they are found today. These data suggest relief of ~3 km at ~80 Ma, i.e., the Sierra Nevada were a large mountain range by Late Cretaceous time. Apatite helium ages were corrected for small variations in vertical and horizontal position to allow comparisons on a fixed elevation traverse at a constant distance from the tilt axis of the range (see House et al. 1998 for correction details).

for a critical evaluation. The coming decade will provide a much clearer view of which applications and techniques "work," and which suffer insurmountable problems.

Beyond the (U-Th)/He method itself, it will become increasingly important to link these new methods to existing techniques. For example, detailed cooling histories through the temperature range ~110 to 45ºC should be possible using combined apatite fission track (Gallagher et al. 1998) and (U-Th)/He methods. Similarly, titanite He ages can logically be used to evaluate and refine cooling histories derived from K-feldspar $^{40}Ar/^{39}Ar$ multidomain modeling (Lovera et al. 1989). Before these linkages can be made it is critical to establish that the various techniques are accurately intercalibrated, an exercise that will likely require both laboratory investigations and analysis of well-understood natural systems.

ACKNOWLEDGMENTS

Many members of my research group, especially M. House, P. Reiners, D. Stockli, and R. Wolf, contributed substantially to the ideas and content of this paper. This work was supported by the National Science Foundation and by a fellowship award from the David and Lucille Packard Foundation. Fin Stuart, Raphael Pik and Tibor Dunai provided helpful reviews.

REFERENCES

Albarède F (1978) The recovery of spatial isotope distributions from step-wise degassing data. Earth Planet Sci Lett 39:387-397

Bahr R, Lippolt HJ, Wernicke, RS (1994) Temperature-induced ^{4}He degassing of specularite and botryoidal hematite: A ^{4}He retentivity study. J Geophys Res 99:17695-17707

Bender M (1973) Helium-Uranium dating of corals. Geochim Cosmochim Acta 37:1229-1247

Coyle D, Wagner, G (1998) Positioning the titanite fission-track partial annealing zone. Chem Geol 149:117-125

Damon PE, Kulp JL (1957) Determination of radiogenic helium in zircon by stable isotope dilution technique. Trans Roy Soc Edinburgh 38:945-953

Damon PE, Green WD (1963) Investigations of the helium age dating method by stable isotope dilution technique, *In* Radioactive Dating. Vienna, IAEA, p 55-69

Dodson MH (1973) Closure temperatures in cooling geological and petrological systems. Contrib Mineral Petrol 40:259-274

Dunai T, Roselieb, K. (1996) Sorption and diffusion of helium in garnet; implications for volatile tracing and dating. Earth Planet Sci Lett 139:411-421

Fanale FP, Kulp JL (1962) The helium method and the age of the Cornwall, Pennsylvania magnetite ore. Econ Geol 57:735-746

Farley KA, Wolf RW, Silver LT (1996) The effects of long alpha-stopping distances on (U-Th)/He ages. Geochim Cosmochim Acta 60:4223-4229

Farley KA (2000) Helium diffusion from apatite: General behavior as illustrated by Durango fluorapatite. J Geophys Res 105:2903-2914

Farley KA, Kohn BP, Pillans B (2002) The effects of secular disequilibrium on (U-Th)/He systematics and dating of quaternary volcanic ziron and apatite. Earth Planet Sci Lett (in press)

Fechtig H, Kalbitzer S (1966) The diffusion of argon in potassium bearing solids, *In* Schaeffer OA, Zähringer J (eds) Potassium-Argon Dating: Heidelberg, Springer, p 68-106

Ferreira M, Macedo R, Reynolds J, Riley J, Rowe M (1975) Rare gas dating. II. Attempted U-He dating of young volcanic rocks from the Madiera Archipelago. Earth Planet Sci Lett 25:142-150

Fitzgerald P, Gleadow A (1990) New approaches in fission track geochronology as a tectonic tool: examples from the Trans-Antarctic mountains. Nuclear Tracks and Radiation Measurement 17:351-357

Gallagher K (1995) Evolving temperature histories from apatite fission-track data. Earth Planet Sci Lett 136: 21-435

Gallagher K, Brown R, Johnson C (1998) Fission track analysis and its applications to geological problems. Ann Rev Earth Planet Sci 26:519-572

Graham DW, Jenkins WJ, Kurz MD, Batiza R (1987) Helium isotope disequilibrium and geochronology of glassy submarine basalts. Nature 326:384-386

Green P, Duddy I, Gleadow A, Tingate P, Laslett G (1986) Thermal annealing of fission tracks in apatite. 1. A qualitative description. Chem Geol 59:237-253

House MA, Wernicke BP, Farley KA, Dumitru T (1997) Cenozoic thermal evolution of the central Sierra Nevada from (U-Th)/He thermochronometry. Earth Planet Sci Lett 151:167-179

House MA, Wernicke BP, Farley KA (1998) Dating topography of the Sierra Nevada, California, using apatite (U-Th)/He ages. Nature 396:66-69

House MA, Farley KA, Kohn BP (1999) An empirical test of helium diffusion in apatite: Borehole data from the Otway basin, Australia. Earth Planet Sci Lett 170:463-474

House MA, Farley KA, Stockli D (2000) Helium chronometry of apatite and titanite using Nd-YAG laser heating. Earth Planet Sci Lett 183:365-368

Hurley PM (1952) Alpha ionization damage as a cause of low He ratios. EOS Trans Am Geophys Union 33:174-183

Hurley PM (1954) The helium age method and the distribution and migration of helium in rocks. *In* Nuclear Geology. Faul H (ed) John Wiley & Sons, New York, p 301-329

Jonckheere R, Mars M, Van den haute P, Rebetez M, Chambaudet A (1993) L'apatite de Durango (Mexique): Analyse d'un minéral standard pour la datation par traces de fission. Chem Geol 103: 141-154

Kelley SP (2002) K-Ar and Ar-Ar dating. Rev Mineral Geochem 47:785-818

Ketcham RA, Donelick, RA, Donelick, MB (2000) AFTSolve: A program for multi-kinetic modeling of apatite fission-track data. Geol Mater Res 2:1-32

Leventhal JS (1975) An evaluation of the U-Th-He method for dating young basalts. J Geophys Res 80:1911-1914

Lippolt HJ, Weigel E (1988) ^{4}He diffusion in ^{40}Ar retentive minerals. Geochim Cosmochim Acta 52: 1449-1458

Lippolt HJ, Wernicke RS, Boschmann W (1993) ^{4}He diffusion in specular hematite: Phys Chem Min 20:415-418

Lippolt HJ, Leitz M, Wernicke RS, Hagedorn B (1994) (U+Th)/He dating of apatite: Experience with samples from different geochemical environments. Chem Geol 112:179-191

Lovera O, Richter F, Harrison T (1989) The ^{40}Ar/^{39}Ar thermochronometry for slowly cooled samples having a distribution of diffusion domain sizes. J Geophys Res 94:17917-17935

McDougall I, Harrison TM (1999) Geochronology and thermochronology by the ^{40}Ar/^{39}Ar method. Oxford University Press, New York.

Mancktelow NS, Grasemann B (1997) Time-dependent effects of heat advection and topography on cooling histories during erosion. Tectonophysics 270:167-195

Niedermann S (2002) Cosmic-ray-produced noble gases in terrestrial rocks as a dating tool for surface processes. Rev Mineral Geochem 47:731-784

Reiners, PW, Farley KA (1999) Helium diffusion and (U-Th)/He thermochronometry of titanite. Geochim Cosmochim Acta 63:3845-3859

Reiners PW, Brady R, Farley KA, Fryxell J, Wernicke BP, Lux, D (2000) Helium and argon thermochronometry of the Gold Butte Block, south Virgin Mountains, Nevada. Earth Planet Sci Lett 178:315-326

Reiners PW, Farley KA (2001a) (U-Th)/He thermochronometry of zircon. Initial results from the Gold Butte Block, Nevada and Fish Canyon Tuff. Tectonophysics (in press)

Reiners PW, Farley KA (2001b) Influence of crystal size on apatite (U-Th)/He thermochronology: an example from the Bighorn mountains, Wyoming. Earth Planet Sci Lett 188:413-420

Schmitz MD, Bowring SA (2001) U-Pb zircon and titanite systematics of the Fish Canyon Tuff: an assessment of high-precision U-Pb geochronology and its application to young volcanic rocks. Geochim Cosmochim Acta 65:2571-2587

Spotila JA, Farley KA, Yule JD, Reiners PW (2001) Near-field convergence along the San Andreas fault zone in southern California, based on exhumation constrained by (U-Th)/He dating. J Geophys Res 106:26731-26746

Stockli DF, Farley KA, Dumitru T (2000) Calibration of the (U-Th)/He thermochronometer on an exhumed fault block, White Mountains, California. Geology 28:983-986

Strutt R (1905) On the radio-active minerals. Proc Roy Soc London 76:88-101

Stuart FM, Persano C (1999) Laser melting of apatite for (U-Th)/He chronology: Progress to date. EOS Trans Am Geophys Union 80:F1169

Stuwe K, White L, Brown R (1994) The influence of eroding topography on steady-state isotherms—application to fission track analysis. Earth Planet Sci Lett 124:63-74

Toyoda S, Ozima M (1988) Investigations of excess ^{4}He and ^{40}Ar in beryl by laser extraction technique. Earth Planet Sci Lett 90:69-76

Trull TW, Kurz MD, Jenkins WJ (1991) Diffusion of cosmogenic ^{3}He in olivine and quartz: implications for surface exposure dating. Earth Planet Sci Lett 103:241-256

Turekian K, Kharkar D, Funkhouser J, Schaeffer O (1970) An evaluation of the U-He method of dating bone. Earth Planet Sci Lett 7:420-424

Warnock AC, Zeitler PK, Wolf RA, Bergman SC (1997) An evaluation of low-temperature apatite U-Th/He thermochronometry. Geochim Cosmochim Acta 61:5371-5377

Watson E, Harrison T, Ryerson F (1985) Diffusion of Sm, Sr, and Pb in fluorapatite. Geochim Cosmochim Acta 49:1813-1823

Wernicke RS, Lippolt HJ (1992) Botryoidal hematite from the Schwarzwald (Germany): heterogeneous uranium distributions and their bearing on the helium dating method. Earth Planet Sci Lett 114: 287-300

Wernicke RS, Lippolt HJ (1994a) ^{4}He age discordance and release behavior of a double shell botryoidal hematite from the Schwarzwald, Germany. Geochim Cosmochim Acta 58:421-429

Wernicke RS, Lippolt HJ (1994b) Dating of vein specularite using internal (U+Th)/^{4}He isochrones. Geophys Res Lett 21:345-347

Wolf RA, Farley KA, Silver LT (1996a) Assessment of (U-Th)/He thermochronometry: the low-temperature history of the San Jacinto Mountains, California. Geology 25:65-68

Wolf RA, Farley KA, Silver LT (1996b) Helium diffusion and low-temperature thermochronometry of apatite. Geochim Cosmochim Acta 60:4231-4240

Wolf RA (1997) Development of the (U-Th)/He Thermochronometer. PhD Dissertation, California Institute of Technology, Pasadena, California

Wolf RA, Farley KA, Kass DM (1998) A sensitivity analysis of the apatite (U-Th)/He thermochronometer. Chem Geol 148:105-114

Yamada R, Tagami T, Nishimura S, Ito H (1995) Annealing kinetics of fission tracks in zircon—an experimental study. Chem Geol 122:249-258

Young E, Myers A, Munson E, Conklin N (1969) Mineralogy and geochemistry of fluorapatite from Cerro de Mercado, Durango, Mexico: U S Geol Surv Prof Paper 650-D:D84-D93

Zeitler PK, Herczig, AL, McDougall I, Honda M (1987) U-Th-He dating of apatite: A potential thermochronometer. Geochim Cosmochim Acta 51:2865-2868

Ziegler JF (1977) Helium: Stopping powers and ranges in all elemental matter. Pergamon, New York

APPENDIX: ANALYTICAL TECHNIQUES

Current techniques for (U-Th)/He dating at Caltech are as follows:

1. For resistance heating, an aliquot of appropriate size (usually a few to a few tens of apatite crystals) is loaded into a stainless steel capsule. A mating lid is spot-welded to the top of the capsule, sealing the grains in but permitting free exchange of gases. The capsule is loaded into a "Christmas tree" arrangement above a resistively-heated vacuum furnace. After pumping and furnace outgassing, a hot blank is run, after which the capsule is dropped into the furnace and held at 950°C for twenty minutes. Numerous experiments show that this time-temperature combination is adequate to completely outgas inclusion-free apatite of less than 200 μm in minimum dimension.

   For laser-heating, the sample, often a single crystal, is placed in a small Pt envelope. After establishing the hot blanket on an empty Pt foil envelope, the sample envelope is lased to reach a temperature of 1000°C (±50°C) for 5 min. The slightly higher temperature was selected to yield a stable pyrometer reading; the shorter duration is allowed because the thermal mass of the system is so small that heating to the desired temperature is essentially instantaneous. For zircon and titanite the sample is lased at 1350°C for 30 min.

2. The evolved gases are spiked with a known amount of 99+% ^{3}He (obtained from Oak Ridge National Laboratory) and cryogenically concentrated on charcoal held at 16 K. After transfer to the charcoal is complete, the charcoal is warmed to 37 K, and the now purified He is admitted into a quadrupole mass spectrometer. The ^{4}He/^{3}He ratio is measured in static mode over a period of about two minutes, after which the entire system is pumped.

3. The *same sample* is again heated to the original temperature for the same holding time, and the measurement procedure is repeated. For inclusion-free samples, this "re-extract" step will yield He at levels indistinguishable from that of the preceding hot blank. However, a substantial amount of He in this step has been found to be the telltale signature of mineral inclusions, at least in the case of apatite. If such He is found, the analysis is rejected as likely to be compromised by inclusions.

4. After He analysis the capsules or Pt envelopes are retrieved from the vacuum chamber. Capsules are opened and inspected to ensure that all grains remain in the capsule (occasionally grains are lost when lids are improperly welded or if the capsule is deformed). Grain loss compromises the age, biasing it toward high values because the parent abundance is measured on a smaller aliquot than the daughter. Grains are transferred to a clean teflon beaker. Pt foil envelopes are not usually opened as this often leads to grain breakage and loss; instead the entire envelope is transferred to a beaker.

5. Apatites are dissolved in concentrated HNO_3, spiked with known amounts of ^{230}Th and ^{235}U, and diluted with water to the desired volume (usually 2 ml). To ensure complete dissolution and sample-spike equilibration the solution is placed in an oven at 90°C for more than one hour. This does nothing to the Pt foil.

   Zircons and titanites are flux-melted at 1100°C in 5 mg of Li or Na borate, again without removal from the Pt foil packet. No Pt dissolves under these conditions. After fusion the glass is dissolved in 10% HNO_3 then spiked and heated like the apatite samples.

6. Th- and U-isotope ratios are measured directly on these solutions by inductively-coupled plasma mass spectrometry.